Advances in Neurobiology

Volume 1

Series Editor
Abel Lajtha

For further volumes, go to
http://www.springer.com/series/8787

John P. Blass
Editor

Neurochemical Mechanisms in Disease

Springer

Editor
John P. Blass
Weil Medical College of Cornell
University
White Plains, NY 10605, USA
jpblass@mail.med.cornell.edu

ISSN 2190-5215 e-ISSN 2190-5223
ISBN 978-1-4419-7103-6 e-ISBN 978-1-4419-7104-3
DOI 10.1007/978-1-4419-7104-3
Springer New York Dordrecht Heidelberg London

Printed on acid-free paper

Springer is part of Springer Science+Business Media (www.springer.com)

Preface

This volume of *Advances in Neurobiology* deals with the neurochemistry of disease. Included are chapters on both human diseases and animal "model" diseases.

***Sources of human tissue*.**
The three main sources of human neural tissues for chemical studies have been:

Brain obtained at autopsy
Brain obtained at biopsy or incidental to neurosurgery
Nonneural tissues containing molecules identical to those in clinically affected brain or nerve

In theory, peripheral nerve is more available than brain, but chemical analyses of peripheral nerves are relatively limited.

A major problem in doing chemistry on brain obtained at autopsy is the possibility of artifacts arising during the process of dying or in the interval between death and chemical analysis: agonal or postmortem artifacts. During the first decades of modern neurochemistry, this problem was considered so severe that few studies were done on autopsy material. Over 30 years ago, Davison and Bowen and their coworkers at Queen Square in London recognized that it was possible to study meaningfully in autopsy brain those molecules that were stable agonally and postmortem (Bowen et al., 1976). Their recognition that one of these proteins was the enzyme choline acetyltransferase allowed them to make major discoveries about the vulnerability of the cholinergic system in Alzheimer disease; that discovery has led to the only available treatments for this common and devastating condition.

Davison and coworkers did extensive control experiments to ensure that they were studying properties of the brain rather than artifacts that arose during the process of dying or after death. Unfortunately subsequent workers have not always adhered to those meticulous standards. It is relatively easy to obtain pieces of human autopsy brain in a hospital or medical school from people who had a variety of diseases of the brain as well as from "controls" free of brain disease detected clinically during life. These are, of course, not truly "healthy controls." They are, after all, dead; they have to have died of something. The ease with which samples of human brain can be obtained has, unfortunately, allowed publication of studies where control of sample quality has been sloppy. Interpretation of neurochemical data obtained from autopsy brain must *always* take into account the possibility of the agonal or postmortem artifacts that so worried earlier neurochemists.

Chemical measurements have also been made on human brain tissue obtained from biopsies or from therapeutic surgery (Smith et al., 1983). Here, agonal and postmortem artifacts are not a problem. However, the tissue available is limited. Patient welfare rather than scientific utility must determine the amounts of tissue available and the anatomical sites from which it comes.

An experimental approach that avoids concerns about the quality of brain tissue is to study in available peripheral tissues genes or gene products that are identical to those in neural tissues. Examples include blood cells and cultured skin fibroblasts, and to a lesser extent biopsies of other tissues such as muscle (Bubber et al., 2005). Most workers assume that genes are identical in all tissues examined from an individual but worry about epigenetic modifications, to DNA as well as to post-translational and posttranscriptional products. However, extensive data indicate that study of proteins in peripheral tissues can often give critical information about those molecules in the brain: the standard A striking example is the use of white blood cells and cultured skin fibroblasts to elucidate enzyme defects in inborn errors of metabolism. A well-known example is Tay–Sachs disease (GM_2–gangliosidosis) (Roe and Shur, 2007).

***Animal tissues*.**

Brain and other tissue from sick experimental animals is as readily available as from healthy animals. That includes transgenic and other animals with "model human diseases." But, it is vital to remember that mice are not men, nor are rats or other experimental animals. For instance, triple transgenic mice have been crafted that develop light microscopic lesions that mimic those of Alzheimer disease (AD). (Pietropaolo et al., 2009). However, direct molecular studies document that such triple gene mutations are not the cause of human AD (Tanzi et al., 1991). Treatments have been identified that benefit "Alzheimer mice" (Sung et al., 2004) but not human patients with this illness. (Petersen et al., 2005; Tabet et al., 2000)

***Disease*.**

Neurochemical studies of illness of the brain typically involve comparing a set of samples classified as "disease" versus a set of samples labeled "control." Clinicians or pathologists do the classification, not chemists. At the extremes of health or illness, it may seem easy to decide who is sick and who is not. In fact, the line is hard to draw. If bizarre and often self-destructive behavior is a sign of mental illness, do we classify adolescence as a form of madness? Are intestinal parasites found in the majority of people in a population "normal" or a form of disease? We have been treating hookworm even though this "germ of laziness" was once endemic in the states of the old Confederacy. Social consensus is particularly important in labeling as "sick" behaviors that are odd but not harmful. Certain sexual variants are considered worth treating in the United States but are thought of as harmless eccentricities in England. (That shocked some of my fellow Americans who went for additional training in psychiatry at the Maudsley Hospital in London.) Soldiers who sacrifice their lives intentionally for their comrades are not classified as suicidally insane. Instead we give them medals.

Specific diseases.

For the last three centuries, it has been conventional to classify sick people as having one or another specific disease. That includes illness of the nervous system, psychiatric as well as neurological. In fact, the concept of specific diseases is a useful but fundamentally unrealistic abstraction. It is one of those approximations that comes out of the English Enlightenment, that are too useful to be discarded even though they do not stand up to close analysis. Grouping patients according to their "disease" helps to provide guidelines for their care, even though in fact every sick person is different from every other sick person. Skilled care requires individualization of care. The British psychiatrist R. E. Kendall has developed the logic of this conundrum with great clarity (Kendall, 1975).

An historical aside may clarify the issues. In the medical tradition that went from the ancients (Hippocrates and Galen) through the Middle Ages until the Enlightenment, physicians basically thought about disease in terms of mechanism. The conventional "theory of humors" was a crude attempt to describe illness in terms of imbalances in body composition, before the invention of modern chemistry and biochemistry.

The modern theory of "specific diseases" was developed in the 1600 s by an English physician, Thomas Sydenham (Haas, 1996). His Latin was too weak for him to study the medical literature of his time, but the professoriat at Oxford granted him a medical degree anyway: his brother was one of Oliver Cromwell's colonels. Came the Restoration, and Sydenham had to make a living. Fortunately, he was a genius. He recognized that specific patterns of signs and symptoms could define clinical entities that typically responded to specific medications. His model was the use of quinine to treat malaria, to treat "tertian and quartan fevers." His concept of specific diseases responding to specific medicines was so powerful that it has come to dominate medicine.

In the later nineteenth and early twentieth century, German-speaking neuropsychiatrists ("alienists") defined neuropsychiatric diseases for which they could not find a neuropathological substratum in terms of the aberrant behaviors. Although sensible enough for the state of knowledge at that time, this approach has been breaking down in recent decades. It is now clear that the same gene mutation can lead to different psychiatric syndromes, to different "diseases" as they are now defined. One classic example is the gene *DISC 1,* which can predispose to "schizophrenia" as well as to "bipolar disease" (manic-depressive psychosis) and "depression" (Chubb et al., 2008).

Perhaps more important, behavioral patterns alone do not predict response to chemicals that act on the nervous system, that is, to medications (Blass, 2006). Thus, behavioral manifestations do not identify specific diseases in the sense originally defined by Sydenham. That is true even of the detailed behavioral classifications created by the committees that write the *Diagnostic and Statistical Manual* of the American Psychiatric Association (the successive versions of *DSM*) (American Psychiatric Association, 2000).

Neurobiology and specifically neurochemistry may—one hopes—give rise to more biologically based and therefore presumably more clinically useful definitions. The editors hope that this volume on the neurochemistry of disease will further that aim.

White Plains, NY John P. Blass

References

American Psychiatric Association (2000) Diagnostic and Statistical Manual of Mental Disorders DSM-IV-TR Fourth Edition (Text Revision). American Psychiatric Association, New york, NY

Blass JP (2006) Concise clinical pharmacology: CNS therapeutics. McGraw-Hill, New york, NY

Bowen DM, Smith CB, White P, Davison AN (1976) Neurotransmitter-related enzymes and indices of hypoxia in senile dementia and other abiotrophies. Brain 99:459–496

Bubber P, Haroutunian V, Fisch G, Blass JP, Gibson GE (2005) Mitochondrial abnormalities in Alzheimer brain: mechanistic implications. Ann Neurol 57:695–703

Chubb JE, Bradshaw NJ, Soares DC, Porteous DJ, Millar JK (2008) The DISC locus in psychiatric illness. Mol Psychiatry 13:36–64

Haas LF (1996) Thomas Sydenham (1624–89). J Neurol Neurosurg Psychiatry 61:465

Kendall RE (1975) The role of diagnosis in psychiatry. Blackwell Scientific Publication, Oxford, Philadelphia, PA (distributed by J. B. Lippincott)

Petersen RC, Thomas RG, Grundman M, Bennett D, Doody R, Ferris S, Galasko D, Jin S, Kaye J, Levey A, Pfeiffer E, Sano M, van Dyck CH, Thal LJ (2005 Jun 9) Alzheimer's Disease Cooperative Study Group. Vitamin E and donepezil for the treatment of mild cognitive impairment. N Engl J Med 352(23):2379–2388

Pietropaolo S, Sun Y, Li R, Brana C, Feldon J, Yee BK (2009) Limited impact of social isolation on Alzheimer-like symptoms in a triple transgenic mouse model. Behav Neurosci 123:181–195

Roe AM, Shur N (2007) From new screens to discovered genes: the successful past and promising present of single gene disorders. Am J Med Genet C Semin Med Genet 145C:77–86

Smith CC, Bowen DM, Sims NR, Neary D, Davison AN (1983) Amino acid release from biopsy samples of temporal neocortex from patients with Alzheimer's disease. Brain Res 264:138–141

Sung S, Yao Y, Uryu K, Yang H, Lee VM, Trojanowski JQ, Praticò D (2004) Early vitamin E supplementation in young but not aged mice reduces Abeta levels and amyloid deposition in a transgenic model of Alzheimer's disease. FASEB J 18:323–325

Tabet N, Birks J, Grimley Evans J (2000) Vitamin E for Alzheimer's disease. Cochrane Database Syst Rev(4):CD002854

Tanzi RE, George-Hyslop PS, Gusella JF (1991) Molecular genetics of Alzheimer disease amyloid. J Biol Chem 266:20579–20582

Contents

Contributors

Lasse K. Bak Faculty of Pharmaceutical Sciences, Department of Pharmacology and Pharmacotherapy, University of Copenhagen, DK-2100, Copenhagen, Denmark, laba@farma.ku.dk

Eric L. Barker Department of Medicinal Chemistry and Molecular Pharmacology, Purdue University, West Lafayette, IN 47906, USA, barkerel@purdue.edu

Nicole Baumann Laboratoire de Neurochimie, Hôpital de la Salpêtrière, F-75013 Paris, France; Université Paris 6, F-75000 Paris, France, nicole.baumann@upmc.fr

Nike Beaubier Department of Pathology, Taub Center for the Study of Alzheimer's Disease and the Aging Brain, Columbia University Medical Center, New York, NY 10032, USA, ntb2101@columbia.edu

Lynn M. Bekris Department of Medicine, University of Washington School of Medicine, Seattle, WA, USA, lbekris@u.washington.edu

Chantal Bémeur Neuroscience Research Unit, CHUM (Hôpital Saint-Luc), Department of Nutrition, University of Montreal, Montreal, QC, Canada, H2X 3J4, chan.chan@videotron.ca

Thomas D. Bird Geriatric Research, Education and Clinical Center, Veterans Affairs Puget Sound Health Care System, Seattle, WA, USA; Department of Neurology, University of Washington School of Medicine, Seattle, WA, USA, tomnroz@u.washington.edu

John P. Blass Department of Medicine, University of Washington School of Medicine, Seattle, WA, USA; Departments of Neurology, Neurosciences, and Medicine, Weill Medical College of Cornell University, New York, NY 10021, USA; Burke Medical Research Institute, White Plains, NY 10605, USA, jpblass@mail.med.cornell.edu

Ivan Bodis-Wollner Department of Neurology, SUNY Downstate Medical Center, Brooklyn, NY 11203, USA; Department of Ophthalmology, SUNY Downstate Medical Center, Brooklyn, NY 11203, USA; Division of Movement Disorders, Department of Neurology, Parkinson's Disease and Related Disorders

Clinic, Center of Excellence NPF, Brooklyn, NY 11203, USA, ivan.bodis-wollner@downstate.edu

D. Allan Butterfield Department of Chemistry, University of Kentucky, Lexington, KY 40506, USA; Sanders-Brown Center on Aging, University of Kentucky, Lexington, KY 40506, USA; Center of Membrane Sciences, University of Kentucky, Lexington, KY 40506, USA, dabcns@uky.edu

Roger F. Butterworth Department of Neurology, University of New Mexico Health Sciences Center, Albuquerque, NM 87131, USA

Eduardo Candelario-Jalil Department of Neurology, University of New Mexico Health Sciences Center, Albuquerque, NM 87131, USA, ecandelario-jalil@salud.unm.edu

Gemma Casadesus Department of Neurosciences, Case Western Reserve University, Cleveland, OH, USA, gemma.casadesus@case.edu

Arthur J.L. Cooper Department of Biochemistry and Molecular Biology, New York Medical College, Valhalla, NY 10595, USA, arthur_cooper@nymc.edu

Frank Culicchia LSU Department of Neurosurgery, Louisiana State University Health Science Center, New Orleans, LA 70112, USA, fculic@lsuhsc.edu

S. Hossein Fatemi Department of Pharmacology, University of Minnesota Medical School, Minneapolis, MN 55455, USA; Department of Neuroscience, University of Minnesota Medical School, Minneapolis, MN 55455, USA; Division of Neuroscience Research, Department of Psychiatry, University of Minnesota Medical School, Minneapolis, MN 55455, USA, fatem002@umn.edu

Timothy D. Folsom Division of Neuroscience Research, Department of Psychiatry, University of Minnesota Medical School, Minneapolis, MN 55455, USA, folso013@umn.edu

Afsaneh Gaillard Institut de Physiologie et Biologie Cellulaires, CNRS 6187, Université de Poitiers, 86022 Poitiers Cedex, France

Francisco García-Sierra Department of Cell Biology, CINVESTAV-IPN, México City, México, fgarcia-sierra@cell.cinvestav.mx

Imad Ghorayeb CNRS 5227, Université de Bordeaux II, 330776 Bordeaux Cedex, France

Cheng-Xin Gong Department of Biochemistry and Molecular Biology, New York Medical College, Valhalla, NY 10595, USA, cxgong@mail.csi.cuny.edu

Lianna Heidt Department of Pathology, Taub Center for the Study of Alzheimer's Disease and the Aging Brain, Columbia University Medical Center, New York, NY 10032, USA, heidtl@iupui.edu

Gyongyi Horvath INSERM U676, Hôpital Robert Debré, F-75019 Paris, France, horvath@phys.szote.u-szeged.hu

Mohamed Jaber Institut de Physiologie et Biologie Cellulaires, CNRS 6187, Université de Poitiers, 86022 Poitiers Cedex, France, mohamed.jaber@unv-poitiers.fr

Nael Nadif Kasri Department of Neurology, SUNY Downstate Medical Center, Brooklyn, NY 11203, USA, n.nadif@donders.ru.nl

Jane Kasten-Jolly New York State Department of Health, Wadsworth Center, Albany, NY 12201, USA

David A. Lawrence New York State Department of Health, Wadsworth Center, Albany, NY 12201, USA, david.lawrence@wadsworth.org

Hyoung-gon Lee Department of Pathology, Case Western Reserve University, Cleveland, OH, USA, hyoung-gon.lee@case.edu

Mark A. Lovell Department of Medicine, University of Washington School of Medicine, Seattle, WA, USA, malove2@uky.edu

Walter J. Lukiw LSU Neuroscience Center of Excellence, Louisiana State University Health Science Center, New Orleans, LA 70112, USA, wlukiw@lsuhsc.edu

Akio Masuda Division of Neurogenetics, Center for Neurological Diseases and Cancer, Nagoya University Graduate School of Medicine, Nagoya, Japan, amasuda@med.nagoya-u.ac.jp

R. Mathew Department of Neurology, Medical College Hospital, Thiruvananthapuram 695011, Kerala, India, robert_mathew90@yahoo.co.in

P.S. Mathuranath Cognition and Behavioural Neurology Section (CBNC), Department of Neurology, Sree Chitra Tirunal Institute for Medical Sciences and Technology (SCTIMST), Thiruvananthapuram 695011, Kerala, India, mathu@sctimst.ac.in

Siddhartha Mondragón-Rodríguez Department of Cell Biology, CINVESTAV-IPN, México City, México, catalisis78@hotmail.com

Jane A. Montgomery Neuroscience Research Unit, CHUM (Hôpital Saint-Luc), Montreal, QC, Canada, H2X 3J4, janemontgomery@sympatico.ca

Herman Moreno Department of Neurology and Physiology/Pharmacology, SUNY Downstate Medical Center, Brooklyn, NY 11203, USA, herman.moreno@downstate.edu

Kinji Ohno Division of Neurogenetics, Center for Neurological Diseases and Cancer, Nagoya University Graduate School of Medicine, Nagoya, Japan, ohnok@med.nagoya-u.ac.jp

Guylène Page Groupe de Recherche sur le Vieillissement Cerebral (GReViC), EA 3808, Université de Poitiers, 86022 Poitiers Cedex, France

Gregory M. Pastores Neurology and Pediatrics, New York University School of Medicine, New York, NY 10016, USA; Neurogenetics Laboratory, New York University School of Medicine, New York, NY 10016, USA, gregory.pastores@med.nyu.edu

George Perry Department of Pathology, Case Western Reserve University, Cleveland, OH, USA; College of Sciences, University of Texas at San Antonio, San Antonio, TX, USA, george.perry@utsa.edu

Robert B. Petersen Department of Pathology, Case Western Reserve University, Cleveland, OH, USA, rbp@case.edu

Danielle Pham-Dinh U676 INSERM, University Denis Diderot, Hopital Robert Debre, 48 Bd Serurier 75019, Paris, France, danielle.pham-dinh@inserm.fr

Elena M. Ribe Department of Pathology, Taub Center for the Study of Alzheimer's Disease and the Aging Brain, Columbia University Medical Center, New York, NY 10032, USA, elena_ribe@hotmail.com

Gary A. Rosenberg Department of Neurology, University of New Mexico Health Sciences Center, Albuquerque, NM 87131, USA; Department of Neurosciences, University of New Mexico Health Sciences Center, Albuquerque, NM 87131, USA; Department of Cell Biology and Physiology, University of New Mexico Health Sciences Center, Albuquerque, NM 87131, USA, grosenberg@salud.unm.edu

Arne Schousboe Department of Neurology, SUNY Downstate Medical Center, Brooklyn, NY 11203, USA, as@farma.ku.dk

Natalie R. Sealover Department of Medicinal Chemistry and Molecular Pharmacology, Purdue University, West Lafayette, IN 47906, USA, nsealove@purdue.edu

Mark A. Smith Department of Pathology, Case Western Reserve University, Cleveland, OH 44106, USA, mark.smith@case.edu

G. Srinivas Department of Biochemistry, Sree Chitra Tirunal Institute for Medical Sciences and Technology (SCTIMST), Thiruvananthapuram 695011, Kerala, India, srinivasg@sctimst.ac.in

Jerry A. Stitzel Neurology and Pediatrics, New York University School of Medicine, New York, NY 10016, USA; Neurogenetics Laboratory, New York University School of Medicine, New York, NY 10016, USA, stitzel@colorado.edu

Rukhsana Sultana Department of Chemistry, University of Kentucky, Lexington, KY 40506, USA; Sanders-Brown Center on Aging, University of Kentucky, Lexington, KY 40506, USA

Saeid Taheri Department of Neurology, University of New Mexico Health Sciences Center, Albuquerque, NM 87131, USA, staheri@salud.unm.edu

Carol M. Troy Departments of Pathology and Neurology, Taub Center for the Study of Alzheimer's Disease and the Aging Brain, Columbia University Medical Center, New York, NY 10032, USA, cmt2@columbia.edu

Debby Tsuang Department of Psychiatry and Behavioral Sciences, University of Washington School of Medicine, Seattle, WA, USA; Mental Illness Research, Education and Clinical Center, Veterans Affairs Puget Sound Health Care System, Seattle, WA, USA, dwt1@u.washington.edu

Linda Van Aelst Department of Neurology, SUNY Downstate Medical Center, Brooklyn, NY 11203, USA, vanaelst@cshl.edu

Helle S. Waagepetersen Faculty of Pharmaceutical Sciences, Department of Pharmacology and Pharmacotherapy, University of Copenhagen, DK-2100, Copenhagen, Denmark, hsw@farma.ku.dk

Chang-En Yu Department of Medicine, University of Washington School of Medicine, Seattle, WA, USA; Geriatric Research, Education and Clinical Center, Veterans Affairs Puget Sound Health Care System, Seattle, WA, USA, changeyu@u.washington.edu

Xiongwei Zhu Department of Pathology, Case Western Reserve University, Cleveland, OH, USA, xiongwei.zhu@case.edu

Mechanisms Versus Diagnoses

John P. Blass

Abstract Science is a branch of scholarship: it provides explanations for material phenomena in terms of matter and energy. Medicine by contrast is a trade: it applies scientific knowledge but also requires nonscientific skills such as empathy. Neurochemistry is the science that deals with the molecules that make up nervous systems and with their interactions. Neurology and psychiatry are the trades of those who try to help people with diseases of the nervous system to heal. Scientists including neurochemists have the luxury of taking the time needed to probe deeply into the phenomena they study. Neurologists and psychiatrists more often face sharp constraints on how long they can take to try to help the sick human beings for whom they care. Examples used to illustrate this distinction include psychoses and demyelinating diseases. The existence of a large and often impressive body of scholarship in neurology and psychiatry can foster the illusion that these are scholarly rather than fundamentally practical activities. For convenience, modern physicians conceptualize the phenomena they see as discrete "diseases." Sometimes their concepts turn out to be scientifically valid. Often, sadly, they do not. The current chapter deals with neurochemical mechanisms rather than listing abnormalities in molecules in clinically defined "diseases." Neurochemical mechanisms in sick people are real-world entities that can be discovered by observation and whenever possible by experimentation. "Diseases" are abstractions constructed by physicians and others to help figure out what is wrong with patients and how to try to help. This chapter is on the chemistry of nervous systems of people whose actions are unusual enough to draw medical attention to them. It does not deal with such nonmaterial concepts as "free will" or "the soul," nor with the relationship of mind to brain. This limitation is intentional and potentially powerful. A neurologist or psychiatrist armed with the array of chemicals that constitute the modern pharmacopeia can do much more than even the most sympathetic and understanding physician or other counselor who is limited to "talk therapy." Sigmund Freud and his fellow alienists in Vienna at the turn of the last century yearned to have such medicines available. In general, the

J.P. Blass (✉)
Dementia Research Service, Burke Medical Research Institute, White Plains, NY 10605, USA
e-mails: jpblass@mail.med.cornell.edu; jbass@burke.org; jpblass@yahoo.com

J.P. Blass (ed.), *Neurochemical Mechanisms in Disease*,
Advances in Neurobiology 1, DOI 10.1007/978-1-4419-7104-3_1,

Viennese-trained psychiatrists who fled to the United States were perfectly willing to use psychotropic medications, although they made sure to talk extensively to their patients as well. Even the psychoanalysts in that group held to the slogan, "There is also a brain." Despite the disputes among "schools of psychoanalysis" that went on with talmudic intensity, the able among those practitioners never forgot that their goal was to aid the troubled individuals who came to them for help.

Keywords Disease · Classification · Mechanism · Tay–Sachs disease · Psychoses

Contents

1 Introduction

1.1 Focus of This Volume

This volume of the *Handbook of Neurochemistry* deals with chemical mechanisms in the nervous systems of sick people. It is intentionally not a catalogue of neurochemical phenomena in specific diseases, as those entities are currently defined. Thus, there are no chapters on the "Neurochemistry of Depression" or "Neurochemistry of Stroke." The choice to focus on mechanisms rather than on diseases deserves explication.

1.1.1 Contrasts Between Science and Clinical Medicine

Science is primarily a scholarly endeavor; the practice of medicine is primarily a practical trade. Scientists try to find out more and more about the world, even if doing so is time-consuming. Medicine, in contrast, depends more on what is useful than on what is "true." That includes the specialties of neurology and psychiatry. The important information for medicine is that which enables medical practicioners to decide how best to try to help individual sick people. Generally those decisions must be made within a limited amount of time in order to be useful. Even the decision to do nothing is a decision that affects patient welfare. When possible, medical decisions are based on scientific information, but where the science is lacking the decisions still have to be made.

Chemists are confident that knowledge about the molecules of which the material world consists can help human beings to understand and deal with that world. We do not doubt that molecules are real things, even if too small to be seen (at least without an electron microscope). We believe that understanding the molecules that make up the brain and how those molecules interact can help us understand our brains and therefore ourselves. Many examples prove that alterations in the molecules that make up human beings can lead to sickness, that is, to signs and symptoms that cause significant disability in patients. We recognize that individual sick people, like molecules, are real things.

"Diseases," in contrast to molecules and to individual patients, are abstractions (Kendell, 1975). The properties of molecules are determined by experimentation; the properties of diseases are determined by consensus conferences. The names and classifications of diseases are inventions of the human mind, developed for the intellectual convenience of those of us who have been trying to take care of sick people.

1.1.2 Utility of the Disease Concept

Physicians have found the concept of specific "diseases" to be a useful intellectual construct, despite the sometimes tenuous relationship of specific diseases to underlying reality. Medical practitioners rapidly experience the truism that every patient differs significantly from every other patient. However, the concept of specific "diseases" allows physicians to classify patients into groups who are likely to react in similar ways and specifically to respond to specific therapies. These intellectual constructs change as more information becomes available. Disease classifications and therefore disease categories will almost certainly continue to change as biology advances. The current rapid increases in understanding the chemistry and molecular genetics of the nervous system are already changing the way dysfunctions of the nervous system are thought of and therefore classified. For instance, the "hereditary ataxias" presented an almost impenetrable forest of erudition as long as differentiation among them depended on clinical signs and symptoms. Now that the responsible genes have been discovered, classification is straightforward and so is differential diagnosis (Duenas et al., 2006; Lodi et al., 2006). Both now depend on laboratory studies of the molecular genetics.

Simply enumerating the chemical abnormalities in entities that are still defined by clinical rather than biological criteria is unlikely to have lasting value, as those categories become increasingly out of date and discarded. On the other hand, mechanisms that have been defined experimentally are not likely to stay scientifically valid, although our understanding of them will, it is hoped, continue to deepen.

1.2 Historical Background

People attempting to heal the sick have almost always used the best knowledge of the universe available to them to do so. For the priests of Aesculapius in ancient

Greece, that was a mixture of mysticism and empiricism. (Aesculapius may originally have been an unusually skilled doctor, whose deification helped maintain a flow of patients to the hospital [shrine] that he founded.) Today, physicians try to use the more objective system of observations and interpretations that make up science, including the science of neurochemistry. (Of course, the "art of medicine" often still requires a certain amount of mumbo-jumbo. Many patients lack confidence in a diagnosis and treatment unless it is described to them in long words of Latin and Greek origin that they do not understand.)

1.2.1 Hippocratic Tradition

The Hippocratic tradition dominated medical practice from ancient times through the middle ages and up to early modern times. This tradition is relatively easy to adapt to modern molecular medicine. First, it is descriptive rather than theoretical. Hippocratic physicians described the things that they could observe going wrong in their patients so clearly and accurately that we can attribute many of the illnesses they describe to agents discovered only within the last century or so. A classic example is the "Plague of Athens," which had major effects on the war between Athens and Sparta. It is now thought to have been an outbreak of adult measles. Hippocratic physicians recognized that groups of people often had the same sorts of things going wrong with them, particularly during epidemics of what we now call infectious diseases. In fact, one of the books in the Hippocratic corpus is entitled "Epidemics."

1.2.2 Theory of Humors

The ancient and medieval tradition was also crudely chemical, although the chemistry of the time was terribly primitive compared to modern knowledge. It emphasized the theory of humors. In health, the humors were in balance. In illness, they were out of balance. Treatment consisted in restoring balance. For instance, a person who had too much moisture needed treatments to help dry out enough to regain balance. The medieval form of this theory of humors depended heavily on the writings of Galen, who had been among other things physician to the emperor Marcus Aurelius. Our chemistry is almost infinitely more complex, but the thrust of modern mechanistic thinking has significant similarities. When a person today retains water and salt, we typically prescribe a diuretic to "dry him out." Our more sophisticated concept of "balance of humors" is "healthy homeostasis."

1.2.3 Sydenham's Conceptualization of Specific Diseases

The modern theory of diseases as distinct entitites was popularized in the 1600s in large part by a great English physician, Thomas Sydenham. He lived during the era of the religious wars in Europe, and his brother was one of Oliver Cromwell's colonels. The faculty at Oxford gave Thomas the medical qualification he wanted even though he was barely literate in Latin, the language in which the medical literature of his time was written. Came the restoration of the Stuart monarchy, and Thomas had to make a living. He did so by practicing the profession whose literature he couldn't really read. Fortunately, he was a genius. Science at that time was heavily a matter of classification, including classification of the new species of plants and animals being discovered in the Americas and other continents previously unexplored by Europeans. Sydenham classified illnesses analogously to the way Linnaeus was classifying plants.

Sydenham used two sets of criteria for his classifications: clinical signs and symptoms and response to specific therapy. His model was what we now recognize to be malaria. This disease typically causes high fevers every third or fourth day, reflecting the life cycle of the parasites which we now recognize to cause it. Sydenham characterized this clinical entity as "tertian fever" or "quartan fever." He recognized the value of treating it with extracts of cinchona bark, whose active principle we now recognize to be quinine. This eminently practical approach—clinical signs and symptoms indicating the need for a particular treatment—proved so useful that it came to dominate medicine, not only in Britain but also in other countries. Because Sydenham's concept included response to a particular treatment (often a medicine containing one or more critical molecules) it was to some extent a chemical classification. However, it is worth noting that Sydenham himself felt that his friend Dalton's ideas about atoms had no significance for clinical medicine. Given the medical ignorance of the time, his conclusion was correct when he made it.

1.2.4 Chemical and Biological Refinements of Sydenham's Concepts

Over the succeeding centuries, developments in chemistry and biology led the concept of what constituted a "disease" to depend less on purely clinical observations and instead on more putatively "scientific" characteristics. The growth of chemistry, especially the development of the chemistry of dyes during the nineteenth century, led to the discovery of chemicals that stained human tissues obtained at autopsy. The increased information that then came from pathology led to the definition of diseases as "clinical-pathological" entities, that is, conditions in which a clincal pattern was associated with a more or less specific anatomic pathology. This approach still dominates neurology textbooks. Confusing clinical entities such as Alzheimer disease are considered to be based on hard scientific definitions, inasmuch as they are associated with characteristic neuropatholgical changes revealed by microscopic examination after staining with appropriate dyes. Psychiatry has been considered a "soft" specialty in part because of the lack of recognized anatomic pathology in the brains of people with such major disorders as schizophrenia and depression. Now that modern imaging techniques are increasingly identifying "objective" abnormalities in the major mental illnesses, psychiatry has been described as becoming more scientific.

Bacteriology and subsequently virology also led to important modifications of Sydenham's concept of diseases. Instead of such general classifications as "pthisis" for inflammation of the lungs, physicians came to recognize more specific entities such as "tuberculosis" or "diplococcus pneumoniae pneumonia." The development of convenient modern techniques for culturing pathogenic infectious agents and determining their sensitivities to specific antibiotics has allowed this biological knowledge to become part of the daily practice of medicine. We recognize an infection with "multiple antibiotic-resistant staphylococcus aureus" (MRSA) as an entity independent of the organ infected and the resulting anatomic pathology. However, it is worth noting that the sensitivities to particular antibiotics of different strains of the same species of bacteria now vary so much that rational treatment of infections still involves direct laboratory studies of the patient being treated, namely culture of the responsible micro-organisms from the specific patient and tests of its sensitivity to specific antibiotics. You and I may both have pneumonia, and both your and my pneumonia may be due to infection with diplococcus pneumoniae, but your pneumonia may respond to and be appropriately treated with penicillin whereas mine needs to be treated with another antibiotic. Knowing that clinically important difference early in the course of our treatment requires laboratory tests.

1.2.5 Molecular Studies and Clinical Specificity

Modern biochemistry and particularly modern nucleic acid chemistry (molecular genetics) are forcing practitioners to re-evaluate their concepts of what constitutes specific diseases. Nowhere is this more evident than in diseases of the nervous system. (Hereditary ataxias are an example discussed above; psychoses are an example discussed below.)

It would have been convenient if abnormalities in specific genes were to have led reliably to specific clinical symptoms and signs, that is, to specific "diseases." Unfortunately, they do not. The general pattern has been that when an abnormal gene is associated with a clinically defined entity, investigators at first assume that it is more or less specifically associated with the "disease" in which it was discovered. Subsequent studies of larger populations with a larger variety of "diseases" typically show that abnormalities of the gene in question turn out to occur in a variety of different diseases and usually even in people who have no clinically significant disability. "Diseases" defined clinically or even by a combination of clinical and laboratory findings generally turn out, on extensive study, to be genetically heterogeneous in two senses: different genes can lead to the same clinical pattern, and abnormalities of a single gene can lead to different clinical syndromes. Stated technically, epidemiologically based studies typically reveal that diseases defined clinically are genetically heterogeneous, and the consequences of mutations in a single gene most often turn out to be clinically heterogeneous. The following paragraphs give examples of these complexities.

1.3 Example 1: Tay–Sachs Disease

A classical "homogeneous" inborn error of metabolism, namely Tay–Sachs disease (GM_2 gangliosidosis), provides a clear example.[6]

1.3.1 Clinical Patterns

This condition was recognized clinically in the nineteenth century in infants of Ashkenazi Jewish heritage, who suffered from a form of severe psychomotor retardation in infancy and early death. These children were hard to distinguish clinically from other infants who had other forms of devastating, early psychomotor retardation with blindness, that is, other forms of "familial amaurotic idiocy." Differential clinical diagnosis depended on such clinical signs as an "exaggerated startle response," that is, an infant with Tay–Sachs disease was supposed to cry even more than usual if startled by something like a loud clap of the physician's hands.

1.3.2 Neuropathology, Neurochemistry, and Molecular Biology

Neuropathological observations subsequently allowed a more biological definition of the subgroup of children with "true" Tay–Sachs disease. Light and then electron microscopy revealed characteristic "whorls" of material stored in their brain. Subsequent neurochemical studies identified that material as GM_2 ganglioside. Enzymatic studies showed that Tay–Sachs disease was due to a lack of a functional form of an enzyme that catalyzes the breakdown of GM_2 ganglioside, namely hexosaminidase A. Molecular genetic studies demonstrated that this lack was due to mutations in the *HEXA* gene that encodes this enzyme. Definitive clinical diagnosis of Tay–Sachs disease now requires molecular genetic confirmation. The clinical overlap among patients with "lipid storage diseases" is so great that a specific diagnosis based on history and physical examination is no more than an informed guess.

Thus neurochemistry and molecular biology appeared to have identified a biologically homogeneous population who suffered from a particular constellation of clinical signs and symptoms due to homozygous recessive inheritance of a mutation-specific gene in a particular ethnic group, that is, from a specific "molecular disease." Clinical applications of these neurochemical discoveries have allowed Ashkenazi Jewish couples to be tested for carrier status even before the woman becomes pregnant. Prenatal testing of cells obtained at amniocentesis from fetuses at risk for this disease has allowed termination of the affected pregnancies in this ethnic group for whom therapeutic abortion is religiously acceptable. This triumph of modern medicine appeared to hold up as long as the chemical analyses were so expensive and tedious that they were done largely in children who fit or were at risk for the expected clinical characteristics. However, as cheaper and more automated procedures were developed that allowed testing of larger populations, the associations between gene and ethnicity and gene and clinical syndrome both broke down.

1.3.3 Genetic Variability

HEXA deficiency has turned out to be neither genetically nor ethnically homogeneous. A variety of different alterations—mutations—in the *HEXA* gene have been associated with classic, infantile Tay–Sachs disease. The existence of different

Table 1 Clinical presentations of *HEXA* deficiency

Psychomotor retardation	Gravel et al. (1995)
Early infantile form	
Late infantile form	
Sandhoff disease	Hendriksz et al. (2004)
Infantile form	
Juvenile form	
Juvenile progressive dystonia	Meek et al. (1984)
Spinocerebellar disease (several clinical syndromes)	Argov and Navon (1984), Oonk et al. (1979), and Rapin et al. (1976)
Motor neuron disease	Argov and Navon (1984) and Drory et al. (2003)
Focal muscular atrophy	Iype et al. (2006)
Dementia	Hammer (1998) and O'Neill et al. (1978)
Depression	Hammer (1998) and Renshaw et al. (1992)
Schizophrenia	Hammer (1998) and Rosebush et al. (1995)
Postpartum psychosis	Lichtenberg et al. (1988)
Asymptomatic	Navon et al. (1973)

This list is not exhaustive, nor is the citation of references. New syndromes associated with *HEXA* deficiency are still appearing in the medical literature.

mutations in a single gene among different individuals and among different populations is, of course, the rule rather than the exception in studies of inherited diseases. Clinically typical, *HEXA*-deficient Tay–Sachs disease occurs in a number of ethnic groups. In some, the mutations tend to differ from those most frequently found in Ashkenazi Jews. For instance, there is a "French Canadian" mutation as well as an "Ashkenazi Jewish" mutation. (See Gravel et al., Table 1, for discussion and references.) However, the overlaps in mutations among ethnic groups are wide. They confirm the principle, well known to human geneticists, that genome studies tell many of us things that we did not know, or want to know, or want our spouses to know. (Genetic counselors are obligated to warn a family for whom molecular genetic studies are recommended that for perhaps 15% of children, the supposed father is not the biological father.)

1.3.4 Variations in Clinical Phenotype

More important for this discussion, abnormalities of the responsible *HEXA* gene have now been associated with a dozen or more clinically distinct patterns (Table 1), including "schizophrenia" and including people with no clinically significant disability. Put technically, "adult onset Tay–Sachs disease" is clinically pleomorphic. A steady stream of reports continues to appear describing variant neurological abnormalities in people with "adult onset Tay–Sachs disease."

Systematic large studies of the incidence of *HEXA* abnormalities among patients in common diagnostic categories such as "schizophrenia" are in short supply. What studies have been done encourage further work (Goodman, 1994). Studies of another inborn error of metabolism associated with "schizophrenia syndromes," metachromatic leukodystrophy, have shown a high incidence of abnormalities in sulfatide metabolism. The abnormalities have been attributed to "pseudosulfatase deficiency" (Alvarez et al., 1995; Galbraith et al., 1989; Herska et al., 1987). Molecular genetic investigations have not been reported. There appears not to have been a systematic study at the molecular genetic level of the incidence of abnormalities in genes responsible for storage disorders such as Tay–Sachs disease and metachromatic leukodystrophy, what have previously been called "Type II schizophrenics." These unfortunate patients suffer from relentless, generally drug-unresponsive, progressive psychoses which sooner or later turn into dementia, and are associated with brain atrophy with enlarged ventricles. Type II schizophrenics appear to have a progressive brain disease. One may speculate that some of them have as yet unelucidated variants of disorders that in other, more severe forms lead to progressive psychomotor failure in infancy or early childhood.

1.4 *Example 2: Psychoses*

The neurochemical and molecular genetic study of psychoses including schizophrenia is beset by problems of clinical definition.

1.4.1 Recognizing Psychosis

The first of these problems is deciding who is psychotic. The difficulty in defining precise criteria for whether someone is crazy is summarized in an old Quaker saying: "All the world is mad but me and Thee, and sometimes I doubt Thee." Whole nations can go mad, for instance, the paranoia of the highly educated German-speaking world during the time of the Nazis. (The review of the movie *Saving Private Ryan* in the *Süddeutsche Zeitung* pointed out that using the Nazis as a horrible example is less controversial than using more up-to-date examples of insane cruelty masking itself as politics.) Sets of diagnostic criteria for psychosis and for specific psychoses continue to be promulgated, not least in the volumes of the *Diagnostic and Statistical Manual of the American Psychiatric Association* (most recently *DSM IV-TR*). Applying these criteria well requires skill and training.

Although the lines between mad, odd, and sane are hard to draw precisely, common sense often allows easy classification. As usual, a clinical anecdote is illustrative.

> A woman suffering from a severe (masked) depression lost her appetite to the point where her body weight fell to a dangerously low level (body mass index of 14.3). The neurologist who saw her immediately arranged admission to a psychiatric hospital. The admitting resident there was concerned about whether the patient met *DSM IV-TR* criteria for depression, and if so of what type. The neurologist responded, somewhat rudely: "Look, this lady has nearly succeeded in starving herself to death. Please admit her and feed her, if necessary through a tube, and treat her for depression. There will be plenty of time to worry about how to classify her mental disease once she is no longer at imminent risk of death from an infection or other complication of starvation."

1.4.2 Classification of Psychoses

Clinical observers have classified forms of madness in different ways over the centuries. The Hippocratic physicians used the single category "delirium" for madness whether an external cause could be recognized or not. In modern terms, they did not distinguish between endogenous and exogenous psychoses, and between endogenous and exogenous neurointoxicants. Instead they described what was happening in individual patients or groups of patients. As knowledge of the molecular bases of psychotic behavior increases, including endogenous chemical imbalances in the brain, we may yet go back to a modernized version of their formulation.

The modern classification of madness goes back about a century, to the German psychiatrist Bleuler. Among the mental disorders he classified that had no neuropathological stigmata at that time were schizophrenia (thought disorder) and depression and manic-depressive disease (mood disorders). This distinction has persisted in psychiatry. It continues to be widely although not universally accepted. However, modern molecular studies are bringing the biology of the distinction between "thought disorder" and "mood disorder" into serious question.

1.4.3 The *DISC1* Locus

DISC1 is an example of a gene that predisposes to both thought disorder and mood disorder (Craddock et al., 2006; Porteus et al., 2006). The association of this gene with schizophrenia was discovered in a family in whom the insanity was associated with a balanced translocation. A number of studies then demonstrated and confirmed that abnormalities in this gene were associated with "typical" schizophrenia. Further studies showed that abnormalities in *DISC1* were also associated with manic-depressive (bipolar) psychosis. This was not too surprising, because the clinical differential diagnosis between schizophrenia and bipolar disease can be difficult, particularly while the sufferers are very crazy. Further studies then showed that abnormalities in *DISC1* were also associated with recurrent unipolar depression, which is relatively easy to tell from schizophrenia and, with careful examination, even from bipolar disorder. Porteus et al. (2006) concluded that: "DISC1 is a generalizable genetic risk factor for psychiatric illness that also influences cognition in healthy subjects."

1.4.4 Other Loci

Other loci also contribute to the risk of schizophrenia as well as other diseases. Mutations in the neuroregulin 1 gene (*NRG-1*) are also associated with both thought disorders and mood disorders (Green et al., 2005). Abnormalities in the 15q13-q14 region of chromosome 15 predispose replicably to the existence of schizophrenia, but also to bipolar disorder, schizoaffective disorder, Prader–Willi syndrome (a developmental disorder associated with psychosis), and some forms of juvenile epilepsy (Leonard and Freedman, 2006). Other loci associated with both "schizophrenia" and "bipolar disease" have been described on chromosome 13 and chromosome 22 and in relation to genes encoding components of myelin.

1.4.5 Modifier Loci

Presumably, the variable clinical presentations of a single genetic abnormality reflect the influences of other genes as well as of varying environmental events. The effect of other parts of an individual's genome have been referred to as actions of "modifier genes" or "genetic background." The effects of specific environmental influences on the clinical expression of a variation in DNA are also being delineated. A striking example is the combination of the interaction of the Val^{158} allele of catechol-O-methyl transferase (COMT) and cannabis use in causing "schizophrenia." Whether the effects of this allele by itself are clinically significant is controversial. However, it is clear that those who carry this allele and also use cannabis during their adolescence have a tenfold increased risk of becoming psychotic, compared to the general population (Caspi et al., 2005). Is their psychosis "schizophrenia" or "cannabis toxicity?" Does it matter? As the late Houston Merritt said about a patient presented to him as a diagnostic problem, "We all know what is wrong with this person; we are just debating what to call it." These semantic problems can be an entertaining exercise in medical erudition, but semantic distinctions should not alter the quality of the care we give to individual patients.

1.4.6 Implications for Research on Mental Illness

The recognition that the same fundamental biological changes can lead to both thought disorders (schizophrenias) and mood disorders (bipolar disease and sometimes unipolar depression) does not contradict clinical experience as much as it brings into question interpretations of neurochemical research on these disorders. Bleuler, one of the psychiatrists most responsible for the distinction between thought disorders and mood disorders, himself recognized that the clinical distinction ("differential diagnosis") between these conditions can be extremely difficult. The standard emergency room pharmacological treatment for a patient with an acute psychosis involves treatment suitable for both conditions. However, researchers have in the past used patients with the diagnosis of "bipolar disorder" as "disease controls" for studies of "schizophrenia," and vice versa. Several metabolites discovered decades ago in the urine of mentally ill people were dismissed as "nonspecific findings" because their excretion was associated with both conditions. In the light of modern molecular genetic discoveries, that interpretation may have been wrong. The patients in the two categories may have had different clinical manifestations of the same biological, neurochemical abnormality. Skolnick (2006), who has extensive experience and expertise in the development of new pharmaceuticals, has proposed that the best way now available for developing innovative treatments for complex illnesses such as psychoses is to define genetic risk factors and then develop innovative new drugs based on the genetic information.

1.5 Example 3: Multiple Sclerosis and Demyelination

1.5.1 Clinical Patterns

The medical school definition of "multiple sclerosis" (MS) is demyelination within the central nervous system that varies in space and time. The term refers to a disorder in which patches of demyelination develop during exacerbations. In the most common forms of MS, exacerbations are separated by varying periods of time in which the disease does not appear to progress. Whether progressive demyelination without clear periods of remission should be considered a form of MS is a matter of definition, about which clinicians specializing in the care of patients with this disorder have argued. Conventional medical nomenclature classifies as distinct entities a number of disorders of myelin that can mimic MS clinically. These include, for instance, the sometimes devastating demyelination localized to the pons or the demyelination that can follow infectious diseases and/or vaccinations.

The precise meaning of the term"multiple sclerosis" is "many scars." The words themselves do not indicate that the scars are even in the nervous system, let alone in myelin. The British terminology, "disseminated sclerosis," is no more descriptive. Charcot, who first distinguished this condition from other disorders of the nervous system including syphilis, coined the slightly more precise French term, "sclerose en plaque."

An inconvenient but more descriptive name for this MS is "intermittent, patchy demyelination." This clumsy term makes clear that "multiple sclerosis" is unlikely to be a single entity in terms of cause (etiology) or disease mechanisms (pathophysiology). In principle, any conditions or combination of conditions that lead to intermittent, patchy demyelination are forms of "multiple sclerosis." If a clear cause can be identified, the condition is by convention not referred to as multiple sclerosis. The disease is therefore by definition of unclear etiology. The neurology literature of the last 100 years contains confident declarations that multiple sclerosis has been proven to be a viral disease, that it has been proven not to be a viral disease, that it has been proven to be an immune disease, that immune mechanisms in multiple sclerosis have been shown to be secondary to the disease process, and so on.

1.5.2 Proposed Mechanisms

Current opinions on cause and mechanism include the possibility that MS is often due to a form of "molecular mimicry," in which an immune response to an infective or other exogenous agent leads to the formation of antibodies and/or cells that cross-react destructively with components of normal myelin. "Molecular mimicry" is well established in certain other disorders of the nervous system (Candler et al., 2006) including paraneoplastic syndromes (Posner, 2003).

1.5.3 Chemistry of Demyelination

In contrast to the confusion about what multiple sclerosis is or are, the chemistry of demyelinating processes is rather well defined. (See Chapters XXX.) Whatever the causes of "multiple sclerosis" eventually turn out to be, they lead to demyelinating processes which will in all likelihood be very similar to or identical at the neurochemical level with the demyelinating processes that have already been elucidated. The information about the mechanisms of demyelination is based on firm, robust, reproducible experimental observations. This information is likely to expand, but what is now proven is unlikely to change. This solid science contrasts with the theoretically rather vague although clinically useful clinical conceptualization of "multiple sclerosis" as an entity. Describing the neurochemical correlates of MS would be an exercise in phenomenology, and unstable phenomenology as the clinical definition of this syndrome changes. Demyelination can, however, be meaningfully discussed in terms of mechanistic neurochemistry.

1.6 Implications

The practice of medicine is a service profession, not a science. It has more in common with a trade than with a branch of scholarship. The physician has the responsibility of trying to keep in mind all the variables that pertain to the person he or she is trying to help, and must choose within a relatively short time to do nothing or to do something practical, such as prescribing a medication or giving advice. The scientist has the responsibility and the luxury of taking the time to think deeply about one aspect of a problem. There is truth to the cliché that scientists are paid to think more and more about less and less (at least until they become senior enough to have "administrative responsibilities" including raising large sums of money).

Science has undoubtedly contributed in major ways to the improvement of human health, as documented by increasing longevity in developed countries. Not least have been improvements in nutrition, in the safety of the water and food supplies, and in maternal and child health including vaccination against communicable plagues of childhood, such as diptheria. The mental health of both patients and practitioners requires that we also believe in the value of curative medicine (as do the profits of pharmaceutical companies). Unfortunately, there are observations which temper that confidence. During a doctors' strike in Israel some decades ago, the death rate fell, with no compensatory blip before or after the time during which everything but emergency medicine and the refilling of medications was suspended. These data do not lead to a recommendation that the treatment of illness be suspended. They do suggest a seemly humility both among those of us who practice medicine and those in the scientific community who provide the information on which those of us who have practiced medicine claim to have based our recommendations.

Medical practitioners should and will continue to adapt the information made available by science to improve their treatment of patients. Scientists using the experimental method and drawing compelling conclusions from their data will continue to add to the body of definitive information on which medical practicitoners can draw. This volume concentrates on the solid scientific studies of neurochemical mechanisms. It leaves to clinical textbooks the discussion of the classification (nosology) of diseases and the discussion of biological abnormalities in those often vaguely defined illnesses, as well as discussions of the interventions now believed to be appropriate.

References

Alvarez LM, Castillo ST, Perez ZJA, Vargas RI, Sepulveda GR, Zuniga CMA (1995) Activity of aryl sulfatase a enzyme in patients with schizophrenic disorders. Rev Invest Clin 47: 387–392

Argov Z&Navon R (1984) Ann Neurol 16:14–20

Candler PM, Dale RC, Griffin S, Church AJ, Wait R, Chapman MD, Keir G, Giovannoni G, Rees JH (2006 Apr) Post-streptococcal opsoclonus-myoclonus syndrome associated with anti-neuroleukin antibodies. J Neurol Neurosurg Psychiatry 77(4):507–512

Caspi A, Moffit TE, Cannon M, McClay J, Murray R, Harrington H et al (2005) Moderation of the effect of adolescent-onset cannabis use on adult psychosis by a functional polymorphism in the catechol-O-methyltransferase gene: longitudinal evidence of a gene X-environment interaction. Biol Psychiatry 57:1117–1127

Craddock N, O'Donovan MC, Owen MJ (2006) Genes for schizopohrenia and bipolar disorder? Implications for psychiatric nosology. Schizophr Bull 32:9–16

Drory VE et al (2003) Muscle Nerve 28:109–112

Duenas AM, Gold R, Giunti P (2006) Molecular pathogenesis of spinocerebellar ataxias. Brain 129(Pt 6):1357–1370, Epub 2006 Apr 13

Galbraith DA, Gordon BA, Feleki V, Gordon N, Cooper AJ (1989) Metachromatic leukodystrophy (MLD) in hospitalized adult schizophrenic patients resistant to drug treatment. Can J Psychiatry 34:299–302

Goodman AB (1994) Medical conditions in Ashkenazi schizophrenic pedigrees. Activity of aryl sulfatase a enzyme in patients with schizophrenic disorders. Schizophr Bull 20:507–517

Gravel RA et al (1995) In: Scriver CR et al (eds) Metabolic and molecular basis of inherited disease, 7th edn. McGraw-Hill, New York, NY, pp 2839–2882

Green EK, Raybould R, MacGregor S, Gordon-Smith K, Heron J, Hyde S et al (2005) Operation of the schizophrenia susceptibility gene, neuregulin 1, across traditional diagnostic bounderies to increase risk for bipolar disorder. Arch Gen Psychiat 62:642–648

Hammer MB (1998) Psychosomatics 39:446–448

Hendriksz CJ et al (2004) J Inherit Metab Dis 27:241–249

Herska M, Moscovich DG, Kalian M, Gottlieb D, Bach G (1987 Mar) Aryl sulfatase a deficiency in psychiatric and neurologic patients. Am J Med Genet 26(3):629–635

Iype M et al (2006) Ann Ind Acad Neurol 9:110–112

Kendell RE (1975) The role of diagnosis in psychiatry. Blackwell, London, See also: Kendell RE (2002) The distinction between personality disorder and mental illness. Br J Psychiat 180: 110–115

Leonard S, Freedman R (2006) Genetics of chromosome 15q13-q14 in schizophrenia. Biol Psychiatry 60:115–122

Lichtenberg P et al (1988) Br J Psychiat 153:387–389

Lodi R, Tonon C, Calabrese V, Schapira AH (2006) Friedreich's ataxia: from disease mechanisms to therapeutic interventions. Antioxid Redox Signal 8:438–443

Meek D et al (1984) Ann Neurol 15:348–352

Navon R et al (1973) Am J Human Genet 25:287–292
O'Neill B et al (1978) Neurology 28:1117–1123
Oonk JGW et al (1979) Neurology 29:380–384
Porteus DJ, Thompson P, Brandon NJ, Millar JK (2006) The genetics and biology of DISC 1 – an emerging role in psychosis and cognition. Biol Psychiatry 60:123–131
Posner JB (2003 Sept) Immunology of paraneoplastic syndromes: overview. Ann N Y Acad Sci 998:178–186
Rapin I et al (1976) Arch Neurol 33:120–130
Renshaw PF et al (1992) Ann Neurol 31:342–344
Rosebush PI et al (1995) J Clin Psychiat 56:347–353
Skolnick EM (2006) Mechanisms of action of medicines for schizophrenia and bipolar illness: status and limitations. Biol Psychiatry 59:1039–1045

Molecular Mechanisms of Neuronal Death

Elena M. Ribe, Lianna Heidt, Nike Beaubier, and Carol M. Troy

Abstract Cellular homeostasis, maintenance of the balance of life and death at the cellular level, is essential for tissue integrity from development through senescence. During development of the nervous system programmmed cell death is responsible for establishing the number of neurons and shaping the nervous system. After development the majority of the postmitotic neurons should live for the life of the organism. Aberrant neuronal death occurs in neurodegenerative diseases and there is still no clear understanding of the mechanisms involved. In this chapter we discuss the molecules and pathways that regulate the life and death of cells and illustrate how these pathways are potentially involved in neurodegenerative diseases. By understanding the molecular mechanisms that regulate cell death we can then begin to identify which pathways are dysregulated in neurodegenerative diseases.

Keywords Neuron death · Caspase · IAP · Smac/DIABLO · TNF · Fas · PIDD · RAIDD · Neurodegenerative disease

Contents

C.M. Troy (✉)
Departments of Pathology and Neurology, Taub Center for the Study of Alzheimer's Disease and the Aging Brain, Columbia University Medical Center, New York, NY 10032, USA
e-mail: cmt2@columbia.edu

J.P. Blass (ed.), *Neurochemical Mechanisms in Disease*,
Advances in Neurobiology 1, DOI 10.1007/978-1-4419-7104-3_2,

1 Introduction

Cellular homeostasis, that is, the balance of life and death at the cellular level, is a requirement for maintaining the integrity of tissues from development through maturity. During development large numbers of superfluous cells are removed by an active process termed *programmed cell death* (PCD) (Burek and Oppenheim, 1996). It was through the genetic studies of developmental death in *C. elegans* that the genes required for PCD were identified (Hengartner and Horvitz, 1994). These gene families are highly conserved from *C. elegans* to humans. Often PCD is used interchangeably with apoptosis; this is not accurate, as PCD refers specifically to developmental death. *Apoptosis* and *necrosis* were described as morphologically distinct processes (Kerr et al., 1972). In apoptosis cellular changes include cell membrane blebbing, cell shrinkage, chromatin condensation, and nuclear fragmentation (Kerr et al., 1972). Eventually the cell disintegrates, generating the so-called apoptotic bodies that will be engulfed via phagocytosis by nearby cells, thus avoiding an inflammatory response in the surrounding tissue. This lack of inflammatory response allows apoptosis to occur without damaging neighboring healthy cells. In contrast, necrosis, in which the cell suffers a major insult leading to rapid swelling, subsequent rupture of the plasma membrane and release of the intracellular contents into the surrounding cellular environment causes a strong inflammatory response. Apoptosis maintains physiological balance and its dysregulation results in pathological conditions, such as neurodegenerative diseases, cancer, and autoimmune disorders. Another mode of cell death is *autophagy*, which is characterized by the formation of large autophagic vacuoles and little inflammation (Levine and Yuan, 2005). Most autophagy does not lead to cell death but is a mechanism by which intracellular components are recycled (Yoshimori, 2007). Although the classification of the different forms of cell death seems to be clear, the boundaries are not so well defined in vivo and crosstalk can occur (Lockshin and Zakeri, 2004). With this idea in mind, we discuss the pathways of apoptotic neuronal death that occur in acute and chronic pathological conditions such as Alzheimer's disease, Parkinson's disease, Huntington's disease, stroke/ischemic disease, and motor neuron diseases.

2 Caspases: Key Players in Apoptosis

Caspases are the main proteins involved in the execution of apoptosis (Troy and Salvesen, 2002). They are a family of cysteine aspartate proteases with a conserved QACXG motif at the active site. To date, 13 mammalian caspases have been identified (Lamkanfi et al., 2002). Synthesized as inactive precursors or zymogens, they can be classified based on their structure, mode of activation, cleavage specificity, and function. According to their function caspases can be subdivided into three groups, shown in Fig. 1:

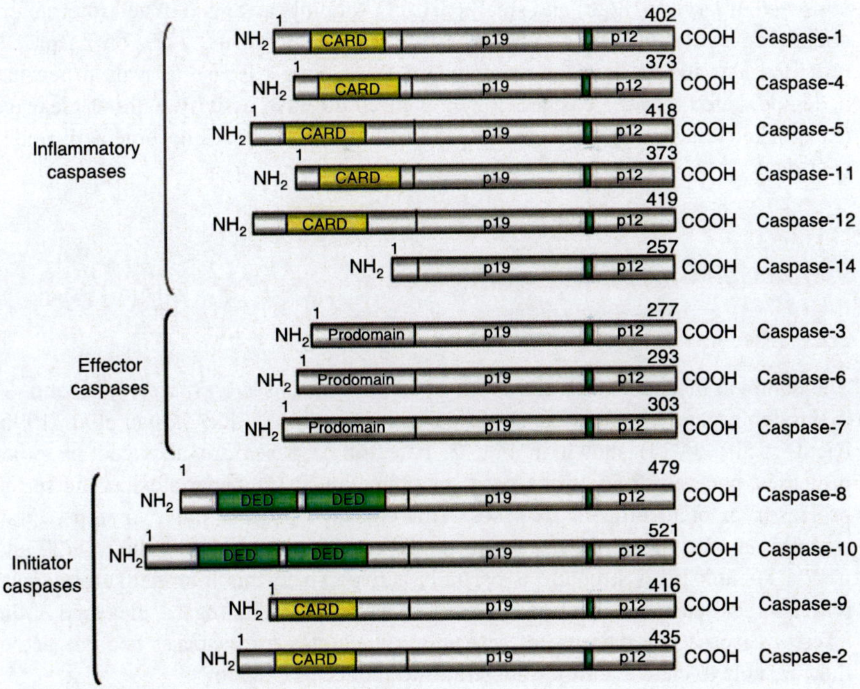

Fig. 1 Mammalian caspases

(1) Inflammatory caspases: caspase-1, -4, -5, -11, -12, and -14.
(2) Initiator caspases: involved in the apoptotic process. These caspases, also known as apical caspases, are structurally characterized by the presence of a long prodomain at the N-terminal region containing different protein–protein interaction motifs such as death effector domain (DED) found in caspase-8 and-10 or caspase recruitment domain (CARD), present in caspase-2 and -9. Via

these domains, caspases establish homotypic interactions with specific adaptor molecules.

(3) Effector caspases: this group of proteases cleave cellular substrates during apoptosis. Due to their function in the apoptotic paradigm they are also known as executioner caspases. They are characterized by the presence of short prodomains. This group contains caspase-3, -6, and -7.

2.1 Caspase Activation

Synthesized as inactive precursors or zymogens, caspases require activation to execute apoptosis (Nicholson and Thornberry, 1997). Early studies suggested that all caspases required proteolytic cleavage for their activation and that mature caspases consisted of large (p18/20) and small (p10/12) subunits arranged in heterotetramers containing two active sites (Walker et al., 1994). However, work on caspase-9 provided new insight into the mechanisms underlying caspase activation because it demonstrated that the caspase-9 zymogen could have activity without cleavage (Stennicke et al., 1999). Thus, the question, "How do caspases become activated?" is critical.

2.2 Mechanisms of Activation

2.2.1 Effector Caspases

The common mechanism of activation of effector caspases (caspases-3, -6, and -7) is through proteolytic cleavage at critical aspartic acid residues (Quan et al., 1996; Riedl et al., 2001a) shown in Fig. 2. Effector caspases are activated by other proteases, generally initiator caspases or granzyme B (an aspartate-specific serine protease), or other effector caspases. This cleavage process has two steps. First, a molecule of zymogen is cleaved at the linker region generating the p18/20 and p10/12 subunits; this structure is partially active. Then, this intermediate interacts with another heterodimer forming the active caspase. In this regard, cleavage of the effector caspase is a measure of activation. Once effector caspases become active they are able to cleave multiple substrates to induce cell death.

2.2.2 Initiator Caspases

Inactive initiator caspases exist as monomers and activation is achieved by proximity-induced dimerization (Boatright and Salvesen, 2003) shown in Fig. 2. Adaptor proteins, which interact with the prodomains of the caspases, bring the caspase molecules into proximity. When initiator caspases dimerize, they undergo conformational changes that result in an active enzyme without a requirement for

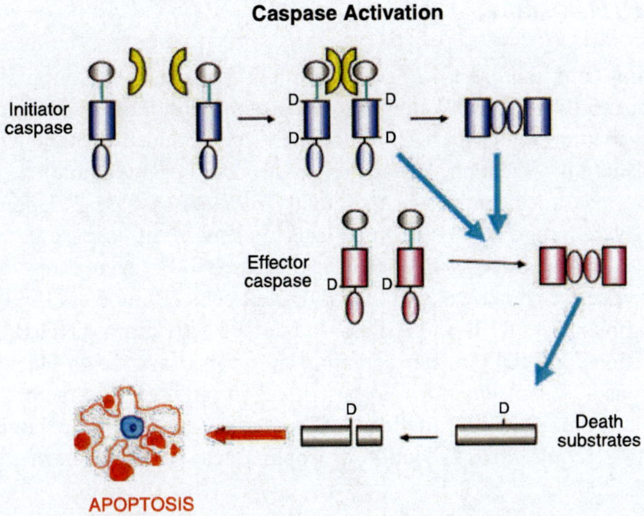

Fig. 2 Caspase activation

cleavage. Thus, cleavage cannot be used as a measure of activation when studying initiator caspases. Because caspase-9 is an initiator caspase that does not require cleavage for its activation, some studies have used the cleavage of caspase-3 as a surrogate measure for caspase-9 activation. However, caspase-8 can also cleave caspase-3. Thus, caspase-3 cleavage/activity is not a specific measurement of caspase-9 activation.

2.3 The Apoptosome

The most widely studied model is caspase-9 activation. Release of cytochrome c from the mitochondria into the cytosol promotes the assembly of the apoptosome, a complex composed of cytochrome c, Apaf-1 (Apoptosis protease-activating factor-1), and caspase-9. The presence of Apaf-1, which is the specific adaptor for caspase-9, recruits procaspase-9 to the apoptosome resulting in caspase-9 activation (Bao and Shi, 2007; Riedl and Salvesen, 2007).

2.4 The DISC

A similar process occurs for caspase-8 activation. In this case, oligomerization of the death adaptor protein Fas-Associated Death Domain (FADD) recruits procaspase-8 into the death-inducing signalling complex (DISC) allowing caspase-8 dimerization and subsequent activation (Shi, 2006).

2.5 The PIDDosome

An activating complex has also been identified for caspase-2, containing RAIDD (RIP-associated ICH-1/CED-3 homologous protein with a death domain), the specific death adaptor for caspase-2, and PIDD (p53-induced protein with a death domain) (Tinel and Tschopp, 2004; Park et al., 2007). This complex, termed the PIDDosome, has not been shown to actually mediate caspase-2 dependent death but rather, overexpression of PIDD can lead to cleavage of caspase-2 which is not necessarily an indication of activation (Tinel et al., 2007). Overexpression of PIDD does lead to death that is blocked in RAIDD-null cells (Berube et al., 2005). PIDD can also complex with RIP1 and NEMO and induce activation of NFκB, suggesting a dual function for PIDD in the regulation of survival and death (Janssens et al., 2005). Caspase-2 has been shown to be critical for both trophic factor deprivation and β-amyloid mediated neuronal death (Troy et al., 2000, 2001), shown in Fig. 3 and RAIDD is required for execution of trophic factor deprivation mediated death (Wang et al., 2006).

Once dimerized in the activating complexes there is often autocleavage of the caspase which, for caspase-2 and -8, has been shown to enhance caspase activity (Chang et al., 2003; Baliga et al., 2004). As death proceeds and effector caspases are activated there is subsequent cleavage of initiator caspases. This cleavage may lead to further enhancement of caspase activity, as in the case of caspase-9 where the initial autocleavage of caspase-9 to the p37 fragment allows XIAP to bind and inhibit activity, and the subsequent cleavage by caspase-3 to the p35 fragment relieves the XIAP inhibition thus enhancing caspase-9 activity (Denault et al., 2007).

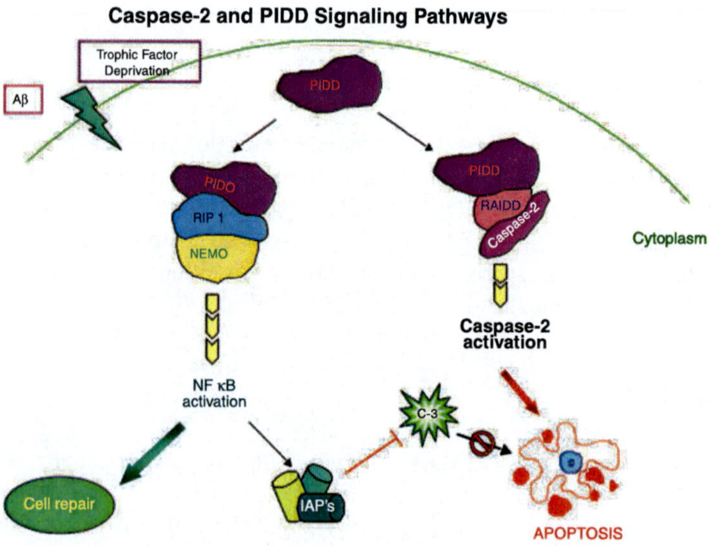

Fig. 3 Caspase-2 activation and PIDD signaling pathways

2.6 *The Inflammasome*

The activation of the inflammatory caspases uses a mechanism resembling that of the initiator caspases. The presence of a complex, known as the inflammasome (Martinon et al., 2002), is required for activation of this set of proteases. The recruitment of caspases into this complex results in their activation. For caspase-1, the adaptor ASC (apoptosis-associated specklike protein containing a CARD) is critical in inflammasome formation in response to a variety of stimuli, whereas involvement of the adaptors Ipaf (ICE-protease-activating factor) and NALP3 is stimulus-dependent (Mariathasan, 2007).

3 Apoptotic Routes: Intrinsic and Extrinsic Pathways

Cells undergoing apoptosis take one of two major pathways: the death receptor (extrinsic) pathway, or the mitochondrial (intrinsic) pathway. Once the cell is dead, the cellular contents form the apoptotic bodies, which are cleared by phagocytosis in a process involving neighboring cells and/or macrophages. This intricate process is tightly regulated so that there is a fine balance between prosurvival and prodeath signals for each route in the apoptotic pathway.

3.1 *The Extrinsic or Receptor-Mediated Pathway*

The extrinsic or receptor-mediated pathway is activated when a death ligand binds to its specific receptor on the cell membrane surface. The main death receptors are all members of the tumor necrosis factor (TNF) superfamily of receptors, which includes TNFR, Fas, p75, and TRAIL. All these receptors are characterized by the presence of domains rich in cysteine, which mediate the binding between ligand and receptor. The receptors are synthesized as transmembrane homotrimers and when they bind to their specific death ligand a DISC is formed. This complex recruits death domain (DD)-containing adaptor proteins that interact with and recruit procaspase-8, leading to caspase-8 activation. Caspase-8 activation results in the cleavage and activation of downstream effector caspases which in turn cleave a plethora of substrates, ultimately leading to cell death. Caspase-10, present only in humans, is also activated in this way.

3.1.1 TNF Pathway

TNF is a proinflammatory cytokine produced mainly by macrophages. There are two main types of receptors, TNF-R1 and TNF-R2. TNF-R2 is primarily found in the immune system and is activated by membrane-bound TNF (Wajant et al., 2003). However, TNF-R1, which is ubiquitously expressed, can be activated by both membrane-bound and soluble TNF. When TNF binds to the TNF receptor, TRADD (TNFRSF1A-associated via the death domain) is able to establish

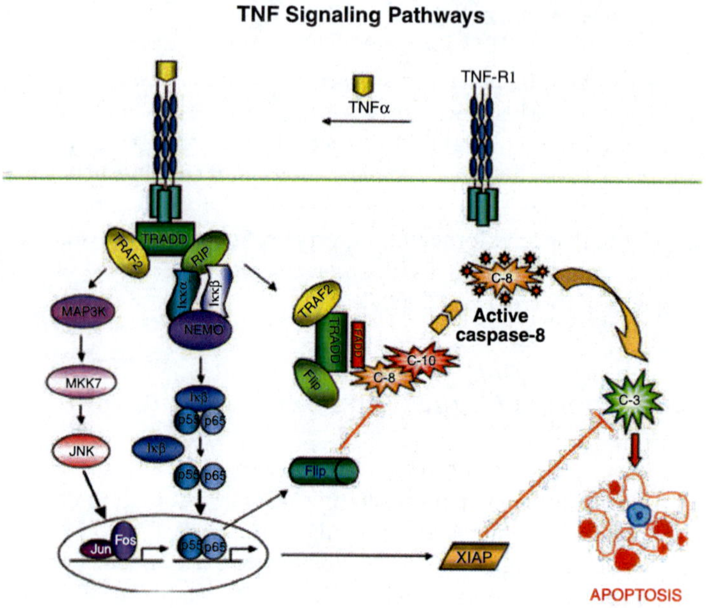

Fig. 4 TNF signaling pathways

homophilic interaction with the DD of the TNF receptor (Hsu et al., 1995), shown in Fig. 4. The binding of TRADD to the TNF receptor–ligand complex facilitates the subsequent binding of TRAF2 (TNF receptor-associated factor 2) and RIP1 (receptor-interacting kinase-1), a DD-containing serine threonine kinase.

When TRAF2 and RIP1 bind to the complex, two sequential pathways are activated: the NFκB pathway and the activated caspase-8 pathway. In the first step of the NFκB pathway, TNF activates the IκBα pathway in a process that depends on the degradation of the inhibitor IκB by the proteasome. The Iκκ complex (IκB kinase) mediates the phosphorylation of the inhibitor IκB. The Iκκ complex is formed by two related IκB kinases, IκBα and IκBβ, and NFκB essential modulator (NEMO), a regulatory protein also known as IκBγ. The roles of TRAF2 and RIP in the Iκκ complex are recruitment and stabilization, respectively (Devin et al., 2003).

In nonstimulated cells, the Iκκ complex remains inactive in the cytoplasm because of the binding of the IκB inhibitor. However, when the complex is recruited to the TNF receptor it becomes active and it is able to phosphorylate the IκB inhibitor which is in turn degraded via the proteasome (Aggarwal, 2003). The degradation of the inhibitor frees the NFκB complex to translocate to the nucleus where it activates transcription of several genes, including XIAP, c-IAP1, and c-IAP2 (Stehlik et al., 1998; Wajant, 2003).

Thus, TNF induces a strong prosurvival signal secondary to NFκB activation. This is the main difference between TNF and Fas or TRAIL, which only mediate apoptosis. TNF can have cytotoxic effects, but only when NFκB activation is inhibited. In the second (caspase-8) part of the pathway, TRADD, which is bound to the TNF receptor, acts as a platform allowing the complex to interact with Fas-associated death domain (Yeh et al., 1998; Thorburn, 2004). Once FADD is bound to the complex, it recruits caspase-8 to form a cytoplasmic DISC protein complex that finally ends with the death of the cell (Micheau and Tschopp, 2003). The TNF receptor can also mediate an alternative pathway through the recruitment of RAIDD, which facilitates the binding of RIP1, establishing homophilic interactions via the DD found in both proteins (Duan and Dixit, 1997). This interaction mediates the recruitment of caspase-2 which in turn leads to apoptosis (Kim et al., 2000). A complex of caspase-2 with TRAF2 and RIP1 has been found that induces NFκB activation independent of caspase-2 enzymatic activity (Lamkanfi et al., 2005).

3.1.2 FAS Pathway

Fas plays a key role in the regulation of apoptosis. The Fas–Fas ligand (FasL) interaction has a special relevance because the initial characterization of the DISC formation was discovered while studying this interaction (Kischkel et al., 1995). When Fas ligand binds to its receptor, Fas, also known as CD95, a structural change takes place facilitating the trimerization of the receptor, which then mediates the recruitment of DD-containing proteins, in this case, FADD. FADD is a molecule with a double nature because it not only contains a DD but also a DED through which it establishes interactions with procaspase-8 (Chinnaiyan et al., 1995). Once procaspase-8 is recruited into the DISC complex, it is autoproteolytically processed by proximity-induced dimerization, which enhances the enzymatic activity (Fig. 5). Another study shows that the DISC complex can also contain caspase-10 but that caspase-10 cannot completely replace the caspase-8 function in apoptosis (Sprick et al., 2002). It appears that caspase-8 and -10 may have some nonredundant

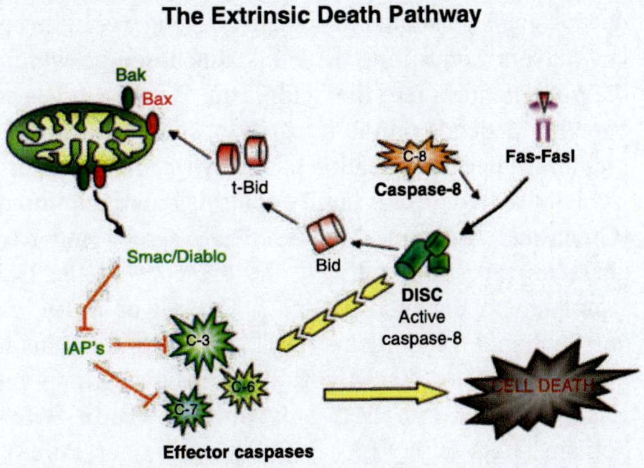

Fig. 5 The extrinsic death pathway

functions. People lacking caspase-10 can develop autoimmunolymphoproliferative syndrome II (Rieux-Laucat et al., 2003).

Modulators of caspase-8 dependent apoptosis, specifically FLIP, have also been identified in the DISC. FLIP is synthesized in two isoforms, short and long. Both have DEDs in tandem and high homology to the N-terminus of caspase-8 (Irmler et al., 1997). When FLIP is recruited into the DISC, it disrupts the complex and acts as an inhibitor of caspase-8 so that caspase-8 cannot become active. This prevents the cell from undergoing apoptosis. However, FLIP can also activate caspase-8 and caspase-10 by forming heterodimers (Boatright et al., 2004).

3.1.3 TRAIL Pathway

Although its physiological role is not completely understood, TRAIL plays a role in apoptosis in blood cells and in the immune system (Thomas and Hersey, 1998). Five TRAIL receptors have been described, which can be divided in two groups: death-inducing receptors and death-inhibitory receptors. As their own names indicate, the first group is actively involved in the apoptotic response and the second group has a defective cytoplasmic DD so they function as competitive inhibitors when they bind to TRAIL. The cascade involving TRAIL is similar to the one induced by Fas. TRAIL binds to its receptor initiating DISC formation and recruitment of caspases-8 and -10 and FLIP. DISC formation generates the active conformation of caspase-8 which in turn activates caspase-3 resulting in cell death. Although there can be an interconnection between this main pathway and the NFκB pathway, TRAIL is a weak inducer of the latter. As with the TNF receptor-mediated pathway, the activation of NFκB is mediated by RIP1 and TRAF2 (Lin et al., 2000). However, the prosurvival signal is completely masked by the strong apoptotic response.

3.2 The Intrinsic Pathway

The intrinsic pathway is the death pathway followed when apoptosis is triggered by death signals generated inside the cell (Fig. 6). In this pathway, mitochondria are the key players, controlling the cell status based on which molecules are released from the mitochondria into the cytoplasm. Because release of molecules from the mitochondria depends on the integrity of the mitochondrial membranes, mitochondrial membrane permeabilization has a key role in the origin and progression of the intrinsic pathway. The Bcl-2 family controls the regulation of mitochondrial permeability (Green and Amarante-Mendes, 1998; Green and Kroemer, 2004). This family is characterized structurally by the presence of the Bcl-2 homology (BH) domain. Family members such as Bcl-2, Bcl-Xl, or Bcl-w can have antiapoptotic effects and contain 4 BH domains (BH1, 2, 3, 4) and a transmembrane domain.

Other proteins from the Bcl-2 family are proapoptotic. The proapoptotic group is subclassified into BH3-only proteins (Bid), BH3-only with a transmembrane domain (Bad, Bim, Bik, Bmf, Hrk, Nox, or Puma), and multi-BH (BH1, 2, 3)

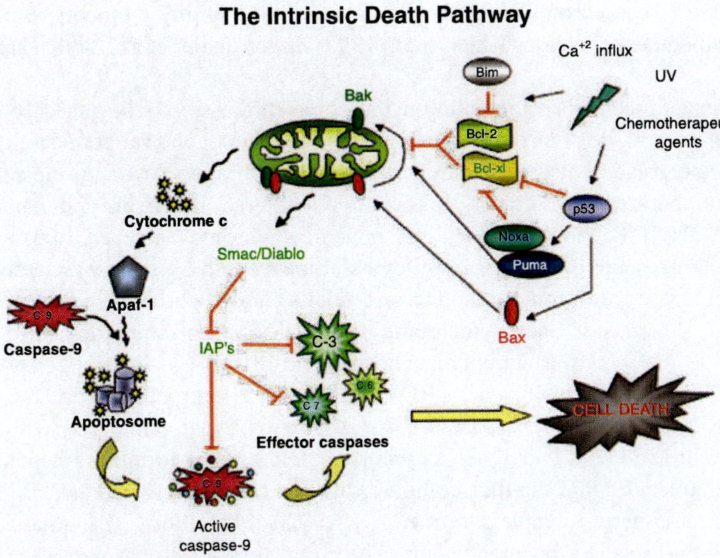

Fig. 6 The intrinsic death pathway

domains with a transmembrane domain (Bax, Bak, Bok) (Adams and Cory, 1998). The BH3-only proteins trigger apoptosis induced by the lack of trophic support or intracellular damage and thus work as damage sensors in the cell (Cheng et al., 2001). Bcl-2, the prototype family member, is found in perinuclear membranes, mitochondria, and endoplasmic reticulum (Korsmeyer et al., 1995). It has important functions in controlling both calcium and mitochondrial membrane homeostasis (Danial and Korsmeyer, 2004).

Following intracellular damage, members of the Bcl-2 family undergo oligomerization and attach to the outer mitochondrial membrane. A good example is the case of Bax and Bak. In healthy cells, Bax is present as a monomer in the cytoplasm but during the apoptotic cascade, it oligomerizes and translocates to the outer mitochondrial membrane. Bak localization seems to be mitochondrial, even in healthy cells, but undergoes conformational changes during apoptosis leading to its aggregation (Danial and Korsmeyer, 2004). Once these proteins are inserted into the outer mitochondrial membrane and become oligomerized, the mitochondrial membrane is disrupted releasing intermembrane proteins, such as cytochrome c, into the cytosol, which compromises cell viability. The involvement of cytochrome c in the apoptotic cascade was initially surprising because cytochrome c is known as an essential component of the respiratory chain. Thus, cytochrome c has a dual role. It promotes the generation of ATP and cell viability while inside the mitochondria, and, when outside the mitochondrial space in the cytosol it promotes cell death. Cytochrome c is found in the mitochondrial interspace and its release is controlled by members of the Bcl-2 family (Green and Amarante-Mendes, 1998; Chipuk et al., 2006). In this context, the antiapoptotic Bcl-2 family members, Bcl-2 and Bcl-XL, will prevent the release of cytochrome c whereas the proapoptotic family members, Bax, Bak, and Bid mediate its release (Kluck et al., 1997; Jurgensmeier et al., 1998; Luo et al., 2005).

The exact mechanism mediating cytochrome c release is still not fully understood. In general, it is believed that a change in the mitochondrial permeability precedes cytochrome c release. However, caspase activation and cytochrome c release can occur before detecting any mitochondrial alteration (Green and Amarante-Mendes, 1998). Because caspases can induce cytochrome c release, it also seems possible that a small initial leakage of cytochrome c could cause caspase activation, which in turn would promote the massive release of cytochrome c from the mitochondria. Either way, once cytochrome c is released into the cytoplasm it binds to Apaf-1, which is the mammalian homologue of the *C. elegans* CED-4 (Zou et al., 1997). Apaf-1 contains a CARD domain at its N-terminus that interacts with the CARD domain of procaspase-9 (Li et al., 1997). Apaf-1 interacts with dATP and cytochrome c and undergoes a conformational change forming a heptamer of APAF-1 molecules that can then complex with pro-caspase-9 (Zou et al., 1999).

This multimeric complex formed by dATP, cytochrome c, Apaf-1, and procaspase-9 is called the apoptosome. The recruitment of procaspase-9 via Apaf-1 into the apoptosome allows the activation of caspase-9 by proximity-induced dimerization. The active caspase-9 is now able to cleave downstream effector caspase-3, -6, and-7, which then cleave myriad cellular substrates involved in DNA metabolism, cytoskeletal and structural proteins, and regulators of the cell cycle, all of which compromise cell integrity and lead to cell death when disrupted (Li et al., 1997). However, cytochrome c is not the only molecule released from mitochondria during the execution of the intrinsic pathway. Smac/Diablo is also released from mitochondria into the cytoplasmic space where it binds to the BIR3 of XIAP, acts as an IAP antagonist and ultimately leads to the activation of caspase-9 and -3 (Chai et al., 2000; Verhagen et al., 2000). Omi/HtrA2 is also released from mitochondria during apoptosis and although it functions, as does Smac/DIABLO, as a competitive inhibitor of the IAPs, it seems to be a more potent inhibitor because Omi/HtrA2 not only binds to and inactivates the IAPs but can also proteolytically process them (Yang et al., 2003). Other pro-apoptotic molecules are released from the mitochondria and although their final consequences are cell disruption and death, these effects are generally considered to be caspase-independent.

4 Natural Inhibitors of Caspase Activity

4.1 The Inhibitor of Apoptosis Proteins

Caspases kill cells by cleaving a broad spectrum of cellular substrates. To ensure that the death pathway is not accidentally activated, caspase activity must be carefully regulated to prevent aberrant caspase activation. Some members of the inhibitor of apoptosis protein (IAP) family can suppress caspase activity thus avoiding unwanted

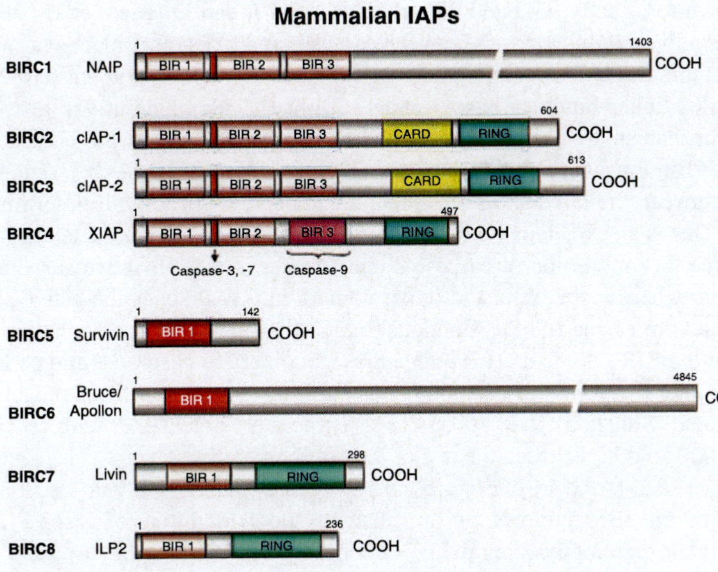

Fig. 7 Mammalian IAPs

apoptosis (Prunell and Troy, 2004). IAPs are phylogenetically highly conserved from *c. elegans* to mammals. There are eight human genes identified that belong to the IAP family (Fig. 7) (Deveraux and Reed, 1999): neuronal-apoptosis-inhibitory protein (NAIP or BIRC1), c-IAP1 (BIRC2), c-IAP2 (BIRC3), XIAP (BIRC4), survivin (BIRC5), Apollon (BRUCE or BIRC6), melanoma-associated IAP (Livin or BIRC7), and hILP-2 (TS-IAP or BIRC8). This family of proteins is characterized by the baculovirus IAP repeat (BIR) domain. The BIR is a 65-amino-acid domain with a high cysteine and histidine content.

There are two types of BIR domains (Salvesen and Duckett, 2002). Type I binds to and inhibits caspases. Type II also binds to caspases, and in addition functions in the cell cycle. The type II BIR domains are found in two mammal IAPs, survivin (BIRC5) and BIRC6. Most of the IAPs also contain a RING domain at the carboxy-terminus region which behaves as an E3 ubiquitin ligase. The RING domain adds ubiquitin residues to target proteins so they will be degraded by the proteasome. IAP-mediated protein ubiquitination has a crucial role in the regulation of apoptosis because it can target the IAP itself and also enhance the antiapoptotic effect by targeting proapoptotic molecules for degradation. In adddition to the RING domain, c-IAP1 and c-IAP2 also contain a CARD domain located in the C-terminal region between the RING domain and BIR3. The function of CARD domains in these two IAPs is not yet known. Usually CARD motifs interact with other CARD-containing proteins, but the classical location for these protein–protein interactions is the N-terminus, not the middle of the structure as in the case of the IAPs.

The best-studied IAP is XIAP, which is the most potent IAP. It is an ubiquitously expressed 56 kDa protein with 3 BIR domains and one RING domain. XIAP has been shown to directly bind and inhibit caspase-3, -7, and -9 (Riedl et al., 2001b). The protein–protein interactions between caspases and IAPs takes place via specific regions within the IAP structure. XIAP–BIR3 domain interacts with caspase-9 and XIAP–BIR2-linker binds caspase-3 and -7. Both BIR domains utilize a two-site binding mechanism to inhibit caspases (Scott et al., 2005). One site has been defined as the IAP-Binding Motif (IBM)–interacting groove. When caspase-3, -7, and -9 are cleaved between the large and small subunits, the new small subunit N-terminus is an IBM. This is an exosite, a functionally important site outside of the active site of the enzyme. For inhibition of caspase-3 and -7 there is also an active-site directed interaction, where a stretch of the linker domain of XIAP spans the active site of the caspase. For caspase-9, the functional inhibitory interaction is via a helix found right after the BIR3 domain. This interaction monomerizes caspase-9 and collapses the active site. Because dimerization is essential for caspase-9 activity the enzyme is inactivated (Shiozaki et al., 2003). XIAP is the most potent IAP with efficiency 100- to 1000-fold higher than the rest of the family members.

c-IAP1 and c-IAP2 are the closest paralogues of XIAP and can also bind to caspases by the IBM grooves but are relatively poor inhibitors of caspase activity. The linker region preceding the BIR2 is not a good inhibitor of caspase-3 or -7 (Eckelman and Salvesen, 2006). The BIR3 domains of cIAPs have only one of the four dimer interface–interacting residues required to inactivate caspase-9 and neither inhibits caspase-9 (Eckelman et al., 2006). IAPs can also be cleaved by caspases that may affect their activity. When XIAP is cleaved between BIR2 and BIR3, the BIR3-RING fragment becomes a more potent inhibitor of caspase-9 activity than the whole molecule (Deveraux et al., 1999). The N-t cleaved fragment of XIAP still has the ability to inhibit caspase-3 and-7, but to a much lesser extent than full-length XIAP.

IAPs have been extensively studied in the context of cancer because of the IAPs' ability to regulate members of the NFκB family and because NFκB activation seems to upregulate expression of IAPs (Stehlik et al., 1998). More recently, IAPs have been implicated in neurodegenerative diseases. In sympathetic neurons deprived of trophic factors XIAP inhibits caspase-3 activity (Troy et al., 2001) (Fig. 3). In motor neurons damaged by sciatic nerve axotomy, there is a significant decrease in the levels of endogenous XIAP and NAIP (Perrelet et al., 2004). Expression of NAIP is increased in AD, whereas that of XIAP is decreased. Treatment with glial-derived neurotrophic factor (GDNF) rescues this effect and promotes motor neuron survival (Perrelet et al., 2002). Inhibition of XIAP or NAIP blocks the neuroprotective effect of GDNF, pointing out a direct effect of IAP activity and motor neuron degeneration. Similar results have been found in the case of ischemic injury where overexpression of XIAP reduced the infarct size, the number of cells exhibiting apoptotic phenotype, and improved neurological activity (Xu et al., 1999). The fact that IAPs are endogenous inhibitors of caspase activity makes them a good therapeutic target for diseases characterized by excessive or premature cell death, such as stroke, AD, PD, and other neurodegenerative disorders. IAPs may also participate in physiological regulation of normal nervous system function. XIAP regulates activated caspase-3 in a songbird model of learning (Huesmann and Clayton, 2006).

4.2 *Natural Inhibitors of the Inhibitor of Apoptosis Proteins: IAP Antagonists*

After discovering that IAPs bind to and inhibit caspase activity, several studies focused on the isolation of endogenous regulators of IAP activity (Crook et al., 1993; Birnbaum et al., 1994). The first molecule identified was the second mitochondria-derived activator of caspases (Smac), also known as DIABLO, an IAP binding protein that in healthy cells is found in mitochondria (Du et al., 2000; Verhagen et al., 2000). This protein contains 239 amino acids. After stimulation, Smac/DIABLO translocates from the mitochondria to the cytosol where it binds to and blocks XIAP activity. This binding is associated with four hydrophobic residues, Ala-Val-Pro-Ile, at the Smac/DIABLO N-terminus which form the IAP-binding motif (Shi, 2002). Smac/DIABLO binds to the BIR3 domain of XIAP at the same site as caspase-9 (Liu et al., 2000; Wu et al., 2000). Therefore, the interaction of Smac/DIABLO with XIAP displaces caspase-9, thus abrogating the inhibitory effect of XIAP on caspase-9 activity.

Smac/DIABLO is not the only regulator of IAP activity. Several studies in mammalian cells have demonstrated the presence of additional molecules that suppress IAP activity in a similar fashion to Smac/DIABLO. The best-studied example is Omi/HtrA2 (Suzuki et al., 2001; Hegde et al., 2002; Martins et al., 2002; van Loo et al., 2002). This protein exhibits, as does Smac/DIABLO, mitochondrial localization with cytoplasmic release upon stimulation.

Apart from IAPs, there are several nonmammalian regulators of caspases, which are active-site specific inhibitors (Callus and Vaux, 2007). One example is a serpin from the cowpox virus, cytokine response modifier A (crmA). CrmA forms a covalent complex with the initiator caspase-1 and -8 resulting in irreversible inhibition of these caspases. It also inhibits caspase-6 but less efficiently (Dobo et al., 2006). The baculoviral protein p35 is a broad spectrum caspase inhibitor that irreversibly inactivates caspases (Bump et al., 1995; Fisher et al., 1999).

4.3 *Phosphorylation*

Phosphorylation is the major form of posttranslational modification. It is important to note that caspase activity differs from caspase activation. Activation refers to the conformational changes that rearrange the caspase molecule leading to the active enzyme. Caspase activity is defined as the ability of a caspase to cleave substrates. Caspase phosphorylation is able to modulate caspase activity. A clear example is the case of human caspase-9 which can be phosphorylated at a consensus sequence by Akt, a serine-threonine kinase implicated in apoptosis suppression (Cardone et al., 1998). Caspase-9 phosphorylation by Akt induces a modification in the caspase structure rendering it unable to form the tetramer required for activity. There is also evidence that phosphorylation may regulate caspase-2 activity (Nutt et al., 2005; Shin et al., 2005).

4.4 Nitrosylation

Caspases can also be modified by nitrosylation. S-nitrosylation of the active site cysteine has been shown to inactivate multiple caspases (Mannick et al., 2001). The physiological relevance of this mechanism is not yet fully understood.

5 ER-Stress

Although mitochondria are the main organelles involved in the intrinsic apoptotic pathway, the endoplasmic reticulum (ER) also plays an important role. The ER is the biggest intracellular reservoir of Ca^{2+} and Ca^{2+} functions as a second messenger interconnecting the mitochondrial pathway with the ER. When a small amount of cytochrome c is released from the mitochondria into the cytosol, there is uptake by the ER which in turn responds by releasing Ca^{2+}. This Ca^{2+} in turn disrupts the resting mitochondrial membrane potential and causes a massive release of cytochrome c that activates caspases and leads to cell death. This apoptotic activation via Ca^{2+} efflux from the ER seems to be important in disorders such as AD and stroke (Rao et al., 2004).

The main function of the ER is to ensure that only those proteins folded properly will be transported through the multivesicular secretory pathway. This property is extremely important in the case of neurodegenerative diseases because most are characterized by the presence of inclusion bodies formed from aberrantly folded protein. Amyloid plaques (β-amyloid aggregates) and neurofibrillary tangles (intracellular inclusions of hyperphosphorylated tau) are the sine qua non of Alzheimer's disease (AD), as are Lewy bodies (α-synuclein inclusions) in Parkinson's disease (PD), Pick's bodies (tau inclusions) in frontotemporal lobar degeneration, and Hirano bodies, cytoplasmatic protein aggregates of actin and actin-associated proteins, which are present in several neurodegenerative disorders such as AD and Creutzfeldt–Jacob disease. When the ER is damaged it cannot correctly regulate the accumulation of unfolded or misfolded proteins. This leads to a reduction in protein synthesis to prevent accumulation and activation of the chaperones that reside in the ER so they can contribute to the proper folding of newly synthesised proteins. There is also an increase in the degradation rate (Breckenridge et al., 2003). However, if these compensatory changes are inadequate, cell integrity will become compromised, leading to death.

Increasing evidence suggests that members of the Bcl-2 family may act not only at the mitochondrial levels but also at the ER level. There is work that suggests that Bak and Bax are involved in controlling Ca^{2+} homeostasis in the ER because double knock-out mice for Bax and Bak exhibit impaired Ca^{2+} efflux from the ER and uptake by the mitochondria; this is correlated with low levels of apoptotic cell death (Nutt et al., 2002b, a). The relevance of these data to human neurodegenerative disorders is not yet clear because so far only caspase-12 has been reported to become activated after ER stress-induced apoptosis. There is evidence showing that both Bax and Bak are required in order to activate caspase-12 (Scorrano et al., 2003; Zong et al., 2003). In this context, both Bcl-2 family members would promote Ca^{2+} efflux from the ER, which in turn would permeabilize the outer mitochondrial membrane. Caspase-12 would then be released from the ER into cytosolic space. However, these studies were done in rodents, and there is no evidence that the caspase-12 protein is expressed in humans, although several studies suggest that human caspase-4 may have redundant functions with rodent caspase-12 (Hitomi et al., 2004).

6 Crosstalk Between the Intrinsic and Extrinsic Pathways

Although apoptosis proceeds through two major pathways in the cell that are initiated through the activation of different caspases forming different multimeric complexes, both pathways converge on the activation of downstream caspases (Danial and Korsmeyer, 2004). TRAIL employs the extrinsic pathway for triggering apoptosis, but there is also involvement of the mitochondrial pathway. In this paradigm, Smac/DIABLO is released from mitochondria, which block the inhibitory effect of XIAP on caspase-3 activity, resulting in the execution of apoptosis mediated by TRAIL. A similar situation occurs in the case of Fas-mediated apoptosis.

It is worth mentioning that in death receptor-mediated apoptosis, cells can be divided into two groups depending on the requirement for mitochondria to induce a complete apoptotic response. Type I cells do not require the mitochondrial pathway because the recruitment of procaspase-8 into the DISC complex is enough to fully activate caspase-8 which then activates effector caspases. However, Type II cells are characterized by an incomplete apoptotic response unless mitochondria are involved (Scaffidi et al., 1999). In this type of cell, efficient activation of effector caspases requires the mitochondrial amplification loop (Fig. 5). Caspase-8 cleaves cytosolic Bid, a BH3-only protein, which when cleaved to tBid is able to translocate to the mitochondria and trigger release of the proapoptotic factors cytochrome c and Smac/DIABLO (Li et al., 1998; Deng et al., 2002). The release of cytochrome c triggers apoptosome formation, subsequent caspase-9 activation, and finally the activation of effector caspases such as caspase-3.

Another positive feedback loop is established after DISC formation because this complex allows caspase-8 autoactivation which in turn cleaves downstream effector caspase-3. The cleavage of one caspase by another must be examined in relation to the timing of the ongoing cellular events in order to understand the relevance of these events. That is, as death proceeds, there is activation of initiator caspases—no cleavage necessary—leading to activation by cleavage of effector caspases. Once activated, the effector caspases may cleave initiator caspases, but this event is not necessary for activity of initiator caspases and may even decrease activity under certain conditions. Thus, as our knowledge of caspase activation increases, the prior assumptions about caspase cascades must be re-evaluated.

7 Neurodegenerative Diseases: An Example of Dysregulated Apoptosis

Because neurons are not normally replaced during the lifespan of an organism, they must possess very robust antiapoptotic mechanisms. If premature death of neurons does occur, it leads to irreversible neurodegenerative diseases. Important examples are Alzheimer's disease, Parkinson's disease, Huntington's disease (HD), and amyotrophic lateral sclerosis (ALS), all of which are characterized by the loss of neurons and the inability of the remaining ones to repopulate depleted areas of the brain. There is still debate about the mechanisms leading to neuronal death in these diseases, however, evidence is mounting that apoptosis is the major pathway (Jellinger and Stadelmann, 2001; Ayala-Grosso et al., 2002; Ugolini et al., 2003; Cribbs et al., 2004; Kermer et al., 2004). But the possibility that the apoptotic pathway coexists with necrosis cannot be excluded (Yuan et al., 2003). A major criticism of the apoptotic neuronal death hypothesis in neurodegenerative diseases is that most of the studies carried out in postmortem human tissue fail to show a significant number of neurons exhibiting the typical apoptotic phenotype. However, considering that the period of time required for neurons to die is on the order of a few hours and that brains from the end-stage of disease were losing neurons for decades, it actually seems reasonable that only a small number of neurons would be found to exhibit the morphological hallmarks of apoptotic death at any given time point.

Many reports correlate the increased expression of caspases and the presence of cleaved caspases with certain types of degenerative diseases but the causal link has not been shown. For example, the increased expression of caspase-1, -2, -3, -5, -6, -7, -8, and -9 have been reported in AD (Chan and Mattson, 1999; LeBlanc et al., 1999; Lu et al., 2000; Pompl et al., 2003), caspases-3, -8 and -9 in PD (Anglade et al., 1997), caspases-1 and -3 in ALS (Pasinelli et al., 1998), and caspases-1 and -8 in HD (Sanchez et al., 1999). Altered expression levels of receptors and death ligands suggest a role for death pathways in these disorders. It has been reported that an increase in Fas expression may be harmful to both neurons and glia, and has been associated with neurodegeneration in diseases such as AD, PD, ALS, and HD (Barone and Parsons, 2000; Ugolini et al., 2003).

Following a similar trend, an up-regulation in the expression levels of TNF receptors has been associated not only with those diseases already mentioned, but also with prion disease and ischemic brain injury. It is not simply the change in the expression levels of death ligands or receptors that is leading to increased apoptosis in these disorders. In certain cases, such as HD, spinocerebellar ataxia, and spinal muscular atrophy, the polyQ expansions introduced into the protein as a result of unstable CAG repeats in the target genes have a tendency to aggregate, forming proteinaceous inclusions in the nuclei of the affected cells ultimately leading to apoptosis (Martin et al., 1999). In these cases, the polyQ repeats are triggering ER stress because the aggregated proteins cannot be properly degraded. It has also has been reported that these aggregates can bind to procaspase-8 and that this binding leads to caspase-8 activation and subsequent cell death (Sanchez et al., 1999). Due to the increasing life expectancy in developed nations, the incidence of neurodegenerative diseases of aging is increasing exponentially. Because AD is the primary cause of dementia among the elderly population and ALS is the most common adult onset disorder of motor neurons, we take a global overview of the molecular mechanisms leading to neuronal cell death in both diseases.

7.1 Alzheimer's Disease (AD)

Alzheimer's disease is characterized by two main histopathological hallmarks, senile plaques, which are extracellular accumulations of amyloid beta peptide (Aβ), and neurofibrillary tangles (NFT), which are intracellular inclusions of hyperphosphorylated tau protein. Accompanying these features is a profound synaptic and neuronal loss in specific vulnerable brain regions including the hippocampus and entorhinal cortex (Terry et al., 1981; Small et al., 1997). Although the pathogenesis of AD is still being debated, it is generally agreed that Aβ peptide, especially the longer 42 amino acid isoform, which is generated by proteolytic cleavage from the amyloid precursor protein (APP), is the key player in the etiopathology of AD (Hardy and Selkoe, 2002). Because the amyloid hypothesis states that the Aβ peptide is highly neurotoxic, both NFT and neuronal death are considered secondary elements caused by an imbalance between Aβ production and clearance (Hardy and Higgins, 1992). This hypothesis has been revised because it originally postulated that the most toxic species were the fibrillar peptides, but new evidence suggests that the soluble oligomeric species may play a more critical role in the pathogenesis and/or progression of the disease inasmuch as they are able to block basal synaptic transmission, alter hippocampal long-term potentiation (LTP), and mediate neuronal death (Lannfelt et al., 1995; Larson et al., 1999; Walsh et al., 2002; Walsh and Selkoe, 2007).

Multiple studies have shown that several caspases are involved in Aβ-induced neuronal cell death (Gervais et al., 1999; Troy et al., 2000; Allen et al., 2001). Experimental evidence shows that the cytoplasmic tail of APP is cleaved by caspases-3, -6, -7, and -8, and that senile plaques as well as degenerating neurons are enriched in caspase-cleaved APP (Gervais et al., 1999; Zhang et al., 2000). Moreover, both mitochondrial and ER dysfunction play an essential role in mediating cell death induced by Aβ peptides (Pereira et al., 1999). Neurons from caspase-2 null and caspase-12 null mice are resistant to Aβ-mediated neuronal cell death (Nakagawa et al., 2000; Troy et al., 2000). Caspase-2 may be involved in mitochondrial permeabilization whereas caspase-12 acts at the level of the ER (Nakagawa et al., 2000; Zhang et al., 2005).

Recent data suggest that the link between amyloid pathology and NFT degeneration may reside at the level of caspases because Aβ can promote the pathological assembly of tau filaments in vitro by triggering the activation of caspases that can cleave tau and contribute to the filament polymerization (Gamblin et al., 2003; Rissman et al., 2004; Cotman et al., 2005). Aβ accumulation also triggers caspase activation through disruption of the secretory pathway, thus generating ER stress. Caspase activation at this level also cleaves tau, which precedes tau hyperphosphorylation, and seems to be an early event in AD tau pathology (Guo et al., 2004; Rissman et al., 2004). The accumulation of Aβ can disrupt proteasomal degradation and lead to activation of caspases (Blandini et al., 2006) which in turn are able to cleave tau, thus contributing to the formation of the NFTs (Chung et al., 2001; Gamblin et al., 2003; Rissman et al., 2004). Moreover, experimental data suggest that when caspases are activated, proteosomal degradation is inhibited in order to fully activate the apoptotic cascade, which provides an amplification loop leading unequivocally to the death of the cell (Sun et al., 2004). In addition, APP and Aβ can activate kinases (GSK-3β, SAPK/JNK, p38) that directly phosphorylate tau at certain residues contributing to tau hyperphosphorylation (Kins et al., 2003; Ferrer et al., 2005). In this context, the proteolytic cleavage of tau provides the link between Aβ and tau pathology. However, it is still unknown whether tau processing is required and causal for neurodegeneration, or is a secondary event related to caspase activation in the degenerating cells. In conclusion, multiple mechanisms coexist in the cell, which, when dysregulated, lead to neuronal degeneration.

7.2 Amyotrophic Lateral Sclerosis (ALS)

Amyotrophic lateral sclerosis is the most prevalent adult onset motor neuron disorder. The hallmark histophatological feature is the progressive loss of upper motor neurons in the motor cortex and lower motor neurons in both the spinal cord and brain stem, first described by Charcot in 1869. Accompanying the cell loss are intracellular inclusions of ubiquitinated proteins and strong reactivity to neurofilament markers in the axons (Ince et al., 1998). This is a multifactorial disorder with a diversity of etiologic mechanisms, such as genetic factors, protein aggregation, and oxidative stress, all contributing to the progression of the disease as well as cell death of the injured motor neurons via apoptotic routes.

Although the vast majority of ALS is sporadic, a small subset of familial ALS has been well studied. About 20% of the autosomal dominant familial cases have mutations in superoxide dismutase 1 (SOD1) (Rosen et al., 1993). Although other causal gene mutations have been identified in ALS, ALS 2 or alsin, ALS 4 or senataxin, and ALS 8 or VAPB, more than 100 mutations have been identified in the SOD1 gene and SOD1 mutations are the most prevalent familial form of the disease (Andersen et al., 2003). SOD1 is a 153 amino-acid-free radical scavenger whose function is to transform superoxide free radicals into hydrogen peroxide. SOD1 is a highly expressed protein representing about the 1% of total brain protein. The reason why motor neurons are susceptible to damage in the presence of SOD1 mutations remains unclear. It is thought that mutations in SOD1 do not generate a loss of function, but on the contrary, may be toxic gain of function mutations. Very recent work suggests that, although the motor neurons are more susceptible to death, the presence of mutant SOD1 in the astrocytes induces death of motor neurons that contain wild-type or mutant SOD1 (Di Giorgio et al., 2007; Nagai et al., 2007). There has been enormous interest in understanding the role of oxidative stress in

ALS because SOD1 encodes for an antioxidant enzyme. Although the relevance of oxidative stress is not fully understood, it is believed that mutations in SOD1 promote a structural change that allows a higher rate of interaction between the substrates and the active site of the enzyme, resulting in increased production of free radical species. However, there are not sufficient experimental data supporting this hypothesis because if SOD1 mutants cause peroxynitrite-dependent cell death in vitro, it would be expected that reduction in the levels of peroxynitrite by inhibition of neuronal nitric oxide synthase (nNOS) would improve the motor neuron outcomes. However, these experiments did not show a decrease in motor neuron damage (Facchinetti et al., 1999; Upton-Rice et al., 1999; Son et al., 2001).

Another possible event leading to ALS is mitochondrial dysfunction (Albers and Beal, 2000; Menzies et al., 2002). Again, several properties converge at this level because mitochondria are able to maintain Ca^{2+} homeostasis and are the source of intracellular ATP. Mitochondria generate intracellular free radicals and can also play a key role as mediators of the apoptotic pathway. Mitochondrial dysfunction has been reported in vitro as well as in vivo. Expression of mutant SOD1 (G93A) in a motor neuron cell line leads to mitochondrial abnormalities, not only at the morphological level, but also at the biochemical level, with impaired activity of complexes II and IV of the respiratory chain leading to the activation of apoptotic mechanisms and subsequent cell death (Menzies et al., 2002; Takeuchi et al., 2002; Fukada et al., 2004). In transgenic mice overexpressing mutant SOD1, mitochondrial vacuolization in motor neurons has been noted as an early event (Wong and Strong, 1998). Impaired activity in several complexes of the respiratory chain and reduced ATP synthesis have also been reported in murine models of the disease (Jung et al., 2002; Mattiazzi et al., 2002). Moreover, translocation of cytochrome c from mitochondria to the cytosolic space, triggering the apoptotic cascade, is a feature of these animals (Guegan et al., 2001; Zhu et al., 2002). Following this line of thought, it has been described that the antiapoptotic protein Bcl-2 can interact with aggregates of SOD1 in the spinal cord, thus decreasing the availability of Bcl-2 to prevent apoptosis (Pasinelli et al., 2004).

Motor neurons can have extremely long axons that travel from the spinal cord all the way to the target muscle. Preserving the morphology of these axons requires the presence of structural proteins, such as neurofilaments. Neurofilaments are the main component of the cytoskeleton in neurons and although their primary role is to maintain cell shape, they are also involved in axonal transport and influence axonal caliber. Inclusions of aberrantly assembled neurofilaments, phosphorylated or not, in the cell bodies and axons of motor neurons is one of the histopathological hallmarks of ALS (Ince et al., 1998). Transgenic mice carrying SOD1 mutations exhibit abnormalities in neurofilament organization, as well as intracellular proteinaceous inclusions, and reduced axonal transport in the ventral root (Tu et al., 1996; Zhang et al., 1997). Moreover, more than 1% of sporadic ALS cases carry deletions or expansion in the neurofilament NF-H gene (Meyer and Potter, 1995; Tomkins et al., 1998).

It is not only NF-H filaments that are involved in the disease. Transgenic mice overexpressing peripherin, an intermediate filament, develop late onset motor neuron degeneration and altered neurofilament assembly (Beaulieu et al., 1999). This alteration in neurofilament structure, together with misfolded SOD1 proteins, may lead to cellular stress, mediated mainly by the ER. This altered situation reduces the ability of the proteasome to mediate protein degradation, thus compromising protein turnover in the cell, which in turn affects surrounding organelles, such as mitochondria, and potentially activates and/or amplifies the apoptotic cascade. Experimental evidence shows that motor neurons die mainly by apoptotic mechanisms (Martin, 1999; Guegan et al., 2001; Sathasivam et al., 2001). The study of cellular models of mutant SOD1 overexpression shows that these cells die via a programmed cell death when exposed to oxidative stress (Cookson and Shaw, 1999). Moreover, the animal models overexpressing mutant SOD1 show an upregulation in expression and activation of caspase-1 and -3 in the spinal cord of symptomatic animals (Li et al., 2000; Vukosavic et al., 2000). Although great strides have been made in understanding the molecular mechanisms underlying the motor neuron degeneration in ALS, the complex interplay among genetic factors, altered axonal transport, oxidative stress, protein aggregation, and mitochondrial dysfunction make this multifactorial disease a very challenging disorder for therapeutic intervention.

8 Dissecting Death Pathways in Vivo

The increasing number of transgenic and knock-out murine models available in the last decade has offered the possibility of studying in vivo those proteins believed to be associated with certain neurodegenerative disorders. These models provide a more accurate view than the cellular models in which the microenvironment is abolished. However, the in vivo models must also be interpreted with caution because the knock-down of certain genes may induce genetic compensation by related family members that could mask the effect of the exogenous genes. Overexpression may be associated with lethality, or can induce artifacts due to the overexpression process and not due to the introduction of the exogenous gene per se. We also have to keep in mind the genetic background of the particular mouse because certain mutants can be lethal on one background but perfectly viable on another. If we take the results generated by these models with caution, understanding that the models try to mimic neurodegenerative disorders but are still far from perfectly reproducing the phenotype of human diseases, the models can contribute to a better understanding of the etiopathology of the disease, help untangle molecular mechanisms triggering the degenerative process, and provide tools for the identification of potential therapeutic targets. The value of culture systems in deciphering mechanisms should not be underestimated. This is well-illustrated by recent studies of the role of astrocytes in motor neuron death which showed that astrocytes expressing the mutant SOD1 protein induced death of motor neurons whether or not the neurons expressed the mutant SOD1(Di Giorgio et al., 2007; Nagai et al., 2007). It is important to remember that any of the model systems under study are approximations of the diseases and each have their own advantages and disadvantages as systems of study.

References

Adams JM, Cory S (1998) The Bcl-2 protein family: arbiters of cell survival. Science 281: 1322–1326

Aggarwal BB (2003) Signalling pathways of the TNF superfamily: a double-edged sword. Nat Rev Immunol 3:745–756

Albers DS, Beal MF (2000) Mitochondrial dysfunction and oxidative stress in aging and neurodegenerative disease. J Neural Transm Suppl 59:133–154

Allen JW, Eldadah BA, Huang X, Knoblach SM, Faden AI (2001) Multiple caspases are involved in beta-amyloid-induced neuronal apoptosis. J Neurosci Res 65:45–53

Andersen PM, Sims KB, Xin WW, Kiely R, O'Neill G, Ravits J, Pioro E, Harati Y, Brower RD, Levine JS, Heinicke HU, Seltzer W, Boss M, Brown RH Jr. (2003) Sixteen novel mutations in the Cu/Zn superoxide dismutase gene in amyotrophic lateral sclerosis: a decade of discoveries, defects and disputes. Amyotroph Lateral Scler Other Motor Neuron Disord 4:62–73

Anglade P, Vyas S, Javoy-Agid F, Herrero MT, Michel PP, Marquez J, Mouatt-Prigent A, Ruberg M, Hirsch EC, Agid Y (1997) Apoptosis and autophagy in nigral neurons of patients with Parkinson's disease. Histol Histopathol 12:25–31

Ayala-Grosso C, Ng G, Roy S, Robertson GS (2002) Caspase-cleaved amyloid precursor protein in Alzheimer's disease. Brain Pathol 12:430–441

Baliga BC, Read SH, Kumar S (2004) The biochemical mechanism of caspase-2 activation. Cell Death Differ 11:1234–1241

Bao Q, Shi Y (2007) Apoptosome: a platform for the activation of initiator caspases. Cell Death Differ 14:56–65

Barone FC, Parsons AA (2000) Therapeutic potential of anti-inflammatory drugs in focal stroke. Expert Opin Investig Drugs 9:2281–2306

Beaulieu JM, Nguyen MD, Julien JP (1999) Late onset of motor neurons in mice overexpressing wild-type peripherin. J Cell Biol 147:531–544

Berube C, Boucher LM, Ma W, Wakeham A, Salmena L, Hakem R, Yeh WC, Mak TW, Benchimol S (2005) Apoptosis caused by p53-induced protein with death domain (PIDD) depends on the death adapter protein RAIDD. Proc Natl Acad Sci USA 102:14314–14320

Birnbaum MJ, Clem RJ, Miller LK (1994) An apoptosis-inhibiting gene from a nuclear polyhedrosis virus encoding a polypeptide with Cys/His sequence motifs. J Virol 68:2521–2528

Blandini F, Sinforiani E, Pacchetti C, Samuele A, Bazzini E, Zangaglia R, Nappi G, Martignoni E (2006) Peripheral proteasome and caspase activity in Parkinson disease and Alzheimer disease. Neurology 66:529–534

Boatright KM, Deis C, Denault JB, Sutherlin DP, Salvesen GS (2004) Activation of caspases-8 and -10 by FLIP(L). Biochem J 382:651–657

Boatright KM, Salvesen GS (2003) Mechanisms of caspase activation. Curr Opin Cell Biol 15:725–731

Breckenridge DG, Germain M, Mathai JP, Nguyen M, Shore GC (2003) Regulation of apoptosis by endoplasmic reticulum pathways. Oncogene 22:8608–8618

Bump NJ, Hackett M, Hugunin M, Seshagiri S, Brady K, Chen P, Ferenz C, Franklin S, Ghayur T, Li P, et al. (1995) Inhibition of ICE family proteases by baculovirus antiapoptotic protein p35. Science 269:1885–1888

Burek MJ, Oppenheim RW (1996) Programmed cell death in the developing nervous system. Brain Pathol 6:427–446

Callus BA, Vaux DL (2007) Caspase inhibitors: viral, cellular and chemical. Cell Death Differ 14:73–78

Cardone MH, Roy N, Stennicke HR, Salvesen GS, Franke TF, Stanbridge E, Frisch S, Reed JC (1998) Regulation of cell death protease caspase-9 by phosphorylation. Science 282: 1318–1321

Chai J, Du C, Wu JW, Kyin S, Wang X, Shi Y (2000) Structural and biochemical basis of apoptotic activation by Smac/DIABLO. Nature 406:855–862

Chan SL, Mattson MP (1999) Caspase and calpain substrates: roles in synaptic plasticity and cell death. J Neurosci Res 58:167–190
Chang DW, Xing Z, Capacio VL, Peter ME, Yang X (2003) Interdimer processing mechanism of procaspase-8 activation. EMBO J 22:4132–4142
Cheng EH, Wei MC, Weiler S, Flavell RA, Mak TW, Lindsten T, Korsmeyer SJ (2001) BCL-2, BCL-X(L) sequester BH3 domain-only molecules preventing BAX- and BAK-mediated mitochondrial apoptosis. Mol Cell 8:705–711
Chinnaiyan AM, O'Rourke K, Tewari M, Dixit VM (1995) FADD, a novel death domain-containing protein, interacts with the death domain of Fas and initiates apoptosis. Cell 81:505–512
Chipuk JE, Bouchier-Hayes L, Green DR (2006) Mitochondrial outer membrane permeabilization during apoptosis: the innocent bystander scenario. Cell Death Differ 13:1396–1402
Chung CW, Song YH, Kim IK, Yoon WJ, Ryu BR, Jo DG, Woo HN, Kwon YK, Kim HH, Gwag BJ, Mook-Jung IH, Jung YK (2001) Proapoptotic effects of tau cleavage product generated by caspase-3. Neurobiol Dis 8:162–172
Cookson MR, Shaw PJ (1999) Oxidative stress and motor neurone disease. Brain Pathol 9:165–186
Cotman CW, Poon WW, Rissman RA, Blurton-Jones M (2005) The role of caspase cleavage of tau in Alzheimer disease neuropathology. J Neuropathol Exp Neurol 64:104–112
Cribbs DH, Poon WW, Rissman RA, Blurton-Jones M (2004) Caspase-mediated degeneration in Alzheimer's disease. Am J Pathol 165:353–355
Crook NE, Clem RJ, Miller LK (1993) An apoptosis-inhibiting baculovirus gene with a zinc finger-like motif. J Virol 67:2168–2174
Danial NN, Korsmeyer SJ (2004) Cell death: critical control points. Cell 116:205–219
Denault JB, Eckelman BP, Shin H, Pop C, Salvesen GS (2007) Caspase-3 attenuates XIAP-mediated inhibition of caspase-9. Biochem J 405(1):11–19
Deng Y, Lin Y, Wu X (2002) TRAIL-induced apoptosis requires Bax-dependent mitochondrial release of Smac/DIABLO. Genes Dev 16:33–45
Deveraux QL, Leo E, Stennicke HR, Welsh K, Salvesen GS, Reed JC (1999) Cleavage of human inhibitor of apoptosis protein XIAP results in fragments with distinct specificities for caspases. EMBO J 18:5242–5251
Deveraux QL, Reed JC (1999) IAP family proteins – suppressors of apoptosis. Genes Dev 13: 239–252
Devin A, Lin Y, Liu ZG (2003) The role of the death-domain kinase RIP in tumour-necrosis-factor-induced activation of mitogen-activated protein kinases. EMBO Rep 4:623–627
Di Giorgio FP, Carrasco MA, Siao MC, Maniatis T, Eggan K (2007) Non-cell autonomous effect of glia on motor neurons in an embryonic stem cell-based ALS model. Nat Neurosci 10:608–614
Dobo J, Swanson R, Salvesen GS, Olson ST, Gettins PG (2006) Cytokine response modifier a inhibition of initiator caspases results in covalent complex formation and dissociation of the caspase tetramer. J Biol Chem 281:38781–38790
Du C, Fang M, Li Y, Li L, Wang X (2000) Smac, a mitochondrial protein that promotes cytochrome c-dependent caspase activation by eliminating IAP inhibition. Cell 102:33–42
Duan H, Dixit VM (1997) RAIDD is a new 'death' adaptor molecule. Nature 385:86–89
Eckelman BP, Salvesen GS (2006) The human anti-apoptotic proteins cIAP1 and cIAP2 bind but do not inhibit caspases. J Biol Chem 281:3254–3260
Eckelman BP, Salvesen GS, Scott FL (2006) Human inhibitor of apoptosis proteins: why XIAP is the black sheep of the family. EMBO Rep 7:988–994
Facchinetti F, Sasaki M, Cutting FB, Zhai P, MacDonald JE, Reif D, Beal MF, Huang PL, Dawson TM, Gurney ME, Dawson VL (1999) Lack of involvement of neuronal nitric oxide synthase in the pathogenesis of a transgenic mouse model of familial amyotrophic lateral sclerosis. Neuroscience 90:1483–1492
Ferrer I, Gomez-Isla T, Puig B, Freixes M, Ribe E, Dalfo E, Avila J (2005) Current advances on different kinases involved in tau phosphorylation, and implications in Alzheimer's disease and tauopathies. Curr Alzheimer Res 2:3–18

Fisher AJ, Cruz W, Zoog SJ, Schneider CL, Friesen PD (1999) Crystal structure of baculovirus P35: role of a novel reactive site loop in apoptotic caspase inhibition. EMBO J 18:2031–2039

Fukada K, Zhang F, Vien A, Cashman NR, Zhu H (2004) Mitochondrial proteomic analysis of a cell line model of familial amyotrophic lateral sclerosis. Mol Cell Proteomics 3: 1211–1223

Gamblin TC, Chen F, Zambrano A, Abraha A, Lagalwar S, Guillozet AL, Lu M, Fu Y, Garcia-Sierra F, LaPointe N, Miller R, Berry RW, Binder LI, Cryns VL (2003) Caspase cleavage of tau: linking amyloid and neurofibrillary tangles in Alzheimer's disease. Proc Natl Acad Sci USA 100:10032–10037

Gervais FG, Xu D, Robertson GS, Vaillancourt JP, Zhu Y, Huang J, LeBlanc A, Smith D, Rigby M, Shearman MS, Clarke EE, Zheng H, Van Der Ploeg LH, Ruffolo SC, Thornberry NA, Xanthoudakis S, Zamboni RJ, Roy S, Nicholson DW (1999) Involvement of caspases in proteolytic cleavage of Alzheimer's amyloid- beta precursor protein and amyloidogenic A beta peptide formation. Cell 97:395–406

Green DR, Amarante-Mendes GP (1998) The point of no return: mitochondria, caspases, and the commitment to cell death. Results Probl Cell Differ 24:45–61

Green DR, Kroemer G (2004) The pathophysiology of mitochondrial cell death. Science 305: 626–629

Guegan C, Vila M, Rosoklija G, Hays AP, Przedborski S (2001) Recruitment of the mitochondrial-dependent apoptotic pathway in amyotrophic lateral sclerosis. J Neurosci 21:6569–6576

Guo H, Albrecht S, Bourdeau M, Petzke T, Bergeron C, LeBlanc AC (2004) Active caspase-6 and caspase-6-cleaved tau in neuropil threads, neuritic plaques, and neurofibrillary tangles of Alzheimer's disease. Am J Pathol 165:523–531

Hardy JA, Higgins GA (1992) Alzheimer's disease: the amyloid cascade hypothesis. Science 256:184–185

Hardy J, Selkoe DJ (2002) The amyloid hypothesis of Alzheimer's disease: progress and problems on the road to therapeutics. Science 297:353–356

Hegde R, Srinivasula SM, Zhang Z, Wassell R, Mukattash R, Cilenti L, DuBois G, Lazebnik Y, Zervos AS, Fernandes-Alnemri T, Alnemri ES (2002) Identification of Omi/HtrA2 as a mitochondrial apoptotic serine protease that disrupts inhibitor of apoptosis protein–caspase interaction. J Biol Chem 277:432–438

Hengartner MO, Horvitz HR (1994) Programmed cell death in *Caenorhabditis elegans*. Curr Opin Genet Dev 4:581–586

Hitomi J, Katayama T, Eguchi Y, Kudo T, Taniguchi M, Koyama Y, Manabe T, Yamagishi S, Bando Y, Imaizumi K, Tsujimoto Y, Tohyama M (2004) Involvement of caspase-4 in endoplasmic reticulum stress-induced apoptosis and Abeta-induced cell death. J Cell Biol 165:347–356

Hsu H, Xiong J, Goeddel DV (1995) The TNF receptor 1-associated protein TRADD signals cell death and NF-kappa B activation. Cell 81:495–504

Huesmann GR, Clayton DF (2006) Dynamic role of postsynaptic caspase-3 and BIRC4 in zebra finch song-response habituation. Neuron 52:1061–1072

Ince PG, Tomkins J, Slade JY, Thatcher NM, Shaw PJ (1998) Amyotrophic lateral sclerosis associated with genetic abnormalities in the gene encoding Cu/Zn superoxide dismutase: molecular pathology of five new cases, and comparison with previous reports and 73 sporadic cases of ALS. J Neuropathol Exp Neurol 57:895–904

Irmler M, Hofmann K, Vaux D, Tschopp J (1997) Direct physical interaction between the *Caenorhabditis elegans* 'death proteins' CED-3 and CED-4. FEBS Lett 406:189–190

Janssens S, Tinel A, Lippens S, Tschopp J (2005) PIDD mediates NF-kappaB activation in response to DNA damage. Cell 123:1079–1092

Jellinger KA, Stadelmann C (2001) Problems of cell death in neurodegeneration and Alzheimer's disease. J Alzheimers Dis 3:31–40

Jung C, Higgins CM, Xu Z (2002) Mitochondrial electron transport chain complex dysfunction in a transgenic mouse model for amyotrophic lateral sclerosis. J Neurochem 83: 535–545

Jurgensmeier JM, Xie Z, Deveraux Q, Ellerby L, Bredesen D, Reed JC (1998) Bax directly induces release of cytochrome c from isolated mitochondria. Proc Natl Acad Sci USA 95: 4997–5002
Kermer P, Liman J, Weishaupt JH, Bahr M (2004) Neuronal apoptosis in neurodegenerative diseases: from basic research to clinical application. Neurodegener Dis 1:9–19
Kerr JF, Wyllie AH, Currie AR (1972) Apoptosis: a basic biological phenomenon with wide-ranging implications in tissue kinetics. Br J Cancer 26:239–257
Kim JW, Choi EJ, Joe CO (2000) Activation of death-inducing signaling complex (DISC) by pro-apoptotic C-terminal fragment of RIP. Oncogene 19:4491–4499
Kins S, Kurosinski P, Nitsch RM, Gotz J (2003) Activation of the ERK and JNK signaling pathways caused by neuron-specific inhibition of PP2A in transgenic mice. Am J Pathol 163: 833–843
Kischkel FC, Hellbardt S, Behrmann I, Germer M, Pawlita M, Krammer PH, Peter ME (1995) Cytotoxicity-dependent APO-1 (Fas/CD95)-associated proteins form a death-inducing signaling complex (DISC) with the receptor. EMBO J 14:5579–5588
Kluck RM, Bossy-Wetzel E, Green DR, Newmeyer DD (1997) The release of cytochrome c from mitochondria: a primary site for Bcl-2 regulation of apoptosis. Science 275:1132–1136
Korsmeyer SJ, Yin XM, Oltvai ZN, Veis-Novack DJ, Linette GP (1995) Reactive oxygen species and the regulation of cell death by the Bcl-2 gene family. Biochim Biophys Acta 1271: 63–66
Lamkanfi M, Declercq W, Kalai M, Saelens X, Vandenabeele P (2002) Alice in caspase land. A phylogenetic analysis of caspases from worm to man. Cell Death Differ 9:358–361
Lamkanfi M, D'Hondt K, Vande Walle L, van Gurp M, Denecker G, Demeulemeester J, Kalai M, Declercq W, Saelens X, Vandenabeele P (2005) A novel caspase-2 complex containing TRAF2 and RIP1. J Biol Chem 280:6923–6932
Lannfelt L, Basun H, Vigo-Pelfrey C, Wahlund LO, Winblad B, Lieberburg I, Schenk D (1995) Amyloid beta-peptide in cerebrospinal fluid in individuals with the Swedish Alzheimer amyloid precursor protein mutation. Neurosci Lett 199:203–206
Larson J, Lynch G, Games D, Seubert P (1999) Alterations in synaptic transmission and long-term potentiation in hippocampal slices from young and aged PDAPP mice. Brain Res 840:23–35
LeBlanc A, Liu H, Goodyer C, Bergeron C, Hammond J (1999) Caspase-6 role in apoptosis of human neurons, amyloidogenesis, and Alzheimer's disease. J Biol Chem 274: 23426–23436
Levine B, Yuan J (2005) Autophagy in cell death an innocent convict? J Clin Invest 115:2679–2688
Li P, Nijhawan D, Budihardjo I, Srinivasula SM, Ahmad M, Alnemri ES, Wang X (1997) Cytochrome c and dATP-dependent formation of Apaf-1/caspase-9 complex initiates an apoptotic protease cascade. Cell 91:479–489
Li M, Ona VO, Guegan C, Chen M, Jackson-Lewis V, Andrews LJ, Olszewski AJ, Stieg PE, Lee JP, Przedborski S, Friedlander RM (2000) Functional role of caspase-1 and caspase-3 in an ALS transgenic mouse model. Science 288:335–339
Li H, Zhu H, Xu CJ, Yuan J (1998) Cleavage of BID by caspase 8 mediates the mitochondrial damage in the Fas pathway of apoptosis. Cell 94:491–501
Lin Y, Devin A, Cook A, Keane MM, Kelliher M, Lipkowitz S, Liu ZG (2000) The death domain kinase RIP is essential for TRAIL (Apo2L)-induced activation of IkappaB kinase and c-Jun N-terminal kinase. Mol Cell Biol 20:6638–6645
Liu Z, Sun C, Olejniczak ET, Meadows RP, Betz SF, Oost T, Herrmann J, Wu JC, Fesik SW (2000) Structural basis for binding of Smac/DIABLO to the XIAP BIR3 domain. Nature 408: 1004–1008
Lockshin RA, Zakeri Z (2004) Apoptosis, autophagy, and more. Int J Biochem Cell Biol 36: 2405–2419
Lu DC, Rabizadeh S, Chandra S, Shayya RF, Ellerby LM, Ye X, Salvesen GS, Koo EH, Bredesen DE (2000) A second cytotoxic proteolytic peptide derived from amyloid beta-protein precursor [see comments]. Nat Med 6:397–404

Luo X, He Q, Huang Y, Sheikh MS (2005) Transcriptional upregulation of PUMA modulates endoplasmic reticulum calcium pool depletion-induced apoptosis via Bax activation. Cell Death Differ 12:1310–1318

Mannick JB, Schonhoff C, Papeta N, Ghafourifar P, Szibor M, Fang K, Gaston B (2001) S-Nitrosylation of mitochondrial caspases. J Cell Biol 154:1111–1116

Mariathasan S (2007) ASC, Ipaf and Cryopyrin/Nalp3: bona fide intracellular adapters of the caspase-1 inflammasome. Microbes Infect 9:664–671

Martin LJ (1999) Neuronal death in amyotrophic lateral sclerosis is apoptosis: possible contribution of a programmed cell death mechanism. J Neuropathol Exp Neurol 58:459–471

Martin J, Van Regemorter N, Del-Favero J, Lofgren A, Van Broeckhoven C (1999) Spinocerebellar ataxia type 7 (SCA7) – correlations between phenotype and genotype in one large Belgian family. J Neurol Sci 168:37–46

Martinon F, Burns K, Tschopp J (2002) The inflammasome: a molecular platform triggering activation of inflammatory caspases and processing of proIL-beta. Mol Cell 10: 417–426

Martins LM, Iaccarino I, Tenev T, Gschmeissner S, Totty NF, Lemoine NR, Savopoulos J, Gray CW, Creasy CL, Dingwall C, Downward J (2002) The serine protease Omi/HtrA2 regulates apoptosis by binding XIAP through a reaper-like motif. J Biol Chem 277:439–444

Mattiazzi M, D'Aurelio M, Gajewski CD, Martushova K, Kiaei M, Beal MF, Manfredi G (2002) Mutated human SOD1 causes dysfunction of oxidative phosphorylation in mitochondria of transgenic mice. J Biol Chem 277:29626–29633

Menzies FM, Cookson MR, Taylor RW, Turnbull DM, Chrzanowska-Lightowlers ZM, Dong L, Figlewicz DA, Shaw PJ (2002) Mitochondrial dysfunction in a cell culture model of familial amyotrophic lateral sclerosis. Brain 125:1522–1533

Meyer MA, Potter NT (1995) Sporadic ALS and chromosome 22: evidence for a possible neurofilament gene defect. Muscle Nerve 18:536–539

Micheau O, Tschopp J (2003) Induction of TNF receptor I-mediated apoptosis via two sequential signaling complexes. Cell 114:181–190

Nagai M, Re DB, Nagata T, Chalazonitis A, Jessell TM, Wichterle H, Przedborski S (2007) Astrocytes expressing ALS-linked mutated SOD1 release factors selectively toxic to motor neurons. Nat Neurosci 10:615–622

Nakagawa T, Zhu H, Morishima N, Li E, Xu J, Yankner BA, Yuan J (2000) Caspase-12 mediates endoplasmic-reticulum-specific apoptosis and cytotoxicity by amyloid-beta. Nature 403: 98–103

Nicholson DW, Thornberry NA (1997) Caspases: killer proteases. Trends Biochem Sci 22:299–306

Nutt LK, Chandra J, Pataer A, Fang B, Roth JA, Swisher SG, O'Neil RG, McConkey DJ (2002b) Bax-mediated Ca2+ mobilization promotes cytochrome c release during apoptosis. J Biol Chem 277:20301–20308

Nutt LK, Margolis SS, Jensen M, Herman CE, Dunphy WG, Rathmell JC, Kornbluth S (2005) Metabolic regulation of oocyte cell death through the CaMKII-mediated phosphorylation of caspase-2. Cell 123:89–103

Nutt LK, Pataer A, Pahler J, Fang B, Roth J, McConkey DJ, Swisher SG (2002a) Bax and Bak promote apoptosis by modulating endoplasmic reticular and mitochondrial Ca2+ stores. J Biol Chem 277:9219–9225

Park HH, Logette E, Raunser S, Cuenin S, Walz T, Tschopp J, Wu H (2007) Death domain assembly mechanism revealed by crystal structure of the oligomeric PIDDosome core complex. Cell 128:533–546

Pasinelli P, Belford ME, Lennon N, Bacskai BJ, Hyman BT, Trotti D, Brown RH Jr. (2004) Amyotrophic lateral sclerosis-associated SOD1 mutant proteins bind and aggregate with Bcl-2 in spinal cord mitochondria. Neuron 43:19–30

Pasinelli P, Borchelt DR, Houseweart MK, Cleveland DW, Brown RH Jr. (1998) Caspase-1 is activated in neural cells and tissue with amyotrophic lateral sclerosis-associated mutations in copper-zinc superoxide dismutase. Proc Natl Acad Sci USA 95:15763–15768

Pereira C, Santos MS, Oliveira C (1999) Involvement of oxidative stress on the impairment of energy metabolism induced by A beta peptides on PC12 cells: protection by antioxidants. Neurobiol Dis 6:209–219

Perrelet D, Ferri A, Liston P, Muzzin P, Korneluk RG, Kato AC (2002) IAPs are essential for GDNF-mediated neuroprotective effects in injured motor neurons in vivo. Nat Cell Biol 4: 175–179

Perrelet D, Perrin FE, Liston P, Korneluk RG, MacKenzie A, Ferrer-Alcon M, Kato AC (2004) Motoneuron resistance to apoptotic cell death in vivo correlates with the ratio between X-linked inhibitor of apoptosis proteins (XIAPs) and its inhibitor, XIAP-associated factor 1. J Neurosci 24:3777–3785

Pompl PN, Yemul S, Xiang Z, Ho L, Haroutunian V, Purohit D, Mohs R, Pasinetti GM (2003) Caspase gene expression in the brain as a function of the clinical progression of Alzheimer disease. Arch Neurol 60:369–376

Prunell GF, Troy CM (2004) Balancing neuronal death. J Neurosci Res 78:1–6

Quan LT, Tewari M, O'Rourke K, Dixit V, Snipas SJ, Poirier GG, Ray C, Pickup DJ, Salvesen GS (1996) Proteolytic activation of the cell death protease Yama/CPP32 by granzyme B. Proc Natl Acad Sci USA 93:1972–1976

Rao RV, Ellerby HM, Bredesen DE (2004) Coupling endoplasmic reticulum stress to the cell death program. Cell Death Differ 11:372–380

Riedl SJ, Fuentes-Prior P, Renatus M, Kairies N, Krapp S, Huber R, Salvesen GS, Bode W (2001a) Structural basis for the activation of human procaspase-7. Proc Natl Acad Sci USA 98: 14790–14795

Riedl SJ, Renatus M, Schwarzenbacher R, Zhou Q, Sun C, Fesik SW, Liddington RC, Salvesen GS (2001b) Structural basis for the inhibition of caspase-3 by XIAP. Cell 104:791–800

Riedl SJ, Salvesen GS (2007) The apoptosome: signalling platform of cell death. Nat Rev Mol Cell Biol 8:405–413

Rieux-Laucat F, Fischer A, Deist FL (2003) Cell-death signaling and human disease. Curr Opin Immunol 15:325–331

Rissman RA, Poon WW, Blurton-Jones M, Oddo S, Torp R, Vitek MP, LaFerla FM, Rohn TT, Cotman CW (2004) Caspase-cleavage of tau is an early event in Alzheimer disease tangle pathology. J Clin Invest 114:121–130

Rosen DR, Siddique T, Patterson D, Figlewicz DA, Sapp P, Hentati A, Donaldson D, Goto J, O'Regan JP, Deng HX, et al. (1993) Mutations in Cu/Zn superoxide dismutase gene are associated with familial amyotrophic lateral sclerosis [published erratum appears in Nature 1993 Jul 22;364(6435):362] [see comments]. Nature 362:59–62

Salvesen GS, Duckett CS (2002) IAP proteins: blocking the road to death's door. Nat Rev Mol Cell Biol 3:401–410

Sanchez I, Xu CJ, Juo P, Kakizaka A, Blenis J, Yuan J (1999) Caspase-8 is required for cell death induced by expanded polyglutamine repeats. Neuron 22:623–633

Sathasivam S, Ince PG, Shaw PJ (2001) Apoptosis in amyotrophic lateral sclerosis: a review of the evidence. Neuropathol Appl Neurobiol 27:257–274

Scaffidi C, Schmitz I, Zha J, Korsmeyer SJ, Krammer PH, Peter ME (1999) Differential modulation of apoptosis sensitivity in CD95 type I and type II cells. J Biol Chem 274:22532–22538

Scorrano L, Oakes SA, Opferman JT, Cheng EH, Sorcinelli MD, Pozzan T, Korsmeyer SJ (2003) BAX and BAK regulation of endoplasmic reticulum Ca2+: a control point for apoptosis. Science 300:135–139

Scott FL, Denault JB, Riedl SJ, Shin H, Renatus M, Salvesen GS (2005) XIAP inhibits caspase-3 and -7 using two binding sites: evolutionarily conserved mechanism of IAPs. EMBO J 24: 645–655

Shi Y (2002) Mechanisms of caspase activation and inhibition during apoptosis. Mol Cell 9: 459–470

Shi Y (2006) Mechanical aspects of apoptosome assembly. Curr Opin Cell Biol 18:677–684

Shin S, Lee Y, Kim W, Ko H, Choi H, Kim K (2005) Caspase-2 primes cancer cells for TRAIL-mediated apoptosis by processing procaspase-8. EMBO J 24:3532–3542

Shiozaki EN, Chai J, Rigotti DJ, Riedl SJ, Li P, Srinivasula SM, Alnemri ES, Fairman R, Shi Y (2003) Mechanism of XIAP-mediated inhibition of caspase-9. Mol Cell 11: 519–527

Small GW, Rabins PV, Barry PP, Buckholtz NS, DeKosky ST, Ferris SH, Finkel SI, Gwyther LP, Khachaturian ZS, Lebowitz BD, McRae TD, Morris JC, Oakley F, Schneider LS, Streim JE, Sunderland T, Teri LA, Tune LE (1997) Diagnosis and treatment of Alzheimer disease and related disorders. Consensus statement of the American Association for Geriatric Psychiatry, the Alzheimer's Association, and the American Geriatrics Society. JAMA 278: 1363–1371

Son M, Fathallah-Shaykh HM, Elliott JL (2001) Survival in a transgenic model of FALS is independent of iNOS expression. Ann Neurol 50:273

Sprick MR, Rieser E, Stahl H, Grosse-Wilde A, Weigand MA, Walczak H (2002) Caspase-10 is recruited to and activated at the native TRAIL and CD95 death-inducing signalling complexes in a FADD-dependent manner but can not functionally substitute caspase-8. EMBO J 21: 4520–4530

Stehlik C, de Martin R, Kumabashiri I, Schmid JA, Binder BR, Lipp J (1998) Nuclear factor (NF)-kappaB-regulated X-chromosome-linked iap gene expression protects endothelial cells from tumor necrosis factor alpha-induced apoptosis. J Exp Med 188:211–216

Stennicke HR, Deveraux QL, Humke EW, Reed JC, Dixit VM, Salvesen GS (1999) Caspase-9 can be activated without proteolytic processing. J Biol Chem 274:8359–8362

Sun XM, Butterworth M, MacFarlane M, Dubiel W, Ciechanover A, Cohen GM (2004) Caspase activation inhibits proteasome function during apoptosis. Mol Cell 14:81–93

Suzuki Y, Imai Y, Nakayama H, Takahashi K, Takio K, Takahashi R (2001) A serine protease, HtrA2, is released from the mitochondria and interacts with XIAP, inducing cell death. Mol Cell 8:613–621

Takeuchi H, Kobayashi Y, Ishigaki S, Doyu M, Sobue G (2002) Mitochondrial localization of mutant superoxide dismutase 1 triggers caspase-dependent cell death in a cellular model of familial amyotrophic lateral sclerosis. J Biol Chem 277:50966–50972

Terry RD, Peck A, DeTeresa R, Schechter R, Horoupian DS (1981) Some morphometric aspects of the brain in senile dementia of the Alzheimer type. Ann Neurol 10:184–192

Thomas WD, Hersey P (1998) TNF-related apoptosis-inducing ligand (TRAIL) induces apoptosis in Fas ligand-resistant melanoma cells and mediates CD4 T cell killing of target cells. J Immunol 161:2195–2200

Thorburn A (2004) Death receptor-induced cell killing. Cell Signal 16:139–144

Tinel A, Janssens S, Lippens S, Cuenin S, Logette E, Jaccard B, Quadroni M, Tschopp J (2007) Autoproteolysis of PIDD marks the bifurcation between pro-death caspase-2 and pro-survival NF-kappaB pathway. EMBO J 26:197–208

Tinel A, Tschopp J (2004) The PIDDosome, a protein complex implicated in activation of caspase-2 in response to genotoxic stress. Science 304:843–846

Tomkins J, Usher P, Slade JY, Ince PG, Curtis A, Bushby K, Shaw PJ (1998) Novel insertion in the KSP region of the neurofilament heavy gene in amyotrophic lateral sclerosis (ALS). Neuroreport 9:3967–3970

Troy CM, Rabacchi SA, Friedman WJ, Frappier TF, Brown K, Shelanski ML (2000) Caspase-2 mediates neuronal cell death induced by beta-amyloid. J Neurosci 20:1386–1392

Troy CM, Rabacchi SA, Hohl JB, Angelastro JM, Greene LA, Shelanski ML (2001) Death in the balance: alternative participation of the caspase-2 and -9 pathways in neuronal death induced by nerve growth factor deprivation. J Neurosci 21:5007–5016

Troy CM, Salvesen GS (2002) Caspases on the brain. J Neurosci Res 69:145–150

Tu PH, Raju P, Robinson KA, Gurney ME, Trojanowski JQ, Lee VM (1996) Transgenic mice carrying a human mutant superoxide dismutase transgene develop neuronal cytoskeletal pathology resembling human amyotrophic lateral sclerosis lesions. Proc Natl Acad Sci USA 93:3155–3160

Ugolini G, Raoul C, Ferri A, Haenggeli C, Yamamoto Y, Salaun D, Henderson CE, Kato AC, Pettmann B, Hueber AO (2003) Fas/tumor necrosis factor receptor death signaling is required for axotomy-induced death of motoneurons in vivo. J Neurosci 23:8526–8531

Upton-Rice MN, Cudkowicz ME, Mathew RK, Reif D, Brown RH Jr. (1999) Administration of nitric oxide synthase inhibitors does not alter disease course of amyotrophic lateral sclerosis SOD1 mutant transgenic mice. Ann Neurol 45:413–414

Verhagen AM, Ekert PG, Pakusch M, Silke J, Connolly LM, Reid GE, Moritz RL, Simpson RJ, Vaux DL (2000) Identification of DIABLO, a mammalian protein that promotes apoptosis by binding to and antagonizing IAP proteins. Cell 102:43–53

Vukosavic S, Stefanis L, Jackson-Lewis V, Guegan C, Romero N, Chen C, Dubois-Dauphin M, Przedborski S (2000) Delaying caspase activation by bcl-2: a clue to disease retardation in a transgenic mouse model of amyotrophic lateral sclerosis. J Neurosci 20:9119–9125

Wajant H (2003) Death receptors. Essays Biochem 39:53–71

Wajant H, Pfizenmaier K, Scheurich P (2003) Tumor necrosis factor signaling. Cell Death Differ 10:45–65

Walker NP, Talanian RV, Brady KD, Dang LC, Bump NJ, Ferenz CR, Franklin S, Ghayur T, Hackett MC, Hammill LD, et al. (1994) Crystal structure of the cysteine protease interleukin-1 beta-converting enzyme: a (p20/p10)2 homodimer. Cell 78:343–352

Walsh DM, Klyubin I, Fadeeva JV, Cullen WK, Anwyl R, Wolfe MS, Rowan MJ, Selkoe DJ (2002) Naturally secreted oligomers of amyloid beta protein potently inhibit hippocampal long-term potentiation in vivo. Nature 416:535–539

Walsh DM, Selkoe DJ (2007) Abeta Oligomers – a decade of discovery. J Neurochem 101: 1172–1184

Wang Q, Maniati M, Jabado O, Pavlaki M, Troy CM, Greene LA, Stefanis L (2006) RAIDD is required for apoptosis of PC12 cells and sympathetic neurons induced by trophic factor withdrawal. Cell Death Differ 13:75–83

Wong NK, Strong MJ (1998) Nitric oxide synthase expression in cervical spinal cord in sporadic amyotrophic lateral sclerosis. Eur J Cell Biol 77:338–343

Wu G, Chai J, Suber TL, Wu JW, Du C, Wang X, Shi Y (2000) Structural basis of IAP recognition by Smac/DIABLO. Nature 408:1008–1012

Xu D, Bureau Y, McIntyre DC, Nicholson DW, Liston P, Zhu Y, Fong WG, Crocker SJ, Korneluk RG, Robertson GS (1999) Attenuation of ischemia-induced cellular and behavioral deficits by X chromosome-linked inhibitor of apoptosis protein overexpression in the rat hippocampus. J Neurosci 19:5026–5033

Yang QH, Church-Hajduk R, Ren J, Newton ML, Du C (2003) Omi/HtrA2 catalytic cleavage of inhibitor of apoptosis (IAP) irreversibly inactivates IAPs and facilitates caspase activity in apoptosis. Genes Dev 17:1487–1496

Yeh WC, Pompa JL, McCurrach ME, Shu HB, Elia AJ, Shahinian A, Ng M, Wakeham A, Khoo W, Mitchell K, El-Deiry WS, Lowe SW, Goeddel DV, Mak TW (1998) FADD: essential for embryo development and signaling from some, but not all, inducers of apoptosis. Science 279: 1954–1958

Yoshimori T (2007) Autophagy: paying Charon's toll. Cell 128:833–836

Yuan J, Lipinski M, Degterev A (2003) Diversity in the mechanisms of neuronal cell death. Neuron 40:401–413

Zhang Y, Goodyer C, LeBlanc A (2000) Selective and protracted apoptosis in human primary neurons microinjected with active caspase-3, -6, -7, and -8. J Neurosci 20:8384–8389

Zhang B, Tu P, Abtahian F, Trojanowski JQ, Lee VM (1997) Neurofilaments and orthograde transport are reduced in ventral root axons of transgenic mice that express human SOD1 with a G93A mutation. J Cell Biol 139:1307–1315

Zhang X, Uthaisang W, Hu L, Ernberg IT, Fadeel B (2005) Epstein-Barr virus-encoded latent membrane protein 1 promotes stress-induced apoptosis upstream of caspase-2-dependent mitochondrial perturbation. Int J Cancer 113:397–405

Zhu S, Stavrovskaya IG, Drozda M, Kim BY, Ona V, Li M, Sarang S, Liu AS, Hartley DM, Wu du C, Gullans S, Ferrante RJ, Przedborski S, Kristal BS, Friedlander RM (2002) Minocycline inhibits cytochrome c release and delays progression of amyotrophic lateral sclerosis in mice. Nature 417:74–78

Zong WX, Li C, Hatzivassiliou G, Lindsten T, Yu QC, Yuan J, Thompson CB (2003) Bax and Bak can localize to the endoplasmic reticulum to initiate apoptosis. J Cell Biol 162:59–69

Zou H, Henzel WJ, Liu X, Lutschg A, Wang X (1997) Apaf-1, a human protein homologous to *C. elegans* CED-4, participates in cytochrome c-dependent activation of caspase-3. Cell 90: 405–413

Zou H, Li Y, Liu X, Wang X (1999) An APAF-1.cytochrome c multimeric complex is a functional apoptosome that activates procaspase-9. J Biol Chem 274:11549–11556

van Loo G, Saelens X, Matthijssens F, Schotte P, Beyaert R, Declercq W, Vandenabeele P (2002) Caspases are not localized in mitochondria during life or death. Cell Death Differ 9:1207–1211

Animal Models of Neurodegenerative Diseases

Imad Ghorayeb, Guylène Page, Afsaneh Gaillard, and Mohamed Jaber

Abstract The use of animals as models of neurodegenerative disorders has allowed the determination of biological targets and biomarkers of several diseases, has yielded new therapeutical perspectives, and is essential before performing novel clinical assays. This review discusses the nature, use, and limits of animal models and how to obtain them for several neurodegenerative disorders such as multiple system atrophy, amyotrophic lateral sclerosis, and Huntington's disease, with a special emphasis on Parkinson's and Alzheimer's diseases. When possible, rodent, invertebrate and primate models are presented and discussed in relation to human disease. Finally, we highlight discrepancies between animal models and human neuropathology leading to question the pertinence of some of these findings to human disorders probably because of the wide spectrum of parameters defining a disease. Another point raised by these studies is the growing necessity to standardize the experimental procedures used to obtain an animal model, housing and breeding conditions, assessments of phenotypes investigated and, ultimately the interpretation of results obtained and their relevance to the pathology.

Keywords Parkinson's disease · Huntington disease · Alzheimer's disease · Amyotrophic lateral sclerosis · Multiple system atrophy · Tauopathies · Nucleotide repeats

Contents

M. Jaber (✉)
Institut de Physiologie et Biologie Cellulaires, CNRS 6187, Université de Poitiers, 86022 Poitiers Cedex, France
e-mail: mohamed.jaber@unv-poitiers.fr

J.P. Blass (ed.), *Neurochemical Mechanisms in Disease*,
Advances in Neurobiology 1, DOI 10.1007/978-1-4419-7104-3_3,

1 Introduction

Given the inherent complexity of neuronal systems and the disease process, animal models have become mandatory in neuroscience research in general and for understanding the pathogenesis of neurodegenerative diseases in particular. Indeed, investigation of human pathologies relies mostly on postmortem human brains and on clinical criteria that neither allow the identification of the causal chains that have led to a disease nor the biological basis of a given pathology. Thus, animal models of neurodegenerative disorders have become a widespread laboratory "tool". Use and housing of these models require animal facilities dedicated to research purposes and that are controlled by specific policies, guidelines, and procedures at local, national, and international levels. Whatever the country and the regulations, the accreditation process is long and difficult to obtain and projects involving animal models are reviewed on a regular basis to ensure the welfare of the animals, the appropriateness of the species used for a given investigation, the adequacy of the experimental procedures with the postulated hypothesis, and that a minimal number of animals is used for a given study.

The use of primates is often dependent on the solidity of previous research performed in lower species such as rodents (often mice and rats) but also worms (*Caenorhabditis elegans*), flies (*Drosophila Melanogaster*), and zebrafish. Primate animal models are still an essential step before reaching human clinical research for obvious and frequently confirmed similarities between the two species, be it behavioral (large clinical repertoire), anatomical, physiological, or genetic. Major constraints of an ethical, practical, and cost nature have limited the use of primates to a few specialized centers. Thus, mainstream research in neurodegenerative disease has focused on rodent animal models that were used to better understand the pathology and the underlying biological mechanisms, develop standardized diagnoses (biological tests, identification of biomarkers, etc.), and search for potential new treatments for these diseases. The use of rodent models was also strengthened given the possibility of performing genetic manipulations to mimic some of the genetic features of the diseases.

Here, we detail several animal models of neurodegenerative disorders with a special focus on the two major ones in humans, namely Parkinson's and Alzheimer's diseases. As detailed in this chapter, some animal models can present spontaneous syndromes, often due to analogous mutations with the human disease but are more generally obtained following toxin injections, physical (mechanical lesions), or genetic manipulations. Some animal models can mimic the behavioral consequences of a given neurodegenerative disease with drawbacks related to specific animal behavior that is only remotely related to humans. A disease gene-based model (also referred to as an "etiological model") can indeed reproduce the etiology of a genetically determined form of a given disease, although adaptations that can occur following a genetic manipulation throughout development can be quite different between the animal and the human. Because manipulation of the mouse genome has become standardized and available at relatively moderate costs, the use of genetically altered mice strains to model neurodegenerative disorders has become increasingly widespread. This use will tend to be generalized following the publication of the assembled mouse genome sequence (Botcherby, 2002). Transgenic mice can be generated to overexpress a gene to reproduce a gain of function mutation, to knock out a gene for a nonexpression mutation, and to mutate a gene to express an altered protein. Many other variances are available where a gene is silenced during development only or expressed/knocked-out in a specific brain area, for instance. These transgenic mice are used to map disease features, determine genetic and environmental factors that can precipitate disease progression, detail behavioral and cellular consequences of altering the expression of a disease-related gene, and test potential therapeutics.

Meticulous gene manipulations have generated a wealth of information regarding the etiology of pathology, the identification of its biological basis and the behavioral consequences of such a manipulation. However, increasing concern is raised as to the adequacy of these animal models to human diseases. Indeed, it is safe to state early in this book chapter that animal models generated so far fail to reproduce faithfully the myriad biochemical, cellular, and behavioral changes reported in a given neurodegenerative disease in humans. The ideal animal model reproducing all hallmarks of a given neurodegenerative disorder is an unattainable aim, as it is expected to develop specific and reproducible behavioral symptoms and biological features related to the disease along with slow onset and selective cell loss. Instead, an animal model is considered acceptable when it demonstrates its usefulness in understanding the pathogenesis of a disease, its behavioral, cellular, and molecular consequences and in exploring potential treatment avenues. This can sometimes be achieved even when animal models show striking differences with the human pathology.

In this line, the general message that can be drawn throughout this review is that we have reached a point in research using animal models where it has become a pressing necessity to standardize not only the experimental procedures used to obtain an animal model, but also the housing and breeding conditions, age and sex of the animals, qualitative and quantitative assessments of phenotypes investigated and, ultimately the interpretation of results obtained and their relevance to the pathology.

2 Alzheimer's Disease (AD)

2.1 The Human Disease

Late-onset Alzheimer's disease (AD) is the most prevalent subtype of age-related dementia accounting for 60% of cases of dementia and with a mean prevalence estimate of 3.4% (Kalaria et al., 2008). If growth in the older population continues, it is projected that the prevalence of AD will nearly quadruple in the next 50 years, by which time approximately 1 in 45 individuals will be afflicted with the disease (Brookmeyer et al., 1998).

In AD, neurodegeneration targets specific brain regions early in its course, especially cholinergic basal forebrain and medial temporal lobe structures. The sequential involvement of the posterior cingulated, temporal, and parietal cortical regions completes the progression of the disease. The neuropathological hallmarks of AD include massive neuronal cell and synapse loss at specific sites and the presence of senile plaques and neurofibrillary tangles (NFTs). The senile plaques are formed from deposits of amyloid-β peptide (Aβ) that is derived from the amyloid precursor protein (APP) whereas the NFTs contain hyperphosphorylated microtubule-associated protein (MAP) tau. Phosphorylation of both APP and tau represents a biochemical link between the two characteristic lesions of AD (Duyckaerts et al., 2008).

Most AD cases occur sporadically (SAD), although inheritance of certain susceptibility genes enhances the risk. In early-onset familial AD (FAD), which accounts for less than 5% of the total number of AD cases, autosomal dominant mutations have been identified in three genes: APP, presenilin 1 (PS-1), and presenilin 2 (PS-2), each of which leads to an overabundance of Aβ (Gotz and Ittner, 2008). The presenilins are components of the proteolytic γ-secretase complex that, together with β-secretase, generates Aβ fragments from the cleavage of APP. Most FAD cases are caused by mutations in PSEN1 and PSEN2, of which over 130 have been identified. In SAD, various susceptibility genes have been identified, including apolipoprotein E (ApoE). It is actually considered that the genetic risk factor that accounts for more cases of AD than any other is the ApoE4 allele located on chromosome 19 (Bertram and Tanzi, 2008).

Because neuropathological confirmation is required for the diagnosis of definite AD, only diagnosis of probable and possible AD can be made in living patients according to the commonly used criteria for AD diagnosis. These include progressive memory loss with cognitive deficits in at least two cognitive domains (McKhann et al., 1984). As the disease progresses, the characteristic clinical features of aphasia, apraxia, and agnosia emerge along with consequent amnesia and personality changes. At present, there are no known curative or preventive measures for AD and current symptomatic treatments of AD are of limited benefit, as they are not directed at the underlying biological basis of the disease.

2.2 Rodent Models

The identification of the genetic defects and mutations that cause FAD has led to the generation of transgenic rodent AD models. Nowadays, mice are the most popular animal models for AD, although rat models are developed as well. Furthermore, invertebrate models of AD have been developed and are presented at the end of this section.

2.2.1 Pharmacological Models of Alzheimer's Disease

Aβ neurotoxicity is studied in rodents (mouse and rat) after intracerebral injections of Aβ peptides ($A\beta_{1-40/42}$, $A\beta_{22/25-35}$) previously fibrillary aggregated by incubation at 37°C for 4 days minimum. Usually, rodents are intracerebroventricularily injected with a dose range between 3 and 9 nmol for a mouse and a dose of 15 nmol for a rat (Maurice et al., 1996; Stepanichev et al., 2003). Some authors injected the aggregated Aβ peptide directly into the hippocampus or into frontal and cingulated cortices uni- or bilaterally (Cetin and Dincer, 2007; Gonzalo-Ruiz et al., 2006). Examination of Congo red-stained tissue sections demonstrated the presence of numerous amyloid deposits throughout the brain areas and a decrease in cresyl violet-stained cells indicating a significant cell loss. Furthermore, Aβ-injected mice showed learning and memory deficits after 1 week postinjection (Fu et al., 2006; Gonzalo-Ruiz et al., 2006; Maurice et al., 1996; Stepanichev et al., 2003). Although these Aβ-injected rodent models did not encompass all of the neuropathological effects observed in AD, they are useful to understand the toxicity of amyloid deposits, in particular in the cholinergic system, and to screen for neuroprotective molecules active on the amyloid process (Fu et al., 2006; Gonzalo-Ruiz et al., 2006).

2.2.2 Transgenic Mouse Models of Alzheimer's Disease

APP Mice

After the first discovery of the mutation in the APP gene by Hardy and Allsop (1991), authors described the first NSEAPP mouse model of AD (Quon et al., 1991). Then, other human APP transgenics were developed: PDAPP, Tg2576, APP23, TgCRND8, and J20. The APP transgene carried one or two mutations at the β-secretase site (Swedish mutation) and/or at the γ-secretase site (London mutation) and was driven by various mouse promoters for gene coding for neuron-specific enolase (NSE), platelet-derived growth factor (PDGF), the prion protein (Prp), or thymus antigen (Thy-1.2). For Thy-1.2, the thymus-specific intronic regulatory element has been removed to target expression specifically to the mouse brain (Andra et al., 1996). Most app transgenes utilize a cDNA encoding the APP_{695} isoform, which is the predominant species expressed in the brain, or the longer APP_{751} species. Except for NSEAPP mice that show diffuse (preamyloid) plaques, all others displayed amyloid plaques which resembled the mature (neuritic) plaques characteristic of AD with positive thioflavin-S staining. These amyloid deposits were observed at 6–12 months according to the model from the hippocampus to cortical and limbic areas in a progressive manner showing regional specificity like that seen in AD pathology. In TgCRND8 mice expressing both the Swedish and London mutations under the Prp promoter, thioflavine-S-positive amyloid deposits became evident by 3 months of age (Chishti et al., 2001). The amyloid plaques were associated with dystrophic neurites, gliosis, and synaptic loss only in PDAPP mice. Despite the extent of amyloid burden, clear neurodegeneration has not been demonstrated except in the hippocampal CA1 region (14% of neuronal loss) of 14–18-month-old APP23 mice with an apparent correlation with senile plaques load (Calhoun et al., 1999). A positive immunoreactivity of phosphorylated tau protein was detected, however, no paired-helicoidal filament (PHFs) was noted in these transgenic APP mice. To date, it seems that the APP23 mice are the only strain to show a cerebral amyloid angiopathy (CAA). Clinically, the Aβ form of CAA is a significant contributor to haemorrhagic stroke, and up to 90% of AD patients may develop CAA over the disease course. Modest cholinergic deficits have also been reported in aged APP23 mice (Boncristiano et al., 2002).

Behavioural studies described age-dependent cognitive deficits assessed by using a Morris water maze. This behavioural test measures spatial reference memory. In these transgenic APP mice, both their acquisition of hidden platform locations and their retention of spatial reference information are affected (Table 1).

APP/PS-1 Mice

Most FAD cases are caused by the mutations in PS-1 and PS-2. Presenilins are polytopic transmembrane proteins which are, in combination with three or other proteins (aph-1, pen-2, and nicastrin), required for an efficient γ-secretase complex and activity to generate amyloid peptides (Edbauer et al., 2003). Although pathogenic mutations in APP and presenilins do not coexist in human AD, it was tempting to cross APP and PS-1 mutant mice and to assess whether mutant PS-1 would cause elevated Aβ levels. Overexpression alone of PS-1 M146L, M146V FAD-associated mutations induced a selective increase of $A\beta_{42}$ production. Crossing APP transgenic mice with PS-1 mutant mice causes an elevation of $A\beta_{42}/A\beta_{40}$ levels and an acceleration of amyloid deposits by 4 months of age in $APP_{SWE}/PS\text{-}1_{dE9}$ mice (Garcia-Alloza et al., 2006), by 6 months of age in PSAPP (Tg2576 mouse × $PS\text{-}1_{M146L}$ mouse), compared to 9 months in Tg2576 mice and by 1 month of age in TgCRND8/PS-1 mice compared to 3 months TgCRND8 mice (Chishti et al., 2001; Holcomb et al., 1998). In various double APP/PS-1 transgenic mice, no clear evidence for neurodegeneration in either frontal cortex or CA1 hippocampus was

Table 1 Amyloid Transgenic Rodent Models of AD Pathology

Name	Gene Mutation	Promoter	Amyloid Deposits (DP, AP)	P-Tau	NFT	Neuronal loss	Memory deficits	Inflammatory reaction
NSEAPP mice	APP_{751}	NSE	DP (8 M)	No	No	No	Yes	Rare (5%)
PDAPP mice	APP_{V717F}	PDGFβ	AP (6–8 M)	Yes (18 M)	No	No	Yes	Yes
Tg2576 mice	$APP_{695(K670N,M671L)}$	PrP	AP (9–11 M)	Yes	No	No	Yes	Yes
APP23 mice	$APP_{751(K670N,M671L)}$	Thy-1	AP (6 M) CAA (6 M)	Yes	No	14% (14–18 M)	Yes	Yes
TgCRND8 mice	$APP_{695(K670N,M671L,\ V717F)}$	PrP	AP (3 M)	NR	No	NR	Yes	Yes
J20 mice	$APP_{K670N,M671L,\ V717F}$	PDGFβ	AP	NR	No	NR	Yes	No
PSAPP mice	$APP_{695(K670N,M671L)}$ $PS\text{-}1_{M146L}$	PrP PDGFβ	AP (6 M)	Yes	NR	No	Yes	Yes
TgCRND8/PS-1 mice	$APP_{695(K670N,M671L,\ V717F)}$ PS-1	PrP	AP (1 M)	Yes	NR	NR	Yes	Yes
$APP_{SWE}/PS\text{-}1_{dE9}$ mice	$APP_{695(K595N,M596L)}$ $PS\text{-}1_{dE9}$	PrP PrP	AP (4–6 M) CAA (6 M)	NR	No	Monoaminergic (50%)	Yes	Yes
$APP_{SL}/PS\text{-}1$ mice	$APP_{751(K670N,M671L\ and\ V717I)}$ $PS\text{-}1_{M146L}$	Thy-1 HMG	AP (6 M)	Yes (4 M)	No	35% (17 M)	Yes	Yes
APP/PS-1 KI mice	$APP_{751(K670N,M671L\ and\ V717I)}$ $PS\text{-}1_{M233T,L235P}$ Knock-in	Thy-1 PS-1 mouse	AP (6 M)	Yes (11 M)	NR	50% (CA1, 6 M) 44% (DG, 12 M)	Yes	Yes
APP rat	$APP_{695(K670N,M671L)}$	PrP	No	No	No	No	Yes	No
PSAPP rat	$APP_{695(K670N,M671L)}$ $PS\text{-}1_{M146V}$	PDGFβ Rat synapsin 1 and human PS-1	AP (9–11 M) CAA ()	Yes (19–22 M)	No	No	Yes	Yes

AP: amyloid neuritic plaques; CAA: cerebral amyloid angiopathy; DG: Dentate gyrus; DP: diffuse (preamyloid) plaques; NFT: neurofibrillar tangles; NR: not reported; P-Tau: hyperphosphorylated MAPT tau protein.

evident except in the model APP$_{SL}$/PS-1$_{M146L}$ mice developed by Blanchard et al. (2003) where a loss (35%) of neurons in the pyramidal cell layer of the hippocampus was seen at 17 months of age (Schmitz et al., 2004). Recently, an intense subcortical monoaminergic neurodegeneration (50% neuronal loss) was observed in APP$_{SWE}$/PS-1$_{dE9}$ (Liu et al., 2008b). It is to be noted that none of these mouse models showed any NFTs. However, many studies described a behavioural phenotype in various APP/PS-1 transgenic mice (Higgins and Jacobsen 2003; Janus and Westaway 2001; Reiserer et al., 2007; Savonenko et al., 2005). In particular, the performance on the Y-maze, that measures spatial working memory, was impaired before amyloid deposits in PSAPP mice (Holcomb et al., 1999).

Taken together, these findings show that it is difficult to obtain a mouse model reproducing AD perfectly, especially with both the neuronal loss and NFTs. However, Casas et al. (2004) produced a new model with many features of AD, the APP$_{SL}$/PS-1 knock-in mice. These transgenic mice have two mutations in the human APP gene at K670N/M671L and V717I sites corresponding to β- and γ-secretase sites, respectively. In addition, their endogenous ps-1 gene carries the M233T and L235P mutations known to be linked to very early onset FAD at 29 and 35 years of age, respectively. These mice displayed a massive neuronal loss (49% in the ten-month-old APP$_{SL}$/PS-1 KI mice) in the CA1 region of the hippocampus with an intense neuronal apoptosis (Casas et al., 2004; Page et al., 2006). This neuronal loss distribution closely parallels the strong intraneuronal Aβ immunostaining and intracellular thioflavine-S-positive material but does not correlate with extracellular deposits (Christensen et al., 2008b). Furthermore, the authors also described a loss of neurons (44%) in the dendate gyrus granule layer (Cotel et al., 2008). The APP$_{SL}$/PS-1 KI mouse model exhibits early robust brain and spinal cord axonal degeneration (Wirths et al., 2007, 2006). At the same time-point, a dramatic age-dependent reduced ability to perform working memory and motor tasks is observed. These mice are smaller and show development of a thoracolumbar kyphosis, together with an incremental loss of body weight (Wirths et al., 2008b). Onset of the observed behavioural alterations correlates well with robust axonal degeneration in brain and spinal cord and with abundant hippocampal CA1 neuron loss (Bayer and Wirths 2008). Although our group detected hyperphosphorylated tau protein in cell bodies of neurons in 11-month-old APP$_{SL}$/PS-1 KI mice (unpublished data obtained by G Page), NFTs were not reported yet.

Contrary to studies showing a minor loss of cholinergic interneurones in the motor cortex of APP$_{SWE}$/PS-1$_{(deltaE9)}$ mice (Perez et al., 2007), the APP$_{SL}$/PS-1 KI mouse model shows a loss of choline acetyl transferase-positive neurons only in the motor nuclei Mo5 (motor trigeminal nucleus) and 7 N (facial motor nucleus) accumulating various intracellular Aβ species (Christensen et al., 2008a). The cholinergic forebrain complex consisting of Ch1-4 showed no Aβ pathology, with neither extracellular Aβ plaque deposition, nor intracellular accumulation of Aβ peptides. These fibres from this region displayed swollen ChAT-positive dystrophic neurites surrounding Aβ plaques in the cortex and hippocampal formation.

Another neuropathological alteration is the inflammatory processes, such as microglial activation and astrocyte reactivity, that occurs early during the course of the disease (Eikelenboom et al., 2006). At the age of six months, the APP_{SL}/PS-1 KI mouse model upregulates different astro- and microglia markers in both brain and spinal cord including GFAP, cathepsin D, members of the Toll-like receptors family, TGFβ-1, and osteopontin (Casas et al., 2004; Damjanac et al., 2007; Wirths et al., 2008a). Another interesting feature is the occurrence of ganglioside alterations and an accumulation of ceramide species in the cerebral cortex of APP_{SL}/PS1 KI mice as it was shown in human AD brains (Barrier et al., 2007, 2008). As early as three months, these lipid alterations were increased and could be linked to the massive neuronal death observed at sixmonths (Table 1).

APP/BACE Mice

The type I transmembrane aspartyl proteinase β-site APP cleaving enzyme (BACE1) was identified as the major β-secretase for generation of Aβ peptides by neurons (Luo et al., 2003). BACE cleaves APP at Asp1 and Glu11, whereas subsequent cleavage by γ-secretase gives rise to the Aβ (1–40/42) and Aβ (11–40/42) amyloid peptides. Deficiency of BACE1 in a double transgenic combination with human mutant APP rescued the early hippocampal memory deficits and correlated with a dramatic reduction in Aβ levels (Ohno et al., 2004). On the other hand, mice overexpressing BACE1 in addition to human wild-type (WT), APP, or mutant APP increased the amyloidogenic processing of APP as revealed by increased levels of the APP metabolites sAPPβ, β-CTF, and Aβ peptides (Willem et al., 2004). No CAA was observed probably due to the higher rate of self-association and fibrillogenic capacity of the shorter and less soluble N-truncated Aβ11–42 peptides that form amyloid deposits in the parenchyma, indicating that BACE1 is in tight control of the balance in amyloid pathology in brain, promoting either parenchyma or vasculature.

APP/ApoE Mice

In epidemiological investigations, it has been found that the ApoE4 allele is genetically associated with sporadic AD with a frequency of 45% compared with 15% in the general population (Corder et al., 1993). The pathological contributions of ApoE to amyloid and tau pathology in AD have been studied in different types of transgenic mice deficient in endogenous murine ApoE and/or overexpressed different ApoE isoforms, including various combinations with mutant human APP and PS-1 (Holtzman 2004). ApoE knock-out mice have significantly decreased synaptophysin and MAP 2 staining, supporting the role of ApoE in the maintenance of synapses and dendrites during aging (Masliah et al., 1995). The finding that ApoE deficiency delayed amyloid plaque deposition in mice, whereas overexpression of human ApoE4 and not ApoE3 by transferring gene promoter accelerated plaque formation in transgenic mice, suggested a gain of function of ApoE4 (Bales et al., 1999; Carter et al., 2001). Authors showed that ApoE4 did not change the balance of amyloidogenic to nonamyloidogenic pathways. Nevertheless, the levels of Aβ42 and Aβ40 increased by ApoE4 overexpression, indicating that ApoE4 acted downstream of the production of amyloid peptides, that is, slowed down the degradation and clearance of Aβ peptides (Van Dooren et al., 2006). Furthermore, the neuron-specific proteolysis of ApoE4 was linked to increased phosphorylation of tau in the brain of ApoE mice (Brecht et al., 2004). Another most interesting finding is the development of CAA in the cortex, hippocampus, and thalamus of APP/ApoE4 and APP/ApoE4/PS-1 mice (Fryer et al., 2003).

APP/ADAM Mice

The endoproteolysis of APP within the Aβ sequence by the α-secretase can preclude the formation of any Aβ peptides. In addition, cleavage by α-secretase releases the N-terminal soluble ectodomain of APP, known as APPα, which has been claimed to exert neurotrophic and neuroprotective properties (Mattson, 1997). Proteinases belonging to the ADAM family (A Desintegrin and Metalloproteinases) were the main candidates as physiologically relevant α-secretases. ADAM10 and ADAM17 single knock-outs have been shown to be lethal embryonically, whereas ADAM9 knock-outs are viable. Transgenic mice are developed with overexpression of ADAM10 or a dominant negative catalytically inactive ADAM10 mutant with human mutant APP in double transgenic mice. Moderate neuronal levels of ADAM10 increased the secretion of APPα concomitantly with a reduction in the production of Aβ peptides, preventing any deposition of amyloid plaques. long-term potentiation (LTP) and cognitive deficits were also improved, suggesting a fundamental rescue of synaptic function via the increased activity of α-secretases (Postina et al., 2004).

Tau and Tau/APP Mice

Although mutations in tau do not lead to AD, they produce dementia such as frontotemporal dementia with Parkinsonism associated with chromosome 17 (FTDP-17). In AD and other tauopathies, the MAP tau protein is abnormally hyperphosphorylated and is accumulated as intraneuronal tangles of PHF in cell bodies of neurons. Furthermore, the number of tangles correlates significantly with the degree of dementia, more so than the amyloid plaque numbers. Tau exists in six isoforms (352–441 amino acids) by alternative splicing of exons 2, 3, and 10, with isoforms containing either three or four C-terminal tandem microtubule-binding domain repeats and either no, one, or two shorter N-terminal domains (Fig. 1). Preparations of PHFs from AD brains reveal only three isoforms of tau corresponding to abnormally phosphorylated tau (Goedert et al., 1992).

Pathogenic mutations in the tau gene that cause FTD and FTDP-17 either reduce the ability of tau to bind to microtubules or alter the splicing of exon 10 resulting in increased 4 repeat tau isoforms. The first transgenic tau models (ALZ7 line) expressing wild-type human tau were generated in 1995 before pathogenic tau mutations had been identified. Overexpressing the longest isoform of human tau (4 repeats) under the human Thy-1 promoter resulted in hyperphosphorylation of tau and somatodendritic localization (Gotz et al., 1995). There are no NFTs, but these mice suffered from a severe axonopathy instead, with progressive paralysis of the hindlimbs, extending to the forelimbs, and age-related increased impairment

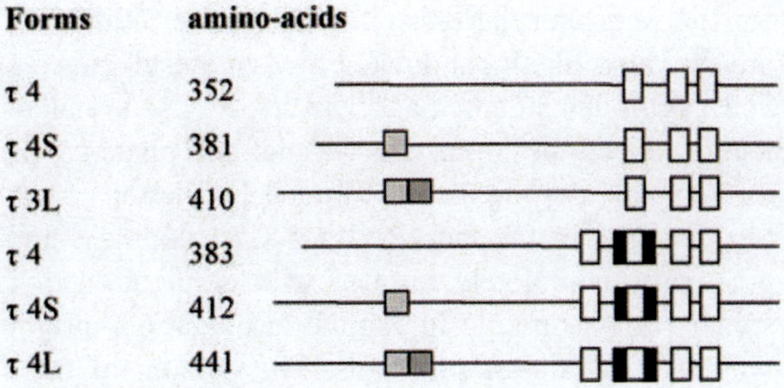

Fig. 1 Schematic representation of six human brain tau isoforms, *White boxes:* three or four tubulin-binding domains, *Grey boxes:* inserts from exon 2 near the N-terminus, *Vertical lines* in boxes: inserts from exon 3 near the N-terminus, and *Black boxes:* inserts from exon 10 near the N-terminus

in the performance of tasks such as beam walking and rotarod (Spittaels et al., 1999). Overexpressing wild-type human 3 repeat tau under the mouse PrP promoter also resulted in hyperphosphorylation of tau and axonopathy in the spinal cord with NFTs in the hippocampus, amygdala, and entorhinal cortex, albeit at very old ages (18–20 months) (Ishihara et al., 2001). Because overexpression of wild-type human tau in mice replicated very limited aspects of tau pathology in AD, many groups turned to the discovered pathogenic tau mutations for use in animal models. In 2000, Lewis et al. (2000) published their JNPL3 mouse model where the transgene contains the most common tau mutation (P301L) associated with FTDT-17 under the mouse PrP promoter. These mice with no amyloid pathology developed NFTs associated with astrogliosis, apoptosis in the spinal cord, and motor and behavioural disturbances.

Before producing a bigenic APP/tau mouse model, some authors injected a synthetic Aβ42 into the somatosensory cortex and contralateral hippocampus of P301L mice resulting in a fivefold increase in NFT numbers in the amygdala, which receives projections from both cortex and hippocampus (Gotz et al., 2001). However, Aβ was not capable of inducing NFT formation in non-NFT-forming WT tau transgenic mice (Gotz et al., 2001). Crossing Tg2576 mice with JNPL3 tau mice resulted in a double transgenic mouse line showing a more than sevenfold increase in NFT numbers at 9–11 months of age compared to JNPL3 mice. However, the presence of tau did not affect amyloid pathology (Lewis et al., 2001). To address the relationship of plaques and NFTs, Oddo et al. (2003) developed a triple transgenic mouse model named 3xTg-AD. These mice harbor mutations of APP_{SWE}, $PS\text{-}1_{M146V}$ KI, and tau_{P301L} and develop senile plaques first in the cortex (around 3 months of age) that spread to the hippocampus by 6 months. Tangles develop after amyloid pathology with hippocampal origin at 12 months of age and extend to the cortex. This regional and temporal development of pathology closely mimics the development of pathology in AD. These mice also exhibit synaptic dysfunction, including LTP deficits that precede senile plaques and tangles formation (Oddo et al., 2003). In this triple transgenic model, cognitive deficits are observed whereas no cell loss is depicted.

In vitro, many kinases can phosphorylate tau, but it is very difficult to establish the equivalent in the brain in vivo and to define exactly which kinases are responsible for the phosphorylation of tau at precise amino acid residues. Two kinases that are the most likely candidates in vivo are glycogen synthase kinase-3β (GSK-3β) and cyclin-dependent kinase 5 (cdk-5). Neuronal overexpression of GSK-3β by itself reduced the brain size without any phenotypic repercussion or development of tauopathy despite increased phosphorylation of tau (Spittaels et al., 2000, 2002). Surprisingly, in the tau-4R × GSK-3β double combination, the axonopathy was practically completely rescued with elimination of axonal dilations in brain and spinal cord, reduction in axonal degeneration and muscular atrophy, and the alleviation of all motor problems. The amount of tau associated with microtubules was reduced by 50% compared to single htau-4R transgenic animals and unbound tau was phosphorylated, leading to the conclusion that hyperphosphorylation of protein tau does not cause tauopathy per se. Recently, two novel bigenic mouse models, APP_L/Tau_{P301L} with amyloid and tau pathology and GSK-3β/Tau_{P301L} with tauopathy only showed remarkable parallels: aggravation of tauopathy, severe cognitive and behavioural defects in young adults before the onset of amyloid deposits or tauopathy and activated GSK-3β with pathological phospho-epitopes of tau (S396/S404, characteristic GSK-3β motif). These findings indicate that Aβ induces tauopathy through activation of GSK-3β (Terwel et al., 2008).

In addition, cdk-5 and its activating subunit p35 or its N-truncated p25 product have been inactivated or expressed in transgenic mouse brain with various degrees of success. The expression of cdk-5 with p35 and tau-4R in triple transgenic mice has yielded no additional new insights to the problem. Then, inducible p25 mice controlled by tetracycline displayed a dramatic neurodegeneration and neuroinflammation. A 30% decrease in brain weight was evident in a three-month observation period after the induction of p25 at the age of 6 weeks (Muyllaert et al., 2008; Table 2).

2.2.3 Transgenic Rat Models of Alzheimer's Disease

Parallel to the generation of transgenic mice, several transgenic rat models have also been produced as rats are a better rodent model for studies involving neurobehavioural testing, cannulation, sampling of cerebrospinal fluid, electrophysiology, neuroimaging, and cell-based transplant manipulations (Abbott 2004). The first transgenic rat line was generated by Flood et al. (2007). Rats have human APP_{SL} and human $PS1_{M146V}$ gene mutations and developed amyloid deposits at 9 months of age. APP_{SWE} rat model was reported but no amyloid pathology was observed except for a low intracellular accumulation of Aβ (Echeverria et al., 2004). Another APP_{SWE} rat model has been generated and produced mild, extracellular Aβ immunostaining and failed to develop compact, mature amyloid plaques by the age of 22 months (Folkesson et al., 2007). Recently, the model of Flood et al. (2007) has been more characterized: from the age of 9 months on, this rat model of AD had amyloid deposits in both diffuse and compact forms associated with activated microglia and reactive astrocytes; two months before the appearance of amyloid plaques, impaired LTP was revealed on hippocampal slices, accompanied by impaired spatial learning and memory in the Morris water maze; a mild

Table 2 Tau Transgenic Mouse Models

Name	Gene Mutation	Promoter	Amyloid Deposits	P-Tau	NFT	Axonopathy	Neuronal Loss	Memory Deficits	Inflammatory reaction
ALZ7, ALZ17	4R tau	hThy-1, mThy-1	No	Yes	No	Yes	No	No	NR
7TauTg	3R tau	PrP	No	Yes	Yes	Yes	NR	NR	NR
JNPL3	4R tau_{P301L}	PrP	No	Yes	Yes	Yes	Yes	Yes	Yes
pR5	4R tau_{P301L}	Thy-1.2	No	Yes	Yes	Yes	Yes	NR	NR
TAPP	$APP_{695(K670N,M671L)}$	PrP	AP (9–11 M)	Yes	Yes	Yes	Yes	Yes	Yes
	4R tau_{P301L}	PrP							
3xTg-AD	APP_{SWE}	Thy-1	AP	Yes	Yes	Yes	NR	Yes	Yes
	PS-1_{M146V} KI	Thy-1							
	Tau_{P301L}	Thy-1							

AP: amyloid neuritic plaques; h: human; m: mouse; NFT: neurofibrillar tangles; NR: not reported; P-Tau: hyperphosphorylated MAPT tau protein.

amyloid angiopathy was also described on the leptomeningeal blood vessels (Liu et al., 2008a).

2.3 Invertebrate Models

Species as diverse as the fly *Drosophila Melanogaster*, the nematode *Caenorhabditis elegans*, and the sea lamprey *Petromyzon marinus* have been employed to provide new insight into the pathogenesis of AD. As we also show in other animal models of neurodegenerative disorders, these lower species offer several advantages compared to rodent models. The sea lamprey was used to study the degenerative changes linked to tau overexpression as it presents six giant neurons in the hindbrain which resemble most large vertebrate neurons and are readily accessible for manipulation (Hall et al., 1997). Flies and worms have other advantages: easy and fast to breed, cheap, no ethical limitations, powerful genetics, and modifier (suppressor and enhancer) screens and drug screenings are possible. These models are useful to understand the normal functions and regulation of APP, PS, and tau genes. Genetic approaches could identify cellular processes that can suppress Aβ- or tau-dependent pathology.

The fly APP homologue, APPL, does not contain the segment of APP cleaved to generate pathogenic amyloid peptides. Therefore, some authors studied the physiological functions of APP and APPL in Drosophila. Both proteins were shown to function as vesicular receptors for kinesin 1, a motor-mediated anterograde vesicle trafficking. Flies lacking APPL or overexpressed of WT and mutant APP have axonal transport defects and only APP overexpression increased cell death in the larval brain (Cauchi and van den Heuvel 2006). Other authors introduced FAD-linked mutations at conserved residues in the *Drosophila* PS gene or overexpressed APP/BACE and showed an increased neurotoxicity in the fly with production of amyloid peptides (Sang and Jackson 2005). Modelling AD in the fly was also attempted by delivering transgenes encoding Aβ40 and Aβ42 peptides. Results with Aβ42 peptides specifically expressed in brain tissue showed a reduced longevity, locomotor deficits, impaired olfactory memory, and neurodegeneration whereas Aβ40 flies were not affected.

As for the fly, *C. elegans* has an APP homologue, APL-1. The RNAi knockdown results in a more severe uncoordinated phenotype and genetic deletion results in embryonic or larval lethality (Link, 2005; Segalat and Neri, 2003). Furthermore, a transgenic *C. elegans* expressing human Aβ has been developed and shown neurodegeneration, amyloid deposits, oxidative stress, and upregulation of many stress-related genes (Wu and Luo, 2005). In contrast to APP, the deletion of the worm tau homologue or the fly tau homologue does not result in any detectable phenotype, probably due to compensation by other MAPs. Authors produced transgenic *Drosophila* and *C. elegans* by introducing either the normal human tau gene or various mutant forms of the human tau gene. Invertebrate animals displayed neurodegeneration, a shortened life span, axonal transport defects, vacuolization in the cortex of the fly, and positive immunostaining for a series of NFT-specific epitopes without insoluble tau fibrils in *Drosophila* contrary to *C. elegans* (Gotz et al., 2004).

2.4 Primate Models

2.4.1 Spontaneous Approaches

Although nonhuman primates do not spontaneously develop AD, age-related behavioural and neurodegenerative changes occur in monkeys. Indeed, it has been shown that nonhuman primates of several species exhibit cerebrovascular and parenchymal Aβ amyloidosis but without or with paucity of tau lesions (Gearing et al., 1994, 1997; Martin et al., 1991; Struble et al., 1985; Walker et al., 1990; Wisniewski et al., 1973). Significant intraneuronal tau pathology was only recently documented in an aged chimpanzee (Rosen et al., 2008). Although the lesion profile in this chimpanzee differed somewhat from that in AD, the occurrence of both tau-immunoreactive paired helical filaments and Aβ-amyloidosis indicates that the molecular mechanisms for the pathogenesis of the two key hallmarks of AD, namely NFTs and senile plaques, are present in aged chimpanzees. In this monkey, although age probably played a role in the pathogeny of tauopathy, additional factors are suggested to be involved because it is unusual to encounter tau-immunoreactive neurons and processes in older animals (Gearing et al., 1994).

Similarly to brain pathology, it was also found that the monkeys undergo an age-related decline in several domains of cognitive function (Bartus et al., 1979, 1978; Lai et al., 1995; Moore et al., 2006; Rapp 1993, 1990; Voytko 1999). However, these changes were not correlated with neuronal loss in memory-related brain regions such as the hippocampus and entorhinal cortex (Peters et al., 1996) but with extensive loss of neurons in subcortical cholinergic basal forebrain regions similar to AD (Smith et al., 1999). The validity of these spontaneous models remains, however, questionable because by contrast to patients with AD in whom severe neuronal cell loss in the hippocampus can be found, the brain of normal aging subjects displayed almost no neuron loss in this region (West et al., 1994). It can be concluded that the neurodegenerative processes associated with normal aging and with AD are qualitatively different and that human AD is not accelerated by aging but is a distinct pathological process. The validity of such models is also weakened by the lack of correlation between the degree of amyloid plaque accumulation and cognitive decline in aged monkeys (Sloane et al., 1997). Therefore, the pathological and cognitive changes observed in the aged nonhuman primates emphasize their value as animal models for studies of human aging but question their relevance to the human AD.

2.4.2 Lesioning Approaches

Lesioned animal models are based upon the assumption that the destruction of basal forebrain cholinergic neurons by injection of a neurotoxin, such as ibotenic acid, is sufficient to reproduce some of the cognitive impairments associated with AD, mainly memory and learning deficits. Indeed, pathology in the basal forebrain cholinergic neurons is a prominent feature of AD and it may be responsible for the severe memory deficits observed in these patients. Impaired learning abilities, visual discrimination, and memory deficit were thus elicited following lesions of the nucleus basalis of Meynert (NBM) (Irle and Markowitsch, 1987; Ridley et al., 1985; Roberts et al., 1990). In other studies, however, large lesions of the NBM did not impair memory or produced only transient mild deficit in visual recognition memory (Aigner et al., 1987, 1991; Voytko et al., 1994). Surprisingly, a cholinesterase inhibitor (physostigmine) produced modest improvement in performance in the control group but not in the experimental animals.

Thus, no consensual conclusion is available about the behavioural and cognitive effects of basal forebrain cholinergic neuron lesions in nonhuman primates. However, one has to keep in mind that AD is not solely a disease of the cholinergic system.

2.4.3 Pharmacological Approaches

Based on evidence of modest improvements in cognitive function in patients with AD and in normal human volunteers with the augmentation of central cholinergic neurotransmission by cholinesterase inhibitors such as physostigmine and tacrine, many animal studies have investigated the effects of systemic administrations of direct or indirect cholinergic modulators. The ability of cholinesterase inhibitors to reverse cognitive impairments induced by the muscarinic antagonist scopolamine has been demonstrated in nonhuman primates and has been the most widely exploited approach used in preclinical animal assays to identify potential therapies for AD (Aigner and Mishkin 1986; Bartus and Johnson 1976; Fitten et al., 1988; Rupniak et al., 1991, 1989; Rupniak et al., 1997; Tang et al., 1997).

Although these models have provided a framework to understand AD and to test the preclinical development of drugs to treat the cognitive symptoms, fundamental questions persist regarding the validity of measures of behavioural function in animals in terms of reflecting clinically relevant measurements of cognition.

2.5 Perspectives

Several animal models of AD have been developed in species ranging from worms to primates. Although none completely recapitulate the disease process, they have proven to be useful models for neuropathological changes. As discussed below for all animal models of neurodegenerative diseases, there is no perfect animal model for AD. It all comes to the question that is to be answered. For instance, for screening purposes of molecules against Aβ aggregation, one can use an Aβ-injected rodent; to explore mechanisms involved in neuronal death, an animal model with amyloid and tau pathology and neuronal loss such as APP_{SL}PS-1 KI or Tau/APP mice are perhaps better suited. However, it is important to note that amyloid plaques are probably not at the origin of neuronal death inasmuch as the active vaccination in AD patients did not rescue the cognitive impairment whereas amyloid plaques were suppressed or reduced in the brain of AD patients (Holmes et al., 2008). Many reports underlined that the intraneuronal accumulation of Aβ instead of extracellular Aβ deposits triggers an early transient pathological event leading to neuronal loss in AD. It will be interesting to study the transcriptome of $APP_{SL}PS$-1 KI from the embryonic to adult life (up six months where a massive neuronal loss is depicted) in order to find new genes involved in the dysfunction of cell life. Furthermore, all findings in these animal AD models open a new field of research to develop an AD animal model: researchers may undertake the biological, molecular, and behavioural knowledge to associate APP/PS-1 for Aβ accumulation with another molecular target involved in neuronal death, cognitive deficits, or NFTs and inflammatory processes induced by intracellular Aβ neurotoxicity.

3 Parkinson's Disease (PD)

3.1 The Human Disease

Parkinson's disease (PD) is the second most common neurodegenerative disorder after AD. Although PD can develop at any age, it begins most commonly in older adults, with a peak age at onset at around 60 years (von Campenhausen et al., 2005). The likelihood of developing PD increases with age, with a lifetime risk of about 2% for men and 1.3% for women (Elbaz et al., 2002).

Most PD cases are sporadic, of unknown aetiology, but rare cases of monogenic mutations show that there are multiple causes for the neuronal degeneration (Fahn, 2003). To date, more than seven genes are known to cause familial PD. Also, 13 genetic loci, PARK1-13, have been suggested for rare forms of the disease such as autosomal dominant and autosomal recessive PD. The pathological hallmarks of PD are the loss of the nigrostriatal dopamine (DA) neurons and the presence of intracellular proteinacious alpha-synuclein-positive inclusions in surviving neurons termed Lewy bodies (LB) and Lewy neurites (LN). A recently proposed staging procedure of PD pathology suggests a premotor period in which typical pathological changes, LB and LN, spread from the olfactory bulb and vagus nerve to lower brainstem regions (stages 1–2), followed by a symptomatic period when pathological changes involve the midbrain including the substantia nigra (stage 3), mesocortex (stage 4), and neocortex (stages 5–6) (Braak et al., 2003).

When PD becomes clinically overt, tremor, rigidity, bradykinesia, and postural instability are considered to be the cardinal signs of the disease. The course of the disease is chronic and progressive, and may be considerably complicated by a wide range of motor and nonmotor features, many of which contribute to increased disability as well as diminished quality of life in patients and caregivers (Schrag et al., 2000).

α-synuclein is a 140-amino-acid protein that is encoded by a gene, SNCA, on chromosome 4 and that is abundantly expressed in many parts of the brain and localized mostly to presynaptic nerve terminals, mainly as an isoform of 140 amino acids. Structurally, α-synuclein is composed of three domains, an N-terminal amphipathic region (residues 1–60), a central hydrophobic region known as the non-β-amyloid component (residues 61–95) and a C-terminal acidic region (residues 96–140). Two categories of mutations causing familiar forms of the PD are known in the *SNCA* gene: point mutations, leading to missense variants in the encoded protein, and whole-locus multiplications leading to severe overexpression of the wild-type protein (Cookson, 2005; Moore et al., 2005; Polymeropoulos et al., 1997; Singleton et al., 2003; Spillantini and Goedert, 2000). Multiplications are rare, perhaps responsible for 1% of the PD families compatible with autosomal dominant inheritance. Point mutations are exceedingly rare: Ala53Thr is found in about 15 families of Greek ancestry; Ala30Pro and Glu46Lys are present in single families of German and Spanish origin, respectively.

Three missense mutations in α-synuclein gene (A53T, A30P, and E46K; Polymeropoulos et al., 1997; Zarranz et al., 2004), in addition to genomic triplications of a region of α-synuclein gene, are associated with autosomal dominant PD (Singleton et al., 2003). α-Synuclein has an increased propensity to aggregate due to its hydrophobic non-β-amyloid component domain. The presence of fibrillar α-synuclein as a major structural component of LB in PD suggests a role of aggregated α-synuclein in disease pathogenesis (Spillantini et al., 1998a). Recent studies provide compelling evidence of non-β-amyloid component domain and a truncated form of α-synuclein in mediating neurodegeneration in vivo.

3.2 Rodent Animal Models

There are both toxin and genetic animal models of PD. Many different toxins are used to generate DA degeneration. The most frequently used toxins in rodent models of PD are 6-hydroxydopamine (6-OHDA) and 1-methyl-4-phenyl-1, 2, 3, 6-tetrahydropyridine (MPTP).

3.2.1 The 6-OHDA Model

6-OHDA shares some structural similarities with DA and norepinephrine, and has a high affinity for several membrane transporters such as the DA (DAT) and norepinephrine transporters (NET) (Bezard et al., 1999; Breese and Traylor 1971; Pifl et al., 1993). 6-OHDA cannot cross the blood–brain barrier and must therefore be injected directly into the brain (Sachs and Jonsson, 1975). Once inside the neurons, it is rapidly oxidized into 6-OHDA-quinone and hydrogen peroxide, both of which are highly toxic (Saner and Thoenen, 1971) as they inhibit the mitochondrial respiratory chain enzyme complex I and IV, thus causing neurodegeneration of DA neurons (Glinka and Youdim, 1995; Ichitani et al., 1991). The extent of loss of DA neurons and their striatal terminals is dependent upon the dose of the toxin injected and the site of toxin injection. The toxin can be injected intrastriatally, into the median forebrain bundle (MFB, that comprises the nigrostriatal tract), or directly into the subsantia nigra (SN). This toxin does not produce LB-like inclusions (Dauer and Przedborski, 2003).

Striatal Injection

6-OHDA delivered into the striatal DA terminals has been widely used to examine neuroprotective strategies. Unilateral delivery of 6-OHDA into the striatum produces a slow and progressive retrograde degeneration of DA neurons. One major advantage of this model is that it damages only DA neurons projecting to the striatum, allowing for examination of neuroprotective strategies. In addition, because in the striatum there are no NE terminals, this allows 6-OHDA to be specific to DA neurons. One drawback to a striatal injection to model PD is that behavioural deficits are more subtle and thus can be difficult to detect. In addition, depending on the degree of DA depletion in the striatum, animals were reported to recover within several days, unless the lesion extends 80%. This recovery is attributed to compensatory mechanisms (increased release, decreased reuptake) of residual DA neurons and to changes in crossed projections from the contralateral hemisphere.

Medial Forebrain Bundle Injection

6-OHDA placed along the MFB produces a rapid degeneration of DA neurons and terminals where a loss of DA levels in the striatum can be detected 24 h after 6-OHDA injection and a significant loss of DA neurons in the SN by 3 days post-6-OHDA. In addition to producing a large cell death to the nigrostriatal pathway, unilateral MFB injections produce reliable, long-lasting behavioural deficits. A major issue regarding placement of 6-OHDA along the MFB is that of specificity. Because 6-OHDA is a catecholamine analogue and not simply an analogue of DA, when placed in the MFB, 6-OHDA can produce damage to NE terminals. In order to create specific damage only to DA neurons, 6-OHDA can be used in conjunction with a NE uptake inhibitor (such as dismethylimipramine), thereby blocking entry of 6-OHDA into NE terminals. Another drawback to the MFB lesion is that it can produce (depending on dose) a rather large and rapid cell death that can sometimes overwhelm potential neuroprotective strategies that may take longer time periods to produce beneficial effects. In addition, because of the speed with which MFB lesions produce death of DA neurons, it does not closely mimic the chronic course of the clinical condition.

Substantia Nigra Injection

The injection of 6-OHDA in the SN destroys the DA cell bodies within a few hours and before degeneration of striatal terminals (Jeon et al., 1995). Injection of 6-OHDA to the ventral midbrain produces a nearly complete destruction of SN neurons and striatal tyrosine hydroxylase (TH)-immunoreactive terminals. Delivery of 6-OHDA into the SN appears to be a more useful approach for testing cell replacement therapies (Hirsch et al., 2003).

Behavioural Impairment Following 6-OHDA Lesions

In the unilateral 6-OHDA model, also known as the "hemiparkinson model," the intact hemisphere serves as an internal control structure (Perese et al., 1989; Schwarting and Huston, 1996). Among the motor tests used following 6-OHDA lesions, the "gold standard" measures the extent of a DA lesion following administration of the DA precursor, L-DOPA, or DA agonists, such as apomorphine (Ungerstedt and Arbuthnott, 1970) and counting the number of rotations. Amphetamines have been termed indirect DA agonists, because they affect DA receptors indirectly by increasing the extracellular availability of endogenous striatal DA (Jones et al., 1998). Amphetamine treatment can induce ipsilateral rotations; the direction of turning is attributed to the release of DA in the unlesioned hemisphere. Apomorphine is a DA receptor agonist which stimulates both classes of DA receptors (D1, D2). Apomorphine treatment can induce contralateral rotations; the direction of turning is attributed to the stimulation of supersensitive D1-receptor and D2-receptor, especially in the lesioned hemisphere.

This approach easily allows the control of the extent of DA lesion and evaluates the power of therapeutic treatments, a major advantage of the 6-OHDA model of PD (Beal, 2001). The other deficit in 6-OHDA lesioned animal models is sensory neglect to visual, tactile, or olfactory stimuli that can be evaluated as the thresholds for leg withdrawal to footshocks. This behavior is believed to be due to damage in the lateral hypothalamus through which the ascending fibres of mesencephalic DA neurons pass. In addition, many researchers use the forepaw usage deficit contralateral to the side of the lesion as a method to evaluate the behavioural consequences of 6-OHDA and the potential efficiency of neuroprotective agents or cell transplantation strategies. Contralateral deficits with massive lesions were also observed in the "staircase" test, where the rat has to reach downwards for food with either only the left or the right paw. Behavioural asymmetries following unilateral 6-HAD lesions were also found in swimming rats tested in circular pools (for a complete review of this issue see Schwarting and Huston 1996).

In summary, the 6-OHDA model does not mimic all pathological and clinical features of human Parkinsonism. It induces DA neuron death, whereas the formation of cytoplasmatic inclusions (LB) does not occur. However, these models are very useful for testing cell replacement therapies or neuroprotective treatments.

3.2.2 The MPTP Model

It was in the late 1970s that a by-product of a synthetic drug, MPTP, was identified as a cause of Parkinsonism in drug addicts (Langston et al., 1983). The subsequent identification of MPTP as a dopaminergic toxin led to it becoming the most widely used toxin to mimic the clinical and pathological hallmarks of PD. MPTP is highly lipophilic and readily crosses the blood–brain barrier. After administration, MPTP is metabolized in astrocytes to its active metabolite 1-methyl-4-phenylpyridinium (MPP+) by the monoamine oxidase B (MAO B), an enzyme involved in monoamine degradation (Nicklas et al., 1985; Przedborski and Vila 2003). MPP+ is selectively taken up by the DAT and is accumulated in mitochondria where it inhibits complex I of the electron transport chain (Langston et al., 1984b; Mizuno et al., 1987; Nicklas et al., 1985). This reduces ATP production and causes an increase in free-radical production. Dopaminergic neurons in SNc are particularly vulnerable to the action of MPTP (Giovanni et al., 1991). In rodents, MPTP is systemically administered, either intraperitonealy or subcutaneously, and with repeated injections. There are marked species differences in susceptibility to the neurotoxic effects of MPTP. For instance, rats are resistant to MPTP toxicity as their catecholamine neurons seem to better cope with, and survive, impaired energy metabolism.

Mice strains vary widely in their sensitivity to the toxin. When administered with multiple high MPTP doses, they exhibit striatal DA reductions, SN neuron loss, and behavioural impairment (Heikkila et al., 1984). However, depending on the endpoint tested, MPTP effects in mice vary with dose, route, number, and timing of injections, as well as gender, age (Jarvis and Wagner 1985), and strain (Tipton and Singer 1993).

The MPP+, the toxic metabolite of MPTP can also be used to obtain animal models of PD. Systemic administration of MPP+ does not damage central DA neurons, because it does not readily cross the blood–brain barrier due to its charge. However, its direct injection into the brain effectively destroys much of the DA nigrostriatal pathway.

The rotarod and open field locomotion tests are used to evaluate the motor deficits following MPTP treatments. These tests are only effective if they are employed shortly after treatment when the mice are still intoxicated by MPTP. Mice tested later show no deficit on the rotarod (Meredith and Kang 2006). More sensitive measures, such as gait analysis, or the pole or grid tests, have been able to detect DA loss as low as 50% (Meredith and Kang 2006). However, motor deficits do not correlate well with the extent of DA neuronal loss, striatal DA levels, or the dose of MPTP (Rousselet et al., 2003).

Today, MPTP represents the most important and most frequently used Parkinsonian toxin applied in animal models (Beal, 2001; Przedborski et al., 2001). The major advantage of the MPTP is that it directly causes a specific intoxication of dopaminergic structures and it induces in humans symptoms virtually identical to PD (Przedborski and Vila, 2003). The major drawback of MPTP is that the cell loss is strain-, age-, and gender-dependent in mice (Smeyne et al., 2005; Sundstrom et al., 1987).

3.2.3 Genetic Rodent Models of PD

PD is generally a sporadic disorder, but in a significant proportion of cases (10–15% in most studies) it runs in families without a clearcut Mendelian pattern. Currently, there have been 13 defined loci identified as associated with high-penetrant autosomal dominant or recessive PD, of which causative mutations in specific genes have been identified. These genes include α-synuclein, parkin, ubiquitin carboxyl-terminal esterase L1 (UCH-L1), PTEN-induced putative kinase 1 (PINK1), DJ-1, and leucine-rich repeat kinase 2 (LRRK2). As outlined in Table 3, most of these mutations can be characterized by an early onset of disease.

PD Caused by Mutations in the α-Synuclein Gene (PARK1)

Overexpression of α-synuclein lacking residues 71–82 failed to aggregate and form oligomeric species in the *Drosophila* model of the disorder resulting in an absence

Table 3 Summary of the Eight Main Mutations Leading to Parkinson's Disease

Locus	Chromosome	Gene	Inheritance	Lewy Bodies
PARK1	4q21	*a-synuclein*	Dominant	Yes
PARK2	6q25	*parkin*	Recessive	No (only 1 case)
PARK3	2p13	Unknown	Dominant	Yes
PARK4	4q21	*a-synuclein*	Dominant	Yes
PARK5	4p14	*UCH-L1*	Dominant	Unknown
PARK6	1p35–36	PINK1	Recessive	Unknown
PARK7	1p36	*DJ-1*	Recessive	Unknown
PARK8	12q12	LRRK2 (dardarian)	Dominant	Variable

of dopaminergic pathology as no loss of tyrosine hydroxylase-positive neurons was observed. The expression of a truncated form of α-synuclein showed an enhanced ability to aggregate into large inclusions bodies, an increased accumulation of high molecular weight alpha-synuclein species, and an enhanced neurotoxicity in vivo (Periquet et al., 2007). To investigate the function of α-synuclein in mice, several transgenic mice lacking α-synuclein or expressing either WT or mutated (A30P, A53T, or both) human α-synuclein were generated.

The first line of α-synuclein knock-out mice displays a reduced level of DA in the striatum (Abeliovich et al., 2000), however, behavioural assessment did not reveal any major impairment. The second line of α-synuclein-null mice generated by Dauer et al. (2002) were completely resistant to MPTP intoxication, likely due to an incapacity of MPP+ to inhibit complex I in these mice. A third line of α-synuclein knock-out mice generated showed a partial protection to MPTP-induced striatal DA loss and an increased methamphetamine-induced DA depletion (Schluter et al., 2003).

Expression of truncated α-synuclein under the TH promoter led to nigrostriatal pathology (Tofaris et al., 2006). Expression of amino acids 1–130 of the human protein with the A53T mutation caused embryonic loss of DA neurons in the SN whereas expression of the full-length protein did not (Wakamatsu et al., 2008). Expression of amino acids 1–120 of the wild-type human protein on a α-synuclein null background only led to decreased striatal DA without loss of DA neurons in SN (Tofaris et al., 2006). Although several α-synuclein-null mice and transgenic overexpression mutations have been created, none exhibited consistent neuronal degeneration of DA terminals.

PD Caused by Mutations in the Parkin Gene (PARK2)

The parkin gene, which maps to chromosome 6, encodes a 465 amino acid protein containing an N-terminal ubiquitin-like domain, a central linker region, and C-terminal RING domain. The parkin protein functions as an E3 ubiquitin protein ligase, and is involved in the degradation of cellular proteins by the proteasomal pathway. The loss of parkin's E3 ligase activity due to mutations leads to autosomal recessive juvenile PD (Kitada et al., 1998; Shimura et al., 2000; Zhang et al.,

2000). Mutations in parkin were first identified in 1998 in Japanese patients with autosomal recessive juvenile Parkinsonism (Kitada et al., 1998). About 50% of the mutations found in parkin are point mutations. The remaining 50% consist of genomic rearrangements. By targeting different exons of the parkin gene, several parkin knock-out mice were generated. In mice, exon 3 deletion did not affect the number of nigral DA neurons. However, the mice exhibited behavioural deficits that are associated with the basal ganglia function and have decreased DA release in response to amphetamine (Goldberg et al., 2003). Similar to exon 3 deletion, exon 7 deletion did not affect the nigral neuron numbers, but decreased TH-producing cells in the locus coeruleus (Von Coelln et al., 2004). Mice with a knock-out of exon 2 exhibited age-related declines in striatal DA and an increase in D1/D2 receptor binding. Behavioural testing and immuno-labeling of dopaminergic nigral neurons revealed no abnormalities compared to WT mice (Sato et al., 2006). Overall, parkin knock-out mice fail to develop a Parkinsonian phenotype, but the different knock-out models generated may provide means to examine the role of parkin in protein turnover, oxidative stress, and mitochondrial dysfunction.

PTEN-Induced Kinase-1 (PINK1) Mutations

The protein PTEN-induced putative kinase 1 (PINK1) was identified to be gene-associated with the PARK6 locus on chromosome 1p36 that is linked to a rare familial form of PD (Valente et al., 2004). Mutations in the PINK1 are a common cause of autosomal recessive PD (Hatano et al., 2004). PINK1 contains 8 exons and encodes a protein of 581 amino acids with a mitochondrial targeting motif and a serine–threonine protein kinase domain. Most reported mutations were distributed throughout the serine–threonine protein kinase domain. Thus, loss of function of kinase activity of PINK1 is the most probable disease mechanism (Silvestri et al., 2005; Valente et al., 2004). To date, no mammalian in vivo studies of PINK1 loss of function have been reported. However, PINK1 loss-of-function mutants in *Drosophila* result in mitochondrial morphological defects in the male germline, muscle, and DA neurons as well as reduced ATP content (Park et al., 2006). These phenotypic effects were attributed to severe mitochondrial dysfunction such as enlargement and fragmentation of christae.

DJ-1 (PARK7) Mutations

The PARK7 locus, localized on chromosome 1p36, has been linked with autosomal recessive early-onset PD. Recent studies have identified mutations in the DJ-1 gene, associated with the PARK7 locus (Bonifati et al., 2003). The first mutations described were a large chromosomal deletion in a Dutch family and a L166P point mutation in an Italian family (Bonifati et al., 2003). DJ-1 is a highly conserved and ubiquitous protein that is widely expressed in both neurons and glia (Bader et al., 2005). DJ-1 knock-out mice show motor impairments and nigrostriatal DA dysfunction associated with reduced DA overflow, resulting in increased reuptake of DA by the DAT (Chen et al., 2005; Goldberg et al., 2005). In agreement with the observations that DJ-1 knock-out mice have enhanced DA reuptake capacity, DJ-1 knock-out mice have enhanced sensitivity to MPTP, which led to increased striatal DA denervation (Kim et al., 2005; Manning-Bog et al., 2007). However, DJ-1 knock-out mice lack SN degeneration, suggesting that loss of DJ-1 function might confer increased susceptibility to Parkinsonism as a result of underlying SN dysfunction.

LRRK2/Dardarin Mutations

Genomewide linkage analysis of a Japanese family with autosomal dominant PD identified a linkage with a genetic locus located on chromosome 12, which has been termed PARK8 (Funayama et al., 2002). Mutations in the leucine-rich repeat kinase 2 gene (the protein has been named dardarin) have been identified in families with autosomal dominant late onset PD (Paisan-Ruiz et al., 2004; Zimprich et al., 2004a, b). The neuropathology associated with LRRK2 mutation consists of nigral neuronal degeneration and gliosis but with variable intraneuronal protein inclusions including LB, tau-positive NFTs, ubiquitin-positive intranuclear and cytoplasmic inclusions, or the absence of distinctive inclusions/aggregates (Funayama et al., 2002; Giasson et al., 2006; Khan et al., 2005; Rajput et al., 2006; Ross et al., 2006; Wszolek et al., 2004). These observations have led to the suggestion that LRRK2 could be a critical central regulator of protein aggregation and deposition relevant to a wide array of neurodegenerative disorders (Taylor et al., 2006). Within the nigrostriatal pathway, LRRK2 is localized at high levels to medium-sized spiny output projection neurons, cholinergic interneurons, and various GABAergic interneuronal subtypes in the caudate putamen, but at markedly lower levels in DA neurons of the SNc (Biskup et al., 2006; Higashi et al., 2007a, b). *Drosophila* LRRK2 mutants displayed reduced female fertility and fecundity, impaired locomotor activity, and a progressive reduction in TH immunostaining and aberrant morphology in certain DA clusters despite normal numbers of DA neurons (Lee et al., 2007). These results suggest that LRRK2 is critical for the integrity of dopaminergic neurons and intact locomotive activity in Drosophila.

3.3 Nonhuman Primate Models

Initial primate models were developed by using toxins that specifically targeted DA neurons, the most successful of which was MPTP (Langston et al., 1984a). In monkeys, MPTP produces an irreversible and severe Parkinsonian syndrome characterized by all of the features of PD, including tremor, rigidity, slowness of movement, postural instability, and freezing. In these animals, the beneficial response to levodopa and development of long-term motor complications to medical therapy, namely dyskinesias, are virtually identical to those seen in PD patients (Bezard et al., 2001; Jenner 2003; Langston et al., 1984a). The findings that the MPTP nonhuman primates exhibit cognitive deficits and autonomic disturbances comparable to patients with PD (Goldstein et al., 2003; Schneider and Pope-Coleman, 1995) bring this model closer to the idiopathic PD.

Mutations in the α-synuclein gene have been shown to cause familiar PD, suggesting that abnormal accumulation of α-synuclein may trigger neurodegeneration (Polymeropoulos et al., 1997). Inasmuch as one of the limitations of the MPTP nonhuman primate model of PD is the absence of the progressive development of the α-synuclein pathology that is the hallmark of idiopathic PD, overexpression of α-synuclein was recently achieved in nonhuman primates. Indeed, unilateral injection of human α-synuclein expressing viral vectors into the SN of adult marmosets caused selective loss of DA neurons accompanied by α-synuclein-positive cytoplasmic inclusions and degenerative changes in TH-positive axons and dendrites as well as motor impairment reminiscent of DA denervation (Kirik et al., 2003). This model did not, however, display the wider clinical Parkinsonian repertoire that can be elicited in the MPTP-lesioned monkey and was not challenged with levodopa to test the reversibility of its motor impairment. In addition to valuably complementing the existing nonhuman primate models, this approach will pave the way for the refining of new therapeutic strategies.

4 Multiple System Atrophy (MSA)

4.1 The Human Disease

Multiple system atrophy (MSA) is a fatal adult-onset neurodegenerative disorder of unknown etiology characterized by autonomic failure and motor impairment resulting from levodopa unresponsive Parkinsonism, cerebellar ataxia, and pyramidal signs. Eighty percent of cases show predominant Parkinsonism (MSA-P) due to underlying striatonigral degeneration (SND), and the remaining 20% develop predominant cerebellar ataxia (MSA-C) associated with olivopontocerebellar atrophy (Wenning et al., 2004). These features result from progressive multisystem neuronal loss that is associated with oligodendroglial α-synuclein inclusions (Lantos 1998). There is a lack of effective therapies particularly for the motor features of MSA. Most patients deteriorate rapidly and survival beyond 10 years after disease onset is unusual. MSA is less common than PD as epidemiological studies estimate a prevalence of 1.9–4.9 people per 100,000 (Chrysostome et al., 2004; Schrag et al., 1999).

Histopathologically, there is variable neuron loss in the striatum, SNc, cerebellum, pons, inferior olives, and intermediolateral column of the spinal cord. Glial pathology includes astrogliosis, microglial activation, and argyrophilic oligodendroglial cytoplasmic inclusions (GCIs) (Papp et al., 1989). In MSA brains, α-synuclein aggregates in the cytoplasm, axons, and nuclei of neurons, and in the nuclei of oligodendroglia (Benarroch 2002; Fearnley and Lees 1990; Lantos 1998; Wenning et al., 1997). Thus, in contrast to neuronal α-synuclein inclusions in PD,

MSA is also characterised by oligodendroglial α-synuclein inclusion pathology, suggesting a unique but poorly understood pathogenic mechanism that could ultimately lead to neuron loss via disturbance of axonal function (Wenning et al., 2008).

4.2 Rodent Animal Models

Inasmuch as the major, although not the only, histopathological feature of MSA-P is nigral and striatal degeneration, the most evident and direct approach to generate animal models of this disease is with double nigral and striatal lesions using specific toxins. This can be achieved by either stereotaxic or systemic lesions. Stereotaxic lesions are essentially performed unilaterally to obtain impairment in paw reaching behaviour and a rotational behaviour induced by either amphetamine or the DA receptor agonist apomorphine. In this case, SNc lesion is done simultaneously or before striatal lesion. For this, DA neurons within the SNc can be stereotaxically lesioned with 6-OHDA applied within the striatum or the MFB. Striatal lesion is usually obtained by stereotaxical injection within the striatum of quinolinic acid (QA). QA is a tryptophan metabolite and a glutamate NMDA agonist with potent excitotoxic effects. Once injected into the striatum QA preferentially induces loss of medium spiny GABAergic neurons, that constitute 90% of the striatal neurons, while sparing most of the remaining interneurons (Figueredo-Cardenas et al., 1998; Foster et al., 1983; Ghorayeb et al., 2001; Stone 1993). This model was first developed by the group of Wenning et al. (1996) that administered 6-OHDA into the left MFB of male Wistar rats, followed 3–4 weeks later by intrastriatal injection of QA into the ipsilateral striatum. The model was used to test the potential efficiency of striatal fetal allografts derived from striatal primordium alone or combined with cografts of ventral mesencephalon. They showed that cografted rats have a reduction in amphetamine-induced rotation but do not improve deficits of more complex behavior. These stereotaxic unilateral double lesion approaches were instrumental in evaluating neuroprotection efficiency and transplantation strategies but they bear several drawbacks as they are invasive, with immediate histological consequences, as opposed to the progressive nature of the disease, and they do not mimic the clinical symptoms observed in the human pathology.

Some of these limitations may be circumvented with systemic lesions that have also been extensively performed to produce animal models of this disorder and that provide a more dynamic approach of the neurodegenerative process and the subsequent behavioural consequences (Fernagut et al., 2004; Stefanova et al., 2003). In these approaches, DA neurons are degenerated following MPTP systemic injection that induces PD-like syndromes in several species including mice and primates (Burns et al., 1983). Selective damage of the striatum is obtained with 3-nitropropionic acid (3-NP), a mycotoxin inhibitor of succinate dehydrogenase (SDH) in most species Przedborski (Alexi et al., 1998; Brouillet et al., 1999), and thus that induces metabolic failure by inhibiting mitochondrial respiration (Alexi et al., 1998; Brouillet et al., 1999; Brouillet and Hantraye 1995; Guyot et al., 1997;

Ludolph et al., 1991). The susceptibility, nature (apoptotic or necrotic), and extent of the striatal damage depend upon the species, animal strain, age, dose administered, and administration schedule (acute versus chronic) (Alexi et al., 1998; Ouary et al., 2000; Pang and Geddes 1997). In mice, 3-NP produces an acute early oxidative stress followed by an apoptotic striatal neuronal death in the following days (Kim and Chan 2001). Mice lesioned with this protocol developed severe and long-lasting motor disorders as assessed with rotarod, pole test, and general locomotor activity measures. Striatal and nigral damage were also evident with significant neuronal loss and astroglial activation (Fernagut et al., 2004).

In general, these double-lesion approaches are considered by many to be too simplistic as they fail to model the MSA pathology closely. For instance, the lesional approach does not induce GCIs inclusions, one of the hallmarks of MSA that is believed to be involved in neurodegeneration (Papp et al., 1989). In this line, the discovery that GCIs contain a significant level of α-synuclein (Spillantini et al., 1998b; Wakabayashi et al., 1998) has led to the development of transgenic animal models overexpressing this protein under the control of the proteolipid-protein (Kahle et al., 2002) or the 20,30-cyclic nucleotide 30-phosphodiesterase promoter (Yazawa et al., 2005). However, none of the generated mice showed a major degeneration in the nigrostriatal pathway although some showed a moderate loss of nigral DA neurons.

Another drawback to the use of the double-lesion models is that neurotoxins can interact, rendering it difficult to control and replicate the extent of the lesion. To overcome these interactions, a model striatonigral degeneration which uses a single unilateral administration of 1-methyl-4-phenylpyridinium ion (MPP^+) into the rat striatum has been developed (Ghorayeb et al., 2002). This resulted in both nigral and striatal degeneration and motor behaviour impairments in relation to this double degeneration.

Researchers also applied 3-NP to α-synuclein transgenic animals hoping to induce striatal degeneration as well. These lesioned transgenic mice showed severe loss of nigral and striatal neurons in addition to astrogliosis and microglial activation reminiscent of the pathology of MSA and thus are considered to be closer to the human disease; they are currently used to test the efficiency of neuroprotective agents (Stefanova et al., 2008).

The fortuitous discovery that transgenic mice overexpressing the α1B-adrenergic receptor bear several features in common with MSA, spurred curiosity among researchers, as implication of the NE transmission in the pathogenesis of MSA was never previously suspected (Zuscik et al., 2000). Although the group that has developed these mice do acknowledge that MSA is not due to a mutated form of this receptor, this transgenic model may nevertheless be useful in dissecting the neurotransmission pathway that might be implicated in this disease. Transgenic mice for this receptor show prominent cerebellum and medulla neurodegeneration as well as moderate to significant degeneration in the basal ganglia, periaquedual gray, spinal cord, thalamus, and cerebral cortex. Brain regions showed positive staining for ubiquitin and α-synuclein, two proteins typically found in inclusion bodies, and caspase-3 expression was documented in the white matter tracts of the striatum and cerebellum. Behaviourally, these transgenic mice had reproductive problems, reduced weight, and reduced locomotor activity that was age related. In addition, these mice showed increased seizures with age and a generalized pattern of brain damage not found in MSA.

4.3 Primate Animal Models

The first effort to model SND as the core neuropathology underlying MSA-P in nonhuman primates was based on the use of selective nigral and striatal neurotoxins, as previously performed to mimic PD and Huntington's disease in monkeys (Brouillet et al., 1999; Langston et al., 1984a). Systemic and sequential chronic administration of the mitochondrial inhibitor 3-NP and MPTP in one nonhuman primate reproduced levodopa-unresponsive Parkinsonism and SND-like pathological changes characteristic of MSA-P (Ghorayeb et al., 2000). Indeed, the administration of MPTP induced a marked levodopa-sensitive Parkinsonian syndrome associated with akinesia, bilateral rigidity, and flexed posture as well as tremor episodes. The subsequent chronic intoxication with 3-NP resulted in a progressive further deterioration of the motor status, and, after the appearance of lower limb dystonia and an abrupt aggravation of Parkinsonism, the dopaminergic responsiveness disappeared except for levodopa-induced orofacial dyskinesias. Histopathologically, this sequential intoxication produced a severe degeneration of the SNc and of the dorsolateral putamen and head of the caudate nucleus comparable with that found in MSA-P. Although this double-lesion primate model of SND may serve as a preclinical testbed for the evaluation of novel therapeutic strategies in MSA-P, its reliability and validity were not tested further.

5 Amyotrophic Lateral Sclerosis (ALS)

5.1 The Human Disease

Amyotrophic lateral sclerosis (ALS) is one of the major forms of motor neuron disease (MND), a heterogeneous group of degenerative disorders causing progressive motor neuron death leading to paralysis and death. Amyotrophic lateral sclerosis is a relatively rare disease with a reported population incidence of between 1.5 and 2.5 per 100,000 per year (Logroscino et al., 2008). This fatal disease results from the degeneration of motor neurons in the motor cortex, brainstem, and spinal cord. The pathogenesis of MND is poorly understood and may include genetic and/or environmental factors, with a common end-stage outcome. There are currently no significant treatments to alter the fatal outcome.

About 10% of ALS cases are familial (FALS), with a Mendelian pattern of inheritance. About 20% of these cases are associated with mutations in the copper/zinc superoxide dismutase 1 gene (SOD1) (Valdmanis and Rouleau 2008). To date, more than 100 different mutations within all exons of the SOD1 gene and its introns have been identified as being involved in the development of chromosome 21q-linked

FALS. The remaining 90% of ALS cases are classified as sporadic (SALS), although there is accumulating evidence that subpopulations of patients with SALS have common inherited susceptibility genes (Greenway et al., 2006).

In SALS, degeneration of the corticospinal tracts in the anterior and lateral columns of the spinal cord is particularly evident. The cytopathology of the affected motor neurons in SALS is characterized by the following two important intracytoplasmic inclusions: the Bunina bodies, which are small eosinophilic intraneuronal inclusions in the remaining lower motor neurons, are generally considered to be a specific pathological hallmark of ALS. Although the nature and significance of Bunina bodies in ALS are not yet clear, the bodies may be abnormal accumulations of unknown proteinous materials (Okamoto et al., 2008). Skeins-like inclusions (SLIs) and round hyaline inclusions (RHIs) in the remaining anterior horn cells are another pathological characteristic finding of ALS. Both inclusions are detected by ubiquitin immunohistochemistry but are negative for phosphorylated neurofilament protein and SOD1.

In FALS, two types can be discriminated by histopathology. One type of FALS is neuropathologically identical to SALS, and frequently contains Bunina bodies. The other form of FALS is that showing posterior column involvement in addition to the pathological features of SALS (Valdmanis and Rouleau, 2008). Neuropathologically, this entity is characterized by the presence of LB-like hyaline inclusions (LBHIs) in the anterior horn cells throughout the spinal cord. It is to be noted that many SOD1-mutated FALS cases are of the posterior column involvement type with neuronal LBHIs and mild corticospinal tract involvement, in contrast to severe degeneration of the lower motor neurons (Kato, 2008).

5.2 Animal Models

Several rodent animal models of ALS have been generated targeting a set of proteins ranging from the SOD1 gene to genes causing neurofilament abnormalities or defects in microtubule-based transport (Cozzolino et al., 2008; Julien and Kriz 2006; Kato, 2008). To date, SOD1 mutants are widely considered as the closest mutants to the human pathology despite the fact that it is still debated how mutations in SOD1 gene may lead to ALS syndromes. In this line, an impressive number of SOD1 transgenic rodents expressing various SOD1 mutations have been generated, most replicating rather efficiently many behavioural and anatomical features of ALS. Because the pathology is believed to be due to a gain of function following SOD1 mutation, the main difference in these lines seems to be the number of copies of SOD1 mRNA expressed. Indeed, the toxicity of the SOD1 mutation does not seem to be related to a decreased enzymatic activity as some mutants actually show an increased activity whereas knock-out animals for SOD1 show almost no motor neuron death (Reaume et al., 1996).

Studies on transgenic mice expressing various SOD1 mutants have generated a wealth of information. Although no clear picture can be drawn, it is now admitted that multiple cascades of events are involved in motor neuron death that are independent of the enzymatic activity involving the copper catalytic site but related to the aggregation of misfolded mutant SOD1 (Chou et al., 1998; Hyun et al., 2003; Jonsson et al., 2004; Takamiya et al., 2003). How this is related to the events leading to neuron death is yet to be determined, especially given the wide variety of biochemical alterations ranging from excitotoxicity through alerted glutamate transmission, oxidative damage, defects in calcium homeostatsis, caspase activation, mitochondrial malfunction, and cytoskeleton alterations (Guegan et al., 2001; Howland et al., 2002; Liu et al., 2004; Swerdlow et al., 1998; Van Den Bosch et al., 2006). The level of expression of SOD1 seems to be proportional to the life span of the animals; that is, the more copies they express, the shorter time they live, with some animals having up to 40 times increase in the mRNA levels of SOD1 (Jonsson et al., 2006). A caveat of this approach is that one can question the validity of such models inasmuch as high levels of SOD1 protein can produce histopathological artefacts such as the formation of vacuoles. Another factor that should be considered regarding the SOD1 protein is its stability and degradation rate especially in the spinal cord so that even with low levels of protein some mutants show significant motor neuron loss (Sato et al., 2005). This further bolsters the protein aggregation hypothesis as a key element in the histopathology of the disease.

Most of the SOD-1 transgenic mice express motor deficits that start with a mild tremor followed by atrophy of hind limb muscles ultimately leading to a complete paralysis where mice can no longer sustain themselves and are thus sacrificed. The early histopathological feature in SOD1 transgenic mice is formation of perikarya, axonal, and dendritic vacuoles (Wong et al., 1995) that appear before neuronal loss and astrocytosis as early as 4–6 weeks of age in the G93A mice where glycine is substituted to alanine at position 93 (Zhang et al., 1997). At this time-point, mice are still asymptomatic as the first symptoms appear at 3 months of age when loss of large motor neurons is observed in the spinal cord with massive vacuolization. At 5 months of age, mice are paralyzed most probably due to the substantial loss of motor neurons accompanied by marked gliosis, intracellular inclusions reminiscent of LB, and phosphorylated neurofilaments filling few motor neurons (for a review of cell death features in ALS see Cleveland, 1999).

Nonneuronal abnormalities are also thought to be involved in ALS (Bruijn et al., 1997; Howland et al., 2002; Lin et al., 1998; Nagai et al., 2007; Rothstein et al., 1995). For instance, altered reuptake of glutamate by astrocytes through glutamate transporter EAAT2 was observed in mice or rats expressing mutant SOD1 (Vermeiren et al., 2006). This should lead to increased extracellular glutamate and thus substantial activation of glutamate receptors and subsequent increase in intracellular Ca++ homeostasis. Increased cytokine levels indicating inflammatory processes through microglial activation were also reported in transgenic SOD1 mutant mice (Hensley and Floyd, 2002) and in human tissues (Henkel et al., 2004) suggesting that motor neuron degeneration implicates the inflammation processes. As mentioned above, motor neurons of ALS patients contain spheroids that are axonal inclusion bodies essentially composed of intermediate filaments.

Neurofilament and peripherin mutations were reported in rare forms of ALS (Gros-Louis et al., 2004; Leung et al., 2004), leading researchers to develop animal models bearing these mutations (Millecamps et al., 2006). Although these mice developed no evident MND, some exhibited moderate sensorimotor and spatial deficits probably due to the observed reduction in conduction velocity.

Patients with an autosomal recessive form of juvenile ALS show deletion mutations in ALS2 gene coding for alsin, a protein that seems to be involved in the Ras transduction pathway (Yang et al., 2001). ALS2 knock-out mice, however, show mild behavioural abnormalities especially in motor coordination accompanied by discrete and age-related loss of cerebellar Purkinke cells (Cai et al., 2008).

Presently, there is no currently available nonhuman primate model of ALS. Transgenic rodent models that exhibit many of the pathological changes in human ALS provide useful tools for drug testing and essays of genetic manipulations as no effective treatment for ALS has yet been found. Animal models with lower gene copy number encoding the mutant SOD1 proteins and with slower and later onset of disease may prove more appropriate to the human pathology. In addition, one might also question the validity of the mutant SOD1 mice as models of gene defects that account for only 2% of ALS cases.

6 Huntington's Disease (HD)

6.1 The Human Disease

Huntington's disease (HD) is an inherited autosomal dominant progressive neurodegenerative disease that is commonly diagnosed at the age of 35–50 years. Typically, onset of symptoms is in middle age, but the disorder can manifest at any time between infancy and senescence. Its prevalence in North America and Europe varies between 0.5 and 10/100,000; it is highest in populations of western European origin and lowest in African and Asian populations (Harper, 1992).

The underlying genetic cause is an expanded trinucleotide CAG repeat of more than 36 units in the IT15 (for "interesting transcript") gene encoding the huntingtin (HTT) protein in chromosome 4 (1993). This will lead to the production of mutant HTT protein with an abnormally long polyglutamine residue (polyQ). The disease occurs when the critical threshold of about 37 polyQ is exceeded. One important characteristic of HD pathology is the vulnerability of a particular brain region, the caudate–putamen, despite similar expression of the mutated HTT protein with expanded polyQ in other brain areas. The ensuing degeneration with atrophy, neuronal loss, and gliosis, initially involves the striatum, then the cerebral cortex, and eventually degeneration may appear throughout the brain as a constellation of the toxic effect of the mutation and the ensuing secondary changes (Albin, 1995; Vonsattel et al., 2008). Interestingly, not all striatal cells are equally affected by the degenerative process. Immunocytochemical studies combined with neurochemical analysis have consistently shown that HD preferentially affects the GABAergic medium-sized spiny neurons, leaving the other subpopulations of striatal cells largely unaffected, at least in the early course of the disease (Cicchetti et al., 1996). Given that these neurons constitute up to 90% of the striatal neurons in total, the consequences of this degeneration are devastating (Jaber et al., 1996).

The actual causative pathway from the HD gene mutation to neuronal dysfunction and loss has not yet been established but two pathogenic processes have been suggested as the basis for neurodegeneration in HD. One process involves interaction of mutant HTT with other proteins to confer a toxic gain of function. Alternatively, mutant HTT might homodimerize or heterodimerize to build a poorly soluble HTT protein that aggregates within ubiquitinated neuronal intranuclear inclusions and dystrophic neuritis in the HD cortex and striatum (DiFiglia et al., 1997).

Clinically, HD is increasingly recognized as a phenotypically heterogeneous disorder. Its motor features can be conceptually divided into positive and negative. Positive motor features are those characterized by excessive movement, such as chorea and dystonia; conversely, negative motor signs describe a poverty of movement, including bradykinesia and apraxia. These motor symptoms, along with personality changes and cognitive decline, form the classic triad of HD symptoms. Myoclonus, tics, and tremor can also occur as part of the clinical spectrum of HD as well as choreoathetotic movements in the oro-bucco-facial regions that progressively interfere with the voluntary control of vocalisation, chewing, and swallowing. General intellectual abilities show a mild diffuse impairment within one year of onset of overt motor signs, but as the disease progresses, a more severe exacerbation of the early impairments produces a general intellectual state that will approach the range of mental retardation. The diagnosis of HD is established on the basis of genetic testing and to date there is no treatment available to modify the natural course of the disease.

6.2 Rodent Animal Models

Mouse models of HD can be classified into several different categories: (1) transgenic mice expressing exon-1 fragments of the human HTT gene containing polyQ mutations in addition to both alleles of murine wild-type huntingtin (Hdh); (2) knock-in mice with pathogenic CAG repeats inserted within the existing murine Hdh gene; and (3) mice that express the full-length human HTT gene in addition to the murine Hdh.

The first reported transgenic HD mouse was the R6 mouse that overexpresses exon 1 of the mutated human HTT gene under the control of the human corresponding promoter (Mangiarini et al., 1996). This inserted gene harbored up to 120–150 CAG-repeats and the transgene is expressed at 31% of endogenous levels. These mice show a slow progression of the disease and limited nuclear inclusions.

Many lines of R6 mice were generated afterwards; they differed mostly by the length of the repeats and by the level of expression of the transgene. To date, the most used mice are probably the R6/2 mice that contain 150 CAG repeats and that express the transgene at 75% of endogenous levels although the R6/1 line, with a lower number of repeats and expression rate, shows a more progressive course of disease. The R6/2 mice show weight loss and progressive and homogeneous motor deficits that start as early as 5–6 weeks and that become overt by 8 weeks (Carter et al., 1999). These behavioural phenotypes include tremor, clasping, convulsions that can be quantified on rotarod tests, grip strength, and general locomotor activity assessment. Life expectancy of these mice is rather short (death occurs at 10–15 weeks of age), probably due to the extensive length of the CAG repeats which lead researchers to draw a parallel with the juvenile form of HD. Survival rates of R6/2 mice were used by researchers in neuroprotective studies and were shown to correlate well with improved motor behaviour (Dedeoglu et al., 2003; Jin et al., 2005). Histologically, these mice show cortical cerebellar and striatal atrophy, but with very little if any cell loss (Turmaine et al., 2000). Protein aggregates and inclusions containing ubiquitine and HTT proteins were also observed but with an extent and distribution beyond what is found in HD (Davies et al., 1997). In addition, the HTT protein was found within the nucleus of cortical and subcortical neurons as also found in postmortem studies of HD patients' brain and other CAG-repeat diseases (DiFiglia et al., 1997; Gutekunst et al., 1999). Interestingly, as shown by the team of A. Hannan, these mice when raised in an enriched environment show marked behavioural recovery and reduced volume loss implicating environmental conditions in this archetypical genetic disorder (reviewed in Laviola et al., 2008). Another line of mice in this category is the N171-82Q mice that harbor a longer N-terminal fragment of HTT than R6/2 mice with 82 polyQ (Schilling et al., 1999). These mice show striatal atrophy and a greater degree of cell loss but with more heterogeneity in the phenotype than R6/2 mice.

Interestingly, a rat model of transgenic HD with a truncated HTT fragment with 51 repeats under the control of the native HTT promoter exhibits adult-onset neurological phenotypes with progressive motor dysfunction and typical pathological alterations in the form of nuclear inclusions in the brain and shrinkage in striatal volume as well as reduced glucose consumption (von Horsten et al., 2003). The distribution of nuclear inclusions is rather limited as they were observed mainly in the striatum and globus pallidus; neuronal loss is moderate. These rats show progressive weight loss and die prematurely.

The second category of mice with insertions of repeats within the mouse HTT gene showed a discrete behavioural phenotype that was evident only when measures were performed during the night cycle, that is, when mice are known to be generally more active (Menalled and Chesselet, 2002). The mice with 111 CAG repeats inserted into the murine HD gene have a progressively developing nuclear phenotype that is specific for striatal neurons (Wheeler et al., 2000). These ubiquinated nuclear inclusions are seldom found in 10–18-month-old mice. Some reactive gliosis was reported but with no cell loss or reduction in the brain volume whatever the region.

Several lines of mice belonging to the third category (i.e, that harbor the full-length IT15 gene) have been generated. The HD48Q and HD89Q transgenic mice have an insertion of the full-length human IT15 gene under the control of the cytomegalovirus promoter (CMV). They show a progressive behavioural phenotype and striatal but also Purkinje neuronal loss, with a small degree of nuclear inclusions (Reddy et al., 1998).

Alternative cloning vectors have been developed; they can be used for genomic fragments of up to 2 mb for the yeast artificial chromosomes (YAC) (Slow et al., 2003) and up to 100 kb for bacterial artificial chromosomes (BAC) (Giraldo and Montoliu, 2001). The YAC transgenic mice expressing the human IT15 gene with 48–128 repeats show a slow disease progression with motor abnormalities that range from initial hyperactivity, to impaired motor coordination and finally to hypokinesia. These behavioural changes are accompanied by almost exclusive striatal cell loss as well as nuclear and neuropil aggregates but in a lesser extension than the R6/2 mice. The BAC mice with 226 CAG repeats show tremor, head bobbing, and curling at 3 months of age followed by hypoactivity at 6 months of age, then death. Selective striatal and cerebral cortex neuronal loss was documented.

6.3 Invertebrate Animal Models

Drosophila and *Caenorhabditis elegans* animal models were also used by researchers for screening purposes of genes and pathways that might be involved in neurodegenerative diseases or that might help manage the disorder. The use of these simple models, that present nevertheless several features of neuronal functions in higher organisms, has increased recently as they offer a unique opportunity to dissect detailed mechanisms related to the development of neurodegenerative disorders. The first reports of polyQ repeat reported insertions of fragments of the human HTT gene that resulted in perinuclear cytoplasmic protein aggregation with repeats up to 150-fold but not with a lower number of repeats (2–95) (Faber et al., 1999; Satyal et al., 2000). This model has been used to identify evolutionary conserved suppressors of polyQ toxicity such as PQE-1 which invalidation exacerbated neurodegeneration and cell death and which overexpression was protective (Faber et al., 1999).

PolyQ insertions in *Drosophila* animal models yielded cell death and aggregate formation. Suppressor screen studies identified protein folding and clearance, RNA maturation, and gene expression as essential steps in HD (Kazemi-Esfarjani and Benzer, 2000). Indeed, two suppressors were identified that contain a chaperone-related J domain. One suppressor gene, dHDJ1, is homologous to human heat shock protein 40/HDJ1 whereas the second, dTPR2, is homologous to the human tetratricopeptide repeat protein 2.

However, caution needs to be exercised when interpreting results obtained in these simple animal models as they do not express the mutant gene in the same cellular phenotype as in humans and intracellular pathways can sometimes be very different from higher model organisms.

6.4 Primate Animal Models

6.4.1 Lesioning Approaches

Earlier studies of HD most often used direct intrastriatal injection of kainaite or QA, a non-NMDA, and NMDA glutamate agonists, to mimic the axon-sparing striatal lesion observed in HD (Ferrante et al., 1993; McGeer and McGeer, 1976). However, as excitotoxic striatal lesions do not elicit persistent spontaneous motor symptoms this has led to the generation of toxin-induced models to study mitochondrial impairment and excitotoxicity-induced cell death, which are both mechanisms of degeneration seen in the HD brain. These models, most of them based on 3-NP lesioning, are often used in HD studies (Brouillet et al., 1999). Interestingly, whereas the neurodegenerative effects were preferentially localized within the striatum, the decrease in SDH activity for a given dose of 3-NP was shown to be homogeneously distributed throughout the brain (Brouillet et al., 1998). The toxic effects of 3-NP in the human were first discovered when farmers from China ingested sugarcane contaminated with the fungus *Arthrinium*. The metabolism of this fungus produces 3-NP which invariably caused cell death in the caudate and putamen with consequent appearance of persistent and severe dystonia in these intoxicated individuals (Ludolph et al., 1991).

Systemic injection of 3-NP in nonhuman primates showed that a partial but prolonged energy impairment induced by the toxin is sufficient to replicate most of the clinical and pathophysiological hallmarks of HD, including spontaneous choreiform and dystonic movements, frontal-type cognitive deficits, and progressive heterogeneous striatal degeneration with preferential degeneration of the medium-sized spiny GABAergic neurons and a relative sparing of interneurons and afferents, as observed in HD striatum (Brouillet et al., 1999).

6.4.2 Genetic Approaches

Genetic approaches using either local transfer of mutated HTT into the monkey striatum (Palfi et al., 2007) or, more interestingly, gene introduction into oocytes (Yang et al., 2008) seem to be the way forward in establishing HD models which closely replicate the pathogenesis of the human disease. By inserting a virus vector carrying part of the mutated human HTT gene, with 84 CAG repeats, into unfertilised monkey egg cells, a transgenic model of HD in a rhesus macaque that expresses polyQ-expanded HTT was developed. Hallmark features of HD, including nuclear inclusions and neuropil aggregates, were observed in the brains of this model. Additionally, the transgenic monkeys showed important clinical features of HD, including dystonia and chorea (Yang et al., 2008). Because the nonhuman primates show neuroanatomical and behavioural characteristics that closely resemble those of humans, a transgenic model in monkeys may prove to be the gold-standard animal model of neurodegenerative diseases and pave the way to generating nonhuman primate models for other neurological conditions that are caused by single-gene mutations, such as familial forms of PD, AD, and ALS.

7 Conclusion

The tremendous amount of research focused on animal models of neurodegenerative diseases, and the impressive amount of data generated, clearly illustrate the significance of their use as a valuable research tool. However, research performed so far has also highlighted discrepancies between models and human neuropathology leading to question the pertinence of some of these findings to human disorders. As detailed above, a given pathology can be mirrored by numerous different animal models and determining which data obtained from these models are relevant to human pathology is problematic. Indeed, a mouse model simply carrying a human mutation or lesion is far from replicating the constellation of clinical symptoms, the pathogenic cascades, and the neuroanatomic and neuropathological changes observed in human pathology. This is especially true when the human pathology has no spontaneous equivalent in animals, which the case for most neurodegenerative disorders.

In addition, the nature of the alteration performed in these models to mimic a neurodegenerative disorder, as well as features inherent to the animal models and their housing conditions, also constitute a drawback. For instance, animal models are often young males that are of an inbred species, thus almost genetically identical, and living in a very standardized environment. This is hardly the case of patients suffering from a neurodegenerative disorder. Given the tremendous amount of data currently available pointing to the implication of gender and gene/environment interactions in modulating brain function, one must use caution before translating findings in these animal models to human disorders (Laviola et al., 2008).

Furthermore, the question addressed and the methodology used in the exploration of animal models are among the main factors of variance between clinical research, mostly performed on human subjects and postmortem brains, and more fundamental research on mouse models. A clear and consensus definition of the criteria needed for a given animal model to be considered adequate is hard to reach among scientists and clinicians even for "straightforward pathologies" such as PD implicating mainly, but not only, degeneration of the nigral DA neurons or for HD due to a well-defined genetic mutation. This is due to the wide spectrum of parameters defining a disease such as its onset, the related behavioural consequences, and the underlying neuropathological features, rendering difficult the quest of generating the ultimate animal model. The challenge of obtaining such an ideal animal model is even greater in psychiatric disorders where the closest model to a human pathology is the drug addiction one, as attempts to model complex illnesses such as schizophrenia or depression remain, at best, unsatisfactory. Animal models are nevertheless still generated, sometimes following exquisite and complex constructions, mainly because of the complexities of the human brain and of disease processes and the inherent technical limitations of exploring the human disease by means other than on postmortem brains. Although medical imagery procedures have gained significant and impressive advances this last decade, they do not provide elements to determine the pathogenesis of a disease or the causal chains involved. Thus, and despite their current limitations, animal models of neurodegenerative diseases are still essential elements in the laborious attempts to determine the etiology of a given disease, understand its progression, and the relationship between the observed clinical phenotypes and the histological features or hallmarks of the disease. A growing need is now acknowledged to combine-join-converge research programs between clinicians and basic researchers who should reach for a consensual language. This should help extrapolate findings obtained in animal models to the human pathology and identify and apply means that will prevent or delay, if not cure, the disease.

Acknowledgments GP wishes to thank AIRMA (Association Internationale pour la Recherche sur la Maladie d'Alzheimer), the LECMA (Ligue Européenne Contre la Maladie d'Alzheimer) and The Poitiers University Hospital. Work on animal models of neurodegenerative disorders in the group of MJ and AG is financed by the Fondation de France, Institut pour la Recherche sur la Moelle Epinière and the CNRS interdisciplinary program "Aging".

References

Abbott A (2004) Laboratory animals: the Renaissance rat. Nature 428:464–466

Abeliovich A, Schmitz Y, Farinas I, Choi-Lundberg D, Ho WH, Castillo PE, Shinsky N, Verdugo JM, Armanini M, Ryan A, Hynes M, Phillips H, Sulzer D, Rosenthal A (2000) Mice lacking alpha-synuclein display functional deficits in the nigrostriatal dopamine system. Neuron 25:239–252

Aigner TG, Mishkin M (1986) The effects of physostigmine and scopolamine on recognition memory in monkeys. Behav Neural Biol 45:81–87

Aigner TG, Mitchell SJ, Aggleton JP, DeLong MR, Struble RG, Price DL, Wenk GL, Mishkin M (1987) Effects of scopolamine and physostigmine on recognition memory in monkeys with ibotenic-acid lesions of the nucleus basalis of Meynert. Psychopharmacology (Berl) 92: 292–300

Aigner TG, Mitchell SJ, Aggleton JP, DeLong MR, Struble RG, Price DL, Wenk GL, Pettigrew KD, Mishkin M (1991) Transient impairment of recognition memory following ibotenic-acid lesions of the basal forebrain in macaques. Exp Brain Res 86:18–26

Albin RL (1995) Selective neurodegeneration in Huntington's disease. Ann Neurol 38: 835–836

Alexi T, Hughes PE, Faull RL, Williams CE (1998) 3-Nitropropionic acid's lethal triplet: cooperative pathways of neurodegeneration. Neuroreport 9:R57–R64

Andra K, Abramowski D, Duke M, Probst A, Wiederhold KH, Burki K, Goedert M, Sommer B, Staufenbiel M (1996) Expression of APP in transgenic mice: a comparison of neuron-specific promoters. Neurobiol Aging 17:183–190

Bader V, Ran Zhu X, Lubbert H, Stichel CC (2005) Expression of DJ-1 in the adult mouse CNS. Brain Res 1041:102–111

Bales KR, Verina T, Cummins DJ, Du Y, Dodel RC, Saura J, Fishman CE, DeLong CA, Piccardo P, Petegnief V, Ghetti B, Paul SM (1999) Apolipoprotein E is essential for amyloid deposition in the APP(V717F) transgenic mouse model of Alzheimer's disease. Proc Natl Acad Sci USA 96:15233–15238

Barrier L, Ingrand S, Damjanac M, Rioux Bilan A, Hugon J, Page G (2007) Genotype-related changes of ganglioside composition in brain regions of transgenic mouse models of Alzheimer's disease. Neurobiol Aging 28:1863–1872

Barrier L, Ingrand S, Fauconneau B, Page G (2008) Gender-dependent accumulation of ceramides in the cerebral cortex of the APP(SL)/PS1Ki mouse model of Alzheimer's disease. Neurobiol Aging

Bartus RT, Dean RL, Fleming DL (1979) Aging in the rhesus monkey: effects on visual discrimination learning and reversal learning. J Gerontol 34:209–219
Bartus RT, Fleming D, Johnson HR (1978) Aging in the rhesus monkey: debilitating effects on short-term memory. J Gerontol 33:858–871
Bartus RT, Johnson HR (1976) Short-term memory in the rhesus monkey: disruption from the anti-cholinergic scopolamine. Pharmacol Biochem Behav 5:39–46
Bayer TA, Wirths O (2008) Review on the APP/PS1KI mouse model: intraneuronal Abeta accumulation triggers axonopathy, neuron loss and working memory impairment. Genes Brain Behav 7(Suppl 1):6–11
Beal MF (2001) Experimental models of Parkinson's disease. Nat Rev Neurosci 2:325–334
Benarroch EE (2002) New findings on the neuropathology of multiple system atrophy. Auton Neurosci 96:59–62
Bertram L, Tanzi RE (2008) Thirty years of Alzheimer's disease genetics: the implications of systematic meta-analyses. Nat Rev Neurosci 9:768–778
Bezard E, Boraud T, Bioulac B, Gross CE (1999) Involvement of the subthalamic nucleus in glutamatergic compensatory mechanisms. Eur J Neurosci 11:2167–2170
Bezard E, Brotchie JM, Gross CE (2001) Pathophysiology of levodopa-induced dyskinesia: potential for new therapies. Nat Rev Neurosci 2:577–588
Biskup S, Moore DJ, Celsi F, Higashi S, West AB, Andrabi SA, Kurkinen K, Yu SW, Savitt JM, Waldvogel HJ, Faull RL, Emson PC, Torp R, Ottersen OP, Dawson TM, Dawson VL (2006) Localization of LRRK2 to membranous and vesicular structures in mammalian brain. Ann Neurol 60:557–569
Blanchard V, Moussaoui S, Czech C, Touchet N, Bonici B, Planche M, Canton T, Jedidi I, Gohin M, Wirths O, Bayer TA, Langui D, Duyckaerts C, Tremp G, Pradier L (2003) Time sequence of maturation of dystrophic neurites associated with Abeta deposits in APP/PS1 transgenic mice. Exp Neurol 184:247–263
Boncristiano S, Calhoun ME, Kelly PH, Pfeifer M, Bondolfi L, Stalder M, Phinney AL, Abramowski D, Sturchler-Pierrat C, Enz A, Sommer B, Staufenbiel M, Jucker M (2002) Cholinergic changes in the APP23 transgenic mouse model of cerebral amyloidosis. J Neurosci 22:3234–3243
Bonifati V, Rizzu P, van Baren MJ, Schaap O, Breedveld GJ, Krieger E, Dekker MC, Squitieri F, Ibanez P, Joosse M, van Dongen JW, Vanacore N, van Swieten JC, Brice A, Meco G, van Duijn CM, Oostra BA, Heutink P (2003) Mutations in the DJ-1 gene associated with autosomal recessive early-onset parkinsonism. Science 299:256–259
Botcherby M (2002) Harvesting the mouse genome. Comp Funct Genomics 3:319–324
Braak H, Del Tredici K, Rub U, de Vos RA, Jansen Steur EN, Braak E (2003) Staging of brain pathology related to sporadic Parkinson's disease. Neurobiol Aging 24: 197–211
Brecht WJ, Harris FM, Chang S, Tesseur I, Yu GQ, Xu Q, Dee Fish J, Wyss-Coray T, Buttini M, Mucke L, Mahley RW, Huang Y (2004) Neuron-specific apolipoprotein e4 proteolysis is associated with increased tau phosphorylation in brains of transgenic mice. J Neurosci 24:2527–2534
Breese GR, Traylor TD (1971) Depletion of brain noradrenaline and dopamine by 6-hydroxydopamine. Br J Pharmacol 42:88–99
Brookmeyer R, Gray S, Kawas C (1998) Projections of Alzheimer's disease in the United States and the public health impact of delaying disease onset. Am J Public Health 88: 1337–1342
Brouillet E, Conde F, Beal MF, Hantraye P (1999) Replicating Huntington's disease phenotype in experimental animals. Prog Neurobiol 59:427–468
Brouillet E, Guyot MC, Mittoux V, Altairac S, Conde F, Palfi S, Hantraye P (1998) Partial inhibition of brain succinate dehydrogenase by 3-nitropropionic acid is sufficient to initiate striatal degeneration in rat. J Neurochem 70:794–805
Brouillet E, Hantraye P (1995) Effects of chronic MPTP and 3-nitropropionic acid in nonhuman primates. Curr Opin Neurol 8:469–473

Bruijn LI, Becher MW, Lee MK, Anderson KL, Jenkins NA, Copeland NG, Sisodia SS, Rothstein JD, Borchelt DR, Price DL, Cleveland DW (1997) ALS-linked SOD1 mutant G85R mediates damage to astrocytes and promotes rapidly progressive disease with SOD1-containing inclusions. Neuron 18:327–338

Burns RS, Chiueh CC, Markey SP, Ebert MH, Jacobowitz DM, Kopin IJ (1983) A primate model of parkinsonism: selective destruction of dopaminergic neurons in the pars compacta of the substantia nigra by N-methyl-4-phenyl-1,2,3,6-tetrahydropyridine. Proc Natl Acad Sci USA 80:4546–4550

Cai H, Shim H, Lai C, Xie C, Lin X, Yang WJ, Chandran J (2008) ALS2/alsin knockout mice and motor neuron diseases. Neurodegener Dis 5:359–366

Calhoun ME, Burgermeister P, Phinney AL, Stalder M, Tolnay M, Wiederhold KH, Abramowski D, Sturchler-Pierrat C, Sommer B, Staufenbiel M, Jucker M (1999) Neuronal overexpression of mutant amyloid precursor protein results in prominent deposition of cerebrovascular amyloid. Proc Natl Acad Sci USA 96:14088–14093

Carter DB, Dunn E, McKinley DD, Stratman NC, Boyle TP, Kuiper SL, Oostveen JA, Weaver RJ, Boller JA, Gurney ME (2001) Human apolipoprotein E4 accelerates beta-amyloid deposition in APPsw transgenic mouse brain. Ann Neurol 50:468–475

Carter RJ, Lione LA, Humby T, Mangiarini L, Mahal A, Bates GP, Dunnett SB, Morton AJ (1999) Characterization of progressive motor deficits in mice transgenic for the human Huntington's disease mutation. J Neurosci 19:3248–3257

Casas C, Sergeant N, Itier JM, Blanchard V, Wirths O, van der Kolk N, Vingtdeux V, van de Steeg E, Ret G, Canton T, Drobecq H, Clark A, Bonici B, Delacourte A, Benavides J, Schmitz C, Tremp G, Bayer TA, Benoit P, Pradier L (2004) Massive CA1/2 neuronal loss with intraneuronal and N-terminal truncated Abeta42 accumulation in a novel Alzheimer transgenic model. Am J Pathol 165:1289–1300

Cauchi RJ, van den Heuvel M (2006) The fly as a model for neurodegenerative diseases: is it worth the jump? Neurodegener Dis 3:338–356

Cetin F, Dincer S (2007) The effect of intrahippocampal beta amyloid (1–42) peptide injection on oxidant and antioxidant status in rat brain. Ann N Y Acad Sci 1100:510–517

Chen L, Cagniard B, Mathews T, Jones S, Koh HC, Ding Y, Carvey PM, Ling Z, Kang UJ, Zhuang X (2005) Age-dependent motor deficits and dopaminergic dysfunction in DJ-1 null mice. J Biol Chem 280:21418–21426

Chishti MA, Yang DS, Janus C, Phinney AL, Horne P, Pearson J, Strome R, Zuker N, Loukides J, French J, Turner S, Lozza G, Grilli M, Kunicki S, Morissette C, Paquette J, Gervais F, Bergeron C, Fraser PE, Carlson GA, George-Hyslop PS, Westaway D (2001) Early-onset amyloid deposition and cognitive deficits in transgenic mice expressing a double mutant form of amyloid precursor protein 695. J Biol Chem 276:21562–21570

Chou SM, Wang HS, Taniguchi A, Bucala R (1998) Advanced glycation endproducts in neurofilament conglomeration of motoneurons in familial and sporadic amyotrophic lateral sclerosis. Mol Med 4:324–332

Christensen DZ, Bayer TA, Wirths O (2008a) Intracellular Abeta triggers neuron loss in the cholinergic system of the APP/PS1KI mouse model of Alzheimer's disease. Neurobiol Aging

Christensen DZ, Kraus SL, Flohr A, Cotel MC, Wirths O, Bayer TA (2008b) Transient intraneuronal A beta rather than extracellular plaque pathology correlates with neuron loss in the frontal cortex of APP/PS1KI mice. Acta Neuropathol 116:647–655

Chrysostome V, Tison F, Yekhlef F, Sourgen C, Baldi I, Dartigues JF (2004) Epidemiology of multiple system atrophy: a prevalence and pilot risk factor study in Aquitaine, France. Neuroepidemiology 23:201–208

Cicchetti F, Gould PV, Parent A (1996) Sparing of striatal neurons coexpressing calretinin and substance P (NK1) receptor in Huntington's disease. Brain Res 730:232–237

Cleveland DW (1999) From Charcot to SOD1: mechanisms of selective motor neuron death in ALS. Neuron 24:515–520

Cookson MR (2005) The biochemistry of Parkinson's disease. Annu Rev Biochem 74:29–52
Corder EH, Saunders AM, Strittmatter WJ, Schmechel DE, Gaskell PC, Small GW, Roses AD, Haines JL, Pericak-Vance MA (1993) Gene dose of apolipoprotein E type 4 allele and the risk of Alzheimer's disease in late onset families. Science 261:921–923
Cotel MC, Bayer TA, Wirths O (2008) Age-dependent loss of dentate gyrus granule cells in APP/PS1KI mice. Brain Res 1222:207–213
Cozzolino M, Ferri A, Carri MT (2008) Amyotrophic lateral sclerosis: from current developments in the laboratory to clinical implications. Antioxid Redox Signal 10:405–443
Damjanac M, Rioux Bilan A, Barrier L, Pontcharraud R, Anne C, Hugon J, Page G (2007) Fluoro-Jade B staining as useful tool to identify activated microglia and astrocytes in a mouse transgenic model of Alzheimer's disease. Brain Res 1128:40–49
Dauer W, Kholodilov N, Vila M, Trillat AC, Goodchild R, Larsen KE, Staal R, Tieu K, Schmitz Y, Yuan CA, Rocha M, Jackson-Lewis V, Hersch S, Sulzer D, Przedborski S, Burke R, Hen R (2002) Resistance of alpha -synuclein null mice to the parkinsonian neurotoxin MPTP. Proc Natl Acad Sci USA 99:14524–14529
Dauer W, Przedborski S (2003) Parkinson's disease: mechanisms and models. Neuron 39: 889–909
Davies SW, Turmaine M, Cozens BA, DiFiglia M, Sharp AH, Ross CA, Scherzinger E, Wanker EE, Mangiarini L, Bates GP (1997) Formation of neuronal intranuclear inclusions underlies the neurological dysfunction in mice transgenic for the HD mutation. Cell 90:537–548
Dedeoglu A, Kubilus JK, Yang L, Ferrante KL, Hersch SM, Beal MF, Ferrante RJ (2003) Creatine therapy provides neuroprotection after onset of clinical symptoms in Huntington's disease transgenic mice. J Neurochem 85:1359–1367
DiFiglia M, Sapp E, Chase KO, Davies SW, Bates GP, Vonsattel JP, Aronin N (1997) Aggregation of huntingtin in neuronal intranuclear inclusions and dystrophic neurites in brain. Science 277:1990–1993
Duyckaerts C, Potier MC, Delatour B (2008) Alzheimer disease models and human neuropathology: similarities and differences. Acta Neuropathol 115:5–38
Echeverria V, Ducatenzeiler A, Alhonen L, Janne J, Grant SM, Wandosell F, Muro A, Baralle F, Li H, Duff K, Szyf M, Cuello AC (2004) Rat transgenic models with a phenotype of intracellular Abeta accumulation in hippocampus and cortex. J Alzheimers Dis 6:209–219
Edbauer D, Winkler E, Regula JT, Pesold B, Steiner H, Haass C (2003) Reconstitution of gamma-secretase activity. Nat Cell Biol 5:486–488
Eikelenboom P, Veerhuis R, Scheper W, Rozemuller AJ, van Gool WA, Hoozemans JJ (2006) The significance of neuroinflammation in understanding Alzheimer's disease. J Neural Transm 113:1685–1695
Elbaz A, Bower JH, Maraganore DM, McDonnell SK, Peterson BJ, Ahlskog JE, Schaid DJ, Rocca WA (2002) Risk tables for parkinsonism and Parkinson's disease. J Clin Epidemiol 55:25–31
Faber PW, Alter JR, MacDonald ME, Hart AC (1999) Polyglutamine-mediated dysfunction and apoptotic death of a Caenorhabditis elegans sensory neuron. Proc Natl Acad Sci USA 96: 179–184
Fahn S (2003) Description of Parkinson's disease as a clinical syndrome. Ann N Y Acad Sci 991:1–14
Fearnley JM, Lees AJ (1990) Striatonigral degeneration. A clinicopathological study. Brain 113(Pt 6):1823–1842
Fernagut PO, Diguet E, Bioulac B, Tison F (2004) MPTP potentiates 3-nitropropionic acid-induced striatal damage in mice: reference to striatonigral degeneration. Exp Neurol 185:47–62
Ferrante RJ, Kowall NW, Cipolloni PB, Storey E, Beal MF (1993) Excitotoxin lesions in primates as a model for Huntington's disease: histopathologic and neurochemical characterization. Exp Neurol 119:46–71
Figueredo-Cardenas G, Harris CL, Anderson KD, Reiner A (1998) Relative resistance of striatal neurons containing calbindin or parvalbumin to quinolinic acid-mediated excitotoxicity compared to other striatal neuron types. Exp Neuro 149:356–372

Fitten LJ, Perryman K, Tachiki K, Kling A (1988) Oral tacrine administration in middle-aged monkeys: effects on discrimination learning. Neurobiol Aging 9:221–224

Flood DG, Lin YG, Lang DM, Trusko SP, Hirsch JD, Savage MJ, Scott RW, Howland DS (2007) A transgenic rat model of Alzheimer's disease with extracellular Abeta deposition. Neurobiol Aging

Folkesson R, Malkiewicz K, Kloskowska E, Nilsson T, Popova E, Bogdanovic N, Ganten U, Ganten D, Bader M, Winblad B, Benedikz E (2007) A transgenic rat expressing human APP with the Swedish Alzheimer's disease mutation. Biochem Biophys Res Commun 358:777–782

Foster AC, Collins JF, Schwarcz R (1983) On the excitotoxic properties of quinolinic acid, 2,3-piperidine dicarboxylic acids and structurally related compounds. Neuropharmacology 22:1331–1342

Fryer JD, Taylor JW, DeMattos RB, Bales KR, Paul SM, Parsadanian M, Holtzman DM (2003) Apolipoprotein E markedly facilitates age-dependent cerebral amyloid angiopathy and spontaneous hemorrhage in amyloid precursor protein transgenic mice. J Neurosci 23: 7889–7896

Fu AL, Dong ZH, Sun MJ (2006) Protective effect of N-acetyl-L-cysteine on amyloid beta-peptide-induced learning and memory deficits in mice. Brain Res 1109:201–206

Funayama M, Hasegawa K, Kowa H, Saito M, Tsuji S, Obata F (2002) A new locus for Parkinson's disease (PARK8) maps to chromosome 12p11.2-q13.1. Ann Neurol 51:296–301

Garcia-Alloza M, Robbins EM, Zhang-Nunes SX, Purcell SM, Betensky RA, Raju S, Prada C, Greenberg SM, Bacskai BJ, Frosch MP (2006) Characterization of amyloid deposition in the APPswe/PS1dE9 mouse model of Alzheimer disease. Neurobiol Dis 24:516–524

Gearing M, Rebeck GW, Hyman BT, Tigges J, Mirra SS (1994) Neuropathology and apolipoprotein E profile of aged chimpanzees: implications for Alzheimer disease. Proc Natl Acad Sci USA 91:9382–9386

Gearing M, Tigges J, Mori H, Mirra SS (1997) beta-Amyloid (A beta) deposition in the brains of aged orangutans. Neurobiol Aging 18:139–146

Ghorayeb I, Fernagut PO, Aubert I, Bezard E, Poewe W, Wenning GK, Tison F (2000) Toward a primate model of L-dopa-unresponsive parkinsonism mimicking striatonigral degeneration. Mov Disord 15:531–536

Ghorayeb I, Fernagut PO, Hervier L, Labattu B, Bioulac B, Tison F (2002) A 'single toxin-double lesion' rat model of striatonigral degeneration by intrastriatal 1-methyl-4-phenylpyridinium ion injection: a motor behavioural analysis. Neuroscience 115:533–546

Ghorayeb I, Puschban Z, Fernagut PO, Scherfler C, Rouland R, Wenning GK, Tison F (2001) Simultaneous intrastriatal 6-hydroxydopamine and quinolinic acid injection: a model of early-stage striatonigral degeneration. Exp Neurol 167:133–147

Giasson BI, Covy JP, Bonini NM, Hurtig HI, Farrer MJ, Trojanowski JQ, Van Deerlin VM (2006) Biochemical and pathological characterization of Lrrk2. Ann Neurol 59:315–322

Giovanni A, Sieber BA, Heikkila RE, Sonsalla PK (1991) Correlation between the neostriatal content of the 1-methyl-4-phenylpyridinium species and dopaminergic neurotoxicity following 1-methyl-4-phenyl-1,2,3,6-tetrahydropyridine administration to several strains of mice. J Pharmacol Exp Ther 257:691–697

Giraldo P, Montoliu L (2001) Size matters: use of YACs, BACs and PACs in transgenic animals. Transgenic Res 10:83–103

Glinka YY, Youdim MB (1995) Inhibition of mitochondrial complexes I and IV by 6-hydroxydopamine. Eur J Pharmacol 292:329–332

Goedert M, Spillantini MG, Cairns NJ, Crowther RA (1992) Tau proteins of Alzheimer paired helical filaments: abnormal phosphorylation of all six brain isoforms. Neuron 8: 159–168

Goldberg MS, Fleming SM, Palacino JJ, Cepeda C, Lam HA, Bhatnagar A, Meloni EG, Wu N, Ackerson LC, Klapstein GJ, Gajendiran M, Roth BL, Chesselet MF, Maidment NT, Levine MS, Shen J (2003) Parkin-deficient mice exhibit nigrostriatal deficits but not loss of dopaminergic neurons. J Biol Chem. 278:43628–43635

Goldberg MS, Pisani A, Haburcak M, Vortherms TA, Kitada T, Costa C, Tong Y, Martella G, Tscherter A, Martins A, Bernardi G, Roth BL, Pothos EN, Calabresi P, Shen J (2005) Nigrostriatal dopaminergic deficits and hypokinesia caused by inactivation of the familial Parkinsonism-linked gene DJ-1. Neuron 45:489–496

Goldstein DS, Li ST, Holmes C, Bankiewicz K (2003) Sympathetic innervation in the 1-methyl-4-phenyl-1,2,3,6-tetrahydropyridine primate model of Parkinson's disease. J Pharmacol Exp Ther 306:855–860

Gonzalo-Ruiz A, Perez JL, Sanz JM, Geula C, Arevalo J (2006) Effects of lipids and aging on the neurotoxicity and neuronal loss caused by intracerebral injections of the amyloid-beta peptide in the rat. Exp Neurol 197:41–55

Gotz J, Chen F, van Dorpe J, Nitsch RM (2001) Formation of neurofibrillary tangles in P301l tau transgenic mice induced by Abeta 42 fibrils. Science 293:1491–1495

Gotz J, Ittner LM (2008) Animal models of Alzheimer's disease and frontotemporal dementia. Nat Rev Neurosci 9:532–544

Gotz J, Probst A, Spillantini MG, Schafer T, Jakes R, Burki K, Goedert M (1995) Somatodendritic localization and hyperphosphorylation of tau protein in transgenic mice expressing the longest human brain tau isoform. EMBO J 14:1304–1313

Gotz J, Streffer JR, David D, Schild A, Hoerndli F, Pennanen L, Kurosinski P, Chen F (2004) Transgenic animal models of Alzheimer's disease and related disorders: histopathology, behavior and therapy. Mol Psychiatry 9:664–683

Greenway MJ, Andersen PM, Russ C, Ennis S, Cashman S, Donaghy C, Patterson V, Swingler R, Kieran D, Prehn J, Morrison KE, Green A, Acharya KR, Brown RH Jr., Hardiman O (2006) ANG mutations segregate with familial and 'sporadic' amyotrophic lateral sclerosis. Nat Genet 38:411–413

Gros-Louis F, Lariviere R, Gowing G, Laurent S, Camu W, Bouchard JP, Meininger V, Rouleau GA, Julien JP (2004) A frameshift deletion in peripherin gene associated with amyotrophic lateral sclerosis. J Biol Chem 279:45951–45956

Guegan C, Vila M, Rosoklija G, Hays AP, Przedborski S (2001) Recruitment of the mitochondrial-dependent apoptotic pathway in amyotrophic lateral sclerosis. J Neurosci 21:6569–6576

Gutekunst CA, Li SH, Yi H, Mulroy JS, Kuemmerle S, Jones R, Rye D, Ferrante RJ, Hersch SM, Li XJ (1999) Nuclear and neuropil aggregates in Huntington's disease: relationship to neuropathology. J Neurosci 19:2522–2534

Guyot MC, Palfi S, Stutzmann JM, Maziere M, Hantraye P, Brouillet E (1997) Riluzole protects from motor deficits and striatal degeneration produced by systemic 3-nitropropionic acid intoxication in rats. Neuroscience 81:141–149

Hall GF, Yao J, Lee G (1997) Human tau becomes phosphorylated and forms filamentous deposits when overexpressed in lamprey central neurons in situ. Proc Natl Acad Sci USA 94:4733–4738

Hardy J, Allsop D (1991) Amyloid deposition as the central event in the aetiology of Alzheimer's disease. Trends Pharmacol Sci 12:383–388

Harper PS (1992) The epidemiology of Huntington's disease. Hum Genet 89:365–376

Hatano Y, Li Y, Sato K, Asakawa S, Yamamura Y, Tomiyama H, Yoshino H, Asahina M, Kobayashi S, Hassin-Baer S, Lu CS, Ng AR, Rosales RL, Shimizu N, Toda T, Mizuno Y, Hattori N (2004) Novel PINK1 mutations in early-onset parkinsonism. Ann Neurol 56:424–427

Heikkila RE, Hess A, Duvoisin RC (1984) Dopaminergic neurotoxicity of 1-methyl-4-phenyl-1,2,5,6-tetrahydropyridine in mice. Science 224:1451–1453

Henkel JS, Engelhardt JI, Siklos L, Simpson EP, Kim SH, Pan T, Goodman JC, Siddique T, Beers DR, Appel SH (2004) Presence of dendritic cells, MCP-1, and activated microglia/macrophages in amyotrophic lateral sclerosis spinal cord tissue. Ann Neurol 55:221–235

Hensley K, Floyd RA (2002) Reactive oxygen species and protein oxidation in aging: a look back, a look ahead. Arch Biochem Biophys 397:377–383

Higashi S, Biskup S, West AB, Trinkaus D, Dawson VL, Faull RL, Waldvogel HJ, Arai H, Dawson TM, Moore DJ, Emson PC (2007a) Localization of Parkinson's disease-associated LRRK2 in normal and pathological human brain. Brain Res 1155:208–219

Higashi S, Moore DJ, Colebrooke RE, Biskup S, Dawson VL, Arai H, Dawson TM, Emson PC (2007b) Expression and localization of Parkinson's disease-associated leucine-rich repeat kinase 2 in the mouse brain. J Neurochem 100:368–381

Higgins GA, Jacobsen H (2003) Transgenic mouse models of Alzheimer's disease: phenotype and application. Behav Pharmacol 14:419–438

Hirsch EC, Hoglinger G, Rousselet E, Breidert T, Parain K, Feger J, Ruberg M, Prigent A, Cohen-Salmon C, Launay JM (2003) Animal models of Parkinson's disease in rodents induced by toxins: an update. J Neural Transm Suppl 65:89–100

Holcomb LA, Gordon MN, Jantzen P, Hsiao K, Duff K, Morgan D (1999) Behavioral changes in transgenic mice expressing both amyloid precursor protein and presenilin-1 mutations: lack of association with amyloid deposits. Behav Genet 29:177–185

Holcomb L, Gordon MN, McGowan E, Yu X, Benkovic S, Jantzen P, Wright K, Saad I, Mueller R, Morgan D, Sanders S, Zehr C, O'Campo K, Hardy J, Prada CM, Eckman C, Younkin S, Hsiao K, Duff K (1998) Accelerated Alzheimer-type phenotype in transgenic mice carrying both mutant amyloid precursor protein and presenilin 1 transgenes. Nat Med 4: 97–100

Holmes C, Boche D, Wilkinson D, Yadegarfar G, Hopkins V, Bayer A, Jones RW, Bullock R, Love S, Neal JW, Zotova E, Nicoll JA (2008) Long-term effects of Abeta42 immunisation in Alzheimer's disease: follow-up of a randomised, placebo-controlled phase I trial. Lancet 372:216–223

Holtzman DM (2004) In vivo effects of ApoE and clusterin on amyloid-beta metabolism and neuropathology. J Mol Neurosci 23:247–254

Howland DS, Liu J, She Y, Goad B, Maragakis NJ, Kim B, Erickson J, Kulik J, DeVito L, Psaltis G, DeGennaro LJ, Cleveland DW, Rothstein JD (2002) Focal loss of the glutamate transporter EAAT2 in a transgenic rat model of SOD1 mutant-mediated amyotrophic lateral sclerosis (ALS). Proc Natl Acad Sci USA 99:1604–1609

Hyun DH, Lee M, Halliwell B, Jenner P (2003) Proteasomal inhibition causes the formation of protein aggregates containing a wide range of proteins, including nitrated proteins. J Neurochem 86:363–373

Ichitani Y, Okamura H, Matsumoto Y, Nagatsu I, Ibata Y (1991) Degeneration of the nigral dopamine neurons after 6-hydroxydopamine injection into the rat striatum. Brain Res 549: 350–353

Irle E, Markowitsch HJ (1987) Basal forebrain-lesioned monkeys are severely impaired in tasks of association and recognition memory. Ann Neurol 22:735–743

Ishihara T, Zhang B, Higuchi M, Yoshiyama Y, Trojanowski JQ, Lee VM (2001) Age-dependent induction of congophilic neurofibrillary tau inclusions in tau transgenic mice. Am J Pathol 158:555–562

Jaber M, Robinson SW, Missale C, Caron MG (1996) Dopamine receptors and brain function. Neuropharmacology 35:1503–1519

Janus C, Westaway D (2001) Transgenic mouse models of Alzheimer's disease. Physiol Behav 73:873–886

Jarvis MF, Wagner GC (1985) Age-dependent effects of 1-methyl-4-phenyl-1,2,5,6-tetrahydropyridine (MPTP). Neuropharmacology 24:581–583

Jenner P (2003) The MPTP-treated primate as a model of motor complications in PD: primate model of motor complications. Neurology 61:S4–S11

Jeon BS, Jackson-Lewis V, Burke RE (1995) 6-Hydroxydopamine lesion of the rat substantia nigra: time course and morphology of cell death. Neurodegeneration 4:131–137

Jin K, LaFevre-Bernt M, Sun Y, Chen S, Gafni J, Crippen D, Logvinova A, Ross CA, Greenberg DA, Ellerby LM (2005) FGF-2 promotes neurogenesis and neuroprotection and prolongs survival in a transgenic mouse model of Huntington's disease. Proc Natl Acad Sci USA 102:18189–18194

Jones SR, Gainetdinov RR, Wightman RM, Caron MG (1998) Mechanisms of amphetamine action revealed in mice lacking the dopamine transporter. J Neurosci 18:1979–1986

Jonsson PA, Ernhill K, Andersen PM, Bergemalm D, Brannstrom T, Gredal O, Nilsson P, Marklund SL (2004) Minute quantities of misfolded mutant superoxide dismutase-1 cause amyotrophic lateral sclerosis. Brain 127:73–88

Jonsson PA, Graffmo KS, Andersen PM, Brannstrom T, Lindberg M, Oliveberg M, Marklund SL (2006) Disulphide-reduced superoxide dismutase-1 in CNS of transgenic amyotrophic lateral sclerosis models. Brain 129:451–464

Julien JP, Kriz J (2006) Transgenic mouse models of amyotrophic lateral sclerosis. Biochim Biophys Acta 1762:1013–1024

Kahle PJ, Neumann M, Ozmen L, Muller V, Jacobsen H, Spooren W, Fuss B, Mallon B, Macklin WB, Fujiwara H, Hasegawa M, Iwatsubo T, Kretzschmar HA, Haass C (2002) Hyperphosphorylation and insolubility of alpha-synuclein in transgenic mouse oligodendrocytes. EMBO Rep 3:583–588

Kalaria RN, Maestre GE, Arizaga R, Friedland RP, Galasko D, Hall K, Luchsinger JA, Ogunniyi A, Perry EK, Potocnik F, Prince M, Stewart R, Wimo A, Zhang ZX, Antuono P (2008) Alzheimer's disease and vascular dementia in developing countries: prevalence, management, and risk factors. Lancet Neurol 7:812–826

Kato S (2008) Amyotrophic lateral sclerosis models and human neuropathology: similarities and differences. Acta Neuropathol 115:97–114

Kazemi-Esfarjani P, Benzer S (2000) Genetic suppression of polyglutamine toxicity in Drosophila. Science 287:1837–1840

Khan NL, Jain S, Lynch JM, Pavese N, Abou-Sleiman P, Holton JL, Healy DG, Gilks WP, Sweeney MG, Ganguly M, Gibbons V, Gandhi S, Vaughan J, Eunson LH, Katzenschlager R, Gayton J, Lennox G, Revesz T, Nicholl D, Bhatia KP, Quinn N, Brooks D, Lees AJ, Davis MB, Piccini P, Singleton AB, Wood NW (2005) Mutations in the gene LRRK2 encoding dardarin (PARK8) cause familial Parkinson's disease: clinical, pathological, olfactory and functional imaging and genetic data. Brain 128:2786–2796

Kim GW, Chan PH (2001) Oxidative stress and neuronal DNA fragmentation mediate age-dependent vulnerability to the mitochondrial toxin, 3-nitropropionic acid, in the mouse striatum. Neurobiol Dis 8:114–126

Kim RH, Smith PD, Aleyasin H, Hayley S, Mount MP, Pownall S, Wakeham A, You-Ten AJ, Kalia SK, Horne P, Westaway D, Lozano AM, Anisman H, Park DS, Mak TW (2005) Hypersensitivity of DJ-1-deficient mice to 1-methyl-4-phenyl-1,2,3,6-tetrahydropyrindine (MPTP) and oxidative stress. Proc Natl Acad Sci USA 102:5215–5220

Kirik D, Annett LE, Burger C, Muzyczka N, Mandel RJ, Bjorklund A (2003) Nigrostriatal alpha-synucleinopathy induced by viral vector-mediated overexpression of human alpha-synuclein: a new primate model of Parkinson's disease. Proc Natl Acad Sci USA 100:2884–2889

Kitada T, Asakawa S, Hattori N, Matsumine H, Yamamura Y, Minoshima S, Yokochi M, Mizuno Y, Shimizu N (1998) Mutations in the parkin gene cause autosomal recessive juvenile parkinsonism. Nature 392:605–608

Lai ZC, Moss MB, Killiany RJ, Rosene DL, Herndon JG (1995) Executive system dysfunction in the aged monkey: spatial and object reversal learning. Neurobiol Aging 16:947–954

Langston JW, Ballard P, Tetrud JW, Irwin I (1983) Chronic Parkinsonism in humans due to a product of meperidine-analog synthesis. Science 219:979–980

Langston JW, Forno LS, Rebert CS, Irwin I (1984a) Selective nigral toxicity after systemic administration of 1-methyl-4-phenyl-1,2,5,6-tetrahydropyrine (MPTP) in the squirrel monkey. Brain Res 292:390–394

Langston JW, Irwin I, Langston EB, Forno LS (1984b) 1-Methyl-4-phenylpyridinium ion (MPP+): identification of a metabolite of MPTP, a toxin selective to the substantia nigra. Neurosci Lett 48:87–92

Lantos PL (1998) The definition of multiple system atrophy: a review of recent developments. J Neuropathol Exp Neurol 57:1099–1111

Laviola G, Hannan AJ, Macri S, Solinas M, Jaber M (2008) Effects of enriched environment on animal models of neurodegenerative diseases and psychiatric disorders. Neurobiol Dis 31: 159–168

Lee SB, Kim W, Lee S, Chung J (2007) Loss of LRRK2/PARK8 induces degeneration of dopaminergic neurons in Drosophila. Biochem Biophys Res Commun 358:534–539

Leung CL, He CZ, Kaufmann P, Chin SS, Naini A, Liem RK, Mitsumoto H, Hays AP (2004) A pathogenic peripherin gene mutation in a patient with amyotrophic lateral sclerosis. Brain Pathol 14:290–296

Lewis J, Dickson DW, Lin WL, Chisholm L, Corral A, Jones G, Yen SH, Sahara N, Skipper L, Yager D, Eckman C, Hardy J, Hutton M, McGowan E (2001) Enhanced neurofibrillary degeneration in transgenic mice expressing mutant tau and APP. Science 293: 1487–1491

Lewis J, McGowan E, Rockwood J, Melrose H, Nacharaju P, Van Slegtenhorst M, Gwinn-Hardy K, Paul Murphy M, Baker M, Yu X, Duff K, Hardy J, Corral A, Lin WL, Yen SH, Dickson DW, Davies P, Hutton M (2000) Neurofibrillary tangles, amyotrophy and progressive motor disturbance in mice expressing mutant (P301L) tau protein. Nat Genet 25:402–405

Lin CL, Bristol LA, Jin L, Dykes-Hoberg M, Crawford T, Clawson L, Rothstein JD (1998) Aberrant RNA processing in a neurodegenerative disease: the cause for absent EAAT2, a glutamate transporter, in amyotrophic lateral sclerosis. Neuron 20:589–602

Link CD (2005) Invertebrate models of Alzheimer's disease. Genes Brain Behav 4:147–156

Liu J, Lillo C, Jonsson PA, Vande Velde C, Ward CM, Miller TM, Subramaniam JR, Rothstein JD, Marklund S, Andersen PM, Brannstrom T, Gredal O, Wong PC, Williams DS, Cleveland DW (2004) Toxicity of familial ALS-linked SOD1 mutants from selective recruitment to spinal mitochondria. Neuron 43:5–17

Liu L, Orozco IJ, Planel E, Wen Y, Bretteville A, Krishnamurthy P, Wang L, Herman M, Figueroa H, Yu WH, Arancio O, Duff K (2008a) A transgenic rat that develops Alzheimer's disease-like amyloid pathology, deficits in synaptic plasticity and cognitive impairment. Neurobiol Dis 31:46–57

Liu Y, Yoo MJ, Savonenko A, Stirling W, Price DL, Borchelt DR, Mamounas L, Lyons WE, Blue ME, Lee MK (2008b) Amyloid pathology is associated with progressive monoaminergic neurodegeneration in a transgenic mouse model of Alzheimer's disease. J Neurosci 28:13805–13814

Logroscino G, Traynor BJ, Hardiman O, Chio A, Couratier P, Mitchell JD, Swingler RJ, Beghi E (2008) Descriptive epidemiology of amyotrophic lateral sclerosis: new evidence and unsolved issues. J Neurol Neurosurg Psychiatry 79:6–11

Ludolph AC, He F, Spencer PS, Hammerstad J, Sabri M (1991) 3-Nitropropionic acid-exogenous animal neurotoxin and possible human striatal toxin. Can J Neurol Sci 18:492–498

Luo Y, Bolon B, Damore MA, Fitzpatrick D, Liu H, Zhang J, Yan Q, Vassar R, Citron M (2003) BACE1 (beta-secretase) knockout mice do not acquire compensatory gene expression changes or develop neural lesions over time. Neurobiol Dis 14:81–88

Mangiarini L, Sathasivam K, Seller M, Cozens B, Harper A, Hetherington C, Lawton M, Trottier Y, Lehrach H, Davies SW, Bates GP (1996) Exon 1 of the HD gene with an expanded CAG repeat is sufficient to cause a progressive neurological phenotype in transgenic mice. Cell 87:493–506

Manning-Bog AB, Caudle WM, Perez XA, Reaney SH, Paletzki R, Isla MZ, Chou VP, McCormack AL, Miller GW, Langston JW, Gerfen CR, Dimonte DA (2007) Increased vulnerability of nigrostriatal terminals in DJ-1-deficient mice is mediated by the dopamine transporter. Neurobiol Dis 27:141–150

Martin LJ, Sisodia SS, Koo EH, Cork LC, Dellovade TL, Weidemann A, Beyreuther K, Masters C, Price DL (1991) Amyloid precursor protein in aged nonhuman primates. Proc Natl Acad Sci USA 88:1461–1465

Masliah E, Mallory M, Ge N, Alford M, Veinbergs I, Roses AD (1995) Neurodegeneration in the central nervous system of apoE-deficient mice. Exp Neurol 136:107–122

Mattson MP (1997) Cellular actions of beta-amyloid precursor protein and its soluble and fibrillogenic derivatives. Physiol Rev 77:1081–1132

Maurice T, Lockhart BP, Privat A (1996) Amnesia induced in mice by centrally administered beta-amyloid peptides involves cholinergic dysfunction. Brain Res 706:181–193

McGeer EG, McGeer PL (1976) Duplication of biochemical changes of Huntington's chorea by intrastriatal injections of glutamic and kainic acids. Nature 263:517–519

McKhann G, Drachman D, Folstein M, Katzman R, Price D, Stadlan EM (1984) Clinical diagnosis of Alzheimer's disease: report of the NINCDS-ADRDA Work Group under the auspices of Department of Health and Human Services Task Force on Alzheimer's Disease. Neurology 34:939–944

Menalled LB, Chesselet MF (2002) Mouse models of Huntington's disease. Trends Pharmacol Sci 23:32–39

Meredith GE, Kang UJ (2006) Behavioral models of Parkinson's disease in rodents: a new look at an old problem. Mov Disord 21:1595–1606

Millecamps S, Robertson J, Lariviere R, Mallet J, Julien JP (2006) Defective axonal transport of neurofilament proteins in neurons overexpressing peripherin. J Neurochem 98:926–938

Mizuno Y, Sone N, Saitoh T (1987) Effects of 1-methyl-4-phenyl-1,2,3,6-tetrahydropyridine and 1-methyl-4-phenylpyridinium ion on activities of the enzymes in the electron transport system in mouse brain. J Neurochem 48:1787–1793

Moore TL, Killiany RJ, Herndon JG, Rosene DL, Moss MB (2006) Executive system dysfunction occurs as early as middle-age in the rhesus monkey. Neurobiol Aging 27:1484–1493

Moore DJ, Zhang L, Troncoso J, Lee MK, Hattori N, Mizuno Y, Dawson TM, Dawson VL (2005) Association of DJ-1 and parkin mediated by pathogenic DJ-1 mutations and oxidative stress. Hum Mol Genet 14:71–84

Muyllaert D, Terwel D, Kremer A, Sennvik K, Borghgraef P, Devijver H, Dewachter I, Van Leuven F (2008) Neurodegeneration and neuroinflammation in cdk5/p25-inducible mice: a model for hippocampal sclerosis and neocortical degeneration. Am J Pathol 172: 470–485

Nagai M, Re DB, Nagata T, Chalazonitis A, Jessell TM, Wichterle H, Przedborski S (2007) Astrocytes expressing ALS-linked mutated SOD1 release factors selectively toxic to motor neurons. Nat Neurosci 10:615–622

Nicklas WJ, Vyas I, Heikkila RE (1985) Inhibition of NADH-linked oxidation in brain mitochondria by 1-methyl-4-phenyl-pyridine, a metabolite of the neurotoxin, 1-methyl-4-phenyl-1,2,5,6-tetrahydropyridine. Life Sci 36:2503–2508

Oddo S, Caccamo A, Shepherd JD, Murphy MP, Golde TE, Kayed R, Metherate R, Mattson MP, Akbari Y, LaFerla FM (2003) Triple-transgenic model of Alzheimer's disease with plaques and tangles: intracellular Abeta and synaptic dysfunction. Neuron 39:409–421

Ohno M, Sametsky EA, Younkin LH, Oakley H, Younkin SG, Citron M, Vassar R, Disterhoft JF (2004) BACE1 deficiency rescues memory deficits and cholinergic dysfunction in a mouse model of Alzheimer's disease. Neuron 41:27–33

Okamoto K, Mizuno Y, Fujita Y (2008) Bunina bodies in amyotrophic lateral sclerosis. Neuropathology 28:109–115

Ouary S, Bizat N, Altairac S, Menetrat H, Mittoux V, Conde F, Hantraye P, Brouillet E (2000) Major strain differences in response to chronic systemic administration of the mitochondrial toxin 3-nitropropionic acid in rats: implications for neuroprotection studies. Neuroscience 97:521–530

Page G, Rioux Bilan A, Ingrand S, Lafay-Chebassier C, Pain S, Perault Pochat MC, Bouras C, Bayer T, Hugon J (2006) Activated double-stranded RNA-dependent protein kinase and neuronal death in models of Alzheimer's disease. Neuroscience 139:1343–1354

Paisan-Ruiz C, Jain S, Evans EW, Gilks WP, Simon J, van der Brug M, Lopez de Munain A, Aparicio S, Gil AM, Khan N, Johnson J, Martinez JR, Nicholl D, Carrera IM, Pena AS, de Silva R, Lees A, Marti-Masso JF, Perez-Tur J, Wood NW, Singleton AB (2004) Cloning of the gene containing mutations that cause PARK8-linked Parkinson's disease. Neuron 44: 595–600

Palfi S, Brouillet E, Jarraya B, Bloch J, Jan C, Shin M, Conde F, Li XJ, Aebischer P, Hantraye P, Deglon N (2007) Expression of mutated huntingtin fragment in the putamen is sufficient to produce abnormal movement in non-human primates. Mol Ther 15:1444–1451

Pang Z, Geddes JW (1997) Mechanisms of cell death induced by the mitochondrial toxin 3-nitropropionic acid: acute excitotoxic necrosis and delayed apoptosis. J Neurosci 17:3064–3073
Papp MI, Kahn JE, Lantos PL (1989) Glial cytoplasmic inclusions in the CNS of patients with multiple system atrophy (striatonigral degeneration, olivopontocerebellar atrophy and Shy-Drager syndrome). J Neurol Sci 94:79–100
Park J, Lee SB, Lee S, Kim Y, Song S, Kim S, Bae E, Kim J, Shong M, Kim JM, Chung J (2006) Mitochondrial dysfunction in Drosophila PINK1 mutants is complemented by parkin. Nature 441:1157–1161
Perese DA, Ulman J, Viola J, Ewing SE, Bankiewicz KS (1989) A 6-hydroxydopamine-induced selective parkinsonian rat model. Brain Res 494:285–293
Perez SE, Dar S, Ikonomovic MD, DeKosky ST, Mufson EJ (2007) Cholinergic forebrain degeneration in the APPswe/PS1DeltaE9 transgenic mouse. Neurobiol Dis 28:3–15
Periquet M, Fulga T, Myllykangas L, Schlossmacher MG, Feany MB (2007) Aggregated alpha-synuclein mediates dopaminergic neurotoxicity in vivo. J Neurosci 27:3338–3346
Peters A, Rosene DL, Moss MB, Kemper TL, Abraham CR, Tigges J, Albert MS (1996) Neurobiological bases of age-related cognitive decline in the rhesus monkey. J Neuropathol Exp Neurol 55:861–874
Pifl C, Giros B, Caron MG (1993) Dopamine transporter expression confers cytotoxicity to low doses of the parkinsonism-inducing neurotoxin 1-methyl-4-phenylpyridinium. J Neurosci 13:4246–4253
Polymeropoulos MH, Lavedan C, Leroy E, Ide SE, Dehejia A, Dutra A, Pike B, Root H, Rubenstein J, Boyer R, Stenroos ES, Chandrasekharappa S, Athanassiadou A, Papapetropoulos T, Johnson WG, Lazzarini AM, Duvoisin RC, Di Iorio G, Golbe LI, Nussbaum RL (1997) Mutation in the alpha-synuclein gene identified in families with Parkinson's disease. Science 276:2045–2047
Postina R, Schroeder A, Dewachter I, Bohl J, Schmitt U, Kojro E, Prinzen C, Endres K, Hiemke C, Blessing M, Flamez P, Dequenne A, Godaux E, van Leuven F, Fahrenholz F (2004) A disintegrin-metalloproteinase prevents amyloid plaque formation and hippocampal defects in an Alzheimer disease mouse model. J Clin Invest 113:1456–1464
Przedborski S, Chen Q, Vila M, Giasson BI, Djaldatti R, Vukosavic S, Souza JM, Jackson-Lewis V, Lee VM, Ischiropoulos H (2001) Oxidative post-translational modifications of alpha-synuclein in the 1-methyl-4-phenyl-1,2,3,6-tetrahydropyridine (MPTP) mouse model of Parkinson's disease. J Neurochem 76:637–640
Przedborski S, Vila M (2003) The 1-methyl-4-phenyl-1,2,3,6-tetrahydropyridine mouse model: a tool to explore the pathogenesis of Parkinson's disease. Ann N Y Acad Sci 991:189–198
Quon D, Wang Y, Catalano R, Scardina JM, Murakami K, Cordell B (1991) Formation of beta-amyloid protein deposits in brains of transgenic mice. Nature 352:239–241
Rajput A, Dickson DW, Robinson CA, Ross OA, Dachsel JC, Lincoln SJ, Cobb SA, Rajput ML, Farrer MJ (2006) Parkinsonism, Lrrk2 G2019S, and tau neuropathology. Neurology 67:1506–1508
Rapp PR (1990) Visual discrimination and reversal learning in the aged monkey (Macaca mulatta). Behav Neurosci 104:876–884
Rapp P (1993) Neuropsychological analysis of learning and memory in the aged nonhuman primate. Neurobiol Aging 14:627–629
Reaume AG, Elliott JL, Hoffman EK, Kowall NW, Ferrante RJ, Siwek DF, Wilcox HM, Flood DG, Beal MF, Brown RH Jr., Scott RW, Snider WD (1996) Motor neurons in Cu/Zn superoxide dismutase-deficient mice develop normally but exhibit enhanced cell death after axonal injury. Nat Genet 13:43–47
Reddy PH, Williams M, Charles V, Garrett L, Pike-Buchanan L, Whetsell WO Jr., Miller G, Tagle DA (1998) Behavioural abnormalities and selective neuronal loss in HD transgenic mice expressing mutated full-length HD cDNA. Nat Genet 20:198–202
Reiserer RS, Harrison FE, Syverud DC, McDonald MP (2007) Impaired spatial learning in the APPSwe + PSEN1DeltaE9 bigenic mouse model of Alzheimer's disease. Genes Brain Behav 6:54–65

Ridley RM, Baker HF, Drewett B, Johnson JA (1985) Effects of ibotenic acid lesions of the basal forebrain on serial reversal learning in marmosets. Psychopharmacology (Berl) 86: 438–443

Roberts AC, Robbins TW, Everitt BJ, Jones GH, Sirkia TE, Wilkinson J, Page K (1990) The effects of excitotoxic lesions of the basal forebrain on the acquisition, retention and serial reversal of visual discriminations in marmosets. Neuroscience 34:311–329

Rosen RF, Farberg AS, Gearing M, Dooyema J, Long PM, Anderson DC, Davis-Turak J, Coppola G, Geschwind DH, Pare JF, Duong TQ, Hopkins WD, Preuss TM, Walker LC (2008) Tauopathy with paired helical filaments in an aged chimpanzee. J Comp Neurol 509:259–270

Ross OA, Toft M, Whittle AJ, Johnson JL, Papapetropoulos S, Mash DC, Litvan I, Gordon MF, Wszolek ZK, Farrer MJ, Dickson DW (2006) Lrrk2 and Lewy body disease. Ann Neurol 59:388–393

Rothstein JD, Van Kammen M, Levey AI, Martin LJ, Kuncl RW (1995) Selective loss of glial glutamate transporter GLT-1 in amyotrophic lateral sclerosis. Ann Neurol 38:73–84

Rousselet E, Joubert C, Callebert J, Parain K, Tremblay L, Orieux G, Launay JM, Cohen-Salmon C, Hirsch EC (2003) Behavioral changes are not directly related to striatal monoamine levels, number of nigral neurons, or dose of parkinsonian toxin MPTP in mice. Neurobiol Dis 14: 218–228

Rupniak NM, Samson NA, Steventon MJ, Iversen SD (1991) Induction of cognitive impairment by scopolamine and noncholinergic agents in rhesus monkeys. Life Sci 48:893–899

Rupniak NM, Steventon MJ, Field MJ, Jennings CA, Iversen SD (1989) Comparison of the effects of four cholinomimetic agents on cognition in primates following disruption by scopolamine or by lists of objects. Psychopharmacology (Berl) 99:189–195

Rupniak NM, Tye SJ, Field MJ (1997) Enhanced performance of spatial and visual recognition memory tasks by the selective acetylcholinesterase inhibitor E2020 in rhesus monkeys. Psychopharmacology (Berl) 131:406–410

Sachs C, Jonsson G (1975) Mechanisms of action of 6-hydroxydopamine. Biochem Pharmacol 24:1–8

Saner A, Thoenen H (1971) Model experiments on the molecular mechanism of action of 6-hydroxydopamine. Mol Pharmacol 7:147–154

Sang TK, Jackson GR (2005) Drosophila models of neurodegenerative disease. NeuroRx 2: 438–446

Sato S, Chiba T, Nishiyama S, Kakiuchi T, Tsukada H, Hatano T, Fukuda T, Yasoshima Y, Kai N, Kobayashi K, Mizuno Y, Tanaka K, Hattori N (2006) Decline of striatal dopamine release in parkin-deficient mice shown by ex vivo autoradiography. J Neurosci Res 84:1350–1357

Sato T, Nakanishi T, Yamamoto Y, Andersen PM, Ogawa Y, Fukada K, Zhou Z, Aoike F, Sugai F, Nagano S, Hirata S, Ogawa M, Nakano R, Ohi T, Kato T, Nakagawa M, Hamasaki T, Shimizu A, Sakoda S (2005) Rapid disease progression correlates with instability of mutant SOD1 in familial ALS. Neurology 65:1954–1957

Satyal SH, Schmidt E, Kitagawa K, Sondheimer N, Lindquist S, Kramer JM, Morimoto RI (2000) Polyglutamine aggregates alter protein folding homeostasis in Caenorhabditis elegans. Proc Natl Acad Sci USA 97:5750–5755

Savonenko A, Xu GM, Melnikova T, Morton JL, Gonzales V, Wong MP, Price DL, Tang F, Markowska AL, Borchelt DR (2005) Episodic-like memory deficits in the APPswe/PS1dE9 mouse model of Alzheimer's disease: relationships to beta-amyloid deposition and neurotransmitter abnormalities. Neurobiol Dis 18:602–617

Schilling G, Becher MW, Sharp AH, Jinnah HA, Duan K, Kotzuk JA, Slunt HH, Ratovitski T, Cooper JK, Jenkins NA, Copeland NG, Price DL, Ross CA, Borchelt DR (1999) Intranuclear inclusions and neuritic aggregates in transgenic mice expressing a mutant N-terminal fragment of huntingtin. Hum Mol Genet 8:397–407

Schluter OM, Fornai F, Alessandri MG, Takamori S, Geppert M, Jahn R, Sudhof TC (2003) Role of alpha-synuclein in 1-methyl-4-phenyl-1,2,3,6-tetrahydropyridine-induced parkinsonism in mice. Neuroscience 118:985–1002

Schmitz C, Rutten BP, Pielen A, Schafer S, Wirths O, Tremp G, Czech C, Blanchard V, Multhaup G, Rezaie P, Korr H, Steinbusch HW, Pradier L, Bayer TA (2004) Hippocampal neuron loss exceeds amyloid plaque load in a transgenic mouse model of Alzheimer's disease. Am J Pathol 164:1495–1502

Schneider JS, Pope-Coleman A (1995) Cognitive deficits precede motor deficits in a slowly progressing model of parkinsonism in the monkey. Neurodegeneration 4:245–255

Schrag A, Ben-Shlomo Y, Quinn NP (1999) Prevalence of progressive supranuclear palsy and multiple system atrophy: a cross-sectional study. Lancet 354:1771–1775

Schrag A, Jahanshahi M, Quinn N (2000) What contributes to quality of life in patients with Parkinson's disease? J Neurol Neurosurg Psychiatry 69:308–312

Schwarting RK, Huston JP (1996) The unilateral 6-hydroxydopamine lesion model in behavioral brain research. Analysis of functional deficits, recovery and treatments. Prog Neurobiol 50:275–331

Segalat L, Neri C (2003) C. elegans as a model for human inherited degenerative diseases. Med Sci (Paris) 19:1218–1225

Shimura H, Hattori N, Kubo S, Mizuno Y, Asakawa S, Minoshima S, Shimizu N, Iwai K, Chiba T, Tanaka K, Suzuki T (2000) Familial Parkinson disease gene product, parkin, is a ubiquitin-protein ligase. Nat Genet 25:302–305

Silvestri L, Caputo V, Bellacchio E, Atorino L, Dallapiccola B, Valente EM, Casari G (2005) Mitochondrial import and enzymatic activity of PINK1 mutants associated to recessive parkinsonism. Hum Mol Genet 14:3477–3492

Singleton AB, Farrer M, Johnson J, Singleton A, Hague S, Kachergus J, Hulihan M, Peuralinna T, Dutra A, Nussbaum R, Lincoln S, Crawley A, Hanson M, Maraganore D, Adler C, Cookson MR, Muenter M, Baptista M, Miller D, Blancato J, Hardy J, Gwinn-Hardy K (2003) alpha-Synuclein locus triplication causes Parkinson's disease. Science 302:841

Sloane JA, Pietropaolo MF, Rosene DL, Moss MB, Peters A, Kemper T, Abraham CR (1997) Lack of correlation between plaque burden and cognition in the aged monkey. Acta Neuropathol 94:471–478

Slow EJ, van Raamsdonk J, Rogers D, Coleman SH, Graham RK, Deng Y, Oh R, Bissada N, Hossain SM, Yang YZ, Li XJ, Simpson EM, Gutekunst CA, Leavitt BR, Hayden MR (2003) Selective striatal neuronal loss in a YAC128 mouse model of Huntington disease. Hum Mol Genet 12:1555–1567

Smeyne M, Jiao Y, Shepherd KR, Smeyne RJ (2005) Glia cell number modulates sensitivity to MPTP in mice. Glia 52:144–152

Smith DE, Roberts J, Gage FH, Tuszynski MH (1999) Age-associated neuronal atrophy occurs in the primate brain and is reversible by growth factor gene therapy. Proc Natl Acad Sci USA 96:10893–10898

Spillantini MG, Crowther RA, Jakes R, Cairns NJ, Lantos PL, Goedert M (1998a) Filamentous alpha-synuclein inclusions link multiple system atrophy with Parkinson's disease and dementia with Lewy bodies. Neurosci Lett 251:205–208

Spillantini MG, Crowther RA, Jakes R, Hasegawa M, Goedert M (1998b) alpha-Synuclein in filamentous inclusions of Lewy bodies from Parkinson's disease and dementia with lewy bodies. Proc Natl Acad Sci USA 95:6469–6473

Spillantini MG, Goedert M (2000) The alpha-synucleinopathies: Parkinson's disease, dementia with Lewy bodies, and multiple system atrophy. Ann N Y Acad Sci 920:16–27

Spittaels K, Van den Haute C, Van Dorpe J, Bruynseels K, Vandezande K, Laenen I, Geerts H, Mercken M, Sciot R, Van Lommel A, Loos R, Van Leuven F (1999) Prominent axonopathy in the brain and spinal cord of transgenic mice overexpressing four-repeat human tau protein. Am J Pathol 155:2153–2165

Spittaels K, Van den Haute C, Van Dorpe J, Geerts H, Mercken M, Bruynseels K, Lasrado R, Vandezande K, Laenen I, Boon T, Van Lint J, Vandenheede J, Moechars D, Loos R, Van Leuven F (2000) Glycogen synthase kinase-3beta phosphorylates protein tau and rescues the axonopathy in the central nervous system of human four-repeat tau transgenic mice. J Biol Chem 275:41340–41349

Spittaels K, Van den Haute C, Van Dorpe J, Terwel D, Vandezande K, Lasrado R, Bruynseels K, Irizarry M, Verhoye M, Van Lint J, Vandenheede JR, Ashton D, Mercken M, Loos R, Hyman B, Van der Linden A, Geerts H, Van Leuven F (2002) Neonatal neuronal overexpression of glycogen synthase kinase-3 beta reduces brain size in transgenic mice. Neuroscience 113: 797–808

Stefanova N, Poewe W, Wenning GK (2008) Rasagiline is neuroprotective in a transgenic model of multiple system atrophy. Exp Neurol 210:421–427

Stefanova N, Puschban Z, Fernagut PO, Brouillet E, Tison F, Reindl M, Jellinger KA, Poewe W, Wenning GK (2003) Neuropathological and behavioral changes induced by various treatment paradigms with MPTP and 3-nitropropionic acid in mice: towards a model of striatonigral degeneration (multiple system atrophy). Acta Neuropathol 106: 157–166

Stepanichev MY, Moiseeva YV, Lazareva NA, Onufriev MV, Gulyaeva NV (2003) Single intracerebroventricular administration of amyloid-beta (25–35) peptide induces impairment in short-term rather than long-term memory in rats. Brain Res Bull 61:197–205

Stone TW (1993) Neuropharmacology of quinolinic and kynurenic acids. Pharmacol Rev 45: 309–379

Struble RG, Price DL Jr., Cork LC, Price DL (1985) Senile plaques in cortex of aged normal monkeys. Brain Res 361:267–275

Sundstrom E, Stromberg I, Tsutsumi T, Olson L, Jonsson G (1987) Studies on the effect of 1-methyl-4-phenyl-1,2,3,6-tetrahydropyridine (MPTP) on central catecholamine neurons in C57BL/6 mice. Comparison with three other strains of mice. Brain Res 405:26–38

Swerdlow RH, Parks JK, Cassarino DS, Trimmer PA, Miller SW, Maguire DJ, Sheehan JP, Maguire RS, Pattee G, Juel VC, Phillips LH, Tuttle JB, Bennett JP Jr., Davis RE, Parker WD Jr. (1998) Mitochondria in sporadic amyotrophic lateral sclerosis. Exp Neurol 153:135–142

Takamiya R, Takahashi M, Myint T, Park YS, Miyazawa N, Endo T, Fujiwara N, Sakiyama H, Misonou Y, Miyamoto Y, Fujii J, Taniguchi N (2003) Glycation proceeds faster in mutated Cu, Zn-superoxide dismutases related to familial amyotrophic lateral sclerosis. FASEB J 17: 938–940

Tang Y, Mishkin M, Aigner TG (1997) Effects of muscarinic blockade in perirhinal cortex during visual recognition. Proc Natl Acad Sci USA 94:12667–12669

Taylor JP, Mata IF, Farrer MJ (2006) LRRK2: a common pathway for parkinsonism, pathogenesis and prevention? Trends Mol Med 12:76–82

Terwel D, Muyllaert D, Dewachter I, Borghgraef P, Croes S, Devijver H, Van Leuven F (2008) Amyloid activates GSK-3beta to aggravate neuronal tauopathy in bigenic mice. Am J Pathol 172:786–798

The Huntington's Disease Collaborative Research Group (1993) A novel gene containing a trinucleotide repeat that is expanded and unstable on Huntington's disease chromosomes. Cell 72:971–983

Tipton KF, Singer TP (1993) Advances in our understanding of the mechanisms of the neurotoxicity of MPTP and related compounds. J Neurochem 61:1191–1206

Tofaris GK, Garcia Reitbock P, Humby T, Lambourne SL, O'Connell M, Ghetti B, Gossage H, Emson PC, Wilkinson LS, Goedert M, Spillantini MG (2006) Pathological changes in dopaminergic nerve cells of the substantia nigra and olfactory bulb in mice transgenic for truncated human alpha-synuclein(1–120): implications for Lewy body disorders. J Neurosci 26:3942–3950

Turmaine M, Raza A, Mahal A, Mangiarini L, Bates GP, Davies SW (2000) Nonapoptotic neurodegeneration in a transgenic mouse model of Huntington's disease. Proc Natl Acad Sci USA 97:8093–8097

Ungerstedt U, Arbuthnott GW (1970) Quantitative recording of rotational behavior in rats after 6-hydroxy-dopamine lesions of the nigrostriatal dopamine system. Brain Res 24:485–493

Valdmanis PN, Rouleau GA (2008) Genetics of familial amyotrophic lateral sclerosis. Neurology 70:144–152

Valente EM, Abou-Sleiman PM, Caputo V, Muqit MM, Harvey K, Gispert S, Ali Z, Del Turco D, Bentivoglio AR, Healy DG, Albanese A, Nussbaum R, Gonzalez-Maldonado R, Deller T, Salvi S, Cortelli P, Gilks WP, Latchman DS, Harvey RJ, Dallapiccola B, Auburger G, Wood NW (2004) Hereditary early-onset Parkinson's disease caused by mutations in PINK1. Science 304:1158–1160

Van Den Bosch L, Van Damme P, Bogaert E, Robberecht W (2006) The role of excitotoxicity in the pathogenesis of amyotrophic lateral sclerosis. Biochim Biophys Acta 1762:1068–1082

Van Dooren T, Muyllaert D, Borghgraef P, Cresens A, Devijver H, Van der Auwera I, Wera S, Dewachter I, Van Leuven F (2006) Neuronal or glial expression of human apolipoprotein e4 affects parenchymal and vascular amyloid pathology differentially in different brain regions of double- and triple-transgenic mice. Am J Pathol 168:245–260

Vermeiren C, Hemptinne I, Vanhoutte N, Tilleux S, Maloteaux JM, Hermans E (2006) Loss of metabotropic glutamate receptor-mediated regulation of glutamate transport in chemically activated astrocytes in a rat model of amyotrophic lateral sclerosis. J Neurochem 96:719–731

Von Coelln R, Thomas B, Savitt JM, Lim KL, Sasaki M, Hess EJ, Dawson VL, Dawson TM (2004) Loss of locus coeruleus neurons and reduced startle in parkin null mice. Proc Natl Acad Sci USA 101:10744–10749

Vonsattel JP, Keller C, Pilar Amaya MD (2008) Neuropathology of Huntington's Disease. Handb Clin Neurol 89:599–618

Voytko ML (1999) Impairments in acquisition and reversals of two-choice discriminations by aged rhesus monkeys. Neurobiol Aging 20:617–627

Voytko ML, Olton DS, Richardson RT, Gorman LK, Tobin JR, Price DL (1994) Basal forebrain lesions in monkeys disrupt attention but not learning and memory. J Neurosci 14:167–186

von Campenhausen S, Bornschein B, Wick R, Botzel K, Sampaio C, Poewe W, Oertel W, Siebert U, Berger K, Dodel R (2005) Prevalence and incidence of Parkinson's disease in Europe. Eur Neuropsychopharmacol 15:473–490

von Horsten S, Schmitt I, Nguyen HP, Holzmann C, Schmidt T, Walther T, Bader M, Pabst R, Kobbe P, Krotova J, Stiller D, Kask A, Vaarmann A, Rathke-Hartlieb S, Schulz JB, Grasshoff U, Bauer I, Vieira-Saecker AM, Paul M, Jones L, Lindenberg KS, Landwehrmeyer B, Bauer A, Li XJ, Riess O (2003) Transgenic rat model of Huntington's disease. Hum Mol Genet 12:617–624

Wakabayashi K, Ikeuchi T, Ishikawa A, Takahashi H (1998) Multiple system atrophy with severe involvement of the motor cortical areas and cerebral white matter. J Neurol Sci 156: 114–117

Wakamatsu M, Ishii A, Iwata S, Sakagami J, Ukai Y, Ono M, Kanbe D, Muramatsu S, Kobayashi K, Iwatsubo T, Yoshimoto M (2008) Selective loss of nigral dopamine neurons induced by overexpression of truncated human alpha-synuclein in mice. Neurobiol Aging 29:574–585

Walker LC, Masters C, Beyreuther K, Price DL (1990) Amyloid in the brains of aged squirrel monkeys. Acta Neuropathol 80:381–387

Wenning GK, Colosimo C, Geser F, Poewe W (2004) Multiple system atrophy. Lancet Neurol 3:93–103

Wenning GK, Granata R, Laboyrie PM, Quinn NP, Jenner P, Marsden CD (1996) Reversal of behavioural abnormalities by fetal allografts in a novel rat model of striatonigral degeneration. Mov Disord 11:522–532

Wenning GK, Stefanova N, Jellinger KA, Poewe W, Schlossmacher MG (2008) Multiple system atrophy: a primary oligodendrogliopathy. Ann Neurol 64:239–246

Wenning GK, Tison F, Ben Shlomo Y, Daniel SE, Quinn NP (1997) Multiple system atrophy: a review of 203 pathologically proven cases. Mov Disord 12:133–147

West MJ, Coleman PD, Flood DG, Troncoso JC (1994) Differences in the pattern of hippocampal neuronal loss in normal ageing and Alzheimer's disease. Lancet 344:769–772

Wheeler VC, White JK, Gutekunst CA, Vrbanac V, Weaver M, Li XJ, Li SH, Yi H, Vonsattel JP, Gusella JF, Hersch S, Auerbach W, Joyner AL, MacDonald ME (2000) Long glutamine tracts cause nuclear localization of a novel form of huntingtin in medium spiny striatal neurons in HdhQ92 and HdhQ111 knock-in mice. Hum Mol Genet 9:503–513

Willem M, Dewachter I, Smyth N, Van Dooren T, Borghgraef P, Haass C, Van Leuven F (2004) beta-site amyloid precursor protein cleaving enzyme 1 increases amyloid deposition in brain parenchyma but reduces cerebrovascular amyloid angiopathy in aging BACE × APP[V717I] double-transgenic mice. Am J Pathol 165:1621–1631

Wirths O, Breyhan H, Marcello A, Cotel MC, Bruck W, Bayer TA (2008a) Inflammatory changes are tightly associated with neurodegeneration in the brain and spinal cord of the APP/PS1KI mouse model of Alzheimer's disease. Neurobiol Aging

Wirths O, Breyhan H, Schafer S, Roth C, Bayer TA (2008b) Deficits in working memory and motor performance in the APP/PS1ki mouse model for Alzheimer's disease. Neurobiol Aging 29:891–901

Wirths O, Weis J, Kayed R, Saido TC, Bayer TA (2007) Age-dependent axonal degeneration in an Alzheimer mouse model. Neurobiol Aging 28:1689–1699

Wirths O, Weis J, Szczygielski J, Multhaup G, Bayer TA (2006) Axonopathy in an APP/PS1 transgenic mouse model of Alzheimer's disease. Acta Neuropathol 111:312–319

Wisniewski HM, Ghetti B, Terry RD (1973) Neuritic (senile) plaques and filamentous changes in aged rhesus monkeys. J Neuropathol Exp Neurol 32:566–584

Wong PC, Pardo CA, Borchelt DR, Lee MK, Copeland NG, Jenkins NA, Sisodia SS, Cleveland DW, Price DL (1995) An adverse property of a familial ALS-linked SOD1 mutation causes motor neuron disease characterized by vacuolar degeneration of mitochondria. Neuron 14:1105–1116

Wszolek ZK, Pfeiffer RF, Tsuboi Y, Uitti RJ, McComb RD, Stoessl AJ, Strongosky AJ, Zimprich A, Muller-Myhsok B, Farrer MJ, Gasser T, Calne DB, Dickson DW (2004) Autosomal dominant parkinsonism associated with variable synuclein and tau pathology. Neurology 62:1619–1622

Wu Y, Luo Y (2005) Transgenic C. elegans as a model in Alzheimer's research. Curr Alzheimer Res 2:37–45

Yang SH, Cheng PH, Banta H, Piotrowska-Nitsche K, Yang JJ, Cheng EC, Snyder B, Larkin K, Liu J, Orkin J, Fang ZH, Smith Y, Bachevalier J, Zola SM, Li SH, Li XJ, Chan AW (2008) Towards a transgenic model of Huntington's disease in a non-human primate. Nature 453: 921–924

Yang Y, Hentati A, Deng HX, Dabbagh O, Sasaki T, Hirano M, Hung WY, Ouahchi K, Yan J, Azim AC, Cole N, Gascon G, Yagmour A, Ben-Hamida M, Pericak-Vance M, Hentati F, Siddique T (2001) The gene encoding alsin, a protein with three guanine-nucleotide exchange factor domains, is mutated in a form of recessive amyotrophic lateral sclerosis. Nat Genet 29: 160–165

Yazawa I, Giasson BI, Sasaki R, Zhang B, Joyce S, Uryu K, Trojanowski JQ, Lee VM (2005) Mouse model of multiple system atrophy alpha-synuclein expression in oligodendrocytes causes glial and neuronal degeneration. Neuron 45:847–859

Zarranz JJ, Alegre J, Gomez-Esteban JC, Lezcano E, Ros R, Ampuero I, Vidal L, Hoenicka J, Rodriguez O, Atares B, Llorens V, Gomez Tortosa E, del Ser T, Munoz DG, de Yebenes JG (2004) The new mutation, E46K, of alpha-synuclein causes Parkinson and Lewy body dementia. Ann Neurol 55:164–173

Zhang Y, Gao J, Chung KK, Huang H, Dawson VL, Dawson TM (2000) Parkin functions as an E2-dependent ubiquitin- protein ligase and promotes the degradation of the synaptic vesicle-associated protein, CDCrel-1. Proc Natl Acad Sci USA 97:13354–13359

Zhang B, Tu P, Abtahian F, Trojanowski JQ, Lee VM (1997) Neurofilaments and orthograde transport are reduced in ventral root axons of transgenic mice that express human SOD1 with a G93A mutation. J Cell Biol 139:1307–1315

Zimprich A, Biskup S, Leitner P, Lichtner P, Farrer M, Lincoln S, Kachergus J, Hulihan M, Uitti RJ, Calne DB, Stoessl AJ, Pfeiffer RF, Patenge N, Carbajal IC, Vieregge P, Asmus F, Muller-Myhsok B, Dickson DW, Meitinger T, Strom TM, Wszolek ZK, Gasser T (2004a) Mutations in LRRK2 cause autosomal-dominant parkinsonism with pleomorphic pathology. Neuron 44:601–607

Zimprich A, Muller-Myhsok B, Farrer M, Leitner P, Sharma M, Hulihan M, Lockhart P, Strongosky A, Kachergus J, Calne DB, Stoessl J, Uitti RJ, Pfeiffer RF, Trenkwalder C, Homann N, Ott E, Wenzel K, Asmus F, Hardy J, Wszolek Z, Gasser T (2004b) The PARK8 locus in autosomal dominant parkinsonism: confirmation of linkage and further delineation of the disease-containing interval. Am J Hum Genet 74:11–19

Zuscik MJ, Sands S, Ross SA, Waugh DJ, Gaivin RJ, Morilak D, Perez DM (2000) Overexpression of the alpha1B-adrenergic receptor causes apoptotic neurodegeneration: multiple system atrophy. Nat Med 6:1388–1394

Vitamins Deficiencies and Brain Function

Chantal Bémeur, Jane A. Montgomery, and Roger F. Butterworth

Keywords Vitamins · Thiamine · Wernicke–Korsakoff syndrome · Pyridoxine · Niacin · Folic acid · Antioxidants

Contents

1 Introduction

The consequences of malnutrition on the central nervous system are diverse and depend to a significant extent on the stage of development or maturity of the brain as well as on the severity of the nutritional deficiency. For example, vitamin deficiencies result in a wide range of neuropathology and neuropsychiatric symptomatology

---

R.F. Butterworth (✉)
Neuroscience Research Unit, CHUM (Hôpital Saint-Luc), Montreal, QC, Canada H2X 3J4
e-mail: roger.butterworth@umontreal.ca

J.P. Blass (ed.), *Neurochemical Mechanisms in Disease*,
Advances in Neurobiology 1, DOI 10.1007/978-1-4419-7104-3_4,

depending upon the nature and extent of the vitamin deficiency. The most common vitamin deficiency disorders are those associated with the group B vitamins, particularly thiamine (vitamin B_1). The likelihood of multiple vitamin deficiencies should be borne in mind.

This review chapter summarizes current knowledge on vitamin deficiencies, examines the role of vitamins in cellular function, and reviews current mechanisms involved in the pathogenesis of brain dysfunction in vitamin deficiencies.

2 Thiamine (Vitamin B_1)

Thiamine is a water-soluble vitamin and is also known as vitamin B_1, or aneurin (Fig. 1). Both the pyrimidine and thiazole moieties are necessary for biological activity, which is maximal when only one methylene group bridges the two moieties.

Fig. 1 Structure of thiamine

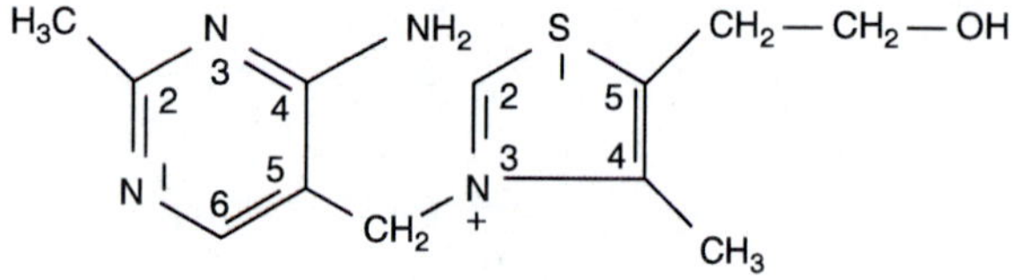

Thiamine status is influenced by the diet and by a variety of other factors, including its bioavailability in food products, ethanol consumption, the presence of antithiamine factors in the diet as well as folate and protein status. Ingested thiamine is fairly well absorbed, rapidly converted to phosphorylated forms, stored poorly, and excreted in the urine in a variety of hydrolyzed and oxidized products (TanPhaichitr et al., 1999).

In developed countries, clinical thiamine deficiency occurs most commonly in alcoholics and in patients with grossly impaired nutritional status associated with, for example, gastrointestinal disease or AIDS (Butterworth, 2006). Thiamine deficiency may result from inadequate dietary intake of the vitamin. Chronic alcohol consumption can result in thiamine deficiency by causing inadequate nutritional thiamine intake, decreased absorption from the gastrointestinal tract, and impaired thiamine utilization in the cells. People differ in their susceptibility to thiamine deficiency and different brain regions may be more or less sensitive to this condition. Thiamine deficiency is more common in developing countries where polished rice is the staple diet. Peripheral nerve damage (neuropathy) is a common consequence of thiamine deficiency. The neuropathy tends to be worse distally than proximally, involves myelin more than axons, and is often painful. The neuropathy is linked to multiple deficiencies of water-soluble vitamins that often occur together in foods and are known as the vitamin B complex.

2.1 Thiamine Deficiency-Related Neurological Disorders

Beriberi (infantile and adult) and Wernicke's encephalopathy (WE) are clinical manifestations attributed to thiamine deficiency. Beriberi is characterized by peripheral neuropathy including sensory, motor, and reflex functions affecting the distal segments of limbs more severely than proximal ones (TanPhaichitr, 1985). WE is a metabolic disease due to thiamine deficiency and is characterized by lesions in the thalamus, hypothalamus (including mammillary nuclei), and cerebellum (Victor et al., 1971; Harper and Butterworth, 1997).

WE is seriously underdiagnosed both in alcoholic and nonalcoholic patients. It has been estimated that in alcoholic patients, the diagnosis of WE is missed in up to 80% of cases (Harper, 1979). Similarly, a review of the literature describing WE patients with HIV–AIDS revealed that 80% of cases had again not been adequately diagnosed clinically during life (Butterworth et al., 1991). The principal reason for its consistent underdiagnosis results from the overuse of the classical textbook definition of WE which requires that a triad of neuropsychiatric symptoms (ophthalmoplegia, ataxia, global confusional state) be present for diagnosis. In practice, it is rare that this triad of symptoms is present; rather, many patients diagnosed subsequently with WE present only with psychomotor slowing or apathy. In the meantime, a definitive diagnosis of WE can nowadays be accurately made using magnetic resonance imaging (MRI) (Charness and DeLaPaz, 1987).

Korsakoff's psychosis is considered by some to represent a progression of WE. It is characterized by a striking loss of working memory with relatively little loss of reference memory. Prompt treatment of Wernicke's syndrome with thiamine is believed to prevent the development of Korsakoff's syndrome, but the latter responds little if at all to treatment with thiamine.

Abnormalities of thiamine-related processes have also been reported in a wide range of neurodegenerative diseases. Brain tissue from patients with Alzheimer Disease (AD) contains decreased concentrations of thiamine diphosphate (TDP) (Héroux et al., 1996) and TDPase activities are reduced by up to 60% in this material (Rao et al., 1993). Furthermore, activities of TDP-dependent enzymes are decreased in AD brains (Gibson et al., 1988; Butterworth and Besnard, 1990) with activities of alpha-ketoglutarate dehydrogenase (αKGDH) showing particularly low levels in patients with both genetic and sporadic forms of the disease. In patients bearing the epsilon 4 allele of the apolipoprotein E gene, the correlation between αKGDH activity and clinical dementia rating is 0.7 (Gibson et al., 1988). Amyloid-β peptide (Aβ) is an important component of senile plaques in AD. There is increasing evidence to suggest that excess Aβ production is the cause of AD and a recent study showed that exposure of isolated brain mitochondria to Aβ caused a significant reduction in activities of the thiamine-dependent enzymes αKGDH and pyruvate dehydrogenase complex (PDHC) (Casley et al., 2002; see Section 2.2.1), suggesting that these changes contribute to neuronal cell death in AD. Reduced activities of αKGDH have also been described in Parkinson's disease (Mizuno et al., 1994) and progressive subnuclear palsy (Albers et al., 2000).

Several thiamine antagonists including oxythiamine, pyrithiamine, and amprolium cause thiamine deficiency in animals. The most extensive studies of thiamine deficiency in laboratory animals have utilized rats and mice. Dietary thiamine deficiency is induced with artificial diets complete in all food stuffs except thiamine. Because thiamine deficiency induces loss of appetite, each control animal must be pair-fed to equal food consumption by the thiamine-deficient group of animals. The thiamine antagonists oxythiamine and pyrithiamine are converted to catalytically inactive pyrophosphates that compete for TDP binding sites on the enzymes. Mice fed with combinations of pyrithiamine and a low-thiamine diet develop abnormal neurological responses within 5–7 days and overt neurological symptoms by day 8 or 9; death often occurs by day 10. In pyrithiamine-treated rats, abnormalities of motor performance occur by day 3, additional neurological symptoms by day 12, and death within 2 weeks (Butterworth, 2006).

2.2 Thiamine and Cell Metabolism/Function

2.2.1 Thiamine as Enzyme Cofactor

Thiamine uptaken into the cell is phosphorylated to TDP by the enzyme thiamine pyrophosphokinase. TDP is then further phosphorylated to thiamine triphosphate (TTP) or is dephosphorylated to thiamine monophosphate (TMP).

Evidence suggests that thiamine phosphorylation/dephosphorylation is a compartmentalized process in the brain. Thiamine phosphate esters are significantly more concentrated in neurons compared to other brain cells (Laforenza et al., 1988). Moreover, TDPase activities are twentyfold higher in neurons whereas TMPase is expressed primarily by glial cells. In nerve terminals, TTP is rapidly synthesized from TDP by the action of TDP phosphoryltransferase but the TTP ester does not accumulate to high concentrations; rather it is rapidly hydrolysed to TDP by the action of TTPase, an enzyme which is also enriched in nerve terminals. Nerve stimulation results in release of thiamine which is mainly in the form of TMP (Cooper and Pincus, 1979). Taken together, these findings suggest that trafficking of thiamine and TMP occurs between neurons and astrocytes in brain as shown in a simplified schematic manner in Fig. 2.

TDP-dependent enzymes include transketolase, an enzyme component of the pentose shunt pathway, pyruvate dehydrogenase complex, and αKGDH a tricarboxylic acid cycle enzyme (Fig. 3). Branched-chain ketoacid dehydrogenases are also TDP-dependent.

Given the mitochondrial localization of pyruvate and α-ketoglutarate dehydrogenase, it is not surprising that thiamine deficiency has multiple metabolic consequences including lactate accumulation (Peters, 1936; Navarro et al., 2005), alanine increases (Butterworth and Héroux, 1989), and reduced synthesis of high-energy phosphates (Aikawa et al., 1984). Inasmuch as an effective tricarboxylic acid cycle is imperative for the synthesis of neurotransmitters (acetylcholine, glutamate, GABA) in brain, thiamine deficiency leads to impairments in their synthesis (Gibson and Blass, 1985; Butterworth and Héroux, 1989; Navarro et al., 2005) (Fig. 3).

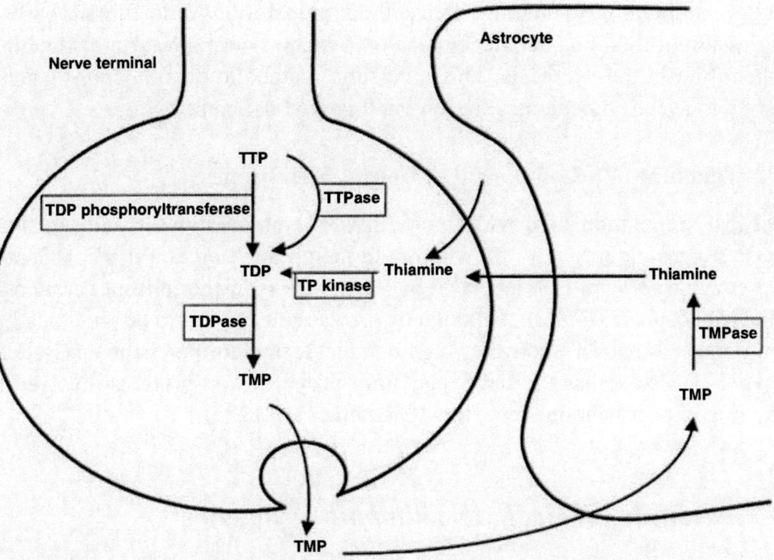

Fig. 2 Intercellular trafficking and thiamine and thiamine esters in brain. TMP: thiamine monophosphate, TDP: thiamine diphosphate, TTP: thiamine triphosphate, TPKinase: thiamine pyrophosphokinase

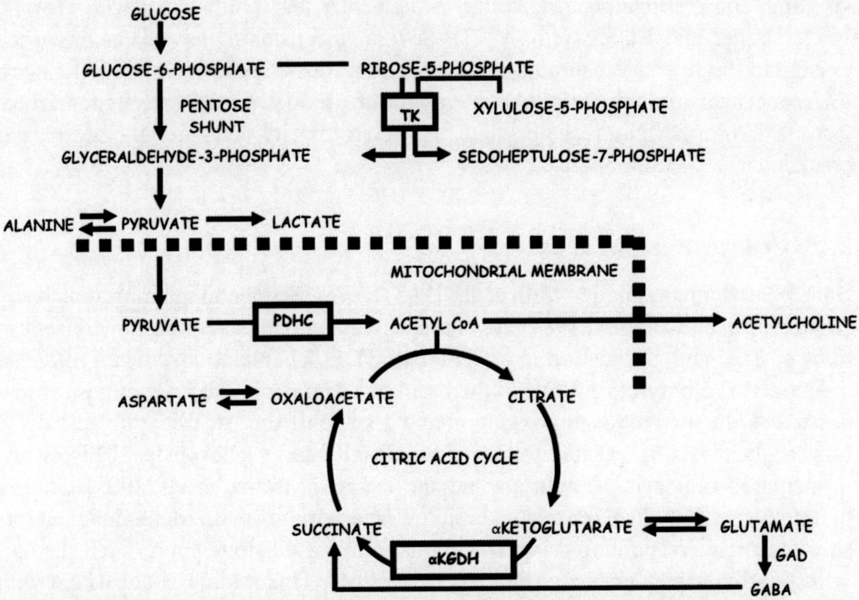

Fig. 3 TDP-dependent enzymes involved in brain glucose oxidation and the pentose shunt pathway. Impairment of TDP-dependent enzymes leads to decreased synthesis of neurotransmitters (acetylcholine, glutamate, GABA, aspartate), cellular energy compromise, and lactate accumulation. PDHC: Pyruvate dehydrogenase complex, αKGDH: α Ketoglurarate dehydrogenase, TK: Transkelotase

Addition of thiamine to thiamine-free cellular preparations or to animals early in the progression of thiamine deficiency results in a rapid normalisation of function and of neurotransmitter synthesis. This reversible metabolic phenomenon is generally referred to as the "biochemical lesion" in thiamine deficiency.

2.2.2 Thiamine as a Component of Neural Membranes

Electrical stimulation of a wide range of nerve preparations results in thiamine release suggesting a role for the vitamin in membrane function that is independent of its enzyme cofactor role mediated by TDP. TDP is further phosphorylated to thiamine triphosphate (Fig. 2). Although its precise role has yet to be elucidated, it has been proposed that TTP activates high-conductance chloride channels (Bettendorf, 1994). TTP also appears to have regulatory properties on proteins involved in the clustering of acetylcholine receptors (Gautam et al., 1995).

2.3 Neuronal Cell Death in Thiamine Deficiency

Chronic thiamine deficiency leads to two distinct types of neuropathological lesions. The first type is characterized by neuronal disintegration, vascular endothelial cell swelling, and sparing of the neuropil. This type of damage is seen in the thalamus and inferior olives. On the other hand, destruction of the neuropil, endothelial cell swelling, and neuronal sparing occur in periventricular brainstem nuclei (Torvik, 1985; Harper and Butterworth, 1997). Several mechanisms have been proposed to explain the selective neuronal cell damage and loss due to thiamine deficiency. These mechanisms include cellular energy failure, oxidative/nitrosative stress, focal lactic acidosis, NMDA receptor-mediated excitotoxicity, and blood–brain barrier breakdown.

2.3.1 Cellular Energy Failure

Both WE in humans (Butterworth et al., 1993) and experimental thiamine deficiency (Butterworth and Héroux, 1989) are characterised by decreases in brain concentrations of TDP and a reduction in activities of TDP-dependent enzymes. Prolonged reduction in activity of αKGDH in the brain due to thiamine deficiency results in a decreased glucose (pyruvate) oxidation and a switch from tricarboxylic acid cycle flux to glycolysis in an attempt to maintain high-energy phosphates. This results in increased synthesis of brain alanine and lactate (Navarro et al., 2005). Studies of oxidative metabolism in isolated brain mitochondria from thiamine-deficient rats show decreased respiration using α-ketoglutarate as a substrate but no such changes in respiration using succinate (Parker et al., 1984). This finding is consistent with decreased activities of αKGDH (see Fig. 3).

Direct measurement of high-energy phosphates in the brains of thiamine-deficient animals reveals early losses of ATP in brainstem (Aikawa et al., 1984). The focal accumulation of lactate in vulnerable brain structures may result in reduced pH (Hakim, 1984), a situation that is exacerbated following glucose loading of thiamine-deficient animals (Navarro et al., 2005).

2.3.2 Oxidative/Nitrosative Stress

Accumulation of reactive oxygen species has been reported in the thiamine-deficient brain (Langlais et al., 1997). Other indicators consistent with oxidative/nitrosative stress in the brain due to thiamine deficiency include reports of early activation of microglia (Todd and Butterworth, 1999; Gibson and Zhang, 2002) and increased expression of inducible nitric oxide synthase leading to increased nitrotyrosine immunoreactivity in vulnerable brain regions (Calingasan et al., 1998) as well as reports of increased expression of hemoxygenase-1 and inter-cellular adhesion molecule-1 (Gibson and Zhang, 2002). There is evidence to suggest that vascular factors also contribute to thiamine deficiency-related brain damage. Such factors include increases of endothelial nitric oxide synthase (eNOS) (Kruse et al., 2004). Moreover, targeted disruption (knock-down) of the eNOS gene attenuates the neuronal cell death in thiamine-deficient mice (Gibson and Zhang, 2002). eNOS knock-down but not knock-down of iNOS or nNOS leads to a reduction in protein tyrosine nitration (Beauchesne et al., 2009), suggesting a major role of eNOS as the source of nitric oxide-related nitrosative stress in thiamine deficiency.

Thiamine-dependent enzymes and processes are modified in the brains of patients with a wide range of neurodegenerative diseases (see Section 2.1) where the decline in enzyme activity is linked to the neuropathology and symptoms of these disorders. In addition to the finding that thiamine deficiency leads to oxidative stress (above), it has been proposed that oxidative stress causes disruption of thiamine-dependent processes (Gibson and Zhang, 2002). These authors proposed that the interaction of thiamine with oxidative processes is part of a cascade of events leading to neurodegeneration and, conversely, the reversal of the effects of thiamine deficiency by antioxidants together with the amelioration of other forms of oxidative stress by thiamine suggest that thiamine acts as a site-directed antioxidant.

2.3.3 NMDA Receptor-Mediated Excitotoxicity

The nature of the brain lesions observed in chronic thiamine deficiency resembles those described in excitotoxic brain injury mediated by the NMDA receptor (Langlais and Mair, 1990). Evidence consistent with a role of excitotoxicity in the pathogenesis of thiamine deficiency-related brain damage includes the finding of increased extracellular glutamate in brain regions that are particularly vulnerable to thiamine deficiency (Hazell et al., 1993) and the report that pretreatment with the NMDA receptor antagonist MK801 leads to significant neuroprotection (Langlais and Mair, 1990). One possible explanation for the increased extracellular brain concentrations of glutamate in thiamine deficiency is the reported loss in expression of high-affinity astrocytic glutamate transporters in vulnerable brain regions (Hazell et al., 2001).

2.3.4 The Blood–Brain Barrier Disruption

Haemorrhagic lesions are characteristic of experimental thiamine deficiency and WE in humans indicative of a breakdown of the blood–brain barrier (BBB). A study using immunoglobulin G (IgG) as an indicator of BBB integrity in thiamine-deficient rats revealed increased IgG immunoreactivity in the inferior colliculus and inferior olive prior to the onset of cell death in these regions (Calingasan et al., 1995). Similar early changes of BBB have been reported in the thiamine-deficient mouse (Harata and Iwasaki, 1995) microglial activation leading to the release of reactive oxygen species and cytokines are early cellular events with the potential to lead to BBB breakdown in thiamine deficiency (Todd and Butterworth, 1999).

3 Pyridoxine (Vitamin B_6)

Vitamin B_6 or pyridoxine participates in over 100 enzymatic reactions as the cofactor, pyridoxal phosphate (PLP). It exists in three forms: the alcohol, the amine, or the aldehyde. Pyridoxal phosphate is an essential cofactor for enzymes involved in the synthesis of many neurotransmitters.

Pyridoxine has been used to counteract nausea during pregnancy. Mothers who use pyridoxine supplements give birth to babies with higher pyridoxine requirements. Pyridoxine dependency has been reported in some newborns with seizures and pyridoxine treatment reverses the seizure activity in these infants (Bernstein, 1990). A number of commonly used drugs are pyridoxine antagonists. These include isoniazid, hydralazine, cycloserine, and penicillamine. Use of these drugs can result in peripheral neuropathy, seizures, and other neurological sequelae. Coadministration of pyridoxine reverses these side effects without affecting the efficacy of the initial treatment. MRI and positron emission tomographic studies in pyridoxine-deficient patients reveal diffuse structural abnormalities with progressive dilatation of the ventricular system and atrophy of the cerebral cortex and white matter (Gospe and Hecht, 1998; Shih et al., 1996). Pyridoxine deficiency in the rat leads to decreased dendritic arborization and reduced numbers of synapses and myelinated axons (Fig. 4) (Gerster, 1996).

Paradoxically, pyridoxine itself can also cause pathology in the central nervous system consisting of necrosis of dorsal root ganglia neurons and a centrifugal axonal atrophy and breakdown of peripheral and central sensory axons (Xu et al., 1989). This may occur at doses as low as 200–500 mg/d. However, in clinical trials using 100–150 mg/d to treat carpal tunnel syndrome, no toxicity was reported, suggesting that this a safe dose in adults. On the other hand, there are insufficient data to recommend long-term use of pyridoxine in children.

Pyridoxine plays a role in (1) the control of the hypothalamo-pituitary end-organ system, (2) melatonin synthesis, and (3) convulsive seizure activity. Neurological deficits resulting from pyridoxine deficiency can largely be explained by decreased activity of glutamic acid decarboxylase, 5-hydroxytryptophan decarboxylase, and ornithine decarboxylase (Dakshinamurti et al., 1990). The products of these

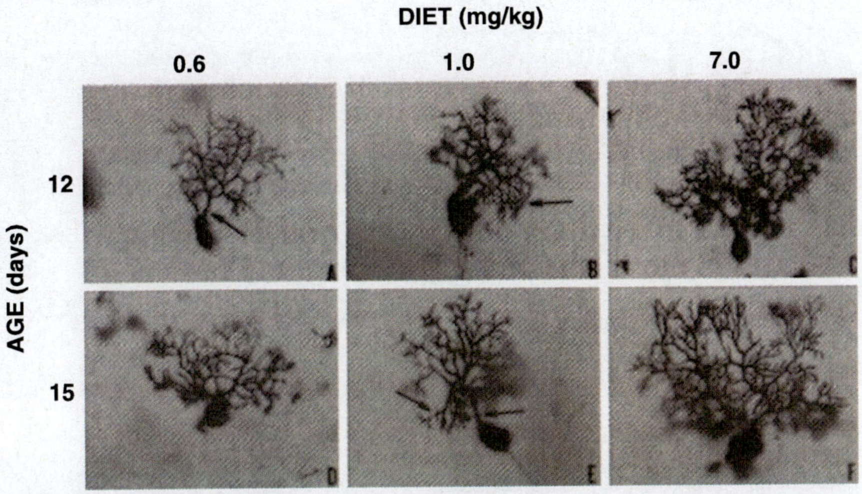

Fig. 4 Decreased dendritic arborisation in pyridoxine deficiency. Figure shows reduced Purkinje cells arborisation at 12 and 15 days in rat pups fed 0.6, 1.0, and 7 mg pyridoxine/Kg diet (modified from Chang et al., 1981)

enzymes are GABA, serotonin, and putrescine, respectively. Putrescine is a precursor of the polyamines, spermidine and spermine. Spermidine and spermine function as allosteric modulators of NMDA receptors, potentiating NMDA currents when glycine and glutamate are saturating. Dihydroxyphenylalanine decarboxylase, which also requires PLP as a cofactor, is less sensitive to pyridoxine deficiency.

The hypothalamus contains high concentrations of the monoamines dopamine and serotonin and these neurotransmitters have inhibitory or excitatory effects, respectively, on the anterior pituitary. For example, thyroid-stimulating hormone (TSH) secretion is increased by serotoninergic and decreased by dopaminergic activation. Pyridoxine deficiency in rats is associated with low levels of PLP in the hypothalamus, with no change in dopamine concentrations, but decreased levels of serotonin (Dakshinamurti et al., 1990). This correlates with decreased thyroid status and decreased pituitary TSH. Treatment with pyridoxine returns these parameters to normal.

Melatonin is produced in the pineal gland from tryptophan in a four-step reaction sequence depicted in Fig. 5. The pineal gland regulates diurnal variation of various physiological processes through the secretion of melatonin. Pyridoxine deficiency results in decreased concentrations of N-acetylserotonin and melatonin in the pineal gland during the dark phase (Dakshinamurti et al., 1990). Melatonin also acts at the level of the hypothalamus, resulting in increased prolactin release. Physiological levels of prolactin result in the initiation of lactation in females. Dopamine has an inhibitory effect resulting in decreased prolactin release. Mild pyridoxine deficiency results in decreased prolactin secretion as dopamine levels are not changed despite decreases in serotonin.

When pyridoxine deficiency is induced in pregnant rats, spontaneous convulsions are seen in the offspring at 3–4 days of age. Seizures are of short duration,

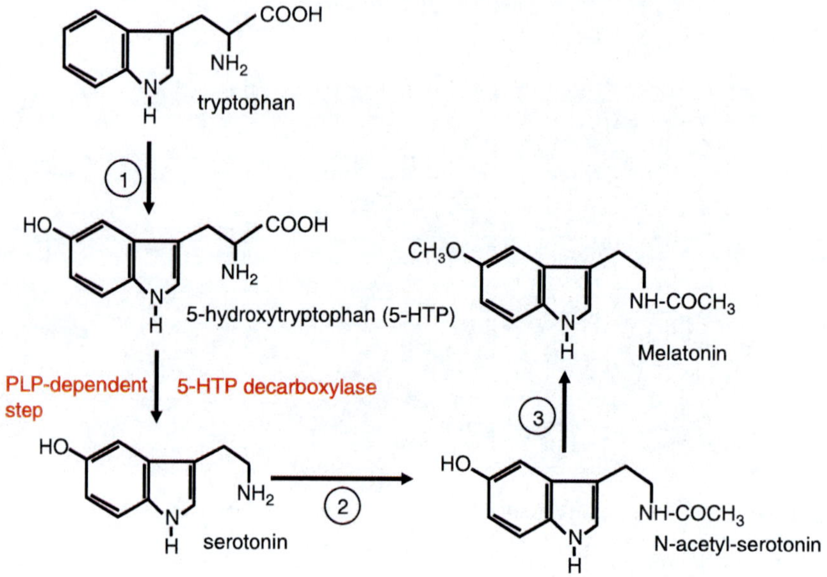

Fig. 5 Tryptophan is converted to 5-hydroxytryptophan by tryptophan hydroxylase (1). The production of serotonin in the next step is catalysed by 5-HTP decarboxylase (a PLP-dependent enzyme). Serotonin is then acetylated to N-acetyl-serotonin by N-acetyltransferase (2). Melatonin is then produced via the action of hydroxyindole-*O*-methyltransferase (3)

but occur at frequent intervals. Brain analysis indicates decreased PLP and glutamic acid decarboxylase (Dakshinamurti et al., 1990). When pyridoxine deficiency is induced in female rats during lactation, the rat pups develop abnormal EEG recordings at 3–5 weeks of age. This is associated with increased ^{3}H-GABA binding to $GABA_a$ receptors and ^{3}H-Baclofen to $GABA_b$ receptors, suggested to be due to increased receptor sensitivity resulting from chronic decreased synaptic GABA. These changes correlated with decreased PLP and GABA in the cerebellum of deficient rats. In another study using pyridoxine-deficient adult male rats, it was demonstrated that picrotoxin, a $GABA_a$ receptor antagonist, injected into the ventro-posterior-lateral thalamic nucleus, resulted in a reduced threshold for seizure activity (Dakshinamurti et al., 1990). The decreased inhibitory effect due to decreased GABA, combined with the accumulation of glutamic acid resulting from decreased decarboxylase activity is a likely explanation for the seizure activity seen in pyridoxine-deficient rats.

4 Cobalamin (Vitamin B_{12})

Vitamin B_{12} or cobalamin is present in meat and dairy products. Following ingestion, it is transformed into either methylcobalamin or adenosyl-cobalamin.

The former is the cytosolic form and is responsible for a number of important methylation reactions such as the conversion of homocysteine to methionine, an important precursor of S-adenosylmethionine which in turn is required for the production of some neurotransmitters (norepinephrine and glutamate), as well as the maintenance of myelin. Adenosyl-cobalamin is a mitochondrial cofactor for methylmalonyl-CoA mutase. Vitamin B_{12} deficiency results in both hematological (pernicious anemia) and neurological changes. Treatment of B_{12} deficiency results in the reversal of the anemia, but may or may not reverse the neurological consequences. To further complicate the picture, B_{12} deficiency is usually diagnosed following a blood test demonstrating megaloblastic changes associated with low serum B_{12}. Unfortunately, in up to 25% of patients, the neurological symptoms precede or are the only signs of B_{12} deficiency. Consequently, the diagnosis is frequently missed (Carmel, 2005).

Vitamin B_{12} deficiency results in subacute combined degeneration of the spinal cord. It is characterized by muscle weakness, parasthesias, various mental problems, and more rarely, visual disturbances. Neuropathological examination demonstrates a spongy appearance in the white matter due to distention of the myelin sheath. In later stages, there is evidence of axonal disintegration. Microscopically, there are multifocal vacuolated and demyelinated lesions in the white matter of the spinal cord affecting the posterior and lateral columns in particular. Early lesions consist of swelling of myelin sheaths; fibres of highest diameter are predominantly affected. Ultrastructural studies are limited to experimental animal models. In nonhuman primates, the neuropathology is indistinguishable topographically and microscopically from that of subacute combined degeneration of the spinal cord in humans. The degeneration of myelin is characterized by separation of myelin lamellae and the formation of intramyelinic vacuoles leading to destruction of the myelin sheath (Agamanolis et al., 1978).

There are only two enzyme reactions that require a cobalamin cofactor, and inhibition of neither of these reactions can easily explain the neurological consequences of B_{12} deficiency. A number of hypotheses have been proposed. For example, it has been suggested that the formation of branched chain fatty acids caused by the accumulation of propionyl-CoA via the inhibition of methylmalonyl-CoA mutase results in abnormal composition of the myelin sheath (Carmel, 2005). Propionyl-CoA can substitute for acetyl-CoA in the acetyl-CoA synthetase reaction, the first step in fatty acid synthesis. However, inherited disorders of cobalamin metabolism which result in much higher accumulations of propionyl-CoA do not result in subacute combined degeneration of the spinal cord. An explanation consistent with methylcobalamin deficiency, is that the lack of methylcobalamin traps methyltetrahydrofolate as shown in Fig. 6. This depletes methylenetetrahydrofolate which is necessary for thymidylate synthesis thus affecting DNA synthesis as well as decreasing the syntheis of S-adenosyl-methionine which is a methyl donor to brain lipids. More recently, the role of homocysteine has been investigated. Homocysteine accumulates as a result of the inhibition of methionine synthase which requires methylcobalamin as a cofactor (Briddon, 2003). It has been suggested that homocysteine is a better functional marker of B_{12} deficiency than serum B_{12} levels (Bates

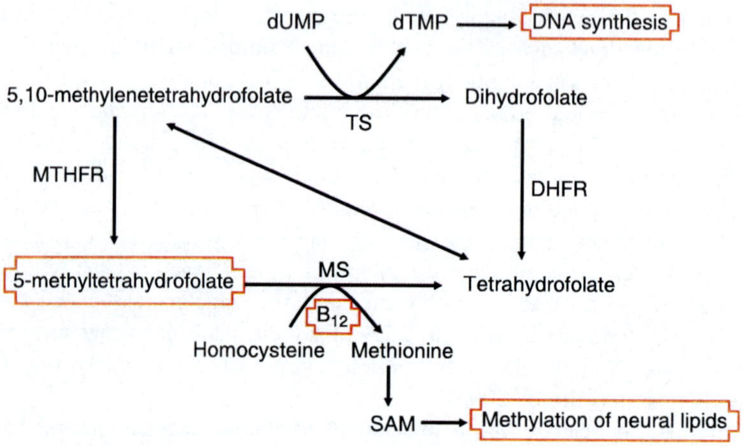

Fig. 6 Methyl trap hypothesis: 5,10-Methylenetetrahydrofolate is reduced to 5-methyltetrahydrofolate in an irreversible reaction. When vitamin B_{12} is deficient, methyl groups are trapped as 5-methyltetrahydrofolate, resulting in decreased substrates for DNA synthesis and neural lipid methylation. MTHFR, methylenetetrahydrofolate reductase; DHFR, dihydrofolate reductase; MS, Methionine synthase; TS, thymidylate synthase; SAM, S-adenosyl-methionine; dUMP, deoxyuridine 5′-monophosphate; dTTP, deoxythymidine 5′-monophosphate

et al., 1997). However, the inherited disorder of homocystinuria with much higher homocysteine levels does not result in subacute combined degeneration.

Vitamin B_{12} deficiency has been linked to increased cytokine production, in particular the myelinolytic tumor necrosis factor-alpha (TNFα), suggesting that inflammation may be the source of the neurological damage (Miller, 2002; Scalabrino et al., 2003, 2005). In a rat model of subacute combined degeneration, increased TNFα production was observed, as well as decreased neurotrophic factors, epidermal growth factor, and interleukin-6 production. This was associated with myelin vacuolation in the central nervous system (Scalabrino et al., 2003, 2005). Although changes in cytokine production are well documented, along with a number of well-defined neurochemical abnormalities, the exact mechanism to explain the selective neuropathological damage caused by B_{12} deficiency is still unknown.

5 Niacin (Vitamin B_3)

Niacin and niacinamide refer to nicotinic acid and its amide. Nicotinic acid is a pyridine derivative synthesized from tryptophan.

Experimental niacin deficiency usually requires a diet high in corn. Zein, the major storage protein of American corn, contains little tryptophan. Ingestion of corn can therefore be expected to raise the ratio, relative to tryptophan, of other long chain neutral amino acids that compete for the same carrier. Corn-fed dogs develop “black tongue,” with prominent abnormalities implicating the gastrointestinal system

(Gibson and Blass, 1985). Also, antiniacin compounds can induce deficiency states within days in rodents on a normal diet. Neurological symptoms are more obvious in rats with antimetabolite-induced niacin deficiencies than in the corn-fed dogs.

Niacin deficiency causes pellagra which includes mental symptoms and was once common in areas where American corn was a dietary staple. Addition of purified niacin to the diet has largely abolished the disorder. Pellagra is associated with the "four Ds": dermatitis, diarrhea, dementia, and death. Dietary niacin deficiency reduces the levels of NAD and NADP coenzymes in the brain. Niacin requirements can be modified by genetic and environmental factors. Hartnup syndrome is a hereditary disorder in which tryptophan transport is impaired and niacin requirements increase. A chronic toxic delirium may be the only clinical abnormality, at least early in the course of the disorder. The delirium may resemble some forms of schizophrenia (Gibson and Blass, 1985). Neuropathological changes seen in pellagra are restricted to neurons and the characteristic finding is chromatolysis. Affected neurons are ballooned, there is loss of Nissl substance, and the nuclei are located eccentrically. Although the issue of chromatolysis has been debated (Harper and Butterworth, 1997) the consensus now is that the brain regions and lesion characteristics are a function of the nature of the underlying cause (dietary, alcohol-related, or isoniazid toxicity).

At present, pellagra is encountered most often in patients with chronic alcoholism, often referred to as alcoholic pellagra encephalopathy (APE). APE patients often show only disturbances of consciousness, but may also manifest myoclonus and ataxia. Administration of niacin is recommended in APE patients showing myoclonus and ataxia even without the classical symptoms found in endemic pellagra patients (Sakai et al., 2006).

Epidemiological studies suggest that niacin may be implicated in the pathogenesis of Parkinson's disease via the following process. NAD produced from niacin releases nicotinamide via poly(ADP-ribosyl)ation which is activated in Parkinson's disease. Released excess nicotinamide is methylated to 1-methylnicotinamide (MNA) in the cytoplasm by nicotinamide N-methyltransferase. MNA destroys several subunits of complex I via superoxide formation. This can destroy complex I subunits either directly or indirectly via mitochondrial DNA damage, and stimulates poly(ADP-ribosyl)ation. It has been proposed that this implicates nicotinamide as a potential causal agent in the development of Parkinson's disease (Fukushima et al., 2004).

6 Folic Acid (Vitamin B_9)

Folate (pteroylglutamic acid) is essential for the synthesis and methylation of DNA during fetal and early postnatal development (Nunn et al., 1986). Folate deficiency may result from poor diet, malabsorption, from treatment with anticonvulsant drugs such as phenytoin or primidone, as well as from antifolate drugs such as methotrexate. Folate deficiency during pregnancy leads to an increased prevalence of fetal malformations such as spina bifida and related neural tube defects. Findings from two multicentre trials confirmed that folate supplements starting periconceptionally and continuing through pregnancy reduce the risk of neural tube defects (Werler et al., 1993; MRC Vitamin Study, 1991).

Studies in experimental animals suggest that folate deficiency during gestation and lactation results in alterations of myelin lipids (Hirono and Wada, 1978). Studies in the developing rat central nervous system suggest that folate uptake and storage depend upon a folate-binding protein (10-formyltetrahydrofolate dehydrogenase) which is preferentially localized in glial cells (Martinasevic et al., 1999).

7 Antioxidant Vitamins

Oxidative stress has been clearly implicated in a wide range of human diseases by an impressive body of scientific evidence. Oxidative stress is an imbalance in the equilibrium status of pro-oxidant/antioxidant systems in cells (Sies, 1985) and comes from external sources such as ionizing radiation, toxins, drugs, chemicals and environmental pollutants, or endogenous sources resulting from (patho)physiological metabolism of the cell. Antioxidant compounds can be classified into several general categories: (1) antioxidant enzymes such as superoxide dismutase, glutathione peroxidase, catalase, and heme-oxygenase, (2) antioxidants such as vitamin E (α-tocopherol), vitamin C (ascorbic acid), and carotenoids, (3) secondary antioxidants including selenium, zinc, riboflavin, and manganese and, finally, (4) antioxidants such as flavonoids, coenzyme Q, lipoic acid, albumin, and bilirubin. Deficiency of any antioxidant vitamin or nutrient has the potential to lead to an imbalance which may to cause oxidative stress.

7.1 α-Tocopherol (Vitamin E)

α-Tocopherol is a lipid-soluble vitamin which is an effective antioxidant. Vitamin E consists of two groups of lipid-soluble compounds: tocopherol and tocotrienol. In humans, α-tocopherol predominates and is considered the more active form of the vitamin. Vitamin E was first isolated as a factor that prevented infertility in rats (Evans and Bishop, 1922). It can inhibit the peroxidation of polyunsaturated fatty acids and it stimulates prostacyclin synthesis which promotes vasodilation and platelet aggregation. Vitamin E also protects membrane structure. Vitamin E deficiency is quite rare. Nevertheless, pure vitamin E deficiency secondary to deficiencies in absorption have been described (Traber, 2006). Vitamin E deficiency can occur with abetalipoproteinemia, cholestatic liver disease, fat malabsorption, celiac disease, cystic fibrosis, and small bowel resection. The typical neurological syndrome in humans is a spinocerebellar degeneration, with loss of reflexes, ataxia, and impaired vibration and position sense.

A severe and chronic deficiency of vitamin E is associated with a characteristic neurological syndrome with typical clinical, neuropathological, and electrophysiological abnormalities in both humans and experimental animals. Chronic vitamin E deficiency (38 weeks) in mice decreases superoxide radical production in multiple regions of male brain (Cuddihy et al., 2004). These results suggest that α-tocopherol can act as a nonclassic uncoupler. Significant impairment in neural and visual function are observed in vitamin E-deficient rats after approximately 8 months (Hayton and Muller, 2004). Low serum vitamin E is associated with demyelinating motor–sensory neuropathy related to spinocerebellar ataxia (Puri et al., 2005). Vitamin E supplementation leads to clinical and electrophysiological recovery of sensory conduction and evoked potentials; motor nerve conduction, however, shows only partial recovery. Vitamin E deficiency has also been associated with an increase in lipid peroxidation and protein oxidation in the rat brain (Jolitha et al., 2006). The concentration of free malondialdehyde (an indicator of lipid peroxidation) is significantly increased in tissues from vitamin E-deficient compared to control animals. This is consistent with a deficiency of α-tocopherol causing increased lipid peroxidation leading to abnormal neural electrophysiology (Hayton and Muller, 2004). A longitudinal study recently showed significant improvements in growth and a number of electrophysiological parameters of both neural and visual function after repletion with vitamin E (Hayton et al., 2006). It was suggested that vitamin E could play a role in hypothalamo–pituitary system regulation. Early vitamin E supplementation may provide considerable improvement of neurological signs and other associated abnormalities (Marzouki et al., 2005).

A wide range of cell culture, animal, and human epidemiological studies are suggestive of a role of vitamin E in brain function and in the prevention of neurodegeneration. It was recently suggested that vitamin E deficiency results in transcriptional alterations in the cerebral cortex of the rat which are consistent with the observed neurological and electrophysiological alterations (Hyland et al., 2006). Vitamin E deficiency was shown to have a strong impact on gene expression in the hippocampus. An important number of genes found to be regulated by vitamin E are associated with hormones and hormone metabolism, and clearance of amyloid-beta and advanced glycated end-products. A protective effect of vitamin E in AD progression has been reported (Rota et al., 2005). A recent study strongly supports the hypothesis of an impairment of lipophilic antioxidant delivery to neuronal cells in AD which could facilitate oxidative stress (Mas et al., 2006). Low-plasma vitamin E concentrations may represent a higher risk of developing dementia in subsequent years (Helmer et al., 2003).

The retention and secretion of vitamin E are regulated by α-tocopherol transfer protein (αTP) in the brain. Dysfunction of αTP results in deficiency of vitamin E in humans and mice, and increased oxidative stress in mouse brain. Ataxia with isolated vitamin E deficiency (AVED) is an autosomal recessive neurodegenerative disorder due to mutations in the αTP protein gene on chromosome 8q13. This genetic disorder is characterized by neurological symptoms often with a striking resemblance to those of Friedrich's ataxia. AVED patients have progressive spinocerebellar symptoms and markedly reduced plasma levels of vitamin E (Mariotti et al., 2004). Vitamin E supplementation therapy allows stabilization of the neurological conditions in most of the patients. However, development of spasticity and retinitis pigmentosa can appear during therapy (Mariotti et al., 2004).

7.2 Ascorbic Acid (Vitamin C)

Vitamin C is used as the generic descriptor for all compounds exhibiting qualitatively the biological activity of ascorbic acid. Ascorbic acid is an unsaturated sugar derivative that is a potent reducing agent. The oxidation of ascorbic acid to dehydroascorbic acid is reversible (Fig. 7). Both forms are biologically active. Because dehydroascorbic acid is readily reduced *in vivo*, it possesses vitamin C (anti-scurvy) activity, whereas diketogulonic acid, a metabolite, has no activity. Vitamin C has many functions in the organism, not least of which is the absorption and metabolism of iron. It is an effective antioxidant. Ascorbic acid participates in neurotransmitter synthesis as well as the synthesis of collagen. Vitamin C is necessary for the synthesis of carnitine and facilitates immune functions. Finally, ascorbic acid participates in the hydroxylation of catecholamines. The uptake of ascorbic acid into synaptosomes requires glucose and oxygen; uptake into the brain appears to be via the cerebrospinal fluid rather than the blood. Fatigue and emotional changes are common in the full-blown deficiency disease scurvy, but diffuse disease of the small blood vessels with small haemorrhages is much more striking.

Ascorbic acid has been implicated in many neurological diseases. There is a strong inverse relation between serum vitamin C concentration and stroke incidence

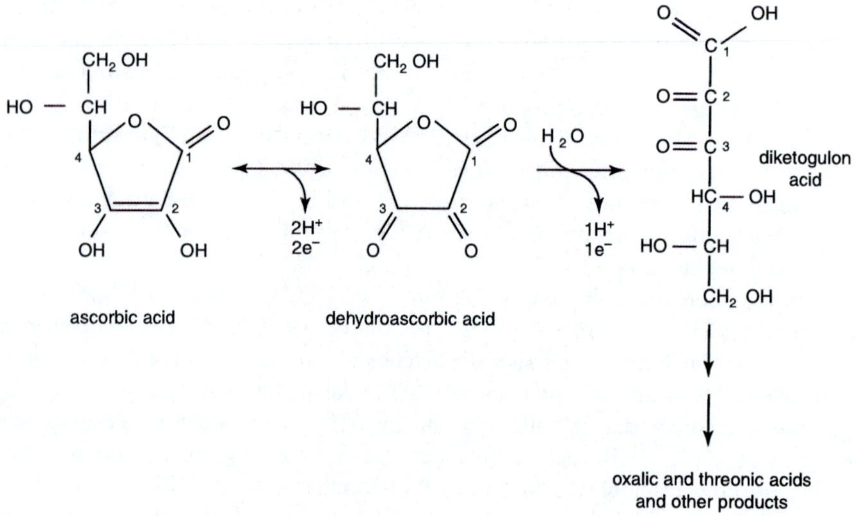

Fig. 7 Structure of ascorbic acid (vitamin C). Oxidation of ascorbic acid to dehydrogenase acid is reversible; both forms are biologically active

(Sanchez-Moreno et al., 2004). Epidemiological evidence links adequate vitamin C ingestion with decreased risk of suffering from a stroke. Inversely, decreased plasma antioxidant status is associated with increased neurological damage following a stroke. It was recently shown that dehydroascorbic acid, the oxidized form of vitamin C which is a superoxide scavenger, normalizes several markers of oxidative stress and inflammation in acute hyperglycemic focal cerebral ischemia in the rat (Bémeur et al., 2005). It was also demonstrated that intraventricular ascorbic acid injection is neuroprotective after hypoxic-ischemic brain injury in rats (Miura et al., 2006). Vitamin C is also able to protect the hypothalamus from oxidative stress induced in rats by an environmental toxicant (Muthuvel et al., 2006). Ascorbic acid confers protection from increased free-radical activity in the brain of spontaneously hypertensive rats by improving total antioxidant and superoxide dismutase status, thus preventing high blood pressure and its complications (Newaz et al., 2005). Also, intravenous cerebroprotective doses of citrate/sorbitol-stabilized DHA are correlated with increased brain ascorbate levels and a suppression of excessive lipid peroxidation (Mack et al., 2006).

A case-control study showed that plasma vitamin C levels were lower in subjects with dementia compared to controls, which was not explained by their dietary vitamin C intakes (Charlton et al., 2004). Low brain ascorbic acid and glutathione levels associated with a perturbation of the dopaminergic system actively participate in the development of some cognitive deficits affecting schizophrenic patients (Castagné et al., 2004). It has been proposed that low ascorbate in striatal extracellular fluid may contribute to Huntington's disease symptoms (Rebec et al., 2006) and evidence suggests that the level of extracellular ascorbate plays a critical role in regulating corticostriatal glutamate transmission (Rebec et al., 2005). A recent study suggested that ascorbate could participate in normalizing neuronal function in Huntington's disease (Rebec et al., 2006).

Antioxidant vitamins, particularly vitamins E and C may act synergistically. A short period of combined deficiency of vitamins E and C causes profound central nervous system dysfunction in guinea pigs (Burk et al., 2006). The damage consists mainly of nerve cell death, axonal degeneration, vascular injury, and associated glial cell responses. These findings suggest that the paralysis and death caused by combined deficiency of vitamins E and C in these animals is caused by severe damage to brainstem and spinal cord. Also, a recent study demonstrated that vitamin E and vitamin C prevented oxidative stress due to maternal hyperphenylalaninemia, (an inborn error of intermediary metabolism) in the brains of rat pups (Martínez-Cruz et al., 2006). Pretreatment with α-tocopherol and ascorbic acid prevents the impairment of energy metabolism caused by hyperargininemia in the cerebellum and hippocampus of rats (Delwing et al., 2006). Vitamin C and E administration, alone or in combination, increases striatal catalase activity in rats subjected to oral dyskinesias, which are implicated in a series of neuropathologies and associated with increased oxidative stress. A beneficial effect of these vitamins on reserpine-induced oral dyskinesia in rats has also been reported (Faria et al., 2005) and a recent study suggested that vitamins C and E hold promise in helping prevent AD (Frank and Gupta, 2005). Finally, low brain glutathione and ascorbic acid levels associated with a perturbation of the dopaminergic system may actively participate in the development of some cognitive deficits affecting schizophrenic patients (Castagné et al., 2004).

7.3 Carotenoids

Carotenoids are plant pigments which constitute more than 600 compounds, most of them being lipid-soluble and which contribute significantly to the nutritional benefits of fruit and vegetable consumption. β-carotene is the most common form of the vitamin and is the precursor of vitamin A. β-cryptoxanthine is another precursor of vitamin A. The latter is a powerful lipid-soluble antioxidant which protects cellular membranes from oxidative stress. Vitamin A is carried into the plasma by retinal binding protein which is synthesized in the liver.

Decreased carotenoid concentrations are associated with increased risk of stroke (Leppälä et al., 1999) and vitamin A levels are decreased in stroke patients (Cherubini et al., 2000). Plasma concentrations of alpha- and beta-carotene are lower in patients with acute ischemic stroke than in healthy controls and are negatively correlated with neurological deficits in stroke patients (Chang et al., 2005).

References

Agamanolis DP, Victor M, Harris JW, Hines JD, Chester EM, Kark JA (1978) An ultrastructural study of subacute combined degeneration of the spinal cord in vitamin B12-deficient rhesus monkeys. J Neuropathol Exp Neurol 37(3):273–299

Aikawa H, Watanabe IS, Furuse T, Iwasaki Y, Satoyoshi E, Sumi T, Moroji T (1984) Low energy levels in thiamine-deficient encephalopathy. J Neuropathol Exp Neurol 43(3):276–287

Albers DS, Augood SJ, Park LC, Browne SE, Martin DM, Adamson J, Hutton M, Standaert DG, Vonsattel JP, Gibson GE, Beal MF (2000) Frontal lobe dysfunction in progressive supranuclear palsy: evidence for oxidative stress and mitochondrial impairment. J Neurochem 74(2):878–881

Bates CJ, Mansoor MA, van der Pols J, Prentice A, Cole TJ, Finch S (1997) Plasma total homocysteine in a representative sample of 972 British men and women aged 65 and over. Eur J Clin Nutr 51(10):691–697

Beauchesne E, Desjardins P, Hazell AS, Butterworth RF (2009) eNOS gene deletion restores blood-brain barrier integrity and attenuates neurodegeneration in the thiamine-deficient mouse brain. J Neurochem 111:452–459

Bémeur C, Ste-Marie L, Desjardins P, Vachon L, Butterworth RF, Hazell AS, Montgomery J (2005) Dehydroascorbic acid normalizes several markers of oxidative stress and inflammation in acute hyperglycemic focal cerebral ischemia in the rat. Neurochem Int 46(5): 399–407

Bernstein AL (1990) Vitamin B_6 in clinical neurology. Ann N Y Acad Sci 585:250–260

Bettendorff L (1994) Thiamine in excitable tissues: reflections on a non-cofactor role. Metab Brain Dis 9(3):183–210

Briddon A (2003) Homocysteine in the context of cobalamin metabolism and deficiency states. Amino Acids 24(1–2):1–12

Burk RF, Christensen JM, Maguire MJ, Austin LM, Whetsell WO Jr, May JM, Hill KE, Ebner FF (2006) A Combined deficiency of vitamins E and C causes severe central nervous system damage in guinea pigs. J Nutr 136(6):1576–1581

Butterworth RF (2006) Metabolic Encephalopathies. In: Siegel GJ, Albers RW, Brady ST, Price DL (eds) Basic neurochemistry, 7th edn. Elsevier, London, pp 593–602

Butterworth RF, Besnard AM (1990) Thiamine-dependent enzyme changes in temporal cortex of patients with Alzheimer's disease. Metab Brain Dis 5(4):179–184

Butterworth RF, Gaudreau C, Vincelette J, Bourgault AM, Lamothe F, Nutini AM (1991) Thiamine deficiency and Wernicke's encephalopathy in AIDS. Metab Brain Dis 6(4):207–212

Butterworth RF, Héroux M (1989) Effect of pyrithiamine treatment and subsequent thiamine rehabilitation on regional cerebral amino acids and thiamine-dependent enzymes. J Neurochem 52(4):1079–1084

Butterworth RF, Kril JJ, Harper CG (1993) Thiamine-dependent enzyme changes in the brains of alcoholics: relationship to the Wernicke-Korsakoff syndrome. Alcohol Clin and Exp Res 17(5):1084–1088

Calingasan NY, Baker H, Sheu KF, Gibson GE (1995) Blood-brain barrier abnormalities in vulnerable brain regions during thiamine deficiency. Exp Neurol 134(1):64–72

Calingasan NY, Park LC, Calo LL, Trifiletti RR, Gandy SE, Gibson GE (1998) Induction of nitric oxide synthase and microglial responses precede selective cell death induced by chronic impairment of oxidative metabolism. Am J Pathol 153(2):599–610

Carmel R (2005) Cobalamin (Vitamin B_{12}). In: Shils ME, Shike M, Ross AC, Caballero B, Cousins RJ (eds) Modern nutrition in health and disease, 10th edn. Lippincott William & Wilkins, New York, NY, pp 482–497

Casley CS, Canevari L, Land JM, Clark JB, Sharpe MA (2002) β-Amyloid inhibits integrated mitochondrial respiration and key enzyme activities. J Neurochem 80(1):91–100

Castagné V, Rougemont M, Cuenod M, Do KQ (2004) Low brain glutathione and ascorbic acid associated with dopamine uptake inhibition during rat's development induce long-term cognitive deficit: relevance to schizophrenia. Neurobiol Dis 15(1):93–105

Chang SJ, Kirksey A, Morré DM (1981) Effects of vitamin B_6 deficiency on morphological changes in dendritic trees of Purkinje cells in developing cerebellum of rats. J Nutr 111(5):848–857

Chang CY, Chen JY, Ke D, Hu ML (2005) Plasma levels of lipophilic antioxidant vitamins in acute ischemic stroke patients: correlation to inflammation markers and neurological deficits. Nutrition 21(10):987–993

Charlton KE, Rabinowitz TL, Geffen LN, Dhansay MA (2004) Lowered plasma vitamin C, but not vitamin E, concentrations in dementia patients. J Nutr Health Aging 8(2): 99–107

Charness ME, DeLaPaz RL (1987) Mamillary body atrophy in Wernicke's encephalopathy: antemortem identification using magnetic resonance imaging. Ann Neurol 22(5):595–600

Cherubini A, Polidori MC, Bregnocchi M, Pezzuto S, Cecchetti R, Ingegni T, Di Iorio A, Senin U, Mecocci P (2000) Antioxidant profile and early outcome in stroke patients. Stroke 31(10): 2295–2300

Cooper JR, Pincus JH (1979) The role of thiamine in nervous tissue. Neurochem Res 4(2):223–239

Cuddihy SL, Musiek ES, Morrow JD, Dugan LL (2004) Long-term vitamin E deficiency in mice decreases superoxide radical production in brain. Ann N Y Acad Sci 1031: 428–431

Dakshinamurti K, Paulose CS, Viswanathan M, Siow YL, Sharma SK, Bolster B (1990) Neurobiology of pyridoxine. Ann N Y Acad Sci 585:128–144

Delwing D, Tagliari B, Chiarani F, Wannmacher CMD, Wajner M, Wyse AT (2006) Alpha-tocopherol and ascorbic acid administration prevents the impairment of brain energy metabolism of hyperargininemic rats. Cell Mol Neurobiol 26(2):177–189

Evans HM, Bishop KS (1922) On the existence of a hitherto unrecognized dietary factor essential for reproduction. Science 56(1458):650–651

Faria RR, Abílio VC, Grassl C, Chinen CC, Negrão LT, de Castro JP, Fukushiro DF, Rodrigues MS, Gomes PH, Registro S, de Carvalho RdeC, D'Almeida V, Silva RH, Ribeiro RdeA, Frussa-Filho R (2005) Beneficial effects of vitamin C and vitamin E on reserpine-induced oral dyskinesia in rats: critical role of striatal catalase activity. Neuropharmacology 48(7):993–1001

Frank B, Gupta S (2005) A review of antioxidants and Alzheimer's disease. Ann Clin Psychiatry 17(4):269–286
Fukushima T, Ohta M, Tanaka K, Kaneko S-Y, Maeda T, Sasaki A (2004) Niacin metabolism and Parkinson's disease. Asia Pac J Clin Nutr 13(Suppl):S176
Gautam M, Noakes PG, Mudd J, Nichol M, Chu GC, Sanes JR, Merlie JP (1995) Failure of postsynaptic specialization to develop at neuromuscular junctions of rapsyn-deficient mice. Nature 377(6546):232–236
Gerster H (1996) The importance of vitamin B6 for development of the infant. Human medical and animal Experimental studies. Z Ernahrungswiss 35(4):309–317
Gibson GE, Blass JP (1985) Oxidative metabolism and acetylcholine synthesis during acetylpyridine treatment. Neurochem Res 10(4):453–467
Gibson GE, Sheu KF, Blass JP, Baker A, Carlson KC, Harding B, Perrino P (1988) Reduced activities of thiamine-dependent enzymes in the brains and peripheral tissues of patients with Alzheimer's disease. Arch Neurol 45(8):836–840
Gibson GE, Zhang H (2002) Interactions of oxidative stress with thiamine homeostasis promote neurodegeneration. Neurochem Int 40(6):493–504
Gospe SM Jr, Hecht ST (1998) Longitudinal MRI findings in pyridoxine-dependent seizures. Neurology 51(1):74–78
Hakim AM (1984) The induction and reversibility of cerebral acidosis in thiamine deficiency. Ann Neurol 16(6):673–679
Harata N, Iwasaki Y (1995) Evidence for early blood-brain barrier breakdown in experimental thiamine deficiency in the mouse. Metab Brain Dis 10(2):159–174
Harper CG (1983) The incidence of Wernicke's encephalopathy in Australia. A neuropathological study of 131 cases. J Neurol Neurosurg Psychiatry 46:593–598
Harper CG, Butterworth RF (1997) Nutritional and metabolic disorders. In: Graham DI, Lantos PL (eds) Greenfield's neuropathology. Arnold, London, pp 601–655
Hayton SM, Kriss T, Wase A, Muller DP (2006) Effects on neural function of repleting vitamin E-deficient rats with alpha-tocopherol. J Neurophysiol 95(4):2553–2559
Hayton SM, Muller DP (2004) Vitamin E in neural and visual function. Ann N Y Acad Sci 1031:263–270
Hazell AS, Butterworth RF, Hakim AM (1993) Cerebral vulnerability is associated with selective increase in extracellular glutamate concentration in experimental thiamine deficiency. J Neurochem 61(3):1155–1158
Hazell AS, Rao KV, Danbolt NC, Pow DV, Butterworth RF (2001) Selective down-regulation of the astrocyte glutamate transporters GLT-1 and GLAST within the medial thalamus in experimental Wernicke's encephalopathy. J Neurochem 78(3):560–568
Helmer C, Peuchant E, Letenneur L, Bourdel-Marchasson I, Larrieu S, Dartigues JF, Dubourg L, Thomas MJ, Barberger-Gateau P (2003) Association between antioxidant nutritional indicators and the incidence of dementia: results form the PAQUID prospective cohort study. Eur J Clin Nutr 57(12):1555–1561
Hirono H, Wada Y (1978) Effects of dietary folate deficiency on developmental increase of myelin lipids in rat brain. J Nutr 108(5):766–772
Hyland S, Muller D, Hayton S, Stoecklin E, Barella L (2006) Cortical gene expression in the vitamin E-deficient rat: possible mechanisms for the electrophysiological abnormalities of visual and neural function. Ann Nutr Metab 50(5):433–441
Héroux M, Raghavendra Rao VL, Lavoie J, Richardson JS, Butterworth RF (1996) Alterations of thiamine phosphorylation and of thiamine-dependent enzymes in Alzheimer's disease. Metab Brain Dis 11(1):81–88
Jolitha AB, Subramanyam MV, Asha Devi S (2006) Modification by vitamin E and exercise of oxidative stress in regions of aging rat brain: studies on superoxide dismutase isoenzymes and protein oxidation status. Exp Gerontol 41(8):753–763
Kruse M, Navarro D, Desjardins P, Butterworth RF (2004) Increased brain endothelial nitric oxide synthase expression in thiamine deficiency: relationship to selective vulnerability. Neurochem Int 45(1):49–56

Laforenza U, Patrini C, Rindi G (1988) Distribution of thiamine, thiamine phosphates, and thiamine metabolizing enzymes in neuronal and glial cell enriched fractions of rat brain. J Neurochem 51(3):730–735

Langlais PJ, Anderson G, Guo SX, Bondy SC (1997) Increased cerebral free radical production during thiamine deficiency. Metab Brain Dis 12(2):137–143

Langlais PJ, Mair RG (1990) Protective effects of the glutamate antagonist MK-801 on pyrithiamine-induced lesions and amino acid changes in rat brain. J Neurosci 10(5): 1664–1674

Leppälä JM, Virtamo J, Fogelholm R, Albanes D, Heinonen OP (1999) Different risk factors for different stroke subtypes: association of blood pressure, cholesterol, and antioxidants. Stroke 30(12):2535–2540

Mack WJ, Mocco J, Ducruet AF, Laufer I, King RG, Zhang Y, Guo W, Pinsky DJ, Connolly ES Jr (2006) A cerebroprotective dose of intravenous citrate/sorbitol-stabilized dehydroascorbic acid is correlated with increased cerebral ascorbic acid and inhibited lipid peroxidation after murine reperfused stroke. Neurosurgery 59(2):383–388

Mariotti C, Gellera C, Rimoldi M, Mineri R, Uziel G, Zorzi G, Pareyson D, Piccolo G, Gambi D, Piacentini S, Squitieri R, Capra R, Castellotti B, Di Donato S (2004) Ataxia with isolated vitamin E deficiency: neurological phenotype, clinical follow-up and novel mutations in TTPA gene in Italian families. Neurol Sci 25(3):130–137

Martinasevic MK, Rios GR, Miller MW, Tephly TR (1999) Folate and folate-dependent enzymes associated with rat CNS development. Dev Neurosci 21(1):29–35

Martínez-Cruz F, Osuna C, Guerrero JM (2006) Mitochondrial damage induced by fetal hyperphenylalaninemia in the rat brain and liver: its prevention by melatonin, Vitamin E, and Vitamin C. Neurosci Lett 392(1–2):1–4

Marzouki N, Benomar A, Yahyaoui M, Birouk N, Elouazzani M, Chkili T, Benlemlih M (2005) Vitamin E deficieny ataxia with (744 del A) mutation on alpha-TTP gene: genetic and clinical peculiarities in Moroccan patients. Eur J Med Genet 48(1):21–28

Mas E, Dupuy AM, Artero S, Portet F, Cristol JP, Ritchie K, Touchon J (2006) Functional vitamin E deficiency in ApoE4 patients with Alzheimer's disease. Dement Geriatr Cogn Disord 21(3):198–204

Miller JW (2002) Vitamin B_{12} deficiency, tumor necrosis factor-alpha, and epidermal growth factor: a novel function for vitamin B_{12}? Nutr Rev 60(5):142–144

Miura S, Ishida A, Nakajima W, Ohmura A, Kawamura M, Takada G (2006) Intraventricular ascorbic acid administrations decreases hypoxic-ischemic brain injury in newborn rats. Brain Res 1095(1):159–166

Mizuno Y, Matuda S, Yoshino H, Mori H, Hattori N, Ikebe S (1994) An immunohistochemical study on alpha-ketoglutarate dehydrogenase complex in Parkinson's disease. Ann Neurol 35(2):204–210

MRC Vitamin Study Research Group (1991) Prevention of neural tube defects: result of the Medical Research Council Vitamin Study. Lancet 338(8760):131–137

Muthuvel R, Venkataraman P, Krishnamoorthy G, Gunadharini DN, Kanagaraj P, Jone Stanley A, Srinivasan N, Balasubramanian K, Aruldhas MM, Arunakaran J (2006) Antioxidant effect of ascorbic acid on PCB (Aroclor 1254) induced oxidative stress in hypothalamus of rats. Clin Chim Acta 365(1–2):297–303

Navarro D, Zwingmann C, Hazell AS, Butterworth RF (2005) Brain lactate synthesis in thiamine deficiency: a re-evaluation using $^{1}H/^{13}C$ nuclear magnetic resonance spectroscopy. J Neurosci Res 79(1–2):33–41

Newaz MA, Yousefipour Z, Nawa NN (2005) Modulation of nitric oxide synthase activity in brain, liver, and blood vessels of spontaneously hypertensive rats by ascorbic acid: protection from free radical injury. Clin Exp Hypertens 27(6):497–508

Nunn JF, Chanarin I, Tanner AG, Owen ER (1986) Megaloblastic bone marrow changes after repeated nitrous oxide anaesthesia. Reversal with folinic acid. Br J Anaesth 58(12):1469–1470

Parker WD Jr, Hass R, Stumpf DA, Parks J, Eguren LA, Jackson C (1984) Brain mitochondrial metabolism in experimental thiamine deficiency. Neurology 34(11):1477–1481

Peters RA (1936) The biochemical lesion in vitamin B_1 deficiency. application of modern biochemical analysis in its diagnosis. Lancet 1:1161–1164

Puri V, Chaudhry N, Tatke M, Prakash V (2005) Isolated vitamin E deficiency with demyelinating neuropathy. Muscle Nerve 32(2):230–235

Rao VL, Richardson JS, Butterworth RF (1993) Decreased activities of thiamine diphosphatase in frontal and temporal cortex in Alzheimer's disease. Brain Res 631(2):334–336

Rebec GV, Conroy SK, Barton SJ (2006) Hyperactive striatal neurons in symptomatic Huntington R6/2 mice: variations with behavioral state and repeated ascorbate treatment. Neuroscience 137(1):327–336

Rebec GV, Witowski SR, Sandstrom MI, Rostand RD, Kennedy RT (2005) Extracellular ascorbate modulates cortically evoked glutamate dynamics in rat striatum. Neurosci Lett 378(3): 166–170

Rota C, Rimbach G, Minihane AM, Stoecklin E, Barella L (2005) Dietary vitamin E modulates differential gene expression in the rat hippocampus: potential implications for its neuroprotective properties. Nutr Neurosci 8(1):21–29

Sakai K, Nakajima T, Fukuhara N (2006) A suspected of alcoholic pellagra encephalopathy with marked response to niacin showing myoclonus and ataxia as chief complaints. No To Shinkei 58(2):141–144

Sanchez-Moreno C, Dashe JF, Scott T, Thaler D, Folstein MF, Martin A (2004) Decreased levels of plasma vitamin C and increased concentrations of inflammatory and oxidative stress markers after stroke. Stroke 35:163–168

Scalabrino G (2005) Cobalamin (vitamin B_{12}) in subacute combined degeneration and beyond tradional interpretations and novel theories. Exp Neurol 192(2):463–479

Scalabrino G, Buccellato FR, Veber D, Mutti E (2003) New basis of the neurotrophic action of vitamin B_{12}. Clin Chem Lab Med 41(11):1435–1437

Shih JJ, Kornblum H, Shewmon DA (1996) Global brain dysfunction in an infant with pyridoxine dependency: evaluation with EEG, evoked potentials, MRI, and PET. Neurology 47(3): 824–826

Sies H (1985) Oxidative stress: introductory remarks.In: Sies H (eds) Oxidative stress. Academic Press, Orlando, pp 1–8

TanPhaichitr V (1985) Epidemiology and clinical assessment of vitamin deficiencies in Thai children. In: Eeckels RE, Ransome-Kuti O, Kroonenberg CC (eds) Child health in the tropics. Martinus Nijhoff Publishers, Dordrecht, pp 157–166

TanPhaichitr V (1999) Thiamin. In: Shils ME, Olsen JA, Shike M et al (eds) Modern nutrition in health and disease, 9th edn. Lippincott Williams & Wilkins, Baltimore, MD

Todd KG, Butterworth RF (1999) Early microglial response in experimental thiamine deficiency: an immunohistochemical analysis. Glia 25(2):190–198

Torvik A (1985) Two types of brain lesions in Wernicke's encephalopathy. Neuropathol Appl Neurobiol 11(3):179–190

Traber MG (2006) How much vitamin E... Just enough! Am J Clin Nutr 84(5):959–960

Victor M, Adams RD, Collins GH (1971) The Wernicke-Korsakoff syndrome. A clinical and pathological study of 245 patients, 82 with post-mortem examinations. Contemp Neurol Ser 7:1–206

Werler MM, Shapiro S, Mitchell AA (1993) Periconceptional folic acid exposure and risk of occurrent neural tube defects. JAMA 269(10):1257–1261

Xu Y, Sladky JT, Brown MJ (1989) Dose-dependent expression of neuronopathy after experimental pyridoxine intoxication. Neurology 39(8):1077–1083

Brain Edema in Neurological Diseases

Eduardo Candelario-Jalil, Saeid Taheri, and Gary A. Rosenberg

Abstract In the brain, the transport of water and solute is precisely regulated. This maintains a stable and unique microenvironment that is critical to brain function. Cerebral edema results from the excess of fluid in the brain's intra- and extracellular spaces. This occurs in response to a wide variety of insults, including cerebral ischemia, hypoxia, infection, brain tumors, and neuroinflammation. Cytotoxic edema leads to intracellular swelling without alterations in vascular permeability. Vasogenic edema is associated with damage to the blood–brain barrier. These types of edema rarely exist in isolation. In most neuropathological conditions, one type of edema predominates, but both coexist. This chapter focuses on the major molecular mechanisms triggering brain edema, including alterations in ion channels and transporters, matrix metalloproteinases, tight junction protein degradation, free radicals, and products of the arachidonic acid metabolism. We review present knowledge of the contribution to brain edema of molecules such as aquaporins, vasopressin, vascular endothelial growth factor, angiopoietins, and bradykinin. We further examine brain imaging modalities that have revolutionized clinical diagnosis of cerebral edema. Finally, we provide a critical evaluation of the current strategies for the treatment of brain edema.

Keywords Vasogenic edema · Neurovascular unit · Matrix metalloproteinases · Aquaporins · Blood–brain barrier · Imaging

Contents

G.A. Rosenberg (✉)
Department of Neurology, University of New Mexico, Albuquerque, NM 87131, USA; Department of Neurosciences, University of New Mexico Health Sciences Center, Albuquerque, NM 87131, USA; Department of Cell Biology and Physiology, University of New Mexico Health Sciences Center, Albuquerque, NM 87131, USA
e-mail: grosenberg@salud.unm.edu

J.P. Blass (ed.), *Neurochemical Mechanisms in Disease*,
Advances in Neurobiology 1, DOI 10.1007/978-1-4419-7104-3_5,

1 Introduction

Cerebral edema occurs in response to a wide variety of insults, including ischemia, hypoxia, infection, and noninfectious inflammation. Shifts in brain water, which is the basis of the cellular swelling, are due to osmotic forces, and result in increases in intra- and extracellular spaces. A reasonable amount of tissue swelling can be tolerated in most parts of the body, however, the restrictions imposed by the rigid tentorium and bony skull cause life-threatening herniation with relatively small increases in the brain compartments. Two early anatomists, Monroe (1733–1817) and Kellie (1758–1829), recognized that increased intracranial pressure due to swelling in the cerebrospinal fluid (CSF), blood, or brain tissue compartments could increase intracranial pressure; the concept of limited expansion capacity of the intracranial contents is called the Monroe–Kellie doctrine.

Brain cell integrity depends on a continuous supply of oxygen and glucose in order to perform chemiosmotic work that maintains the cell membranes. Loss of ATP causes failure of the ATPase-mediated electrolyte pumps that remove sodium in exchange for potassium with the result that osmotic pressure builds up within the cell and *cytotoxic* edema occurs. If the cellular membrane remains intact, the swollen cell can survive. Once the membranes break down, however, preservation of function is no longer possible. Another type of edema occurs with damage to blood vessels by trauma, ischemia, hypertension, or infections, which disrupt the endothelial tight junctions allowing fluid and toxins to cross the blood–brain barrier (BBB) and enter the brain. Linearly arranged white matter tracts serve as conduits for fluid to move from one place to another within the brain. Leaky blood vessels cause fluid to be transported between white matter tracts, which is referred to as *vasogenic* edema. A third type of edema results from the transependymal flow of fluid into the brain; this type of extracellular edema is referred to as *interstitial edema* (Klatzo, 1967; Fishman, 1975; Kimelberg, 1995; Rosenberg and Yang, 2007).

Knowledge of the mechanisms of edema formation have expanded dramatically in the past few years with discovery of important molecular mechanisms involved in water movement across membranes and degradation of tight junction proteins in endothelial cells. Many of the concepts regarding brain edema are well established and based on a large body of work that was done in the past 50 years, and has been summarized in several monographs (Katzman and Pappius, 1973; Rapoport, 1976; Rosenberg, 1990; Fishman, 1992). Our goal is to refer to that work, but to emphasize the more recent studies based on advances in molecular biology and brain imaging that will lead to novel therapies (Rabinstein, 2006; Bardutzky and Schwab, 2007; Zador et al., 2007).

2 Water Homeostasis in the Brain: Physiology of the Brain Fluids

Water flow in the central nervous system (CNS) is unique, and regulation of water balance is of paramount importance for brain functions. For most tissues, there is a conveying water flow into the tissue at the interface of the endothelial cells of blood vessels, bringing in hydrophilic substances and electrolytes (Kimelberg, 2004). Selectively permeable membranes keep plasma proteins and charged substances within the vasculature. It is estimated that 93% of plasma proteins are retained in the vascular space, which create an osmotic driving force in the venous capillaries for the return of fluid to the blood (Kimelberg, 2004). However, due to the presence of the BBB, water flow occurs differently in the CNS, maintaining a stable and unique microenvironment for the normal function of neurons and other cells.

Brain and blood interactions occur at three interfaces: endothelial cells, which form the major site of the BBB, the choroid plexus ependymal cell lining, and the arachnoid granulations. These sites are key in regulating the exchange of substances between brain and blood, thus maintaining the composition of brain electrolytes, as well as the content of proteins and other substances.

Brain vascular endothelial cells are linked by tight junction proteins creating high-resistance junctions between cells that effectively prevent the movement of hydrophilic substances, including electrolytes, such as Na^+ and K^+. Water moves across the lipid bilayer of endothelial cells through simple diffusion and vesicular transport (Tait et al., 2008). However, specialized water channels are formed by molecules called aquaporins (AQPs), which are highly expressed in blood–brain interfaces to facilitate the transport of water across cell membranes.

Transport of water into the cerebrospinal fluid (CSF) and interstitial fluid (ISF) forms the source of the CSF that fills the cerebral ventricles and the subarachnoid spaces around the brain and spinal cord. Early studies by Weed and Cushing identified the CSF as a "Third Circulation," functioning along with the fluid between the cells, the ISF, as the lymph of the brain (Weed, 1935, 1938). The ISF circulates between the cells and drains into the CSF; it is formed osmotically by the extrusion of sodium by the sodium–potassium ATPase pump in endothelial cell membranes (Abbott, 2004). The CSF is secreted by the choroid plexi into the ventricles and is also derived from the ISF produced mainly by the brain capillaries. The ISF communicate with the CSF through gaps between cells forming the ependymal lining of the ventricles. The CSF is eventually drained across the arachnoid granulations that protrude into the sagital sinus, thus completing the circulation of the CSF and ISF (Klatzo, 1994; Abbott, 2004). An alternative model of CSF circulation has been proposed, implying that the main absorption of CSF occurs through the brain capillaries and bulk flow through the arachnoid granulations is only a complementary outlet route for CSF (Greitz and Hannerz, 1996; Greitz et al., 1997).

Water transport from the vasculature into the ventricle is facilitated by aquaporin-1 (AQP1) highly expressed in the apical (ventricular-facing) membrane of the choroid plexus, and via AQP4 in the ependymal lining of the ventricles (Zador et al., 2007). Deletion of AQP1 reduces by fivefold osmotically induced water transport in the choroid plexus (Oshio et al., 2005). CSF production is significantly reduced in AQP1-deficient mice, but only by 20–25%, indicating a substantial contribution of extrachoroidal fluid production by the brain parenchyma (Zador et al., 2007). Fluid from the subarachnoid space is drained through the arachnoid granulations into the low-pressure venous sinus that exit the cranium. Astrocytic processes lining the pial membrane heavily express AQP4 which facilitate water flux into the subarachnoid space (Zador et al., 2007). Excess ISF is also eliminated by a transcapillary (AQP4-rich) route into the blood (Greitz et al., 1997; Tait et al., 2008).

3 The Neurovascular Unit and Tight Junction Proteins

Normal function of the brain depends on the BBB, which provides a highly selective barrier between the blood and the brain parenchyma that creates a special microenvironment crucial for brain homeostasis. Endothelial cells, astrocytes, perivascular microglia, neurons, and pericytes comprise the neurovascular unit (Fig. 1) (Ballabh

Fig. 1 Cellular and molecular constituents of the neurovascular unit. The blood–brain barrier (BBB) is formed by endothelial cells, which are surrounded by the basal lamina and the astrocytic end-feet processes. The perivascular astrocytes provide the connection between the neurons and the BBB. Astrocytic processes heavily express aquaporin 4 (AQP4). Within the basal lamina reside the pericytes, which are important in BBB stability. The basal lamina provides structural integrity to the capillaries and is mainly composed of type IV collagen, fibronectin, heparin sulfate, laminin, and entactin. Perivascular microglial cells make contact with cerebral microvessels and modulate the functioning of the BBB. The tight junctions and adherent junctions connect brain endothelial cells, and confer the low paracellular permeability of the BBB. The tight junction proteins (TJPs) form an intricate complex of proteins linked to the actin cytoskeleton. Claudins and occludin have four transmembrane domains with two extracellular loops, which are important in forming the "seal" between two adjacent endothelial cells. These proteins associate with the cytoskeleton via accessory proteins such as zona occludens ZO-1, ZO-2, AF6, and cingulin. The junctional adhesion molecule (JAM) family forms part of the TJPs, and mediates attachment of cell membranes via homophilic interactions. The most important components of the adherent junctions are vascular endothelial (VE)-cadherin and platelet endothelial cell adhesion molecule-1 (PECAM-1). VE-cadherin is linked to the actin cytoskeleton via catenins

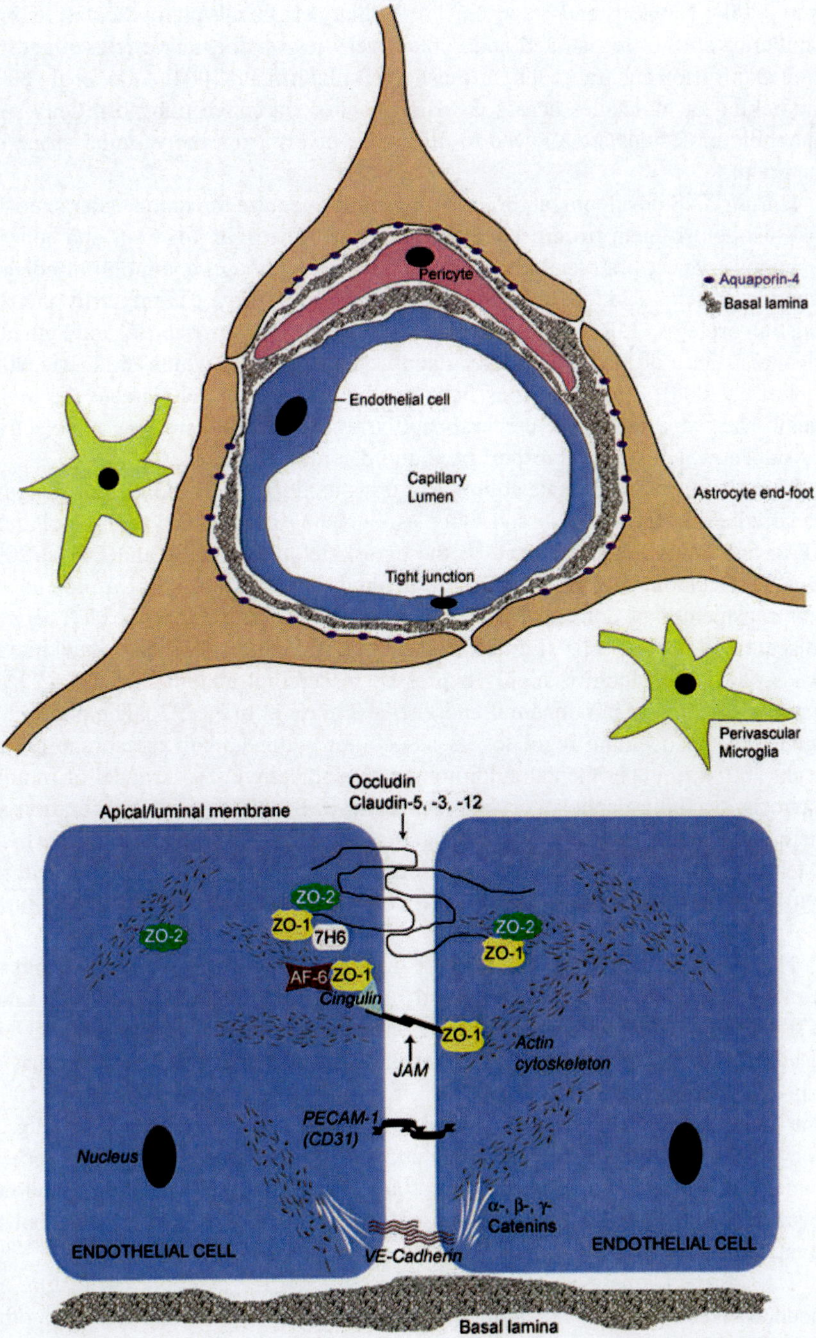

Fig. 1 (Continued)

et al., 2004; Hawkins and Davis, 2005). Unlike peripheral microvasculature, brain capillaries are not fenestrated and contain very few endocytic vesicles suggesting limited diffusion and transcellular transport (Ballabh et al., 2004; Zador et al., 2007). Trafficking of molecules across the BBB occurs via active transport. Only small lipophilic molecules are allowed to diffuse passively from the vascular space into the brain.

During CNS development, brain blood vessels acquire the unique characteristics that distinguish them from peripheral capillaries. The tight junctions and adherent junctions connect brain endothelial cells (Fig. 1). Although disruption of adherent junction proteins can lead to increased BBB permeability, it is primarily the tight junction proteins (TJPs) that confer the low paracellular permeability and high electrical resistance of the BBB (Bazzoni and Dejana, 2004; Hawkins and Davis, 2005; Zlokovic, 2008). Tight junctions between the endothelial cells create the unique membrane properties of the cerebral capillaries by greatly increasing the electrical resistance, which blocks transport of nonlipid soluble substances.

The TJPs form an intricate complex of transmembrane (occludin, claudins, junctional adhesion molecule-1) and cytoplasmic (zona occludens-1 and -2, cingulin, AF-6, and 7H6) proteins linked to the cytoskeleton (Hawkins and Davis, 2005) (Fig. 1). Occludin was the first TJP discovered. It is a 60- to 65-kDa protein with four transmembrane domains and two extracellular loops that span the cleft between adjacent endothelial cells (Furuse et al., 1993; Hirase et al., 1997; Hawkins and Davis, 2005). Occludin is highly expressed in cerebral endothelium (Fig. 2) and sparsely distributed in nonneural endothelia (Hirase et al., 1997). The phosphorylation state of occludin regulates its association with the cell membrane (Hirase et al., 2001). In experimental autoimmune encephalomyelitis, a model of multiple sclerosis, occludin dephosphorylation precedes the neurological deterioration and increased leakage of plasma proteins across the BBB (Morgan et al., 2007). The C-terminal cytoplasmic domain of occludin is involved in its association with the cytoskeleton via accessory proteins such as zona occludens ZO-1 and ZO-2 (Furuse et al., 1993).

The claudins are a large family of at least 24 members. Claudin-5, -3, and -12 are localized at the BBB (Wolburg and Lippoldt, 2002; Nitta et al., 2003) and it is still debatable whether claudin-1 is present at the BBB. The extracellular tails of claudins from adjacent cells self-assemble to form the tight junctions that are "zip-locked" together (Nitta et al., 2003; Krause et al., 2008; Piontek et al., 2008). The junctional adhesion molecule (JAM) family forms part of the TJPs (Fig. 1). They are believed to mediate the early attachment of adjacent cell membranes via homophilic interactions (Dejana et al., 2000; Bazzoni and Dejana, 2004) and may regulate transendothelial leukocyte migration (Del Maschio et al., 1999), but the function of JAM in the mature BBB is largely undefined.

The adherent junctions are ubiquitously found in the cerebral vasculature and mediate several functions, including the adhesion of endothelial cells to each other, contact inhibition during remodeling, and growth of the vasculature, and mediate in part the regulation of paracellular permeability (Hawkins and Davis, 2005). The most important components of the adherent junctions are vascular endothelial

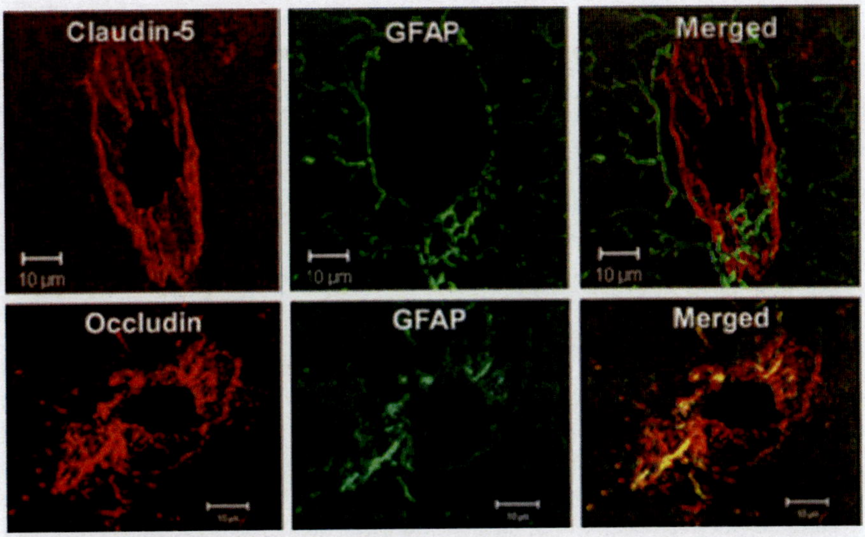

Fig. 2 Confocal micrographs showing the immunoreactivity for the tight junction proteins, claudin-5 (**a**) and occludin (**b**) in rat brain microvessels. Claudin-5 (*red*) in blood vessels is separated from the astrocytes (glial fibrillary acidic protein, GFAP, in *green*) surrounding them. The merged images show that claudin-5 and GFAP staining are separate (**Panel (a)**, far *right*). (**b**) Occludin is highly expressed in the cerebral endothelium. Occludin expression and GFAP staining are co-localized around blood vessels in the rat brain. *Scale bars* indicate 10 μm. Modified from Yang et al. (2007)

(VE)-cadherin and platelet endothelial cell adhesion molecule-1 (PECAM-1). VE-cadherin is an endothelial-specific Ca^{2+}-regulated protein that is linked to the cytoskeleton via catenins (Fig. 1). PECAM-1, also known as CD31, is a key participant in the migration of blood-borne cells across the BBB. Changes in the adherent junction proteins can lead to increased paracellular permeability (Abbruscato and Davis, 1999) and leukocyte trafficking in the CNS (Newman, 1994; Garrido-Urbani et al., 2008).

On the abluminal surface of the endothelial cells is a thin layer of basal lamina composed mainly of type IV collagen, fibronectin, heparan sulfate, laminin, and entactin. Entactin (also termed nidogen) is a basement membrane glycoprotein that connects type IV collagen and laminin to add a structural element to the capillary, and plays a role in cell interactions with the extracellular matrix. Fibronectin from the cells joins the basal lamina to the endothelium. Basal lamina provide structure through type IV collagen, charge barriers by heparan sulfate, and binding sites on the laminin and fibronectin molecules (Zlokovic, 2008).

Within the basal lamina reside the pericytes (Fig. 1). Mesenchymal in origin, pericytes form an incomplete envelopment around the endothelial cells and within the microvascular basement membrane of capillaries and postcapillary venules. Cell bodies and cytoplasmic processes of pericytes, as well as the endothelial cells, are enveloped by the same basal lamina, except for where they make direct contact with each other (Diaz-Flores et al., 1991). They are important in BBB stability as well as angiogenesis. They have been implicated in blood flow regulation at the capillary level (Hirschi and D'Amore, 1996). Their expression of smooth muscle actin (SMA) and desmin, two proteins found in smooth muscle cells, and their adherence to the endovascular cells make them very strong candidates for blood flow regulators in the microvasculature (Hirschi and D'Amore, 1996). Pericytes are contractile and seem to serve as a smooth muscle equivalent in the brain capillaries. They also display several macrophage properties including phagocytosis and antigen presentation (Thomas, 1999). Interaction between pericytes and endothelial cells is important for the maturation, remodeling, and maintenance of the vascular system via the secretion of growth factors or modulation of the extracellular matrix (Lai and Kuo, 2005). There is also evidence that pericytes are involved in the transport across the BBB and the regulation of vascular permeability (Hirschi and D'Amore, 1996; Thomas, 1999; Dore-Duffy, 2008).

Surrounding the endothelia and basal lamina are the astrocytic foot processes, which have multiple ion transporters and channels, and heavily express AQP4, suggesting that these processes facilitate ion and water transport across the BBB (Zador et al., 2007) (Fig. 1). Neurons and perivascular microglia are the other cellular components of the neurovascular unit. In the adult brain neurons, which are not in direct contact with endothelial cells, probably exert an influence indirectly. However, astrocytes directly mediate the neurovascular connections by enwrapping their foot processes around brain microvessels (Kim et al., 2006; Kaur and Ling, 2008). Neuronal activity modulates cerebral blood flow, and astrocytes mediate this process (Anderson and Nedergaard, 2003; Schipke and Kettenmann, 2004). Astrocytes by releasing vasoactive molecules mediate the neuron–astrocyte–endothelial signaling pathway and play a profound role in coupling blood flow to neuronal activity (Jakovcevic and Harder, 2007; Koehler et al., 2009).

Perivascular microglia make contact with cerebral microvessels and modulate the functioning of the neurovascular unit (Kaur and Ling, 2008). There are two important sources of microglia in the brain. During development, leptomeningeal mesenchymal cells enter the brain and transform into microglia (Bechmann et al., 2007). Circulating monocytes provide another major source of brain microglia (Bechmann et al., 2005, 2007). Perivascular microglial cells, which are bone marrow derived, continuously turn over in the CNS, and are immunoregulatory cells that connect the CNS with the peripheral immune system (Williams et al., 2001). Microglia are phagocytic cells with the capability of antigen presentation. They rapidly respond to a wide variety of stimuli including inflammation and hypoxia/ischemia (Block et al., 2007; del Zoppo et al., 2007). Activated microglia release several inflammatory factors, which modulate the permeability properties of the neurovascular unit (Stoll and Jander, 1999; Block et al., 2007).

There are complex interactions among the different cellular components of the neurovascular unit and the extracellular matrix, determining its permeability properties during both physiological and pathological conditions. This highlights the severe limitations of cell culture-based models to mimic neurological diseases associated with BBB disruption. Transwell culture systems of endothelial cells alone rarely achieve adequate transendothelial electrical resistance (TEER). Cocultures of astrocytes with endothelial cells reach higher levels of TEER. The incorporation of luminal structures into the coculture system provides a flow component that most closely mimics the in vivo situation (Krizanac-Bengez et al., 2006).

4 Cytotoxic Brain Edema

Brain edema is defined as an abnormal accumulation of fluid associated with volumetric enlargement of the brain (Klatzo, 1967). Excess fluid can accumulate in the intracellular or extracellular spaces. Two types of brain edema have been defined based on the site of damage and where the fluid accumulates. Cytotoxic edema results in intracellular swelling without alterations in vascular permeability. Vasogenic edema is associated with damage to the BBB leading to flow of water and plasma constituents into the brain. These types of edema rarely exist in isolation; typically, one type of edema dominates the other, but both co-exist.

Cytotoxic edema, which results from pathological processes that interfere with the normal function of cell membranes, constricts the extracellular space, constraining movement of fluid between the cells. The swelling is predominantly localized to the glial processes around brain capillaries with sparing of the neurons (Kimelberg, 2004; Zador et al., 2007). The main reason for this is the presence of a high density of AQP4 in the astrocytic foot processes that make astrocytes swell rapidly in response to an osmotic gradient.

The forces driving water flow to form cytotoxic edema are osmotic, generated in brain injury conditions (ischemia, trauma, hypoxia) by disturbances in ionic homeostasis due to failure of the Na^+/K^+ ATPase and/or dramatic influx of Na^+ and Ca^{2+} via ionotropic glutamate receptors (excitotoxicity) and other ionic channels. These pathological alterations in cellular ionic homeostasis result in Na^+ and water flow from the intravascular and extracellular space into the intracellular compartment.

5 Vasogenic Brain Edema

The key feature of vasogenic edema is the breakdown of the BBB and subsequent leakage of the intravascular fluid into the extracellular space of the brain parenchyma resulting in expansion of the extracellular space. Vasogenic edema moves more readily in between the linearly arranged fibers that form the white matter. The gray matter restricts water movement because of the dense nature of the neuropil, whereas the more loosely connected long fiber tracts can be separated to allow edema fluid to flow. Because of the lack of cell damage in vasogenic edema, once the damage to the blood vessel resolves, there may be a return to normal in the edematous tissue. This is generally not the case in cytotoxic edema, which is due to direct injury to the cells. White matter fiber tracts provide conduits for the bulk flow of vasogenic edema (Cserr and Ostrach, 1974; Rosenberg et al., 1980). Characteristic patterns of increased water in the projections of the white matter beneath the cortical ribbon can be readily observed in certain MRI pulse sequences (Fig. 3).

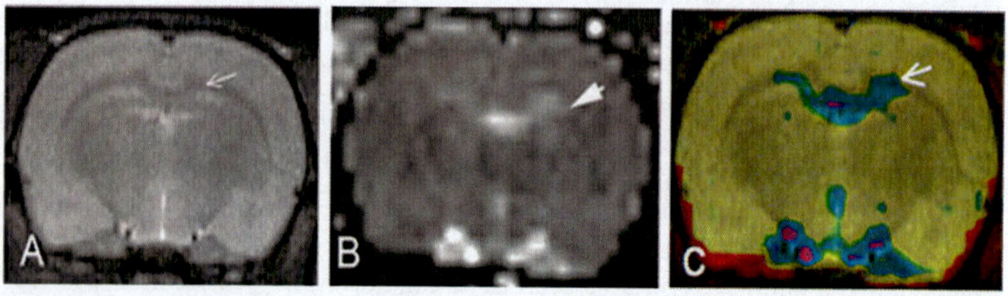

Fig. 3 (**a**) T2-weighted MR image with anatomical location of the WM damage in a rat model of chronic hypoperfusion induced by permanent ligation of the bilateral common carotid arteries. WM damage is seen as a subtle region of diffuse signal loss in the medial corpus callosum (*arrow*). The region of damage is variable between rats and varies from medial WM loss in some rats to lateral in others or a more generalized bilateral signal loss. (**b**) ADC map reconstructed from raw DWI MRI data slice matched to the T2w images. The ADC map shows bilateral regions of increase ADC values (hyperintense region on the map) with the region of higher signal more pronounced on the *right side* (*arrow*). (**c**) Image generated by overlaying converted multicolor ADC map over structural T2w image, showing the increased ADC in the white matter (corpus callosum—*white arrow*). Taken from Sood et al. (2009)

6 Role of Aquaporins in Brain Edema

Aquaporins (AQPs) are a family of at least 13 members of small membrane-spanning proteins that assemble in cell membranes as homotetramers (Verkman and Mitra, 2000; Agre et al., 2002; Verkman, 2005). Each monomer is approximately 30 kDa and six α-helical domains with cytosolically oriented amino- and carboxy-termini surround the water pore (Verkman and Mitra, 2000). AQPs can transport water in both directions (Tait et al., 2008). Early experiments demonstrating that erythrocyte membranes are more permeable to water than expected from water diffusion through a lipid bilayer provided the first experimental evidence of the existence of AQPs (Sidel and Solomon, 1957).

The principal AQP in mammalian brain is AQP4. Brain AQP4 is heavily expressed at the borders between brain parenchyma and major fluid compartments including astrocytic foot processes, glia limitans, ependymal cells, and subependymal astrocytes (Nielsen et al., 1997; Rash et al., 1998; Badaut et al., 2002). This distribution pattern indicates that AQP4 controls water flow into and out of the brain (Tait et al., 2008). AQP1 is expressed in the apical membrane of the choroid plexus and plays an important role in CSF formation (Boassa et al., 2006; Zador et al., 2007; Tait et al., 2008). There is controversy about whether AQP9 is expressed in the brain (Zador et al., 2007; Tait et al., 2008). However, a recent study using mice with targeted deletion of the AQP9 gene provides conclusive evidence for expression of AQP9 in neurons (Mylonakou et al., 2009).

Water moving from the blood into the brain through an intact BBB has to cross three membranes: luminal and abluminal endothelial cell membranes, and the membrane of the astrocyte foot processes (Kimelberg, 2004; Tait et al., 2008). High density of AQP4 is present in the vascular-facing astrocytic membranes. Although at a lower density, AQP4 is also expressed in endothelial cell membranes. Because of the close apposition of the astrocytic foot processes and their high density of AQP4, water that crosses the BBB will rapidly and preferentially end up in the perivascular astrocyte (Kimelberg, 2004).

AQP4 is likely to be one of the most abundant molecules at the brain–blood interface and has been shown to play an important role in edema associated with many brain pathologies (Badaut et al., 2002; Zeynalov et al., 2008). In a clinically relevant model of ischemic stroke, AQP4 knock-out mice had decreased cerebral edema and improved outcome. In AQP4-deficient mice, brain tissue water content and swelling of pericapillary astrocytic foot processes were significantly reduced in comparison with wild-type controls (Manley et al., 2000). Similarly, in a model of water intoxication, AQP4-null mice display a decreased brain water content and a significant improvement in survival (Manley et al., 2000; Zador et al., 2007). Significantly reduced brain edema after cerebral ischemia and water intoxication has been reported in α-syntrophin-deficient mice, which have reduced AQP4 expression in astrocyte foot processes (Amiry-Moghaddam et al., 2003, 2004). Transgenic mice overexpressing endothelin-1 in astrocytes showed more BBB disruption with increased water accumulation and brain edema possibly because of elevated AQP4 expression in astrocytic end-feet following temporary focal cerebral ischemia (Lo et al., 2005).

Deletion of AQP4 reduces edema in models in which cytotoxic edema is the predominant pathophysiological mechanism. However, in conditions in which vasogenic edema is significant, AQP4 deletion exacerbates brain edema (Zador et al., 2007). AQP4 function has been demonstrated to be of great importance in the clearance of extracellular fluid and resolution of vasogenic edema (Papadopoulos et al., 2004; Zador et al., 2007). AQP4 deletion results in increased brain swelling in vasogenic edema because of impaired removal of excess brain water through glial limitans and ependymal barriers. AQP4-deficient mice have higher intracranial pressure (ICP) and brain water content after continuous intraparenchymal fluid infusion. In a freeze-injury model of vasogenic brain edema, AQP4-deficient mice had remarkably worse clinical outcome, higher ICP, and greater brain water content. Similarly, in a brain tumor edema model involving stereotactic implantation of melanoma cells, tumor growth was comparable in wild-type and AQP4-deficient mice. However, AQP4-deficient mice had higher ICP and corresponding accelerated neurological deterioration (Papadopoulos et al., 2004; Zador et al., 2007). Results from these studies indicate that AQP4-mediated transcellular water movement is crucial for fluid clearance in vasogenic brain edema.

Together, these studies emphasize the importance of AQPs in water flux and brain edema formation and suggest that AQPs are potential targets for drug development. In addition to controlling brain water balance, AQPs participate in cell migration and neuronal excitability (Papadopoulos and Verkman, 2008; Tait et al., 2008). The complex involvement of AQPs in multiple aspects of brain function, and the opposite role of AQPs in cytotoxic and vasogenic edema, will require greater understanding before AQPs can be considered targets of therapy.

7 Injury Cascade in Brain Edema: Molecular Mechanisms

Following brain injury, a cascade of highly interconnected pathological events triggers edema (Fig. 4). Alteration of ionic homeostasis due to metabolic failure is the main process resulting in cytotoxic edema. In addition to the injury occurring at the cell membrane, there are other molecular mechanisms involving the extracellular matrix and endothelial cells that lead to the breakdown of the BBB associated with vasogenic edema. Although the number of molecules involved in cell death is large, the timing of expression results in a cascade effect that evolves over days and weeks. The initial events result in loss of energy stores, fall in ATP, and rise in extracellular glutamate. Excitation of membrane glutamate channels allows Ca^{2+} to enter the cell, triggering mitochondrial damage and induction of cytokines, proteases, and free radicals. The final common pathways of cell death involve acid hydrolases and neutral proteases, such as plasminogen activator/plasmin and matrix metalloproteinases (MMPs). Free radicals of nitrogen and oxygen add to the damage. Other molecules that are involved in these cascades include vasopressin V1a receptors, bradykinin, and prostaglandins.

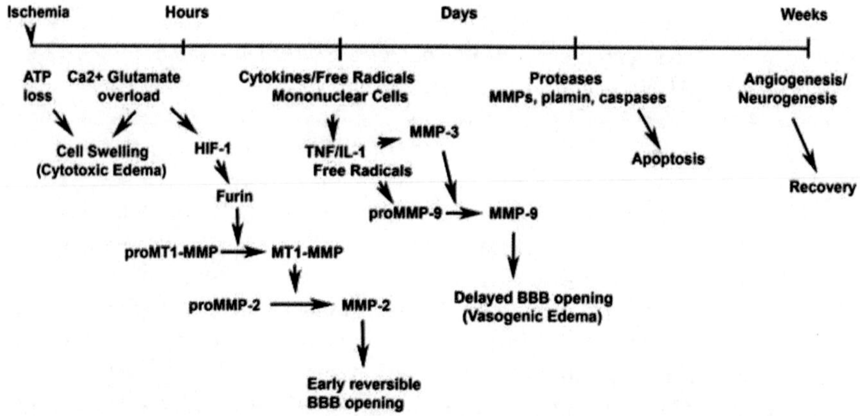

Fig. 4 Molecular cascade involved in cerebral edema. The time course of events is depicted at the top of the drawing beginning with the initiating ischemic event and progressing over several weeks. In the first hours, there is energy failure with Ca^{2+} and glutamate entering the cells. The cell swelling produces cytotoxic edema. HIF-1α is formed. Furin, an intracellular convertase that activates membrane type metalloproteinase (MT1-MMP), is formed. The activated MT1-MMP activates the constitutively expressed MMP-2. Reversible opening of the BBB occurs. After 24–48 h there is formation of a second group of molecules that turn on cassettes of genes. The cytokines include TNF-α and IL-1β, which activate transcription factors to induce MMP-3 and MMP-9. The second wave of MMPs leads to irreversible damage to the BBB with delayed vasogenic edema. Induction of caspases occurs in the nucleus and apoptosis takes place. Finally, angiogenesis and neurogenesis participate in the repair process

7.1 Cation Channels Involved in Cytotoxic Edema

The Na^+/K^+ ATPase is the main active transport mechanism responsible for maintaining ionic homeostasis, and this process involves continuous expenditure of ATP. Normal cell volume depends on the constant extrusion of intracellular Na^+ by the Na^+/K^+ ATPase. Ischemia/hypoxia results in abrogation of mitochondrial oxidative phosphorylation, and a rapid loss of ATP compromising the cellular ionic homeostasis. Sodium ion influx drives Cl^- influx via chloride channels, resulting in an increased osmolarity inside the cell that drives inflow of water mainly via AQP channels (Badaut et al., 2002; Amiry-Moghaddam and Ottersen, 2003; Liang et al., 2007). Membrane blebbing is a characteristic morphological alteration of cytotoxic edema. Oncosis (from the Greek *oncos*, meaning swelling) describes the cell death induced by cytotoxic edema. No significant alterations in the BBB are seen in the initial stages of cytotoxic edema, and fluid movement from the extracellular space into the cell does not lead to any change in total brain volume (Liang et al., 2007).

Shortly after middle cerebral artery occlusion (MCAO), cytotoxic edema occurs. Swelling of astrocytes is more prominent than neuronal swelling. Astrocytes are highly vulnerable to cytotoxic edema because they are involved in clearance of K^+ and glutamate, which cause high osmolarity and promote water inflow. Moreover, the expression of high levels of AQP4 in astrocytes makes them selectively vulnerable to pathological swelling following ischemia/hypoxia (Liang et al., 2007; Zador et al., 2007).

Cerebral tissue acidosis following ischemia or traumatic brain injury contributes to cytotoxic brain edema formation. In vitro lactacidosis induces swelling of glial cells by intracellular Na^+- and Cl^- accumulation by the Na^+/H^+-antiporter, Cl^-/HCO_3^- antiporters, and the Na^+–K^+–$2Cl^-$ cotransport (Staub et al., 1990; Ringel et al., 2006a).

Many studies have shown that pharmacological blockade of ion channels, including nonselective cation channels, reduces cytotoxic edema and ischemic brain injury in animal models of focal ischemia (Hoehn-Berlage et al., 1997; Miller, 2004; Liang et al., 2007). The following cation channels have been shown to participate in the development of cytotoxic edema following brain injury: NMDA receptor channel, acid-sensing ion channels (ASIC), sulfonylurea receptor 1 (SUR1)-regulated $NC_{Ca\text{-}ATP}$ channels, TRP channels, and the electroneutral cotransporter NKCC channel.

The SUR1-regulated $NC_{Ca\text{-}ATP}$ channel has recently received much attention due to growing evidence from preclinical and clinical studies demonstrating the therapeutic potential of blocking SUR1 by sulfonylureas such as glibenclamide (glyburide) in conditions associated with cytotoxic edema, such as ischemic stroke and spinal cord injury (Kunte et al., 2007; Simard et al., 2007, 2008; Simard et al., 2009b, a). The SUR1-regulated $NC_{Ca\text{-}ATP}$ channel is not constitutively expressed in the CNS, but is strongly upregulated under conditions of hypoxia or injury in all members of the neurovascular unit. The SUR1-regulated $NC_{Ca\text{-}ATP}$ channel conducts all inorganic monovalent cations and opening of this channel induces a strong inward current that depolarizes the cell completely and leads to oncotic cell swelling (Simard et al., 2008). In a rodent model of massive ischemic stroke with malignant cerebral edema, pharmacological blockade of SUR1-regulated $NC_{Ca\text{-}ATP}$ channel with glibenclamide reduced mortality and cerebral edema by half (Simard et al., 2006).

7.2 Role of MMPs in the Formation of Vasogenic Edema

Several recent reviews have been published on MMPs in brain injury (Ning et al., 2008; Candelario-Jalil et al., 2009; Rosenberg, 2009). This section focuses only on their role in BBB disruption and formation of vasogenic edema.

Proteases contribute to the inflammatory response to injury, forming a final common pathway that leads to BBB breakdown, hemorrhage, and cell death. After traumatic and ischemic injuries, there is a buildup of lactate, which is increased with hyperglycemia. Acidosis leads to release of acid hydrolases, which are destructive enzymes that attack cellular components, including membranes, resulting in cell necrosis. In situations where the pH remains neutral, increases in intracellular calcium and cytokines cause induction of neutral proteases. The main neutral proteases are the extracellular matrix-degrading MMPs, plasminogen activator/plasmin, and caspases.

Matrix metalloproteinases are a gene family of 26 zinc-dependent proteases that act on the extracellular matrix during injury and repair (Yong, 2005). Normally they contribute to the remodeling of extracellular matrix, angiogenesis, and neurogenesis (Wang et al., 2006). The MMPs are produced in a latent form and remain inactive until they are activated by other proteases or free radicals (Cunningham et al., 2005; Liu and Rosenberg, 2005). During an inflammatory response as part of an injury, inducible MMPs with AP-1 and NF-κB sites in their gene promoter regions, are induced by cytokines, such as tumor necrosis factor-α (TNF-α) and interleukin-1β (IL-1β) (Rosenberg, 2002). MMPs break down the basal lamina and tight junction proteins, opening the BBB and leading to hemorrhage (Yang et al., 2007). The plasminogen activator/plasmin system contributes to the vascular injury directly and indirectly by activating the MMPs (Cuzner and Opdenakker, 1999).

As part of the neuroinflammatory response in brain hypoxia–ischemia, multiple sclerosis, and CNS infections, MMPs mediate the increased permeability of the BBB, which leads to vasogenic edema. MMPs attack proteins of the extracellular matrix including collagen type IV, laminin, fibronectin, and proteoglycans (Asahi et al., 2001; Rosenberg, 2002; Gu et al., 2005). Degradation of basal lamina components by MMPs compromises the structural integrity of capillaries. Proteolytic weakening of the vessel walls may also increase risks of rupture and hemorrhage (Mun-Bryce and Rosenberg, 1998). In addition, tight junction proteins in endothelial cells are susceptible to MMP proteolysis. Occludin, claudin-5, and ZO-1 are vulnerable to attack by MMPs in ischemic brain injury (Asahi et al., 2001; Rosenberg and Yang, 2007; Yang et al., 2007; McColl et al., 2008; Liu et al.,

2009), neuroinflammation (Gurney et al., 2006; Reijerkerk et al., 2006), and gliomas (Ishihara et al., 2008).

The plasminogen/plasmin system acts in synergy with the MMPs. Plasmin is an activator of several MMPs. This is important in the treatment of acute stroke with tissue plasminogen activator (tPA). The risk of hemorrhagic conversion after tPA treatment is increased when the time between stroke onset and the injection of the drug is over 1 h; as the time increases, the risk of BBB disruption with hemorrhage leading to death, also increases. The tPA lyses the fibrin clot in the blood vessel and restores flow to the ischemic brain tissue. When the BBB is opened, as occurs in the early stages of a stroke due to the production of MMP-2, the tPA escapes into the brain tissue where it activates MMP-9. Inhibitors to MMPs block the opening of the BBB and reduce the risk of hemorrhage after tPA treatment (Lapchak et al., 2000; Pfefferkorn and Rosenberg, 2003; Wang et al., 2004; Murata et al., 2008).

Cerebral ischemia activates latent MMPs and induces de novo synthesis and release of MMPs (Rosenberg et al., 1996a; Romanic et al., 1998; Rosenberg et al., 1998; Heo et al., 1999; Asahi et al., 2000). MMP inhibitors significantly reduce vasogenic brain edema following ischemia (Rosenberg et al., 1998; Lapchak et al., 2000; Gasche et al., 2001; Copin et al., 2008).

Pharmacological blockade of MMPs using broad-spectrum inhibitors significantly reduces brain edema following intracerebral hemorrhage (Rosenberg and Navratil, 1997), cortical impact injury (Shigemori et al., 2006), and bacterial meningitis (Paul et al., 1998; Leib et al., 2000, 2001). This body of experimental evidence emphasizes the key role of MMPs in BBB damage and edema formation in a wide range of neuropathological conditions.

7.3 Oxidative Stress and Brain Edema

Free radical formation is an important contributor to cell death and brain injury in many neurological diseases. Shortly after brain damage by hypoxia–ischemia, hemorrhage, or trauma, excessive reactive oxygen species (ROS) production occurs, and at the same time, there is an impairment of antioxidant protective mechanisms, which leads to oxidative stress (Heo et al., 2005).

During cerebral ischemia, ROS contribute to cytotoxic edema by perturbing the functioning of plasma membrane ion transport systems such as Na^+–K^+-ATPase, Ca^{2+}-ATPase and Na^+–Ca^{2+} exchanger. The proposed mechanisms underlying ion transport modulation by ROS include the peroxidation of membrane phospholipids, the oxidation of sulfhydryl groups located on the ion transport proteins, and oxidative protein modification (Kourie, 1998). Oxidative stress triggers the release of mediators known to be responsible for cytotoxic cell swelling, such as K^+ ions, glutamate, and lactic and arachidonic acid (Ringel et al., 2006b).

Oxidative stress damages endothelial cells of the BBB and contributes to vasogenic edema (Chan et al., 1984; Chan, 2001). Incubation of endothelial cells with ROS-generating systems increases the permeability of endothelial monolayers (Imaizumi et al., 1996; Lagrange et al., 1999; Fischer et al., 2005). The superoxide radical ($O_2^{-\bullet}$) has been identified as the primary ROS involved in increased vascular permeability and edema formation in global and focal cerebral ischemia, cold brain injury, and brain tumors (Heo et al., 2005). Scavenging $O_2^{-\bullet}$ radicals using recombinant superoxide dismutase (SOD) or polyethylene glycol-SOD reduces ischemia-induced BBB injury and vasogenic edema (Armstead et al., 1992; Schleien et al., 1994). Treatment of ischemic rats with encapsulated SOD in biodegradable poly (D,L-lactide co-glycolide) nanoparticles (SOD-NPs) maintained BBB integrity, thereby preventing edema, reduced oxidative injury following reperfusion, and protected neurons from undergoing apoptosis (Reddy and Labhasetwar, 2009). Further evidence emphasizing the important role of ROS formation in brain edema development comes from transgenic animals overexpressing antioxidant enzymes. Brain water content and infarct size are significantly reduced after transient focal cerebral ischemia in transgenic mice overexpressing the human Cu,Zn-SOD (SOD1) compared with nontransgenic controls (Kinouchi et al., 1991; Yang et al., 1994; Kokubo et al., 2002). These SOD1-overexpressing mice also have reduced vasogenic edema and infarction after cold-trauma brain injury (Chan et al., 1991). Conversely, mice bearing a disruption of the SOD1 gene had increased infarct volume and brain swelling after temporary focal cerebral ischemia (Kondo et al., 1997), but not following permanent focal ischemia where there is no reperfusion injury (Fujimura et al., 2001).

Hyperglycemia increases oxidative stress and MMP-9 expression/activity, exacerbating BBB breakdown and dramatically increasing edema formation after ischemia-reperfusion injury in the rat (Kamada et al., 2007). Heterozygous *SOD1* transgenic rats, carrying human *SOD1* genes with a four- to sixfold increase in Cu/Zn SOD activity, showed a significant reduction in hyperglycemia-induced Evans blue leakage, vasogenic edema, and MMP-9 activation after experimental ischemia compared with control nontransgenic rats (Kamada et al., 2007).

Transgenic mice overexpressing the intracellular form of glutathione peroxidase (GPx1) displayed reduced infarct size and edema formation compared with nontransgenic mice at 24 h of reperfusion following 1 h of middle cerebral artery occlusion (Weisbrot-Lefkowitz et al., 1998; Ishibashi et al., 2002). Absence of GPx1 exacerbates cerebral ischemia-reperfusion injury as shown by larger infarct volumes, increased activation of MMP-9, and a dramatic disruption of the BBB in GPx1-null mice compared with wild-type controls (Wong et al., 2008).

The gp^{91phox} (Nox2) containing NADPH oxidase is an important source of ROS during cerebral ischemia (Kunz et al., 2007). It has been demonstrated that genetic deletion of gp^{91phox} confers protection against ischemic stroke in mice (Walder et al., 1997). In gp^{91phox} deficient mice, ischemic stroke-induced BBB breakdown, brain edema, and lesion volume were largely attenuated compared with those in wild-type mice (Kahles et al., 2007). In another study, intracerebral injection of collagenase produced less bleeding in gp^{91phox} null mice than wild-type animals. Brain edema formation, neurological deficit and a high mortality rate were observed in wild-type, but not in gp^{91phox} knock-out mice (Tang et al., 2005). These studies suggest that formation of ROS by NADPH oxidase plays a central role in BBB injury and edema in stroke and intracerebral hemorrhage.

Nitric oxide (NO) is a free radical that has both beneficial and deleterious actions during ischemia/reperfusion depending on the cell type in which it is generated (Gursoy-Ozdemir et al., 2004). Excessive NO generation by neuronal nitric oxide synthase (nNOS) is cytotoxic (Huang et al., 1994). On the contrary, endothelial nitric oxide synthase (eNOS) knock-outs develop larger infarcts because NO of endothelial origin promotes survival by improving blood flow during ischemia (Huang et al., 1996; Gursoy-Ozdemir et al., 2004). However, excessive production of NO by eNOS during reperfusion may contribute to ischemic brain injury via peroxynitrite formation, which is the product of the reaction between NO and superoxide radical (Gursoy-Ozdemir et al., 2000, 2004; Han et al., 2006). In a model of transient focal ischemia in the mouse, superoxide and peroxynitrite formation was particularly intense in microvessels and astrocytic end-foot processes surrounding them. There was colocalization of sites with peroxynitrite formation and vascular injury, as shown by increased Evans blue leakage and MMP-9 labeling, suggesting an association between peroxynitrite and microvascular injury (Gursoy-Ozdemir et al., 2004). Nonselective NOS inhibition has been shown to significantly reduce brain edema, BBB disruption, and infarct size in experimental stroke (Nagafuji et al., 1992; Kozniewska et al., 1995).

Many studies have shown that synthetic antioxidant compounds significantly reduce brain edema formation in experimental models of hemorrhage (Nakamura et al., 2004, 2008) and ischemia (Ding-Zhou et al., 2003b; Ginsberg et al., 2003; Suda et al., 2007).

7.4 Involvement of Vasopressin in Cerebral Edema

Arginine vasopressin (AVP) is a neuropeptide that is synthesized in the hypothalamus and transported to the neurohypophysis, from where it is released into the blood. AVP is commonly known as the antidiuretic hormone because it increases water reabsorption by the kidney.

Centrally released AVP contributes to brain capillary water permeability, brain ionic homeostasis, and the regulation of CSF production (Rosenberg et al., 1990; Niermann et al., 2001; Bhardwaj, 2006). AVP mediates its action through three G-protein coupled receptors: V1a, V1b, and V2. Unlike V2 receptors, V1a and V1b are widely expressed in the brain. There is a causative role for centrally formed AVP in brain edema formation following cerebral damage including trauma and ischemia (Doczi et al., 1984; Dickinson and Betz, 1992; Shuaib et al., 2002). AVP, through a V1 receptor- and [Ca^{2+}]-dependent mechanism, stimulates the BBB Na-K-Cl cotransporter to participate in ischemia-induced edema formation (O'Donnell et al., 2005). Antagonists of vasopressin V1 receptors confer significant protection against brain edema and neuronal cell death induced by ischemia (Ikeda et al., 1997a; Laszlo et al., 1999; Shuaib et al., 2002; Vakili et al., 2005; Molnar et al., 2008a, b), hemorrhage (Rosenberg et al., 1992), traumatic brain injury (Szmydynger-Chodobska et al., 2004; Pascale et al., 2006; Trabold et al., 2008), subarachnoid hemorrhage

(Doczi et al., 1984; Laszlo et al., 1999), and cold-induced vasogenic edema (Ikeda et al., 1997b; Bemana and Nagao, 1999). There is a relationship between AVP and AQP in the kidney, but the interaction between AVP and water channels in the brain remains to be elucidated.

7.5 Vascular Endothelial Growth Factor and Angiopoietins

Vascular endothelial growth factor (VEGF) and angiopoietins are families of vascular-specific growth factors that regulate blood vessel growth, maturation, and function (Thurston, 2002). VEGF, the predominant angiogenic growth factor, triggers endothelial cell proliferation, migration, and increased vascular permeability due to the formation of nascent vessels, which essentially consist of immature endothelium with few pericytes and little mature matrix (Carmeliet, 2003; Ferrara et al., 2003; Ballabh et al., 2007). By acting as a capillary permeability-enhancing agent, VEGF also affects the integrity of the BBB.

The angiopoietins, Ang-1 and Ang-2, differently modulate the actions of VEGF in angiogenesis (Zhu et al., 2005). In particular, Ang-1 and its endothelial receptor, Tie2, mediate the maturation and stabilization of VEGF-induced vasculature by promoting the recruitment of smooth muscle cells (pericytes) to the abluminal surface of the newly generated vascular bed, promoting the structural integrity of blood vessels (Yancopoulos et al., 2000; Ballabh et al., 2007). In contrast, Ang-2, a natural antagonist of Ang-1, is associated with both initial angiogenesis and capillary destabilization. An increase in the expression of Ang-2 in the presence of VEGF promotes vessel sprouting and increased vascular permeability (Carmeliet, 2003; Roviezzo et al., 2005).

VEGF is an angiogenic factor that induces increased permeability of the BBB leading to the formation of edema following ischemia–hypoxia (Mayhan, 1999; Schoch et al., 2002; Kaur and Ling, 2008). VEGF is associated with endothelial proliferation and neovascularization, suggesting that VEGF promotes angiogenesis and repair following stroke (Zhang et al., 2002). However, new vessels lack a fully mature BBB, and are consequently leaky (Zhang and Chopp, 2002). VEGF also directly increases the permeability of the BBB via the synthesis/release of nitric oxide and subsequent activation of soluble guanylate cyclase (Mayhan, 1999). Furthermore, VEGF may increase BBB permeability by inducing alterations in endothelial TJP. It has been shown in vitro that VEGF significantly reduces occludin and ZO-1 expression and disrupts the molecular organization of both proteins, which leads to tight junction disassembly (Wang et al., 2001; Fischer et al., 2002).

Increased VEGF production following ischemia has been shown to contribute to BBB disruption and vasogenic edema (Zhang et al., 2000). Astrocytes are the main cell type expressing VEGF following brain ischemia (Kaur et al., 2006). Antagonism of VEGF using a fusion protein, mFlt (1–3)-IgG, which sequesters VEGF, reduces ischemia/reperfusion-related brain edema and injury (van Bruggen et al., 1999). Inhibition of endogenous VEGF by topical application of an anti-VEGF antibody in the ischemic cortex decreased extravasation of ^{14}C-AIB, which suggests that endogenous VEGF is in part responsible for the BBB breakdown during the early stage of focal cerebral ischemia (Chi et al., 2007).

In cerebral ischemia, Ang-1 is able to antagonize VEGF-mediated BBB disruption, in association with inhibition of MMP-9 activity (Valable et al., 2005). In ischemic animals, administration of VEGF leads to increased BBB permeability and to an induction of MMP-9 activity. Conversely, the coadministration of Ang-1 and VEGF blocks the BBB disruption and reduces MMP-9 activity, resulting in a dramatic reduction in edema volume (Valable et al., 2005). On the contrary, the combined administration of VEGF and Ang-2 leads to an increase in MMP-9 activity and BBB disruption (Zhu et al., 2005).

7.6 Bradykinin

Bradykinin is an endogenous inflammatory substance that increases vascular permeability and produces tissue edema. The kallikrein–kinin system is very rapidly activated following brain injury resulting in the activation of kallikrein that cleaves kininogen to produce bradykinin. The effects of bradykinin are mediated by two different receptors: B1 and B2. Very low levels of B1 are found under normal conditions. In contrast, the B2 receptor is constitutively expressed in a wide variety of tissues including the brain and mediates the majority of bradykinin effects (Couture et al., 2001).

Bradykinin promotes edema as shown in numerous models of brain injury including bacterial meningitis (Lorenzl et al., 1996), traumatic brain and spinal cord injury (Plesnila et al., 2001; Hellal et al., 2003; Ivashkova et al., 2006), and cerebral ischemia (Lehmberg et al., 2003; Groger et al., 2005; Klasner et al., 2006; Lumenta et al., 2006).

There are conflicting data on the specific role of B1 versus B2 receptors in bradykinin-induced edema following focal cerebral ischemia. A large number of studies indicate that blockade of the B2 receptor using pharmacological agents dramatically reduces edema and infarct size, and improves neurological function in animal models (Zausinger et al., 2002; Ding-Zhou et al., 2003a; Klasner et al., 2006; Lumenta et al., 2006). Kinin B2-deficient mice had improved motor function, smaller infarct volumes, and developed less brain edema than wild-type controls after focal cerebral ischemia (Groger et al., 2005). Contrary to these data, it has been found that postischemic brain injury is dramatically exacerbated in B2-null mice following temporary middle cerebral artery occlusion (Xia et al., 2006). Compared with wild-type controls, mice lacking the bradykinin B2 receptor displayed a higher mortality rate and neurological deficit scores, larger infarct volumes, more apoptosis, and increased neutrophil infiltration after ischemic stroke (Xia et al., 2006), suggesting that the B2 receptor promotes survival and suppresses apoptosis and inflammation after cerebral ischemia. Adding even more to the controversy on the role of bradykinin receptors in ischemic brain injury, a recent study has found that B2 deficiency did not confer neuroprotection and had no effect on the development of brain edema in a mouse model of focal ischemia (Austinat et al., 2009). Interestingly, B1 receptor knock-out mice developed smaller brain infarction, and fewer neurological deficits compared with wild-type controls. This was accompanied by a significant reduction in edema and endothelin-1 expression, as well as less neuroinflammation (Austinat et al., 2009).

7.7 Arachidonic Acid and Brain Edema

Arachidonic acid is a polyunsaturated fatty acid that is released from membrane phospholipids by the action of phospholipase A_2 (Bosetti, 2007). Large amounts of arachidonic acid are released following brain ischemia and trauma (Phillis and O'Regan, 2004; Phillis et al., 2006). Arachidonic acid has been implicated in vasogenic cerebral edema (Chan and Fishman, 1984; Staub et al., 1994). The deleterious effects of arachidonic acid, which may contribute to cerebral edema, include enhanced production of prostanoids and free radicals via its metabolism by cyclooxygenase (COX) and lipoxygenase (LOX) enzymes. However, it was found in a previous in vitro study using C6 cells that arachidonic acid-induced glial swelling is not due to formation of prostaglandins and leukotrienes (Winkler et al., 2000). The authors speculated on the possible mechanism, but it remains to be determined how arachidonic acid directly induces glia cell swelling in this in vitro model.

It has been shown that arachidonic acid metabolism could contribute to the pathogenesis of cerebral edema. Treatment with indomethacin, a COX inhibitor, nordihydroguaiaretic acid, a LOX inhibitor, or their combination significantly reduced vasogenic edema induced by freezing lesions (Yen and Lee, 1987). 5-LOX inhibitors significantly decreased vascular permeability both within the tumors and in brain adjacent to tumor, suggesting that capillary permeability is influenced by endogenous leukotrienes, which play an important role in brain tumor edema (Baba et al., 1992). Similarly, in transient focal cerebral ischemia, leakage of immunoglobulin G into the brain parenchyma was significantly reduced in 12/15-LOX knock-out mice as well as wild-type mice treated with baicalein, a LOX inhibitor. Likewise, brain edema was significantly ameliorated in 12/15-LOX null mice and baicalein-treated wild-type animals (Jin et al., 2008).

Experimental evidence indicates that COX modulates BBB permeability in neuroinflammatory conditions, ischemia, and hemorrhage. The COX inhibitor, KBT-3022, prevented brain edema induced by bilateral carotid occlusion and recirculation in gerbils (Yamamoto et al., 1996). In the collagenase model of intracerebral hemorrhage, the brain water content of rats treated with the COX-2 inhibitor, celecoxib, decreased both in lesioned and nonlesioned hemispheres in a dose-dependent manner, which was accompanied with reduced perihematomal cell death (Chu et al., 2004). Delayed damage to the BBB and vasogenic edema, which follow ischemic stroke, were significantly diminished by administration of the COX-2 inhibitor nimesulide (Candelario-Jalil et al., 2007a). Inhibition of COX activity with indomethacin prevented BBB damage following intracerebral administration of TNF-α in the rat. Indomethacin significantly reduced TNF-α-induced MMP-9 and MMP-3 expression and activity and attenuated free radical formation (Candelario-Jalil et al., 2007b).

8 Imaging Brain Edema

Technological advances in brain imaging have revolutionized diagnosis of cerebral edema. This occurred through the development of tomographic imaging, beginning with computed tomography (CT) and complemented with magnetic resonance imaging (MRI). These methods essentially image water in brain tissue either through the loss of tissue density that blocked X-rays in CT or the gain of water that is the main source of signal in MRI. The theories related to shifts in water that result in the imbalance associated with edema have been described above. The manner in which images are generated in either method depends strongly on the movement of water between the intra- and extracellular compartments and the movement of water into the brain across an injured blood vessel. Current concepts relate the changes seen in images are due to changes in the intracellular–extracellular water ratio secondary to disruption of intracellular energy metabolism and loss of ionic gradients. Cellular swelling reduces extracellular space and increases tortuosity of extracellular space pathways; there is a restriction of water movement between cells in cytotoxic edema that rapidly affects the diffusion of water and results in a decrease in the apparent diffusion coefficient (ADC) of water. Conversely, when the extracellular space expands as water crosses into the brain and enters the extracellular space, there is an increase in water diffusion that appears as an increase in the ADC. When there is an ischemic insult, cells swell and water enters the intracellular compartment without an increase in total water content in the affected zone. Fig. 5 schematically shows the hypothetical changes in water diffusion within different compartments as results of edema.

8.1 Imaging by CT

Brain tissue water content is inversely correlated with X-ray attenuation and can thus be measured with CT (Rieth et al., 1980; Unger et al., 1988; Dzialowski et al., 2004). Hypoattenuated areas on CT represent an increase in the net water content of the involved brain parenchyma. Lowering of CT attenuation allows for quantification and localization of edema, which is the result of net change in water content of the area of interest. This increase in water content could be readily related to the vasogenic edema. Based on this physics, different CT techniques have been developed to monitor edema: (1) noncontrast-enhanced CT (NECT), (2) perfusion CT (PCT), and (3) CT angiography (CTA).

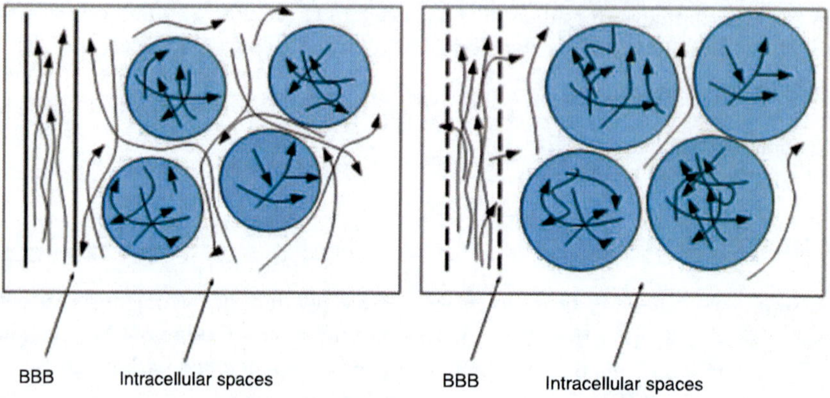

Fig. 5 Diffusion of molecules can be restricted in closed spaces, such as cells. Diffusion might also be hindered by obstacles that result in tortuous pathways. Exchange between compartments also slows down molecular displacements. Left panel shows a model of the movement of fluids (diffusion and bulk flow) with three compartments of healthy brain tissues. Bulk flow is seen in vascular compartment while the diffusion happens in the interstitial spaces and cell compartments. *Right panel* shows the changes in diffusion of water as a result of edema. Molecular displacement between compartments increases as a result of BBB breakdown in vasogenic edema. Tortuosity will decrease as intracellular space is reduced

8.1.1 Noncontrast-Enhanced CT (NECT)

It has been shown that an increase by 1% of tissue water results in a decrease of X-ray attenuation of 2.6 Hausfield units (HU) (Rieth et al., 1980; von Kummer et al., 2001). For example, during MCA occlusion, attenuation decreased to 69.3 HU after 1 h, 66.6 HU after 2 h, 65.4 HU after 3 h, and 64.1 HU after 4 h. After reperfusion, attenuation remained stable in the 1-h occlusion group but further and steadily declined in the 2-, 3-, and 4-h occlusion groups. Attenuation during reperfusion in the 1-h occlusion group differed significantly from that in the 2-, 3-, and 4-h occlusion groups (Dzialowski et al., 2004, 2006, 2007). In comparison with ADC measured by MRI, it was shown that CT measurements continue to decrease linearly at a rate of 0.4 HU/h, whereas the decrease in ADC was almost completed after 1.5 h (Kucinski et al., 2002; Doczi and Schwarcz, 2003). Therefore, there should be different causality for these observations. With NECT ischemic changes cannot be observed before any morphological changes can be observed (von Kummer et al., 2001); X-ray hypoattenuation at CT is highly specific for irreversible ischemic brain damage (von Kummer et al., 2001). Therefore, NECT is not able to identify edema before the appearance of vasogenic edema.

8.1.2 Perfusion CT

Two different methods have been used to achieve CT perfusion: xenon CT and intravenous contrast-enhanced CT perfusion. Xenon CT provides an accurate quantitative measure of cerebral blood flow, but its use in an emergency setting is limited. Contrast-enhanced perfusion CT studies are done by monitoring the first pass of an iodinated contrast agent bolus through the cerebral vasculature (Hoeffner et al., 2004; Wintermark et al., 2008). Any increase in Hausfield units is directly proportional to the iodine concentration. Dynamic sequential acquisition of data follows, and then data are analyzed to generate parameters of interest (e.g., cerebral blood volume, CBV; cerebral blood flow, CBF; and time to peak, TTP). PCT has shown promise as a method for rapid assessment of cerebral hemodynamics. The lesion size calculated by PCT was not different from the one calculated by perfusion-weighted imaging (PWI) (Schramm et al., 2004) and DWI (Roberts et al., 2001; Eastwood et al., 2003). Fig. 6 shows the superiority of PCT over NECT in detecting the edema following ischemic stroke (Dhamija and Donnan, 2008).

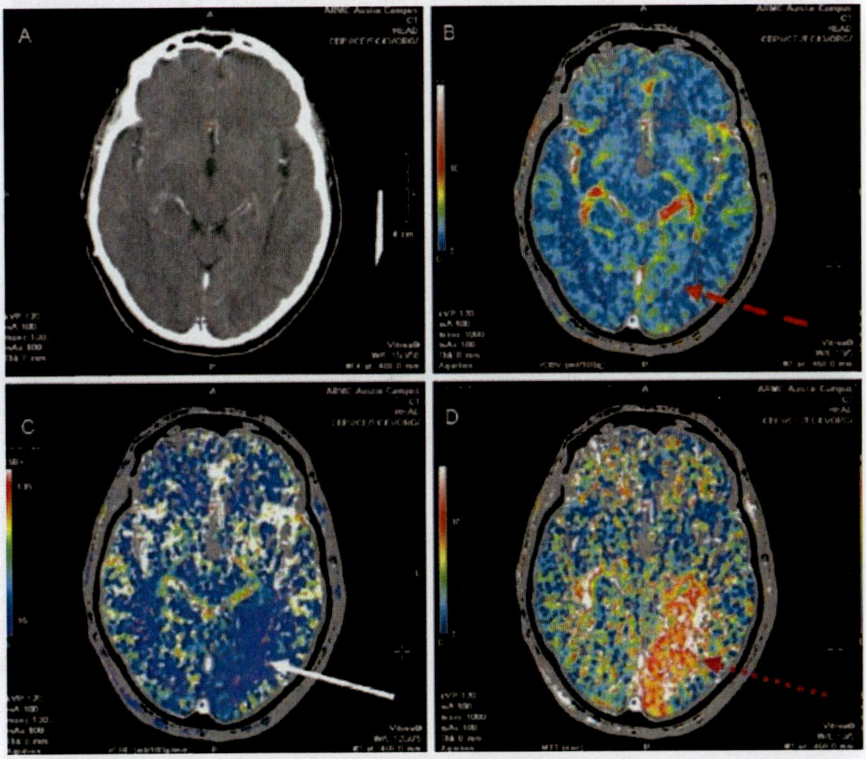

Fig. 6 CT Perfusion: noncontrast-enhanced CT (NECT) and CT perfusion maps of a patient presenting within 1 h of stroke onset. NECT was normal and CT perfusion revealed large penumbra in form of increased mean transient time (MTT), normal cerebral blood volume (CBV), and reduced cerebral blood flow (CBF). **(a)** Normal NECT. **(b)** CT perfusion showing normal CBV. **(c)** CT perfusion showing reduced CBF. **(d)** CT perfusion showing increased MTT. Taken from Dhamija and Donnan (2008). Available at the website of the journal *Annals of Indian Academy of Neurology.* http://www.annalsofian.org/temp/AnnIndianAcadNeurol11512-4945421_134414.pdf

8.2 Imaging by MRI

Common MR imaging is based on proton imaging; clinical proton imaging is largely water imaging. MR can be adapted to provide noninvasive measurements of water mobility and content in biological tissues. Several different techniques in MR imaging have been developed to measure abnormalities in water mobility and water content of tissues. Diffusion-weighted imaging (DWI) and diffusion tensor imaging (DTI) have been developed to measure the changes in water mobility in cerebral tissues. DWI measures one-dimensional distribution of water diffusion. The result of DWI imaging is represented as an ADC map. On the other hand, DTI measures three-dimensional distribution of water diffusion. Two important parameters, fractional anisotropy (FA) and mean diffusivity (MD) are calculated from DTI images to represent the diffusion abnormalities.

T2-weighted imaging is used to measure the net change in water content of underlying tissue. Similar to NECT, T2 weighted images detect vasogenic edema, but are not sufficient to detect cytotoxic edema. The newest technique, susceptibility weighted imaging (SWI), measures vasculature changes within the edema region.

8.2.1 T2-Weighted Imaging

T2 is the transverse relaxation time and shows how long transverse magnetization would last in a uniform magnetic field. T2 relaxation depends on the presence of static internal fields in the substances. These are generally due to protons on large molecules. An important event in the pathophysiological cascade that leads to infarction following ischemia is the net movement of water from the extracellular space into the intracellular compartment without an increase in total water content in the affected zone. Because of the lack of a change in water, the T2-weighted image remains normal at this stage. When the BBB breaks down, leading to vasogenic edema, there is an increase in total water content, which produces the bright signal on the T2-weighted image. The intense appearance of the vasogenic edema on T2-weighted MR images is because the motion of protons in vasogenic edema is not so slow. Therefore, T2 remains long. It was suggested that true infarct extent on T2-weighted can probably only be assessed on scans obtained beyond seven weeks after stroke (O'Brien et al., 2004).

8.2.2 Diffusion-Weighted Imaging (DWI)

Diffusion-weighted MRI measures water self-diffusion and depends on: (1) diffusion distance within the cells, (2) tortuosity of the interstitial spaces (Helmer et al., 1995), and (3) transport through the cell membrane. Diffusion-weighted imaging assesses microscopic mobility of water. The rate of water diffusion within tissues measured by conventional MR methods is found to be significantly lower than for free solutions, and measurements are often summarized in terms of an ADC, which is a measure of the effective distance over which water can migrate within the tissue within a specified time. The ADC differs from the intrinsic diffusion coefficient in a manner that is dependent on the microstructure and composition of the tissue.

By obtaining images with gradients of differing strengths (i.e., differing b values), an ADC can be calculated, providing a quantitative measurement of water translational motion independent of magnetic field strength and gradient strength. To determine the ADC, at least two b values are needed. A number of pathological conditions, such as ischemic stroke and prolonged seizures, produce significant changes in the ADC compared with healthy tissues. Moreover, having ADC values allows the in vivo monitoring of changes in the ratio of extracellular to intracellular volume and the development of cellular swelling or shrinkage by measuring the ADC of the tissue water. In DWI images, regions with a high diffusion constant, for example, ventricles, tend to be darker and those with low diffusion constant brighter. The contrast of the ADC map is the inverse of DWI. DWI is more sensitive than CT in the identification of acute ischemia and can visualize major ischemia more easily than CT.

DWI has been adopted to evaluate the development of edema in clinical and research applications in a variety of neurological disorders including stroke (Provenzale and Sorensen, 1999; Neumann-Haefelin et al., 2000a, c; Neumann-Haefelin et al., 2000b; Chan et al., 2002; Chen et al., 2006b; Taguchi et al., 2007), head trauma (Marmarou et al., 2000b, a; Barzo et al., 2002), and metabolic disturbances such as systemic hyponatremia. Fig. 7 represents DWI images of a stroke patient.

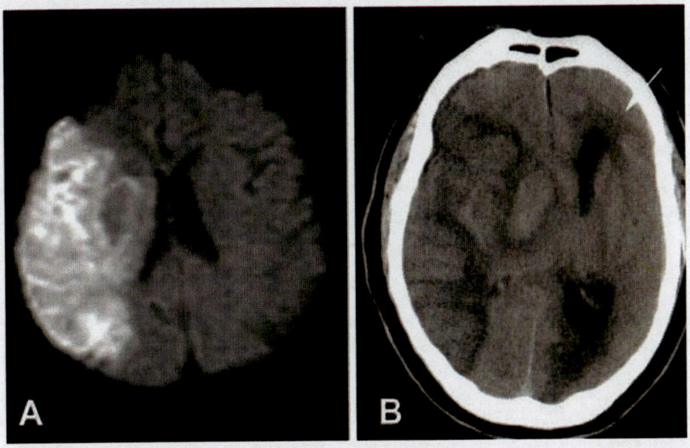

Fig. 7 **(a)** Diffusion-weighted MRI of a patient with large middle cerebral artery stroke. There is involvement of the entire vessel territory with possibly some hemorrhage in the basal ganglia. The image was made within hours of the infarct and there is minimal compression of the ventricles. **(b)** CT days after the infarct shows the massive shift of the midline structures away from the evolving mass lesion. Compression of the CSF outflow tracts causes the hydrocephalus with interstitial edema in the white matter adjacent to the ventricles

8.2.3 Diffusion Tensor Imaging (DTI)

When the diffusion is isotropic, the probability of finding a water molecule after a certain time is spherical, which can be described by one parameter. Tissue water diffusion is affected by the presence and orientation of barriers to translational motion (Kremer et al., 2007; Rollins, 2007). The measured ADC values can vary depending on the direction in which the diffusion-sensitive gradients are applied. ADC is direction-dependent especially in the WM area. In this case, we can assume that the diffusion process leads to an elliptical shape of the probability with the longest axis aligned along the fiber direction. In order to fully characterize the diffusion ellipsoid six parameters are needed. These parameters are organized in a tensor, called the diffusion tensor. Six diffusion constants along six independent axes are measured. Having a tensor data DTI measures the diffusion properties: (1) magnitude, (2) direction, and (3) anisotropy of water molecule in tissues. DTI was used in a mouse model of traumatic brain injury. At every time-point, DTI was more sensitive to injury than conventional magnetic resonance imaging, and relative anisotropy distinguished injured from control mice with no overlap between groups. DTI changes predicted the approximate time since trauma (Mac Donald et al., 2007). DTI has been used in carbon monoxide poisoning to follow recovery (Terajima et al., 2008).

8.2.4 Susceptibility-Weighted Imaging (SWI)

Susceptibility differences between tissues have been used as a new type of MR contrast by SWI sequences (Haacke et al., 2004; Haacke, 2006; Hu et al., 2008; Haacke et al., 2009). SWI is a fully velocity-compensated high-resolution 3-D gradient-echo sequence that uses magnitude and filtered-phase information, both separately and in combination with each other, to create new sources of contrast (Mittal et al., 2009). In SWI, there is a kind of mixture of spin density, T1, T2*, CSF suppression, and susceptibility sensitivity. SWI images reveal regions of edema identical to FLAIR images because of short TR and comparatively longer TE, however, SWI does not reveal a low signal in CSF because of a low flip angle. DWI highlights the edematous regions affected by stroke, whereas SWI shows changes in oxygen saturation along with other sources of susceptibility. Therefore, SWI demonstrates the affected vascular territory in stroke. The hypothesis is that the deoxyhemoglobin content of small vessels is increased over their normal values due to slower or restricted flow, making these vessels visible (Haacke et al., 2004; Haacke, 2006; Hu et al., 2008; Haacke et al., 2009).

9 Clinical Conditions Associated with Brain Edema

The consequences of brain edema depend on the amount of tissue involved, the effect by intracranial pressure, and the threat of herniation. Small lesions such as limited edema around a metastatic lesion or an early abscess may have little clinical impact. On the other hand, a large middle cerebral artery stroke with massive edema may block CSF flow, resulting in unilateral hydrocephalus and herniation. When the edema is generalized and the intracranial pressure massively increased as can occur with head trauma, there is a threat of secondary ischemia due to loss of cerebral blood flow.

Brain tumors cause brain edema through several mechanisms. Highly vascular tumors often have vessels with leaky BBB and both the mass lesion and the vasogenic edema produce the pathological changes. In the case of metastatic tumors, which can act as a foreign object, there is swelling in the tissue around the mass from disruption of the BBB and cellular function. The resulting edema around the metastatic tissue fans out into the white matter in fingerlike projections. Generally, there are multiple masses due to metastatic lesions and one lesion in primary tumors. With some tumors such as low-grade astrocytomas, which have tissue characteristics close to normal brain tissue, relatively little edema accompanies the mass.

A different pattern is seen in the cerebral edema occurring with ischemia/hypoxia. Lesions evolve over time as described above. The early energy failure causes cellular swelling with cytotoxic edema. This can occur within minutes as shown in DWI studies in animals. The cell swelling compresses the extracellular space, constricting water diffusion, which appears on a diffusion-weighted MRI as a hyperintense region with a corresponding dark area on the ADC scan (Fig. 7a). Soon after the ischemic event, there is a transient opening of the BBB in reperfused brain. A second more severe opening is seen at 24–48 h in experimental animals. These openings are associated with the expression of MMPs. Large infarcts cause life-threatening edema because of compromise of blood flow and herniation (Fig. 7b).

Purely vasogenic edema is uncommon in vascular disease. When there is a sudden rise in blood pressure and the autoregulatory range of normal blood pressure control is exceeded, an acute hypertensive crisis causes a pure form of vasogenic edema. This is best illustrated in the young pregnant woman who has a sudden rise in blood pressure during eclampsia (Fig. 8). The level of the blood pressure is less important than the change. In a young person with normal blood vessels and a low blood pressure, a marked increase, which may remain under what would be considered a normal range, could result in damage to the blood vessels. On the other hand, a person with long-standing hypertension may tolerate a further rise without developing vasogenic edema. In the hypertensive crisis, there is a predilection for the posterior circulation to be involved more dramatically than the anterior circulation. The vasogenic edema expands the extracellular space and fluid accumulated in the white matter of the posterior regions, producing a characteristic pattern (Fig. 8). The key to diagnosis lies in the MRI, where the lack of changes on DWI, with extensive white matter edema on T2 and FLAIR images, indicates that an ischemic injury has not occurred and without an ischemic/hypoxic injury, recovery generally occurs over a period of several weeks.

Another pattern of edema is seen with inflammatory and infectious disease processes. With infections, there is an upregulation of adhesion molecules on the inner surface of the blood vessel. White blood cells cross the BBB and release proteases

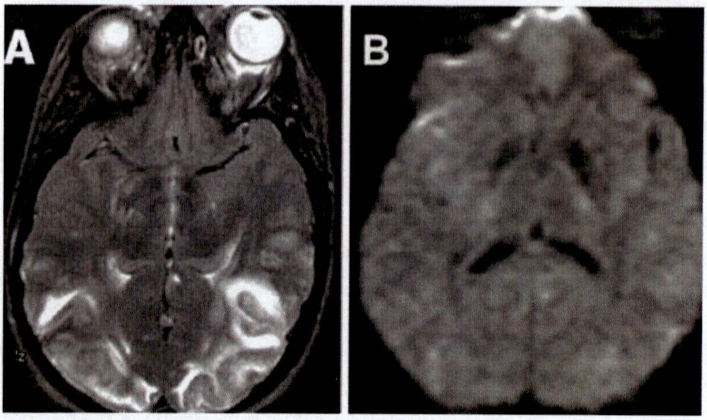

Fig. 8 Patient with hypertensive encephalopathy secondary to eclampsia with the HELLP (hemolysis, elevated liver enzymes, and low platelets) syndrome. **(a)** A T2-weighted MRI showing the extensive cerebral edema in the posterior white matter regions with less involvement of the gray matter. **(b)** Diffusion-weighted images with only one small area of involvement. The lack of DWI changes is consistent with this being a vasogenic type of edema, and the patient had a good recovery without residual

and free radicals intended to fight the infection, but the resulting inflammatory response can damage normal tissues. Bacterial and viral meningitis is by definition limited to the meninges and does not lead to brain edema. However, in some individuals there is penetration of the organisms into the brain along the Virchow–Robin spaces. When there is meningoencephalitis, there is brain edema in the adjacent brain. Occasionally the inflammation around the blood vessels penetrating the brain causes a stroke further aggravating the injury region. When the parenchyma is involved, the infection leads to a cerebritis, which eventually walls off, becoming an abscess. The tissue around the abscess becomes edematous with vasogenic edema, forming a ring around the outside of the lesion.

Another form of inflammatory response occurs in autoimmune processes, such as multiple sclerosis, which involves infiltration primarily by T cells. The site of the inflammation is the venules particularly in the white matter. A series of veins in the region of the corpus callosum are vulnerable, producing enhancing lesions that fan out from the corpus callosum. The myelinated fibers are the site of most of the injury, however, recent evidence suggests that eventually the axons are damaged in multiple sclerosis (Trapp et al., 1998). Loss of myelin leads to expression of excess numbers of sodium channels. Glutamate channels are activated with calcium overload. The denuded axons with excess sodium and glutamate channels are more vulnerable to minor forms of hypoxia, making it possible that edema as part of an hypoxia-related injury occurs in the white matter (Trapp and Stys, 2009).

Interstitial edema is seen in the periventricular regions in patients with hydrocephalus (Fig. 9). The widened extracellular space is the site of transependymal flow of CSF. The movement of ISF into the frontal white matter leads to difficulty walking and incontinence in the syndrome of normal pressure hydrocephalus.

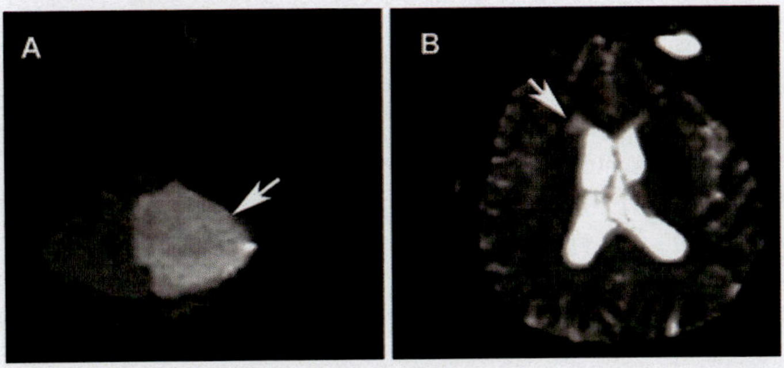

Fig. 9 Cerebellar infarct with secondary hydrocephalus and transependymal fluid movement (interstitial edema). (**a**) Initial diffusion-weighted image with cerebellar infarct in the territory of the left posterior inferior cerebellar artery. (**b**) Echo-planar T_2 axial image shows enlargement of the ventricles prior to surgery for hydrocephalus. *Arrow* shows transependymal movement of fluid

Identification of patients with adult-onset hydrocephalus that will respond to a ventriculoperitoneal shunt is challenging. Criteria have been established, but the rate of response remains low (Boon et al., 2000; Kahlon et al., 2005; Marmarou et al., 2005).

10 Treatment of Brain Edema

A large number of studies in animals have tested potential treatments for cerebral edema. Although many have been shown to work in animal studies, treatment of cerebral edema in humans has been extremely difficult to study, and in spite of multiple studies, convincing evidence of efficacy is lacking for many of the currently used treatments. In a recent review of several decades of studies, no agent met vigorous criteria for efficacy. There was some enthusiasm for decompressive surgery in massive ischemic edema, but this conclusion was reached on the basis of several uncontrolled studies (Rabinstein, 2006). Why have the treatment efforts lagged so far behind the rapid advances in understanding the underlying molecular mechanisms and successes in the treatment of animal models of brain edema? One obvious reason is the difficulty in identifying patients with similar lesions that can be entered into controlled studies. Obtaining consents for experimental treatments in poorly responsive patients raises ethical questions about patient protection. Another is that numbers of patients with severe edema seen at any one center are generally too few to conduct a randomized study, making costly multicenter studies necessary. Finally, long-term follow-up is necessary to adequately test a new treatment, and many of the studies are short-term.

Current practice has dictated the treatment of cerebral edema in patients. The two treatments most commonly used are osmotic agents and steroids. The key to the treatment of cerebral edema, which is still empirical, is the accurate identification of the type of injury. In cytotoxic edema, for example, mannitol and hypertonic saline provide short-term relief to control life-threatening increased intracranial pressure. Another example is the use of a short-term course of high-dose steroids to reduce the inflammatory response and reduce the vasogenic edema. A common mistake is to use steroids for treatment of cytotoxic edema; a large number of studies have documented the futility of steroid treatment in stroke. They have been less effective in cytotoxic edema, however, and are contraindicated in the treatment of edema secondary to stroke or hemorrhage. In fact, systemic complications of corticosteroids can worsen the patient's condition in the treatment of patients with intracerebral hemorrhage (Qureshi et al., 2001).

Hypertonic solutions are used to reduce the water content of brain tissue; initially urea was used, but it entered the brain and caused a rebound in CSF pressure (Pappius and Dayes, 1965). Presently, hypertonic solutions of mannitol and saline are used to reduce brain volume, lower CSF production, and improve cerebral blood flow. Earlier studies employed 3 g/kg of mannitol, which had a dramatic effect on the serum electrolytes, and permitted only one or two doses to be given. More recently, it was found that low doses of mannitol (0.25–1.0 g/kg) are as effective as the higher doses without affecting the electrolytes. Noninfarcted regions are mainly affected by the hypertonic solutions rather than in the infarcted hemisphere (Videen et al., 2001). Mannitol also changes the rheological characteristics of the blood and may have an antioxidant effect. Prolonged administration of mannitol results in an electrolyte imbalance that may override its benefit and that must be carefully monitored. More recently, hypertonic saline has been advocated for use in treatment of cerebral edema (Zeynalov et al., 2008). Studies in animals have shown that it lowers intracranial pressure, and studies in humans are being done (Chen et al., 2006a).

Most treatments have been directed at controlling the secondary consequences of brain edema rather than treating the underlying causes. Although not directly aimed at the edema itself, reducing the blood and CSF volumes is used to lower the intracranial pressure. Blood volume can be reduced with hyperventilation, which lowers carbon dioxide. However, excessive hyperventilation can cause vasoconstriction and ischemia. Reduction of CSF volume can be done mechanically by placing a drainage catheter into one of the ventricles, which may be difficult if the cerebral ventricles are compressed by the edema. Agents that reduce the production of CSF, such as acetazolamide or diuretics, may be used, but are of marginal benefit.

Edema surrounding brain tumors, particularly metastatic brain tumors, responds dramatically to treatment with high doses of dexamethasone. The corticosteroid closes the BBB rapidly. Hence, it is important to obtain contrast-enhanced MRI or computed tomographic scans before treatment with corticosteroids; otherwise, enhancement of the lesion may be missed. High doses of corticosteroids have been shown to be effective in brain edema secondary to inflammation in multiple sclerosis; the steroids act by closing the BBB, which can be seen on contrast-enhanced MRI (Noseworthy et al., 2000). The opening of the BBB is associated with elevated levels of the proinflammatory cytokine, TNF-α. Inflammatory lesions, such as those that occur in acute attacks of multiple sclerosis, respond well to high-dose methylprednisolone. Treatment with 1 g/day of methylprednisolone for 3–5 days reduces the inflammatory changes in the blood vessels during an acute exacerbation. Dramatic reduction in enhancement on MRI may be seen after treatment. However, the effect is lost after several months. High-dose steroids reduce the MMP-9 in the brain as reflected in the CSF, preserving the integrity of the BBB (Rosenberg et al., 1996b).

Treatment of edema surrounding an intracerebral hemorrhage has recently been intensively studied because of the side-effect of hemorrhagic transformation in patients treated with tPA. As in studies of edema secondary to ischemia, a large number of animal studies have documented the use of various agents to reduce the edema secondary to the hemorrhage. One promising study using recombinant activated factor VII reduced growth of the hemorrhage in an initial study (Mayer et al., 2005). A subsequent study failed to confirm the results of the first because of a high rate of thrombotic complications (Mayer et al., 2008). Several other studies are underway, including stereotactic removal of the clot that has been dissolved with intraventricular tPA, but results are not available for this study.

11 Conclusions

Cerebral edema is common in many neurological disorders. When the metabolism of the cell fails, the cells are unable to remove sodium and the cells swell. This is seen on MRI as an area of restricted diffusion. Alternatively, the blood vessels may be damaged as part of an injury, infection, or autoimmune process, resulting in vasogenic edema, which has a predilection for the white matter. Common mechanisms involved in cellular swelling include failure of membrane ion pumps, retention of sodium inside the cell, and increases in intracellular calcium, initiating an inflammatory response. Proteases and free radicals degrade extracellular matrix and tight junction proteins leading to vascular edema and hemorrhage. An important advance in our understanding of brain edema was made with the discovery of AQPs, which were shown to be located in astrocyte endfeet and water-controlling surfaces of the brain. Although knowledge of the mechanisms of brain edema continues to grow, the challenge remains the translation of that information into treatments.

Acknowledgments This work was supported by grants from the National Institutes of Health (NIH) to GAR (R01 NS045847 and R01 NS052305).

References

Abbott NJ (2004) Evidence for bulk flow of brain interstitial fluid: significance for physiology and pathology. Neurochem Int 45:545–552

Abbruscato TJ, Davis TP (1999) Protein expression of brain endothelial cell E-cadherin after hypoxia/aglycemia: influence of astrocyte contact. Brain Res 842:277–286

Agre P, King LS, Yasui M, Guggino WB, Ottersen OP, Fujiyoshi Y, Engel A, Nielsen S (2002) Aquaporin water channels–from atomic structure to clinical medicine. J Physiol 542:3–16

Amiry-Moghaddam M, Otsuka T, Hurn PD, Traystman RJ, Haug FM, Froehner SC, Adams ME, Neely JD, Agre P, Ottersen OP, Bhardwaj A (2003) An alpha-syntrophin-dependent pool of AQP4 in astroglial end-feet confers bidirectional water flow between blood and brain. Proc Natl Acad Sci USA 100:2106–2111

Amiry-Moghaddam M, Ottersen OP (2003) The molecular basis of water transport in the brain. Nat Rev Neurosci 4:991–1001

Amiry-Moghaddam M, Xue R, Haug FM, Neely JD, Bhardwaj A, Agre P, Adams ME, Froehner SC, Mori S, Ottersen OP (2004) Alpha-syntrophin deletion removes the perivascular but not endothelial pool of aquaporin-4 at the blood-brain barrier and delays the development of brain edema in an experimental model of acute hyponatremia. FASEB J 18:542–544

Anderson CM, Nedergaard M (2003) Astrocyte-mediated control of cerebral microcirculation. Trends Neurosci 26:340–345

Armstead WM, Mirro R, Thelin OP, Shibata M, Zuckerman SL, Shanklin DR, Busija DW, Leffler CW (1992) Polyethylene glycol superoxide dismutase and catalase attenuate increased blood-brain barrier permeability after ischemia in piglets. Stroke 23:755–762

Asahi M, Asahi K, Jung JC, del Zoppo GJ, Fini ME, Lo EH (2000) Role for matrix metalloproteinase 9 after focal cerebral ischemia: effects of gene knockout and enzyme inhibition with BB-94. J Cereb Blood Flow Metab 20:1681–1689

Asahi M, Wang X, Mori T, Sumii T, Jung JC, Moskowitz MA, Fini ME, Lo EH (2001) Effects of matrix metalloproteinase-9 gene knock-out on the proteolysis of blood-brain barrier and white matter components after cerebral ischemia. J Neurosci 21:7724–7732

Austinat M, Braeuninger S, Pesquero JB, Brede M, Bader M, Stoll G, Renne T, Kleinschnitz C (2009) Blockade of bradykinin receptor B1 but not bradykinin receptor B2 provides protection from cerebral infarction and brain edema. Stroke 40:285–293

Baba T, Chio CC, Black KL (1992) The effect of 5-lipoxygenase inhibition on blood-brain barrier permeability in experimental brain tumors. J Neurosurg 77:403–406

Badaut J, Lasbennes F, Magistretti PJ, Regli L (2002) Aquaporins in brain: distribution, physiology, and pathophysiology. J Cereb Blood Flow Metab 22:367–378

Ballabh P, Braun A, Nedergaard M (2004) The blood-brain barrier: an overview: structure, regulation, and clinical implications. Neurobiol Dis 16:1–13

Ballabh P, Xu H, Hu F, Braun A, Smith K, Rivera A, Lou N, Ungvari Z, Goldman SA, Csiszar A, Nedergaard M (2007) Angiogenic inhibition reduces germinal matrix hemorrhage. Nat Med 13:477–485

Bardutzky J, Schwab S (2007) Antiedema therapy in ischemic stroke. Stroke 38: 3084–3094

Barzo P, Marmarou A, Fatouros P, Portella G, Czigner A, Bullock R, Young H (2002) [Cerebral edema and changes of cerebral blood volume in patients with head injuries]. Orv Hetil 143:1625–1634

Bazzoni G, Dejana E (2004) Endothelial cell-to-cell junctions: molecular organization and role in vascular homeostasis. Physiol Rev 84:869–901

Bechmann I, Galea I,, Perry VH (2007) What is the blood-brain barrier (not)? Trends Immunol 28:5–11

Bechmann I, Goldmann J, Kovac AD, Kwidzinski E, Simburger E, Naftolin F, Dirnagl U, Nitsch R, Priller J (2005) Circulating monocytic cells infiltrate layers of anterograde axonal degeneration where they transform into microglia. FASEB J 19:647–649

Bemana I, Nagao S (1999) Treatment of brain edema with a nonpeptide arginine vasopressin V1 receptor antagonist OPC-21268 in rats. Neurosurgery 44:148–155

Bhardwaj A (2006) Neurological impact of vasopressin dysregulation and hyponatremia. Ann Neurol 59:229–236

Block ML, Zecca L, Hong JS (2007) Microglia-mediated neurotoxicity: uncovering the molecular mechanisms. Nat Rev Neurosci 8:57–69

Boassa D, Stamer WD, Yool AJ (2006) Ion channel function of aquaporin-1 natively expressed in choroid plexus. J Neurosci 26:7811–7819

Boon AJ, Tans JT, Delwel EJ, Egeler-Peerdeman SM, Hanlo PW, Wurzer HA, Hermans J (2000) The Dutch normal-pressure hydrocephalus study. How to select patients for shunting? An analysis of four diagnostic criteria. Surg Neurol 53:201–207

Bosetti F (2007) Arachidonic acid metabolism in brain physiology and pathology: lessons from genetically altered mouse models. J Neurochem 102:577–586

Candelario-Jalil E, Gonzalez-Falcon A, Garcia-Cabrera M, Leon OS, Fiebich BL (2007a) Post-ischaemic treatment with the cyclooxygenase-2 inhibitor nimesulide reduces blood-brain barrier disruption and leukocyte infiltration following transient focal cerebral ischaemia in rats. J Neurochem 100:1108–1120

Candelario-Jalil E, Taheri S, Yang Y, Sood R, Grossetete M, Estrada EY, Fiebich BL, Rosenberg GA (2007b) Cyclooxygenase inhibition limits blood-brain barrier disruption following intracerebral injection of tumor necrosis factor-alpha in the rat. J Pharmacol Exp Ther 323:488–498

Candelario-Jalil E, Yang Y, Rosenberg GA (2009) Diverse roles of matrix metalloproteinases and tissue inhibitors of metalloproteinases in neuroinflammation and cerebral ischemia. Neuroscience 158:983–994

Carmeliet P (2003) Angiogenesis in health and disease. Nat Med 9:653–660

Chan PH (2001) Reactive oxygen radicals in signaling and damage in the ischemic brain. J Cereb Blood Flow Metab 21:2–14

Chan PH, Fishman RA (1984) The role of arachidonic acid in vasogenic brain edema. Fed Proc 43:210–213

Chan LL, Khoo JB, Thng CH, Lim WE, Tay KH, Tan EK, Chang HM, Chen C, Wong MC, Tan KP (2002) Diffusion weighted MR imaging in acute stroke: the SGH experience. Singapore Med J 43:118–123

Chan PH, Schmidley JW, Fishman RA, Longar SM (1984) Brain injury, edema, and vascular permeability changes induced by oxygen-derived free radicals. Neurology 34:315–320

Chan PH, Yang GY, Chen SF, Carlson E, Epstein CJ (1991) Cold-induced brain edema and infarction are reduced in transgenic mice overexpressing CuZn-superoxide dismutase. Ann Neurol 29:482–486

Chen PE, Simon JE, Hill MD, Sohn CH, Dickhoff P, Morrish WF, Sevick RJ, Frayne R (2006b) Acute ischemic stroke: accuracy of diffusion-weighted MR imaging–effects of b value and cerebrospinal fluid suppression. Radiology 238:232–239

Chen CH, Toung TJ, Sapirstein A, Bhardwaj A (2006a) Effect of duration of osmotherapy on blood-brain barrier disruption and regional cerebral edema after experimental stroke. J Cereb Blood Flow Metab 26:951–958

Chi OZ, Hunter C, Liu X, Weiss HR (2007) Effects of anti-VEGF antibody on blood-brain barrier disruption in focal cerebral ischemia. Exp Neurol 204:283–287

Chu K, Jeong SW, Jung KH, Han SY, Lee ST, Kim M, Roh JK (2004) Celecoxib induces functional recovery after intracerebral hemorrhage with reduction of brain edema and perihematomal cell death. J Cereb Blood Flow Metab 24:926–933

Copin JC, Merlani P, Sugawara T, Chan PH, Gasche Y (2008) Delayed matrix metalloproteinase inhibition reduces intracerebral hemorrhage after embolic stroke in rats. Exp Neurol 213: 196–201

Couture R, Harrisson M, Vianna RM, Cloutier F (2001) Kinin receptors in pain and inflammation. Eur J Pharmacol 429:161–176

Cserr HF, Ostrach LH (1974) Bulk flow of interstitial fluid after intracranial injection of blue dextran 2000. Exp Neurol 45:50–60

Cunningham LA, Wetzel M, Rosenberg GA (2005) Multiple roles for MMPs and TIMPs in cerebral ischemia. Glia 50:329–339

Cuzner ML, Opdenakker G (1999) Plasminogen activators and matrix metalloproteases, mediators of extracellular proteolysis in inflammatory demyelination of the central nervous system. J Neuroimmunol 94:1–14

Dejana E, Lampugnani MG, Martinez-Estrada O, Bazzoni G (2000) The molecular organization of endothelial junctions and their functional role in vascular morphogenesis and permeability. Int J Dev Biol 44:743–748

Del Maschio A, De Luigi A, Martin-Padura I, Brockhaus M, Bartfai T, Fruscella P, Adorini L, Martino G, Furlan R, De Simoni MG, Dejana E (1999) Leukocyte recruitment in the cerebrospinal fluid of mice with experimental meningitis is inhibited by an antibody to junctional adhesion molecule (JAM). J Exp Med 190:1351–1356

del Zoppo GJ, Milner R, Mabuchi T, Hung S, Wang X, Berg GI, Koziol JA (2007) Microglial activation and matrix protease generation during focal cerebral ischemia. Stroke 38: 646–651

Dhamija RK, Donnan GA (2008) The role of neuroimaging in acute stroke. Ann Indian Acad Neurol 11:S12–S23

Diaz-Flores L, Gutierrez R, Varela H, Rancel N, Valladares F (1991) Microvascular pericytes: a review of their morphological and functional characteristics. Histol Histopathol 6: 269–286

Dickinson LD, Betz AL (1992) Attenuated development of ischemic brain edema in vasopressin-deficient rats. J Cereb Blood Flow Metab 12:681–690

Ding-Zhou L, Marchand-Verrecchia C, Palmier B, Croci N, Chabrier PE, Plotkine M, Margaill I (2003b) Neuroprotective effects of (S)-N-[4-[4-[(3,4-Dihydro-6-hydroxy-2,5,7,8-tetramethyl-2H-1-benzopyran-2-yl) carbonyl]-1-piperazinyl]phenyl]-2-thiophenecarboximid-amide (BN 80933), an inhibitor of neuronal nitric-oxide synthase and an antioxidant, in model of transient focal cerebral ischemia in mice. J Pharmacol Exp Ther 306:588–594

Ding-Zhou L, Margaill I, Palmier B, Pruneau D, Plotkine M, Marchand-Verrecchia C (2003a) LF 16-0687 Ms, a bradykinin B2 receptor antagonist, reduces ischemic brain injury in a murine model of transient focal cerebral ischemia. Br J Pharmacol 139:1539–1547

Doczi T, Laszlo FA, Szerdahelyi P, Joo F (1984) Involvement of vasopressin in brain edema formation: further evidence obtained from the Brattleboro diabetes insipidus rat with experimental subarachnoid hemorrhage. Neurosurgery 14:436–441

Doczi T, Schwarcz A (2003) Correlation of apparent diffusion coefficient and computed tomography density in acute ischemic stroke. Stroke 34:e

Dore-Duffy P (2008) Pericytes: pluripotent cells of the blood brain barrier. Curr Pharm Des 14:1581–1593

Dzialowski I, Hill MD, Coutts SB, Demchuk AM, Kent DM, Wunderlich O, von Kummer R (2006) Extent of early ischemic changes on computed tomography (CT) before thrombolysis: prognostic value of the Alberta Stroke Program Early CT Score in ECASS II. Stroke 37:973–978

Dzialowski I, Klotz E, Goericke S, Doerfler A, Forsting M, von Kummer R (2007) Ischemic brain tissue water content: CT monitoring during middle cerebral artery occlusion and reperfusion in rats. Radiology 243:720–726

Dzialowski I, Weber J, Doerfler A, Forsting M, von Kummer R (2004) Brain tissue water uptake after middle cerebral artery occlusion assessed with CT. J Neuroimaging 14:42–48

Eastwood JD, Lev MH, Wintermark M, Fitzek C, Barboriak DP, Delong DM, Lee TY, Azhari T, Herzau M, Chilukuri VR, Provenzale JM (2003) Correlation of early dynamic CT perfusion imaging with whole-brain MR diffusion and perfusion imaging in acute hemispheric stroke. AJNR Am J Neuroradiol 24:1869–1875

Ferrara N, Gerber HP, LeCouter J (2003) The biology of VEGF and its receptors. Nat Med 9: 669–676

Fischer S, Wiesnet M, Renz D, Schaper W (2005) H2O2 induces paracellular permeability of porcine brain-derived microvascular endothelial cells by activation of the p44/42 MAP kinase pathway. Eur J Cell Biol 84:687–697

Fischer S, Wobben M, Marti HH, Renz D, Schaper W (2002) Hypoxia-induced hyperpermeability in brain microvessel endothelial cells involves VEGF-mediated changes in the expression of zonula occludens-1. Microvasc Res 63:70–80

Fishman RA (1975) Brain edema. N Engl J Med 293:706–711

Fishman RA (1992) Cerebrospinal fluid in diseases of the nervous system. W.B. Saunders Co, Philadelphia, PA

Fujimura M, Morita-Fujimura Y, Copin J, Yoshimoto T, Chan PH (2001) Reduction of copper, zinc-superoxide dismutase in knockout mice does not affect edema or infarction volumes and the early release of mitochondrial cytochrome c after permanent focal cerebral ischemia. Brain Res 889:208–213

Furuse M, Hirase T, Itoh M, Nagafuchi A, Yonemura S, Tsukita S (1993) Occludin: a novel integral membrane protein localizing at tight junctions. J Cell Biol 123:1777–1788

Garrido-Urbani S, Bradfield PF, Lee BP, Imhof BA (2008) Vascular and epithelial junctions: a barrier for leucocyte migration. Biochem Soc Trans 36:203–211

Gasche Y, Copin JC, Sugawara T, Fujimura M, Chan PH (2001) Matrix metalloproteinase inhibition prevents oxidative stress-associated blood-brain barrier disruption after transient focal cerebral ischemia. J Cereb Blood Flow Metab 21:1393–1400

Ginsberg MD, Becker DA, Busto R, Belayev A, Zhang Y, Khoutorova L, Ley JJ, Zhao W, Belayev L (2003) Stilbazulenyl nitrone, a novel antioxidant, is highly neuroprotective in focal ischemia. Ann Neurol 54:330–342

Greitz D, Greitz T, Hindmarsh T (1997) A new view on the CSF-circulation with the potential for pharmacological treatment of childhood hydrocephalus. Acta Paediatr 86:125–132

Greitz D, Hannerz J (1996) A proposed model of cerebrospinal fluid circulation: observations with radionuclide cisternography. AJNR Am J Neuroradiol 17:431–438

Groger M, Lebesgue D, Pruneau D, Relton J, Kim SW, Nussberger J, Plesnila N (2005) Release of bradykinin and expression of kinin B2 receptors in the brain: role for cell death and brain edema formation after focal cerebral ischemia in mice. J Cereb Blood Flow Metab 25:978–989

Gu Z, Cui J, Brown S, Fridman R, Mobashery S, Strongin AY, Lipton SA (2005) A highly specific inhibitor of matrix metalloproteinase-9 rescues laminin from proteolysis and neurons from apoptosis in transient focal cerebral ischemia. J Neurosci 25:6401–6408

Gurney KJ, Estrada EY, Rosenberg GA (2006) Blood-brain barrier disruption by stromelysin-1 facilitates neutrophil infiltration in neuroinflammation. Neurobiol Dis 23:87–96

Gursoy-Ozdemir Y, Bolay H, Saribas O, Dalkara T (2000) Role of endothelial nitric oxide generation and peroxynitrite formation in reperfusion injury after focal cerebral ischemia. Stroke 31:1974–1981

Gursoy-Ozdemir Y, Can A, Dalkara T (2004) Reperfusion-induced oxidative/nitrative injury to neurovascular unit after focal cerebral ischemia. Stroke 35:1449–1453

Haacke EM (2006) Susceptibility weighted imaging (SWI). Z Med Phys 16:237

Haacke EM, Mittal S, Wu Z, Neelavalli J, Cheng YC (2009) Susceptibility-weighted imaging: technical aspects and clinical applications, part 1. AJNR Am J Neuroradiol 30:19–30

Haacke EM, Xu Y, Cheng YC, Reichenbach JR (2004) Susceptibility weighted imaging (SWI). Magn Reson Med 52:612–618

Han F, Shirasaki Y, Fukunaga K (2006) Microsphere embolism-induced endothelial nitric oxide synthase expression mediates disruption of the blood-brain barrier in rat brain. J Neurochem 99:97–106

Hawkins BT, Davis TP (2005) The blood-brain barrier/neurovascular unit in health and disease. Pharmacol Rev 57:173–185

Hellal F, Pruneau D, Palmier B, Faye P, Croci N, Plotkine M, Marchand-Verrecchia C (2003) Detrimental role of bradykinin B2 receptor in a murine model of diffuse brain injury. J Neurotrauma 20:841–851

Helmer KG, Dardzinski BJ, Sotak CH (1995) The application of porous-media theory to the investigation of time-dependent diffusion in in vivo systems. NMR Biomed 8:297–306

Heo JH, Han SW, Lee SK (2005) Free radicals as triggers of brain edema formation after stroke. Free Radic Biol Med 39:51–70

Heo JH, Lucero J, Abumiya T, Koziol JA, Copeland BR, del Zoppo GJ (1999) Matrix metalloproteinases increase very early during experimental focal cerebral ischemia. J Cereb Blood Flow Metab 19:624–633

Hirase T, Kawashima S, Wong EY, Ueyama T, Rikitake Y, Tsukita S, Yokoyama M, Staddon JM (2001) Regulation of tight junction permeability and occludin phosphorylation by Rhoa-p160ROCK-dependent and -independent mechanisms. J Biol Chem 276:10423–10431

Hirase T, Staddon JM, Saitou M, Ando-Akatsuka Y, Itoh M, Furuse M, Fujimoto K, Tsukita S, Rubin LL (1997) Occludin as a possible determinant of tight junction permeability in endothelial cells. J Cell Sci 110(14):1603–1613

Hirschi KK, D'Amore PA (1996) Pericytes in the microvasculature. Cardiovasc Res 32:687–698

Hoeffner EG, Case I, Jain R, Gujar SK, Shah GV, Deveikis JP, Carlos RC, Thompson BG, Harrigan MR, Mukherji SK (2004) Cerebral perfusion CT: technique and clinical applications. Radiology 231:632–644

Hoehn-Berlage M, Hossmann KA, Busch E, Eis M, Schmitz B, Gyngell ML (1997) Inhibition of nonselective cation channels reduces focal ischemic injury of rat brain. J Cereb Blood Flow Metab 17:534–542

Hu J, Yu Y, Juhasz C, Kou Z, Xuan Y, Latif Z, Kudo K, Chugani HT, Haacke EM (2008) MR susceptibility weighted imaging (SWI) complements conventional contrast enhanced T1 weighted MRI in characterizing brain abnormalities of Sturge-Weber Syndrome. J Magn Reson Imaging 28:300–307

Huang Z, Huang PL, Ma J, Meng W, Ayata C, Fishman MC, Moskowitz MA (1996) Enlarged infarcts in endothelial nitric oxide synthase knockout mice are attenuated by nitro-L-arginine. J Cereb Blood Flow Metab 16:981–987

Huang Z, Huang PL, Panahian N, Dalkara T, Fishman MC, Moskowitz MA (1994) Effects of cerebral ischemia in mice deficient in neuronal nitric oxide synthase. Science 265:1883–1885

Ikeda Y, Teramoto A, Nakagawa Y, Ishibashi Y, Yoshii T (1997b) Attenuation of cryogenic induced brain oedema by arginine vasopressin release inhibitor RU51599. Acta Neurochir (Wien) 139:1173–1180

Ikeda Y, Toda S, Kawamoto T, Teramoto A (1997a) Arginine vasopressin release inhibitor RU51599 attenuates brain oedema following transient forebrain ischaemia in rats. Acta Neurochir (Wien) 139:1166–1172

Imaizumi S, Kondo T, Deli MA, Gobbel G, Joo F, Epstein CJ, Yoshimoto T, Chan PH (1996) The influence of oxygen free radicals on the permeability of the monolayer of cultured brain endothelial cells. Neurochem Int 29:205–211

Ishibashi N, Prokopenko O, Weisbrot-Lefkowitz M, Reuhl KR, Mirochnitchenko O (2002) Glutathione peroxidase inhibits cell death and glial activation following experimental stroke. Brain Res Mol Brain Res 109:34–44

Ishihara H, Kubota H, Lindberg RL, Leppert D, Gloor SM, Errede M, Virgintino D, Fontana A, Yonekawa Y, Frei K (2008) Endothelial cell barrier impairment induced by glioblastomas and transforming growth factor beta2 involves matrix metalloproteinases and tight junction proteins. J Neuropathol Exp Neurol 67:435–448

Ivashkova Y, Svetnitsky A, Mayzler O, Pruneau D, Benifla M, Fuxman Y, Cohen A, Artru AA, Shapira Y (2006) Bradykinin B2 receptor antagonism with LF 18-1505T reduces brain edema and improves neurological outcome after closed head trauma in rats. J Trauma 61:879–885

Jakovcevic D, Harder DR (2007) Role of astrocytes in matching blood flow to neuronal activity. Curr Top Dev Biol 79:75–97

Jin G, Arai K, Murata Y, Wang S, Stins MF, Lo EH, van Leyen K (2008) Protecting against cerebrovascular injury: contributions of 12/15-lipoxygenase to edema formation after transient focal ischemia. Stroke 39:2538–2543

Kahles T, Luedike P, Endres M, Galla HJ, Steinmetz H, Busse R, Neumann-Haefelin T, Brandes RP (2007) NADPH oxidase plays a central role in blood-brain barrier damage in experimental stroke. Stroke 38:3000–3006

Kahlon B, Sundbarg G, Rehncrona S (2005) Lumbar infusion test in normal pressure hydrocephalus. Acta Neurol Scand 111:379–384

Kamada H, Yu F, Nito C, Chan PH (2007) Influence of hyperglycemia on oxidative stress and matrix metalloproteinase-9 activation after focal cerebral ischemia/reperfusion in rats: relation to blood-brain barrier dysfunction. Stroke 38:1044–1049

Katzman R, Pappius HM (1973) Brain electrolytes and fluid metabolism. Williams & Wilkins Co, Baltimore, MD

Kaur C, Ling EA (2008) Blood brain barrier in hypoxic-ischemic conditions. Curr Neurovasc Res 5:71–81
Kaur C, Sivakumar V, Zhang Y, Ling EA (2006) Hypoxia-induced astrocytic reaction and increased vascular permeability in the rat cerebellum. Glia 54:826–839
Kim JH, Park JA, Lee SW, Kim WJ, Yu YS, Kim KW (2006) Blood-neural barrier: intercellular communication at glio-vascular interface. J Biochem Mol Biol 39:339–345
Kimelberg HK (1995) Current concepts of brain edema. Review of laboratory investigations. J Neurosurg 83:1051–1059
Kimelberg HK (2004) Water homeostasis in the brain: basic concepts. Neuroscience 129:851–860
Kinouchi H, Epstein CJ, Mizui T, Carlson E, Chen SF, Chan PH (1991) Attenuation of focal cerebral ischemic injury in transgenic mice overexpressing CuZn superoxide dismutase. Proc Natl Acad Sci USA 88:11158–11162
Klasner B, Lumenta DB, Pruneau D, Zausinger S, Plesnila N (2006) Therapeutic window of bradykinin B2 receptor inhibition after focal cerebral ischemia in rats. Neurochem Int 49:442–447
Klatzo I (1967) Presidental address. Neuropathological aspects of brain edema. J Neuropathol Exp Neurol 26:1–14
Klatzo I (1994) Evolution of brain edema concepts. Acta Neurochir Suppl (Wien) 60:3–6
Koehler RC, Roman RJ, Harder DR (2009) Astrocytes and the regulation of cerebral blood flow. Trends Neurosci 32:160–169
Kokubo Y, Matson GB, Derugin N, Hill T, Mancuso A, Chan PH, Weinstein PR (2002) Transgenic mice expressing human copper-zinc superoxide dismutase exhibit attenuated apparent diffusion coefficient reduction during reperfusion following focal cerebral ischemia. Brain Res 947:1–8
Kondo T, Reaume AG, Huang TT, Carlson E, Murakami K, Chen SF, Hoffman EK, Scott RW, Epstein CJ, Chan PH (1997) Reduction of CuZn-superoxide dismutase activity exacerbates neuronal cell injury and edema formation after transient focal cerebral ischemia. J Neurosci 17:4180–4189
Kourie JI (1998) Interaction of reactive oxygen species with ion transport mechanisms. Am J Physiol 275:C1–C24
Kozniewska E, Roberts TP, Tsuura M, Mintorovitch J, Moseley ME, Kucharczyk J (1995) NG-nitro-L-arginine delays the development of brain injury during focal ischemia in rats. Stroke 26:282–289
Krause G, Winkler L, Mueller SL, Haseloff RF, Piontek J, Blasig IE (2008) Structure and function of claudins. Biochim Biophys Acta 1778:631–645
Kremer S, Oppenheim C, Schmitt E, Dietemann JL (2007) [Diffusion MRI: technique and clinical applications]. J Radiol 88:428–443
Krizanac-Bengez L, Hossain M, Fazio V, Mayberg M, Janigro D (2006) Loss of flow induces leukocyte-mediated MMP/TIMP imbalance in dynamic in vitro blood-brain barrier model: role of pro-inflammatory cytokines. Am J Physiol Cell Physiol 291:C740–C749
Kucinski T, Vaterlein O, Glauche V, Fiehler J, Klotz E, Eckert B, Koch C, Rother J, Zeumer H (2002) Correlation of apparent diffusion coefficient and computed tomography density in acute ischemic stroke. Stroke 33:1786–1791
Kunte H, Schmidt S, Eliasziw M, del Zoppo GJ, Simard JM, Masuhr F, Weih M, Dirnagl U (2007) Sulfonylureas improve outcome in patients with type 2 diabetes and acute ischemic stroke. Stroke 38:2526–2530
Kunz A, Anrather J, Zhou P, Orio M, Iadecola C (2007) Cyclooxygenase-2 does not contribute to postischemic production of reactive oxygen species. J Cereb Blood Flow Metab 27:545–551
Lagrange P, Romero IA, Minn A, Revest PA (1999) Transendothelial permeability changes induced by free radicals in an in vitro model of the blood-brain barrier. Free Radic Biol Med 27:667–672
Lai CH, Kuo KH (2005) The critical component to establish in vitro BBB model: pericyte. Brain Res Brain Res Rev 50:258–265
Lapchak PA, Chapman DF, Zivin JA (2000) Metalloproteinase inhibition reduces thrombolytic (tissue plasminogen activator)-induced hemorrhage after thromboembolic stroke. Stroke 31:3034–3040

Laszlo FA, Varga C, Nakamura S (1999) Vasopressin receptor antagonist OPC-31260 prevents cerebral oedema after subarachnoid haemorrhage. Eur J Pharmacol 364:115–122

Lehmberg J, Beck J, Baethmann A, Uhl E (2003) Bradykinin antagonists reduce leukocyte-endothelium interactions after global cerebral ischemia. J Cereb Blood Flow Metab 23:441–448

Leib SL, Clements JM, Lindberg RL, Heimgartner C, Loeffler JM, Pfister LA, Tauber MG, Leppert D (2001) Inhibition of matrix metalloproteinases and tumour necrosis factor alpha converting enzyme as adjuvant therapy in pneumococcal meningitis. Brain 124:1734–1742

Leib SL, Leppert D, Clements J, Tauber MG (2000) Matrix metalloproteinases contribute to brain damage in experimental pneumococcal meningitis. Infect Immun 68:615–620

Liang D, Bhatta S, Gerzanich V, Simard JM (2007) Cytotoxic edema: mechanisms of pathological cell swelling. Neurosurg Focus 22:E2

Liu W, Hendren J, Qin XJ, Shen J, Liu KJ (2009) Normobaric hyperoxia attenuates early blood-brain barrier disruption by inhibiting MMP-9-mediated occludin degradation in focal cerebral ischemia. J Neurochem 108:811–820

Liu KJ, Rosenberg GA (2005) Matrix metalloproteinases and free radicals in cerebral ischemia. Free Radic Biol Med 39:71–80

Lo AC, Chen AY, Hung VK, Yaw LP, Fung MK, Ho MC, Tsang MC, Chung SS, Chung SK (2005) Endothelin-1 overexpression leads to further water accumulation and brain edema after middle cerebral artery occlusion via aquaporin 4 expression in astrocytic end-feet. J Cereb Blood Flow Metab 25:998–1011

Lorenzl S, Kodel U, Frei K, Pfister HW (1996) Effect of the bradykinin B2 receptor antagonist Hoe140 in experimental pneumococcal meningitis in the rat. Eur J Pharmacol 308:335–341

Lumenta DB, Plesnila N, Klasner B, Baethmann A, Pruneau D, Schmid-Elsaesser R, Zausinger S (2006) Neuroprotective effects of a postischemic treatment with a bradykinin B2 receptor antagonist in a rat model of temporary focal cerebral ischemia. Brain Res 1069:227–234

Mac Donald CL, Dikranian K, Bayly P, Holtzman D, Brody D (2007) Diffusion tensor imaging reliably detects experimental traumatic axonal injury and indicates approximate time of injury. J Neurosci 27:11869–11876

Manley GT, Fujimura M, Ma T, Noshita N, Filiz F, Bollen AW, Chan P, Verkman AS (2000) Aquaporin-4 deletion in mice reduces brain edema after acute water intoxication and ischemic stroke. Nat Med 6:159–163

Marmarou A, Black P, Bergsneider M, Klinge P, Relkin N (2005) Guidelines for management of idiopathic normal pressure hydrocephalus: progress to date. Acta Neurochir Suppl 95: 237–240

Marmarou A, Fatouros PP, Barzo P, Portella G, Yoshihara M, Tsuji O, Yamamoto T, Laine F, Signoretti S, Ward JD, Bullock MR, Young HF (2000b) Contribution of edema and cerebral blood volume to traumatic brain swelling in head-injured patients. J Neurosurg 93:183–193

Marmarou A, Portella G, Barzo P, Signoretti S, Fatouros P, Beaumont A, Jiang T, Bullock R (2000a) Distinguishing between cellular and vasogenic edema in head injured patients with focal lesions using magnetic resonance imaging. Acta Neurochir Suppl 76:349–351

Mayer SA, Brun NC, Begtrup K, Broderick J, Davis S, Diringer MN, Skolnick BE, Steiner T (2005) Recombinant activated factor VII for acute intracerebral hemorrhage. N Engl J Med 352:777–785

Mayer SA, Brun NC, Begtrup K, Broderick J, Davis S, Diringer MN, Skolnick BE, Steiner T (2008) Efficacy and safety of recombinant activated factor VII for acute intracerebral hemorrhage. N Engl J Med 358:2127–2137

Mayhan WG (1999) VEGF increases permeability of the blood-brain barrier via a nitric oxide synthase/cGMP-dependent pathway. Am J Physiol 276:C1148–C1153

McColl BW, Rothwell NJ, Allan SM (2008) Systemic inflammation alters the kinetics of cerebrovascular tight junction disruption after experimental stroke in mice. J Neurosci 28:9451–9462

Miller BA (2004) Inhibition of TRPM2 function by PARP inhibitors protects cells from oxidative stress-induced death. Br J Pharmacol 143:515–516

Mittal S, Wu Z, Neelavalli J, Haacke EM (2009) Susceptibility-weighted imaging: technical aspects and clinical applications, part 2. AJNR Am J Neuroradiol 30:232–252

Molnar AH, Varga C, Berko A, Rojik I, Parducz A, Laszlo F, Laszlo FA (2008a) Prevention of hypoxic brain oedema by the administration of vasopressin receptor antagonist OPC-31260. Prog Brain Res 170:519–525

Molnar AH, Varga C, Berko A, Rojik I, Parducz A, Laszlo F, Laszlo FA (2008b) Inhibitory effect of vasopressin receptor antagonist OPC-31260 on experimental brain oedema induced by global cerebral ischaemia. Acta Neurochir (Wien) 150:265–271

Morgan L, Shah B, Rivers LE, Barden L, Groom AJ, Chung R, Higazi D, Desmond H, Smith T, Staddon JM (2007) Inflammation and dephosphorylation of the tight junction protein occludin in an experimental model of multiple sclerosis. Neuroscience 147:664–673

Mun-Bryce S, Rosenberg GA (1998) Matrix metalloproteinases in cerebrovascular disease. J Cereb Blood Flow Metab 18:1163–1172

Murata Y, Rosell A, Scannevin RH, Rhodes KJ, Wang X, Lo EH (2008) Extension of the thrombolytic time window with minocycline in experimental stroke. Stroke 39:3372–3377

Mylonakou MN, Petersen PH, Rinvik E, Rojek A, Valdimarsdottir E, Zelenin S, Zeuthen T, Nielsen S, Ottersen OP, Amiry-Moghaddam M (2009) Analysis of mice with targeted deletion of AQP9 gene provides conclusive evidence for expression of AQP9 in neurons. J Neurosci Res 87: 1310–1322

Nagafuji T, Matsui T, Koide T, Asano T (1992) Blockade of nitric oxide formation by N omega-nitro-L-arginine mitigates ischemic brain edema and subsequent cerebral infarction in rats. Neurosci Lett 147:159–162

Nakamura T, Keep RF, Hua Y, Schallert T, Hoff JT, Xi G (2004) Deferoxamine-induced attenuation of brain edema and neurological deficits in a rat model of intracerebral hemorrhage. J Neurosurg 100:672–678

Nakamura T, Kuroda Y, Yamashita S, Zhang X, Miyamoto O, Tamiya T, Nagao S, Xi G, Keep RF, Itano T (2008) Edaravone attenuates brain edema and neurologic deficits in a rat model of acute intracerebral hemorrhage. Stroke 39:463–469

Neumann-Haefelin T, Moseley ME, Albers GW (2000a) New magnetic resonance imaging methods for cerebrovascular disease: emerging clinical applications. Ann Neurol 47:559–570

Neumann-Haefelin T, Wittsack HJ, Fink GR, Wenserski F, Li TQ, Seitz RJ, Siebler M, Modder U, Freund HJ (2000c) Diffusion- and perfusion-weighted MRI: influence of severe carotid artery stenosis on the DWI/PWI mismatch in acute stroke. Stroke 31:1311–1317

Neumann-Haefelin T, Wittsack HJ, Wenserski F, Li TQ, Moseley ME, Siebler M, Freund HJ (2000b) Diffusion- and perfusion-weighted MRI in a patient with a prolonged reversible ischaemic neurological deficit. Neuroradiology 42:444–447

Newman PJ (1994) The role of PECAM-1 in vascular cell biology. Ann N Y Acad Sci 714:165–174

Nielsen S, Nagelhus EA, Amiry-Moghaddam M, Bourque C, Agre P, Ottersen OP (1997) Specialized membrane domains for water transport in glial cells: high-resolution immunogold cytochemistry of aquaporin-4 in rat brain. J Neurosci 17:171–180

Niermann H, Amiry-Moghaddam M, Holthoff K, Witte OW, Ottersen OP (2001) A novel role of vasopressin in the brain: modulation of activity-dependent water flux in the neocortex. J Neurosci 21:3045–3051

Ning M, Wang X, Lo EH (2008) Chapter 6 Reperfusion injury after stroke: neurovascular proteases and the blood-brain barrier. Handb Clin Neurol 92:117–136

Nitta T, Hata M, Gotoh S, Seo Y, Sasaki H, Hashimoto N, Furuse M, Tsukita S (2003) Size-selective loosening of the blood-brain barrier in claudin-5-deficient mice. J Cell Biol 161: 653–660

Noseworthy JH, Lucchinetti C, Rodriguez M, Weinshenker BG (2000) Multiple sclerosis. N Engl J Med 343:938–952

O'Brien P, Sellar RJ, Wardlaw JM (2004) Fogging on T2-weighted MR after acute ischaemic stroke: how often might this occur and what are the implications? Neuroradiology 46:635–641

O'Donnell ME, Duong V, Suvatne J, Foroutan S, Johnson DM (2005) Arginine vasopressin stimulation of cerebral microvascular endothelial cell Na-K-Cl cotransporter activity is V1 receptor and [Ca] dependent. Am J Physiol Cell Physiol 289:C283–C292
Oshio K, Watanabe H, Song Y, Verkman AS, Manley GT (2005) Reduced cerebrospinal fluid production and intracranial pressure in mice lacking choroid plexus water channel Aquaporin-1. FASEB J 19:76–78
Papadopoulos MC, Manley GT, Krishna S, Verkman AS (2004) Aquaporin-4 facilitates reabsorption of excess fluid in vasogenic brain edema. FASEB J 18:1291–1293
Papadopoulos MC, Verkman AS (2008) Potential utility of aquaporin modulators for therapy of brain disorders. Prog Brain Res 170:589–601
Pappius HM, Dayes LA (1965) Hypertonic urea. Its effect on the distribution of water and electrolytes in normal and edematous brain tissues. Arch Neurol 13:395–402
Pascale CL, Szmydynger-Chodobska J, Sarri JE, Chodobski A (2006) Traumatic brain injury results in a concomitant increase in neocortical expression of vasopressin and its V1a receptor. J Physiol Pharmacol 57(Suppl 11):161–167
Paul R, Lorenzl S, Koedel U, Sporer B, Vogel U, Frosch M, Pfister HW (1998) Matrix metalloproteinases contribute to the blood-brain barrier disruption during bacterial meningitis. Ann Neurol 44:592–600
Pfefferkorn T, Rosenberg GA (2003) Closure of the blood-brain barrier by matrix metalloproteinase inhibition reduces rtPA-mediated mortality in cerebral ischemia with delayed reperfusion. Stroke 34:2025–2030
Phillis JW, Horrocks LA, Farooqui AA (2006) Cyclooxygenases, lipoxygenases, and epoxygenases in CNS: their role and involvement in neurological disorders. Brain Res Rev 52:201–243
Phillis JW, O'Regan MH (2004) A potentially critical role of phospholipases in central nervous system ischemic, traumatic, and neurodegenerative disorders. Brain Res Rev 44:13–47
Piontek J, Winkler L, Wolburg H, Muller SL, Zuleger N, Piehl C, Wiesner B, Krause G, Blasig IE (2008) Formation of tight junction: determinants of homophilic interaction between classic claudins. FASEB J 22:146–158
Plesnila N, Schulz J, Stoffel M, Eriskat J, Pruneau D, Baethmann A (2001) Role of bradykinin B2 receptors in the formation of vasogenic brain edema in rats. J Neurotrauma 18:1049–1058
Provenzale JM, Sorensen AG (1999) Diffusion-weighted MR imaging in acute stroke: theoretic considerations and clinical applications. AJR Am J Roentgenol 173:1459–1467
Qureshi AI, Tuhrim S, Broderick JP, Batjer HH, Hondo H, Hanley DF (2001) Spontaneous intracerebral hemorrhage. N Engl J Med 344:1450–1460
Rabinstein AA (2006) Treatment of cerebral edema. Neurologist 12:59–73
Rapoport SI (1976) Blood-brain barrier in physiology and medicine. Raven Press, New York, NY
Rash JE, Yasumura T, Hudson CS, Agre P, Nielsen S (1998) Direct immunogold labeling of aquaporin-4 in square arrays of astrocyte and ependymocyte plasma membranes in rat brain and spinal cord. Proc Natl Acad Sci USA 95:11981–11986
Reddy MK, Labhasetwar V (2009) Nanoparticle-mediated delivery of superoxide dismutase to the brain: an effective strategy to reduce ischemia-reperfusion injury. FASEB J 23:1384–1395
Reijerkerk A, Kooij G, van der Pol SM, Khazen S, Dijkstra CD, de Vries HE (2006) Diapedesis of monocytes is associated with MMP-mediated occludin disappearance in brain endothelial cells. FASEB J 20:2550–2552
Rieth KG, Fujiwara K, Di Chiro G, Klatzo I, Brooks RA, Johnston GS, O'Connor CM, Mitchell LG (1980) Serial measurements of CT attenuation and specific gravity in experimental cerebral edema. Radiology 135:343–348
Ringel F, Baethmann A, Plesnila N (2006a) Lactacidosis-induced glial cell swelling depends on extracellular Ca2+. Neurosci Lett 398:306–309
Ringel F, Bieringer F, Baethmann A, Plesnila N (2006b) Effect of oxidative stress on glial cell volume. J Neurotrauma 23:1693–1704
Roberts HC, Roberts TP, Smith WS, Lee TJ, Fischbein NJ, Dillon WP (2001) Multisection dynamic CT perfusion for acute cerebral ischemia: the "toggling-table" technique. AJNR Am J Neuroradiol 22:1077–1080

Rollins NK (2007) Clinical applications of diffusion tensor imaging and tractography in children. Pediatr Radiol 37:769–780

Romanic AM, White RF, Arleth AJ, Ohlstein EH, Barone FC (1998) Matrix metalloproteinase expression increases after cerebral focal ischemia in rats: inhibition of matrix metalloproteinase-9 reduces infarct size. Stroke 29:1020–1030

Rosenberg GA (1990) Brain fluids and metabolism. Oxford University Press, New York, NY

Rosenberg GA (2002) Matrix metalloproteinases in neuroinflammation. Glia 39:279–291

Rosenberg GA (2009) Matrix metalloproteinases and their multiple roles in neurodegenerative diseases. Lancet Neurol 8:205–216

Rosenberg GA, Dencoff JE, Correa N Jr., Reiners M, Ford CC (1996b) Effect of steroids on CSF matrix metalloproteinases in multiple sclerosis: relation to blood-brain barrier injury. Neurology 46:1626–1632

Rosenberg GA, Estrada EY, Dencoff JE (1998) Matrix metalloproteinases and TIMPs are associated with blood-brain barrier opening after reperfusion in rat brain. Stroke 29: 2189–2195

Rosenberg GA, Estrada E, Kyner WT (1990) Vasopressin-induced brain edema is mediated by the V1 receptor. Adv Neurol 52:149–154

Rosenberg GA, Kyner WT, Estrada E (1980) Bulk flow of brain interstitial fluid under normal and hyperosmolar conditions. Am J Physiol 238:F42–F49

Rosenberg GA, Navratil M (1997) Metalloproteinase inhibition blocks edema in intracerebral hemorrhage in the rat. Neurology 48:921–926

Rosenberg GA, Navratil M, Barone F, Feuerstein G (1996a) Proteolytic cascade enzymes increase in focal cerebral ischemia in rat. J Cereb Blood Flow Metab 16:360–366

Rosenberg GA, Scremin O, Estrada E, Kyner WT (1992) Arginine vasopressin V1-antagonist and atrial natriuretic peptide reduce hemorrhagic brain edema in rats. Stroke 23: 1767–1774

Rosenberg GA, Yang Y (2007) Vasogenic edema due to tight junction disruption by matrix metalloproteinases in cerebral ischemia. Neurosurg Focus 22:E4

Roviezzo F, Tsigkos S, Kotanidou A, Bucci M, Brancaleone V, Cirino G, Papapetropoulos A (2005) Angiopoietin-2 causes inflammation in vivo by promoting vascular leakage. J Pharmacol Exp Ther 314:738–744

Schipke CG, Kettenmann H (2004) Astrocyte responses to neuronal activity. Glia 47: 226–232

Schleien CL, Eberle B, Shaffner DH, Koehler RC, Traystman RJ (1994) Reduced blood-brain barrier permeability after cardiac arrest by conjugated superoxide dismutase and catalase in piglets. Stroke 25:1830–1835

Schoch HJ, Fischer S, Marti HH (2002) Hypoxia-induced vascular endothelial growth factor expression causes vascular leakage in the brain. Brain 125:2549–2557

Schramm P, Schellinger PD, Klotz E, Kallenberg K, Fiebach JB, Kulkens S, Heiland S, Knauth M, Sartor K (2004) Comparison of perfusion computed tomography and computed tomography angiography source images with perfusion-weighted imaging and diffusion-weighted imaging in patients with acute stroke of less than 6 hours' duration. Stroke 35:1652–1658

Shigemori Y, Katayama Y, Mori T, Maeda T, Kawamata T (2006) Matrix metalloproteinase-9 is associated with blood-brain barrier opening and brain edema formation after cortical contusion in rats. Acta Neurochir Suppl 96:130–133

Shuaib A, Xu WC, Yang T, Noor R (2002) Effects of nonpeptide V(1) vasopressin receptor antagonist SR-49059 on infarction volume and recovery of function in a focal embolic stroke model. Stroke 33:3033–3037

Sidel VW, Solomon AK (1957) Entrance of water into human red cells under an osmotic pressure gradient. J Gen Physiol 41:243–257

Simard JM, Chen M, Tarasov KV, Bhatta S, Ivanova S, Melnitchenko L, Tsymbalyuk N, West GA, Gerzanich V (2006) Newly expressed SUR1-regulated NC(Ca-ATP) channel mediates cerebral edema after ischemic stroke. Nat Med 12:433–440

Simard JM, Geng Z, Woo SK, Ivanova S, Tosun C, Melnichenko L, Gerzanich V (2009b) Glibenclamide reduces inflammation, vasogenic edema, and caspase-3 activation after subarachnoid hemorrhage. J Cereb Blood Flow Metab 29:317–330
Simard JM, Tsymbalyuk O, Ivanov A, Ivanova S, Bhatta S, Geng Z, Woo SK, Gerzanich V (2007) Endothelial sulfonylurea receptor 1-regulated NC Ca-ATP channels mediate progressive hemorrhagic necrosis following spinal cord injury. J Clin Invest 117:2105–2113
Simard JM, Woo SK, Bhatta S, Gerzanich V (2008) Drugs acting on SUR1 to treat CNS ischemia and trauma. Curr Opin Pharmacol 8:42–49
Simard JM, Yurovsky V, Tsymbalyuk N, Melnichenko L, Ivanova S, Gerzanich V (2009a) Protective effect of delayed treatment with low-dose glibenclamide in three models of ischemic stroke. Stroke 40:604–609
Sood R, Yang Y, Taheri S, Candelario-Jalil E, Estrada EY, Walker EJ, Thompson J, Rosenberg GA (2009) Increased apparent diffusion coefficients on MRI linked with matrix metalloproteinases and edema in white matter after bilateral carotid artery occlusion in rats. J Cereb Blood Flow Metab 29:308–316
Staub F, Baethmann A, Peters J, Weigt H, Kempski O (1990) Effects of lactacidosis on glial cell volume and viability. J Cereb Blood Flow Metab 10:866–876
Staub F, Winkler A, Peters J, Kempski O, Kachel V, Baethmann A (1994) Swelling, acidosis, and irreversible damage of glial cells from exposure to arachidonic acid in vitro. J Cereb Blood Flow Metab 14:1030–1039
Stoll G, Jander S (1999) The role of microglia and macrophages in the pathophysiology of the CNS. Prog Neurobiol 58:233–247
Suda S, Igarashi H, Arai Y, Andou J, Chishiki T, Katayama Y (2007) Effect of edaravone, a free radical scavenger, on ischemic cerebral edema assessed by magnetic resonance imaging. Neurol Med Chir (Tokyo) 47:197–202
Szmydynger-Chodobska J, Chung I, Kozniewska E, Tran B, Harrington FJ, Duncan JA, Chodobski A (2004) Increased expression of vasopressin v1a receptors after traumatic brain injury. J Neurotrauma 21:1090–1102
Taguchi Y, Takashima S, Hirano K, Dougu N, Toyoda S, Tanaka K (2007) Cerebral microbleeds identified by diffusion weighted MR imaging in a case of acute ischemic stroke. Intern Med 46:1275–1276
Tait MJ, Saadoun S, Bell BA, Papadopoulos MC (2008) Water movements in the brain: role of aquaporins. Trends Neurosci 31:37–43
Tang J, Liu J, Zhou C, Ostanin D, Grisham MB, Neil Granger D, Zhang JH (2005) Role of NADPH oxidase in the brain injury of intracerebral hemorrhage. J Neurochem 94:1342–1350
Terajima K, Igarashi H, Hirose M, Matsuzawa H, Nishizawa M, Nakada T (2008) Serial assessments of delayed encephalopathy after carbon monoxide poisoning using magnetic resonance spectroscopy and diffusion tensor imaging on 3.0T system. Eur Neurol 59:55–61
Thomas WE (1999) Brain macrophages: on the role of pericytes and perivascular cells. Brain Res Brain Res Rev 31:42–57
Thurston G (2002) Complementary actions of VEGF and angiopoietin-1 on blood vessel growth and leakage. J Anat 200:575–580
Trabold R, Krieg S, Scholler K, Plesnila N (2008) Role of vasopressin V(1a) and V(2) receptors for the development of secondary brain damage after traumatic brain injury in mice. J Neurotrauma 25:1459–1465
Trapp BD, Peterson J, Ransohoff RM, Rudick R, Mork S, Bo L (1998) Axonal transection in the lesions of multiple sclerosis. N Engl J Med 338:278–285
Trapp BD, Stys PK (2009) Virtual hypoxia and chronic necrosis of demyelinated axons in multiple sclerosis. Lancet Neurol 8:280–291
Unger E, Littlefield J, Gado M (1988) Water content and water structure in CT and MR signal changes: possible influence in detection of early stroke. AJNR Am J Neuroradiol 9:687–691
Vakili A, Kataoka H, Plesnila N (2005) Role of arginine vasopressin V1 and V2 receptors for brain damage after transient focal cerebral ischemia. J Cereb Blood Flow Metab 25:1012–1019

Valable S, Montaner J, Bellail A, Berezowski V, Brillault J, Cecchelli R, Divoux D, Mackenzie ET, Bernaudin M, Roussel S, Petit E (2005) VEGF-induced BBB permeability is associated with an MMP-9 activity increase in cerebral ischemia: both effects decreased by Ang-1. J Cereb Blood Flow Metab 25:1491–1504

van Bruggen N, Thibodeaux H, Palmer JT, Lee WP, Fu L, Cairns B, Tumas D, Gerlai R, Williams SP, van Lookeren CM, Ferrara N (1999) VEGF antagonism reduces edema formation and tissue damage after ischemia/reperfusion injury in the mouse brain. J Clin Invest 104:1613–1620

Verkman AS (2005) More than just water channels: unexpected cellular roles of aquaporins. J Cell Sci 118:3225–3232

Verkman AS, Mitra AK (2000) Structure and function of aquaporin water channels. Am J Physiol Renal Physiol 278:F13–F28

Videen TO, Zazulia AR, Manno EM, Derdeyn CP, Adams RE, Diringer MN, Powers WJ (2001) Mannitol bolus preferentially shrinks non-infarcted brain in patients with ischemic stroke. Neurology 57:2120–2122

von Kummer R, Bourquain H, Bastianello S, Bozzao L, Manelfe C, Meier D, Hacke W (2001) Early prediction of irreversible brain damage after ischemic stroke at CT. Radiology 219: 95–100

Walder CE, Green SP, Darbonne WC, Mathias J, Rae J, Dinauer MC, Curnutte JT, Thomas GR (1997) Ischemic stroke injury is reduced in mice lacking a functional NADPH oxidase. Stroke 28:2252–2258

Wang W, Dentler WL, Borchardt RT (2001) VEGF increases BMEC monolayer permeability by affecting occludin expression and tight junction assembly. Am J Physiol Heart Circ Physiol 280:H434–H440

Wang X, Tsuji K, Lee SR, Ning M, Furie KL, Buchan AM, Lo EH (2004) Mechanisms of hemorrhagic transformation after tissue plasminogen activator reperfusion therapy for ischemic stroke. Stroke 35:2726–2730

Wang L, Zhang ZG, Zhang RL, Gregg SR, Hozeska-Solgot A, LeTourneau Y, Wang Y, Chopp M (2006) Matrix metalloproteinase 2 (MMP2) and MMP9 secreted by erythropoietin-activated endothelial cells promote neural progenitor cell migration. J Neurosci 26:5996–6003

Weed LH (1935) Certain anatomical and physiological aspects of the meninges and cerebrospinal fluid. Brain 58:383–397

Weed LH (1938) Meninges and Cerebrospinal Fluid. J Anat 72:181–215

Weisbrot-Lefkowitz M, Reuhl K, Perry B, Chan PH, Inouye M, Mirochnitchenko O (1998) Overexpression of human glutathione peroxidase protects transgenic mice against focal cerebral ischemia/reperfusion damage. Brain Res Mol Brain Res 53:333–338

Williams K, Alvarez X, Lackner AA (2001) Central nervous system perivascular cells are immunoregulatory cells that connect the CNS with the peripheral immune system. Glia 36:156–164

Winkler AS, Baethmann A, Peters J, Kempski O, Staub F (2000) Mechanisms of arachidonic acid induced glial swelling. Brain Res Mol Brain Res 76:419–423

Wintermark M et al (2008) Acute stroke imaging research roadmap. Stroke 39:1621–1628

Wolburg H, Lippoldt A (2002) Tight junctions of the blood-brain barrier: development, composition and regulation. Vascul Pharmacol 38:323–337

Wong CH, Bozinovski S, Hertzog PJ, Hickey MJ, Crack PJ (2008) Absence of glutathione peroxidase-1 exacerbates cerebral ischemia-reperfusion injury by reducing post-ischemic microvascular perfusion. J Neurochem 107:241–252

Xia CF, Smith RS Jr., Shen B, Yang ZR, Borlongan CV, Chao L, Chao J (2006) Postischemic brain injury is exacerbated in mice lacking the kinin B2 receptor. Hypertension 47:752–761

Yamamoto N, Yokota K, Yamashita A, Oda M (1996) Effect of KBT-3022, a new cyclooxygenase inhibitor, on experimental brain edema in vitro and in vivo. Eur J Pharmacol 297: 225–231

Yancopoulos GD, Davis S, Gale NW, Rudge JS, Wiegand SJ, Holash J (2000) Vascular-specific growth factors and blood vessel formation. Nature 407:242–248

Yang G, Chan PH, Chen J, Carlson E, Chen SF, Weinstein P, Epstein CJ, Kamii H (1994) Human copper-zinc superoxide dismutase transgenic mice are highly resistant to reperfusion injury after focal cerebral ischemia. Stroke 25:165–170
Yang Y, Estrada EY, Thompson JF, Liu W, Rosenberg GA (2007) Matrix metalloproteinase-mediated disruption of tight junction proteins in cerebral vessels is reversed by synthetic matrix metalloproteinase inhibitor in focal ischemia in rat. J Cereb Blood Flow Metab 27:697–709
Yen MH, Lee SH (1987) Effects of cyclooxygenase and lipoxygenase inhibitors on cerebral edema induced by freezing lesions in rats. Eur J Pharmacol 144:369–373
Yong VW (2005) Metalloproteinases: mediators of pathology and regeneration in the CNS. Nat Rev Neurosci 6:931–944
Zador Z, Bloch O, Yao X, Manley GT (2007) Aquaporins: role in cerebral edema and brain water balance. Prog Brain Res 161:185–194
Zausinger S, Lumenta DB, Pruneau D, Schmid-Elsaesser R, Plesnila N, Baethmann A (2002) Effects of LF 16-0687 Ms, a bradykinin B(2) receptor antagonist, on brain edema formation and tissue damage in a rat model of temporary focal cerebral ischemia. Brain Res 950:268–278
Zeynalov E, Chen CH, Froehner SC, Adams ME, Ottersen OP, Amiry-Moghaddam M, Bhardwaj A (2008) The perivascular pool of aquaporin-4 mediates the effect of osmotherapy in postischemic cerebral edema. Crit Care Med 36:2634–2640
Zhang Z, Chopp M (2002) Vascular endothelial growth factor and angiopoietins in focal cerebral ischemia. Trends Cardiovasc Med 12:62–66
Zhang ZG, Zhang L, Jiang Q, Zhang R, Davies K, Powers C, Bruggen N, Chopp M (2000) VEGF enhances angiogenesis and promotes blood-brain barrier leakage in the ischemic brain. J Clin Invest 106:829–838
Zhang ZG, Zhang L, Tsang W, Soltanian-Zadeh H, Morris D, Zhang R, Goussev A, Powers C, Yeich T, Chopp M (2002) Correlation of VEGF and angiopoietin expression with disruption of blood-brain barrier and angiogenesis after focal cerebral ischemia. J Cereb Blood Flow Metab 22:379–392
Zhu Y, Lee C, Shen F, Du R, Young WL, Yang GY (2005) Angiopoietin-2 facilitates vascular endothelial growth factor-induced angiogenesis in the mature mouse brain. Stroke 36: 1533–1537
Zlokovic BV (2008) The blood-brain barrier in health and chronic neurodegenerative disorders. Neuron 57:178–201

Monoamine Transporter Pathologies

Natalie R. Sealover and Eric L. Barker

Abstract The monoamine neurotransmitters control a variety of functions including movement, appetite, mood, reward, and memory. The monoamine transporters are responsible for the termination of synaptic signaling by removing neurotransmitters from the synaptic cleft. Altered monoaminergic transporter function has been implicated in the pathology of disease states such as depression, anxiety, addiction, autism, Parkinson's disease, and attention deficit hyperactivity disorder (ADHD). This review considers the mechanism of transporter action and regulation of transporter function. The implications of transporter polymorphisms are also addressed. Finally, a brief overview is presented that highlights important findings as well as existing problems that need to be considered in future studies.

Keywords Dopamine transporter · Norepinephrine transporter · Serotonin transporter · Polymorphism · Attention deficit hyperactivity disorder · Parkinson's disease · Addiction · Anxiety · Depression · Autism

Abbreviations

ADHD	Attention deficit hyperactivity disorder
β-CIT	2β-carbomethoxy-3β-(4-iodophenyl) tropane
nor-β-CIT	N-(2-fluoroethyl)-2β-carbomethoxy-3β-(4-iodophenyl)-nortropane
DAT	Dopamine transporter
GAD	Generalized anxiety disorder
K_m	Substrate affinity
LeuT	Leucine transporter
$LeuT_{Aa}$	*Aquifex aeolicus* leucine transporter

E.L. Barker (✉)
Department of Medicinal Chemistry and Molecular Pharmacology, Purdue University, West Lafayette, IN 47906, USA
e-mail: barkerel@purdue.edu

J.P. Blass (ed.), *Neurochemical Mechanisms in Disease*,
Advances in Neurobiology 1, DOI 10.1007/978-1-4419-7104-3_6,

MAOI	Monoamine oxidase inhibitor
MAPK	Mitogen-activated protein kinase
MDMA, "ecstasy"	3,4-methylenedioxymethamphetamine
MPP^+	1-methyl-4-phenylpyridinium
NET	Norepinephrine transporter
NSS	Neurotransmitter/sodium symporter
OCD	Obsessive compulsive disorder
PD	Parkinson's disease
PKC	Protein kinase C
β-PMA	Phorbol 12-myristate13-acetate
PP2A	Protein phosphatase 2A
SERT	Serotonin transporter
SNP	Single nucleotide polymorphism
SPECT	Single-photon emission computed tomography
SSRI	Selective serotonin reuptake inhibitor
TAAR1	Trace amine-associated receptor 1
TCA	Tricyclic antidepressant
TMH	Transmembrane helice
VMAT	Vesicular monoamine transporter
V_{max}	Transport capacity
VNTR	Variable number tandem repeat

Contents

1 Introduction to Monoamine Transporters

1.1 The Monoamine Transporter Family

Synaptic transmission requires the release of neurotransmitters into the extracellular space to bind pre-or postsynaptic receptors, conveying a chemical message to nerve cells (Torres et al., 2003a). Termination of this signaling occurs rapidly by uptake of the released neurotransmitter into the presynaptic cell by high-affinity neurotransmitter transporters. The clearance of the monoamines dopamine, norepinephrine, and serotonin occurs via the dopamine transporter (DAT), norepinephrine transporter (NET), and serotonin transporter (SERT), respectively (Torres et al., 2003a)

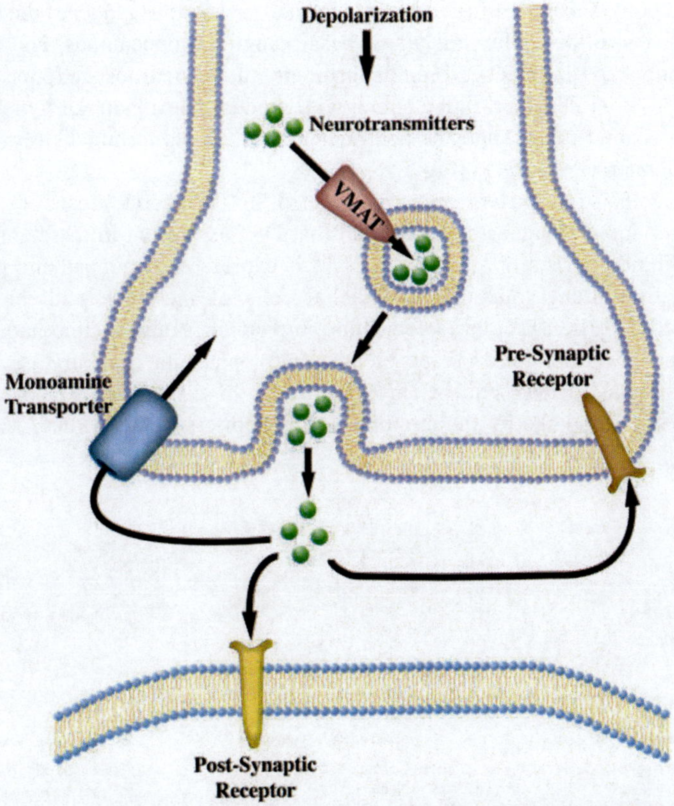

Fig. 1 General model of the release of vesicular neurotransmitter stores in response to cellular depolarization and the reuptake of the neurotransmitters by the monoamine transporters. Cytosolic neurotransmitters are taken into vesicles by VMAT and stored until the cell becomes depolarized, causing these vesicular stores to fuse with the plasma membrane and release the neurotransmitters into the synaptic cleft. Neurotransmitters in the synaptic cleft are available to bind pre- or postsynaptic receptors. Termination of signaling occurs when the neurotransmitters are taken back into the presynaptic cell by the monoamine transporters

(Fig. 1). These monoamine transporters belong to the SLC6 gene family of Na^{+}-Cl^{-}-coupled neurotransmitter transporters that is also referred to as the neurotransmitter sodium symporter (NSS) family (Chen et al., 2004). In addition to the monoamine transporters, the NSS family includes subfamilies of transporters for GABA, amino acids, creatine, and the osmolytes betaine and taurine (Chen et al., 2004).

1.2 Neuroanatomy

In the brain, monoamine transporters are found on neurons that contain their respective neurotransmitter (Torres et al., 2003a). For example, neuronal cells that produce dopamine are localized in the substantia nigra, ventral tegmental area, and hypothalamus (Lin and Madras, 2006). The processes of dopaminergic neurons extend into the caudate nucleus, putamen, nucleus accumbens, and prefrontal cortex (Lin and Madras, 2006). Serotonergic neurons are located in the raphe nuclei of the brainstem and project into the cortex, thalamus, basal ganglia, hippocampus, and amygdala (Jacobs and Azmitia, 1992). Norepinephrine-producing neurons are found primarily in the locus coeruleus and raphe nuclei with moderate levels in the hypothalamus, midline thalamic nuclei, and the bed nucleus of the stria terminalis (Torres et al., 2003a; Donnan et al., 1991) (Fig. 2).

Monoamine transporters are also located in peripheral areas of the body. Eisenhofer and colleagues demonstrated that DAT is present in the stomach, pancreas, and kidney (Eisenhofer, 2001). NET is expressed in sympathetic peripheral neurons, the adrenal medulla, endothelial cells of the lung, and the placenta (Eisenhofer, 2001). SERT has been found in platelets (Talvenheimo and Rudnick, 1980), the intestinal tract (Wade et al., 1996), placenta (Padbury et al., 1997; Balkovetz et al., 1989), and in chromaffin cells of the adrenal gland (Schroeter et al., 1997). Reuptake by the monoamine transporters is the primary mechanism

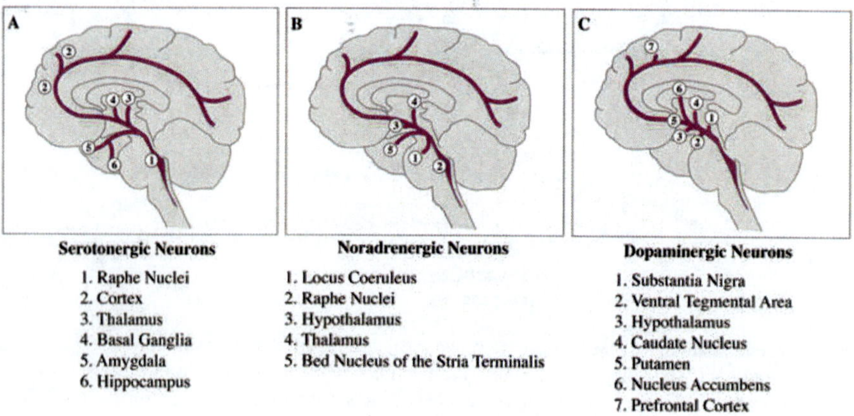

Fig. 2 (**a**) Location of serotonergic neurons and their projections in the human brain. (**b**) Location of noradrenergic neurons and their projections in the human brain. (**c**) Location of dopaminergic neurons and their projections in the human brain

of terminating monoaminergic neurotransmitter signaling in the central nervous system and periphery.

1.3 Physiological Functions

The monoamine transporters are involved in the regulation of many physiological functions. DAT has been implicated in addiction and reward response, movement, cognition, and memory (Greengard, 2001). Altered dopaminergic regulation is involved in depression, suicide, anxiety, aggression, schizophrenia, attention deficit hyperactivity disorder (ADHD), and Parkinson's disease (Jayanthi and Ramamoorthy, 2005; Gainetdinov and Caron, 2003). SERT is involved in the regulation of appetite, libido, mood, anxiety, fear, reward, aggression, and memory (Barnes and Sharp, 1999). Disrupted serotonergic function has been implicated in depression, suicide, impulsive violence, autism, and alcoholism (Jayanthi and Ramamoorthy, 2005). NET plays an important role in arousal, mood, aggression, addiction, and attention, as well as in thermal and cardiac regulation (Jayanthi and Ramamoorthy, 2005; Howell and Kimmel, 2008). Alteration of the noradrengeric system can result in cardiac disease and psychiatric disorders including depression and anxiety (Jayanthi and Ramamoorthy, 2005).

1.4 Structure and Transport Mechanism

The monoamine transporters contain 12 alpha helical transmembrane helices (TMHs) with a putative large extracellular loop between TMHs III and IV with potential glycosylation sites (Melikian et al., 1996, 1994) (Fig. 3). The amino and carboxy termini are located intracellularly and contain putative phosphorylation sites (Torres et al., 2003a). Uptake of the monoamines by their respective transporters utilizes an ion gradient generated by the plasma membrane Na^+/K^+ ATPase (Torres et al., 2003a). NET and SERT are thought to translocate one Na^+ ion and one Cl^- ion with the substrate per transport cycle, whereas DAT is predicted to transport two Na^+ ions and one Cl^- ion with its substrate (Torres et al., 2003a).

The alternating access transport model has been used to describe the mechanism by which substrates are transported across the membrane via the monoamine transporters (Forrest et al., 2008). This model postulates that the transporter can exist in at least two conformations. These conformations include an extracellularly facing form that is open to the extracellular environment and can bind substrate and Na^+ and Cl^- ions (Forrest et al., 2008). An intracellularly facing form allows the release of substrate into the cell and the binding of the countertransported K^+ ion to reverse the conformation of the transporter (Forrest et al., 2008). The alternating access model is supported by recent crystal structures of other transporters (Weyand et al., 2008; Faham et al., 2008). Two additional conformations of these transporters have also been described. A closed–closed conformation is predicted that prevents accessibility of substrate and ions from either side of the transporter. This closed–closed conformation was observed in the crystal structure of a leucine transporter

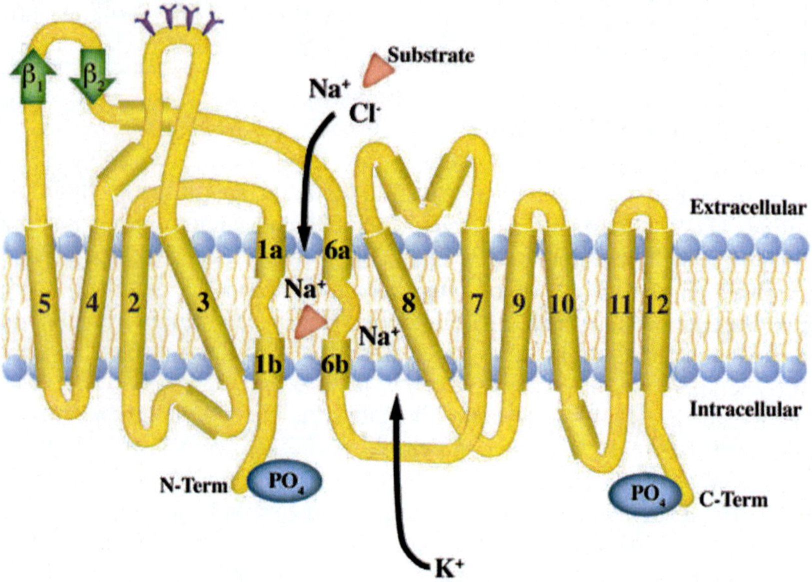

Fig. 3 Schematic representation of the predicted topology of the monoamine transporters based on the crystallization of $LeuT_{Aa}$ (Yamashita et al., 2005). The representation demonstrates how extracellular Na^+, Cl^-, and substrate are exchanged for intracellular K^+. The putative phosphorylation sites on the N-terminus and C-terminus are shown along with predicted glycosylation sites between TMH III and TMH IV. This figure was adapted from Yamashita et al. (2005)

from *Aquifex aeolicus* ($LeuT_{Aa}$), a bacterial homologue of the NSS transporter family (Yamashita et al., 2005). The closed–closed conformation has closed intra- and extracellular gates and may serve as an intermediate between the extracellularly and intracellularly facing states. Another conformation is predicted to have open intra- and extracellular gates. In this conformation the transporter is predicted to operate in a channel mode, allowing substrate molecules and ions to pass through the transporter quickly without an opening and closing of the gates for each transport cycle (Torres et al., 2003a).

Comprehensive understanding of the mechanism of monoamine transport has been hampered by the lack of a crystal structure of these membrane transporters. As mentioned above, in 2005, $LeuT_{Aa}$ was crystallized (Yamashita et al., 2005). This structure and the cocrystallization of $LeuT_{Aa}$ with the tricyclic antidepressants (TCAs) have provided several clues about the putative structure of the monoamine transporters (Singh et al., 2007; Yamashita et al., 2005; Zhou et al., 2007). The $LeuT_{Aa}$ structures reveal binding sites for substrate and Na^+ ions located about halfway through the pore of the protein, interacting with TMHs III and VIII and the unwound regions of TMHs I and VI (Singh et al., 2007). The protein structure shows TMHs I through V are related to VI through X by a pseudo twofold axis in the membrane plane (Yamashita et al., 2005). Cocrystallization studies with the TCAs have identified a putative binding pocket in $LeuT_{Aa}$ that places the TCA binding site just above the extracellular gate (Singh et al., 2007; Zhou et al., 2007). This structure may reveal a similar binding site for the mammalian monoamine transporters, although some have questioned if the TCA binding site of $LeuT_{Aa}$ is likely to be reflective of such a site in the monoamine transporters (Henry et al., 2007; Rudnick,2007).

1.5 Vesicular Monoamine Transporters

Although plasma membrane monoamine transporters are responsible for the reuptake of neurotransmitters from the synapse, vesicular monoamine transporters (VMAT) sequester monoamines into synaptic vesicles in preparation for fusion with the plasma membrane and release into the synapse (Schuldiner et al., 1995). Vesicular uptake is coupled to a proton gradient across the vesicle membrane rather than the sodium gradient used with the plasma membrane transporters (Schuldiner et al., 1995). These vesicular transporters are not neurotransmitter-specific; rather, they transport the monoamines nonselectively (Johnson, Jr., 1988; Henry et al., 1998).

VMAT is predicted to have similar membrane topology to the plasma membrane monoamine transporters, although they do not share homologous sequences (Erickson et al., 1992). Hydrophobicity studies predict 12 TMHs with amino and carboxy termini located in the cytoplasm (Erickson et al., 1992). The large extracellular loop between TMHs III and IV of the plasma membrane transporters is located between TMH I and II in VMAT (Erickson et al., 1992). VMAT1 is located in the neuroendocrine cells of the adrenal medulla and intestinal tract, whereas VMAT2 is found in monoaminergic neurons of the central nervous system (Erickson et al., 1996).

Because VMAT regulates the level of cytosolic monoamines, researchers have examined a role for VMAT in disease states. Although no direct pathological links to aberrant VMAT function have been described, altered dopamine regulation can lead to drug addiction, Parkinson's disease, and schizophrenia (Mazei-Robison et al., 2008). Psychostimulants have been demonstrated to affect dopaminergic signaling by altering DAT and VMAT function (Fleckenstein et al., 2009). Such alterations can be neurotoxic and may provide a role for the monoamine transporters in Parkinson's disease (Fleckenstein et al., 2009).

2 Regulation of Plasma Membrane Monoamine Transporters

Plasma membrane monoamine transporters serve an important regulatory role in maintaining appropriate levels of monoamines in the synapse (Torres et al., 2003a). Aberrant regulation of transporter expression and function has been implicated in several disease states (Howell and Kimmel, 2008). The monoamine transporters are regulated by interaction with a number of substrates and antagonists with varying affinities for the transporters at the plasma membrane (Sulzer et al., 1995; Gutman and Owens, 2006; Fleckenstein et al., 2007). In addition to transporting their respective neurotransmitters, the monoamine transporters can lose substrate selectivity under certain conditions. DAT and NET can each transport dopamine and norepinephrine (Giros et al., 1994), and SERT displays an increased preference for dopamine at elevated temperatures (Saldana and Barker, 2004). Amphetamines such as methamphetamine and 3,4-methylenedioxymethamphetamine (MDMA, "ecstasy") are also substrates of the monoamine transporters, as are some neurotoxins such as 1-methyl-4-phenylpyridinium (MPP^+) (Torres et al., 2003a). In addition, the monoamine transporters are influenced by several classes of antagonists, including cocaine and antidepressants (Torres et al., 2003a).

Monoamine transporter regulation can occur by altering transporter surface expression. Monoamine transporters contain sites for potential phosphorylation in the cytoplasmic loops and the carboxy terminal region (Jayanthi and Ramamoorthy, 2005). Samuvel and colleagues demonstrated that p38 mitogen-activated protein kinase (MAPK) regulates SERT by inhibiting cell surface expression (Samuvel et al., 2005). Treatment of cells and synaptosomes with the PKC activator, phorbol 12-myristate13-acetate (β-PMA) reduces monoamine transport capacity (V_{max}) without altering substrate affinity (K_m) (Samuvel et al., 2005). Other agents that maintain the phosphorylated state of the monoamine transporters such as phosphatase inhibitors also reduce V_{max} (Vaughan et al., 1997; Ramamoorthy et al., 1998; Jayanthi et al., 2004; Apparsundaram et al., 1998b, a). The phosphatase inhibitor, okadaic acid, downregulates DAT, NET, and SERT activity (Ramamoorthy et al., 1998). These studies suggest that phosphorylation of monoamine transporters impairs plasma membrane expression. SERT and protein phosphatase 2A (PP2A) form a complex that is regulated by p38 MAPK activation (Zhu et al., 2005). This complex is inhibited by PP2A inhibitors and PKC activators (Bauman et al., 2000). This complex is stabilized in the presence of the substrate 5-HT (Bauman et al., 2000). These studies provide a mechanism for the regulation of transporter function through the interaction of SERT with PP2A.

Monoamine transporter function is also regulated by glycosylation. The large extracellular loop between TMH III and TMH IV of the monoamine transporters contains consensus sites for glycosylation (Melikian et al., 1994, 1996). The glycosylated form of the transporter is the mature form that undergoes insertion into the plasma membrane (Sitte et al., 2004).

Functional monoamine transporters are predicted to form oligomers (Milner et al., 1994; Jess et al., 1996; Kilic and Rudnick, 2000). One study reports the existence of a dimer of dimers (Kilic and Rudnick, 2000). This tetramer is proposed to be the functional form that exists in the plasma membrane (Kilic and Rudnick, 2000). A leucine heptad repeat in TMH II and a glycophorin-like motif in TMH VI are thought to play a role in stabilizing the oligomeric form of DAT (Torres et al., 2003b). The formation of SERT dimers results from a putative interaction involving TMH XI and TMH XII (Just et al., 2004).

The monoamine transporters are also regulated by a feedback mechanism that involves monoamine autoreceptors located on the presynaptic cell membrane (Hjorth et al., 2000; Schmitz et al., 2002; Garcia et al., 2004). These autoreceptors detect the levels of various monoamines in the synapse and modulate the release of monoamines to keep appropriate levels of neurotransmitter in the synapse (Hjorth et al., 2000; Schmitz et al., 2002; Garcia et al., 2004). The exact mechanism of this feedback loop is unknown (Hjorth et al., 2000; Schmitz et al., 2002; Garcia et al., 2004). The autoreceptors are the D2 short isoform (D2 s), α_{2A}, and 5-HT_{1B} receptors for dopamine, norepinephrine, and serotonin, respectively (Xie et al., 2008). The feedback loop may also be controlled by the trace amine-associated receptor 1 (TAAR1). TAAR1 is a G protein-coupled receptor that is activated by the biogenic monoamines, trace amines, and psychostimulants (Borowsky et al., 2001). Xie and colleagues demonstrated the regulation of DAT by TAAR1 and the regulation of TAAR1 signaling by D2 s (Xie and Miller, 2007; Xie et al., 2007). Similar studies have been conducted with NET and SERT to show the regulation of these transporters by TAAR1 and monoamine autoreceptors (Xie et al., 2008).

3 Transporter Gene Polymorphisms

Several genetic polymorphisms have been identified for the genes encoding the monoamine transporters. A brief review of these genetic variations and possible associations with disease states is presented below and in Table 1. A comprehensive review by Hahn and Blakely examines the impact of genetic variations of the SLC6 gene family (Hahn and Blakely, 2007).

Table 1 Summary of identified polymorphisms for NET, DAT, and SERT

Polymorphism	Effect of Polymorphism	Possible Pathological Associations
NET		
A457P	Impaired transport, decreased cell surface expression	Orthostatic intolerance, increased heart rate
F528C	Elevated transport, decreased TCA potency	High blood pressure
–3801 (A/T)	Transcription factor-based repression of NET expression	ADHD
DAT		
A559V	Increased Na^+ sensitivity, spontaneous DA efflux	ADHD, bipolar disorder
3′ untranslated VNTR (480 bp)	Unknown	ADHD
SERT		
I425V	Increased transport, increased V_{max}, decreased K_m	OCD, Asperger's syndrome
5-HTTLPR (s)	Reduced gene transcription	OCD, ADHD, depression
VNTR	Regulates transcription	No known links

3.1 NET

A number of nonsynonymous single nucleotide polymorphisms (SNPs) that result in single amino acid substitutions have been identified for the monoamine transporters.

The hNET SNP A457P was discovered in a familial form of orthostatic intolerance (Hahn et al., 2003; Shannon et al., 2000). The A457P allele was found to be associated with increased heart rate and plasma norepinephrine levels (Hahn et al., 2003). Molecular studies demonstrate that hNET A457P has severely impaired transport function and decreased cell surface expression, revealing a mechanism for impaired hNET function and cardiovascular disease (Hahn et al., 2003). Approximately 20 more coding region SNPs have been identified for hNET, primarily associated with altered psychiatric and cardiovascular phenotypes (Hahn et al., 2005). The precise functional role of many of these variants remains largely undefined. However, the hNET variant F528C was discovered in patients with high blood pressure (Hahn et al., 2005). Hahn and colleagues found the hNET variant to have elevated transport levels, decreased tricyclic antidepressant potency, and an insensitivity to PKC downregulation by β-PMA (Hahn et al., 2005).

In addition to the potential significance of coding region SNPs, variations in the hNET promoter region have also been identified (Kim et al., 2006). The substitution of adenine to thymine at –3081 has been linked to ADHD (Kim et al., 2006). The thymine substitution establishes a palindromic E2-box motif that binds the neural-expressed repressors of transcription, Slug and Scratch (Kim et al., 2006). Slug and Scratch bind the E2-box motif and repress *SLC6A2* promoter activity only when the thymine substitution is present. These data suggest that the –3081(A/T) polymorphism, resulting in transcription factor-based repression of *SLC6A2*, may increase the risk of ADHD development (Kim et al., 2006).

3.2 DAT

The presence of SNPs is not unique to NET. Studies have revealed variants of DAT and SERT as well. A rare DAT coding SNP, A559V, has been identified in two male children diagnosed with ADHD (Mazei-Robison et al., 2008) and a female with bipolar disorder (Grunhage et al., 2000). Cellular studies have demonstrated an increased sensitivity to intracellular sodium and increased DA efflux for hDAT A559V in the absence of efflux-inducing amphetamines (Mazei-Robison et al., 2008). Homology modeling based on $LeuT_{Aa}$, places A559 at the extracellular end of TMH 12 (Mazei-Robison et al., 2008). Studies have implicated TMHs 11 and 12 as forming the interface for monoamine transporter dimerization (Just et al., 2004). Dimerization is known to be important for serotonin efflux (Seidel et al., 2005). The increased dopamine efflux observed for A559V may be due to impairment of transporter dimerization (Mazei-Robison et al., 2008). Interestingly, Chen and colleagues demonstrated that mutating S528 to alanine in DAT TMH 11 results in increased dopamine efflux (Chen and Justice, 2000). These findings suggest a mechanism by which altered DA efflux may be linked to disease states.

The DAT gene is located on chromosome 5 and contains a variable number tandem repeat (VNTR) polymorphism in the 3′-untranslated region. This VNTR is composed of 40 bp repeats that commonly contain nine or ten copies. Multiple investigations have found a link between ADHD and the 480 bp VNTR (Barr et al., 2001; Chen et al., 2003; Cook, Jr. et al., 1995; Curran et al., 2001; Daly et al., 1999; Gill et al., 1997; Waldman et al., 1998). Researchers are unclear if the number of repeats in the 3′ untranslated VNTR directly controls the expression level of the DAT gene or if the allele containing this VNTR is in linkage disequilibrium with functional DNA variants that contribute to the ADHD phenotype (Barr et al., 2001).

3.3 SERT

SERT gene variants have been implicated in neuropsychiatric disorders. Ozaki and colleagues identified the presence of an I425V coding region SNP in some individuals affected with obsessive compulsive disorder (OCD) and Asperger's syndrome (Ozaki et al., 2003). Studies in cultured cells found the I425V mutation to cause an increased rate of transporter activity with an increase in V_{max} and decrease in K_m (Kilic et al., 2003). Cell surface expression was unchanged for the mutant. The elevated transport is thought to be caused by altered cGMP-dependent protein kinase activity (PKG). The I425V mutation results in constitutive activation of SERT similarly to the way nitric oxide stimulates wild-type SERT via a PKG-dependent pathway (Kilic et al., 2003). The stimulation of SERT by cGMP is disrupted in the I425V mutant, although the exact mechanism by which this occurs remains unknown. Thr 276 is predicted to be in the second intracellular loop, between TMH IV and V, and is the site of PKG phosphorylation on SERT (Ramamoorthy et al., 2007). Ile 425 is predicted to reside in the middle of TMH VIII near putative substrate and inhibitor binding sites. It is unclear if the I425V mutation activates the transporter in a manner that makes Thr 276 phosphorylation irrelevant, or if this mutation indirectly increases the level of Thr 276 phosphorylation by interfering with the activity of a phosphatase (Zhang et al., 2007).

Two common polymorphisms have also been reported in the promoter region of the SERT gene. The first is the insertion or deletion of a 44 bp sequence that results in a long (L) or short (S) allele termed the 5-HTTLPR (Lesch et al., 1996). The S variant displays threefold reduced gene transcription, leading to decreased transporter expression and 5-HT uptake (Lesch et al., 1996). Patients with major depression who are homozygous for the long allele (L/L) or heterozygous (L/S) respond better to treatment with the SSRIs fluvoxamine and paroxetine than those homozygous for the short allele (S/S) (Lesch et al., 1996; Zanardi et al., 2000). The S allele has been associated with an increased risk of depression, obsessive-compulsive disorder, and ADHD (Torres et al., 2003a).

The second type of promoter polymorphism is a VNTR in the second intron composed of 17-bp repeats (Ogilvie et al., 1996). Ten and twelve sets of repeats are most common (Lesch et al., 1994). Studies with embryonic stem cells and transgenic embryos implicate the VNTR as playing a role in the regulation of transcription, although no definitive links are known between this VNTR and disease states (Torres et al., 2003a).

4 Addiction

4.1 Psychostimulant Addiction

The rewarding and reinforcing effects of psychostimulants appear to rely primarily on the dopamine system, although studies have demonstrated the ability of serotonin and norepinephrine systems to produce behavioral and neurochemical effects in response to psychostimulants (Howell and Kimmel, 2008). DAT and VMAT2 are critical players in the regulation of dopamine levels in the synapse and cytosol, respectively. The GABAergic system can regulate dopaminergic signaling by controlling the firing rate of dopamine neurons (Churchill et al., 1992; Steffensen et al., 1998). Psychostimulants exert their effects by increasing levels of extracellular neurotransmitter. Psychostimulants are classified as uptake inhibitors or releasers. Cocaine is an example of an uptake inhibitor (Table 2). Cocaine exerts its effects by binding to DAT, NET, or SERT. This binding prevents the transport of neurotransmitter, resulting in increased synaptic neurotransmitter levels. Amphetamines such as MDMA are classified as releasers. They are substrates of the monoamine transporters. Releasers reverse the direction of transport from inward to outward, leading to an increase in the levels of neurotransmitter in the synaptic cleft (Fleckenstein et al., 2007; Rothman and Baumann, 2003).

Repeated exposure to psychostimulants can modify neurotransmitter systems and result in tolerance or increased sensitivity. This exposure alters the effects of

Table 2 Structures of psychostimulants and K_i values in μmol/L for inhibition of [^{3}H] 5-HT, [^{3}H] NE, and [^{3}H] DA uptake at hSERT, hNET, and hDAT, respectively

Structure	Name	hSET	hNET	hDAT
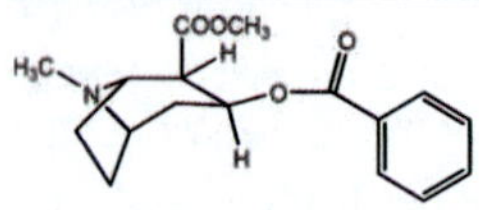	Cocaine	0.74± 0.03	0.48 ± 0.05	0.23 ± 0.03
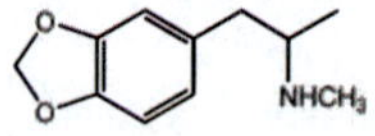	MDMA	2.41 ± 0.73	1.19 ± 0.13	8.29 ± 1.67
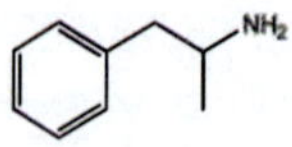	Amphetamine	38.46 ± 3.84	0.07 ± 0.01	0.64 ± 0.14
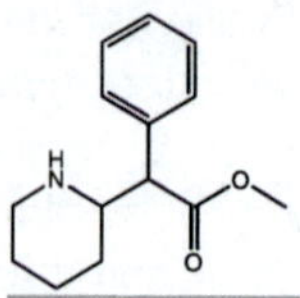	Methylphenidate	132.43 ± 10.71	0.10 ± 0.01	0.06 ± 0.01

Data were obtained in Intestine 407 cells transfected with hSERT, hNET, or hDAT (Han and Gu, 2006).

drugs on brain neurochemistry and behavior, ultimately disrupting the neurobiological regulation of functions related to addiction (Howell and Kimmel, 2008). Interestingly, conflicting studies have reported that cocaine administration in rodents may result in increased, decreased, or unaltered DAT, D1, and D2 receptor levels (Pilotte et al., 1994; Wilson et al., 1994; Claye et al., 1995; Boulay et al., 1996; Tella et al., 1996; Letchworth et al., 1997; Letchworth et al., 1999). Repeated cocaine use has been shown to increase DAT activity in humans (Mash et al., 2002). The initial increase in extracellular dopamine after cocaine administration is thought to result in increased DAT function as a compensatory mechanism. Increased DAT function in turn leads to reduced levels of extracellular dopamine even in the absence of cocaine. This cycle of altered synaptic dopamine levels and DAT function is thought to contribute to the addictive properties of psychostimulants such as cocaine.

Despite significant public health concerns surrounding psychostimulant abuse, currently no effective pharmacotherapies exist (Howell and Kimmel, 2008). To date, treatment for cocaine addiction has been the most widely studied of all psychostimulants. Researchers have examined potential benefits of antidepressants and dopamine receptor agonists and antagonists for cocaine addiction with little success. The TCA desipramine was reported to be an effective treatment in outpatient clinical trials (Levin and Lehman, 1991). Further clinical trials were not able to confirm this effectiveness (Arndt et al., 1992; Campbell et al., 1994). Similarly, treatment with the selective serotonin reuptake inhibitor (SSRI) fluoxetine, appeared initially promising (Walsh et al., 1994), but further studies were unable to demonstrate effectiveness over placebo controls (Batki et al., 1996; Grabowski et al., 1995). Clinical studies with the D2-like receptor agonist bromocriptine have yielded inconclusive results (Gorelick, 1992). Studies targeting GABAergic transmission have shown recent promise (Sofuoglu and Kosten, 2005). Treatment with baclofen, an antispasticity medication and $GABA_B$ receptor agonist, has resulted in increased cocaine abstinence in cocaine-addicted patients (Shoptaw et al., 2003). Similarly, the GABA transporter inhibitor tiagabine that is used for treating epilepsy has been shown to reduce cocaine dependence (Gonzalez et al., 2003). These studies implicate the GABAergic system as a promising target for the development of useful pharmacotherapies for the treatment of cocaine addiction.

4.2 Alcoholism

Alcoholism is characterized by the development of tolerance, craving, and withdrawal (Heinz et al., 2004). Repeated exposure to alcohol results in neuroadaptive changes in the central dopaminergic and serotonergic systems (Heinz et al., 2004). Several studies have directly implicated DAT and SERT in alcoholism. A reduction in SERT expression was found in a sample of alcoholic patients (Heinz et al., 1998) and a high frequency of the short 5-HTTLPR was observed in alcoholic patients (Hammoumi et al., 1999). Studies in rodents demonstrated an ethanol-induced release of dopamine that reinforced the mesolimbic reward system (Mereu et al., 1984; Di Chiara and Imperato, 1988). A recent study by Hillemacher and colleagues found significant hypermethylation of the DAT promoter in alcohol-dependent patients compared to healthy control subjects (Hillemacher et al., 2009). They proposed that ethanol consumption in alcoholics may lead to reduced craving due to hypermethylation-induced downregulation of genes including DAT (Hillemacher et al., 2009). Hypermethylation of the DAT promoter is thought to inhibit gene transcription, leading to reduced DAT expression and increased levels of synaptic dopamine. The mechanism for DAT promoter methylation in response to ethanol consumption is unknown, although the long-term regulation of gene expression by epigenetic mechanisms such as DNA methylation has been suggested as playing a role in the pathophysiology of several psychiatric disorders (Hillemacher et al., 2009).

5 Anxiety and Depression

Alterations in the serotonergic and noradrenergic systems are well established in the pathophysiology of mood disorders, including anxiety and depression. Studies have demonstrated a linkage between the short 5-HTTLPR and psychiatric conditions (Olivier et al., 2008). Anxiety disorders include panic, phobias, obsessive compulsive disorder, and generalized anxiety disorder (GAD) (Keller et al., 2006). These disorders are often treated by blocking NET with compounds such as reboxetine or atomoxetine (Morilak and Frazer, 2004). Chronic treatment with reboxetine or desipramine in rats has been shown to decrease NET binding sites (Gould et al., 2003; Frazer and Benmansour, 2002). Anxiety disorders carry periods of high emotional distress accompanied by physiological hyperarousal (Keller et al., 2006). Keller and colleagues demonstrated that NET-deficient mice respond to stress-inducing environments with heightened autonomic cardiovascular response (Keller et al., 2006). This cardiovascular response is consistent with a NET deficiency linked to increased blood pressure and heart rate due to anxiety and fear-inducing stimuli (Keller et al., 2006).

Several theories have attempted to explain the pathology of depression. One of these theories is the monoamine theory of depression (Heninger et al., 1996). This theory proposes that impaired monoaminergic function is the central basis behind depression. Serotonin and norepinephrine are the two monoamines that have been primarily implicated in the disease. Pharmacological treatment of depression has focused on increasing synaptic levels of these two neurotransmitters (Table 3).

The first class of antidepressants was developed in the early 1950s with the discovery of an antitubercular drug iproniazid that possesses mood-elevating properties (Nutt, 2002). Iproniazid is a monoamine oxidase inhibitor (MAOI). Monoamine oxidase is the enzyme that breaks down serotonin, dopamine, and norepinephrine. The inhibition of monoamine oxidase increases levels of monoamines in the synapse.

Table 3 Structures of antidepressants and K_i values in nmol/L for [^{3}H] 5-HT or [^{3}H] NE inhibition at hSERT and hNET, respectively

Structure	Name	hSERT	hNET
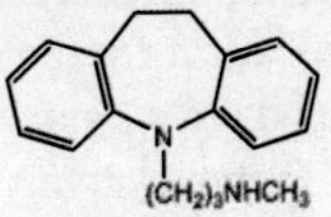	Desipramine	163 ± 5	3.5 ± 0.6
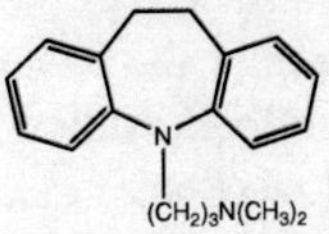	Imipramine	20 ± 2	142 ± 8
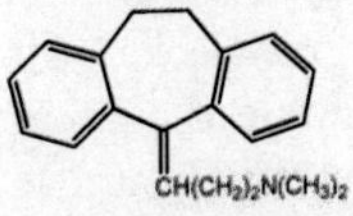	Amitriptyline	36 ± 1	102 ± 9
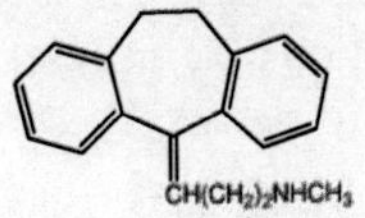	Nortriptyline	279 ± 20	21 ± 0.77
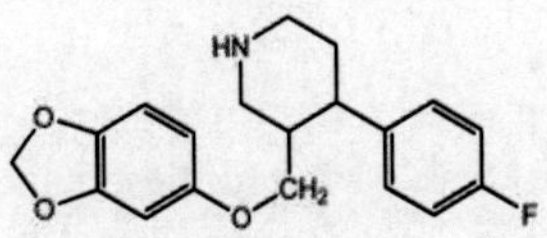	Paroxetine	0.83 ± 0.06	328 ± 25
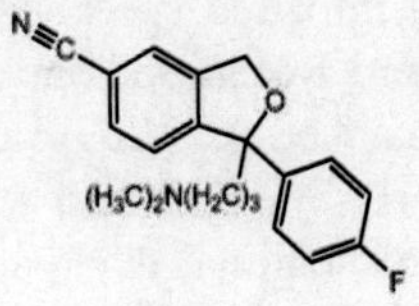	Citalopram	8.9 ± 0.7	30,285 ± 1600
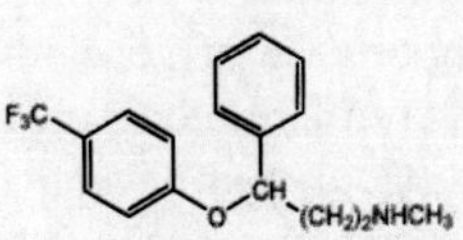	Fluoxetine	20 ± 2	2186 ± 142
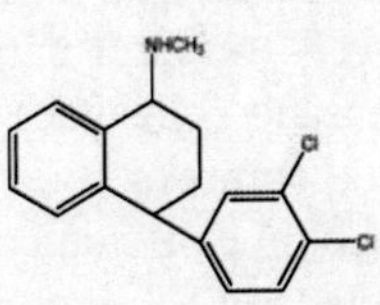	Sertraline	3.3 ± 0.4	1716 ± 151
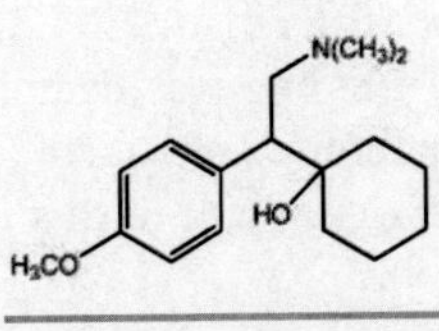	Venlafaxine	102 ± 9	1644 ± 84

Data were obtained in HEK-293 cells transfected with hSERT or hNET (Owens et al., 1997).

Potentially life-threatening interactions with foods containing tyramine and tryptophan led to the disuse of these drugs and the development of different classes of antidepressants. Since the discovery of the first MAOI, TCAs, SSRIs, serotonin–norepinephrine reuptake inhibitors (SNRIs), and some atypical antidepressants such as buproprion have been used in the treatment of depression. Except for the atypical antidepressants, the aforementioned classes of antidepressants increase synaptic levels of serotonin or norepinephrine by inhibiting SERT or NET, respectively (White et al., 2005). White and colleagues provide a comprehensive review of the antidepressants in each of these classes (White et al., 2005). Despite the development of new classes of antidepressants, the effectiveness of these therapeutics remains no better than the MAOIs and patient compliance remains low (Song et al., 1993). Inhibiting SERT and NET rapidly increases synaptic neurotransmitter levels, but the maximal clinical effect is not observed until after several weeks of treatment (Gelenberg and Chesen, 2000). As with chronic administration of NET inhibitors, long-term exposure to SERT inhibitors results in decreased SERT surface expression (Benmansour et al., 1999, 2002). These decreased SERT and NET levels may help to explain the lapse in time from the initial administration of antidepressants to their maximum clinical efficacy.

6 Autism

Autism is a neurodevelopmental disorder that appears in early childhood and results in severely impaired behavioral functions (Folstein and Rosen-Sheidley, 2001). Children with autism display poor social interactions, impaired speech development, and an interest in repetitive activities (Folstein and Rosen-Sheidley, 2001). Autism is recognized as a heritable disorder (Macdonald et al., 1989), although twin-based studies indicate that the disorder is not always inherited (Murphy et al., 2000). Research indicates that autism is linked to neuronal disorganization and the disarrangement of neurotransmitter pathways (Pardo and Eberhart, 2007). The serotonin hypothesis of autism describes the importance of genes that regulate the serotonin system. In particular, genes that control serotonin metabolism and neurotransmission have received much attention (Cook and Leventhal, 1996; Buitelaar and Willemsen-Swinkels, 2000). The serotonin hypothesis of autism is supported by an improvement in behavioral functions in autistic patients receiving treatment with SSRIs (Hollander et al., 2003) or 5-HT2 receptor antagonists (Pardo and Eberhart, 2007). A recent study by Makkonen and colleagues demonstrates reduced SERT binding capacity in the medial frontal cortex of children with autism (Makkonen et al., 2008). This study used single-photon emission computed tomography (SPECT) to analyze the binding of [^{123}I] labeled N-(2-fluoroethyl)-2β-carbomethoxy-3β-(4-iodophenyl)-nortropane, ([^{123}I] nor-β-CIT) to SERT and DAT. A significant decrease in SERT binding, but not DAT binding was demonstrated. Whereas a number of factors likely contribute to the pathology of autism, a significant amount of data indicates a role for the serotonergic system in this complex disorder.

7 Parkinson's Disease (PD)

Parkinson's disease (PD) is characterized by rigidity in movement, resting tremor, bradykinesia, and difficulty in maintaining postural stability (Gelb et al., 1999). PD is a neurodegenerative disease marked by Lewy bodies or abnormal protein aggregates and the loss of dopaminergic cells in the substantia nigra (Gelb et al., 1999). These dopaminergic neurons project into the striatum. Disruption of these neural circuits inhibits pathways responsible for controlling movement. The cause of PD is generally unknown. However, the progressive loss of dopaminergic neurons has led researchers to examine the role of the dopamine system in PD.

Postmortem studies show a correlation between PD and DAT concentrations in the striatum (Niznik et al., 1991). Molecular imaging techniques have been used to determine DAT levels using DAT radiotracers such as 2β-carbomethoxy-3β-(4-iodophenyl) tropane ([^{123}I] β-CIT) and [^{11}C] cocaine (Shih et al., 2006). These techniques are useful for the evaluation and diagnosis of patients with PD and to monitor the progression of the disease (Shih et al., 2006). The delivery of DA to surviving nerve terminals by treatment with L-DOPA helps alleviate some of the early symptoms of PD (Nutt, 2002). Unfortunately, the ameliorating effects of L-DOPA fade and within five years of treatment approximately half of patients display involuntary, sometimes debilitating dyskinesias (Bezard et al., 2001). This reduced ability of L-DOPA to treat patients is thought to result from a failing ability to store synthesized DA (Lee et al., 2008). The reduction in the ability to store DA is thought to result from the immediate release of all DA synthesized from L-DOPA (Lee et al., 2008). Lee and colleagues demonstrated that the sprouting of DA terminals and decreased DAT function help to prevent the appearance of PD symptoms until approximately 60% of dopaminergic neurons in the substantia nigra are lost (Lee et al., 2008). They proposed that the DA terminal sprouting and decreased DAT function helped contribute to the DA release responsible for the reduced therapeutic benefits of L-DOPA (Lee et al., 2008).

8 Important Findings and the Need for Future Studies

Significant advances have been made in understanding the role of the monoamine transporters in the pathology of disease states. Despite these advances, studies are needed to further elucidate monoamine transporter structure and function. In recent years, genetic animal models exhibiting disruption of the monoamine transporters have provided a unique method for investigating the role of these transporters in physiology and pathology. Studies with DAT, SERT, and NET knock-out mice are able to reveal subtle interplay among these transporters (Gainetdinov and Caron, 2003). Knock-out of individual transporters exposes secondary functions and compensatory mechanisms of the remaining monoamine transporters.

The development of improved pharmacological agents targeting the monoamine transporters has been hindered by a lack of structural information. Although numerous biochemical and molecular studies have provided insight into the structural and mechanistic aspects of monoamine transporter function, to date no crystal structure exists. Yamashita and colleagues provided a significant contribution to the field with the crystallization of $LeuT_{Aa}$ (Yamashita et al., 2005) and further information was revealed with the subsequent cocrystallization of $LeuT_{Aa}$ with the TCAs (Singh et al., 2007; Zhou et al., 2007). These structures have allowed for homology modeling of the monoamine transporters and have revitalized the structural studies of these transporters (Indarte et al., 2008; Jorgensen et al., 2007a, b; Celik et al., 2008; Forrest et al., 2007; White et al., 2006).

Important discoveries have been made in the findings of transporter gene polymorphisms. The identification of promoter polymorphisms and coding region polymorphisms in individuals has helped to link these variations with disease states. Further studies are needed to understand the genetic and environmental contributions of these variants to disease. Additionally, a number of regulatory mechanisms and interacting proteins have been identified for the monoamine transporters. Future studies will need to identify additional players in this complex network of interactions. A greater understanding of this network will be vital in understanding and treating diseases such as addiction, ADHD, autism, and PD.

Acknowledgments Special thanks to David Allen for all of his help with figure construction.

References

Apparsundaram S, Galli A, DeFelice LJ, Hartzell HC, Blakely RD (1998a) Acute regulation of norepinephrine transport: I. Protein kinase C-linked muscarinic receptors influence transport capacity and transporter density in SK-N-SH cells. J Pharmacol Exp Ther 287: 733–743

Apparsundaram S, Schroeter S, Giovanetti E, Blakely RD (1998b) Acute regulation of norepinephrine transport: II. PKC-modulated surface expression of human norepinephrine transporter proteins. J Pharmacol Exp Ther 287:744–751

Arndt IO, Dorozynsky L, Woody GE, McLellan AT, O'Brien CP (1992) Desipramine treatment of cocaine dependence in methadone-maintained patients. Arch Gen Psychiatry 49:888–893

Balkovetz DF, Tiruppathi C, Leibach FH, Mahesh VB, Ganapathy V (1989) Evidence for an imipramine-sensitive serotonin transporter in human placental brush-border membranes. J Biol Chem 264:2195–2198

Barnes NM, Sharp T (1999) A review of central 5-HT receptors and their function. Neuropharmacology 38:1083–1152

Barr CL, Xu C, Kroft J, Feng Y, Wigg K, Zai G, Tannock R, Schachar R, Malone M, Roberts W, Nothen MM, Grunhage F, Vandenbergh DJ, Uhl G, Sunohara G, King N, Kennedy JL (2001) Haplotype study of three polymorphisms at the dopamine transporter locus confirm linkage to attention-deficit/hyperactivity disorder. Biol Psychiatry 49:333–339

Batki SL, Washburn AM, Delucchi K, Jones RT (1996) A controlled trial of fluoxetine in crack cocaine dependence. Drug Alcohol Depend 41:137–142

Bauman AL, Apparsundaram S, Ramamoorthy S, Wadzinski BE, Vaughan RA, Blakely RD (2000) Cocaine and antidepressant-sensitive biogenic amine transporters exist in regulated complexes with protein phosphatase 2A. J Neurosci 20:7571–7578

Benmansour S, Cecchi M, Morilak DA, Gerhardt GA, Javors MA, Gould GG, Frazer A (1999) Effects of chronic antidepressant treatments on serotonin transporter function, density, and mRNA level. J Neurosci 19:10494–10501

Benmansour S, Owens WA, Cecchi M, Morilak DA, Frazer A (2002) Serotonin clearance in vivo is altered to a greater extent by antidepressant-induced downregulation of the serotonin transporter than by acute blockade of this transporter. J Neurosci 22:6766–6772

Bezard E, Brotchie JM, Gross CE (2001) Pathophysiology of levodopa-induced dyskinesia: potential for new therapies. Nat Rev Neurosci 2:577–588

Borowsky B, Adham N, Jones KA, Raddatz R, Artymyshyn R, Ogozalek KL, Durkin MM, Lakhlani PP, Bonini JA, Pathirana S, Boyle N, Pu X, Kouranova E, Lichtblau H, Ochoa FY, Branchek TA, Gerald C (2001) Trace amines: identification of a family of mammalian G protein-coupled receptors. Proc Natl Acad Sci USA 98:8966–8971

Boulay D, Duterte-Boucher D, Leroux-Nicollet I, Naudon L, Costentin J (1996) Locomotor sensitization and decrease in [3H]mazindol binding to the dopamine transporter in the nucleus accumbens are delayed after chronic treatments by GBR12783 or cocaine. J Pharmacol Exp Ther 278:330–337

Buitelaar JK, Willemsen-Swinkels SH (2000) Autism: current theories regarding its pathogenesis and implications for rational pharmacotherapy. Paediatr Drugs 2:67–81

Campbell JL, Thomas HM, Gabrielli W, Liskow BI, Powell BJ (1994) Impact of desipramine or carbamazepine on patient retention in outpatient cocaine treatment: preliminary findings. J Addict Dis 13:191–199

Celik L, Sinning S, Severinsen K, Hansen CG, Moller MS, Bols M, Wiborg O, Schiott B (2008) Binding of serotonin to the human serotonin transporter. Molecular modeling and experimental validation. J Am Chem Soc 130:3853–3865

Chen CK, Chen SL, Mill J, Huang YS, Lin SK, Curran S, Purcell S, Sham P, Asherson P (2003) The dopamine transporter gene is associated with attention deficit hyperactivity disorder in a Taiwanese sample. Mol Psychiatry 8:393–396

Chen N, Justice JB (2000) Differential effect of structural modification of human dopamine transporter on the inward and outward transport of dopamine. Mol Brain Res 75: 208–215

Chen NH, Reith ME, Quick MW (2004) Synaptic uptake and beyond: the sodium-and chloride-dependent neurotransmitter transporter family SLC6. Pflugers Arch 447:519–531

Churchill L, Dilts RP, Kalivas PW (1992) Autoradiographic localization of gamma-aminobutyric acidA receptors within the ventral tegmental area. Neurochem Res 17:101–106

Claye LH, Akunne HC, Davis MD, DeMattos S, Soliman KF (1995) Behavioral and neurochemical changes in the dopaminergic system after repeated cocaine administration. Mol Neurobiol 11:55–66

Cook EH, Leventhal BL (1996) The serotonin system in autism. Curr Opin Pediatr 8:348–354

Cook EH Jr., Stein MA, Krasowski MD, Cox NJ, Olkon DM, Kieffer JE, Leventhal BL (1995) Association of attention-deficit disorder and the dopamine transporter gene. Am J Hum Genet 56:993–998

Curran S, Mill J, Tahir E, Kent L, Richards S, Gould A, Huckett L, Sharp J, Batten C, Fernando S, Ozbay F, Yazgan Y, Simonoff E, Thompson M, Taylor E, Asherson P (2001) Association study of a dopamine transporter polymorphism and attention deficit hyperactivity disorder in UK and Turkish samples. Mol Psychiatry 6:425–428

Daly G, Hawi Z, Fitzgerald M, Gill M (1999) Mapping susceptibility loci in attention deficit hyperactivity disorder: preferential transmission of parental alleles at DAT1, DBH and DRD5 to affected children. Mol Psychiatry 4:192–196

Di Chiara G, Imperato A (1988) Drugs abused by humans preferentially increase synaptic dopamine concentrations in the mesolimbic system of freely moving rats. Proc Natl Acad Sci USA 85:5274–5278

Donnan GA, Kaczmarczyk SJ, Paxinos G, Chilco PJ, Kalnins RM, Woodhouse DG, Mendelsohn FA (1991) Distribution of catecholamine uptake sites in human brain as determined by quantitative [3H] mazindol autoradiography. J Comp Neurol 304:419–434

Eisenhofer G (2001) The role of neuronal and extraneuronal plasma membrane transporters in the inactivation of peripheral catecholamines. Pharmacol Ther 91:35–62

Erickson JD, Eiden LE, Hoffman BJ (1992) Expression cloning of a reserpine-sensitive vesicular monoamine transporter. Proc Natl Acad Sci USA 89:10993–10997

Erickson JD, Schafer MK, Bonner TI, Eiden LE, Weihe E (1996) Distinct pharmacological properties and distribution in neurons and endocrine cells of two isoforms of the human vesicular monoamine transporter. Proc Natl Acad Sci USA 93:5166–5171

Faham S, Watanabe A, Besserer GM, Cascio D, Specht A, Hirayama BA, Wright EM, Abramson J (2008) The crystal structure of a sodium galactose transporter reveals mechanistic insights into Na+/sugar symport. Science 321:810–814

Fleckenstein AE, Volz TJ, Hanson GR (2009) Psychostimulant-induced alterations in vesicular monoamine transporter-2 function: Neurotoxic and therapeutic implications. Neuropharmacology 56:133–138

Fleckenstein AE, Volz TJ, Riddle EL, Gibb JW, Hanson GR (2007) New insights into the mechanism of action of amphetamines. Annu Rev Pharmacol Toxicol 47:681–698

Folstein SE, Rosen-Sheidley B (2001) Genetics of autism: complex aetiology for a heterogeneous disorder. Nat Rev Genet 2:943–955

Forrest LR, Tavoulari S, Zhang YW, Rudnick G, Honig B (2007) Identification of a chloride ion binding site in Na+/Cl-dependent transporters. Proc Natl Acad Sci USA 104:12761–12766

Forrest LR, Zhang YW, Jacobs MT, Gesmonde J, Xie L, Honig BH, Rudnick G (2008) Mechanism for alternating access in neurotransmitter transporters. Proc Natl Acad Sci USA 105: 10338–10343

Frazer A, Benmansour S (2002) Delayed pharmacological effects of antidepressants. Mol Psychiatry 7(Suppl 1):S23–S28

Gainetdinov RR, Caron MG (2003) Monoamine transporters: from genes to behavior. Annu Rev Pharmacol Toxicol 43:261–284. Epub;%2002 Sep 17

Garcia AS, Barrera G, Burke TF, Ma S, Hensler JG, Morilak DA (2004) Autoreceptor-mediated inhibition of norepinephrine release in rat medial prefrontal cortex is maintained after chronic desipramine treatment. J Neurochem 91:683–693

Gelb DJ, Oliver E, Gilman S (1999) Diagnostic criteria for Parkinson disease. Arch Neurol 56: 33–39

Gelenberg AJ, Chesen CL (2000) How fast are antidepressants? J Clin Psychiatry 61:712–721

Gill M, Daly G, Heron S, Hawi Z, Fitzgerald M (1997) Confirmation of association between attention deficit hyperactivity disorder and a dopamine transporter polymorphism. Mol Psychiatry 2:311–313

Giros B, Wang YM, Suter S, McLeskey SB, Pifl C, Caron MG (1994) Delineation of discrete domains for substrate, cocaine, and tricyclic antidepressant interactions using chimeric dopamine-norepinephrine transporters. J Biol Chem 269:15985–15988

Gonzalez G, Sevarino K, Sofuoglu M, Poling J, Oliveto A, Gonsai K, George TP, Kosten TR (2003) Tiagabine increases cocaine-free urines in cocaine-dependent methadone-treated patients: results of a randomized pilot study. Addiction 98:1625–1632

Gorelick DA (1992) Alcohol and cocaine. Clinical and pharmacological interactions. Recent Dev Alcohol 10:37–56

Gould GG, Pardon MC, Morilak DA, Frazer A (2003) Regulatory effects of reboxetine treatment alone, or following paroxetine treatment, on brain noradrenergic and serotonergic systems. Neuropsychopharmacology 28:1633–1641

Grabowski J, Rhoades H, Elk R, Schmitz J, Davis C, Creson D, Kirby K (1995) Fluoxetine is ineffective for treatment of cocaine dependence or concurrent opiate and cocaine dependence: two placebo-controlled double-blind trials. J Clin Psychopharmacol 15:163–174

Greengard P (2001) The neurobiology of dopamine signaling. Biosci Rep 21:247–269

Grunhage F, Schulze TG, Muller DJ, Lanczik M, Franzek E, Albus M, Borrmann-Hassenbach M, Knapp M, Cichon S, Maier W, Rietschel M, Propping P, Nothen MM (2000) Systematic screening for DNA sequence variation in the coding region of the human dopamine transporter gene (DAT1). Mol Psychiatry 5:275–282

Gutman DA, Owens MJ (2006) Serotonin and norepinephrine transporter binding profile of SSRIs. Essent Psychopharmacol 7:35–41

Hahn MK, Blakely RD (2007) The functional impact of SLC6 transporter genetic variation. Annu Rev Pharmacol Toxicol 47:401–441

Hahn MK, Mazei-Robison MS, Blakely RD (2005) Single nucleotide polymorphisms in the human norepinephrine transporter gene affect expression, trafficking, antidepressant interaction, and protein kinase C regulation. Mol Pharmacol 68:457–466

Hahn MK, Robertson D, Blakely RD (2003) A mutation in the human norepinephrine transporter gene (SLC6A2) associated with orthostatic intolerance disrupts surface expression of mutant and wild-type transporters. J Neurosci 23:4470–4478

Hammoumi S, Payen A, Favre JD, Balmes JL, Benard JY, Husson M, Ferrand JP, Martin JP, Daoust M (1999) Does the short variant of the serotonin transporter linked polymorphic region constitute a marker of alcohol dependence? Alcohol 17:107–112

Han DD, Gu HH (2006) Comparison of the monoamine transporters from human and mouse in their sensitivities to psychostimulant drugs. BMC Pharmacol 6(6):6

Heinz A, Goldman D, Gallinat J, Schumann G, Puls I (2004) Pharmacogenetic insights to monoaminergic dysfunction in alcohol dependence. Psychopharmacology (Berl) 174:561–570

Heinz A, Ragan P, Jones DW, Hommer D, Williams W, Knable MB, Gorey JG, Doty L, Geyer C, Lee KS, Coppola R, Weinberger DR, Linnoila M (1998) Reduced central serotonin transporters in alcoholism. Am J Psychiatry 155:1544–1549

Heninger GR, Delgado PL, Charney DS (1996) The revised monoamine theory of depression: a modulatory role for monoamines, based on new findings from monoamine depletion experiments in humans. Pharmacopsychiatry 29:2–11

Henry LK, Meiler J, Blakely RD (2007) Bound to be different: neurotransmitter transporters meet their bacterial cousins. Mol Interv 7:306–309

Henry JP, Sagne C, Bedet C, Gasnier B (1998) The vesicular monoamine transporter: from chromaffin granule to brain. Neurochem Int 32:227–246

Hillemacher T, Frieling H, Hartl T, Wilhelm J, Kornhuber J, Bleich S (2009) Promoter specific methylation of the dopamine transporter gene is altered in alcohol dependence and associated with craving. J Psychiatr Res 43:388–392

Hjorth S, Bengtsson HJ, Kullberg A, Carlzon D, Peilot H, Auerbach SB (2000) Serotonin autoreceptor function and antidepressant drug action. J Psychopharmacol 14:177–185

Hollander E, Phillips AT, Yeh CC (2003) Targeted treatments for symptom domains in child and adolescent autism. Lancet 362:732–734

Howell LL, Kimmel HL (2008) Monoamine transporters and psychostimulant addiction. Biochem Pharmacol 75:196–217

Indarte M, Madura JD, Surratt CK (2008) Dopamine transporter comparative molecular modeling and binding site prediction using the LeuT(Aa) leucine transporter as a template. Proteins 70:1033–1046

Jacobs BL, Azmitia EC (1992) Structure and function of the brain serotonin system. Physiol Rev 72:165–229

Jayanthi LD, Ramamoorthy S (2005) Regulation of monoamine transporters: influence of psychostimulants and therapeutic antidepressants. AAPS J 7:E728–E738

Jayanthi LD, Samuvel DJ, Ramamoorthy S (2004) Regulated internalization and phosphorylation of the native norepinephrine transporter in response to phorbol esters. Evidence for localization in lipid rafts and lipid raft-mediated internalization. J Biol Chem 279:19315–19326

Jess U, Betz H, Schloss P (1996) The membrane-bound rat serotonin transporter, SERT1, is an oligomeric protein. FEBS Letters 394:44–46

Johnson RG Jr. (1988) Accumulation of biological amines into chromaffin granules: a model for hormone and neurotransmitter transport. Physiol Rev 68:232–307

Jorgensen AM, Tagmose L, Jorgensen AM, Bogeso KP, Peters GH (2007a) Molecular dynamics simulations of Na+/Cl(–)-dependent neurotransmitter transporters in a membrane-aqueous system. ChemMedChem 2:827–840

Jorgensen AM, Tagmose L, Jorgensen AM, Topiol S, Sabio M, Gundertofte K, Bogeso KP, Peters GH (2007b) Homology modeling of the serotonin transporter: insights into the primary escitalopram-binding site. ChemMedChem 2:815–826

Just H, Sitte HH, Schmid JA, Freissmuth M, Kudlacek O (2004) Identification of an additional interaction domain in transmembrane domains 11 and 12 that supports oligomer formation in the human serotonin transporter. J Biol Chem 279:6650–6657

Keller NR, Diedrich A, Appalsamy M, Miller LC, Caron MG, McDonald MP, Shelton RC, Blakely RD, Robertson D (2006) Norepinephrine transporter-deficient mice respond to anxiety producing and fearful environments with bradycardia and hypotension. Neuroscience 139:931–946

Kilic F, Murphy DL, Rudnick G (2003) A human serotonin transporter mutation causes constitutive activation of transport activity. Mol Pharmacol 64:440–446

Kilic F, Rudnick G (2000) Oligomerization of serotonin transporter and its functional consequences. Proc Natl Acad Sci USA 97:3106–3111

Kim CH, Hahn MK, Joung Y, Anderson SL, Steele AH, Mazei-Robinson MS, Gizer I, Teicher MH, Cohen BM, Robertson D, Waldman ID, Blakely RD, Kim KS (2006) A polymorphism in the norepinephrine transporter gene alters promoter activity and is associated with attention-deficit hyperactivity disorder. Proc Natl Acad Sci USA 103:19164–19169

Lee J, Zhu WM, Stanic D, Finkelstein DI, Horne MH, Henderson J, Lawrence AJ, O'Connor L, Tomas D, Drago J, Horne MK (2008) Sprouting of dopamine terminals and altered dopamine release and uptake in Parkinsonian dyskinaesia. Brain 131:1574–1587

Lesch KP, Balling U, Gross J, Strauss K, Wolozin BL, Murphy DL, Riederer P (1994) Organization of the human serotonin transporter gene. J Neural Transm Gen Sect 95: 157–162

Lesch KP, Bengel D, Heils A, Sabol SZ, Greenberg BD, Petri S, Benjamin J, Muller CR, Hamer DH, Murphy DL (1996) Association of anxiety-related traits with a polymorphism in the serotonin transporter gene regulatory region. Science 274:1527–1531

Letchworth SR, Daunais JB, Hedgecock AA, Porrino LJ (1997) Effects of chronic cocaine administration on dopamine transporter mRNA and protein in the rat. Brain Res 750: 214–222

Letchworth SR, Sexton T, Childers SR, Vrana KE, Vaughan RA, Davies HM, Porrino LJ (1999) Regulation of rat dopamine transporter mRNA and protein by chronic cocaine administration. J Neurochem 73:1982–1989

Levin FR, Lehman AF (1991) Meta-analysis of desipramine as an adjunct in the treatment of cocaine addiction. J Clin Psychopharmacol 11:374–378

Lin Z, Madras BK (2006) Human genetics and pharmacology of neurotransmitter transporters. Handb Exp Pharmacol 175:327–371

Macdonald H, Rutter M, Howlin P, Rios P, Le Conteur A, Evered C, Folstein S (1989) Recognition and expression of emotional cues by autistic and normal adults. J Child Psychol Psychiatry 30:865–877

Makkonen I, Riikonen R, Kokki H, Airaksinen MM, Kuikka JT (2008) Serotonin and dopamine transporter binding in children with autism determined by SPECT. Dev Med Child Neurol 50:593–597

Mash DC, Pablo J, Ouyang Q, Hearn WL, Izenwasser S (2002) Dopamine transport function is elevated in cocaine users. J Neurochem 81:292–300

Mazei-Robison MS, Bowton E, Holy M, Schmudermaier M, Freissmuth M, Sitte HH, Galli A, Blakely RD (2008) Anomalous dopamine release associated with a human dopamine transporter coding variant. J Neurosci 28:7040–7046

Melikian HE, McDonald JK, Gu H, Rudnick G, Moore KR, Blakely RD (1994) Human norepinephrine transporter. Biosynthetic studies using a site-directed polyclonal antibody. J Biol Chem 269:12290–12297

Melikian HE, Ramamoorthy S, Tate CG, Blakely RD (1996) Inability to N-glycosylate the human norepinephrine transporter reduces protein stability, surface trafficking, and transport activity but not ligand recognition. Mol Pharmacol 50:266–276

Mereu G, Fadda F, Gessa GL (1984) Ethanol stimulates the firing rate of nigral dopaminergic neurons in unanesthetized rats. Brain Res 292:63–69

Milner HE, Bēliveau R, Jarvis SM (1994) The in situ size of the dopamine transporter is a tetramer as estimated by radiation inactivation. Biochimica et Biophysica Acta (BBA) Biomembranes 1190:185–187

Morilak DA, Frazer A (2004) Antidepressants and brain monoaminergic systems: a dimensional approach to understanding their behavioural effects in depression and anxiety disorders. Int J Neuropsychopharmacol 7:193–218

Murphy M, Bolton PF, Pickles A, Fombonne E, Piven J, Rutter M (2000) Personality traits of the relatives of autistic probands. Psychol Med 30:1411–1424

Niznik HB, Fogel EF, Fassos FF, Seeman P (1991) The dopamine transporter is absent in parkinsonian putamen and reduced in the caudate nucleus. J Neurochem 56:192–198

Nutt DJ (2002) The neuropharmacology of serotonin and noradrenaline in depression. Int Clin Psychopharmacol 17(Suppl 1):S1–S12

Ogilvie AD, Battersby S, Bubb VJ, Fink G, Harmar AJ, Goodwim GM, Smith CA (1996) Polymorphism in serotonin transporter gene associated with susceptibility to major depression. Lancet 347:731–733

Olivier JD, Van Der Hart MG, Van Swelm RP, Dederen PJ, Homberg JR, Cremers T, Deen PM, Cuppen E, Cools AR, Ellenbroek BA (2008) A study in male and female 5-HT transporter knockout rats: an animal model for anxiety and depression disorders. Neuroscience 152: 573–584

Owens MJ, Morgan WN, Plott SJ, Nemeroff CB (1997) Neurotransmitter receptor and transporter binding profile of antidepressants and their metabolites. J Pharmacol Exp Ther 283:1305–1322

Ozaki N, Goldman D, Kaye WH, Plotnicov K, Greenberg BD, Lappalainen J, Rudnick G, Murphy DL (2003) Serotonin transporter missense mutation associated with a complex neuropsychiatric phenotype. Mol Psychiatry 8:933–936

Padbury JF, Tseng YT, McGonnigal B, Penado K, Stephan M, Rudnick G (1997) Placental biogenic amine transporters: cloning and expression. Brain Res Mol Brain Res 45:163–168

Pardo CA, Eberhart CG (2007) The neurobiology of autism. Brain Pathol 17:434–447

Pilotte NS, Sharpe LG, Kuhar MJ (1994) Withdrawal of repeated intravenous infusions of cocaine persistently reduces binding to dopamine transporters in the nucleus accumbens of Lewis rats. J Pharmacol Exp Ther 269:963–969

Ramamoorthy S, Giovanetti E, Qian Y, Blakely RD (1998) Phosphorylation and regulation of antidepressant-sensitive serotonin transporters. J Biol Chem 273:2458–2466

Ramamoorthy S, Samuvel DJ, Buck ER, Rudnick G, Jayanthi LD (2007) Phosphorylation of threonine residue 276 is required for acute regulation of serotonin transporter by cyclic GMP. J Biol Chem 20(282):11639–11647

Rothman RB, Baumann MH (2003) Monoamine transporters and psychostimulant drugs. Eur J Pharmacol 479:23–40

Rudnick G (2007) What is an antidepressant binding site doing in a bacterial transporter? ACS Chem Biol 2:606–609

Saldana SN, Barker EL (2004) Temperature and 3,4-methylenedioxymethamphetamine alter human serotonin transporter-mediated dopamine uptake. Neurosci Lett 354:209–212

Samuvel DJ, Jayanthi LD, Bhat NR, Ramamoorthy S (2005) A role for p38 mitogen-activated protein kinase in the regulation of the serotonin transporter: evidence for distinct cellular mechanisms involved in transporter surface expression. J Neurosci 25:29–41

Schmitz Y, Schmauss C, Sulzer D (2002) Altered dopamine release and uptake kinetics in mice lacking D2 receptors. J Neurosci 22:8002–8009

Schroeter S, Levey AI, Blakely RD (1997) Polarized expression of the antidepressant-sensitive serotonin transporter in epinephrine-synthesizing chromaffin cells of the rat adrenal gland. Mol Cell Neurosci 9:170–184

Schuldiner S, Shirvan A, Linial M (1995) Vesicular neurotransmitter transporters: from bacteria to humans. Physiol Rev 75:369–392

Seidel S, Singer EA, Just H, Farhan H, Scholze P, Kudlacek O, Holy M, Koppatz K, Krivanek P, Freissmuth M, Sitte HH (2005) Amphetamines take two to tango: an oligomer-based counter-transport model of neurotransmitter transport explores the amphetamine action. Mol Pharmacol 67:140–151

Shannon JR, Flattem NL, Jordan J, Jacob G, Black BK, Biaggioni I, Blakely RD, Robertson D (2000) Orthostatic intolerance and tachycardia associated with norepinephrine-transporter deficiency. N Engl J Med 342:541–549

Shih MC, Hoexter MQ, Andrade LA, Bressan RA (2006) Parkinson's disease and dopamine transporter neuroimaging: a critical review. Sao Paulo Med J 124:168–175

Shoptaw S, Yang X, Rotheram-Fuller EJ, Hsieh YC, Kintaudi PC, Charuvastra VC, Ling W (2003) Randomized placebo-controlled trial of baclofen for cocaine dependence: preliminary effects for individuals with chronic patterns of cocaine use. J Clin Psychiatry 64:1440–1448

Singh SK, Yamashita A, Gouaux E (2007) Antidepressant binding site in a bacterial homologue of neurotransmitter transporters. Nature 448:952–956

Sitte HH, Farhan H, Javitch JA (2004) Sodium-dependent neurotransmitter TRANSPORTERS: OLIGOMERIZATION as a determinant of transporter function and trafficking. Mol Interv 4:38–47

Sofuoglu M, Kosten TR (2005) Novel approaches to the treatment of cocaine addiction. CNS Drugs 19:13–25

Song F, Freemantle N, Sheldon TA, House A, Watson P, Long A, Mason J (1993) Selective serotonin reuptake inhibitors: meta-analysis of efficacy and acceptability. BMJ 306: 683–687

Steffensen SC, Svingos AL, Pickel VM, Henriksen SJ (1998) Electrophysiological characterization of GABAergic neurons in the ventral tegmental area. J Neurosci 18:8003–8015

Sulzer D, Chen TK, Lau YY, Kristensen H, Rayport S, Ewing A (1995) Amphetamine redistributes dopamine from synaptic vesicles to the cytosol and promotes reverse transport. J Neurosci 15:4102–4108

Talvenheimo J, Rudnick G (1980) Solubilization of the platelet plasma membrane serotonin transporter in an active form. J Biol Chem 255:8606–8611

Tella SR, Ladenheim B, Andrews AM, Goldberg SR, Cadet JL (1996) Differential reinforcing effects of cocaine and GBR-12909: biochemical evidence for divergent neuroadaptive changes in the mesolimbic dopaminergic system. J Neurosci 16:7416–7427

Torres GE, Carneiro A, Seamans K, Fiorentini C, Sweeney A, Yao WD, Caron MG (2003b) Oligomerization and trafficking of the human dopamine transporter. Mutational analysis identifies critical domains important for the functional expression of the transporter. J Biol Chem 278:2731–2739

Torres GE, Gainetdinov RR, Caron MG (2003a) Plasma membrane monoamine transporters: structure, regulation and function. Nat Rev Neurosci 4:13–25

Vaughan RA, Huff RA, Uhl GR, Kuhar MJ (1997) Protein kinase C-mediated phosphorylation and functional regulation of dopamine transporters in striatal synaptosomes. J Biol Chem 272:15541–15546

Wade PR, Chen J, Jaffe B, Kassem IS, Blakely RD, Gershon MD (1996) Localization and function of a 5-HT transporter in crypt epithelia of the gastrointestinal tract. J Neurosci 16: 2352–2364

Waldman ID, Rowe DC, Abramowitz A, Kozel ST, Mohr JH, Sherman SL, Cleveland HH, Sanders ML, Gard JM, Stever C (1998) Association and linkage of the dopamine transporter gene and attention-deficit hyperactivity disorder in children: heterogeneity owing to diagnostic subtype and severity. Am J Hum Genet 63:1767–1776

Walsh SL, Preston KL, Sullivan JT, Fromme R, Bigelow GE (1994) Fluoxetine alters the effects of intravenous cocaine in humans. J Clin Psychopharmacol 14:396–407

Weyand S, Shimamura T, Yajima S, Suzuki S, Mirza O, Krusong K, Carpenter EP, Rutherford NG, Hadden JM, O'Reilly J, Ma P, Saidijam M, Patching SG, Hope RJ, Norbertczak HT, Roach PC, Iwata S, Henderson PJ, Cameron AD (2008) Structure and molecular mechanism of a nucleobase-cation-symport-1 family transporter. Science 322:709–713

White KJ, Kiser PD, Nichols DE, Barker EL (2006) Engineered zinc-binding sites confirm proximity and orientation of transmembrane helices I and III in the human serotonin transporter. Protein Sci 15:2411–2422

White KJ, Walline CC, Barker EL (2005) Serotonin transporters: implications for antidepressant drug development. AAPS J 7:E421–E433

Wilson JM, Nobrega JN, Corrigall WA, Coen KM, Shannak K, Kish SJ (1994) Amygdala dopamine levels are markedly elevated after self – but not passive-administration of cocaine. Brain Res 668:39–45

Xie Z, Miller GM (2007) Trace amine-associated receptor 1 is a modulator of the dopamine transporter. J Pharmacol Exp Ther 321:128–136

Xie Z, Westmoreland SV, Bahn ME, Chen GL, Yang H, Vallender EJ, Yao WD, Madras BK, Miller GM (2007) Rhesus monkey trace amine-associated receptor 1 signaling: enhancement by monoamine transporters and attenuation by the D2 autoreceptor in vitro. J Pharmacol Exp Ther 321:116–127

Xie Z, Westmoreland SV, Miller GM (2008) Modulation of monoamine transporters by common biogenic amines via trace amine-associated receptor 1 and monoamine autoreceptors in human embryonic kidney 293 cells and brain synaptosomes. J Pharmacol Exp Ther 325: 629–640

Yamashita A, Singh SK, Kawate T, Jin Y, Gouaux E (2005) Crystal structure of a bacterial homologue of Na+/Cl-dependent neurotransmitter transporters. Nature 437:215–223

Zanardi R, Benedetti F, Di Bella D, Catalano M, Smeraldi E (2000) Efficacy of paroxetine in depression is influenced by a functional polymorphism within the promoter of the serotonin transporter gene. J Clin Psychopharmacol 20:105–107

Zhang YW, Gesmonde J, Ramamoorthy S, Rudnick G (2007) Serotonin transporter phosphorylation by cGMP-dependent protein kinase is altered by a mutation associated with obsessive compulsive disorder. J Neurosci 27:10878–10886

Zhou Z, Zhen J, Karpowich NK, Goetz RM, Law CJ, Reith ME, Wang DN (2007) LeuT-desipramine structure reveals how antidepressants block neurotransmitter reuptake. Science 317:1390–1393

Zhu CB, Carneiro AM, Dostmann WR, Hewlett WA, Blakely RD (2005) p38 MAPK activation elevates serotonin transport activity via a trafficking-independent, protein phosphatase 2A-dependent process. J Biol Chem 280:15649–15658

Glutamate and Glutamine in Brain Disorders

Lasse K. Bak, Arne Schousboe, and Helle S. Waagepetersen

Abstract Diseases of the brain account for much human suffering and place a huge burden on the health care systems. Thus, research into the pathology of brain diseases and improved pharmacotherapy is of significant value. In this respect, knowledge on malfunctions of metabolic homeostasis related to the neurotransmission process is still limited. As evident from this chapter, failure of the metabolic homeostasis of the two amino acids of major importance, namely glutamate and glutamine, is a hallmark of a wide range of both neurological and psychiatric diseases. This chapter deals with representative brain diseases as well as the methodology of research related to metabolism. In addition, future need for research and potential new targets for pharmacotherapy are discussed.

Keywords Epilepsy · Glutamate–glutamine cycle · Ischemia · Metabolism · Neurodegenerative disorder · NMRS · Psychiatric disorder

Abbreviations

EAAT	Excitatory amino acid transporter
GABA	γ-Aminobutyric acid
GDH	Glutamate dehydrogenase
GS	Glutamine synthetase
MCAO	Middle cerebral artery occlusion
MS	Mass spectrometry
MSA	Multiple system atrophy
NMDA	*N*-methyl-D-aspartate
NMRS	Nuclear magnetic resonance spectroscopy
3-NPA	3-Nitropropionic acid
PAG	Phosphate-activated glutaminase

A. Schousboe (✉)
Faculty of Pharmaceutical Sciences, Department of Pharmacology and Pharmacotherapy, University of Copenhagen, DK-2100, Copenhagen, Denmark
e-mail: as@farma.ku.dk

J.P. Blass (ed.), *Neurochemical Mechanisms in Disease*,
Advances in Neurobiology 1, DOI 10.1007/978-1-4419-7104-3_7,

TCA Tricarboxylic acid
VGLUT Vesicular glutamate transporter

Contents

1 Introduction

Together, neurological and psychiatric disorders account for the vast majority of suffering from chronic illnesses (Cowan and Kandel, 2001). Thus, such disorders are important to understand at all levels, ranging from primary care of patients to the cellular and molecular levels. The focus of this chapter is on metabolic aspects of brain disorders related to malfunction of glutamate and glutamine homeostasis. The following sections deal with representative pathologies discussing selected recent as well as future need for research and potential targets for drug treatment.

Apart from being one of 20 amino acids used for protein synthesis, glutamate is the most abundant excitatory neurotransmitter in the mammalian brain. In addition, glutamate is the immediate precursor for γ-aminobutyric acid (GABA), the most abundant inhibitory neurotransmitter. Moreover, glutamate serves an important function in intermediary metabolism as donor of the amino group in transaminations of α-keto acids to form other α-amino acids. The cognate keto acid of glutamate, α-ketoglutarate, is a tricarboxylic acid (TCA) cycle intermediate and glutamate can be oxidized in the TCA cycle thereby acting as an energy substrate. Glutamine, produced from glutamate via the astrocytic glutamine synthetase (GS) reaction serves as a precursor for neuronal transmitter glutamate (see below). In addition, the GS reaction is considered the most important reaction for fixating ammonia in the brain playing a fundamental role for detoxifying ammonia in hyperammonemic conditions. Paradoxically, glutamate is a potent excitotoxin as well which means that tight homeostatic control of glutamate is of vital importance. This involves a number of both cytosolic and mitochondrial enzymes as well as transporters located in the plasma, vesicular, and mitochondrial membranes (see Waagepetersen et al., 2007).

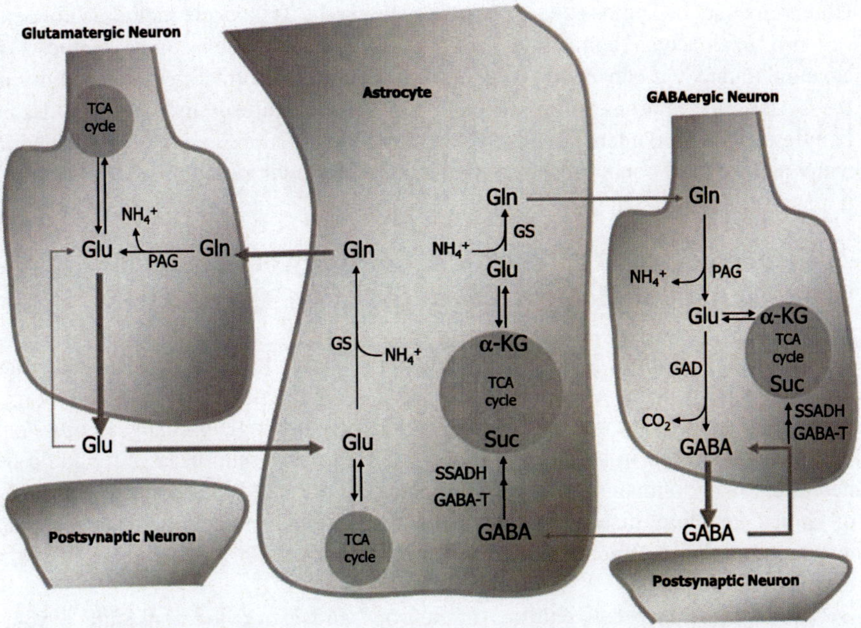

Fig. 1 Schematic representation of the glutamate/GABA-glutamine cycle. See text for details. The present dogma dictates that all (or most) of the released glutamate (Glu) is taken up into the glial compartment, whereas released GABA is primarily accumulated into the presynaptic neuron by neuronal reuptake (as indicated by the size of the arrows); this view may change in the future. Abbreviations: GABA, γ-aminobutyric acid; GABA-T, GABA transaminase; GAD, glutamate decarboxylase; Glu, glutamate; Gln, glutamine; GS, glutamine synthetase; α-KG, α-ketoglutarate; PAG, phosphate-activated glutaminase; SSADH, succinate semialdehyde dehydrogenase; TCA, tricarboxylic acid

For the following discussions, one vital component of glutamate homeostasis is particularly important to bring to mind, namely the glutamate–glutamine cycle. A more comprehensive discussion of brain glutamate and glutamine homeostasis is available elsewhere, therefore only a brief introduction is provided here (Bak et al., 2006; Waagepetersen et al., 2007). Based on discoveries of intercellular compartmentalization of glutamine and glutamate pools, related to astrocytes and neurons, respectively, a glutamate–glutamine cycle working between neurons and astrocytes was suggested more than three decades ago (van den Berg and Garfinkel, 1971; Benjamin and Quastel, 1972; Berl and Clarke, 1983; Ottersen et al., 1992). The cycle for a glutamatergic synapse is outlined in Fig. 1 (left part) in which released neurotransmitter is taken up into surrounding astrocytes, transformed into glutamine by the astrocyte-specific enzyme GS (Norenberg and Martinez-Hernandez, 1979) and released into the extracellular space from which it is taken up into neurons and transformed back to glutamate by phosphate-activated glutaminase (PAG; Kvamme et al., 2001). In the GABAergic synapse (Fig. 1; right part), GABA is taken up into astrocytes and catabolized to the TCA cycle intermediate succinate via the concerted action of GABA transaminase and succinate semialdehyde dehydrogenase.

Glutamine may be synthesized from succinate via the TCA cycle including condensation of oxaloacetate and acetyl-CoA forming citrate and subsequent synthesis of α-ketoglutarate and conversion to glutamate. Glutamate formed by PAG activity in the GABAergic neurons is the precursor for GABA catalyzed by glutamate decarboxylase. This fundamental neuronal–astrocytic interaction seems to be affected in many pathological conditions, as is shown in the following sections.

2 Methodological Approaches in the Study of Glutamate and Glutamine Homeostasis in the Brain

Animal models of brain disorders are typically generated by pharmacological treatment, surgical procedures, or more recently by techniques of molecular biology. Such treatments might give rise to symptoms or pathological changes consistent with those observed in humans suffering from a given brain disease. It should be noted that many human diseases are not naturally observed in rodents, the animals of choice for many disease models. Still, these animal models constitute a vital tool for studying malfunctions and potential treatments of a wide range of brain disorders.

Although primary cell cultures of neurons and astrocytes constitute a valuable tool for studying cell-specific metabolism, their use is limited in the context of studying brain pathology as they do not constitute a native biological system. For this reason, most work done using cell cultures has been focused on mimicking pathologies such as ischemia (hypoxia and glucose-deprivation), a condition fairly easy to create experimentally. However, pathological conditions with a more complex pathogenesis, such as epilepsy, have typically been studied in more intact systems, for example, in vivo or ex vivo studies on animal models or acute preparations of native brain tissue.

The techniques used to study metabolism in brain (pathology) include labeling of precursors with either radioactive (14C, 3H) or stable isotopes (13C, 15 N) and subsequently analyzing the incorporation of these isotopes into metabolites under different experimental conditions. The use of radio-labeled precursors is a classical technique still widely used; however, even though still a valid method, the amount of information gained using radio-labeled precursors is limited compared to employing stable isotopes in combination with more advanced analytical techniques such as nuclear magnetic resonance spectroscopy (NMRS) or mass spectrometry (MS). The major difference between NMRS and MS is that the NMRS experiment reveals the location of the labeled atoms within the molecule of interest whereas MS data provide information regarding only the number of labeled atoms in a given molecule. However, MS analysis is much more sensitive, faster, and inexpensive compared to NMRS.

No doubt the most valuable in vivo technique for studying metabolism is NMRS. Although in vivo NMRS is a powerful technique providing real-time data, it is hampered by high cost, problems with sensitivity, low resolution, the need for anesthetizing the animal (of course this is not necessary for human subjects), and lack of specific methods for differentiating between neuronal and glial compartments. Thus, in vivo NMRS relies on a number of assumptions as well as results from in vitro (cell culture) work combined with mathematical modeling to make sense of the data obtained. However, despite these shortcomings, refinement of this technique promises to bring about an increasing amount of valuable data in the coming decades.

To date, the best way to probe glial and neuronal metabolism is the use of labeled glucose and acetate. The underlying principle is that acetate is specifically taken up and metabolized by astrocytes because of specific uptake into this cellular compartment (Waniewski and Martin, 1998). It is then presumed that oxidative metabolism of glucose primarily takes place in neurons (Taylor et al., 1996; Qu et al., 2000). A number of studies take advantage of this by employing differentially labeled glucose and acetate. Thus, combined injection of [1-13C]glucose and [1,2-13C]acetate into an experimental animal produces different labeling patterns in metabolites, that is, mono-labeled from [1-13C]glucose (neuronal compartment) and double-labeled from [1,2-13C]acetate (astrocytic compartment; see Fig. 2 for detailed explanation).

3 Glutamate and Glutamine Homeostasis in Selected Brain Disorders

3.1 Epilepsy

More than 13 decades ago, an epileptic seizure was defined as an occasional, sudden, and excessive discharge of grey matter (Jackson, 1873). However, seizures are merely symptoms of underlying brain pathologies although in most cases no causative brain disorder can be identified and a purely descriptive diagnosis is made. Current drugs for treating epilepsy target either sodium channels or neurotransmitter metabolic enzymes, transporters, or receptors; although the mechanisms of action are not always clear, most drugs seem to either inhibit the glutamatergic (excitatory) system or potentiate the GABAergic (inhibitory) system. The concept of regarding epilepsy as an imbalance between excitation and inhibition seems to implicate that glutamatergic or GABAergic systems may be involved in the pathology. Indeed, many studies have revealed that the levels of several amino acids including glutamate, GABA, and glutamine are altered in epilepsy in animal models as well as in humans.

Biopsies taken from human subjects suffering from temporal lobe epilepsy revealed decreased glutamate–glutamine cycling in sclerotic hippocampal tissue, as evidenced by labeling in glutamate and glutamine from [2-13C]glucose infused prior to resection of the hippocampal tissue (Petroff et al., 2002). In fact, glutamate–glutamine cycling seems to be a mechanism commonly affected in epilepsy. Accordingly, inhibiting astrocytic GS (and thus glutamine transfer to neurons for transmitter glutamate synthesis) in hippocampal neuronal/astrocytic cocultures or hippocampal slices (in which epileptiform activity was induced by GABAA

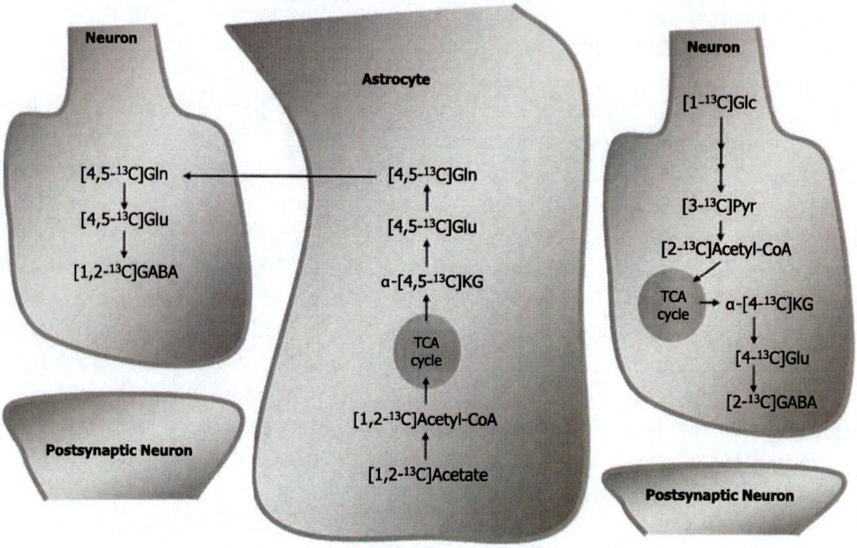

Fig. 2 Schematic representation of synapses showing how differentially labeled [1-13C]glucose ([1-13C]Glc) and [1,2-13C]acetate produces different labeling patterns in major metabolites. [1,2-13C]Acetate is specifically taken up into astrocytes (see text for references) and metabolized in the TCA cycle to α-[4,5-13C]ketoglutarate (α-[4,5-13C]KG) which is converted to [4,5-13C]glutamate ([4,5-13C]Glu) and [4,5-13C]glutamine ([4,5-13C]Gln) which is subsequently transferred to neurons. In neurons, [4,5-13C]glutamine is converted to [4,5-13C]glutamate and subsequently to [1,2-13C]GABA in GABAergic neurons. [1-13C]glucose is primarily metabolized in the neuronal compartment (see text for references). Glycolytic processing of [1-13C]glucose leads to [3-13C]pyruvate ([3-13C]Pyr) which is metabolized to α-[4-13C]ketoglutarate and subsequently to [4-13C]glutamate and [2-13C]GABA in GABAergic neurons. The depicted scheme is the simplest possible labeling patterns produced when α-[4,5-13C]ketoglutarate leaves the TCA cycle in the first turn. In addition, [4,5-13C]glutamate formed from [4,5-13C]glutamine in the neuronal compartment may to a large degree be converted to α-[4,5-13C]ketoglutarate and metabolized in the TCA cycle thus forming additional labeling patterns in glutamate and GABA. Such cycling of astrocyte-derived glutamine has been suggested to be substantial in GABAergic neurons (Waagepetersen et al., 1999)

receptor block) reduced spontaneous epileptiform spiking activity, indicating that the reduced flow of glutamine to the neurons reduced neuronal activity (Bacci et al., 2002). The same effect on spike activity was observed in cultured hippocampal neurons in which glutamine transport had been blocked (Bacci et al., 2002). Furthermore, intraperitoneal injection of leucine or its cognate keto acid, α-ketoisocaproate, augmented the occurrence of absence seizures in genetic absence epilepsy rats from Strasburg (GAERS; Dufour et al., 2001a, b). This was argued to be mediated by a decrease in the amount of glutamate available for neurotransmission, which may correlate to seizure activity in this animal model of nonconvulsive absence epilepsy (Danober et al., 1998). However, a more direct approach to study the metabolic disturbances involved in cortical and thalamic brain regions in the GAERS model was performed by combined injection of [1-13C]glucose and

[1,2-13C]acetate, showing increased glutamate–glutamine cycling in the cortex but not in the thalamus in conjunction with a decreased amount of cortical GABA (Melø et al., 2006). Hence, increased glutamatergic input from the cortex to the thalamus may affect thalamic filter function, thus playing a role in inducing absence seizures in the GAERS model.

In general, epilepsy animal models show metabolic disturbances of both neurons and astrocytes. However, it has been argued that the initial or primary change might take place in only one of these cell types (e.g., Sonnewald and Kondziella, 2003; Kondziella et al., 2003). A recent study in a rat model of lithium–pilocarpine-induced temporal lobe epilepsy showed the same extent of [1,2-13C]acetate metabolism in controls as in epileptic animals, implying that astrocytic metabolism is not compromised in these animals (Melø et al., 2005). However, glutamate labeling from [1-13C]glucose was reduced, suggesting that the metabolic malfunction in this epileptic model is in the neuronal compartment.

Although glutamate and glutamine metabolism is certainly affected in epileptic conditions, it does not seem clear whether this is what is causing the seizure activity or vice versa; basically a chicken and egg dilemma that may be best resolved by mapping the pathogenesis taking all initial and persistent changes into account. However, knowledge of metabolic malfunctions is very useful in the context of developing novel drugs for symptomatic treatment of epileptic disorders that target metabolic changes (see Section 4 for a further discussion on this matter).

3.2 Ischemic Conditions

Cerebral ischemia, (i.e., absolute or relative shortage of blood supply to a part of the brain such as following an ischemic stroke) is a serious condition that can lead to physical impairment or death. Stroke is a leading cause of death and adult disability in the industrialized part of the world (Thom et al., 2006). The outcome of a cerebral ischemic episode is greatly influenced by the duration, that is, the time from onset of ischemia until reperfusion; treatment is directed at the cause of the impaired blood supply, that is, thrombolysis or surgical intervention (e.g., see the review by Juttler et al., 2006).

The cellular and molecular events taking place during an ischemic episode have been studied extensively in animal models as well as in tissue preparations and cell cultures. The lack of oxygen and glucose causes a dramatic chain of events ultimately leading to cell death and necrosis of the affected tissue. It is generally thought that excitotoxic insults are key elements in the pathology, as neurons might succumb to excitotoxic mechanisms rather than energy deprivation per se, especially in the penumbral zone (Huang et al., 1997). One of the initial events in ischemia may be impairment or reversal of astrocytic glutamate uptake (Rossi et al., 2000), resulting in increased extracellular levels of glutamate and concomitant dysfunction of the glutamate–glutamine cycle. Interestingly, the eventual demise of astrocytes in ischemia might be due to lack of intracellular glutamate (caused by a reversal of the glutamate transporter) for synthesis of glutathione and associated oxidative damage, a mechanism possibly involving the lipoxygenase pathway (Re et al., 2006). However, it is clear from cell culture studies that astrocytes are much more resistant to deprivation of oxygen and glucose than neurons (e.g., Almeida et al., 2002). A widely employed rodent model of cerebral ischemia is middle cerebral artery occlusion (MCAO), a surgical procedure in which the middle cerebral artery is temporarily or permanently occluded resulting in infarction of only one hemisphere (Longa et al., 1989). As discussed below, the MCAO model or variations thereof have been employed in a number of studies. These studies indicate malfunctions in cellular metabolism and impairment of neuronal–astrocytic interactions both during the ischemic episode and following reperfusion.

After 1 h of reperfusion following 2 or 3 h of MCAO in conscious rats, regional as well as cellular differences were found with regard to incorporation of 14C into glutamate and glutamine from [14C]glucose or [14C]acetate (Thoren et al., 2005, 2006). It was found that labeling of glutamine from [14C]acetate was reduced compared to the nonaffected hemisphere in striatum, but not in focal or perifocal cortical tissue (Thoren et al., 2005). This signifies the regional diversity in susceptibility to ischemic insults apparently based on differences in properties of glial cells. In contrast, when [14C]glucose was employed, labeling in both glutamate and glutamine decreased in both striatum and cortex (Thoren et al., 2006). This indicates that neuronal glucose metabolism in both regions is significantly affected which is consistent with a reported decrease in 2-deoxy-glucose utilization in a similar model (Belayev et al., 1997). Surprisingly, for both brain regions the ATP: ADP ratio and phosphocreatine level were found to be maintained following reperfusion; however, increases in [14C]lactate was manifest throughout both regions (Thoren et al., 2006). The cellular origin of this lactate may be both neuronal and astrocytic, probably reflecting increased anaerobic glycolysis secondary to lack of O_2 for oxidative metabolism.

It should be noted that astrocytes contain the only supply of glycogen in the brain, although it is of limited size (McKenna et al., 2006). Glycogen may support anaerobic glycolysis in astrocytes during ischemia producing lactate which is released into the extracellular space; however, the functional role of brain glycogen during normal or ischemic conditions is not known in detail (McKenna et al., 2006). In this respect, lactate has been suggested to function as an (obligatory) energy substrate for neurons recovering from hypoxia or aglycemia (Fowler, 1993; Schurr et al., 1997a, b, c; Cater et al., 2001), although these observations have been argued to be an artifact of the preparation procedures of the experimental models employed (Okada and Lipton, 2007). In line with this, exogenous pyruvate may serve a neuroprotective role in the postischemic brain (e.g., Desagher et al., 1997) being primarily a neuronal substrate as evidenced in mice employing injections of [3-13C]pyruvate (Gonzalez et al., 2005); however, as noted by these authors, the clinical use may be hampered by the risk of seizures induced by high doses of exogenous pyruvate.

In a series of studies in rats by Håberg et al. (1998, 2001, 2006), different time periods of MCAO (ranging from 30 to 240 min) were studied ex vivo employing injection of [1-13C]glucose and [1,2-13C]acetate and subsequent analysis by NMRS. Already in the early stage of ischemia (30 min of MCAO) astrocytic metabolism was compromised in the ischemic core (lateral caudoputamen and lower parietal cortex) as evidenced by decreased synthesis of [4,5-13C]glutamine from [1,2-13C]acetate (Håberg et al., 2001). In addition, decreased neuronal formation of [4,5-13C]glutamate from the [4,5-13C]glutamine formed in astrocytes was evident, suggesting impaired transfer of glutamine from astrocytes to neurons (Håberg et al., 2001). Even though neurons can maintain glutamate synthesis throughout 240 min of MCAO, the total amount of glutamate in the affected tissue is decreased after an ischemic episode (Håberg et al., 2001), underlining the serious metabolic deficiencies occurring under such conditions. Interestingly, [1,2-13C]GABA was already absent from ischemic brain regions after 30 min of MCAO (Håberg et al., 2001). This may indicate that (i) at the GABAergic synapse, astrocyte-to-neuron trafficking of [4,5-13C]glutamine was completely abolished or (ii) that the [4,5-13C]glutamate derived from [4,5-13C]glutamine is metabolized in the neuronal TCA cycle rather than being used directly for synthesis of GABA.

As indicated above, primary astrocytic malfunction may cause subsequent harm to the neurons but certainly astrocytes play an important role for neuronal survival following reperfusion as well. In rats, after 120 min of MCAO followed by 120 min of reperfusion interesting metabolic differences were found between the ischemic core (lateral caudoputamen and lower parietal cortex) and the penumbral zone (frontoparietal cortex; Håberg et al., 2006). Following reperfusion, astrocytic metabolism was significantly improved in the penumbral zone, as evidenced by increased metabolism of [1,2-13C]acetate whereas in the ischemic core the situation was the reverse, showing decreased [1,2-13C]acetate metabolism compared to 120 min of MCAO alone (Håberg et al., 2006); this was most likely caused by decreased activity of acetyl-CoA synthetase, catalyzing the conversion of [1,2-13C]acetate to [1,2-13C]acetyl-CoA. In the penumbra, label from [1-13C]glucose into [4-13C]glutamate was not affected by 120 min reperfusion compared to after 120 min of MCAO alone, which was in contrast to the ischemic core where almost no [4-13C]glutamate was present at this point (Håberg et al., 2006). This suggests that after 120 min of MCAO, neuronal metabolism in the ischemic core is not rescued by 120 min of reperfusion whereas neuronal integrity is preserved in the penumbral zone.

The MCAO animal model seems to be a robust system in which to investigate the metabolic changes occurring in ischemia. Not surprising, rather severe dysfunctional homeostasis of glutamate and glutamine metabolism seem to be present in ischemia, including interference with neuronal–astrocytic interactions. However, the severity of these dysfunctions seemed to vary among brain regions, possibly due to differential properties of the local glial cells.

3.3 Neurodegenerative Disorders

Neurodegeneration is a progressive, fatal deterioration of neuronal function, playing a part in many brain pathologies including Alzheimer's, Huntington's, and Parkinson's diseases and olivopontocerebellar atrophy, which are commonly referred to as neurodegenerative diseases. Olivopontocerebellar atrophy and Huntington's disease are part of the group of trinucleotide repeat or polyglutamine disorders, featuring excessive repeats of genomic CAG sequences on different chromosomes affecting gene function. Generally, only few studies have been conducted on glutamate and glutamine metabolism in neurodegenerative disorders. Some studies have determined levels of amino acids in blood and brain or cerebrospinal fluid (CSF) and found some differences; however, levels of metabolites only offer limited information on dysfunctional metabolism. Interestingly, there seems to be an increased interest in employing NMRS in studying these pathologies and representative studies are discussed below.

Glutamate metabolism and glutamate–glutamine cycling might be compromised in Alzheimer's disease, inasmuch as aberrant expression of glutamate metabolizing enzymes including glutamate dehydrogenase (GDH), GS, and PAG has been reported (Robinson, 2000, 2001; Burbaeva et al., 2005). In addition, a study employing in vivo NMR spectroscopy suggested decreased glutamatergic neurotransmission activity and TCA cycling rate in patients suffering from Alzheimer's disease, as suggested by labeling patterns in glutamate and glutamine from infused [1-13C]glucose (Lin et al., 2003).

Neurodegeneration related to GDH malfunction and subsequent glutamate toxicity may constitute part of the pathology of some forms of multiple system atrophies (MSAs) known as olivopontocerebellar atrophy or spinocerebellar ataxia affecting the cerebellum, pons, and inferior olives (review, Plaitakis et al., 1993); symptoms include Parkinsonism, ataxia, and autonomic dysfunction. GDH deficiencies were found in patients suffering from cerebellar degeneration (Plaitakis et al., 1979, 1980) and abnormal levels of brain glutamate have been observed in patients suffering from MSA (Perry et al., 1981). In addition, brain TCA cycle metabolism may also be compromised as the activity of α-ketoglutarate dehydrogenase, the rate-limiting enzyme in the TCA cycle, seems to be decreased in postmortem biopsies from MSA patients (Mastrogiacomo and Kish, 1994). Apparently, no studies employing stable isotope-labeled substrates and NMRS or MS seem to have been performed in either MSA patients or any animal models of cerebellar degeneration. One problem may be that no valid animal model is available that displays the right combination of pathological changes compared to patients suffering from GDH-deficient forms of MSA (Miret-Duvaux et al., 1990), although systemic administration to rodents of 3-acetylpyridine has been argued to produce lesions corresponding to those observed in patients (see the review by Plaitakis et al., 1993). The elucidation of metabolic defects in MSAs would be of significant value for understanding these debilitating neurodegenerative diseases.

The levels of glutamine and glutamate were found to increase and decrease, respectively, in CSF of patients suffering from Parkinson's disease (Mally et al., 1997), however, other studies employing in vivo NMRS found that the [glutamate+glutamine]/[creatine] ratio in the basal ganglia and striatum of Parkinson's disease patients was no different from healthy control subjects (Clarke et al., 1997; Taylor-Robinson et al., 1999). On the other hand, a study employing 1-methyl-4-phenyl-1,2,3,6-tetrahydropyridine (MPTP; a drug found to induce striatal lesions of dopaminergic pathways and cause Parkinsonism) treatment of cats and NMRS analysis found lower [glutamate+glutamine]/[creatine] ratio than in control animals (Podell et al., 2003).

Chronic treatment with the mitochondrial toxin 3-nitropropionic acid (3-NPA; inhibits succinate dehydrogenase, a TCA cycle enzyme) induces lesions in the striatum of animals leading to symptoms similar to Huntington's disease in humans (Brouillet et al., 1998). Initial, acute intoxication with 3-NPA in mice selectively targeted GABAergic neurons inhibiting their TCA cycle, whereas glutamatergic pathways, glial cells, and glutamate–glutamine cycle function were unaffected (Hassel and Sonnewald, 1995). However, this selective inhibition of (GABAergic) neuronal function in the striatum is still somewhat of a mystery, as 3-NPA affects other brain regions as well (Brouillet et al., 1998) showing a general 18% decrease in TCA cycle flux in one in vivo NMRS study of rat brain (Henry et al., 2002).

In conclusion, glutamate and glutamine homeostasis seem to play a role in neurodegenerative pathologies, although the picture is not consistent. Much more research is needed to elucidate the potential role of malfunction of glutamate and glutamine homeostasis in neurodegenerative disorders. Especially, ex and in vivo NMRS employing animal models as well as patients should be employed to this end, as such approaches provide the most precise information.

3.4 Psychiatric Disorders

The psychiatric disorders represent a duality of suffering for the patients, as they not only have to combat the disease but also have to endure the stigmatization of being mentally ill. Thus, not only is there a need for effective pharmacotherapy and research into the pathogenesis, there is also a need for changing the public view on these disorders.

There is increasing evidence that glutamate/glutamine homeostasis is disturbed in a number of psychiatric disorders. Thus, in the anterior cingulate cortex, GDH and GS were found to be expressed to a lower extent in patients suffering from bipolar disorder and major depressive disorder, respectively (Beasley et al., 2006). Conversely, it was found that transcripts (i.e., mRNA) for PAG and GS are present to a higher extent in the thalamus of patients suffering from schizophrenia and, in addition, increased transcripts for the glial glutamate transporters (EAAT1, EAAT2) as well as a vesicular glutamate transporter (VGLUT2) have been reported (Smith et al., 2001a, b). Furthermore, one study suggests that EAAT2 protein may actually be increased in prefrontal cortex of schizophrenic patients and that atypical antipsychotic pharmacotherapy (e.g., clozapine) may normalize the expression of EAAT2 protein (Matute et al., 2005). In this respect, it should be noted that such "hyperactive" astrocytic glutamate transport may constitute a target for novel pharmacotherapeutic approaches, as discussed by Nanitsos et al. (2005). As always, mRNA levels may not reflect actual formation of functional protein and, moreover, when using patients there is always the risk that treated and nontreated patients as well as the stage of the disease may affect the results. The last point is illustrated by the study of Ohrmann et al. (2005) in which only chronic but not first-episode patients showed significantly lower glutamate/glutamine levels in the prefrontal cortex. Interestingly, another study published the same year by another group did not find such differences and, quite disturbingly, they found the opposite effect on the glutamate and glutamine levels, that is, increased concentrations in both prefrontal cortex and hippocampus (van Elst et al., 2005). This may illustrate the complexity of such diseases, and one way to investigate this under more controlled conditions is to turn to animal models.

In general, animal models of psychiatric disease are somewhat of a conundrum, as symptoms are usually subjective; for example, in schizophrenia, symptoms consist of so-called positive (thought disorder; delusions) and negative symptoms (emotional disorder). One potential high-quality animal model of schizophrenia producing both negative and positive symptoms is based on the glutamate hypofunction hypothesis and may be induced by treating rats with the noncompetitive, activity-dependent *N*-methyl-D-aspartate (NMDA) receptor blocker MK-801 (or dizocilpine; the systematic IUPAC name is [5R,10S]-[+]-5-methyl-10,11-dihydro-5H-dibenzo[a,d]cyclohepten-5,10-imine; Carlsson et al., 2001). MK-801 was originally being developed by the drug company Merck & Co. as a neuroprotective agent but was discontinued because of the above ability to induce cellular lesions, eventually leading to schizophrenic symptoms (Olney et al., 1989). Eyjolfsson et al. (2006) argues, that repeated "low" doses (0.1 mg/kg) of MK-801 mimics some of the typical behavioral changes observed in schizophrenic patients (e.g., hyperlocomotion) whereas neurochemical changes consistent with observations in so-called first-episode patients (Kondziella et al., 2006) were only present after repeated "high" doses (0.5 mg/kg).

A single dose (0.5 mg/kg) of MK-801 given to rats that were subsequently injected with [1-13C]glucose and [1,2-13C]acetate in combination, caused an increase in total glutamine content as well as in [4-13C]glutamine synthesized from [1-13C]glucose (Brenner et al., 2005); the mechanism may involve decreased NO-mediated inhibition of GS (Kosenko et al., 1995, 2003). In the study by Brenner et al. (2005), no concomitant increase in [4-13C]glutamate was observed which indicates that the glutamate–glutamine cycle may be affected in the neuron-to-astrocyte direction; however, labeling from [1,2-13C]acetate in glutamate, glutamine, and GABA were not altered, suggesting that the flow of glutamine from astrocytes to neurons was not influenced by a single injection of MK-801. In contrast, another study by the same group employing repetitive injections of MK-801 (0.5 mg/kg; rats; every day for 6 days) showed decreased synthesis of [4,5-13C]glutamate and [4,5-13C]glutamine from [1,2-13C]acetate in the prefrontal cortex (Kondziella et al., 2006) implying that the glutamate–glutamine cycle is impaired in the astrocyte-to-neuron direction in this model. As suggested by Kondziella et al. (2006) as well as Eyjolfsson et al. (2006), repeated injections rather than a single injection of MK-801 may be a better animal model, as the neurochemical and behavioral changes are more in keeping with the changes observed in patients.

In conclusion, psychiatric disorders seem to involve a significant component of disruption of glutamate/glutamine homeostasis. In addition, a high-quality animal model of schizophrenia based on repetitive injections of MK-801 has been established and might prove to be of significant value in future research.

4 Potential Drug Targets Related to Glutamate and Glutamine Homeostasis

Clearly, interfering with glutamate/glutamine homeostasis seems an attractive goal to pursue for symptomatic treatment of a number of brain disorders. In the treatment of epilepsy, drugs targeting GABA transporters (tiagabin) and GABA-transaminase (vigabatrin) have been on the market for a number of years, providing proof of principle for the neurotransmitter cycling systems as pharmacological targets (Sarup et al., 2003). However, with regard to glutamate homeostasis no such drugs have been marketed, although one effect of atypical antipsychotic pharmacotherapy may be a reduction in astrocytic EAAT2 protein (Matute et al., 2005) and it has been suggested that a novel approach in treating psychotic disease should be directed specifically at regulating astrocytic glutamate transport "hyperfunction" (Nanitsos et al., 2005).

One concern with targeting the glutamatergic neurotransmitter system might be that glutamatergic synapses are so abundant and that glutamate is an important metabolite in intermediary metabolism, making interference with glutamate homeostasis a potential nightmare with regard to adverse effects. Thus, most drug development directed at the glutamatergic system seems to have been focused on ionotropic glutamate receptors as pharmacological targets, although G-protein coupled receptors have been attracting increased attention. Recent work by Rae et al. (2005) shows that metabotropic glutamate receptors might constitute attractive drug targets for regulating the metabolism associated with the glutamate–glutamine cycle. Agonists and antagonists of metabotropic glutamate receptors of groups I and II (coupled to the phosphoinositide/Ca2+ and the cyclic AMP second messenger systems, respectively) affected TCA cycle activity as well as the glutamate–glutamine cycling rate in Guinea pig slices. It was observed that group I agonists/antagonists affected TCA cycle activity, whereas group II agonists/antagonists influenced the glutamate–glutamine cycling rate.

Thus, metabolic homeostasis of the glutamate/glutamine system may be a fertile avenue to pursue in order to identify novel targets for pharmacotherapy directed at a number of diseases of the brain.

5 Concluding Remarks

It seems clear from this discussion that metabolic changes related to glutamate and glutamine homeostasis are an integral part of many pathologies of the brain and that interfering with the glutamate/glutamine system may be an attractive target for drug treatment. However, much work still needs to be performed in order to elucidate the many aspects of metabolic dysfunction in neurological and psychiatric diseases. Especially, increased use of in vivo MR spectroscopy will be important in this respect to elucidate the complex metabolic changes in patients and animal models of brain disease. However, although employing 13C-labeling and spectroscopy techniques may aid in mapping metabolic dysfunctions, the exact mechanisms underlying these disturbances need to be addressed as well. The best way to bring this about is to employ a different set of techniques including proteomics methodology to determine changes in the related proteome. Furthermore, refinement of existing, and development of new, animal models of brain diseases are an important issue. This may become easier as knowledge of potential genetic and/or developmental causes of brain disease becomes available, making genetically modified animals the model of choice for these investigations. No doubt, the future will bring an increased appreciation of the metabolic disturbances involved leading to an improved understanding of the pathology of brain disorders.

References

Almeida A, Delgado-Esteban M, Bolanos JP, Medina JM (2002) Oxygen and glucose deprivation induces mitochondrial dysfunction and oxidative stress in neurones but not in astrocytes in primary culture. J Neurochem 81:207–217

Bacci A, Sancini G, Verderio C, Armano S, Pravettoni E, Fesce R, Franceschetti S, Matteoli M (2002) Block of glutamate-glutamine cycle between astrocytes and neurons inhibits epileptiform activity in hippocampus. J Neurophysiol 88:2302–2310

Bak LK, Schousboe A, Waagepetersen HS (2006) The glutamate/GABA-glutamine cycle: aspects of transport, neurotransmitter homeostasis and ammonia transfer. J Neurochem 98: 641–653

Beasley CL, Pennington K, Behan A, Wait R, Dunn MJ, Cotter D (2006) Proteomic analysis of the anterior cingulate cortex in the major psychiatric disorders: evidence for disease-associated changes. Proteomics 6:3414–3425

Belayev L, Zhao W, Busto R, Ginsberg MD (1997) Transient middle cerebral artery occlusion by intraluminal suture: I. Three-dimensional autoradiographic image-analysis of local cerebral glucose metabolism-blood flow interrelationships during ischemia and early recirculation. J Cereb Blood Flow Metab 17:1266–1280

Benjamin AM, Quastel JH (1972) Locations of amino acids in brain slices from the rat. Tetrodotoxin-sensitive release of amino acids. Biochem J 128:631–646

Berl S, Clarke DD (1983) The metabolic compartmentation concept. In: Hertz L, Kvamme E, McGeer EG, Schousboe A (eds) Glutamine, glutamate and GABA in the central nervous system. Liss, New York, NY, pp 205–217

Brenner E, Kondziella D, Håberg A, Sonnewald U (2005) Impaired glutamine metabolism in NMDA receptor hypofunction induced by MK801. J Neurochem 94:1594–1603

Brouillet E, Guyot MC, Mittoux V, Altairac S, Conde F, Palfi S, Hantraye P (1998) Partial inhibition of brain succinate dehydrogenase by 3-nitropropionic acid is sufficient to initiate striatal degeneration in rat. J Neurochem 70:794–805

Burbaeva GS, Boksha IS, Tereshkina EB, Savushkina OK, Starodubtseva LI, Turishcheva MS (2005) Glutamate metabolizing enzymes in prefrontal cortex of Alzheimer's disease patients. Neurochem Res 30:1443–1451

Carlsson A, Waters N, Holm-Waters S, Tedroff J, Nilsson M, Carlsson ML (2001) Interactions between monoamines, glutamate, and GABA in schizophrenia: new evidence. Annu Rev Pharmacol Toxicol 41:237–260

Cater HL, Benham CD, Sundstrom LE (2001) Neuroprotective role of monocarboxylate transport during glucose deprivation in slice cultures of rat hippocampus. J Physiol 531:459–466

Clarke CE, Lowry M, Horsman A (1997) Unchanged basal ganglia N-acetylaspartate and glutamate in idiopathic Parkinson's disease measured by proton magnetic resonance spectroscopy. Mov Disord 12:297–301

Cowan WM, Kandel ER (2001) Prospects for neurology and psychiatry. JAMA 285:594–600

Danober L, Deransart C, Depaulis A, Vergnes M, Marescaux C (1998) Pathophysiological mechanisms of genetic absence epilepsy in the rat. Prog Neurobiol 55:27–57

Desagher S, Glowinski J, Premont J (1997) Pyruvate protects neurons against hydrogen peroxide-induced toxicity. J Neurosci 17:9060–9067

Dufour F, Nalecz KA, Nalecz MJ, Nehlig A (2001a) Metabolic approach of absence seizures in a genetic model of absence epilepsy, the GAERS: study of the leucine-glutamate cycle. J Neurosci Res 66:923–930

Dufour F, Nalecz KA, Nalecz MJ, Nehlig A (2001b) Modulation of absence seizures by branched-chain amino acids: correlation with brain amino acid concentrations. Neurosci Res 40:255–263

Eyjolfsson EM, Brenner E, Kondziella D, Sonnewald U (2006) Repeated injection of MK801: an animal model of schizophrenia? Neurochem Int 48:541–546

Fowler JC (1993) Glucose deprivation results in a lactate preventable increase in adenosine and depression of synaptic transmission in rat hippocampal slices. J Neurochem 60:572–576

Gonzalez SV, Nguyen NH, Rise F, Hassel B (2005) Brain metabolism of exogenous pyruvate. J Neurochem 95:284–293

Hassel B, Sonnewald U (1995) Selective inhibition of the tricarboxylic acid cycle of GABAergic neurons with 3-nitropropionic acid in vivo. J Neurochem 65:1184–1191

Henry PG, Lebon V, Vaufrey F, Brouillet E, Hantraye P, Bloch G (2002) Decreased TCA cycle rate in the rat brain after acute 3-NPA treatment measured by in vivo 1H-[13C] NMR spectroscopy. J Neurochem 82:857–866

Huang R, Sochocka E, Hertz L (1997) Cell culture studies of the role of elevated extracellular glutamate and K+ in neuronal cell death during and after anoxia/ischemia. Neurosci Biobehav Rev 21:129–134

Håberg A, Qu H, Haraldseth O, Unsgard G, Sonnewald U (1998) In vivo injection of [1-13C]glucose and [1,2-13C]acetate combined with ex vivo 13C nuclear magnetic resonance spectroscopy: a novel approach to the study of middle cerebral artery occlusion in the rat. J Cereb Blood Flow Metab 18:1223–1232

Håberg A, Qu H, Saether O, Unsgard G, Haraldseth O, Sonnewald U (2001) Differences in neurotransmitter synthesis and intermediary metabolism between glutamatergic and GABAergic neurons during 4 hours of middle cerebral artery occlusion in the rat: the role of astrocytes in neuronal survival. J Cereb Blood Flow Metab 21:1451–1463

Håberg A, Qu H, Sonnewald U (2006) Glutamate and GABA metabolism in transient and permanent middle cerebral artery occlusion in rat: importance of astrocytes for neuronal survival. Neurochem Int 48:531–540

Jackson JH (1873) On the anatomical, physiological and pathological investigation of epilepsies. West Riding Lunatic Asylum Med Rep 3:315–339

Juttler E, Kohrmann M, Schellinger PD (2006) Therapy for early reperfusion after stroke. Nat Clin Pract Cardiovasc Med 3:656–663

Kondziella D, Brenner E, Eyjolfsson EM, Markinhuhta KR, Carlsson ML, Sonnewald U (2006) Glial-neuronal interactions are impaired in the schizophrenia model of repeated MK801 exposure. Neuropsychopharmacology 31:1880–1887

Kondziella D, Hammer J, Sletvold O, Sonnewald U (2003) The pentylenetetrazole-kindling model of epilepsy in SAMP8 mice: glial-neuronal metabolic interactions. Neurochem Int 43:629–637

Kosenko E, Kaminsky Y, Grau E, Minana MD, Grisolia S, Felipo V (1995) Nitroarginine, an inhibitor of nitric oxide synthetase, attenuates ammonia toxicity and ammonia-induced alterations in brain metabolism. Neurochem Res 20:451–456

Kosenko E, Llansola M, Montoliu C, Monfort P, Rodrigo R, Hernandez-Viadel M, Erceg S, Sanchez-Perez AM, Felipo V (2003) Glutamine synthetase activity and glutamine content in brain: modulation by NMDA receptors and nitric oxide. Neurochem Int 43:493–499

Kvamme E, Torgner IA, Roberg B (2001) Kinetics and localization of brain phosphate activated glutaminase. J Neurosci Res 66:951–958
Lin AP, Shic F, Enriquez C, Ross BD (2003) Reduced glutamate neurotransmission in patients with Alzheimer's disease – an in vivo 13C magnetic resonance spectroscopy study. MAGMA 16:29–42
Longa EZ, Weinstein PR, Carlson S, Cummins R (1989) Reversible middle cerebral artery occlusion without craniectomy in rats. Stroke 20:84–91
Mally J, Szalai G, Stone TW (1997) Changes in the concentration of amino acids in serum and cerebrospinal fluid of patients with Parkinson's disease. J Neurol Sci 151:159–162
Mastrogiacomo F, Kish SJ (1994) Cerebellar Alpha-Ketoglutarate Dehydrogenase-Activity Is Reduced in Spinocerebellar Ataxia Type-1. Ann Neurol 35:624–626
Matute C, Meløne M, Vallejo-Illarramendi A, Conti F (2005) Increased expression of the astrocytic glutamate transporter GLT-1 in the prefrontal cortex of schizophrenics. Glia 49: 451–455
McKenna MC, Gruetter R, Sonnewald U, Waagepetersen HS, Schousboe A (2006) Energy metabolism of the brain. In: Siegel GJ, Albers RW, Brady ST, Price DL (eds) Basic neurochemistry, 7th edn. Elsevier, Amsterdam, pp 531–557
Melø TM, Nehlig A, Sonnewald U (2005) Metabolism is normal in astrocytes in chronically epileptic rats: a 13C NMR study of neuronal-glial interactions in a model of temporal lobe epilepsy. J Cereb Blood Flow Metab 25:1254–1264
Melø TM, Sonnewald U, Touret M, Nehlig A (2006) Cortical glutamate metabolism is enhanced in a genetic model of absence epilepsy. J Cereb Blood Flow Metab 26:1496–1506
Miret-Duvaux O, Frederic F, Simon D, Guenet JL, Hanauer A, Delhaye-Bouchaud N, Mariani J (1990) Glutamate dehydrogenase in cerebellar mutant mice: gene localization and enzyme activity in different tissues. J Neurochem 54:23–29
Nanitsos EK, Nguyen KT, St'astny F, Balcar VJ (2005) Glutamatergic hypothesis of schizophrenia: involvement of Na+/K+-dependent glutamate transport. J Biomed Sci 12:975–984
Norenberg MD, Martinez-Hernandez A (1979) Fine structural localization of glutamine synthetase in astrocytes of rat brain. Brain Res 161:303–310
Ohrmann P, Siegmund A, Suslow T, Spitzberg K, Kersting A, Arolt V, Heindel W, Pfleiderer B (2005) Evidence for glutamatergic neuronal dysfunction in the prefrontal cortex in chronic but not in first-episode patients with schizophrenia: a proton magnetic resonance spectroscopy study. Schizophr Res 73:153–157
Okada Y, Lipton P (2007) Glucose, oxidative energy metabolism, and neuronal function in brain slices-glycolysis plays a key role in neural activity. In: Gibson GE, Dienel G (eds) Handbook of neurochemistry and molecular neurobiology Springer-Verlag, Heidelburg, Germany, vol 5. pp 17–39
Olney JW, Labruyere J, Price MT (1989) Pathological changes induced in cerebrocortical neurons by phencyclidine and related drugs. Science 244:1360–1362
Ottersen OP, Zhang N, Walberg F (1992) Metabolic compartmentation of glutamate and glutamine: morphological evidence obtained by quantitative immunocytochemistry in rat cerebellum. Neuroscience 46:519–534
Perry TL, Kish SJ, Hansen S, Currier RD (1981) Neurotransmitter amino acids in dominantly inherited cerebellar disorders. Neurology 31:237–242
Petroff OA, Errante LD, Rothman DL, Kim JH, Spencer DD (2002) Glutamate-glutamine cycling in the epileptic human hippocampus. Epilepsia 43:703–710
Plaitakis A, Flessas P, Natsiou AB, Shashidharan P (1993) Glutamate dehydrogenase deficiency in cerebellar degenerations: clinical, biochemical and molecular genetic aspects. Can J Neurol Sci 20(Suppl 3):S109–S116
Plaitakis A, Nicklas WJ, Desnick RJ (1979) Glutamate dehydrogenase deficiency in three patients with spinocerebellar ataxia: a new enzymatic defect? Trans Am Neurol Assoc 104:54–57
Plaitakis A, Nicklas WJ, Desnick RJ (1980) Glutamate dehydrogenase deficiency in three patients with spinocerebellar syndrome. Ann Neurol 7:297–303

Podell M, Hadjiconstantinou M, Smith MA, Neff NH (2003) Proton magnetic resonance imaging and spectroscopy identify metabolic changes in the striatum in the MPTP feline model of parkinsonism. Exp Neurol 179:159–166

Qu H, Håberg A, Haraldseth O, Unsgard G, Sonnewald U (2000) 13C MR spectroscopy study of lactate as substrate for rat brain. Dev Neurosci 22:429–436

Rae C, Moussa C, Griffin JL, Bubb WA, Wallis T, Balcar VJ (2005) Group I and II metabotropic glutamate receptors alter brain cortical metabolic and glutamate/glutamine cycle activity: a 13C NMR spectroscopy and metabolomic study. J Neurochem 92:405–416

Re DB, Nafia I, Meløn C, Shimamoto K, Goff LK, Had-Aissouni L (2006) Glutamate leakage from a compartmentalized intracellular metabolic pool and activation of the lipoxygenase pathway mediate oxidative astrocyte death by reversed glutamate transport. Glia 54:47–57

Robinson SR (2000) Neuronal expression of glutamine synthetase in Alzheimer's disease indicates a profound impairment of metabolic interactions with astrocytes. Neurochem Int 36:471–482

Robinson SR (2001) Changes in the cellular distribution of glutamine synthetase in Alzheimer's disease. J Neurosci Res 66:972–980

Rossi DJ, Oshima T, Attwell D (2000) Glutamate release in severe brain ischaemia is mainly by reversed uptake. Nature 403:316–321

Sarup A, Larsson OM, Schousboe A (2003) GABA transporters and GABA-transaminase as drug targets. Curr Drug Targets CNS Neurol Disord 2:269–277

Schurr A, Payne RS, Miller JJ, Rigor BM (1997a) Brain lactate is an obligatory aerobic energy substrate for functional recovery after hypoxia: further in vitro validation. J Neurochem 69:423–426

Schurr A, Payne RS, Miller JJ, Rigor BM (1997b) Brain lactate, not glucose, fuels the recovery of synaptic function from hypoxia upon reoxygenation: an in vitro study. Brain Res 744:105–111

Schurr A, Payne RS, Miller JJ, Rigor BM (1997c) Glia are the main source of lactate utilized by neurons for recovery of function posthypoxia. Brain Res 774:221–224

Smith RE, Haroutunian V, Davis KL, Meador-Woodruff JH (2001a) Expression of excitatory amino acid transporter transcripts in the thalamus of subjects with schizophrenia. Am J Psychiatry 158:1393–1399

Smith RE, Haroutunian V, Davis KL, Meador-Woodruff JH (2001b) Vesicular glutamate transporter transcript expression in the thalamus in schizophrenia. Neuroreport 12:2885–2887

Sonnewald U, Kondziella D (2003) Neuronal glial interaction in different neurological diseases studied by ex vivo 13C NMR spectroscopy. NMR Biomed 16:424–429

Taylor A, McLean M, Morris P, Bachelard H (1996) Approaches to studies on neuronal/glial relationships by 13C-NMRS analysis. Dev Neurosci 18:434–442

Taylor-Robinson SD, Turjanski N, Bhattacharya S, Seery JP, Sargentoni J, Brooks DJ, Bryant DJ, Cox IJ (1999) A proton magnetic resonance spectroscopy study of the striatum and cerebral cortex in Parkinson's disease. Metab Brain Dis 14:45–55

Thom T, Haase N, Rosamond W, Howard VJ, Rumsfeld J, Manolio T, Zheng ZJ, Flegal K, O'Donnell C, Kittner S, Lloyd-Jones D, Goff DC Jr., Hong Y, Adams R, Friday G, Furie K, Gorelick P, Kissela B, Marler J, Meigs J, Roger V, Sidney S, Sorlie P, Steinberger J, Wasserthiel-Smoller S, Wilson M, Wolf P (2006) Heart disease and stroke statistics – 2006 update: a report from the American Heart Association Statistics Committee and Stroke Statistics Subcommittee. Circulation 113:e85–e151

Thoren AE, Helps SC, Nilsson M, Sims NR (2005) Astrocytic function assessed from 1-14C-acetate metabolism after temporary focal cerebral ischemia in rats. J Cereb Blood Flow Metab 25:440–450

Thoren AE, Helps SC, Nilsson M, Sims NR (2006) The metabolism of 14C-glucose by neurons and astrocytes in brain subregions following focal cerebral ischemia in rats. J Neurochem 97: 968–978

Waagepetersen HS, Schousboe A, Sonnewald U (2007) Glutamate, glutamine and GABA: metabolic aspects. In: Oja SS, Schousboe A, Saransaari P (eds) Handbook of neurochemistry and molecular neurobiology Springer-Verlag, Heidelburg, Germany, vol 6. pp 1–21

Waagepetersen HS, Sonnewald U, Larsson OM, Schousboe A (1999) Synthesis of vesicular GABA from glutamine involves TCA cycle metabolism in neocortical neurons. J Neurosci Res 57: 342–349

Waniewski RA, Martin DL (1998) Preferential utilization of acetate by astrocytes is attributable to transport. J Neurosci 18:5225–5233

van den Berg CJ, Garfinkel D (1971) A simulation study of brain compartments. Metabolism of glutamate and related substances in mouse brain. Biochem J 123:211–218

van Elst LT, Valerius G, Buchert M, Thiel T, Rusch N, Bubl E, Hennig J, Ebert D, Olbrich HM (2005) Increased prefrontal and hippocampal glutamate concentration in schizophrenia: evidence from a magnetic resonance spectroscopy study. Biol Psychiatry 58:724–730

Rho-Linked Mental Retardation Genes

Nael Nadif Kasri and Linda Van Aelst

Abstract Mental retardation (MR) is generally defined as a global reduction in cognitive abilities, which manifests before the age of 18. The causes of MR are extremely heterogeneous, including environmental factors as well as genetic changes, such as chromosomal abnormalities and single-gene mutations. Great progress has been made in recent years towards the identification of MR genes, particularly X-linked MR genes. A largely remaining challenge, however, is to connect the genetic causes of MR to processes that establish and/or modify neuronal circuit function. Several of the currently identified genes are associated with MR code for regulators and effectors of the Rho subfamily of GTP-binding proteins, which are key regulators of the actin cytoskeleton. The identification and characterization of Rho-linked genes associated with different forms of MR have shed light on our current understanding as to how defective cellular signaling can result in abnormal neuronal connectivity, which can give rise to impaired information processing underlying cognitive function. Aberrations in defined Rho-mediated signaling pathways have been linked to defects in the formation and remodeling of dendritic spines and/or the maturation and activity-dependent modification of the efficacy of synapses. In this review, we focus on the role of Rho GTPases and their associated signaling molecules in the control of spine structure and synaptic function, and highlight their involvement in MR resulting from a variety of genetic mutations within regulators and effectors of these molecules.

Keywords Mental retardation (MR) · Dentritic spines · Synaptic structure and function · Actin cytoskeleton · Rho GTPases · Rho-linked MR genes · Nonsyndromic and syndromic X-linked MR · Autosomal syndromic MR

Contents

L. Van Aelst (✉)
Cold Spring Harbor Laboratory, Cold Spring Harbor, New York, NY 11724, USA
e-mail: vanaelst@cshl.edu

J.P. Blass (ed.), *Neurochemical Mechanisms in Disease*,
Advances in Neurobiology 1, DOI 10.1007/978-1-4419-7104-3_8,

1 Mental Retardation

1.1 Definition, Causes, and Classification

Mental retardation (MR) is generally defined as a global reduction in cognitive and intellectual abilities, which manifests before the age of 18, and is estimated to affect 1 to 3% of the population (Chelly et al., 2006). Intellectual functioning and its severity is commonly based on the evaluation of the Full Scale Intelligence Quotient (FSIQ), and MR is represented by an intelligence quotient (IQ) lower than 70. Based on the IQ, MR is commonly classified in two main groups: severe MR with an IQ below 50, and mild MR with an IQ between 50 and 70. The causes of MR are extremely heterogeneous and include nongenetic factors such as premature birth, infectious disease, and fetal alcohol syndrome, as well as genetic changes that include chromosomal abnormalities and single-gene mutations (Mandel and Chelly, 2004; Ropers and Hamel, 2005; Vaillend et al., 2008). Conventionally, genetic forms of MR have been subdivided into syndromic and nonsyndromic forms; with syndromic MR being characterized by associated clinical, radiological, metabolic, or biological features whereas in the case of nonsyndromic or nonspecific MR, cognitive impairment represents the only manifestation of the disease (Chelly and Mandel, 2001; Ropers and Hamel, 2005). It should be noted, however, that more recent genotype/phenotype studies and clinical re-evaluations of patients indicate that the boundaries between syndromic and nonsyndromic MR are vanishing. Moreover, several of the MR-related genes emerged as being involved in both forms of MR (Frints et al., 2002; Chelly et al., 2006).

Great progress has been made in recent years towards the identification of MR-related genes, resulting in a list of approximately 300 genes. A complete list of MR- and associated syndromes-related genes has been rigorously reviewed by Inlow and Restifo (Inlow and Restifo, 2004). Among these genes, several are associated with severe brain abnormalities, such as neuronal heterotopia, lissencephaly, and microcephaly (Chelly et al., 2006). In these cases, MR is likely to be a secondary symptom inasmuch as the involved gene products are likely to play a role in proper development of the CNS. A vast number of other genes have been associated with MR disorders with no apparent/gross abnormalities in brain structure and architecture, and, as discussed below, current efforts are geared towards unraveling the cellular bases for these MR conditions. Whereas the initial excess in identifying genes mutated in syndromic and nonsyndromic forms of X-linked MR led to the hypothesis that a disproportionally high density of genes influencing cognitive abilities reside on the X chromosome, recent estimates suggest a downturn in X-linked MR prevalence (Mandel and Chelly, 2004; Poirier et al., 2006). This is further supported by the emerging high frequency of pathogenic autosomal copy number variations.

1.2 *Mental Retardation Is Associated with Abberations in Spine Structure and Synaptic Function*

A major challenge has been to connect the genetic causes of MR to the relevant cellular processes and pathways underlying the pathophysiology of human cognitive disorders. One consistent feature of neurons in patients with MR is alterations in the size, shape, stability, and/or number of dendritic spines, which are highly specialized and dynamic structures on the dendrites on which most excitatory synapses in the brain are located (Kaufmann and Moser, 2000; Fiala et al., 2002; Newey et al., 2005; Tada and Sheng, 2006). More than three decades ago, studies using Golgi impregnations of postmortem material from normal and mentally retarded children demonstrated that changes in dendritic spine density and shape were associated with MR (Huttenlocher, 1970; Marin-Padilla, 1972; Huttenlocher, 1974; Purpura, 1974; Kaufmann and Moser, 2000). For instance, Purpura reported a loss of spines with an absence of short thick spines and a predominance of abnormally long thin spines (resembling immature filopodia) in patients with MR of unknown etiology (Purpura, 1974). Similar alterations with the presence of long tortuous spines have been observed in other defined classes of MR, such as fragile-X syndrome (Rudelli et al., 1985; Hinton et al., 1991; Wisniewski et al., 1991; Irwin et al., 2001). The latter is caused by a trinucleotide CGG repeat expansion and hypermethylation in the 5′ untranslated region of the *FMR1* gene, which is the most frequent single-gene cause of MR (see further Section 2.2.4).

In the case of Down's syndrome, which results from trisomy of chromosome 21, a reduction in spine density in the neocortex and hippocampus is a common feature; however, short thin spines and spines with large heads and thin necks have also been described (Marin-Padilla, 1972, 1976; Suetsugu and Mehraein, 1980; Takashima et al., 1981; Ferrer and Gullotta, 1990; Takashima et al., 1994). Subsequent studies using mouse models with genetically generated MR (e.g., *FMR1* knock-out mouse and partially trisomic mouse, Ts65Dn) consistently reported defects in dendritic spine morphology (Comery et al., 1997; Nimchinsky et al., 2001; Belichenko et al., 2002; Bakker and Oostra, 2003; Dierssen et al., 2003; Galdzicki and Siarey, 2003; Grossman et al., 2006). Thus, alterations in the shape, stability, and/or number of dendritic spines are likely to be a contributive factor to MR.

What are the functional implications of alterations in spine morphology and how are they linked to MR? There are numerous observations indicating that spine size, which can range over two orders of magnitude, is of physiological importance. For instance, larger spines can greatly outlast small spines (months compared with hours) (Holtmaat et al., 2005). It is important to note that large spines contain large synapses (Harris et al., 1992) with more glutamate-sensitive AMPA receptors (AMPARs), the principal receptors for fast excitatory neurotransmission in the mammalian central nervous system (Baude et al., 1995; Nusser et al., 1998; Kharazia and Weinberg, 1999; Takumi et al., 1999), and thus are functionally stronger than small spines. This strong positive correlation between spine size and synaptic strength is maintained in the face of plasticity (Matsuzaki et al., 2004; Kopec et al., 2006). Indeed, increasing evidence indicates that synaptic plasticity is associated with changes in spine morphology. These morphological changes of spines depend on NMDAR activation and are thought to contribute to activity-dependent formation and elimination of synaptic connections.

Two forms of synaptic plasticity, which are considered to be major cellular mechanisms underlying learning and memory, are long-term potentiation (LTP) and long-term depression (LTD) (reviewed in Citri and Malenka, 2008). LTP-inducing stimuli, associated with the addition of AMPARs at the postsynaptic site, cause the formation of new spines and/or the enlargement of existing spines, whereas LTD-inducing stimuli, associated with internalization of AMPARs, lead to shrinkage and/or retraction of spines (reviewed in Malinow and Malenka, 2002; Bredt and Nicoll, 2003; Shepherd and Huganir, 2007; Citri and Malenka, 2008). Given that learning deficit is a constant feature of patients with MR, alterations in synaptic structure and function are thought to be attributed to some of the MR conditions. Consistent with this notion, increasing evidence suggests that impairments in synaptogenesis and synaptic plasticity contribute to mental and neurological disorders, including MR (Fiala et al., 2002; Bagni and Greenough, 2005; Halpain et al., 2005; Chahrour and Zoghbi, 2007; Dolen and Bear, 2008) (see also Section 2.2). Hence, it is not surprising that large efforts have been devoted towards unraveling the molecular and cellular mechanisms underlying synaptic structure, function, and plasticity (reviewed in Shepherd and Huganir, 2007; Citri and Malenka, 2008).

Ample evidence points to an active contribution of actin to the modulation of spine morphology and the efficacy of pre- and postsynaptic terminals (reviewed in Cingolani and Goda, 2008). Actin filaments form the main cytoskeleton of dendritic spines, which are remarkably dynamic. It is widely believed that the regulated polymerization and/or depolymerization of actin underlie spine motility, growth, and shape (Tada and Sheng, 2006; Cingolani and Goda, 2008). Moreover, several observations support the view that dynamic actin filaments are a prerequisite for synapse formation. For instance, activity-dependent synaptogenesis is blocked by actin depolymerizing agents; disruption of signaling pathways implicated in synaptic actin reorganization results in synaptogenesis defects; and finally, actin plays a part in the development of dendritic spines, thus linking the synapse with actin (reviewed in Cingolani and Goda, 2008). The actin network is also directly involved in synaptic regulation at mature synapses, such as LTP and LTD. Actin-GFP FRET experiments demonstrated that changes in actin polymerization/depolymerization occur in response to different patterns of synaptic stimulation. In particular, these studies showed that tetanic stimulation causes a shift of actin equilibrium towards filamentous actin (F-actin), whereas prolonged low-frequency stimulation causes a shift in actin equilibrium towards G-actin, resulting in a loss of postsynaptic actin

(Okamoto et al., 2004). Furthermore, F-actin has been shown to be required for stable LTP, suggesting that nascent actin filaments stabilize synaptically delivered AMPARs (Kim and Lisman, 1999; Krucker et al., 2000; Fukazawa et al., 2003; Chen et al., 2004; Okamoto et al., 2004; Lin et al., 2005; Matus, 2005; Honkura et al., 2008). Interestingly, a more recent study showed that synaptic insertion of the AMPAR subunit GluR1, independent of its role of increasing synaptic strength, is required for stable spine enlargement after plasticity-inducing stimuli (Kopec et al., 2007). These findings suggest that AMPARs and nascent actin filaments are interdependent and mutually stabilizing. Together, these studies point to an important role for actin in functional and structural plasticity.

The central role of actin in the regulation of synaptic structure and function pointed to Rho GTPase family members as central contributors, inasmuch as they are key regulators of actin dynamics and organization (Van Aelst and D'Souza-Schorey, 1997). Indeed, Rho proteins emerged as key regulators of spine morphogenesis, and more recently have been implicated in synapse formation and synaptic plasticity. Furthermore, and significantly, mutations in regulators and effectors of the Rho GTPases have been found to underlie various forms of MR. In the remainder of this review, we first briefly discuss the role of Rho GTPases in spine and synapse formation, and subsequently describe in more detail some of the Rho GTPase signaling pathways involved in different forms of MR.

2 Rho GTPases

2.1 Rho GTPases Control Synaptic Structure and Function

The Rho family of small GTPases are low-molecular-weight guanine nucleotide-binding proteins, which act as molecular switches cycling between an active GTP-bound form and an inactive GDP-bound form (see Fig. 1). Their activity is tightly controlled by dedicated guanine nucleotide exchange factors (GEFs), which promote GTP-loading, GTPase activating proteins (GAPs), which enhance hydrolysis of the bound GTP, and guanine-nucleotide-dissociation-inhibitors (GDI), which prevent the exchange of GDP for GTP (Van Aelst and D'Souza-Schorey, 1997). Activated GTP-bound Rho GTPases interact with specific effector molecules to mediate their cellular actions. Of the Rho GTPase family members, RhoA, Rac1, and Cdc42 have been characterized most extensively. These GTPases are best known for their effects on the actin cytoskeleton, and, hence, it is not surprising that they emerged as critical regulators of spine formation and/or maintenance (Govek et al., 2005) (Fig. 1).

Several lines of evidence pointed to a role for Rac in spine formation and/or maintenance and the control of spine morphology in different model organisms (Govek et al., 2005). These studies largely relied on imaging of individual, fluorescently labeled neurons expressing constitutively active (CA) and dominant-negative (DN) mutant forms of Rac. Expression of CA Rac1 in hippocampal brain slices resulted in the formation of multiple small spines (Nakayama et al., 2000; Tashiro et al.,

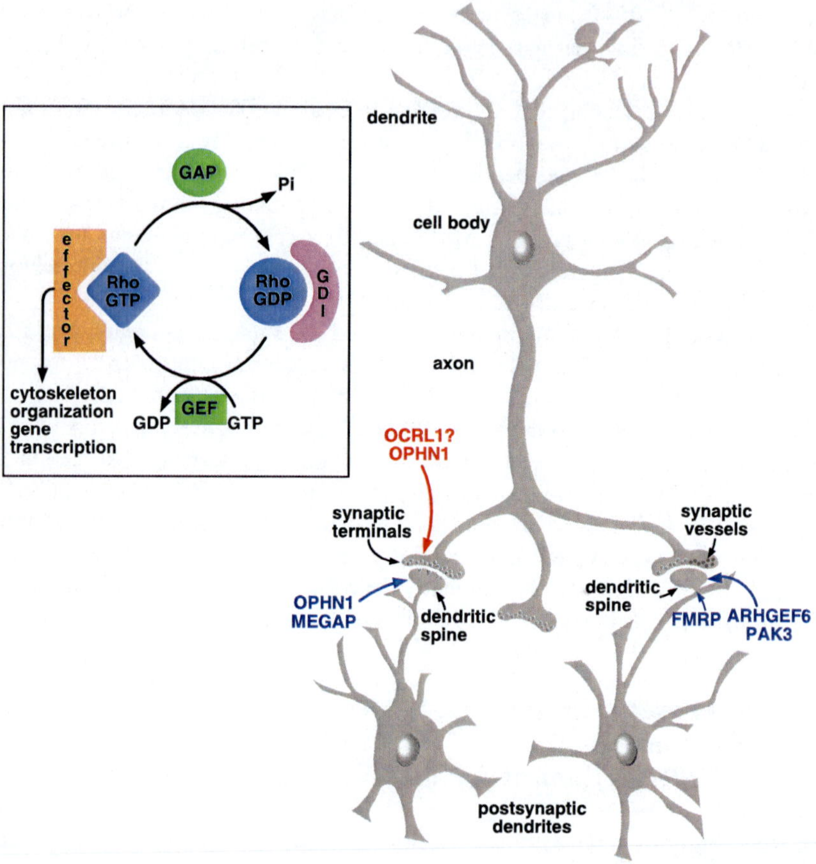

Fig. 1 Regulatory cycle for the activation and inactivation of Rho GTPases and their involvement in synapse development and maturation. *Left panel*: Rho GTPases cycle between an inactive GDP and an active GTP bound form. Their activity is tightly controlled by dedicated guanine nucleotide exchange factors (GEFs), which promote GTP-loading; GTPase activating proteins (GAPs), which enhance their intrinsic rate of GTP hydrolysis; and guanine nucleotide dissociation inhibitors (GDIs), which prevent exchange of GDP for GTP and inhibit the intrinsic GTPase activity of GTP-bound GTPases. Only in their active state, Rho GTPases bind to their downstream effectors and exert their effects on various important biological activities. *Right panel*: Rho GTPases have been implicated in various aspects of neuronal development, including spine/synapse development and maturation. A number of Rho-associated MR gene products are indicated at the appropriate positions. Blue label indicates postsynaptic localization and red label presynaptic localization. Abbreviations: ARHGEF6, Rho guanine nucleotide exchange factor 6; FMRP: fragile-X mental retardation-1 protein; MEGAP, Mental disorder-associated GAP protein; OCRL1, the oculocerebrorenal syndrome of Lowe protein 1; OPHN1, oligophrenin-1; PAK3, p21-activated kinase 3

2000; Pilpel and Segal, 2004). This spine phenotype was also observed in transgenic mice expressing CA Rac1 in Purkinje cells (Luo et al., 1996). Such spines appear to be often engaged in multiple synaptic contacts, which is rarely seen in normal animals (Luo et al., 1996). On the other hand, expression of a DN Rac1 mutant in mouse and rat hippocampal slices caused a reduction in spine density and a corresponding reduction in synapse formation (Tashiro et al., 2000; Tashiro and Yuste, 2004). Notably, the spines of DN Rac transfected neurons were in general significantly longer than control spines, and detailed analysis revealed that blockade of Rac transforms a subset of existing spines into long, thin filopodia-like protrusions. Furthermore, inhibition of Rac1 reduces spine head growth (particularly in mature neurons), morphological changes, and spine stability (Tashiro and Yuste, 2004).

Interestingly, a recent study examining single and double *Rac1* and *Rac3* (which encodes the closely related, neuron-specific, Rac3 family member) knock-out mice demonstrated that spine formation is strongly hampered only in hippocampal neurons lacking both Rac1 and Rac3, implying that Rac1 and Rac3 play complementary roles during late stages of neuronal development. This study additionally showed that the double knock-out mice displayed neurological abnormalities (Corbetta et al., 2009). Recent studies have also coupled Rac1 function to synaptic activity. Wiens et al. found that overexpression of wild-type or CA Rac1 enhances excitatory synaptic transmission and induces clustering of AMPARs in both pre-existing and newly formed dendritic spines, demonstrating that Rac1 can regulate the function of excitatory synapses (Wiens et al., 2005). These findings indicate that Rac1 is not only important for spine morphology and motility, but is also directly coupled to synaptic function. Positive regulators of Rac examined in the context of dendritic spine morphogenesis and/or synaptic function include Tiam1, Kalirin, and α- and β-PIX (Penzes et al., 2008); see Section 2.2).

Evidence has been provided that the Rac-GEF Tiam1 acts as a critical mediator of *N*-methyl-D-aspartate receptor (NMDAR)-dependent spine development (Tolias et al., 2005). Tolias et al. showed that Tiam1 is necessary for spine and synapse development and that it interacts with the NMDAR. Following glutamate application, they observed that NMDAR-mediated increases in intracellular calcium causes phosphorylation of Tiam1, with a concomitant increase in Rac1 activity required for spine remodeling (Tolias et al., 2005). In a subsequent study, the same group examined EphB receptors, as they are known to form a complex with NMDARs and positively modulate their function. They found that Tiam1 also mediates EphB receptor-dependent dendritic spine development, and proposed a model in which Tiam1 by functioning downstream of both EphB and NMDAR may act as a convergence point to help integrate activity-dependent and -independent signaling pathways during the development and remodeling of synaptic connections (Tolias et al., 2007). Recent work also indicates that the Rac-GEF kalirin-7 is a key component in coupling NMDAR activation to Rac activation and structural plasticity in mature cortical neurons.

Xie et al. found that activation of the calcium/calmodulin-dependent kinase II family member (CaMKII) following NMDAR activation directly phosphorylates kalirin-7 on its N-terminus, thereby stimulating its GEF activity (Xie et al., 2007). Knock-down of kalirin-7 levels reduces basal spine density and the frequency of miniature excitatory postsynaptic currents (mEPSCs) (Xie et al., 2007). Also, kalirin has been linked to the EphB receptor. Activation of the EphB receptor by ephrin-B has been shown to translocate kalirin-7 to synapses where it locally activates Rac1 and its effector PAK, which presumably regulates the actin cytoskeleton to contribute thereby to proper dendritic spine development (Penzes et al., 2003). Finally, the Rac-GEF βPIX has been shown to be regulated by NMDAR activation and to be critical for activity-dependent synaptogenesis. In particular, Saneyoshi et al. demonstrated that CaMK kinase (CaMKK)/CaMKI and βPIX form a signaling complex in spines, in which CaMKK/CaMKI phosphorylates and stimulates the GEF activity of βPIX to enhance Rac activity and promote formation/stabilization of mushroom-shaped spines (Saneyoshi et al., 2008). Key effectors mediating the effects of Rac on spine morphogenesis and potentially synaptic function include group 1 PAK kinases and the WAVE proteins, and are further discussed in Section 2.2.

As with other cellular functions, RhoA appears to work in an opposite fashion to Rac in the regulation of spine structure. In general, increased RhoA activity has been coupled to reduced spine length, size, and density (Tashiro et al., 2000; Ryan et al., 2005; Elia et al., 2006; Sfakianos et al., 2007; Zhang and Macara, 2008), whereas, conversely, low levels of RhoA have been associated with the maintenance of dendritic and spine structures (Nakayama et al., 2000; Van Aelst and Cline, 2004; Sfakianos et al., 2007). Interestingly, a few groups reported a decrease in endogenous RhoA activity upon glutamate receptor activation (Van Aelst and Cline, 2004; Schubert et al., 2006), suggesting a link between synaptic input and regulation of endogenous RhoA activity. Recent studies corroborated this and provided further insight into potential molecular links between synaptic activity and RhoA signaling.

Kang et al. found a complex formation between the RhoA-GEF, Lfc/GEF-H1, and AMPARs and showed that these proteins colocalize in spines (Kang et al., 2009). Furthermore, they demonstrated that Lfc/GEF-H1 activity negatively regulates spine density and length through a RhoA signaling cascade, and that AMPAR-dependent changes in spine development were eliminated by downregulation of Lfc/GEF-H1. Thus, these data suggest that Lfc/GEF-H1 is a key mediator of AMPAR activity-dependent structural plasticity in hippocampal neurons. Nadif Kasri et al. found that the Rho-GAP, Oligophrenin-1, is regulated by synaptic activity and NMDAR activation, and, significantly, that oligophrenin-1 in turn controls synapse maturation and plasticity at the hippocampal CA3-CA1 synapse by stabilizing AMPARs (see Section 2.2.1). Finally, the p190RhoGAP has been implicated in the regulation of hippocampal synapse stability by regulating Rho activity in the dendritic spine (Sfakianos et al., 2007).

The effects of RhoA activity on spine number and morphology are mediated, at least in part, by the RhoA effector, Rho kinase (Nakayama et al., 2000; Tashiro and Yuste, 2004; Yuste and Bonhoeffer, 2004). Different targets of Rho-kinase have been identified, such as LIMK, myosin light chain (MLC), and MLC phosphatase. Rho-kinase phosphorylates and activates LIMK, which in turn phosphorylates and inactivates the actin depolymerization factor (ADF) cofilin (Maekawa et al., 1999; Sumi et al., 1999; Ohashi et al., 2000; Amano et al., 2001). Phosphorylation of MLC by Rho-kinase results in the stimulation of myosin–actin interactions (Amano et al., 1996). Rho-kinase can also regulate the amount of phosphorylated MLC by phosphorylating and inactivating MLC phosphatase (Kimura et al., 1996). Significantly, a recent study has demonstrated that myosinIIB, which binds and contracts actin filaments, is essential for spine morphology and dynamics, as well as synaptic function (Ryu et al., 2006).

The role for Cdc42 in spine morphogenesis is less well defined. In hippocampal pyramidal neurons in organotypic slices, expression of a CA- or DN-Cdc42 mutant did not have a significant effect on spine density or length (Tashiro et al., 2000). However, Cdc42 has been demonstrated to affect spine formation in other systems. Loss of function of Cdc42 in vertical system (VS) neurons in the *Drosophila* visual system leads to a reduction in the density of spinelike structures (Scott et al., 2003), and reduced Cdc42 protein expression is associated with reduced cortical pyramidal neuron spine density and synapses in insulinlike growth factor 1 *(Igf1)–/–* brains (Cheng et al., 2003). Furthermore, Cdc42 has been shown to mediate the effects of upstream activators such as the EphB receptor and the Cdc42-specific GEF, intersectin-1, on spine morphogenesis in rat hippocampal neurons (Irie and Yamaguchi, 2002). The presence of both the Cdc42 effector N-Wasp and the EphB receptor had a synergizing effect on the GEF activity of intersectin-1, resulting in high levels of Cdc42-GTP; whereas DN mutant forms of intersectin-1, N-Wasp and Cdc42 interfered with spine formation (Irie and Yamaguchi, 2002).

These findings have led to a model in which the EphB receptor, in a complex with intersectin-1 and N-Wasp, triggers the activation of Cdc42 to promote actin polymerization via N-Wasp and the Arp2/3 complex, leading to spherical expansion of dendritic spine heads. A more recent study identified Numb as an intersectin-1 binding protein (Nishimura et al., 2006). They found that Numb enhanced intersectin-1's GEF activity for filopodia formation, and demonstrated a role for Numb in spine development. Moreover, they found that Numb forms a complex with the EphB2 receptor and NMDA-type glutamate receptors at the postsynapse together with intersectin, which potentially links Numb to EphB and glutamate receptor signaling for synaptic development. In addition to N-Wasp, the insulin receptor substrate 53 (IRSp53) and PAK3 have also been shown to mediate the effects of Cdc42 on spine morphogenesis. Of note, whereas IRSp53 seems to bind equally to Cdc42 and Rac1, Pak3 preferentially binds to Cdc42 (Choi et al., 2005; Kreis et al., 2007; see also Section 2.2).

Taken together, these studies clearly implicate Rho GTPase signaling in the structural remodeling of dendritic spines. Emerging evidence also points to a critical role for Rho GTPase signaling in the regulation of synaptic function and plasticity. Notably, additional regulators and effectors of Rho GTPases implicated in spine morphogenesis (that are not discussed here) have been reported; for a more detailed description of regulators and effectors of Rho GTPases, see reviews: Govek et al. (2005) and van Galen and Ramakers (2005).

2.2 *Mutations in Regulators and Effectors of Rho GTPases Underlie Various Forms of Mental Retardation*

As discussed above, MR has been associated with abnormalities in spine structure and function, and Rho GTPases have been implicated in the regulation of these processes. It is thus not surprising that mutations in several regulators (GEFs and

GAPs) and effectors of the Rho GTPases have been found to underlie or contribute to various forms of MR. These include syndromic and nonsyndromic X-linked forms of MR, as well as autosomal syndromic MR. Below, we discuss several examples demonstrating the involvement of Rho GTPase signaling in the etiology of different forms of MR. These examples also tackle the emerging view of how mutations in Rho-linked genes could result in MR, that is, by disrupting the normal development, structure, and/or plasticity of neuronal networks via perturbations in the regulation of the actin cytoskeleton and gene expression (see also Fig. 2).

2.2.1 Oligophrenin-1 (OPHN1)

OPHN1 was the first identified Rho-linked MR gene (Billuart et al., 1998). It encodes the protein OPHN1 that contains a BAR (Bin, amphiphysin, Rvs) and PH domain at its N-terminus, and a GAP domain shown to negatively regulate Rho family members at its C-terminus (Fauchereau et al., 2003; Govek et al., 2004). *OPHN1* was initially identified by the analysis of a balanced translocation t(X;12) observed in a female patient with mild MR (Bienvenu et al., 1997). Subsequent studies have revealed the presence of *OPHN1* mutations in families with MR associated with cerebellar hypoplasia and lateral ventricle enlargement (Tentler et al., 1999; Bergmann et al., 2003; Philip et al., 2003; des Portes et al., 2004; Zanni et al., 2005). Abnormal behavior, impaired language skills, and motor development delays were described for several of the patients (Tentler et al., 1999; Bergmann et al., 2003; Philip et al., 2003). All *OPHN1* mutations identified to date have been shown, or predicted, to result in OPHN1 loss of function (Zanni et al., 2005), and, interestingly, inactivation of *ophn1* in mice has recently been demonstrated to recapitulate some of the human phenotypes, such as behavioral, social, and cognitive impairments (Khelfaoui et al., 2007).

The OPHN1 protein is expressed in multiple tissues, although with highest levels in the brain, where it is found in neurons of all major regions, including hippocampus and cortex, and is present in axons, dendrites, and spines (Govek et al., 2004). Thus, OPHN1 is present both pre- and postsynaptically in neurons. Recent studies have begun to unveil how mutations in *OPHN1* may affect neuronal function. In a first study, it was found that knock-down of *OPHN1*, by using RNA interference (RNAi), in CA1 pyramidal neurons in hippocampal slices results in a significant decrease in dendritic spine length (Govek et al., 2004). This phenotype was mimicked using a constitutive active (CA) RhoA mutant and was rescued by inhibiting a key effector of RhoA, termed Rho-kinase (Govek et al., 2004). As discussed above, Rho-kinase can influence the actin cytoskeleton by acting on LIM kinase (LIMK), myosin light chain, and/or MLC phosphatase (see also Fig. 2).

These findings support a model in which loss of OPHN1 causes aberrations in spine morphology during development as a result of changes in the actin cytoskeleton triggered upon elevation of RhoA and Rho-kinase activities. More recently, mice lacking the *Ophn1* gene were generated, and analysis of these mice showed a decrease in mature spines (Khelfaoui et al., 2007). Surprisingly, in this mouse knock-out model, no obvious deficits in synaptic transmission or plasticity were

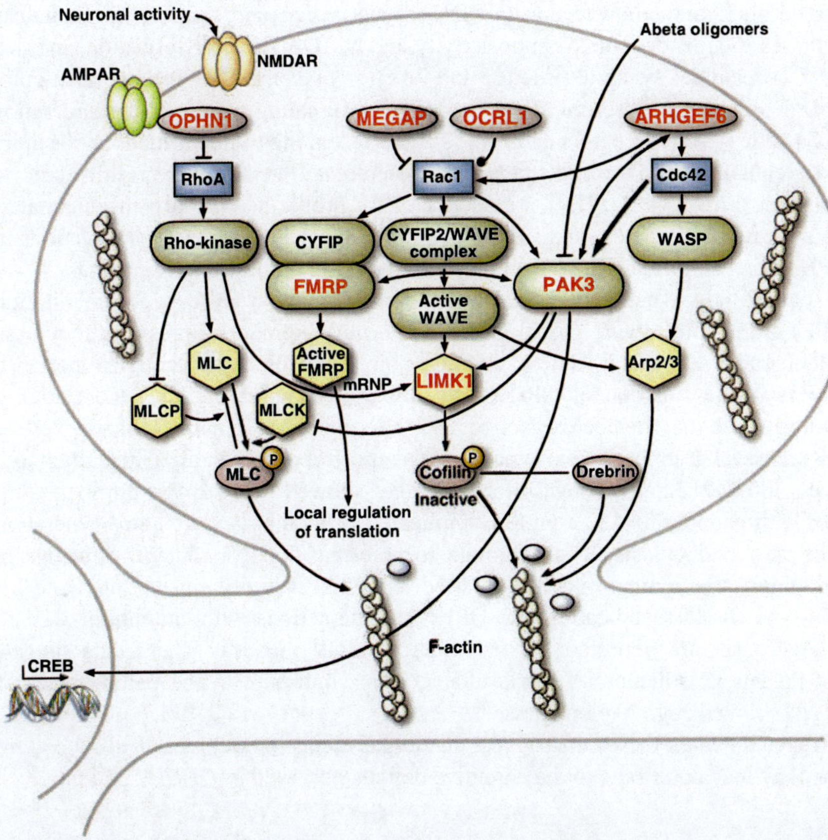

Fig. 2 Rho-linked mental retardation proteins and effector pathways connecting Rho GTPases to actin dynamics. Proteins encoded by Rho-linked genes involved in different forms of MR are highlighted in *red* text. See main text for explanation. Abbreviations: AMPAR, α-amino-3-hydroxy-5-methyl-4-isoxazolepropionic acid receptor; ARHGEF6, a Rho guanine nucleotide exchange factor 6; Arp2/3, actin-related proteins 2 and 3; Aß, amyloid ß; CREB, cAMP-responsive element-binding protein; CYFIP, cytoplasmic FMR1 interacting protein; FMRP, fragile X syndrome protein; LIM, Lin-11, Isl-1 and Mec-3 kinase; MEGAP, Mental disorder-associated GAP protein; MLC, myosin light chain; MLCK, myosin light chain kinase; MLCP, myosin light chain phosphatase; NMDAR, *N*-methyl D-aspartate receptor; OCRL1, oculocerebrorenal syndrome of Lowe; OPHN1, oligophrenin-1; PAK, p21-activated kinases; WASP, Wiskott-Aldrich-syndrome protein; WAVE, WASP family Verprolin-homologous protein

observed. The interpretation of these data is, however, complicated by the fact that OPHN1 is absent both pre- and postsynaptically in global *ophn1* knock-out mice, and there could also be compensatory adaptations during development in the *ophn1* knock-out mice. Indeed, by temporally and spatially manipulating *OPHN1* gene expression, Nadif Kasri et al. recently demonstrated that postsynaptic OPHN1 plays a key role in activity-dependent maturation and plasticity of excitatory synapses by regulating their structural and functional stability (Nadif Kasri et al., 2009).

Furthermore, they showed that OPHN1's localization and function in excitatory synapses is dependent on synaptic activity and NMDA receptor activation, and that OPHN1 regulates synaptic structure and function by controlling the stabilization of AMPA receptors. Therefore, defective OPHN1 signaling results in destabilization of synaptic AMPA receptors and spine structure, leading to impairment in plasticity and eventually loss of spines and NMDA receptors. Together, these results indicate that critical levels of OPHN1 are necessary for proper activity-driven glutamatergic synapse development and suggest a cellular mechanism by which mutations in *OPHN1* can contribute to the cognitive deficits observed in *OPHN1* patients.

Interestingly, Khelfaoui et al. did report a decrease in paired-pulse facilitation (PPF) in the adult *ophn1* global knock-out, a measurement of presynaptic release (Khelfaoui et al., 2007). These findings point towards the potential importance of presynaptic function of OPHN1 signaling. In support of this, recent studies demonstrated that reduced/defective OPHN1 signaling impairs synaptic vesicle (SV) retrieval at hippocampal synapses (Nakano-Kobayashi et al., 2009; Khelfaoni et al., 2009). Nakano-Kobayashi et al. further showed that OPHN1 forms a complex with endophilin A1, a protein implicated in membrane curvature generation during SV endocytosis. It is important to note that OPHN1 mutants defective in endophilin A1 binding, or with impaired Rho-GAP activity, fail to substitute for wild-type OPHN1, indicating that OPHN1's interactions with endophilin A1 and Rho GTPases are important for its function in SV retrieval. These data suggest that defects in efficient SV retrieval may also contribute to the pathogenesis of *OPHN1*-linked cognitive impairment.

Taken together, these data suggest that impairments in both long- and short-term plasticity may contribute to the cognitive deficits observed in *OPHN1* patients.

2.2.2 p21-Activated Kinase 3 (PAK3)

The second Rho-linked MR gene identified is *PAK3*. Mutations in *PAK3* were found to be the cause of nonsyndromic X-linked MR (see below). *PAK3* encodes a member of the group I p21-activated serine/threonine kinases (PAK) (Dan et al., 2001a). The group I PAK proteins (including PAK1, PAK2, and PAK3) function as effectors of the Rac1 and Cdc42 GTPases, and have been demonstrated to mediate their effects on the actin cytoskeleton and gene expression (Jaffer and Chernoff, 2002; Bokoch, 2003). One mechanism by which PAKs affect the actin cytoskeleton involves phosphorylation and activation of LIMK (Stanyon and Bernard, 1999), which in turn phosphorylates and inhibits cofilin, an actin filament depolymerizing/severing factor, thereby stabilizing actin filaments and promoting actin polymerization (Yang et al., 1998; Edwards et al., 1999; Dan et al., 2001). The regulation of myosins is likely to be another component of PAK-mediated cytoskeletal signaling. There is evidence that PAK1 can interfere with myosin light chain function via direct phosphorylation and inhibition of myosin light chain kinase (MLCK) (Sanders et al., 1999; Bokoch, 2003). This action of PAK may assist in the disassembly of actin stress fibers triggered by PAK (see also Fig. 2). The group I PAK kinases exist in a dormant state in the cytoplasm as a result of an N-terminal autoinhibitory region, which assumes a configuration that prevents the activation of the C-terminal kinase domain. Upon binding to Rac-GTP or Cdc42-GTP, the autoinhibition is alleviated, resulting in activation of the PAK proteins and their autophosphorylation (Jaffer and Chernoff, 2002; Bokoch, 2003). Noteworthy is a recent study that reported PAK3 binds significantly more Cdc42 than Rac1, and is selectively activated by endogenous Cdc42, suggesting that PAK3 is a selective effector of Cdc42 (Kreis et al., 2007). Among the PAK proteins, PAK1 and PAK3 are highly expressed in the brain. Both proteins are present in the hippocampus and cortex, with PAK3 being particularly highly expressed in postmitotic neurons of the dentate gyrus and cortical layers II/III and V (Kreis and Barnier, 2009). In neurons, PAK3 shows a diffuse distribution throughout the soma and proximal dendrites and is present in dendritic spines (Boda et al., 2004).

As discussed below, both PAK1 and PAK3 proteins have been implicated in spine morphogenesis, however, as of this writing, only mutations in PAK3 have been identified that are associated with nonsyndromic MR. In particular, five different mutations in the *PAK3* gene have been identified in several X-linked MR pedigrees. The first *PAK3* mutation, R419X, found in family MRX30, introduced a premature stop codon that abolishes the kinase activity of the truncated product (Allen et al., 1998). Since then four additional mutations have been identified in MR patients. These include the R67C and the A365E mutations located in the p21-binding domain and in the kinase domain, respectively; the W446S mutation located in the catalytic domain; and, finally, a splice mutation located at the 5′ end of intron 6 leading to a disruption of the reading frame with a premature stop codon at position 128 (Bienvenu et al., 2000; Gedeon et al., 2003; Peippo et al., 2007; Rejeb et al., 2008). Biochemical analysis demonstrated that PAK3 proteins harboring the R419X and A365E mutations, and presumably also the W446S mutation, are devoid of kinase activity, whereas the PAK3 protein with the R67C mutation has a functional kinase domain but displays a decrease in binding to Cdc42 and a decrease in its activation by this GTPase (Kreis et al., 2007).

Several lines of evidence have demonstrated a role for PAK3 (as well as PAK1) in the regulation of dendritic spine morphogenesis, synapse formation, and/or synaptic plasticity. First, a study using transgenic mice in which the catalytic activity of the PAK family members, PAK1 and PAK3, is inhibited by expression of the PAK-autoinhibitory domain (AID-PAK) revealed that cortical neurons of these mice have fewer spines than control animals and show a shift in the overall spine population towards shorter spines with larger heads and postsynaptic densities. Interestingly, these mice also show enhanced LTP and reduced LTD in the cortex, as well as specific deficits in the consolidation phase of hippocampus-dependent memory, suggesting a role for PAK in memory retention (Hayashi et al., 2004).

Secondly, Boda et al. observed that RNAi-mediated suppression of PAK3, or expression of a dominant negative, kinase-dead, PAK3 mutant (R419X), in rat hippocampal organotypic slice cultures results in the formation of abnormal elongated dendritic spines and filopodia-like protrusions, as well as a decrease in mature spine synapses. They observed that these defects were associated with reduced expression of AMPARs at the synapse and defective LTP (Boda et al., 2004). Interestingly, a more recent study compared the effects of three different PAK3 mutants (R67C, A365E and R419X) on spine morphogenesis and observed that these mutant proteins affect spinogenesis differentially (Kreis et al., 2007). Specifically, they found that expression of the PAK3 kinase-dead mutants, A365E and R419X, in CA1 neurons of hippocampal brain slices profoundly altered spine morphology without affecting spine density, whereas expression of the PAK3 R67C mutant drastically decreased spine density. Based on these data, a model was proposed in which PAK3 may act at two different steps during spine formation, namely at (1) the initiation of spines and (2) at spine maturation (Kreis et al., 2007).

Finally, mice lacking the *PAK3* gene have been generated, and analysis of these mice showed selective impairment in late-phase hippocampal LTP, a distinct form of long-term synaptic plasticity involving new gene expression (Meng et al., 2005). Surprisingly, in this mouse knock-out model, no obvious deficits in spine morphology or density were observed. The differences seen with regard to spine morphology between the knock-out and RNAi studies could potentially reflect differences between a homogeneous and a heterogeneous cell population, respectively, or could be attributed to compensatory mechanisms (e.g., PAK1 or PAK2) in the knock-out mice. Indeed, it has recently been shown that expression of active PAK1 can revert the long spine phenotype induced by RNAi-mediated suppression of PAK3 (Boda et al., 2008), although it should be noted that PAK1 and PAK3 also seem to have distinct roles in spine morphogenesis (Boda et al., 2008). The *PAK3* knock-out mice did, however, show a dramatic decrease in the levels of the phosphorylated/active form of cAMP-responsive element-binding protein (CREB) in the hippocampus, whereas no changes in the total CREB protein levels were observed (Meng et al., 2005). Several studies have shown that CREB function is important for synaptic plasticity and memory formation in mice (Kandel, 2001; Lonze and Ginty, 2002). Therefore, the reduced CREB function may be responsible for the impairment in late-phase hippocampal LTP in these mice.

Together, these studies indicate that mutations in PAK3, which are associated with nonsyndromic MR, cause aberrant spine structure and/or function as a result of altered actin dynamics and/or transcriptional regulation (see also Fig. 2). Interestingly, defects in PAK signaling not only result in MR, but recently have also been associated with Alzheimer disease (AD) (reviewed in Kreis and Barnier, 2009). This may not be that surprising considering the analogy between AD and MR; that is, both conditions share in common spine loss or spine alterations. AD is defined clinically as a gradual loss of cognitive performance with the onset of a slowly progressive impairment of memory during mid-to-late adult life. The neuropathological hallmarks include amyloid deposits (Aß), neurofibrillary tangles, and reductions in the number of neurons and synapses in many areas of the brain, but especially in the cerebral cortex and the hippocampus (LaFerla and Oddo, 2005).

Of particular interest is that the Aß oligomers implicated in AD were shown to reduce PAK1 and PAK3 expression levels and activities in the hippocampus and temporal cortex, resulting in a loss of drebrin from the spines and synaptic dysfunctions (Zhao et al., 2006; Ma et al., 2008). Drebrin is localized at spines in adult brains and is required for active clustering and synaptic targeting of PSD95

(Takahashi et al., 2003). Expression of active PAK in hippocampal neurons could prevent the effects induced by Aß oligomers, and significantly, pharmacological PAK inhibition in adult mice was sufficient to cause drebrin loss and memory impairment (Zhao et al., 2006). Thus, these findings indicate that loss of PAK3 and/or PAK1 is involved in both developmental-dependent and age-dependent cognitive deficits, such as observed in AD.

2.2.3 Rho Guanine Nucleotide Exchange Factor 6 (ARHGEF6)

The *ARHGEF6* gene, also known as *αPIX* or *Cool-2*, is another Rho GTPase-linked gene shown to be involved in nonsyndromic X-linked MR (Kutsche et al., 2000). It codes for a Cdc42/Rac1 GEF, which harbors a number of interesting motifs implicated in protein–protein interactions (Bagrodia et al., 1998; Manser et al., 1998; Kutsche et al., 2000; Koh et al., 2001; Feng et al., 2002). Besides the Dbl homology (DH) and plextrin homology (PH) domains, it contains an N-terminally located calponin homology (CH) domain, an SH3 domain, a GIT binding domain, and a C-terminally located leucine zipper that mediates the formation of homo- and heterodimers. The dimeric form of ARHGEF6/Cool-2/αPIX was found to act as a specific GEF for Rac1, whereas the monomeric form as a GEF for Cdc42 as well as Rac (Feng et al., 2004). Significantly, ARHGEF6 has been shown to directly interact with group I PAK kinases, as well as with the synaptic adaptor protein GIT1 (G-protein coupled receptor kinase-interacting protein1) (Bagrodia et al., 1998; Manser et al., 1998; Daniels et al., 1999; Feng et al., 2002; Zhang et al., 2003). The latter protein has been shown to be crucial for spine formation; its loss of expression significantly decreases the number of spines (Zhang et al., 2005). Furthermore, GIT1 has been found to be important for the localization of the closely related family member ßPIX to dendritic spines and to activate Rac1 and its downstream effector PAK locally (Zhang et al., 2005).

The first mutation in *ARHGEF6* associated with nonsyndromic X-linked MR was identified in a male carrying a reciprocal X;21 translocation breakpoint located between exons 10 and 11 of the *ARHGEF6* gene (Kutsche et al., 2000). Subsequently, additional mutations have been identified in the first intron of the gene that result in preferential skipping of exon 2 and a predicted protein product lacking the first 28 amino acids in affected males in a large MRX family (MRX46) (Kutsche et al., 2000). A recent study demonstrated that the ARHGEF6 protein is present in CA3 and CA1 neurons of the hippocampus and that expression of epitope-tagged ARHGEF6 in hippocampal slice cultures shows a punctate staining in dendritic spines that colocalizes with PSD-95 and other synaptic proteins (Node-Langlois et al., 2006). The same study also revealed a requirement for ARHGEF6 in spine morphogenesis. Whereas overexpression of ARHGEF6 did not alter spine morphology, RNAi-mediated knock-down of ARHGEF6 resulted in abnormalities in spine morphology similar to those reported for knock-down of PAK3: a decrease of large mushroom-type spines and an increase of elongated spines and filopodia-like protrusions (Node-Langlois et al., 2006).

Consistent with a role for ARHGEF6 in the regulation of spine morphogenesis, the *Drosophila* homologue, *dPIX*, was also shown to play a major role in regulating postsynaptic structures and protein localization at the glutamatergic neuromuscular junction (Parnas et al., 2001). It is important to note that the defect in spine structure in ARHGEF6 RNAi-treated neurons could be rescued by coexpression of a constitutively active PAK3 protein, but not with wild-type PAK3 (Node-Langlois et al., 2006). By contrast, the phenotype caused by knock-down of PAK3 could not be rescued by overexpression of ARHGEF6. Together, these results indicate that ARHGEF6 is involved in the same signaling pathway as PAK3, thereby controlling spine morphogenesis and plasticity of synaptic networks. Hence, similar mechanisms are likely to underlie cognitive deficits associated with mutations in *ARHGEF6* and *PAK3*. Interestingly, a recent study focusing on the closely related family member βPIX suggested a potential mechanism by which the PIX proteins are regulated in the synapse. As discussed above, Saneyoshi et al. identified a signaling pathway upstream of βPIX by which NMDAR activation during neuronal development or plasticity can modulate spinogenesis. They found that CaMKK/CaMKI interacts with βPIX/GIT1 and mediates phosphorylation of Ser516 in βPIX to enhance Rac activity and promote formation/stabilization of mushroom-shaped spines (Saneyoshi et al., 2008).

2.2.4 CYFIP/Rac/PAK and Fragile X Syndrome

Fragile X syndrome (FXS) is the most common inherited cause of MR with approximately 1 in 4000 males affected. In the vast majority of cases, this X-linked disorder is caused by an unstable expansion of the CGG trinucleotide repeat and hypermethylation of CpG dinucleotides in the 5′ untranslated region of the *FMR1* gene, which results in transcriptional silencing of *FMR1*. The first clinical indication of FXS is often delayed developmental milestones, such as mild motor delays and/or language delays. Autistic-like behaviors such as hand flapping, poor eye contact, and hand biting may be observed. The average IQ in adult men with the completely methylated full mutation is approximately 40. Less affected males, which typically have incomplete methylation and thus resulting in an incomplete activation of FMR1, may have an IQ in the borderline to low normal range. Physical features may include macroorchidism that is apparent just before puberty and those related to a connective tissue dysplasia, which include a long, narrow face, prominent ears, joint hypermobility, and flat feet (reviewed in Garber et al., 2006; Bassell and Warren, 2008; Garber et al., 2008)

FMR1 encodes a selective RNA-binding protein (FMRP) that regulates the local translation of a subset of mRNAs at synapses in response to activation of metabotropic glutamate receptors (mGluRs) and possibly other receptors. In the absence of FMRP, increased and dysregulated mRNA translation is believed to contribute to altered spine morphology, synaptic function, and loss of protein synthesis-dependent plasticity (reviewed in Bear et al., 2004; Bagni and Greenough, 2005; Bassell and Warren, 2008). As mentioned before, the shape and density of dendritic spines are altered in patients and in *FMR1*-deficient mice brains. A few reports suggested that FMRP could affect spine morphogenesis through regulation of "cargo" mRNAs, such as Map1B and profilin mRNAs (Lu et al., 2004; Reeve et al., 2005). More recent studies, mainly performed in *Drosophila*, have linked FMRP's effect on spine morphology to the Rac1 GTPase signaling pathway. One group demonstrated that the mRNA encoding Rac1 is present in Fmr1-messenger ribonucleoprotein complexes (Lee et al., 2003). Furthermore, evidence was provided that *Fmr1* and *Rac1* genetically interact, and that Rac1 mediates at least in part the effects of Fmr1 (*Drosophila* fragile X-related protein) on dendritic branching (Lee et al., 2003). An independent study demonstrated a biochemical association between the Fmr1-interacting protein dCYFIP and dRac1 (Schenck et al., 2003). Phenotypic analyses and genetic interaction experiments placed dRac, CYFIP, and dFMRP in a common pathway controlling axonogenesis and synaptogenesis. Furthermore, evidence was presented that Rac1 negatively regulates CYFIP, which in turn negatively regulates Fmr1, with the net result that dRac1 positively regulates dFMR1 action on neuronal morphogenesis (Schenck et al., 2003). Together with the above findings, these data suggest that there is a feedback loop between Rac1 and Fmr1 functions in vivo.

The mammalian homologues of *Drosophila* CYPIP, CYFIP1, and CYPIP2, have also been shown to interact with FMRP. In mammals, CYFIP1 (also known as p140/Sra-1) was initially identified as a target of Rac1 (Kobayashi et al., 1998), whereas CYFIP2 (also termed PIR121) was found to be part of the WAVE protein complex, which mediates actin nucleation by Rac (Eden et al., 2002). In its inactive state, this complex contains WAVE and four other proteins: HSPC300, Nap125, Abi2, and PIR121. When active Rac1 is added, the complex dissociates, freeing WAVE and HSPC300, thereby allowing WAVE to activate the actin-related protein 2/3 (Arp2/3) complex to induce actin polymerization (see Fig. 2). In analogy to the mechanism of WAVE activation, a model was proposed in which CYFIP dissociates from FMRP/Fmr1 upon interaction with activated Rac1, allowing released FMRP/Fmr1 to regulate local protein translation. A recent study also reported an interaction between PAK1 and FMRP and demonstrated that inhibition of group I PAK kinases rescued symptoms of knock-out (KO) *FMR1* mice (Hayashi et al., 2007). Specifically, the spine abnormalities observed in *FMR1* KO mice were partially restored by postnatal expression of a dominant negative PAK transgene (AID-PAK). Furthermore, the reduced cortical long-term potentiation was fully restored and several of the behavioral abnormalities associated with *FMR1* KO mice were ameliorated by the PAK-AID transgene. Whereas the precise underpinnings of the PAK1/FMRP interaction remain to be established, it is tempting to speculate (analogous to the CYFIP/WAVE complex) that FMRP and PAK1 could inhibit each other to form an inactive complex. Activation of PAK by GTPases would then trigger the dissociation of the two proteins allowing FMRP to regulate protein translation.

Together, these data suggest a model in which FMRP, Rac1, CYFIP, and/or PAK act together in a dynamic signaling complex(es) to regulate actin dynamics and control local protein translation, processes that are key to neuronal morphogenesis and connectivity.

2.2.5 Oculocerebrorenal Syndrome of Lowe Protein 1 (OCRL1)

Oculocerebrorenal syndrome of Lowe (OCRL) or Lowe syndrome is a rare X-linked developmental disorder characterized by MR, congenital cataracts, and renal Fanconi syndrome (Attree et al., 1992). The gene responsible for OCRL was initially identified by positional cloning of X chromosome breakpoints and encodes a protein termed OCRL1, an inositol polyphosphate-5-phosphatase (Attree et al., 1992; Lowe, 2005). In addition to the central polyphosphate-5-phosphatase domain, which uses PI(4,5)P_2 and PI(3,4,5)P_3 as the preferred substrates (Zhang et al., 1995; Schmid et al., 2004), the protein also contains at its C-terminus an ASH (ASPM, SPD2, Hydin) domain (Ponting, 2006) and Rho-GAP-like domain.

OCRL1 was initially localized to the Golgi complex (Olivos-Glander et al., 1995; Dressman et al., 2000), and it is recruited to membrane ruffles in response to growth factor stimulation and Rac activation (Faucherre et al., 2003). The GAP domain of OCRL1 has been shown to interact with Rac1, however, it does not appear to possess appreciable GAP activity towards Rac1 (Faucherre et al., 2003). More recent studies showed that OCRL1 is also present on endosomes and is important at early steps of the endocytic pathway (Erdmann et al., 2007), including clathrin-coated pits, which is consistent with the ability of OCRL1 to bind to clathrin, the endocytic clathrin adaptor AP-2, and the endosomal protein Rab5 (Ungewickell et al., 2004; Choudhury et al., 2005; Hyvola et al., 2006). In addition, OCRL1 was also found to bind the Rab5 effector APPL1 on peripheral endosomes; this interaction is mediated by the ASH–RhoGAP-like domains of OCRL1 (Erdmann et al., 2007).

Mutations that cause Lowe syndrome have been mapped exclusively to the *OCRL1* gene. The overwhelming majority of missense mutations are localized to the 5-phosphatase domain, underscoring the importance of the 5-phosphatase activity of this protein (McCrea et al., 2008). A small number of missense mutations are also located in the ASH and RhoGAP-like domains (McCrea et al., 2008), raising the question as to whether Rac and/or APPL1 interaction may play a role in the disease. The observation that *OCRL1*-deficient fibroblasts derived from Lowe-syndrome patients, in addition to increased PI(4,5)P_2 levels, also had alterations in the actin cytoskeleton, an increased sensitivity to actin depolymerizing agents, and mislocalization of the actin-binding proteins α-actinin and gelsolin (Suchy and Nussbaum, 2002), led initially to the postulation that abnormal cytoskeleton may contribute to the disease process, thus possibly involving Rho GTPase signaling. However, a more recent study showed that although all six known disease-causing missense mutations in the ASH and Rho-GAP domains abolished binding to APPL1, some of these mutations preserved the ability to bind Rac (McCrea et al., 2008).

Thus far, APPL1 is the only protein whose binding is consistently disrupted by patient missense mutations in the C-terminal region of OCRL. The same group also demonstrated that APPL1 helps localize OCRL1 to specific cellular sites, and a model was proposed in which disruption of OCRL1 binding to APPL1 would impair the proper localization of OCRL1 as well as disconnect OCRL1 from a protein network potentially linked to the disease phenotype (McCrea et al., 2008). Future studies will, however, be required to further unravel the signaling networks involved.

Surprisingly, *OCRL1* knock-out mice do not develop Lowe syndrome. A potential explanation for this observation is that the OCRL1 loss of function is compensated by the phosphatase, Inpp5b, which shares high homology with OCRL1, and which is more expressed in mice than in humans (Jefferson and Majerus, 1995; Janne et al., 1998; Hellsten et al., 2001; Astle et al., 2006).

2.2.6 Mental-Disorder-Associated GAP (MEGAP)

Mutations in Rho-linked genes that give rise to mental retardation are not only found on the X-chromosome, but have also been identified on autosomes. For example, the *MEGAP* (mental-disorder-associated GAP) gene was identified by positional cloning, as the only gene disrupted with a balanced de novo translocation of chromosome t(X;3)(p11.2;p25) (Endris et al., 2002). This patient exhibited severe MR and locomotor impairments that are associated with 3p-syndrome. Whereas other genes have also been implicated (Angeloni et al., 1999; Sotgia et al., 1999), 11 patients with 3p-syndrome MR displayed loss of heterozygosity for MEGAP, supporting the notion that reduced levels of this protein are causally linked to this form of MR. Notably, the MEGAP gene product had previously been identified as a WAVE-associated protein (WRP) (Soderling et al., 2002), and as a ROBO interacting protein (srGAP3) (Wong et al., 2001). The mRNA transcript of the MEGAP/WRP/srGAP3 gene is predominantly expressed in fetal and adult brain, and is enriched in the neurons of the hippocampus, cortex, and amygdala (Endris et al., 2002).

Biochemical studies showed that MEGAP/WRP/srGAP3 strongly enhances the intrinsic hydrolytic activity of Rac1 and to a significant lesser extent of Cdc42 (Endris et al., 2002). Together with the observation that MEGAP/WRP/srGAP3 directly binds to WAVE-1, a model was proposed in which MEGAP/WRP/srGAP3 functions in a negative feedback loop that inactivates Rac1 associated with WAVE-1, thereby controlling actin dynamics and spine morphogenesis. Significantly, Soderling et al. generated and characterized WAVE-1 knock-out mice and reported that WAVE-1 knock-out mice exhibited defects in balance and coordination, reduced anxiety, and deficits in learning and memory (Soderling et al., 2003). Interestingly, these phenotypes are strikingly similar to those observed in $3p^-$ syndrome patients. Morphological analysis of neurons in both the CA1 region of the hippocampus and the outer layer of the cortex of WAVE-1 knock-out mice revealed a reduction in spine density and abnormal spine morphology. Furthermore, electrophysiological recordings from hippocampal slices showed that WAVE-1 knock-out mice exhibit increased LTP and reduced LTD (Soderling et al., 2007). To determine whether the MEGAP/WRP/srGAP3's interaction with WAVE-1 contribute to WAVE's effect on spine density, synaptic plasticity, and memory, Soderling et al. generated mice that express WAVE-1 without the MEGAP/WRP/srGAP3 binding site. They observed that these WAVE-1 knock-in mice have reduced spine density and altered synaptic plasticity, as well as specific deficits in memory retention (Soderling et al., 2007). Thus, MEGAP/WRP/srGAP3's interaction with WAVE is important for WAVE's function in neural plasticity and cognitive behavior.

Together these findings imply that signaling through MEGAP/WRP/srGAP3 and WAVE-1 to the actin cytoskeleton is important for normal neuronal function and connectivity and that alteration of this pathway (e.g., upon loss or reduced expression of MEGAP/WRP/srGAP3) affects the expression of normal behaviors, including learning and memory.

3 Conclusions

Rho GTPase mediated signaling pathways modulate actin cytoskeleton dynamics and gene expression, which are critical for structural and functional plasticity in the developing and mature nervous system. Such synaptic remodeling and plasticity are thought to underlie the anatomic basis for learning and memory formation and normal cognitive function. Consistent with this are the findings demonstrating an association between various MR conditions and mutations in Rho-linked genes. The current view of how mutations in Rho-linked genes contribute to MR is by disrupting the normal development, structure, and/or plasticity of neuronal networks via perturbations in the actin cytoskeleton and gene regulation networks. Evidence supporting such a view has come from MR patients, mouse models of MR, and RNAi studies in hippocampal and cortical slices. Further elucidation of the molecular and cellular mechanisms by which Rho signaling contributes to the above disorders will not only shed light on the epidemiology of these diseases, but also on basic mechanisms of neuronal development and function and may provide candidates for therapeutic intervention.

Acknowledgments Because of space limitations, we are not able to cite the work of many of our colleagues who have made valuable contributions to this field. LVA is supported by NSF and NIH grants. NNK is a postdoctoral fellow from the Fund for Scientific Research Flanders and is supported by the Human Frontiers Science Program.

References

Allen KM, Gleeson JG, Bagrodia S, Partington MW, MacMillan JC, Cerione RA, Mulley JC, Walsh CA (1998) PAK3 mutation in nonsyndromic X-linked mental retardation. Nat Genet 20:25–30

Amano M, Ito M, Kimura K, Fukata Y, Chihara K, Nakano T, Matsuura Y, Kaibuchi K (1996) Phosphorylation and activation of myosin by Rho-associated kinase (Rho-kinase). J Biol Chem 271:20246–20249

Amano T, Tanabe K, Eto T, Narumiya S, Mizuno K (2001) LIM-kinase 2 induces formation of stress fibres, focal adhesions and membrane blebs, dependent on its activation by Rho-associated kinase-catalysed phosphorylation at threonine-505. Biochem J 354: 149–159

Angeloni D, Lindor NM, Pack S, Latif F, Wei MH, Lerman MI (1999) CALL gene is haploinsufficient in a 3p– syndrome patient. Am J Med Genet 86:482–485

Astle MV, Seaton G, Davies EM, Fedele CG, Rahman P, Arsala L, Mitchell CA (2006) Regulation of phosphoinositide signaling by the inositol polyphosphate 5-phosphatases. IUBMB Life 58:451–456

Attree O, Olivos IM, Okabe I, Bailey LC, Nelson DL, Lewis RA, McInnes RR, Nussbaum RL (1992) The Lowe's oculocerebrorenal syndrome gene encodes a protein highly homologous to inositol polyphosphate-5-phosphatase. Nature 358:239–242

Bagni C, Greenough WT (2005) From mRNP trafficking to spine dysmorphogenesis: the roots of fragile X syndrome. Nat Rev Neurosci 6:376–387

Bagrodia S, Taylor SJ, Jordon KA, Van Aelst L, Cerione RA (1998) A novel regulator of p21-activated kinases. J Biol Chem 273:23633–23636

Bakker CE, Oostra BA (2003) Understanding fragile X syndrome: insights from animal models. Cytogenet Genome Res 100:111–123

Bassell GJ, Warren ST (2008) Fragile X syndrome: loss of local mRNA regulation alters synaptic development and function. Neuron 60:201–214

Baude A, Nusser Z, Molnar E, McIlhinney RA, Somogyi P (1995) High-resolution immunogold localization of AMPA type glutamate receptor subunits at synaptic and non-synaptic sites in rat hippocampus. Neuroscience 69:1031–1055

Bear MF, Huber KM, Warren ST (2004) The mGluR theory of fragile X mental retardation. Trends Neurosci 27:370–377

Belichenko PVKA, Masliah E, Villar AJ, Epstein CJ, Mobley WC (2002) Excitatory-inhibitory relationships in fascia dentata in Ts65Dn mice. Soc Neurosci Abstract. p 5

Bergmann C, Zerres K, Senderek J, Rudnik-Schoneborn S, Eggermann T, Hausler M, Mull M, Ramaekers VT (2003) Oligophrenin 1 (OPHN1) gene mutation causes syndromic X-linked mental retardation with epilepsy, rostral ventricular enlargement and cerebellar hypoplasia. Brain 126:1537–1544

Bienvenu T, Der-Sarkissian H, Billuart P, Tissot M, des Portes V, Bruls T, Chabrolle JP, Chauveau P, Cherry M, Kahn A, Cohen D, Beldjord C, Chelly J, Cherif D (1997) Mapping of the X-breakpoint involved in a balanced X;12 translocation in a female with mild mental retardation. Eur J Hum Genet 5:105–109

Bienvenu T, des Portes V, McDonell N, Carrie A, Zemni R, Couvert P, Ropers HH, Moraine C, van Bokhoven H, Fryns JP, Allen K, Walsh CA, Boue J, Kahn A, Chelly J, Beldjord C (2000) Missense mutation in PAK3, R67C, causes X-linked nonspecific mental retardation. Am J Med Genet 93:294–298

Billuart P, Bienvenu T, Ronce N, des Portes V, Vinet MC, Zemni R, Roest Crollius H, Carrie A, Fauchereau F, Cherry M, Briault S, Hamel B, Fryns JP, Beldjord C, Kahn A, Moraine C, Chelly J (1998) Oligophrenin-1 encodes a rhoGAP protein involved in X-linked mental retardation. Nature 392:923–926

Boda B, Alberi S, Nikonenko I, Node-Langlois R, Jourdain P, Moosmayer M, Parisi-Jourdain L, Muller D (2004) The mental retardation protein PAK3 contributes to synapse formation and plasticity in hippocampus. J Neurosci 24:10816–10825

Boda B, Jourdain L, Muller D (2008) Distinct, but compensatory roles of PAK1 and PAK3 in spine morphogenesis. Hippocampus 18:857–861

Bokoch GM (2003) Biology of the p21-activated kinases. Annu Rev Biochem 72:743–781

Bredt DS, Nicoll RA (2003) AMPA receptor trafficking at excitatory synapses. Neuron 40:361–379

Chahrour M, Zoghbi HY (2007) The story of Rett syndrome: from clinic to neurobiology. Neuron 56:422–437

Chelly J, Khelfaoui M, Francis F, Cherif B, Bienvenu T (2006) Genetics and pathophysiology of mental retardation. Eur J Hum Genet 14:701–713

Chelly J, Mandel JL (2001) Monogenic causes of X-linked mental retardation. Nat Rev Genet 2:669–680

Chen Y, Bourne J, Pieribone VA, Fitzsimonds RM (2004) The role of actin in the regulation of dendritic spine morphology and bidirectional synaptic plasticity. Neuroreport 15:829–832

Cheng CM, Mervis RF, Niu SL, Salem N Jr., Witters LA, Tseng V, Reinhardt R, Bondy CA (2003) Insulin-like growth factor 1 is essential for normal dendritic growth. J Neurosci Res 73:1–9

Choi J, Ko J, Racz B, Burette A, Lee JR, Kim S, Na M, Lee HW, Kim K, Weinberg RJ, Kim E (2005) Regulation of dendritic spine morphogenesis by insulin receptor substrate 53, a downstream effector of Rac1 and Cdc42 small GTPases. J Neurosci 25:869–879

Choudhury R, Diao A, Zhang F, Eisenberg E, Saint-Pol A, Williams C, Konstantakopoulos A, Lucocq J, Johannes L, Rabouille C, Greene LE, Lowe M (2005) Lowe syndrome protein OCRL1 interacts with clathrin and regulates protein trafficking between endosomes and the trans-Golgi network. Mol Biol Cell 16:3467–3479

Cingolani LA, Goda Y (2008) Actin in action: the interplay between the actin cytoskeleton and synaptic efficacy. Nat Rev Neurosci 9:344–356

Citri A, Malenka RC (2008) Synaptic plasticity: multiple forms, functions, and mechanisms. Neuropsychopharmacology 33:18–41

Comery TA, Harris JB, Willems PJ, Oostra BA, Irwin SA, Weiler IJ, Greenough WT (1997) Abnormal dendritic spines in fragile X knockout mice: maturation and pruning deficits. Proc Natl Acad Sci USA 94:5401–5404

Corbetta S, Gualdoni S, Ciceri G, Monari M, Zuccaro E, Tybulewicz VL, de Curtis I (2009) Essential role of Rac1 and Rac3 GTPases in neuronal development. FASEB J 23:1347–1357

Dan C, Kelly A, Bernard O, Minden A (2001a) Cytoskeletal changes regulated by the PAK4 serine/threonine kinase are mediated by LIM kinase 1 and cofilin. J Biol Chem 276:32115–32121

Dan I, Watanabe NM, Kusumi A (2001b) The Ste20 group kinases as regulators of MAP kinase cascades. Trends Cell Biol 11:220–230

Daniels RH, Zenke FT, Bokoch GM (1999) Alphapix stimulates p21-activated kinase activity through exchange factor-dependent and -independent mechanisms. J Biol Chem 274: 6047–6050

des Portes V, Boddaert N, Sacco S, Briault S, Maincent K, Bahi N, Gomot M, Ronce N, Bursztyn J, Adamsbaum C, Zilbovicius M, Chelly J, Moraine C (2004) Specific clinical and brain MRI features in mentally retarded patients with mutations in the Oligophrenin-1 gene. Am J Med Genet A 124:364–371

Dierssen M, Benavides-Piccione R, Martinez-Cue C, Estivill X, Florez J, Elston GN, DeFelipe J (2003) Alterations of neocortical pyramidal cell phenotype in the Ts65Dn mouse model of Down syndrome: effects of environmental enrichment. Cereb Cortex 13:758–764

Dolen G, Bear MF (2008) Role for metabotropic glutamate receptor 5 (mGluR5) in the pathogenesis of fragile X syndrome. J Physiol 586:1503–1508

Dressman MA, Olivos-Glander IM, Nussbaum RL, Suchy SF (2000) OCRL1, a PtdIns(4,5)P(2) 5-phosphatase, is localized to the trans-Golgi network of fibroblasts and epithelial cells. J Histochem Cytochem 48:179–190

Eden S, Rohatgi R, Podtelejnikov AV, Mann M, Kirschner MW (2002) Mechanism of regulation of WAVE1-induced actin nucleation by Rac1 and Nck. Nature 418:790–793

Edwards DC, Sanders LC, Bokoch GM, Gill GN (1999) Activation of LIM-kinase by Pak1 couples Rac/Cdc42 GTPase signalling to actin cytoskeletal dynamics. Nat Cell Biol 1:253–259

Elia LP, Yamamoto M, Zang K, Reichardt LF (2006) p120 catenin regulates dendritic spine and synapse development through Rho-family GTPases and cadherins. Neuron 51:43–56

Endris V, Wogatzky B, Leimer U, Bartsch D, Zatyka M, Latif F, Maher ER, Tariverdian G, Kirsch S, Karch D, Rappold GA (2002) The novel Rho-GTPase activating gene MEGAP/srGAP3 has a putative role in severe mental retardation. Proc Natl Acad Sci USA 99:11754–11759

Erdmann KS, Mao Y, McCrea HJ, Zoncu R, Lee S, Paradise S, Modregger J, Biemesderfer D, Toomre D, De Camilli P (2007) A role of the Lowe syndrome protein OCRL in early steps of the endocytic pathway. Dev Cell 13:377–390

Fauchereau F, Herbrand U, Chafey P, Eberth A, Koulakoff A, Vinet MC, Ahmadian MR, Chelly J, Billuart P (2003) The RhoGAP activity of OPHN1, a new F-actin-binding protein, is negatively controlled by its amino-terminal domain. Mol Cell Neurosci 23:574–586

Faucherre A, Desbois P, Satre V, Lunardi J, Dorseuil O, Gacon G (2003) Lowe syndrome protein OCRL1 interacts with Rac GTPase in the trans-Golgi network. Hum Mol Genet 12:2449–2456

Feng Q, Albeck JG, Cerione RA, Yang W (2002) Regulation of the Cool/Pix proteins: key binding partners of the Cdc42/Rac targets, the p21-activated kinases. J Biol Chem 277:5644–5650

Feng Q, Baird D, Cerione RA (2004) Novel regulatory mechanisms for the Dbl family guanine nucleotide exchange factor Cool-2/alpha-Pix. EMBO J 23:3492–3504

Ferrer I, Gullotta F (1990) Down's syndrome and Alzheimer's disease: dendritic spine counts in the hippocampus. Acta Neuropathol (Berl) 79:680–685

Fiala JC, Spacek J, Harris KM (2002) Dendritic spine pathology: cause or consequence of neurological disorders? Brain Res Brain Res Rev 39:29–54

Frints SG, Froyen G, Marynen P, Fryns JP (2002) X-linked mental retardation: vanishing boundaries between non-specific (MRX) and syndromic (MRXS) forms. Clin Genet 62: 423–432

Fukazawa Y, Saitoh Y, Ozawa F, Ohta Y, Mizuno K, Inokuchi K (2003) Hippocampal LTP is accompanied by enhanced F-actin content within the dendritic spine that is essential for late LTP maintenance in vivo. Neuron 38:447–460

Galdzicki Z, Siarey RJ (2003) Understanding mental retardation in Down's syndrome using trisomy 16 mouse models. Genes Brain Behav 2:167–178

Garber K, Smith KT, Reines D, Warren ST (2006) Transcription, translation and fragile X syndrome. Curr Opin Genet Dev 16:270–275

Garber KB, Visootsak J, Warren ST (2008) Fragile X syndrome. Eur J Hum Genet 16:666–672

Gedeon AK, Nelson J, Gecz J, Mulley JC (2003) X-linked mild non-syndromic mental retardation with neuropsychiatric problems and the missense mutation A365E in PAK3. Am J Med Genet 120A:509–517

Govek EE, Newey SE, Akerman CJ, Cross JR, Van der Veken L, Van Aelst L (2004) The X-linked mental retardation protein oligophrenin-1 is required for dendritic spine morphogenesis. Nat Neurosci 7:364–372

Govek EE, Newey SE, Van Aelst L (2005) The role of the Rho GTPases in neuronal development. Genes Dev 19:1–49

Grossman AW, Elisseou NM, McKinney BC, Greenough WT (2006) Hippocampal pyramidal cells in adult Fmr1 knockout mice exhibit an immature-appearing profile of dendritic spines. Brain Res 1084:158–164

Halpain S, Spencer K, Graber S (2005) Dynamics and pathology of dendritic spines. Prog Brain Res 147:29–37

Harris KM, Jensen FE, Tsao B (1992) Three-dimensional structure of dendritic spines and synapses in rat hippocampus (CA1) at postnatal day 15 and adult ages: implications for the maturation of synaptic physiology and long-term potentiation. J Neurosci 12:2685–2705

Hayashi ML, Choi SY, Rao BS, Jung HY, Lee HK, Zhang D, Chattarji S, Kirkwood A, Tonegawa S (2004) Altered cortical synaptic morphology and impaired memory consolidation in forebrain-specific dominant-negative PAK transgenic mice. Neuron 42:773–787

Hayashi ML, Rao BS, Seo JS, Choi HS, Dolan BM, Choi SY, Chattarji S, Tonegawa S (2007) Inhibition of p21-activated kinase rescues symptoms of fragile X syndrome in mice. Proc Natl Acad Sci USA 104:11489–11494

Hellsten E, Evans JP, Bernard DJ, Janne PA, Nussbaum RL (2001) Disrupted sperm function and fertilin beta processing in mice deficient in the inositol polyphosphate 5-phosphatase Inpp5b. Dev Biol 240:641–653

Hinton VJ, Brown WT, Wisniewski K, Rudelli RD (1991) Analysis of neocortex in three males with the fragile X syndrome. Am J Med Genet 41:289–294

Holtmaat AJ, Trachtenberg JT, Wilbrecht L, Shepherd GM, Zhang X, Knott GW, Svoboda K (2005) Transient and persistent dendritic spines in the neocortex in vivo. Neuron 45:279–291

Honkura N, Matsuzaki M, Noguchi J, Ellis-Davies GC, Kasai H (2008) The subspine organization of actin fibers regulates the structure and plasticity of dendritic spines. Neuron 57:719–729

Huttenlocher PR (1970) Dendritic development and mental defect. Neurology 20:381

Huttenlocher PR (1974) Dendritic development in neocortex of children with mental defect and infantile spasms. Neurology 24:203–210

Hyvola N, Diao A, McKenzie E, Skippen A, Cockcroft S, Lowe M (2006) Membrane targeting and activation of the Lowe syndrome protein OCRL1 by rab GTPases. EMBO J 25:3750–3761

Inlow JK, Restifo LL (2004) Molecular and comparative genetics of mental retardation. Genetics 166:835–881

Irie F, Yamaguchi Y (2002) EphB receptors regulate dendritic spine development via intersectin, Cdc42 and N-WASP. Nat Neurosci 5:1117–1118

Irwin SA, Patel B, Idupulapati M, Harris JB, Crisostomo RA, Larsen BP, Kooy F, Willems PJ, Cras P, Kozlowski PB, Swain RA, Weiler IJ, Greenough WT (2001) Abnormal dendritic spine characteristics in the temporal and visual cortices of patients with fragile-X syndrome: a quantitative examination. Am J Med Genet 98:161–167

Jaffer ZM, Chernoff J (2002) p21-activated kinases: three more join the Pak. Int J Biochem Cell Biol 34:713–717

Janne PA, Suchy SF, Bernard D, MacDonald M, Crawley J, Grinberg A, Wynshaw-Boris A, Westphal H, Nussbaum RL (1998) Functional overlap between murine Inpp5b and OCRL1 may explain why deficiency of the murine ortholog for OCRL1 does not cause Lowe syndrome in mice. J Clin Invest 101:2042–2053

Jefferson AB, Majerus PW (1995) Properties of type II inositol polyphosphate 5-phosphatase. J Biol Chem 270:9370–9377

Kandel ER (2001) The molecular biology of memory storage: a dialogue between genes and synapses. Science 294:1030–1038

Kang MG, Guo Y, Huganir RL (2009) AMPA receptor and GEF-H1/Lfc complex regulates dendritic spine development through RhoA signaling cascade. Proc Natl Acad Sci USA 106:3549–3554

Kaufmann WE, Moser HW (2000) Dendritic anomalies in disorders associated with mental retardation. Cereb Cortex 10:981–991

Kharazia VN, Weinberg RJ (1999) Immunogold localization of AMPA and NMDA receptors in somatic sensory cortex of albino rat. J Comp Neurol 412:292–302

Khelfaoui M, Denis C, van Galen E, de Bock F, Schmitt A, Houbron C, Morice E, Giros B, Ramakers G, Fagni L, Chelly J, Nosten-Bertrand M, Billuart P (2007) Loss of X-linked mental retardation gene oligophrenin1 in mice impairs spatial memory and leads to ventricular enlargement and dendritic spine immaturity. J Neurosci 27:9439–9450

Khelfaoui M, Pavlowsky A, Powell AD, Valnegri P, Cheong KW, Blandin Y, Passafaro M, Jefferys JG, Chelly J, Billuart P (2009) Inhibition of RhoA pathway rescues the endocytosis defects in Oligophrenin-1 mouse model of mental retardation. Hum Mol Genet 18: 2575–2583

Kim CH, Lisman JE (1999) A role of actin filament in synaptic transmission and long-term potentiation. J Neurosci 19:4314–4324

Kimura K, Ito M, Amano M, Chihara K, Fukata Y, Nakafuku M, Yamamori B, Feng J, Nakano T, Okawa K, Iwamatsu A, Kaibuchi K (1996) Regulation of myosin phosphatase by Rho and Rho-associated kinase (Rho-kinase). Science 273:245–248

Kobayashi K, Kuroda S, Fukata M, Nakamura T, Nagase T, Nomura N, Matsuura Y, Yoshida-Kubomura N, Iwamatsu A, Kaibuchi K (1998) p140Sra-1 (specifically Rac1-associated protein) is a novel specific target for Rac1 small GTPase. J Biol Chem 273:291–295

Koh CG, Manser E, Zhao ZS, Ng CP, Lim L (2001) Beta1PIX, the PAK-interacting exchange factor, requires localization via a coiled-coil region to promote microvillus-like structures and membrane ruffles. J Cell Sci 114:4239–4251

Kopec CD, Li B, Wei W, Boehm J, Malinow R (2006) Glutamate receptor exocytosis and spine enlargement during chemically induced long-term potentiation. J Neurosci 26:2000–2009

Kopec CD, Real E, Kessels HW, Malinow R (2007) GluR1 links structural and functional plasticity at excitatory synapses. J Neurosci 27:13706–13718

Kreis P, Barnier JV (2009) PAK signalling in neuronal physiology. Cell Signal 21:384–393

Kreis P, Thevenot E, Rousseau V, Boda B, Muller D, Barnier JV (2007) The p21-activated kinase 3 implicated in mental retardation regulates spine morphogenesis through a Cdc42-dependent pathway. J Biol Chem 282:21497–21506

Krucker T, Siggins GR, Halpain S (2000) Dynamic actin filaments are required for stable long-term potentiation (LTP) in area CA1 of the hippocampus. Proc Natl Acad Sci USA 97: 6856–6861

Kutsche K, Yntema H, Brandt A, Jantke I, Nothwang HG, Orth U, Boavida MG, David D, Chelly J, Fryns JP, Moraine C, Ropers HH, Hamel BC, van Bokhoven H, Gal A (2000) Mutations in ARHGEF6, encoding a guanine nucleotide exchange factor for Rho GTPases, in patients with X-linked mental retardation. Nat Genet 26:247–250

LaFerla FM, Oddo S (2005) Alzheimer's disease: abeta, tau and synaptic dysfunction. Trends Mol Med 11:170–176

Lee A, Li W, Xu K, Bogert BA, Su K, Gao FB (2003) Control of dendritic development by the Drosophila fragile X-related gene involves the small GTPase Rac1. Development 130: 5543–5552

Lin B, Kramar EA, Bi X, Brucher FA, Gall CM, Lynch G (2005) Theta stimulation polymerizes actin in dendritic spines of hippocampus. J Neurosci 25:2062–2069

Lonze BE, Ginty DD (2002) Function and regulation of CREB family transcription factors in the nervous system. Neuron 35:605–623

Lowe M (2005) Structure and function of the Lowe syndrome protein OCRL1. Traffic 6: 711–719

Lu R, Wang H, Liang Z, Ku L, O'Donnell WT, Li W, Warren ST, Feng Y (2004) The fragile X protein controls microtubule-associated protein 1B translation and microtubule stability in brain neuron development. Proc Natl Acad Sci USA 101:15201–15206

Luo L, Hensch TK, Ackerman L, Barbel S, Jan LY, Jan YN (1996) Differential effects of the Rac GTPase on Purkinje cell axons and dendritic trunks and spines. Nature 379:837–840

Ma QL, Yang F, Calon F, Ubeda OJ, Hansen JE, Weisbart RH, Beech W, Frautschy SA, Cole GM (2008) p21-activated kinase-aberrant activation and translocation in Alzheimer disease pathogenesis. J Biol Chem 283:14132–14143

Maekawa M, Ishizaki T, Boku S, Watanabe N, Fujita A, Iwamatsu A, Obinata T, Ohashi K, Mizuno K, Narumiya S (1999) Signaling from Rho to the actin cytoskeleton through protein kinases ROCK and LIM-kinase. Science 285:895–898

Malinow R, Malenka RC (2002) AMPA receptor trafficking and synaptic plasticity. Annu Rev Neurosci 25:103–126

Mandel JL, Chelly J (2004) Monogenic X-linked mental retardation: is it as frequent as currently estimated? The paradox of the ARX (Aristaless X) mutations. Eur J Hum Genet 12:689–693

Manser E, Loo TH, Koh CG, Zhao ZS, Chen XQ, Tan L, Tan I, Leung T, Lim L (1998) PAK kinases are directly coupled to the PIX family of nucleotide exchange factors. Mol Cell 1:183–192

Marin-Padilla M (1972) Structural abnormalities of the cerebral cortex in human chromosomal aberrations: a Golgi study. Brain Res 44:625–629

Marin-Padilla M (1976) Pyramidal cell abnormalities in the motor cortex of a child with Down's syndrome. A Golgi study. J Comp Neurol 167:63–81

Matsuzaki M, Honkura N, Ellis-Davies GC, Kasai H (2004) Structural basis of long-term potentiation in single dendritic spines. Nature 429:761–766

Matus A (2005) Growth of dendritic spines: a continuing story. Curr Opin Neurobiol 15:67–72

McCrea HJ, Paradise S, Tomasini L, Addis M, Melis MA, De Matteis MA, De Camilli P (2008) All known patient mutations in the ASH-RhoGAP domains of OCRL affect targeting and APPL1 binding. Biochem Biophys Res Commun 369:493–499

Meng J, Meng Y, Hanna A, Janus C, Jia Z (2005) Abnormal long-lasting synaptic plasticity and cognition in mice lacking the mental retardation gene Pak3. J Neurosci 25:6641–6650

Nadif Kasri N, Nakano-Kobayashi A, Malinow R, Li B, Van Aelst L (2009) The Rho-linked mental retardation protein oligophrenin-1 controls synapse maturation and plasticity by stabilizing AMPA receptors. Genes Dev 23:1289–1302

Nakano-Kobayashi A, Nadif Kasri N, Newey SE, Van Aelst L (2009) The Rho-linked mental retardation protein OPHN1 controls synaptic vesicle endocytosis via endophilin A1. Curr Biol 19:1133–1139

Nakayama AY, Harms MB, Luo L (2000) Small GTPases Rac and Rho in the maintenance of dendritic spines and branches in hippocampal pyramidal neurons. J Neurosci 20: 5329–5338

Newey SE, Velamoor V, Govek EE, Van Aelst L (2005) Rho GTPases, dendritic structure, and mental retardation. J Neurobiol 64:58–74

Nimchinsky EA, Oberlander AM, Svoboda K (2001) Abnormal development of dendritic spines in FMR1 knock-out mice. J Neurosci 21:5139–5146

Nishimura T, Yamaguchi T, Tokunaga A, Hara A, Hamaguchi T, Kato K, Iwamatsu A, Okano H, Kaibuchi K (2006) Role of numb in dendritic spine development with a Cdc42 GEF intersectin and EphB2. Mol Biol Cell 17:1273–1285

Node-Langlois R, Muller D, Boda B (2006) Sequential implication of the mental retardation proteins ARHGEF6 and PAK3 in spine morphogenesis. J Cell Sci 119:4986–4993

Nusser Z, Lujan R, Laube G, Roberts JD, Molnar E, Somogyi P (1998) Cell type and pathway dependence of synaptic AMPA receptor number and variability in the hippocampus. Neuron 21:545–559

Ohashi K, Nagata K, Maekawa M, Ishizaki T, Narumiya S, Mizuno K (2000) Rho-associated kinase ROCK activates LIM-kinase 1 by phosphorylation at threonine 508 within the activation loop. J Biol Chem 275:3577–3582

Okamoto K, Nagai T, Miyawaki A, Hayashi Y (2004) Rapid and persistent modulation of actin dynamics regulates postsynaptic reorganization underlying bidirectional plasticity. Nat Neurosci 7:1104–1112

Olivos-Glander IM, Janne PA, Nussbaum RL (1995) The oculocerebrorenal syndrome gene product is a 105-kD protein localized to the Golgi complex. Am J Hum Genet 57: 817–823

Parnas D, Haghighi AP, Fetter RD, Kim SW, Goodman CS (2001) Regulation of postsynaptic structure and protein localization by the Rho-type guanine nucleotide exchange factor dPix. Neuron 32:415–424

Peippo M, Koivisto AM, Sarkamo T, Sipponen M, von Koskull H, Ylisaukko-oja T, Rehnstrom K, Froyen G, Ignatius J, Jarvela I (2007) PAK3 related mental disability: further characterization of the phenotype. Am J Med Genet A 143A:2406–2416

Penzes P, Beeser A, Chernoff J, Schiller MR, Eipper BA, Mains RE, Huganir RL (2003) Rapid induction of dendritic spine morphogenesis by trans-synaptic ephrinB-EphB receptor activation of the Rho-GEF kalirin. Neuron 37:263–274

Penzes P, Cahill ME, Jones KA, Srivastava DP (2008) Convergent CaMK and RacGEF signals control dendritic structure and function. Trends Cell Biol 18:405–413

Philip N, Chabrol B, Lossi AM, Cardoso C, Guerrini R, Dobyns WB, Raybaud C, Villard L (2003) Mutations in the oligophrenin-1 gene (OPHN1) cause X linked congenital cerebellar hypoplasia. J Med Genet 40:441–446

Pilpel Y, Segal M (2004) Activation of PKC induces rapid morphological plasticity in dendrites of hippocampal neurons via Rac and Rho-dependent mechanisms. Eur J Neurosci 19: 3151–3164

Poirier K, Lacombe D, Gilbert-Dussardier B, Raynaud M, Desportes V, de Brouwer AP, Moraine C, Fryns JP, Ropers HH, Beldjord C, Chelly J, Bienvenu T (2006) Screening of ARX in mental retardation families: consequences for the strategy of molecular diagnosis. Neurogenetics 7:39–46

Ponting CP (2006) A novel domain suggests a ciliary function for ASPM, a brain size determining gene. Bioinformatics 22:1031–1035

Purpura DP (1974) Dendritic spine "dysgenesis" and mental retardation. Science 186: 1126–1128

Reeve SP, Bassetto L, Genova GK, Kleyner Y, Leyssen M, Jackson FR, Hassan BA (2005) The Drosophila fragile X mental retardation protein controls actin dynamics by directly regulating profilin in the brain. Curr Biol 15:1156–1163

Rejeb I, Saillour Y, Castelnau L, Julien C, Bienvenu T, Taga P, Chaabouni H, Chelly J, Ben Jemaa L, Bahi-Buisson N (2008) A novel splice mutation in PAK3 gene underlying mental retardation with neuropsychiatric features. Eur J Hum Genet 16:1358–1363

Ropers HH, Hamel BC (2005) X-linked mental retardation. Nat Rev Genet 6:46–57

Rudelli RD, Brown WT, Wisniewski K, Jenkins EC, Laure-Kamionowska M, Connell F, Wisniewski HM (1985) Adult fragile X syndrome. Clinico-neuropathologic findings. . Acta Neuropathol (Berl) 67:289–295

Ryan XP, Alldritt J, Svenningsson P, Allen PB, Wu GY, Nairn AC, Greengard P (2005) The Rho-specific GEF Lfc interacts with neurabin and spinophilin to regulate dendritic spine morphology. Neuron 47:85–100

Ryu J, Liu L, Wong TP, Wu DC, Burette A, Weinberg R, Wang YT, Sheng M (2006) A critical role for myosin IIb in dendritic spine morphology and synaptic function. Neuron 49: 175–182

Sanders LC, Matsumura F, Bokoch GM, de Lanerolle P (1999) Inhibition of myosin light chain kinase by p21-activated kinase. Science 283:2083–2085

Saneyoshi T, Wayman G, Fortin D, Davare M, Hoshi N, Nozaki N, Natsume T, Soderling TR (2008) Activity-dependent synaptogenesis: regulation by a CaM-kinase kinase/CaM-kinase I/betaPIX signaling complex. Neuron 57:94–107

Schenck A, Bardoni B, Langmann C, Harden N, Mandel JL, Giangrande A (2003) CYFIP/Sra-1 controls neuronal connectivity in Drosophila and links the Rac1 GTPase pathway to the fragile X protein. Neuron 38:887–898

Schmid AC, Wise HM, Mitchell CA, Nussbaum R, Woscholski R (2004) Type II phosphoinositide 5-phosphatases have unique sensitivities towards fatty acid composition and head group phosphorylation. FEBS Lett 576:9–13

Schubert V, Da Silva JS, Dotti CG (2006) Localized recruitment and activation of RhoA underlies dendritic spine morphology in a glutamate receptor-dependent manner. J Cell Biol 172: 453–467

Scott EK, Reuter JE, Luo L (2003) Small GTPase Cdc42 is required for multiple aspects of dendritic morphogenesis. J Neurosci 23:3118–3123

Sfakianos MK, Eisman A, Gourley SL, Bradley WD, Scheetz AJ, Settleman J, Taylor JR, Greer CA, Williamson A, Koleske AJ (2007) Inhibition of Rho via Arg and p190RhoGAP in the postnatal mouse hippocampus regulates dendritic spine maturation, synapse and dendrite stability, and behavior. J Neurosci 27:10982–10992

Shepherd JD, Huganir RL (2007) The cell biology of synaptic plasticity: AMPA receptor trafficking. Annu Rev Cell Dev Biol 23:613–643

Soderling SH, Binns KL, Wayman GA, Davee SM, Ong SH, Pawson T, Scott JD (2002) The WRP component of the WAVE-1 complex attenuates Rac-mediated signalling. Nat Cell Biol 4:970–975

Soderling SH, Guire ES, Kaech S, White J, Zhang F, Schutz K, Langeberg LK, Banker G, Raber J, Scott JD (2007) A WAVE-1 and WRP signaling complex regulates spine density, synaptic plasticity, and memory. J Neurosci 27:355–365

Soderling SH, Langeberg LK, Soderling JA, Davee SM, Simerly R, Raber J, Scott JD (2003) Loss of WAVE-1 causes sensorimotor retardation and reduced learning and memory in mice. Proc Natl Acad Sci USA 100:1723–1728

Sotgia F, Minetti C, Lisanti MP (1999) Localization of the human caveolin-3 gene to the D3S18/D3S4163/D3S4539 locus (3p25), in close proximity to the human oxytocin receptor gene. Identification of the caveolin-3 gene as a candidate for deletion in 3p-syndrome. FEBS Lett 452:177–180

Stanyon CA, Bernard O (1999) LIM-kinase1. Int J Biochem Cell Biol 31:389–394

Suchy SF, Nussbaum RL (2002) The deficiency of PIP2 5-phosphatase in Lowe syndrome affects actin polymerization. Am J Hum Genet 71:1420–1427

Suetsugu M, Mehraein P (1980) Spine distribution along the apical dendrites of the pyramidal neurons in Down's syndrome. A quantitative Golgi study. Acta Neuropathol (Berl) 50:207–210

Sumi T, Matsumoto K, Takai Y, Nakamura T (1999) Cofilin phosphorylation and actin cytoskeletal dynamics regulated by rho- and Cdc42-activated LIM-kinase 2. J Cell Biol 147: 1519–1532

Tada T, Sheng M (2006) Molecular mechanisms of dendritic spine morphogenesis. Curr Opin Neurobiol 16:95–101

Takahashi H, Sekino Y, Tanaka S, Mizui T, Kishi S, Shirao T (2003) Drebrin-dependent actin clustering in dendritic filopodia governs synaptic targeting of postsynaptic density-95 and dendritic spine morphogenesis. J Neurosci 23:6586–6595

Takashima S, Becker LE, Armstrong DL, Chan F (1981) Abnormal neuronal development in the visual cortex of the human fetus and infant with down's syndrome. A quantitative and qualitative Golgi study. Brain Res 225:1–21

Takashima S, Iida K, Mito T, Arima M (1994) Dendritic and histochemical development and ageing in patients with Down's syndrome. J Intellect Disabil Res 38(Pt 3):265–273

Takumi Y, Ramirez-Leon V, Laake P, Rinvik E, Ottersen OP (1999) Different modes of expression of AMPA and NMDA receptors in hippocampal synapses. Nat Neurosci 2:618–624

Tashiro A, Minden A, Yuste R (2000) Regulation of dendritic spine morphology by the rho family of small GTPases: antagonistic roles of Rac and Rho. Cereb Cortex 10:927–938

Tashiro A, Yuste R (2004) Regulation of dendritic spine motility and stability by Rac1 and Rho kinase: evidence for two forms of spine motility. Mol Cell Neurosci 26:429–440

Tentler D, Gustavsson P, Leisti J, Schueler M, Chelly J, Timonen E, Anneren G, Willard HF, Dahl N (1999) Deletion including the oligophrenin-1 gene associated with enlarged cerebral ventricles, cerebellar hypoplasia, seizures and ataxia. Eur J Hum Genet 7:541–548

Tolias KF, Bikoff JB, Burette A, Paradis S, Harrar D, Tavazoie S, Weinberg RJ, Greenberg ME (2005) The Rac1-GEF Tiam1 couples the NMDA receptor to the activity-dependent development of dendritic arbors and spines. Neuron 45:525–538

Tolias KF, Bikoff JB, Kane CG, Tolias CS, Hu L, Greenberg ME (2007) The Rac1 guanine nucleotide exchange factor Tiam1 mediates EphB receptor-dependent dendritic spine development. Proc Natl Acad Sci USA 104:7265–7270

Ungewickell A, Ward ME, Ungewickell E, Majerus PW (2004) The inositol polyphosphate 5-phosphatase Ocrl associates with endosomes that are partially coated with clathrin. Proc Natl Acad Sci USA 101:13501–13506

Vaillend C, Poirier R, Laroche S (2008) Genes, plasticity and mental retardation. Behav Brain Res 192(1):88–105

Van Aelst L, Cline HT (2004) Rho GTPases and activity-dependent dendrite development. Curr Opin Neurobiol 14:297–304

Van Aelst L, D'Souza-Schorey C (1997) Rho GTPases and signaling networks. Genes Dev 11:2295–2322

van Galen EJ, Ramakers GJ (2005) Rho proteins, mental retardation and the neurobiological basis of intelligence. Prog Brain Res 147:295–317

Wiens KM, Lin H, Liao D (2005) Rac1 induces the clustering of AMPA receptors during spinogenesis. J Neurosci 25:10627–10636

Wisniewski KE, Segan SM, Miezejeski CM, Sersen EA, Rudelli RD (1991) The Fra(X) syndrome: neurological, electrophysiological, and neuropathological abnormalities. Am J Med Genet 38:476–480

Wong K, Ren XR, Huang YZ, Xie Y, Liu G, Saito H, Tang H, Wen L, Brady-Kalnay SM, Mei L, Wu JY, Xiong WC, Rao Y (2001) Signal transduction in neuronal migration: roles of GTPase activating proteins and the small GTPase Cdc42 in the Slit-Robo pathway. Cell 107:209–221

Xie Z, Srivastava DP, Photowala H, Kai L, Cahill ME, Woolfrey KM, Shum CY, Surmeier DJ, Penzes P (2007) Kalirin-7 controls activity-dependent structural and functional plasticity of dendritic spines. Neuron 56:640–656

Yang N, Higuchi O, Ohashi K, Nagata K, Wada A, Kangawa K, Nishida E, Mizuno K (1998) Cofilin phosphorylation by LIM-kinase 1 and its role in Rac-mediated actin reorganization. Nature 393:809–812

Yuste R, Bonhoeffer T (2004) Genesis of dendritic spines: insights from ultrastructural and imaging studies. Nat Rev Neurosci 5:24–34

Zanni G, Saillour Y, Nagara M, Billuart P, Castelnau L, Moraine C, Faivre L, Bertini E, Durr A, Guichet A, Rodriguez D, des Portes V, Beldjord C, Chelly J (2005) Oligophrenin 1 mutations frequently cause X-linked mental retardation with cerebellar hypoplasia. Neurology 65: 1364–1369

Zhang X, Jefferson AB, Auethavekiat V, Majerus PW (1995) The protein deficient in Lowe syndrome is a phosphatidylinositol-4,5-bisphosphate 5-phosphatase. Proc Natl Acad Sci USA 92:4853–4856

Zhang H, Macara IG (2008) The PAR-6 polarity protein regulates dendritic spine morphogenesis through p190 RhoGAP and the Rho GTPase. Dev Cell 14:216–226

Zhang H, Webb DJ, Asmussen H, Horwitz AF (2003) Synapse formation is regulated by the signaling adaptor GIT1. J Cell Biol 161:131–142

Zhang H, Webb DJ, Asmussen H, Niu S, Horwitz AF (2005) A GIT1/PIX/Rac/PAK signaling module regulates spine morphogenesis and synapse formation through MLC. J Neurosci 25:3379–3388

Zhao L, Ma QL, Calon F, Harris-White ME, Yang F, Lim GP, Morihara T, Ubeda OJ, Ambegaokar S, Hansen JE, Weisbart RH, Teter B, Frautschy SA, Cole GM (2006) Role of p21-activated kinase pathway defects in the cognitive deficits of Alzheimer disease. Nat Neurosci 9:234–242

Cognitive Deficits in Neurodegenerative Disorders: Parkinson's Disease and Alzheimer's Disease

Ivan Bodis-Wollner and Herman Moreno

Abstract Among the most important neurodegenerative disorders affecting aging adults around the world are Alzheimer's disease (AD), which affects around 4.5 million people in the United States, and Parkinson's disease (PD), which has a prevalence of about 160 cases per 100,000 people and an incidence of about 20 cases per 100,000 people per year. In both disorders prevalence and incidence increase with age. AD is the main dementing disorder, whereas cognitive dysfunction and dementia eventually occur in 20–40% of patients with PD. In this chapter the neurobiology of these diseases is reviewed. Classical, anatomically defined local circuits are summarized. Data obtained using advanced imaging techniques, such as SPECT, and functional MRI, and electrophysiological recordings, are highlighted. The main emphasis for both PD and AD is on cognitive deficits from the perspective of brain circuits and synaptic physiological abnormalities as well as on their biochemical correlates. In particular, among nonmotor defects in Parkinson's disease sensory deficits are also emphasized in relation to visuocogntive and attentive dysfunction. The main neurotransmitter systems involved are dopamine (in PD) and acetylcholine (both in PD and AD). The logic role of dopamine in the retinal circuitry is discussed in relation to sensory (visual) dysfunction in PD. The contribution of neurotransmitter/modulators beyond the dopaminergic and cholinergic systems in the basal ganglia and in several cortical areas is reviewed. This involves glutamate, adenosine, and GABA. The cognitive effect of genetic variability of catechol-o-methyltranferase, in the prefrontal cortex is summarized. Although advances in the understanding of AD and PD pathophysiology have been significant, fundamental issues remain unsolved. The powerful neuropathological arguments concerning the progression of PD based on alpha synuclein predict late involvement of cortical

---

I. Bodis-Wollner (✉)
Department of Neurology, SUNY Downstate Medical Center, Brooklyn, NY 11203, USA; Department of Ophthalmology, SUNY Downstate Medical Center, Brooklyn, NY 11203, USA; Division of Movement Disorders, Department of Neurology, Parkinson's Disease and Related Disorders Clinic, Center of Excellence NPF, Brooklyn, NY 11203, USA
e-mail: ivan.bodis-wollner@downstate.edu

J.P. Blass (ed.), *Neurochemical Mechanisms in Disease*,
Advances in Neurobiology 1, DOI 10.1007/978-1-4419-7104-3_9,

circuits, presumably responsible for cognitive changes. The role of acetylcholine and diverse cholinergic receptors in cognitive dysfunction in both AD and PD will need further studies. Future studies may potentially lead to a bridging theory of cognitive impairment in both AD and PD.

Keywords Dopamine · Basal ganglia · Frontal cortex · Striatum · Vision · D1 and D2 dopamine receptors · Retina · Visual cognition · GABA · Subthalamic nucleus · Cholinergic mechanisms · Glutamate · Adenosine · Thalamocortical processing · Alzheimer's disease (AD) · Mild cognitive impairment (MCI) · Amyloid precursor protein (APP) · Amyloid beta (Aβ) tau · Apolipoprotein E ε4 (APOE4) · Cerebral blood volume (CBV) · Positron emission tomography (PET) · Magnetic resonance imaging (MRI)

Contents

1 Introduction

Parkinson's disease (PD) is the second most common neurodegenerative disorder after Alzheimer's disease. With increased longevity and improved health care, our society is experiencing an unprecedented challenge posed by these neurodegenerative disorders. Alzheimer's disease alone is now the third most expensive disease to treat in the United States, costing close to $100 billion annually.

The availability of genetically modified mice has advanced our understanding of several neurodegenerative processes. Cellular neurobiology experiments have informed us about mechanisms of neuronal dysfunction in AD and PD mouse models. For instance, recent studies have identified that synaptic transmission is one of the earliest events in the cognitive abnormalities that characterize AD and PD. The integration of this information with data-based circuits modeling, in which neuronal electrical properties, synaptic transmission parameters, and brain oscillations can now be evaluated and it has been recently addressed in PD and AD.

The recent availability of in vivo studies such as functional magnetic resonance imaging (fMRI) that can noninvasively obtain information about the metabolic state of the brain thus allows the use of longitudinal and cross-species studies designs in both neurodegenerative mouse models and patients. These types of work carry our knowledge from basic neuroscience to clinical neurology and back, to combine basic science with clinically used methods to address clinical issues in the neurobiology of aging and neurodegenerative diseases. Knowledge of basic aspects of cellular physiological changes induced by specific pathologies, that is, synuclein-related in PD or amyloid-related in AD is essential to identify potential therapeutic targets. The present chapter discusses data that exemplify relevant findings on synaptic and circuit abnormalities in relation to PD and AD and clinical information that may correlate with such abnormalities or complement it.

2 Parkinson's Disease: An Overview

Parkinson's disease is a progressive neurological disorder that affects critical domains of daily living. Since its first description as a "shaking palsy" it has been recognized that it affects many nonmotor functions. PD causes autonomic, sensory, cognitive, and behavioral problems all of which can significantly impair quality of life. The original observations by Carlsson and his collegues (1957) that 3-4-Dihydroxyphenylalanine reverses reserpine-induced akinesia in rabbits led to the first clinical use of levodopa by Birkmayer and Hornykiewicz (1961).

Cognitive impairment, labeled as moderate to marked "dementia" was first reported in a large number of patients to occur tenfold more commonly than in controls (Lieberman et al., 1979). Demented PD patients in this study responded less well to levodopa therapy and it was suggested that PD with dementia may represent a different disorder from PD without dementia. We distinguish select cognitive impairments from dementia in PD and discuss neurochemical and molecular mechanisms of cognition in PD.

Cognitive impairment in PD is traditionally specified with neuropsychological testing, performed with clinically validated methods. In the recent decade, however, considerable knowledge was gained from imaging (PET and fMRI) and EEG and "event"-related potentials (ERP) studies. They use specific experimental cognitive paradigms to tax select cognitive operations, such as executive functions, short-term memory, and visuospatial orientation.

In the last decades, administration of new forms of therapy in addition to levodopa, has led to a better understanding of the role of dopaminergic and non-dopaminergic circuits in PD. Many scientists are engaged in trying to specify the neurochemical building blocks of PD and develop a rational pharmacotherapy of the whole disease and not just its motor manifestations. There are several promising avenues of applying so-called "dirty drugs," that is, medications that have more than one effect in the CNS. The circuitry responsible for cognitive and some other manifestations of the disease is still based on the core abnormality of dopaminergic deficiency in the basal ganglia, however, our understanding of the basic basal ganglia circuitry has undergone revision and refinement. One contribution to a better neuroanatomical understanding is based on the effects of nondopaminergic manipulations, in particular transcranial magnetic stimulation (TMS) and deep brain stimulation (DBS).

DBS as a treatment for PD was introduced nearly two decades ago. The electrophysiological results obtained with DBS have contributed considerable new knowledge regarding the basal ganglia/cortex circuits responsible for the manifestations of PD. Human intraoperative monitoring and electrophysiological recordings combined with microdialysis have yielded biochemical evidence on the role of GABA and the subthalamic and the pedunculopontine nucleus in the PD.

We discuss the evolution of the development of PD based on the neuropathological studies of Braak and collegues. Based on the distribution of Lewy bodies

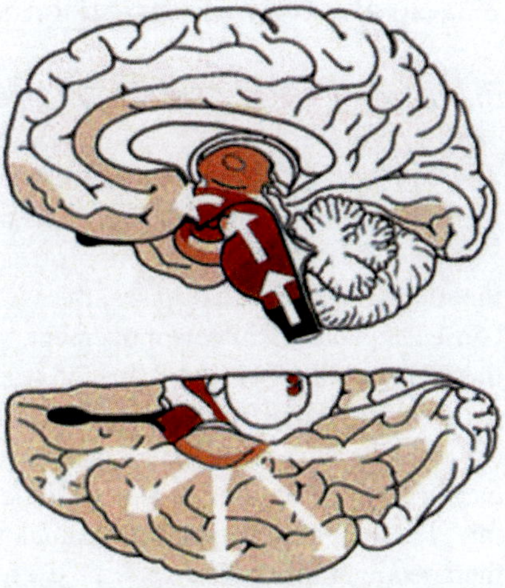

Fig. 1 In the CNS, the pathological process of PD commences in the dorsal motor nucleus of the vagal nerve (medulla oblongata) from where it follows an ascending course, affecting additional nuclei in the lower brainstem, in basal portions of the mid- and forebrain, and eventually reaching the cerebral cortex (Braak et al., 2004). However, as summarized, several neuronal circuits that are relevant to cognitive/emotional changes in PD are also involved. In these circuits not all DA neurons have long axons

at different stages of PD, Braak et al. (2002) suggested that PD progresses from peripheral to central long axon projecting neurons in a caudal-cranial direction (Fig. 1). Braak's neuroanatomical model predicts relatively late occurrence of cognitive changes in PD, but not all clinical data and observations fit neatly into the model.

One defined dopaminergic circuitry affected in PD is the retina. The retinal dopaminergic circuit is of interest from several points of view. For one, it provides an opportunity to study the logic role of diverse dopamine receptors. The retina is a multilayered structure with a single output line and three layers of neurons with lateral and feedback connections. Second, many patients with PD suffer from visual hallucinations and vision in PD can be to a great degree attributed to specifically impaired retinal processing. Anatomical and electrophysiological data show both in humans and in MPTP-treated monkeys that retinal dopaminergic deficiency may be one reason for visual dysfunction in PD. Neuropharmacological manipulations in the monkey model of PD have led to a better understanding of the "antagonistic" role of D1 and D2 type recptors. The retina has not been subjected to studies for evaluating the presence of alpha synuclein. On the other hand, animal and human studies suggest a correlation of motor disease progression with visual dysfunction. Affected retinal dopaminergic neurons do not have long projections. Recent imaging data, however, show neuronal thinning involving both inner nuclear layers and retinal ganglion cells. Retinal ganglion cells are among neurons having the longest axons in the CNS. Whether PD retinopathy starts in a subset of ganglion cells, before (consistent with the postulates of Braak) the process attacks dopaminergic neurons, is unknown at present.

3 Neurobiology of Parkinson's Disease

3.1 Etiology and Molecular Progression of PD

The etiology of PD still eludes us; however, an understanding of cognitive impairment, linked to the classical concepts of PD, is emerging.

Since the original description of Parkinson's disease in 1817, as shaking palsy that does not affect the senses, the last decades have witnessed a slowly emerging consensus that PD affects movement, sensation, cognition, and mood. Furthermore these noncardinal features do not respond well to dopaminergic therapy. Recent pathological data (Braak et al., 2002) suggest a caudo-cranial evolution of the disease progressing from nondopaminergic and nonmotor extracerebral neurons to basal ganglia to cortical neurons. Hence some nonmotor symptoms may be heralding PD. The possibility of precardinal and premotor diagnosis is challenging from a therapeutic point of view

Although the MPTP model is nearly perfect once developed, it is a toxic model that occurs suddenly after the introduction of the toxin, whereas PD is a progressive disease. Nevertheless the elucidation of the action of MPTP on the mitochondrial respiratory chain, more precisely on the step between I and II has led to the search to the search for agents with neuroprotective abilities in the mitochondrial respiratory chain.

3.2 PD as a Synucleinopathy

The degenerative process of idiopathic PD (iPD) is associated with the anatomical progression of CEB1-synuclein into select neurons. It is an aggregate of the misfolded protein and appears in dendrites and axons, Lewy neurites (LNs), as well as punctuate structures and/or Lewy bodies (LBs) in the somata of involved nerve cells. It is thought that only projection neurons with a long axon become involved, whereas short-axoned cells resist the pathology. It has been proposed that vulnerable brain regions in PD are anatomically interconnected.

Before the era of synuclein immunocytochemistry, Qualman et al. (1984) observed in a postmortem study LBs in the esophageal Auerbach plexus of two dysphagic PD patients but not in Meissner's plexus. Subsequently, Wakabayashi et al. (1988) reported LBs and LNs in both plexuses of clinically diagnosed PD patients and asymptomatic incidental cases. In the gut, the bulk of the proteins were observed in cellular processes and cell bodies of vasoactive intestinal polypeptide (VIPergic) neurons (Wakabayashi et al., 1988, 1993).

Braak and collegues (2006) suggested that the process starts in the neuronal plexus of the GI tract. They used immunocytochemisty to investigate gastric myenteric and submucosal plexuses in five autopsy individuals, whose brains were also staged for Parkinson-associated synucleinopathy. CEB1-synuclein immunoreactive inclusions were found in neurons of the submucosal Meissner plexus, whose axons project into the gastric mucosa and terminate in direct proximity to fundic glands. They suggested that a yet to be identified environmental pathogen capable of passing the gastric epithelial lining might induce CEB1-synuclein misfolding and aggregation in specific cell types of the submucosal plexus and reach the brain via a consecutive series of projection neurons. Some of the axonal aggregations occurred directly beneath the epithelial lining of the stomach, therefore they suggested that alpha synuclein pathology of the submucosal plexus of Meissner could represent the beginning of an "uninterrupted series of projection cells that ultimately link the Enteric Nervous System with the cerebral cortex." In the CNS, the process appears to commence in the dorsal motor nucleus of the vagal nerve (medulla oblongata) and in the olfactory bulb from where it follows an ascending course, affecting additional nuclei in the lower brainstem, in basal portions of the mid- and forebrain, and eventually reaching the cerebral cortex (Braak et al., 2004). However, as summarized in the following, several neuronal circuits that are relevant to cognitive/emotional changes in PD are also involved. In these circuits not all DA neurons have long axons.

It should be noted that the Braak scheme is based on alpha synucleinopathy. More understanding is needed before PD is accepted as generalized alpha synucleinopathy and in the Braak scheme a specific vulnerability of dopaminergic neurons, unless they have long axons. There is little knowledge at this point about what alpha-synuclein does in and to the cell and it is not yet clear why some long axon dopaminergic neurons are particularly affected in PD.

4 Basal Ganglia Circuit

4.1 Central Role of Dopamine in PD

Parkinson disease is a clinical diagnosis based on four essential cardinal symptoms. These are rigidity, tremor, bradykinesia, and loss of postural reflexes. The clinical diagnosis is about 80% consistent with postmortem histopathological diagnosis. Clinically a number of other features are considered for postmortem diagnosis, but the essential pathology is the depigmentation and reduced number of tyrosine hydroxylase (TH)-labeled dopaminergic neurons in the area known as the substantia nigra. This deficit in turn causes impaired transmission between presynaptic dopaminergic neurons and postsynaptic dopamine receptors. This essential feature, a specific deficiency of dopaminergic neurons, is consistent with the spectacular clinical success of rational pharmacotherapy. This consists of treating patients with medications that promote presynaptic dopamine content in releasable synaptic vesicles and with ligands that directly bind to postsynaptic dopamine receptors. In the early 1980s the chance observation was made that methyl-phenyl tetrahydropyridine (MPTP) causes in man and monkeys a syndrome phenotypically almost identical to Parkinson disease. MPTP selectively affects dopaminergic neurons via the physiological high-affinity uptake system of TH-labeled neurons. These observations led to pathological studies showing degeneration of dopaminergic neurons in the substantia nigra as in PD and that the monkeys and humans affected by MPTP do respond to dopaminergic treatment. The selective effect of MPTP is well explained by the high-affinity uptake system of dopaminergic neurons; in this case MPTP finds its targets via the same system.

In this chapter we review the classical model of PD and summarize the two "direct" and "indirect" loops of the dopaminergic circuit of the striato-caudal system and the "antagonistic" effects of D1 and D2 type receptors.

The retina contains both types of receptors in humans and in the monkey model of PD. We summarize the retinopathy of PD, the retinal dopaminergic circuit, and the functional significance of D1 and D2 antagonism in the retina. Then we review some newly described roles of adenosine in the classic basal ganglia circuit and the molecular mechanisms associated with adenosine in the circuit. The antagonistic effects of D1 and D2 type receptors in neuronal responses have been considered an enigma for a long time. Recent biochemical studies revealed a cascade of molecular events that explain in part the role of D1 and D2 receptors in cognitive dysfunction in PD.

4.2 The Classical Basal Ganglia Circuit

The classical model of the cortico–basal ganglia–cortical circuit, with its indirect and direct pathways, was developed to explain the phenomenon of hypokinesia in Parkinson's disease (Albin et al., 1989; Alexander and Crutcher, 1990; DeLong, 1990). The most researched cortico–subcortical circuit is the "motor circuit" because of its importance for movement disorders. The motor circuit is composed of several subcircuits that originate from the motor cortex and several premotor areas (Fig. 2). In a general sense, tonic output from this circuit, arising in motor portions of the GPi and SNr, may regulate the overall amount of movement. Increased basal ganglia output could translate into less movement through inhibition of thalamocortical projection neurons, whereas reduced basal ganglia output could translate into increased movement because of disinhibition of these neurons. Although no direct evidence is available, it has been proposed that the combined action of information traveling via the direct and indirect pathways may scale or focus movements. To achieve scaling of movement parameters or termination of movements, striatal output would initially inhibit specific neuronal populations in the GPi and SNr via the direct pathway, hence facilitating movement, followed by disinhibition of the same GPi and SNr neuron via input over the indirect pathway, thus inhibiting ongoing movement.

In the alternative focusing model, inhibition of relevant pallidal and nigral neurons via the direct pathway would allow intended movements to proceed, whereas unintended movements would be suppressed by concomitant increased excitatory input to other GPi and SNr neurons via the indirect pathway. The balance between direct and indirect pathways is regulated by the differential actions of dopamine on striatal neurons from terminals of neurons in the substantia nigra pars compacta. Release of dopamine in the striatum increases activity along the direct pathway (acting on D1 receptors in striatal neurons) and reduces activity along the indirect pathway (acting on D2 receptors). Together these actions result in a net reduction

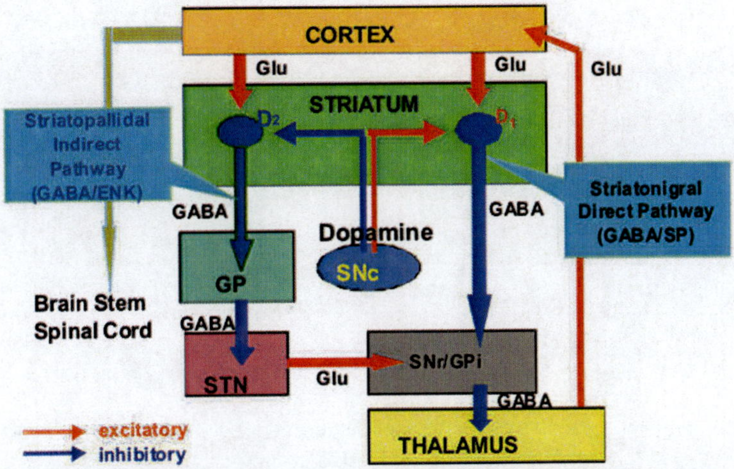

Fig. 2 A simplified diagram, based on Wichmann and DeLong (2003). The cortical motor areas give rise to a specific motor subcircuit. *Red arrows:* inhibitory (γ-aminobutyric acid [GABA]–ergic) connections; *green arrows:* excitatory (glutamatergic) connections. GP the globus, substantia nigra pars compacta; SNr, substantia nigra pars reticulata; STN, The sketch is only a skeleton of the much more intricate interconnected system of the basal ganglia with the cortex. These principal basal ganglia–thalamocortical circuits establish a balance between excitatory and inhibitory neurotransmission. The categorical division between the so-called direct (D1-receptor linked) and the indirect (D2-receptor linked) striatal output pathways may be complicated though (not shown here) by pre-synaptic receptor mechanisms. Presynsaptic dopamine receptors have higher affinity to dopaminergic ligands than do postsynaptic receptors

in GPi and SNr activity. Conversely, a decrease in striatal dopamine release would result in an increase in GPi and SNr activity. The two pathways regulating basal ganglia output via a balance of D1 and D2 receptors bears a certain logic similarity to their role in the retina regulating its output neurons, the retinal ganglion cells (Fig. 3).

Increased understanding of the anatomy and function of the basal ganglia and their role in motor and nonmotor disorders (Bodis-Wollner et al., 1983) now posits the basal ganglia at the core of cortical connections. The basal ganglia are now seen (DeLong and Wichmann, 2007) as components of parallel, re-entrant cortico–subcortical circuits, which originate from individual cortical areas, traverse the basal ganglia and thalamus, and connect to a number of separate neuronal groups terminating in the frontal cortices. Dopamine (DA) is a powerful neuromodulator for a wide variety of behaviors and some sensory processing.

The working hypothesis of cognitive deficits in PD is that they result from impairment of specific cortico–subcortical circuits. These circuits in part depend on dopamine-linked synapses. Although the original model (see Fig. 1) has been modified, the essential role of dopamine deficiency of the basal ganglia remains at the core of cognitive deficiencies in PD involving the frontal cortices.

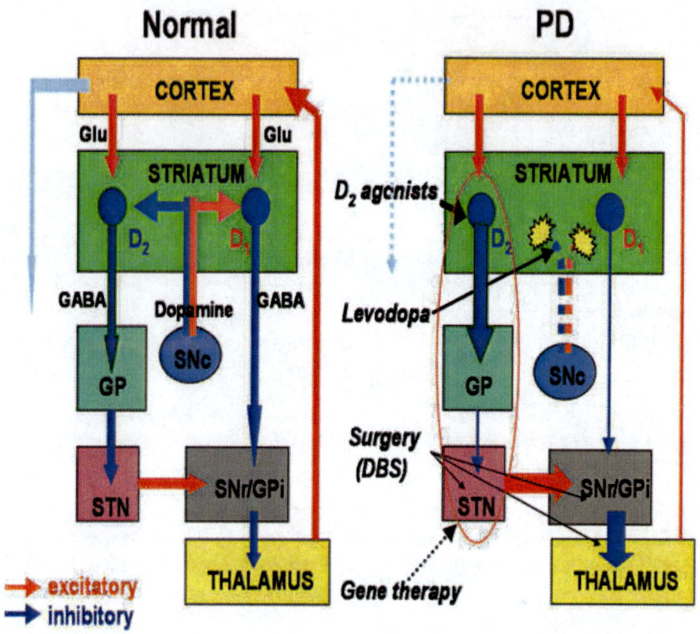

Fig. 3 The imbalance of striatal GABAergic output pathways in PD (after Alexander and Crutcher, 1990. *Blue arrows:* inhibitory (γ-aminobutyric acid [GABA]–ergic) connections; *red arrows:* excitatory (glutamatergic) connections. GP, globus pallidus; SNc, substantia nigra pars compacta; SNr, substantia nigra pars reticulate) and presumed sites of commonly applied therapies (courtesy of A. Mori)

5 Frontal Cortices, Striatum, and Cognition in PD

5.1 Fontostriatal Circuits in PD

The frontostriatal circuits connect the basal ganglia with cortical areas that are involved in cognitive, motor, and emotional processes. Furthermore, a correlation between cortical DA innervation and expression of cognitive capacities, including learning, has been shown by a number of studies (Nieoullon, 2002). Therefore a role of dopamine in impaired cognitive processing in PD is not surprising, Considerable evidence accumulated from rodent and monkey experiments over the last two decades suggests that DA activity in the frontal cortex is reciprocally linked to that in functionally related basal ganglia (BG) structures. However, the functional importance of this in humans is still unknown. To address this issue, we measured endogenous DA release using positron emission tomography in 15 healthy subjects as they practiced the first training session of a finger sequence learning task. Significant results were observed not only in striatal areas but also in extrastriatal "motor" regions, bilaterally. Faster learning was specifically coupled to lower DA release in the sensorimotor part of the globus pallidus pars interna (GPi) contralateral to the moving hand, which was paralleled by a higher increase in DA levels in the presupplementary motor area (pre-SMA). This finding provides original evidence supporting a motor-learning-related interaction between DA release in left GPi and pre-SMA, a mechanism that may also apply to other anatomically and functionally interconnected BG and frontal cortical areas as a function of behavior.

Cortical activity during executive processing in PD depends on striatal mechanisms even in early stages of the disease, as shown with functional imaging (Monchi et al., 2007). They have developed a new card-sorting task that is known to require frontal involvement and represent executive functions including planning and set-shifting. They have first shown that in young healthy adults, the caudate nucleus is specifically required when a set-shift must be planned. They studied a group of early-stage Parkinson's disease patients (seven right-handed patients at Hoehn and Yahr stages 1 and 2; mean age 62 years, range 56–70) and matched control subjects. Decreased cortical activation was observed in the patient group in the condition significantly involving the caudate nucleus. Their study showed a pattern of either reduced or increased activation depending on whether the caudate nucleus was involved in the task. This activation pattern included not only the prefrontal regions but also posterior cortical areas in the parietal and prestriate cortex. These findings are not in agreement with the traditional model, which proposes that the nigrostriatal dopamine depletion results in decreased cortical activity. These observations provide further evidence in favor of the hypothesis that not only the nigrostriatal and frontal mechanisms are affected in PD executive dysfunction.

5.2 *Impaired Memory in PD: Thalamocortical Circuitry*

Aggleton and Brown (1999) proposed two parallel brain systems with qualitatively different contributions to memory. The proposed functional-anatomical division is of relevance to cognition in PD. Hay et al. (2002) investigated memory performance in patients with either mild Parkinson's disease, moderate Parkinson's disease, or amnesia using measures of habit (automatic memory) and conscious recollection (intentional memory). Patients with amnesia displayed the expected dissociation between (intact) habit learning and (deficient) conscious recollection, patients with moderate PD were impaired on both measures whereas the mild PD patients had no abnormalities on either one. Hay et al. (2002) attributed the deficiencies in habit learning to striatal dysfunction, whereas conscious recollection is thought to result from disruption of prefrontal cortical processing.

Evidence of executive dysfunction Parkinson's disease patients is also consistent with proposals that frontostriatal circuit damage produces widespread prefrontal dysfunction (e.g., Buytenhuijs et al., 1994; Leplow et al., 1997; Owen et al., 1998; Dujardin et al., 2001).

It has been recognized recently that the thalamic dopaminergic system degenerates in PD. In addition to the GABAergic thalamic input to the cortex it has now been recognized that the nigrothalamic connection is affected in humans

(Garcia-Cabezas et al., 2007a) and in MPTP monkeys (Sanchez-Gonzalaez et al., 2005; Garcia-Cabezas et al., 2007b). However, the contribution of the dopaminergic thalamocortical system to cognition has not been elucidated.

Deep brain stimulation (DBS) has yielded quantitative information on the diversity of different functional loops between cerebral cortex and the subthalamic area in PD. These studies suggest that functional subloops between the subthalamic area and cerebral cortical motor regions can be distinguished by their frequency, cortical topography, and temporal relationships (Fogelson et al., 2006). Tuning to distinct frequencies may provide a means of marking and segregating related processing, over and above any anatomical segregation of processing streams.

They recorded EEG and local field potentials (LFPs) from macroelectrodes inserted into the subthalamic nucleus area in nine awake patients following functional neurosurgery for PD. Patients were studied after overnight withdrawal of medication. Coherence between EEG and SA LFPs was apparent in the theta (3–7 Hz), alpha (8–13 Hz), lower beta (14–20 Hz), and upper beta (21–32 Hz) bands, although activity in the alpha and upper beta bands dominated. Theta coherence predominantly involved mesial and lateral areas, alpha and lower beta coherence the mesial and ipsilateral motor areas, and upper beta coherence the midline cortex. SA LFPs led EEG in the theta band. In contrast, EEG led the depth LFP in the lower and upper beta bands. SA LFP activity in the alpha band could either lead or lag EEG.

In PD patients undergoing surgical procedures one may record enhanced beta range oscillatory rhythms in the BG and effective levodopa medication attenuates beta activity. The dominant cortical synaptic input to the BG and to the ST is glutamatergic. Given this fact, it is likely that local BG oscillatory rhythms derive from cortical inputs, but not dopamine. Dopamine is not known to regulate quick synaptic currents via phasic ionotropic mechanisms (Traub et al., 2008) hypothesize that abnormal oscillations which occur in PD arise at the cortex, even though recorded deeper.

Trottenberg et al. (2006) recorded oscillatory activity in the gamma frequency (60–100 Hz) band in local field potentials (LFPs) recorded from the region of the subthalamic nucleus (STN) in PD patients. Spike-triggered averages of LFP activity suggested that the discharges of neurons in this region were locked to gamma oscillations in the LFP. They suggested that gamma band oscillations in the LFP are likely to represent local neuronal discharge.

Gamma activity (Gray and Singer, 1989; Singer, 1993; Singer, 1999) reflects synchronization of thalamo-cortical neuronal groups in order for them to act together, a necessary prerequisite for voluntary motor action and for forming a coherent percept (Joliot et al., 1994; Tallon-Baudry et al., 1997). Fixation on visual stimuli that are optimal for foveal processing results in a time-locked increase in the power of the gamma component of the human EEG (Tzelepi et al., 2000; Bodis-Wollner et al., 2001).

When saccades are performed either in the presence of visual stimuli or in the dark, gamma range activity is enhanced in the intrasaccadic period over the parietal and occipital cortices (Bodis-Wollner and Tzelepi, 2002; Forgacs et al., 2008).

Gamma power increase starts at the beginning of the movement, prior to achieving the final eye position and then returns to baseline following the end of the saccade. Power is higher in the absence of visual input. The perisaccadic modulation of gamma power we call "phasic" gamma. Conceive of a person sleeping in the dark, awakened and quickly looking for an invisible visual target. Our intention to see even in the absence of visual stimulus, is a prerequisite for the saccades to start. Perisaccadic phasic gamma in the absence of visual stimulation may represent intrasaccadic cerebral processes we called "pre-emptive" (Bodis-Wollner, 2008). The term "pre-emptive" represents that the saccades are directed and not random; they may emerge from a repertoire of an infinite number of trajectories by suppressing unwanted actions.

5.3 Genetic Variability of Catechol-O-Methyltranferase, Prefrontal Cortex, and Cognition

Dopamine receptors, in particular D1 receptors, are abundant in the prefrontal cortex (Goldman-Rakic et al., 2000). Cathecol-O-amine transferase (COMT) is one of the metabolic enzymes of catecholamines in the tissues, including the brain (Axelrod, 1966). In the last two decades COMT inhibitors have been introduced into clinical practice to enhance the effectiveness of levodopa therapy, without the need to administer higher doses of levodopa. COMT alters dopamine levels in the prefrontal cortex (PFC) dopamine system. The COMT gene contains a functional polymorphism (Val^{158}Met) that has been associated with variation in PFC function, including "prefrontal tests" of cognition in PD (Malholtra et al., 2002). The COMT inhibitor medication tolcapone, which easily crosses the blood–brain barrier, improved cognition in eight advanced PD patients: in particular improvement was noted in the attentional task, auditory verbal short-term memory, visuospatial recall, constructional praxia, and motoric (Gaspirini et al., 1997).

In the dopamine-depleted awake rat model (Tunbridge et al., 2004) tolcapone significantly and specifically improved extradimensional (ED) set-shifting performance, originally described by Teuber and Proctor (1964) in PD. Microdialysis showed that tolcapone significantly potentiated the increase in extracellular dopamine (DA) elicited by either local administration of the depolarizing agent potassium chloride or systemic administration of the antipsychotic agent clozapine. Although extracellular norepinephrine (NE) was also elevated by local depolarization and clozapine, the increase was not enhanced by tolcapone. Apparently COMT activity specifically affects ED set-shifting and is a significant modulator of mPFC DA but not NE under conditions of increased catecholaminergic transmission. The interaction between clozapine and tolcapone may have implications for the treatment of schizophrenia (Inada et al., 2003).

Schott et al. (2006) measured dopaminergic midbrain functions in a human episodic memory task. They quantified responses in 51 young, healthy adults using functional magnetic resonance imaging. Their specific question was how polymorphisms in dopamine clearance pathways affect encoding-related brain activity. Successful episodic encoding was associated with activation of the substantia nigra. This midbrain activation was modulated by the number of tandem repeat (VNTR) polymorphisms in the dopamine transporter (DAT1) gene. Despite no differences in memory performance between genotype groups, carriers of the (low-expressing) 9-repeat allele of the DAT1 VNTR showed relatively higher midbrain activation when compared with subjects homozygous for the 10-repeat allele, who express DAT1 at higher levels. The catechol-O-methyl transferase Val108/158Met polymorphism, known to modulate enzyme activity, affected encoding-related activity in the right prefrontal cortex and in occipital brain regions but not in the midbrain. Moreover, subjects homozygous for the (low-activity) metallele showed stronger functional coupling between the PFC and the hippocampus during encoding. Their study provides strong support for a role of dopaminergic neuromodulation in human episodic memory formation. It also supports the hypothesis of anatomically and functionally distinct roles for DAT1 and COMT in dopamine metabolism, with DAT1 modulating rapid, phasic midbrain activity and COMT being particularly involved in prefrontal dopamine clearance.

6 Vision and Visual Cognition

6.1 Short-Term Memory for Visual Stimuli and Spatial Orientation in PD

In the definition of dementia memory is one of the three most important defining characteristics by *DSM-IV* criteria. In the commonly applied experimental paradigms to elicit event-related potentials for a successful response the target stimulus has to be stored in the active working memory. The paradigms require a comparison between the stored stimulus and a subsequently presented one for same–different decision making. There has been an attempt to relate neuropsychological deficits to discrete neuranatomical brain areas.

Prior to the era of in vivo brain functional imaging, this was based on human and animal lesion studies, with remarkable foresight. Spatial orientation deficits (Teuber and Proctor, 1964) are thought to reflect deficits of the posterior cortices and set-shifting impairment has been thought to reflect mostly frontal functions (Taylor et al., 1986).

Aggleton and Brown (1999) proposed two parallel brain systems with qualitatively different contributions to memory. Pathological anatomical studies in the monkey support the concept of this division of memory systems. Following bilateral symmetrical frontal ablations Macaque monkeys are impaired in object–reward association memory (Gaffan and Parker, 2000) and object RM (Kowalska et al., 1991). Thus, the select deficiencies in recall-aware memory reported by Hay et al. and others (Levin et al., 1989; Buytenhuijs et al., 1994; Hugdahl et al., 1991; Daum et al., 1995; Leplow et al., 1997; Knoke et al., 1998; Owen et al., 1998; Dujardin et al., 2001) appear to arise as a consequence of a breakdown in frontally mediated strategic memory processes implicated in intentional and effortful memory processes.

Spatial orientation deficit and set-shifting impairment have been noted in neuropsychological studies in PD (Raskin et al., 1990). One part of the working memory system, the visuospatial sketchpad, is devoted to the maintenance of visual information. The visuospatial sketchpad shows a specific selective impairment in PD: even when the visual subsystem responsible for the object-related visual analysis seems to be spared until the later stages of the disease, the visual processing of spatial location, motion, and three-dimensional properties is impaired (Moreaud et al., 1997; Owen et al., 1993, 1997; Postle et al., 1997; Lee et al., 1999). Cognitive changes in PD may be independent or precede global dementia. Kuroiwa and his collegues (Wang et al., 1999a, b, 2000; Li et al., 2003, 2005) introduced an S1–S2 paradigm (task S) as well as oddball paradigm (task O) visual event-related potentials under different interstimulus intervals (ISIs). In several studies they have shown that PD patients have particular difficulty with longer interstimulus intervals. ERP measurements were correlated with motor disability, WAIS-R, and regional cerebral blood flow ((99)Tc-ECD SPECT) examination. In advanced PD patients, P300 latency to S2-same and reaction time was significantly prolonged, whereas rCBF at bilateral frontal, temporal, and the right parietal lobes was decreased. P300 latency to S2-same was significantly correlated with the rCBF at bilateral temporal lobes. Reaction time was significantly correlated with the rCBF at the right frontal and parietal lobes, as well as the temporal and occipital lobes. They suggested that P300 changes in nondemented PD in the late stage is related to temporal lobe dysfunction, suggesting the importance of a memory task-dependent subdivision of cortico–basal ganglia circuits in PD.

Based on the results in humans it is likely that the P300 abnormalities predominantly reflect working memory impairment in PD. Kemps et al. (2004) compared the visuospatial sketch pad and central executive components of working memory as potential cognitive mechanisms of visuospatial dysfunction in PD. Patients performed more poorly in both concurrent task conditions, implicating a reduction in both visuospatial sketch pad and central executive resources. The impact of the concurrent tasks varied with disease severity, with the central executive deficit prominent at disease onset, but the visuospatial sketch pad deficit became apparent only in the moderate stages of the illness. Studies that examined P300 amplitude in PD are few (Wang et al., 1999a). The first positive "hump," the anterior P3a, is attenuated in amplitude in patients with PD without dementia (Lagopoulos et al., 1998). This component possibly reflects passive or automatic math/mismatch processing and elicited by the more infrequently appearing stimulus regardless of its target or nontarget status. The reduction of the classical P3b was also found and correlated with a poor performance in the Wisconsin Card Sorting Test (WCST; Tsuchiya et al., 2000). These results suggest that the passive and active orienting responses of PD patients to novel events is impaired and that recording P300 might provide a neurophysiological and quantitative measure of attentional and cognitive deficits linked to the frontal lobe in nondemented PD. Furthermore, the amplitudes of the NoGo-P300 and NoGo-N200 (a negative component appearing around 200 ms after the stimulus onset) were also significantly smaller in the PD group than in the control group (Bokura et al., 2005). The NoGo-P300 amplitude was significantly correlated with the WCST and the verbal fluency test scores, as well as with the number of commission errors. These data suggest that there is also an impairment of inhibitory function in PD and that this deficit may be related to impaired inhibitory executive function in the frontal lobe.

Visual perceptual categorization and ERPs were studied by Antal et al. (2000). Classifying visual targets is often difficult for PD patients. Antal et al. (2002) tested 16 de novo, early PD patients. They were shown, one after another, color photographs of either natural scenes or animals, each different. Their task was to categorize each picture as "scene" or "animal". There was a significant poststimulus amplitude difference around 200 ms in control, but not in PD patients, suggesting a visual categorization deficit in PD out of 1000 randomly presented

Delayed-response tests are well suited to test the spatial location of objects. PD patients even with mild symptoms have difficulty in maintaining, even briefly, the memory trace of spatial locations of irregular polygons, whereas they successfully keep online the shapes of the same stimuli (Postle et al., 1997). However, errors in this kind of task (usually errors in pointing movements to remembered visual targets) can be attributed to various factors, such as the misperception of the target position, errors in spatial memory, errors in the transformation from visual information to an appropriate motor command, or to a deficit in proprioceptive information processing of the arm. A recent study reports that pointing movements in PD are impaired due to a deficit in processing of proprioceptive information, which appears early in the course of the disease, and by a visual feedback problem, which emerges in later stages of the disease (Keijsers et al., 2005).

Although some studies have suggested that the visual subsystem responsible for the object-related visual analysis seems to be spared until the later stages of PD (Lee et al., 1999; Moreaud et al., 1997; Owen et al., 1997; Postle et al., 1997; Amick et al., 2003), others have found that it is not always the case (Antal et al., 2002). Rather, attention-biased object-related weighting and selecting processes can be dysfunctional even in young PD patients. During a visual categorization process a diminished differential N1 component was observed in de novo and also in treated PD patients. This component represents the basic visual feature encoding and initiating stages of perceptual categorization in the first 200 ms poststimulus period (Thorpe et al., 1996). It is hypothesized that the neostriatum may mediate feature weighting and extraction processes and the differential N1 may refer to this function. In PD, this is possibly dysfunctional, as reflected by the diminished differential N1 (Antal et al., 2002). In agreement with this hypothesis, in patients with AD, in which the cortico–cortical pathways mediating feedforward mechanisms are impaired, this component was not diminished compared to the controls but appeared later.

6.2 Aging and Cognitive Event-Related Potentials

It has been suggested that with age the dopaminergic system progressively weakens in various animal species. However, the generalization of these observations from observing motor dysfunction to the suggestion that PD represents an accelerated form of aging is not generally accepted. Evidence from various sources, including patients, animals, the effects of experimental pharmacological intervention, and molecular genetics shows that DA is also critically implicated in select higher-order cognitive functioning. It remains to be seen whether DA-dependent select cognitive deficits in PD also characterize the aging process (Bäckman et al., 2006).

A delayed P300 in PD is not due to aging per se (Tachibana et al., 1997) and there is a significant inverse relationship of delayed P300 and score of the Mini Mental State (Maeshima et al., 2002). The P300–P100 latency difference calculated from the concurrently obtained visual ERP is significantly longer in younger PD patients and differentiates them from controls (Antal et al., 1996; Sagliocco et al., 1997).

6.3 Neurotransmitters and Cognitive ERP-S in PD

Does the P300 abnormality represent only dopaminergic dysfunction? Electrophysiological evidence shows that DA receptors are involved in visual working memory in the prefrontal area (for a review see Goldman-Rakic, 1998), which was also identified as one of the generators of P300 (Halgren et al., 1998). Indeed, levodopa treatment shortens the latency of P300 in some PD patients (Stanzione et al., 1991; Sohn et al., 1998). Contrary to these findings, some investigators have described a prolonged P300 latency in medicated patients (Hansch et al., 1982; Prasher and Findley, 1991). However, medicated patients are more severely affected and the delayed P300 might also correlate with disease severity. In the animal model of PD, in behaving MPTP monkeys the visual P300 is beneficially affected by levodopa treatment (Glover et al., 1988) and in the healthy monkey D2 receptor blockade impairs the latency of the visual ERP but surprisingly it enhances its amplitude (Antal et al., 1997). It is known that task difficulty prolongs ERP latency while enhancing its amplitude. Based on this explanation the D2 receptor blockade may perhaps induce increased noise in the thalamocortical cognitive loop.

The modulation of P300 by nondopaminergic agents such as cholinergic substances has been studied in monkeys (Antal et al., 1994) and in healthy subjects (Dierks et al., 1994; Frodl-Bauch et al., 1999). Delayed P300 improved in PD patients following treatment with amantidine, a low-affinity uncompetitive NMDA receptor antagonist (Bandini et al., 2002). In this study amantidine's effect was noticeable not only as a monotherapy, but also in patients treated with levodopa. It is suggested that amantadine has DA-mimetic properties and it cannot be therefore excluded that amantidine improves cognitive ERP-s in PD as a DA-mimetic agent.

6.4 Dopamine in Visual Processing in the Retina

The retina is multilayered, with distinct neural elements in each layer. Receptors in the outer layers convert and use light energy to produce electrical currents, affecting subsequent neurons. There are three major forward synapses before the ganglion cells receive signals in the innermost layer of neurons. In between there are several lateral and feedback connections. Amacrine cells, including dopamine amacrine cells, are located in the inner layer close to ganglion cells (Fig. 4). The inner retina (IRL) includes the nerve fiber layer, the ganglion cell layer, and the inner-plexiform layer whereas the outer retina (ORL) consists of layers starting from inner nuclear layer up to and including the retinal pigment epithelium.

Fig. 4 A TH labeled (dopaminergic) amacrine cell of the rat retina. Note that the receptor layer is on the *top* and the faintly visible ganglion cell layer is on the *bottom* (from Mytileneou and Bodis-Wollner, Department of Neurology, The Mt. Sinai School of Medicine, around 1978; unpublished data)

Quantitative morphology of gross retinal histology in humans can be measured in vivo using optical coherence tomography (OCT) (Shulman et al., 1996) (Fig. 5). The electrophysiological measure obtained with corneal electrodes in the intact eye, using patterned visual stimuli, such as sinusoidal gratings, the so-called PERG is an average response of foveally located retinal ganglion cells. The amplitude of the pattern electroretinogram as a function of spatial frequency of the sinusoidal grating stimulus also shows the inverted U-shaped function (in both man and monkey) as the CS curve. This PERG output function is the result of the massed averaged response of central retinal ganglion cells (Maffei et al., 1989). The bandpass function is changed to a lowpass function in PD and in the monkey using systemic MPTP (Ghilardi et al., 1988a, b), intraocular 6-Hydroxydopamine (Ghilardi et al., 1989), and D2 receptor blockade (Tagliati et al., 1994; Fig. 6).

These results lend themselves to the functional interpretation of retinal ganglion cell disorganization in the dopamine-deficient PD retina once the contrast transfer function is understood as the sum (difference) of center and surround responses (Enroth-Cugell and Robson 1966; Fig. 7). Based on electrophysiological and functional studies in primates, one may accept the existence of two classes of ganglion cells: one center dominated with strong surround, having narrow spatial tuning

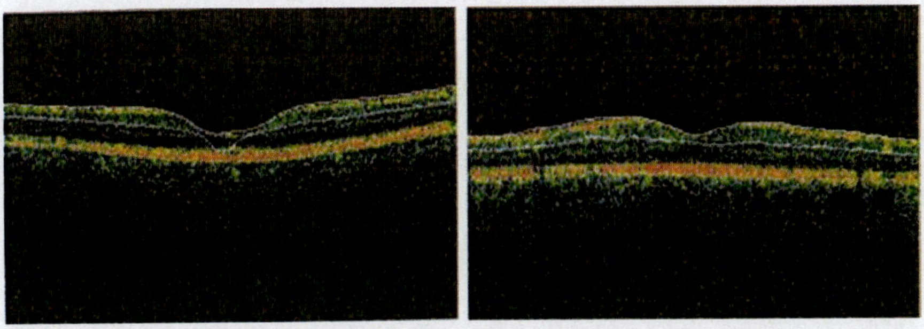

Fig. 5 *Left*: OCT image of 6 mm of the macular retinal layers of a 46-year-old healthy subject with an IOP 15 mmHg and Snellen visual acuity 20/20. *Right*: OCT image of 6 mm of the macular retinal layers of a 50-year-old, moderately advanced (Hoehn and Yahr staging 2.5) PD patient (Unified Parkinsons Disease Rating System motor score 17) prior to any anti-Parkinsonian treatment. IOP was 14 mmHg and Snellen visual acuity 20/25 (from Hajee et al., 2009)

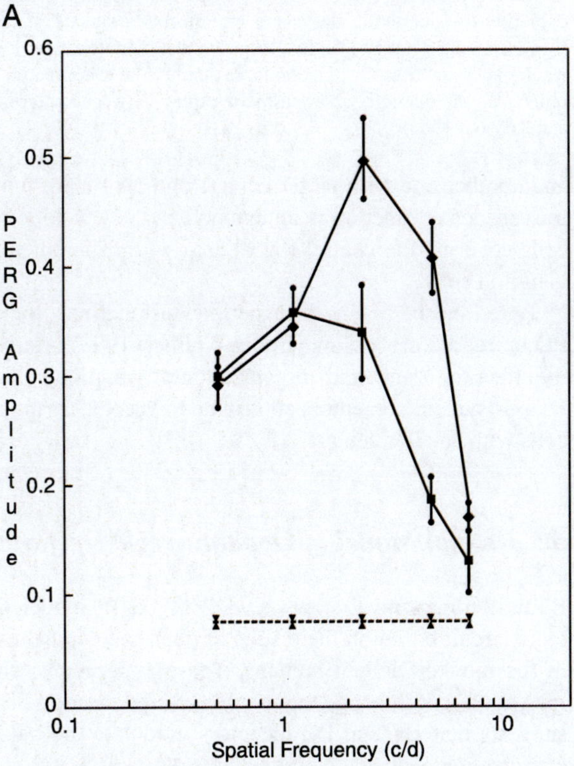

Fig. 6 The pattern ERG PERG in PD. The normally strongly bandpass PERG amplitude function (*top curve*) shows lowpass shape in levodopa-treated (*middle curve*) and untreated PD patients (after Tagliati et al., 1996). Note that the treatment slope on the right side of the curve, representing summation in the center mechanism, is attenuated as a result of D2 antagonist treatment

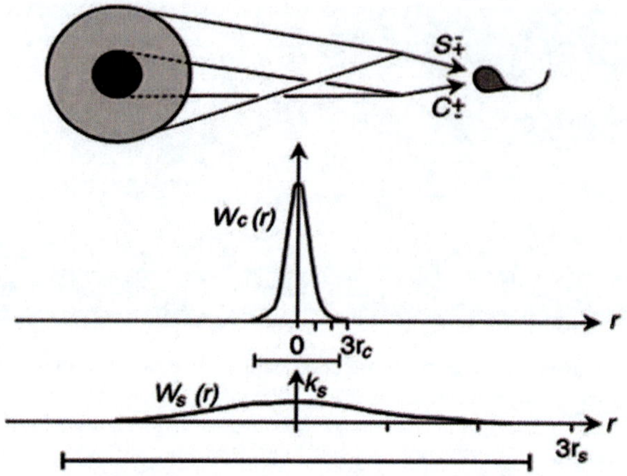

Fig. 7 Antagonistic center/surround interaction is the basic model of a foveal retinal ganglion cell receptive field. Note that the center mechanism is characterized by a narrow and tall response (sensitivity) profile, whereas the surround is broad and has a low, spatially extended profile. The response profiles are established by preganglionic circuitry. The ganglion cell performs the linear operation of subtracting the center and surround signals. If the surround mechanism is selectively attenuated it may lead to a response that monotonically grows with center stimulation. As a result the spatial transfer function loses tuning. The exact spatial frequency at which tuning occurs reflects on the diameter and optimal interplay between center and surround (after Enroth-Cugell and Robson, 1966)

and another one with larger centers and less sharp tuning. The PERG spatial contrast response function is understood as the envelope output of all retinal ganglion cells covering the central foveal area with different weights for the two classes of ganglion cells.

Based on the results of experimental pharmacological studies and the effect of PD on retinal processing (Bodis-Wollner, 1990) inferred that D1 receptors primarily affect the "surround" organization of ganglion cells with large centers, whereas D2 postsynaptic receptors contribute to "center" response amplification of ganglion cells with smaller centers.

6.5 Retinal Model of Dopaminergic Dysfunction in PD

Bodis-Wollner and Tzelepi (1998, 2002, 2005) modeled the preganglionic dopaminergic circuit based on the results of pharmacological experimental data (see above) in the monkey and in humans. These experiments showed that select dopamine receptor ligands change the spatial transfer function of the retina in a manner which suggests that D1 and D2 receptors modulate the balance of center and surround organization of foveal ganglion cells of the primate.

Based on results in vertebrates Bodis-Wollner (1990) assumed that in primates the surround organization of the retinal ganglion cell is under D1 receptor control. D1 receptor activation causes disjunction of horizontal cells, otherwise coupled in an extended chain (Piccolino et al., 1987). In other words, when D1 activity is present, the horizontal cell signal is more concentrated in a smaller area; it does not get diffused. As a consequence under D1 activity the surround becomes smaller

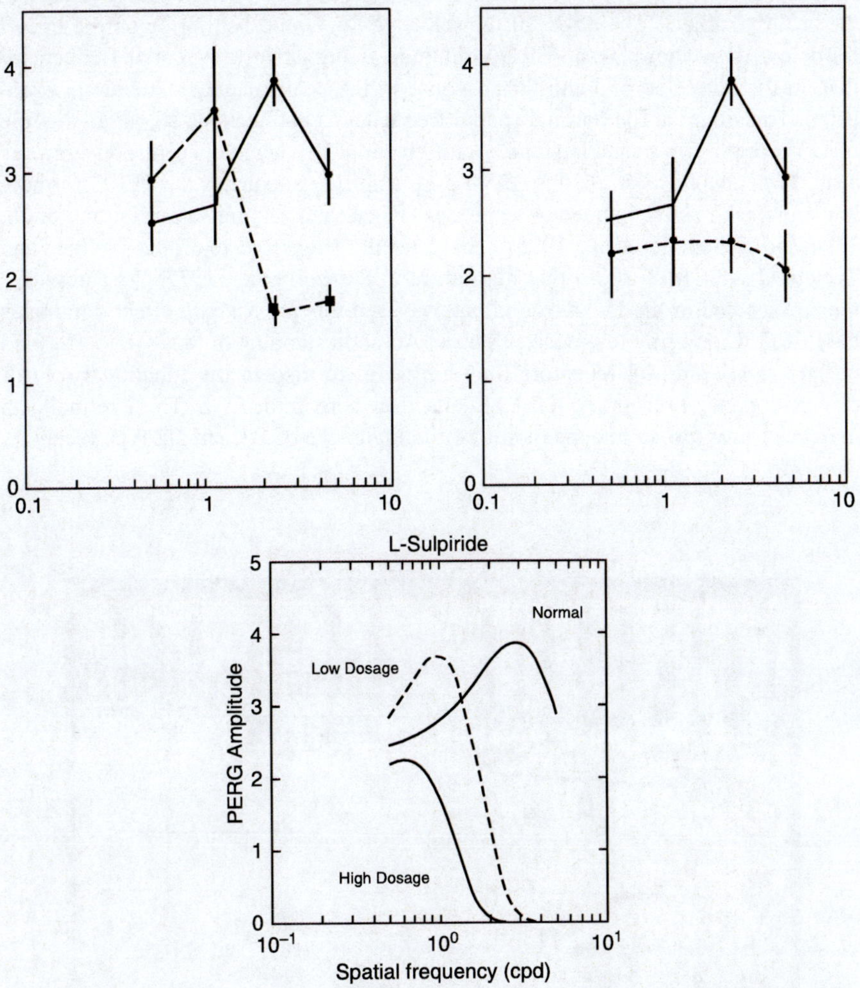

Fig. 8 The effect of the selective D2 ligand l-sulpiride on the PERG of a low-dose monkey. *Upper left*: Note that for a low-dose sulpiride PERG amplitude is higher at low spatial frequencies than in the untreated one and the response to the peak spatial frequency is attenuated. This effect at the optimal spatial frequency is not unexpected, given the role of D2 receptors in center response amplification: the heightened response at low spatial frequencies is surprising. *Upper right*: The effect of high-dose sulpiride on the PERG: both low and peak spatial frequency responses are attenuated. (Both figures after Tagliati et al., 1994 and Stanzione et al., 1995.) *Bottom*: a model of the dose and spatial frequency-dependent effect of D2 blocking on the PERG according to the subtractive Gaussian center-surround interactive mechanisms. (Based on the experimental results of Tagliati et al., 1994 and Stanzione et al., 1995, after Bodis-Wollner and Tzelepi, 2002, 2005.)

but stronger. D2 receptors promote coupling between rods and cones in the xenopus (Krizaj et al., 1998). In our model we assumed that 2–3 neighboring receptors are coupled, thereby amplifying the center strength. Consequently complete lack of D2 activation should lead to a loss of center sensitivity by a factor of 2–3. Tagliati et al. (1994) and Stanzione et al. (1995) found a seemingly paradoxical effect when the effects of a low-dose and a high-dose sulpiride were compared in the healthy subject. The effect of the selective D2 ligand l-sulpiride on the PERG for a low dose shows that PERG amplitude is higher at low spatial frequencies than in the untreated one and the response to the peak spatial frequency is attenuated. This effect at the optimal spatial frequency is not unexpected, given the role of D2 receptors in center response amplification: the heightened response at low spatial frequencies is surprising. High-dose sulpiride attenuates the PERG for both low and peak spatial frequency responses (Fig. 8; both figures after Tagliati et al., 1994 and Stanzione et al., 1995). These results suggested to Bodis-Wollner and Tzelepi (1998, 2002, 2005) that high-affinity (Skirboll et al., 1977) D2 autoreceptors are located in the D1 surround pathway and this presynaptic effect dominates the PERG when using low-dose sulpiride. An understanding of the logic performed by retinal D1 and D2 receptors may be useful to discern the functional role of diverse dopamine receptors in DA circuits elsewhere in the CNS. These retinal data may be relevant to an understanding of the logic role of D1 and D2 type receptors (Fig. 9).

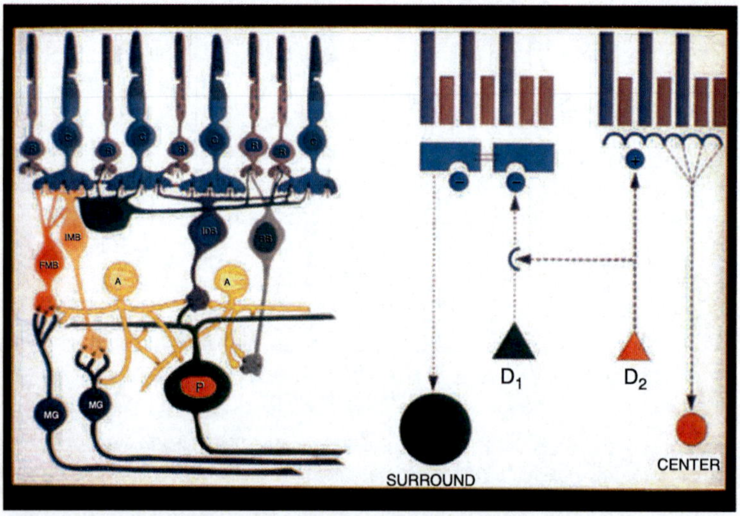

Fig. 9 *Left*: a simplified diagram of the retina with principal neurons and interconnections. On top is the receptor layer and the bottom shows the ganglion cells with their axons ("nerve fiber") which can be seen on ophthalmoscopic examination and quantified using modern retinal imaging in vivo. *Right:* a schematic representation of the preganglionic dopaminergic connections including pre-synaptic D2 receptor connection

7 Nondopaminergic Signals and Cognition in PD

7.1 GABA and the Subthalamic Nucleus

Deep brain stimulation was introduced nearly two decades ago into the treatment of PD. Over the last years it has become a rather widespread treatment modality, with some rather impressive results in carefully selected patients. The effects are seen as motor improvement. However, adverse mood and cognitive effects have been also described and preoperative selection criteria have to be adhered to minimize the chances for adverse effects (Lang et al., 2006). Some of the results also raised questions concerning our neuroanatomical and neuropharmacological concepts of the classical basal ganglia circuitry. One of the evaluating techniques is intraoperative microdialysis which was developed by Stanzione and his collegues (Fedele et al., 2001; Mazzone et al., 2005) as a technique into human intraoperative electrophysiological monitoring and recordings. Their and some other studies have yielded biochemical evidence of the functional role of the subthalamic nucleus in the PD relevant basal ganglia circuitry. It was noted that lesions of the subthalamic nucleus may reverse Parkinsonism (Bergman et al., 1990).

Deep brain stimulation (DBS) of the subthalamic nucleus in Parkinson's disease patients augments STN-driven excitation of the internal globus pallidus (Stefani et al., 2005). Stefani et al. (2006) quantified the biochemical effects of STN-DBS in two basal ganglia mechanisms (putamen, PUT, and GPi) and in a thalamic relay nucleus, the anteroventral thalamus (VA). In six advanced PD patients undergoing surgery, microdialysis samples were collected from GPi, PUT, and VA before, during, and after 1 h of STN-DBS. cGMP was measured in the GPi and PUT as an index of glutamatergic transmission, whereas GABA ([gamma]-aminobutyric acid) was measured in the VA. During clinically effective STN–DBS, a significant decrease in GABA extracellular concentrations in the VA (–25%) was achieved. Simultaneously, cGMP extracellular concentrations were enhanced in the PUT (+200%) and GPi (+481%). The results suggest that DBS differentially affects fibers crossing the STN area: it activates the STN–GPi pathway while inhibiting the GPi–VA one. These findings support a thalamic disinhibition, as the main factor responsible for the clinical effect of STN–DBS. Inhibitory interneuron play is essential for regulating neuronal circuits and behavior by controlling spike timing and neuronal rhythms (see later). However, there is great diversity of GABAergic interneurons and GABAA receptor subtypes. There are indications that region- and domain-specific location of these receptor subtypes are differentially involved in several nonmotor manifestations of PD such as regulation of sleep, anxiety, memory, and sensorimotor processing, in addition to postnatal developmental plasticity (Mohler, 2007).

The pedunculopomtine nucleus (PPN) is reciprocally connected to the BG. It input is from the globus pallidus interna, substantia nigra pars reticulata, and STN. Its output is glutamatergic to STN, GPOi, and SNC. It has been reported that in PD about 50% of its intrinsic cholinergic neurons are lost. Combined deep brain stimulation of the subthalamic and pedunculopontine (PPN) nuclei has been recently proposed as surgical treatment of advanced PD. STN stimulation alone has been shown to provide selective improvement of the grammatical aspect of language. In five advanced PD patients combined deep brain stimulation (STN + PPN) did not change the overall cognitive profile, however, language was affected. There was a trend towards reduction of ungrammatical errors (particularly substitution of free and inflectional morphemes) when stimulating the PPN, even when the STN was switched off (Zanini et al., 2009). The PPN reaches the motor cortex via thalamic nuclei, however, its connections to other cortices are less well explored.

7.2 Cholinergic Mechanisms

"Dementia" has been recognized for many decades as one of the nonmotor features of PD. Research proceeded along several approaches to understand whether dementia in PD represents an overlap syndrome with primarily dementing disorders, such as Alzeheimer disease, diffuse Lewy body disease, and frontotemporal dementia, just to name a few. Both anatomical and neurochemical studies addressed dementia in PD. Dementia, by the definition of *Diagnostic Manual (DSM-IV)* criteria includes memory dysfunction and one of the following: aphasia, apraxia, agnosia, or disturbance of executive functions. By this standard PD patients rarely have dementia. They have select and specific cognitive impairments such as "logical memory" (Pang et al., 1990; Bodis-Wollner et al., 1995) which correlate with modality dependent changes in event-related potentials.

The earliest cognitive impairment in PD is evident for executive abilities, visuospatial orientation, and memory. However, many studies of cognition in PD use the term "dementia" without defining the specific deficits. With this in mind, it is still worth reviewing the contrast between Alzheimer disease and PD, in particular with reference to cholinergic mechanisms.

It was suggested that cognitive dysfunction in PD, similarly to AD results from loss of cholinergic neurons in the nucleus basalis Meynert (Whitehouse, 1981). However, it was reported by Perry et al. (1985) that "dementia" in this disease usually occurs in the absence of substantial Alzheimer type changes in the cortex and may be related to abnormalities in the cortical cholinergic system.

Thus, in Parkinsonian patients with dementia there are extensive reductions of choline acetyltransferase and less extensive reductions of acetylcholinesterase in all four cortical lobes. Choline acetyltransferase reductions in temporal neocortex correlated with the degree of mental impairment assessed by a memory and information test but not with the extent of plaque or tangle formation. In PD but not in Alzheimer's disease the decrease in neocortical (particularly temporal) choline acetyltransferase correlates with the number of neurons in the nucleus of Meynert suggesting that primary degeneration of these cholinergic neurons may be related, directly or indirectly, to declining cognitive function in Parkinson's disease. In fact the correlation of cholinergic deficits in PD is more evident and is more severe than in AD (Bohnen et al., 2003). In 18 of 22 patients who were diagnosed with dementia Aarsland et al. (2005) reported on the presence of limbic and neocortical Lewy bodies-associated cholinergic deficits without the presence of changes typical of AD.

Thus the evidence suggests that acetylcholine may have an important role in nondopaminergic cognitive changes. If so, some cognitive defects could possibly be treated with choline-esterase inhibitors. Although the ACh inibitor Rivastigmine has been reported to have some benefits in PD (Emre, 2004), up to now there has been a dissonance between the scientific evidence for cholinergic deficiency in PD brains and the clinical use of cholinergic agents in treating cognitive impairment in PD.

One reason may be the concern for a potential negative motor effect of cholinergic medication in PD. Treatment of PD with anticholinergic medications precedes the l-dopa era and is still used in some clinical centers Cholinergic interneurons play an essential role in striatal mechanisms and increased acetylcholine release in the striatum is pathognomic for PD (Cragg, 2006).

ACh interneurons interact with the dopaminergic system in several ways. Dopamine-dependent long-term potentiation (LTD) is mediated by D2 receptors (in the indirect pathway) but D2 receptors are also reported on cholinergic interneurons. Acetylcholine release activates mainly through M1 muscarinic receptors at glutamatergic axo-spinous synapses (Hersch et al., 1994).

It is possible therefore that selective receptor agonist therapy would benefit cognition in PD without causing negative motor effects. Establishing the relationship between select cognitive deficits and nicotinic versus muscarinic neurotransmission may lay the foundation for rational pharmacotherapy of cognitive dysfunction in PD.

7.3 Glutamate, Thalamocortical Processing, and D1 and D2 Dopamine Receptors

The results of several imaging studies have shown correction of abnormal motor, but not cognitive, network activity by treatment with dopaminergic therapy and deep brain stimulation. Some nondopaminergic elements of the circuitry, however, are known to contact diverse postsynaptic dopamine receptors.

Select psychomotor functions are under the influence of dopamine in modulating cortical and thalamic glutamatergic signals impinging upon principal medium spiny neurons (MSNs) of the striatum. Dopamine D1 receptor signaling enhances dendritic excitability and glutamatergic signaling in striatonigral MSNs, whereas D2 receptor signaling exerts the opposite effect in striatopallidal MSNs. The functional antagonism between these two major striatal dopamine receptors also extends to the regulation of synaptic plasticity. Furthermore these studies have also shown that long-term alterations in dopamine signaling produce profound and cell-type-specific reshaping of corticostriatal connectivity and function. However, at this point little is known of the effects of selective dopamine receptor ligands on memory performance in PD.

7.4 Adenosine

The role of adenosine, originally described by Greengard (1972), is being increasingly emphasized and studied from the molecular to the behavioral level. Furthermore, anecdotical clinical evidence concerning cognitive effects of alpha adenosine A2 type receptor blockers is consistent with the results of animal studies (Takahashi et al., 2010).

The molecular role of adenosine is primarily in the postsynaptic phosphorylation cascade in the action of dopamine (Nishi et al., 2008). Phosphodiesterase (PDE) regulates cellular cAMP/protein kinase A (PKA) signaling. The effect of dopamine is largely mediated through the cAMP/PKA signaling cascade, and therefore is controlled by PDE activity. PDEs with different substrate specificities and subcellular localizations are expressed in neurons. PDE4 and PDE10A have different roles in the regulation of cAMP/PKA signaling in the striatum. In vitro and in vivo biochemical techniques Nishi et al. (2008) used selective PDE inhibitors to regulate phosphorylation of presynaptic (e.g., tyrosine hydroxylase (TH)) and postsynaptic (e.g., dopamine- and cAMP-regulated phosphoprotein of M(r) 32 kDa (DARPP-32)) substrates for PKA. The PDE4 inhibitor, rolipram, induced a large increase in TH Ser40 phosphorylation at dopaminergic terminals that was associated with a commensurate increase in dopamine synthesis and turnover in striatum in vivo. Rolipram induced a small increase in DARPP-32 Thr34 phosphorylation preferentially in striatopallidal neurons by activating adenosine A(2A) receptor signaling in striatum. In contrast, the PDE10A inhibitor, papaverine, had no effect on TH phosphorylation or dopamine turnover, but instead robustly increased DARPP-32 Thr34 and GluR1 Ser845 phosphorylation in striatal neurons. Inhibition of PDE10A by papaverine-activated cAMP/PKA signaling in both striatonigral and striatopallidal neurons, resulted in potentiation of dopamine D(1) receptor signaling and inhibition of dopamine D(2) receptor signaling. These biochemical results are supported by immunohistochemical data demonstrating differential localizations of PDE10A and PDE4 in striatum.

Over the last decade, adenosine receptors in the central nervous system have been implicated in the modulation of cognitive functions. Despite the general view that endogenous adenosine modulates cognition through the activation of adenosine A1 receptors, evidence is now emerging on a possible role of A2A receptors in learning and memory. Takahashi et al. (2010) reviewed studies using diverse animal models to provide a comprehensive picture of the recent evidence of a relationship between adenosinergic function and memory deficits. They conclude that caffeine (a nonselective adenosine receptor antagonist) and selective adenosine A2A receptor antagonists can improve memory performance in rodents evaluated through different tasks. Their review also suggests that caffeine and selective antagonists may also afford protection against memory dysfunction elicited in experimental models of aging, Alzheimer's disease, PD, and, in spontaneously hypertensive rats (SHR), a putative genetic model of attention deficit hyperactivity disorder (ADHD) (Fig. 10).

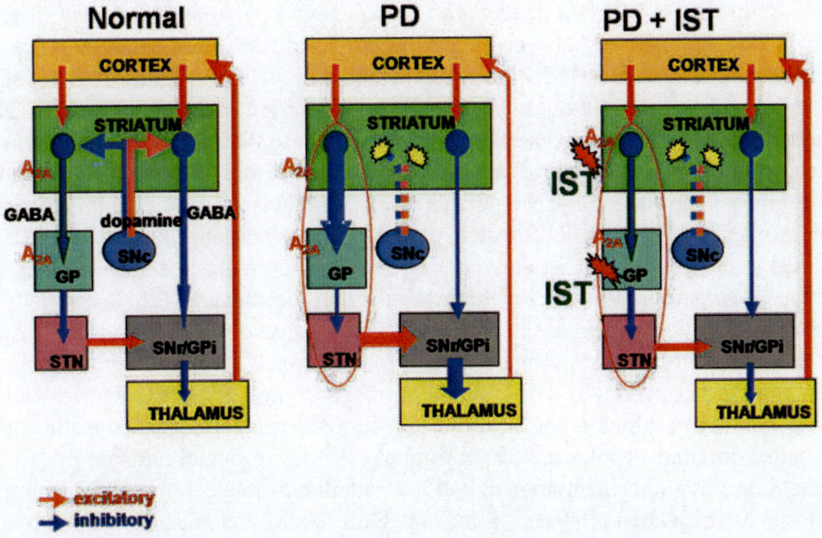

Fig. 10 Regulation of the striatopallidal indirect GABAergic pathway: A_{2A} receptor-mediated dual excitatory modulation of the indirect pathway. Presumed action of the nondopaminergic adenosine receptor ligand in the basal ganglia circuit. Adenosine A(2A) receptors are localized to the indirect striatal output function (courtesy of Mori and Shindou, 2003)

From the clinical point of view the development of nondopaminergic therapy is highly attractive as direct dopaminergic therapy is associated with a number of complications. Whether medications that affect alpha adenosine a-2 type receptors are effective and whether they bypass some of the complications associated with direct dopaminergic therapy, remains to be seen. Current research also suggests that cyclic-nucleotide PDE isoforms could be targets for developing novel therapies for neuropsychiatric and neurodegenerative disorders affecting dopamine neurotransmission.

8 The Alzheimer's Disease Case: An Overview

Listen, stranger; this was myself: this was I. (W. Faulkner, very last sentence in "The Jail")

Alzheimer's disease has prevalence estimates of approximately 10% in individuals over age 65 and 30% in individuals over age 85 in the United States. Clinically AD presents as a progressive deterioration of selective cognitive domains, with initial symptoms indicating a decline in memory function, particularly a loss of episodic memory, which is considered a subcategory of declarative memory. But it is also well documented that a large number of elderly people have poorer memory performances, with prevalence of up to 40% in individuals over 60 years (Hanninen et al., 1996) and in many cases other cognitive deficits (di Carlo et al., 2007). Prospective studies show that elderly subjects who exhibit mild cognitive impairment, a proposed transitional stage between "normal aging" and "dementia" go on to develop dementia at a rate of 10–15% per year, which is 5–7 times higher than for age-matched individuals without such impairment (Petersen and Negash, 2008). The challenges in the diagnosis, predictors to conversion to AD, and possible modifiers such as diet and education level are briefly presented in this chapter.

From a neuropathology perspective AD is characterized by accumulation of senile plaques (beta amyloid-related pathology) and neurofibrillary tangles (tau-related pathology). Until recently it has been proposed that beta amyloid is an extracellular pathology, and tau is intracellular, but recent studies challenge both of these statements, as discussed in detail below. The cause of the disease still remains unknown but involves abnormal cleavage of a neuronal membrane protein called amyloid precursor protein (APP) and abnormal accumulation of a fragment called B-amyloid (Aβ), which is the substrate for the senile plaques. This constitutes the so-called amyloid hypothesis, which implies a causative model initiated by Aβ. A more recent hypothesis linking Aβ and tau pathologies to a common upstream initiator(s) has also been proposed (Small and Duff, 2008). The mechanism of Aβ and tau-induced changes in neuronal activity and their relationship with cognitive dysfunction are also topics of the present chapter. Thus, the cognitive deficits in AD are covered from both the clinical and basic science perspectives.

9 Cognitive Decline in the Elderly; Is It "Aging", MCI, or Early AD

9.1 Normal Aging

Advanced aging is accompanied by cognitive decline even in the absence of disease. Several theories posit that cognitive deficits arise from alterations in functional properties of co-ordinated brain systems or from subtle anatomical disconnection between brain regions that ordinarily function together, most likely due to white matter abnormalities (O'Sullivan et al., 2001; Pfefferbaum et al., 2005). Based on the structural observation of age-associated white matter degeneration, O'Sullivan proposed the "disconnection" hypothesis: decline in normal aging emerges from changes in functional integration between systems of brain areas in addition to dysfunction of specific gray matter areas. An indirect method based on analysis of spontaneous fluctuations within brain systems has been proposed to detect system integrity (Greicius et al., 2003 in 4). The basis of this technique is that functional MRI (fMRI) detects the spontaneous low-frequency fluctuations that are coherent within large-scale systems, such as motor (Biswal et al., 1995) and sensory (De Luca et al., 2005). Recently this technique was used to measure the integrity of a large-scale system involving frontal and posterior brain regions; this system is often referred to as the "default network" (Raichle et al., 2001) and is associated with the internally directed mental states including memory, planning, and related cognitive processes. This study demonstrated an age-dependent reduction in the correlation between the anterior and posterior systems, in elderly individuals free of AD. This correlated with poor performance in the following psychometric tests, in order of strength: memory, processing speed, and executive function (Andrews-Hanna et al., 2007). One could propose that this disconnection ultimately leads to corticosubcortical dysrrhythmias, such as thalamo-cortical abnormalities, that could potentially explain cognitive deficits (Llinás et al., 2005).

In contrast, other researchers proposed that age-related processes, some of which underlie cognitive decline, do not target cortical regions equally, suggesting that the effect of aging is not cognitively diffuse (Small, 2001). Using a different MR imaging technique, in which basal metabolism is measured by cerebral blood volume (CBV), this group (Small et al., 2000) has identified the hippocampal dentate gyrus region as the most sensitive structure to the aging process, which correlated with the memory decline observed in the elderly. These proposed mechanisms—disconnection and site-specific vulnerability—are not mutually exclusive. We can envision a situation in which deafferentation produces specific gray matter dysfunction or neuronal dysfunction may lead to axonal abnormalities in specific circuits. Of note, almost all cases of non-AD causes of memory decline in humans and nonhuman mammalian species are hippocampal-based. The exact cause of non-AD age-dependent memory decline is a matter of debate; many mechanisms have been proposed including: adrenal and gestational hormonal levels, changes in cerebrovascular supply, oxidative stress, and disrupted neuronal calcium homeostasis. All these different abnormalities are present at different levels in the elderly, but the extent to which they contribute to memory failure is still unknown.

What else to take into account before considering MCI or AD diagnosis: The fact that different life exposures including education, occupation, and leisure, impart a reserve against Alzheimer's disease in epidemiological studies, raised the possibility of a different brain response to the aging process or to neurodegenerative entities in general. This term is referred to as cognitive reserve (CR). It has been proposed that the neuronal implementation of CR may involve two major components: neuronal reserve and neuronal compensation (Stern, 2006). Neuronal reserve refers to CNS networks or cognitive paradigms that due to their activity become less susceptible to disruption. Using this type of CR would be a normal process that is constantly used in healthy individuals. However, such networks may also help an individual cope with brain pathology. Neuronal compensation refers to the process by which subjects suffering from brain pathology use brain structures or networks not normally used by healthy individuals. These two hypothesized mechanisms have been supported by several fMRI studies (reviewed in Stern, 2006). It is plausible to propose that an individual's CR may be amenable to change upon specific exposures or interventions. This could potentially be used as the basis for a behavioral therapy to treat AD.

9.2 Mild Cognitive Impairment (MCI)

MCI is not a disease per se, but rather it has been defined as a condition of intermediate symptomatology between the cognitive changes of aging and very early dementia (Petersen and Negash, 2008). In fact, it can be viewed as a cognitive decline at the normal tail end of a continuum. The rationale for the study of MCI is derived from the idea that the earlier one intervenes in a neurodegenerative process, the more likely the damage done to the CNS can be prevented. The concept of MCI has evolved considerably over the years. The term MCI was initially used by Reisberg to describe individuals with a global deterioration scale (GDS) of 3. Others have used the clinical dementia rating scale (CDR) of 0.5. But these cutoff values do not necessarily correspond to specific diagnoses. A patient with CDR0.5 can meet the criteria for MCI, mild dementia, or AD. Recently MCI has emerged to represent a stage of impairment beyond what is considered normal for age, but not of sufficient magnitude to warrant the diagnosis of dementia or AD. Originally MCI was defined by memory complaint, memory impairment for age (adjusted for education and socioeconomic background), preserved general cognitive function, and intact activities of daily living. Recently the criteria have been expanded to include two subtypes: amnesic (original criteria) and nonamnesic (nonmemory cognitive domain impaired; Winblad et al., 2004). It has been suggested that nonamnesic MCI patients may have underlying brain pathology different from AD, and the amnesic MCI patients are more likely to be diagnosed with AD over time (Devanand et al., 2008a). Using these criteria it is rather subjective to diagnose MCI versus normal aging; in most cases the appropriate diagnosis becomes clear only with time.

9.3 Predicting Conversion from "Normal Aging" to MCI and from MCI to AD?

The clue to this transition seems to be found in the sense of olfaction. Early in the course of AD, degeneration occurs in the entorhinal cortex–hippocampal–subicular complex (Price and Morris, 1999). The olfactory bulb, particularly the anterior olfactory nucleus, shows numerous neurofibrillary tangles (NFTs). Odor identification deficits during life may be associated with NFTs in the hippocampus (Wilson et al., 2007).

Clinically, AD patients consistently show deficits in odor identification compared to controls (Doty et al., 1991). These deficits have been shown to be a true decline in odor identification ability that cannot be explained by lexical difficulty in interpreting written words in the multiple choice test formats. The University of Pennsylvania has developed a smell identification test (UPSIT), with ranges 0–40, which is widely used in clinical settings. Recent studies have shown that odor identification deficits predict conversion from normal to MCI, particularly decline of verbal memory (Wilson et al., 2007). Other series (Devanand et al., 2008a) of studies have demonstrated that olfaction has a strong predictive power of MCI to AD in both types of MCI. Many other biomarkers have been used to predict AD conversion, including neuropsychological tests, cerebrospinal fluid markers (Beta amyloid, hyperphosphorylated tau and isoprostane), and MRI entorhinal cortex volume. Combining several of these markers (i.e., olfactory measures, selective remaining test immediate recall (verbal memory), MRI-hippocampus/entorhinal cortex volumes and functional activities questionnaire values) strongly predicted conversion to AD (Devanand et al., 2008b).

9.4 Diagnosis of AD

AD is a genetically heterogeneous disorder. Four genes have been identified (Preselinin 1-PSEN1, Preselinin 2-PSEN2, Amyloid precursor protein-APP, and Apolipoprotein E ε4-APOE4) and additional chromosomal regions and genes are being investigated. The autosomal-dominant genes cause early-onset before 60 years of age. Mutations in these causative genes account for less than 5% of all cases of AD. PSEN1 is the most common causative gene; its mutation is associated with the earliest age at onset, seizures, myoclonus, and language deficits. APP mutations cause dementia typical of AD. Mutations in PSEN2 have been identified only in one family.

In most cases the cause of the disease is believed to be complex, resulting from a combination of susceptibility genes interacting with each other as well as with environmental factors. APOE4, the most common of the known susceptibility genes, is distinct from the causative genes, because it is neither sufficient nor necessary to cause AD. Genetic testing for diagnosis and predictive purposes is available for early-onset AD, and appropriate genetic counseling is strongly suggested, almost mandatory.

Clinically AD is characterized by a gradual progressive decline in intellectual function, problem solving, language, and perception. Patients manifest characteristic cognitive and behavioral findings. The most common presenting cognitive symptom is short-term memory impairment (mostly episodic memory) and forgetfulness. As the disease progresses long-term memory is also affected. In addition to memory impairment, diagnosis requires impairment of at least one other cognitive domain, including judgment, abstract reasoning, language (primarily word finding difficulties or anomia, common symptom) orientation, praxis, and attention. It should be emphasized that some patients with AD may develop memory deficits only very late in the course of the disease. Noncognitive behavioral manifestations include changes in personality and mood or in behavior (paranoia, delusions, anger, aggression, restlessness agitation, wandering, sleep–wake cycle disturbance, hallucinations, and illusions). The rest of the neurological examination is relatively normal. Extrapyramidal signs such as rigidity and bradykinesia may portend a more rapid decline.

Standardized instruments are available for staging of AD, these include the Clinical Dementia Rating scale (CDR) and the Global Deterioration Scale (GDS). Staging is useful mainly for following disease progression and management. The CDR evaluates memory, orientation, judgment, problem solving, community affairs, home and hobbies, and personal care. This is used to classify AD into categories: mild, moderate, and severe. Many comorbidities, such as dyslipedemias and diabetes, seem to worsen AD progression, making the approach to this pathology multimodal, and requiring metabolic, neurological, behavioral, and psychosocial interventions

9.5 Diet in AD

Diet may play an important role in the causation and prevention of AD (Luchsinger and Mayeux, 2004) but the results may be conflictive. Higher intake of vitamins C, E, and B12, flavonoids, unsaturated fatty acids, fish, and folate, moderate ethanol, and lower total fats, have been related to lower risk of AD. Other studies failed to find association between intake of vitamins C, E, B12, carotenes, fats, or levels of vitamin B12 and AD risk. There is a new and perhaps more ecological approach to this situation, which is to study the effect of dietary patterns (rather than individual foods or nutrients) on the risk for AD. One such dietary pattern is the Mediterranean diet (MeDi), which is characterized by high intake of vegetables, legumes, fruits, cereals, unsaturated fatty acids (mostly in the form of olive oil) but low intake of saturated fatty acids and moderately high intake of fish and low to moderate intake of dairy products (mostly cheese and yogurt) and low intake of meat and poultry, and a regular but moderate amount of ethanol, primarily in the form of wine and generally during meals (Trichopoulou et al., 1995). There are several lines of evidence showing that MeDi is related to lower risk of cardiovascular disease, several forms of cancer, and overall mortality. Recently it was reported that higher adherence to the MeDi is associated with a reduced risk for AD, in a cohort of nondemented individuals at baseline where AD was prospectively assessed. The association observed between MeDi and risk for AD was not mediated by vascular comorbidity (Scarmeas et al., 2006a, b). This constitutes another potential nonpharmacological intervention to treat and/or prevent AD and suggests that there are multiple metabolic processes related to AD pathophysiology.

10 Imaging AD

10.1 Can Neuronal Dysfunction Be Visualized Before Cell Death?

As part of the initial assessment of a patient with dementia, brain imaging is required. With imaging one can get information on structure or function, or a combination of the two. In AD, clinical structural MRI findings are not specific, with reports of neocortical, hippocampal, or global atrophy and white matter hyperintensities and/or associated mircovascular disease. The main reason to perform a MRI is to rule out the existence of another entity, which may be treated differently. In a clinical setting, functional imaging is basically limited to positron emission tomography (PET), where two modalities are currently used: [18F]fluorodeoxy glucose (FDG) to evaluate brain glucose metabolism (Gonzalez et al., 1995) and more recently imaging of amyloid-beta, using Aβ selective carbon-11 labeled thioflavin-T ([C-11]PIB) or stilbene ([C-11]SB-11) compounds. FDG-PET may show hypometabolism in temporoparietal and posterior cingulate regions, or more extensive abnormalities depending on AD stage. PIB/SB-PET studies have proven a valuable method to confirm AD diagnosis or to evaluate response to treatment, nevertheless its use is still limited, because PET compounds have a very short half-life, making the distribution to PET facilities difficult. So far in the United States there are only about 40 PET radiochemistry facilities that can provide this service. Research MR-imaging groups have developed several structural analysis programs that allow the quantitative evaluation of specific brain regions. Using these methods, it has been found that atrophy rates of entorhinal cortex best correlate with cognitive deficits in AD (Du et al., 2003). Others have used a more visual score to evaluate white matter lesions, the Scheltens score, which also correlates with cognitive decline in AD (Brickman et al., 2008).

AD researchers have also extensively used functional imaging. Almost all in vivo techniques that measure brain metabolism are based on Fick's principle postulated in 1870, which has been used to measure blood flow to different organs. This principle describes a relationship between oxidative metabolism and hemodynamic variables—cerebral blood flow, cerebral blood volume, and deoxyhemoglobin—to assess metabolism in the living brain. These techniques include: near-infrared spectroscopy (NIRS), contrast-enhanced computerized tomography (CT) to evaluate cerebral blood volume (CBV), PET and single-photon emission tomography (SPECT) measurements of cerebral blood flow (CBF), and magnetic resonance imaging (MRI) measurements of CBF, CBV, and deoxyhemoglobin (BOLD signals). To varying degrees, all these approaches have proven capable of detecting AD-related metabolic changes (El Fakhri et al., 2003; Small et al., 2000; Dixon et al., 2002). It is important to note that these techniques can detect metabolic changes caused by diseases in which there is a relative absence of cell loss, including a range of psychiatric illnesses (Costa et al., 1999) as well as in aging (Noda et al., 2002), establishing that imaging correlates of metabolism can detect cell dysfunction. In a recent study (Moreno et al., 2007) high spatial resolution CBV maps performed in both humans with AD and AD mouse models showed early entorhinal cortex hypometabolism which then extended to the other hippocampal subregions. This correlated with cognitive symptoms. This method was able to detect AD related MRI-changes before Aβ plaques developed in the mouse model (Fig. 11).

New imaging approaches applied to AD are under development. One such approach is based on the ability to detect Aβ plaques with MRI by using ^{19}F and ^{1}H containing amyloidophilic Congo red-type compounds (Higuchi et al., 2005). This would allow a better spatial resolution in plaque location and earlier detection. The development of longer half-life F-18 label-PET radioligands have advanced

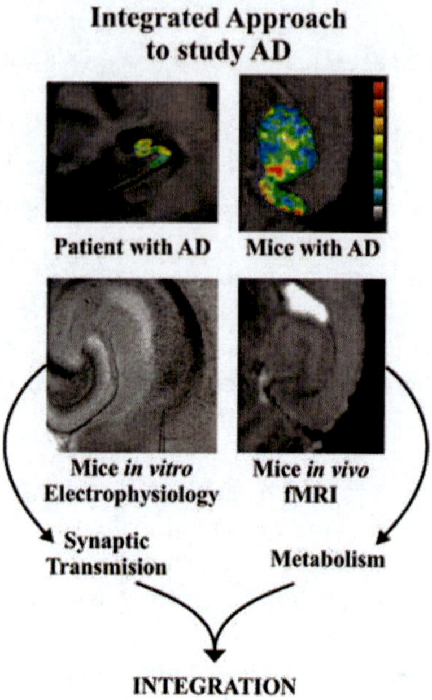

Fig. 11 Integrated approach to study neurodegeneration. In the upper panel are shown hippocampal cerebral blood volume (CBV) maps generated in mice (*right*) and humans (*left*). In the lower panel an in vitro horizontal brain slice (*left*) at the same position where the MRI generated images (*right*). These types of preparations exemplified the techniques that can be used crosssspecies (*upper panel*) and techniques that complement each other (*lower panel*)

significantly too. In contrast, efforts to develop a SPECT Aβ imaging agent have been largely unsuccessful. To conclude, it seems possible, although still at early stages, that neuronal metabolic changes preceding cell death can indeed be detected by imaging methods. This is a significant advancement, which allows for very early diagnosis and intervention.

11 APP Processing and Its Relation to Cognition

11.1 Amyloid Hypothesis

A strong genetic association exists between early onset familial forms of AD (FAD) and the 42 amino acid species of the Aβ peptide (Hutton et al., 1998; Younkin, 1998). It has been determined that autosomal dominant mutations in the genes for amyloid precursor protein (APP), presenilin 1 (PS1), and presenilin 2 (PS2) increase production of Aβ42 and correlate strongly with the FAD syndrome. In addition, the ε4 isoform of the apolipoprotein E gene, which is the major risk for late-onset disease, affects the rate of Aβ aggregation (Bales et al., 1999). Aβ has been the central point of AD research for over a decade and is generally considered as the upstream causative factor. This has been the basis of the amyloid hypothesis. Recently this hypothesis has been challenged or at least reconsidered by several investigators (Ash, Duff, etc.); some propose a dual hypothesis (Small and Duff, 2008), which implies an upstream event that initiates both Aβ- and tau-related pathologies. Aβ is a cleaved product of APP via the sequential action of two protease complexes, the β and γ secretases. (Selkoe, 2001). β secretase cleaves APP at the N-terminus producing the membrane-bound moiety C99 and the secreted APPs β segment. Subsequently, C99 is cleaved by the γ secretase to generate the C-terminus of Aβ, resulting in a series of 28 to 43 amino acid length Aβ peptides. Under normal conditions such events result in a higher proportion of Aβ40 than Aβ42 moieties. Under altered conditions, such as in transgenic mice harboring human APP mutations, an increased production of Aβ42 develops, followed by many pathophysiological features of AD including amyloid plaques, dystrophic neuritis, and synaptic dysfunction. Nevertheless none of the amyloidogenic mice or even mice that develop Aβ- and tau-related pathology have significant neuronal loss.

11.2 Aβ Extra- or Intracellular and in Which Compartment?

Amyloidogenic mouse models have shown that overproduction of Aβ leads to dystrophic axons and dendrites around amyloid plaques (LaFerla et al., 2007). It is also clear that anterograde axonal transport delivers Aβ peptide into plaques (Stokin et al., 2005). Substantial controversy remains over the sites of APP processing and Aβ release (Lee et al., 2005). Some studies implicate the axon as a site of Aβ production. Consistent with the amyloid deposition hypothesis is the fact that plaque formation increases if poor axonal transport delays the progress of APP and its processing enzymes through the axon (Stokin et al., 2005). Other reports failed to reproduce parts of this model, in which APP and its processing enzymes are cotransported (Lazarov et al., 2005). Some Aβ release occurs at the synapse in an activity-dependent manner, but Aβ can also be released from more proximal sites (Adalbert et al., 2007). A study that evaluated Aβ dynamics in human brain interstitial fluid (ISF) after trauma, reported that Aβ concentrations increased as neurological status improved and decreased when neurological status declined. Brain ISF Aβ concentrations were also lower when cerebral hypometabolism was present, reflecting depressed neuronal function (Brody et al., 2008); interestingly most of the normally secreted Aβ detected in this study did not appear to be either Aβ40 or Aβ42.

Extracellular accumulation of Aβ represents the foundation of the amyloid cascade hypothesis. The importance of intracellular Aβ accumulation in the pathogenesis of AD has emerged as a possibility in recent studies (LaFerla et al., 2007). These studies, implemented in human and mouse brains were made possible by the development of antibodies that could differentiate Aβ40 and Aβ42 from the transmembrane amyloid precursor protein from which they derived (Gouras et al., 2005). Other studies using transgenic mice harboring constructs that target Aβ either intracellularly or extracellularly showed that only transgenic mice producing the intracellular Aβ developed neurodegeneration (LaFerla et al., 1995). Some experimental evidence suggests that intracellular Aβ accumulates because a portion of the Aβ by not being secreted remains in the cytosol. Given that the vast majority of Aβ is normally secreted, such results indicate that Aβ is predominantly cleaved at or near the plasmalemmal inner surface or as part of the secretory pathway (Laferla et al., 2007). Another possible mechanism that explains intracellular Aβ accumulation involves Aβ endocytosis (D'andrea et al 2001).

There is strong evidence that Aβ42 is responsible for memory decline in AD, however, in humans the extent of Aβ accumulation correlates poorly with memory abnormalities (Giannakopoulos et al., 2003). Indeed, a specific challenge in addressing Aβ in AD concerned the role of specific pools of Aβ (e.g., extracellular, intracellular, membrane associated, or insoluble) in the genesis of the pathology. Newly reported pathological findings, using sequential brain extraction procedures demonstrated that only the intracellular Aβ42 levels and membrane-bound compartments were significant higher in the neocortex of AD cases than controls and correlated with neurological deficit, whereas Aβ40 levels were similar in patients with AD and in controls (Steinerman et al., 2008).

The relevance of extracellular Aβ toxicity has also been fully documented, with many reports emphasizing two issues: there is no clear relationship between amyloid plaque number and AD clinical status in humans or behavioral deficit in humans or mouse models, and cognitive deficits occur much before plaque deposit in AD mouse models, and are related to the appearance of oligomeric forms of Aβ, mainly Aβ42. The main Aβ oligomeric subspecies identified have been (1) dimeric forms in CSF from AD patients (Klyubin et al., 2008), and (2) 56-kDa specie, a potential dodecameric Aβ42 assembly in an AD mouse model, that when injected into young rats produced similar deficits in memory as those seen in the AD mouse (Lesné et al., 2006). Finally, it should be pointed out that Aβ40 and Aβ42 are not the only toxic proteolytic products of APP. Indeed, several groups have proposed that other C-terminal fractions (C99 and C88) or CTF50 are also related to AD pathophysiology.

12 Revisiting the Unforgettable Tau

12.1 Aβ and Tau Interaction

For more than two decades neurobiologists have known that both Aβ and tau are prominent in the CNS structures targeted by AD. As described above, Aβ hypothesis has been favored due to its genetic links. Indeed, no mutations in the tau gene have ever been linked to the disease and even today many tau experts concede that Aβ-related toxicity initiates neuronal dysfunction. Based on these premises several Aβ reducing agents have been or are in the process of being tested (tarenflurbil, tramiposate, active, and passive Aβ immunizations). However, to date these human trials have been largely disappointing. Furthermore, a recent study documenting the long-term effect of Aβ immunization in patients that had been immunized in September 2000 (of note, phase I of the trial was halted because of lethal complications in a small number of patients) who deceased, reports that the immunization with Aβ42 resulted in a near complete reduction of Aβ pathology, but this clearance did not prevent progressive neurodegeneration. All these patients had severe neurocognitive deficits and advanced NFT pathology, even involving primary cortex (Holmes et al., 2008). These two sets of findings—the failure of drugs that reduce Aβ load and the neuropathological study—not only suggest that the focus of AD treatment must also consider tau physiopathology in the equation, but has been the basis for proposing the dual pathway hypothesis, specifically referring to late onset AD (Small and Duff, 2008). This group hypothesizes that Aβ and tau pathologies are driven by single upstream molecular events: potential candidate molecules include Apolipoprotein E ε4-(APO-E4), the glycogen synthase kinase 3 (GSK3), and the retromer complex.

Focusing on the interaction tau-Aβ, mice expressing disease-causing double-mutations in APP and varying the number of the mouse tau gene (either, none, one or the normal two copies) have been generated. These mice developed brain amyloid pathology due to the APP mutation, but as usual in AD mouse models none had NFTs or neuronal loss. As the mice aged, those with two tau gene copies became impaired in spatial memory tests. Animals with one tau copy performed slightly better and mice with no tau gene had normal memory scores (Roberson et al., 2007). These results suggest that tau reduction somehow can block Aβ mediated neuronal dysfunction. Consistent with these reports, mice bearing a mutant tau that is linked to hereditary tauopathy that can be suppressed with doxycycline developed age-dependent NFTs, where the tau gene is expressed (SantaCruz et al., 2005). These mice developed brain atrophy and abnormal spatial memory. Suppression of the mutant tau gene with doxycycline improved the mice's spatial memory but did not affect NFT accumulation. Some have even proposed that NFT serves as protective role against cellular toxicity, because phosphorylated tau can sequester redox-active heavy metals (Castellani et al., 2008). This leaves by default one viable hypothesis of a "toxic tau intermediate" yet to be discovered.

13 Synaptic Dysfunction in AD

13.1 Is AD a Neuronal Disconnection Syndrome?

In neuropathological studies of synaptic damage, several groups have identified synaptic loss in the hippocampus ~50%, neocortex 25–30%, whereas in the cerebellum, an area not affected in AD, there are no changes in synapse number (Bertoni-Freddari et al., 1990; Gylys et al., 2004; Almeida et al., 2005). These changes are accompanied by decreased levels of pre- (synaptophysin) and post-synaptic (synaptopodin and PSD95) proteins in the samples from the AD brains compared to non-AD age-matched controls. The loss of synapses and the loss of synaptic proteins are confined to the brain regions known to be affected in AD.

There are several lines of evidence from experiments in vitro and in vivo that soluble oligomeric Aβ is responsible for a decrease in long-term potentiation in several synapses, mainly in the hippocampus, and disruption of neuronal synaptic plasticity. Further experiments have identified that dimeric and trimeric subspecies of Aβ inhibit LTP (Cleary et al., 2005; Townsend et al., 2006; Puzzo et al., 2005). Thus, although there is evidence for oligomeric Aβ (oβ)-induced synaptic dysfunction, the pre- or postsynaptic sites of action and the specific mechanism responsible for such dysfunction have not been established. Experiments addressing these questions were performed in the squid giant synapse, demonstrating that although intra-axonal oAβ42 peptide produced failure of synaptic transmission, intra-axonal oAβ40 peptide produced no significant changes in synaptic transmission. The effect of oAβ42 peptide is mediated by a cascade of events involving caseine kinase activity, abnormal fast axonal transport, and the rapid clathrin-independent endocytosis pathway. This set of events resulted in reduction of transmitter release (Moreno et al., 2009). This suggests that a dying-back phenomenon may be occurring in AD, causing synaptic loss, as has been recently proposed. (Pigino et al., 2009). In addition, other factors, such as Aβ-mediated decreased expression of sodium channels in hippocampal GABAergic interneurons, which leads to a hyperactive sate of the hippocampall circuit, have also been proposed (Palop et al., 2008). In agreement with this "exitotoxic event" acute extracellular exposure of hippocampus subicular pyramidal neurons to nanomolar concentrations of oAβ42 produced an increase of spontaneous synaptic events and larger basal dendritic calcium levels, and intracellular oAβ42 produced a synaptic failure (Angulo et al., 2008). Interestingly

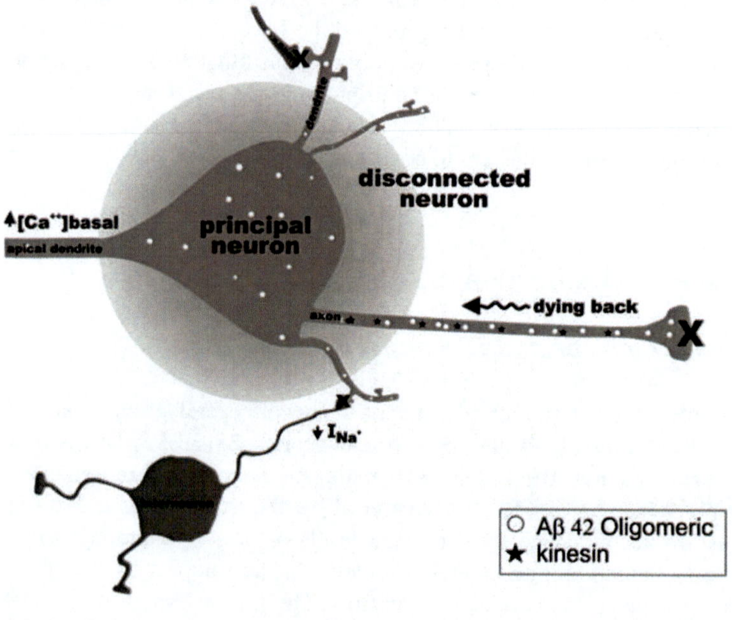

Fig. 12 Model of a "disconnected neuron." Shown are the events known to occur in different compartments of the neuron due to amyloid pathology. Note that both the presynaptic and postsynaptic elements are dysfunctional in AD

an upregulation of several neurotransmitter receptors (cholinergic, GABAergic, and glutamatergic) in young AD mouse models, with a subsequent age-dependent decline in their expression levels has also been reported (Bell et al., 2006).

The role of tau in synaptic dysfunction has not been as well documented as Aβ. Tau has been reported to change intracellular calcium levels, probably via the interaction of extracellular tau with muscarinic receptors M1 and M3 (Gómez-Ramos et al., 2008). Also a spatially association of tau modifications with intraneuronal Aβ, in which both pathologies co-occur at synaspses has been reported (Takahashi et al., 2010). Finally, using two-photon calcium imaging in layer 2/3 of an AD mouse model cortical neurons, an increase in the frequency of spontaneous calcium transients in the vicinity of amyloid plaques was seen (Busche et al., 2008).

Taking this information together with the imaging data (MRI-CBV) maps demonstrating a hypermetabolic phase in very young AD and Down-syndrome mice and a later age-dependent hypometabolism (Moreno et al., 2006, 2007), it seems plausible to propose that this series of events leads to a biphasic neuronal disconnection syndrome, with an initial hyperactive state followed by subsequent synaptic failure phase (Fig. 12).

14 Future Perspectives

Although the advances in the understanding of AD and PD pathophysiology have been significant, fundamental issues remain unsolved.

The powerful neuropathological arguments concerning the progression of PD based on alpha synuclein predicts late involvement of cortical circuits, presumably responsible for cognitive changes. This needs to be established with a multivariate analysis of longitudinal studies in a large number of demographically well-defined populations. The relationship of PD to frontotemporal dementia (Kertesz et al., 2005) and other neurodegenerative disease remains to be defined. The search for specific protein aggregates characterizing or perhaps defining each diagnostic entity appears promising. More attention to the sensitivity of pre- and postsynaptic dopamine receptors may help further development of rational pharmacotherapy in PD.

The retina is the most accessible part of the central nervous system and its morphology and functional properties have been well characterized. Future neuropharmacological/ therapeutical research may take advantage of the retina in PD. Although the inner retina has been shown to be significantly thinned in over two thirds of PD patients, the presence of alpha synuclein has not been established.

There is a demonstrable paucity of rational pharmacotherapeutical agents in the treatment of cognitive dysfunction in PD. The evidence suggests that multiple circuits and multiple neurotransmitter systems are involved: developing therapy will be a challenging task. Potentially targeting specific cognitive dysfunctions and specific pathochemical mechanisms may be rewarding.

The possible future directions of AD research may include: (a) more detailed information in the synaptic dysfunction (i.e., localizing the pre- and postsynaptic events, identification of the "toxic tau subspecies" and its molecular pathway); (b) identification of the molecular/functional basis for the regional sensitivity/specificity: why is entorhinal cortex affected initially in AD and how does it progress? What confers "resistance" to the cerebellum; (c) reconsider therapeutic strategies; we may need a molecular step up; (d) the use of in vivo techniques that allow a better understanding of brain circuitry disarray in AD such as magnetoencephalography (MEG); (e) the use of data-based circuits modeling, in which brain oscillations can be investigated and used in conjunction with MEG data, to be able to propose the mechanisms of the cognitive deficits observed more realistically.

To finish, I refer to Faulkner. "The past is never dead," Gavin Stevens says in *Requiem for a Nun*, and he adds, "It is not even past." Such fundamental considerations may not remain the same under AD pathological processes, a rather devastating new reality, or perhaps quite the opposite; in any case we must understand it, if an appropriate treatment is to be developed.

Acknowledgments I thank Yafell Serulle and Michaelangelo Fuortes for critical reading of the manuscript and Sergio Angulo for assistance with graphics design. HM was supported by NIH grant AG27476.

References

Aarsland D, Perry R, Brown A, Larsen JP, Ballard C (2005) Neuropathology of dementia in Parkinson's disease: a prospective community based study. Ann Neurol 58:773–776

Adalbert R, Gilley J, Coleman MP (2007) Abeta, tau and ApoE4 in Alzheimer's disease: the axonal connection. Trends Mol Med 13(4):135–142

Aggleton JP, Brown MW (1999) Episodic memory, amnesia, and the hippocampal-anterior thalamic axis. Behav Brain Sci 22:425–489

Albin RL, Young AB, Penney JB (1989) The functional anatomy of basal ganglia disorders. Trends Neurosci 12:366–375

Alexander GE, Crutcher MD (1990) Functional architecture of basal ganglia circuits: neural substrates of parallel processing. Trends Neurosci 13:266–271

Almeida CG, Tampellini D, Takahashi RH, Greengard P, Lin MT, Snyder EM, Gouras GK (2005) Beta-amyloid accumulation in APP mutant neurons reduces PSD-95 and GluR1 in synapses. Neurobiol Dis 20(2):187–198

Amick MM, Cronin-Golomb A, Gilmore GC (2003) Visual processing of rapidly presented stimuli is normalized in Parkinson's disease when proximal stimulus strength is enhanced. Vision Res 43:2827–2835

Andrews-Hanna JR, Snyder AZ, Vincent JL, Lustig C, Head D, Raichle ME, Buckner RL (2007) Disruption of large-scale brain systems in advanced aging. Neuron 56(5):924–935

Angulo S, Vela D, Bianchi R, Small S, Moreno H (2008) Acute effects of oligomeric ABeta 1–42 on hippocampal subicular neurons. SFN abstracts 829.12

Antal A, Keri S, Bodis-Wollner I (1997) Dopamine D2 receptor blockade alters the primary and cognitive components of visual evoked potentials in the monkey, Macaca fascicularis. Neurosci Lett 232:179–181

Antal A, Keri S, Dibo G, et al. (2002) Electrophysiological correlates of visual categorization: evidence for cognitive dysfunctions in early Parkinson's disease. Brain Res Cogn Brain Res 13:153–158

Antal A, Keri S, Kovacs G, Janka Z, Benedek G (2000) Early and late components of visual categorization: an event related potential study. Cogn Brain Res 9:117–119

Antal A, Kovanecz I, Bodis-Wollner I (1994) Visual discrimination and P300 are affected in parallel by cholinergic agents in the behaving monkey. Physiol Behav 56:161–166

Antal A, Pfeiffer R, Bodis-Wollner I (1996) Simultaneously evoked primary and cognitive visual evoked potentials distinguish younger and older patients with Parkinson's disease. J Neural Trans 103:1053–1067

Axelrod J (1966) Methylation reactions in the formation and metabolism of catecholamines and other biogenic amines. Pharmacol Rev 18(1):95–113

Bäckman L, Nyberg L, Lindenberger U, Shu-Chen Li, Farde L (2006) The correlative triad among aging, dopamine, and cognition: current status and future prospects. Neurosci Biobehav Rev 30:791–807

Bales KR, Verina T, Cummins DJ, Du Y, Dodel RC, Saura J, Fishman CE, DeLong CA, Piccardo P, Petegnief V, Ghetti B, Paul SM (1999) Apolipoprotein E is essential for amyloid deposition in the APP(V717F) transgenic mouse model of Alzheimer's disease. Proc Natl Acad Sci USA 96(26):15233–15238

Bandini F, Pierantozzi M, Bodis-Wollner I (2002) The visuo-cognitive and motor effect of amantidine in non-Caucasian patients with Parkinson's disease. A clinical and electrophysiological study. J Neural Transm 109:41–51

Bell KF, Ducatenzeiler A, Ribeiro-da-Silva A, Duff K, Bennett DA, Cuello AC (2006) The amyloid pathology progresses in a neurotransmitter-specific manner. Neurobiol Aging 27(11): 1644–1657

Bergman H, Wichmann T, DeLong M (1990) Reversal of experimental parkinsonism by lesions of the subthalamic nucleus. Science 249:1436–1438

Bertoni-Freddari C, Fattoretti P, Casoli T, Meier-Ruge W, Ulrich J (1990) Morphological adaptive response of the synaptic junctional zones in the human dentate gyrus during aging and Alzheimer's disease. Brain Res 517(1–2):69–75

Birkmayer W, Hornykiewicz O (1961) Der L-3,4-dioxyphenalalanin (DOPA)- Effekt bei der Parkinson Akinesie. Wien Klin Wochenschr 73:787–788

Biswal B, Yetkin FZ, Haughton VM, Hyde JS (1995) Functional connectivity in the motor cortex of resting human brain using echo-planar MRI. Magn Reson Med 34(4):537–541

Bodis-Wollner I (1990) Visual deficits related to dopamine deficiency in experimental animals and Parkinson's disease patients. Trends Neurosci 13:296–302

Bodis-Wollner I (2008) Pre-emptive perception. Perception 3:263–278

Bodis-Wollner I, Borod JC, Cicero B, Haywood CS, Raskin S, Mylin L, Sliwinski M, Falk A, Yahr MD (1995) Modality dependent changes in event related potentials correlate with specific cognitive function in non-demented patients with Parkinson's disease. J Neural Trans 9: 197–209

Bodis-Wollner I, Davis J, Tzelepi A, Bezerianos T (2001) Wavelet transform of the EEG reveals differences in low and high gamma responses to elementary visual stimuli. Clin Electroencephalogr 32:139–144

Bodis-Wollner I, Tzelepi A (1998) The push-pull action of dopamine on spatial tuning of the monkey retina: the effects of dopaminergic deficiency and selective Dl and D2 receptor ligands on the pattern electroretinogram. Vision Res 38:1479–1487

Bodis-Wollner I, Tzelepi A (2002) Push-pull model of dopamine's action in the retina, Chapter 5. In: Hung GK, Ciuffreda KJ (eds) Models of the visual system. Kluwer Academic/Plenum Publishers, New York, NY, Boston, MA, Dordrecht, London and Moscow, pp 191–214

Bodis-Wollner I, Tzelepi A (2005) The effect of diverse dopamine receptors on spatial processing in the central retina: a model, Chapter 17. In: Jenkin M, Harris L (eds) Seeing spatial form. Oxford University Press, New York, NY

Bodis-Wollner I, Yahr MD, Mylin LH (1983) Non-motor functions of the basal ganglia. In: Hassler RG, Christ RG (eds) Advances in neurology. Parkinson-specific motor and mental disorders, vol 40. Raven Press, New York, pp 289–298

Bohnen NI, Kaufer DI, Ivanco LS, Lopresti B, Koeppe RA, Davis JG, Mathis CA, Moore RY, DeKosky ST (2003) Cortical cholinergic function is more severely affected in parkinsonian dementia than in Alzheimer disease: an in vivo positron emission tomographic study. Arch Neurol 60:1745–1748

Bokura H, Yamaguchi S, Kobayashi S (2005) Event-related potentials for response inhibition in Parkinson's disease. Neurophyschologia 43(6):967–975

Braak H, Del Tredici K, Bratzke H, Hamm-Clement J, Sandmann-Keil D, Rub U (2002) Staging of the intracerebral inclusion body pathology associated with idiopathic Parkinson's disease (preclinical and clinical stages). J Neurol 249(Suppl 3:III):1–5

Braak H, Ghebremedhin E, Rüb U, Bratzke H, Del Tredici K (2004) Stages in the development of Parkinson's disease-related pathology. Cell Tissue Res 318(1):121–134

Braak H, de Vos R, Bohl BC, Del Tredici K (2006) Gastric CEB1-synuclein immunoreactive inclusions in Meissner's and Auerbach's plexuses in cases staged for Parkinson's disease-related brain pathology. Neurosci Lett 396:67–72

Brickman AM, Honig LS, Scarmeas N, Tatarina O, Sanders L, Albert MS, Brandt J, Blacker D, Stern Y (2008) Measuring cerebral atrophy and white matter hyperintensity burden to predict the rate of cognitive decline in Alzheimer disease. Arch Neurol 65(9):1202–1208

Brody DL, Magnoni S, Schwetye KE, Spinner ML, Esparza TJ, Stocchetti N, Zipfel GJ, Holtzman DM (2008) Amyloid-beta dynamics correlate with neurological status in the injured human brain. Science 321(5893):1221–1224

Busche MA, Eichhoff G, Adelsberger H, Abramowski D, Wiederhold KH, Haass C, Staufenbiel M, Konnerth A, Garaschuk O (2008) Clusters of hyperactive neurons near amyloid plaques in a mouse model of Alzheimer's disease. Science 321(5896):1686–1689

Buytenhuijs EL, Berger HJ, Van Spaendonck KP, Horstink MW, Borm GF, Cools AR (1994) Memory and learning strategies in patients with Parkinson's disease. Neuropsychologia 32:335–342

Carlsson A, Lindquist M, Magnusson T (1957) 3-4 Dihydroxyphenylalanine and 5-hydroxytryptophanas reserpine antagonists. Nature 180:1200–12001

Castellani RJ, Nunomura A, Lee HG, Perry G, Smith MA (2008) Phosphorylated tau: toxic, protective, or none of the above. J Alzheimers Dis 14(4):377–383

Cleary JP, Walsh DM, Hofmeister JJ, Shankar GM, Kuskowski MA, Selkoe DJ, Ashe KH (2005) Natural oligomers of the amyloid-beta protein specifically disrupt cognitive function. Nat Neurosci 8(1):79–84

Costa DC, Pilowsky LS, Ell PJ (1999) Nuclear medicine in neurology and psychiatry. Lancet 354(9184):1107–1111

Cragg SJ (2006) Meaningful silences; how dopamine listens to the Ach pause. Trends Neurosci 29:125–131

D'Andrea MR, Nagele RG, Wang HY, Peterson PA, Lee DH (2001) Evidence that neurones accumulating amyloid can undergo lysis to form amyloid plaques in Alzheimer's disease. Histopathology 38(2):120–134

Daum I, Schugens MM, Spieker s (1995) Memory and skill acquisition in Parkinson's disease and frontal lobe dysfunction. Cortex 31:413–432

De Luca M, Smith S, De Stefano N, Federico A, Matthews PM (2005) Blood oxygenation level dependent contrast resting state networks are relevant to functional activity in the neocortical sensorimotor system. Exp Brain Res 167(4):587–594

DeLong MR (1990) Primate models of movement disorders of basal ganglia origin. Trends Neurosci 13:281–285

DeLong MR, Wichmann T (2007) Circuits and circuit disorders of the basal ganglia. Arch Neurol 64(1):20–24

Devanand DP, Liu X, Tabert MH, Pradhaban G, Cuasay K, Bell K, de Leon MJ, Doty RL, Stern Y, Pelton GH (2008b) Combining early markers strongly predicts conversion from mild cognitive impairment to Alzheimer's disease. Biol Psychiatry 64(10):871–879

Devanand DP, Tabert MH, Cuasay K, Manly JJ, Schupf N, Brickman AM, Andrews H, Brown TR, Decarli C, Mayeux R (2008a) Olfactory identification deficits and MCI in a multi-ethnic elderly community sample. Neurobiol Aging 9:1593–1600

Di Carlo A, Lamassa M, Baldereschi M, Inzitari M, Scafato E, Farchi G, Inzitari D (2007) CIND and MCI in the Italian elderly: frequency, vascular risk factors, progression to dementia. Neurology 68(22):1909–1916

Dierks T, Frolich L, Ihi R, Maurer K (1994) Event-related potentials and psychopharmacology. Cholinergic modulation of P300. Pharmacopsychiatry 27:72–74

Dixon RM, Bradley KM, Budge MM, Styles P, Smith AD (2002) Longitudinal quantitative proton magnetic resonance spectroscopy of the hippocampus in Alzheimer's disease. Brain 125(Pt 10):2332–2341

Doty RL, Perl DP, Steele JC, Chen KM, Pierce JD Jr, Reyes P, Kurland LT (1991) Olfactory dysfunction in three neurodegenerative diseases. Geriatrics 46(Suppl 1):47–51

Du AT, Schuff N, Zhu XP, Jagust WJ, Miller BL, Reed BR, Kramer JH, Mungas D, Yaffe K, Chui HC, Weiner MW (2003) Atrophy rates of entorhinal cortex in AD and normal aging. Neurology 60(3):481–486

Dujardin K, Defebvre L, Grunberg C (2001) Memory and executive function in sporadic and familial Parkinson's disease. Brain 124:389–398

El Fakhri G, Kijewski MF, Johnson KA, Syrkin G, Killiany RJ, Becker JA, Zimmerman RE, Albert MS (2003) MRI-guided SPECT perfusion measures and volumetric MRI in prodromal Alzheimer disease. Arch Neurol 60(8):1066–1072

Emre M (2004, Aug) Dementia in Parkinson's disease: cause and treatment. Curr Opin Neurol 17(4):399–404. Review

Enroth-Cugell C, Robson JG (1966, Dec) The contrast sensitivity of retinal ganglion cells of the cat. J Physiol 187(3):517–552

Fedele E, Mazzone P, Stefani A, Bassi A, Ansaldo MA, Raiteri M, Altibrandi MG, Pierantozzi M, Giacomini P, Bernardi G, Stanzione P (2001) Microdialysis in Parkinsonian patient basal ganglia: acute apomorphine-induced clinical and electrophysiological effects not paralleled by changes in the release of neuroactive amino acids. Exp Neurol 167(2):356–365

Fogelson N, Williams D, Tijssen M, van Bruggen G, Speelman H, Brown P (2006) Different functional loops between cerebral cortex and the subthalmic area in Parkinson's disease. Cerebral Cortex 16:64–75

Forgacs P, Bodis-Wollner I (2004) Nicotinic receptors and cognition in Parkinson's disease: the importance of neuronal synchrony. J Neural Transm 111:1317–1331

Forgacs P, von Gizycki H, Selesnick I, Syed N, Ebrahim K, Avitable M, Amassian V, Lytton W, Bodis-Wollner I (2008) Perisaccadic parietal and occipital gamma power in light and in complete darkness. Perception 3:419–432

Frodl-Bauch T, Bottlender R, Hegerl U (1999) Neurochemical substrates and neuroanatomical generators of event-related P300. Neuropsychobiology 40:86–94

Gaffan D, Parker A (2000, Apr) Mediodorsal thalamic function in scene memory in rhesus monkeys. Brain 123(Pt 4):816–827

Garcia-Cabezas MA, Marinez-Sanchez P, Sanchez-Gonzalez M, Garzon M, Cavada C (2007a) The thalamic dopaminergic system is expanded in humans compared to rodents. Soc Neurosci Abstr 38:12

Garcia-Cabezas MA, Rico B, Sanchez-Gonzalez M, Cavada C (2007b) Distribution of the dopamine innervation in the macaque and human thalamus. Neuroimage 34:965–985

Gaspirini M, Fabrizio E, Bonifati V, Meco G (1997) Cognitive improvement during Tolcapone treatment in Parkinson's disease. J Neural Transm 104:887–894

Ghilardi MF, Bodis-Wollner I, Onofrj MC, Marx MS, Glover AA (1988a) Spatial frequency dependent abnormalities of the pattern electroretinogram and visual evoked potentials in a parkinsonian monkey model. Brain 111:131–184

Ghilardi MF, Chung E, Bodis-Wollner I, Dvorzniak M, Glover A, Onofrj M (1988b) Systemic 1-methyl,4-phenyl,1-2-3-6-tetrahydropyridine (MPTP) administration decreases retinal dopamine content in primates. Life Sci 43:255–262

Ghilardi MF, Marx MS, Bodis-Wollner I, Camras CB, Glover AA (1989) The effect of intraocular 6-hydroxydopamine on retinal processing of primates. Ann Neurol 25:357–364

Giannakopoulos P, Herrmann FR, Bussière T, Bouras C, Kövari E, Perl DP, Morrison JH, Gold G, Hof PR (2003) Tangle and neuron numbers, but not amyloid load, predict cognitive status in Alzheimer's disease. Neurology 60(9):1495–1500

Glover A, Ghilardi MF, Bodis-Wollner I, Onofrj M (1988) Alterations in event-related potentials (ERPs) of MPTP-treated monkeys. Electroencephalogr Clin Neurophysiol 7 1(46):1–468

Goldman-Rakic P (1998) The cortical dopamine system: role in memory and cognition. Adv Pharmacol 42:707–711

Goldman-Rakic PS, Muly EC, Williams GV (2000) D_1 receptors in prefrontal cells and circuits. Brain Res Rev 31:295–301

Gómez-Ramos A, Díaz-Hernández M, Rubio A, Miras-Portugal MT, Avila J (2008) Extracellular tau promotes intracellular calcium increase through M1 and M3 muscarinic receptors in neuronal cells. Mol Cell Neurosci 37(4):673–681

Gonzalez RG, Fischman AJ, Guimaraes AR, Carr CA, Stern CE, Halpern EF, Growdon JH, Rosen BR (1995) Functional MR in the evaluation of dementia: correlation of abnormal dynamic cerebral blood volume measurements with changes in cerebral metabolism on positron emission tomography with fludeoxyglucose F 18 AJNR. Am J Neuroradiol 16(9): 1763–1770

Gouras GK, Almeida CG, Takahashi RH (2005) Intraneuronal Abeta accumulation and origin of plaques in Alzheimer's disease. Neurobiol Aging 26(9):1235–1244

Gray CM, Singer W (1989) Stimulus-specific neuronal oscillations in orientation columns of cat visual cortex. Proc Natl Acad Sci USA 86:1698–1702

Greengard P (1972) Adenosine 3′:5′-cyclic monophosphate as a mediator in the action of neurohumoral agents. Biochem J 128:75–80

Greicius MD, Krasnow B, Reiss AL, Menon V (2003) Functional connectivity in the resting brain: a network analysis of the default mode hypothesis. Proc Natl Acad Sci USA 100(1):253–258

Gylys KH, Fein JA, Yang F, Wiley DJ, Miller CA, Cole GM (2004) Synaptic changes in Alzheimer's disease: increased amyloid-beta and gliosis in surviving terminals is accompanied by decreased PSD-95 fluorescence. Am J Pathol 165(5):1809–1817

Hajee M, March W, Lazzaro D, Wolintz A, Shrier E, Glazman S, Bodis-Wollner I (2009) Inner retinal layer thinning in Parkinson's disease. Arch Ophthalmol 127(6):737–741

Halgren E, Marinkovic K, Chauvel P (1998) Generators of the late cognitive potentials in auditory and visual oddball tasks. Electroencephalogr Clin Neurophysiol 106:156–164

Hanninen T, Koivisto K, Reinikainen KJ, Helkala EL, Soininen H, Mykkänen L, Laakso M, Riekkinen PJ (1996) Prevalence of ageing-associated cognitive decline in an elderly population. Age Ageing 25(3):201–205

Hansch EC, Syndulko K, Cohen SN, Goldberg ZI, Potvin AR, Tourtellotte WW (1982) Cognition in Parkinson disease: an event-related potential perspective. Ann Neurol 11:599–607

Hay JF, Moscovitch M, Levine B (2002) Dissociating habit and recollection: evidence from Parkinson's disease, amnesia and focal lesion patients. Neuropsychologia 40:1324–1334

Hersch SM, Hutekunst CA, Rees HD, Heilman CJ, Levey AI (1994) Distribution of m1–m4 muscarinic receptor proteins in the rat striatum: light and electron microscopic immunocytochemistry using subtype-specific antibodies. J Neurosci 14:3351–3363

Higuchi M, Iwata N, Matsuba Y, Sato K, Sasamoto K, Saido TC (2005) 19F and 1H MRI detection of amyloid beta plaques in vivo. Nat Neurosci 8(4):527–533

Hoehn MM, Yahr MD (1967) Parkinsonism: onset, progression and mortality. Neurology 17: 427–442

Holmes C, Boche D, Wilkinson D, Yadegarfar G, Hopkins V, Bayer A, Jones RW, Bullock R, Love S, Neal JW, Zotova E, Nicoll JA (2008) Long-term effects of Abeta42 immunisation in Alzheimer's disease: follow-up of a randomised, placebo-controlled phase I trial. Lancet 372(9634):216–223

Hugdahl K, Wester K, Asbjørnsen A (1991, Oct) Auditory neglect after right frontal lobe and right pulvinar thalamic lesions. Brain Lang 41(3):465–473

Hutton M, Pérez-Tur J, Hardy J (1998) Genetics of Alzheimer's disease. Essays Biochem 33: 117–131

Inada T, Nakamura A, Iijima Y (2003) Relationship between catechol-O-methyltransferase polymorphism and treatment-resistant schizophrenia. Am J Med Genet 120B:35–39

Joliot M, Ribary U, Llinás R (1994) Neuromagnetic coherent oscillatory activity in the vicinity of 40 Hz coexists with cognitive temporal binding in the human brain. Proc Natl Acad Sci USA 91:6339–6343

Keijsers NLW, Admiraal MA, Cools AR, Bloem BR, Gielen CCAM (2005) Differential progression of proprioceptive and visual information processing deficits in Parkinson's disease. Eur J Neurosci 21:239–248

Kemps E, Szmalec A, Vandierendonck A, Crevits L (2004) Visuo-spatial processing in Parkinson's disease: evidence for diminished visuo-spatial sketch pad and central executive resources. Parkinsonism Relat Disord 11:181–186

Kertesz A, McMonagle P, Blair M, Davidson W, Munoz D (2005) The evolution and pathology of frontotemporal dementia. Brain 128(9):1996–2005

Klyubin I, Betts V, Welzel AT, Blennow K, Zetterberg H, Wallin A, Lemere CA, Cullen WK, Peng Y, Wisniewski T, Selkoe DJ, Anwyl R, Walsh DM, Rowan MJ (2008) Amyloid beta protein dimer-containing human CSF disrupts synaptic plasticity: prevention by systemic passive immunization. J Neurosci 28(16):4231–4237

Knoke D, Taylor AE, Saint-Cyr JA (1998) The differential effects of cueing on recall in Parkinson's disease and normal subjects. Brain Cogn 38:261–274

Kowalska DM, Bachevalier J, Mishkin M (1991) The role of the inferior prefrontal convexity in performance of delayed nonmatching-to-sample. Neuropsychologia 29(6):583–600

Krizaj D, Gábriel R, Owen WG, Witkovsky P (1998, Sep 7) Dopamine D2 receptor-mediated modulation of rod-cone coupling in the Xenopus retina. J Comp Neurol 398(4):529–538

LaFerla FM, Green KN, Oddo S (2007) Intracellular amyloid-beta in Alzheimer's disease. Nat Rev Neurosci 8(7):499–509

LaFerla FM, Tinkle BT, Bieberich CJ, Haudenschild CC, Jay G (1995) The Alzheimer's A beta peptide induces neurodegeneration and apoptotic cell death in transgenic mice. Nat Genet 9(1):21–30

Lagopoulos J, Gordon E, Barhamali H et al (1998) Dysfunctions of automatic (P300a) and controlled (P300b) processing in Parkinson's disease. Neurol Res 20:5–10

Lang AE, Houeto JL, Krack P, Kubu C, Lyons KE, Moro E, Ondo W, Pahwa R, Poewe W, Tröster AI, Uitti R, Voon V (2006) Deep brain stimulation: preoperative issues. Mov Disord 21(Suppl 14):S171–S196

Lazarov O, Morfini GA, Lee EB, Farah MH, Szodorai A, DeBoer SR, Koliatsos VE, Kins S, Lee VM, Wong PC, Price DL, Brady ST, Sisodia SS (2005) Axonal transport, amyloid precursor protein, kinesin-1, and the processing apparatus: revisited. J Neurosci 25(9):2386–2395

Lee AC, Harris JP (1999) Problems with the perception of space in Parkinson's disease. Neuro-ophthalmology 22:1–15

Lee EB, Zhang B, Liu K, Greenbaum EA, Doms RW, Trojanowski JQ, Lee VM (2005) BACE overexpression alters the subcellular processing of APP and inhibits Abeta deposition in vivo. J Cell Biol 168(2):291–302

Leplow B, Dierks C, Herrmann P, Pieper N, Annecke R, Ulm G (1997) Remote memory in Parkinson's disease and senile dementia. Neuropsychologia 35:547–555

Lesné S, Koh MT, Kotilinek L, Kayed R, Glabe CG, Yang A, Gallagher M, Ashe KH (2006) A specific amyloid-beta protein assembly in the brain impairs memory. Nature 440(7082): 352–357

Levin BE, Llabre MM, Weiner WJ (1989) Cognitive impairments associated with early Parkinson's disease. Neurology 39:557–561

Li M, Kuroiwa Y, Wang L, Kamitani T, Omoto S, Hayashi E, Takahashi T, Suzuki Y, Koyano S, Ikegami T, Matsubara S (2005) Visual event-related potentials under different interstimulus intervals in Parkinson's disease: relation to motor disability, WAIS-R, and regional cerebral blood flow. Parkinsonism Relat Disord 11(4):209–219

Li M, Kuroiwa Y, Wang L, Kamitani T, Takahashi T, Suzuki Y, Omoto S (2003) Early sensory information processes are enhanced on visual oddball and S1–S2 tasks in Parkinson's disease: a visual event-related potentials study. Parkinsonism Relat Disord 9(6):329–340

Lieberman A, Dziatolowski M, Kupersmith M, Serby M, Goodgold A, Korein J, Goldstein M (1979) Dementia in Parkinson disease. Ann Neurol Oct 6(4):355–359

Llinás R, Urbano FJ, Leznik E, Ramírez RR, van Marle HJ (2005) Rhythmic and dysrhythmic thalamocortical dynamics: GABA systems and the edge effect. Trends Neuros 28(6):325–333

Luchsinger JA, Mayeux R (2004) Dietary factors and Alzheimer's disease. Lancet Neurol 3(10):579–587

Maeshima S, Itakura T, Komai N, Matsumoto T, Ueyoksi A (2002) Relationships between event-related potentials (P300) and activities of daily living in Parkinson's disease. Brain Inj 16:1–8

Maffei L, Fiorentini A, Bisti S, Hollander H (1989) Pattern ERG in the monkey after section of the optic nerve. Exp Brain Res 59:423–425

Malholtra AK, Kestler LJ, Mazzanti C, Bates JA, Goldberg T, Goldman D (2002) A functional polymorphism in the COMT gene and performance on a test of prefrontal cognition. Am J Psychiatry 159:652–654

Mazzone P, Lozano A, Stanzione P, Galati S, Scarnati E, Peppe A, Stefani A (2005, Nov 28) Implantation of human pedunculopontine nucleus: a safe and clinically relevant target in Parkinson's disease. Neuroreport 16(17):1877–1881

Mohler H (2007) Molecular regulation of cognitive functions and developmental plasticity: impact of GABAA receptors. J Neurochem 102(1):1–12

Monchi O, Petrides M, Mejia-Constain B, Strafella A (2007) Cortical activity in Parkinson's disease during executive processing depends on striatal involvement. Brain 130(1):233–244

Moreaud O, Fournet N, Roulin JL, Naegele B, Pellat J (1997) The phonological loop in medicated patients with Parkinson's disease: presence of phonological similarity and word length effects. J Neurol Neurosurg Psychiatry 62:609–611

Moreno H, Lee T, Yu R, Hua H, Small S (2006) fMRI studies of hippocampal dysfunction and therapeutic response in mouse models of Alzheimer's disease. Proceedings of the 10th international conference on Alzheimer. lzheimer's&Dementia:2;3:S676

Moreno H, Wu WE, Lee T, Brickman A, Mayeux R, Brown TR, Small SA (2007) Imaging the Abeta-related neurotoxicity of Alzheimer disease. Arch Neurol 64(10):1467–1477

Moreno H, Yu E, Pigino G, Hernandez I, Kim N, Moreira J, Sugimori M, Llinas R (2009) Direct inhibition of synaptic transmission by pre-synaptic injection of oligomeric amyloid beta. Proc Natl Acad Sci USA 106(14):5901–5906

Mori A, Shindou T (2003, Dec 9) Modulation of GABAergic transmission in the striatopallidal system by adenosine A2A receptors: a potential mechanism for the antiparkinsonian effects of A2A antagonists. Neurology 61(11 Suppl 6):S44–S48

Nieoullon A (2002) Dopamine and the regulation of cognition and attention. Prog Neurobiol 67:53–83

Nishi A, Kuroiwa M, Miller DB, O'Callaghan JP, Bateup HS, Shute T, Sotogaku N, Fukuda T, Heints N, Greengard P, Snyder GL (2008) Distinct roles of PDE4 and PDE10A in the regulation of cAMP/PKA signaling in the striatum. J Neurosci 28:10460–10471

Noda A, Ohba H, Kakiuchi T, Futatsubashi M, Tsukada H, Nishimura S (2002) Age-related changes in cerebral blood flow and glucose metabolism in conscious rhesus monkeys. Brain Res 936(1–2):76–81

O'Sullivan M, Jones DK, Summers PE, Morris RG, Williams SC, Markus HS (2001) Evidence for cortical "disconnection" as a mechanism of age-related cognitive decline. Neurology 57(4):632–638

Owen AM, Dayan J, Dagher A (1998) Abnormal basal ganglia outflow in Parkinson's disease identified with PET. Implications for higher cortical functions. Brain 121:949–965

Owen AM, Iddon JL, Hodges JR, Summers BA, Robbins TW (1997) Spatial and non-spatial working memory at different stages of Parkinson's disease. Neuropsychologia 35:519–532

Owen AM, Roberts AC, Hodges JR (1993a) Contrasting mechanisms of impaired attentional set-shifting in patients with Parkinson's disease. Brain 116:1159–1175

Palop J, Ho K, Thwin T, Mucke L (2008) Depletion of voltage-gated sodium channels in transgenic mouse models of Alzheimer's disease: a potential cause of Aβ-induced epileptiform activity. SFN abstracts 638.18

Pang S, Borod JC, Hernandez A, Bodis-Wollner I, Raskin S, Mylin L, Coscia L, Yahr MD (1990) The auditory P300 correlates with specific cognitive deficits in Parkinson's disease. J Neural Trans 2:249–264

Perry E, Curtis M, Dick D, John T (1985) Cholinergic correlates of cognitive impairment in Parkinson' s disease: comparisons with Alzheimer' s disease. J Neurol Neurosurg Psychiatry 48:413–421

Petersen RC, Negash S (2008) Mild cognitive impairment: an overview. CNS Spectr 13(1):45–53

Pfefferbaum A, Adalsteinsson E, Sullivan EV (2005) Frontal circuitry degradation marks healthy adult aging: evidence from diffusion tensor imaging. Neuroimage 26(3):891–899

Piccolino M, Witkovsky P, Trimarchi C (1987) Dopaminergic mechanisms underlying the reduction of electrical coupling between horizontal cells of the turtle retina induced by d-amphetamine, bicuculline, and veratridine. J Neurosci 7(8):2273–2284

Pigino G, Morfini G, Atagi Y, Deshpande A, Yu C, LaDu J, Busciglio J, Brady S (2009) Novel pathogenic mechanism for intraneuronal amyloid beta: disruption of fast axonal transport. Proc Natl Acad Sci USA 106(14):5907–5912

Postle BR, Jonides J, Smith EE, Corkin S, Growdon JH (1997) Spatial, but not obkect, delayed response is impaired in early Parkinson's disease. Neuropsychology 11:171–179

Prasher D, Findley L (1991) Dopaminergic induced changes in cognitive and motor processing in Parkinson's disease: an electrophysiological investigation. J Neurol Neurosurg Psychiatry 54:603–609

Price JL, Morris JC (1999) Tangles and plaques in nondemented aging and "preclinical" Alzheimer's disease. Ann Neurol 45(3):358–368

Puzzo D, Vitolo O, Trinchese F, Jacob JP, Palmeri A, Arancio O (2005) Amyloid-beta peptide inhibits activation of the nitric oxide/cGMP/cAMP-responsive element-binding protein pathway during hippocampal synaptic plasticity. J Neurosci 25(29):6887–6897

Qualman SJ, Haupt HM, Yang P, Hamilton SR (1984) Esophageal Lewy bodies associated with ganglion cell loss in achalasia. Similarity to Parkinson's disease. Gastroenterology 87(4): 848–856

Raichle ME, MacLeod AM, Snyder AZ, Powers WJ, Gusnard DA, Shulman GL (2001) A default mode of brain function. Proc Natl Acad Sci USA 98(2):676–682

Raskin S, Borod J, Tweedy J (1990) Neuropsychological aspects of Parkinson's disease. Neuropsychol Rev 1:185–221

Rieger JW, Kim A, Argyelan M, Farber M, Glazman S, Liebeskind M, Bodis-Wollner I (2008) Cortical control of voluntary saccades in Parkinson's disease. Electroencephalogr Clin Neurosci 39:169–174

Roberson ED, Scearce-Levie K, Palop JJ, Yan F, Cheng IH, Wu T, Gerstein H, Yu GQ, Mucke L (2007) Reducing endogenous tau ameliorates amyloid beta-induced deficits in an Alzheimer's disease mouse model. Science 316(5825):750–754

Sagliocco L, Bandini F, Pierantozzi M, Mari Z, Tzelepi A, Ko C, Gulzar J, Bodis-Wollner I (1997) Electrophysiological evidence for visuocognitive dysfunction in younger non-Caucasian patients with Parkinson's disease. J Neural Trans 104:427–439

Sanchez-Gonzalaez MA, Garcia-Cabezas MA, Rico B, Cavada C (2005) The primate thalamus is akey target for brain dopamine. J Neurosci 25:6076–6083

Santacruz K, Lewis J, Spires T, Paulson J, Kotilinek L, Ingelsson M, Guimaraes A, DeTure M, Ramsden M, McGowan E, Forster C, Yue M, Orne J, Janus C, Mariash A, Kuskowski M, Hyman B, Hutton M, Ashe KH (2005) Tau suppression in a neurodegenerative mouse model improves memory function. Science 309(5733):476–481

Scarmeas N, Stern Y, Mayeux R, Luchsinger JA (2006b) Mediterranean diet, Alzheimer disease, and vascular mediation. Arch Neurol 63(12):1709–1717

Scarmeas N, Stern Y, Tang MX, Mayeux R, Luchsinger JA (2006a) Mediterranean diet and risk for Alzheimer's disease. Ann Neurol 59(6):912–921

Schott B, Seidenbecher ConstanzeI, Fenker I, Lauer D, Bunzeck C, Bernstein H-G N, Tischmeyer W, Gundelfinger ED, Heinze HJ, Duzel E (2006) The dopaminergic midbrain participates in human episodic memory formation: evidence from genetic imaging. J Neurosci 26(5): 1407–1417

Schulman JS, Pedut-Kloizman T, Hertzmark E, et al. (1996) Reproducibility of nerve fiber layer thickness measurements using optical coherence tomography. Ophthalmology 103:1889–1898

Selkoe DJ (2001) Alzheimer's disease: genes, proteins, and therapy. Physiol Rev 81(2):741–766

Singer W (1993) Synchronisation of cortical activity and its putative role in information processing and learning. Annu Rev Physiol 55:349–374

Singer W (1999, Feb 4) Neurobiology. Striving for coherence. Nature 397(6718):391, 393. Comment on: Nature 1999 Feb 4;397(6718):434–436. Nature 1999 Feb 4;397(6718):430–433

Skirboll LR, Grace AA, Bunney BS (1977) Dopamine auto- and postsynaptic receptors: electrophysiological evidence for differential sensitivity to dopamine agonists. Science 206:80–82

Small SA (2001) Age-related memory decline: current concepts and future directions. Arch Neurol 58(3):360–364

Small SA, Duff K (2008) Linking Abeta and tau in late-onset Alzheimer's disease: a dual pathway hypothesis. Neuron 60(4):534–542

Small GW, Ercoli LM, Silverman DH, Huang SC, Komo S, Bookheimer SY, Lavretsky H, Miller K, Siddarth P, Rasgon NL, Mazziotta JC, Saxena S, Wu HM, Mega MS, Cummings JL, Saunders AM, Pericak-Vance MA, Roses AD, Barrio JR, Phelps ME (2000) Cerebral metabolic and cognitive decline in persons at genetic risk for Alzheimer's disease. Proc Natl Acad Sci USA 97(11):6037–6042

Sohn YH, Kim GW, Huh K, Kim JS (1998) Dopaminergic influences on the P300 abnormality in Parkinson's disease. J Neurol Sci 158:83–87

Stanzione P, Fattapposta F, Giunti P, et al. (1991) P300 variations in parkinsonian patients before and during dopaminergic monotherapy: a suggested dopamine component in P300. Electroencephalogr Clin Neurophysiol 80:446–453

Stanzione P, Pierantozzi M, Semprini R, Tagliati M, Traversa R, Peppe A, Pierelli F, Bernardi G (1995, Sep) Increasing doses of l-sulpiride reveal dose- and spatial frequency-dependent effects of D2 selective blockade in the human electroretinogram. Vision Res 35(18):2659–2664

Stefani A, Fedele E, Galati S, Pepicelli O, Frasca S, Pierantozzi M, Peppe A, Brusa L, Orlacchio A, Hainsworth AH, Gattoni G, Stanzione P, Bernardi G, Raiteri M, Mazzone P (2005, Mar) Subthalamic stimulation activates internal pallidus: evidence from cGMP microdialysis in PD patients. Ann Neurol 57(3):448–452

Stefani A, Fedele E, Galati S, Raiteri M, Pepicelli O, Brusa L, Pierantozzi M, Peppe A, Pisani A, Gattoni G, Hainsworth AH, Bernardi G, Stanzione P, Mazzone P (2006) Deep brain stimulation in Parkinson's disease patients: biochemical evidence. J Neural Transm Suppl 70:401–408

Steinerman JR, Irizarry M, Scarmeas N, Raju S, Brandt J, Albert M, Blacker D, Hyman B, Stern Y (2008) Distinct pools of beta-amyloid in Alzheimer disease-affected brain: a clinicopathologic study. Arch Neurol 65(7):906–912

Stern Y (2006) Cognitive reserve and Alzheimer disease. Alzheimer Dis Assoc Disord 20(2): 112–117

Stokin GB, Lillo C, Falzone TL, Brusch RG, Rockenstein E, Mount SL, Raman R, Davies P, Masliah E, Williams DS, Goldstein LS (2005) Axonopathy and transport deficits early in the pathogenesis of Alzheimer's disease. Science 307(5713):1282–1288

Tachibana H, Aragane K, Kawabata K, Sugita M (1997) P3 latency change in aging and Parkinson's disease. Arch Neurol 54:296–302

Tagliati M, Bodis-Wollner I, Kovanecz I, Stanzione P (1994) Spatial frequency tuning of the monkey pattern ERG depends on D2 receptor-linked action of dopamine. Vision Res 34:2051–2057

Tagliati M, Bodis-Wollner I, Yahr M (1996) The pattern electroretinogram in Parkinson's disease reveals lack of retinal spatial tuning. Electroencephalogr Clin Neurophysiol 100:1–11

Takahashi RH, Capetillo-Zarate E, Lin MT, Milner TA, Gouras GK (2010) Co-occurrence of Alzheimer's disease beta-amyloid and tau pathologies at synapses. Neurobiol Aging 31(7):1145–1152

Takahashi RN, Pamplona FA, Prediger RD (2008) Adenosine receptor antagonists for cognitive dysfunction: a review of animal studies. Front Biosci 13:2614–2632

Tallon-Baudry C, Bertrand O, Delpuech C, Permier J (1997) Oscillatory gamma-band (30–70 Hz) activity induced by a visual search task in humans. J Neurosci 17:722–734

Taylor A, Saint-Cyr A, Lang A (1986) Frontal lobe dysfunction in Parkinson's disease: the cortical focus of neostriatal outflow. Brain 109:845–883

Teuber H, Proctor F (1964) Some effects of basal ganglia lesions in subhuman primates and man. Neuropsychologia 2:85–93

Thorpe S, Fize D, Marlot C (1996) Speed of processing in the human visual system. Nature 381:520–522

Townsend M, Shankar GM, Mehta T, Walsh DM, Selkoe DJ (2006) Effects of secreted oligomers of amyloid beta-protein on hippocampal synaptic plasticity: a potent role for trimers. J Physiol 572(Pt 2):477–492

Traub RD, Miles R, Schulman LS (2008) Neuronal networks of the hippocampus. Cambridge University Press, Cambridge, ISBN:9780521063319

Trichopoulou A, Kouris-Blazos A, Wahlqvist ML, Gnardellis C, Lagiou P, Polychronopoulos E, Vassilakou T, Lipworth L, Trichopoulos D (1995) Diet and overall survival in elderly people. BMJ 311(7018):1457–1460

Trottenberg T, Fogelson N, Kühn AA, Kivi A, Kupsch A, Schneider GH, Brown P (2006) Subthalamic gamma activity in patients with Parkinson's disease. Exp Neurol 200(1):56–65

Tsuchiya H, Yamaguchi S, Kobayashi S (2000) Impaired novelty detection and frontal lobe dysfunction in Parkinson disease. Neuropsychologia 38:645–654

Tunbridge EM, Bannerman DM, Sharp T, Harrison PJ (2004) Catechol-o-methyltransferase inhibition improves set-shifting performance and elevates stimulated dopamine release in the rat prefrontal cortex. J Neurosci 24:5331–5335

Tzelepi A, Bezerianos T, Bodis-Wollner I (2000) Functional properties of sub-bands of oscillatory brain waves to pattern visual stimulation in man. Clin Neurophysiol 111:259–269

Wakabayashi K, Takahashi H, Ohama E, Takeda S, Ikuta F (1993) Lewy bodies in the visceral autonomic nervous system in Parkinson's disease. Adv Neurol 60:609–612

Wakabayashi K, Takahashi H, Takeda S, Ohama E, Ikuta F (1988) Parkinson's disease: the presence of Lewy bodies in Auerbach's and Meissner's plexuses. Acta Neuropathol 76(3):217–221

Wang L, Kuroiwa Y, Kamitani T (1999a) Visual event-related potential changes at two different tasks in nondemented Parkinson's disease. J Neurol Sci 164(2):139–147

Wang L, Kuroiwa Y, Kamitani T, Takahashi T, Suzuki Y, Hasegawa O (1999b) Effect of interstimulus interval on visual P300 in Parkinson's disease. J Neurol Neurosurg Psychiatry 67(4):497–503

Wang L, Kuroiwa Y, Li M, Kamitani T, Wang J, Takahashi T, Suzuki Y, Ikegami T, Matsubara S (2000) The correlation between P300 alterations and regional cerebral blood flow in nondemented Parkinson's disease. Neurosci Lett 282(3):133–136

Whitehouse PJ, Price DL, Clark AW, Coyle JT, DeLong MR (1981, Aug) Alzheimer disease: evidence for selective loss of cholinergic neurons in the nucleus basalis. Ann Neurol 10(2):122–126

Wichmann T, DeLong MR (2003) Functional neuroanatomy of the basal ganglia in Parkinson's disease. Adv Neurol 91:9–18

Wilson RS, Arnold SE, Schneider JA, Tang Y, Bennett DA (2007) The relationship between cerebral Alzheimer's disease pathology and odour identification in old age. J Neurol Neurosurg Psychiatry 78(1):30–35

Winblad B, Palmer K, Kivipelto M, Jelic V, Fratiglioni L, Wahlund LO, Nordberg A, Bäckman L, Albert M, Almkvist O, Arai H, Basun H, Blennow K, de Leon M, DeCarli C, Erkinjuntti T, Giacobini E, Graff C, Hardy J, Jack C, Jorm A, Ritchie K, van Duijn C, Visser P, Petersen RC (2004) Mild cognitive impairment – beyond controversies, towards a consensus: report of the International Working Group on Mild Cognitive Impairment. J Intern Med 256(3):240–246

Younkin SG (1998) The role of A beta 42 in Alzheimer's disease. J Physiol Paris 92(3–4):289–292

Zanini S, Moschella V, Stefani A, Peppe A, Pierantozzi M, Galati S, Costa A, Mazzone P, Stanzione P (2009) Grammar improvement following deep brain stimulation of the subthalamic and the pedunculopontine nuclei in advanced Parkinson's disease: a pilot study. Parkinsonism Relat Disord 15(8):606–609

NF-κB in Brain Diseases

Cheng-Xin Gong

Abstract Nuclear factor κB (NF-κB) is a family of major transcription factors that play various important roles under physiological and pathological conditions. NF-κB transcription factors are ubiquitously expressed, including neurons and glial cells of the central nervous system (CNS). The roles of NF-κB family in the CNS, both as mediators of transcriptional response to synaptic activity and in behavioral paradigms of learning and memory, are the focus of recent studies. In this chapter, the general structure, major functions, and the regulation of NF-κB signaling are first described concisely. Advances made in understanding of the roles of NF-κB in the CNS and in brain diseases are then reviewed in more detail. The NF-κB signaling pathway as a potential therapeutic target of brain diseases is discussed at the end.

Keywords Alzheimer disease · Brain · Cytokines · Glial cells · Huntington's disease · Iκ · IκB kinase · Ischemic and traumatic brain injury · Learning · Memory · Multiple sclerosis · Neuronal Plasticity · Neuroprotection · NF-κB · Nuclear factor · Parkinson's disease · Rel family · Seizures · Synaptic transmission · Therapeutic target · Transcription factor

Abbreviations

AD	Alzheimer disease
CNS	Central nervous system
HD	Huntington's disease
IκB	Inhibitor of NF-κB
IKK	IκB kinase
IL	Interleukin
LTP	Long-term potentiation
NES	Nuclear export signal

C.-X. Gong (✉)
Department of Neurochemistry, New York State Institute for Basic Research in Developmental Disabilities, Staten Island, NY 10314-6399, USA
e-mail: cxgong@mail.csi.cuny.edu

J.P. Blass (ed.), *Neurochemical Mechanisms in Disease*,
Advances in Neurobiology 1, DOI 10.1007/978-1-4419-7104-3_10,

NF-κB	Nuclear factor κB
NIK	NF-κB–inducing kinase
NGF	Nerve growth factor
PD	Parkinson's disease
SOD	Superoxide dismutase
SRD	Signal response domain
TNF	Tumor necrosis factor

Contents

1 Introduction

Nuclear factor κB (NF-κB) is the prototype of a family of major transcription factors that play an essential role in several aspects of physiological and pathological conditions. More than two decades ago, Sen and Baltimore (1986) discovered it as a nuclear factor that, when activated by agents such as bacterial lipopolysaccharide, binds to a 10-bp sequence in the enhancer region of the gene encoding the κ light chain of antibody molecules in B cells (hence, κB). Because of the growing biomedical importance of nuclear factors, studies on NF-κB and its implications have been a major area of research in the recent years.

NF-κB transcription factors are expressed ubiquitously in mammalian cells. Expression of multiple NF-κB family members has been reported in different cells, including neurons and glial cells, of the central nervous system (CNS). However, the role of the NF-κB family in the nervous system, both as mediators of transcriptional response to synaptic activity and in behavioral paradigms of learning and memory, was found only recently. Investigating the functions of NF-κB transcription factors in the CNS is now a new frontier, both for the general field of NF-κB research and for understanding of transcriptional regulation in the brain.

2 Structure of NF-κB and IκB Family

NF-κB transcription factors are highly conserved across species. In mammals, the NF-κB family consists of five members: p50 (product of the NF-κB1 gene), p52 (product of the NF-κB2 gene), p65 (also known as RelA), c-Rel, and RelB. They all share a Rel homology domain (~300 amino acids in length) and, thus, the NF-κB family is also known as the Rel family, in reference to c-Rel that was first discovered as a proto-oncogene. This Rel domain contains the crucial functional region for DNA binding, dimerization, nuclear localization, and interaction with the inhibitors of NF-κB (IκB) (Fig. 1, top). The NF-κB family members function as dimers, and the five subunits can homodimerize or heterodimerize. Many, but not all, of the possible homodimer and heterodimer combinations have been observed in cells.

The NF-κB family can be classified into two subgroups, based on the presence or absence of a transcriptional activation domain. p50 and p52 do not contain distinct transcription activation domain and are, therefore, categorized class I. The homodimers of p50 and p52, and the p50/p52 heterodimer may occupy the NF-κB–binding sites of DNA and, thus, function as repressors of gene transcription (Franzoso et al., 1992). The three other NF-κB family members, p65, c-Rel, and RelB, constitute the class II subgroup. NF-κB dimers containing one or two of these polypeptides act as activators of transcription by virtue of the presence of at least one transcription activation domain. The two most abundant and biologically well-characterized NF-κB dimers are p50 homodimer and p50/p65 heterodimer.

NF-κB exists in the cytoplasm in an inactive form via association with the inhibitor of NF-κB (IκB) proteins. Six IκB members have been characterized (Fig. 1, middle): IκBα, IκBβ, IκBε, Bcl-3, IκBζ, and IκBγ. The most prominent ones are IκBα, IκBβ, and IκBε. This group of proteins contains either six or seven ankyrin repeats, a 33-aa motif that mediates protein–protein interactions. The ankyrin repeats in IκBα, IκBβ, and IκBε are flanked by two segments, the amino-terminal signal response domain (SRD) and the carboxyl-terminal acidic region, which is rich in prolines, glutamates, serines, and threonines (PEST). The SRD and the PEST sequence have been shown to be essential for interactions with NF-κB dimers (Ernst et al., 1995; Malek et al., 1998). The N-terminus of IκB also contains the nuclear export signal (NES) that functions to constantly expel the NF-κB/IκB complex from the nucleus (Huang and Miyamoto, 2001).

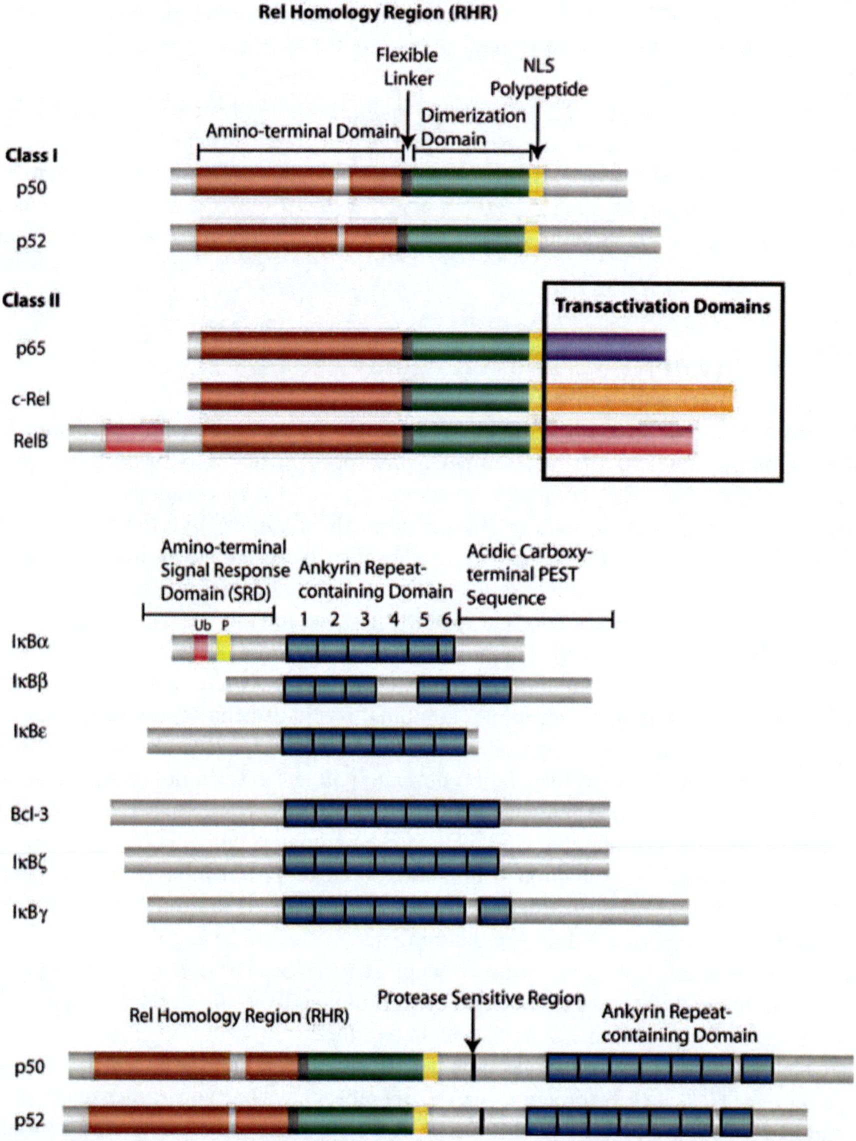

Fig. 1 Domains of the NF-κB and IκB families. *Top*, the NF-κB transcription factors are divided into two subgroups, Class I and Class II, depending on the presence or absence of transcription activation domains. The Rel homology region is indicated with the amino-terminal domain in *red* and dimerization domain in *green*. Other structural elements of interest such as nuclear localization sequence (NLS) are also labeled. *Middle*, the IκB family members are aligned according to ankyrin repeats. The amino-terminal signal response domain (SRD) and the carboxyl-terminal acidic region rich in prolines, glutamates, serines, and threonines (PEST) are indicated. P and Ub indicate the sites for inducible phosphorylation and ubiquitination, respectively. *Bottom*, p105 and p100 proteins contain p50 and p52, respectively, in the amino-terminal half and ankyrin repeats in the carboxyl-terminal half. The carboxyl-terminal half of p105 is homologous to IκBγ. (Modified from Huxford et al., 1999, with permission from Cold Spring Harbor Laboratory Press)

Interestingly, two larger proteins, p105 (also called NF-κB1) and p100 (NF-κB2), contain the Rel homology region of p50 and p52 in their amino-terminal half and the ankyrin repeats in their carboxyl-terminal half (Fig. 1, bottom). Evidence suggests that p50 and p52 are actually derived from p105 and p100, respectively, by proteolytic processing, so that p105 and p100 are sometimes called precursors of p50 and p52. IκBγ can also be generated by proteolytic processing from p105. Full-length p105 and p100 act as IκB as well (Naumann et al., 1993).

3 General Biological Role of NF-κB

NF-κB transcription factors promote the expression of over 200 genes involved in a variety of cellular processes, indicating that they play important roles in multiple aspects of biology. In addition, many more genes have the NF-κB–binding sequence in their promoters, which have not yet clearly been shown to be controlled by NF-κB. These NF-κB–regulating and potentially NF-κB–regulating genes and their original references are listed comprehensively on the website of Dr. T. D. Gilmore of Boston University (http://people.bu.edu/gilmore/nf-kb/target/index.html). These genes can be divided into the following groups: (1) cytokines/chemokines and their modulators; (2) immunoreceptors; (3) proteins involved in antigen presentation; (4) cell adhesion molecules; (5) acute phase proteins; (6) stress response genes; (7) cell surface receptors; (8) apoptosis regulators; (9) growth factors, ligands, and their modulators; (10) early response genes; (11) transcription factors; (12) viruses; (13) enzymes; and (14) miscellaneous genes not fitting into the groups above. Although the functionally important NF-κB–binding sites are located in the promoter/enhancer region of all these genes, the transcription of individual genes and the amount of transcribed product after NF-κB activation under specific circumstances depend on many factors, including the composition of NF-κB dimers, the nature of the NF-κB activating stimulus, and the number of consensus sites in the target gene. In addition, NF-κB works in cooperation with other transcription factors, especially activator protein-1 (Karin et al., 2001; Zhou et al., 2001).

Studies using gene knock-out animal models have revealed both specific and redundant functions of each NF-κB family member in regulation of cell survival and immune responses. For instance, the deletion of the p65 (RelA) gene in mice causes embryonic lethality due to extensive apoptosis in the liver (Beg et al., 1995), indicating that the function of p65 cannot be compensated for by other NF-κB family proteins and is essential for the survival of the mouse embryo. On the other hand, mice lacking p50 or RelB are immunodeficient but otherwise develop normally to adulthood (Burkly et al., 1995; Sha et al., 1995; Weih et al., 1995). The knockout of multiple members of the NF-κB family results in more severe phenotypes, which suggests that there is some functional redundancy between the NF-κB family members.

NF-κB is essential for normal functioning of the immune system. It plays key roles in regulating the expression of many cytokines, which are critical mediators of the immune system and are crucial for immune cell communication and effector functions during an active immune response. Studies on c-Rel–deficient mice have demonstrated that c-Rel is essential for IL-2, IL-3, GM-CSF, and γ-IFN expression in T lymphocytes; IL-6 expression in B cells; TNF-α expression in macrophages; and IL-12 expression in dendritic cells (Gerondakis et al., 1996; Liou et al., 1999; Sanjabi et al., 2000; Weinmann et al., 2001). When p105/p50 is knocked out, functional defects in the immune system appear despite otherwise normal development and phenotype (Sha et al., 1995). P105/p50 is essential for the survival of nonactivated B cells but not for all B cell-activated pathways (Snapper et al., 1996; Grumont et al., 1999). C-Rel knock-out mice show normal development but have B and T cell deficiencies (Kontgen et al., 1995). Mice deficient in the NF-κB2 gene (p100/p52) mainly have defects in lymph nodes and splenic architecture, although development is normal (Caamano et al., 1998). All these studies demonstrate the vital role of NF-κB in normal development and functioning of the immune system.

The activation of NF-κB is a double-edged sword. Although needed for proper development and immune system function, NF-κB, if inappropriately overactivated, can mediate inflammation and tumorigenesis. That duality is especially striking in relation to cancer, a proinflammatory disease. Most inflammatory agents mediate their effects through the activation of NF-κB, and the latter is suppressed by anti-inflammatory agents. Similarly, most carcinogens and tumor promoters activate NF-κB, whereas chemopreventive agents suppress it, suggesting its strong linkage with cancer. Paradoxically, most agents, including cytokines, chemotherapeutic agents, and radiation that induce apoptosis, also activate NF-κB (Beg and Baltimore, 1996). Thus, NF-κB is a part of the cells' autodefense mechanism and may mediate desensitization, chemoresistance, and radioresistance (Wang et al., 1999).

The most studied role of NF-κB to date is its role in malignant transformation and hyperplasia, in the control of apoptosis, and in immune functions. Some NF-κB proteins act as oncogenes. C-Rel consistently transforms cells in culture, is itself activated by a retroviral promoter insertion in an avian B cell lymphoma, and is frequently amplified in Hodgkin's lymphoma, diffuse large B cell lymphomas, and some follicular and mediastinal B cell lymphomas (Gilmore et al., 2004). Several oncogenes mediate their effects by activating NF-κB. Among them are oncogenic Ras (Mayo et al., 2001; Kim et al., 2002) and c-myc (Kim et al., 2000). It has been shown that NF-κB activation induces cellular transformation, proliferation, invasion, and angiogenesis, and mediates metastasis (reviewed by Aggarwal, 2004). On the other hand, functional blockage of NF-κB in transgenic murine and human epidermis produced hyperplastic epithelium in vivo (Seitz et al., 1998). Selective inhibition of NF-κB signaling in murine skin resulted in the spontaneous development of squamous cell carcinomas (Seitz et al., 1998; van Hogerlinden et al., 1999). NF-κB blockage also triggered invasive human epidermal neoplasia (Dajee et al., 2003). It appears that either overactivation or inactivation of NF-κB may lead to tumorigenesis, depending on circumstances. It is possible that NF-κB has different roles in different cell types. The role of NF-κB in tumorigenesis has been reviewed elsewhere recently (Aggarwal, 2004).

NF-κB also has a dual effect on controlling apoptosis. Because NF-κB regulates the expression of many genes involved in apoptosis, the regulation of apoptosis by NF-κB is very complicated and is far from being elucidated. The NF-κB signaling pathway has emerged as a critical regulator of the apoptotic response. In most circumstances, NF-κB is antiapoptotic, by activating the expression of antiapoptotic genes, but it can also promote apoptosis in response to certain death-inducing signals in certain cell types reviewed by (Kucharczak et al., 2003).

4 Regulation of NF-κB Signaling

NF-κB transcription factors are expressed ubiquitously in all cell types, but normally they are present as an inactive complex in the cytoplasm via their noncovalent interaction with IκB. In response to varieties of stimuli, including cytokines, viral and bacterial pathogens, and stress-inducing agents, the latent cytoplasmic NF-κB/IκB complex is activated by phosphorylation on the conserved serine residues in the amino-terminal portion of IκB. Phosphorylation targets IκB for ubiquitination, which leads to degradation of the IκB by the 26S proteasome. Degradation of IκB releases NF-κB by unmasking the nuclear localization signal present in the Rel-family polypeptides that permits translocation to the nucleus and binds to its cognate DNA-binding site (5′-GGGRNNYYCC-3′) in the promoter/enhancer regions of specific genes (Fig. 2).

Phosphorylation of IκB is catalyzed by a multimeric complex referred to as IκB kinase (IKK), which is activated by various stimuli. The IKK complex consists of two catalytic subunits (IKK-α and IKK-β) and a regulatory subunit IKK-γ or NF-κB essential modulator (NEMO) (DiDonato et al., 1997; Zandi et al., 1997; Rothwarf et al., 1998; Yamaoka et al., 1998). The activated IKK complex recruits IκB proteins and phosphorylates them at serine residues (in the case of IκB-α, Ser32 and Ser36 are phosphorylated). IKK can be phosphorylated and activated by another kinase called NF-κB–inducing kinase (NIK), which may be involved in an NF-κB–inducing signaling cascade induced by tumor necrosis factor (TNF) (Malinin et al., 1997).

The activation of IKK is considered a major mechanism of NF-κB activation in the classical pathway. However, in certain cases, such as in response to shortwave UV light (Li and Karin, 1998; Kato et al., 2003), pervanadate (Mukhopadhyay et al., 2000), H_2O_2 (Takada et al., 2003), hypoxia/reoxygenation (Fan et al., 2003), nerve growth factor (NGF) (Bui et al., 2001), erythropoietin (Digicaylioglu and Lipton, 2001), and Her-2 (Pianetti et al., 2001), the activation of NF-κB does not seem to involve phosphorylation of IκB by IKK or even IκB degradation.

Studies have shown that NF-κB signaling is also regulated by phosphorylation of Rel proteins themselves. RelA (p65), the most dominant NF-κB protein, is phosphorylated by cAMP-dependent protein kinase at Ser276, which enhances its transcription function (Zhong et al., 1997). RelA phosphorylation can occur before IκB degradation, which creates a moiety that is primed and "sitting on ready." In some cases, the phosphorylation of RelA can occur after the dissociation from IκB

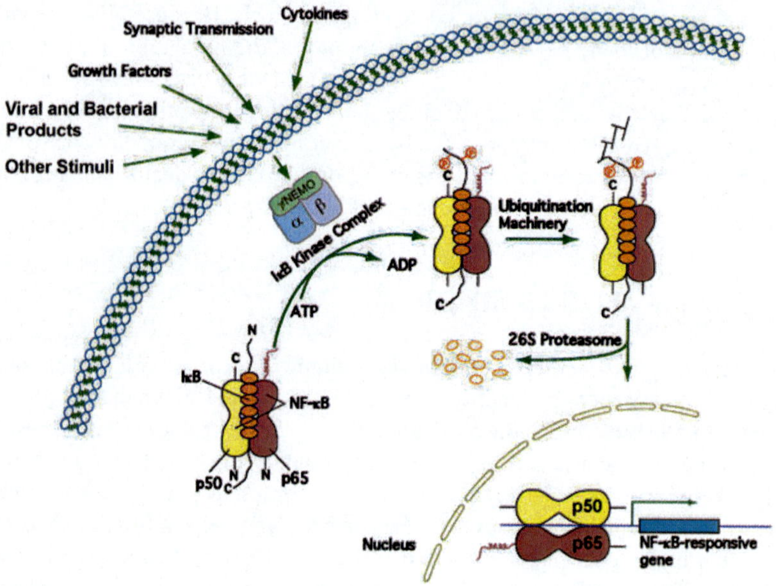

Fig. 2 Major pathway of NF-κB activation. NF-κB dimer (e.g., p50/p65 dimer) is normally in a latent form as a result of the association with the inhibitory protein IκB, which masks the nuclear localization signal and DNA-binding domains of NF-κB and retains NF-κB in the cytosol. Inducing stimuli (e.g., cytokines, synaptic transmission, and growth factors) activate the IκB kinase complex, which rapidly phosphorylates IκB, leading to its degradation through the ubiquitin-dependent proteasome pathway. Degradation of IκB by the 26S proteasome then allows NF-κB to translocate to the nucleus and activate transcription of target genes. (Reproduced from Meffert and Baltimore, 2005 with permission from Elsevier)

(Zhong et al., 1997; Wang et al., 2000). The requirement for both phosphorylation of RelA and degradation of IκB may set up a two-step mechanism that could serve to integrate disparate signals. It could also be a failsafe algorithm to place more stringent control on a powerful cellular agent to prevent inadvertent activation. In addition, ceramide has been reported to activate NF-κB via activating atypical protein kinase C that, in turn, phosphorylates Ser311 of RelA (Duran et al., 2003). The p38 MAPK was also shown to be required for NF-κB–dependent gene expression (Carter et al., 1999).

In the CNS, IκB-α is the most prominent member of its class, but IκB-β and IκB-γ also play supporting roles. IκB-γ is actually either encoded by alternative mRNA species derived from the gene for p105 or produced by proteolytic cleavage of p105 protein (Inoue et al., 1992; Heron et al., 1995). Full-length p105 and p100 can act as IκB as well (Naumann et al., 1993). Some of the previously characterized NF-κB activators have unique roles in the CNS that might be relevant in determining the function of neuronal NF-κB transcription factors. For instance, the cytokine TNFα could be a mediator of neuronal plasticity in noninflammatory settings of the hippocampus (Beattie et al., 2002). Although the free radical nitric oxide is involved in cell-mediated killing in the immune system, its alternative CNS role in regulating synaptic efficacy is well documented; see the review by (Schuman and Madison, 1994). NF-κB can also be activated by stimuli specific to the nervous system, such as β-amyloid, NGF, and neurotransmission. However, current knowledge of NF-κB activation in the CNS is very limited.

5 Role of NF-κB Signaling in the CNS

5.1 NF-κB in the CNS

The key functions of the CNS are information transmission, processing, and storage. Neurons communicate with each other via synapses, which are specialized cellular compartments consisting of presynaptic (sending) and postsynaptic (receiving) parts. Neuronal function is supported and assisted by glial cells. NF-κB transcription factors are present in both neurons and glial cells, including synapses of the neurons. The p50/p65 dimer is the major NF-κB in the CNS, which is either constitutively active or forms a complex with IκB. In addition, there are other NF-κB–binding proteins in the brain, such as brain-specific transcription factor specifically detected in the gray matter (Korner et al., 1989), developing brain factors enriched highly in developing cortex (Cauley and Verma, 1994), and neuronal NF-κB–binding factor with different target sequence requirements (Moerman et al., 1999). These NF-κB binding factors were not assigned to specific genes, nor could they be tested directly in reporter gene assays. It appears that an additional level of complexity is added by overlapping mutually exclusive or synergistically acting binding sites of NF-κB for other transcription factors in the CNS.

5.2 Activators and Inhibitors of NF-κB in the CNS

Many activators of NF-κB have been identified (see the review by Kaltschmidt et al., 2005), some of which are only seen in the nervous system, such as glutamate acting as an NF-κB activator via the main ionotropic glutamate receptors or NGF via the p75 receptor (Carter et al., 1996). In microglia, all neurotrophins activate NF-κB (Nakajima et al., 1998). Some molecules, such as TNF, can both activate and repress NF-κB in neurons, depending on the cell types and circumstances (Kaltschmidt et al., 1999).

Several anti-inflammatory cytokines that are known from the immune system inhibit NF-κB in the nervous system, but how these molecules act in the neuron is not well understood. One possibility might be the induction of IκB transcription. Some molecules have dual activity in regulating NF-κB, depending on their concentrations. A recent review has summarized the inhibitors of NF-κB in the nervous system (Kaltschmidt et al., 2005).

5.3 NF-κB–Regulating Genes in the CNS

As discussed above, there are many genes whose expression is regulated by NF-κB, and this list of genes is still growing. Most of the NF-κB–regulating genes are identified from studies in nonneuronal cells. Only a limited number of genes that are regulated by brain NF-κB and with direct relevance for the nervous system have been described. They include neural cell adhesion molecule (Simpson and Morris, 2000; Liu et al., 2003), inducible nitric oxide synthase (Madrigal et al., 2001), amyloid precursor protein (Grilli et al., 1995), μ-opioid receptors (Kraus et al., 2003), brain-derived neurotrophic factor (Lipsky et al., 2001), inducible cyclooxygenase-2 (Kaltschmidt et al., 2002), Ca^{2+}/calmodulin-dependent protein kinase IIδ (Kassed et al., 2004), galanin receptor (Lorimer et al., 1997), neuropeptide Y-Y1 receptor (Musso et al., 1997), and myelin basic protein (Paez et al., 2006). With the increasing interest in studies of the role of NF-κB in the CSN, there is no doubt that more genes will be found to be regulated by NF-κB in the brain. Many NF-κB–regulating genes that were observed in nonneuronal cells are probably also regulated by NF-κB in the brain.

5.4 Role of NF-κB in Synaptic Transmission and Neuronal Plasticity

Neuronal plasticity is essential for the transfer and storage of information by neurons. Recent studies have suggested that NF-κB may be crucially involved in the important process of neuronal plasticity. The capability of NF-κB in transmitting information from active synapses to the nucleus is supported by several studies demonstrating the presence of NF-κB in synapses (Kaltschmidt et al., 1993; Meberg et al., 1996; Meffert et al., 2003). A robust increase of p65 mRNA was observed after long-term potentiation in vivo (Meberg et al., 1996). This increase might be part of a feedforward mechanism leading to increased DNA-binding to the κB elements during long-term potentiation (LTP). NF-κB is activated in neurons by glutamate, an important synaptic neurotransmitter, and depolarization (Guerrini et al., 1995; Kaltschmidt et al., 1995). Memory consolidation in crabs also involves the activation of NF-κB–like activity (Freudenthal et al., 1998). In rat, traumatic brain injury first activates axonal NF-κB, and later, the activated NF-κB in the nuclei is detected, and this activation could be detected as long as one year after brain injury (Nonaka et al., 1999). In glutamate-stimulated hippocampal neurons, the return of p65 from neuritic to nuclear distribution was observed (Wellmann et al., 2001; Meffert et al., 2003).

Unlike most other cells, NF-κB is constitutively activated, which appears to be the result of synaptic activity (Kaltschmidt et al., 1994). The basal constitutive NF-κB activity in neurons could be repressed by specific inhibitors of action potential generation, glutamate receptors, and L-type Ca^{2+} channels (Meffert et al., 2003). This suggests that extracellular influx of Ca^{2+}, either through NMDA receptor or L-type Ca^{2+} channels, could activate NF-κB (Kaltschmidt et al., 2005). A pivotal role for NF-κB in synaptic plasticity was further supported by studies in which blockade of NF-κB activation impairs synaptic plasticity (Albensi and Mattson, 2000). A recent study reported that p50 knock-out mice exhibit impaired learning ability (Kassed et al., 2002). Taken together, NF-κB may act as a signal transducer to transmit information from active synapses to the nucleus in addition to its well-known role as a transcription factor, which transduces a synaptic signal into a transcription event.

5.5 Role of NF-κB in Learning and Memory

NF-κB appears to be involved in translating short-term signals from distant sites in neurites into long-term changes in gene expression, which may play a key role in plasticity, development, and survival. Double knock-out of p65 and TNF-RI results in a severe learning deficit (Meffert et al., 2003). A modulation of learning and memory was also observed in two other transgenic mouse models where NF-κB was repressed by tetracycline-regulated expression of dominant negative IκBα (Meffert and Baltimore, 2005). This nondegradable IκB remains bound to NF-κB and suppresses all NF-κB activation. In a recent study, repression of NF-κB by IκB in neurons resulted in behavioral deficits and a reduction in LTP and LTD induction (Kaltschmidt et al., 2006), which occurs via protein kinase A/CREB signaling. On the other hand, IκB expression driven by the prion promoter resulted in enhanced learning in older animals (Meffert and Baltimore, 2005). It is possible that neuronal NF-κB at physiological levels is needed for learning, but inhibition of pathological NF-κB hyperactivation in elder age might enhance learning. Overall, a consensus of behavioral studies in both crabs and mice suggests that, in most settings, NF-κB plays a positive role in learning and memory and that general or p65-subunit–specific inhibition of NF-κB function can lead to deficits in a variety of learning paradigms.

5.6 Role of NF-κB in Neuroprotection

Because NF-κB generally has antiapoptosis action, its activation is critical to neuronal survival. Several neurotrophic factors and cytokines may promote neuronal survival by NF-κB activation. Trophic factor deprivation has been found to result in a rapid and sustained increase in the level of IκB-α and IκB-β in cultured cerebellar granule neurons and to lead to sustained inhibition of NF-κB (Kovacs et al., 2004). A peptide inhibitor of NF-κB blocked the ability of NGF to prevent death of cultured sympathetic neurons (Maggirwar et al., 1998), suggesting a role of NF-κB in the control of neuronal death during development of the nervous system. When NF-κB is inhibited, cultured PC12 cells go to apoptosis, and NGF is unable to prevent it (Taglialatela et al., 1997). The cytokine-transforming growth factor-β1 (TGF-β1) may prevent neuronal apoptosis via an NF-κB-mediated mechanism, because this ability is blocked by NF-κB decoy DNA (Zhu et al., 2004).

The molecular mechanism by which NF-κB enhances neuronal viability is far from fully understood. The majority of the studies have focused on the direct binding and activating of prosurvival genes by NF-κB. These genes fall into three broad classes: antiapoptotic proteins, antioxidants, and neurotrophic factors. In addition to the conventional transcriptional inductions of prosurvival genes, NF-κB appears to have other viability-enhancing means at its disposal. For example, glucocorticoid receptor agonists have neurodegenerative effects through inhibition of glucose transport (Sapolsky, 1996). RelA inhibits the transcriptional activity of glucocorticoid receptors through direct binding (Ray and Prefontaine, 1994).

The roles of NF-κB in neuronal survival are complex. Mice lacking the p50 subunit of NF-κB exhibit increased neurotoxin-induced damage to neuronal cells as compared to wild-type mice (Yu et al., 1999), but decreased damage following a focal ischemic stroke (Nurmi et al., 2004). It was observed that activation of NF-κB is correlated with neurotoxicity in other paradigms. It is possible that the trophic or toxic dichotomy of NF-κB could be attributed to the specific subunits of the activated complex and the differential regulation in different types of cells.

5.7 Role of NF-κB in Glial Cells

Regardless of any direct role of NF-κB in neuronal cells, its activity in glia could dramatically influence neuroplasticity. Astrocytes play important roles in the modulation of synaptic activity. Hence, changes in gene expression in astrocytes could alter these processes, and NF-κB is as likely as any other factor to effect such changes. In fact, NF-κB is involved in an elevation of astrocytic major glutamate transporters (EAAT2) by epidermal growth factor (EGF) (Zelenaia et al., 2000; Su et al., 2003), brain-derived neurotrophic factor (BDNF), and subtoxic dose of Aβ (Rodriguez-Kern et al., 2003).

As the CNS representatives of the monocytic cell lineage, microglia undergo an inflammatory type of activation in response to brain injury and stress. Among the products of microglia activated by inflammatory signals is nitric oxide, which is produced by the exquisitely NF-κB–sensitive, inducible nitric oxide synthase. This enzyme can be elevated in an NF-κB-dependent manner in astrocytes (Akama et al., 1998). Microglial activation is associated with a marked increase in expression of cyclooxygenase-2, an oxyradical-generating enzyme, and agents that inhibit NF-κB can suppress lipopolysaccharide-induced cyclooxygenase-2 expression, suggesting an important role of NF-κB in microglial activation and oxyradical production.

The role of microglial NF-κB in neuronal injury is complicated by elevated production of neurotrophic factors by the activated microglia. Activated microglia produce NGF, basic fibroblast growth factor (bFGF), and TNF, each of which has been shown to prevent neuronal death in various experimental models of neurodegenerative diseases. A potent inducer of NF-κB activation in astrocytes is bradykinin, an inflammatory mediator produced in the brain in response to ischemia and trauma (Schwaninger et al., 1999). Acting through an NF-κB–mediated pathway, bradykinin induces IL-6 production in astroctyes, which stimulates production of several inflammation-related cytokines.

6 Role of NF-κB in Brain Diseases

Because NF-κB plays many important roles, it is not surprising that dysregulation of NF-κB signaling is involved in the pathogenesis of a number of human diseases. Except those resulting from mutations affecting components of the NF-κB signaling pathway reviewed by (Courtois and Smahi, 2006), the mechanisms by which NF-κB is involved in disease pathogenesis appear to be complicated, requiring future investigation. The diseases in which abnormal NF-κB regulation has been reported to play significant roles include atherosclerosis, AIDS, tumors, diabetes, heart diseases, muscular dystrophy, rheumatoid arthritis, inflammatory bowel diseases, bone resorption, and some neurodegenerative diseases reviewed by (Kumar et al., 2004). Investigation of the roles of NF-κB in brain diseases is just beginning, focusing mainly on acute and chronic neurodegeneration. Although the exact roles of NF-κB in these brain diseases are not yet known, these roles certainly deserve thorough investigation in the near future. Hence, these studies are reviewed in this section.

6.1 Role of NF-κB in Ischemic and Traumatic Brain Injury

NF-κB is dramatically activated in brain tissue in rodent models of stroke or cardiac arrest. Transient global forebrain ischemia causes NF-κB activation in CA1 hippocampal neurons (Clemens et al., 1997). A delayed increase in NF-κB activation in association with reactive glial cells was also observed several days after focal ischemia/reperfusion (Gabriel et al., 1999). Studies of mice lacking the p50 subunit of NF-κB suggest that, overall, NF-κB activation enhances ischemic neuronal death, but its effects differ among different cell types (Schneider et al., 1999). NF-κB activation in microglia promotes ischemic neuronal degeneration, whereas activation of NF-κB in neurons may increase their survival after a stroke. In cultured neuronal cells, activation of NF-κB protects them against excitotoxic and metabolic insults relevant to the pathogenesis of stroke, including glucose deprivation and exposure to glutamate (Cheng et al., 1994; Yu et al., 1999). The cortical and striatal neurons of mice that fail to induce the κB-responsive Mn-superoxide dismutase (SOD) gene due to lack of TNF-α receptors are more vulnerable to focal ischemic injury (Schmidt-Ullrich et al., 1996). The neuroprotective effect of endogenous TNF-α is likely mediated by NF-κB activation in neurons, because mice lacking p50 and mice treated with κB decoy DNA exhibit increased vulnerability of hippocampal neurons to excitotoxicity (Yu et al., 1999). The IκB kinase complex (IKK) is also activated in a mouse model of stroke and appears to play a key function in ischemic brain damage (Herrmann et al., 2005). Inhibition of neuronal IKK activity in transgenic mice that either lack IKK2 or express a dominant inhibitor of IKK reduces infarct size markedly. In contrast, constitutive activation of IKK2 enlarges the infarct size (Herrmann et al., 2005). The postischemic inflammatory response is critical to the consequence of stroke, and this response is mainly mediated by NF-κB signaling (reviewed by (Zheng and Yenari, 2004)). Therefore, NF-κB may be a potential molecular target for ischemic stroke therapy.

NF-κB activation also occurs in the cerebral cortex within hours of traumatic brain injury in rats, and this activation becomes maximal within the first 24 h (Nonaka et al., 1999; Sanz et al., 2002). Immunohistochemical staining indicates an increase of p65 level in the axons first and, subsequently, in neuronal cell bodies. The increased p65 level also occurs in the neighboring microglia and astrocytes. This increase in p65 immunoreactivity persists for many months, especially in the margins of the progressively enlarging ventricle, suggesting a role for NF-κB in a prolonged inflammatory process. In addition, expression of IκBα is also observed in astrocytes and microglial cells of the corpus callosum in traumatic brain injury at the time of NF-κB activation (Sanz et al., 2002).

6.2 Role of NF-κB in Seizures

In laboratory animals, NF-κB activity is rapidly increased in hippocampal neurons within 4–6 h after kainate-induced seizures, which is followed by a delayed and sustained NF-κB activation in glial cells (Yu et al., 1999). Intraventricular infusion of κB decoy DNA prior to administration of kainate causes a significant increase in the extent of neuronal death, suggesting an excitoprotective role for seizure-induced neuronal NF-κB activation. In mice lacking the p50 subunit of NF-κB, which is required for the vast majority of κB DNA-binding activity in the hippocampus, seizure-induced neuronal degeneration is greater than in control mice (Yu et al., 1999). Cultured hippocampal neurons from p50-deficient mice exhibit enhanced elevation of intracellular calcium levels upon exposure to glutamate and are more vulnerable to excitotoxicity as compared with neurons from wild-type mice. These studies suggest that the p50 subunit of NF-κB plays a major role in protecting neurons against excitotoxicity.

Excitotoxic and ischemic injury to neurons is partially mediated by dysregulation of cellular calcium homeostasis, resulting in a prolonged elevation of intracellular calcium levels. Neuronal NF-κB activation can stabilize intracellular calcium concentrations under ischemia-like conditions (Barger et al., 1995; Barger and Mattson, 1996), possibly via induction of several different genes, including those encoding calcium-binding proteins and glutamate receptor (Cheng et al., 1994; Furukawa and Mattson, 1998; Gary et al., 2000).

6.3 Role of NF-κB in Alzheimer Disease (AD)

Recent studies suggest that dysregulation of NF-κB signaling might be involved in the pathogenesis of AD. NF-κB immunoreactivity is found especially in and around the early senile plaques in AD brain, whereas mature plaques show mainly reduced NF-κB activity (Kaltschmidt et al., 1999). Several reports suggest that amyloid β (Aβ) peptide can activate NF-κB in neurons, suggesting a plausible mechanism by which Aβ may act in AD (Barger et al., 1995). Actually, elevation and activation of p65 and p50 subunits of NF-κB have been observed in AD brain (Yan et al., 1995;

Boissiere et al., 1997; Kaltschmidt et al., 1997). Activation of NF-κB protects hippocampal neurons against oxidative stress-induced apoptosis (Mattson et al., 1997). On the other hand, inhibition of NF-κB potentiates Aβ-mediated neuronal apoptosis (Kaltschmidt et al., 1999). The proapoptotic protein prostate–apoptosis response-4 (Par-4), which is implicated in AD, kills neurons partially by inhibiting NF-κB activity (Guo et al., 1998a). Interestingly, expression of IκB-α, IκB-γ and its precursor, p105, are also increased in AD brain (Yoshiyama et al., 2001; Huang et al., 2005), and the increased IκBα expression is in a distribution that corresponds to the neurofibrillary pathology of AD (Yoshiyama et al., 2001).

It is interesting that a low dose (0.1 μM) of Aβ is able to activate NF-κB and to protect against a high cytotoxic dose (10 μM) of Aβ (Kaltschmidt et al., 1999). This finding actually led to the discovery of an essential role for NF-κB in preconditioning (Blondeau et al., 2001; Ravati et al., 2001). The underlying mechanisms might be similar to a process described by Baltimore (1988) as intracellular immunization against virus infection. Overexpression of transdominant IκB-α completely abolishes the preconditioning effect of NF-κB. General evidence suggests that constitutive NF-κB activity is essential for neuronal survival (Bhakar et al., 2002). This protective role might be perturbed in AD brain, for example, by oxidative stress.

Activation of neuronal NF-κB in AD may be a neuroprotective response, but activation of NF-κB in glial cells may mediate the production of proinflammatory cytokine and nitric oxide associated with the amyloid and neurofibrillary pathology in AD (Chen et al., 2005; Ho et al., 2005). NF-κB might also play a role in amyloidogenesis of AD, because the enhancer region 5′ to the APP gene contains NF-κB–binding sites, and expression of APP can be induced by NF-κB (Grilli et al., 1996). A recent report suggests that NF-κB activation may also mediate sAPPα release (Choi et al., 2006).

Mutations of the presenilin-1 gene are the major cause of inherited early-onset AD. Presenilin-1 mutations impair the ability of neurons to induce NF-κB activation under conditions of oxidative stress in the pathogenesis of AD (Guo et al., 1998b). An abnormal NF-κB response occurs in neurons expressing mutant presenilin-1, such that it is activated rapidly but then drops to a very low level for a prolonged period. Transgenic mice with presenilin mutation exhibit impaired NF-κB activation in response to exposure to trimethyltin (Kassed et al., 2003).

6.4 Role of NF-κB in Parkinson's Disease (PD) and Huntington's Disease (HD)

PD and HD are age-related movement disorders that involve degeneration of dopaminergic neurons in the substantia nigra and medium spiny neurons in the striatum, respectively. It is striking that there is a seventyfold increase in the percentage of dopaminergic neurons with nuclear immunoreactive NF-κB p65, which indicates NF-κB activation, in the substantia nigra of PD patients as compared to age-matched controls (Hunot et al., 1997). This observation suggests a role of NF-κB activation in PD. Increased levels of oxidative stress and mitochondrial dysfunction are implicated in the pathogenesis of both PD and HD (Rao and Balachandran, 2002; Jenner, 2003). NF-κB activity increased in affected neurons in the substantia nigra and striatum may represent an early protective response to ongoing oxidative stress and mitochondrial dysfunction (Browne et al., 1999; Jenner, 2003). Consistent with this is that an NF-κB inhibitor increases the vulnerability of dopaminergic neurons to Parkinsonian neurotoxin 6-hydroxydopamine (Park et al., 2004). Mice lacking the p50 subunit of NF-κB exhibit increased damage to striatal neurons and worsened motor dysfunction after administration of the mitochondrial toxin 3-nitropropionic acid in an HD animal model (Yu et al., 2000). Levels of Mn-SOD are increased in response to 3-nitropropionic acid in striatal cells of wild-type mice, but not in striatal cells of mice lacking p50, suggesting a pivotal role of NF-κB in this neuroprotective response. However, NF-κB activation may also promote the death of neurons under conditions such as oxidative and metabolic stress that often occur in neurodegenerative diseases (Schneider et al., 1999; Gill and Windebank, 2000). In a neuronal cell line, mutant huntingtin is found to activate NF-κB, and blockage of the NF-κB activation reduces the toxicity of the mutant huntingtin (Khoshnan et al., 2004). What determines whether NF-κB activation is beneficial or detrimental for neurons in the context of neurodegenerative disorders is barely understood, but it likely involves regulatory elements that determine whether NF-κB increases the expression of pro- or antiapoptotic genes. Microglial activation has been shown to contribute to neuronal death in PD, and this activation may be mediated by the NF-κB/p38 MAPK pathway (Wilms et al., 2003).

6.5 Role of NF-κB in Multiple Sclerosis

Multiple sclerosis is a chronic autoimmune disease of the CNS, in which myelin and myelin-forming oligodendrocytes become the target of an inflammatory response, leading to their depletion. Although the molecular mechanism of oligodendrocyte depletion is not well understood, increased levels of TNF-α and IL-1β transcripts and activation of NF-κB have been observed in active multiple sclerosis lesions (Gveric et al., 1998; Bonetti et al., 1999). Both TNF-α and IL-1β are NF-κB–regulated proinflammatory cytokines that also cause apoptosis of oligodendrocytes (Selmaj and Raine, 1988). In CNS glial cells treated with proinflammatory cytokine, inhibition of NF-κB transactivation by IL-4 protects differentiating oligodendrocyte progenitors (Paintlia et al., 2006). This observation further supports a role of NF-κB in the pathogenesis of multiple sclerosis. Theiler's virus infection in the CNS induces a demyelinating disease very similar to multiple sclerosis. This infection directly induces proinflammatory cytokines in primary astrocytes via NF-κB activation (Palma et al., 2003), suggesting that NF-κB is critical for the development of immune-mediated demyelination. Genetic studies demonstrate that inhibitors of the NF-κB cascade comprise prime candidate genes predisposing to multiple sclerosis (Miterski et al., 2002). NF-κB also regulates transcription of myelin basic protein gene in oligodendroglioma cells (Huang et al., 2002).

7 NF-κB Signaling Pathway as a Potential Therapeutic Target

The involvement of NF-κB in several vital biological functions and in the pathogenesis of many human diseases suggests that it could be an important target for therapeutic intervention. The first evidence that NF-κB pathways could be inhibited came from studies of IκB-α mutant that could not be phosphorylated by IKK and thus not degraded by proteasome (Ghosh et al., 1998). This IκB-α mutant sequesters NF-κB in the cytoplasm and thus prevents the induction of specific NF-κB target genes. Delivering this IκB-α suppressor mutant by adenoviral vectors has been effective in rheumatoid arthritis models (Bondeson et al., 1999) and in reducing the resistance of tumors to chemotherapy in a mouse model (Wang et al., 1999). Targeting NF-κB for treating diseases has recently been reviewed elsewhere (Monaco and Paleolog, 2004; Panwalker et al., 2004; Verma, 2004).

The emerging data described above suggest that NF-κB plays important roles in cellular response to injury of the CNS in both acute and chronic neurodegenerative conditions. Therefore, the NF-κB pathway is no doubt a potential important target for therapeutic intervention of neurological disorders. Drugs targeting NF-κB in the CNS of animal models of neurodegenerative conditions are just beginning to be tested. In a rat model of embolic focal cerebral ischemia, bortezomib (a potent and selective inhibitor of proteasome) was found to reduce adverse cerebrovascular events, including secondary thrombosis, inflammatory response and blood–brain barrier, and hence reduce infarct volume and neurological functional deficits when administrated within 4 h after stroke onset. These protective actions are mediated by blocking endothelial NF-κB (Zhang et al., 2006). In a mouse model of stroke, a selective small molecule inhibitor of IKK reduces the infarct volume and cell death in a therapeutic window of 4.5 h (Herrmann et al., 2005). A natural green tea constituent, (–)-epigallocatechin-5-gallate, can limit brain inflammation and reduce neuronal damage via inhibiting NF-κB overactivation in an animal model of autoimmune encephalomyelitis, which opens a new therapeutic avenue for inflammatory brain diseases (Aktas et al., 2004).

Recent studies have demonstrated that a number of pharmacological agents act via their activities to inhibit NF-κB. The immunosuppressive and anti-inflammatory actions of glucocorticosteroids are mediated in part by the induction of IκB-α synthesis (Yamamoto and Gaynor, 2001). Nonsteroidal anti-inflammatory drugs also inhibit endotoxin- and cytokine-induced nuclear translocation of NF-κB by preventing IκB-α phosphorylation and degradation. Some naturally occurring and synthetic inhibitors of ubiquitin-proteasome also block NF-κB activation by preventing IκB degradation (Adams et al., 2000). Several pharmaceutical companies are now developing novel specific inhibitors of IKK (Haefner, 2002).

Because NF-κB is involved in a variety of neuronal functions and memory processing, use of any agents targeting the NF-κB pathway in brain diseases is complicated and warrants extensive studies. In general, activation of NF-κB in neurons protects them against degeneration, but activation of NF-κB in microglia promotes neuronal degeneration. Hence, ideal agents to target NF-κB should be cell type-selective in their actions. For example, inhibitors of NF-κB that selectively target microglial cells may suppress damaging neural inflammation without affecting the normal functions of NF-κB in neurons. Selecting such cell type-selective agents will be a major focus of future research.

Acknowledgments The author is grateful to Drs. K. Iqbal and I. Grundke-Iqbal of our institute and Dr. X. Zhu of the Institute of Pathology, Case Western Reserve University, Cleveland, OH, for critical reading of the manuscript, to Ms. M. Marlow of our institute for help in editing this manuscript, and to Ms. J. Murphy and Ms. S. Warren for secretarial assistance. Studies in my laboratory are supported in part by funds from the New York State Office of Mental Retardation and Developmental Disabilities, and grants from the NIH (R01 AG16760, R01 AG027429), the Alzheimer's Association, Chicago, IL (IIRG-05-13095), and the Li Foundation, Inc., New York, NY.

References

Adams J, Palombella VJ, Elliott PJ (2000) Proteasome inhibition: a new strategy in cancer treatment. Invest New Drugs 18:109–121

Aggarwal BB (2004) Nuclear factor-kappaB: the enemy within. Cancer Cell 6:203–208

Akama KT, Albanese C, Pestell RG, Van Eldik LJ (1998) Amyloid beta-peptide stimulates nitric oxide production in astrocytes through an NFkappaB-dependent mechanism. Proc Natl Acad Sci USA 95:5795–5800

Aktas O, Prozorovski T, Smorodchenko A, Savaskan NE, Lauster R et al (2004) Green tea epigallocatechin-3-gallate mediates T cellular NF-kappa B inhibition and exerts neuroprotection in autoimmune encephalomyelitis. J Immunol 173:5794–5800

Albensi BC, Mattson MP (2000) Evidence for the involvement of TNF and NF-kappaB in hippocampal synaptic plasticity. Synapse 35:151–159

Baltimore D (1988) Gene therapy intracellular immunization. Nature 335:395–396

Barger SW, Horster D, Furukawa K, Goodman Y, Krieglstein J et al (1995) Tumor necrosis factors alpha and beta protect neurons against amyloid beta-peptide toxicity: evidence for involvement of a kappa B-binding factor and attenuation of peroxide and Ca2+ accumulation. Proc Natl Acad Sci USA 92:9328–9332

Barger SW, Mattson MP (1996) Induction of neuroprotective kappa B-dependent transcription by secreted forms of the Alzheimer's beta-amyloid precursor. Brain Res Mol Brain Res 40: 116–126

Beattie EC, Stellwagen D, Morishita W, Bresnahan JC, Ha BK et al (2002) Control of synaptic strength by glial TNFalpha. Science 2955:2282–2285

Beg AA, Baltimore D (1996) An essential role for NF-kappaB in preventing TNF-alpha-induced cell death. Science 274:782–784

Beg AA, Sha WC, Bronson RT, Ghosh S, Baltimore D (1995) Embryonic lethality and liver degeneration in mice lacking the RelA component of NF-kappa B. Nature 376: 167–170

Bhakar AL, Tannis LL, Zeindler C, Russo MP, Jobin C et al (2002) Constitutive nuclear factor-kappa B activity is required for central neuron survival. J Neurosci 22:8466–8475

Blondeau N, Widmann C, Lazdunski M, Heurteaux C (2001) Activation of the nuclear factor-kappaB is a key event in brain tolerance. J Neurosci 21:4668–4677

Boissiere F, Hunot S, Faucheux B, Duyckaerts C, Hauw JJ, Agid Y et al (1997) Nuclear translocation of NF-kappaB in cholinergic neurons of patients with Alzheimer's disease. Neuroreport 8:2849–2852

Bondeson J, Foxwell B, Brennan F, Feldmann M (1999) Defining therapeutic targets by using adenovirus: blocking NF-kappaB inhibits both inflammatory and destructive mechanisms in rheumatoid synovium but spares anti-inflammatory mediators. Proc Natl Acad Sci USA 96:5668–5673

Bonetti B, Stegagno C, Cannella B, Rizzuto N, Moretto G et al (1999) Activation of NF-kappaB and c-jun transcription factors in multiple sclerosis lesions. Implications for oligodendrocyte pathology. Am J Pathol 155:1433–1438

Browne SE, Ferrante RJ, Beal MF (1999) Oxidative stress in Huntington's disease. Brain Pathol 9:147–163

Bui NT, Livolsi A, Peyron JF, Prehn JH (2001) Activation of nuclear factor kappaB and Bcl-x survival gene expression by nerve growth factor requires tyrosine phosphorylation of IkappaBalpha. J Cell Biol 152:753–764

Burkly L, Hession C, Ogata L, Reilly C, Marconi LA et al (1995) Expression of relB is required for the development of thymic medulla and dendritic cells. Nature 373:531–536

Caamano JH, Rizzo CA, Durham SK, Barton DS, Raventos-Suarez C et al (1998) Nuclear factor (NF)-kappa B2 (p100/p52) is required for normal splenic microarchitecture and B cell-mediated immune responses. J Exp Med 187:185–196

Carter BD, Kaltschmidt C, Kaltschmidt B, Offenhauser N, Bohm-Matthaei R et al (1996) Selective activation of NF-kappa B by nerve growth factor through the neurotrophin receptor p75. Science 272:542–545

Carter AB, Knudtson KL, Monick MM, Hunninghake GW (1999) The p38 mitogen-activated protein kinase is required for NF-kappaB-dependent gene expression. The role of TATA-binding protein (TBP). J Biol Chem 274:30858–30863

Cauley K, Verma IM (1994) Kappa B enhancer-binding complexes that do not contain NF-kappa B are developmentally regulated in mammalian brain. Proc Natl Acad Sci USA 91:390–394

Chen J, Zhou Y, Mueller-Steiner S, Chen LF, Kwon H, Yi S et al (2005) SIRT1 protects against microglia-dependent amyloid-beta toxicity through inhibiting NF-kappaB signaling. J Biol Chem 280:40364–40374

Cheng B, Christakos S, Mattson MP (1994) Tumor necrosis factors protect neurons against metabolic-excitotoxic insults and promote maintenance of calcium homeostasis. Neuron 12:139–153

Choi S, Kim JH, Roh EJ, Ko MJ, Jung JE et al (2006) Nuclear factor-kappa B activated by capacitative Ca2+ entry enhances muscarinic receptor-mediated sAPPalpha release in SH-SY5Y cells. J Biol Chem 281:12722–12728

Clemens JA, Stephenson DT, Smalstig EB, Dixon EP, Little SP (1997) Global ischemia activates nuclear factor-kappa B in forebrain neurons of rats. Stroke 28:1073–1080. discussion 1080–1081

Courtois G, Smahi A (2006) NF-kappaB-related genetic diseases. Cell Death Differ 13:843–851

Dajee M, Lazarov M, Zhang JY, Cai T, Green CL et al (2003) NF-kappaB blockade and oncogenic Ras trigger invasive human epidermal neoplasia. Nature 421:639–643

DiDonato JA, Hayakawa M, Rothwarf DM, Zandi E, Karin M (1997) A cytokine-responsive IkappaB kinase that activates the transcription factor NF-kappaB. Nature 388:548–554

Digicaylioglu M, Lipton SA (2001) Erythropoietin-mediated neuroprotection involves cross-talk between Jak2 and NF-kappaB signalling cascades. Nature 412:641–647

Duran A, Diaz-Meco MT, Moscat J (2003) Essential role of RelA Ser311 phosphorylation by zetaPKC in NF-kappaB transcriptional activation. EMBO J 22:3910–3918

Ernst MK, Dunn LL, Rice NR (1995) The PEST-like sequence of I kappa B alpha is responsible for inhibition of DNA binding but not for cytoplasmic retention of c-Rel or RelA homodimers. Mol Cell Biol 15:872–882

Fan C, Li Q, Ross D, Engelhardt JF (2003) Tyrosine phosphorylation of I kappa B alpha activates NF kappa B through a redox-regulated and c-Src-dependent mechanism following hypoxia/reoxygenation. J Biol Chem 278:2072–2080

Franzoso G, Bours V, Park S, Tomita-Yamaguchi M, Kelly K et al (1992) The candidate oncoprotein Bcl-3 is an antagonist of p50/NF-kappa B-mediated inhibition. Nature 359:339–342

Freudenthal R, Locatelli F, Hermitte G, Maldonado H, Lafourcade C et al (1998) Kappa-B like DNA-binding activity is enhanced after spaced training that induces long-term memory in the crab Chasmagnathus. Neurosci Lett 242:143–146

Furukawa K, Mattson MP (1998) The transcription factor NF-kappaB mediates increases in calcium currents and decreases in NMDA- and AMPA/kainate-induced currents induced by tumor necrosis factor-alpha in hippocampal neurons. J Neurochem 70:1876–1886
Gabriel C, Justicia C, Camins A, Planas AM (1999) Activation of nuclear factor-kappaB in the rat brain after transient focal ischemia. Brain Res Mol Brain Res 65:61–69
Gary DS, Sooy K, Chan SL, Christakos S, Mattson MP (2000) Concentration- and cell type-specific effects of calbindin D28k on vulnerability of hippocampal neurons to seizure-induced injury. Brain Res Mol Brain Res 75:89–95
Gerondakis S, Strasser A, Metcalf D, Grigoriadis G, Scheerlinck JY (1996) Rel-deficient T cells exhibit defects in production of interleukin 3 and granulocyte-macrophage colony-stimulating factor. Proc Natl Acad Sci USA 93:3405–3409
Ghosh S, May MJ, Kopp EB (1998) NF-kappa B and Rel proteins: evolutionarily conserved mediators of immune responses. Annu Rev Immunol 16:225–260
Gill JS, Windebank AJ (2000) Ceramide initiates NFkappaB-mediated caspase activation in neuronal apoptosis. Neurobiol Dis 7:448–461
Gilmore TD, Kalaitzidis D, Liang MC, Starczynowski DT (2004) The c-Rel transcription factor and B-cell proliferation: a deal with the devil. Oncogene 23:2275–2286
Grilli M, Goffi F, Memo M, Spano P (1996) Interleukin-1beta and glutamate activate the NF-kappaB/Rel binding site from the regulatory region of the amyloid precursor protein gene in primary neuronal cultures. J Biol Chem 271:15002–15007
Grilli M, Ribola M, Alberici A, Valerio A, Memo M et al (1995) Identification and characterization of a kappa B/Rel binding site in the regulatory region of the amyloid precursor protein gene. J Biol Chem 270:26774–26777
Grumont RJ, Rourke IJ, Gerondakis S (1999) Rel-dependent induction of A1 transcription is required to protect B cells from antigen receptor ligation-induced apoptosis. Genes Dev 13:400–411
Guerrini L, Blasi F, Denis-Donini S (1995) Synaptic activation of NF-kappa B by glutamate in cerebellar granule neurons in vitro. Proc Natl Acad Sci USA 92:9077–9081
Guo Q, Fu W, Xie J, Luo H, Sells SF et al (1998a) Par-4 is a mediator of neuronal degeneration associated with the pathogenesis of Alzheimer disease. Nat Med 4:957–962
Guo Q, Robinson N, Mattson MP (1998b) Secreted beta-amyloid precursor protein counteracts the proapoptotic action of mutant presenilin-1 by activation of NF-kappaB and stabilization of calcium homeostasis. J Biol Chem 273:12341–12351
Gveric D, Kaltschmidt C, Cuzner ML, Newcombe J (1998) Transcription factor NF-kappaB and inhibitor I kappaBalpha are localized in macrophages in active multiple sclerosis lesions. J Neuropathol Exp Neurol 57:168–178
Haefner B (2002) NF-kappa B: arresting a major culprit in cancer. Drug Discov Today 7:653–663
Heron E, Deloukas P, van Loon AP (1995) The complete exon-intron structure of the 156-kb human gene NFKB1, which encodes the p105 and p50 proteins of transcription factors NF-kappa B and I kappa B-gamma: implications for NF-kappa B-mediated signal transduction. Genomics 30:493–505
Herrmann O, Baumann B, de Lorenzi R, Muhammad S, Zhang W et al (2005) IKK mediates ischemia-induced neuronal death. Nat Med 11:1322–1329
Ho GJ, Drego R, Hakimian E, Masliah E (2005) Mechanisms of cell signaling and inflammation in Alzheimer's disease. Curr Drug Targets Inflamm Allergy 4:247–256
Huang Y, Liu F, Grundke-Iqbal I, Iqbal K, Gong CX (2005) NF-kappaB precursor, p105, and NF-kappaB inhibitor, IkappaBgamma, are both elevated in Alzheimer disease brain. Neurosci Lett 373:115–118
Huang TT, Miyamoto S (2001) Postrepression activation of NF-kappaB requires the amino-terminal nuclear export signal specific to IkappaBalpha. Mol Cell Biol 21:4737–4747
Huang CJ, Nazarian R, Lee J, Zhao PM, Espinosa-Jeffrey A et al (2002) Tumor necrosis factor modulates transcription of myelin basic protein gene through nuclear factor kappa B in a human oligodendroglioma cell line. Int J Dev Neurosci 20:289–296

Hunot S, Brugg B, Ricard D, Michel PP, Muriel MP et al (1997) Nuclear translocation of NF-kappaB is increased in dopaminergic neurons of patients with parkinson disease. Proc Natl Acad Sci USA 94:7531–7536

Huxford T, Malek S, Ghosh G (1999) Structure and mechanism in NF-kappaB/IkappaB signalling. Cold Spring Harb Symp Quant Biol LXIV:533–540

Inoue J, Kerr LD, Kakizuka A, Verma IM (1992) I kappa B gamma, a 70 kd protein identical to the C-terminal half of p110 NF-kappa B: a new member of the I kappa B family. Cell 68: 1109–1120

Jenner P (2003) Oxidative stress in Parkinson's disease. Ann Neurol 53(Suppl 3):S26–S36. discussion S36–38

Kaltschmidt C, Kaltschmidt B, Baeuerle PA (1993) Brain synapses contain inducible forms of the transcription factor NF-kappa B. Mech Dev 43:135–147

Kaltschmidt C, Kaltschmidt B, Baeuerle PA (1995) Stimulation of ionotropic glutamate receptors activates transcription factor NF-kappa B in primary neurons. Proc Natl Acad Sci USA 92:9618–9622

Kaltschmidt C, Kaltschmidt B, Neumann H, Wekerle H, Baeuerle PA (1994) Constitutive NF-kappa B activity in neurons. Mol Cell Biol 14:3981–3992

Kaltschmidt B, Linker RA, Deng J, Kaltschmidt C (2002) Cyclooxygenase-2 is a neuronal target gene of NF-kappaB. BMC Mol Biol 3:16

Kaltschmidt B, Ndiaye D, Korte M, Pothion S, Arbibe L et al (2006) NF-kappaB regulates spatial memory formation and synaptic plasticity through protein kinase A/CREB signaling. Mol Cell Biol 26:2936–2946

Kaltschmidt B, Uherek M, Volk B, Baeuerle PA, Kaltschmidt C (1997) Transcription factor NF-kappaB is activated in primary neurons by amyloid beta peptides and in neurons surrounding early plaques from patients with Alzheimer disease. Proc Natl Acad Sci USA 94:2642–2647

Kaltschmidt B, Uherek M, Wellmann H, Volk B, Kaltschmidt C (1999) Inhibition of NF-kappaB potentiates amyloid beta-mediated neuronal apoptosis. Proc Natl Acad Sci USA 96:9409–9414

Kaltschmidt B, Widera D, Kaltschmidt C (2005) Signaling via NF-kappaB in the nervous system. Biochim Biophys Acta 1745:287–299

Karin M, Takahashi T, Kapahi P, Delhase M, Chen Y et al (2001) Oxidative stress and gene expression: the AP-1 and NF-kappaB connections. Biofactors 15:87–89

Kassed CA, Butler TL, Navidomskis MT, Gordon MN, Morgan D et al (2003) Mice expressing human mutant presenilin-1 exhibit decreased activation of NF-kappaB p50 in hippocampal neurons after injury. Brain Res Mol Brain Res 110:152–157

Kassed CA, Butler TL, Patton GW, Demesquita DD, Navidomskis MT et al (2004) Injury-induced NF-kappaB activation in the hippocampus: implications for neuronal survival. FASEB J 18:723–724

Kassed CA, Willing AE, Garbuzova-Davis S, Sanberg PR, Pennypacker KR (2002) Lack of NF-kappaB p50 exacerbates degeneration of hippocampal neurons after chemical exposure and impairs learning. Exp Neurol 176:277–288

Kato T Jr, Delhase M, Hoffmann A, Karin M (2003) CK2 Is a C-Terminal IkappaB Kinase Responsible for NF-kappaB Activation during the UV Response. Mol Cell 12:829–839

Khoshnan A, Ko J, Watkin EE, Paige LA, Reinhart PH et al (2004) Activation of the IkappaB kinase complex and nuclear factor-kappaB contributes to mutant huntingtin neurotoxicity. J Neurosci 24:7999–8008

Kim BY, Gaynor RB, Song K, Dritschilo A, Jung M (2002) Constitutive activation of NF-kappaB in Ki-ras-transformed prostate epithelial cells. Oncogene 21:4490–4497

Kim DW, Gazourian L, Quadri SA, Romieu-Mourez R, Sherr DH et al (2000) The RelA NF-kappaB subunit and the aryl hydrocarbon receptor (AhR) cooperate to transactivate the c-myc promoter in mammary cells. Oncogene 19:5498–5506

Kontgen F, Grumont RJ, Strasser A, Metcalf D, Li R et al (1995) Mice lacking the c-rel proto-oncogene exhibit defects in lymphocyte proliferation, humoral immunity, and interleukin-2 expression. Genes Dev 9:1965–1977

Korner M, Rattner A, Mauxion F, Sen R, Citri Y (1989) A brain-specific transcription activator. Neuron 3:563–572

Kovacs AD, Chakraborty-Sett S, Ramirez SH, Sniderhan LF, Williamson AL et al (2004) Mechanism of NF-kappaB inactivation induced by survival signal withdrawal in cerebellar granule neurons. Eur J Neurosci 20:345–352

Kraus J, Borner C, Giannini E, Hollt V (2003) The role of nuclear factor kappaB in tumor necrosis factor-regulated transcription of the human mu-opioid receptor gene. Mol Pharmacol 64: 876–884

Kucharczak J, Simmons MJ, Fan Y, Gelinas C (2003) To be, or not to be: NF-kappaB is the answer–role of Rel/NF-kappaB in the regulation of apoptosis. Oncogene 22:8961–8982. Erratum in: Oncogene. 2004 23:8858

Kumar A, Takada Y, Boriek AM, Aggarwal BB (2004) Nuclear factor-kappaB: its role in health and disease. J Mol Med 82:434–448

Li N, Karin M (1998) Ionizing radiation and short wavelength UV activate NF-kappaB through two distinct mechanisms. Proc Natl Acad Sci USA 95:13012–13017

Liou HC, Jin Z, Tumang J, Andjelic S, Smith KA et al (1999) c-Rel is crucial for lymphocyte proliferation but dispensable for T cell effector function. Int Immunol 11:361–371

Lipsky RH, Xu K, Zhu D, Kelly C, Terhakopian A et al (2001) Nuclear factor kappaB is a critical determinant in N-methyl-D-aspartate receptor-mediated neuroprotection. J Neurochem 78:254–264

Liu H, Yu W, Liou LY, Rice AP (2003) Isolation and characterization of the human DC-SIGN and DC-SIGNR promoters. Gene 313:149–159

Lorimer DD, Matkowskj K, Benya RV (1997) Cloning, chromosomal location, and transcriptional regulation of the human galanin-1 receptor gene (GALN1R). Biochem Biophys Res Commun 241:558–564

Madrigal JL, Moro MA, Lizasoain I, Lorenzo P, Castrillo A et al (2001) Inducible nitric oxide synthase expression in brain cortex after acute restraint stress is regulated by nuclear factor kappaB-mediated mechanisms. J Neurochem 76:532–538

Maggirwar SB, Sarmiere PD, Dewhurst S, Freeman RS (1998) Nerve growth factor-dependent activation of NF-kappaB contributes to survival of sympathetic neurons. J Neurosci 18: 10356–10365

Malek S, Huxford T, Ghosh G (1998) Ikappa Balpha functions through direct contacts with the nuclear localization signals and the DNA binding sequences of NF-kappaB. J Biol Chem 273:25427–25435

Malinin NL, Boldin MP, Kovalenko AV, Wallach D (1997) MAP3K-related kinase involved in NF-kappaB induction by TNF, CD95 and IL-1. Nature 385:540–544

Mattson MP, Goodman Y, Luo H, Fu W, Furukawa K (1997) Activation of NF-kappaB protects hippocampal neurons against oxidative stress-induced apoptosis: evidence for induction of manganese superoxide dismutase and suppression of peroxynitrite production and protein tyrosine nitration. J Neurosci Res 49:681–697

Mayo MW, Norris JL, Baldwin AS (2001) Ras regulation of NF-kappa B and apoptosis. Methods Enzymol 333:73–87

Meberg PJ, Kinney WR, Valcourt EG, Routtenberg A (1996) Gene expression of the transcription factor NF-kappa B in hippocampus: regulation by synaptic activity. Brain Res Mol Brain Res 38:179–190

Meffert MK, Baltimore D (2005) Physiological functions for brain NF-kappaB. Trends Neurosci 28:37–43

Meffert MK, Chang JM, Wiltgen BJ, Fanselow MS, Baltimore D (2003) NF-kappa B functions in synaptic signaling and behavior. Nat Neurosci 6:1072–1078

Miterski B, Bohringer S, Klein W, Sindern E, Haupts M et al (2002) Inhibitors in the NFkappaB cascade comprise prime candidate genes predisposing to multiple sclerosis, especially in selected combinations. Genes Immun 3:211–219

Moerman AM, Mao X, Lucas MM, Barger SW (1999) Characterization of a neuronal kappaB-binding factor distinct from NF-kappaB. Brain Res Mol Brain Res 67:303–315

Monaco C, Paleolog E (2004) Nuclear factor kappaB: a potential therapeutic target in atherosclerosis and thrombosis. Cardiovasc Res 61:671–682

Mukhopadhyay A, Manna SK, Aggarwal BB (2000) Pervanadate-induced nuclear factor-kappaB activation requires tyrosine phosphorylation and degradation of IkappaBalpha. Comparison with tumor necrosis factor-alpha. J Biol Chem 275:8549–8555

Musso R, Grilli M, Oberto A, Gamalero SR, Eva C (1997) Regulation of mouse neuropeptide Y Y1 receptor gene transcription: a potential role for nuclear factor-kappa B/Rel proteins. Mol Pharmacol 51:27–35

Nakajima K, Kikuchi Y, Ikoma E, Honda S, Ishikawa M et al (1998) Neurotrophins regulate the function of cultured microglia. Glia 24:272–289

Naumann M, Nieters A, Hatada EN, Scheidereit C (1993) NF-kappa B precursor p100 inhibits nuclear translocation and DNA binding of NF-kappa B/rel-factors. Oncogene 8:2275–2281

Nonaka M, Chen XH, Pierce JE, Leoni MJ, McIntosh TK et al (1999) Prolonged activation of NF-kappaB following traumatic brain injury in rats. J Neurotrauma 16:1023–1034

Nurmi A, Lindsberg PJ, Koistinaho M, Zhang W, Juettler E et al (2004) Nuclear factor-kappaB contributes to infarction after permanent focal ischemia. Stroke 35:987–991

Paez PM, Garcia CI, Pasquini JM (2006) Expression of myelin basic protein in two oligodendroglial cell lines is modulated by apotransferrin through different transcription factors. J Neurosci Res 83:606–618

Paintlia AS, Paintlia MK, Singh I, Singh AK (2006) IL-4-induced peroxisome proliferator-activated receptor gamma activation inhibits NF-kappaB trans activation in central nervous system (CNS) glial cells and protects oligodendrocyte progenitors under neuroinflammatory disease conditions: implication for CNS-demyelinating diseases. J Immunol 176: 4385–4398

Palma JP, Kwon D, Clipstone NA, Kim BS (2003) Infection with Theiler's murine encephalomyelitis virus directly induces pro-inflammatory cytokines in primary astrocytes via NF-kappaB activation: potential role for the initiation of demyelinating disease. J Virol 77:6322–6331

Panwalkar A, Verstovsek S, Giles F (2004) Nuclear factor-kappaB modulation as a therapeutic approach in hematologic malignancies. Cancer 100:1578–1589

Park SH, Choi WS, Yoon SY, Ahn YS, Oh YJ (2004) Activation of NF-kappaB is involved in 6-hydroxydopamine-but not MPP+ -induced dopaminergic neuronal cell death: its potential role as a survival determinant. Biochem Biophys Res Commun 322:727–733

Pianetti S, Arsura M, Romieu-Mourez R, Coffey RJ, Sonenshein GE (2001) Her-2/neu overexpression induces NF-kappaB via a PI3-kinase/Akt pathway involving calpain-mediated degradation of IkappaB-alpha that can be inhibited by the tumor suppressor PTEN. Oncogene 20:1287–1299

Rao AV, Balachandran B (2002) Role of oxidative stress and antioxidants in neurodegenerative diseases. Nutr Neurosci 5:291–309

Ravati A, Ahlemeyer B, Becker A, Klumpp S, Krieglstein J (2001) Preconditioning-induced neuroprotection is mediated by reactive oxygen species and activation of the transcription factor nuclear factor-kappaB. J Neurochem 78:909–919

Ray A, Prefontaine KE (1994) Physical association and functional antagonism between the p65 subunit of transcription factor NF-kappa B and the glucocorticoid receptor. Proc Natl Acad Sci USA 91:752–756

Rodriguez-Kern A, Gegelashvili M, Schousboe A, Zhang J, Sung L et al (2003) Beta-amyloid and brain-derived neurotrophic factor, BDNF, up-regulate the expression of glutamate transporter GLT-1/EAAT2 via different signaling pathways utilizing transcription factor NF-kappaB. Neurochem Int 43:363–370

Rothwarf DM, Zandi E, Natoli G, Karin M (1998) IKK-gamma is an essential regulatory subunit of the IkappaB kinase complex. Nature 395:297–300

Sanjabi S, Hoffmann A, Liou HC, Baltimore D, Smale ST (2000) Selective requirement for c-Rel during IL-12 P40 gene induction in macrophages. Proc Natl Acad Sci USA 97: 12705–12710

Sanz O, Acarin L, Gonzalez B, Castellano B (2002) NF-kappaB and IkappaBalpha expression following traumatic brain injury to the immature rat brain. J Neurosci Res 67:772–780
Sapolsky RM (1996) Stress, glucocorticoids, and damage to the nervous system: the current state of confusion. Stress 1:1–19
Schmidt-Ullrich R, Memet S, Lilienbaum A, Feuillard J, Raphael M et al (1996) NF-kappaB activity in transgenic mice: developmental regulation and tissue specificity. Development 122:2117–2128
Schneider A, Martin-Villalba A, Weih F, Vogel J, Wirth T (1999) NF-kappaB is activated and promotes cell death in focal cerebral ischemia. Nat Med 5:554–559
Schuman EM, Madison DV (1994) Nitric oxide and synaptic function. Annu Rev Neurosci 17: 153–183
Schwaninger M, Sallmann S, Petersen N, Schneider A, Prinz S et al (1999) Bradykinin induces interleukin-6 expression in astrocytes through activation of nuclear factor-kappaB. J Neurochem 73:1461–1466
Seitz CS, Lin Q, Deng H, Khavari PA (1998) Alterations in NF-kappaB function in transgenic epithelial tissue demonstrate a growth inhibitory role for NF-kappaB. Proc Natl Acad Sci USA 95:2307–2312
Selmaj KW, Raine CS (1988) Tumor necrosis factor mediates myelin and oligodendrocyte damage in vitro. Ann Neurol 23:339–346
Sen R, Baltimore D (1986) Multiple nuclear factors interact with the immunoglobulin enhancer sequences. Cell 46:705–706
Sha WC, Liou HC, Tuomanen EI, Baltimore D (1995) Targeted disruption of the p50 subunit of NF-kappa B leads to multifocal defects in immune responses. Cell 80:321–330
Simpson CS, Morris. BJ (2000) Regulation of neuronal cell adhesion molecule expression by NF-kappa B. J Biol Chem 275:16879–16884
Snapper CM, Zelazowski P, Rosas FR, Kehry MR, Tian M et al (1996) B cells from p50/NF-kappa B knockout mice have selective defects in proliferation, differentiation, germ-line CH transcription, and Ig class switching. J Immunol 156:183–191
Su ZZ, Leszczyniecka M, Kang DC, Sarkar D, Chao W et al (2003) Insights into glutamate transport regulation in human astrocytes: cloning of the promoter for excitatory amino acid transporter 2 (EAAT2). Proc Natl Acad Sci USA 100:1955–1960
Taglialatela G, Robinson R, Perez-Polo JR (1997) Inhibition of nuclear factor kappa B (NFkappaB) activity induces nerve growth factor-resistant apoptosis in PC12 cells. J Neurosci Res 47: 155–162
Takada Y, Mukhopadhyay A, Kundu GC, Mahabeleshwar GH, Singh S (2003) Hydrogen peroxide activates NF-kappa B through tyrosine phosphorylation of I kappa B alpha and serine phosphorylation of p65: evidence for the involvement of I kappa B alpha kinase and Syk protein-tyrosine kinase. J Biol Chem 278:24233–24241
van Hogerlinden M, Rozell BL, Ahrlund-Richter L, Toftgard R (1999) Squamous cell carcinomas and increased apoptosis in skin with inhibited Rel/nuclear factor-kappaB signaling. Cancer Res 59:3299–3303
Verma IM (2004) Nuclear factor (NF)-kappaB proteins: therapeutic targets. Ann Rheum Dis 63(Suppl 2):ii57–ii61
Wang CY, Cusack JC Jr, Liu R, Baldwin AS Jr (1999) Control of inducible chemoresistance: enhanced anti-tumor therapy through increased apoptosis by inhibition of NF-kappaB. Nat Med 5:412–417
Wang D, Westerheide SD, Hanson JL, Baldwin AS Jr (2000) Tumor necrosis factor alpha-induced phosphorylation of RelA/p65 on Ser529 is controlled by casein kinase II. J Biol Chem 275:32592–32597
Weih F, Carrasco D, Durham SK, Barton DS, Rizzo CA et al (1995) Multiorgan inflammation and hematopoietic abnormalities in mice with a targeted disruption of RelB, a member of the NF-kappa B/Rel family. Cell 80:331–340

Weinmann AS, Mitchell DM, Sanjabi S, Bradley MN, Hoffmann A et al (2001) Nucleosome remodeling at the IL-12 p40 promoter is a TLR-dependent, Rel-independent event. Nat Immunol 2:51–57
Wellmann H, Kaltschmidt B, Kaltschmidt C (2001) Retrograde transport of transcription factor NF-kappa B in living neurons. J Biol Chem 276:11821–11829
Wilms H, Rosenstiel P, Sievers J, Deuschl G, Zecca L (2003) Activation of microglia by human neuromelanin is NF-kappaB dependent and involves p38 mitogen-activated protein kinase: implications for Parkinson's disease. FASEB J 17:500–502
Yamamoto Y, Gaynor RB (2001) Therapeutic potential of inhibition of the NF-kappaB pathway in the treatment of inflammation and cancer. J Clin Invest 107:135–142
Yamaoka S, Courtois G, Bessia C, Whiteside ST, Weil R et al (1998) Complementation cloning of NEMO, a component of the IkappaB kinase complex essential for NF-kappaB activation. Cell 93:1231–1240
Yan SD, Yan SF, Chen X, Fu J, Chen M et al (1995) Non-enzymatically glycated tau in Alzheimer's disease induces neuronal oxidant stress resulting in cytokine gene expression and release of amyloid beta-peptide. Nat Med 1:693–699
Yoshiyama Y, Arai K, Hattori T (2001) Enhanced expression of I-kappaB with neurofibrillary pathology in Alzheimer's disease. Neuroreport 12:2641–2645
Yu Z, Zhou D, Bruce-Keller AJ, Kindy MS, Mattson MP (1999) Lack of the p50 subunit of nuclear factor-kappaB increases the vulnerability of hippocampal neurons to excitotoxic injury. J Neurosci 19:8856–8865
Yu Z, Zhou D, Cheng G, Mattson MP (2000) Neuroprotective role for the p50 subunit of NF-kappaB in an experimental model of Huntington's disease. J Mol Neurosci 15:31–44
Zandi E, Rothwarf DM, Delhase M, Hayakawa M, Karin M (1997) The IkappaB kinase complex (IKK) contains two kinase subunits, IKKalpha and IKKbeta, necessary for IkappaB phosphorylation and NF-kappaB activation. Cell 91:243–252
Zelenaia O, Schlag BD, Gochenauer GE, Ganel R, Song W et al (2000) Epidermal growth factor receptor agonists increase expression of glutamate transporter GLT-1 in astrocytes through pathways dependent on phosphatidylinositol 3-kinase and transcription factor NF-kappaB. Mol Pharmacol 57:667–678
Zhang L, Zhang ZG, Liu X, Hozeska A, Stagliano N (2006) Treatment of embolic stroke in rats with bortezomib and recombinant human tissue plasminogen activator. Thromb Haemost 95:166–173
Zheng Z, Yenari MA (2004) Post-ischemic inflammation: molecular mechanisms and therapeutic implications. Neurol Res 26:884–892
Zhong H, SuYang H, Erdjument-Bromage H, Tempst P, Ghosh S (1997) The transcriptional activity of NF-kappaB is regulated by the IkappaB-associated PKAc subunit through a cyclic AMP-independent mechanism. Cell 89:413–424
Zhou LZ, Johnson AP, Rando TA (2001) NF kappa B and AP-1 mediate transcriptional responses to oxidative stress in skeletal muscle cells. Free Radic Biol Med 31:1405–1416
Zhu Y, Culmsee C, Klumpp S, Krieglstein J (2004) Neuroprotection by transforming growth factor-beta1 involves activation of nuclear factor-kappaB through phosphatidylinositol-3-OH kinase/Akt and mitogen-activated protein kinase-extracellular-signal regulated kinase1,2 signaling pathways. Neuroscience 123:897–906

Trinucleotide-Expansion Diseases

Arthur J. L. Cooper and John P. Blass

Abstract Many diseases, most with a strong neurodegenerative component, are now known to result from an expansion of a trinucleotide repeat sequence within the genome. In many cases, the longer the repeat the earlier the onset, and the more rapid and severe is the disease progression. Almost all of these diseases may be divided into three groups. In the first group, the expansion is either in an untranslated region/intron of the gene (e.g., fragile X syndrome, myotonic dystrophy type 1 (MD1), Friedreich ataxia, or spinocerebellar ataxia type 12 (SCA12)), or is in a DNA stretch that does not code for a protein (spinocerebellar ataxia type 8 (SCA8)). In group 1, with the exception of SCA12, the repeat triplet is not CAG. Generally, the mutations in group 1 result in low or absent levels of the protein corresponding to the affected gene and/or altered message RNA metabolism. In the second group, caused by $(CAG)_n$ expansions, the mutation is in an exon of the gene and each protein is expressed with an expanded polyglutamine (Q_n) domain. At least nine neurodegenerative diseases belong in this second group. The most common of these diseases is Huntington disease (HD). Others are dentatorubralpallidoluysian atrophy (DPLA; Haw River syndrome), spinobulbar muscular atrophy (SBMA; Kennedy disease) and seven forms of spinocerebellar ataxia [SCA1, SCA2, SCA3 (Machado–Joseph disease), SCA6, SCA7, and SCA17]. The mutated genes appear to be unrelated except for the fact that each possesses a $(CAG)_n/Q_n$ expansion. These diseases are characterized by insoluble protein aggregates in the affected areas. The aggregates contain the mutated protein. The CAG-expansions are widely thought to confer a pathological gain of function to the mutated protein, although in some cases a pathological decrease of function may also contribute. In eight of the Q_n-expansion disorders, the disease phenotype occurs when n is greater than about 36. Disease expansions may result in n values up to about 80, but larger values may sometimes occur. In the third group, the nucleotide expansion is in a coding exon and gives rise to an elongation of a polyalanine (A_n) stretch in the mutated expressed protein. At least nine diseases have been shown to be due to an

A.J.L. Cooper (✉)
Department of Biochemistry and Molecular Biology, New York Medical College, Valhalla, NY 10595, USA
e-mail: arthur_cooper@nymc.edu

J.P. Blass (ed.), *Neurochemical Mechanisms in Disease*,
Advances in Neurobiology 1, DOI 10.1007/978-1-4419-7104-3_11,

A_n expansion. Eight of the mutations are in transcription factors, and in one case the mutation is in the polyadenylate-binding protein. The disease phenotypes variably include mental retardation and malformations of the brain, genitourinary tract, skull, and digits. Both the normal size of the amino acid repeat and the pathological length of the repeat tend to be smaller in the A_n-expansion diseases than in the Q_n-expansion diseases. Although many of the trinucleotide-expansion diseases are rare (some exceedingly rare), they offer insights into pathophysiological processes that may pertain to the more common neurodegenerative diseases such as Alzheimer disease (AD) and Parkinson disease (PD).

Keywords Cerebellar ataxias · Dentatorubral pallidoluysian atrophy · Fragile X syndrome · Friedreich ataxia · Huntington disease · Myotonic dystrophy · Spinobulbar muscular atrophy · Trinucleotide expansion diseases

Abbreviations

AD	Alzheimer disease
ARX	Aristaless-related homeobox
BDNF	Brain-derived neurotrophic factor
BPES	Blepharophimosis, ptosis and epicanthus inversus
CBP	CREB binding protein
CCD	Cleidocranial dysplasia
CCHS	Congenital central hypoventilation syndrome
CREB	cAMP response-element-binding
DM	Myotonic dystrophy
DM1	Myotonic dystrophy type 1
DM2	Myotonic dystrophy type 2
DMPK	Dystrophia myotonica protein kinase
DRPLA	Dentatorubral-pallidoluysian atrophy
FMRP	Fragile mental retardation protein 1
FRAXE	Fragile XE syndrome
FRDA	Friedreich ataxia
TXTAS	Fragile X tremor/ataxia syndrome
GST	Glutathione *S*-transferase
HD	Huntington disease
HFGS	Hand–foot–genital syndrome
HPE	Holoprosencephaly
Htt	Huntingtin
ISSX	X-linked infantile spasm syndrome
MBNL	Muscleblind-like
MRX	Sex-linked mental retardation
NMDAR	*N*-methyl-D-aspartate receptor
NRSE	Neuronal restrictive silencer element
OPMD	Oculopharyngeal muscular dystrophy
PD	Parkinson disease

PRTS	Partington syndrome
PT	Permeability transition
SBMA	Spinobulbar muscular atrophy
SCA	Spinocerebellar ataxia
SPD	Synpolydactyly type II
TBP	TATA-binding protein
TG	Transglutaminase
UTR	5′-Untranslated region
WS	West syndrome

Contents

1 Introduction

The presence of trinucleotide repeats in the genome has been known for many years. It was originally thought that these repeats were inconsequential. However, beginning in the early 1990s it became apparent that many diseases were caused by pathological expansions of trinucleotide repeats within the genome.

Initially, almost all trinucleotide-repeat diseases were broadly assigned to two groups based on the position of the trinucleotide repeat within the gene (Cummings and Zoghbi, 2000). The first group is based on the positioning of the trinucleotide expansion in a noncoding region of the gene/genome. By contrast, the second group is characterized by an expansion in a coding region. All members of this group contain expanded CAG repeats in the affected gene and an expanded polyglutamine (Q_n) domain in the mutated expressed protein (Cummings and Zoghbi, 2000). In the last decade a third group of trinucleotide-expansion disease has been described. It has become apparent that trinucleotide expansions in a coding region can give rise to proteins containing polyalanine (A_n) expansions. Nine diseases are presently known to be caused by an A_n expansion.

Diseases in groups 1 and 2 share several defining characteristics: (1) The mutant repeats often show somatic and germline instability, expanding more often than contracting during succeeding generational transmissions; (2) subsequent generations often exhibit earlier age of onset and more rapid progression of the disease (anticipation), which correlates with the size of the expansion; and (3) paternal transmission often carries a greater risk of trinucleotide expansion (imprinting) (Cummings and Zoghbi, 2000).

2 Diseases Due to a Noncoding Trinucleotide Expansion

These diseases are typically characterized by variable and often very large trinucleotide expansions in the genome, which result in multiple organ dysfunction. The disease phenotype is frequently variable, perhaps due to somatic heterogeneity (Cummings and Zoghbi, 2000). The larger mutations are often transmitted from clinically silent intermediate size expansions termed premutations. Generally, the mutations produce instability of the message resulting in diminished or loss of protein (loss of function), or to aberrant alternative splicing giving rise to "toxic" mRNAs (loss and/or toxic gain of function) (Cummings and Zoghbi, 2000). In one case (SCA 8), the expansion is not in a gene coding for a protein. Currently known noncoding trinucleotide expansion diseases are listed in Table 1.

Table 1 Diseases resulting from expansions of noncoding trinucleotide repeats

Disease	Gene/Locus	Trinucleotide Expansion	Affected Protein
Fragile X syndrome	*FMR1* (FRAXA) Xq27.3	GCC	FMR-1 protein(FMRP)
Myotonic dystrophy (type 1)	*DMPK* 19q13	CTG	Dystrophia myotonica protein kinase (*DMPK*)
Friedreich ataxia	*X25* (*frataxin*) 9q13	GAA	Frataxin
Spinocerebellar ataxia type 8 (SCA 8)[a]	*SCA8* 13q21	CTG	–
Spinocerebellar ataxia type 12 (SCA 12)	*SCA12* 5q31-33	CAG	PP2A-PR55β

[a]The mutation is not associated with a gene coding for a protein.

2.1 Fragile X Syndrome (FRAXA)

At least nine "fragile" sites are present in human chromosomes. The "fragility" refers to chromosomal sites that break easily in the presence of certain compounds such as aphidocolin, methotrexate, and high doses of caffeine (Sutherland, 1979a, b). These sites are generally $(GCC)_n$ stretches and in most cases are protected if folate, thymidine, or folinic acid is in the medium (Sutherland, 1979a, b, 2003).

The first disease recognized to be due to a trinucleotide expansion was fragile X syndrome, which was shown to result from an increase of GCC repeats in the 5′-untranslated region (UTR) of the affected gene (*FMR1*) (Kremer et al., 1991). The disease is characterized by mental retardation, macro-orchidism, dysmorphic features, and aberrant behavior (including attention deficit/hyperactivity disorder, anxiety, epilepsy, and autism) (D'Huist and Kooy, 2009). The disease is X-linked and both sexes can be affected. However, females account for only a third of cases, and females on average have milder symptoms. The mutation maps to chromosome Xq27.3, resulting in loss of the fragile mental retardation protein 1 (FMRP). Mutations in the *FMR1* gene account for 15–25% of all X-linked forms of mental retardation (Strelnikov et al., 1999).

FMRP is an RNA-binding protein that shuttles between the nucleus and cytosol and binds to several mRNAs including its own (Zalfa and Bagni, 2004). FMRP is located in synapses and loss of FMRP affects synaptic plasticity (Zalfa and Bagni, 2004). Fragile X syndrome is second only to Down syndrome as a cause of mental retardation in males (Kremer et al., 1991; Rosenberg, 1996). The *FMR1* gene in normal individuals contains a $(GCC)_n$ domain, where $n = 5$–54. In carriers, the repeat number may be 60–230 (premutations), whereas in affected males the GCC repeat number is from 230 to 4000. The full mutation is accompanied by DNA hypermethylation of the GCC expansion and also of a nearby CpG island, resulting in loss of the expression of the *FMR1* gene (Rosenberg, 1996; Zalfa and Bagni, 2004). Hypermethylation of the CpG island is accompanied by histone deacetylation. Thus, amplification of the CGG repeat results in a change of the chromatin structure to a very condensed, transcriptionally inactive structure (D'Huist and Kooy, 2009). It was originally thought that an n of 60–230 is benign, but such expansions are now known to sometimes cause a clinically distinct syndrome, fragile X tremor/ataxia syndrome (FXTAS) in elderly males. The syndrome may result from aberrant RNA interactions rather than decreased FMRP (Hagerman et al., 2001).

2.2 Other "Fragile" Syndromes

A second, less common, form of fragile X syndrome (FRAXE) is associated with a mutation at the Xq28 locus of the *FMR2* gene and is also caused by a GCC expansion (Gécz et al., 1996; Frints et al., 2002; Lesca et al., 2003; Gu and Nelson, 2003). FMR2 belongs to the ALF (AF4/LAF4/FMR2) family of transcription factors (Bitoun and Davies, 2009). FRAXE patients generally have mild mental retardation and variable behavioral problems. Hypermethylation leads to transcriptional silencing of the *FMR2* gene and to a subsequent loss of the gene product FMR2. The FMR2 protein is normally expressed at high levels in the amygdala and hippocampus (Chakrabarti et al., 1998), suggesting that its loss could lead to neurological abnormalities.

In addition to FRXA and FRXE, five other folate-sensitive fragile sites (FRA10A, FRA11B, FRA12A, FRA16A, and FRAXF) and two nonfolate-sensitive fragile sites (FRA10B and FRA16B) have been identified in the human genome and have been molecularly characterized (Gécz et al., 1996; Strelnikov et al., 1999; Sarafidou et al., 2004; Lukusa and Fryns, 2008). These sites represent expanded DNA repeat sequences resulting from a dynamic mutation involving the normally occurring polymorphic CCG/CGG trinucleotide repeats at the folate-sensitive and AT-rich minisatellite repeats at the nonfolate-sensitive fragile sites (Lukusa and Fryns, 2008). Mutations in FRA11B are associated with Jacobsen syndrome (Lukusa and Fryns, 2009). The possible association of the common fragile sites FRA10A, FRA12A, FRA16A, and FRAXF with neuropsychiatric and developmental disorders is still poorly understood (Sutherland, 2003; Sarafidou et al., 2004; Lukusa and Fryns, 2009).

2.3 Myotonic Dystrophy Type 1 (DM1)

Myotonic dystrophy is a dominantly inherited disorder that is the most common form of muscular dystrophy worldwide, affecting ~1 in 8500 adults (Lee and Cooper, 2009). Shortly after the discovery of the trinucleotide repeat expansion in FRAXA, it was found that myotonic dystrophy type 1 (DM1) is due to a CTG expansion at the UTR of a gene (*DMPK*) at chromosome 19q13.2 encoding a member of the cAMP-dependent protein kinase family (DMPK; dystrophia myotonica protein kinase) (Shelbourne and Johnson, 1992; Brook et al., 1992; Lee and Cooper, 2009). In normal individuals, the CTG repeat ranges in size from 5 to 37. In mildly affected, adult-onset patients CTG repeats expand to 50–1000, whereas in severely affected early onset patients the expansion is greater than 1000 repeats (Korade-Mirnics et al., 1998). The clinical phenotype is extremely variable and strong anticipation in subsequent generations is a feature of this disease. Myotonia and progressive muscle weakness are characteristics of the adult-onset disease. Developmental and mental abnormalities together with hypotonia and respiratory distress are characteristics of the more severe congenital disease. It has been suggested that the CTG expansion may alter DMPK protein levels by interfering with *DMPK* gene transcription, with RNA processing, and/or with translation. The results would be abnormal phosphorylation of downstream substrates (discussed by Cummings and Zoghbi, 2000). Other possibilities include alterations in expression of nearby genes or sequestration of RNA-binding proteins leading to abnormal RNA processing (Cummings and Zoghbi, 2000). DM1 transcripts have been shown to accumulate in the nuclei of muscle cells (Jiang et al., 2004). DMPK mRNA is widely expressed in cortical and subcortical neurons and the mutant transcripts accumulate in discrete foci within the neuronal nuclei. Human EXP (expansion RNA-binding) proteins are homologous to muscleblind proteins that are critical for terminal differentiation of embryonic pharyngeal, visceral, and somatic muscle bodies, and for eye formation in *Drosophila*. In DM1, proteins in the muscleblind family are recruited to these foci in the nuclei depleting their levels in the cytosol (Jiang et al., 2004). Additionally, pre-mRNAs show abnormal regulation of alternative splicing. These findings suggest a toxic gain of function of DMPK mRNA, which may contribute to the muscle and neurological symptoms (Jiang et al., 2004; Lee and Cooper, 2009).

2.4 *Friedreich Ataxia (FRDA)*

FRDA is the most common inherited ataxia with a prevalence of about 1 in 50,000 in the general population. FRDA is characterized by ataxia, dysarthria, diminished reflexes, cardiomyopathy, diabetes, and degeneration of spinal cord, dorsal root ganglia, and several peripheral systems (Cummings and Zoghbi, 2000). The disease is inherited in an autosomal recessive fashion. The affected gene (*X25; frataxin*) is located at chromosome 9q13-q21.1. The mutation is due to an expansion of a GAA repeat in a noncoding region (intron 1) of a gene that codes for a highly conserved nuclear-encoded protein named frataxin that binds to the inner mitochondrial membrane (Dürr et al., 1996; Campuzano et al., 1996). The mutation appears to alter mRNA processing, resulting in low levels of frataxin. In normal individuals the GAA sequence is usually repeated 7–22 times, but in homozygous patients with Friedreich ataxia, the nucleotide repeat number may be 66–1700 (Sharma et al., 2004). Most Friedrich ataxia patients have expansions in both alleles. The longer the repeats, the lower the level of expressed frataxin and the more severe is the disease phenotype (Gatchel and Zoghbi, 2005). Two mild cases of Friedreich ataxia have been reported in heterozygotes for large expansions and an allele for a 44 and 66 triplet repeat, respectively. Due to somatic instability, 15% (GAA-44) and 75% (GAA-66) of cells contained alleles with $\geq$66 triplet repeats, suggesting a plausible mechanism for the mild phenotype. No such instability was noted in a sibling who

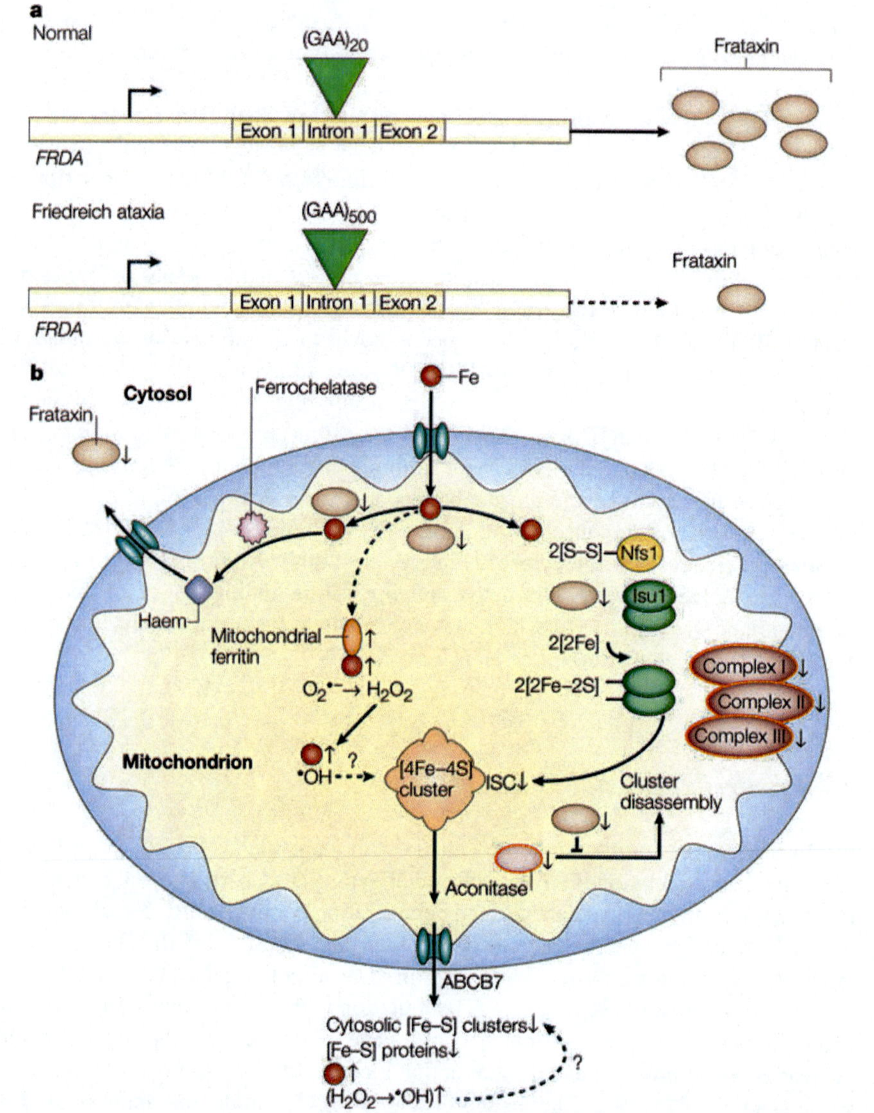

Fig. 1 Molecular and biochemical basis of Friedreich's ataxia (FRDA). (a) A GAA-repeat expansion in the first intron of the *FRDA* gene results in decreased levels of frataxin as a result of inhibition of transcriptional elongation. (b) Alterations in mitochondrial biochemistry that are associated with reduced frataxin levels. Proposed functions for frataxin include iron binding, protection and synthesis of Fe–S clusters, providing a binding partner for ferrochetalase in heme (haem) metabolism, and providing a metabolic switch between heme metabolism and Fe–S cluster biosynthesis. In FRDA, reduction of frataxin results in lowered levels of aconitase and respiratory complexes I, II, and III. Cytosolic proteins that contain Fe–S clusters may also be affected. Inability to form Fe–S clusters leads to an accumulation of iron, which leads to increased free radical formation (Fenton chemistry) in these organelles. Increased free radical formation may feed back to further decrease levels of Fe–S clusters, which are known to be sensitive to oxidative stress.

possessed an expanded GAA allele and a GAA-37 allele and was clinically normal (Sharma et al., 2004).

Frataxin is normally targeted to mitochondria, and mitochondrial respiratory chain dysfunction, mitochondrial iron accumulation, and oxidative stress are important components of the FRDA disease mechanism. It is possible that frataxin is normally involved in mitochondrial iron metabolism, including the formation of iron–sulfur centers (Lane and Richardson, 2010). As discussed by Lane and Richardson (2010) the metabolic defect in FRDA leads to mitochondrial iron loading that results from dysregulation of mitochondrial iron metabolism/iron–sulfur cluster biosynthesis, heme biogenesis, and mitochondrial iron storage. Additionally, the abnormal mitochondrial iron deposits may promote Fenton chemistry (hydroxyl radical formation) and thus contribute to toxic free-radical production that compromises cellular dysfunction, eventually leading to cell death (Pandolfo and Pastore, 2009). In FRDA patients with vitamin E and CoQ_{10} deficiency, supplementation with these antioxidants appears to be beneficial to heart muscle mitochondria, and, to a lesser extent, skeletal muscle mitochondria. The antioxidant idebenone also appears to be beneficial to heart muscle in FRDA patients (reviewed by Cooper and Schapira, 2003).

A mechanism, as proposed by Gatchel and Zoghbi (2005), linking mitochondrial biochemical defects to loss of frataxin is shown in Fig. 1.

2.5 *Spinocerebellar Ataxia Type 8 (SCA8)*

SCA8 is a rare neurodegenerative disease caused by an expansion of a CTG repeat on chromosome 13q21 (Koob et al., 1999). The disease is characterized by progressive ataxia with cerebellar atrophy, decreased vibration sense, and brisk reflexes (Cummings and Zoghbi, 2000). The normal allele contains 15–50 triplet repeats. By contrast, 71–800 repeats have been observed in SCA8 patients. SCA8 is interesting because it shows a complex inheritance pattern with extremes of incomplete penetration, in which only one or two affected individuals are found in a given family (Ikeda et al., 2004). As noted above, SCA8 is unique among the trinucleotide repeat disorders in that the predicted gene product is not a protein, but rather is a noncoding RNA. It was suggested that the expressed RNA may be an antisense RNA that regulates the expression of other genes (Mutsuddi et al., 2004). Later it was found that the mutation produces toxic RNAs that alter RNA splicing activities of MBNL (Muscleblind-like) and CELF (CUGBP and ETR-3-like) proteins (Daughters et al., 2009).

Fig. 1 (continued) *NFS1* is a gene that encodes mitochondrial cysteine desulfurase. This enzyme catalyzes the conversion of cysteine to alanine plus sulfane sulfur (S^0). The S^0 is incorporated into the Fe–S clusters. *Isu1* is a gene that encodes a scaffold protein on which the Fe–S clusters are assembled. ABCB7, ATP-binding cassette, sub-family B, member 7 (ABC transporter 7 protein); ISC, Fe–S cluster. From Gatchel and Zoghbi (2005) with permission

Interestingly, a small subset of patients with major psychoses, but without ataxia, appears to possess the SCA8 mutation (Vincent et al., 2000). Thus, additional genetic and/or environmental factors may play an important role in the expression of SCA 8 ataxia disease phenotype (Ikeda et al., 2004) or psychoses (Vincent et al., 2000).

2.6 Spinocerebellar Ataxia Type 12 (SCA12)

This is an extremely rare disorder, characterized by a variety of abnormalities of movement as well as dementia in older subjects (Holmes et al., 1999). The disease is caused by a CAG expansion in the UTR of the *PPP2R2B* gene (Holmes et al., 1999; Bahl et al., 2005). The expanded allele length ranges from 55 to 69 repeats. SCA12 is the only known disease caused by an expansion of a CAG repeat in a noncoding region of a gene. *PPP2R2B* encodes a brain-specific regulatory subunit (PR55β) of protein phosphatase 2A (PP2A). PP2A has been implicated in modulation of the cell cycle progression, tau phosphorylation, and apoptosis. Holmes et al. (1999) suggested that the CAG expansion may affect PR55β expression, perhaps by altering PP2A function in the brain. Interestingly, Holmes et al. (2003) suggested that the mutation might result in an overproduction of the PR55β mRNA.

3 Diseases Due to a Coding Trinucleotide Expansion—Polyglutamine (Q_n)-Expansion Diseases

Ten CAG-expansion diseases are currently known. In nine of these diseases (listed in Table 2), the mutation occurs in an exon (coding region) and results in an expansion of a Q_n domain in the expressed protein. These nine diseases may therefore conveniently be referred to as Q_n-expansion diseases. Some noncoding, non-CAG trinucleotide-expansion diseases may have expansions numbering in the thousands. By contrast, maximal expansions in the coding CAG-expansion diseases only occasionally exceed 100 trinucleotide repeats. In most cases, the protein is expressed normally, which suggests that the expanded Q_n domain exerts a toxic gain of function to the mutated protein. Nevertheless, limited loss of function probably occurs in some cases, as discussed below. In general, all nine Q_n-expansion diseases are characterized by progressive neuronal dysfunction, most often beginning in adult life. However, only selected neurons are affected. All the Q_n-expansion diseases are characterized by protein aggregates in the affected regions, which may by cytosolic, nuclear, or both cytosolic and nuclear (Table 2).

None of the mutated proteins in the CAG-expansion diseases are related, except for the possession of an expanded Q_n domain. In every case, except SCA6, the mutated protein is widely expressed, not only throughout the brain, but also throughout the body. This wide expression raises important questions. What accounts for the restriction of the major disease phenotype to nervous tissue? And what accounts

Table 2 $(CAG)_n/Q_n$-expansion diseases[a]

Disease	Gene/Locus	Affected Protein	Normal	Disease	Aggregate Location
Spinobulbar muscular atrophy (SBMA, Kennedy disease)	*AR* Xq13-21	Androgen receptor (AR)	9–36	38–62	n,c
Huntington disease (HD)	*Hd* 4p16.3	Huntingtin	6–35	36–121	n,c
Dentatorubralpallidoluysian atrophy (DRPLA, Haw River syndrome)	*DRPLA* 12p13.31	Atrophin-1	6–35	49–88	n,c
Spinocerebellar ataxia type 1 (SCA1)	*SCA1* 6p23	Ataxin-1	6–44[b]	39–82	n
Spinocerebellar ataxia type 2 (SCA2)	*SCA2* 12q24.1	Ataxin-2	15–31	36–63	c
Spinocerebellar ataxia type 3 (SCA3, Machado–Joseph disease)	*SCA3* (*MJD1*) 14q32.1	Ataxin-3	12–40[c]	45–84	n,c
Spinocerebellar ataxia type 6 (SCA6)	*SCA6* (*CACNA1A*) 19p13	α_{1A}-Voltage-dependent channel subunit	4–18	21–33	c
Spinocerebellar ataxia type 7 (SCA7)	*SCA7* 13p12-13	Ataxin-7	4–35	37–460	n
Spinocerebellar ataxia type 17 (SCA17)[d]	*SCA17* 6q27	TATA-binding protein (TBP)	34≤43	45–55	n

[a]From Cummings and Zoghbi (2000) updated to include Spinocerebellar ataxia type 17 and new information on CAG repeat size in some of the diseases. Subcellular location of brain protein aggregates typical of the Q_n-expansion diseases are as follows: n, nuclear; c, cytosolic.
[b]Alleles with 21 or more repeats are interrupted by 1–3 CAT trinucleotides; disease alleles contain pure CAG repeats.
[c]There are occasional variant triplets at the third (CAA), fourth (AAG), and sixth (CAA) positions of the CAG repeat in the human gene (Limprasert et al., 1996). CAA is a redundant triplet code for glutamine. AAG codes for lysine.
[d]The CAG stretch in the gene is interspersed with a few CAA trinucleotides.

for the often subtle differences in neuropathology among the different Q_n-expansion diseases? In other words, by what mechanism does the selective vulnerability among the different Q_n-expansion diseases occur? Finally, what mechanism can explain the incredibly sharp demarcation between pathological and nonpathological n values of n in the Q_n domains?

To date the literature is replete with studies characterizing Q_n disease, particularly Huntington disease (HD). What follows is a brief description of the discovery of each Q_n-expansion disease, and some key references, followed by a discussion of current theories on the mechanisms by which expanded Q_n domains exert their neurotoxicity. Because of the huge volume of literature on Q_n-expansion diseases, we have had to be selective in the references quoted. The diseases are listed in approximately the chronological order in which the mutation was discovered.

3.1 Spinobulbar Muscular Atrophy (SBMA; Kennedy Disease)

Shortly after the discovery of the repeat disorder in FRAXA, it was shown that SBMA is caused by an expansion of a CAG repeat at chromosome Wq11-112 (La Spada et al., 1991, 1992). The mutation is caused by an expansion of a CAG repeat within the first exon of the *AR* gene coding for AR (the androgen receptor) (La Spada et al., 1991, 1992; Amato et al., 1993). Unlike all other known Q_n-expansion diseases, which are inherited in an autosomal dominant fashion, SBMA inheritance is X-linked. Carrier females are usually clinically normal, although some may have mild symptoms. Affected males exhibit mild hypogonadism and gynecomastasia, although they are fertile. These features suggest mild loss of function in the affected gene. The main symptoms of SBMA are slowly progressive muscle weakness and atrophy of bulbar, facial, and limb muscles (Suzuki et al., 2009). The key histopathological findings of SBMA are an extensive loss of lower motor neurons in the anterior horn of the spinal cord as well as in brainstem motor nuclei and intranuclear accumulations of mutant AR protein in the residual motor neurons (ARs are found on lower motor neurons as well as on other CNS neurons) (Suzuki et al., 2009). These findings suggest a toxic gain of function in the mutated SBMA.

Androgens are important in regulating sexually dimorphic neurons in the rat brain. They are also important in signaling pathways in the motor neurons. As pointed out by Cary and La Spada (2008) a thorough understanding of androgen receptor signaling in motor neurons should provide important inroads toward the development of effective treatments for SBMA and a variety of other devastating motor neuron diseases. Suppression of disease progression by castration or by administration of leuprorelin acetate (a luteinizing hormone-releasing hormone antagonist) in a mouse model of SBMA has been reported (Suzuki et al., 2009). Some efficacy of leuporelin has also been demonstrated in a phase 2 clinical trial (Banno et al., 2009).

3.2 Huntington Disease (HD)

In 1993, the Huntington Disease Collaborative Research Group reported that the mutation in HD is due to a CAG expansion in the gene at chromosome 4p16.3 (The Huntington Disease Collaborative Research Group, 1993). The affected protein, which contains an expanded Q_n domain near the N-terminus, was named huntingtin (Htt). Htt is a large protein ($M_r \sim 350{,}000$). Homozygous Htt knock-out mice embryos die in utero, and it has been difficult to assign a biological function to Htt. Nevertheless, some evidence suggests that Htt may be an iron-regulated protein essential for normal nuclear and perinuclear organelles (Hilditch-Maguire et al., 2000). More recent evidence suggests a role in intracellular vesicular trafficking (Caviston and Holzbaur, 2009). HD is the most common of the $(CAG)_n/Q_n$-expansion diseases, despite the fact that new expansion mutational expansions in the *Htt* gene are believed to be exceedingly rare. The incidence of HD worldwide is about 5–10 per 100,000 individuals. Japan has a very low rate (0.1–0.5 per 100,000), whereas in the Lake Maracaibo region of Venezuela the incidence exceeds 100 per 100,000. In the United States in the 1980s, it was estimated that 25,000 persons had HD (Conneally, 1984). HD is typically fully penetrant and is characterized by movement disorders (chorea), psychiatric and behavioral disorders, and cognitive decline. Symptoms usually begin in adulthood (~30–40 years of age) and inexorably worsen over a period of 10–25 years. In early onset cases (age <20; ~5–10% of cases) survival time after onset is shorter and symptoms include rigidity, bradykinesia, and tremor. Seizures may also occur. The cause of death is related to debility and immobility, weight loss, and trouble swallowing. Pneumonia is the most common cause of death (Greenamyre and Shoulson, 1994). Some mild cortical atrophy may occur in end-stage disease, but the most striking change in brain morphology occurs in the caudate nucleus. The caudate may be reduced to a thin rim of tissue resulting in greatly enlarged ventricles. There is also atrophy of the putamen and globus pallidus. Microscopically, medium-sized spiny GABAergic projection neurons in the striatum are most vulnerable, whereas medium-sized and large aspiny interneurons are less affected (Greenamyre and Shoulson, 1994). Although fully penetrant, genetic and environmental factors may modulate the age of onset of HD (The US–Venezuela Collaborative Research Project and Wexler, 2004).

3.3 Spinocerebellar Ataxia Type 1 (SCA1)

SCA1 is due to a $(CAG)_n$ expansion toward the N-terminus of the gene *ATXN-1* that maps to chromosome 6p22-p23 and codes for the protein ataxin-1 (Orr et al., 1993; Zoghbi and Orr, 2009). The disease is characterized by degeneration of the cerebellum, spinal cord, and brainstem. The disease phenotype appears to be due to a toxic gain of function in the mutated protein and perhaps to some loss of function (Zoghbi and Orr, 2009). Protein aggregates are particularly prominent in the nuclei of Purkinje neurons. Ataxin-1 can be phosphorylated at serine 776. Mutation of this residue to an alanine greatly reduces the disease phenotype in CSA1 transgenic mice, despite the fact that the mutated protein (with a Q_{82} domain) accumulates in the Purkinje cell nuclei (Emamian et al., 2003). This finding suggests an effect of the expanded Q_n domain on the properties of the protein at a residue (serine) distal to the mutation. Studies with DNA microarrays have suggested that the presence of the disease protein in SCA1-transgenic mice results in major changes in the expression of nine genes during disease progression. Interestingly, five of these genes centered on glutamate signaling in Purkinje cells (Serra et al., 2004).

3.4 Spinocerebellar Ataxia Type 2 (SCA 2)

SCA2 is caused by a $(CAG)_n$ expansion in exon 1 of the *ATXN2* gene (coding for the cytosolic protein ataxin 2 (Atx2)) located in chromosome 12q24.1. Cerebellar Purkinje cells are targeted in this disease. Mutant Atx2-58Q, but not wild-type Atx2-22Q, specifically associates with the cytosolic C-terminal region of type 1 inositol 1,4,5-trisphosphate receptor (InsP(3)R1), an intracellular Ca^{2+} release channel (Liu et al., 2009). The studies of Liu et al. (2009) suggest that disturbed Ca^{2+} signaling may play an important role in SCA2 neuropathology. The authors suggested that the ryanodine receptor (RyanR) may be a potential therapeutic target to treat SCA2 patients. It is of interest that in a cohort of Central European subjects the n in the Q_n domains of Atx2 was remarkably consistent. Q_{22} represented 92% of the alleles and Q_{23} represented 5–7% (Figueroa et al., 2009). The finding of such a constant number suggests evolutionary pressure to maintain this number in the Central European cohort. The finding with normal Atx2 contrasts with that noted for Htt, for example, where the wild-type (normal) Htt may have an n in the Q_n domain ranging from 11 to 34 (Koroshetz and Martin, 1997). Recent studies suggest that an expansion of Q_n domains in Atx2 beyond $n = 22$ may be associated with disruption of energy metabolism and severe obesity in some children (Figueroa et al., 2009).

3.5 Spinocerebellar Ataxia Type 3 (SCA3; Machado–Joseph Disease)

SCA3 is characterized by progressive ataxia and external ophthalmoplegia. Recent studies have also suggested that auditory, vestibular, and ingestion-related dopaminergic and cholinergic systems may also be compromised (Rüb et al., 2008). Mental faculties in SCA3 patients usually remain intact. The disease phenotype is quite variable, but the major symptom with all patients is difficulty in walking. When the disease presents before age 20 it is usually characterized by marked spasticity, akinesia, and dystonia-like posturing. When the disease presents after age fifty, it is usually characterized by amyotrophic polyneuropathy, with fasciculations accompanying the ataxia. Other cases fall between the two (Uitti, 1994). The disease occurs in some families of Portuguese descent, and "hotspots" occur in Japan and parts of India.

The gene for SCA 3, which maps to chromosome 14q24.3-q32, was shown to be due to a CAG expansion in exon 10 and to a corresponding increase in the Q_n domain in the affected protein (ataxin-3) (Kawaguchi et al., 1994). There are occasional variant triplets at the third (CAA), fourth (AAG), and sixth (CAA) positions of the CAG repeat in the human gene (Limprasert et al., 1996). The nucleotide following the last trinucleotide in the CAG repeat in the human gene is a G in all cases studied where the number of CAG repeats is less than 20. In 55% of the genes containing a CAG repeat of between 27 and 40 trinucleotides, a C nucleotide followed the CAG repeat; and in all cases of expanded (pathological) CAG repeats, a C nucleotide followed the CAG. The authors suggested that the C variant may be associated with CAG repeat instability (Limprasert et al., 1996). Ataxin-3 is a 42-kDa protein in which the Q_n domain is positioned toward the C-terminus. The N-terminus, which is whimsically termed Josephin, is highly conserved from nematodes to humans (Chow et al., 2004). There are two ubiquitin-binding sites between Josephin and the Q_n domain. Josephin, which is globular and monomeric, possesses ubiquitin protease activity (Chow et al., 2004). The expanded Q_n domain may destabilize the Josephin domain perhaps resulting in a loss of function (Chow et al., 2004).

3.6 Dentatorubral Pallidoluysian Atrophy (DRPLA; Haw River Syndrome)

This disease, which is characterized by progressive ataxia, choreoathetosis, dystonia, seizures, myoclonus, and dementia, maps to chromosome 12p12-ter (Koide et al., 1994; Nagafuchi et al., 1994). The disease is fairly common in Japan, but very rare in Caucasians. The mutated protein, which has an M_r of about 140,000, contains a Q_n domain in the middle of the protein with a serine-rich region in front of this region (Margolis et al., 1996). The mutated protein was originally named atrophin, but is now named atrophin-1 because there is also another closely related protein in humans (atrophin-2). Evidence suggests that the two proteins may act as transcriptional corepressors during embryonic development (Zoltewicz et al., 2004).

3.7 Spinocerebellar Ataxia Type 6 (SCA 6)

This disease is a late onset disorder of the cerebellum characterized by selective and progressive loss of Purkinje cells. Initially the disease was thought to be confined to the cerebellar cortex, dentate nucleus, and inferior olives, but more recent studies suggest more widespread cerebellar involvement (Seidel et al., 2009; Wang et al., 2010). The disease is caused by a CAG expansion in the *SCA6* gene at chromosome 19p13, which codes for the α_{1A}-voltage-dependent calcium channel subunit (Zhuchenko et al., 1997). Six isoforms of this protein have been described. The CAG repeat is within the open reading frame and is predicted to encode Q_n domains in three of the isoforms (Zhuchenko et al., 1997). The protein is abundantly expressed in Purkinje cell bodies and dendrites (Restituito et al., 2000). The disease is unique among the Q_n-expansion diseases in that the pathological number of repeats is much lower than in the other diseases in this group and the maximal expansion noted thus far is small (Table 2). Unlike the other diseases in this group, the affected protein in SCA6 is a membrane-spanning protein. The mutation apparently shifts the voltage dependence of channel activation and rate of inactivation, and impairs normal G-protein regulation of P/Q channels (Restituito et al., 2000). This may be regarded as a loss of function. In addition, the mutated SCA6 may impair normal proteasome function (Seidel et al., 2009) and prevent cell death (Matsuyama et al., 2004).

A point mutation in the SCA6 gene gives rise to a different disease phenotype, namely familial hemiplegic migraine (Ophoff et al., 1996). Inasmuch as patients with familial hemiplegic migraine do not exhibit cerebellar ataxia, the ataxia in SCA6 is presumably a pathological gain of function resulting from the Q_n expansion. Thus, SCA6 seems to exhibit features of both loss and gain of function for the mutated protein.

Two episodic ataxias have been described in the literature. It is now clear that episodic ataxia type 2 (EA2) is identical with SCA6. The other (EA1), which is characterized by attacks of generalized ataxia and by continuous myokymia (irregular twitching), is due to a point mutation in the voltage-gated potassium channel gene *KCNA1* (Cusimano et al., 2004). Clearly, EA1 and EA2/SCA-6 are both examples of channelopathies.

3.8 Spinocerebellar Ataxia Type 7 (SCA7)

SCA7 is characterized by late-onset neuronal loss in the cerebellum, brainstem, and retina (Miller et al., 2009). It is the only Q_n-expansion disease in which the retina is affected. The disease is caused by a CAG expansion in the gene at chromosome 13p12-13 (David et al., 1997). The affected gene has an M_r of about 100,000. Because the biological function of the protein was unclear, it was initially named ataxin-7. The SCA7 gene product, ataxin-7, is now known to be a subunit of a transcriptional coactivator complex (STAGA or TFTC) that has histone acetyltransferase activity (Helmlinger et al., 2004, 2006a, b). Thus, transcriptional regulation seems to be altered in SCA-7 (and possibly other Q_n-expansion diseases).

3.9 Spinocerebellar Ataxia Type 17 (SCA17)

This was the last of the SCAs to be shown to arise from a Q_n expansion in the mutated protein. Koide et al. (1999) identified a CAG expansion in the transcription factor TATA-binding factor protein *TBP* gene in a patient with short stature, pyramidal signs, and mental retardation. Since then this extremely rare disease has been identified in a few European and Japanese families (Rolfs et al., 2003). Cerebral proteinaceous deposits are especially prominent in this disease (Rolfs et al., 2003). The disease is interesting because some patients present first with exclusively pure psychiatric symptoms while having no signs of ataxia or movement disorders (Rolfs et al., 2003). The authors point out that SCA17 represents one of the very few psychiatric diseases for which there is a known monogenic cause (Rolfs et al., 2003). SCA17 also appears to be a risk factor for Parkinson-like symptoms (Lee et al., 2009).

4 Possible Factors Contributing to Neurodegeneration in the $(CAG)_n/Q_n$-Expansion Diseases

As noted above, the neurodegenerative disease phenotypes in the Q_n-expansion diseases are probably caused mostly by a toxic gain of function. This is suggested by the autosomal dominant mode of inheritance (except BMA) and by most experiments with cell and animal models (Ross, 2002). HD is common enough in the Lake Maracaibo area where HD homozygotes are sometimes encountered (The US–Venezuela Collaborative Research Project and Wexler, 2004). In agreement with the gain of function hypothesis, these individuals have an almost identical disease phenotype to heterozygotes. However, as also noted above, some loss of normal function may also occur in some cases (Ross, 2002). It is also possible that the aberrant protein expressed by the mutant allele interacts with the normal protein expressed by the other allele. This dominant-negative interaction might lead to a partial loss of function in some cases (Ross, 2002).

Although the expanded Q_n domains impart a toxic gain of function, this does not explain the selective vulnerability in the various Q_n-expansion diseases. In all cases, the mutated Q_n domain is expressed throughout the brain. Therefore, the selective vulnerability must reside in a toxic gain of function that somehow involves the non-mutated part of the protein (Cummings and Zoghbi, 2001; Zoghbi and Orr, 2009).

Many excellent reviews and discussions on possible mechanisms contributing to Q_n-expansion diseases have been published (e.g., Gatchel and Zoghbi, 2005; Di Prospero and Fischbeck, 2005; Pearson et al., 2005). Possible mechanisms are discussed below. Because much of the work on Q_n-expansion diseases has been directed to HD, much of the following discussion is heavily weighted toward HD.

4.1 Toxic Protein Aggregates

A characteristic feature of all the Q_n-expansion diseases is the presence of insoluble protein aggregates (inclusion bodies) in the affected brain regions. Some authors have suggested that these insoluble aggregates are toxic and thereby contribute to the disease process (e.g., Bates, 2003). However, other studies indicate that the insoluble aggregates per se may not be toxic (e.g., Saudou et al., 1998; Kuemmerle et al., 1999; Kaytor et al., 2004; Mitra et al., 2009), and that their concentrations do not necessarily correlate with damaged regions of the brain. The insoluble protein aggregates may represent the end-stage of a cascade of previous events that relate more directly to toxicity than the aggregates themselves (Ross and Poirier, 2004). Nevertheless, an understanding of the mechanism of aggregate formation may provide clues as to pathological mechanisms.

Two theories have been proposed concerning the origin of aggregates in the Q_n-expansion diseases. The first was put forward by Max Perutz and is known as the "polar zipper hypothesis" (e.g., Perutz and Windle, 2001). Proteins/polypeptides containing Q_n domains in vitro form hydrogen bonds that link the glutaminyl side chains of the Q_n domain to the peptide backbone in an adjacent protein/polypeptide resulting in highly ordered β-pleated sheets that are often extremely insoluble. β-Pleated sheets occur in bacteria overexpressing proteins containing Q_n domains (Scherzinger et al., 1999). However, the aggregates in HD brain appear to be less ordered (Karpuj et al., 1999).

The second theory involves the action of the enzyme family transglutaminases (TGs). Q_n domains are excellent substrates of TGs (Kahlem et al., 1998; Cooper et al., 2002), which catalyze calcium-dependent protein cross-linking between glutaminyl (Q) and lysyl (K) residues. Often, but not always, such cross-linked proteins are insoluble. In HD mice lacking TG 2 (the major TG in brain; also known as tissue TG) (R6/2 TGase $2^{-/-}$) exhibit as many, and possibly more, insoluble inclusions in the brain than their R/6 TGase $2^{+/+}$ littermates (Mastroberardino et al., 2002). The R6/2 TGase $2^{-/-}$ mice live longer than the R6/2 TGase $2^{+/+}$ mice, suggesting that insoluble aggregates are not the only toxic manifestation of Q_n domains. The finding also suggests that TG 2 may not be responsible for most of the aggregate formation in HD mouse brain. It should be noted, however, that the brain also contains TGs 1 and 3 and possibly other TGs that may also contribute to protein cross-linking (Zainelli et al., 2005). A possible explanation for the findings of increased insoluble cerebral proteinaceous aggregates in the R6/2 TGase $2^{-/-}$ mice compared to the R6/2 TGase $2^{+/+}$ mice relates to the work of Lai et al. (2004). These authors found that a thioredoxin fusion protein containing a Q_{62} polypeptide has a tendency in vitro to aggregate into insoluble polymers. However, this aggregation was arrested by TG2, which cross-linked the Q_{62} fusion protein into soluble high-M_r polymers. TG2-catalyzed conversion of the Q_{62} fusion protein into soluble aggregates was decreased in the presence of amine substrates (Lai et al., 2004). The authors suggested that TGs may contribute to the pathogenicity of mutant Htt by catalyzing the formation of toxic, soluble Q_n-containing fragments.

Most probably, the aggregates in the Q_n-expansion disease are formed by a combination of (a) noncovalent ordered interactions (polar zippers), (b) noncovalent interactions of disordered misfolded proteins, and (c) covalent modifications (TG-catalyzed cross-linking).

We suggest the following mechanism implicating TGs in the neuropathology of HD and other Q-expansion diseases. The cross-linking activity of brain TGs is normally very low or quiescent in vivo. However, with aging calcium dysregulation begins to occur (Foster and Kumar, 2002) and inherent TG activity is increased

(Park et al., 1999). Both factors will lead to increased protein cross-linking in the aging brain. Misfolded mutant Htt has a tendency with time to generate insoluble protein aggregates. However, activation of TG competes with this process, such that a threshold is reached later in life in which toxic soluble, cross-linked mutant Htt fragments begin to accumulate. With time, excessive removal of "normal" Q_n-containing protein, including transcription factors through both noncovalent interactions and covalent cross-linking may occur. Possibly, protein synthesis can keep up with "lost" proteins early in life, but with aging the process of replenishment may be less efficient. This would lead to loss of function through loss of biologically active Q_n-containing proteins.

4.2 Disrupted Proteasome Function

The aggregates in each of the Q_n-expansion diseases (except in the case of SCA6) are immunopositive for ubiquitin (reviewed by Cummings and Zoghbi, 2000). In addition, the aggregates often contain parts of the proteasome machinery and chaperones (reviewed by Cooper et al., 2002). The possibility therefore exists that the ubiquitin proteasome pathway is disrupted in HD and other Q_n-expansion diseases (Cooper et al., 2002; Mandrusiak et al., 2003; Wang et al., 2008). Clogging of the proteasome might occur if the aggregates contain polar zippers or TG-catalyzed cross-links that cannot be "unzipped." Clogging of proteasomes could prevent the removal of damaged and misfolded proteins. Such proteins, including those containing the expanded Q_n domain, might therefore accumulate and exert toxicity, perhaps by interacting aberrantly with other proteins/polypeptides. In support of this hypothesis, interference with the ubiquitin-tagging pathway, or interference of clearance by the proteasome machinery, either singly or in combination would be expected to increase the toxicity of mutant Htt in cell culture (Saudou et al., 1998). Moreover, SCA1 transgenic mice, with a block in the ubiquitin pathway, had markedly fewer intranuclear aggregates, but markedly worse SCA1 pathology (Cummings et al., 1998; Cummings and Zoghbi, 2001).

Inasmuch as chaperones help fold proteins into their "correct" configurations, one might infer that chaperones would help in lessening the toxicity associated with misfolded proteins containing Q_n expansions (Cummings and Zoghbi, 2000). Indeed, there is some evidence that this is the case both for cell culture models of Q_n-expansion diseases and for a *Drosophila* retina model of HD (reviewed by Cummings and Zoghbi, 2002). However, chaperones do not appear to mitigate disease phenotype in at least one animal model of a Q_n-expansion disease (Helmlinger et al., 2004). As in human SCA7, transgenic SCA7 mice develop retinopathy. Helmlinger et al. (2004) developed transgenic mice, which specifically overexpress Hsp70 and HDJ2 along with the Q_n-containing protein. Although coexpression prevented aggregate formation in a cell model it did not prevent either neuronal toxicity or aggregate formation in intact mice. Moreover, protein aggregates in SCA7 mice contained cleaved mutant ataxin-7, whereas in the transfected cells the aggregates contained full-length ataxin-7. Thus, the possibility that disrupted proteasome dysfunction contributes to the neuropathology of Q_n-expansion diseases remains controversial. Moreover, in vitro-generated Q_n aggregates failed to inhibit purified proteasomes, whereas filamentous Htt aggregates isolated from mouse brain resulted in inhibition (Ortega et al., 2007). Perhaps formation of inclusion bodies is a protective mechanism to remove potentially harmful aggregates from solution (Mitra et al., 2009). However, this mechanism may eventually fail in HD brain. Indeed the activity of the proteasome machinery is significantly lower in postmortem HD brain tissue (Ortega et al., 2007).

Macroautophagy (sometimes more simply referred to as autophagy) is an additional mechanism for degrading damaged or misfolded cellular proteins (Renna et al., 2010). It has been suggested that autophagy may be especially useful in degrading mutant Htt-containing fragments (Renna et al., 2010). If this hypothesis is correct then small molecule stimulators of autophagy (e.g., rapamycin, rilmenidine) might be useful in treating HD and other Q_n-expansion diseases (Renna et al., 2010; Rose et al., 2010). In this regard, rilmenidine may be especially efficacious as it has a long clinical safety use (Rose et al., 2010).

4.3 Interference with Gene Expression

Many transcription factors, such as CBP (CREB binding protein) and TBP (TATA binding protein), contain Q_n domains (Perutz et al., 1994; Schaffar et al., 2004), which may assist in the assembly of the transcriptosome. Therefore, it is conceivable that the pathological gain of function in Q_n-expansion diseases is due at least in part to aberrant interaction among the mutated protein and various transcription factors (e.g., Sugars and Rubinsztein, 2003; Li and Li, 2004). Indeed, proteins containing aberrant Q_n domains have been shown to interact with various transcription factors including CREB, CBP, $TAF_{\mu}130$, SP1, and TP53, some of which have been shown to be present in the protein aggregates in affected brain regions (Gatchel and Zoghbi, 2005). Moreover, one of the Q_n-expansion diseases is due to a mutation within a transcription factor itself (TBP in SCA17; Table 2). In the case of HD, N-terminal fragments of mutated Htt (containing the expanded Q_n domain) accumulate in the nucleus (Zainelli et al., 2003), apparently as a result of interference with the nuclear export machinery (Cornett et al., 2005). Moreover, nuclear-targeting of mutant Htt fragments produces a Huntington-disease-like phenotype in HD transgenic mice (Schilling et al., 2004). Thus, it is possible that one of the toxic effects of expanded Q_n domains is the alteration of transcription factor interactions (e.g., Schaffar et al., 2004). It has been shown that mutant Htt binds to CBP and p53. The latter protein regulates transcription of various mitochondrial proteins (Sawa, 2001).

Wild-type, but not mutant Htt, stimulates transcription of brain-derived neurotrophic factor (BDNF), and neuronal restrictive silencer element (NRSE) is the target of wild-type Htt activity on the BDNF promoter II (Zuccato et al., 2003).

Moreover, it was shown that mutant Htt in a mouse model of HD facilitates CRE-dependent transcription (Obrietan and Hoyt, 2004). Thus, mutated Htt may cause either increases (Obrietan and Hoyt, 2004) or decreases (Zuccato et al., 2003) in transcriptional regulation. Such alterations may contribute to the pathological response in HD and other CAG-expansion diseases. A particularly intriguing Q_n-containing protein is PQBP-1, which binds to both Q_n expansions and to brain-specific transcription factor Brn-2 (Waragai et al., 1999). Thus, aberrant interactions between an expanded Q_n domain and PQBP-1 may, in turn, result in aberrant transcription of Brn-2 and neuropathology.

Glutamine-rich transcription factor Sp1 is readily cross-linked by TG 2 (Han and Park, 2000). Inasmuch as some TG 2 is present in the nucleus, and Q_n domains are excellent substrates, it is possible that TGs may modulate the activity of at least some transcription factors in vivo. Because TG activity is increased in HD brain, and because the expanded Q_n domain of Htt is an excellent TG substrate, the possibility exists that TGs play a critical role in altered transcription level and properties in Q_n-expansion diseases.

Recent work has suggested that REST [RE1 (repressor element 1)-silencing transcription factor] function is disrupted in HD brain (Bithell et al., 2009). As discussed by Bithell et al. (2009), REST is a master regulator of many neuronal genes, including BDNF. In addition, recent work suggests that REST regulates transcription of regulatory miRNAs (microRNAs), many of which are involved in neuronal protein expression. Thus, mutant Htt not only appears to directly dysregulate target genes of REST, but also to indirectly dysregulate neuronal transcription (Bithell et al., 2009).

Polyalanine (A_n) expansions also give rise to disease phenotypes (Section 5). In eight of these diseases, the expansion is in a transcription factor and the disease phenotype is evident at birth. If transcription factor dysfunction contributes to Q_n-expansion diseases, then a hypothesis explaining the disease phenotype must account for the fact that the disease phenotype is present at birth in the A_n-expansion diseases, but is typically adult onset in Q_n-expansion diseases.

4.4 Interference with Mitochondrial Function

Marked interference with mitochondrial function is a feature of HD brain (Browne and Beal, 2004; Browne et al., 1997; Browne, 2008; Nicholls, 2009; Reddy et al., 2009; Quintanilla and Johnson, 2009; Su et al., 2010). For example, Browne et al. (1997) showed that citrate synthase-corrected complex II–III activity is markedly reduced in both HD caudate (–29%) and putamen (–67%), and complex IV specific activity is reduced in HD putamen (–62%). Tabrizi et al. (1999) reported that aconitase specific activity is reduced to 8, 27, and 52% of control activities in HD caudate, putamen, and cerebral cortex, respectively. Tabrizi et al. (2000) also reported that aconitase and complex IV activities are decreased in the striatum of 12-wk HD transgenic (R6/2) mice, and complex IV activity is decreased in cerebral cortex. As noted previously for human HD, oxidative stress indicators (increased inducible NO synthase and nitrotyrosine) were detected in brains of HD-transgenic mice (Tabrizi et al., 2000). Deficits in energy metabolism in human HD brain also occur in human HD muscle. Thus, Lodi et al. (2000) demonstrated by means of in vivo magnetic resonance spectroscopy (MRS) a decreased ATP/(PCr + P_i) ratio in skeletal muscle of HD patients. During recovery from exercise the maximal rate of ATP synthesis was decreased by 44% in symptomatic patients and by 35% in presymptomatic HD carriers compared to controls (Lodi et al., 2000). Gårseth et al. (2000) reported small but significant decreases in lactate and citrate in the CSF of HD patients as assessed by MRS. Dietary supplementation with 2% creatine significantly improved survival and improved symptoms in HD-transgenic mouse models, possibly through redress in part of the energy deficits (Andreassen et al., 2001).

Panov et al. (2002) reported that lymphoblast mitochondria from patients with HD have a lower membrane potential and depolarize at lower calcium loads than do mitochondria from controls. These authors also showed a defect in brain mitochondria similar to those isolated from HD-transgenic mice. This defect preceded behavioral and pathological abnormalities. Panov et al. (2003) also showed that GST (glutathione *S*-transferase) constructs with a pathological-length Q_n insert induced a small but significant reduction in membrane potential (State 4) of mitochondria isolated from normal rat liver and normal human lymphoblasts. With successive increments of Ca^{2+} aliquots, mitochondria exposed to pathological-length Q_n domains depolarized much earlier and to a greater extent than did mitochondria exposed to nonpathological constructs (Panov et al., 2003). The striatum is particularly ("selectively") vulnerable in HD, and mitochondria isolated from striatum of rat brain may be more susceptible to the Ca^{2+}-induced permeability transition (PT) than are cortical mitochondria. The susceptibility of striatal mitochondria has been demonstrated by measurements of depolarization, swelling, Ca^{2+} uptake, reactive oxygen species, and respiration (Brustovetsky et al., 2003).

For a recent review of mitochondrial calcium function and dysfunction in neurodegenerative diseases (including HD) see Nicholls (2009).

4.5 Aberrant Caspase Activity

Several studies have shown that cystamine is beneficial in mouse models of HD (e.g., Karpuj et al., 2002; Dedeoglu et al., 2002; Bailey and Johnson, 2006; Van Raamsdonk et al., 2005). In one study, brain aggregates were reduced by cystamine treatment in the HD mice (Dedeoglu et al., 2002). Cystamine is an in vitro inhibitor of Ca^{2+}-dependent TGs and is an inhibitor of caspases in cells in culture (Lesort et al., 2003). Some studies have suggested a role for caspases in HD and that inhibition of these enzymes may be beneficial in HD mouse models (e.g., Ona et al., 1999; Chen et al., 2000; Lesort et al., 2003). Caspase activity has also been implicated in SCA-3 (Shoesmith Berke et al., 2004). Because caspases contain a crucial cysteine residue at the active site they are expected to be sensitive to inhibition by cystamine through disulfide interchange reactions (Lesort et al., 2003). However, cystamine does not accumulate to any great extent in the brains of mice treated with pharmacological levels of cystamine (Pinto et al., 2005). Moreover, the magnitude of the protective effect of cystamine is similar in R6/2 TGase $2^{+/+}$ mice to that in R6/2 TGase $2^{-/-}$ mice (Bailey and Johnson, 2006). Thus, the mechanism by which cystamine exerts its beneficial effects in the intact HD transgenic mice, especially in regard to caspase and TG activity, must await further study.

Nevertheless, recent evidence does suggest a possible pivotal role for caspases in HD neuropathology. For example, there is some evidence that certain proteolytic fragments generated from Htt are neurotoxic (Ratovitski et al., 2009, and references cited therein). Htt undergoes proteolysis by calpains and caspases at the N-terminus between amino acid residues 460 and 470 (Ratovitski et al., 2009). Evidently the proteolytic cleavage is heterogeneous. Htt can be phosphorylated at serine-421 (S421) by the prosurvival signaling kinases Akt and SGY (Warby et al., 2009). Interestingly, within the brain, phosphorylation of Htt is lowest in the striatum. Caspase 6-cleavage of Htt at amino acid 586 appears to be a crucial factor in Htt neurotoxicity (Warby et al., 2009). Phosphorylation of Htt reduces the nuclear accumulation of caspase-6-generated Htt fragments by reducing caspase-6 cleavage. Inasmuch as different cells contain different complements of calpains, caspases, and phosphorylation/dephosphorylation machinery, it is possible that production of toxic fragments will be cell-specific (Ratovitski et al., 2009; Warby et al., 2009) and may explain in part the remarkable selectivity of different neuronal populations in the various Q_n-expansion diseases.

4.6 Increased Excitotoxicity/Oxidative Stress

Quinolinate has been known for more than 25 years to produce HD-like pathology in rodents (e.g., Beal et al., 1986). Thus, it has been suggested that the genetic defect in HD may result in heightened neuronal susceptibility to excitotoxic injury. Guidetti et al. (2004) have shown that the levels of quinolinate (an endogenous neuroactive metabolite of the kynurenine pathway of tryptophan metabolism) and 3-hydroxykynurenate (a free radical generator and additional metabolite of the kynurenine pathway) are elevated three- to fourfold in low-grade HD brain (grade 0/1) in the neocortex and neostriatum, but not in the cerebellum. In contrast, levels of these compounds tended to decrease in HD brain in advanced grades (grades 2–4). NAD(P)H oxidase has been suggested to contribute to neurotoxicity in an excitotoxic/pro-oxidant model of HD in rats (intrastriatal injection of quinolinate) (Maldonado et al., 2010)

Calcineurin is a calcium-dependent serine/threonine phosphatase involved in the regulation of glutamate receptor signaling (Xifró et al., 2009). It has been suggested that reduction of calcineurin A (the catalytic subunit of the calcineurin heterodimer) activity may contribute to the pathophysiology of HD (Xifró et al., 2009). Some evidence suggests that the excitoxicity associated with overstimulation of the *N*-methyl-D-aspartate receptor (NMDAR) is associated with the pathogenesis of HD (Milnerwood et al., 2010, and references cited therein). Milnerwood et al. (2010) cite evidence that synaptic NMDAR transmission drives neuroprotective gene expression, whereas extrasynaptic gene expression promotes cell death. The authors suggest that elevated extrasynaptic NMDAR activity may contribute to the neurodegeneration of HD.

4.7 Defects in Axonal Transport

Several authors have provided evidence that pathological-length Q_n repeats promote aberrant protein interactions that cause defects in axonal transport (Gunawardena and Goldstein, 2005; Smith et al., 2009; Schweitzer et al., 2009; Wu et al., 2010). Gunawardena and Goldstein (2005) have aptly described the phenomenon as "deadly traffic jams along the neuronal highway." These jams would be particularly troublesome in long-narrow caliber axons.

4.8 Integration of Mechanisms

Any theory that attempts to unify all the competing mechanisms that have been proposed to account for the toxicity of expanded Q_n domains must account for the following observations: (1) Q_n-expansion diseases typically become manifest in adulthood, and (2) different brain regions are selectively vulnerable in the various Q_n-expansion diseases despite widespread expression of mutated protein throughout the brain and body. Evidently, proteins containing pathological-length Q_n expansions exhibit normal (or near normal) biological functions early in life. Typically, only in adulthood does a pathological gain in function become prominent. This situation contrasts dramatically with the A_n-expansion diseases, where pathology is evident at birth, and disruption of normal protein function is present even in utero (Section 5).

It is now becoming clear that differences in disease phenotype among the different Q_n-expansion diseases are not only dictated by the length of *n*, but also in part by the intrinsic function of the disease-causing mutation (Gatchel and Zoghbi, 2005). For example, modifications outside the Q_n domain, such as phosphorylation of ataxin-1 at a crucial serine residue (Emamian et al., 2003) and SUMOylation of Htt (Steffan et al., 2004) are important determinants of toxicity. Moreover, Boat (brother of ataxin-1) was shown to interact with ataxin-1 at multiple sites, and altered expression of Boat in Purkinje cells may contribute to the neurodegeneration in SCA1 (Mizutani et al., 2005). Differences in susceptibility to TG-catalyzed cross-linking among the various mutated Q_n-containing proteins may also contribute to the selectivity.

A summary of many of the pathological mechanisms postulated to occur in Q_n-expansion diseases and how they might be interrelated is shown in Fig. 2 (Steps 1–8). The Q_n expansion causes aberrant protein conformation, which in turn leads to altered Ca^{2+} homeostasis (1), mitochondrial dysfunction (2), altered energy metabolism (3), and excitoxicity/oxidative stress (4). The protein containing the

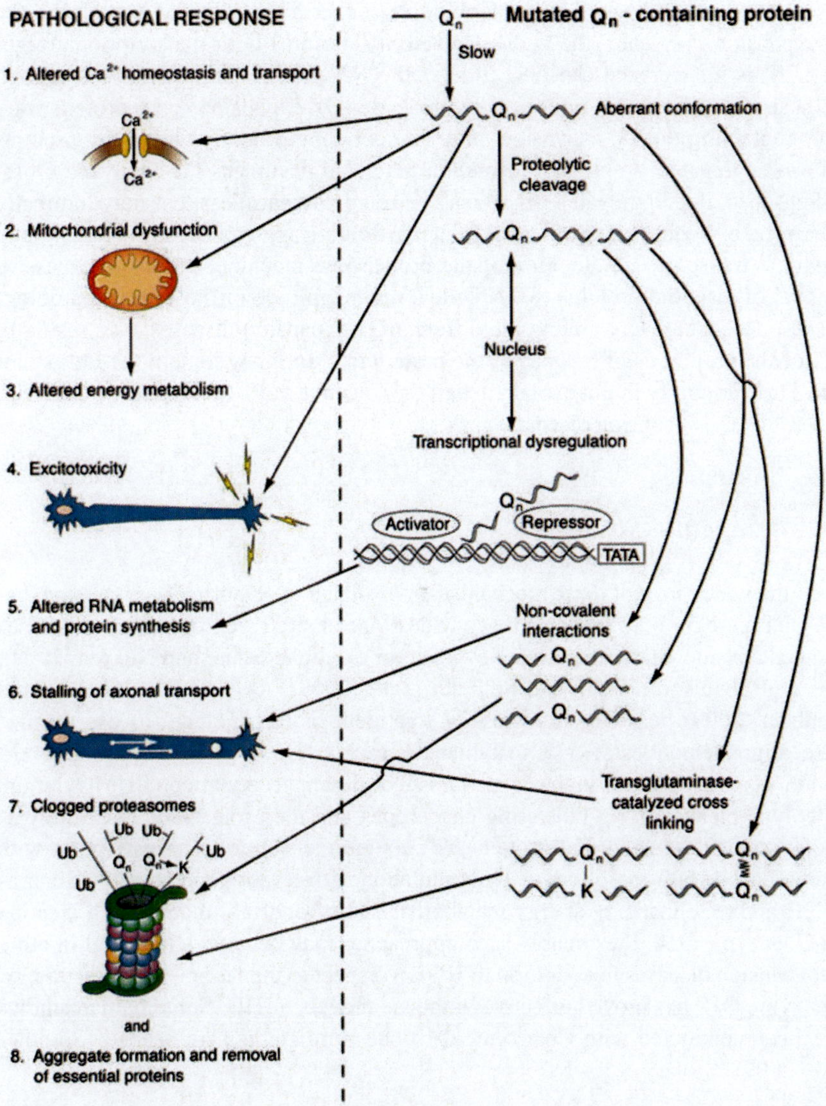

Fig. 2 Proposed pathological responses in polyglutamine (Q_n)-expansion diseases. Pathological responses include (*1*) altered calcium homeostasis, (*2*) mitochondrial dysfunction, (*3*) altered energy metabolism, (*4*) excitoxicity, (*5*) altered transciptio/translation, (*6*) stalling of axonal transport, (*7*) proteasome malfunction, and (*8*) removal of essential proteins/factors in insoluble protein aggregates (formed via non-covalent and/or covalent cross linking). For a more detailed discussion of the various pathological responses see the text. K, lysine residue; Q, glutamine residue; Ub, ubiquitin. The figure is adapted from Gatchel and Zoghbi (2005) but with considerable modification

Q_n expansion is subjected to proteolytic cleavage. The fragment containing the Q_n expansion may enter the nucleus where it interferes with transcriptional regulation. This interference can lead to altered RNA metabolism and altered protein synthesis (5). The intact protein containing the Q_n expansion, or a protein fragment containing the Q_n expansion, may interact noncovalently with other proteins to form aggregates. Aggregate formation may also occur via TG-catalyzed cross-linking. Insoluble aggregates per se may be relatively harmless, but may indirectly be harmful by sequestering essential proteins. Soluble aggregates may cause stalling of axonal transport (6), clogging of the proteasome machinery (7), and excessive removal of essential proteins (8). Note that the magnitude of the various pathological responses may vary among the different Q_n-expansion diseases, as a result, in part, of the properties of the part of the protein that does not contain the Q_n expansion. This variability might explain in part the selective vulnerability associated with the various Q_n-expansion diseases.

4.9 Therapeutic Strategies

Given the wide range of toxic mechanisms postulated to occur in Q_n-expansion diseases, it is perhaps not surprising that a wide range of strategies to treat such diseases has been considered. For recent reviews see for example, Bauer and Nukina (2009), Ross and Shoulson (2009), and Spindler et al. (2009). As discussed by Ross and Shoulson (2009) possible strategies for treatment of HD include (a) use of antisense oligonucleotides/SiRNA to target the mutant *Htt*, (b) alteration of the Htt protein posttranslationally (by, e.g., phosphorylation, acetylation, SUMOylation, proteolytic cleavage), (c) bolstering chaperones and the proteasome machinery as a defenses against abnormal proteins, (d) countering abnormal transcription with histone deacetylase inhibitors or by stimulation of relevant gene products such as BDNF, and (e) enhancing energy metabolism/mitochondrial function with creatine and Coenzyme Q10. Presumably such approaches may be generally useful in other Q_n-expansion diseases in adddition to HD. In respect to the last-mentioned strategy, Coenzyme Q10 has proved effective in mouse models of HD. Some human studies have been conducted with Coenzyme Q10 but more studies are needed (Spindler et al., 2009).

5 Diseases Due to a Coding Trinucleotide Expansion—Polyalanine (A_n)-Expansion Diseases

5.1 General Description

Much of the following discussion on A_n-expansion diseases is from excellent reviews by Brown and Brown (2004) and Messaed and Rouleau (2009). Nine diseases are currently known to be associated with an expansion of an A_n domain in the affected protein (Table 3). Synpolydactyly type II (SPD), cleidocranial dysplasia

Table 3 A_n-expansion diseases

Disease	Gene	Affected Protein	A_n Expansion
Synpolydactyly type II (SPD)	*HOXD13*	Transcription factor	(15 → 22–29)
Cleidocranial dysplasia (CCD)	*RUNX2* (*CBFA1*)	Transcription factor	(17 → 27)
Oculopharyngeal muscular dystrophy (OPMD)	*PABPN1*	Polyadenylate-binding protein	(10 → 11–17)
Holoprosencephaly (HPE5)	*ZIC2*	Transcription factor	(15 → 25)
Hand–foot–genital syndrome (HFGS)	*HOXA13*	Transcription factor	(Tract 1; 14 → 22) (Tract 2; 12 → 18) (Tract 3; 18 → 24–30)
Blepharophimosis, ptosis, and epicanthus inversus (BPES)	*FOXL2*	Transcription factor	(14 → 22–24)
X-linked mental retardation with hypopituitarism (XH)	(MRX) *SOX3*	Transcription factor	(15 → 26)
X-linked infantile spasm syndrome (ISSX; West syndrome, WS) Partington syndrome (PRTS) X-linked lissencephaly with ambiguous genitalia (XLAG) X-linked mental retardation 36 and 54 (MRX)	*ARX*	Transcription factor	(Tract 1; 16 → 23) (Tract 2; 12 → 20)
Congenital central hypoventilation syndrome (CCHS) Ondine's curse	*PMX2B* (*PHOX2B*)	Transcription factor	(20 → 25–33)

Modified from Brown and Brown (2004) and Messaed and Rouleau (2009).

(CCD), oculopharyngeal muscular dystrophy (OPMD), hand–foot–genital syndrome (HFGS), and blepharophimosis, ptosis, and epicanthus inversus (BPES) are inherited in an autosomal dominant fashion. Holoprosencephaly (HPE) is a severe developmental disease resulting in death in infancy and the causative mutation presumably arises de novo. Congenital central hypoventilation syndrome (CCHS; Ondine curse) cases also appear to be due to spontaneous mutations. Two A_n-expansion diseases are X-linked. A mutation in the *SOX3* gene, giving rise to sex-linked mental retardation (MRX) with growth hormone deficiency, has been described in one large family (Brown and Brown, 2004). A_n-Expansion mutations can occur in two different tracts in the expressed protein of the *ARX* gene, giving rise to nonsyndromic MRX and two syndromic conditions (X-linked infantile spasm syndrome (ISSX) also known as West syndrome (WS) and Partington syndrome (PRTS)). A_n-Expansion mutations can occur in three different tracts in the *HOXA13* gene.

5.2 *Comparison of A_n-Expansion Diseases with Q_n-Expansion Diseases*

Although expansions in Q_n and A_n domains give rise to disease phenotypes, the characteristics of the diseases are remarkably dissimilar. (1) The number of alanine repeats is fixed at a single value in the normal (nonmutated) A_n-containing proteins. This number varies between 10 and 18, depending on the protein (Table 3). In contrast, although 10–20 glutamine repeats occur in normal (nonmutated) Q_n-containing proteins, the number is not fixed and may vary considerably in the nonmutated protein among the general population. (2) The expansions giving rise to disease phenotypes in the A_n-expansion diseases are small, with a maximum extension of 14 alanine residues. In contrast, pathological extensions of Q_n domains can be much larger (compare Tables 2 and 3). (3) The A_n expansion appears to be meiotically and somatically stable, whereas this is not the case with the Q_n expansions. (4) The disease phenotypes of the Q_n-expansion diseases are almost entirely restricted to neural tissue. In contrast, although severe neurological defects can occur in the A_n-expansion diseases, all give rise to extraneural disease phenotypes. (5) The A_n-expansion disease phenotypes are present at birth whereas the onset of Q_n-expansion diseases is typically in adulthood. (6) The diseases caused by Q_n expansions are thought to arise predominantly (but not exclusively) via a toxic gain of function (Section 5). By contrast, the diseases caused by A_n expansions are thought to arise by loss of function, gain of function, or by a dominant-negative effect, depending on the disease. (7) All but one of the A_n-expansion diseases is associated directly with a transcription factor, whereas only one of the Q_n-expansion diseases is associated directly with a known transcription factor. (8) The Q_n-expansion disease proteins are usually coded in the genome by "perfect" runs of CAG repeats, and thus slippage mechanisms may account for the meiotic and somatic instabilities of the Q_n domain. On the other hand, the A_n-expansion disease proteins are usually coded by "imperfect" runs containing any of the four triplets that code for alanine (CCG, GCA, GCT, and GCC). These differences suggest that Q_n tracts are easier to expand than A_n tracts (Messaed and Rouleau, 2009). Increases in the length of the A_n domain are thought to arise through unequal recombination. However, slippage may occur in BPES and possibly ISSX (Brown and Brown, 2004).

6 Possible Mechanisms Contributing to A_n-Expansion Diseases

Disruption of transcription factor function will result in altered expression of downstream target genes and in abnormal development (Brown and Brown, 2004). But how are the functions of the transcription factors altered by an A_n expansion? It has been suggested that the A_n domains may (1) have a role in repression, (2) act as spacers or hinges, or (3) play important roles in correct protein–protein interactions and protein–DNA interactions during transcription. Clearly much work needs to be done to elucidate the pathological mechanisms associated with A_n-expansion diseases (reviewed by Brown and Brown, 2004). A_n polypeptides form fibrils through β-sheet formation (Nguyen and Hall, 2004), and it has been shown that transfection of cells with an A_n expansion in the aristaless-related homeobox (ARX) protein results in nuclear protein aggregation, filamentous nuclear inclusions, and increased cell death (Nasrallah et al., 2004). It was suggested that nuclear protein aggregation likely underlies the pathogenesis of diseases caused by A_n expansions in the ARX protein and possibly in other A_n-containing transcription factors (Nasrallah et al., 2004). The aggregates themselves may be toxic or toxicity may result in loss of transcription function.

A summary of possible mechanisms relating to the A_n-expansion diseases has been provided by Messaed and Rouleau (2009). These authors point out that in mammalian cells chaperones bind to misfolded proteins produced during translation or later during their aggregation (in the cytosol and/or nuclear compartment) in an attempt to correctly refold/solubilize them. Unsuccessful folding and prolonged association of the misfolded protein with chaperones can stimulate ubiquitination and targeting to the proteasome machinery. However, in the A_n-expansion diseases this quality-control mechanism may be insufficient. Depending on the disease, aggregates may form in the cytosol, nucleus, or both compartments. These aggregates may sequester essential cellular factors preventing them from reaching their targets. In the case of SPD, HFGS, and XH the essential cellular factor recruited into the deposits may be the wild-type A_n-containing protein which may lead to a dominant-negative effect. In addition, expanded A_n tracts may result in (a) decreased binding efficiency to DNA, (b) pathological competition with the wild-type protein for binding to DNA, or (c) interfere with cofactors important for DNA binding of the wild-type protein.

As summarized by Messaed and Rouleau (2009), in order to delineate the cellular pathways involved in pathogenesis, future work requires an understanding of the mechanism relating to (a) the selective tissue vulnerability of expanded A_n domains, (b) the nature of the targeted genes, and (c) the nature of the targeted interaction partners.

7 Other Nucleotide-Expansion Diseases

Examples of diseases caused by tetra- and pentanucleotide expansions are known. For example, myotonic dystrophy type 2 (DM2) is caused by a tetranucleotide expansion in the affected gene on chromosome 19q13.3 (Meola and Moxley, 2004; Lee and Cooper, 2009). In this case, an expanded CCTG repeat occurs in intron 1 of the zinc finger 9 (*ZFN9*) gene (Meola and Moxley, 2004). The expanded repeat in transcribed RNA forms nuclear inclusions in both types of myotonic dystrophy (Mankodi, 2008). The aberrrant RNA sequesters muscleblind-like protein 1 (MBNL1), a splice regulator protein, and depletes MBNL1 in the nucleus. Loss of MBNL1 results in altered splicing of ClC-1 (chloride channel 1) mRNA, inactive ClC-1 and loss of chloride conductance in muscle membranes (Mankodi, 2008).

Spinocerebellar ataxia 10 (SCA 10) is caused by a pentanucleotide (ATTCT) expansion in intron 9 of the *SCA10* (*ATXN10*) gene at chromosome 22q13 encoding an approximately 55-kd protein (ataxin 10) of unknown function (Matsuura et al., 2000; Wakamiya et al., 2006). In experiments with HEK293 cells in culture, it was shown that the SCA10 protein is essential for neuronal survival (März et al., 2004).

Spinocerebellar ataxia type 31 is associated with pentanucleotide repeat [$(TGGAA)_n$] expansion at chromosme 16g22.1 in introns of the *TK2* (thymidine kinase) and *BEAN* (brain expressed, associated with Nedd4) genes (Sato et al., 2009). The length of the pentanucleotide repeat correlates inversely with the age of onset of symptoms. Purkinje cells in the cerebellum are affected. Aberrant RNA foci may result in disrupted splicing factors in these cells (Sato et al., 2009).

A mutation in the *junctophilin-3* (*JPH3*) gene at chromosome 16q24.3 in a variably spliced exon gives rise to an autosomal dominantly inherited disease that is clinically indistinguishable from HD (Margolis et al., 2004). The disease appears to be restricted largely, if not exclusively, to families of African descent, and is referred to as Huntington disease-like 2 (HDL2) (Margolis et al., 2004; Rodrigues et al., 2008). The mutation results from an expansion in a CTG/CAG repeat. Although the expansion is in an exon (exon 2A), junctophilin-3 protein containing a domain with increased Q residues does not appear to be produced, suggesting that the pathological phenotype is due to loss of JPH3 protein and/or to aberrant RNA processing/metabolism (Margolis et al., 2004).

8 Conclusions

Before the advent of modern genetic analyses, precise classification of many inherited neurodegenerative diseases, especially among those that exhibited closely related disease phenotypes, was difficult and often resulted in contentious debate. Beginning in the early 1990s, the genetic basis of a large number of inherited neurodegenerative diseases began to be elucidated. Many of the inherited neurodegenerative diseases were found to be caused by trinucleotide expansions either in a noncoding region of the affected gene or in a coding region (Q_n-expansion diseases; A_n-expansion diseases). This genetic underpinning has essentially brought order out of anarchy and chaos (Margolis, 2002). Many inherited neurodegenerative diseases can now be classified on a rational nosology based on well-defined genetic mutations. A_n-expansions in mutated proteins result in neurological damage, but these diseases are invariably accompanied by cranial and somatic morphological defects. There are still many inherited neurodegenerative diseases for which a mutation has not yet been described. For example, there are many inherited SCAs whose genetic basis has not yet been determined. Possibly, at least some of these neurodegenerative diseases will be shown to be due to nucleotide-expansions. It will be interesting to determine whether some forms of purely psychiatric disease also can be firmly assigned to trinucleotide-expansion diseases.

Among the trinucleotide-expansion diseases, it is still not yet clear what accounts for the sensitivity of nervous tissue to the altered genotype and what accounts for the selective vulnerability of different brain regions among the various diseases. However, some progress is being made. A feature of the Q_n-expansion diseases and A_n-expansion diseases, is the presence of (or propensity to form) aberrant protein deposits in affected brain regions. In this respect, these diseases are similar to the more common neurodegenerative diseases such as PD and AD, which also exhibit aberrant protein deposits. To what extent these deposits play a role in the pathogenesis of AD and PD is still under debate. An understanding of the origin of aberrant protein deposits in the Q_n- and A_n-expansion diseases may suggest possible therapeutic strategies not only for these diseases, but also for the more common AD and PD.

Acknowledgments We thank Dr. John T. Pinto for helpful suggestions. Part of the work cited from the authors' laboratory was supported by NIH grant 2P01 AG14930.

References

Amato AA, Prior TW, Barohn RJ, Snyder P, Papp A, Mendell JR (1993) Kennedy's disease: a clinicopathologic correlation with mutations in the androgen receptor gene. Neurology 43: 791–794

Andreassen OA, Dedeoglu A, Ferrante RJ, Jenkins BG, Ferrante KL, Thomas M, Friedlich A, Browne SE, Schilling G, Borchelt DR, Hersch SM, Ross CA, Beal MF (2001) Creatine increases survival and delays motor symptoms in a transgenic mouse model of Huntington's disease. Neurobiol Dis 8:479–491

Bahl S, Virdi K, Mittal U, Sachdeva MP, Kalla AK, Holmes SE, O'Hearn E, Margolis RL, Jain S, Srivastava AK, Mukerji M (2005) Evidence of a common founder for SCA12 in the Indian population. Ann Hum Genet 69:528–534

Bailey CDC, Johnson GVW (2006) The protective effects of cystamine in the R6/2 Huntington's disease mouse involve mechanisms other than the inhibition of transglutaminase. Neurobiol Aging 27:871–879

Banno H, Katsuno M, Suzuki K, Takeuchi Y, Kawashima M, Suga N, Takamori M, Ito M, Nakamura T, Matsuo K, Yamada S, Oki Y, Adachi H, Minamiyama M, Waza M, Atsuta N, Watanabe H, Fujimoto Y, Nakashima T, Tanaka F, Doyu M, Sobue G (2009) Phase 2 trial of leuprorelin in patients with spinal and bulbar muscular atrophy. Ann Neurol 65:140–150

Bates G (2003) Huntingtin aggregation and toxicity in Huntington's disease. Lancet 361: 1642–1644

Bauer PO, Nukina N (2009) The pathogenic mechanisms of polyglutamine diseases and current therapeutic strategies. J Neurochem 110:1737–1765

Beal MF, Kowall NW, Ellison DW, Mazurek MF, Swartz KJ, Martin JB (1986) Replication of the neurochemical characteristics of Huntington's disease by quinolinic acid. Nature 321:168–171

Bithell A, Johnson R, Buckley NJ (2009) Transcriptional dysregulation of coding and non-coding genes in cellular models of Huntington's disease. Biochem Soc Trans 37:1270–1275

Bitoun E, Davies KE (2009) The robotic mouse: understanding the role of AF4, a cofactor of transcriptional elongation and chromatin remodelling, in Purkinje cell function. Cerebellum 8:175–183

Brook JD, McCurrach ME, Harley HG, Buckler AJ, Church D, Aburatani H, Stanton VP, Thirion JP, Hudson T, et al (1992) Molecular basis of myotonic dystrophy: expansion of a trinucleotide (CTG) repeat at the 3′ end of a transcript encoding a protein kinase family member. Cell 68:799–808

Brown LY, Brown SA (2004) Alanine tracts: the expanding story of human illness and trinucleotide repeats. Trends Genet 20:51–58

Browne SE (2008) Mitochondria and Huntington's disease pathogenesis: insight from genetic and chemical models. Ann N Y Acad Sci 1147:358–382

Browne SE, Beal MF (2004) The energetics of Huntington's disease. Neurochem Res 29:531–546

Browne SE, Bowling AC, MacGarvey U, Baik MJ, Berger SC, Muqit MMK, Bird ED, Beal MF (1997) Oxidative damage and metabolic dysfunction in Huntington's disease: selective vulnerability of the basal ganglia. Ann Neurol 41:646–653

Brustovetsky N, Brustovetsky T, Purl KJ, Capano M, Crompton M, Dubinsky JM (2003) Increased susceptibility of striatal mitochondria to calcium-induced permeability transition. J Neurosci 23:4858–4867

Campuzano A, Montermini L, Moltò MD, Pianese L, Cossée M, Cavalcanti F, Monros E, Rodius F, Duclos F, Monticelli A, Zara F, Cañizares J, Koutnikova H, Bidichandani SI, Gellera C, Brice A, Trouillas P, De Michele G, Filla A, De Frutos R, Palau F, Patel PI, Di Donato S, Mandel JL, Cocozza S, Koenig M, Pandolfo M (1996) Friedreich's ataxia: autosomal recessive disease caused by an intronic GAA triplet repeat expansion. Science 271:1423–1427

Cary GA, La Spada AR (2008) Androgen receptor function in motor neuron survival and degeneration. Phys Med Rehabil Clin N Am 19:479–494, viii

Caviston JP, Holzbaur EL (2009) Huntingtin as an essential integrator of intracellular vesicular trafficking. Trends Cell Biol 19:147–155

Chakrabarti L, Bristulf J, Foss GS, Davies KE (1998) Expression of the murine homologue of *FMR2* in mouse brain and during development. Hum Mol Genet 7:441–448

Chen M, Ona VO, Li M, Ferrante RJ, Fink KB, Zhu S, Bian J, Guo L, Farrell LA, Hersch SM, Hobbs W, Vonsattel J-P, Cha JH, Friedlander RM (2000) Minocycline inhibits caspase-1 and caspase-3 expression and delays mortality in a transgenic mouse model of Huntington disease. Nat Med 6:797–801

Chow MK, Mackay JP, Whisstock JC, Scanlon MJ, Bottomley SP (2004) Structural analysis of the Josephin domain of the polyglutamine protein ataxin-3. Biochem Biophys Res Commun 322:387–394

Conneally PM (1984) Huntington's disease: genetics and epidemiology. Am J Hum Genet 36: 506–526

Cooper AJL, Jeitner TM, Gentile V, Blass JP (2002) Cross linking of polyglutamine domains catalyzed by tissue transglutaminase is greatly favored with pathological length repeats: does transglutaminase activity play a role in $(CAG)_n/Q_n$-expansion diseases? Neurochem Int 40: 53–67

Cooper JM, Schapira AH (2003) Friedreich's ataxia: disease mechanisms, antioxidants and coenzyme Q_{10} therapy. Biofactors 18:163–171

Cornett J, Cao F, Wang CE, Ross CA, Bates GP, Li SH, Li XJ (2005) Polyglutamine expansion of huntingtin impairs nuclear export. Nat Genet 37:198–204

Cummings CJ, Reinstein E, Sun Y, Antalffy B, Jiang Y, Ciechanover A, Orr HT, Beaudet AL, Zoghbi HY (1998) Mutations of the E6-AP ubiquitin ligase reduces nuclear inclusion frequency while accelerating polyglutamine-induced pathology in SCA1 mice. Neuron 24: 879–892

Cummings CJ, Zoghbi HY (2000) Fourteen and counting: unraveling trinucleotide repeat disorders. Human Mol Genet 9:909–916

Cummings CJ, Zoghbi HY (2001) Trinucleotide repeats: mechanisms and pathophysiology. Annu Rev Genomics Hum Genet 1:281–328

Cusimano A, D'Adamo MC, Pessia M (2004) An episodic ataxia type-1 mutation in the S1 segment sensitizes the hKv1.1 potassium channel to extracellular Zn^{2+}. FEBS Lett 576: 237–244

Daughters RS, Tuttle DL, Gao W, Ikeda Y, Moseley ML, Ebner TJ, Swanson MS, Ranum LP (2009) RNA gain-of-function in spinocerebellar ataxia type 8. PLoS Genet 5(8):e1000600. Epub 2009 Aug 14

David G, Abbas N, Stevanin G, Dürr A, Yvert G, Cancel G, Weber C, Imbert G, Saudou F, Antoniou E, Drabkin H, Gemmill R, Giunti P, Benomar A, Wood N, Ruberg M, Agid Y, Mandel JL, Brice A (1997) Cloning of the SCA7 gene reveals a highly unstable CAG repeat expansion. Nat Genet 17:65–70

Dedeoglu AD, Kubilus JK, Jeitner TM, Matson SA, Bogdanov M, Kowall NW, Matson WR, Cooper AJL, Ratan RR, Beal MF, Hersch SM, Ferrante RJ (2002) Therapeutic effects of the transglutaminase inhibitor, cystamine, in a murine model of Huntington's disease. J Neurosci 22:8942–8950

D'Huist C, Kooy RF (2009) Fragile X syndrome: from molecular genetics to therapy. J Med Genet 46:577–584

Di Prospero NA, Fischbeck KH (2005) Therapeutic development for triplet repeat expansion diseases. Nat Rev Genet 6:756–765

Dürr A, Cossee M, Agid Y, Campuzano V, Mignard C, Penet C, Mandel JL, Brice A, Koenig M (1996) Clinical and genetic abnormalities in patients with Friedreich's ataxia. N Eng J Med 335:1169–1175

Emamian ES, Kaytor MD, Duvick LA, Zu T, Tousey SK, Zoghbi HY, Clark HB, Orr HT (2003) Serine 776 of ataxin-1 is critical for polyglutamine-induced disease in SCA1 transgenic mice. Neuron 38:375–387

Figueroa KP, Farooqi S, Harrup K, Frank J, O'Rahilly S, Pulst SM (2009) Genetic variance in the spinocerebellar ataxia type 2 (ATXN2) gene in children with severe early onset obesity. PLoS One 4:e8280

Foster TC, Kumar A (2002) Calcium dysregulation in the aging brain. Neuroscientist 8:297–301

Frints SG, Froyen G, Marynen P, Fryns JP (2002) X-linked mental retardation: vanishing boundaries between non-specific (MRX) and syndromic (MRXS) forms. Clin Genet 62:423–432

Gårseth M, Sonnewald U, White LR, Rød M, Zwart JA, Nygaard O, Aasly J (2000) Proton magnetic resonance spectroscopy of cerebrospinal fluid in neurodegenerative disease: indication of glial energy impairment in Huntington chorea, but not Parkinson disease. J Neurosci Res 60:779–782

Gatchel JR, Zoghbi HY (2005) Diseases of unstable repeat expansions. Nature Rev Genet 6: 743–755

Gécz J, Gedeon AK, Sutherland GR, Mulley JC (1996) Identification of the gene *FMR2*, associated with FRAXE mental retardation. Nat Genet 13:105–108

Greenamyre JT, Shoulson I (1994) Huntington's disease. In neurodegenerative diseases. In: Calne DB (ed) Neurodegenerative diseases. W. B. Saunders Company, New York, NY, pp 685–704

Gu Y, Nelson DL (2003) *FMR2* function: insight from a mouse knockout model. Cytogenet Genome Res 100:129–139

Guidetti P, Luthi-Carter RE, Augood SJ, Schwarcz R (2004) Neostriatal and cortical quinolinate levels are increased in early grade Huntington's disease. Neurobiol Dis 17:455–461

Gunawardena S, Goldstein LSB (2005) Polyglutamine diseases and transport problems. Deadly traffic jams and neuronal highways. Arch Neurol 62:46–51

Hagerman RJ, Leehey M, Heinrichs W, Tassone F, Wilson R, Hills J, Grigsby J, Gage B, Hagerman PJ (2001) Intention tremor, parkinsonism, and generalized brain atrophy in male carriers of fragile X. Neurology 57:127–130

Han JA, Park SC (2000) Transglutaminase-dependent modulation of transcription factor Sp1 activity. Mol Cells 10:612–618

Helmlinger D, Hardy S, Eberlin A, Devys D, Tora L (2006b) Both normal and polyglutamine-expanded ataxin-7 are components of TFTC-type GCN5 histone acetyltransferase-containing complexes. Biochem Soc Symp 73:155–163

Helmlinger D, Hardy S, Sasorith S, Klein F, Robert F, Weber C, Miguet L, Potier N, van-Dorsselaer A, Wurtz JM, Mandel JL, Tora L, Devys D (2004) Ataxin-7 is a subunit of GCN5 histone acetyltransferase-containing complexes. Hum Mol Genet 13:1257–1267

Helmlinger D, Tora L, Devys D (2006a) Transcriptional alterations and chromatin remodeling in polyglutamine diseases. Trends Genet 22:562–570

Hilditch-Maguire. P, Trettel F, Passsani LA, Auerbach A, Persichetti F, MacDonald ME (2000) Huntingtin: an iron-regulated protein essential for normal nuclear and perinuclear organelles. Hum Mol Genet 9:2789–2797

Holmes SE, O'Hearn E, Margolis RL (2003) Why is SCA12 different from other SCAs? Cytogenet Genome Res 100:189–197

Holmes SE, O'Hearn EE, McInnis MG, Gorelick-Feldman DA, Kleiderlein JJ, Callahan C, Kwak NG, Ingersoll-Ashworth RG, Sherr M, Sumner AJ, Sharp AH, Ananth U, Seltzer WK, Boss MA, Vieria-Saecker AM, Epplen JT, Riess O, Ross CA, Margolis RL (1999) Expansion of a novel CAG trinucleotide repeat in the 5′ region of *PPP2R2B* is associated with SCA12. Nat Genet 23:391–392

Ikeda Y, Dalton JC, Moseley ML, Gardner KL, Bird TD, Ashizawa T, Seltzer WK, Pandolfo M, Milunsky A, Potter NT, Shoji M, Vincent JB, Day JW, Ranum LP (2004) Spinocrebellar ataxia type 8: molecular genetic comparisons and haplotype analysis of 37 families with ataxia. Am J Hum Genet 75:3–16

Jiang H, Mankodi A, Swanson MS, Moxley RT, Thornton CA (2004) Myotonic dystrophy type I associated with nuclear foci of mutant RNA, sequestration of muscleblind proteins, and deregulated alternative splicing in neurons. Hum Mol Genet 13:3079–3088

Kahlem P, Green H, Djian P (1998) Transglutaminase action imitates Huntington's disease: selective polymerization of Huntingtin containing expanded polyglutamine. Mol Cell 1:595–601

Karpuj MV, Becher MW, Springer JE, Chabas D, Youssef S, Pedotti R, Mitchell D, Steinman L (2002) Prolonged survival and decreased abnormal movements in transgenic model of Huntington disease, with administration of the transglutaminase inhibitor cystamine. Nat Med 8:143–149

Karpuj MV, Garren H, Slunt H, Price DL, Gusella J, Becher MW, Steinman L (1999) Transglutaminase aggregates huntingtin into nonamyloidogenic polymers, and its enzymatic activity increases in Huntington's disease nuclei. Proc Natl Acad Sci USA 96:7388–7393

Kawaguchi Y, Okamoto T, Taniwaki M, Aizawa M, Inoue M, Katayama S, Kawakami H, Nakamura S, Nishimura M, Akiguchi I, et al (1994) CAG expansions in a novel gene for Machado-Joseph disease at chromosome 14q32.1. Nat Genet 8:221–228

Kaytor MD, Wilkinson KD, Warren ST (2004) Modulating huntingtin half-life alters polyglutamine-dependent aggregate formation and cell toxicity. J Neurochem 89:962–973

Koide R, Ikeuchi T, Onodera O, Tanaka H, Igarishi S, Endo K, Takahashi H, Kondo R, Ishikawa A, Hayashi T (1994) Unstable expansion of CAG repeat in hereditary dentatorubral-pallidoluysian atrophy (DRPLA). Nat Genet 6:9–13

Koide R, Kobayashi S, Shimohata T, Ikeuchi T, Maruyama M, Saito M, Yamada M, Takahashi H, Tsuji S (1999) A neurological disease caused by an expanded CAG trinucleotide repeat in the TATA-binding protein gene: a new polyglutamine disease? Hum Mol Genet 8: 2047–2053

Koob MD, Moseley ML, Schut LJ, Benzow KA, Bird TD, Day JW, Ranum LP (1999) An untranslated CTG expansion causes a novel form of spinocerebellar ataxia (SCA8). Nat Genet 21:379–384

Korade-Mirnics Z, Babitzke P, Hoffman E (1998) Myotonic dystrophy: molecular windows on a complex etiology. Nucleic Acids Res 26:1363–1368

Koroshetz WJ, Martin JB (1997) Huntington's disease. In: Rosenberg RN, Prusiner SB, DiMauro S, Barchi RL (eds) The molecular and genetic basis of neurological disease. Butterworth-Heinemann, Boston, MA, pp 545–564

Kremer EJ, Pritchard M, Lynch M, Yu S, Holman K, Baker E, Warren ST, Schlessinger D, Sutherland GR, Richards RI (1991) Mapping of DNA instability at the fragile X to a trinucleotide repeat sequence $p(CCG)_n$. Science 252:1711–1714

Kuemmerle S, Gutekanst C-A, Klein AM, Li X-J, Li SH, Beal MF, Hersch SM, Ferrante RJ (1999) Huntingtin aggregates may not predict neuronal death in Huntington's disease. Ann Neurol 46:842–849

La Spada AR, Roling DB, Harding AE, Warner CL, Spiegel R, Hausmanowa-Petrusewicz I, Yee WC, Fischbeck KH (1992) Meiotic stability and genotype-phenotype correlation of the trinucleotide repeat in X-linked spinal and bulbar muscular atrophy. Nat Genet 2: 301–304

La Spada AR, Wilson EM, Lubahn DB, Harding AE, Fischbeck KH (1991) Androgen receptor gene mutations in X-linked spinal and bulbar muscular atrophy. Nature 352:77–79

Lai TS, Tucker T, Burke JR, Strittmatter WJ, Greenberg CS (2004) Effect of tissue transglutaminase on the solubility of proteins containing expanded polyglutamine repeats. J Neurochem 88:1253–1260

Lane DJ, Richardson DR (2010) Frataxin, a molecule of mystery: trading stability for function in its iron-binding site. Biochem J 426:e1–3

Lee JE, Cooper TA (2009) Pathogenic mechanisms of myotonic dystrophy. Biochem Soc Trans 37:1281–1286

Lee WW, Kim SY, Kim JY, Kim HJ, Park SS, Jeon BS (2009) Extrapyramidal signs are a common feature of spinocerebellar ataxia type 17. Neurology 73:1708–1709

Lesca G, Biancalana V, Brunel MJ, Quack B, Calender A, Lespinasse J (2003) Clinical, cytogenetic, and molecular description of a FRAXE French family. Psychiatr Genet 13: 43–46

Lesort M, Lee M, Tucholski J, Johnson GVW (2003) Cystamine inhibits caspase activity. Implications for the treatment of polyglutamine disorders. J Biol Chem 278:3825–3830

Li SH, Li XJ (2004) Huntingtin-protein interactions and the pathogenesis of Huntington's disease. Trends Genet 20:146–154

Limprasert P, Nouri N, Heyman RA, Nopparatana C, Kamonslip M, Deininger PL, Keats BJ (1996) Analysis of CAG repeat of the Machado-Joseph gene in human, chimpanzee and monkey populations: a variant nucleotide is associated with the number of CAG repeats. Hum Mol Genet 5:207–213

Liu J, Tang TS, Tu H, Nelson O, Herndon E, Huynh DP, Pulst SM, Bezprozvanny I (2009) Deranged calcium signaling and neurodegeneration in spinocerebellar ataxia type 2. J Neurosci 29:9148–9962

Lodi R, Schapira AH, Manners D, Styles P, Wood NW, Taylor DJ, Warner TT (2000) Abnormal in vivo skeletal muscle energy metabolism in Huntington's disease and dentatorubralpallidoluysian atrophy. Ann Neurol 48:72–76

Lukusa T, Fryns JP (2008) Human chromosome fragility. Biochim Biophys Acta 1779:3–16

Maldonado PD, Molina-Jijón E, Villeda-Hernández J, Galván-Arzate S, Santamaría A, Pedraza-Chaverrí J (2010) NAD(P)H oxidase contributes to neurotoxicity in an excitotoxic/prooxidant model of Huntington's disease in rats: protective role of apocynin. J Neurosci Res 88: 620–629

Mandrusiak LM, Beitel LK, Wang X, Scanlon TC, Chevalier-Larsen E, Merry DE, Trifiro MA (2003) Transglutaminase potentiates ligand-dependent proteasome dysfunction induced by polyglutamine-expanded androgen receptor. Hum Mol Genet 12:1497–1506

Mankodi A (2008) Myotonic dystrophies. Neurol India 56:298–304

Margolis RL (2002) The spinocerebellar ataxias: order emerges from chaos. Curr Neurol Sci Rep 2:447–456

Margolis RL, Holmes SE, Rosenblatt A, Gourley L, O'Hearn E, Ross CA, Seltzer WK, Walker RH, Ashizawa T, Rasmussen A, Hayden M, Almqvist EW, Harris J, Fahn S, Macdonald ME, Mysore J, Shimohata T, Tsuji S, Potter N, Nakaso K, Adachi Y, Nakashima K, Bird T, Krause A, Greenstein P (2004) Huntington's disease like 2 (HDL-2) in North America and Japan. Ann Neurol 56:670–674

Margolis RL, Li SH, Young WS, Wagster MV, Stine OC, Kidwai AS, Ashworth RG, Ross CA (1996) DRPLA gene (atrophin-1) sequence and mRNA expression in human brain. Brain Res Mol Brain Res 36:219–226

März P, Probst A, Lang S, Schwager M, Rose-John S, Otten U, Ozbek S (2004) Ataxin-10, the spinocerebellar ataxia type 10 neurodegenerative disorder protein, is essential for survival of cerebellar neurons. J Biol Chem 279:35542–35550

Mastroberardino PG, Iannocola C, Nardacci R, Bernassola F, De Lurenzi V, Melino G, Moreno S, Pavone F, Oliverio S, Fésüs L, Piacentini M (2002) 'Tissue' transglutaminase ablation reduces neuronal death and prolongs survival in a mouse model of Huntington's disease. Cell Death Differ 9:873–880

Matsuura T, Yamagata T, Burgess DL, Rasmussen A, Grewal RP, Watase K, Khajavi M, McCall AE, Davis CF, Zu L, Achari M, Pulst SM, Alonso E, Noebels JL, Nelson DL, Zoghbi HY, Ashizawa T (2000) Large expansion of the ATTCT pentanucleotide repeat in spinocerebellar ataxia type 10. Nat Genet 26:191–194

Matsuyama Z, Yanagisawa NK, Aoki Y, Black JL 3rd, Lennon VA, Mori Y, Imoto K, Inuzuka T (2004) Polyglutamine repeats of spinocerebellar ataxia 6 impair the cell-death-preventing effect of CaV2.1 Ca^{2+} channel – loss of-function cellular model of SCA6. Neurobiol Dis 17: 198–204

Meola G, Moxley RT 3rd (2004) Myotonic dystrophy type 2 and related myotonic disorders. J Neurol 251:1173–1182

Messaed C, Rouleau GA (2009) Molecular mechanisms underlying polyalanine diseases. Neurobiol Dis 34:397–405

Miller RC, Tewari A, Miller JA, Garbern J, Van Stavern GP (2009) Neuro-ophthalmologic features of spinocerebellar ataxia type 7. J Neuroophthalmol 29:180–186

Milnerwood AJ, Gladding CM, Pouladi MA, Kaufman AM, Hines RM, Boyd JD, Ko RW, Vasuta OC, Graham RK, Hayden MR, Murphy TH, Raymond LA (2010) Early increase in extrasynaptic NMDA receptor signaling and expression contributes to phenotype onset in Huntington's disease mice. Neuron 65:178–190

Mitra S, Tsvetkov AS, Kinkbeiner S (2009) Protein turnover and inclusion body formation. Autophagy 5:1037–1038

Mizutani A, Wang L, Rajan H, Vig PJ, Alaynick WA, Thaler JP, Tsai CC (2005) Boat, an AXH domain protein, suppresses the cytotoxicity of mutant ataxin-1. EMBO J 24:3339–3351

Mutsuddi M, Marshall CM, Benzow KA, Koob MD, Rebay I (2004) The Spinocerebellar ataxia 8 noncoding RNA causes neurodegeneration and associates with stauffen in *Drosphila*. Curr Biol 14:302–308

Nagafuchi S, Yanagisawa H, Sato K, Shirayama T, Ohsaki E, Bundo M, Takeda T, Tadokoro K, Kondo I, Murayama N, et al (1994) Dentatorubral and pallidoluysian atrophy expansion of an unstable CAG trinucleotide on chromosome 12p. Nat Genet 6:14–18

Nasrallah IM, Minarcik JC, Golden JA (2004) A polyalanine tract expansion in Arx forms intranuclear inclusions and results in increased cell death. J Cell Biol 167:411–416

Nguyen HD, Hall CK (2004) Molecular dynamics of spontaneous fibril formation by random-coil peptides. Proc Natl Acad Sci USA 101:16180–16185

Nicholls DG (2009) Mitochondrial calcium function and dysfunction in the central nervous system. Biochim Biophys Acta 1787:1416–1424

Obrietan K, Hoyt KR (2004) CRE-mediated transcription is increased in Huntington's disease transgenic mice. J Neurosci 24:791–796

Ona VO, Li M, Vonsatell J-P, Andrews LJ, Khan SQ, Chung WM, Frey AS, Menon AS, Li XJ, Stieg PE, Yuan J, Penney JB, Young AB, Cha JH, Friedlander RM (1999) Inhibition of caspase-1 slows disease progression in a mouse model of Huntington's disease. Nature 399: 263–267

Ophoff RA, Terwindt GM, Vergouwe MN, van Eijk R, Oefner PJ, Hoffman SM, Lamerdin JE, Mohrenweiser HW, Bulman DE, Ferrari M, Haan J, Lindhout D, van Ommen GJ, Hofker MH, Ferrari MD, Frants RR (1996) Familial hemiplegic migraine and episodic ataxia type-2 are caused by mutations in the Ca^{2+} channel gene *CACNL1A4*. Cell 87:543–552

Orr HT, Chung MY, Banfi S, Kwiatkowski TJ, Jr Servadio A, Beaudet AL, McCall AE, Duvick LA, Ranum LP, Zoghbi HY (1993) Expansion of an unstable trinucleotide CAG repeat in spinocerebellar ataxia type 1. Nat Genet 4:221–226

Ortega Z, Díaz-Hernández M, Lucas JJ (2007) Is the ubiquitin-proteasome system impaired in Huntington's disease? Cell Mol Life Sci 64:2245–2257

Pandolfo M, Pastore A (2009) The pathogenesis of Friedreich ataxia and the structure and function of frataxin. J Neurol 256(Suppl 1):9–17

Panov AV, Burke JR, Strittmatter WJ, Greenamayre JT (2003) In vitro effects of polyglutamine tracts on Ca^{2+}-dependent depolarization of rat and human mitochondria: relevance to Huntington's disease. Arch Biochem Biophys 410:1–6

Panov AV, Gutekunst CA, Leavitt BR, Hayden MR, Burke JR, Strittmatter WJ, Greenamayre JT (2002) Early mitochondrial calcium defects in Huntington's disease are a direct effect of polyglutamines. Nat Neurosci 5:731–736

Park SC, Yeo EJ, Han JA, Hwang YC, Choi JY, Park JS, Park YH, Kim KO, Kim IG, Seong SC, Kwak SJ (1999) Aging process is accompanied by increase in transglutaminase C. J Gerentol A Biol Sci Med 54:B78–B83

Pearson CE, Nichol Edamura K, Cleary JD (2005) Repeat instability: mechanisms of dynamic mutations. Nature Rev Genet 6:729–742

Perutz MF, Johnson T, Suzuki M, Finch JT (1994) Glutamine repeats as polar zippers: their possible role in inherited neurodegenerative diseases. Proc Natl Acad Sci USA 91: 5355–5358

Perutz MF, Windle AH (2001) Cause of neural death in neurodegenerative diseases attributable to expansion of glutamine repeats. Nature 412:143–144

Pinto JT, Van Raamsdonk J, Leavitt BR, Hayden MR, Thaler HT, Jeitner TM, Krasnikov BF, Cooper AJL (2005) Treatment of YAC128 mice with cystamine does not lead to its accumulation in plasma or brain: implications for the treatment of Huntington disease. J Neurochem 94:1087–1101

Quintanilla RA, Johnson GVW (2009) Role of mitochondrial dysfunction in the pathogenesis of Huntington's disease. Brain Res Bull 80:242–247

Ratovitski T, Gucek M, Jiang H, Chighladze E, Waldron E, D'Ambola J, Hou Z, Liang Y, Poirier MA, Hirschhorn RR, Graham R, Hayden MR, Cole RN, Ross CA (2009) Mutant huntingtin N-terminal fragments of specific size mediate aggregation and toxicity in neuronal cells. J Biol Chem 284:10855–10867

Reddy PH, Mao P, Manczak M (2009) Mitochondrial structural and functional dynamics in Huntington's disease. Brain Res Rev 61:33–48

Renna M, Jimenez-Sanchez M, Sarkar S, Rubinsztein DC (2010) Chemical inducers of autophagy that enhance the clearance of mutant proteins in neurodegenerative diseases. J Biol Chem 285:11061–11067

Restituito S, Thompson RM, Eliet J, Raike RS, Riedl M, Charnet P, Gomez CM (2000) The polyglutamine expansion in spinocerebellar ataxia type 6 causes a β subunit specific enhanced activation of P/Q-type calcium channels in *Xenopus* oocytes. J Neurosci 20:6394–6403

Rodrigues GG, Walker RH, Brice A, Cazeneuve C, Russaouen O, Teive HA, Munhoz RP, Becker N, Raskin S, Werneck LC, Junior WM, Tumas V (2008) Huntington's disease-like 2 in Brazil – report of 4 patients. Mov Disord 23:2244–2247

Rolfs A, Koeppen AH, Bauer I, Bauer P, Buhlmann S, Topka H, Schöls L, Riess O (2003) Clinical features and neuropathology of autosomal dominant spinocerebellar ataxia (SCA17). Ann Neurol 54:367–373

Rose C, Menzies FM, Renna M, Acevedo-Arozena A, Corrochano S, Sadiq O, Brown SD, Rubinsztein DC (2010) Rilmenidine attenuates toxicity of polyglutamine expansions in a mouse model of Huntington's disease. Hum Mol Genet 2010 Mar 17. [Epub ahead of print]

Rosenberg RN (1996) DNA-triplet repeats and neurologic disease. New Engl J Med 335: 1222–1224

Ross CA (2002) Polyglutamine pathogenesis: emergence of unifying mechanisms for Huntington's disease and related disorders. Neuron 35:819–822

Ross CA, Poirier MA (2004) Protein aggregation and neurodegenerative disease. Nat Med 10: S10–S17

Ross CA, Shoulson I (2009) Huntington disease. pathogenesis, biomarkers, and approaches to experimental therapeutics. Parkinsonism Rel Disord 15:S135–S138

Rüb U, Brunt ER, Deller T (2008) New insights into the pathoanatomy of spinocerebellar ataxia type 3 (Machado-Joseph disease). Curr Opin Neurol 21:111–116

Sarafidou T, Kahl C, Martinez-Garay I, Mangelsdorf M, Gesk S, Baker E, Kokkinaki M, Talley P, Maltby EL, French L, Harder L, Hinzmann B, Nobile C, Richkind K, Finnis M, Deloukas P, Sutherland GR, Kutsche K, Moschonas NK, Siebert R, Gécz J, and European Collaborative Consortium for the Study of ADLTE (2004) Folate-sensitive fragile site FRA10A is due to an expansion of a CGG repeat in a novel gene FRA10AC1, encoding a nuclear protein. Genomics 84:69–81

Sato N, Amino T, Kobayashi K, Asakawa S, Ishiguro T, Tsunemi T, Takahashi M, Matsuura T, Flanigan KM, Iwasaki S, Ishino F, Saito Y, Murayama S, Yoshida M, Hashizume Y, Takahashi Y, Tsuji S, Shimizu N, Toda T, Ishikawa K, Mizusawa H (2009) Spinocerebellar ataxia type 31 is associated with "inserted" penta-nucleotide repeats containing $(TGGAA)_n$. Am J Hum Genet 85:544–557

Saudou F, Finkbeiner S, Devys D, Greenberg ME (1998) Huntingtin acts in the nucleus to induce apoptosis but death does not correlate with the formation of intranuclear inclusions. Cell 95:55–66

Sawa A (2001) Mechanisms for neuronal cell death and dysfunction in Huntington's disease: pathological cross-talk between the nucleus and the mitochondria? J Mol Med 79:375–381

Schaffar G, Breuer P, Boteva R, Behrends C, Tzvetkov N, Strippel N, Sakahira H, Siegers K, Hayer-Hartl M, Hartl FU (2004) Cellular toxicity of polyglutamine expansion proteins: mechanism of transcription factor deactivation. Mol Cell 15:95–105

Scherzinger E, Sittler A, Schweiger K, Heiser V, Lurz R, Hasenbank R, Bates GP, Lehrach H, Wanker EE (1999) Self-assembly of polyglutamine-containing huntingtin fragments into amyloid-like fibrils: implications for Huntington's disease pathology. Proc Natl Acad Sci USA 96:4604–4609

Schilling G, Savonenko AV, Klevytska A, Morton JL, Tucker SM, Poirier M, Gale A, Chan N, Gonzales V, Slunt HH, Coonfield ML, Jenkins NA, Copeland NG, Ross CA, Borchelt DR (2004) Nuclear-targeting of mutant huntingtin fragments produces Huntington's disease-like phenotypes in transgenic mice. Hum Mol Genet 13:1599–1610

Schweitzer JK, Krivda JP, D'Souza-Schorey C (2009) Neurodegeneration in Niemann-Pick Type C disease and Huntington's disease: impact of defects in membrane trafficking. Curr Drug Targets 10:653–665

Seidel K, Brunt ER, de Vos RA, Dijk F, van der Want HJ, Rüb U, den Dunnen WF (2009) The p62 antibody reveals various cytoplasmic protein aggregates in spinocerebellar ataxia type 6. Clin Neuropathol 28:344–349

Serra HG, Byam CE, Lande JD, Tousey SK, Zoghbi HY, Orr HT (2004) Gene profiling links SCA1 pathophysiology to glutamate signaling in Purkinje cells of transgenic mice. Hum Mol Genet 13:2535–2543

Sharma R, De Biase I, Gómez M, Delatycki MB, Ashizawa T, Bidichandani SI (2004) Friedreich ataxia in carriers of unstable borderline GAA triplet-repeat alleles. Ann Neurol 56:898–901

Shelbourne P, Johnson K (1992) Myotonic dystrophy: another case of too many repeats? Hum Mutat 1:183–189

Shoesmith Berke SJ, Flores Schmied FA, Brunt ER, Ellerby LM, Paulson HL (2004) Caspase-mediated proteolysis of the polyglutamine disease protein ataxin-3. J Neurochem 89:908–918

Smith R, Bacos K, Fedele V, Soulet D, Walz HA, Obermüller S, Lindqvist A, Björkqvist M, Klein P, Onnerfjord P, Brundin P, Mulder H, Li JY (2009) Mutant huntingtin interacts with β-tubulin and disrupts vesicular transport and insulin secretion. Hum Mol Genet 18:3942–3954

Spindler M, Beal MF, Henchcliffe C (2009) Coenzyme Q10 effects in neurodegenerative diseases. Neuropsychiatr Dis Treat 5:597–610

Steffan JS, Arawal N, Pallos J, Rockabrand E, Trotman LC, Slepko N, Illes K, Lukacsovich T, Zhu YZ, Cattaneo E, Pandolfi PP, Thompson LM, Marsh JL (2004) SUMO modification and Huntington's disease pathology. Science 304:100–104

Strelnikov V, Nemtsova M, Chesnokova G, Kuleshov N, Zaletayev D (1999) A simple multiplex FRAXA FRAXE, and FRAXF PCR assay convenient for wide screening programs. Hum Mutat 13:166–169

Su B, Wang X, Zheng L, Perry G, Smith MA, Zhu X (2010) Abnormal mitochondrial dynamics and neurodegenerative diseases. Biochim Biophys Acta 1892:135–142

Sugars KL, Rubinsztein DC (2003) Transcriptional abnormalities in Huntington disease. Trends Genet 19:233–238

Sutherland GR (1979a) Heritable fragile sites on human chromosomes. I. Factors affecting expression in lymphocyte culture. Am J Hum Genet 31:125–135

Sutherland GR (1979b) Heritable fragile sites on human chromosomes. II. Distribution, phenotypic effects, and cytogenetics. Am J Hum Genet 31:136–148

Sutherland GR (2003) Rare fragile sites. Cytogenet Genome Res 100:77–84

Suzuki K, Kastuno M, Banno H, Sobue G (2009) Pathogenesis-targeting therapeutics for spinal and bulbar muscular atrophy (SBMA). Neuropathology 29:509–516

Tabrizi SJ, Cleeter MW, Xuereb J, Taanman JW, Cooper JM, Schapira AH (1999) Biochemical abnormalities and excitotoxicity in Huntington's disease brain. Ann Neurol 45:25–32

Tabrizi SJ, Workman J, Hart PE, Mangiarini L, Mahal A, Bates G, Cooper JM, Schapira AH (2000) Mitochondrial dysfunction and free radical damage in the Huntington R6/2 transgenic mouse. Ann Neurol 47:80–86

The Huntington Disease Collaborative Research Group. (1993) A novel gene containing a trinucleotide repeat that is expanded and unstable on Huntington disease chromosome. Cell 72:971–983

Uitti RJ (1994) Cerebellar degenerations. In neurodegenerative diseases. In: Calne DB (ed) Neurodegenerative diseases. W. B. Saunders Company, New York, NY, pp 787–806

Van Raamsdonk JM, Pearson J, Bailey CDC, Rogers DA, Johnson GVW, Hayden MR, Leavitt BR (2005) Cystamine treatment is neuroprotective in the YAC128 mouse model of Huntington disease. J Neurochem 95:210–220

Vincent JB, Yuan QP, Schalling M, Adolfsson R, Azevedo MH, Macedo A, Bauer A, DallaTorre C, Medeiros HM, Pato MT, Pato CN, Bowen T, Guy CA, Owen MJ, O'Donovan MC, Paterson AD, Petronis A, Kennedy JL (2000) Long repeat tracts in major psychosis. Am J Med Genet (Neuropsychiatric Genetics) 96:873–876

Wakamiya M, Matsuura T, Liu Y, Schuster GC, Gao R, Xu W, Sarkar PS, Lin X, Ashizawa T (2006) The role of ataxin 10 in the pathogenesis of spinocerebellar ataxia type 10. Neurology 67:607–613

Wang J, Wang CE, Orr A, Tydlacka S, Li SH, Li XJ (2008) Impaired ubiquitin-proteasome system activity in the synapses of Huntington's disease mice. J Cell Biol 180:1177–1189

Wang X, Wang H, Xia Y, Jiang H, Shen L, Wang S, Shen R, Xu Q, Luo X, Tang B (2010) Spinocerebellar ataxia type 6: systematic patho-anatomical study reveals different phylogenetically defined regions of the cerebellum and neural pathways undergo different evolutions of the degenerative process. Neuropathology 2010 Jan 26. [Epub ahead of print]

Waragai M, Lammers CH, Takeuchi S, Imafuku I, Udagawa Y, Kanazawa I, Kawabata M, Mouradian MM, Okazawa H (1999) PQBP-1, a novel polyglutamine tract-binding protein, inhibits transcription activation by BRn-2 and affects cell survival. Hum Mol Gent 8: 977–987

Warby SC, Doty CN, Graham RK, Shively J, Singaraja RR, Hayden MR (2009) Phosphorylation of huntingtin reduces the accumulation of its nuclear fragments. Mol Cell Neurosci 40: 121–127

The US – Venezuela Collaborative Research Project and Wexler NS (2004) Venezuelan kindreds reveal that genetic and environmental factors modulate Huntington's disease age of onset. Proc Natl Acad Sci USA 101:3498–3503

Wu LL, Fan Y, Li S, Li XJ, Zhou XF (2010) Huntingtin-associated protein-1 interacts with pro-brain-derived neurotrophic factor and mediates its transport and release. J Biol Chem 285:5614–5623

Xifró X, Giralt A, Saavedra A, García-Martínez JM, Díaz-Hernández M, Lucas JJ, Alberch J, Pérez-Navarro E (2009) Reduced calcineurin protein levels and activity in exon-1 mouse models of Huntington's disease: role in excitotoxicity. Neurobiol Dis 36:461–469

Zainelli GM, Dudeck NL, Ross CA, Kim SY, Muma NA (2005) Mutant huntingtin protein: a substrate for transglutaminase 1, 2, and 3. J Neuropathol Exp Neurol 64:58–65

Zainelli GM, Ross CA, Troncos JC, Muma NA (2003) Transglutaminase cross-links in intranuclear inclusions in Huntington disease. J Neuropathol Exp Neurol 62:14–24

Zalfa F, Bagni C (2004) Molecular insights into mental retardation: multiple functions for the fragile X mental retardation protein? Curr Issues Mol Biol 6:73–88

Zhuchenko O, Bailey J, Bonnen P, Ashizawa T, Stockton DW, Amos C, Dobyns WB, Subramony SH, Zoghbi HY, Lee CC (1997) Autosomal dominant cerebellar ataxia (SCA6) associated with small polyglutamine expansions in the α_{1A}-voltage-dependent calcium channel. Nat Genet 15:62–69

Zoghbi HY, Orr HT (2009) Pathogenic mechanisms of a polyglutamine-mediated neurodegenerative disease, spinocerebellar ataxia type 1. J Biol Chem 284:7425–7429

Zoltewicz JS, Stewart NJ, Leung R, Peterson AS (2004) Atrophin 2 recruits histone deacylase and is required for the function of multiple signaling centers during mouse embryogenesis. Development 131:3–14

Zuccato C, Tartari M, Crotti A, Goffredo D, Valenza M, Coni L, Cataudella T, Leavitt BR, Hayden MR, Timmusk T, Rigamonti D, Cattaneo E (2003) Huntingtin interacts with REST/NRSF to modulate the transcription of NRSE-controlled neuronal genes. Nat Gen 35:76–83

CNS Cytokines

Jane Kasten-Jolly and David A. Lawrence

Abstract The first description of the inflammatory process appeared as early as the first century AD. Among the first things learned about inflammation is that vascular permeability is increased and leukocyte extravasation occurs. It is now realized that the central nervous system (CNS) is not as devoid of immune cell entrance as once believed and that neuroinflammation can occur. Even in the CNS absence of peripheral immune cells, cytokines from the periphery can influence glial activation in response to endogenous or exogenous stresses. Activated glial cells will secrete proinflammatory cytokines among other factors. The presence of relatively high concentrations of proinflammatory cytokines, such as IL-1, IL-6, and TNF-α, in the brain produces sickness behavior. Neuroinflammation is not only caused by viral or bacterial infection, but can also be the result of physical injury or neurodegenerative diseases, such as Alzheimer's disease, Parkinson's disease, multiple sclerosis, and cerebral palsy. This chronic neuroinflammation is associated with a number of common factors; most notable among these is the increased concentration of proinflammatory cytokines. In addition to the ones listed above, others have been detected including, IL-18, IL-33, and HMGB1. Although TGF-β1 functions most often as an anti-inflammatory cytokine, under certain circumstances it, too, can have proinflammatory activity. Other common features of neuroinflammation include increased production of reactive oxygen species (ROS) and nitric oxide (NO), which function to increase apoptosis and promote neuronal damage. Activation of astrocytes is detected by elevated GFAP expression. Activated astrocytes promote chemokine expression causing permeability of the blood–brain barrier (BBB), thus allowing leukocytes to enter the brain tissue. The heavy metal Pb accumulates in glial cells and in doing so can potentiate cytokine and glutamate-mediated increases in the BBB permeability, as well as cause chronic glial cell activation. Pb's ability to promote gliosis and deficiencies in chaperone protein function has prompted a comparison of Pb toxicity to certain neurodegenerative disorders, such as Alzheimer's

D.A. Lawrence (✉)
New York State Department of Health, Wadsworth Center, Albany, NY 12201, USA
e-mail: david.lawrence@wadsworth.org

J.P. Blass (ed.), *Neurochemical Mechanisms in Disease*,
Advances in Neurobiology 1, DOI 10.1007/978-1-4419-7104-3_12,

and Parkinson's diseases. Toxicity of other metals, such as, Al, Cu, Cd, Zn, and Hg was also found to share common features with Alzheimer's disease.

Keywords Alzheimer's disease · Astrocytes · Blood–brain barrier (BBB) · Central nervous system (CNS) · Chaperone proteins · Chemokines · Cytokines · Heavy metals · Inflammation · Lead (Pb) · Map kinases · Microglia · Neurons · Nitric oxide (NO) · Reactive oxygen species (ROS)

Contents

1 Introduction

The inflammatory process was first described by Cornelius Celsus in the first century AD when he listed the four signs of inflammation: *rubor et tumor cum calore et dolore* (redness and swelling with heat and pain). A fifth sign of inflammation has been annexed to the cardinal four, *functio laesa* (disruption of normal function); although its origin has been questioned (Rather, 1971), it is the disturbance of normal function that may lead to the potential pathological consequences of inflammation. The original four signs of inflammation were based mainly on peripheral skin observations. It wasn't until the nineteenth and twentieth centuries that the cellular aspects of inflammation began to be understood. Julius Cohnheim observed that in the inflamed tissue there was increased vasodilation along with vascular permeability and leukocyte extravasation. Sir Thomas Lewis noted that the vascular changes were mediated by chemical substances present in serum (Cotran et al., 1994). It was not realized until much later that inflammation could occur in hidden places, such as the central nervous system (CNS) and that inflammation could be initiated by endogenous or exogenous agents. With advancements in molecular and cellular methodology during the past few decades, much has been learned about the mediators of the inflammation process. Much of this knowledge has been obtained through the study of disease-related inflammation (Sheng et al., 1996; Akiyama et al., 2000; Combs et al., 2000; Cagnin et al., 2002; Rosales-Corral et al., 2004; Eikekenboom et al., 2006; Shepherd et al., 2006; Zipp and Aktas, 2006; Lassmann, 2007), injury, and animal models (Jafarian-Tehrani and Sternberg, 1999;

Kielian and Hickey, 2000). Although inflammation usually has positive influences in defense against pathogens, the processes can cause cellular damage and initiate or exacerbate various pathologies, especially autoimmune diseases and neurobehavioral diseases in susceptible humans and animals. Inflammation is closely associated with oxidative stress, and chronic inflammation along with its accompanying oxidant damage can lead to an acceleration of aging-related changes, which includes cytotoxicity (Norris et al., 2005). The present chapter is restricted to discussion of neuroinflammation, including information about the roles of cytokine, chemokines, MAP kinases, glia, and reactive oxygen species. Closing paragraphs present a discussion of metal neurotoxicity, with emphasis on the immune- and neurotoxicant lead (Pb), and the neuroinflammatory aspects of this toxicity.

1.1 Cytokines and Neuroinflammation

1.1.1 Interleukin (IL)-1, IL-6, and TNF-α

The proinflammatory cytokines, IL-6, IL-1β, and TNF-α, are associated with sickness behavior and their levels increase in the CNS during viral or bacterial infections (Quan et al., 1999; Pollmacher et al., 2002; Anisman, 2004; Dantzer et al., 2007; 2008). The sickness behavior paradigm of lethargy (malaise and lack of mobility), loss of appetite and drinking, and altered body temperature (fever) is due to the signals of proinflammation cytokines delivered to the hypothalamus (Woiciechowsky et al., 1999). The brain monitors peripheral innate immune responses by several methods (Quan and Banks, 2007). One pathway involves activation of afferent nerves by locally produced cytokines and pathogen-associated molecular patterns (PAMPs), such as vagal nerve, during an abdominal infection. A second mechanism involves the humoral (antibody-mediated) pathway and Toll-like receptors (TLRs) located on macrophage-like cells residing in the circumventricular organs and the choroid plexus (Rao et al., 2005). These cells respond to immune complexes of antibody:antigen and PAMPs by producing proinflammatory cytokines. These cytokines then enter the brain by volume diffusion. Third, the overflow of cytokines in the systemic circulation can gain entry to the brain via cytokine transporters at the blood–brain barrier (Matyszak, 1998; Karman et al., 2006; Niederkorn, 2006). A fourth pathway involves activation of IL-1 receptors located on perivascular macrophages and endothelial cells of brain venules resulting in production of prostaglandin E2 by upregulation of cyclooxygenase (COX) enzyme synthesis (Pasinetti, 1998; Lacroix and Rivest, 1998; Li et al., 2001; Konsman et al., 2004; Sapirstein et al., 2005; Inoue et al., 2006). These prostaglandins will diffuse to the brain targets where they will alter the setpoint for various regulatory processes (Dantzer et al., 2007). However, it was noted that not all of the effects of prostaglandin E_2 were proinflammatory (Zhang and Rivest, 2001). Increased amounts of proinflammatory cytokines from the periphery lead to production of proinflammatory cytokines by microglia cells within the brain (Eikelenboom and Veerhuis., 1996; Hull et al., 1996; Becher et al., 2000; Calvo et al., 2005; Dantzer et al., 2007). The visible response to the increased proinflammatory cytokines in the brain is sickness behavior (Horai et al., 1998; Cartmell et al., 1999).

Administration of LPS systemically to mice increases expression, both mRNA and protein, of IL-1β and other proinflammatory cytokines in the brain (Iwai et al., 2006). Also, administration of IL-1β or TNF-α to mice resulted in decreased motor activity, social withdrawal, reduced food and water intake, increased slow-wave sleep, and altered cognition (Campbell et al., 2007; Dantzer et al., 2008). IL-6 has been associated with fever and hippocampus associated cognitive impairment (Smith et al., 2007). Moreover, overexpression of IL-6 in a transgenic mouse model has demonstrated that increased IL-6 will cause astrocytosis and neurodegeneration (Campbell et al., 1993; Steffensen et al., 1994; Jafarian-Tehrani and Sternberg, 1999). IL-6 signals through the JAK1/STAT3 pathway, and it has been shown that LPS-induced plasma levels of IL-6 cause nuclear translocation of the transcription factor STAT3 in certain brain structures including the area postrema, the vascular organs of the lamina terminalis, and the subfornical organ, as well as the hypothalamic supraoptic nucleus (Rummel et al., 2004).

The predominant CNS source of IL-6 is the activated astrocyte. IL-6 expression in astrocytes is regulated by proinflammatory factors (such as IL-1β and TNF-α), neurotransmitters, and second messengers (Van Wagoner and Benveniste, 1999). Expression of the proinflammatory cytokines is promoted through activation of NF-kB which is present in inactive form in the cytoplasm by its association with IκBα (Yabe et al., 2005). Proinflammatory cytokines may downregulate their own expression by increasing the expression of IκB or decreasing its proteolysis in certain cells of the brain (Laflamme and Rivest, 1999). Inhibition of NF-κB activation and inhibition of IκBα degradation occurs via a mechanism involving α-melanocyte stimulating hormone (αMSH), which is a pro-opiomelanocortin (POMC) derivative (Ichiyama et al., 1999).

In the CNS, increases in proinflammatory cytokines lead to increased formation of reactive oxygen species (ROS) and upregulation of genes that produce toxic products, such as reactive nitrogen species (RNS) (Floyd, 1999; Patel et al., 2003). Synthesis of nitric oxide (NO), can be induced in the brain by mediators of inflammation present in the cerebrospinal fluid (Kong et al., 2000). Several reports have indicated that IL-1 can induce nitric oxide synthase (iNOS) gene expression and thus promote the formation of NO through regulation by interleukin-1 converting enzyme (ICE, caspase-1) (Jones et al., 2005; Juttler et al., 2007). In summary, activated glial cells release NO through increased expression of iNOS, upregulated by the presence of high concentrations of proinflammatory cytokines (Kifle et al., 1996; Stasiolek et al., 2000). Formation of ROS and NRS moieties can alter protein, DNA, RNA, lipid, and carbohydrate structures. Thus, unregulated inflammation can culminate in pathological impairment of normal functions; with disregulated oxidative stress, inhibition of mitochrondrial respiration can occur resulting in cytotoxicity (Brown and Bal-Price, 2003). It has been found that nitric oxide production can be inhibited by β- and γ-melanocortin in the mouse brain (Muceniece et al., 2004).

1.1.2 IL-18

Data from human and rodent studies have shown that IL-18 (previously referred to as interferon-gamma inducing factor, IGIF) expression can be associated with neuropathology in infection, autoimmune disease, ischemia, or closed head injury (Felderhoff-Mueser et al., 2005). IL-18 is a member of the IL-1 family of cytokines. Like IL-1β, it is synthesized as an inactive precursor protein (24 kDa) that is subsequently cleaved to the active18 kDa protein by caspase-1(ICE) (Nhan et al., 2006). The active form of IL-18 induces signal transduction by binding to its receptor, IL-18α/β receptor (IL1Rrp/IL1RAPL) expressed by diverse cell types, including neurons and glia cells. In adult brains of untreated BALB/c mice, IL-18 is constitutively the most highly expressed cytokine (Fig. 1). In the developing brain, IL-18 has been found in association with hypoxic-ischemic brain injury. Mice lacking IL-18 expression had smaller infarct size and a lesser extent of subcortical white matter injury (Felderhoff-Mueser et al., 2005). In a hypoxia model, IL-18 was associated with increased neuronal apoptosis. Therefore, IL-18 can exhibit neuropathology with respect to neuroinflammation and neurodegeneration. In experiments performed on the rat dentate gyrus, in vitro, IL-18 was found to

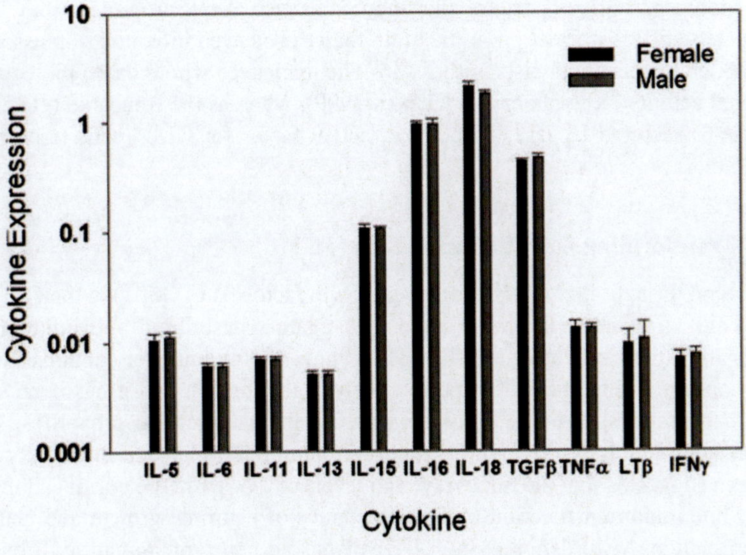

Fig. 1 Constitutive expression of CNS cytokines. RNA from brains of male and female BALB/c mouse pups at 21 days of age was quantified by real-time RT-PCR. Whole-brain RNA was isolated using the Qiagen Lipid Tissue Midi RNA isolation kit. Brains were pooled by gender within each litter. Cytokine mRNA quantity was normalized to endogenous control GAPDH. Each bar represents mean ± S.D. for *N* of 3 L. All of the cytokine levels significantly ($p < 0.05$) differed from each other except the following: IL-6:TNFα, IL-13:IFNγ, and TNFα:IFNγ; LT-β from IL-5, IL-6, IL-11, IL-13, TNFα, and IFNγ

impair the induction of long-term-potentiation (LTP) in NMDA receptor expressing neurons (Curran and O'Connor, 2001). Increased levels of IL-18 have now been reported to be present in the neurodegenerative disorder, Alzheimer's disease (AD) (Bossu et al., 2008) and have been found associated with stress activation of brain microglia (Sugama et al., 2007).

Human and mouse studies have shown that IL-1β and IL-18 are key players in fundamental inflammatory processes that increase during aging (Bodles and Barger, 2004; Joseph et al., 2005; Dinarello, 2006). Further evidence of the role of IL-18 in neuroinflammation is the finding that caspase-1 deficiency reduces inflammation-mediated transcription in the brain (Mastronardi et al., 2007). In normal brain tissue caspase-1 is activated within molecular platforms called inflammasomes. Key proteins of inflammasomes are proteins containing caspase recruitment domains (CARDS), or pyrin domains (PYDS). CARD-only proteins are termed COPs and PYD-only proteins are termed POPs. These proteins modulate the inflammasome activity in response to pathogen infection and tissue destruction (Stehlik and Dorfleutner, 2007). For example, caspase-1 activation can be blocked by COPs, Iceberg, and COP1/Pseudo-ICE, with a CARD similar in sequence to caspase-1. Expression of Iceberg in monocytes abrogates the secretion of IL-1β in response to LPS stimulation. Because IL-18 is constitutively expressed at a high level in the adult mouse brain, it is possible that the noncleaved IL-18 proprotein may have a completely different function in the brain than the cleaved inflammation associated form. Such a precedent exists with IL-16. The uncleaved whole IL-16 molecule has neuronal activity (Kurschner and Yuzaki, 1999), whereas the truncated (cleaved by caspase-3) secreted IL-16 is a chemoattractant factor for $CD4^+$ cells (Cruikshank et al., 2000).

1.1.3 Transforming Growth Factor-Beta (TGF-β)

Reports on the activities of transforming growth factor-β1 (TGFβ1) in the CNS have been conflicting in nature with some investigators describing anti-inflammatory effects and others indicating that TGFβ1 can have proinflammatory actions. Normal CNS concentrations of TGFβ1 are relatively high (Fig. 1), and it has been shown to be a potent neurotrophic cytokine with immunosuppressive properties. In the healthy adult brain, TGFβ1 inhibits proliferation of microglial and astrocyte cells. It has been suggested that the relatively high levels of TGFβ1 in the normal adult brain have some important function(s) in maintenance of neuronal growth and neuroimmune function. Mice deficient in TGFβ1 displayed neuroinflammation throughout the brain, excessive astrogliosis, and proliferating microglia displaying a phagocytic, deramified, and abnormally activated phenotype (Makwana et al., 2007). Ultrastructural features of TGFβ1 deficiency showed focal blockade of axonal transport, perinodal damming of axonal organelles, focal demyelination, and myelin debris in granule-rich phagocytic microglia cells. In a ME7 model of a murine prion disease, removal of TGFβ1 resulted in severe cerebral inflammation, increased expression of iNOS, and acute neuronal death in diseased animals. The data indicate a critical role for TFGβ1 in regulation of microglia cells and minimization of brain inflammation in order to avoid further brain tissue damage (Boche et al., 2006).

However, TGFβ1 can also behave in a proinflammatory manner (Grammas and Ovase, 2002). With murine autoimmune encephalomyelitis, overproduction of TGFβ1 locally in the brain led to a more severe and earlier onset of the disease with increased infiltration of mononuclear cells in the brain (Luo et al., 2007). TGFβ1 along with IL-6 are known to promote IL-17 producing T cells, which enhance inflammation. The encephalomyelitis model seems to indicate that the brain will signal to the immune system in certain circumstances as was suggested by increased plasma levels of IL-1, IL-6, and α-1-antichymotrypsin in patients with Alzheimer's disease (Licastro et al., 2000). In astrocyte cultures, TGFβ1, in the presence of LPS and IFN-γ, increased the expression of iNOS (NOS-2). It was determined that the increase in NOS-2 was due to enhanced proliferation of astrocytes producing NOS-2 (Hamby et al., 2006). In a mouse model for cerebral palsy, it was observed that neuronal-derived TGFβ1 mediated in some way the IL-9/mast cell interaction and exacerbated excitotoxicity in newborn mice (Patkai et al., 2001; Mesples et al., 2005). Further evidence of TGFβ1 involvement in mast cell activity comes from in vitro experiments with the mast cell line, D-36. In D-36 cultures, addition of TGFβ1 increased media histamine concentrations. It is the increase in the extracellular histamine concentrations that promotes the excitotoxic neuronal damage in cerebral palsy. Therefore, in neonatal cerebral palsy, the action of TGFβ1 on mast cells is proinflammatory in nature.

1.1.4 IL-33 and HMGB1

The cytokines IL-33 and HMGB1 are both present in the brain at relatively high concentration within the nucleus of cells, astrocytes in the case of IL-33 and neurons and oligo-dendrocyte–like cells for HMGB1(Hudson et al., 2008; Kim et al., 2008). HMGB1 is a nonhistone DNA-binding protein of the nucleoprotein complex, and it regulates TNF-α expression. HMGB1 also is able to increase proinflammatory cytokine levels by enhancing the signaling of TLR9 (Ivanov et al., 2007). Both proinflammatory proteins are released into the cytoplasm and subsequently secreted upon stimulation through signals stemming from infection or injury. Like IL-1β and IL-18, IL-33 is released by inflammasome activation of caspase-1 (Martinon et al., 2007); the central components of inflammasomes detect cytosolic microbial components and "danger" signals, such as ATP and toxins. Glial and astrocyte-enhanced cultures stimulated with PAMPs and ATP produced elevated levels of IL-1β and IL-33 in the culture media. Culture supernatants from these astrocyte enhanced cultures were able to induce the secretion of IL-6, IL-13, and MCP-1 from a mast cell line, MC/9 in culture, in a similar fashion to the addition of IL-33 alone (Hudson et al., 2008). It was also noted that IL-33 levels were increased in brains of mice infected with Theiler's murine encephalomyelitis virus with the promotion of various innate immune effectors on glial cells (Hudson et al., 2008).

The proinflammatory effects of HMGB1 were studied in the postischemic brain of rats. Ischemia injury in the brain proceeds with excitotoxicity-induced acute neuronal cell death in the ischemic core, followed by delayed damage to the penumbra (Lee et al., 2000b). It was observed that HMGB1 was immediately released into the extracellular space after ischemia and subsequently promoted neuroinflammation by microglial activation. Downregulation of HMGB1 expression via (sh) RNA decreased infarct size, microglial activation, and proinflammatory marker induction (Kim et al., 2006). After 1 h of middle cerebral artery occlusion HMGB1 was immediately translocated from the neuron nuclei into the extracellular space during the excitotoxicity-induced acute damage process. About two days after reperfusion HMGB1 was notably induced in activated microglia, astrocytes, and microvascular structures. This induction of HMGB1 expression was sustained for several days. The results emphasize the paracrine and autocrine function of HMGB1 in the postischemic brain (Kim et al., 2008).

1.1.5 IL-10 and IL-13

Interleukins 10 and 13 are anti-inflammatory, because they inhibit IL-1β bioactivity. IL-10 is capable of directly blocking IL-1 expression (Wong et al., 1997), whereas IL-13 induces the synthesis of the IL-1 receptor antagonist (IL-1ra). These interleukins are considered $CD4^+$ T-helper subset type 2 (Th2) and function to counteract the effects of $CD4^+$ T-helper subset type 1 (Th1) proinflammatory activities by inhibiting their action (Martino et al., 2000). In general, cytokines produced by each Th cell subset are inhibitory for the opposite subset. Production of these Th2 cytokines during neuroinflammation may be through the appearance of brain-derived heat shock protein (HSP) peptide complexes in the periphery (Galazka et al., 2006). HSP-induced downregulation of immune responses may involve the generation of immune cell subsets, such as Ag-specific Th2 cells, which secrete cytokines such as IL-10 that inhibit the proinflammatory process. Inhibition of inflammatory processes by IL-10 and IL-13 has been connected with ceramide production. Proinflammatory cytokines promote ceramide production through hydrolysis of plasma membrane sphingomyelin in brain cells. Ceramide plays an important role in coordinating cellular responses to stress, growth suppression, and apoptosis. The anti-inflammatory cytokines IL-10 and IL-13 are capable of blocking ceramide production through a mechanism involving activation of phosphatidylinositol-3-kinase. By blocking ceramide production IL-10 and IL-13 are able to inhibit apoptosis caused by the actions of proinflammatory cytokines, IL-1 and TNF-α (Pahan et al., 2000).

1.2 MAP Kinases and Stress Kinases

Stress stimuli come in a variety of forms, such as deprivation of trophic factors, ionizing radiation, free radicals (e.g., peroxynitrite), hypoxia, ischemia, heat shock, lipid second messengers (such as ceramide) (Singh et al., 1998), TNF-α, or Fas-ligand. In the brain, neurons are especially susceptible to stress stimuli; these stimuli lead to activation of intracellular pathways that either promote apoptosis or defense-adaptation mechanisms. At least three such pathways have been well studied. These pathways lead to the activation of c-Jun N-terminal kinases (JNKs), p38 kinases,

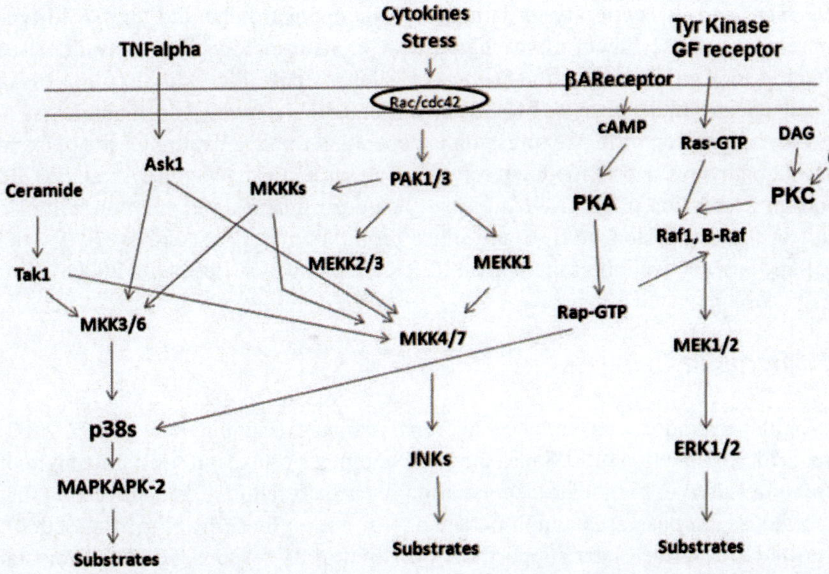

Fig. 2 Outline of signal transduction pathways for stress kinases, JNKs, and for Map kinases, p38 and ERK1/2. Shown is the cross-talk between the various pathways. Map kinases MEKK1/2/3 activate the MAP kinases MKK4 (JNKK1) and MKK7 (JNKK2). MKK4/7 can also be activated by the Map kinases TAK1 and Ask1, which can also activate MKK3/6 of the p38 s pathway. Activation of Rap-GTP is capable of activating the p38 and ERK1/2 pathways. As indicated, ceramide can activate the p38 pathway, and, also, has been suggested to play a role in activation of the ERK1/2 pathway

and extracellular signal-regulated kinases (ERKs) (Mielke and Herdegen, 2000). The cascades for these pathways are overlapping in places, as shown in Fig. 2. Note that each pathway can be stimulated by TNF-α and both p38 and JNK cascades can be stimulated by ceramide. Activation of JNK or p38 kinases leads to upregulation or activation of a number of factors, including transcription factors (ATF2, CREB, ELK1, MEF2C, CHOP), translation factors (eIF4E), MAP kinase-associated proteins, heat shock protein (Hsp27), and phosphorylation of tau at position Ser422 (phosphorylated in neurofibrillary tangles, but not in normal brain). Stimulation of the JNK pathway leads to inactivation of NFAT4, glucocorticoid receptor, and Bcl2, an inhibitor of apoptosis.

Involvement of JNK, p38, and ERK1/2 MAP kinases in neurodegeneration is further supported by studies demonstrating that inhibition of their activation reduces brain inflammation and neuron damage (Barone et al., 2001; Angstwurm et al., 2004). Moreover, modulation of the purinergic $P2X_7$ receptor by oxidized ATP (oxATP), during LPS activation of microglia, led to attenuation of inflammatory mediators, resulting in inactivation of the p38 and NF-kB pathways and increased neuronal survival (Ferrari et al., 2006; Choi et al., 2007). Neuronal damage via the p38 pathway can be modulated by the tyrosine phosphatase SHP2 (Chong et al., 2003). It was observed that neurons from mice deficient in SHP2 showed more increased susceptibility to damage by NO exposure than did their wild-type counterparts. SHP2 function was necessary for neuronal survival only after the induction of signal transduction pathways, such as p38, that would culminate in the cell's death. In the absence of p38 activation, SHP2 remained dormant. In astrocytes, the regulation of IL-6 expression is dependent upon activation of the p38 and ERK1/2 pathways and is modulated by oncostatin M (Van Wagoner et al., 2000). Aberrant expression of ICAM-1 on astrocytes during neuroinflammation will also result in the expression of IL-6 and other proinflammatory cytokines, IL-1α and IL-1β, through activation of the p38 and ERK1/2 pathways (Lee et al., 2000a).

1.3 Microglia Cells

Microglial activation can enhance neuronal damage (Cunningham et al., 2002; Rivest, 2003). Injection of LPS into the hippocampus, cortex, and substantia nigra of rat brain resulted in high neurodegeneration in the substantia nigra, suggesting that this brain region possesses a high percentage of microglia cells (Kim et al., 2000); experiments described later support this distribution. In response to environmental toxins and certain endogenous proteins, microglia will become overactivated and will release reactive oxygen species (ROS), which can then cause neuronal damage (Block et al., 2007; Galea et al., 2003). Microglia cells are phagocytes and express a diverse array of membrane receptors (PRRs) that recognize a wide variety of molecular determinates, including phosphatidylserine (Li et al., 2003). Microglia receptors are constitutively expressed and bind PAMPS during the innate immune response. Prominent among these PRRs are the TLRs, of which microglia express TLRs 1–9. Other PRRs on microglia, include scavenger receptors, MAC1 receptors, and complexes of these receptors. The detection of extracellular superoxide is a common result of ligand recognition by PRRs and oxidative stress is a primary cause of neurodegeneration (Kifle et al., 1996).

Through the activity of NADPH oxidase microglia become a robust source of free radicals, both extracellular and intracellular. Intracellular ROS provide a mechanism for proinflammatory signaling. Microglial NADPH oxidase has been associated with a number of neurodegenerative disorders, including Alzheimer's disease and Parkinson's disease (Kalaria et al., 1996). In addition to the extracellular effects, NADPH oxidase is crucial to microglia intracellular signaling. As an example, gangliosides activate microglia through a protein kinase C and NADPH oxidase mechanism (Farooqui et al., 2007). Gangliosides are able to bind to microglia cells via TLR4 (Jou et al., 2006). This binding signals through a JAK/STAT pathway and induces phosphorylation of STAT1 and STAT3 (Kim et al., 2002; Lee et al., 2005), which factors upregulate transcription of inflammation-associated genes (Heese et al., 1998; Pinteaux et al., 2007), such as iNOS, ICAM-1, and MCP-1.

In general, the higher the intracellular ROS concentration is, the higher the inflammatory response will be in the microglia cells. However, prolonged ROS exposure will induce cumulative events harmful to the cell's survival, such as lipid peroxidation and modification of proteins. Eventually the activated microglia will become apoptotic themselves, an outcome that was shown to be enhanced by expression of B cell translocation gene-1 (BTG1) (Lee et al., 2003). Although the deleterious effects of activated and proliferating microglia cells are clear, there is some experimental evidence that suggests a protective role for these cells. In a mouse model for cerebral ischemia, ablation of proliferating resident microglia resulted in increased proinflammatory cytokine expression, increased infarct size, and a 2.7-fold increase in apoptotic cells, mostly neurons (Lalancette-Hebert et al., 2007).

1.4 Astrocytes

Astrocytes, the most abundant glial cell of the CNS, can display innate immune responses triggered by a variety of insults (Xiao and Link, 1999; Becher et al., 2006). These cells possess a wide array of receptors including TLRs, nucleotide-binding oligomerization domains, double-stranded RNA-dependent protein kinase, scavenger receptors, mannose receptor, and components of the complement system (Bugno et al., 1999; Farina et al., 2007). Upon stimulation through one or more of these receptors, the cell will produce cytokines, IL-6, TGFβ, IFNβ, GM-CSF, BAFF, IL-1β, and TNF; chemokines, CCL2, CCL5, CCL20, CXCL-10, CXCL12, CXCL1, CXCL2, and CX3CL1; and neurotrophic factors, NGF, CNTF, BDNF, VEGF, IGF1, and LIF (Croll et al., 2004; Cotman et al., 2007; Krasowska-Zoladek et al., 2007). Two types of events result from astrocyte stimulation: activation of neighboring cells and further amplification of local innate immune responses and modification of BBB permeability and attraction of immune cells from the blood into the neural tissue (Fitch and Silver, 1997; Andjelkovic and Pachter, 1998; Anthony et al., 1998; Stolp et al., 2005; Milner and Campbell, 2006; Andras et al., 2008). The latter activity is mediated by chemokine expression (Mennicken et al., 1999; Biber et al., 2002; Hsieh et al., 2006). Migration of the astrocytes themselves can be induced by the presence of stromal cell-derived factor-1 (SDF-1) and upregulation of astrocyte CXCR4 by IL-6 plus cAMP (Odemis et al., 2002). Further evidence of the innate immune response activity of astrocytes stems from the cell's ability to secrete active α_1-antichymotrypsin (Kanemaru et al., 1996) and expression of syncytin-1 in MS leading to the upregulation of iNOS through an old astrocyte specifically induced substance (OASIS) mechanism (Antony et al., 2007). Protective effects of astrocytes are mediated through the cell's ability to release purines, adenosine and adenosine triphosphate, and guanosine and guanosine triphosphate (Ciccarelli et al., 2001).

1.5 Neuroinflammatory Aspects of Pb Toxicity

1.5.1 Pb Effects on Glial Cells

In the CNS, Pb accumulates preferentially in glial cells rather than in neurons (Tiffany-Castiglioni et al., 1989; Lindahl et al., 1999), and Pb's presence in these cells produces activities that are proinflammatory in nature. As indicated earlier, astrocytes may have a role in increasing permeability of the BBB during inflammation. It was observed that Pb was able to potentiate proinflammatory cytokine and glutamate-mediated increases in permeability of the BBB in mice (Dyatlov et al., 1998). In a study performed on Pb-exposed young (15–30-day-old) rats, it was observed that increased Pb levels resulted in astrocyte cell activation and proliferation, as indicated by elevated GFAP and S-100β in all brain regions examined (Struzynska et al., 2007). Results also showed increased production of IL-6 in the forebrain with a concomitant decrease in levels of the axonal markers synapsin-1 and synaptophysin. The study concluded that Pb caused chronic glial cell activation with coexisting inflammatory and neurodegenerative features.

A more recent study using proton magnetic resonance to study the relative levels of certain metabolites in human brain regions indicated that Pb increased the myoinositol to creatine ratio (mI/Cr) in the hippocampus (Weisskopf et al., 2007). An increased mI/Cr is a distinctive aspect of Alzheimer's disease and is thought to be indicative of gliosis. Therefore, the increased mI/Cr associated with increased bone Pb levels in humans is also suggestive of a neuroinflammatory aspect to Pb toxicity. The study concluded that the glial effects observed might be the more sensitive indicators of the adverse effects of cumulative Pb exposure and these changes are similar to those seen in the early stages of Alzheimer's disease. A further connection between Pb activity and neurodegenerative diseases is the ability of Pb to produce a deficiency in chaperone protein function which then compromises protein secretion, exacerbates protein aggregation, and increases sensitivity to oxidative stress. Alzheimer's disease and Parkinson's disease are characterized by a deficiency in the function of the chaperone protein GRP78 (Bip, HSPA5). In the absence of a diseased state GRP78 facilitates the maturation of the amyloid precursor protein and reduces or prevents the formation of extracellular amyloid deposits. Pb binds to GRP78 and can inhibit its function as a chaperone protein (White et al., 2007). Furthermore, Pb exposure of rodents from pnd0-pnd20 gave a transient increase in amyloid precursor protein mRNA synthesis (White et al., 2007). Therefore, Pb neurotoxicity shares several features of neurodegenerative disorders.

1.5.2 Pb Effects on Cytokines in the CNS

It has been observed that exposure to the heavy metal, lead (Pb), can increase susceptibility to infectious agents (Lawrence, 1981). Moreover, neonatal Pb exposure exacerbated sickness behavior in pups infected with *Listeria monocytogenes*; such sickness behavior was documented as loss of appetite and drinking, decreased bodyweight gain, and lack of mobility (Dyatlov and Lawrence, 2002). These results suggested that Pb might modulate expression of proinflammatory cytokines in the brain.

Gene expression of 14 cytokines was measured by real-time RT-PCR in the perfused brain tissue of male and female 21-day-old mouse pups. As shown earlier, there are no significant differences in the expression levels between males and females. Messenger RNA for IL-4, IL-10, and IL-12p40 was not detectable in the

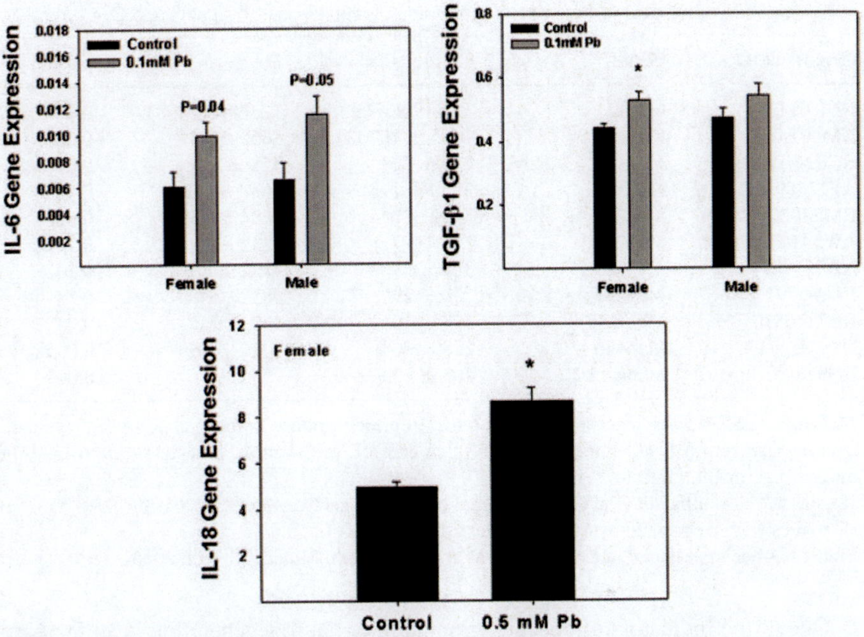

Fig. 3 Pb effect on expression of cytokines IL-6, TGF-β1, and IL-18 in the brain. Cytokine mRNA from the brains of female and male mouse pups at pnd21 was quantified by real-time RT-PCR. Whole-brain RNA was isolated using a Qiagen Lipid Tissue kit. Mouse pup brains from each litter were pooled according to gender. All cytokine RNA quantitation results were normalized to endogenous GAPDH. Each bar represents mean ± S.D. for an *N* of 3 L. Significance, indicated by an asterisk, *, was determined by the Student's *t*-test, $p < 0.05$. The p value for both female and male IL-6 gene expression ±Pb is 0.056 and the p value for male TGF-β1 gene expression ±Pb is 0.20 (Data for IL-6 and TGF-beta has been published in Kasten-Jolly et al., J. Biochem. Molec. Toxicol. 2010)

brains of these pnd21 mouse pups. Among the most abundant cytokine transcripts in the brain are those coding for IL-16, IL-15, IL-18, and TGF-β1. Exposure of the mouse pups to 0.1 mM Pb acetate from gd8 to pnd21 via the dam's drinking water resulted in enhanced expression of IL-6 and TGF-β1 (Fig. 3). If Pb exposure was increased to 0.5 mM an increase in expression of IL-18 in whole brain tissue was observed for the female mouse pups at pnd21 (Fig. 3).

Associated with the upregulation of proinflammatory cytokine gene expression and generation of ROSs is the activation of stress kinases and MAP kinases. Microarray data from whole-brain RNA of 0.1 mM Pb-exposed (gd8 to pnd21) and unexposed control mice indicated upregulation of p38 and MAP kinases within the p38 cascade, such as MAPKAPK-2 (Table 1). The microarray data are in agreement with previous reports (Cordova et al., 2003; Leal et al., 2006) indicating that Pb exposure induces activation of the p38 and ERK1/2 MAP kinase pathways. The mechanism of Pb activation of these pathways is not yet completely understood, but it may occur through the generation of ROSs.

Table 1 Effect of Pb on gene expression of map kinases[a,d] signal[b] ± S.D.

GeneBank ID	Gene	Control	Pb	*p*-value[c]
NM_011951	Mapk1	12,119 ± 719	12,998 ± 1,083	0.15
NM_015806	Mapk6	1485 ± 107	1655 ± 73	0.04
BC024684	Mapk11(p38)	256 ± 32	358 ± 52	0.02
AF128892	Mapk14	3350 ± 173	3638 ± 171	0.06
BM240207	Map2k4	5412 ± 244	5894 ± 358	0.06
AW541674	Map2k7	1157 ± 77	1232 ± 69	0.14
AA929089	Map3k7	3832 ± 167	4152 ± 401	0.14
BF166991	Map4k2	500 ± 29	583 ± 80	0.08
BB734681	Map4k5	363 ± 50	490 ± 90	0.05
NM_016713	Map4k6	2589 ± 64	3086 ± 86	0.005
BG918951	Mapkapk2	783 ± 20	994 ± 91	0.009

[a]Affymetrix MG430A GeneChip data. The data represent mouse brain total RNA from female mouse pups from 3 L of untreated control mice and 3 L of 0.1 mM Pb acetate (gd8 to pnd21) treated experimental mice.
[b]Signal was normalized to the GAPDH signal on each respective GeneChip.
[c]Statistics were performed by Student's *t*-test, significant at $p < 0.05$.
[d]Table has been published in Kasten-Jolly et al., J. Biochem. Molec. Toxicol, 2010.

One of the indicators of neuroinflammation or neurodegeneration is an increase in GFAP expression due to astrocyte activation and proliferation (Hauss-Wegrzyniak et al., 1998; Norris et al., 2005; O'Callaghan and Sriram, 2005; Pannu et al., 2005). Because Pb exposure seemed to enhance gene expression of IL-6 and TGF-β1, it was postulated that Pb might also increase the expression of GFAP. Both IL-6 and TGF-β signaling are needed to promote transcription of the GFAP gene (Taga and Fukuda, 2005). Microarray data for pnd21 female mouse pups exposed to 0.1 mM Pb (gd8-pnd21) showed that Pb did significantly increase gene expression of GFAP (Fig. 4), indicating that Pb may be promoting astrocyte activation.

As indicated above, αMSH has an anti-inflammatory effect in that it can inhibit the degradation of IκBα; thereby blocking the activation of NFκB (a transcription factor for several proinflammatory cytokines). αMSH is a product of the POMC1 gene, which also codes for several other peptide hormones. The microarray data suggested that Pb could decrease POMC1 gene expression. Results obtained by real-time RT-PCR supported this finding (Fig. 5). Shown here are POMC1 RNA levels for untreated controls and Pb-treated females at pnd21 using two different concentrations of Pb, 0.1 and 0.5 mM. As indicated, Pb dampened POMC1 gene expression at each concentration. Therefore, Pb may increase inflammation by decreasing the expression of αMSH through downregulation of the POMC1 gene.

1.6 Neuroinflammatory Effects of Metals Other than Pb

Neurotoxicity of aluminum and copper are associated with the upregulation of stress-related gene expression patterns. Whole human genome microarray data

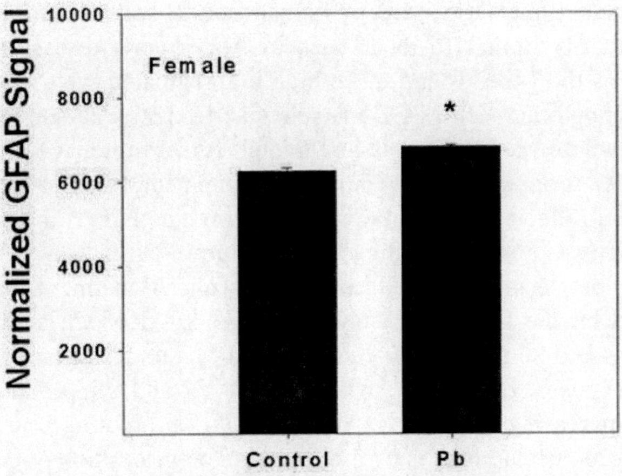

Fig. 4 Pb exposure enhances CNS expression of GFAP. Female BALB/c mice from 3 L treated or untreated with 0.1 mM Pb acetate from gd8 to pd21 via the dam's drinking water, were sacrificed on day 21, and whole-brain RNA was isolated. GFAP expression was measured by Affymetrix GeneChip (MG 430A). Signal for GFAP was normalized to the signal for GAPDH on the same chip. Data were gathered from 3 control litters (distilled water) and 3 Pb-treated litters, that is, 6 GeneChips. The * indicates a $p < 0.05$ for the difference in normalized signal between control and Pb-treated female mice

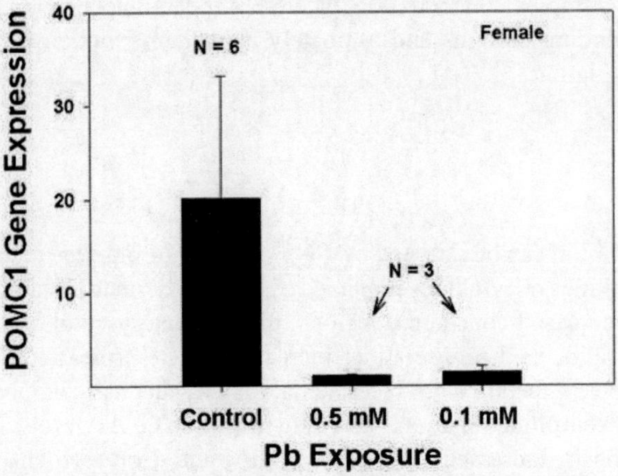

Fig. 5 Pb effect on gene expression of POMC1. POMC1 mRNA from the brains of female mouse pups at pnd21 was quantified by real-time RT-PCR. Whole-brain RNA was isolated using a Qiagen Lipid Tissue kit. Female mouse pup brains from each litter were pooled during isolation of the RNA. POMC1 transcript quantitation was normalized against endogenous GAPDH. Each bar represents the mean ± SEM for an N of 6 L for untreated controls and $N = 3$ L for each experimental group

indicated that aluminum exposure of human neural cells in culture resulted in expression patterns similar to those seen in Alzheimer's disease (Lukiw et al., 2005). Seven of the genes found to be highly upregulated were proinflammatory and were proapoptotic, such as NF-κB subunits, IL-1β precursor, cytosolic phosphlipase A_2, cyclooxygenase-2, beta-amyloid precursor protein (APP), and DAXX. Many of the genes upregulated by aluminum contained promoter binding sites for NF-κB or stress-inducible transcription factors, such as HIF-1, thereby suggesting a role for these two promoter binding factors in proinflammatory gene expression. An association between Al-induced neurotoxicity and Alzheimer's disease was further supported by the finding that mice exposed to Al or Cu showed enhanced oxidative stress and an accumulation of amyloid β peptides (Becaria et al., 2006). Concomitant exposure of the mice to both metals, Al and Cu, produced a cooperative effect on increasing APP levels. Another way of promoting neurodegeneration and altered organ development is by modifying the transcription of genes. This can be achieved by blocking transcription factor binding to the DNA. Sp1 is among the transcription factors affected by metal toxicity. Pb, Zn, and Cd modulate the binding of Sp1 to its DNA target sequences. It was suggested that exposure to heavy metals could alter developmental gene expression in the brain through their interference with binding of Sp1 to its promoter sites (Zawia et al., 1998). A study of how Pb and other heavy metals inhibit transcription factors, like Sp1, was performed using synthetic peptides (Razmiafshari and Zawia, 2000). Here it was found that Pb and other metals, Zn, Cd, and Hg, formed complexes with the peptides that bound the double-stranded DNA with high affinity and did not allow Sp1 DNA-binding. This inhibition of Sp1 DNA binding occurred in a manner dependent on the metal/peptide complex concentration. Therefore, heavy metals can alter the activity of DNA binding proteins and ultimately alter their function in terms of gene expression regulation.

2 Summary

Neuroinflammation can be initiated by CNS entrance of immune cells, peripherally generated cytokines or cytokines produced by glial cells; neuroinflammation is characterized by increased concentrations of proinflammatory cytokines, chemokines, glial cell activation and proliferation, increased BBB permeability, and neuronal damage. The heavy metal Pb produces neural toxicity that has a number of factors in common with neuroinflammation, including increased GFAP levels, increased IL-6 expression, gliosis, increased BBB permeability, and decreased chaperone protein function. Other metals produce neurotoxicity similar to Pb with parallels to features of neurodegenerative disorders, such as Alzheimer's and Parkinson's diseases.

Acknowledgments This work was supported by NIH grant # RO1 ES11135. We thank Michael Ryan and the late Robin Pietropaolo from the Wadsworth Center Microarray Core facility for their assistance in performing the microarray analysis. We thank Donghong Gao for her technical assistance with Pb acetate treating the mice through the dam's drinking water and for obtaining the intact brain tissue for us.

References

Akiyama H, Barger S, Barnum S, Bradt B, Bauer J, Cole GM, Cooper NR, Eikelenboom P, Emmerling M, Fiebich BL, Finch CE, Frautschy S et al (2000) Inflammation and Alzheimer's disease. Neurobiol of Aging 21:383–421

Andjelkovic AV, Pachter J (1998) Central nervous system endothelium in neuroinflammatory, neuroinfectious, and neurodegenerative disease. J Neurosci Res 51:423–430

Andras I, Rha GB, Huang W, Eum SY, Couraud P-O, Romero IA, Hennig B, Toborek M (2008) Simvastatin protects against amyloid β and HIV-1 Tat-induced promoter activities of inflammatory genes in brain endothelial cells. Mol Pharmacol 73:1424–1433

Angstwurm K, Hanisch U-K, Gassemi T, Bille MB, Prinz M, Dirnagl U, Kettenmann H, Weber JR (2004) Tyrosine kinase inhibition reduces inflammation in the acute stage of experimental pneumococcal meningitis. Infect Immun 72:3294–3298

Anisman H (2004) Considering cytokine panels. Brain Behav Immun 18:221–222

Anthony D, Dempster R, Fearn S, Clements J, Wells G, Perry H, Walker K (1998) CXC chemokines generate age-related increases in neutrophil-mediated brain inflammation and blood-brain barrier breakdown. Curr Biol 8:923–926

Antony JM, Ellestad KK, Hammond R, Imaizumi K, Mallot F, Warren KG, Power C (2007) The human endogenous retrovirus envelope glycoprotein, syncytin-1, regulates neuroinflammation and its receptor expression in multiple sclerosis: A role for endoplasmic reticulum chaperones in astrocytes. J Immunol 179:1210–1224

Barone FC, Irving EA, Ray AM, Lee JC, Kassis S, Kumar S, Badger AM, White RF, McVey MJ, Legos JJ, Erhardt JA, Nelson AH, Ohlstein EH, Hunter AJ, Ward K, Smith BR, Adams JL, Parsons AA (2001) SB 239063, a second-generation p38 mitogen-activated protein kinase inhibitor, reduces brain injury and neurological deficits in cerebral focal ischemia. J Pharmacol Exp Ther 296:312–321

Becaria A, Lahiri DK, Bondy SC, Chen D, Hamadeh A, Li H, Taylor R, Campbell A (2006) Aluminum and copper in drinking water enhance inflammatory or oxidative events specifically in the brain. J Neuroimmunol 176:16–23

Becher B, Bechmann I, Greter M (2006) Antigen presentation in autoimmunity and CNS inflammation: how T lymphocytes recognize the brain. J Mol Med 84:532–543

Becher B, Prat A, Antel JP (2000) Brain-immune connection: Immune-regulatory properties of CNS-resident cells. GLIA 29:293–304

Biber K, Zuurman MW, Dijkstra IM, Boddeke HWGM (2002) Chemokines in the brain: neuroimmunology and beyond. Curr Opin Pharmacol 2:63–68

Block ML, Zecca L, Hong J-S (2007) Microglia-mediated neurotoxicity: uncovering the molecular mechanism. Nat Rev Neurosci 8:57–69

Boche D, Cunningham C, Docagne F, Scott H, Perry VH (2006) TGFβ1 regulates the inflammatory response during chronic neurodegeneration. Neurobiol Dis 22:638–650

Bodles AM, Barger SW (2004) Cytokines and the aging brain – what we don't know might help us. Trends Neurosci 27:621–626

Bossu P, Ciaramella A, Salani F, Bizzoni F, Varsi E, De Iulio F, Giubilei F, Gianni W, Trequattrini A, Moro ML, Bernardini S, Caltagirone C, Spalletta G (2008) Interleukin-18 produced by peripheral blood cells is increased in Alzheimers disease and correlates with cognitive impairment. Brain Behav Immun 22:487–492

Brown GC, Bal-Price A (2003) Inflammatory neurodegeneration mediated by nitric oxide, glutamate, and mitochondria. Mol Neurobiol 27:325–355

Bugno M, Witek B, Bereta J, Bereta M, Edwards DR, Kordula T (1999) Reprogramming of TIMP-1 and TIMP-3 expression profiles in brain microvascular endothelial cells and astrocytes in response to proinflammatory cytokines. FEBS Lett 448:9–14

Cagnin A, Gerhard A, Banati RB (2002) In vivo imaging of neuroinflammation. Eur Neuropsycopharm 12:581–586

Calvo C-F, Amigou E, Desaymard C, Glowinski J (2005) A pro- and an anti-inflammatory cytokine are synthesized in distinct brain macrophage cells during innate activation. J Neuroimmunol 170:21–30

Campbell IL, Abraham CR, Masliah E, Kemper P, Inglis JD, Oldstone MB, Mucke L (1993) Neurologic disease induced in transgenic mice by cerebral overexpression of interleukin-6. Proc Natl Acad Sci USA 90:10061–10065

Campbell SJ, Deacon RMJ, Jiang Y, Ferrari C, Pitossi FJ, Anthony DC (2007) Overexpression of IL-1β by adenoviral-mediated gene transfer in the rat brain cause a prolonged hepatic chemokine response, axonal injury and the suppression of spontaneous behaviour. Neurobiol Dis 27:151–163

Cartmell T, Luheshi GN, Rothwell NJ (1999) Brain sites of action of endogenous interleukin-1 in the febrile response to localized inflammation in the rat. J Physiol 518:585–594

Choi HB, Ryu JK, Kim SU, McLarnon JG (2007) Modulation of the puinergic $P2X_7$ receptor attenuates lipopolysaccharide-mediated microglial activation and neuronal damage in inflamed brain. J Neurosci 27:4957–4968

Chong ZZ, Lin S-H, Kang J-Q, Maiese K (2003) The tyrosine phosphatase SHP2 modulates MAP kinase-38 and caspase 1 and 3 to foster neuronal survival. Cell Mol Neurobiol 23: 561–578

Ciccarelli R, Ballerini P, Sabatino G, Rathbone MP, D'Onofrio M, Caciagli F, DiIorio P (2001) Involvement of astrocytes in purine-mediated reparative processes in the brain. Dev Neurosci 19:395–414

Combs CK, Johnson DE, Karlo JC, Cannady SB, Landreth GE (2000) Inflammatory mechanisms in Alzheimer's disease: inhibition of β-amyloid-stimulated proinflammatory responses and neurotoxicity by PPARγ agonists. J Neurosci 20:558–567

Cordova FM, Rodrigues ALS, Giacomelli MBO, Olivera CS, Posser T, Dunkley PR, Leal RB (2003) Lead stimulates ERK 1/2 and $p38^{MAPK}$ phosphorylation in the hippocampus of immature rats. Brain Res 998:65–72

Cotman CW, Berchtold NC, Christie L-A (2007) Exercise builds brain health: key roles of growth factor cascades and inflammation. Trends Neurosci 30:464–472

Cotran RS, Kumar V, Robbins SL (eds) (1994) Pathologic basis of disease. W.B. Saunders, Philadelphia, PA, pp 39–86

Croll SD, Ransohoff RM, Cai N, Zhang Q, Martin FJ, Wei T, Kasselman LJ, Kintner J, Murphy AJ, Yancopoulos GD, Wiegand SJ (2004) VEGF-mediated inflammation precedes angiogenesis in adult brain. Exp Neurol 187:388–402

Cruikshank WW, Kornfeld H, Center DM (2000) Interleukin-16. J Leukocyte Biol 67:757–766

Cunningham C, Konsman J-P, Cartmell T (2002) Cytokines and the ageing brain. Trends Neurosci 25:546–547

Curran B, O'Connor JJ (2001) The pro-inflammatory cytokine interleukin-18 impairs long-term potentiation and NMDA receptor-mediated transmission in the rat hippocampus in vitro. Neuroscience 108:83–90

Dantzer R, Bluthe R-M, Castanon N, Kelley KW, Konsman J-P, Laye S, Lestage J, Parnet P (2007) Cytokines, sickness behavior, and depression. In: Ader R (ed) Psychoneuroimmunology, 4th edn, vol 1. Elsevier, Inc., London, pp 281–318

Dantzer R, O'Conner JC, Freund GG, Johnson RW, Kelley KW (2008) From inflammation to sickness and depression: when the immune system subjugates the brain. Nat Rev Neurosci 9:46–57

Dinarello CA (2006) Interleukin 1 and interleukin 18 as mediators of inflammation and the aging process. Am J Clin Nutr 83(Suppl):447S–455S

Dyatlov VA, Lawrence DA (2002) Neonatal lead exposure potentiates sickness behavior induced by *Listeria monocytogenes* infection of mice. Brain Behav Immun 16(4):477–492

Dyatlov V, Platoshin AV, Lawrence DA, Carpenter DO (1998) Lead potentiates cytokine- and glutamate-mediated increases in permeability of the blood-brain barrier. Neurotoxicology 19:283–292

Eikelenboom E, Veerhuis R (1996) The role of complement and activated microglia in the pathogenesis of Alzheimer's disease. Neurobiol Aging 17:673–680

Eikelenboom P, Veerhuls R, Scheper W, Rozemuller AJM, van Gool WA, Hoozemans JJM (2006) The significance of neuroinflammation in understanding Alzheimer's disease. J Neural Trans 113:1685–1695

Farina C, Aloisi F, Meinl E (2007) Astrocytes are active players in cerebral innate immunity. Trends Immunol 28:138–144

Farooqui AA, Horrocks LA, Farooqui T (2007) Modulation of inflammation in brain: a matter of fat. J Neurochem 101:577–599

Felderhoff-Mueser U, Schmidt OI, Oberholzer A, Buhrer C, Stahel PF (2005) IL-18: a key player in neuroinflammation and neurodegeneration. Trends Neurosci 28:487–493

Ferrari D, Pizzirani C, Adinolfi E, Lemoli RM, Curti A, Idzko M, Panther E, di Virgilio F (2006) The $P2X_7$ receptor: a key player in IL-1 processing and release. J Immunol 176: 3877–3883

Fitch MT, Silver J (1997) Activated macrophages and the blood-brain barrier: inflammation after CNS injury leads to increases in putative inhibitory molecules. Exp Neurobiol 148: 587–603

Floyd RA (1999) Neuroinflammatory processes are important in neurodegenerative diseases: an hypothesis to explain the increased formation of reactive oxygen and nitrogen species as major factors involved in neurodegenerative disease development. Free Radic Biol Med 26: 1346–1355

Galazka G, Stasiolek M, Walczak A, Jurewicz A, Zylicz A, Brosnan CF, Raine CS, Selmaj KW (2006) Brain-derived heat shock protein 70-peptide complexes induce NK cell-dependent tolerance to experimental autoimmune encephalomyelitis. J Immunol 176:1588–1599

Galea E, Heneka MT, Russo CD, Feinstein DL (2003) Intrinsic regulations of brain inflammatory responses. Cell Mol Neurobiol 23:625–635

Grammas P, Ovase R (2002) Cerebrovascular transforming growth factor-β contributes to inflammation in the Alzheimer's disease brain. Am J Pathol 160:1583–1587

Hamby ME, Hewett JA, Hewett SJ (2006) TGF-β1 potentiates astrocytic nitric oxide production by expanding the population of astrocytes that express NOS-2. GLIA 54:566–577

Hauss-Wegrzyniak B, Dobrzanski P, Stoehr JD, Wenk GL (1998) Chronic neuroinflammation in rats reproduces components of the neurobiology of Alzheimer's disease. Brain Res 780: 294–303

Heese K, Hock C, Otten U (1998) Inflammatory signals induce neurotrophin expression in human microglial cells. J Neurochem 70:699–707

Horai BR, Asano M, Sudo K, Kanuka H, Suzuki M, Nishihara M, Takahashi M, Iwakura Y (1998) Production of mice deficient in genes for interleukin (IL)-1α, IL-1β, IL-1α/β, and IL-1 receptor antagonist shows that IL-1β is crucial in turpentine-induced fever development and glucocorticoid secretion. J Exp Med 187:1463–1475

Hsieh M-F, Lai S-L, Chen J-P, Sung J-M, Lin Y-L, Wu-Hsieh BA, Gerard C, Luster A, Liao F (2006) Both CXCR3 and CXCL10/IFN-inducible protein 10 are required for resistance to primary infection by Dengue virus. J Immunol 177:1855–1863

Hudson CA, Christophi GP, Gruber RC, Wilmore JR, Lawrence DA, Massa PT (2008) Induction of IL-33 expression and activity in central nervous system glia. J Leukoc Biol 84: 631–643

Hull M, Fiebich BL, Lieb K, Strauss S, Berger M, Volk B, Bauer J (1996) Interleukin-6-associated inflammatory processes in Alzheimer's disease: new therapeutic options. Neurobiol Aging 17:795–800

Ichiyama T, Zhao H, Catania A, Furukawa S, Lipton JM (1999) α-melanocyte-stimulating hormone inhibits NF-κB activation and IκBα degradation in human glioma cells and in experimental brain inflammation. Exp Neurol 157:359–365

Inoue W, Poole S, Bristow AF, Luheshi GN (2006) Leptin induces cyclooxygenase-2 via an interaction with interleukin-1β in the rat brain. Eur J Neurosci 24:2233–2245

Ivanov S, Dragoi A-M, Wang X, Dallacosta C, Louten J, Musco G, Sitia G, Yap GS, Wan Y, Biron CA, Bianchi ME, Wang H, Chu W-M (2007) A novel role for HMGB1 in TLR9-mediated inflammatory responses to CpG-DNA. Blood 110:1970–1981

Iwai T, Ito S, Tanimitsu K, Udagawa S, Oka J-I (2006) Glucagon-like peptide-1 inhibits LPS-induced IL-1β production in cultured rat astrocytes. Neurosci Res 55:352–360

Jafarian-Tehrani M, Sternberg EM (1999) Animal models of neuroimmune interactions in inflammatory diseases. J Neuroimmunol 100:13–20

Jones NC, Prior MJW, Burden-The E, Marsden CA, Morris P, Murphy S (2005) Antagonism of the interleukin-1 receptor following traumatic brain injury in the mouse reduces the number of nitric oxide synthase-2-positive cells and improves anatomical and functional outcomes. Eur J Neurosci 22:72–78

Joseph JA, Shukitt-Hale B, Casadesus G, Fisher D (2005) Oxidative stress and inflammation in brain aging: nutritional considerations. Neurochem Res 30:927–935

Jou I, Lee JH, Park SY, Yoon HJ, Joe E-H, Park EJ (2006) Gangliosides trigger inflammatory responses via TLR4 in brain glia. Am J Pathol 168:1619–1630

Juttler E, Bonmann E, Spranger M, Kolb-Bachofen V, Suschek CV (2007) A novel role of interleukin-1-converting enzyme in cytokine-mediated inducible nitric oxide synthase gene expression: Implications for neuroinflammatory diseases. Mol Cell Neurosci 34:612–620

Kalaria RN, Cohen DL, Premkumar DRD (1996) Cellular aspects of the inflammatory response in Alzheimer's disease. Neurodegeneration 5:497–503

Kanemaru K, Meckelein B, Marshall CL, Sipe JD, Abraham CR (1996) Synthesis and secretion of active α_1-antichymotrypsin by murine primary astrocytes. Neurobiol Aging 17: 767–771

Karman J, Chu HH, Co DO, Seroogy CM, Sandor M, Fabry Z (2006) Dendritic cells amplify T cell-mediated immune responses in the central nervous system. J Immunol 177:7750–7760

Kielian T, Hickey WF (2000) Proinflammatory cytokine, chemokine, and cellular adhesion molecule expression during the acute phase of experimental brain abscess development. Am J Pathol 157:647–658

Kifle Y, Monnier J, Chesrown SE, Raizada MK, Nick HS (1996) Regulation of the manganese superoxide dismutase and inducible nitric oxide synthase gene in rat neuronal and glial cells. J Neurochem 66:2128–2135

Kim J-B, Choi JS, Yu Y-M, Nam K, Piao C-S, Kim S-W, Lee M-H, Han P-L, Park J-S, Lee J-K (2006) HMGB1, a novel cytokine-like mediator linking acute neuronal death and delayed neuroinflammation in the postischemic brain. J Neurosci 26:6413–6421

Kim J-B, Lim C-M, Yu Y-M, Lee J-K (2008) Induction and subcellular localization of high-mobility group box-1 (HMGB1) in the postischemic rat brain. J Neurosci Res 86:1125–1131

Kim W-G, Mohney RP, Wilson B, Jeohn G-H, Liu B, Hong J-S (2000) Regional difference in susceptibility to lipopolysaccharide-induced neurotoxicity in the rat brain: role of microglia. J Neurosci 20:6309–6316

Kim OS, Park EJ, Joe E-H, Jou I (2002) JAK-STAT signaling mediates gangliosides-induced inflammatory responses in brain microglial cells. J Biol Chem 277:40594–40601

Kong G-Y, Peng Z-C, Costanzo C, Kristensson K, Bentivoglio M (2000) Inducible nitric oxide synthase expression elicited in the mouse brain by inflammatory mediators circulating in the cerebrospinal fluid. Brain Res 878:105–118

Konsman JP, Vigues S, Mackerlova LM, Bristow A, Blomqvist A (2004) Rat brain vascular distribution of interleukin-1 type-1 receptor immunoreactivity: relationship to patterns of inducible cyclooxygenase expression by peripheral inflammatory stimuli. J Comp Neurol 472: 113–129

Krasowska-Zoladek A, Banaszewska M, Kraszpulski M, Konat GW (2007) Kinetics of inflammatory response of astrocytes induced by TLR3 and TLR4 ligation. J Neurosci Res 85: 203–212

Kurschner C, Yuzaki M (1999) Neuronal interleukin-16 (NIL-16): a dual function PDZ domain protein. J Neurosci 19:7770–7780

Lacroix S, Rivest S (1998) Effect of acute systemic inflammatory response and cytokines on the transcription of the genes encoding cyclooxygenase enzymes (COX-1 and COX-2) in the rat brain. J Neurochem 70:452–466

Laflamme N, Rivest S (1999) Effects of systemic immunogenic insults and circulating proinflammatory cytokines on the transcription of the inhibitory factor κBα within specific cellular populations of the rat brain. J Neurochem 73:309–321

Lalancette-Hebert M, Gowing G, Simard A, Weng YC, Kriz J (2007) Selective ablation of proliferating microglial cells exacerbates ischemic injury in the brain. J Neurosci 27: 2596–2605

Lassmann H (2007) Multiple sclerosis: is there neurodegeneration independent from inflammation? J Neurological Sci 259:3–6

Lawrence DA (1981) In vivo and in vitro effects of lead on humoral and cell-mediated immunity. Infect Immun 31:36–143

Leal RB, Ribeiro SJ, Posser T, Cordova FM, Rigon AP, Filho EZ, Bainy ACD (2006) Modulation of ERK1/2 and $p38^{MAPK}$ by lead in the cerebellum of Brazilian catfish *Rhamdia quelen*. Aquat Toxicol 77:98–104

Lee H, Cha S, Lee M-S, Cho G-J, Choi WS, Suk K (2003) Role of antiproliferative B cell translation gene-1 as an apoptotic sensitizer in activation-induced cell death of brain microglia. J Immunol 171:5802–5811

Lee SJ, Drabik K, Van Wagoner NJ, Lee S, Choi C, Dong Y, Benveniste B (2000a) ICAM-1-induced expression of proinflammatory cytokines in astrocytes: involvement of extracellular signal-regulated kinase and p38 mitogen-activated protein kinase pathways. J Immunol 165:4658–4666

Lee J-M, Grabb MC, Zipfel GJ, Choi DW (2000b) Brain tissue responses to ischemia. J Clin Invest 106:723–731

Lee JH, Park EJ, Kim OS, Kim HY, Joe E-H, Jou I (2005) Double-stranded RNA-activated protein kinase is required for the LPS-induced activation of STAT1 inflammatory signaling in rat brain glial cells. GLIA 50:66–76

Li S, Ballou LR, Morham SG, Blatteis CM (2001) Cyclooxygenase-2 mediates the febrile response of mice to interleukin-1β. Brain Res 910:163–173

Li MO, Sarkisian MR, Mehal WZ, Rakic P, Flavell RA (2003) Phosphatidylserine receptor is required for clearance of apoptotic cells. Science 302:1560–1563

Licastro F, Pedrini S, Caputo L, Annoni G, Davis LJ, Ferri C, Casadei V, Grimaldi LME (2000) Increased plasma levels of interleukin-1, interleukin-6 and α-1-antichymotrypsin in patients with Alzheimer's disease: peripheral inflammation or signals from the brain. J Neuroimmunol 103:97–102

Lindahl LS, Bird L, Legare ME, Mikeska G, Bratton GR, Tiffany-Castiglioni E (1999) Differential ability of astroglia and neuronal cells to accumulate lead: dependence on cell type and on degree of differentiation. Toxicol Sci 50:236–243

Lukiw WJ, Percy ME, Kruck TP (2005) Nanomolar aluminum induces pro-inflammatory and pro-apoptotic gene expression in human brain cells in primary culture. J Inorg Biochem 99: 1895–1898

Luo J, Ho PP, Buckwalter MS, Hsu T, Lee LY, Zhang H, Kim D-K, Kim S-J, Gambhir SS, Steinman L, Wyss-Coray T (2007) Glia-dependent TGF-β signaling, acting independently of the TH17 pathway, is critical for initiation of murine autoimmune encephalomyelitis. J Clin Invest 117:3306–3315

Makwana M, Jones LL, Cuthill D, Heuer H, Bohatschek M, Hristova M, Friedrichsen S, Ormsby I, Bueringer D, Koppius A, Bauer K, Doetschman T, Raivich G (2007) Endogenous transforming growth factor β1 suppresses inflammation and promotes survival in adult CNS. J Neurosci 27:11201–11213

Martino G, Furlan R, Brambilla E, Bergami A, Ruffini F, Gironi M, Poliani PL, Grimaldi LME, Comi G (2000) Cytokines and immunity in multiple sclerosis: the dual signal hypothesis. J Neuroimmunol 109:3–9

Martinon F, Gaide O, Pétrilli V, Mayor A, Tschopp J (2007) NALP inflammasomes: a central role in innate immunity. Semin Immunopathol 29:213–229

Mastronardi C, Whelan F, Yildiz OA, Hannestad J, Elashoff D, McCann SM, Licinia J, Wong M-L (2007) Caspase-1 deficiency reduces inflammation-induced brain transcription. Proc Natl Acad Sci 104:7205–7210

Matyszak MK (1998) Inflammation in the CNS: balance between immunological privilege and immune responses. Prog Neurobiol 56:19–35

Mennicken F, Maki R, de Souza EB, Quirion R (1999) Chemokinases and chemokine receptors in the CNS: a possible role in neuroinflammation and patterning. TiPS 20:73–78

Mesples B, Fontaine RH, Lelievre V, Launay J-M, Gressens P (2005) Neuronal TGF-β_1 mediates IL-9/mast cell interaction and exacerbates excitotoxicity in newborn mice. Neurobiol Dis 18:193–205

Mielke K, Herdegen T (2000) JNK and p38 stresskinases – degenerative effectors of signal-transduction-cascades in the nervous system. Prog Neurobiol 61:45–60

Milner R, Campbell IL (2006) Increased expression of the β4 and α5 integrin subunits in cerebral blood vessels of transgenic mice chronically producing the pro-inflammatory cytokines IL-6 or IFN-α in the central nervous system. Mol Cell Neurosci 33:429–440

Muceniece R, Zvejniece L, Kirjanova O, Liepinsh E, Krigere L, Baumane L, Kalvinish I, Wikberg JES, Dambrova M (2004) β- and γ-melanocortins inhibit lipopolysaccharide induced nitric oxide production in mice brain. Brain Res 995:7–13

Nhan TQ, Liles C, Schwartz SM (2006) Physiological functions of caspases beyond cell death. Am J Pathol 169:729–737

Niederkorn JY (2006) See no evil, hear no evil, do no evil: the lessons of immune privilege. Nat Immunol 7:354–359

Norris CM, Kadish I, Blalock EM, Chen K-C, Thibault V, Porter NM, Landfield PW, Kraner SD (2005) Calcineurin triggers reactive/inflammatory processes in astrocytes and is upregulated in aging and Alzheimer's models. J Neurosci 25:4649–4658

O'Callaghan JP, Sriram K (2005) Glial fibrillary acidic protein and related glial proteins as biomarkers of neurotoxicity. Expert Opin Drug Saf 4:433–442

Odemis V, Moepps B, Gierschik P, Engele J (2002) Interleukin-6 and cAMP induce stromal cell-derived factor-1 chemotaxis in astroglia by up-regulating CXCR4 cell surface expression. J Biol Chem 277:39801–39808

Pahan K, Khan M, Singh I (2000) Interleukin-10 and interleukin-13 inhibit proinflammatory cytokine-induced ceramide production through the activation of phosphatidylinositol 3-kinase. J Neurochem 75:576–582

Pannu R, Singh AK, Singh I (2005) A novel role of lactosylceramide in the regulation of tumor necrosis factorα-mediated proliferation of rat primary astrocytes. J Biol Chem 280: 13742–13751

Pasinetti GM (1998) Cyclooxygenase and inflammation in Alzheimer's disease: experimental approaches and clinical interventions. J Neurosci Res 54:1–6

Patel HC, Boutin H, Allan SM (2003) Interleukin-1 in the brain: mechanism of action in acute neurodegeneration. Ann N Y Acad Sci 992:39–47

Patkai J, Mesples B, Dommergues M-A, Fromont G, Thornton EN, Renauld J-C, Evrard P, Gressens P (2001) Deleterious effects of IL-9-activated mast cells and neuroprotection by antihistamine drugs in the developing mouse brain. Pediatr Res 50:222–230

Pinteaux E, Inoue W, Schmidt L, Molina-Holgado F, Rothwell NJ, Luheshi GN (2007) Leptin induces interleukin-1β release from rat microglial cells through a caspase 1 independent mechanism. J Neurochem 102:826–833

Pollmacher T, Haack M, Schuld A, Reichenberg A, Yirmiya R (2002) Low levels of circulating inflammatory cytokines – Do they affect human brain functions? Brain Behav Immun 16: 525–532

Quan N, Banks WA (2007) Brain-immune communication pathways. Brain Behav Immun 21: 727–735

Quan N, Mhlanga JDM, Whiteside MB, McCoy AN, Kristensson K, Herkenham M (1999) Chronic overexpression of proinfammatory cytokines and histopathology in the brains of rats infected with *Trypanosoma brucei*. J Comp Neurobiol 414:114–130

Rao N, Nguyen S, Ngo K, Fung-Leung W-P (2005) A novel splice variant of interlekin-1 receptor (IL-1R)-associated kinase 1 plays a negative regulatory role in Toll/IL-1R-induced inflammatory signaling. Mol Cell Biol 25:6521–6532

Rather LJ (1971) Disturbance of function (functio laesa): the legendary fifth cardinal sign of inflammation, added by Galen to the four cardinal signs of Celsus. Bull N Y Acad Med 47:303–322

Razmiafshari M, Zawia NH (2000) Utilization of a synthetic peptide as a tool to study the interaction of heavy metals with the zinc finger domain of proteins critical for gene expression in the developing brain. Toxicol Appl Pharmacol 166:1–12

Rivest S (2003) Molecular insights on the cerebral innate immune system. Brain Behav Immun 17:13–19

Rosales-Corral S, Tan D-X, Reiter RJ, Valdivia-Velazquez M, Acosta-Martinez JP, Ortiz GG (2004) Kinetics of the neuroinflammation-oxidative stress correlation in rat brain following the injection of fibrillar amyloid-β onto the hippocampus in vivo. J Neuroimmunol 150:20–28

Rummel C, Hubschle T, Gerstberger R, Roth J (2004) Nuclear translocation of the transcription factor STAT3 in the guinea pig brain during systemic or localized inflammation. J Physiol 557:671–687

Sapirstein A, Saito H, Texel SJ, Samad TA, O'Leary E, Bonventre JV (2005) Cytosolic phospholipase $A_2\alpha$ regulates induction of brain cyclooxygenase-2 in a mouse model of inflammation. Am J Physiol Regul Integr Comp Physiol 288:R1774–R1782

Sheng JG, Ito K, Skinner RD, Mrak RE, Rovnaghi CR, van Eldik LJ, Griffen WST (1996) In vivo and in vitro evidence supporting a role for the inflammatory cytokine interleukin-1 as a driving force in Alzheimer pathogenesis. Neurbiol Aging 17:761–766

Shepherd CE, Goyette J, Utter V, Rahimi F, Yang Z, Geczy CL, Halliday GM (2006) Inflammatory S100A9 and S100A12 proteins in Alzheimer's disease. Neurobiol Aging 27:1554–1563

Singh I, Pahan K, Khan M, Singh AK (1998) Cytokine-mediated induction of ceramide production is redox-sensitive. J Biol Chem 273:20354–20362

Smith SEP, Li J, Garbett K, Mirnics K, Patterson PH (2007) Maternal immune activation alters fetal brain development through interleukin-6. J Neurosci 27:10695–10702

Stasiolek M, Gavrilyuk V, Sharp A, Horvath P, Selmaj K, Feinstein DL (2000) Inhibitory and stimulatory effects of lactacystin on expression of nitric oxide synthase type 2 in brain glial cells. J Biol Chem 275:24847–24856

Steffensen SC, Campbell IL, Henriksen SJ (1994) Site-specific hippocampal pathophysiology due to cerebral overexpression of interleukin-6 in transgenic mice. Brain Res 652:149–153

Stehlit C, Dorfleutner A (2007) COPs and POPs: modulators of inflammasome activity. J Immunol 179:7993–7998

Stolp HB, Dziegielewska KM, Ek CJ, Habgood MD, Lane MA, Potter AM, Saunders NR (2005) Breakdown of the blood-brain barrier to proteins in white matter of the developing brain following systemic inflammation. Cell Tissue Res 320:369–378

Struzynska L, Dabrowska-Bouta B, Koza K, Sulkowski G (2007) Inflammation-like glial response in lead-exposed immature rat brain. Toxicol Sci 95:156–162

Sugama S, Fujita M, Hashimoto M, Conti B (2007) Stress induced morphological microglial activation in the rodent brain: involvement of interleukin-18. Neuroscience 146:1388–1399

Taga T, Fukuda S (2005) Role of IL-6 in the neural stem cell differentiation. Clin Rev Allergy Immunol 28:249–256

Tiffany-Castiglioni E, Sierra EM, Wu JN, Rowles TK (1989) Lead toxicity in neuroglia. Neurotoxicology 10:417–443

Van Wagoner NJ, Benveniste EN (1999) Interleukin-6 expression and regulation in astrocytes. J Neuroimmunol 100:124–139

Van Wagoner NJ, Choi C, Repovic P, Benveniste EN (2000) Oncostatin M regulation of interleukin-6 expression in astrocytes: biphasic regulation involving the mitogen-activated protein kinases ERK1/2 and p38. J Neurochem 75:563–575

Weisskopf MG, Hu H, Sparrow D, Lenkinski RE, Wright RO (2007) Proton magnetic resonance spectroscopic evidence of glial effects of cumulative lead exposure in the adult human hippocampus. Environ Health Perspect 115:519–523

White LD, Cory-Slechta DA, Gilbert ME, Tiffany-Castiglioni E, Zawia NH, Virgolini M, Rossi-George A, Lasley SM, Qian YC, Basha MR (2007) New and evolving concepts in the neurotoxicology of lead. Toxicol Appl Pharmacol 225:1–27

Woiciechowsky C, Schoning B, Daberkow N, Asche K, Stoltenburg G, Lanksch WR, Volk H-D (1999) Brain-IL-1β induces local inflammation but systemic anti-inflammatory response through stimulation of both hypothalamic-pituitary-adrenal axis and sympathetic nervous system. Brain Res 816:563–571

Wong M-L, Bongiorno PB, Rettori V, McCann SM, Lici J (1997) Interleukin (IL) 1β, IL-1 receptor antagonist, IL-10, and IL-13 gene expression in central nervous system and anterior pituitary during systemic inflammation with pathophysiological implications. Proc Natl Acad Sci USA 94:227–232

Xiao B-G, Link H (1999) Is there a balance between microglia and astrocytes in regulating Th1/Th2-cell responses and neuropathologies? Trends Immunol 20:477–479

Yabe T, Sanagi T, Schwartz JP, Yamada H (2005) Pigment epithelium-derived factor induces pro-inflammatory genes in neonatal astrocytes through activation of NF-κB and CREB. GLIA 50:223–234

Zawia NH, Sharan R, Brydie M, Oyama T, Crumpton T (1998) Sp1 as a target site for metal-induced perturbation of transcriptional regulation of developmental brain gene expression. Dev Brain Res 107:291–298

Zhang J, Rivest S (2001) Anti-inflammatory effects of prostaglanding E_2 in the central nervous system in response to brain injury and circulating lipopolysaccharide. J Neurochem 76:855–864

Zipp F, Aktas O (2006) The brain as a target of inflammation: common pathways link inflammatory and neurodegenerative diseases. Trends Neurosci 29:518–527

Neurochemistry of Autism

Timothy D. Folsom and S. Hossein Fatemi

Abstract Autism is a neurodevelopmental disorder characterized by presence of social deficits, language abnormalities, stereotypies, and repetitive behavior. Brain pathology is extensive, suggesting widespread dysfunction of neurotransmitter systems. Genetic, biochemical, and gene association studies have shown that a number of neurotransmitters including serotonin, dopamine, oxytocin, GABA and glutamate, and acetylcholine contribute to the pathology of autism. Pharmacological treatment of autism has focused on reduction of symptoms and atypical antipsychotics, antidepressants, mood stabilizers, and anticonvulsants have been shown to successfully reduce many symptoms of autism. In this review we discuss the contributions of neurotransmitter systems to the pathology of autism and pharmacological treatment of autistic symptoms associated with neurochemical dysfunction.

Keywords Serotonin · GABA · Neurotransmitter · Neuropeptide · Pharmacotherapy · Genes

Contents

S.H. Fatemi (✉)
Division of Neuroscience Research, Department of Psychiatry, University of Minnesota Medical School, Minneapolis, MN 55455, USA
e-mail: fatem002@umn.edu

J.P. Blass (ed.), *Neurochemical Mechanisms in Disease*,
Advances in Neurobiology 1, DOI 10.1007/978-1-4419-7104-3_13,

1 Introduction

Autism is a debilitating neurodevelopmental disorder with heritability of >90% (Bailey et al., 1996) and characterized by presence of social deficits, language abnormalities, stereotypies, and repetitive behavior (APA, 1994). There is pervasive brain pathology encompassing different neurotransmitters and brain proteins (Bauman and Kemper, 1994, 2005; Acosta and Pearl, 2003; Palmen et al., 2004). Biochemical reports show involvement of several genes and proteins involved with neurotransmission implicating the GABAergic system (Blatt et al., 2001; Fatemi et al., 2002a, 2009a, b; Fatemi, 2008), cholinergic system (Perry et al., 2001; Lee et al., 2002), serotonergic system (Anderson, 2005), dopaminergic system (Gillberg et al., 1983; Gillberg and Svennerholm, 1987; de Krom et al., 2008), and the neuropeptide oxytocin (Waterhouse et al., 1996; Modahl et al., 1998) with the neuropathology of autism. Pharmacotherapy has focused on reduction of symptoms of autism including aggression, hyperactivity, self-injury, repetitive behavior, and anxiety. Antipsychotics, antidepressants, mood stabilizers, and anticonvulsants, among other drugs have been shown to successfully reduce autistic symptoms. In the current review, we examine the possible contributions of the serotonin, dopamine, acetylcholine, GABA and glutamate, and oxytocin to the pathology of autism and pharmacological treatment strategies that have shown efficacy in reducing symptoms of autism.

2 Serotonin

Serotonin (5-hydroxytrypamine) is an indolamine that is derived from tryptophan. A depletion in dietary tryptophan is known to lead to a worsening of autistic symptoms (McDougle et al., 1996a). Serotonin regulates a host of functions including mood, body temperature, arousal, hormone release, and eating (reviewed by Berger et al., 2009). In the mature brain, serotonin acts as a neurotransmitter, in the developing brain serotonin contributes to the development of serotonergic neurons and brain regions targeted by serotonergic neurons such as the prefrontal cortex (PFC) and hippocampus (Whitaker-Azmitia, 2001). Increased serotonin levels during early development may lead to a number of consequences contributing to brain pathology in autism (Whitaker-Azmita, 2005). Studies have shown that in utero exposure to drugs such as cocaine, that increase serotonin levels, result in higher rates of autism (Davis et al., 1992; Kramer et al., 1994). Conversely, low serotonin may also have negative effects on development and pose a risk factor for autism. Studies using an animal model have shown that when pregnant rats were treated with serotonin depletors it led to altered hippocampal and cortical development (Butkevich et al., 2003), abnormalities in levels of serotonin receptors in brain (Whitaker-Azmita et al., 1987), and behavioral abnormalities including passive avoidance (Shemer et al., 1988). Moreover, a murine model of prenatal viral infection that results in autistic-like behavior in offspring (Shi et al., 2003), has demonstrated that infection on embryonic day 16 (E16) and E18 which correspond to mid- and late second trimester, respectively, results in reduced levels of serotonin in cerebella of exposed offspring (Fatemi et al., 2008; Winter et al., 2008).

Hyperserotonemia has been a consistent finding in subjects with autism, which may be due to activity of serotonin-associated platelet proteins (Hranilović et al., 2008, 2009). Interestingly, 99% of blood serotonin is contained in platelets (Anderson et al., 1987) and studies have shown that there is an approximate 50% increase in blood-levels of serotonin in subjects with autism vs. controls (McBride et al., 1998). Hypotheses for increased serotonin include increased synthesis of serotonin by tryptophan hydroxylase (TPH1), increased uptake of serotonin into platelets via serotonin transporters (5-HTT), diminished release of serotonin from platelets via serotonin 2A receptor, and decreased breakdown of serotonin by monoamine oxidase (MAOA) (Hranilovic et al., 2008). A study by Hranilovic et al. (2008) identified polymorphisms of tryptophan hydroxylase and MAOA with increased serum serotonin levels. Similarly, haplotype analysis has shown a significant association between polymorphisms of TPH1 and increased serotonin in whole blood (Cross et al., 2008).

The serotonin transporter gene 5-HTT (also known as SLC6A4) has been the focus of much research as a potential candidate gene for autism (Cook and Leventhal, 1996; Ozaki et al., 2003). 5-HTT modulates serotonergic neurotransmission by active reuptake of serotonin from the synaptic cleft (Amara and Pacholczyk, 1991). Of the over 20 polymorphisms of 5-HTT, there are two that are of interest due to their functional effects: (1) 5-HTTLPR which has a deletion/insertion at the 5′-flanking regulatory region that results in a long variant (L) and a short variant (S) (Heils et al., 1996). The short variant reduces the efficiency of the 5-HT gene promoter and results in lower gene expression and serotonin uptake ability (Heils et al., 1996); and (2) STin2 which has a variable number of tandem repeats in the second intron and results in three common alleles: STin2.9, STin2.10, and STin2.12 indicating 9, 10, and 12 repeats, respectively (Lesch et al., 1996). A recent meta-analysis of family-based and population-based association studies for 5-HTT found no significant association between 5-HTTLPR and STin2 variants and autism (Huang and Santangelo, 2008).

Selective serotonin reuptake inhibitors (SSRIs) inhibit the neuronal reuptake of serotonin in the central nervous system and have shown mixed efficacy in the treatment of autistic symptoms (Moore et al., 2004). A number of studies have shown reductions in repetitive behaviors, lethargy, inappropriate speech, and improvements in the ability to relate to others, cognition, language improvement with fluoxetine (DeLong et al., 1998; Fatemi et al., 1998; Peral et al., 1999), fluvoxamine (McDougle et al., 1996b), and sertraline (Steingard et al., 1997). However, other studies have shown a lack of response with fluvoxamine (Martin et al., 2003) and citalopram (Couturier and Nicolson, 2002).

3 Dopamine

Dopamine (DA) is a catecholamine synthesized from the amino acid tyrosine and is thought to affect a wide range of behaviors and functions including cognition, motor function, selective attention, and brain-stimulation reward mechanisms (Beninger and Banasikowski, 2008; Boulougouris and Tsaltas, 2008; Cools, 2008; Fox et al., 2008). Dopamine is produced when tyrosine is hydroxylated into L-dihydroxyphenylalanine (L-DOPA), which is in turn converted to dopamine by DOPA decarboxylase (DDC). In the brain several important dopamanergic systems are of importance to autism: (1) the nigrostriatal system in which dopamanergic axons project from the substantia nigra to the neostriatum (Carlson, 2001); (2) the mesolimbic system in which dopamanergic axons project from the ventral tegmental area to the nucleus accumbens, amygdala, and hippocampus (Carlson, 2001); and (3) the mesocortical system in which dopamanergic axons project from the ventral tegmental area to the prefrontal cortex (Carlson, 2001).

Despite the importance of these three systems on regulating a number of behaviors that are known to be impaired in autism, studies of DA levels in CSF, blood, and urine in subjects with autism have been inconsistent. A study of plasma and urine has revealed no differences in levels of DA or its metabolites homovanillic acid (HVA) or 3,4-dihydroxyphenylacetic acid (DOPAC) in subjects with autism when compared with controls (Minderaa et al., 1989). Two studies of HVA in CSF have found elevated levels (Gillberg et al., 1983; Gillberg and Svennerholm, 1987), while others have found no difference (Cohen et al., 1974, 1977; Winsberg et al., 1980; Ross et al., 1985; Narayan et al., 1993). Despite these findings, antipsychotic drugs, which generally function as dopamine blocking agents, have been found efficacious in treatment of autistic symptoms such as repetitive stereotyped behaviors, hyperactivity, and aggression (reviewed by Canitano and Scandurra, 2008; McDougle et al., 2008; Posey et al., 2008).

Genetic studies have investigated linkages between enzymes responsible for the production of dopamine (tyrosine hydroxylase and DDC) and with dopamine receptors with autism. A study of 90 parent–offspring trios recruited in Europe found no evidence of linkage disequilibrium between two polymorphisms of DDC and autism (Lauritsen et al., 2002). Nor did they find linkage disequilibrium between haplotypes of the variants and autism (Lauritsen et al., 2002). Three studies have similarly found no link between tyrosine hydroxylase and autism (Martineau et al., 1994; Comings et al., 1995; Philippe et al., 2002). Thus far, it does not appear as though tyrosine hydroxylase or DDC are candidate genes for autism. Dopamine receptor D3 (DRD3) has also been investigated as a potential autism candidate gene. One study of autistic children found no linkage between DRD3 and autism (Martineau et al., 1994), however, in a more recent study a single nucleotide polymorphism (SNP) for DRD3 was associated with autism (de Krom et al., 2008). As the DRD3 receptor has been shown to be related to obsessive–compulsive behavior (Light et al., 2006), and liability to side effects of antipsychotic medication (Campbell et al., 1997), further study is needed to elucidate what role, if any, this receptor has in autism.

4 Acetylcholine

Acetylcholine (ACh) is a neurotransmitter found in both the central and peripheral nervous systems. In the peripheral nervous system, ACh activates muscles and in the central nervous system ACh is a neuromodulator contributing to functions including learning and memory (Kandel et al., 1995). Postmortem studies of subjects with autism have revealed abnormalities in the basal forebrain of children and adults with autism, with children having larger and more numerous cholinergic neurons and adults having smaller and less numerous cholinergic neurons (Bauman and Kemper, 1994). Cholinergic receptor (muscarinic and nicotinic) abnormalities have also been identified in brains of subjects with autism (Perry et al., 2001; Lee et al., 2002; Martin-Ruiz et al., 2004). Perry et al. (2001) found reduced [^{3}H]Pirenzepine binding to muscarinic M(1) receptors in the parietal cortex of subjects with autism and reduced [^{3}H]epibatidine binding to the α_4 and β_2 nicotinic receptor subunits in both the frontal and parietal cortices (Perry et al., 2001). Moreover, immunocytochemical analysis showed reduced levels of the α_4 and β_2 nicotinic receptor subunits in the parietal cortex, verifying the binding studies (Perry et al., 2001). Similarly, a separate study showed reduced [^{3}H]epibatidine binding in the granule cell, Purkinje, and molecular layers in cerebella of subjects with autism compared with controls that was accompanied by significantly reduced α_4 subunit protein (Lee et al., 2002). In contrast, in the same regions, there was an increase in α-bungarotoxin binding to the α_7 subunit whereas there were no significant changes in muscarinic receptor subunits (Lee et al., 2002). Finally, a study by Martin-Ruiz et al. (2004) verified some of these earlier results by demonstrating reduced [^{3}H]epibatidine binding to α_4 and β_2 receptor subunits and reduced α_4 subunit mRNA in parietal cortex and increased α_7 binding and reduced α_4 subunit protein in cerebellum.

Although studies have shown no differences in cholinergic enzyme markers acetylcholinesterase (Perry et al., 2001) or acetyltransferase (Perry et al., 2001; Lee et al., 2002) activity in subjects with autism, acetylcholinesterase inhibitors including donepezil, rivastigmine, and galantamine have shown some promise in treating symptoms of autism. Interestingly, both donepezil and galantamine are effective in improving prepulse inhibition of the acoustic startle response in mice, suggesting that they may act as cognitive enhancers (Hohnadel et al., 2007). A pilot study using donepezil to treat children and adolescents with autism found that 50% of the subjects demonstrated significant improvement in hyperactivity and irritability (Hardan and Handen, 2002). Treatment with rivastigmine in a 12-week open-label study resulted in increases in expressive speech and overall autistic behavior in subjects with autism (Chez et al., 2004). Treatment with galantamine has been shown to increase verbal fluency (Hertzman, 2003), improve emotional lability and inattention (Nicholson et al., 2006), and reduce anger, social withdrawal, and parent-rated irritability (Nicholson et al., 2006). These results suggest that inhibition of the breakdown of acetylcholine by acetylcholinesterase is efficacious in treating a number of symptoms of autism.

5 GABA and Glutamate

Glutamate is the primary excitatory transmitter substance in brain and spinal cord, and gamma-aminobutyric acid (GABA) is responsible for the majority of inhibitory neurotransmission in the brain (Lam et al., 2006; Carlson, 2001; Kandel et al., 1995). There are few, if any, areas in the brain that are not affected by these two substances (Lam et al., 2006; Carlson, 2001). Several reports have demonstrated abnormalities involving the glutamatergic and GABAergic systems of subjects with autism (Blatt et al., 2001; Dhossche et al., 2002; Fatemi, 2008; Fatemi et al., 2009a, b).

Glutamic acid decarboxylase (GAD) is the rate-limiting enzyme that is responsible for conversion of glutamate to GABA. In the adult brain, GAD exists in two major isoforms: GAD 65 and GAD 67 kDa proteins (Erlander et al., 1991). GAD 65 is a membrane-bound protein largely localized to axon terminals and is involved in vesicular synthesis of GABA (Laprade and Soghomonian, 1999). GAD 67 is a cytoplasmic protein primarily localized to interneurons and is involved in nonvesicular GABA release (Reetz et al., 1991).

Our laboratory has demonstrated that brain levels of GAD 65 and 67 kDa proteins were significantly decreased in cerebellum (GAD65) and parietal cortex (GAD67) in subjects with autism (Fatemi et al., 2002a). Yip et al. (2007) reported a significant decrease in GAD67 mRNA in autistic cerebellum, confirming our previous findings (Fatemi et al., 2002a). The major deficiencies in levels of GAD 65 and 67 kDa proteins in two important brain areas in autism may subserve deficiency in availability of GABA affecting important biological functions such as learning, locomotor activity, reproduction, and circadian rhythms (Soghomonian and Martin, 1998). Additionally, decreases in levels of GAD 65 and 67 kDa proteins in the autistic brain will negatively affect normal processing of visual, somatic, locomotor, and memory information processing, and could also explain the observations of increased blood, platelet, and CSF glutamate levels in the autistic patients (Pan et al., 1999; Moreno-Fuenmayor et al., 1996; Moreno et al., 1992). Moreover, deficiency in GABA due to decreased conversion of glutamate could account for the fact that up to one third of autistic subjects suffer from seizure disorders.

Binding of GABA to its receptors transduces various signals underlying various inhibitory transmissions in the brain. There are three main classes of receptors, $GABA_A$, $GABA_B$, and $GABA_C$ (Guidotti et al., 2005). $GABA_A$ receptors are ligand-gated ion channels that mediate GABA's fast inhibitory action (Brandon et al., 2000). $GABA_A$ receptors are divided into multiple subunits, for example: $\alpha1$–$\alpha6$, $\beta1$–$\beta4$, $\gamma1$–$\gamma4$, δ, ε, π, θ, and $\rho1$–$\rho2$, which combine to form multiple $GABA_A$ receptors (Ma et al., 2005; Brandon et al., 2000). $GABA_B$ receptors are heterodimeric, composed of the $GABA_B$ receptor 1 (GABBR1) and $GABA_B$ receptor 2 (GABBR2) subunits (Jones et al., 1998). $GABA_B$ receptors facilitate the release of neurtransmitters presynaptically generating inhibitory potentials postsynaptically (Bowery, 2000; Kuriyama et al., 2000). $GABA_C$ receptors are ionotropic, similar to $GABA_A$ receptors (Johnsoton et al., 2003) although they exhibit high GABA sensitivity and slow activation and deactivation kinetics (Qian and Ripps, 2008). They are composed of GABA ρ subunits of which there are three (ρ_1–ρ_3) and are expressed primarily in the retina although they are present in other regions of the CNS (Qian and Ripps, 2008).

Our laboratory has demonstrated significanlty reduced levels of $GABA_A$ and $GABA_B$ receptor protein in cerebella (GABRA1, GABBR3, GABBR1, GABBR2), parietal cortex (GABRA1, GABRA2, GABRA3, GABRA5, GABRB3, GABBR1), and prefrontal cortex (GABRA1, GABRA5, GABRA5, GABRB1, GABBR1) of subjects with autism (Fatemi et al., 2009a,b, 2010). All three brain areas have previously been implicated in the pathogenesis of autism (Bauman and Kemper, 1994, 2005). Alterations in all GABA receptors may partially explain the seizure disorders associated with autism. The occurrence of seizure disorders comorbid with autism has been estimated from 4 to 44% (Tuchman and Rapin, 2002). The presence of epileptiform activity may explain cognitive deficits common to children with autism and epilepsy (Binnie, 1993); phenomena that may also occur in children with autism and epilepsy. Multiple laboratories have demonstrated altered expression of GABBR1 and GABBR2 in animal models for seizure disorders (Straessle et al., 2003; Princivalle et al., 2003a; Han et al., 2006). Moreover, expression for GABBR1A, GABBR1B, and GABBR2 are altered in the hippocampus of subjects with temporal lobe epilepsy (Princivalle et al., 2003b).

6 Oxytocin

Oxytocin is a neuropeptide synthesized in the paraventricular and supraoptic nucleus of the hypothalamus. Oxytocin is released from axon terminals of the posterior pituitary into the bloodstream. It is also distributed to the central nervous system and oxytocin binding sites are found throughout, especially in the limbic system (Insel and Young, 2000). Oxytocin has been linked to affiliative behavior, social memory, and behavior, all of which are impaired in autism (Insel et al., 1999). It has been hypothesized that dysfunction of oxytocin and vasopresin contributes to social impairment in autism (Waterhouse et al., 1996).

Animal models have shown that oxytocin plays a role in social recognition (Popik et al., 1992) and that oxytocin antagonists disrupt social memory (Engelmann et al., 1998). Oxytocin knock-out mice have been shown to be unable to recognize conspecifics and this lack of recognition is not due to general disruptions of olfaction, learning, or memory (Ferguson et al., 2000; Choleris et al., 2003). However, a single injection of oxytocin prior to the first encounter with the conspecific allowed for social memory acquisition (Ferguson et al., 2001).

Plasma oxytocin has been shown to be reduced in autistic children and moreover, levels of oxytocin were correlated with social impairment (Modahl et al., 1998). A follow-up study using the same subjects found that the autistic children had higher levels of the precursor of oxytocin when compared with controls, suggesting that reduced plasma oxytocin in autistic children may be related to how oxytocin is processed (Green et al., 2001). Preliminary studies have demonstrated that infusion with oxytocin can reduce repetitive behaviors such as need to know, repeating, self-injury, and touching (Hollander et al., 2003), and increase affective speech comprehension (Hollander et al., 2007) in subjects with autism.

A number of recent studies have linked the oxytocin receptor gene (OXTR) to autism (Wu et al., 2005; Ylisaukko-oja et al., 2006; Jacob et al., 2007; Lerer et al., 2008; Yrigollen et al., 2008). OXTR expression is enhanced in brain regions associated with social behavior including the amygdala and lateral septum (Ferguson et al., 2000). OXTR knock-out mice display defects in social discrimination and demonstrate more aggression than normal mice (Takayanagi et al., 2005). Taken together, these studies suggested a potential role of OXTR in social deficits related to autism. In a study of 314 Finnish autism families a number of loci were identified as potentially conferring susceptibility to autism including 3p24-26 which includes OXTR (Ylisaukko-oja et al., 2006). A study of 57 Caucasian autism trios identified a significant association between an SNP (rs2254298) of OXTR previously associated with autism in a Han Chinese population sample (Wu et al., 2005; Jacob et al., 2007). A third study showed a significant association between SNPs (including rs2254298) and autism in an Israeli sample (Lerer et al., 2008). Moreover, an association between OXTR SNPs and IQ and Vineland Adaptive Behavior Scales (VABS) suggested that OXTR affects cognition and daily living skills in subjects with autism (Lerer et al., 2008). Finally, a link between the OXTR gene and affiliative behaviors, which are impaired in autism, has been identified (Yrigollen et al., 2008).

7 Reelin in Autism

Reelin is a secreted extracellular matrix protein with serine protease activity (DeBergeyck et al., 1998) that is critically involved in guiding brain development in an orderly fashion. Changes in the level of this protein, its receptors, or downstream proteins may cause abnormal corticogenesis. Reelin binds several proteins as likely receptors, including apolipoprotein E receptor 2 (ApoER2), very-low-density lipoprotein receptor (VLDLR), and $\alpha 3\beta 1$ integrin protein (D'Arcangelo et al., 1999; Hiesberger et al., 1999; Dulabon et al., 2000). Reelin binding to ApoER2 and VLDLR receptors induces clustering of the latter receptors, causing dimerization/oligomerization of the adaptor protein, disabled-1 (Dab-1), on the cytosolic aspect of the plasma membrane (Strasser et al., 2004) leading to tyrosine phosphorylation of Dab-1 (Cooper and Howell, 1999), resulting in the transduction of signaling pathway from the Reelin-producing cells. Embryologically, Reelin guides neurons and radial glial cells to their correct positions in the developing brain (Forster et al., 2002; Luque et al., 2003). In adults Reelin may play a role in neurotransmission as a report has indicated that Reelin has a direct effect on enhancement of long-term potentiation (LTP) in the hippocampus (Weeber et al., 2002).

Two studies have demonstrated associations of polymorphisms of the RELN gene with autism (Persico et al., 2001; Zhang et al., 2002). However, four other studies have not found an association (Krebs et al., 2002; Bonora et al., 2003; Devlin et al., 2004; Li et al., 2004). Despite the lack of a clear genetic association between

RELN and autism, protein levels of Reelin have been observed to be reduced in cerebella (Fatemi et al., 2001, 2005), frontal cortex (Fatemi et al., 2005), and blood (Fatemi et al., 2002; Lugli et al., 2003) of subjects with autism. Reduction of Reelin in frontal cortex and cerebella of subjects with autism was verified by qRT-PCR (Fatemi et al., 2005). Moreover, Reelin receptor ApoER2 mRNA was increased in frontal cortex and cerebella of subjects with autism and downstream signaling molecule Dab-1 mRNA was decreased in the same brain areas (Fatemi et al., 2005) suggesting impairments in the Reelin signaling system. These impairments may be partly responsible for the structural and cognitive deficits observed in autism.

8 Conclusion

Autism is a heterogeneous disorder with no definitive etiology. Brain pathology, gene expression, and neurochemical dysfunction of various neurotransmitter signaling systems including serotonin, dopamine, acetylcholine, GABA and glutamate, and oxytocin suggest a role of neurotransmitter systems in the pathology of autism. Pharmacological treatments focus on reduction of various symptoms of autism and SSRIs, antiacetylcholinesterases, and infusions of oxytocin, have all shown some efficacy.

Acknowledgments Grant support by National Institute of Child Health and Human Development (#5R01HD052074-01A2 and #3R01KD052074-03S1) to SHF is gratefully acknowledged

References

Acosta MT, Pearl PL (2003) The neurobiology of autism: new pieces of the puzzle. Curr Neurol Neurosci Rep 3:149–156
Amara SG, Pacholczyk T (1991) Sodium-dependent neurotransmitter reuptake systems. Curr Opin Neurobiol 1:84–90
American Psychiatric Association (1994) Diagnostic and Statistical Manual of Mental Disorders, 4th edn. APA Press, Washington, DC
Anderson GM (2005) Serotonin in autism. In: Bauman ML, Kemper TL (eds) The Neurobiology of Autism. Johns Hopkins University Press, Baltimore, MD, pp 303–318
Anderson GM, Freedman DX, Cohen DJ, Volkmar FR, Hoder EL, McPhedran P, Minderaa RB, Hansen CR, Young JG (1987) Whole blood serotonin in autistic and normal subjects. J Child Psychol Psychiatry 28:885–900
Bailey A, Phillips W, Rutter M (1996) Autism: towards an integration of clinical, genetic, neuropsychological and neurobiological perspectives. J Child Psychol Psychiatry 37: 89–126
Bauman ML, Kemper TL (1994) Neuroanatomic observations of the brain in autism. In: Bauman ML, Kemper TL (eds) The neurobiology of autism. Johns Hopkins University Press, Baltimore, MD, pp 19–145
Bauman ML, Kemper TL (2005) Structural brain anatomy in autism: what is the evidence? In: Bauman ML, Kemper TL (eds) The neurobiology of autism. Johns Hopkins University Press, Baltimore, MD, pp 121–135
Beninger RJ, Banasikowski TJ (2008) Dopaminergic mechanism of reward-related incentive learning: focus on the dopamine D(3) receptor. Neurotox Res 14:57–70

Berger M, Gray JA, Roth BL (2009) The expanded biology of serotonin. Annu Rev Med 60: 355–366

Binnie CD (1993) Significance and management of transitory cognitive impairment due to subclinical EEG discharges in children. Brain Dev 15:23–30

Blatt GJ, Fitzgerald CM, Guptill JT, Booker AB, Kemper TL, Bauman ML (2001) Density and distribution of hippocampal neurotransmitter receptors in autism: an autoradiographic study. J Autism Dev Disord 31:537–544

Bonora E, Beyer KS, Lamb JA, Parr JR, Klauck SM, Benner A, Paolucci M, Abbott A, Ragoussis I, Poustka A, Bailey AJ, Monaco AP (2003) International Molecular Genetic Study of Autism (IMGSAC). Analysis of reelin as a candidate gene for autism. Mol Psychiatry 10:885–892

Boulougouris V, Tsaltas. E (2008) Serotonergic and dopaminergic modulation of attentional processes. Prog Brain Res 172:517–542

Bowery NG (2000) $GABA_B$ receptors structure and function. In: Martin DL, Olsen RW (eds) GABA in the nervous system: the view at fifty years. Lippincott Williams and Wilkins, Philadelphia, PA, pp 233–244

Brandon NJ, Smart TG, Moss SJ (2000) Regulation of GABAA Receptors by protein phosphorylation. In: Martin DL, Olsen RW (eds) GABA in the nervous system: the view at fifty years. Lippincott Williams and Wilkins, Philadelphia, PA, pp 191–206

Butkevich IP, Khozhai LI, Mikhailenko VA, Otellin VA (2003) Decreased serotonin level during pregnancy alters morphological and functional characteristics of tonic nociceptive system in juvenile offspring of the rat. Reprod Biol Endocrinol 1:96

Campbell M, Armenteros JL, Malone RP, Adams PB, Eisenberg ZW, Overall JE (1997) Neuroleptic-related dyskinesias in autistic children: a prospective, longitudinal study. J Am Acad Child Adolesc Psychiatry 36:835–843

Canitano R, Scandurra V (2008) Risperidone in the treatment of behavioral disorders associated with autism in children and adolescents. Neuropsychiatr Dis Treat 4:723–730

Carlson NR (2001) Physiology of behavior. Allyn and Bacon, Boston, MA

Chez MG, Aimonovitch M, Buchanan T, Mrazek S, Tremb RJ (2004) Treating autistic spectrum disorders in children: utility of the cholinesterase inhibitor rivastigmine tartrate. J Child Neurol 19:165–169

Choleris E, Gustafsson JA, Korach KS, Muglia LJ, Pfaff DW, Ogawa S (2003) An estrogen-dependent four-gene micronet regulating social recognition: a study with oxytocin and estrogen receptor-alpha and -beta knockout mice. Proc Natl Acad Sci U S A 100:6192–6197

Cohen DJ, Caprulo BK, Shaywitz BA, Bowers MB Jr (1977) Dopamine and serotonin metabolism in neuropsychiatrically disturbed children: CSF homovanillic acid and 5-hydroxyindoleacetic acid. Arch Gen Psychiatry 34:545–550

Cohen DJ, Shaywitz BA, Johnson WT, Bowers MB Jr (1974) Biogenic amines in autistic and atypical children: cerebrospinal fluid measures of homovillic acid and 5-hydroxyindoleacetic acid. Arch Gen Psychiatry 31:845–853

Comings DE, Gade R, Muhleman D, Sverd J (1995) No association of a tyrosine hydroxylase gene tetranucleotide repeat polymorphism in autism, Tourette syndrome, or ADHD. Biol Psychiatry 37:484–486

Cook EH, Leventhal BL (1996) The serotonin system in autism. Curr Opin Pediatr 8:348–354

Cools R (2008) Role of dopamine in the motivational and cognitive control of behavior. Neuroscientist 14:381–395

Cooper J, Howell BW (1999) Lipoprotein receptors: signaling function in the brain? Cell 97: 671–674

Couturier JL, Nicolson R (2002) A retrospective assessment of citalopram in children and adolescents with pervasive developmental disorders. J Child Adolesc Psychopharmacol 12:243–248

Cross S, Kim SJ, Weiss LA, Delahanty RJ, Sutcliffe JS, Leventhal BL, Cook EH Jr, Veenstra-Vanderweele J (2008) Molecular genetics of the platelet serotonin system in first-degree relatives of patients with autism. Neuropsychopharmacology 33:353–360

D'Arcangelo G, Homayouni R, Keshvara L, Rice DS, Sheldon M, Curran T (1999) Reelin is a ligand for lipoprotein receptors. Neuron 24:471–479

Davis E, Fennoy I, Laraque D, Kanem N, Brown G, Mitchell J (1992) Autism and developmental abnormalities in children with perinatal cocaine exposure. J Natl Med Assoc 84:315–319

DeBergeyck V, Naerhuyzen B, Goffinet AM, deRouvroit C (1998) A panel of monoclonal antibodies against Reelin, the extracellular matrix protein defective in reeler mutant mice. J Neurosci Methods 82:17–24

de Krom M, Staal WG, Ophoff RA, Hendriks J, Buitelaar J, Franke B, de Jonge MV, Bolton P, Collier D, Curran S, van Engeland H, van Ree JM (2008) A common variant in DRD3 receptor is associated with Autism spectrum disorder. Biol Psychiatry 65(7):625–630

DeLong GR, Teague LA, McSwain Kamran M (1998) Effects of fluoxetine treatment in young children with idiopathic autism. Dev Med Child Neurol 40:551–562

Devlin B, Bennett P, Dawson G, Figlewicz DA, Grigorenko EL, McMahon W, Minshew N, Pauls D, Smith M, Spence MA, Rodier PM, Stodgell C, Schellenberg GD; CPEA Genetics Network (2004) Alleles of a reelin CGG repeat do not convey liability to autism in a sample from the CPEA network. Am J Med Genet 126B:46–50

Dhossche D, Applegate H, Abraham A, Maertens P, Bland L, Bencsath A, Martinez J (2002) Elevated plasma gama-aminobutyric acid (GABA) levels in autistic youngsters: stimulus for GABA hypothesis of autism. Med Sci Mon 8:PR1–PR6

Dulabon L, Olson EC, Taglienti MG, Eisenhuth S, McGrath B, Walsh CA, Kreidberg JA, Anton ES (2000) Reelin binds alpha 3 beta 1 integrin and inhibits neuronal migration. Neuron 27:33–44

Engelmann M, Ebner K, Wotjak CT, Landgraf R (1998) Endogenous oxytocin is involved in short-term olfactory memory in female rats. Behav Brain Res 90:89–94

Erlander MG, Tillakaratne NJK, Feldblum S, Patel N, Tobin A (1991) Two genes encode distinct glutamate deoxycarbolases. Neuron 7:97–100

Fatemi SH (2008) The hyperglutamatergic hypothesis of autism. Prog Neuropsychopharmacol Biol Psychiatry 32:911

Fatemi SH, Folsom TD, Reutiman TJ, Thuras PD (2009a) Expression of GABA(B) receptors is altered in brains of subjects with autism. Cerebellum 8(1):64–69

Fatemi SH, Halt A, Stary J, Kanodia R, Schulz SC, Realmuto G (2002a) Glutamic acid decarboxylase 65 and 67 kDa proteins are reduced in parietal and cerebellar cortices of autistic subjects. Biol Psychiatry 52:805–810

Fatemi SH, Realmuto GM, Khan L, Thuras P (1998) Fluoxetine in treatment of adolescent patients with autism: a longitudinal open trial. J Autism Dev Disord 28:303–307

Fatemi SH, Reutiman TJ, Folsom TD, Huang H, Oishi K, Mori S, Smee DF, Pearce DA, Winter C, Sohr R, Juckel G (2008) Maternal infection leads to abnormal gene regulation and brain atrophy in mouse offspring: implications for genesis of neurodevelopmental disorders. Schizophr Res 99:56–70

Fatemi SH, Reutiman TJ, Folsom TD, Rooney RJ, Patel DH, Thuras P (2010) mRNA and protein levels for GabaAalpha4, alpha5, beta1, and GABABR1 receptors are altered in brains from subjects with autism. J Autism Dev Disord 40:743–750

Fatemi SH, Reutiman TJ, Folsom TD, Thuras PD (2009b) GABA(A) Receptor Downregulation in Brains of Subjects with Autism. J Autism Dev Disord 39:223–230

Fatemi SH, Snow AV, Stary JM, Araghi-Niknam M, Brooks AI, Pearce DA, Reutiman TJ, Lee S (2005) Reelin signaling is impaired in autism. Biol Psychiatry 57:777–787

Fatemi SH, Stary JM, Egan EA (2002b) Defective Reelin production as a vulnerability factor in genesis of autism. Cell Mol Neurobiol 22:139–152

Fatemi SH, Stary JM, Halt AR, Realmuto GR (2001) Dysregulation of Reelin and Bcl-2 proteins in autistic cerebellum. J Autism Dev Disord 6:529–535

Ferguson JN, Aldag JM, Insel TR, Young LJ (2001) Oxytocin in the medial amygdala is essential for social recognition in the mouse. J Neurosci 21:8278–8285

Ferguson JN, Young LJ, Hearn EF, Matzuk MM, Insel TR, Winslow JT (2000) Social amnesia in mice lacking the oxytocin gene. Nat Genet 25:284–288

Forster E, Tielsch A, Saum B, Weiss KH, Johanssen C, Graus-Porta D, Muller U, Frotscher M (2002) Reelin, disabled 1, and beta 1 integrins are required for the formation of the radial glial scaffold in the hippocampus. Proc Natl Acad Sci U S A 99:13178–13183

Fox SH, Chuang R, Brotchie JM (2008) Parkinson's disease–opportunities for novel therapeutics to reduce the problems of levadopa therapy. Prog Brain Res 172:479–494

Gillberg C, Svennerholm L (1987) CSF monoamines in autistic syndromes and other pervasive developmental disorders of early childhood. Br J Psychiatry 151:89–94

Gillberg C, Svennerholm L, Hamilton-Hellberg C (1983) Childhood psychosis and monoamine metabolites in spinal fluid. J Autism Dev Disord 13:383–396

Green L, Fein D, Modahl C, Feinstein C, Waterhouse L, Morris M (2001) Oxytocin and autistic disorder: alterations in peptide forms. Biol Psychiatry 50:609–613

Guidotti A, Auta J, Davis JM, Dong E, Grayson DR, Veldic M, Zhang X, Costa E (2005) GABAergic dysfunction in schizophrenia: new treatment strategies on the horizon. Psychopharmacology (Berl) 180:191–205

Han Y, Qin J, Bu DF, Chang XZ, Yang ZX (2006) Successive alterations of hippocampal gamma-aminobutyric acid B receptor subunits in a rat model of febrile seizure. Life Sci 8:2944–2952

Hardan AY, Handen BL (2002) A retrospective open trial of adjunctive donepezil in children and adolescents with autistic disorder. J Child Adolesc Psychopharmacol 12:237–241

Heils A, Teufel A, Petri S, Stober G, Riederer P, Bengel D, Lesch KP (1996) Allelic variation of human serotonin transporter gene expression. J Neurochem 66:2621–2624

Hertzman M (2003) Galantamine in the treatment of adult autism: a report of three clinical cases. Int J Psychiatry Med 33:395–398

Hiesberger T, Trommsdorff M, Howell BW, Goffinet A, Mumby MC, Cooper JA, Herz J (1999) Direct biding of Reelin to VLDL receptor and ApoE receptor 2 induces tyroene phosphorylation of disabled-1 and modulates tau phosphorylation. Neuron 24: 481–489

Hohnadel E, Bouchard K, Terry AV Jr (2007) Galantamine and donepezil attenuate pharmacologically induced deficits in prepulse inhibition in rats. Neuropharmacology 52:542–551

Hollander E, Bartz J, Chaplin W, Phillips A, Sumner J, Soorya L, Anagnostou E, Wasserman S (2007) Oxytocin increases retention of social cognition in autism. Biol Psychiatry 61:498–503

Hollander E, Novotny S, Hanratty M, Yaffe R, DeCaria CM, Aronowitz BR, Mosovich S (2003) Oxytocin infusion reduces repetitive behaviors in adults with autistic and Asperger's disorders. Neuropsychopharmacology 28:193–198

Hranilović D, Bujas-Petković Z, Tomicić M, Bordukalo-Niksić T, Blazević S, Cicin-Sain L (2009) Hyperserotonemia in autism: activity of 5-HT-associated platelet proteins. J Neural Transm 116:493–501

Hranilović D, Novak R, Babić M, Novokmet M, Bujas-Petković Z, Jernej B (2008) Hyperserotonemia in autism: the potential role of 5HT-related gene variants. Coll Antropol 32:75–80

Huang CH, Santangelo SL (2008) Autism and serotonin transporter gene polymorphisms: a systematic review and meta-analysis. Am J Med Genet B Neuropsychiatr Genet 147B:903–913

Insel TR, O'Brien DJ, Leckman JF (1999) Oxytocin, vasopressin, and autism: is there a connection? Biol. Psychiatry 45:145–157

Insel TR, Young LJ (2000) Neuropeptides and the evolution of social behavior. Curr Opin Neurobiol 10:784–789

Jacob S, Brune CW, Carter CS, Leventhal BL, Lord C, Cook EH Jr (2007) Association of the oxytocin receptor gene (OXTR) in Caucasian children and adolescents with autism. Neurosci Lett 417:6–9

Johnsoton GA, Chebib M, Hanrahan JR, Mewett KN (2003) GABA(C) receptors as drug targets. Curr Drug Targets CNS Neurol Disord 2:260–268

Jones KA, Borowsky B, Tamm JA, Craig DA, Durkin MM, Dai M et al (1998) GABA(B) receptors function as a heteromeric assembly of the subunits GABA(B)R1 and GABA(B)R2. Nature 396:674–679

Kandel, ER, Schwartz, JH, Jessell, TH (eds) (1995) Essentials of neuroscience and behavior. Appleton and Lange, Norwalk, CT

Kramer K, Azmitia EC, Whitaker-Azmitia PM (1994) In vitro release of [3H]5-hydroxytryptamine from fetal and maternal brain by drugs of abuse. Brain Res Dev Brain Res 78:142–146

Krebs MO, Betancur C, Leroy S, Bourdel MC, Gillberg C, Leboyer M (2002) Paris Autism Research International Sibpair (PARIS) study. Absence of association between a polymorphic GGC repeat in the 5′ untranslated region of the reelin gene and autism. Mol Psychiatry 7:801–804

Kuriyama K, Hirouchi M, Kimura H (2000) Neurochemical and molecular pharmacological aspects of the GABA(B) receptor. Neurochem Res 25:1233–1239

Lam KS, Aman MG, Arnold LE (2006) Neurochemical correlates of autistic disorder: a review of the literature. Res Dev Disabil 27:254–289

Laprade N, Soghomonian JJ (1999) Gene expression of the GAD67 and GAD65 isoforms of glutamate decarboxylase is differently altered in subpopulations of striatal neurons in adult rats lesioned with 6-OHDA as neonates. Synapse 33:36–48

Lauritsen MB, Børglum AD, Betancur C, Philippe A, Kruse TA, Leboyer M, Ewald H (2002) Investigation of two variants in the DOPA decarboxylase gene in patients with autism. Am J Med Genet 114:466–470

Lee M, Martin-Ruiz C, Graham A, Court J, Jaros E, Perry R, Iversen P, Bauman M, Perry E (2002) Nicotinic receptor abnormalities in the cerebellar cortex in autism. Brain 125:1483–1495

Lerer E, Levi S, Salomon S, Darvasi A, Yirmiya N, Ebstein RP (2008) Association between the oxytocin receptor (OXTR) gene and autism: relationship to Vineland adaptive behavior scales and cognition. Mol Psychiatry 13:980–988

Lesch KP, Bengel D, Heils A, Sabol SZ, Greenberg BD, Petri S, Benjamin J, Muller CR, Hamer DH, Murphy DL (1996) Association of anxiety related traits with a polymorphism in the serotonin transporter gene regulatory region. Science 274:1527–1531

Li J, Nguyen L, Gleason C, Lotspeich L, Spiker D, Risch N, Myers RM (2004) Lack of evidence for an association between WNT2 and RELN polymorphisms and autism. Am J Med Genet 126B:51–57

Light KJ, Joyce PR, Luty SE, Mulder RT, Frampton CM, Joyce LR, Miller AL, Kennedy MA (2006) Preliminary evidence for an association between a dopamine D3 receptor gene variant and obsessive-compulsive personality disorder in patients with major depression. Am J Med Genet B Neuropsychiatr Genet 141:409–413

Lugli G, Krueger JM, Davis JM, Persico AM, Keller F, Smalheiser NR (2003) Methodological factors influencing measurement and processing of plasma reelin in humans. BMC Biochem 4:9

Luque JM, Morante-Oria J, Fairen A (2003) Localization of ApoER2, VLDLR and Dab-1 in radial glia: groundwork for a new model of Reelin action during cortical development. Dev Brain Res 140:195–203

Ma DQ, Whitehead PL, Menold MM, Martin ER, Ashley-Koch AE, Mei H, Ritchie MD, Delong GR, Abramson RK, Wright HH, Cuccaro ML, Hussman JP, Gilbert JR, Pericak-Vance MA (2005) Identification of significant association and gene-gene interaction of GABA receptor subunit genes in autism. Am J Hum Genet 77:377–388

Martin A, Koenig K, Anderson GM, Scahill L (2003) Low-dose fluvoxamine treatment of children and adolescents with pervasive developmental disorders: a prospective, open-label study. J Autism Dev Disord 33:77–85

Martin-Ruiz CM, Lee M, Perry RH, Baumann M, Court JA, Perry EK (2004) Molecular analysis of nicotinic receptor expression in autism. Brain Res Mol Brain Res 123:81–90

Martineau J, Hérault J, Petit E, Guérin P, Hameury L, Perrot A, Mallet J, Sauvage D, Lelord G, Müh JP (1994) Catecholaminergic metabolism and autism. Dev Med Child Neurol 36: 688–697

McBride PA, Anderson GM, Hertzig ME, Snow ME, Thompson SM, Khait VD, Shapiro T, Cohen DJ (1998) Effects of diagnosis, race, and puberty on platelet serotonin levels in autism and mental retardation. J Am Acad Child Adolesc Psychiatry 37:767–776

McDougle CJ, Naylor ST, Cohen DJ, Aghajanian GK, Heninger GR, Price LH (1996a) Effects of tryptophan depletion in drug-free adults with autistic disorder. Arch Gen Psychiatry 53: 993–1000
McDougle CJ, Naylor ST, Cohen DJ, Volkmar FR, Heninger GR, Price LH (1996b) A double-blind, placebo-controlled study of fluvoxamine in adults with autistic disorder. Arch Gen Psychiatry 53:1001–1008
McDougle CJ, Stigler KA, Erickson CA, Posey DJ (2008) Atypical antipsychotics in children and adolescents with autistic and other pervasive developmental disorders. J Clin Psychiatry 69:15–20
Minderaa RB, Anderson GM, Volkmar FR, Akkerhuis GW, Cohen DJ (1989) Neurochemical study of dopamine functioning in autistic and normal subjects. J Am Acad Child Adolesc Psychiatry 28:190–194
Modahl C, Green L, Fein D, Morris M, Waterhouse L, Feinstein C, Levin H (1998) Plasma oxytocin levels in autistic children. Biol Psychiatry 43:270–277
Moore ML, Eichner SF, Jones JR (2004) Treating functional impairment of autism with selective serotonin-reuptake inhibitors. Ann Pharmacother 38:1515–1519
Moreno H, Borjas L, Arrieta A, Salz L, Prassad A, Estevez J, Bonilla E (1992) Clinical heterogeneity of the autistic syndrome: a study of 60 families. Invest Clin 33:13–31
Moreno-Fuenmayor H, Borjas L, Arrieta A, Valera V, Socorro-Candanoza L (1996) Plasma excitatory amino acids in autism. Invest Clin 37:113–128
Narayan M, Srinath S, Anderson GM, Meundi DB (1993) Cerebrospinal fluid levels of homovanillic acid and 5-hydroxyindoleacetic acid in autism. Biol Psychiatry 33:630–635
Nicholson R, Craven-Thuss B, Smith J (2006) A prospective, open-label trial of galantamine in autistic disorder. J Child Adolesc Psychopharmacol 16:621–629
Ozaki N, Goldman D, Kaye WH, Plotnicov K, Greenberg BD, Lappalainen J, Rudnick G, Murphy DL (2003) Serotonin transporter missense mutation associated with a complex neuropsychiatric phenotype. Mol Psychiatry 8(895):933–936
Palmen SJ, van Engeland H, Hof PR, Schmitz C (2004) Neuropathological findings in autism. Brain 127:2572–2583
Pan JW, Lane JB, Hetherington H, Percy AK (1999) Rett Syndrome: H-1 spectroscopic imaging at + 1 Tesla. J Child Neurol 14:524–528
Peral M, Alcami M, Gilaberte I (1999) Fluoxetine in children with autism. J Am Acad Child Adolesc Psychiatry 38:1472–1473
Perry EK, Lee ML, Martin-Ruiz CM, Court JA, Volsen SG, Merrit J, Folly E, Iversen PE, Bauman ML, Perry RH, Wenk GL (2001) Cholinergic activity in autism: abnormalities in the cerebral cortex and basal forebrain. Am J Psychiatry 158:1058–1066
Persico AM, D'Agruma L, Maiorano N, Totaro A, Militerni R, Bravaccio C, Wassink TH, Schneider C, Melmed R, Trillo S, Montecchi F, Palermo M, Pascucci T, Puglisi-Allegra S, Reichelt KL, Conciatori M, Marino R, Quattrocchi CC, Baldi A, Zelante L, Gasparini P, Keller F; Collaborative Linkage Study of Autism (2001) Reelin gene alleles and haplotypes as a factor predisposing to autistic disorder. Mol Psychiatry 6:150–159
Philippe A, Guilloud-Bataille M, Martinez M, Gillberg C, Råstam M, Sponheim E, Coleman M, Zappella M, Aschauer H, Penet C, Feingold J, Brice A, Leboyer M; Paris Autism Research International Sibpair Study (2002) Analysis of ten candidate genes in autism by association and linkage. Am J Med Genet 114:125–128
Popik P, Vetulani J, van Ree JM (1992) Low doses of oxytocin facilitate social recognition in rats. Psychopharmacology (Berl) 106:71–74
Posey DJ, Erickson CA, McDougle CJ (2008) Developing drugs for core social and communication impairment in autism. Child Adolesc Psychiatr Clin N Am 17:787–801
Princivalle AP, Duncan JS, Thom M, Bowery NG (2003b) GABA(B1a), GABA(B1b) and GABA(B2) mRNA variants expression in hippocampus resected from patients with temporal lobe epilepsy. Neuroscience 122:975–984

Princivalle AP, Richards DA, Duncan JS, Spreafico R, Bowery NG (2003a) Modification of GABA(B1) and GABA(B2) receptor subunits in the somatosensory cerebral cortex and thalamus of rats with absence seizures (GAERS). Epilepsy Res 55:39–51

Qian H, Ripps H (2008) Focus on molecules: the GABA(C) receptor. Exp Eye Res 88(6): 1002–1003

Reetz A, Solimena M, Matteoli M, Folli F, Takei K, De Camilli P (1991) GABA and pancreatic beta cells: colocalization of glutamic acid decarboxylase (GAD) and GABA with synaptic-like microvesicles suggests their role in GABA storage and secretion. EMBO J 10:1275–1284

Ross DL, Klykylo WM, Anderson GM (1985) Cerebrospinal fluid levels of homovanillic acid and 5-hydroxyindoleacetic acid in autism. Ann Neurol 18:394

Shemer A, Whitaker-Azmitia PM, Azmitia EC (1988) Effects of prenatal 5-methoxytryptamine and parachlorophenylalanine on serotonergic uptake and behavior in the neonatal rat. Pharmacol Biochem Behav 30:847–851

Shi L, Fatemi SH, Sidwell RW, Patterson PH (2003) Maternal influenza infection causes marked behavioral and pharmacological changes in the offspring. J Neurosci 23:297–302

Soghomonian JJ, Martin DL (1998) Two isoforms of glutamate decarboxylase: why? Trends Pharmacol Sci 19:500–505

Steingard RJ, Zimnitzky B, DeMaso DR, Bauman ML, Bucci JP (1997) Sertraline treatment of transition-associated anxiety and agitation in children with autistic disorder. J Child Adolesc Psychopharmacol 7:9–15

Straessle A, Loup F, Arabadzisz D, Ohning GV, Fritschy JM (2003) Rapid and long-term alterations of hippocampal GABAB receptors in a mouse model of temporal lobe epilepsy. Eur J Neurosci 18:2213–2226

Strasser V, Fasching D, Hauser C, Mayer H, Bock HH, Hiesberger T, Herz J, Weeber EJ, Sweatt JD, Pramatarova A, Howell B, Schneider WJ, Nimpf J (2004) Receptor clustering is involved in reelin signaling. Mol Cell Biol 24:1378–1386

Takayanagi Y, Yoshida M, Bielsky IF, Ross HE, Kawamata M, Onaka T, Yanagisawa T, Kimura T, Matzuk MM, Young LJ, Nishimori K (2005) Pervasive social deficits, but normal parturition, in oxytocin receptor-deficient mice. Proc Natl Acad Sci U S A 102:16096–16101

Tuchman R, Rapin I (2002) Epilepsy in autism. Lancet Neurol 1:352–358

Waterhouse L, Fein D, Modahl C (1996) Neurofunctional mechanisms in autism. Psychol Rev 103:457–489

Weeber EJ, Beffert U, Jones C, Christian JM, Forster E, Sweatt JD, Herz J (2002) Reelin and ApoE receptors cooperate to enhance hippocampal synaptic plasticity and learning. J Biol Chem 277:39944–39952

Whitaker-Azmitia PM (2001) Serotonin and brain development: role in human developmental diseases. Brain Res Bull 56:479–485

Whitaker-Azmitia PM (2005) Behavioral and cellular consequences of increasing serotonergic activity during brain development: a role in autism? Int J Dev Neurosci 23:75–83

Whitaker-Azmitia PM, Lauder JM, Shemmer A, Azmitia EC (1987) Postnatal changes in serotonin receptors following prenatal alterations in serotonin levels: further evidence for functional fetal serotonin receptors. Brain Res 430:285–289

Winsberg BG, Sverd J, Castells S, Hurwic M, Perel JM (1980) Estimation of monoamine and cyclic-AMP turnover and amino acid concentrations of spinal fluid in autistic children. Neuropediatrics 11:250–255

Winter C, Reutiman TJ, Folsom TD, Sohr R, Wolf RJ, Juckel G, Fatemi SH (2008) Dopamine and serotonin levels following prenatal viral infection in mouse–implications for psychiatric disorders such as schizophrenia and autism. Eur Neuropsychopharmacol 18:712–716

Wu S, Jia M, Ruan Y, Liu J, Guo Y, Shuang M, Gong X, Zhang Y, Yang X, Zhang D (2005) Positive association of the oxytocin receptor gene (OXTR) with autism in the Chinese Han population. Biol Psychiatry 58:74–77

Yip J, Soghomonian JJ, Blatt GJ (2007) Decreased GAD67 mRNA levels in cerebellar Purkinje cells in autism: pathophysiological implications. Acta Neuropathol 113:559–568

Ylisaukko-oja T, Alarcón M, Cantor RM, Auranen M, Vanhala R, Kempas E, von Wendt L, Järvelä I, Geschwind DH, Peltonen L (2006) Search for autism loci by combined analysis of autism genetic resource exchange and finnish families. Ann Neurol 59:145–155
Yrigollen CM, Han SS, Kochetkova A, Babitz T, Chang JT, Volkmar FR, Leckman JF, Grigorenko EL (2008) Genes controlling affiliative behavior as candidate genes for autism. Biol Psychiatry 63:911–916
Zhang H, Liu X, Zhang C, Mundo E, Macciardi F, Grayson DR, Guidotti AR, Holden JJ (2002) Reelin gene alleles and susceptibility to autism spectrum disorders. Mol Psychiatry 7: 1012–1017

RNA Pathologies in Neurological Disorders

Kinji Ohno and Akio Masuda

Abstract RNA is not a simple intermediate linking DNA and protein. RNA is widely transcribed from a variety of genomic regions, and extensive studies on the functional roles and regulations of noncoding RNAs including antisense RNAs and small RNAs are in progress. In addition, the human genome project revealed that we humans carry as few as ~22,000 genes. Humans exploit tissue-specific and developmental stage-specific alternative splicing to generate a large variety of molecules in specific cells at specific developmental stages. Neurological disorders are also subject to aberrations of the splicing mechanisms. This review focuses mostly on splicing abnormalities due to pathological alterations of splicing *cis*-elements and *trans*-factors. Pathomechanisms associated with disrupted splicing *cis*-elements can be applied to any human diseases, and we did not restrict the descriptions to neurological diseases. On the other hand, we limited the descriptions of dysregulated splicing *trans*-factors to neurological disorders. Neurological diseases covered in this review include congenital myasthenic syndromes, spinal muscular atrophy, myotonic dystrophy, Alzheimer's disease, frontotemporal dementia with Parkinsonism linked to chromosome 17, facioscapulohumeral muscular dystrophy, fragile X-associated tremor/ataxia syndrome, Prader–Willi syndrome, Rett syndrome, spinocerebellar atrophy type 8, and paraneoplastic neurological disorders.

Keywords The RNA world · Pre-mRNA splicing · Splicing *cis*-elements · Splicing *trans*-factors · Branch point sequence (BPS) · Exonic splicing enhancer (ESE) · Exonic splicing silencer (ESS) · Intronic splicing enhancer (ISE) · Intronic splicing silencer (ISS) · Nonsense-mediated mRNA decay (NMD) · Nonsense-associated skipping of a remote exon (NASRE) · Congenital myasthenic syndromes · Spinal muscular atrophy (SMA) · Myotonic dystrophy (DM1, DM2) · Alzheimer's disease · Frontotemporal dementia with Parkinsonism linked to chromosome 17 (FTDP-17) ·

K. Ohno (✉)
Division of Neurogenetics, Center for Neurological Diseases and Cancer, Nagoya University Graduate School of Medicine, Nagoya, Japan
e-mail: ohnok@med.nagoya-u.ac.jp

J.P. Blass (ed.), *Neurochemical Mechanisms in Disease*,
Advances in Neurobiology 1, DOI 10.1007/978-1-4419-7104-3_14,

Facioscapulohumeral muscular dystrophy (FSHD) · Fragile X-associated tremor/ataxia syndrome (FXTAS) · Prader–Willi syndrome, Rett syndrome · Spinocerebellar atrophy type 8 (SCA8) · Paraneoplastic neurological disorders (PND)

Contents

1 Introduction

The central dogma first enunciated by Francis Crick depicts RNA as an intermediate that links DNA and protein (Crick, 1970). The beginning of life, however, was the RNA world where there were no DNA or proteins (Gilbert, 1986). In the RNA world, RNA was the only carrier of genetic information that DNA currently serves as, and the only functional molecule that proteins currently serve as. Although the RNA transmits no genetic information to progeny and constitutes a limited number of functional molecules in our human body, the RNA world is still in effect in our body. Humans transcribe more than half of our entire genome including noncoding regions. The transcripts work as *antisense RNAs, microRNAs, and snoRNAs*. Researchers are now working to disclose the functional significance of these noncoding RNAs.

The human genome project and the subsequent annotation efforts revealed that we humans carry as few as 22,000 genes. Tissue-specific and developmental stage-specific splicing enables to us to generate more than 100,000 molecules from a limited number of genes (Black, 2003; Licatalosi and Darnell, 2006). Small RNA molecules and RNA splicing mechanisms potentially become targets of neurological diseases (Ranum and Cooper, 2006). This review focuses mostly on splicing aberrations associated with neurological disorders.

2 Physiology of Splicing Mechanisms

In higher eukaryotes, pre-mRNA splicing is mediated by degenerative splicing *cis*-elements comprised of the branch point sequence (BPS), the polypyrimidine tract (PPT), the 5′ and 3′ splice sites, and exonic/intronic splicing enhancers/silencers (Fig. 1). Stepwise assembly of the spliceosome starts from recruitment of *U1 snRNP* to the 5′ splice site, *SF1* to the BPS, *U2AF65* to the PPT, and *U2AF35* to the 3′ end of an intron to form a spliceosome complex E (Sperling et al., 2008). SF1, a 75 kDa protein, is a mammalian homologue of yeast BBP (branch point-binding protein). U2AF65 and U2AF35 bring *U2 snRNP* to the BPS in place of SF1 (Wu et al., 1999; Zorio and Blumenthal, 1999). The BPS establishes base pairing interactions with a stretch of "GUAGUA" of U2 snRNA (Arning et al., 1996; Abovich and Rosbash, 1997), which then bulges out the branch site nucleotide, usually an adenosine to form a spliceosome complex A (Query et al., 1994). Thereafter, pre-mRNAs are spliced in two sequential transesterification reactions mediated by the spliceosome. In the first step, the 2′-OH moiety of the branch site nucleotide carries out a nucleophilic attack against a phosphate at the 5′ splice site, generating a free upstream exon, as well as a lariat carrying the intron and the downstream exon. In the second step, the 3′-OH moiety of the upstream exon attacks the 3′ splice site of the

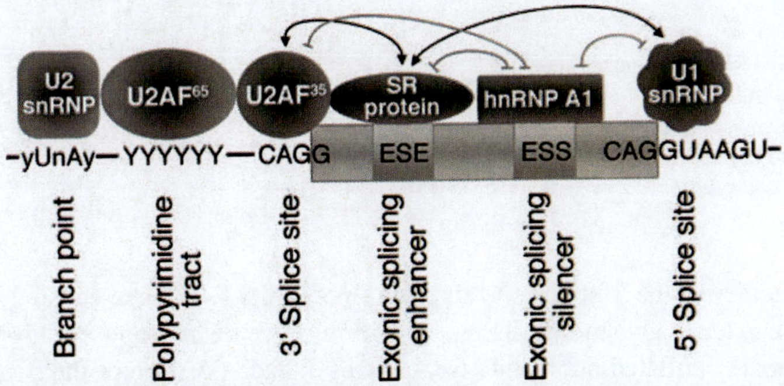

Fig. 1 Representative splicing *cis*-elements and *trans*-factors. Tissue-specific and developmental stage-specific expressions of splicing *trans*-factors including SR proteins and hnRNP A1 enable precise regulations of alternative splicing. ISE and ISS have similar activities as ESE and ESS, but are omitted from the figure

lariat leading to intron excision and ligation of the upstream and downstream exons (Query et al., 1996).

In addition to the "classical" spliceosomal mechanisms, splicing is modulated by exonic/intronic splicing enhancers/silencers (ESE, ISE, ESS, ISS). The *trans*-factors for the splicing enhancers/silencers carry repeats of arginine and serine are accordingly called SR proteins. Tissue-specific and developmental stage-specific expressions of the splicing *trans*-factors enable precise spatial and temporal regulations of the gene expressions. In addition, the splicing *trans*-factors also work on constitutively spliced exons to compensate for highly degenerative "classical" splicing *cis*-elements.

3 Disorders Associated with Disruption of Splicing *Cis*-Elements

3.1 Aberrations of the 5′ Splice Sites

Mutations disrupting the 5′ splice sites have been most frequently reported. U1 snRNA recognizes three nucleotides at the end of an exon and six nucleotides at the beginning of an intron (Fig. 2). The completely matched nucleotides to U1 snRNA are CAG|GTAAGT, where the vertical line represents the exon/intron boundary. The completely matched sequence is observed at 1597 sites out of the entire 189,249 5′ splice sites in the human genome (Sahashi et al., 2007), which is the tenth most common sequence. The completely matched 5′ splice site is rather avoided because, in the second stage of splicing, U1 snRNA is substituted for U5 snRNA. If U1 snRNA is tightly bound to the 5′ splice site, it hinders binding of U5 snRNA.

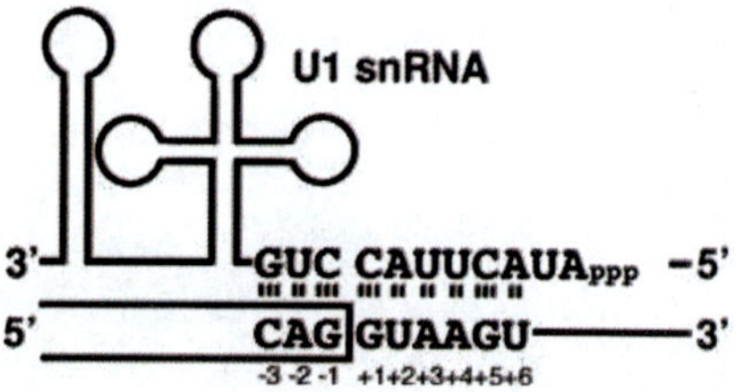

Fig. 2 U1 snRNA recognizes three nucleotides at the 3′ end of an exon and six nucleotides at the 5′end of an intron

Degeneracy of the 5′ splice site and its vulnerability to disease-causing mutations have been extensively studied. Three algorithms have been proposed. First, Shapiro and Senapathy collated nucleotide frequencies at each position of the 5′ splice site. They assumed that nucleotide frequencies at each position of the 5′ splice site represent the splicing signal intensity. They thus constructed a linear regression model so that the most preferred 5′ splice site becomes 1.0 and the most unfavorable 5′ splice site becomes 0.0 (Shapiro and Senapathy, 1987). Second, Rogan and Schneider invented the information contents, R*i*. For example, at a specific position, if a single nucleotide is exclusively used, the information content at this position becomes–$\log_2(1/4) = 2$ bits. Similarly, if two nucleotides are equally used, the information content becomes $-\log_2(2/4) = 1$ bit. In R*i*, the similarity to the consensus sequence is represented by the sum of information bits (Rogan and Schneider, 1995; O'Neill et al., 1998). Third, we found that a new parameter, the SD-Score, which represents a common logarithm of the frequency of a specific 5′ splice site in the human genome, efficiently predicts the splicing signal intensity (Sahashi et al., 2007).

Our algorithm predicts the splicing consequences of mutations with the sensitivity of 97.1% and the specificity of 94.7%. Simulation of all the possible mutations in the human genome using the SD-score algorithm predicts high frequencies of splicing mutations from exon –3 to intron +6 (Table 1). Especially at exon position –3, about one third of mutations are predicted to cause aberrant splicing. Using our algorithm, we predicted and proved that *DYSF* G1842D in Miyoshi myopathy, *ABCD1* R545W in adrenoleucodystrophy, *GLA* Q333X in Fabry disease, and *DMD* Q119X and Q1144X in Duchenne muscular dystrophy are not missense or nonsense mutations but are splicing mutations. Algorithms by us and by others all point to the notion that aberrant splicing caused by mutations at the 5′ splice sites is likely to be underestimated.

Table 1 Predicted ratios of exonic and intronic splicing mutations

Position	–3	–2	–1	+1	+2	+3	+4	+5	+6
Complementary nucleotide	C (%)	A (%)	G (%)	G	T	A (%)	A (%)	G (%)	T (%)
A	1.8	–	93.7	–	–	–	–	93.9	56.9
C	–	89.6	99.7	–	–	99.9	94.4	98.6	75.4
G	35.0	90.5	–	–	–	48.7	96.2	–	56.7
T	76.7	86.2	97.1	–	–	99.9	94.3	97.0	–
All mutations	37.8	88.8	96.8	–	–	82.8	95.0	96.5	63.0

3.2 *Human Branch Point Consensus Sequence*

In an effort to seek an algorithm to predict the position of the branch point sequence (BPS) in humans, we sequenced 367 clones of lariat RT-PCR products arising from 52 introns of 20 human housekeeping genes and identified that the human consensus BPS is simply yUnAy, where "y" represents U or C (Gao et al., 2008) (Fig. 3). The consensus BPS was more degenerative than we had expected and we failed to construct a dependable algorithm that predicts the position of the BPS. Sixteen disease-causing mutations and a polymorphism, however, have been reported to date that disrupt a BPS and cause aberrant splicing (Gao et al., 2008). Among these, eight mutates U at position –2, whereas nine affects A at position 0, which also supports the notion that U at –2 and A at 0 are essential nucleotides.

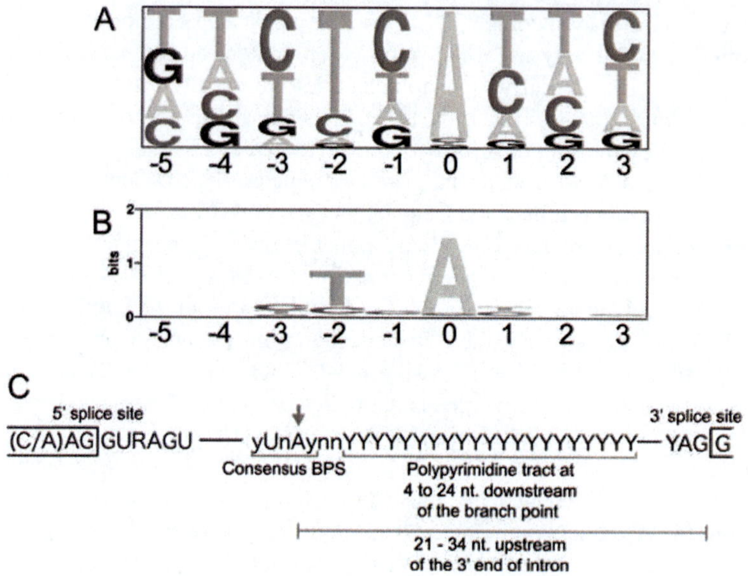

Fig. 3 Human consensus BPS. (**a**) Pictogram and (**b**) WebLogo presentations of BPS. Position 0 represents the branch point. (**c**) Representative sequences and positions of splicing *cis*-elements

3.3 Ectopic AG Dinucleotide Abrogates the AG-Scanning Mechanism

The 3′ end of an intron and the 5′ end of an exon carry a consensus sequence of CAG|G, where the vertical line represents the intron/exon boundary. The AG dinucleotide is scanned from the branch point and the first AG is recognized as the 3′ end of the intron (Chen et al., 2000). In a patient with congenital myasthenic syndrome, we identified duplication of a 16-nt segment comprised of 8 intronic and 8 exonic nucleotides at the intron 10/exon 10 boundary of *CHRNE* encoding the acetylcholine receptor epsilon subunit (Ohno et al., 2005). We found that the upstream AG of the duplicated segment is exclusively used for splicing and that one or two mutations in the upstream BPS had no effect whereas complete deletion of the upstream BPS partially activated the downstream AG. Similar exclusive activation of the upstream AG is reported in *HEXB* (Dlott et al., 1990) and *SLC4A1* (Bianchi et al., 1997). Creation of a cryptic AG dinucleotide close to the 3′ end of an intron should be carefully scrutinized in mutation analysis.

3.4 Mutations That Disrupt ESE *and* ESS

Gorlov and colleagues predicted that more than 16–20% of missense mutations are splicing mutations that disrupt an ESE (Gorlov et al., 2003). According to our own experience, their estimates are likely to be too high. Most ESE/ESS-disrupting mutations, however, are likely to be underestimated, because the positions and sequences of ESE/ESS are highly degenerative.

Four Web services provide valuable information to locate ESE and ESS. First, the ESE Finder (http://rulai.cshl.org/ESE/) calculates the similarity of a given nucleotide sequence to the consensus sequences of four splicing *trans*-factors, SF2/ASF, SC35, SRp40, and SRp55 (Cartegni et al., 2003; Smith et al., 2006). Second, the RESCUE-ESE Web server (http://genes.mit.edu/burgelab/rescue-ese/) shows the similarity of a given sequence to ESE elements of unidentified splicing *trans*-factors (Fairbrother et al., 2002). The same group also provides the FAS-ESS Web service to screen for ESS elements (http://genes.mit.edu/fas-ess/) (Wang et al., 2004). Third, the PESX Web server (http://cubweb.biology.columbia.edu/pesx/) indicates an RNA octamer with putative exonic splicing enhancing or silencing activities (Zhang and Chasin, 2004; Zhang et al., 2005). Fourth, the ESRsearch Web server (http://ast.bioinfo.tau.ac.il/) shows 285 candidate ESE/ESS sequences (Goren et al., 2006), as well as ESE/ESS elements indicated by the RESCUE-ESE, FAS-ESS, and PESX services.

In patients with congenital myasthenic syndromes, we identified that *CHRNE* E154X and EF157V (Ohno et al., 2003), as well as *COLQ* E415G (Kimbell et al., 2004), disrupt an ESE and cause aberrant splicing. The ESE/ESS servers above indicate disruption of candidate splicing *cis*-elements for all three mutations, but we frequently obtain false positives and we cannot simply rely on the servers. Analysis of patient mRNA or analysis using a minigene is generally expected.

3.5 *Mutations That Disrupt* ISE *and* ISS

Identification of mutations disrupting intronic splicing *cis*-elements is more challenging than that of exonic mutations, because introns are longer than exons and splicing mutations can be anywhere in the introns, and because we do not have a dependable algorithm to predict ISE/ISS. The ESRsearch Web server described above is able to indicate consensus sequences recognized by a variety of splicing *trans*-factors including intronic ones.

In a patient with congenital myasthenic syndrome, we identified that *CHRNA1* IVS3-8G>A attenuates binding of *hnRNP H* ~100-fold and causes exclusive inclusion of the downstream exon P3A (Masuda et al., 2008) (Fig. 4). We also identified that polypyrimidine tract binding protein (PTB) silences recognition of exon P3A and tannic acid facilitates the expression of PTB by activating its promoter region (Gao et al., 2009).

3.6 *Spinal Muscular Atrophy (SMA)*

SMA is an autosomal recessive disorder characterized by degeneration of the anterior horn cells of the spinal cord, which causes muscular weakness and atrophy. SMA is caused by loss-of-function mutations including deletion of the *SMN1* gene

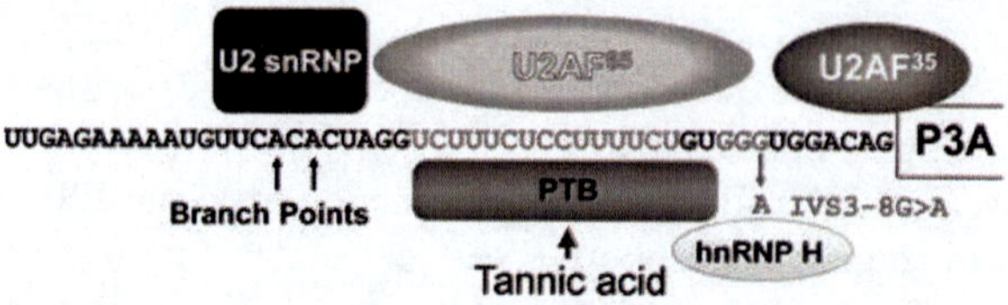

Fig. 4 *CHRNA1* carries a 75-nt exon P3A. Its inclusion generates a nonfunctional alpha subunit of the acetylcholine receptor. hnRNP H and PTB silence recognition of exon P3A and induce its skipping. The IVS3-8G>A mutation identified in a patient with congenital myasthenic syndrome weakens the binding of hnRNP H and causes inclusion of exon P3A. Tannic acid facilitates the expression of PTB and partially ameliorates aberrant splicing due to IVS3-8G>A

that encodes the survival of motor neuron 1. Humans carry almost identical *SMN1* and *SMN2* genes both on chromosome 5q13. *SMN2* carries a C-to-T transition at position 6 of exon 7 compared to *SMN1*, which results in loss of an SF2/ASF-dependent ESE activity (Cartegni et al., 2006). In addition, *SMN2* carries an A-to-G transition at position +100 of intron 7, which creates a high-affinity hnRNP A1-binding site and promotes skipping of exon 7 (Kashima et al., 2007). Skipping of exon 7 in *SMN2* can be ameliorated by therapeutic doses of valproic acid (Brichta et al., 2003, 2006) and of salbutamol (Angelozzi et al., 2008).

4 Skipping of Multiple Exons Caused by a Single Splicing Mutation

4.1 Skipping of Multiple Contiguous Exons

A mutation disrupting a splicing *cis*-element generally affects splicing of a single exon or intron, but sometimes generates aberrant transcripts affecting multiple neighboring exons. Skipping of multiple contiguous exons is accounted for by ordered removal of introns and consequent clustering of neighboring exons (Schwarze et al., 1999; Takahara et al., 2002).

4.2 Nonsense-Associated Skipping of a Remote Exon (NASRE)

A single mutation infrequently causes skipping of a remote exon. In a patient with congenital myasthenic syndrome, we found that a 7-nt deletion in exon 7 of *CHRNE* causes complete skipping of the preceding exon 6. *CHRNE* exon 6 is composed of 101 nucleotides. It carries weak splicing signals and is partially skipped even in normal subjects. The exon 6-skipped transcript, however, is removed by the nonsense-mediated mRNA decay (NMD) mechanism. The 7-nt deletion in exon 7 restores the open reading frame of the exon 6-skipped transcript and renders it immune to NMD. On the other hand, the normally spliced transcript carries a

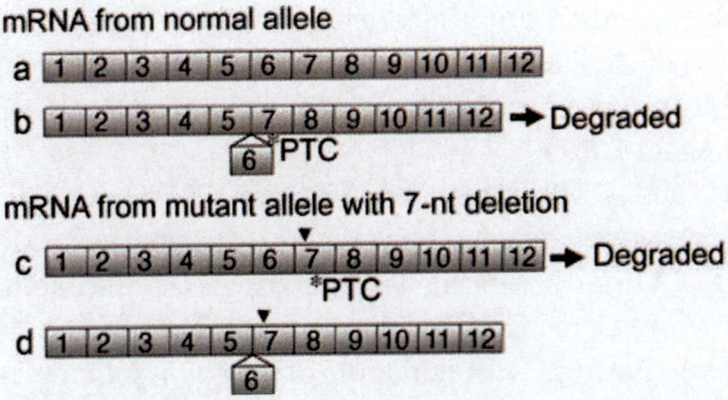

Fig. 5 NASRE. Wild-type *CHRNE* generates the normally spliced transcript (**a**) and the exon 6-skipped transcript (**b**), because exon 6 carries weak splicing signals. The exon-skipped transcript carries a premature termination codon (PTC) and is degraded by NMD. A 7-nt deletion (*arrowhead*) in exon 7 generates a PTC in the normally spliced transcript (**c**) and is degraded by NMD. The deletion resumes the open reading frame from the exon 6-skipped transcript, and the transcript escapes NMD (**d**)

premature stop codon (PTC) after the 7-nt deletion, and is degraded by NMD[1] (Fig. 5). We dubbed this mechanism NASRE, and found that it is in effect in *SLC25A20* (Hsu et al., 2001), *DBT* (Fisher et al., 1993), *BTK* (Haire et al., 1997), and *MLH1* (Clarke et al., 2000).

5 Disorders Associated with Dysregulation of Splicing *Trans*-Factors

5.1 *Myotonic Dystrophy*

Myotonic dystrophy is an autosomal dominant multisystem disorder affecting skeletal muscles, eye, heart, endocrine system, and central nervous system. The clinical symptoms include variable degrees of muscle weakness and wasting, myotonia, cataract, insulin resistance, hypogonadism, cardiac conduction defects, frontal balding, and intellectual disabilities (Harper and Monckton, 2004). Myotonic dystrophy is caused by abnormally expanded CTG repeats in the 3′ untranslated region of the *DMPK* gene encoding the dystrophia myotonica protein kinase on chromosome 19q13 (myotonic dystrophy type 1, DM1) (Brook et al., 1992) or by abnormally expanded CCTG repeats in intron 1 of the *ZNF9* gene encoding the zinc finger protein 9 on chromosome 3q21 (myotonic dystrophy type 2, DM2) (Liquori et al., 2001). In DM1, normal individuals have 5–30 repeats, mildly affected patients

[1]Nonsense-mediated mRNA decay (NMD). NMD is a quality-assurance mechanism that degrades mRNAs harboring a premature termination codon (PTC) (Chang et al., 2007). Proteins translated from mRNAs harboring PTCs potentially have dominant-negative or deleterious activities. In pre-mRNA splicing, an exon–junction complex (EJC) is deposited 20–24 nucleotides upstream of each exon–exon junction. Ribosomes remove EJCs, but, in the presence of a PTC, EJCs stay on the transcript and trigger the NMD pathway in the cytoplasm.

have 50–80 repeats, and severely affected individuals have 2000 or more copies of CTG (Gharehbaghi-Schnell et al., 1998). In DM2, the size of expanded repeats is extremely variable, ranging from 75 to 11,000 repeats, with a mean of 5000 CCTG repeats (Liquori et al., 2001).

In both DM1 and DM2, expanded CTG or CCTG repeats in the noncoding regions sequestrate a splicing *trans*-factor muscleblind encoded by *MBNL1* to intranuclear RNA foci harboring the mutant RNA, and somehow upregulate another splicing *trans*-factor CUG-binding protein encoded by *CUGBP1* (Ranum and Cooper, 2006) (Fig. 6). Dysregulation of the two splicing *trans*-factors then causes aberrant splicing of their target genes. The aberrantly spliced genes identified to date in skeletal and cardiac muscles include *ATP2A1* (*SERCA1*) exon 22, *ATP2A2* (*SERCA2*) intron 19, *CAPN3* exon 16, *CLCN1* intron 2 and exons 6b/7a, *DMD* exons 71 and 78, *DTNA* exons 11A and 12, *FHOD1* (*FHOS*) exon 11a, *GFPT1* (*GFAT1*) exon 10, *INSR* exon 11, *KCNAB1* exons 2b/2c, *LDB3* (*ZASP*) exon 11 (189-nt exon 7 according to RefSeq Build 36.3), *MBNL1* exon 7 (54-nt exon 6 according to RefSeq), *MBNL2* exon 7 (54 nt, no exonic annotation in RefSeq), *MTMR1* exons 2.1 and 2.2, *NRAP* exon 12, *PDLIM3* (*ALP*) exons 5a/5b, *RYR1* exon 70, *TNNT2* exon 5, *TNNT3* fetal exon, *TTN* exons Zr4 and Zr5 (138-nt exon 11 and 138-nt exon 12 according to RefSeq), and *TTN* exon Mex5 (303-nt exon 315 according to RefSeq) (Philips et al., 1998; Savkur et al., 2001; Kimura et al., 2005; Lin et al., 2006). Lin and colleagues report that alternative transcripts observed in myotonic dystrophy are all fetal isoforms (Lin et al., 2006). Muscleblind normally translocates

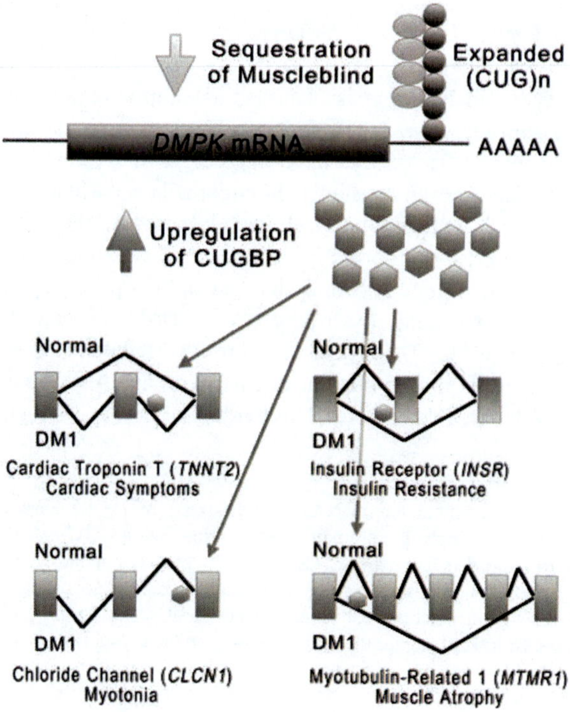

Fig. 6 In DM1, expanded CUG repeats in the 3′ UTR of DMPK sequestrate muscleblind and upregulates CUG-binding protein. Dysregulation of these splicing *trans*-factors causes aberrant splicing of their inherent target genes. Four representative target genes are indicated

from cytoplasm to nucleus in the postnatal period to induce adult-type splicings, and lack of muscleblind in nucleus due to sequestration to RNA foci recapitulates fetal splicing patterns.

5.2 Alzheimer's Disease (AD) and Frontotemporal Dementia with Parkinsonism Linked to Chromosome 17 (FTDP-17)

AD is the most common neurodegenerative disease representing dementia. It is characterized by intracellular neurofibrillary tangles (NFTs) and extracellular amyloid plaques. NFTs are composed of aggregates of the hyperphosphorylated tau protein encoded by *MAPT*. The amyloid plaques are composed of amyloid β peptide (Aβ) that originates from enzymatic cleavage of the amyloid precursor protein (*APP*) by β-secretase followed by γ-secretase (LaFerla et al., 2007). The γ-secretase is an enzyme complex composed of presenilin-1 (*PS1*) or presenilin-2 (*PS2*), as well as nicastrin, anterior pharynx defective (APH-1), and presenilin enhancer 2 (PEN-2) (Takasugi et al., 2003). Autosomal dominant forms of AD constitute ~5% of AD and are caused by mutations in *APP*, *PS1*, or *PS2* (Bertram and Tanzi, 2008).

Although the pathomechanisms underlying sporadic AD remain mostly unknown, *PS2* exon 5 is exclusively skipped in brains of sporadic AD, which is mediated by overexpression of a splicing *trans*-factor, *HMGA1a* (Sato et al., 1999; Manabe et al., 2003). As hypoxia induces the overexpression of HMGA1a, the upregulation of HMGA1a in sporadic AD may or may not represent an agonal state of AD, in which respiratory insufficiency possibly associated with pneumonia frequently becomes the cause of death.

Mutations in *MAPT* are not observed in AD, but are present in FTDP-17. *MAPT* exon 10 is alternatively spliced in normal brain. N279K, K280del, and L284L mutations on exon 10 provoke aberrant splicing of exon 10 by disrupting or enhancing exonic splicing *cis*-elements, and cause FTDP-17 (D'Souza et al., 1999) (Fig. 7). The splicing *trans*-factors for these *cis*-elements are also identified (Jiang et al., 2004; Kondo et al., 2004).

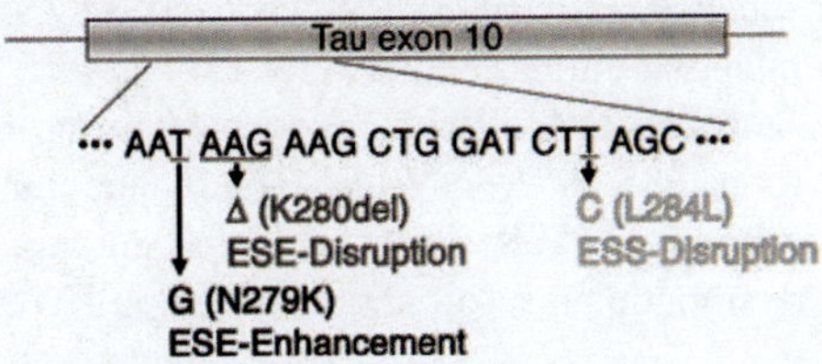

Fig. 7 Mutations on *MAPT* exon 10 cause excessive skipping (N279K and L284L) or inclusion (K280del) of exon 10

5.3 Facioscapulohumeral Muscular Dystrophy (FSHD)

FSHD is the third most common hereditary muscular dystrophy after Duchenne muscular dystrophy and myotonic dystrophy. As its name represents, the disease predominantly affects the face, the scapulae, and the proximal arm muscles. In

FSHD, the number of a 3.3 kb repeat in the subtelomeric region of 4q (4q35), designated *D4Z4*, are abnormally reduced (Wijmenga et al., 1992). Loss of *D4Z4* causes upregulation of FRG1 located upstream of *D4Z4* (Gabellini et al., 2002). FRG1 is a splicing *trans*-factor, and its overexpression causes aberrant splicing of *TNNT3* encoding the troponin T type 3 of fast skeletal muscle and *MTMR1* encoding the myotubularin-related protein 1 (Gabellini et al., 2006). The reported splicing aberrations in FSHD, however, have not been confirmed by us (unpublished data) or by the other groups (personal communications).

5.4 Fragile X-Associated Tremor/Ataxia Syndrome (FXTAS)

Fragile X mental retardation syndrome is caused by abnormal expansion of a CGG repeat in the 5′ untranslated region of *FMR1*, which culminates in hypermethylation of *FMR1* and silences its expression (Kremer et al., 1991). On the other hand, moderate expansion of the CGG repeat in *FMR1* causes FXTAS, which is characterized by intention tremor, Parkinsonism, cognitive decline, and neuropathy (Hagerman and Hagerman, 2004). In FXTAS, CGG-binding proteins including *hnRNP A2* and muscleblind are excessively bound to the expanded CGG repeats of *FMR1* and are depleted from the cellular pool (Iwahashi et al., 2006), which results in the loss their functions in other regulatory processes (Jacquemont et al., 2007).

5.5 Prader–Willi Syndrome (PWS)

PWS is an autosomal dominant disorder characterized by obesity, muscular hypotonia and weakness, mental retardation, short stature, hypogonadotropic hypogonadism, and small distal extremities. The proximal long arm of chromosome 15 (15q11-q13) is normally imprinted in order to achieve parent-specific monoallelic gene expressions. Some genes in this region are expressed only from the maternal allele, and some others are only from the paternal allele. Lack of a functional paternal copy of 15q11-13 causes PWS, whereas lack of a functional maternal copy of *UBE3A* in the same region results in *Angelman syndrome* (Horsthemke and Wagstaff, 2008). PWS is caused by a deletion of the paternal 15q11-q13 or by maternal uniparental disomy 15.

A *snoRNA HBII-52* is located in the defective region of PWS. HBII-52 binds to an ESS in exon Vb of *HTR2C* encoding the serotonin receptor 2C, and its disruption in PWS causes aberrant splicing of *HTR2C* and potentially accounts for dysfunctional serotonergic system in PWS (Kishore and Stamm, 2006).

5.6 Rett Syndrome

Rett syndrome is a neurodevelopmental disorder in females, which is characterized by loss of speech, stereotypical movements of hands, microcephaly, seizures, and mental retardation. Rett syndrome is caused by a mutation in *MECP2* encoding the metyl-CpG-binding protein 2 (Amir et al., 1999). MeCP2 binds to a splicing *trans*-factor *YB-1* and the abnormal regulation of YB-1 causes aberrant splicing of its target genes (Young et al., 2005).

5.7 Spinocerebellar Ataxia Type 8 (SCA8)

SCA8 is caused by an abnormal expansion of CTA/CTG repeats in the protein-noncoding *ATXN8OS*, which represents the *ATXN8* opposite strand (Ikeda et al., 2008). Expanded CUG repeats on the *ATXN8OS* transcript potentially bind to and sequestrate CUG-binding proteins, as we observe in myotonic dystrophy (Mutsuddi and Rebay, 2005). In addition, *ATXN8* on the opposite strand of *ATXN8OS* encodes the Kelch-like 1, and the expanded CAG repeats on *ATXN8* give rise to a polyglutamine tract that forms a cytotoxic aggregate in neuronal cells (Moseley et al., 2006). Furthermore, expression of *ATXN8OS* is colocalized with that of *ATXN8* (Chen et al., 2008). *ATXN8OS* thus potentially serves as an antisense RNA for *ATXN8*, and the abnormal CTA/CTG expansion in *ATXN8OS* may dysregulate the expression of *ATXN8* (Fig. 8).

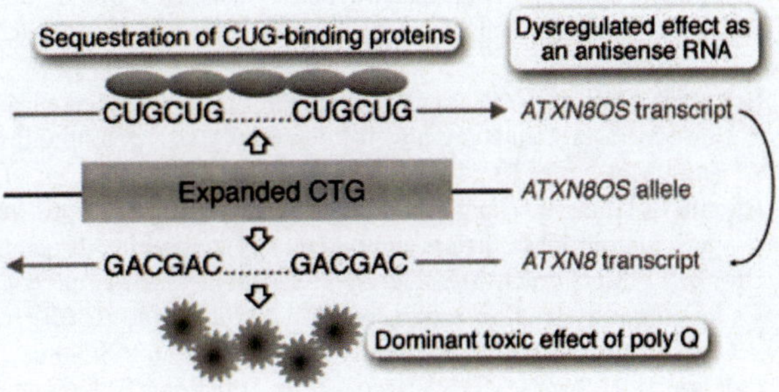

Fig. 8 Expanded CTG on *ATXN8OS* exerts three toxic effects on the bidirectional transcripts

5.8 Paraneoplastic Neurological Disorders (PND)

In PND, tumors outside of the nervous system excrete humoral factors such as hormones and cytokines, or provoke an immune response against specific molecules expressed in tumors, and cause a wide range of neurological symptoms. In paraneoplastic opsoclonus myoclonus ataxia (POMA), autoantibodies are raised against the Nova family of neuron-specific splicing *trans*-factor (Jensen et al., 2000; Ule et al., 2003, 2006; Licatalosi et al., 2008). In paraneoplastic encephalomyelitis and sensory neuropathy (PEN/SN or Hu syndrome), autoantibodies recognize the Hu family of RNA-binding protein (Szabo et al., 1991), a human homologue of the *Drosophila* splicing *trans*-factor *Elav* (Koushika et al., 2000; Soller and White, 2003). In both disorders, autoantibodies downregulate the splicing *trans*-factors and cause aberrant splicing in neuronal cells.

Acknowledgments Works from the authors' laboratories have been supported by Grants-in-Aid from the Ministry of Education, Culture, Sports, Science, and Technology of Japan, and from the Ministry of Health, Labor, and Welfare of Japan.

References

Abovich N, Rosbash M (1997) Cross-intron bridging interactions in the yeast commitment complex are conserved in mammals. Cell 89:403–412

Amir RE, Van den Veyver IB, Wan M, Tran CQ, Francke U et al (1999) Rett syndrome is caused by mutations in X-linked MECP2, encoding methyl-CpG-binding protein 2. Nat Genet 23: 185–188

Angelozzi C, Borgo F, Tiziano FD, Martella A, Neri G et al (2008) Salbutamol increases SMN mRNA and protein levels in spinal muscular atrophy cells. J Med Genet 45:29–31

Arning S, Gruter P, Bilbe G, Kramer A (1996) Mammalian splicing factor SF1 is encoded by variant cDNAs and binds to RNA. RNA 2:794–810

Bertram L, Tanzi RE (2008) Thirty years of Alzheimer's disease genetics: the implications of systematic meta-analyses. Nat Rev Neurosci 9:768–778

Bian Y, Masuda A, Matsuura T, Ito M, Okushin K et al (2009) Tannic acid facilitates expression of the polypyrimidine tract binding protein and alleviates deleterious inclusion of *CHRNA1* exon P3A due to an hnRNP H-disrupting mutation in congenital myasthenic syndrome. Hum Mol Genet 18:1229–1237

Bianchi P, Zanella A, Alloisio N, Barosi G, Bredi E et al (1997) A variant of the EPB3 gene of the anti-Lepore type in hereditary spherocytosis. Br J Haematol 98:283–288

Black DL (2003) Mechanisms of alternative pre-messenger RNA splicing. Annu Rev Biochem 72:291–336

Brichta L, Hofmann Y, Hahnen E, Siebzehnrubl FA, Raschke H et al (2003) Valproic acid increases the SMN2 protein level: a well-known drug as a potential therapy for spinal muscular atrophy. Hum Mol Genet 12:2481–2489

Brichta L, Holker I, Haug K, Klockgether T, Wirth B (2006) In vivo activation of SMN in spinal muscular atrophy carriers and patients treated with valproate. Ann Neurol 59:970–975

Brook JD, McCurrach ME, Harley HG, Buckler AJ, Church D et al (1992) Molecular basis of myotonic dystrophy: expansion of a trinucleotide (CTG) repeat at the 3′ end of a transcript encoding a protein kinase family member. Cell 68:799–808

Cartegni L, Hastings ML, Calarco JA, de Stanchina E, Krainer AR (2006) Determinants of exon 7 splicing in the spinal muscular atrophy genes, SMN1 and SMN2. Am J Hum Genet 78: 63–77

Cartegni L, Wang J, Zhu Z, Zhang MQ, Krainer AR (2003) ESEfinder: a web resource to identify exonic splicing enhancers. Nucleic Acids Res 31:3568–3571

Chang YF, Imam JS, Wilkinson MF (2007) The nonsense-mediated decay RNA surveillance pathway. Annu Rev Biochem 76:51–74

Chen S, Anderson K, Moore MJ (2000) Evidence for a linear search in bimolecular 3′ splice site AG selection. Proc Natl Acad Sci U S A 97:593–598

Chen WL, Lin JW, Huang HJ, Wang SM, Su MT et al (2008) SCA8 mRNA expression suggests an antisense regulation of KLHL1 and correlates to SCA8 pathology. Brain Res 1233: 176–184

Clarke LA, Veiga I, Isidro G, Jordan P, Ramos JS et al (2000) Pathological exon skipping in an HNPCC proband with MLH1 splice acceptor site mutation. Genes Chromosomes Cancer 29:367–370

Crick F (1970) Central dogma of molecular biology. Nature 227:561–563

Dlott B, d'Azzo A, Quon DV, Neufeld EF (1990) Two mutations produce intron insertion in mRNA and elongated beta-subunit of human beta-hexosaminidase. J Biol Chem 265: 17921–17927

D'Souza I, Poorkaj P, Hong M, Nochlin D, Lee VM et al (1999) Missense and silent tau gene mutations cause frontotemporal dementia with parkinsonism-chromosome 17 type, by affecting multiple alternative RNA splicing regulatory elements. Proc Natl Acad Sci U S A 96: 5598–5603

Fairbrother WG, Yeh RF, Sharp PA, Burge CB (2002) Predictive identification of exonic splicing enhancers in human genes. Science 297:1007–1013

Fisher CW, Fisher CR, Chuang JL, Lau KS, Chuang DT et al (1993) Occurrence of a 2-bp (AT) deletion allele and a nonsense (G-to-T) mutant allele at the E2 (DBT) locus of six patients with maple syrup urine disease: multiple-exon skipping as a secondary effect of the mutations. Am J Hum Genet 52:414–424

Gabellini D, D'Antona G, Moggio M, Prelle A, Zecca C et al (2006) Facioscapulohumeral muscular dystrophy in mice overexpressing FRG1. Nature 439:973–977

Gabellini D, Green MR, Tupler R (2002) Inappropriate gene activation in FSHD: a repressor complex binds a chromosomal repeat deleted in dystrophic muscle. Cell 110:339–348

Gao K, Masuda A, Matsuura T, Ohno K (2008) Human branch point consensus sequence is yUnAy. Nucleic Acids Res 36:2257–2267

Gharehbaghi-Schnell EB, Finsterer J, Korschineck I, Mamoli B, Binder BR (1998) Genotype-phenotype correlation in myotonic dystrophy. Clin Genet 53:20–26

Gilbert W (1986) Origin of life: the RNA world. Nature 319:618

Goren A, Ram O, Amit M, Keren H, Lev-Maor G et al (2006) Comparative analysis identifies exonic splicing regulatory sequences – The complex definition of enhancers and silencers. Mol Cell 22:769–781

Gorlov IP, Gorlova OY, Frazier ML, Amos CI (2003) Missense mutations in hMLH1 and hMSH2 are associated with exonic splicing enhancers. Am J Hum Genet 73:1157–1161

Hagerman PJ, Hagerman RJ (2004) The fragile-X premutation: a maturing perspective. Am J Hum Genet 74:805–816

Haire RN, Ohta Y, Strong SJ, Litman RT, Liu YY et al (1997) Unusual patterns of exon skipping in bruton tyrosine kinase are associated with mutations involving the intron 17 3′ splice site. Am J Hum Genet 60:798–807

Harper PS, Monckton DG (2004) Myotonic dystrophy. In: Engel AG (ed) Myology, 3rd edn. McGraw-Hill, New York, NY, pp 1039–1076

Horsthemke B, Wagstaff J (2008) Mechanisms of imprinting of the Prader–Willi/Angelman region. Am J Med Genet A 146A:2041–2052

Hsu BY, Iacobazzi V, Wang Z, Harvie H, Chalmers RA et al (2001) Aberrant mRNA splicing associated with coding region mutations in children with carnitine-acylcarnitine translocase deficiency. Mol Genet Metab 74:248–255

Ikeda Y, Daughters RS, Ranum LP (2008) Bidirectional expression of the SCA8 expansion mutation: one mutation, two genes. Cerebellum 7:150–158

Iwahashi CK, Yasui DH, An HJ, Greco CM, Tassone F et al (2006) Protein composition of the intranuclear inclusions of FXTAS. Brain 129:256–271

Jacquemont S, Hagerman RJ, Hagerman PJ, Leehey MA (2007) Fragile-X syndrome and fragile X-associated tremor/ataxia syndrome: two faces of FMR1. Lancet Neurol 6:45–55

Jensen KB, Dredge BK, Stefani G, Zhong R, Buckanovich RJ et al (2000) Nova-1 regulates neuron-specific alternative splicing and is essential for neuronal viability. Neuron 25:359–371

Jiang H, Mankodi A, Swanson MS, Moxley RT, Thornton CA (2004) Myotonic dystrophy type 1 is associated with nuclear foci of mutant RNA, sequestration of muscleblind proteins and deregulated alternative splicing in neurons. Hum Mol Genet 13:3079–3088

Kashima T, Rao N, Manley JL (2007) An intronic element contributes to splicing repression in spinal muscular atrophy. Proc Natl Acad Sci U S A 104:3426–3431

Kimbell LM, Ohno K, Engel AG, Rotundo RL (2004) C-terminal and heparin-binding domains of collagenic tail subunit are both essential for anchoring acetylcholinesterase at the synapse. J Biol Chem 279:10997–11005

Kimura T, Nakamori M, Lueck JD, Pouliquin P, Aoike F et al (2005) Altered mRNA splicing of the skeletal muscle ryanodine receptor and sarcoplasmic/endoplasmic reticulum Ca2+-ATPase in myotonic dystrophy type 1. Hum Mol Genet 14:2189–2200

Kishore S, Stamm S (2006) The snoRNA HBII-52 regulates alternative splicing of the serotonin receptor 2C. Science 311:230–232

Kondo S, Yamamoto N, Murakami T, Okumura M, Mayeda A et al (2004) Tra2 beta, SF2/ASF and SRp30c modulate the function of an exonic splicing enhancer in exon 10 of tau pre-mRNA. Genes Cells 9:121–130

Koushika SP, Soller M, White K (2000) The neuron-enriched splicing pattern of Drosophila erect wing is dependent on the presence of ELAV protein. Mol Cell Biol 20:1836–1845

Kremer EJ, Pritchard M, Lynch M, Yu S, Holman K et al (1991) Mapping of DNA instability at the fragile X to a trinucleotide repeat sequence p(CCG)n. Science 252:1711–1714

LaFerla FM, Green KN, Oddo S (2007) Intracellular amyloid-beta in Alzheimer's disease. Nat Rev Neurosci 8:499–509

Licatalosi DD, Darnell RB (2006) Splicing regulation in neurologic disease. Neuron 52:93–101

Licatalosi DD, Mele A, Fak JJ, Ule J, Kayikci M et al (2008) HITS-CLIP yields genome-wide insights into brain alternative RNA processing. Nature 456:464–469

Lin X, Miller JW, Mankodi A, Kanadia RN, Yuan Y et al (2006) Failure of MBNL1-dependent post-natal splicing transitions in myotonic dystrophy. Hum Mol Genet 15:2087–2097

Liquori CL, Ricker K, Moseley ML, Jacobsen JF, Kress W et al (2001) Myotonic dystrophy type 2 caused by a CCTG expansion in intron 1 of ZNF9. Science 293:864–867

Manabe T, Katayama T, Sato N, Gomi F, Hitomi J et al (2003) Induced HMGA1a expression causes aberrant splicing of Presenilin-2 pre-mRNA in sporadic Alzheimer's disease. Cell Death Differ 10:698–708

Masuda A, Shen XM, Ito M, Matsuura T, Engel AG et al (2008) hnRNP H enhances skipping of a nonfunctional exon P3A in CHRNA1 and a mutation disrupting its binding causes congenital myasthenic syndrome. Hum Mol Genet 17:4022–4035

Moseley ML, Zu T, Ikeda Y, Gao W, Mosemiller AK et al (2006) Bidirectional expression of CUG and CAG expansion transcripts and intranuclear polyglutamine inclusions in spinocerebellar ataxia type 8. Nat Genet 38:758–769

Mutsuddi M, Rebay I (2005) Molecular genetics of spinocerebellar ataxia type 8 (SCA8). RNA Biol 2:49–52

Ohno K, Milone M, Shen X-M, Engel AG (2003) A frameshifting mutation in *CHRNE* unmasks skipping of the preceding exon. Hum Mol Genet 12:3055–3066

Ohno K, Tsujino A, Shen X-M, Milone M, Engel AG (2005) Spectrum of splicing errors caused by CHRNE mutations affecting introns and intron/exon boundaries. J Med Genet 42:e53

O'Neill JP, Rogan PK, Cariello N, Nicklas JA (1998) Mutations that alter RNA splicing of the human HPRT gene: a review of the spectrum. Mutat Res 411:179–214

Philips AV, Timchenko LT, Cooper TA (1998) Disruption of splicing regulated by a CUG-binding protein in myotonic dystrophy. Science 280:737–741

Query CC, Moore MJ, Sharp PA (1994) Branch nucleophile selection in pre-mRNA splicing: evidence for the bulged duplex model. Genes Dev 8:587–597

Query CC, Strobel SA, Sharp PA (1996) Three recognition events at the branch-site adenine. EMBO J 15:1392–1402

Ranum LP, Cooper TA (2006) RNA-mediated neuromuscular disorders. Annu Rev Neurosci 29:259–277

Rogan PK, Schneider TD (1995) Using information content and base frequencies to distinguish mutations from genetic polymorphisms in splice junction recognition sites. Hum Mutat 6:74–76

Sahashi K, Masuda A, Matsuura T, Shinmi J, Zhang Z et al (2007) In vitro and *in silico* analysis reveals an efficient algorithm to predict the splicing consequences of mutations at the 5′ splice sites. Nucleic Acids Res 35:5995–6003

Sato N, Hori O, Yamaguchi A, Lambert JC, Chartier-Harlin MC et al (1999) A novel presenilin-2 splice variant in human Alzheimer's disease brain tissue. J Neurochem 72:2498–2505

Savkur RS, Philips AV, Cooper TA (2001) Aberrant regulation of insulin receptor alternative splicing is associated with insulin resistance in myotonic dystrophy. Nat Genet 29:40–47

Schwarze U, Starman BJ, Byers PH (1999) Redefinition of exon 7 in the COL1A1 gene of type I collagen by an intron 8 splice-donor-site mutation in a form of osteogenesis imperfecta: influence of intron splice order on outcome of splice-site mutation. Am J Hum Genet 65:336–344

Shapiro MB, Senapathy P (1987) RNA splice junctions of different classes of eukaryotes: sequence statistics and functional implications in gene expression. Nucleic Acids Res 15:7155–7174

Smith PJ, Zhang C, Wang J, Chew SL, Zhang MQ et al (2006) An increased specificity score matrix for the prediction of SF2/ASF-specific exonic splicing enhancers. Hum Mol Genet 15:2490–2508

Soller M, White K (2003) ELAV inhibits 3′-end processing to promote neural splicing of ewg pre-mRNA. Genes Dev 17:2526–2538

Sperling J, Azubel M, Sperling R (2008) Structure and function of the Pre-mRNA splicing machine. Structure 16:1605–1615

Szabo A, Dalmau J, Manley G, Rosenfeld M, Wong E et al (1991) HuD, a paraneoplastic encephalomyelitis antigen, contains RNA-binding domains and is homologous to Elav and Sex-lethal. Cell 67:325–333

Takahara K, Schwarze U, Imamura Y, Hoffman GG, Toriello H et al (2002) Order of intron removal influences multiple splice outcomes, including a two-exon skip, in a COL5A1 acceptor-site mutation that results in abnormal pro-alpha1(V) N-propeptides and Ehlers-Danlos syndrome type I. Am J Hum Genet 71:451–465

Takasugi N, Tomita T, Hayashi I, Tsuruoka M, Niimura M et al (2003) The role of presenilin cofactors in the gamma-secretase complex. Nature 422:438–441

Ule J, Jensen KB, Ruggiu M, Mele A, Ule A et al (2003) CLIP identifies Nova-regulated RNA networks in the brain. Science 302:1212–1215

Ule J, Stefani G, Mele A, Ruggiu M, Wang X et al (2006) An RNA map predicting Nova-dependent splicing regulation. Nature 444:580–586

Wang Z, Rolish ME, Yeo G, Tung V, Mawson M et al (2004) Systematic identification and analysis of exonic splicing silencers. Cell 119:831–845

Wijmenga C, Hewitt JE, Sandkuijl LA, Clark LN, Wright TJ et al (1992) Chromosome 4q DNA rearrangements associated with facioscapulohumeral muscular dystrophy. Nat Genet 2:26–30

Wu S, Romfo CM, Nilsen TW, Green MR (1999) Functional recognition of the 3′ splice site AG by the splicing factor U2AF35. Nature 402:832–835

Young JI, Hong EP, Castle JC, Crespo-Barreto J, Bowman AB et al (2005) Regulation of RNA splicing by the methylation-dependent transcriptional repressor methyl-CpG binding protein 2. Proc Natl Acad Sci U S A 102:17551–17558

Zhang XH, Chasin LA (2004) Computational definition of sequence motifs governing constitutive exon splicing. Genes Dev 18:1241–1250

Zhang XH, Kangsamaksin T, Chao MS, Banerjee JK, Chasin LA (2005) Exon inclusion is dependent on predictable exonic splicing enhancers. Mol Cell Biol 25:7323–7332

Zorio DA, Blumenthal T (1999) Both subunits of U2AF recognize the 3′ splice site in Caenorhabditis elegans. Nature 402:835–838

Neurochemistry of Endogenous Antinociception

Gyongyi Horvath

Abstract It is well known that a multitude of ligands and receptors are involved in the nociceptive system, and some of them increae, whereas others inhibit the pain sensation both peripherally and centrally. These substances, including neurotransmitters, neuromodulators, hormones, cytokines, and the like, may modify the activity of nerves involved in the pain pathways. The organism itself can express very effective antinociception under different circumstances (e.g., stress), and during such situations the levels of various endogenous ligands change. Accordingly, a very exciting field of pain research relates to the roles of endogenous ligands. This chapter provides a comprehensive overview of the endogenous ligands that can produce antinociception, discusses their effects on different receptors and focuses on their action in different parts of the pain pathways. The results show that the net effect of a ligand is determined by the activation/inhibition of the different types of receptors and the location of these receptors, however, only a part of the endogenous substances has been characterized extensively in this respect.

Keywords Pain · Receptor · Endogenous · Antinociception

Contents

G. Horvath (✉)
Department of Physiology, Faculty of Medicine, University of Szeged, H-6701 Szeged, Hungary
e-mail: horvath@phys.szote.u-szeged.hu

J.P. Blass (ed.), *Neurochemical Mechanisms in Disease*,
Advances in Neurobiology 1, DOI 10.1007/978-1-4419-7104-3_15,

1 Introduction

Suffering from pain is a major medical, social, and economic burden worldwide, however, the ideal solution for effective pain-relief remains elusive. Understanding the neurochemistry of antinociception has advanced considerably in recent years. Pain is a dynamic phenomenon resulting from the activity of both excitatory and inhibitory endogenous modulation systems. It is well known that a multitude of substances and receptors are involved in the nociceptive system; some of them increase, and others inhibit the pain sensation both peripherally and centrally (Furst, 1999; Sandkuhler, 1996). Virtually no ligands/receptors are to be found that have not been investigated in this respect. These substances, which include neurotransmitters, neuromodulators, hormones, cytokines, and the like, can modify the activity of nerves involved in the pain pathways. One of the physiological functions of the endogenous system is to tonically regulate nociceptive transmission; therefore the ratio of the pronociceptive and antinociceptive ligands determines the pain sensitivity. The balance between these actions ensures effective modulation of acute pain, whereas during chronic pain the pronociceptive effects appear to prevail. It is also well known that the organism can express very effective antinociception in different circumstances, and during such situations the levels of various endogenous ligands change. Thus, endogenous antinociceptive mechanisms play an important role in the regulation of behavior under stressful circumstances. One of the first explicit notations of stress-induced analgesia (SIA) came from observations of soldiers' behavior in World War II (Beecher, 1957). Endogenous opioid peptides have been associated with SIA as its chemical mediators but other, nonopioid, mediators of SIA are known to exist as well (Ortiz et al., 2008). Not only stress may influence pain sensitivity, but several psychiatric diseases can also change it. Thus, the pain threshold is increased in schizophrenic patients and animal models, and depression can result in increased pain sensitivity or increased analgesic requirement (Becker et al., 2006; Blumensohn et al., 2002; Dworkin, 1994; Jackson and St Onge, 2003; Tuboly et al., 2009). Furthermore, migraine disease and other chronic pain syndromes are also based on the imbalance between pro- and antinociceptive endogenous ligands (Gagnier, 2001).

The endogenous ligands can produce their effects at both peripheral and central (spinal and supraspinal) levels. The first relay in pain pathways activated by Aδ- and C-nociceptors is the spinal dorsal horn (SDH) and, as such, this represents an important site for the modulation of the pain signal. The activation of several pathways is involved in the production of analgesia including pathways that project from the amygdala, hypothalamus (arcuate nucleus: ARC, and lateral area of anterior hypothalamus: LAAH), the somatosensory cortex and the anterior cingulated cortex (ACC) to the midbrain periaqueductal grey matter (PAG) (Millan, 2002; Pilcher et al., 1988). ACC and amygdala are particularly related to the affective component of pain and ACC is also implicated in the cognitive processing of pain (Fields, 2004; Ji and Neugebauer, 2008; Neugebauer et al., 2004; Rainville et al., 1997). The hypothalamus is known to be one of the key structures involved in pain modulation and transmission (Dafny et al., 1996), and the hypothalamic fibers containing opioid neurons terminate in PAG (Pilcher et al., 1988). The LAAH has the capacity to differentially modulate components of the pain signal (i.e., activation of this nucleus inhibits the responses to unmyelinated C-fiber activation) and not change the activity of Aδ fibers (Simpson et al., 2008). The overall effect of this would be to safeguard sensory-discriminative information that could direct motivational behaviors and, at the same time, filter out those components of the pain signal that are less relevant to emergency situations. The thalamus contributes to the emotional component of pain and in particular, the intralaminar parafascicular nucleus receives nociceptive information from the spinal cord by both the spinothalamic and spinopontothalamic tracts and its output is to the ACC. PAG represents the mechanisms whereby cortical and other inputs act to control the nociceptive "gate" in the dorsal horn of the spinal cord. PAG projects rostrally to the medial thalamus and orbital frontal cortex, and also interacts with several brainstem structures to modulate nociception including the rostroventral medulla (RVM) (Jensen and Yaksh, 1989; Sandkuhler, 1996; Smith et al., 1988; Zhao et al., 2007). RVM is considered an important source of descending control of spinal nociceptive neurons (Fields and Basbaum, 1999). RVM is the principal relay in the integration of ascending nociceptive inputs with descending outputs from rostral sites (Fields and Basbaum, 1999), as well as the major source of bulbospinal projections that terminate in laminas I, II, and V of the SDH, mostly via OFF (antinociceptive) and ON (pronociceptive) cells. Descending control of spinal nociception, which originates from the locus ceruleus (LC), is another major determinant of pain sensitivity in different behavioral and emotional states (Willis and Westlund, 1997). These descending modulations are exerted by three main neurochemical systems – noradrenergic, serotonergic and opioidergic – which interact in an intricate manner (Millan, 2002).

A very exciting and rapidly developing field of pain research relates to the roles of different endogenous ligands. The endogenous antinociceptive ligands may have potentially advantageous features: their synthesizing and breakdown enzymes (or the mechanism of their excretion) are available in the body; thus, in general they have short half-lives and they may have lower toxicity. On the other hand, most of the endogenous ligands exhibit lower specificity and affinity for their receptors as compared with exogenous drugs, and/or they exert their effects at several types of receptors at different parts of the body. Therefore, the net effect depends on the localization of the ligands/receptors, and on which receptors and where they will be influenced by a ligand. Accordingly, their effectiveness might be lower than that of synthetic drugs, suggesting that these ligands alone would not be ideal drugs for pain therapy.

This chapter provides a comprehensive overview of endogenous ligands with antinociceptive potential, discussing their effects on different receptors and focusing on their action at distinct levels of neural axis.

2 Small-Molecules

2.1 Class I. Acetylcholine (ACh)

Acetylcholine, the first neurotransmitter to be identified, is an ester of acetic acid and choline with the name 2-acetoxy-N,N,N-trimethylethanaminium. It plays pivotal roles in a diverse array of physiological processes, and its activity is controlled through enzymatic degradation by acetylcholinesterase. The effects of ACh receptor (AChR) agonists and enzyme inhibitors, collectively termed cholinomimetics, in antinociception/analgesia are widely investigated. These compounds successfully inhibit pain signaling in both humans and animals, and are efficacious in a number of different preclinical and clinical pain models, suggesting a broad therapeutic potential (Jones and Dunlop, 2007; Wess et al., 2007). Both peripheral and central cholinergic components may be involved in the antinociception. For example, cholinergic stimulation of lateral hypothalamus increases the pain threshold by activating the descending inhibitory pathways (Holden et al., 2002). However, a major site of action of ACh is the spinal cord (Xu et al., 2000; Zhuo and Gebhart, 1991). Intrathecal cholinergic agents cause antinociception by mimicking the release of ACh from the spinal cholinergic nerves, whereas the inhibition it effects decreases the pain threshold suggesting a tonic activity of these neurons (Hood et al., 1997; Krukowski et al., 1997; Pan et al., 2008). Dorsal root ganglion (DRG) neurons express several markers for cholinergic neurons, and it seems that ACh is synthesized both in unmyelinated and myelinated DRG neurons (Khan et al., 2003; Matsumoto et al., 2007; Sann et al., 1995; Takeda et al., 2003; Tata et al., 2004; Vincler and Eisenach, 2004). Other data have shown that painful stimuli increase ACh level in the spinal cord releasing from the cholinergic interneurons in the SDH, and these neurons are activated by the inhibitory descending noradrenergic and serotoninergicic pain modulatory pathways (Detweiler et al., 1993; Eisenach et al., 1996b; Jones and Dunlop, 2007; Zhuo and Gebhart, 1990).

ACh exerts its physiological actions by binding to and activating two structurally and functionally distinct families of cell-surface receptors, the nicotinic ACh receptors (nAChRs) and the muscarinic ACh receptors (mAChRs). The nAChRs function as ACh-gated cation channels, whereas the mAChRs are members of the superfamily of G-protein-coupled receptors (GPCRs). The fast actions of ACh are mediated by its interaction with nAChRs, a family of pentameric ligand-gated ion channels composed of 1 or more of 17 different subunits, and these receptors are distributed widely throughout the central nervous system (CNS) and the periphery (Kalamida et al., 2007). These subunits are divided into muscle-type and neuronal-type. In the CNS the predominant nAChRs are the homomeric α7 and the heteromeric α4/β2 receptor. These receptors have been widely referred to as the neuronal nAChRs, and they mediate synaptic transmission of ACh by gating inward flux of Na^+ and Ca^{2+} at diverse synapses. Molecular-cloning studies have revealed the existence of five molecularly distinct mammalian mAChR subtypes, M1–M5 (Wess et al., 2007). The M1–M5 receptors can be subdivided into two major functional classes according to their G-protein coupling preference. The M1, M3, and M5 receptors selectively couple to G-proteins of the G_q/G_{11} family, whereas the M2 and M4 receptors preferentially activate G_i/G_o-type G-proteins (Wess et al., 2007). Thus, M1, M3, and M5 are linked to phospholipase-C (PLC), and their stimulation leads to formation of inositol phosphates (inositol-triphosphate: IP3 and diacylglycerol: DAG) and a consequent increase in intracellular calcium, whereas M2 and M4 receptor activation inhibits formation of cyclic adenosine monophosphate (cAMP) through inhibition of adenylate cyclase (AC) (Jones and Dunlop, 2007). Agonist-induced activation of mAChRs leads to a wide range of biochemical and electrophysiological responses, and the precise nature of these responses and the resulting physiological effects primarily depend on the location and the molecular identity of the activated mAChR subtypes (Wess, 1996). Each of the five mAChR subtypes exhibits a distinct pattern of distribution; they are expressed in many regions of the CNS (in both neurons and glial cells) and in various peripheral tissues (Wess et al., 2007). The M1, M4, and M5 receptors are predominantly expressed in the CNS, whereas the M2 and M3 receptor subtypes are widely distributed both in the CNS and in peripheral tissues.

Initial observations that nicotine might have an analgesic activity dates back to 1932 {9808}. The concept of nicotinic analgesia being superior to opioids led the songwriter, Paul Simon to memorialize the event in a song (Arneric et al., 2007). It has been found that nAchRs play a role in modulating pain transmission both centrally and peripherally, however, the results are controversial. Multiple populations of nACh receptors at both spinal and supraspinal level can modulate the transmission of nociceptive stimuli (Damaj et al., 2000; Guimaraes et al., 2000; Jones and Dunlop, 2007; Matsumoto et al., 2007). As regards the activation of nAChRs supraspinally, the ACh administered in the dorsal PAG increased the spinally organized pain threshold, and this effect was inhibited by nAChR antagonists, suggesting an nAChR mediated descending pain control (Guimaraes et al., 2000). Furthermore, the activation of neuronal nAchR in the nucleus raphe magnus (NRM) produces an antinociceptive effect as well (Bannon et al., 1998; Jones and Dunlop, 2007). Stimulation of spinal nAChRs may produce both pronociceptive and antinociceptive behaviors via stimulation of separable populations of nAChRs (Khan et al., 1998; Li and Eisenach, 2002). However, most studies found that intrathecal (IT) administration of nicotine produced antinociception, and the inhibition of nAChRs produced hyperalgesia (Li and Eisenach, 2002; Matsumoto et al., 2007; Rashid et al., 2006; Rashid and Ueda, 2002; Vincler and Eisenach, 2004; Young et al., 2008). Activation of nAChRs may enhance the inhibitory GABAergic (γ-aminobutyric acid) and glycinergic activities in the SDH (Genzen and McGehee, 2005; Kiyosawa et al., 2001; Takeda et al., 2003). It has been proposed that the increased expression of $\alpha 3$ and $\alpha 4$ subunits may contribute to the neuropathic pain, whereas inhibition of $\alpha 3/\beta 2$ subunits produces a pronociceptive effect (Vincler and Eisenach, 2004; Young et al., 2008). The peripheral stimulation of nAChR excites or sensitizes peripheral sensory nerve fibres, but can also mediate cholinergic antinociception (Bernardini et al., 2001; Gilbert et al., 2001). NAChR stimulates nitric oxide synthase (NOS) in DRG neurons, and the synthesized nitric oxide (NO) is able to block ion channels in DRG (Haberberger et al., 2004; Renganathan et al., 2002), however, other studies suggest the activation of calcium channels by NO (Bernardini et al., 2001; Haberberger et al., 2004). There is controversy about the role of $\alpha 7$ nAChR at the periphery, inasmuch as Haberberger et al. (2004) found these receptors on all nociceptive neurons and activation of $\alpha 7$-nAChR elicited antinociceptive effects in an inflammatory pain model by peripheral mechanism (Wang et al., 2005b). However, Lang et al. (2003) could not detect these receptors peripherally, and the deficiency in this receptor did not influence pain sensitivity (Rashid et al., 2006).

Centrally active muscarinic agonists are known to induce robust analgesic effects via activation of spinal and supraspinal mAChRs (Gomeza et al., 1999b; Iwamoto and Marion, 1993; Wess et al., 2007). Perhaps the clearest indication of the role of the individual mAChR subtype in antinociception has been provided by receptor knock-out (KO) mice, but plastic changes can mask the real role of the receptors/ligands (Wess et al., 2003). These data suggest that almost all the mAChRs play a significant role in the decrease of pain sensitivity (Jones and Dunlop, 2007; Wess et al., 2007). Independent of the route of administration, the analgesic efficacy of mAChR agonists was greatly reduced, but not abolished, in $M2R^{-/-}$ mice (Duttaroy et al., 2002; Gomeza et al., 1999a). On the other hand, in $M2R^{-/-}/M4R^{-/-}$ double-knock-out mice, the agonist was virtually devoid of analgesic activity. These findings suggest that the M2 receptor is the predominant mAChR mediating muscarinic antinociception at the spinal and the supraspinal level, but M4 receptors also contribute to the analgesic activity (Chen et al., 2005b). Some data suggest that muscarinic antinociception is mediated by M1/M2 receptors or M1/M3 receptors in rats (Naguib and Yaksh, 1997), whereas others have failed to demonstrate the role of these receptors in antinociception (Velligan et al., 2002). It seems that M5 receptors do not play a significant role in acute pain sensitivity (Wang et al., 2004).

As regards the activation of mAChRs supraspinally, painful stimuli have been reported to increase the neuronal activity in the thalamus and ACh through the activation of M1 AChRs can inhibit this effect (Harte et al., 2004; Jones and Dunlop, 2007). Furthermore, M1 receptors located in the nuclei of RVM are involved in the opioid-induced antinociception (Abe et al., 2003). Data suggest that about 90% of all mAChRs in the spinal cord represent M2Rs, which provides a molecular basis for the predominant functional role of M2Rs at the spinal level (Duttaroy et al., 2002). Presynaptic M2R activation inhibits the glutamate release in the spinal cord and ACh induces a presynaptic stimulatory effect on the release of GABA by activating M1R/M2R, whereas the release of glycine (Gly) from spinal cord interneurons was increased through the activation of M3Rs (Jones and Dunlop, 2007; Li et al., 2002a; Wang et al., 2006c). The spinal cholinergic system also plays a role in the actions of opiates because spinally administered atropine can reduce the analgesic effects of systemically administered morphine in rats, and inhibition of ACh-esterase by neostigmine enhances the ability of morphine to reduce pain sensitivity (Chiang and Zhuo, 1989; Eisenach and Gebhart, 1995; Pan et al., 2008). Thus, there is strong evidence that activation of mAChR in the spinal cord results in an increased release of inhibitory transmitters along with a decrease in the release of excitatory transmitters, and this may mediate their antinociceptive effects.

Evidence is also accumulated for a peripheral site of action for mAChRs in antinociception (Bernardini et al., 2001; Wess et al., 2007; Wess et al., 2003). Electrophysiological and neurochemical studies using skin and skin-saphenous nerve preparations demonstrated that muscarine-induced peripheral antinociception was abolished in $M2R^{-/-}$ mice, indicating that this activity is mediated by the M2R subtype (Bernardini et al., 2002; Pan et al., 2008). Stimulation of peripheral mAChRs reduces the heat-stimulated release of calcitonin gene-related peptide (CGRP) from mouse skin, this effect being absent in tissue taken from M2R knock-out mice (Bernardini et al., 2002; Wess et al., 2003). In summary, both the muscarinic and nicotinic receptor activations are very important in the antinociceptive effects of ACh, and the cholinergic system may offer a number of tractable targets for the development of pain therapeutics.

2.1.1 Choline

Choline, a quaternary saturated amine (and a precursor of ACh), is generally thought of as a relatively inactive molecule, although several studies have shown that it can have direct effects on various biological systems and signal transduction pathways. Choline interacts with both nAChRs and mAChRs as a full and selective agonist at α7-containing nAChRs, and at the M1Rs as well (Alkondon et al., 1997; Carriere and El-Fakahani, 2000). The systemic administration of choline did not change the acute heat pain latency (HP), but decreased the inflammatory pain (Wang et al., 2005b). Both intracerebroventricular (ICV) and intrathecal (IT) administrations of choline produces antinociception in acute heat pain tests, and its effects are blocked by the α7-receptor antagonist, but not by atropine or naloxone (Damaj et al., 2000; Wang et al., 2005b).

2.2 Class II. Amines

2.2.1 Epinephrine (E)/Norepinephrine (NE)

Norepinephrine (4-(2-amino-1-hydroxyethyl) benzene-1,2-diol) and epinephrine ((*R*)-4-(1-hydroxy-2-(methylamino) ethyl) benzene-1,2-diol) are monoamines originating from the adrenal medulla, sympathetic nerve terminals and the CNS. In the CNS, adrenergic cells can be found primarily in the brainstem (RVM, LC) (Milner et al., 2002; Stone et al., 2003). NE is synthesized from tyrosine as a precursor, and E is synthesized via methylation of NE. E/NE perform their actions on the target cell by binding to and activating adrenergic receptors (α_1, α_2, β_1, β_2, and β_3). They are widely distributed both centrally and peripherally, including most motor, sensory, autonomic, and neuroendocrine-related areas (Delfs et al., 2000; Egan et al., 1983; Nicholson et al., 2005; Stone et al., 2003). All of these receptors are GPCRs; α_1-adrenoceptors couple to G_q, which results in increased intracellular Ca^{2+}. α_2-Receptors, couple to G_i and decrease the level of cAMP, whereas β-receptors couple to G_s, and increase intracellular cAMP activity. Several data suggest that the α_2-receptors play the most important role in pain mechanisms, primarily through the inhibition of transmitter release presynaptically, but they also inhibit the projecting neurons (Willis, Jr., 1988). In vitro studies have suggested that α_2-adrenoceptor agonists decrease glutamate release, and inhibit glutamate-mediated neuronal activation, and this action can also contribute to their antinociceptive potency (Faber et al., 1998; Li and Eisenach, 2001). There is conflicting evidence concerning the role of α_2-receptors located supraspinally (Mansikka et al., 1996; Mansikka and Pertovaara, 1995; Ossipov and Gebhart, 1983).

The noradrenergic innervation of the spinal cord arises from noradrenergic nuclei in the brainstem, including the A6 (locus ceruleus), the A5 (lateral reticular nucleus: LRN), and A7 (in the dorsolateral pontine tegmentum) nuclei (Guo et al., 1996; Kwiat and Basbaum, 1992; Proudfit and Clark, 1991). The activity of these neurons can be modulated by ligands acting at α_2-adrenoceptors, changing the descending noradrenergic effects (Aghajanian and Vandermaelen, 1982; Andrade and Aghajanian, 1982; Mansikka and Pertovaara, 1995). Stimulation of α_2-adrenoceptors in the LRN did not influence the mechanical hyperalgesia, whereas α_2-adrenoceptor antagonist reversed the central hyperalgesia induced by mustard oil, without having any effects on nocifensive withdrawal thresholds of an intact limb (Mansikka et al., 1996). The RVM does not contain noradrenergic cells, but receives a dense noradrenergic projection from the A5 and A7 neurons, and these inputs affect pain modulation by RVM neurons (Fields and Basbaum, 1999).

It has been shown that excitatory α_1-adrenoceptor is present on both On and Off cells, but the inhibitory α_2-adrenoceptor is present only on the Off cells, and the activation of On cells can be involved in the increased pain sensitivity during opioid withdrawal (Bie et al., 2003). Thus, activation of α_2-adrenoceptors in NRM may induce antinociceptive effects (Haws et al., 1990; Proudfit, 1988). The activation of LC neurons by α_2-adrenoceptor agonists also produces antinociception, and α_2-adrenoceptor activation might contribute to the antinociceptive effects of amygdala activation (Guo et al., 1996; Ortiz et al., 2008). Some data suggest the role of α_1-receptors in antinociception at supraspinal level, because ICV administration of an α_1-receptor antagonist inhibits the antinociceptive potency of monoamine-reuptake inhibitors (Yokogawa et al., 2002). Activation of supraspinal β_2-adenergic receptors also produces inhibition in nociceptive transmission (Fukui et al., 2004). The role of the noradrenergic system in the control of the activity of spinal neurons involved in the transmission of sensory messages to supraspinal relays is well documented (Eisenach et al., 1996a; Pertovaara, 2006; Skagerberg and Lindvall, 1985; Weil-Fugazza and Godefroy, 1993). NE is released from the descending inhibitory pathways in the spinal cord, and the activation of α_2-adrenergic receptors plays the most important role in this respect, inasmuch as a very effective antinociception can be reached by their activation (Eisenach et al., 1996a; Ganong et al., 1983; Horvath et al., 1994; Kalso et al., 1991; Kuraishi et al., 1985; Reimann et al., 1999).

Peripheral mechanisms might also significantly contribute to their pain-influencing effects, because topical administration of α_2-receptor agonist compounds produces effective antinociception, whereas E produces hyperalgesia via β_2-adrenergic receptors (Ansah and Pertovaara, 2007; Chen and Levine, 2005; Khasar et al., 1999; Moon et al., 1999). The activation of peripheral α_2-adrenoceptors might decrease pain by the inhibition of the activity of C-fibers (Gaumann et al., 1992; Pertovaara, 2006; Yagi and Sumino, 1998). In addition, α_2-adrenoceptor activation can produce peripheral antinociception via action on the immune system by altering the balance of pro- and anti-inflammatory cytokines, and by inducing a release of endogenous opioids from immune cells (Binder et al., 2004; Romero-Sandoval and Eisenach, 2007).

2.2.2 Dopamine (DA)

Dopamine (4-(2-aminoethyl) benzene-1,2-diol) constitutes about 80% of the catecholamine content in the brain (Pivonello et al., 2007; Vallone et al., 2000). Projections originating from brain areas that synthesize this neurotransmitter give rise to four axonal pathways: nigro-striatal, mesolimbic, mesocortical, and tuberoinfundibular. Dopamine receptors (DARs: GPCRs) are widely distributed in the CNS, mainly localized in the striatum, the limbic system, the brain cortex, and the infundibulum, where they mediate the effect of DA on cognition, emotion, regulation of hunger and satiety, locomotor activity, pain, and on the endocrine system (Missale et al., 1998). DARs are widely distributed in the periphery as well, primarily at the level of the cardiovascular system, kidneys and adrenal glands, beyond the peripheral nervous system (PNS). Five distinct DARs receptors have been isolated, and subdivided into two subfamilies, D1- and D2-like, on the basis of their biochemical and pharmacological properties. The D1-like subfamily comprises D1 and D5-R, whereas the D2-like includes D2-, D3-, and D4-R (Brucke et al., 1991). D2-like receptors have a presynaptic location, and D1-like receptors are exclusively postsynaptic (Vallone et al., 2000). The signal transduction pathways activated by DARs are numerous, but the best-described effects are the activation or inhibition of the cAMP pathway and modulation of Ca^{2+} signaling. Receptors of the D1-like subtype are positive regulators of cAMP, whereas the inhibition of AC activity seems to be a general property of D2-like receptors. DARs are also able to activate other mechanisms of signal transduction, including the modulation of the activity of PLC or phospholipase D (PLD) leading to the release of arachidonic acid, as well as the activity of the calcium and potassium channels (Pivonello et al., 2007; Senogles, 2000). Moreover, DARs also seem to modulate the activity of Na^{+}/H^{+} exchangers and the Na^{+}/K^{+}-ATPase (Missale et al., 1998).

Several data have demonstrated a control for dopaminergic neurotransmission in modulating pain perception and natural analgesia within supraspinal regions, including basal ganglia, insula, ACC, thalamus, and PAG (Wood, 2008). The mesolimbic dopaminergic system plays important roles in the suppression of persistent pain, and studies have provided direct evidence that the nucleus accumbens plays a major role in this mechanism (Altier and Stewart, 1999; Carta et al., 1999; Gear et al., 1999; Taylor et al., 2003). The descending dopaminergic system is also involved in pain control. Both DA and its metabolites are present in the spinal cord (Bjorklund and Skagerberg, 1979; Commissiong et al., 1978; Fleetwood-Walker et al., 1988; Jensen and Yaksh, 1984; Takada et al., 1988). The dopaminergic fibers are predominantly localized in the superficial layers of the SDH, and they arise primarily from the hypothalamic areas and from the caudal thalamus, but they can also originate from the substantia nigra (Commissiong et al., 1978). Some dopaminergic sensory neurons in the DRG may innervate the spinal cord, and dopaminergic cell bodies may also be a source for dopamine in the spinal cord (Mouchet et al., 1986; Price and Mudge, 1983). IT administration of the DA agonist apomorphine produces analgesia, and morphine induces an increase in the metabolism of DA in the SDH suggesting that the descending dopaminergic system is involved in the modulation of the activity of the nociceptive neurons induced by morphine (Jensen and Yaksh, 1984; Weil-Fugazza and Godefroy, 1993). Both D1-like and D2-like receptors are found in the SDH, and both D1, D2 and D5 receptors are involved in DA-mediated antinociception (Altier and Stewart, 1998; Dubois et al., 1986; Karper et al., 2000; Morgan and Franklin, 1991). As regards its effect peripherally, it has been shown that the local administration of dopamine causes hyperalgesia by activating primary sensory neurons directly (Steiner et al., 2001).

2.2.3 Serotonin (5-Hydroxi-Tryptamine, 5-HT)

Serotonin was discovered as a potent vasotonic ligand. It plays a role in the inflammatory chemical milieu and is released from platelets, mast cells, and basophils in injured or inflamed tissues as a critical factor in the control of nociceptive transmission (Doak and Sawynok, 1997; Dray, 1995; Tokunaga et al., 1998; Zeitz et al., 2002). Serotoninergic neurons are found in the raphe nuclei in the midbrain, pons, and medulla and serotoninergic fibres project to the several brain regions and the spinal cord. Molecular cloning studies have confirmed the existence of at least 14 subtypes of 5-HT receptors, each encoded by distinct genes (Raymond et al., 2001).

The 5-HT receptors have been divided into seven subfamilies. All of them except 5-HT3 receptors are GPCRs, whereas 5-HT3 receptors are ion channels. There is multiplicity of coupling mechanisms for each 5-HT receptor subtype (Raymond et al., 2001).

As regards the antinociceptive potency of 5-HT at supraspinal level, the results are controversial. Serotoninergicic deficiency is a common factor both in mental depression and chronic pain. It is well known that antidepressants, including the selective serotonin reuptake inhibitors, have antinociceptive effects after systemic or ICV administration (Nayebi et al., 2001; Singh et al., 2001). Increased levels of 5-HT in synaptic clefts are therefore presumed to lead to changes in pain thresholds and induce antinociception. A recent study, by using the formalin test in rats, attempted to determine the identity and possible localization of the receptor subtypes predominantly involved in the antinociceptive effects of antidepressants. Thus, it has been shown that ICV administration of 5-HT2 and 5-HT3-receptor antagonists inhibited the antinociceptive potency of serotonin-reuptake inhibitors (Yokogawa et al., 2002).

The major site of the antinociceptive action of serotonin seems to be the spinal cord, and various studies have identified several types of serotonin receptors in the SDH (Coggeshall and Carlton, 1997; Fields et al., 1991; Furst, 1999). Data suggest that distinct 5-HT receptors subtypes are employed to generate the 5-HT-induced antiallodynic and antinociceptive effects. Serotonin is released from the descending inhibitory pathways in the spinal cord, and the activation of these pathways lead to antinociception (Millan, 2002; Reimann et al., 1999; Van Steenwinckel et al., 2008; Willis, Jr., 1988). IT administered 5-HT has an antinociceptive effect in acute pain models, but it has lower potency and efficacy in models of persistent pain. (Bardin et al., 2000a, b; Kuraishi et al., 1985). It has been suggested that predominantly 5-HT3 receptors are involved in the antinociception by evoking GABA and enkephalin (ENK) release (Huang et al., 2008; Kesim et al., 2005; Li et al., 2000; Wang et al., 2003b). However, some data suggest that 5-HT3 receptors contribute to the maintenance of chronic pain, because the 5-HT3 receptor antagonist ondenasetron reduces mechanical allodynia, and activation of deep SDH neurons that develops following nerve injury (Hamon and Bourgoin, 1999; Oatway et al., 2004; Suzuki et al., 2004). Endogenous 5-HT shows the highest affinity for 5-HT2A receptors subtype, and this subtype is also able to exert antinociceptive action (Hamon and Bourgoin, 1999). Other data have shown that the antinociceptive effects of IT 5-HT or serotonin reuptake inhibitors were blocked by 5-HT1A, B, 2A, 2C, 3, and 4 antagonists, whereas antagonists at 5-HT1D did not influence them (Honda et al., 2006; Jeong et al., 2004; Van Steenwinckel et al., 2008). 5-HT depletion in the SDH antagonizes the analgesic action of morphine (Murphy and Zemlan, 1990), and selective blockade of 5-HT7, but not of 5-HT1A and 5-HT2 receptors attenuated morphine analgesia (Dogrul and Seyrek, 2006). 5-HT produces an algesic response as a component of the inflammatory process at peripheral level (Giordano and Rogers, 1989; Taiwo and Levine, 1992), but activation of 5-HT3 receptors can blunt the pronociceptive effects on 5-HT2 and 5-HT1A receptors (Kesim et al., 2005).

2.2.4 Histamine

The biogenic amine histamine (2-(3H-imidazol-4-yl) ethanamine) is involved in local immune responses, and it is also regarded as a neurotransmitter or modulator in the mammalian brain (Prell and Green, 1986; Schwartz et al., 1991). Histamine is derived from the decarboxylation of the amino acid histidine, a reaction catalyzed by the enzyme L-histidine decarboxylase. In the CNS histamine mostly originates from two cell types, neurons and mast cell. The cell body of histaminergic neurons is localized in the tuberomammillary nucleus of the posterior hypothalamus, and histamine-immunoreactive nerve fibres project widely to the various brain regions and the SDH (Haas and Panula, 2003). Histamine mediates its effects through four histamine receptors (GPCRs) that have been discovered and are designated H1 through H4.

As regards the effect of histamine on the pain threshold, it depends on the site of application and the type of the activated receptor. The H1 receptor knock-out animals or systemic injection H1- receptor antagonist drugs show increased pain threshold (Farzin et al., 2002; Mobarakeh et al., 2000; Sakurada et al., 2002; Yanai et al., 2003; Zamfirova et al., 2007). Activation of H2 receptors induced an increase in the mechanical pain threshold, whereas antagonism of H2 receptors can induce either antinociception or hypernociception (Lamberti et al., 1996; Oluyomi and Hart, 1991). ICV administration of low doses of histamine elicits hyperalgesia, and high doses of histamine produce antinociception (Chung et al., 1984; Parolaro et al., 1989). The injection of histamine into the dorsal raphe nucleus and PAG region produces an antinociception, whereas its injection into the median raphe nucleus causes hyperalgesia (Glick and Crane, 1978; Thoburn et al., 1994). Some data suggest that activation of opioid receptors can increase histamine release in PAG (Barke and Hough, 1993).

The opposite effects of histamine on the pain threshold may be mediated through different subtypes of receptors (Lamberti et al., 1996; Malmberg-Aiello et al., 1994; Thoburn et al., 1994). Thus, ICV injection of histamine H1 receptor agonist produced hypernociception in hot plate (HP) and writhing tests, and H1 receptor antagonists produce antinociceptive effects (Malmberg-Aiello et al., 1998). However, other reports found that H1 antagonist antagonized the histamine-induced antinociception (Parolaro et al., 1989). The ICV injection of either H2 agonists or antagonists raised the pain threshold (Farzin et al., 2002). Moreover, a series of H2 receptor antagonists reduced the antinociceptive effects of H2 receptor agonist (Netti et al., 1988). However, other data suggest that both H1 and H2 receptor activation inhibit the morphine-induced antinociception at both spinal and supraspinal levels (Mobarakeh et al., 2000; Mobarakeh et al., 2002; Mobarakeh et al., 2006). It seems that H3 receptor activation may also decrease the pain threshold, because receptor antagonists have analgesic properties, as these compounds block presynaptic autoreceptors and increase the release of neuronal histamine (Farzin et al., 2002). Spinal administration of histamine produces nociceptive behavior, and a recent study has suggested that this effect is partially mediated by the activation of N-methyl-D-aspartate (NMDA) receptors at polyamine binding sites (Sakurada et al., 2002; Yanai et al., 2003). However, the activation of H3 receptors located on spinal terminals increases the pain threshold by inhibiting the release of excitatory neurotransmitters (Cannon et al., 2003; Cannon et al., 2007). It is well known that the peripherally released histamine is a very effective pain-inducing ligand, however, the activation of H3 receptors at peripheral level can also inhibit the pain sensation (Cannon et al., 2007).

2.2.5 Melatonin (MT)

Melatonin (5-methoxy-N-acetyltryptamine), a pineal neurohormone and a derivative of serotonin, is critically involved in the regulation of important biological functions including circadian rhythms, sleep, mood, and pain (El Shenawy et al., 2002; Sugden, 1983; Zeng et al., 2008). MT and its receptors (MT1 and MT2) are located in the spinal cord and various brain regions (Morgan et al., 1994; Vitte et al., 1990; Zahn et al., 2003). These receptors are GPCRs, and they are linked to activation of multiple signaling pathways, with the inhibition of cAMP formation being the most common (Dubocovich et al., 2003; Reppert et al., 1996). The action of MT may be mediated through an interaction with NMDA receptors and the NOS pathway too (Hernandez-Pacheco et al., 2008; Mantovani et al., 2003; Tu et al., 2004). It can also inhibit calcium influx, and it may exert its central effects by modulating GABAA receptors, therefore, the inhibitory mechanisms of MT might be complex and are yet to be elucidated in detail (Vanecek, 1998; Wu et al., 1999a). Clinical studies have shown that migraine patients have lower nocturnal plasma MT levels than controls, and migraine patients with superimposed depression exhibit the greatest decrease of MT (Gagnier, 2001; Reiter, 1991). Furthermore, MT administration improved the symptoms, and it could be due to a number of the actions of MT: resetting the biological rhythm, relieving anxiety and insomnia, inhibiting both protaglandin and NO synthesis, depressing calcium uptake, or directly affecting cerebral blood vessels. MT can reduce cluster headache, irritable bowel syndrome, and fibromyalgia, although the relationship between depression and chronic pain was not specifically examined in these clinical reports (Citera et al., 2000; Leone et al., 1996; Song et al., 2005).

Systemic administration of MT produced dose-dependent antinociception in HP, and visceral and inflammatory pain tests primarily by supraspinal MT2 receptor activation (El Shenawy et al., 2002; Li et al., 2005b; Sugden, 1983; Tu et al., 2004; Yu et al., 2000a; Zeng et al., 2008). On the other hand, others have shown that systemic MT did not influence the normal pain threshold, but inhibited the development of morphine tolerance (Raghavendra et al., 2000; Raghavendra and Kulkarni, 1999; Raghavendra and Kulkarni, 2000). In contrast, light-induced MT suppression can decrease arthritic pain (Burk, 2008). As regards the activation of the central melatoninergic system, ICV MT produced a significant increase in acute heat pain latency, and reversed the nociception or neuropathy-induced hyperalgesic effects (Li et al., 2005b; Sakurada et al., 2002; Ulugol et al., 2006; Wang et al., 2006b). Its effect can be reversed by naloxone and MT can specifically enhance the antinociception induced by δ-(DOR) , but not by μ-(MOR) opioid agonists (Li et al., 2005b; Yu et al., 2000b).

Intra-ACC administration of MT attenuated mechanical allodynia and improved depression-like behavior without changing the nociceptive response in normal rats, and depressive animals exhibited a lower level of plasma MT concentration and intra-ACC MT receptor expression (Zeng et al., 2008). These results indicate that there is a reciprocal relationship between depression-like behavior, and nociceptive behavior and the melatoninergic system within ACC could play a significant role in this relationship. IT administration of MT did not influence the pain threshold in a postoperative pain model, but potentiated the effect of morphine, and it effectively decreased the capsaicin-induced pain behavior and neuropathic allodynia by activation of MT2 receptors (Ambriz-Tututi and Granados-Soto, 2007; Tu et al., 2004; Zahn et al., 2003). The results suggest that the endogenous MT system in the spinal cord can reduce the generation, development, and maintenance of central sensitization, with a resultant inhibition of hyperalgesia, allodynia. Peripheral administration of MT can also decrease formalin- and glutamate-induced behavior via the activation of the NO-cyclic guanosine monophosphate (GMP)-K^+-channel opening (Hernandez-Pacheco et al., 2008; Mantovani et al., 2006).

2.2.6 Agmatine (AGM)

AGM (decarboxylated arginine), an endogenous amine derived from arginine and its biosynthetic enzyme (arginine decarboxylase), is broadly distributed in the CNS, including the SDH (Li et al., 1994a; Raasch et al., 1995; Reis and Regunathan, 2000). The distribution of AGM-containing neurons is concentrated in regions of the brain that subserve visceral and neuroendocrine control, the processing of emotions, pain perception, and cognition (Reis and Regunathan, 2000). The concentration of AGM in the brain is comparable to that of norepinephrine or dopamine (Li et al., 1994a). AGM possesses modest (micromolar) affinity for α_2-adrenoceptors, and for imidazoline-binding sites (I1 and I2) (Li et al., 1994a; Raasch et al., 1995). Features complicating the interpretation of its influence upon nociceptive processing are that AGM behaves as an inhibitor of NOS, expresses antagonist properties at NMDA receptors and blocks the nAChR cation channels (Fairbanks et al., 2000; Gibson et al., 2002; Reis and Regunathan, 2000; Yang and Reis, 1999). Systemic administration of AGM significantly reversed inflammatory hyperalgesia and neuropathic allodynia; furthermore, it potentiated morphine-induced analgesia (Kolesnikov et al., 1996; Paszcuk et al., 2007). ICV administration of AGM had no antinociceptive potency by itself, but potentiated the effects of morphine through activation of both α_2-adrenoceptors and I2-receptors (Roerig, 2003; Sanchez-Blazquez et al., 2000). AGM suppresses the transmission of nociceptive inputs at the spinal level, primarily through the activation of I-receptors (Auguet et al., 1995; Bradley and Headley, 1997; Hou et al., 2003; Kolesnikov et al., 1996; Pinthong et al., 1995). The single or continuous IT administration of AGM could restore injured hypersensitive animals to normal levels of sensation, but did not influence normal pain sensitivity (Fairbanks et al., 2000; Kekesi et al., 2004). No data are available on the possible effects of AGM at the peripheral level.

2.3 *Class III. Amino Acids and Derivatives*

2.3.1 Glutamate

The excitatory amino acid glutamate plays a key role in the modulation of nociceptive processing by acting through two distinct types of receptors: excitatory ionotropic (tetrameric Ca^{2+}/Na^{+}-channels: NMDA, α-amino-3-hydroxyl-5-methyl-4-isoxazole-propionate: AMPA and kainate) and metabotropic glutamate receptors (mGluRs) (Bleakman et al., 2006). Eight mGluRs have been identified and divided into three groups (I–III) based on their sequence similarity, pharmacology, and G-protein coupling (Conn and Pin, 1997), Group I receptors including mGlu1 and mGlu5 are coupled via G_q to PLC. Group II (mGlu2 and mGlu3) and Group III (mGlu4,6,7,8) receptors activate G_i and inhibit cAMP formation. Group I mGluRs are primarily located postsynaptically on neurons and contribute to biphasic regulation of glutamate synaptic transmission. Group II and III mGluRs are found to contribute to presynaptic regulation of glutamate and GABA transmission. All three groups are distributed throughout the CNS, and several data proved the antinociceptive effect of activation of group II and III receptors (Goudet et al., 2008; Kim et al., 2002; Pan et al., 2008).

Glutamate can produce antinociception at the supraspinal level, because central NMDA receptor activation can lead to the release of endogenous opioid peptides (Bach and Yaksh, 1995; Goudet et al., 2008; Kim et al., 2002; Starowicz et al., 2007). The ICV administration of glutamate enhanced the morphine-induced antinociception, indicating the analgesic interaction between the NMDA and MOR (Hunter et al., 1994; Jacquet, 1988). Neurons immunoreactive for the NMDA receptors and glutamate were identified in the PAG, and a subset of these projects to the RVM (Commons et al., 1999; Ito et al., 2008; Wiklund et al., 1988). In an animal model of inflammatory hyperalgesia, intra-RVM injection of NMDA produced facilitation at lower doses, and inhibition at higher doses, whereas AMPA receptor activation produced dose-dependent inhibition (Guan et al., 2002). Thus, activation of both AMPA and NMDA receptors are involved in the descending modulation after inflammatory hyperalgesia. It has been suggested that glutamate release in the RVM activates Off antinociceptive neurons, and its leads to antinociception (Bleakman et al., 2006; Guan et al., 2002; Starowicz et al., 2007). Activation of mGlu receptors in the brainstem can also produce antinociceptive effects (Bleakman et al., 2006; Kim et al., 2002; Oja and Saransaari, 2000). Glutamate (and aspartate) is a well-known excitatory neurotransmitter and pain-inducing substance at spinal and peripheral levels by the activation of the ionotropic receptors (Bleakman et al., 2006). However, data suggest the involvement of spinal group III mGluR in the modulation of acute, inflammatory, and neuropathic pain (Goudet et al., 2008). The selective activation of group III mGluR at the spinal level inhibited the nociceptive behavior of rats submitted to the formalin test and the mechanical hyperalgesia associated with inflammatory or neuropathic pain. This study provides new evidence for supporting the role of spinal group III mGluRs in the modulation of pain perception in different pathological pain states of various etiologies but not in normal conditions.

2.3.2 γ-Amino-butyric Acid (GABA)

GABA is the major inhibitory neurotransmitter. Organisms synthesize GABA from glutamate using the enzyme L-glutamic acid decarboxylase and pyridoxal phosphate (which is the active form of vitamin B6) as a cofactor. This process converts the principal excitatory neurotransmitter (glutamate) into the principal inhibitory one (GABA). GABA receptors can be classified as GABAA and GABAC receptors, which are ionotropic receptors (pentameric chloride channels), and GABAB receptors, which are metabotropic receptors (Alger and Le Beau, 2001). GABA and its receptors are widely distributed throughout the neuraxis; their concentration in the brain and spinal cord is relatively high (Enna and McCarson, 2006; Willis, Jr., 1988). The activation of GABAA and GABAC receptors increases the neuronal concentration of chloride ion leading to hyperpolarization of the cells. Stimulation of GABAB receptors modifies the level of cAMP, decreases Ca^{2+}, and increases K^+ membrane conductance, leading to cellular hyperpolarization. The function of GABA in the modulation of nociception is crucial and complex. Enna McCarson provided an excellent review of the role of GABA and its receptors in pain transmission, and the results suggest that GABA provides the main neurochemical substrate for local modulation of pain control in different central areas (Enna and McCarson, 2006).

With regard to higher brain regions, there are GABAergic projections from the ventral tegmental area and substantia nigra to the PAG and NRM (Kirouac et al., 2004; Williams and Beitz, 1990). GABAA receptors are located on inhibitory neurons projecting from the RVM to the dorsal horn (Gilbert and Franklin, 2001). Thus, local injection of a GABA agonist into this region facilitates transmission of a pain impulse through the spinal cord (Ito et al., 2008). In contrast, central stimulation of GABAA and GABAB receptors induces antinociception in the formalin test, and this effect may be mediated partly through supraspinal opioid receptor mechanisms (Mahmoudi and Zarrindast, 2002). Furthermore, an increase in overall GABAergic activity in the insular cortex induces analgesia by enhancing the descending inhibition of the spinal cord (Jasmin et al., 2003). Activation of GABAA receptors in the amygdala produced a robust reversal of escape/avoidance behavior, and reduced mechanical hypersensitivity in a neuropathic pain model (Pedersen et al., 2007). These data suggest that GABAA receptor activation increases output from amygdala to brainstem and forebrain areas and this process might selectively attenuate affective nociceptive processing. In the spinal cord GABA is a widespread transmitter and GABA receptors are located in the SDH on pre- and postsynaptic sites in the region of the Aδ- and C-fiber synapses (Huang et al., 2008; Yang et al., 2002). In the SDH GABA and ENK are colocalized in a large population of neurons, and these neurons may represent local inhibitory interneurons which modulate pain transmission (Todd et al., 1992). Neuropathies cause a loss of GABAergic neurons and GABA transports in the rat spinal cord contributing to the pain syndrome (Drew et al., 2004; Lever et al., 2003; McCarson et al., 2006; Moore et al., 2002).

Both GABAA and GABAB receptor activations display antinociceptive activity (Dirig and Yaksh, 1995; Franek et al., 2004; Hwang and Yaksh, 1997; Malan et al., 2002; Patel et al., 2001; Vaught et al., 1985). It has been demonstrated that the GABAergic system contributed to spinal serotonin-mediated antinociception, inasmuch as GABA release may underlie the antinociceptive effects of the descending serotoninergic pathway (Huang et al., 2008; Kawamata et al., 2002). The analgesic response to GABAB agonist is thought to be mediated, in part, by the activation of spinal cord presynaptic receptors that regulate the release of tachykinins, and the inflammatory pain modifies GABAB receptor expression in the DRG and SDH (Riley et al., 2001). Less is known about the involvement of the GABAC receptors in pain. Zheng et al. (2003) localized the GABAC receptors on lamina I and II of DH and DRG, crucial sites for pain transmission, but no data are available on the effect of GABAC agonist at spinal level. Peripheral activations of both GABAA and GABAC receptors produce antinociception (Carlton et al., 1999; Da Motta et al., 2004; Reis and Duarte, 2007). It is supposed that activation of coupled chloride channels causes a hyperpolarization of peripheral terminals of primary afferents, leading to a decrease in action potential generation.

2.3.3 Glycine (Gly)

Glycine, the smallest amino acid, is known to be an inhibitory neurotransmitter, but only a few studies investigated its role in pain modulation (Webb and Lynch, 2007). Initially, Gly was described to be restricted to the mammalian spinal cord, but subsequently it has been detected supraspinally as well (Legendre, 2001). Gly receptors (GlyRs) belong to the superfamily of receptor channels, which are generally composed of five subunits (α1–4, β) (Webb and Lynch, 2007). The different α- and β-subunits are differently localized. The GlyR is a pentameric chloride channel, and it is classically known for mediating inhibitory synaptic transmission between interneurons and motor neurons in reflex circuits of the spinal cord, but they are also found presynaptically, where they modulate neurotransmitter release (Lynch, 2009; Webb and Lynch, 2007). The picture is complicated by the fact that Gly also binds to and activates NMDA receptors, therefore, it can influence the pain threshold by this action as well (see above, Section 2.3.1) (Zeilhofer, 2005).

Changes in glycinergic neurotransmission in the spinal cord dorsal horn are critically involved in the development of pathological pain, and GlyR blockade produces allodynia in "normal" animals and enhances nociceptive responses (Cronin et al., 2004; Sherman and Loomis, 1996; Yaksh, 1989). Immunocytochemical and electrophysiological evidence implicates α3β GlyRs as important mediators of glycinergic inhibitory neurotransmission in nociceptive sensory neuronal circuits in the peripheral laminae of the SDH (Lynch, 2009; Webb and Lynch, 2007). Because α3 subunits are targets for prostaglandin modulation in spinal nociceptive neurons, antinociceptive drugs targeting the GlyR should ideally be specific for this subtype. Thus, inflammation-induced decrease of lamina II glycinergic inhibitory postsynaptic current (IPSC) was found to be abolished in the α3β GlyR knock-out mice, and chronic inflammation did not produce pain sensitisation, but these mice responded normally to acute inflammatory pain stimuli (Harvey et al., 2004). Furthermore, the inhibition of glycine uptake at the spinal level produced antinociception in acute pain tests and in different models of neuropathy (Hermanns et al., 2008; Tanabe et al., 2008). No data are available on the glycine effect at both supraspinal and peripheral levels, but see Section 2.3.5.

2.3.4 D-serine

D-serine is one of the recently identified neurotransmitter candidates, and has attracted extensive attention because of its multiple roles in physiological and pathophysiological conditions (Boehning and Snyder, 2003). D-amino acid oxidase degrades D-serine physiologically, whereas serine racemase directly converts L-serine to D-serine. Both enzymes and D-serine can be found in the brain in the highest concentrations in the forebrain, where the NMDA receptors for glutamate are also highly concentrated. Most strikingly, D-serine occurs selectively in protoplasmic astrocytes, which ensheath synapses in grey matter, whereas most astrocytes are enriched in white matter. The occurrence of D-serine in astrocytes in close proximity to NMDA receptors and its release by glutamate suggest that D-serine is an endogenous ligand for the NMDA receptors. It is quite potent in stimulating the Gly site of the NMDA receptor, and it acts as an endogenous and obligatory coagonist for this receptor (Danysz and Parsons, 1998). Some observations have suggested that the activation of the supraspinal NMDA receptors by D-serine may lead to an increased pain threshold. Accordingly, ICV application of D-serine alone produces a dose-dependent antinociception, and potentiates the antinociception of morphine in the tail-flick (TF) and formalin tests (Hunter et al., 1994; Yoshikawa et al., 2007). Other data have shown that D-serine-induced antinociception was attenuated by the ICV application of a GABAA receptor agonist (Ito et al., 2008). These data suggest functional interactions among the GABAA, NMDA receptors, and MOR in the regulation of the antinociception at the supraspinal level.

2.3.5 Taurine

Taurine (2-aminoethanesulfonic acid) is a phylogenetically ancient nonessential amino acid; one of the most widespread ligands throughout the CNS (Oja and Saransaari, 2000; Zeilhofer, 2005). Taurine differs from most other amino acids in being a sulfonic acid and a β-amino acid. Taurine has been proposed as a possible inhibitory neurotransmitter in several loci of the CNS through the activation of GlyRs (Frizzo et al., 2003; Legendre, 2001; Mathers et al., 1989; Xu et al., 2004a). Taurine induces hyperpolarization and inhibits firing of neurons; it acts as a modulator of synaptic activity in the brain (Oja and Saransaari, 2000). The increase in extracellular taurine upon excessive stimulation of glutamate receptors and under cell-damaging conditions may serve as an important protective mechanism against excitotoxicity. An increase in oral taurine uptake diminishes chronic nociception, but decreases the normal heat–pain threshold (Belfer et al., 1998). As regards its effect in the brain, intra-ACC injection of taurine effectively reduced neuropathic nociception acting on GlyRs (Pellicer et al., 2007). Both IT and intraperitoneal (IP) administration of taurine relieves nociceptive stimulation effects, and pain stimulus releases taurine in the spinal cord (Hornfeldt et al., 1992; Ishikawa et al., 2000; Legendre, 2001; Serrano et al., 1998; Skilling et al., 1990; Smullin et al., 1990).

2.3.6 Kynurenic Acid (KYNA)

Degradation of the essential amino acid tryptophan along the kynurenine pathway yields several neuroactive intermediates, including kynurenic acid (4-oxo-1H-quinoline-2-carboxylic acid) (Moroni et al., 1988; Schwarcz and Pellicciari, 2002; Vecsei and Beal, 1991). This is found both centrally and peripherally in low concentrations (10–150 nM) and is synthesized in the CNS, predominantly by glial cells (Moroni et al., 1988; Pawlak et al., 2000; Schwarcz and Pellicciari, 2002; Turski and Schwarcz, 1988; Urbanska et al., 2000). KYNA at high, nonphysiological concentrations is a broad-spectrum antagonist of ionotropic excitatory amino acid receptors, acting at the Gly (half-maximal inhibitory concentration: $IC_{50} \sim 20\ \mu M$) and the NMDA recognition sites ($IC_{50} \sim 200\ \mu M$) of the NMDA receptor complex (Carpenedo et al., 2001; Ganong et al., 1983; Stone, 1993). In higher concentrations (0.1–1 mM), it also antagonizes the AMPA and kainate receptors, and KYNA is a potent noncompetitive antagonist of α7 nAchRs ($IC_{50} \sim 7\ \mu M$) too (Hilmas et al., 2001; Stone, 2000). Thus, direct support for its physiological role in glutamatergic and cholinergic neurotransmission has been reported (Carpenedo et al., 2001; Nemeth et al., 2005; Schwarcz and Pellicciari, 2002).

A recent study has shown that GPR35, a previously orphan GPCR, functions as a receptor for kynurenic acid (Wang et al., 2006a). KYNA elicits calcium mobilization and IP3 production in a GPR35-dependent manner, and it also induces the internalization of this receptor. GPR35 is predominantly detected in immune cells and the gastrointestinal tract, but it has also been found in the DRG on small- to medium-diameter neurons (Ohshiro et al., 2008). The results suggest that GPR35 may modulate nociception and a continued study of this receptor will provide additional insight into the role of KYNA in pain perception, inasmuch as no in vivo data are available regarding the role of GPR35 in the effects of KYNA. Systemic administration of KYNA produced antinociception in acute heat pain tests and attenuated the development of tolerance (Heinricher and McGaraughty, 1998; Marek et al., 1991). Intracisternally administered KYNA effectively inhibited the capsaicin-induced pain behavior (Hajos and Engberg, 1990). Intra-RVM infusion of KYNA inhibited the opioid-induced antinociception, although the baseline pain threshold was unaffected (Heinricher and McGaraughty, 1998; Heinricher et al., 1999). IT administration of KYNA produces antinociception in different models (Raigorodsky and Urca, 1990; Yaksh, 1989; Yamamoto and Yaksh, 1992; Zhang et al., 2003b), and enhanced the effects of EM-1 and AGM (Horvath et al., 2007; Horvath and Kekesi, 2006; Kekesi et al., 2002). Its peripheral administration also produced antinociception in a joint inflammatory model with low potency (Mecs et al., 2009).

3 Purines

3.1 Adenosine

The endogenous purine mediator adenosine, originating from adenosine 5-triphosphate (ATP), is a widely distributed neuromodulator with complex effects (Sawynok, 1998; Sawynok and Liu, 2003). Four adenosine receptors have been identified and are termed A1, A2A, A2B, and A3 (Fredholm et al., 2001). They are all GPCRs and couple to classical second messenger pathways; A1 and A3 receptor activation decreases the level of cAMP, A2 increases it, whereas A2B receptor stimulates PLC (Sawynok, 1998; Sawynok and Liu, 2003).

It has a complex influence on nociceptive pathways because its effects depend on the receptor subtype activated (Sawynok, 1998; Sawynok and Liu, 2003). The activation of the A1 and A3 receptors produces analgesia by both peripheral and central mechanisms, and a variety of molecules is being developed to provide analgesia through this nonopioid mechanism (Poon and Sawynok, 1998; Sawynok, 1998; Sawynok and Liu, 2003). It is very important that adenosine has a significant role in opioid-induced antinociception (Sawynok, 1998). Only a few data support its antinociceptive potency at the supraspinal level. Thus caffeine (adenosine antagonist) decreased the analgesic effects of ICV-administered opioids, and MOR and DOR (but not κ-opioid receptor [KOR]) agonists (Pham et al., 2003). Furthermore, both A1 and A2A agonists produced antinociception in acute pain models (Pham et al., 2003; Regaya et al., 2004).

Several reports suggest the antinociceptive effect of synthetic adenosine derivatives or adenosine kinase inhibitors in different pain tests at spinal level (Poon and Sawynok, 1998; Sawynok, 1998), however, only few laboratories (those of Sollevi and Eisenach) have investigated adenosine in this regard (Belfrage et al., 1999; Chiari and Eisenach, 1999; Eisenach et al., 2002; Lavand'homme and Eisenach, 1999; Rane et al., 2000; Von Heijne et al., 1999). Most studies have observed effective antinociception in neuropathic pain states, but slight or no effects on normal or inflammatory pain sensitivity have been found (Kekesi et al., 2004; Lavand'homme and Eisenach, 1999; Rane et al., 2000; Sawynok and Liu, 2003). This ineffectivity might be due to uptake and metabolic degradation of adenosine (Sawynok and Liu, 2003).

It seems that antinociceptive effects of adenosine are particularly related to the activation of A1 receptors in the spinal cord where inhibition of intrinsic and primary sensory neurons may contribute to this actions (Sawynok and Liu, 2003; Schulte et al., 2003; Sollevi et al., 1995). Moreover, adenosine agonists produce analgesia largely by interacting with the descending inhibitory noradrenergic system, and the effect of adenosine is blocked by α_2-adrenergic antagonists (Gomes et al., 1999; Sweeney et al., 1987). A component of the antinociceptive action of morphine is also due to the local release of adenosine within the spinal cord (Sawynok et al., 1989; Sweeney et al., 1987). Adenosine acting at its A2A receptor is thought to be pronociceptive, and this effect has been proposed to result from the increase in cAMP levels and activation of NMDA receptors (Hussey et al., 2007; Khasar et al., 1995). As regards the effects of adenosine at peripheral level, the A1 receptor is the predominant receptor subtype mediating antinociception at peripheral level too, whereas the A2A, A2B, and A3 receptors mediate nociception peripherally (Sawynok and Liu, 2003). Thus it has been shown that the activation of A1 receptors produced a significant antihyperalgesic effect in the inflammatory pain model (Vuckovic et al., 2006). In contrast, A2A receptor knock-out mice have a higher nociceptive threshold and this has been suggested to be attributable to the lack of peripheral adenosine A2A receptors (Ledent et al., 1997).

3.2 Nucleotides

Nucleotides are molecules which comprise the structural units of RNA and DNA. Additionally, nucleotides play central roles in the metabolism. These nucleotides might be adenine-(adenosine di- and triphosphate: ADP and ATP), guanine-(guanosine di- and triphosphate: GDP, GTP), or pyrimidine-nucleotides (uridine di- and triphosphate: UDP, UTP). It has been established that these molecules mediate diverse biological effects via P2 purinoceptors (P2Xn: ionotropic and P2Yn: metabotropic receptors) in both the PNS and CNS (Tsuda et al., 2005; Wirkner et al., 2007). The P2X receptors expressed in the brain are primarily distributed throughout the rat hindbrain, including the RVM and LC (Kanjhan et al., 1999; Wirkner et al., 2007). ATP acting on P2X receptors at the supraspinal level produces mechanical and thermal antinociception in rats through the activation of $P2X_3$ receptors (Fukui et al., 2004; Fukui et al., 2006; Wirkner et al., 2007). It is conceivable that the ascending noradrenergic neurons arising from the LC are involved in the supraspinal antinociception by a P2X receptor agonist (Fukui et al., 2004). At the spinal level the expression of the $P2X_3$ receptors appears selective for a subpopulation of small diameter DRG neurons, which are probably associated with nociception (Inoue et al., 2007; Wirkner et al., 2007). As regards the role of P2Y receptors, the activation of some of these receptors by UTP (acting on $P2Y_{2,4,6,14}$) in the brain had no effect on the mechanical nociceptive threshold (Fukui et al., 2001). Only a few data suggest that the expression of these receptors increased in the microglia at the spinal level during neuropathy, and blocking or lack of this receptor produced antinociception (Gerevich and Illes, 2004; Inoue et al., 2007; Tozaki-Saitoh et al., 2008). Thus, the activation of these receptors contributes to an acute nociceptive behavior, hyperalgesia, and allodynia.

In contrast, some data suggest that P2Y receptor agonists can inhibit cytokine release from activated spinal cord microglia (Gerevich and Illes, 2004). This process could interrupt chronic pain development and continuation. Thus, UTP and UDP were shown to be analgesic in the neuropathic pain model (Okada et al., 2002). IT pyrimidine nucleotides elevated the nociceptive threshold in the paw pressure and

TF tests, whereas adenine nucleotides (activate: $P2Y_{1,2,11,12,13}$) lowered it and produced allodynia (Gerevich and Illes, 2004). The ADP analogue ADP-β-S (acting on $P2Y_{1,12,13}$) has also been found to cause analgesia in TF test. As regards the action mechanism, it seems that both pyrimidine nucleotides and ADP-β-S produce antinociception by the activation of P2Y receptor causing the inhibition of the voltage gated N-type Ca^{2+} channels, and decrease the transmitter release in the spinal cord (Gerevich et al., 2004). However, the specific antagonist of $P2Y_{12}$ receptor decreased neuropathic pain, suggesting that this type of purinergic receptors may be critical in the pathogenesis of neuropathic pain (Tozaki-Saitoh et al., 2008). These ligands, mainly the ATP play a facilitatory role in pain transmission at peripheral level (Wirkner et al., 2007).

Extracellular guanine-based purines (GBP), namely the nucleotide guanosine monophosphate (GMP) and the nucleoside guanosine also exert biological effects. Such actions are unrelated to direct G-protein modulation, but GBPs induce modulation of the glutamatergic system by the inhibition of the binding of glutamate and analogues, and prevent cell responses to excitatory amino acids (Burgos et al., 1998; Morciano et al., 2004). Both neurons and astrocytes release guanosine under basal and toxic conditions (Ciccarelli et al., 2001). ICV administration of guanosine or its prosubstance, GTP, produced dose-dependent antinociceptive effects in several different acute nociceptive tests (HP, TF, intraplantar: IPL capsaicin or glutamate) (Schmidt et al., 2008). The action mechanism of this effect is unknown, but it was not inhibited by the adenosine- or opioid receptor antagonists, and guanosine did not increase the adenosine level in the brain. (Traversa et al., 2003). However, guanosine significantly stimulates glutamate uptake, thereby preventing glutamate toxicity, therefore, it is tempting to suppose that the in vivo antinociceptive effect of guanosine can result from its effect on glutamate removal from the synaptic cleft, leading to less activation of glutamate receptors (Frizzo et al., 2003).

4 Other Nonpeptide Molecules

4.1 Ouabain

Endogenous cardiac glycoside inhibitors of Na^+/K^+-ATPase with structures similar to that of plant-derived ouabain (1β,3β,5β,11α,14,19-hexahydroxycard-20(22)-enolide 3-(6-deoxy-α-L-mannopyranoside)) have been isolated from several tissues, including the adrenal cortex and the brain (Van Huysse and Leenen, 1998). Ouabain, through the inhibition of Na^+,K^+-ATPase, may produce several effects including modulation of neural activity and neurotransmitter release. These effects might be related to the pain mechanism; but only few studies investigated its role in this context. ICV-injected ouabain in relatively high doses (μg) exerts an antinociceptive effect and potentiates the analgesic activity of morphine (Calcutt et al., 1971). However, lower doses (ng) of ouabain were able to antagonize the antinociception induced by morphine (Masocha et al., 2003). IT administration of ouabain in high doses also produced analgesia and enhanced the potency of morphine (Zeng et al.,

1999). Another study found that low doses of ouabain did not modify the acute heat pain latency, but higher doses caused excitation and motor impairment, suggesting that ouabain does not produce a pronounced effect on the pain threshold (Horvath et al., 2003). It is supposed that the controversial results might be due to the fact that ouabain produces its effect on all of the cells, inasmuch as all have Na^+,K^+-ATPase, and its net effects on transmitter releases might be dose-dependent.

5 Peptides

A growing number of peptides have been identified in the CNS and the periphery that relate to pain modulation (Palkovits, 1984). They can originate from neurons, endocrine cells, immunocytes, fat and muscle cells, and so on. There is no clear classification for peptides, and there is some overlap between the different groups. The first group of them contains the hormones, however, now it is well established that most of the hormones can originate from neurons, and they can modify the neuronal functions as well. The second group comprises neuropeptides consisting of short chains of amino acids, with some functioning as neurotransmitters others as hormones; they are often localized in axon terminals at synapses and are classified as putative neurotransmitters. They include endorphins, ENKs, and others. Cytokines (the third group) are a special category of signaling molecules that, like hormones and neurotransmitters, are used extensively in cellular communication. Anatomical and structural distinctions between cytokines and classic hormones are fading as we learn more about each. Classic protein hormones circulate in nanomolar (10^{-9}) concentrations that usually vary by less than one order of magnitude. In contrast, some cytokines (such as IL-6) circulate in picomolar (10^{-12}) concentrations that can increase up to 1000-fold during a trauma or infection.

5.1 Peptide Hormones

5.1.1 Oxytocin (OT)

Oxytocin (nonapeptide) is mainly synthesized together with arginin-vasopressin (AVP) in magnocellular neurons of the paraventricular (PVN) and supraoptic (SO) nuclei of the hypothalamus, and acts as a neurohormone during parturition and the milk ejection reflex (Gimpl and Fahrenholz, 2001). However, it can subserve a neurotransmitter/neuromodulator function as well. OT exerts its actions via the OT receptor, a GPCR receptor ($G_{q/11}$ class, stimulates PLC activity), which is localized in many parts of the CNS and PNS (Tribollet et al., 1992). Descending OTergic pathways extend from the hypothalamus to the thalamus and brainstem (Sawchenko and Swanson, 1982), thus OTergic terminals and high-affinity binding sites for OT are present in regions involved in pain perception (Tribollet et al., 1992). High concentrations of OT are found in raphe nuclei, and OT modulates serotonin turnover in the brain (Kovacs, 1986). It has been suggested that a loop exists between the LC and the hypothalamus, a pathway that may be involved in the regulation of the release of OT (and AVP) following painful stimulation (Swanson and McKellar, 1979). At least 25% of OTergic neurons in the hypothalamus project to the SDH, and these projection sites match well OT binding sites in the superficial layers of the SDH and in the autonomic regions (Rousselot et al., 1990).

Although some conflicting results have been reported in the literature, analgesic effects of OT have been proved in most studies after systemic and/or central administration. The ICV injection of OT produced a significant antinociceptive effect in different acute pain models (Arletti et al., 1993; Gao and Yu, 2004; Zubrzycka and Janecka, 2008). An OT-sensitive antinociception can be induced by massage-like stimulation, swim stress, and electrical stimulation the PVN in both naive and neuropathic rats, suggesting the involvement of an endogenous OT receptor-dependent analgesic system. The OT-induced antinociception might be mediated by MOR and KOR activation, which suggests the release of endogenous opioids after OT receptor activation (Zubrzycka et al., 2005). The stimulation of PVN causes OT release in the spinal cord and can influence spinal nociceptive processing (Condes-Lara et al., 2005; Yang, 1994). This OT-specific stimulation of neurons allows the recruitment of GABA-ergic interneurons in lamina II, which produces a generalized elevation of local inhibition.

5.1.2 Vasopressin (Arginine Vasopressin: AVP)

AVP (nonapeptide), the other posterior pituitary hormone, is mainly synthesized in the PVN of hypothalamus. Similarly to OT, descending AVPergic pathways extend from the hypothalamus to the thalamus, medulla oblongata, and the substantia gelatinosa of the SDH (Sawchenko and Swanson, 1982). Especially PAG contains many AVP-containing fibers (Pittman et al., 1981). Three subtypes of AVP receptors (GPCRs), V1, V2, and V3, have been identified, mediating vasoconstriction, water reabsorption, and central nervous system effects, respectively (Holmes et al., 2003). Functionally, V1R activates G-proteins of the $G_{q/11}$, whereas V2R stimulates the Gs proteins. A variety of signaling pathways is associated with V1R including the activation of calcium influx, PLA2, PLC, and PLD; in contrast, V2R activates cAMP. More than one G-protein appears to participate in signal transduction pathways linked to V3Rs, depending on the level of receptor expression and the concentration of AVP. Many experiments discovered that AVP is related to pain modulation, and pain stimuli elevate AVP concentration in different brain areas (NRM, caudate nucleus, and PVN), furthermore microinjection of AVP into these centers raised pain thresholds (Yang et al., 2007a; Zubrzycka and Janecka, 2007). Stimulation of PVN caused antinociception, which was antagonized by anti-AVP (Yang et al., 2007a). Central injection of AVP (ICV, intra-PAG) increased pain threshold and the level of endogenous opioids, thus its effects were reversed by naloxone suggesting that the release of endogenous opioids plays a significant role in its antinociceptive effect (Yang et al., 2007a, b; Zubrzycka et al., 2005; Zubrzycka and Janecka, 2008). The analgesic effect of AVP in PAG can be reversed by a V2 antagonist and V2 antagonist also reduced the basal pain threshold, suggesting an inhibitory tone of these fibres (Yang et al., 2006). It seems that the antinociceptive effect of AVP is limited to the brain nuclei, not to the spinal cord and peripheral organs, because IT or intravenous (IV) injection of AVP or anti-AVP serum did not change the pain threshold (Yang et al., 2007a).

5.1.3 Calcitonin (CT)/Parathyroid Hormone Fragments (PTH)/Tuberoinfundibular Peptide of 39 Residues (TIP39)

Parathormone (PTH) (84 amino acids) acts in concert with calcitonin (CT; 32-amino acids) to maintain the serum calcium level acting on GPCR receptors. TIP39 was purified based on parathyroid hormone-2 receptor (PTHR2) activation, and it is an endogenous ligand for the PTHR2, a GPCR receptor increasing cAMP level (Usdin et al., 2003). Both ligands and their binding sites can be found in regions that are involved in processing pain-related information (Harvey and Hayer, 1993; Olgiati et al., 1983; Usdin et al., 2003). The ability of CT to produce antinociception has led to the suggestion that CT may serve as a neuromodulator in the CNS. Systemic injection of CT produced antinociception in inflammatory pain models, and this was inhibited by ICV but not by IT application of serotonin-antagonists, and neither IT nor ICV administration of α-adrenoceptor antagonist and systemic injection of opioid antagonist influenced the antinociceptive effects (Yamazaki et al., 1999).

ICV administration also was effective in acute heat- and visceral pain models (Welch et al., 1986). One study has investigated the antinociceptive interaction of fragments of the PTH with CT after their ICV administration (Welch and Dewey, 1990). It has been suggested that CT and some PTH fragments interact in the modulation of nonopiate antinociception, possibly via actions on the calcium level in the brain. CT is not an effective antinociceptive ligand at the spinal level (Wiesenfeld-Hallin and Persson, 1984). Only few data support the role of TIP39 in the pain processes. ICV administration of TIP39 partially reversed tactile withdrawal hypersensitivity following carrageenan administration, but did not change the HP latency or the formalin-induced behavioral responses (LaBuda and Usdin, 2004). TIP39 also decreased the aversiveness of paw stimulation, suggesting that it may modulate an effective component of nociception within the brain, and its IT administration produced pronociceptive action (Dobolyi et al., 2002; LaBuda and Usdin, 2004).

5.1.4 Insulin

Insulin (51 amino acids) is produced in the islets of Langerhans in the pancreas. Insulin regulates not only the blood sugar level, but also various CNS functions acting on its tyrosine-kinase receptor (TrK). A few early data suggest that insulin can induce antinociception, and diabetic rats are less sensitive to the antinociceptive effect of morphine (Bodnar et al., 1979; Simon and Dewey, 1981). Both systemic and ICV insulin decreased the formalin-evoked behavior, and the insulin-induced antinociception was independent of hypoglycemic effects, but it could be due to the activation of endogenous dopaminergic, serotoninergic, and opioidergic systems (Anuradha et al., 2004; Takeshita and Yamaguchi, 1997). Insulin inhibits neuronal firing in the hippocampus and hypothalamus, and insulin-induced antinociception could possibly involve all these centers, whereas the IT administration of insulin did not influence the pain threshold (Bitar et al., 1996).

5.1.5 Renin-Angiotensin System (RAS)

Angiotensin II (Ang II; octapeptide) is a key regulator of the cardiovascular system, and it is the main effector of RAS. The final steps of its biosynthesis involve consecutive proteolytic cleavages of its inactive precursors, angiotensinogen (AO) and angiotensin I (Ang I), by renin and angiotensin converting enzyme (ACE), respectively (Suarez et al., 2002). Ang II-containing neurons and fibers were identified in the brain, especially in the hypothalamic regions, the nucleus of the solitary tract, and in the PAG (Sofroniew et al., 1981). Two receptor subtypes Ang II type 1 (AT1) and Ang II type 2 (AT2) (GPCRs: increase PLC and/or cAMP formation) have been identified, and the main biological functions exerted by Ang II are mediated by the AT1 receptor subtype. Ang III (metabolite of Ang II) binds primarily to AT2 subtypes, and the biological function of AT2 receptors is still controversial (McKinley et al., 2003; Suarez et al., 2002). The localization of Ang II and its receptors in the PAG reinforces the suggestion that endogenous Ang II participates in an AT-receptor-mediated modulation of nociception (Pelegrini-da-Silva et al., 2005).

AT2 receptor-deficient mice have increased sensitivity to pain and decreased the levels of brain β-endorphin (Sakagawa et al., 2000). Antinociception following ICV administration of Ang II and Ang III has been demonstrated in several rodent pain models (Pelegrini-da-Silva et al., 2005; Raghavendra et al., 1999). Administration of Ang III into the rat nucleus reticularis gigantocellularis also evokes antinociception in TF test (Yien et al., 1993). Microinjection of angiotensinogen, Ang I, Ang II, or Ang III into the PAG produces a dose-dependent antinociceptive effect in the rat TF-test, which was inhibited by Ang- and opioid receptor antagonists (Pelegrini-da-Silva et al., 2005; Prado et al., 2003; Adams et al., 1986). Additional evidence of the involvement of RAS peptides in nociception includes the reduction of the SIA by Ang receptor antagonists (Haulica et al., 1986; Raghavendra et al., 1999). Nerve terminals containing Ang II-like immunoreactivity (LI) have been identified in the primary sensory neurons and in the SDH as well (Buck et al., 1982), and IT administration of Ang II increased the TF latency (Thurston et al., 1992), whereas others did not find this effect (Cridland and Henry, 1988a).

5.1.6 Melanocortin System (MC)

Amongst the wide range of modulators, the melanocortin system represents a relatively new, intriguing, potential target for pain control (Bertorelli et al., 2005). MCs are a family of endogenous peptides generated by enzymatic cleavage of a common precursor molecule, proopiomelanocortin (POMC). Main members of the MC family are α-, β-, γ-melanocyte-stimulating hormones (MSH: containing 16, 22, 12 amino acids, respectively) and the adrenocorticotropic hormone (ACTH; 39 amino acids). MCs exert their actions through activation of at least five subtypes of receptors (MC1–MC5), which are GPCRs, and each of them is positively coupled to AC. A further peculiarity of the system is that, in addition to endogenous agonists, there are also endogenous antagonists such as agouti protein, which binds preferentially to MC1 receptors and is expressed mainly in the skin, and the agouti-related protein (AgRP), which is an inverse agonist of both MC3 and MC4 receptors and is mainly expressed in the brain (Dinulescu and Cone, 2000; Nijenhuis et al., 2001). Thus, both endogenous agonists and antagonists can be detected in the CNS including the spinal cord (Bertorelli et al., 2005), and the MC system could be under the control not only of an excitatory but also an inhibitory system.

MCs and their receptors are mainly present in the periphery where they can be found primarily on melanoma cells and melanocytes (Wikberg and Mutulis, 2008). In the CNS, the MC1 receptor is present on neurons in the PAG of the midbrain, where it is thought to have a role in pain control (Mogil et al., 2003; Palkovits et al., 1987). Furthermore, mainly MC3 and MC4 receptors are found in the spinal cord. A possible link between MCs and nociception was first postulated by pioneering studies in late the 1970s, early 1980s, showing that ICV administration of α-MSH and ACTH causes hyperalgesia, and reverses the analgesic effects of morphine and β-endorphin (Sandman and Kastin, 1981; Williams et al., 1986). Thus, these ligands produce mainly hyperalgesia, whereas the antagonists produce antinociception in inflammatory and nerve injury models (Bellasio et al., 2003; Bertorelli et al., 2005; Mogil et al., 2003; Sandman and Kastin, 1981; Starowicz et al., 2002; Vrinten et al., 2000; Vrinten et al., 2001). However, MSH has an anti-inflammatory potency as well, and it antagonizes the interleukin-1β-induced hyperalgesia in the PVN (Ceriani et al., 1994; Macaluso et al., 1994). Furthermore, MC1 mediates KOR-mediated analgesia in female mice (Mogil et al., 2003). Inasmuch as agouti protein and AgRP are endogenous antagonists of MC receptors, the inhibition of these receptors by these endogenous ligands can produce effective antinociception, as was shown after their IT administration (Bellasio et al., 2003; Bertorelli et al., 2005).

5.1.7 Corticotropin-Releasing Factor (CRF) and Related Peptides

CRF (41 amino acids) is best known as the major physiological regulator of pituitary ACTH secretion. CRF not only mediates stress responses but also acts as a neuromodulator of synaptic transmission outside of the hypothalamic–pituitary–adrenocortical axis (Ji and Neugebauer, 2008). In addition to CRF, the CRF family encompasses three novel CRF-related mammalian ligands, urocortin 1 (Ucn1), Ucn2, and Ucn3 (Martinez et al., 2004; Perrin and Vale, 1999). Ucn1 contains 40 amino acids, whereas Ucn2 and Ucn3 are composed of 38 amino acids. Each type of Ucns is found in different brain areas including the PAG, but they can be identified peripherally as well (Martinez et al., 2004). These ligands mediate their actions through interaction with two distinct receptor subtypes, CRF1 and CRF2, and both receptors can couple to similar signal transduction pathways (AC and PKA). CRF has preferential affinity for CRF1, Ucn1 binds with equal high affinity to both CRF receptors, and Ucn2 and Ucn3 exhibit high selectivity towards CRF2 receptors (Dautzenberg and Hauger, 2002; De Souza et al., 1985; Korosi et al., 2007). The presence of Ucn1 and Ucn3 immunoreactive nerve terminals in association with

CRF2 receptors in the PAG and spinal cord suggests a modulatory influence of these receptors in pain (Korosi et al., 2007).

Accumulating evidence suggests that peripheral and central CRF are important pain modulators, but the literature on pain-related CRF functions in the CNS is very controversial (Million et al., 2006). CRF, injected ICV, mimicked the effects of stress-induced visceral hyperalgesia, and CRF1/CRF2 antagonists blocked this effect (Gue et al., 1997). However, other data suggest that both CRF1 and CRF2 are involved in SIA (Ji and Neugebauer, 2008; Korosi et al., 2007). A recent study has shown antinociceptive effects of CRF (ICV) in acute and inflammatory pain models, but CRF has also increased pain-related vocalizations and the number of Fos immunopositive spinal neurons (Vit et al., 2006). It seems that the amygdala might be an important site in this respect, because CRF2 receptors could activate inhibitory circuits in the amygdala, whereas CRF1 receptors regulate excitatory processes (Ji and Neugebauer, 2008; Neugebauer et al., 2004). The opposing effects of CRF on nociceptive processing may be mediated through different receptors. Low concentrations of CRF facilitate nociception through CRF1, whereas higher concentrations have inhibitory effects through CRF2 receptors, and this would be consistent with the higher affinity of CRF for CRF1 than CRF2 receptors (Dautzenberg and Hauger, 2002; Ji and Neugebauer, 2008). CRF, Ucn1, and CRF receptors occur in the spinal cord as well, mainly in laminae VII and X, and occasionally in lamina IX, whereas the receptors have not been identified in the superficial laminae of the dorsal horn (Korosi et al., 2007). It has been suggested that these receptors are also involved in stress adaptation processes, such as modulation of SIA and the mediation of visceral nociceptive information at spinal level (Korosi et al., 2007; Million et al., 2006; Robbins and Ness, 2008). Thus, IT-administered CRF produces antinociception, which is reversible by the CRF2 receptor antagonist (Nijsen et al., 2005). CRF and CRF2 receptor expressions were detected in the periphery as well, and the peripheral CRF2 may also be involved in visceral sensitivity (Ayesta and Nikolarakis, 1989; Kiang and Wei, 1985; Million et al., 2006; Schafer et al., 1996). Intra-arterial injection of Uc2 reduced visceral hyperalgesia in vitro, and this effect was inhibited by a selective CRF2 antagonist, suggesting that the peripheral activation of CRF2 receptors has a significant role in visceral antinociception (Million et al., 2006).

5.1.8 Thyrotropin-Releasing Hormone (TRH)

TRH (tripeptide: Glu-His-Pro-NH_2), discovered originally as a hypothalamic hormone, is widely distributed in the CNS, and it also coexists with substance P (SP) and 5-HT in the neurons of the medulla oblongata which projects into the spinal cord (Johansson et al., 1981). TRH exerts a variety of CNS effects, through stimulation of TRHR1 and TRHR2 receptors belonging to GPCR. TRHR1 receptors are expressed in the pituitary gland to release thyrod stimulating hormone, whereas TRHR2 has been shown in the brainstem nuclei, which are involved in descending pain modulation (Cao et al., 1998). Both systemic and supraspinal activations of TRH receptors produce antinociceptive effects in acute heat, mechanical, and visceral pain tests, whereas IT administration has been ineffective (Tanabe et al., 2007; Webster et al., 1983). Thus, injection of TRH into the lateral ventricle, PAG,

NRM, or amygdala generates antinociception in the acute pain test (Reny-Palasse et al., 1989; Webster et al., 1983). ICV administration of TRH has induced analgesic effect with similar or higher potency than morphine in mechanical but not thermal pain tests (Boschi et al., 1983). Its antinociceptive effects are mediated via activation of both descending monoaminergic and serotoninergic pathways (Tanabe et al., 2007), and others have found that TRH activity was resistant to modifications of NE, dopamine, and 5-HT systems (Boschi et al., 1983). The TRH effect was not antagonized by naloxone, but TRH at a nonanalgesic dose prevented the hyperalgesia induced by naloxone (Boschi et al., 1983).

5.1.9 Somatostatin (SST)

Somatostatin (in 14 and 28 amino-acid-containing forms) was originally described as a hypothalamic polypeptide that inhibits the secretion of pituitary growth hormone. It exerts a wide range of effects such as modulation of hormone and neurotransmitter release, cognitive and behavioral processes, the gastrointestinal tract, the cardiovascular system and tumour cell proliferation, but it also has an important neuromodulator function (Gamse et al., 1981; Pan et al., 2008; Pinter et al., 2006). SST is synthesised and stored in capsaicin-sensitive transient receptor potential vanilloid 1 (TRPV1) receptor expressing nociceptive afferents, but it also has been identified in SDH neurons (Pan et al., 2008; Willis, 1988). The effects of SST are mediated via five different GPCR subtypes which can be divided into two main groups on the basis of their sequence similarities and their binding profile towards synthetic somatostatin analogues. SST2, 3, and 5 mediate the endocrine and antiproliferative effects of SST, whereas SST1 and 4 may be responsible for the anti-inflammatory and antinociceptive actions (Helyes et al., 2000; Pinter et al., 2006; Sandor et al., 2006). Exogenously administered SST has been shown to inhibit neurogenic inflammation and nociception in several experimental assessments, and it is effective in the treatment of patients with certain pain conditions, including different types of headaches (Pan et al., 2008; Yu et al., 2004). Central administration of SST (ICV or into the caudate nucleus) increases the pain threshold, suggesting an antinociceptive role of SST at the supraspinal level (Tashev et al., 2001; Zheng and Li, 1995). IT injection of SST has failed to influence TF latency at low doses, however, higher doses have caused motor impairments (Cridland and Henry, 1988a). Much progress has been made in the past ten years in the understanding of the important roles of SST in the regulation of pain transmission at the peripheral level (Helyes et al., 2000; Sandor et al., 2006). Somatostatin released from the activated capsaicin-sensitive sensory nerve terminals reaches the circulation, and it is able to elicit systemic anti-inflammatory and antinociceptive actions. This endogenous counterregulatory mechanism of neurally derived somatostatin has been termed as its "sensocrine function" (Szolcsanyi, 2004; Than et al., 2000).

5.1.10 Prolactin

Prolactin is a polypeptide hormone (199 amino acids) whose major biological actions are related to normal lactation and reproduction. After hormone binding, signal transduction occurs via the cytokine receptor superfamily. It is well known that prolactin level increases during painful stimuli, but only a few data suggest its antinociceptive role. Thus it has been shown that systemic administration of prolactin induced antinociception in visceral pain model, and it can contribute to postictal antinociception as well (Portugal-Santana et al., 2004; Ramaswamy et al., 1985). No other data are available in this context.

5.1.11 Ghrelin

Ghrelin (28 amino acids), a gastric-derived hormone, was discovered as a ligand for the growth hormone secretagogue receptor (GHSR). It has gained increasing attention as a brain–gut hormone with GH-releasing and appetite-inducing functions (Arora and Anubhuti, 2006). GHSR is a GPCR and two isoforms (type 1a and 1b) have been detected (Kojima and Kangawa, 2005). Recent studies have reported that, in addition to the stomach, ghrelin and its receptors are expressed in various peripheral tissues and in the brain, including the pituitary, hypothalamus, pons medulla, oblongata, and SDH, the regions implicated in the control of pain transmission (Kojima and Kangawa, 2005; Vergnano et al., 2008). It is very important in this context that ghrelin specifically inhibits the expression of the proinflammatory cytokines; therefore it may attenuate proinflammatory cytokine-mediated neuropathic pain (Dixit et al., 2004; Guneli et al., 2007; Li et al., 2004). IP and ICV administration of ghrelin reduced the inflammatory hyperalgesia and edema in a naloxone-reversible manner (Sibilia et al., 2006). Ghrelin is able to stimulate the neural activity in the hypothalamic ARC, where it increases the endogenous opioid synthesis and/or activity (Sibilia et al., 2006). Ghrelin increased the inhibitory postsynaptic currents and prevented the capsaicin-induced increase of Fos-LI in the deep SDH (Vergnano et al., 2008). These data suggest that the effect of the ghrelin is mainly due to an action potential-dependent presynaptic release of inhibitory neurotransmitters, and it may be tonically active in the spinal cord. Ghrelin promotes neuronal release of neuropeptide Y, which is another antinociceptive ligand (see below Section 5.2.8; Cowley et al., 2003). IPL administration of ghrelin increased the inflammatory pain threshold, suggesting a peripheral role of this ligand, too (Sibilia et al., 2006).

5.1.12 Orexins

Orexins, also called hypocretins, are the common names given to a pair of highly excitatory neuropeptide hormones that were discovered in rat brain, and they are implicated in body mass regulation (Arora and Anubhuti, 2006; Trivedi et al., 1998). The two related peptides (orexin-A: 33 and orexin-B: 28 amino acids), with approximately 50% sequence identity, are produced by cleavage of a single precursor protein. Orexin-A has two intrachain disulfide bonds and has greater biological importance, and orexin-B is a linear peptide (Smart, 1999). Although these peptides are produced by a very small population of cells in the lateral and posterior hypothalamus, they send projections throughout the brain and to the spinal cord (Marcus et al., 2001; Van Den Pol, 1999; Yamamoto et al., 2002a). The orexin peptides bind to the GPCR orexin receptors (OX-1 and OX-2), which are widely distributed in the CNS (Kukkonen et al., 2002; Trivedi et al., 1998). The deficiency in orexin induced an increased hyperalgesia and less SIA (Watanabe et al., 2004), and ICV administration of orexin was effective on acute and neuropathic pain tests (Mobarakeh et al., 2005; Yamamoto et al., 2003). At the spinal level orexin was also effective in different pain models (visceral, formalin, postoperative, and neuropathic), and the inhibition of the NMDA receptors might play a significant role in this process (Cheng et al., 2003; Kajiyama et al., 2005; Peng et al., 2008; Yamamoto et al., 2002a; Yamamoto et al., 2003). Its peripheral administration inhibits the neuronal vasodilation and this effect can also contribute to a decreased pain sensitivity (Holland et al., 2005).

5.1.13 Bombesin-Related Peptides

Bombesin (BN; amidated tetradecapeptide) was isolated from frog skin. Subsequently, in mammals two BN-like peptides were identified: gastrin-releasing peptide (GRP; 27 amino acids) and neuromedin B (NMB; 10 amino acids) (Jensen et al., 2008). On the basis of the preceding molecular studies, three classes of mammalian bombesin receptors (BB1-3, GPCRs acting primarily through PLC system) were proposed. The BB1 is an NMB-preferring receptor, the BB2 is a GRP-preferring receptor, and the BB3 has low affinity for these peptides. Studies of GRP and NMB immunoreactivity as well as mRNA studies have demonstrated that these peptides and their receptors are widely distributed in mammals in both the nervous system and peripheral tissues, especially in the gastrointestinal tract (Jensen et al., 2008). Only a few studies suggest the role of these peptides in the nociception. GRP or BB1 receptor-deficient mice did not show any impairment in pain threshold (Sun and Chen, 2007; Yamada et al., 2003), but intra-PAG injection of bombesin produced antinociception in the HP and TF tests (Pert et al., 1980; Yu et al., 2004). Furthermore, the IT administration of bombesin and neuromedin B produced nocifensive behavior (Cridland and Henry, 1992).

5.2 Neuropeptides

5.2.1 Opioid-Related Peptides

Morphine, the main alkaloid of opium, is utilized for the treatment of severe pain, and is the gold standard to which all analgesics are compared. Early efforts to understand the endogenous targets of opiate drugs led to the identification of receptor sites. Binding studies suggested four main classes of opioid receptors, named μ- (MOR), δ- (DOR), κ- (KOR), and opioid receptor-like (ORL1) receptors. Opioid receptors comprise a subfamily of structurally homologous GPCRs. Activation of these receptors inhibits the formation of cAMP, close voltage-gated Ca^{2+}-channels and opens inwardly rectifying potassium channels (Dhawan et al., 1996; Jordan et al., 2000; Lambert, 2008). The net effect of these cellular actions is to reduce neuronal excitability and neurotransmitter release.

Opioid receptors and their endogenous ligands are widely distributed in the organism, thus both central and peripheral activation of this system might lead to effective antinociception (Akil et al., 1984; Basbaum and Fields, 1984; Bodnar, 2008; Menetrey and Basbaum, 1987; Palkovits, 2000; Pan et al., 2008). A high dose of naloxone (opioid antagonist) produces hyperalgesia, suggesting a significant role of endogenously released opioids in the development of normal pain sensitivity (Boschi et al., 1983). For example, the distribution of the endomorphins (EMs) along the nociceptive pathway implicates them as particularly important for the modulation of pain (Horvath, 2000). Thus, the EMs have been found unequally in the brain; they are stored in neurons and axon terminals with heterogenous distribution and they are released from synaptosomes by depolarization (Horvath, 2000; Zadina et al., 1997). Nociceptin is also widely distributed in central structures involved in sensory, emotional, and cognitive processing, and in the periphery including the immune cells (Lambert, 2008; Reinscheid et al., 2000). Furthermore, nocistatin (the other opioid-related peptide) is also present in the brain and the spinal cord (Boom et al., 1999; Lee et al., 1999; Okuda-Ashitaka et al., 1998), and its distribution appears to be almost identical to that of nociceptin (Okuda-Ashitaka and Ito, 2000).

As regards the actions of opioids at the supraspinal level, several centers are involved in this process. Some of the analgesic actions of opioids may be due to modulation of the descending pathways to reduce nociceptive transmission in the spinal dorsal horn (Anderson et al., 1977; Basbaum and Fields, 1984). Thus, spinally projecting RVM neurons expressing opioid receptors can mediate the opioid analgesia triggered from the PAG (Anderson et al., 1977; Basbaum and Fields, 1984; Fields and Basbaum, 1999; Millan, 2002), and microinjection of MOR agonists into the RVM elicits analgesia because opioids can reduce synaptic GABA release to spinally projecting neurons (Connor et al., 1999; Fields et al., 1991; Fields and Basbaum, 1999; Hurley et al., 2003). In addition, through presynaptic inhibition of GABA release, activation of opioid receptors may disinhibit spinally projecting noradrenergic neurons in the LC (Pan et al., 2002). It is well known that opioids reduce the sensory discriminative and affective component of pain as well. Thus, microinjection of morphine into the ACC decreases the affective component of pain processing, and activation of presynaptic MOR attenuates GABAergic synaptic input in the amygdala (Finnegan et al., 2005; Finnegan et al., 2006; LaGraize et al., 2006). Furthermore, both the prefrontal cortex and thalamic nuclei are involved in the actions of opioids (Zhao et al., 2007).

It is well known that opioids produce very effective antinociception at spinal levels as well. Opioid receptors in the DRG of sensory neurons undergo axonal transport to reach peripheral nerve terminals, and inflammation induces increases in MOR binding within DRG leading to an improved antinociceptive potency in these circumstances (Endres-Becker et al., 2007; Mousa et al., 2007; Zollner et al., 2003). The endogenous opioid ligands can induce antinociception at peripheral levels as well. During inflammation of the peripheral tissues leukocytes are the important source of the endogenous opioid peptides, and β-endorphin, Met-ENK, dynorphins, and endomorphins are produced and released by these cells (Labuz et al., 2006; Mousa et al., 2002; Rittner et al., 2008).

β-Endorphin

Since the discovery and characterization of β-endorphin (31 amino acids) as an opioid peptide in 1976, the opinion has been widely held that this peptide has a role in the control of pain (Akil et al., 1984; Basbaum and Fields, 1984; Loh et al., 1976; Rossier et al., 1977). POMC-derived β-endorphin is considered to be a key component of the endogenous antinociceptive system attenuating the stress- and inflammation-induced hyperalgesia (Rossier et al., 1977; Stein et al., 1990; Sun et al., 2003). It binds with high affinity to both MOR and DOR (Akil et al., 1984). Pain stimulation induces PAG release of β-endorphin and the ICV administration of β-endorphin produces analgesia (Akil et al., 1984). Similarly, both spinal and peripheral administration of β-endorphin evokes antinociceptive effects in different pain models (Chung et al., 1994; Stein et al., 1990; Suh et al., 1994; Suh et al., 1996).

Leu-enkephalin and Met-enkephalin (Leu-ENK, Met-ENK)

Methionine-enkephalin (Tyr-Gly-Gly-Phe-Met) and leucine-enkephalin (Tyr-Gly-Gly-Phe-Leu) were isolated and characterized as the first endogenous peptidic ligands for DOR receptors (Hughes et al., 1975). ENKs, synthesizing from pre-proenkephalin, possess antinociceptive activity at both spinal and supraspinal levels (Lee et al., 1980; Maldonado et al., 1994; Takemori and Portoghese, 1993; Yu et al., 2004). For example, pain stimulation induces PAG release of Leu-ENK and Met-ENK, and their antinociceptive effects are mediated in the brain through interactions mainly at DOR1 (Yang et al., 2006). Leu-ENK inhibited the nociceptin-induced allodynia in a dose-dependent manner at the spinal level (Honda et al., 2001). In the spinal cord, the ENKs interact with DOR2 receptors, and the release of ENKs in the SDH inhibit the projecting neurons (Mizoguchi et al., 1997; Willis, 1988). Furthermore, ENKergic neurons in the rat SDH are innervated by serotonin terminals, and 5-HT3A receptors colocalized with ENK, thus it seems that the activation of these neurons might be involved in 5-HT-induced antinociception (Huang et al., 2008). As regards their peripheral action, the results showed that clonidine (an α_2-adrenoceptor agonist) can induce peripheral antinociception by the local release of ENKs (Nakamura and Ferreira, 1988).

Dynorphins

Dynorphin A_{1-17}, a heptadecapeptide, and dynorphin A_{1-13}, the N-terminal tridecapeptide of dynorphin A, were isolated from porcine pituitary tissue, and they are produced from preprodynorphin. Both dynorphins possess high affinity for the KOR (Szeto, 2003). The high density of KORs in the spinal cord, medulla, amygdala, hypothalamus, and periphery suggests a possible involvement of dynorphins in the regulation of pain mechanisms (Lai et al., 2008; Menetrey and Basbaum, 1987; Palkovits, 2000; Szeto, 2003). Several reports indicate hyperalgesic or allodynic effects of dynorphins (Fujimoto et al., 1990; Lai et al., 2008; Qu and Isaac, 1993; Rady and Fujimoto, 2002; Wang et al., 2001b; Wen et al., 1985), however, KOR deletion significantly exacerbated mechanical and thermal inflammatory hypersensitivity but there was no change in the formalin-induced pain behavior (Schepers et al., 2008; Xu et al., 2004b). ICV administration of dynorphins can produce antinociception, which is reversed by the KOR antagonist (Fujimoto et al., 1990; Shukla et al., 1992).

As regards their effects at the spinal level, the results are inconsistent. Dynorphins can induce nocifensive behavior or hyperalgesia, and arthritic rats displayed a pronounced rise in immunoreactive dynorphins in the lumbosacral spinal cord, which correlated both with the intensity and time-course of hyperalgesia (Arcaya et al., 1999; Gardell et al., 2004; Millan et al., 1985; Vanderah et al., 1996; Wang et al., 2001b). However, other data suggest that spinal KOR activation is involved in the antinociceptive effects of some opioids, and the selective blockade of KORs increases the formalin-induced nocifensive behavior (Ossipov et al., 1996; Tseng and Collins, 1993). As regards the explanation for these opposing data, increasing evidence suggests that the dynorphin-induced antinociception is KOR-mediated, whereas its pronociceptive effects are elicited by binding of its enzymatic degradation peptide fragments to nonopioid receptors. Therefore, the pronociceptive effecs of dynorphins is mediated by activation of NMDA and/or bradykinin receptors leading to the release of SP and CGRP from primary sensory neurons (Arcaya et al., 1999; Lai et al., 2008; Wang et al., 2001b). However, peripheral application of dynorphins A_{1-17} produced antiallodynia, and this effect was reversed by KOR antagonists (Ko et al., 2000).

Endomorphin-1 and Endomorphin-2 (EM1, EM2)

More than ten years ago, a new group of MOR agonists was discovered and named endomorphins (EMs) by Zadina et al. (1997). Endomorphin-1 (EM-1: Tyr-Pro-Trp-Phe-NH_2) and endomorphin-2 (EM-2: Tyr-Pro-Phe-Phe-NH_2) differ from conventional endogenous opioid receptor ligands in their N-terminal sequence, peptide length, and C-terminal amidation. The pathway for their synthesis is unknown, but they are converted enzymatically by endopeptidases (Horvath, 2000; Zadina et al., 1997). They interact specifically and with high affinity with MOR (Horvath, 2000; Zadina et al., 1997), and they possess partial rather than full agonist properties at MOR (Sim et al., 1998). EM-1 and EM-2 produce their effects through different subtypes of MOR, EM-1 affecting predominantly the MOR2 receptor and EM-2 the MOR1 (Sakurada et al., 2000). The administration of EMs elicits short-lasting antinociception, and tolerance was also observed (Csullog et al., 2001; Horvath et al., 1999; Tseng et al., 2000; Yu et al., 2004; Zadina et al., 1997).

The antinociceptive effects are produced by peripheral, spinal and supraspinal levels as well (Przewlocki et al., 1999). ICV or intrathalamic administration of EMs produced antinociception in both acute and chronic pain models (Zadina et al., 1997; Zhao et al., 2007; Zubrzycka et al., 2005; Zubrzycka and Janecka, 2008). The EMs displayed lower potencies in the mechanical (paw pressure) test than in the heat-pain (TF) test in rats after IT administration (Horvath et al., 1999; Przewlocka et al., 1999), but they exerted high analgesic potency in different inflammatory pain models as well (Csullog et al., 2001; Hao et al., 1999; Przewlocka et al., 1999; Przewlocki et al., 1999; Wang et al., 1999). Neuropathic pain has been assumed to be resistant to treatment with opioids, therefore it is of particular interest that the EMs have high potency in decreasing neuropathic pain (Przewlocka et al., 1999). EM-1, but not EM-2, dose-relatedly reduced the Aβ-fibre evoked responses, therefore, spinal EM-2 exerts selective effects on noxious responses, whereas EM-1 is nonselective (Chapman et al., 1997). IPL administration of EM1 dose-dependently decreased the mechanical allodynia and the thermal hypersensitivity in neuropathic and inflammatory pain models (Labuz et al., 2006; Obara et al., 2004; Mecs et al., 2009).

Tyr-MIF Peptides

The Tyr-MIF (melanocyte-inhibiting factor) family of neuropeptides includes MIF-1 (Pro-Leu-Gly-NH2), Tyr-MIF-1 (Tyr-Pro-Leu-Gyl-NH2), Tyr-W-MIF-1 (Tyr-Pro-Trp-Gly-NH2), and Tyr-K-MIF-1 (Tyr-Pro-Lys-Gly-NH2). All have been isolated from bovine hypothalamus and the cortex of human brain (Zadina et al., 1992; Zadina et al., 1994). They bind to MORs, but Tyr-K-MIF-1 primarily interacts with specific Tyr-MIF-1 binding sites (Zadina et al., 1992; Zadina et al., 1994; Zamfirova et al., 2007). Both ICV and IT administrations of Tyr-W-MIF-1 and/or Tyr-MIF-1 induce prolonged, naloxone-reversible analgesia (Bell et al., 1999; Gergen et al., 1996; Zadina et al., 1993; Zamfirova et al., 2007; Yu et al., 2004). However, the spinal effect is about 75 times stronger than the supraspinal one (Zadina et al., 1996). IP administration of Tyr-K-MIF-1 also produced antinociception by the activation of MORs and stimulated the histaminergic system, too (Zamfirova et al., 2007). However, others have shown that the peripherally and systemically applied Tyr-MIF-1 acts as an opioid antagonist in the TF test (Kastin et al., 1984; Kavaliers, 1987).

Hemorphins

Hemorphins are endogenous peptides belonging to the family of "nonclassical" or "atypical" opioid peptides, derived from hemoglobin (Nyberg et al., 1997). The hemorphin family member peptides vary in size from 4 to 10 amino acids, and they have been identified in the brain, plasma, and cerebrospinal fluid (Nyberg et al., 1997). These peptides include: hemorphin-4 (Tyr-Pro-Trp-Thr), hemorphin-5 (Tyr-Pro-Trp-Thr-Gln), hemorphin-6 (Tyr-Pro-Trp-Thr-Gln-Arg), hemorphin-7 (Tyr-Pro-Trp-Thr-Gln-Arg-Phe), LVV-hemorphin-4 (Leu-Val-Val-Tyr-Pro-Trp-Thr; spinorphin), LVV-hemorphin-6 (Leu-Val-Val-Tyr-Pro-Trp-Thr-Gln-Arg), and LVV-hemorphin-7 (Leu-Val-Val-Tyr-Pro-Trp-Thr-Gln-Arg-Phe). These peptides display affinities for MOR, DOR, and KOR (Davis et al., 1989; Liebmann et al., 1989; Zadina et al., 1992), except for spinorphin, which has an ENK-degrading activity (Nishimura and Hazato, 1993). Only a few studies investigated their role in pain mechanisms. Thus, the ICV administration of hemorphin-4 and hemorphin-5 showed potent antinociceptive effects in the acute pain model in a naloxone-reversible manner, but they did not influence formalin-induced pain behavior (Davis et al., 1989). Peripheral administration of hemorphin-7 decreased the acute inflammation. which may also contribute to its antinociceptive effect (Sanderson et al., 1998).

Spinorphin has antinociceptive potency, and its effect may be due to the inhibition of the degradation of endogenous opioids (Honda et al., 2001; Maldonado et al., 1994; Schmidt et al., 1991; Yamamoto et al., 2002b; Yu et al., 2004). Spinorphin administered ICV did not influence acute pain sensitivity, but potentiated the effects of Leu-ENK. It inhibited the nociceptin-induced allodynia in a dose-dependent manner after IT administration, which was reversed by naloxone (Honda et al., 2001). Thus, spinorphin is not a real endogenous opioid neurotransmitter, but it might enhance the effect of Leu-ENK through inhibition of the degradation of ENK.

Nociceptin

Shortly after the cloning of the three known opioid receptors, a fourth member of this family was identified, the opioid-like receptor (ORL-1), which was found not to bind any of the known natural or synthetic opioid ligands (Reinscheid et al., 2000). In 1995, the natural ligand for this receptor was isolated and named orphanin FQ or nociceptin (Reinscheid et al., 1995). It is a 17-amino acid peptide, the amino terminus of which displays a striking similarity to the known mammalian opioid peptides. It is derived from pronociceptin together with another peptide, nocistatin. Nociceptin has been reported to be an active ligand at multiple sites of nociceptive transmission, ranging from peripheral nociceptors to nociceptive centers in the brain. Pharmacologically, the actions of nociceptin are complex and contradictory. ICV administration of this peptide exerts a pronociceptive action. The neuroanatomical site underlying the pronociceptive actions of nociceptin might be the RVM, where it inhibits the actions of OFF cells (Lambert, 2008). However, its effect at the spinal level depends on the dose applied; that is, a low dose produces nociception, whereas higher doses result in antinociception in acute and neuropathic pain models (Calo et al., 2000; Ma et al., 2003; Mogil and Pasternak, 2001; Yu et al., 2004). It is suggested that low doses may increase the release of SP, whereas high doses inhibit the glutamate release (Lambert, 2008). ORL-1 receptors and nociceptin can be found peripherally as well, and their activation can lead to peripheral antinociception (Lambert, 2008; Obara et al., 2005), whereas other data suggest that nociceptin has pain-inducing effects (McDougall and Larson, 2006).

Nocistatin

A further endogenous peptide that has been implicated in the modulation of pain transmission is the heptadecapeptide nocistatin, produced by the proteolytic cleavage of prepronociceptin (Lee et al., 1999; Okuda-Ashitaka et al., 1998). It has been detected in different parts of the body and exerts its effect through the activation of an unknown GPCR (Joseph et al., 2007; Zeilhofer et al., 2000). Some data have shown that its ICV administration increased the acute and inflammatory pain threshold (Nakagawa et al., 1999; Zhao et al., 1999), whereas other data suggest that it does not influence the normal pain latency, but prevented the nociceptin-induced hyperalgesia (Liu et al., 2006; Nakagawa et al., 1999; Scoto et al., 2005). It presumably acts as a neuromodulator in pain processing at the spinal level as well, because it blocks the hyperalgesia and allodynia induced by nociceptin (its name originates from this observation) or prostaglandin-E2 (Ito et al., 2001; Okuda-Ashitaka et al., 1998; Okuda-Ashitaka and Ito, 2000). It does not influence the TF latency (Zeilhofer et al., 2000) and can decrease the neuropathic pain (in low doses) (Muth-Selbach et al., 2004), but higher doses could increase hyperalgesia or block nociceptin-induced analgesia in neuropathic rats (Ma et al., 2003; Muth-Selbach et al., 2004). However, it increases the flexor reflex response (Xu et al., 1999b), and inconsistent data are available on its effect on the formalin test (Nakano et al., 2000; Yamamoto and Sakashita, 1999; Zeilhofer et al., 2000). Cumulative administration of its C-terminal octapeptide, BPNP-3-8P, also significantly decreased heat hyperalgesia but it did not change the paw withdrawal (PWD) latency on the normal side (Csullog et al., 2001). As regards its peripheral effects, its pronociceptive potency has been reported (Inoue et al., 2003).

5.2.2 Kyotorphin

Kyotorphin (Tyr-Arg), was isolated from bovine brain by Takagi et al. (1979). It is formed by kyotorphin synthase in the presence of ATP and Mg^{2+} in the brain (Kawabata et al., 1995). High concentrations of the dipeptide were found in the brainstem and SDH, and several studies have demonstrated its analgesic properties (Ueda et al., 1980). Kyotorphin binds to its specific receptors (kyotorphin receptor; GPCR) and activates PLC (Lopes et al., 2006; Ueda et al., 1989). ICV administration of kyotorphin produced antinociception in the acute pain test (Kawabata et al., 1994b). Kyotorphin excites cortical neurons directly, and it also exerts indirect opioid action to produce analgesia via the release of Met-ENK (Shiomi et al., 1981). Systemic administration of a kyotorphin receptor agonist leads to antinociception in the acute pain test, and this effect has been antagonized by IT injection of kyotorphin receptor antagonist, but not by ICV application (Ochi et al., 2000). Furthermore, IT kyotorphin also produced antinociception in an acute mechanical pain model (Ochi et al., 2002). Kyotorphin has a nonopioid analgesic effect at peripheral level, which makes it quite appealing for chronic pain treatment (Inoue et al., 1997).

5.2.3 Tachykinins

The mammalian tachykinins are a family of evolutionary conserved peptides that share the common C-terminal motif. Until recently, the family consisted of three peptides: substance P (SP; undecapeptide), neurokinin A (10 amino acids), and neurokinin B (10 amino acids). Since the discovery of a third preprotachykinin gene (*TAC4*), the number of tachykinins has more than doubled to reveal several species-divergent peptides. This group includes hemokinin-1 (HK-1) in mouse and rat, endokinin-1 (EK-1) in rabbit, and EKA, EKB in humans (Page, 2004). Their peripheral expression has led to the proposal that they are the endogenous peripheral SP-like endocrine/paracrine agonists where SP is not expressed. Additionally, three orphan tachykinin gene-related peptides are identified, in rabbit, endokinin-2 (EK-2), and in humans, EKC and EKD (Page, 2004). The biological actions of tachykinins are mediated by at least three different transmembrane GPCRs, namely NK1, NK2, and NK3. SP is preferential, but not exclusive, for NK1, NKA for NK2, and NKB for NK3, therefore, each ligand can interact with all receptors (Maggi et al., 1993; Morteau et al., 2001; Patacchini and Maggi, 2001). The NK1 receptor is widely expressed in both CNS and PNS, and the effects of NK1 receptor involvement in nociceptive transmission are very complex (Quartara and Maggi, 1998). NK1 knock-out mice have substantial impairments of endogenous pain control mechanisms (Bester et al., 2001). At the supraspinal level, the dorsal raphe nucleus and PAG have numerous NK1 and GABA double-labelled neurons (Ma and Bleasdale, 2002).

SP and HK1, acting upon NK1 receptors, might be relevant in descending pain control, because their administration in different brain areas increased the pain threshold, and this effect was reversed by opioid and GABAA receptor antagonists (Altier and Stewart, 1993; Altier and Stewart, 1998; Altier and Stewart, 1999; Fu et al., 2008; Holden and Pizzi, 2008; Rosen et al., 2004; Yu et al., 2004). Furthermore, the injection of SP into the ventrolateral PAG has induced analgesia, and morphine has increased the SP release in this area (Rosen et al., 2004). The activation of SP-containing neurons in the lateral hypothalamus also increases the pain threshold by activating NK-1 receptors in the RVM (Holden and Pizzi, 2008). Therefore, SP may activate the descending antinociceptive pathways through activation of NK1 receptors. Several data suggest that SP, HK1, and EKA/B decrease the pain threshold both spinally and peripherally (Abbadie et al., 1996; Afrah et al., 2001; Beyer et al., 1991; Cridland and Henry, 1988b; Dirig and Yaksh, 1999; Donnerer et al., 1992). However, NK1 receptor activation may also increase the inhibitory neurotransmission by activating inhibitory interneurons in the SDH (Vergnano et al., 2004). Furthermore, both induction of scratching behavior and thermal hyperalgesia by IT administration of SP and EKA/B as well as enhancement of c-Fos-LI following noxious thermal stimulation are suppressed by pretreatment with EKC/D, suggesting that the EKC/D peptide is an antagonist of the NK1 receptor (Naono et al., 2007).

5.2.4 Calcitonin Gene-Related Peptide (CGRP)

Calcitonin gene-related peptide (CGRP; 37 amino acids) expresses predominantly in the nervous system and it influences multiple physiological activities. As regards its action mechanism, molecular correlates for discrete CGRP receptor types are still lacking. Functional CGRP receptors represent a multiprotein entity composed of at least three discrete proteins, that is, the seven-transmembrane receptor calcitonin-receptor-like receptor (CRLR), receptor-activity-modifying protein 1 (RAMP1), and the cytoplasmic receptor component protein (RCP) (Wu et al., 2002). CGRP is involved in many stages of the transmission of nociceptive information, because CGRP-LI has been found to colocalize with that of SP-LI in capsaicin-sensitive nerve terminals in the periphery and SDH, but it has been identified in the DRG and several brain areas as well (Ballet et al., 1998; Brain and Cox, 2006; Olgiati et al., 1983; Skofitsch and Jacobowitz, 1985a). Thus, CGRP-Li fibers and CGRP receptors distribute densely in the amygdala, and these fibers originate from parabrachial nucleus and the thalamic nuclei (Oliver and Keyvan-Fouladi, 2000; Shimada et al., 1989). ICV, intra-PAG and intra-amygdala injection of CGRP induce antinociception (Candeletti and Ferri, 1990; Xu et al., 2003; Yu et al., 2003). It is probable that in the amygdala CGRP-containing terminals activate ENKergic neurons, which project to the PAG releasing ENK (Palkovits, 2000). However, other data have shown that ICV administration of CGRP did not modify pain sensitivity in the TF test and did not affect the antinociceptive action of morphine (Azarov et al., 1995). As regards its effects at spinal and peripheral levels, it is well known that it has a facilitation role in the nociceptive information, and this effect may be mediated via SP mechanism (Ballet et al., 1998; Li et al., 2008; Morton et al., 1991; Nahin and Byers, 1994; Santicioli et al., 1993).

5.2.5 Pituitary Adenylate Cyclase-Activating Polypeptide (PACAP)

Pituitary adenylate cyclase-activating polypeptide-38 (PACAP-38) is a member of the vasoactive intestinal peptide family (VIP) that was originally isolated from ovine hypothalamus (Miyata et al., 1989). Two forms of PACAP have been identified, containing 27 and C-terminally extended 38 amino acids. PACAP acts via GPCRs (activation of AC and PLC): the PAC1 receptors, which specifically bind PACAP and the VPAC1/VPAC2 receptors, which have a similar binding affinity for PACAP and VIP (see Section 5.2.6). Both PACAP and its receptors have been widely described on central and peripheral neurons, smooth muscle cells, and several inflammatory cells (Bartho et al., 2000; Delgado et al., 2003; Dickinson and Fleetwood-Walker, 1999; Narita et al., 1996). On the basis of the morphological and molecular biological results, PACAP has been suggested to be involved in pain transmission, but very few functional data are available to support this theory. Decreased response to different pain stimuli has been observed in PACAP or PAC1 receptor deficiencies (Mabuchi et al., 2004), and PAC1 receptor activation can lead to stimulation of NMDA receptors and synthesis of brain-derived neurotrophic factor (BDNF), and these processes can lead to enhanced nociception (Jongsma et al., 2001a; Mabuchi et al., 2004; Martin et al., 2003).

ICV administration of PACAP did not influence the acute heat pain threshold (TF) in mice, but significantly decreased morphine-induced analgesia (Macsai et al., 2002). As regards its effects at the spinal level, it can also facilitate spinal nociceptive flexor reflexes, and pronociceptive effects were observed, whereas PAC1-antagonists potently reduced mechanical allodynia in a neuropathic nerve model and were also effective in reducing thermal hyperalgesia in the carrageenan model (Davis-Taber et al., 2008; Narita et al., 1996; Ohsawa et al., 2002; Shimizu et al., 2004). However, other studies have shown that IT PACAP 27 or 38 were analgesic in inflammatory and neuropathic pain models (Narita et al., 1996; Yamamoto and Tatsuno, 1995; Zhang et al., 1996). PACAP inhibits the release of proinflammatory/pronociceptive sensory neuropeptides: SP and CGRP from peripheral terminals of capsaicin-sensitive nerves, and it also inhibited acute neurogenic and nonneurogenic inflammatory processes in both mice and rats (Helyes et al., 2007; Nemeth et al., 2006). IPL PACAP-38 did not alter basal mechanical or heat thresholds, but it inhibited the carrageenan- or heat injury-induced hypersensitivities, as well as nocifensive behaviors in the early and late phase of the formalin test (Sandor et al., 2009). In mice, it significantly diminished acetic acid-induced abdominal contractions but exerted no effect on neuropathic mechanical hyperalgesia. In contrast, it markedly increased rotation-induced firing of afferent fibres in the inflamed rat knee joint, clearly demonstrating a peripheral sensitization in this organ (Sandor et al., 2009).

5.2.6 Vasoactive Intestinal Peptide (VIP)

VIP contains 28 amino acids, and it was first isolated from the porcine small intestine, but it has also been detected in the PNS and CNS (Fuji et al., 1983; Jancso et al., 1981). Both VAPC1 and VPAC2 receptors bind VIP with high affinity, therefore, similarly to PACAP, VIP can also regulate different aspects of pain. Thus, its ICV administration elicits analgesia in acute heat pain tests, but impairs the antinociceptive effect of morphine (Macsai et al., 1998). Bilateral application of VIP into the amygdala persistently suppressed the heat-evoked reflexes by exiting amygdala-originating neurons that innervate the PAG antinociceptive cells (Shin, 2005). VIP can produce both analgesia and hyperalgesia at the spinal level, depending on its molecular conformation (Yeomans et al., 2003). However, most of the data have shown that IT VIP induces nocifensive behavior, potentiates the effects of SP and decreased the pain threshold in acute pain models (Beyer et al., 1991; Cridland and Henry, 1988a; Wiesenfeld-Hallin, 1987; Xu and Wiesenfeldhallin, 1991; Yu et al., 2004). As regards its peripheral effect, intra-articular injection of VIP caused increased allodynia in rats with osteoarthritis probably by the activation of PKA (McDougall et al., 2006).

5.2.7 Galanin (GAL)

Galanin is a neuropeptide consisting of 29 or 30 (in humans) amino acids and was originally isolated from porcine intestine (Bartfai et al., 1992). GAL exerts its biological effects by interacting with three high-affinity cell surface receptors GALR1-3, which all belong to the family of GPCRs (Branchek et al., 2000). All three receptors couple to Gi/o and inhibit AC, although GalR2 also signals via Gq/11 to activate PLC and PKC (Wittau et al., 2000). GAL and its receptors are widely distributed in the nervous system and have been implicated in a number of important body functions, including feeding, cognition, endocrine modulation, and nociception (Holmberg et al., 2005; Wiesenfeld-Hallin et al., 1992). Most neurons of the hypothalamic PVN and SO also contain GAL, colocalized with AVP, OT, and opioids (Zubrzycka and Janecka, 2008). In the rat, cell bodies and fibres containing GAL-LI and GAL receptors have been identified in DRG and in laminae I, II, VII, and X of SDH (Rokaeus et al., 1984; Skofitsch and Jacobowitz, 1985b; Wiesenfeld-Hallin et al., 2005). As GAL-LI and receptor numbers in the SDH are decreased following dorsal rhizotomy or capsaicin treatment, it has been suggested that one source of GAL is derived primarily from unmyelinated primary afferent fibres (Xu et al., 1996b). In most systems the effect of GAL on nociception appears to be predominantly inhibitory, mediated at least partially by GALR1 s (Bacon et al., 2002; Blakeman et al., 2003; Grass et al., 2003b; Ji et al., 1994a; Xu et al., 1998). Mice overexpressing GAL or deficient in GAL have become available, and have provided a genetic approach to analyze the role of GAL in nociception. GAL knock-out mice have a lower nociceptive threshold to heat stimulation, and have changed neuropathic pain behaviors (Blakeman et al., 2003; Grass et al., 2003a; Kerr et al., 2000; Malkmus et al., 2005). Furthermore, mice deficient in GALR1 or GALR2 have impaired pain-like behavior (Blakeman et al., 2003; Grass et al., 2003b; Hobson et al., 2006).

Mice overexpressing GAL exhibit significant elevation of the nociceptive threshold to thermal stimulation, but no change in response to mechanical or cold stimulation was seen (Blakeman et al., 2001). ICV perfusion of GAL concentration dependently inhibited pain-induced responses, and its effect was blocked by GAL receptor and by MOR antagonists, whereas it was potentiated by EM2, AVP, and OT (Zubrzycka and Janecka, 2008). Similarly, the analgesic effect of GAL administered in the PAG or arcuate nucleus was also attenuated by naloxone (Sun and Yu, 2005; Wang et al., 2000). These data suggest that GAL can induce antinociception at the supraspinal level by release of the endogenous opioid ligands. The antinociceptive role of GAL at the spinal level has been extensively studied (Wiesenfeld-Hallin et al., 2005). IT administration of GAL produced a dose-dependent increase in the TF latency, but surprisingly, it lowered the threshold to von Frey stimulus (Cridland and Henry, 1988a). The effect of exogenous GAL is predominantly inhibitory under normal conditions, and due to blocking the excitatory effect of SP and CGRP (Hua et al., 2004; Xu et al., 1998; Yu et al., 2001), but GAL can also modulate the release of endogenous opioid ligands (Zhang et al., 2000a). However, other reports proposed that GAL produces a biphasic, dose-dependent effect on nociception through activation of antinociceptive (inhibitory) GALR1 or pronociceptive (excitatory) GALR2 receptors, thus endogenous GAL can potentiate nociceptive processing during inflammation (Kerr et al., 2001; Liu and Hokfelt, 2002). Some results suggest that peripheral GAL has an excitatory role in inflammatory pain, likely mediated by peripheral GALR2 and that GAL can modulate TRPV1 function, whereas activation of peripheral GalR1 results in antinociception (Jimenez-Andrade et al., 2004; Jimenez-Andrade et al., 2006).

5.2.8 Neuropeptide Y (NPY)

NPY is an abundant neuroactive peptide (containing 36 amino acids) that exerts numerous physiological actions, including pain modulation. NPY is expressed in the CNS and PNS, and can be released from sensory, enteric, and sympathetic neurons but also from glial cells (Arora and Anubhuti, 2006; Chronwall and Zukowska, 2004; Dumont et al., 1992). There are six receptors (Y1-6) of NPY which are GPCRs. They exhibit dynamic alterations in signalling pathways, leading to neuronal excitatory or inhibitory effects after receptor activation (Lin et al., 2004). NPY and Y1 and Y2 (most prevalent) receptors are located at key pain signalling centers throughout the nervous systems, particularly the SDH, and their effects on nociceptive modulation has been extensively studied (Allen et al., 1984; Bannon et al., 2000; Gibbs and Hargreaves, 2008; Ji et al., 1994b; Shi et al., 1998; Smith et al., 2007). In contrast to the cellular localization of Y1 receptors, spinal Y2 receptors are located on primary afferent terminals (Brumovsky et al., 2005). Y1 receptor deletion increased acute heat, inflammatory, and neuropathic pain sensitivities (Kuphal et al., 2008). Furthermore, lack of Y1 receptor or antagonism at this receptor inhibited the antinociceptive potency of NPY at the spinal level. These data suggest that the Y1-receptor system exerts tonic inhibitory control and it mediates the antiallodynic effects of NPY during inflammatory and neuropathic pain syndromes.

Several studies have proved the antinociceptive potential of supraspinal NPY (Illes et al., 1993; Li et al., 2005a). After the microinjection of NPY into the RVM, the PAG, the ARC, or in the nucleus accumbens, withdrawal reflexes to noxious heat or tactile stimuli are decreased (Li et al., 2005a; Li et al., 2002b; Wang et al., 2001a; Zhang et al., 2000b). However, intracranial administration of anti-NPY antiserum or Y1 receptor antagonist into the nucleus gracilis reversed nerve injury-mediated mechanical allodynia (Ossipov et al., 2002). IT administration of NPY inhibits transient, inflammatory, and neuropathic pain (Hua et al., 1991; Intondi et al., 2008; Mahinda and Taylor, 2004; Smith et al., 2007; Taiwo and Taylor, 2002). Evidence from pharmacological studies suggests that both Y1 and Y2 agonists can attenuate the flexor reflex in axotomized animals (Xu et al., 1999a). The neurophysiological mechanism of this antinociception involves inhibition of pronociceptive, excitatory neurotransmitter release from primary afferent neurons through activation of Y2 receptors, whereas Y1 receptor activation inhibits GABAergic inhibition on the substantia gelatinosa (Martire et al., 2000; Smith et al., 2007). NPY is colocalized with TRPV1 receptors, and it can inhibit the excitatory transmitter release from the capsaicin-sensitive primary sensory neurons leading to peripheral antinociception (Gibbs et al., 2006a; Gibbs and Hargreaves, 2008).

5.2.9 RFamide Neuropeptides

There is a new family of mammalian neuropeptides, that is, RF (Arg-Phe) amide neuropeptides including neuropeptide FF (NPFF, octapeptide), prolactin-releasing peptide (PrRP, 31 amino acids), RF-amide related peptides (RFRP); RFRP1: Leu-Pro-Leu-Arg-Phe-amide; RFRP2: gonadotropin-inhibitory hormone:dodecapeptide and RFRP3: Leu-Pro-Gln-Arg-Phe-amide, kisspeptins (10, 13 and 14 amino acids), and the 26Rfa (RF[Arg-Phe]amide family 26-amino acid peptide, also known as P518) (Fukusumi et al., 2006). RF-amides represent a group of peptides sharing a C-terminal RF-terminus, and they are involved in many regulatory functions in the body by action on different GPCRs. Both NPFF and RF-amide related peptides produce their effects by activation of NPFF receptors (1 and 2). PrRP is a ligand for an orphan receptor, the GPR10-like receptor, but it can also activate NPFF2 receptors (Engstrom et al., 2003). Kisspeptins are the products of the gene Kiss1 and they are ligands for GPR54, whereas 26Rfa produces its effects by activation of GPR103, a receptor that is widely distributed in the spinal cord (Bruzzone et al., 2007). These peptides and their binding sites are expressed in the CNS (Engstrom et al., 2003; Pertovaara et al., 2005; Sullivan et al., 1991; Yang and Iadarola, 2006). NPFF has both potent pro-opioid antinociceptive and antiopioid-like effects, depending on the sites of administration and the dose (Frances et al., 2001; Panula et al., 1996; Roumy and Zajac, 1998; Wei et al., 1998; Yang et al., 1985). It produces antinociception after ICV or intra-PAG administration in neuropathic and inflammatory models in a naloxone reversible manner (Altier et al., 2000; Kalliomaki et al., 2004; Wei et al., 1998; Wei et al., 2001). On the other hand, NPFF either decreases or does not influence the acute mechanical and thermal pain sensitivities and can antagonize the effect of morphine (Altier et al., 2000; Wei et al., 2001; Yang et al., 1985).

As regards its role at the spinal level, several reports have proved its effectivities in neuropathic and inflammatory pain models; it can potentiate the effect of morphine, but the results about the effects on acute pain tests are inconsistent (Altier et al., 2000; Gouarderes et al., 2000; Kontinen and Kalso, 1995; Wei et al., 2001; Yamamoto et al., 2008; Yang and Iadarola, 2006). Peripheral administration of an NPFF analogue did not produce antinociceptive effect (Wei et al., 2001). The potential role of PrRP in pain was addressed by its ICV and IT injections in both neuropathic and normal rats (Kalliomaki et al., 2004). It was ineffective at the spinal level, but with administration in the dorsal medulla, PrRP produced significant antinociception in normal rats and an antiallodynic effect in neuropathic rats. The PrRP-induced antinociception is not mediated by MOR because it is not reversible by naloxone. However, other data have shown that ICV administration of PrRP promoted hyperalgesia, and it reversed the morphine-induced antinociception (Laurent et al., 2005). Furthermore, PrRP knock-out animals showed increased pain threshold. As regards the RF-amide-related peptides, their injection into the brain (nucleus of solitary tract) inhibited mechanical hyperalgesia, whereas IT administration significantly decreased the acute heat pain sensitivity and the tactile allodynia in a neuropathic pain model (Pertovaara et al., 2005). Spinally applied 26Rfa also significantly decreased the nocifensive behavior in the formalin test, and attenuated the level of mechanical allodynia in a carrageenan-induced inflammatory pain model, but it did not influence the normal heat and mechanical pain sensitivity (Yamamoto et al., 2008). The relation of kisspeptin to the pain has been suggested as well. Thus, a marked elevation in the levels of kisspeptin and GPR54 mRNA as well as protein was observed in the SDH and DRG during inflammation, indicating a possible involvement of the kisspeptin/GPR54 system in chronic inflammatory pain (Mi et al., 2009).

5.2.10 Neurotensin (NeT)

Another endogenous peptide which has been implicated in pain transmission and the central integration of pain responses is neurotensin (NeT) (Dobner, 2006; Gui et al., 2004; Pettibone et al., 2002). NeT is a brain–gut tridecapeptide that fulfils a dual function: as a neurotransmitter/neuromodulator in the nervous system, and as a paracrine and circulating hormone at the periphery. Three NeT receptors, NTR1,

NTR2, and NTR3, have been cloned to date (Dubuc et al., 1999). NTR1 and NTR2 belong to the GPCR family, whereas NTR3 is a single transmembrane domain protein that belongs to a recently identified family of sorting receptors (Mazella, 2001; Mazella and Vincent, 2006). Most of the known peripheral and central effects of NeT are mediated through NTR1. NTR2 may possibly take part in the analgesic response elicited by the central administration of NeT; the biological roles of NTR3 are yet to be discovered in detail (Dubuc et al., 1999; Kitabgi, 2002; Mazella and Vincent, 2006). Various in vivo data support its modulatory role in pain transmission (Dobner, 2006; Yu et al., 2004). NeT normally facilitates visceral nociception, whereas an increased NeT expression can be observed under high stress conditions, and this ligand is required for SIA (Gui et al., 2004). The results indicate that the supraspinal antinociceptive effect of NeT is largely MOR independent (Osbahr et al., 1981). ICV, intra-PAG or intra-RVM administrations of NeT produce analgesic effects in different pain models (HP, TF, writhing, colorectal distension tests) by NTR1 and/or NTR2 activations (Behbehani and Pert, 1984; Dobner, 2006; Dubuc et al., 1999; Nemeroff et al., 1979; Osbahr et al., 1981; Pettibone et al., 2002; Urban et al., 1999). It seems that NeT can excite PAG neurons, which leads to activation of the descending pain inhibitory system (Behbehani and Pert, 1984; Dobner, 2006). NeT also induces analgesia through stimulation of NTR1 and NTR2 at spinal level, as was shown in acute and inflammatory pain models (Roussy et al., 2008; Sarret et al., 2005).

5.2.11 Neurotrophic Factors

Neurotrophic factors are a unique family of polypeptide growth factors that influence the proliferation, differentiation, survival, and death of neuronal and nonneuronal cells. Neurotrophic factors are synthesized as high-molecular-weight precursors and their release from cells is constitutive as well as activity dependent. The nerve growth factor (NGF), brain-derived neurotrophic factor (BDNF), neurotrophin-3 (NT-3), neurotrophin-4 (NT-4), glial cell line-derived neurothrophic factor (GDNF), and insulin-like growth factor-1 (IGF-1) are essential for the health and well being of the nervous system, and also mediate additional higher-order activities, such as learning, memory, and behaviour. Alterations in their levels have been implicated in neurodegenerative disorders, such as Alzheimer's disease and Huntington's disease, as well as psychiatric disorders, including depression and substance abuse. Most of the neurotrophic factors interact with two types of cell-surface receptors, the low-affinity p75 receptor and TrK family of high-affinity tyrosine kinase receptors (TrKA, B, and C). Whereas all neurotrophins (NGF, BDNF, NT-3, and NT-4) bind the p75 receptor with similar affinity, NGF bind TrKA receptors, BDNF and NT-4 bind TrKB receptors, and NT-3 preferentially binds TrKC receptors, and to a lesser extent, TrKA receptors (Huang and Reichardt, 2003).

There is now strong evidence that two neurotrophins, NGF and BDNF act as important mediators and modulators of pain in a variety of circumstances. Particularly, NGF can promote the sensitization and activation of nociceptors (Pezet et al., 2002; Zhao et al., 2006). NGF is produced in the periphery and taken up by SP and BDNF containing sensory nerve terminals, it binds to the TrKA receptors and it is retrogradely transported to their cell bodies within DRG (Delcroix et al., 2003). The altered retrograde supply of NGF contributes to the neuronal response to injury and inflammation by the modulation of SP and CGRP release from the primary sensory neurons (Donnerer et al., 1992; Woolf and Costigan, 1999). Furthermore, NGF acts on mast cells to induce release of 5-HT, which sensitizes nociceptors as well (Theodosiou et al., 1999). Thus, endoneural injection of NGF is sufficient to produce transient histological and behavioral effects like those seen in neuropathic pain models, whereas sequestration of NGF prevents hyperalgesia, which normally accompanies inflammation (Dmitrieva et al., 1997; Lewin et al., 1994; Ruiz et al., 2004). NGF can also increase opioid binding sites, and it enhances the opioid susceptibility of sensory neurons towards better pain control by an upregulation in the number and efficacy of sensory neuron MOR (Chen et al., 1997; Inoue and Hatanaka, 1982; Mousa et al., 2007). Furthermore, NGF may increase the amount of PACAP in sensory neurons, and this effect can also decrease its pain-inducing effects (Jongsma et al., 2001b). This suggests its therapeutic potential for pathological conditions with a reduced susceptibility to opioids such as certain neuropathic pain states.

Among the neurotrophins, BDNF is the most abundant and widely distributed in the CNS. In the sensory system BDNF is constitutively produced by nociceptive-sensitive primary sensory neurons in the DRG, then it is transported anterogradely to the central terminals of sensory neurons in the SDH, where it can be released (together with SP) (Malcangio and Lessmann, 2003; Michael et al., 1997). In sensory neurons the concentration of BDNF and SP depends on the availability of NGF, therefore, the increase in NGF concentration in peripheral tissues that follows an inflammatory insult enhances the expression of these ligands (Malcangio and Lessmann, 2003). BDNF may act as a regulator of neuronal excitability and modulator of synaptic plasticity, playing an important role in pain pathways (Kerr et al., 1999; Malcangio and Lessmann, 2003; Thompson et al., 1999). Conditional BDNF knock-out (in the primary sensory neurons) mice showed several alterations, including a reduced baseline thermal threshold and a decreased inflammatory hyperalgesia, whereas neuropathic pain behavior developed normally (Zhao et al., 2006). IT grafts of BDNF-secreting neurons or overexpression of BDNF in the spinal cord have been shown to alleviate chronic neuropathic pain, and other data suggest an ineffectivity after IT administration of BDNF (Boucher et al., 2000; Cejas et al., 2000; Eaton et al., 1999; Eaton et al., 2002). BDNF can depress sensory neurone transmission in the SDH by an indirect mechanism that requires the release of GABA from interneurons and the activation of GABAB receptors located in the terminals of sensory neurons. Therefore, BDNF decreased neuropathic pain by increasing the GABA release in SDH (Lever et al., 2003). However, other reports have shown that the expression of BDNF is dramatically upregulated in models of inflammatory pain, it enhances the NMDA-receptor-mediated responses and TrKB antagonism significantly reduced the inflammatory pain (Garraway et al., 2005; Kerr et al., 1999; Pezet et al., 2002). Thus, BDNF released from nociceptors along with SP and glutamate appears necessary for the full activation of second-order DH neurons (Pezet et al., 2002).

The role of NT-3 in pain transmission has not been fully worked out. NT-3 significantly attenuates neuronal expression of voltage-gated sodium channels and elevated levels of galanin and PACAP in DRG neurons, and all these effects can influence the pain sensitivity (Wilson-Gerwing and Verge, 2006). It is upregulated in models of neuropathic pain, and can contribute to the mechanical hyperalgesia (Zhou et al., 2000). However, IT administration NT-3 suppressed thermal hyperalgesia associated with neuropathic pain, although others did not find any effects in a similar model (Boucher et al., 2000; Wilson-Gerwing et al., 2005).

Glial cell line-derived neurothrophic factor (GDNF) binds to its high-affinity receptor, glial cell line-derived neurotropic factor receptor α-1 (GFRα1), and a subpopulation of DRG neurons expresses GFRα1 (Lindfors et al., 2006). The effect of GDNF on nociception is still a matter of debate, but it seems that GDNF expression decreases in neuropathic pain states (Nagano et al., 2003). IT GDNF exerts potent analgesic effects on hyperalgesia as well as protective effects against the development of neuropathic pain (Boucher et al., 2000; Sakai et al., 2008; Wang et al., 2003a). In association with the analgesic effects of GDNF, several molecules, including sodium channels, purinergic receptors, and neuropeptides, have been reported to exhibit changes in expression (Boucher et al., 2000; Issa et al., 2001; Wang et al., 2003a). Recently, GDNF was shown to bind to neural cell adhesion molecule (NCAM) via GFRα1, and NCAM signalling plays a role in mediating the analgesic effect of GDNF in rats with nerve injury (Sakai et al., 2008). IPL injection of GDNF induces thermal hyperalgesia, and inflammatory hyperalgesia is attenuated by treatment with an antibody against GDNF (Malin et al., 2006).

Insulin-like growth factor-1 (IGF-1) is a 70-amino acid polypeptide that exerts effects on peripheral growth, differentiation, and survival in a variety of cells and tissues (Daftary and Gore, 2005). IGF-1 is an important regulator of synaptic plasticity and neuronal survival in response to injury. The secretion of growth hormone (GH) stimulates the production of peripheral IGF-1 from its primary target, the liver, as well as from secondary targets such as lungs, thymus, heart, neurons and glia. Immunoreactivity and mRNA transcripts for IGF-1 and its receptor have been reported to exist in the brain and spinal cord, and IGF-1-LI is also transported retrogradely and anterogradely in axons of the peripheral nerve (Bitar et al., 1996; Daftary and Gore, 2005; Hansson et al., 1987). IGF-1 produces its effects through activation of its tyrosine kinase receptor, TrK IGF-1R that signals through the phosphoinositol-3 kinase and mitogen-activated protein kinase (MAPK) signalling cascade (Daftary and Gore, 2005). Only few studies suggest its role in antinociception. Thus, IT administration of IGF-1 produces a dose-dependent antinociception, and the increased IGF-1 level may play a significant role in the serotoninergicic antinociception at the spinal level (Bitar et al., 1996; Bonnefont et al., 2007)

5.3 *Other Peptides*

5.3.1 Endothelins (ET)

It is now firmly established that endothelins, a family of 21-amino-acid residue peptides produced by many cell types, can exert multiple and important actions in many tissues and systems, including those implicated in nociceptive signaling functions (Kedzierski and Yanagisawa, 2001). The potent and widespread actions of ET-1 and other isopeptides of the family (ET-2 and ET-3) are mediated by specific GPCR receptors (ETA and ETB), and the signal transduction pathways involve increases in intracellular calcium levels (Rubanyi and Polokoff, 1994). ET-1-LI and specific binding sites have been found in the SDH and in different brain areas (Gulati and Srimal, 1992; Rubanyi and Polokoff, 1994). Some data imply that endothelin might have a role in neurotransmission that is important for an animal's pro- and antinociception. Injection of ET-1 into the PAG reduces pain response in mice subjected to the HP paradigm (D'Amico et al., 1996). Furthermore, astrocytes produce endogenous cannabinoids (CBs) in response to treatment with ET1 through ETA receptor activation, and these effects can contribute to its antinociceptive potency (Walter et al., 2002; Walter and Stella, 2003). Intrathecal administration of ET1 also induced antinociception in the acute heat pain model (Kamei et al., 1993b; Kamei et al., 1993a). It is suggested that ET-1 induces SP release, which provokes endogenous opioid release in the SDH (Kamei et al., 1993b). The peripheral administration of ET produces pronociceptive effects primarily through ETA receptor activation, but ETB receptor also can contribute to the effect (Da Cunha et al., 2004; Daher et al., 2004; Piovezan et al., 1997; Raffa et al., 1996). One study has shown that ETB receptors normally can display antihyperalgesic and antinociceptive functions in the rat knee-joint incapacitation test (Daher et al., 2004)

5.3.2 Hemopressin

Hemopressin is a nonapeptide derived from the α1-chain of hemoglobin, which was originally isolated from rat brain homogenates (Rioli et al., 2003). It is the first endogenous peptide ligand for cannabinoid-1 (CB1) receptors, and it behaves as an inverse agonist (Dale et al., 2005; Heimann et al., 2007; Lippton et al., 2006; Rioli et al., 2003). Hemopressin causes hypotension in anaesthetised rats and is metabolised in vivo and in vitro by endopeptidase 24.15 (EP24.15), neurolysin (EP24.16), and ACE (Lippton et al., 2006; Rioli et al., 2003). There are no data available on its distribution in the organism including the brain. This peptide selectively binds to CB1 receptors, but did not affect on CB2, MOR, DOR, α2- and β2-adrenergic, AT1 and AT2 and bradykinin B2 receptors (Heimann et al., 2007). The only in vivo experiment showed that IPL, IT and oral hemopressin reduced inflammatory pain sensitivity (Heimann et al., 2007). IP hemopressin also decreased the visceral nociception. Because a large body of evidence has clearly demonstrated the antinociceptive action of cannabinoid CB1 receptor agonists, a possible explanation for these paradoxical effects could be that after CB1 receptor blockade by the antagonist, the released endocannabinoids could induce antinociception by affecting another pain transmission mechanism (see below Section 6.1).

5.3.3 Annexin-A1

Annexin-A1 (37 kDa glucocorticoid-regulated protein), formerly known as lipocortin-1, is a member of the annexin family of calcium and phospholipids-binding proteins (Buckingham and Flower, 1997). It is widely distributed in different tissues including the CNS, and annexin-A1 mediates the anti-inflammatory actions of glucocorticoids (Yang et al., 2004). Few studies have addressed the question of whether annexin-A1 is involved in the activation/modulation of pain pathways. Systemic administration of annexin-A1 peptidomimetics produced analgesia in inflamed rat paws and neutralizing antisera to annexin-A1 prevented the antihyperalgesic activity of glucocorticoids (Ferreira et al., 1997). Annexin-A1 knock-out mice were more susceptible to visceral pain stimulus compared with wild-type, and increased levels of prostaglandin E2 (PGE2) in the spinal cord of knock-out compared with normal mice suggest that annexin-A1 modulates nociceptive processing at the spinal level by downregulating PGE2 spinal nociceptive facilitation (Ayoub et al., 2008). Inhibition of the formalin-induced nociceptive behavior by annexin-A1, administered centrally (ICV) or locally (IPL), is dependent on activation of the receptors of the formylated peptide family, which is a GPCR family (Pieretti et al., 2004). IPL administration of annexin-A1 also significantly decreases the intensity of hyperalgesia by inhibition of neutrophil accumulation (Ferreira et al., 1997; Pieretti et al., 2004). Furthermore, using a rat model of C-fibre modulated bradykinin-induced plasma extravasation, Green et al. (1998) suggest that the inhibitory action of a glucocorticoid on C-fibre activation is mediated by the release of annexin-A1. These results may provide a possible mechanism for the analgesic action of the glucocorticoids, which are routinely given to patients with postoperative pain.

5.4 Cytokines

The term cytokine encompasses a large and diverse family of polypeptide regulators that are produced widely throughout the body by cells of diverse embryological origin. Historically, the term "cytokine" has been used to refer to the immunomodulating agents (interleukins, interferons, etc.). Virtually all nucleated cells, but especially endo/epithelial cells and resident macrophages are potent producers of different cytokines (e.g., IL-1, IL-6, and TNF-α) (Cannon, 2000). The action of cytokines, similarly to hormones, may be autocrine, paracrine, and endocrine. Cytokines have been classified as lymphokines, interleukins, and chemokines, based on their presumed function, cell of secretion, or target of action. Because cytokines are characterized by considerable redundancy and pleiotropism, such distinctions, allowing for exceptions, are obsolete. A classification that proves useful in clinical and experimental practice divides immunological cytokines into those that enhance cytokine responses, type 1 (IFN-γ, TGF-β etc.), and type 2 (IL-4, IL-10, IL-13, etc.), which favour antibody responses. Cytokines are critical to the development and functioning of both the innate and adaptive immune response, although not limited to just the immune system. There is increasing evidence that a number of cytokines and their receptors are involved in the processes that lead to the development and maintenance of pain states. A diverse range of cytokines and other inflammatory mediators are known to be secreted by activated glia, many of which have been shown to modulate nociception/allodynia (Scholz and Woolf, 2007). These include the pronociceptive cytokines: IL-1β, IL-12, IL-18, IFNγ, TNF-α and the antinociceptive cytokines: IL-2, IL-4, and IL-10 (Vale et al., 2003; Yao et al., 2002a).

IL-2 produces analgesic effect in both CNS and PNS (Jiang et al., 2000b; Yao et al., 2002a). Thus, microinjection of IL-2 in ICV, hippocampus, or LC increases the pain threshold, and the antinociceptive effect was related to the increase of Leu-ENK or SP (Guo and Zhao, 2000; Jiang et al., 2000a; Wu et al., 1999b). IT delivery of IL-2 or IL-2 gene also inhibits nociceptive responses by the activation of opioid receptors or the decrease of SP release in the spinal cord and the reduction of Fos protein in superficial SDH (Guo and Zhao, 2000; Song and Zhao, 2000; Wang et al., 1996; Wu et al., 1999b; Yao et al., 2002b; Yao et al., 2002a).

The class of anti-inflammatory cytokines includes IL-4, IL-10, and IL-13 and transforming growth factor-β (Callard et al., 1996; Fiorentino et al., 1991; Hart et al., 1989). These cytokines are produced by several cell types, including T-helper2 lymphocytes, monocytes, macrophages, and mast cells. They are believed to play a role in inhibiting hypersensitivity reactions of macrophage functions, the synthesis of proinflammatory cytokines and the expression of cyclooxygenase-2 (COX2) and inducible nitric oxide synthase (iNOS) (Fernandes et al., 2002). Neither of these cytokines affected the acute pain sensitivity, but they inhibit the inflammatory mechanical hyperalgesia (after systemic administration) (Poole et al., 1995; Vale et al., 2003). This analgesic effect could be related to a peripheral mechanism, probably via the inhibition of the release of the proinflammatory cytokine by resident peritoneal macrophages.

Many studies examined the hyperalgesic action of chemokines, but recent evidence has also pointed towards their antinociceptive effects (Rittner et al., 2008). In early inflammation, granulocytes are activated by chemokines, and their activation induces opioid release leading to antinociception. Thus, chemokines may play an important role in the trafficking of opioid-containing cells to injured tissues and in the release of opioid peptides in inflamed tissue. Levels of CX3CR1 (the receptor for the chemokine fractalkine) mRNA, but not in the levels of fractalkine mRNA, in lumbar DRG significantly increase in neuropathic pain models (Holmes et al., 2008). IT or intra-PAG administration of fractalkine to rats produces pain facilitation (Chen et al., 2007; Johnston et al., 2004). The number of CX3CR1-positive macrophages and the expression of CX3CR1 in macrophages are markedly increased in the nerve proximal to the site of the injury, and intraneural injection of fractalkine significantly delays the development of allodynia, whereas CX3CR1 knock-out mice display an increase in allodynia (Holmes et al., 2008). Thus, fractalkine may play opposing and site-dependent nociceptive roles, although the summation of the two seems to be inhibitory, at least in mice, on the basis of the increased allodynia in the CX3CR1 knock-out animals.

6 Lipids

Lipids are a diverse group of compounds, and they may be divided into eight categories: fatty acyls, glycerolipids, glycerophospholipids, sphingolipids, saccharolipids, polyketides, sterol lipids, and prenol lipids. Once viewed primarily as structural constituents of cell membranes and energy storage sources, lipids are now recognized as also serving as valuable signalling functions. As do other chemical transmitters, lipids bind and activate specific protein receptors to produce their biological effects. In the early 1990s, Mechoulam's group opened the door to a new class of fatty acid derivatives that serve naturally to modulate pain (Devane et al., 1992; Martin et al., 1999).

6.1 Endocannabinoid System and Related Fatty Acid Derivatives

Cannabinoids (CBs, e.g., Δ^9-tetrahydrocannabinol) are a distinct class of psychoactive compounds, which produce a wide array of effects on specific receptors (CB1 and CB2) (Calignano et al., 1998; Hohmann, 2002; Martin et al., 1999). The endogenous cannabinoids are lipid derivaties (except hemopressin; see Section 5.3.2), and much has ben written about their signalling mechanisms and their role in physiological regulation. A feature that distinguishes endocannabinoids from many other neuromodulators is that they are not synthesized in advance and stored in vesicles. Rather, their precursors exist in cell membranes (lipids) and are cleaved by specific enzymes on demand, and endocannabinoids release generally postsynaptically, and they act presynaptically (Walker and Hohmann, 2005). The endocannabinoid system consists of endogenous cannabinoids, cannabinoid receptors and the enzymes responsible for synthesis and degradation of endocannabinoids. The first endocannabinoid identified was arachidonoyl-ethanolamine (anandamide: AEA), and the second one was 2-arachidonoyl-glycerol (2-AG). The concentration of 2-AG in the brain is 50–500-fold higher than the concentration of AEA (Sugiura et al., 2006). Other putative endogenous ligands of cannabinoid receptors are palmityl-ethanolamide (PEA) and virodhamine (O-arachidonoyl-ethanolamine), a derivative of anandamide (Di Marzo et al., 1998; Porter et al., 2002; Walker et al., 2005). The derivative of 2-AG, 2-arachidonyl-glycerylether or noladin ether was also suggested to be an endocannabinoid (Hanus et al., 2001). A new subgroup contains molecules that consist of a lipid moiety conjugated to an amino acid and have been termed lipoamino acids; they are likely to serve a variety of regulatory functions in the brain and other tissues. (Huang et al., 2001). Several endogenous lipoamino acids were detected in a variety of tissues in the rat, that is, N-arachidonoyl-glycine (NAGly),

N-arachidonoyl-alanine (NAAla), N-arachidonoyl-serine (NASer), N-arachidonoyl-taurine (NaTau), and N-arachidonoyl-GABA (NAGABA) (De Petrocellis et al., 2004; Devane et al., 1992; Huang et al., 2001; Huang et al., 2002; Walker et al., 2005). NAGly has been intensively studied (see Section 6.1.3), but only one report has investigated the antinociceptive potency of NAGABA and NAAla. It has been found that their IT administration does not produce an increase in mechanical and thermal pain threshold in the inflammatory pain model (Succar et al., 2007). N-arachidonoyl-dopamine (NADA), oleamide, N-oleyl-dopamine (OLDA), and N-palmitoyl-glycine (PalGly) are also fatty acid derivates, and they have also been identified as endogenous lipids (Huang et al., 2002). All of these ligands constitute a family of ubiquitous endogenous lipids present in varying levels throughout the body, and several of them produce their effects through modulation of CB receptors, whereas other receptor activation/inhibition has also been suggested.

Cannabinoid receptors (CB1 and CB2) are among the most abundant GPCRs (Pan et al., 2008). The CB1 receptor is widely distributed in the CNS and PNS; its density is especially high in the brain, and it preferentially presents on axons and their terminals. CB2 receptors are expressed predominantly peripherally, where they are localized extensively to cells of the immune system, but they can be found on the peripheral nerve terminals as well (Guindon and Hohmann, 2007; Szabo, 2008). As regards its expression in the CNS, neural CB2 receptor expression is very low under normal conditions, but it can be induced in nonneuronal cells under pathological conditions (Guindon and Hohmann, 2007; Pan et al., 2008; Szabo, 2008; Van Sickle et al., 2005; Zhang et al., 2003a). Both CB1 and CB2 receptors primarily signal through the inhibitory GPCR proteins (Gi/o), however, under certain conditions and with certain agonists, coupling via Gs and Gq/11 has also been demonstrated (Mackie, 2008; Pertwee, 2001). Stimulation of CB1 receptors leads to the inhibition of AC, the inhibition of certain voltage-gated calcium channels, and the activation of G protein-linked inwardly rectifying potassium channels and these effects are associated with depression of neuronal excitability and transmitter release. The complexity of the actions of CB2 agonists on neuronal and nonneuronal cells and their signalling properties are only beginning to be explored. Activation of CB2 receptors inhibits AC and in contrast to CB1 receptors, CB2 receptors do not couple to ion channels, but both receptors can activate the MAPK signaling cascade (Howlett et al., 2004).

Considerable progress has been made in understanding the physiological functions of the endocannabinoids, and their corresponding potential pathological implications. Most of the above-mentioned ligands are now recognized as potent modulators of pain and inflammation (Hohmann, 2002; Hohmann et al., 2005; Pertwee, 2001; Quartilho et al., 2003). Cannabinoids induce antinociceptive effects at several levels, and they can mediate the opioid-independent SIA as well (Hohmann et al., 2005). Recent studies have demonstrated the antinociceptive efficacy of cannabinoids in several pain models acting primarily at the central CB receptors (Guindon and Hohmann, 2007; Walker et al., 2005; Walker and Hohmann, 2005). However, several data suggest the antinociceptive potential of peripherally acting cannabinoid agonists (Agarwal et al., 2007; Dogrul et al., 2003; Yesilyurt et al., 2003). Cannabinoids can reduce the production and release of proinflammatory signalling molecules and enhance the release of anti-inflammatory cytokines; moreover, CB2 receptor activation may stimulate the local release of endorphins from cells such as keratinocytes (Ibrahim et al., 2005; Walter and Stella, 2004). Cannabinoids inhibit the release of calcitonin gene-related peptide (CGRP) in isolated skin preparations, suggesting that one mechanism by which these drugs may modulate pain is the inhibition of neuropeptide release from peripheral sensory terminals (Ellington et al., 2002). Thus, nonneuronal substrates as well as neuronal substrates may be responsible for the ability of CB2-selective agonists to influence pain sensitivity. The peripheral action may possibly be extremely important, because low doses of these endogenous ligands may reduce pain without disphoric side effects, and without the abused potential typical of centrally acting cannabimimetic drugs.

As mentioned above, several of the fatty acid derivates can also interact with other GPCRs and ion channels. Thus, they modulate several types of potassium channels, α7-nAChRs, 5-HT3 receptors, and some orphan receptors (GPR55, GPR92, and GPR18) (Demuth and Molleman, 2006; Kohno et al., 2006; Oh et al., 2008; Pertwee, 2007). The orphan GPR55 is an especially serious candidate to become an additional cannabinoid receptor (Pertwee, 2007). As these ligands are lipophilic, they may partition into the cell membrane, where they may reach high local concentrations and thereby influence the actions of membrane proteins via so-called "receptor-independent" mechanisms (Oz, 2006). The best known and characterized ion channel interaction is the activation of TRPV1 channels. TRPV1 is a ligand-gated nonselective cation channel that is considered to be an important integrator of various pain stimuli such as capsaicin, heat, and low pH. Several endogenous lipids represent "chimeric" ligands (AEA, OLDA, and NADA) acting on both cannabinoid and TRPV1 receptors (Starowicz et al., 2008). Because CBs and TRPV1 receptors show coexpression in brain neurons, their coactivations can lead to a cross-talk between them. The role of peripheral TRPV1 receptor in pain has been the subject of several detailed studies (Jancso and Jancso-Gabor, 1980; Nagy et al., 2004; Starowicz et al., 2008; Szolcsanyi, 2000; Szolcsanyi, 2004), and its crucial role in nociception and hyperalgesia has been confirmed in the TRPV1 knock-out mice as well, in which impaired nociception and reduced sensitivity to painful heat in behavioral tests have been reported (Barton et al., 2006; Bolcskei et al., 2005; Caterina et al., 2000;Davis et al., 2000). The TRPV1 receptor activation at spinal level by capsaicin or AEA causes temporary painful behavior and a prolonged antinociception (Di Marzo et al., 2000a; Horvath et al., 2008b; Yaksh et al., 1979). However, TRPV1 antagonists effectively reduce thermal hyperalgesia and mechanical allodynia through both spinal and peripheral mechanisms (Cui et al., 2006).

The expression of TRPV1 in supraspinal structures such as PAG, RVM, the LC, and thalamus suggests its involvement in descending and ascending supraspinal pain processing (Cristino et al., 2006; Maione et al., 2006; Mezey et al., 2000; Starowicz et al., 2008). Microinjection of capsaicin into the PAG increases the latency to thermal nociceptive responses, an effect blocked by NMDA and metabotropic glutamate receptor antagonists (Palazzo et al., 2002; Starowicz et al., 2008). These data suggest that TRPV1 activation in the PAG increases glutamate release, and this leads to activation of postsynaptic glutamate receptors. However, the response to intra-PAG injected capsaicin depends on the location of the injection. Its injection into the dorsolateral-PAG decreases the pain threshold, whereas the capsaicin administration into the ventrolateral-PAG produces antinociception by the increased glutamate release in the RVM, which leads to enhanced activity of antinociceptive OFF cells, and decreased firing of pronociceptive ON cells. Furthermore, capsaicin-induced excitation of LC neurons might also be involved, in part, in its analgesic properties (Hajos et al., 1987). TRPV1 is located presynaptically on afferents to the LC, and its activation may serve to potentiate the release of glutamate and norepinephrine in this brain region (Marinelli et al., 2002). Nociceptive neurons of the medial thalamus also respond to capsaicin, in agreement with the high density of TRPV1 in this area (Cristino et al., 2006). TRPV1 activation evokes glutamate release from the hypothalamus and cerebral cortex as well (Sasamura and Kuraishi, 1999). Both regions send their projections to PAG (Millan, 2002), and these data also suggest a mechanism by which TRPV1 activation may modulate neuronal activity in these central areas.

6.1.1 N-Arachidonoyl-Ethanolamine (Anandamide; AEA)

Anandamide, the first identified and best-studied endocannabinoid, can be found both centrally and peripherally (Calignano et al., 1998; Devane et al., 1992; Walker et al., 2005). It is principally formed from glycerophospholipid by two successive enzymatic reactions: N-acylation of phosphatidyl-ethanolamine to generate N-acylphosphatidyl-ethanolamine (NAPE) by Ca^{2+}-dependent N-acyltransferase, and release of AEA from NAPE by a phosphodiesterase of the PLD type (NAPE-PLD) (Okamoto et al., 2007). It has been hypothesized that AEA could be recycled by the cell to form new endocannabinoid molecules and released into the extracellular space (Placzek et al., 2008). AEA is extremely short-lived, being rapidly inactivated by the enzymes fatty acid amide hydrolase (FAAH) (Cravatt et al., 1996). Termination of AEA signalling appears to involve a two-step process that begins with transport across the plasma membrane, followed by enzymatic hydrolysis into arachidonic acid and ethanolamine. AEA binds to both CB1 and CB2 receptors, behaving as a partial agonist, but it also activates TRPV1 receptors, and it has CB receptor-independent G protein-coupled antinociceptive potency through the activation of the GPR55 (Di Marzo et al., 2000b; Pertwee, 2007; Ryberg et al., 2007). Furthermore, AEA may directly affect the GlyRs and functionally antagonizes the transient receptor potential melastatin 8 (TRPM8) receptor-mediated responses (De Petrocellis et al., 2007; Hejazi et al., 2006; Lozovaya et al., 2005). Furthermore, AEA targets potassium channels, T-type calcium channels, and gap junctions. It is a substrate for COX2 giving rise to amino acid conjugates of the prostaglandins, and induces the expression of COX2 enzyme as well (Chemin et al., 2001; Chen et al., 2005a; Maingret et al., 2001).

Only a few studies have investigated the antinociceptive potency of AEA in vivo, and it seems that the CB1 receptor is the predominant target mediating anandamide's antinociceptive effect (Wise et al., 2007). Systemic administration of AEA is able to attenuate the visceral hyperreflexia induced by inflammation of the urinary bladder and to reduce the second phase response to formalin (Jaggar et al., 1998). FAAH inhibitor significantly increases the potency of anandamide in a mild thermal injury model (Palmer et al., 2008). A recent study has shown that AEA (IV) produces a cannabinoid receptor-independent antinociception, and its effects were inhibited by 5-HT3 receptor antagonist suggesting that the activation of these receptors contributes to the anandamide-induced analgesia (Racz et al., 2008). The ICV administration of AEA also induces dose-related antinociception in the TF-test, and this effect is reduced by pertussin toxin, but not cholera toxin (Calignano et al., 1998; Raffa et al., 1999). Furthermore, the inhibition of FAAH at PAG level also decreases pain sensitivity which was reversed by CB1 and TRPV1 antagonist (De Novellis et al., 2008; Maione et al., 2006; Suplita et al., 2005). AEA is a potent short-lasting antinociceptive ligand at the spinal level in acute and inflammatory pain models (Horvath et al., 2008b; Smith et al., 1994; Welch et al., 1998). The effect of AEA is inhbited by CB1 antagonist, but also by TRPV1 antagonist (Horvath et al., 2008b; Welch et al., 1998). Local administration of anandamide significantly decreases the formalin-induced pain behavior but not the paw edema (Calignano et al., 1998; Guindon et al., 2006a,b). Furthermore, it also inhibits the TRPV1 receptor activation-induced drop in HP latency by activation CB1 receptors (Almasi et al., 2008). Thus, anandamide may activate cannabinoid CB1 receptors located on capsaicin-sensitive primary afferents, resulting in the decreased responsiveness of these afferents to noxious stimuli. However, others have shown that locally administered anandamide activates nociceptors in normal and arthritic rat by stimulating TRPV1 receptors on primary sensory neurons, suggesting a pain-inducing potential of anandamide at this level (Gauldie et al., 2001).

6.1.2 2-Arachidonoyl-Glycerol (2-AG)

2-AG is a 2-acyl-glycerol ester, and its concentration in the brain is 50–500-fold higher than the concentration of anandamide (Mackie, 2008; Sugiura et al., 2006). 2-AG is also short-lived, being rapidly inactivated mainly by the enzyme monoglyceride lipase (MAGL), but it might also be metabolized by FAAH (Bisogno, 2008; Cravatt et al., 1996; Dinh et al., 2002; Saario and Laitinen, 2007; Sugiura et al., 2006). It is a full agonist for CB1 and CB2 receptors with no direct binding to the TRPV1 receptor (Mechoulam et al., 1995; Pertwee, 2001; Sugiura et al., 2006). It is also a substrate for COX2, and 2-AG is capable of suppressing elevation of COX2 expression by activating the CB1 receptors (Bleakman et al., 2006; Zhang and Chen, 2008).

As regards its antinociceptive potency, only a few data are available in this respect; systemic administration of 2-AG has produced antinociception in TF test, and its in vivo potency is similar to anandamide (Mechoulam et al., 1995). 2-AG (IP) does not decrease hyperalgesia after mild thermal injury, but it is effective if it is administered together with a FAAH inhibitor (Palmer et al., 2008). Furthermore, the coadministration of 2-AG (in an ineffective dose IP) with two other endogenous lipids (2-linoleoyl-glycerol and 2-palmitoyl-glycerol) increases the potency of 2-AG, although these lipids are ineffective by themselves (Ben Shabat et al., 1998). This entourage effect might be due to the decreased inactivation of 2-AG. MAGL inhibitor has induced a CB1-mediated enhancement in endocannabinoid-mediated SIA following local administration into either the PAG or SDH (Hohmann et al., 2005; Hohmann, 2007; Suplita II et al., 2006; Suplita et al., 2005). This effect is associated with a profound increase in levels of 2-AG, but not anandamide, suggesting a physiological role for 2-AG in the suppression of pain sensitivity. Topical (IPL) administration of 2-AG and MAGL inhibitor decreased the pain behavior in the late phase of the formalin test and produced antihyperalgesic and antiallodynic effects in a neuropathy pain model (Desroches et al., 2008; Guindon et al., 2007; Hohmann, 2007). Moreover, the antinociceptive effects of 2-AG are prevented by a selective CB2 receptor antagonist, but not by a CB1 receptor antagonist in the formalin test, whereas both antagonists inhibit the antiallodynic and antihyperalgesic effects of 2-AG (Desroches et al., 2008; Guindon et al., 2007). However, local administration of CB1 and CB2 antagonists by themselves failed to induce hyperalgesia, suggesting that the endocannabinoids do not act tonically in the periphery to dampen senstivitiy to pain (Guindon et al., 2007).

6.1.3 N-Arachidonoyl-Glycine (NAGly)

N-Arachidonoyl-glycine (NAGly) was first synthesized as a structural analogue of the AEA. It is expressed within the CNS, with particularly high levels within the spinal cord, but it can be detected in the skin as well (Burstein, 1999; Huang et al., 2001; Rimmerman et al., 2008). NAGly is formed via oxidation of AEA and by conjugation of glycine with arachidonic acid by arachidonyl-CoA, and being rapidly inactivated by FAAH (Burstein, 1999; Huang et al., 2001). The pharmacology of NAGly is still poorly understood, however, several targets for NAGly are emerging. NAGly has no affinity for the CB1 and TRPV1 receptors, although it can activate CB2 binding sites (Devane et al., 1992; Huang et al., 2001; Sipe et al., 2005). It is also a substrate for COX2 giving rise to amino acid conjugates of the prostaglandins, and it inhibits activation of COX2 and 5-lipoxygenase enzymes (Burstein, 1999; Prusakiewicz et al., 2002). Thus, it has a complex effect on prostaglandin synthesis, and a role for COX2 cannot be excluded in its antinociceptive effect (Burstein et al., 2007). Furthermore, NAGly inhibits FAAH, and it is a substrate for this enzyme, therefore, NAGly can regulate the levels of AEA in tissues (Grazia Cascio et al., 2004; Huang et al., 2001). NAGly is a ligand for the orphan receptors GPR18, and it activates this receptor in a pertussis toxin-sensitive manner (Kohno et al., 2006).

NAGly has also been shown to stimulate another orphan receptor GPR92, which is highly expressed in DRG and colocalized with TRPV1 receptors and has been postulated to play a role in sensory perception (Oh et al., 2008). Alternatively, the coexpression of GPR92 and TRPV1 in the DRG raises the possibility that NAGly can exert its pain suppressive effects through GPR92 in the sensory nervous system.

In addition, NAGly inhibits the glycine transporter GLYT2, but it can also influence the GlyRs, thus GlyRs could also mediate some of the analgesic effects of NAGly (Wiles et al., 2006; Yang et al., 2009). As regards its effects after systemic (SC, IP, or oral) administration, the results are inconsistent. Thus, NAGly produces analgesia administered in acute pain models including the HP and formalin tests and has anti-inflammatory activity (Burstein et al., 2007; Huang et al., 2001). However, systemic administration of NAGly, at a dose similar to that used IT, is without effect, and NAGly does not produce antinociception in mild thermal injury model (Palmer et al., 2008; Vuong et al., 2008). IT administration of NAGly reduces the mechanical allodynia and thermal hyperalgesia, and its effect is not influenced by CB1 and CB2 antagonists (Succar et al., 2007; Vuong et al., 2008). In addition, NAGly does not produce the motor side effects associated with exogenous cannabinoid receptor agonists. Consistent with its high levels in skin, NAGly produces analgesia administered peripherally in HP and formalin tests, and it also has anti-inflammatory activity (Burstein et al., 2007; Huang et al., 2001).

6.1.4 2-Arachidonyl-Glycerylether (Noladin Ether)

Mechoulam's laboratory in 2001 has identified noladin ether, the derivative of 2-AG, from porcine brain (Hanus et al., 2001). This compound is more stable compared to 2-AG and anandamide. It has higher affinity to CB1 than CB2 receptors (Hanus et al., 2001), but another study has shown that it has high affinity for CB2 receptors as well, and it is a full agonist at this receptor (Shoemaker et al., 2005). Newer studies have investigated the activity of noladin ether on TRPV1 receptors, and it has been shown that its effects are not connected with this receptor (Duncan et al., 2004). Noladin ether inhibits the CGRP-induced vasorelaxation, and this effect is unaffected by both CB1 and CB2 antagonists, but is inhibited by pertussis toxin (Duncan et al., 2004). These results suggest that noladin ether produces its effects through non-CB1/CB2 GPCR activation, but noladin ether can decrease the MOR expression by acting on CB2 receptors (Paldyova et al., 2008). Only one study has investigated its antinociceptive potency after systemic (IP) administration and it has been shown that noladin ether produced antinociception on HP test in mice (Hanus et al., 2001).

6.1.5 N-Arachidonoyl-Dopamine (NADA)

NADA was first synthesized, and then identified in the brain at the beginning of this century (Bisogno et al., 2000; Huang et al., 2002; Yang et al., 2007a). It is synthesized through a condensation reaction between arachidonic acid and dopamine or between arachidonic acid and tyrosine, and then converted to NADA (Huang et al., 2002). NADA can be inactivated by conversion into the less active 3-O-methyl-NADA by catechol O-methyltransferase, and it is slowly hydrolysed by FAAH as well (Huang et al., 2002). As regards its action mechanism, NADA can activate either TRPV1 or CB1 receptors depending on the location and circumstance (Bisogno et al., 2000; Bisogno, 2008; De Petrocellis et al., 2000; Huang et al., 2002; Mackie, 2008). Thus, NADA activates DRG neurons, and increases the intracellular calcium concentration and the CGRP release by activation of TRPV1 receptors, but it has lower potency than capsaicin (Huang et al., 2002; McDonald et al., 2008; Medvedeva et al., 2008). NADA also functionally antagonizes the TRPM8-mediated responses (De Petrocellis et al., 2007). It elicits analgesia following systemic administration in acute heat pain test (Bisogno et al., 2000). NADA produced slight allodynia after IT administration, and a dose-dependent antihyperalgesic effect was also observed; this effect was inhibited both by CB1 and TRPV1 antagonists (Horvath et al., 2008a; Pitcher et al., 2007). In addition, it causes nocifensive behavior and hyperalgesia when administered peripherally (Huang et al., 2002; Price et al., 2004).

6.1.6 N-Palmitoyl-Ethanolamide (PEA) and N-Palmitoyl-Glycine (PalGly)

Although the subfamily of arachidonoyl amides has received considerable attention, much less is known about the presence and activity of their saturated counterparts (Rimmerman et al., 2008). The most studied member of the saturated acyl amides is N-palmitoyl-ethanolamide (PEA) (Di Marzo et al., 1998; Walker et al., 2005). PEA, found in neural and nonneural tissues, inhibits mast-cell activation and reduces inflammatory responses by a mechanism that may involve binding to CB2 receptors (Calignano et al., 1998; Martin et al., 1999). However, because PEA does not produce an effective activation of cannabinoid receptors, it is generally classified as a cannabimemic compound. Furthermore, PEA is an agonist at the peroxisome proliferator-activated receptor α (PPARα), and at the orphan receptor GPR55 (Lo Verme et al., 2005; Ryberg et al., 2007). An "entourage" effect on anandamide-mediated action may be due to the PEA-induced inhibition of FAAH that leads to an increase of tissue levels of AEA (Costa et al., 2008). Thus, a recent study demonstrated that CB1, PPARα, and TRPV1 receptors mediate the antinociception induced by systemic PEA in the neuropathic pain model, and repeated PEA treatment significantly decreased the enhanced NGF, GDNF, and NT-3 levels in the spinal cord (Costa et al., 2008). Orally administered PEA also reduced inflammatory hyperalgesia and edema by inhibiting mast cell degranulation (Mazzari et al., 1996). It can also attenuate the visceral pain sensitivity and the second phase response to formalin (Jaggar et al., 1998; Lo Verme et al., 2006). Its ICV administration was ineffective in acute heat pain test, whereas it showed marked antinociceptive properties at the peripheral level (Calignano et al., 1998; Calignano et al., 2001; Lo Verme et al., 2006). Local administration of PEA produced antinociception in the formalin test that was blocked by a CB2 receptor-selective antagonist (Calignano et al., 1998). Another study has proved the role of PPARα receptor activation in this respect (Lo Verme et al., 2006). It is noteworthy that local coadministration of PEA with exogenous anandamide produced a synergistic analgesic effect in both phases of the formalin test through a mechanism that involves both CB1 and CB2 receptor subtypes (Calignano et al., 1998; Calignano et al., 2001).

PalGly also has been identified in rat brain, skin, and spinal cord (Rimmerman et al., 2008). It induces transient calcium influx in native adult DRG cells, and stimulates the NO production. PalGly potently inhibits heat-evoked firing of nociceptive neurons in rat SDH, and its effects are not inhibited by CB1 and CB2 antagonists, but are blocked by TRP channel antagonists. No in vivo studies have been performed in this respect yet.

6.1.7 N-Oleoyl-Ethanolamide (OEA), N-Oleoyl-Dopamine (OLDA), and Oleamide

All these substances are derivatives of oleic acid (monounsaturated omega-9 fatty acid). N-oleoyl-ethanolamide (OEA) is an endogenous regulator of food intake, and may have some potential as an antiobesity drug, however, some studies investigated its effects on sensory neurons as well (Fu et al., 2003; Hansen and Artmann, 2008). It does not bind to CB1 and CB2 receptors, but it is an endogenous agonist of TRPV1 and PPARα, however, there are some contradictions in this respect (Ahern, 2003; Almasi et al., 2008; Fu et al., 2003; Lo Verme et al., 2006; Wang et al., 2005a). Only a few inconsistent results suggest its role in the pain. Thus, IP administration of OEA has decreased the pain behavior in formalin and visceral models, and this effect was independent from PPARα activation, although high dose causes writhing behavior (Suardiaz et al., 2007; Wang et al., 2005a). Its IPL administration does not change the acute heat-pain latency, but reverses the thermal hyperalgesia (Almasi et al., 2008). Another study has found nocifensive behavior after its local injection, which could not be observed in TRPV1 knock-out animals (Lo Verme et al., 2006).

The endogenous presence of OLDA has recently been confirmed in the mammalian brain (Chu et al., 2003; Huang et al., 2002). The in vivo pathways of OLDA synthesis are unsettled. The most probable pathway seems to be *N*-acylation of tyrosine by a fatty acid, with tyrosine entering then the normal pathway of dopamine synthesis to form *N*-acyl-dopamine, and it is inactivated by FAAH (Chu et al., 2003). However, it can weakly activate CB1 receptors (Bisogno et al., 2000; Chu et al., 2003). OLDA possesses activity at TRPV1 receptors with potency similar to that of capsaicin, and produces long-lasting nocifensive behavior and thermal hyperalgesia, which is blocked by TRPV1 antagonists (Chu et al., 2003; Szolcsanyi et al., 2004; Walker et al., 2005). Therefore, OLDA may function as either a peripheral or central mediator of TRPV1 activation.

Oleamide (cis-9,10-octadecenoamide or oleic acid amide) originally was found in the cerebrospinal fluid (CSF) of sleep-deprived cats, and has received much attention due to its sleep-inducing properties in mammals (Cravatt et al., 1995; Farrell and Merkler, 2008). The primary site of action of oleamide in the central nervous system remains unclear. It does not interact directly with CB1 receptors, but it interacts with other neurotransmitter–receptor systems (GABAergic, dopaminergic, and serotonergic transmission) (Boring et al., 1996; Walker et al., 2005). Many of oleamide's behavioral effects are consistent with its being an indirect cannabimimetic, increasing either the levels or activity of endogenous cannabinoids (e.g., AEA) (Fedorova et al., 2001). The mechanism by which this occurs remains unclear and may include the suppression of AEA uptake, although it also shares with AEA the same degradatory enzyme, FAAH (Mechoulam et al., 1997). Hence the cannabimimetic effects have largely been attributed to the indirect entourage effect on the endocannabinoid system (Mechoulam et al., 1998). Only one study has reported its effects on pain sensation. Systemic administration of oleamide induces cannabimimetic effects, and it produces relatively long-lasting antinociceptive effects (HP and TF tests), but repeated administration of oleamide causes tolerance as well (Fedorova et al., 2001).

6.2 Eicosanoids

The derivatives of arachidonic acid are eicosanoids, which have four families: the prostaglandins, the prostacyclins, the thromboxanes, and the leukotrienes. Prostanoid is the term used to describe a subclass of eicosanoids consisting of the prostaglandins, the thromboxanes, and the prostacyclins. They are important lipid mediators involved in the transmission of nociceptive pain. Their synthesis is initiated by the generation of arachidonic acid by phospholipase A2 (PLA2), which is metabolized by COX1 enzymes (1–3) to generate short-lived mediators that act as precursors for the synthesis of the biologically active prostanoids: prostaglandin-E2 (PGE2), PGD2, PGI2, PGF2α, or thromboxane A2. In the spinal cord both COX1 and COX2 are expressed. The major prostaglandins produced in the spinal cord are PGE2 and PGD2. The role of PGE2 is to enhance synaptic transmission and increase spinal responses to peripheral stimulation, and it thus plays a major role in the induction of hyperexcitability during peripheral inflammation (Ahmadi et al., 2002; Vanegas and Schaible, 2001; Vasquez et al., 2001). By contrast, much less is known about the role of PGD2. In general, PGD2 is the most produced postanoid in the CNS of mammals. A profound basal release of PGD2 in the spinal cord has been reported, and peripheral nociceptive stimulation and systemic inflammation increases spinal PGD2 biosynthesis (Grill et al., 2008; Willingale et al., 1997). As regards its action mechanism, PGD2 activates two GPCRs, the DP1 and DP2 receptors. Activation of DP1 stimulates AC and increases cAMP concentration, whereas DP2 receptors couple to inhibitor G-proteins and decrease cAMP concentration (Hata et al., 2003; Kostenis and Ulven, 2006). Both DP1 and DP2 receptors are localized in neurons of all laminae within the ventral and dorsal horn (Grill et al., 2008). As regards its action on the pain threshold, IT application of PGD2 evokes hyperalgesia and allodynia (Minami et al., 1994), and allodynia cannot be elicited in mice lacking the PGD synthase (Eguchi et al., 1999), suggesting a pronociceptive role of PGD2. A recent study has shown that spinally administered PGD2 does not change the responses to mechanical stimulation in normal animals, and neither DP1 nor DP2 receptor agonists influences this reflex (Telleria-Diaz et al., 2008). However, IT PGD2 can also inhibit the PGE2-induced allodynia supporting an antinociceptive effect of PGD2 as well (Minami et al., 1996). Furthermore, either PGD2 or a DP1 receptor agonist decreases responses to mechanical stimulation in rats with inflamed joints and the facilitatory effects of PGE2, and the inhibitory effect of PGD2 could have been resulted from the activation of GABAergic inhibitory interneurons (Eguchi et al., 1999; Minami et al., 1997; Telleria-Diaz et al., 2008). Another prostaglandin, which can produce antinociception is the prostaglandin J, which is dominant during the resolution of an inflammatory condition (Burstein et al., 2007). Thus, compounds that promote the synthesis of this PG without significantly raising the level of the other prostaglandins could be considered as good candidates for the treatment of inflammation. Thus, NAGly produces a favorable prostaglandin ratio and is effective in reducing in vivo responses to proinflammatory agensts (Burstein et al., 2007). Cytochrome P450 genes catalyze formation of epoxyeicosatrienoic acids (EETs) from arachidonic acid. The effects of 5,6EET, 8,9EET, 11,12EET, and 14,15EET microinjected into the ventrolateral PAG on the thermally produced TF-response have been studied in rats (Terashvili et al., 2008). 14,15EET dose-dependently increased the TF-latency, whereas other EETs were inactive. The effect of 14,15EET has been blocked by antiserum against β-endorphin or Met-ENK, suggesting that this ligand evokes β-endorphin and Met-ENK releases.

6.3 Gangliosides

Gangliosides are glycosphingolipids that occur in nearly all cellular membranes and are particularly concentrated in the nervous tissue (Zeller and Marchase, 1992). Gangliosides include all sialic acid-containing glycosphingolipids possessing a specific sequence of neutral sugars (in different numbers). Members of the ganglioside family are designated by the capital letter G and are defined by the characteristic neutral sugar chain sequence. The sialic acid content of a ganglioside is designated by a capital letter: A (asialo), M (monosialo), or D (disialo). These agents have been shown to be effective in treating features of a variety of diabetic and toxic peripheral neuropathies. The mechanisms of ganglioside action in peripheral nerves include the enhancement of mean sprouting length, and increase of the number of regenerating axons (Zeller and Marchase, 1992). The protection afforded by gangliosides may be attributable to their ability to attenuate the neural injury induced by glutamate and/or to block the translocation of PKC (Vaccarino et al., 1987; Vorwerk et al., 1999; Zeller and Marchase, 1992). Furthermore, GM_1 ganglioside displays a broad spectrum of neurotrophic effects in vivo and in vitro. Some articles have reported the antinociceptive potency of GM_1 and the pronociceptive effect of anti-GD antibody (Fromm et al., 1993; Goettl et al., 2000; Mao et al., 1992; Sorkin et al., 2002), but opposite data have also been published (Crain and Shen, 1992; Crain and Shen, 2000). The systemic or IT administrations of gangliosides to control animals have no effect on sensory thresholds, but they suppress thermal and mechanical hyperalgesia and also spontaneous pain behavior in neuropathic pain models (Fromm et al., 1993; Mao et al., 1992). The chronic administration of GM_1 to aged rats partially restored the pain responses, but it had no effect on any sensory modality tested in young rats, suggesting a normalgesic effect (Goettl et al., 2000). The coadministration of a highly purified bovine brain ganglioside mixture (without GM_1) with pure GM_1 produced a potentiated antinociceptive effect in a model of peripheral mononeuropathy (Hayes et al., 1992).

6.4 Steroids

Steroids are a large group of compounds, any of which have important biological actions. These effects are typically brought about by steroid binding to nuclear receptors and subsequent changes in gene expression (Beato and Sanchez-Pacheco, 1996). However, steroids can also exert faster effects by activating membrane surface receptors (Losel and Wehling, 2003). Some of the best-characterised membrane surface steroid receptors in mammals are ion channels. For example, GlyRs and GABA receptors display different sensitivities to many neuroactive steroids (Webb and Lynch, 2007). However, most neurosteroids are active at a wide range of receptors so their potential as therapeutic agents seems limited.

6.4.1 Neurosteroids

Neurosteroids are steroid hormones synthesized in the brain that can modulate neuronal function through both gene expression and by direct modulation of neuronal excitability (Rupprecht, 2003). Nongenomic rapid effects of neurosteroids are particularly efficient. Two metabolites of progesterone, allopregnanolone (3α5αTHP: 3α5α-tetrahydro-progesterone) and pregnanolone (3α5βTHP: 3α5β-tetrahydroprogesterone), and sulfated steroids as well, act on GABAA receptors and potentiate their inhibitory function in the CNS (Covey et al., 2000; Keller et al., 2004; Majewska, 1992; Schlichter et al., 2006). Apart from GABAA, pregnanolone significantly reduced GlyR function as well (Jiang et al., 2006). Pregnenolone sulphate (PES) has been shown to modulate the activity of NMDA receptors and a variety of other ionotropic receptors, therefore PES increases neurotransmitter release from a variety of preparations and it affects the strength of synaptic transmission; the chronic IP administration of PES prevents the development of morphine tolerance (Gibbs et al., 2006b; Mameli et al., 2005; Reddy and Kulkarni, 1997). Both PES and the androgen dehydroepiandrosterone sulfate (DHEAS; see below in the section "Androgens") bind to sigma receptors (sigma1 and 2), which are nonopioid, nonphencyclidine receptors (Monnet et al., 1995). The sigma1 receptor has been cloned and its sequence does not resemble that of any mammalian protein, whereas sigma2 receptors have not been cloned. Sigma1 agonists, although having no effects by themselves, caused the amplification of signal transductions incurred upon the stimulation of the glutamatergic, dopaminergic, IP3-related metabotropic, or nerve growth factor-related systems. Inasmuch as this receptor has a significant role in the pain mechanisms, we may not exclude the role of its activation by these neurosteroids in their antinociceptive effects (Guitart et al., 2004), but the activation of this receptor at spinal level induces mechanical allodynia by the activation of spinal NMDA receptors (Roh et al., 2008). ICV administration of allopregnanolone significantly and dose-dependently increases the pain thresholds to heat stimulus, an effect which was mediated by GABAARs (Kavaliers and Wiebe, 1987). Endogenous neurosteroids are produced in the SDH and the elevated concentration of neurosteroids seen in inflammatory pain states significantly reduces thermal heat hyperalgesia (Poisbeau et al., 2005). It has been proposed that neurosteroids could be part of an endogenous modulatory/compensatory mechanism in response to a strong and/or sustained activation of the spinal nociceptive system (Vergnano et al., 2007). The IT administration of allopregnanolone effectively decreased both the mechanical and thermal hyperalgesia, whereas pregnanolone was only efficient on mechanical allodynia and had no effect on thermal heat hyperalgesia (Charlet et al., 2009).

6.4.2 Sexual Hormones

Estrogen and Progesterone

It is well known that women are more sensitive to several types of pain than men and functional bowel disorders are 2–3 times more prevalent in women (Berkley, 1997). The severity of pain symptoms fluctuates with the menstrual cycle suggesting female gonadal hormones modulate pain processing (Bartok and Craft, 1997; Kayser et al., 1996; Houghton et al., 2002; Hucho et al., 2006). Furthermore, pregnancy and parturition are associated with an opioid-mediated maternal analgesia (Dawson-Basoa and Gintzler, 1997). Estrogen has a permissive rather than a modulating function in this respect, and progesterone seems to specifically inhibit GlyRs (Mogil et al., 2003; Webb and Lynch, 2007). The change in estrogen status alone is sufficient to modify the processing of noxious sensory input to the CNS. The classical estrogen receptor exists as two subtypes, ERα and ERβ, and is expressed in the primary sensory neurons, sensory ganglia, dorsal horn, and supraspinal brain regions associated with pain modulation (Bereiter et al., 2005; Merchenthaler et al., 2004; Okamoto et al., 2008). Estrogen receptor signalling dramatically affects uterine cervical structure, and may also enhance pain responses at this level (Ji et al., 2005; Yan et al., 2007). Compared to intact rats, ovariectomy reduces the magnitude of the visceromotor responses and the response of SDH neurons, which is reversed by estradiol replacement (Ren et al., 2000; Tang et al., 2008). The data suggest that estrogens play an important role in modulating visceral nociceptive processing by increasing the spinal NMDA receptor expression and activation (Tang et al., 2008). Furthermore, chronic estrogen treatment increases spontaneous activity of afferents that innervate the uterine cervix, and enhances afferent firing in response to cervical distension, and TRPV1 receptor function is important for estrogen-induced sensitization (Yan et al., 2007).

The effects of progesterone are mediated by two distinct nuclear receptor proteins, PRA and PRB. Some studies reported antinociceptive effects of progesterone. Lactating rats with a high level of progesterone demonstrated significantly less hyperalgesia and progesterone replacement in ovariectomised rats significantly attenuated inflammation-induced hyperalgesia (Ji et al., 2005; Ren et al., 2000). It is thought that progesterone's antihyperalgesic effects include suppression of NMDA receptor activation at the level of the spinal cord (Ren et al., 2000). Because the increase in estradiol and that in progesterone coincide during the estrous cycle, the pronociceptive effect of estradiol and the antinociceptive effect of progesterone may obscure each other, reducing fluctuations during the course of the estrous cycle. Dawson-Basoa and Gintzler (1996, 1997, 1998) have performed studies of the interaction of β-estradiol and progesterone and their potential mechanism of action in respect of pain sensitivity. Simulation of the pregnancy blood profile of β-estradiol and progesterone in nonpregnant, ovariectomised rats have resulted in a statistically significant elevation of the pain threshold in the electric foot shock test, suggesting that the entire pregnancy profile of steroid hormones is responsible for the manifestation of analgesia. As regards the mechanism, it is proposed that the analgesia during pregnancy may result from direct effects of estrogen and progesterone on the CNS. The activation of the receptors caused a 52% increase in the opioid receptor binding density and in the concentration of β-endorphin in the preoptic area (Bridges and Ronsheim, 1987). Additionally, estrogen has been shown to positively regulate pro-ENK mRNA levels in the ventrolateral aspect of the ventromedial hypothalamic nucleus (Romano et al., 1989). These observations may suggest that the opioid systems mediating the analgesic effects of estrogen and progesterone are modulated in a synergistic fashion; accordingly, pregnancy provides a special case of antinociceptive interaction between these endogenous ligands.

Androgens

Men are typically reported to have higher pain thresholds than women, and gonadal hormones, particularly testosterone for males, contribute to this effect. Gonadectomy in adult male rats enhanced inflammation-induced sensitivity to mechanical stimulation and the effects was reversed by testosterone, and the antihyperalgesic potential of morphine decreased in neonatally gonadectomised male animals (Cicero et al., 2002). According to another study, castration reduces both opioid and nonopioid SIA in rats, which was reversed by testosterone replacement (Romero et al., 1988). However, other reports suggest pronociceptive potential of testosterone, because castration induces analgesia in the late phase of the formalin test, which correlates with increased 5-HT level in the SDH (Nayebi and Ahmadiani, 1999; Nayebi and Rezazadeh, 2004). IT administration of testosterone caused analgesia in neuropathic rats (Kibaly et al., 2008). As regards the weak androgens originated primarily from the adrenal cortex (i.e. dehydroepiandrosterone (DHEA) and dehydroepiandrosterone sulfate (DHEAS)), they are abundant in the brain even after adrenalectomy and gonadectomy (Corpechot et al., 1985). Both DHEA and DHEAS can inhibit the GABA- and Gly-induced current, and the chronic systemic (IP) administration of DHEA or DHEAS elevates the basal nociceptive threshold, and prevents the development of morphine tolerance (Majewska, 1992; Ren et al., 2004; Webb and Lynch, 2007). As regards the role of DHEA at the spinal level, its level drops in neuropathy in the SDH (Kibaly et al., 2008). Behavioral analysis shows a rapid pronociceptive and a delayed antinociceptive action of acute DHEA treatment, and the inhibition of its synthesis evokes analgesia. In contrast, the peripheral administration of DHEAS has significant hyperalgesic and vasodilatory actions through histamine release (Uchida et al., 2003).

6.4.3 Glucocorticoids

Glucocorticoids (mainly cortisol) are considered to be essential stress hormones and their levels increase immediately after injury, pain, and the like. The basal level of glucocorticoids is critical for the expression of analgesia, playing a permissive role in this process and they have an important role in the SIA (Panocka et al., 1987; Sutton et al., 1994). The glucocorticoid receptors are widely distributed both centrally and peripherally, and similarly to other steroids, the glucocorticoids also possess genomic and nongenomic actions (DeLeon et al., 1994; Joels, 1997). At the genomic level, the glucocorticoids repress transcription of a number of proinflammatory gene products that include the proinflammatory cytokines and enzymes such as PLA2, COX2, and inducible NOS (Niederberger et al., 2007; Sorrells and Sapolsky, 2007). Glucocorticoids are also involved in the induction of anti-inflammatory products such as interleukin-10 and annexin-A1 (Ayoub et al., 2008; Sorrells and Sapolsky, 2007). The nongenomic actions of the glucocorticoids are very rapid, occurring within minutes, and are dependent on the cytoplasmic glucocorticoid receptor (Buckingham et al., 2006). Glucocorticoids reduce PGE2 biosynthesis by either inhibition of PLA2 activity or inhibition of COX2 protein induction, the former effect considered to be a nongenomic, and the latter a genomic, effect. All of these actions decrease the release of pronociceptive ligands. Therefore, systemic, local, and IT administrations of glucocorticoids decrease the hyperalgesia, and cortisols are routinely given to patients with different pain syndroms (Ferreira et al., 1997; Taguchi et al., 2007). However, other studies have shown that chronic stress induces a long-lasting hyperalgesia, which is inhibited by glucocorticoid receptor antagonists (Khasar et al., 2008). Furthermore, the ICV administration of glucocorticoids decreases both the opioid- and clonidine-induced antinociception (Capasso et al., 1992; Capasso and Loizzo, 2001). Thus, the glucocorticoids influence both the antinociceptive and pronociceptive processes, and the net effects may depend on the qualities of activating factors and their duration.

7 Gases

7.1 Nitric Oxide (NO)

It has been recognized that NO serves as an important intracellular and intercellular messenger molecule in the PNS and CNS, and functions in a variety of physiological and pathophysiological processes (Mizutani and Layon, 1996; Wu and Morris, 1998). NO, a free radical gas, is not stored in synaptic vesicles and released by exocytosis. Three different NOSs are responsible for NO synthesis: neuronal (nNOS or NOS1), endothelial (eNOS or NOS2), and inducible NOS (iNOS or NOS-3) (Boehning and Snyder, 2003). Modulation of nNOS activity by multiple signalling cascades permits the regulated production of NO in response to neuronal stimulation. Unreacted NO has been assumed to simply diffuse away from target areas, but recent studies have suggested an enzymatic inactivation system, thus, myeloperoxidase, an enzyme highly enriched in leukocytes, also regulates NO bioavailability. Once NO is synthesized from L-arginine, it quickly diffuses from one neuron to another neuron and acts on the soluble guanylyl cyclase (sGC) to stimulate the formation of cGMP. This appears to be the principal mode for mediating the effects of NO in the neurons. However, NO also acts by modifying the transition metal centers of a wide variety of proteins. It can function by selectively and reversibly S-nitrosylating cysteine residues on a wide variety of proteins with precise spatial and temporal resolution. These proteins may be ion channels, pumps, or metabolic enzymes; for example, S-nitrosylation activates L-type Ca^{2+} channels, Ca^{2+}-activated K^{+}-channels, and GABAA receptors, but it inhibits NMDA receptors and several classes of Na^{+}-channels.

The role of NO in the nociceptive processes is very controversial. Several lines of evidence have shown that inhibition of NO production reduces pain hypersensitivity; however, other studies suggest the opposite role of NO. Systemic administration of NOS inhibitors have reversed pain hypersensitivity (Chu et al., 2005; Crosby et al., 1995; Mabuchi et al., 2003). Disruption of nNOS partially reduced inflammation-induced mechanical pain hypersensitivity (but not thermal hyperalgesia) (Chu et al., 2005), whereas others have not found this effect in other pain models (Crosby et al., 1995; Tao et al., 2004). Findings support the view that a condition of chronic stress can enhance hyperalgesia induced by systemic nitroglycerin administration (Costa et al., 2005). These observations may be relevant to pain disorder, and particularly to migraine, because nitroglycerin is able to induce spontaneous-like pain attacks in humans. NO appears to play a promoting role in supraspinal pain transmission, inasmuch as a number of NOS inhibitors exhibit potent antinociceptive activities on systemic or ICV administration (Pelligrino et al., 1996). In contrast, an antinociceptive property of brain NO has also been reported in acute mechanical and heat pain tests (Kawabata et al., 1992). Considerable evidence has demonstrated that NO and its enzymes are involved in the central mechanisms of pain at the spinal cord level (Chu et al., 2005; Malmberg and Yaksh, 1993; Meller et al., 1992a; Tao et al., 2003). IT NOS inhibitors have decreased the neuropathic and inflammatory pain sensitivity, and significantly decreased the C-fiber-evoked activity (Chu et al., 2005; Meller et al., 1992b; Meller et al., 1994b; Meller et al., 1994a; Zhang et al., 2005). Moreover, many of the effects of NMDA receptors in inflammatory hyperalgesia appear to be mediated through the production of NO (Chu et al., 2005; Li et al., 1994b; Mabuchi et al., 2003; Meller et al., 1992a; Tao et al., 2004). Given the link between NMDA receptor activation and production of nitric oxide, it is not surprising that activation of NOS and subsequent production of NO is one of the signal transduction systems shown to be involved in the behavioral response to pain stimuli (Meller et al., 1994c).

All of these data indicate that the spinal NOS is essential for pain hypersensitivity. However, the effects of NO on neuronal firing differ among cell types, and NO production can lead to inhibition of neuronal firing and transmission at spinal level (Ma and Eisenach, 2007; Song et al., 1998; Xu et al., 1996a). Thus, spinal NO is directly involved in the analgesic effects of morphine and clonidine and acetylcholine (Lauretti et al., 2000; Ma and Eisenach, 2007; Pan et al., 1998; Song et al., 1998; Xu et al., 1996a; Xu et al., 1997; Xu et al., 2000). As regards the role of NO at the peripheral level, locally released NO also plays a dual role in nociceptive modulation, inducing either nociceptive or antinociceptive responses depending on its amount and tissue level (Kawabata et al., 1994a). Noxious heat induces NO generation causing SP release from the peripheral endings of small-diameter primary afferent neurons (Yonehara and Yoshimura, 1999). Inflammation also enhances peripheral release of NO, which may contribute to edema and hyperalgesia (Lawand et al., 1997; Omote et al., 2001). NO may play an important role in neurogenic inflammation through enhancement of the release of neuropeptides by activating small-diameter primary afferent neurons (Yonehara and Yoshimura, 1999). In contrast, the local increase in NO level can decrease the mechanical hyperalgesia through an increase in cGMP (Steiner et al., 2001). Furthermore, the peripheral MT-induced antinociception is inhibited by NOS inhibitor, because NO production can lead to opening K^+-channels (Hernandez-Pacheco et al., 2008). Several other observations also indicate that NO donors inhibit the ongoing mechanical nociceptor supersensitivity, and NOS inhibitors enhance the hypersensitivity (Ferreira et al., 1991; Ferreira et al., 1992; Lorenzetti and Ferreira, 1996). The simplest explanation for these conflicting observations may be that the role and importance of the pathway varies among the groups of primary sensory neurons mobilized by different types of nociceptive stimuli.

7.2 Carbon Monoxide (CO)

It has been long recognized that the gaseous compound CO is noxious and harmful. CO is generated by haeme oxygenase (HO) that degrades haeme in aging red blood cells giving rise to biliverdin, iron, and CO (Boehning and Snyder, 2003). HO activity can be induced in almost all cell types by cellular stressors. Molecular cloning has revealed three types of HOs, the highly inducible isoform termed HO1, and two constitutive expressed isoforms termed HO2 and HO3. HO2 is selectively concentrated in the brain and testes, and it is colocalized with sGC throughout multiple brain regions. CO activates sGC to generate cGMP, but it can also directly activate different K^+-channels (Boehning and Snyder, 2003). Since the beginning of the 1990s, a growing body of evidence has given support to the physiological actions of CO as a vasoactive substance and a neurotransmitter/modulator. Thus, as with NO, CO is also a labile gaseous messenger in the nervous system (Boehning and Snyder, 2003; Verma et al., 1993). Because the nociceptor activity and excitability may be modulated by intracellular cGMP, CO can influence the pain sensitivity in a cGMP-dependent manner (Duarte et al., 1992; Sousa and Prado, 2001). Various studies have suggested that CO regulates nociception and a part of them indicate pronociceptive effects. Thus, the lack of HO2 enzyme has not changed the normal heat and mechanical threshold, but it reduced the hyperalgesia in inflammatory or nerve injury models; it has not modified the potency of morphine, but it has prevented morphine tolerance (Li and Clark, 2002; Li and Clark, 2003; Liang et al., 2003; Liang et al., 2004). Furthermore, the IT administration of HO inhibitors has not influenced acute heat pain latency, but inhibited pain-related behaviors and increased the potency of morphine (Li and Clark, 2001; Li and Clark, 2002). Thus, the production of CO at the spinal level is important in the behavioral expression of acute mechanical hyperalgesia, but is not involved in thermal hyperalgesia (Meller et al., 1994c). However, another study has found opposite results, that is, IT HO inhibitor increased, whereas CO substrate decreased the formalin-induced behavior (Nascimento and Branco, 2008). IPL administration of an HO inhibitor potentiated mechanical hyperalgesia and the formalin-induced behavior, and the increased CO level decreased the hypersensitivity by increasing the intracellular level of sGC (Nascimento and Branco, 2007; Rosa et al., 2008; Steiner et al., 2001). The effect of CO has been prevented by the NOS blocker, suggesting that the effect of CO depends on the integrity of the NO pathway. In conclusion, our knowledge of the role of CO in the nociceptive processes is incomplete, therefore further studies are required to reveal its role at different levels.

7.3 Hydrogen Sulfide (H_2S)

H_2S is now considered a novel gasotransmitter in peripheral tissues and the CNS (Boehning and Snyder, 2003; Qu et al., 2008). Similarly to CO and NO, H_2S also exists in the brain in relatively high concentrations (50–160 μM), and there is a cross-talk between H_2S and NO. H_2S is formed from cysteine by cystathione β-synthase (CBS) and cystathione γ-lyase (CSE) (Szabo, 2007). CBS transcript levels are high in the brain, with little or no CSE. CBS is activated by stimulation of ionotropic glutamate receptors in the presence of Ca^{2+}. Much progress has been made in the past decade in elucidating the roles of H_2S in physiological and pathological conditions at the cellular level (Bhatia et al., 2005; Qu et al., 2008; Szabo, 2007). It increases cAMP level, which stimulates PKA to phosphorylate and activate postsynaptic NMDA receptors. H_2S also upregulates GABAB receptors, therefore, H_2S can act presynaptically to inhibit neurotransmission (Boehning and Snyder, 2003). However, H_2S directly activates T-type Ca^{2+} channels as well, and H_2S may play a part in maintaining the excitation/inhibition balance (Kawabata et al., 2007). Systemic administration of H_2S donors inhibits acute and inflammatory visceral nociception by opening ATP-sensitive potassium channels (Distrutti et al., 2006b; Distrutti et al., 2006a). However, the H_2S may play a dual role in inflammatory hypernociception (Cunha et al., 2008). Production of endogenous H_2S during inflammation mediates the induction of mechanical hypernociception. The pronociceptive role of H_2S seems to be closely associated with upregulation of neutrophil migration to the inflammatory site, and the activation of T-type Ca^{2+} channel activity. On the other hand, the direct action of H_2S on peripheral nociceptive neurons can produce antinociception by the activation of peripheral K^+ATP channels. Only one study demonstrated the effect of H_2S at the spinal level (Maeda et al., 2009). In isolated DRG neurons, H_2S donor (sodium hydrosulfide: NaHS) facilitated T-type calcium channel-dependent currents, and caused hyperalgesia, and this effect was blocked by a T-type channel inhibitor. More results are available about the peripheral effect of H_2S. H_2S biosynthesis is increased following IPL injection of carrageenan, local administration of NaHS induces a mechanical hypernociception, and the hyperalgesia was decreased by H_2S synthesis inhibitor or by T-type calcium channel blockers (Bhatia et al., 2005; Cunha et al., 2008; Kawabata et al., 2007; Maeda et al., 2009). NaHS also excites capsaicin-sensitive primary afferents and evokes a peripheral release of neurokinins (Patacchini et al., 2004). Intracolonic administration of NaHS caused visceral pain-like nociceptive behavior and referred abdominal hyperalgesia (Matsunami et al., 2009). Retrograde injection of NaHS into the pancreatic duct induced expression of Fos-protein in the superficial layers of the SDH, and the pancreatitis-induced referred pain was decreased by inhibition of the H_2S enzyme (Nishimura et al., 2009). All of these experiments suggest a role of H_2S as a nociceptive messenger in the periphery. The mechanisms of H_2S action in these processes are dependent on the direct modulation of T-type Ca^{2+} channel activity in nociceptors and independent of K^+ATP channels (Maeda et al., 2009). However, another study provides evidence suggesting a nociceptive-intensity-dependent role for peripheral H_2S in nociception. Topical administration of H_2S donor increases the nociceptive behavior of formalin, whereas the H_2S level decreases in the spinal cord with hind paw injection of formalin (Lee et al., 2008). Because H_2S inhibits microglia production of proinflammatory cytokines and nitric oxide (Hu et al., 2007), a decrease in spinal H_2S is pronociceptive in the formalin test by virtue of disinhibition of microglial function. All these preliminary reports suggest a complex role of H_2S in the pain mechanism, and further studies are required to reveal the central role of H_2S in these process.

8 Conclusions

Knowledge of the pathophysiology of pain has evolved substantially inasmuch as the more current hypotheses incorporate gene–environment interactions, endocrine, immunological, and metabolic mediators, and cellular, molecular, and epigenetic factors of plasticity. However, enormous gaps in the knowledge of pain and its treatment persist. The data reveal that the actions of the endogenous substances are very different depending on the type of ligands, the pain tests, and the route of administrations. Activation of the different receptors may produce anti- or pronociception depending on the types and localization of the binding sites. In an ideal case the ligand induces analgesia by the presynaptic inhibition of excitatory neurotransmitter release, and the postsynaptic inhibition of the effects of excitatory neurotransmitters or increase of the release of endogenous inhibitory transmitters from the neurons. Accordingly, the simultaneous engagement of pre- and postsynaptic mechanisms by a combination of drugs may magnify the effects produced by either drug acting at one site independently. Furthermore, both the coordination and plasticity of cellular responses to different receptor activation could be influenced by variables such as the types and numbers of receptors present in each cell type, physical or functional compartmentalisation of the signalling components, and differential and/or overlapping sensitivities to various ligands and/or costimulation with other receptor types. Another crucial factor that underlies the efficacy of a drug is the "robustness" of the network that the compound targets, because the ligands could express their effects at several levels of the pathways. Moreover, all of these variables need to be recognized as dynamic. Thus the specific response of a cell to a ligand will be determined as the sum of many variables, and net effect of different ligands depends upon all of these factors. That is why signal transduction studies performed in transfected cells ultimately must be validated contextually in cells, organs, and intact animals in which the different receptors are endogenously expressed. Not all of the endogenous ligands have been discovered, and even fewer data are available about their interactions, therefore researchers need to continue identifying key cellular and molecular factors affected in sensory pathways during different pain syndromes to characterize potential targets for new drugs and drug combinations.

Acknowledgments This work was supported by a grant of the National Research and Development Office, Hungary (OMFB-0066/2005/DNT) and a Hungarian Research Grant (OTKA, K60278). The author thanks to Dr. Gabriella Kekesi, Dr. Gabor Tuboly and Agnes Tandari for their technical assistance.

References

Abbadie C, Brown JL, Mantyh PW, Basbaum AI (1996) Spinal cord substance P receptor immunoreactivity increases in both inflammatory and nerve injury models of persistent pain. Neuroscience 70:201–209

Abe K, Taguchi K, Kato M, Utsunomiya I, Chikuma T, Hojyo H, Miyatake T (2003) Characterization of muscarinic receptor subtypes in the rostral ventrolateral medulla and effects on morphine-induced antinociception in rats. Eur J Pharmacol 465:237–249

Adams ML, Brase DA, Welch SP, Dewey WL (1986) The role of endogenous peptides in the action of opioid analgesics. Ann Emerg Med 15:1030–1035

Afrah AW, Stiller CO, Olgart L, Brodin E, Gustafsson H (2001) Involvement of spinal N-methyl-D-aspartate receptors in capsaicin-induced in vivo release of substance P in the rat dorsal horn. Neurosci Lett 316:83–86

Agarwal N, Pacher P, Tegeder I, Amaya F, Constantin CE, Brenner GJ, Rubino T, Michalski CW, Marsicano G, Monory K, Mackie K, Marian C, Batkai S, Parolaro D, Fischer MJ, Reeh P, Kunos G, Kress M, Lutz B, Woolf CJ, Kuner R (2007) Cannabinoids mediate analgesia largely via peripheral type 1 cannabinoid receptors in nociceptors. Nat Neurosci 10:870–879

Aghajanian GK, Vandermaelen CP (1982) Alpha 2-adrenoceptor-mediated hyperpolarization of locus coeruleus neurons: intracellular studies in vivo. Science 215:1394–1396

Ahern GP (2003) Activation of TRPV1 by the satiety factor oleoylethanolamide. Nat Neurosci 278:30429–30434

Ahmadi S, Lippross S, Neuhuber WL, Zeilhofer HU (2002) PGE_2 selectively blocks inhibitory glycinergic neurotransmission onto rat superficial dorsal horn neurons. Nat Neurosci 5:34–40

Akil H, Watson SJ, Young E, Lewis ME, Khachaturian H, Walker JM (1984) Endogenous opioids: biology and function. Annu Rev Neurosci 7:223–255

Alger B, Le Beau F (2001) Physiology of the GABA and glycine systems. Springer-Verlag GmbH and Co. KG, New York, NY

Alkondon M, Pereira EFR, Cortes WS, Maelicke A, Albuquerque EX (1997) Choline is a selective agonist of alpha 7 nicotinic acetylcholine receptors in the rat brain neurons. Eur J Neurosci 9:2734–2742

Allen JM, Gibson SJ, Adrian TE, Polak JM, Bloom SR (1984) Neuropeptide-Y in human spinal-cord. Brain Res 308:145–148

Almasi R, Szoke E, Bolcskei K, Varga A, Riedl Z, Sandor Z, Szolcsanyi J, Petho G (2008) Actions of 3-methyl-N-oleoyldopamine, 4-methyl-N-oleoyldopamine and N-oleoylethanolamide on the rat TRPV1 receptor in vitro and in vivo. Life Sci 82:644–651

Altier N, Dray A, Menard D, Henry JL (2000) Neuropeptide FF attenuates allodynia in models of chronic inflammation and neuropathy following intrathecal or intracerebroventricular administration. Eur J Pharmacol 407:245–255

Altier N, Stewart J (1993) Intra-VTA infusions of the substance P analogue, DiMe-C7, and intra-accumbens infusions of amphetamine induce analgesia in the formalin test for tonic pain. Brain Res 628:279–285

Altier N, Stewart J (1998) Dopamine receptor antagonists in the nucleus accumbens attenuate analgesia induced by ventral tegmental area substance P or morphine and by nucleus accumbens amphetamine. J Pharmacol Exp Ther 285:208–215

Altier N, Stewart J (1999) The role of dopamine in the nucleus accumbens in analgesia. Life Sci 65:2269–2287

Ambriz-Tututi M, Granados-Soto V (2007) Oral and spinal melatonin reduces tactile allodynia in rats via activation of MT2 and opioid receptors. Pain 132:273–280

Anderson SD, Basbaum AI, Fields HL (1977) Response of medullary raphe neurons to peripheral stimulation and to systemic opiates. Brain Res 123:363–368

Andrade R, Aghajanian GK (1982) Single cell activity in the noradrenergic A-5 region: responses to drugs and peripheral manipulations of blood pressure. Brain Res 242:125–135

Ansah OB, Pertovaara A (2007) Peripheral suppression of arthritic pain by intraarticular fadolmidine, an alpha 2-adrenoceptor agonist, in the rat. Anesth Analg 105:245–250

Anuradha K, Hota D, Pandhi P (2004) Possible mechanisms of insulin antinociception. Methods Find Exp Clin Pharmacol 26:5–8

Arcaya J-L, Cano G, Gomez G, Maixner W, Suarez-Roca H (1999) Dynorphin A increases substance P release from trigeminal primary afferent C-fibers. Eur J Pharmacol 366:27–34

Arletti R, Benelli A, Bertolini A (1993) Influence of oxytocin on nociception and morphine antinociception. Neuropeptides 24:125–129

Arneric SP, Holladay M, Williams M (2007) Neuronal nicotinic receptors: a perspective on two decades of drug discovery research. Biochem Pharmacol 74:1092–1101

Arora S, Anubhuti (2006) Role of neuropeptides in appetite regulation and obesity – A review. Neuropeptides 40:375–401

Auguet M, Viossat I, Marin J-G, Chabrier P-E (1995) Selective inhibition of inducible nitric oxide synthase by agmatine. Jpn J Pharmacol 69:285–287

Ayesta FJ, Nikolarakis KE (1989) Peripheral but not intracerebroventricular corticotropin- releasing hormone (CRH) produces antinociception which is not opioid mediated. Brain Res 503:219–224

Ayoub SS, Yazid S, Flower RJ (2008) Increased susceptibility of annexin-A1 null mice to nociceptive pain is indicative of a spinal antinociceptive action of annexin-A1. Br J Pharmacol 154:1135–1142

Azarov AV, Szabo G, Czako L, Telegdy G (1995) Effects of calcitonin gene-related peptide on acute and chronic effects of morphine. Pharmacol Biochem Behav 52:595–599

Bach FW, Yaksh TL (1995) Release of beta-endorphin immunoreactivity from brain by activation of a hypothalamic N-methyl-D-aspartate receptor. Neuroscience 65:775–783

Bacon A, Holmes FE, Small CJ, Ghatei M, Mahoney S, Bloom S, Wynick D (2002) Transgenic over-expression of galanin in injured primary sensory neurons. NeuroReport 13: 2129–2132

Ballet S, Mauborgne A, Benoliel JJ, Bourgoin S, Hamon M, Cesselin F, Collin E (1998) Polyarthritis-associated changes in the opioid control of spinal CGRP release in the rat. Brain Res 796:198–208

Bannon AW, Decker MW, Holladay MW, Curzon P, Donnelly-Roberts D, Puttfarcken PS, Bitner RS, Diaz A, Dickenson AH, Porsolt RD, Williams M, Arneric SP (1998) Broad-spectrum, non-opioid analgesic activity by selective modulation of neuronal nicotinic acetylcholine receptors. Science 279:77–81

Bannon AW, Seda J, Carmouche M, Francis JM, Norman MH, Karbon B, McCaleb ML (2000) Behavioral characterization of neuropeptide Y knockout mice. Brain Res 868:79–87

Bardin L, Schmidt J, Alloui A, Eschalier A (2000a) Effect of intrathecal administration of serotonin in chronic pain models in rats. Eur J Pharmacol 409:37–43

Bardin L, Lavarenne J, Eschalier A (2000b) Serotonin receptor subtypes involved in the spinal antinociceptive effect of 5-HT in rats. Pain 86:11–18

Barke KE, Hough LB (1993) Simultaneous measurement of opiate-induced histamine release in the periaqueductal gray and opiate antinociception: an in vivo microdialysis study. J Pharmacol Exp Ther 266:934–942

Bartfai T, Fisone G, Langel U (1992) Galanin and galanin antagonists: molecular and biochemical perspectives. Trends Pharmacol Sci 13:312–317

Bartho L, Lazar Z, Lenard L Jr, Benko R, Toth G, Penke B, Szolcsanyi J, Maggi CA (2000) Evidence for the involvement of ATP, but not of VIP/PACAP or nitric oxide, in the excitatory effect of capsaicin in the small intestine. Eur J Pharmacol 392:183–188

Bartok RE, Craft RM (1997) Sex differences in opioid antinociception. J Pharmacol Exp Ther 282:769–778

Barton NJ, McQueen DS, Thomson D, Gauldie SD, Wilson AW, Salter DM, Chessell IP (2006) Attenuation of experimental arthritis in TRPV1R knockout mice. Exp Mol Pathol 81:166–170

Basbaum AI, Fields HL (1984) Endogenous pain control systems: brainstem spinal pathways and endorphin circuitry. Annu Rev Neurosci 7:309–338

Beato M, Sanchez-Pacheco A (1996) Interaction of steroid hormone receptors with the transcription initiation complex. Endocr Rev 17:587–609

Becker A, Grecksch G, Schroder H (2006) Pain sensitivity is altered in animals after subchronic ketamine treatment. Psychopharmacology 189:237–247

Beecher HK (1957) The measurement of pain: prototype for the quantitative study of subjective responses. Pharmacol Rev 9:59–209

Behbehani MM, Pert A (1984) A mechanism for the analgesic effect of neurotensin as revealed by behavioral and electrophysiological techniques. Brain Res 324:35–42

Belfer I, Davidson E, Ratner A, Beery E, Shir Y, Seltzer Z (1998) Dietary supplementation with the inhibitory amino acid taurine suppresses autotomy in HA rats. NeuroReport 9:3103–3107

Belfrage M, Segerdahl M, Arner S, Sollevi A (1999) The safety and efficacy of intrathecal adenosine in patients with chronic neuropathic pain. Anesth Analg 89:136–142

Bell RL, Vaccarino AL, Olson RD, Olson GA, Nores WL, Cambre JG, Zadina JE, Kastin AJ (1999) Tolerance and morphine-induced cross-tolerance are not shown to Tyr-W-MIF-1 analgesia. Peptides 20:971–978

Bellasio S, Nicolussi E, Bertorelli R, Reggiani A (2003) Melanocortin receptor agonists and antagonists modulate nociceptive sensitivity in the mouse formalin test. Eur J Pharmacol 482:127–132

Ben Shabat S, Fride E, Sheskin T, Tamiri T, Rhee MH, Vogel Z, Bisogno T, De Petrocellis L, Di Marzo V, Mechoulam R (1998) An entourage effect: inactive endogenous fatty acid glycerol esters enhance 2-arachidonoyl-glycerol cannabinoid activity. Eur J Pharmacol 353:23–31

Bereiter DA, Cioffi JL, Bereiter DF (2005) Oestrogen receptor-immunoreactive neurons in the trigeminal sensory system of male and cycling female rats. Arch Oral Biol 50:971–979

Berkley KJ (1997) Sex differences in pain. Behav Brain Sci 20:371–380

Bernardini N, Roza C, Sauer SK, Gomeza J, Wess J, Reeh PW (2002) Muscarinic M2 receptors on peripheral nerve endings: a molecular target of antinociception. J Neurosci 22 :RC229

Bernardini N, Sauer SK, Haberberger R, Fischer MJM, Reeh PW (2001) Excitatory nicotinic and desensitizing muscarinic (M2) effects on C-nociceptors in isolated rat skin. J Neurosci 21:3295–3302

Bertorelli R, Fredduzzi S, Tarozzo G, Campanella M, Grundy R, Beltramo M, Reggiani A (2005) Endogenous and exogenous melanocortin antagonists induce anti-allodynic effects in a model of rat neuropathic pain. Behav Brain Res 157:55–62

Bester H, De Felipe C, Hunt SP (2001) The NK1 receptor is essential for the full expression of noxious inhibitory controls in the mouse. J Neurosci 21:1039–1046

Beyer C, Caba M, Banas C, Komisaruk BR (1991) Vasoactive intestinal polypeptide (VIP) potentiates the behavioral-effect of substance-P intrathecal administration. Pharmacol Biochem Behav 39:695–698

Bhatia M, Sidhapuriwala J, Moochhala SM, Moore PK (2005) Hydrogen sulphide is a mediator of carrageenan-induced hindpaw oedema in the rat. Br J Pharmacol 145:141–144

Bie B, Fields HL, Williams JT, Pan ZZ (2003) Roles of alpha1- and alpha2-adrenoceptors in the nucleus raphe magnus in opioid analgesia and opioid abstinence-induced hyperalgesia. J Neurosci 23:7950–7957

Binder W, Mousa SA, Sitte N, Kaiser M, Stein C, Schafer M (2004) Sympathetic activation triggers endogenous opioid release and analgesia within peripheral inflamed tissue. Eur J Neurosci 20:92–100

Bisogno T (2008) Endogenous cannabinoids: structure and metabolism. J Neuroendocrinol 20:1–9

Bisogno T, Melck D, Bobrov MY, Gretskaya NM, Bezuglov VV, De Petrocellis L, Di Marzo V (2000) N-acyl-dopamines: novel synthetic CB1 cannabinoid-receptor ligands and inhibitors of anandamide inactivation with cannabimemetic activity in vitro and in vivo. Biochem J 351: 817–824

Bitar MS, Al-Bustan M, Nehme CL, Pilcher CWT (1996) Antinociceptive action of intrathecally admistered IGF-I and the expression of its receptor in rat spinal cord. Brain Res 737:292–294

Bjorklund A, Skagerberg G (1979) Evidence for a major spinal cord projection from the diencephalic A11 dopamine cell group in the rat using transmitter-specific fluorescent retrograde tracing. Brain Res 177:170–175

Blakeman KH, Hao J-X, Xu X-J, Jacoby AS, Shine J, Crawley JN, Iismaa T, Wiesenfeld-Hallin Z (2003) Hyperalgesia and increased neuropathic pain-like response in mice lacking galanin receptor 1 receptors. Neuroscience 117:221–227

Blakeman KH, Holmberg K, Hao JX, Xu XJ, Kahl U, Lendahl U, Bartfai T, Wiesenfeld-Hallin Z, Hokfelt T (2001) Mice over-expressing galanin have elevated heat nociceptive threshold. NeuroReport 12:423–425

Bleakman D, Alt A, Nisenbaum ES (2006) Glutamate receptors and pain. Semin Cell Dev Biol 17:592–604

Blumensohn R, Ringler D, Eli I (2002) Pain perception in patients with schizophrenia. J Nerv Ment Dis 190:481–483

Bodnar RJ, Kelly DD, Mansour A, Glusman M (1979) Differential effects of hypophysectomy upon analgesia induced by two glucoprivic stressors and morphine. Pharmacol Biochem Behav 11:303–308

Bodnar RJ (2008) Endogenous opiates and behavior: 2007. Peptides 29:2292–2375

Boehning D, Snyder SH (2003) Novel neuronal modulators. Annu Rev Neurosci 26:105–131

Bolcskei K, Helyes Z, Szabo A, Sandor K, Elekes K, Nemeth J, Almasi R, Pinter E, Petho G, Szolcsanyi J (2005) Investigation of the role of TRPV1 receptors in acute and chronic nociceptive processes using gene-deficient mice. Pain 117:368–376

Bonnefont J, Daulhac L, Etienne M, Chapuy E, Mallet C, Ouchchane L, Deval C, Courade JP, Ferrara M, Eschalier A, Clottes E (2007) Acetaminophen recruits spinal p42/p44 MAPKs and GH/IGF-1 receptors to produce analgesia via the serotonergic system. Mol Pharmacol 71: 407–415

Boom A, Mollereau C, Meunier JC, Vassart G, Parmentier M, Vanderhaeghen J-J, Schiffmann SN (1999) Distribution of the nociceptin and nocistatin precursor transcript in the mouse central nervous system. Neuroscience 91:991–1007

Boring DL, Berglund BA, Howlett AC (1996) Cerebrodiene, arachidonylethanolamide, and hybrid structures: potential for interaction with brain cannabinoid receptors. Prostaglandins Leukot Essent Fatty Acids 55:207–210

Boschi G, Desiles M, Reny V, Rips R, Wrigglesworth S (1983) Antinociceptive properties of thyrotropin releasing hormone in mice - comparison with morphine. Br J Pharmacol 79:85–92

Boucher TJ, Okuse K, Bennett DLH, Munson JB, Wood JN, McMahon SB (2000) Potent analgesic effects of GDNF in neuropathic pain states. Science 290:124–127

Bradley KJ, Headley PM (1997) Effect of agmatine on spinal nociceptive reflex: lack of interaction with alpha2-adrenoceptor or mu-opioid receptor mechanisms. Eur J Pharmacol 331:133–138

Brain SD, Cox HM (2006) Neuropeptides and their receptors: innovative science providing novel therapeutic targets. Br J Pharmacol 147:S202–S211

Branchek TA, Smith KE, Gerald C, Walker MW (2000) Galanin receptor subtypes. Trends Pharmacol Sci 21:109–117

Bridges RS, Ronsheim PM (1987) Immunoreactive beta-endorphin concentrations in brain and plasma during pregancy in rats – possible modulation by progesterone and estradiol. Neuroendocrinology 45:381–388

Brucke T, Wenger S, Podreka I, Asenbaum S (1991) Dopamine receptor classification, neuroanatomical distribution and in vivo imaging. Wien Klin Wochenschr 103:639–646

Brumovsky P, Stanic D, Shuster S, Herzog H, Villar M, Hokfelt T (2005) Neuropeptide Y2 receptor protein is present in peptidergic and nonpeptidergic primary sensory neurons of the mouse. J Comp Neurol 489:328–348

Bruzzone F, Lectez B, Alexandre D, Jegou S, Mounien L, Tollemer H, Chatenet D, Leprince J, Vallarino M, Vaudry H, Chartrel N (2007) Distribution of 26RFa binding sites and GPR103 mRNA in the central nervous system of the rat. J Comp Neurol 503:573–591

Buck SH, Walsh JH, Yamamura HI, Burks TF (1982) Neuropeptides in sensory neurons. Life Sci 30:1857–1866

Buckingham JC, Flower RJ (1997) Lipocortin 1: a second messenger of glucocorticoid action in the hypothalamo-pituitary-adrenocortical axis. Mol Med Today 3:296–302

Buckingham JC, John CD, Solito E, Tierney T, Flower RJ, Christian H, Morris J (2006) Annexin 1, glucocorticoids, and the neuroendocrine-immune interface. Ann N Y Acad Sci 1088:396–409

Burgos JS, Barat A, Souza DO, Ramirez G (1998) Guanine nucleotides protect against kainate toxicity in an *ex vivo* chick retinal preparation. FEBS Lett 430:176–180

Burk R (2008) Aromatase inhibitor-induced joint pain: melatonin's role. Med Hypotheses 71: 862–867

Burstein SH (1999) The cannabinoid acids: nonpsychoactive derivatives with therapeutic potential. Pharmacol Ther 82:87–96

Burstein SH, Adams JK, Bradshaw HB, Fraioli C, Rossetti RG, Salmonsen RA, Shaw JW, Walker JM, Zipkin RE, Zurier RB (2007) Potential anti-inflammatory actions of the elmiric (lipoamino) acids. Bioorg Med Chem 15:3345–3355

Calcutt CR, Doggett NS, Spencer PSJ (1971) Modification of the anti-nociceptive activity of morphine by centrally administered ouabain and dopamine. Psychopharmacologia 21: 111–117

Calignano A, La Rana G, Giuffrida A, Piomelli D (1998) Control of pain initiation by endogenous cannabinoids. Nature 394:277–281

Calignano A, La Rana G, Piomelli D (2001) Antinociceptive activity of the endogenous fatty acid amide, palmitylethanolamide. Eur J Pharmacol 419:191–198

Callard RE, Matthews DJ, Hibbert L (1996) IL-4 and IL-13 receptors: are they one and the same? Immunol Today 17:108–110

Calo G, Guerrini R, Rizzi A, Salvadori S, Regoli D (2000) Pharmacology of nociceptin and its receptor: a novel therapeutic target. Br J Pharmacol 129:1261–1283

Candeletti S, Ferri S (1990) Antinociceptive profile of intracerebroventricular salmon-calcitonin and calcitonin gene-related peptide in the mouse formalin test. Neuropeptides 17:93–98

Cannon JG (2000) Inflammatory cytokines in nonpathological states. News Physiol Sci 15: 298–303

Cannon KE, Leurs R, Hough LB (2007) Activation of peripheral and spinal histamine H3 receptors inhibits formalin-induced inflammation and nociception, respectively. Pharmacol Biochem Behav 88:122–129

Cannon KE, Nalwalk JW, Stadel R, Ge P, Lawson D, Silos-Santiago I, Hough LB (2003) Activation of spinal histamine H3 receptors inhibits mechanical nociception. Eur J Pharmacol 470: 139–147

Cao J, O'Donnell D, Vu H, Payza K, Pou C, Godbout C, Jakob A, Pelletier M, Lembo P, Ahmad S, Walker P (1998) Cloning and characterization of a cDNA encoding a novel subtype of rat thyrotropin-releasing hormone receptor. J Biol Chem 273:32281–32287

Capasso A, Digiannuario A, Loizzo A, Pieretti S, Sorrentino L (1992) Central interaction of dexamethasone and Ru-38486 on morphine antinociception in mice. Life Sci 51:L139–L143

Capasso A, Loizzo A (2001) Clonidine-induced antinociception and locomotor hypoactivity are reduced by dexamethasone in mice. J Pharm Pharmacol 53:351–360

Carlton SM, Zhou S, Coggeshall RE (1999) Peripheral GABAA receptors: evidence for peripheral primary afferent depolarization. Neuroscience 93:713–722

Carpenedo R, Pittaluga A, Cozzi A, Attucci S, Galli A, Raiteri M, Moroni F (2001) Presynaptic kynurenate-sensitive receptors inhibit glutamate release. Eur J Neurosci 13:2141–2147

Carriere JL, El-Fakahani EE (2000) Choline is a full agonist in inducing activation of neuronal nitric oxide synthase via the muscarinic M_1 receptor. Pharmacology 60:82–89

Carta G, Gessa GL, Nava F (1999) Dopamine D(2) receptor antagonists prevent delta9-tetrahydrocannabinol-induced antinociception in rats. Eur J Pharmacol 384:153–156

Caterina MJ, Leffler A, Malmberg AB, Martin WJ, Trafton J, Petersen-Zeitz KR, Koltzenburg M, Basbaum AI, Julius D (2000) Impaired nociception and pain sensitisation in mice lacking the capsaicin receptor. Science 288:306–313

Cejas PJ, Martinez M, Karmally S, McKillop M, McKillop JP, Lunkett JA, Oudega M, Eaton MJ (2000) Lumbar transplant of neurons genetically modified to secrete brain-derived neurotrophic factor attenuates allodynia and hyperalgesia after sciatic nerve constriction. Pain 86: 195–210

Ceriani G, Macaluso A, Catania A, Lipton JM (1994) Central neurogenic antiinflammatory action of alpha-MSH: modulation of peripheral inflammation induced by cytokines and other mediators of inflammation. Neuroendocrinology 59:138–143

Chapman V, Diaz A, Dickenson AH (1997) Distinct inhibitory effects of spinal endomorphin-1 and endomorphin-2 on evoked dorsal horn neuronal responses in the rat. Br J Pharmacol 122:1537–1539

Charlet A, Lasbennes F, Darbon P, Poisbeau P (2009) Fast non-genomic effects of progesterone-derived neurosteroids on nociceptive thresholds and pain symptoms. Pain 139(3): 603–609

Chemin J, Monteil A, Perez-Reyes E, Nargeot J, Lory P (2001) Direct inhibition of T-type calcium channels by the endogenous cannabinoid anandamide. EMBO J 20:7033–7040

Chen JJ, Dymshitz J, Vasko MR (1997) Regulation of opioid receptors in rat sensory neurons in culture. Mol Pharmacol 51:666–673

Chen P, Hu S, Yao J, Moore SA, Spector AA, Fang X (2005a) Induction of cyclooxygenase-2 by anandamide in cerebral microvascular endothelium. Microvasc Res 69:28–35

Chen SR, Wess J, Pan HL (2005b) Functional activity of the M2 and M4 receptor subtypes in the spinal cord studied with muscarinic acetylcholine receptor knockout mice. J Pharmacol Exp Ther 313:765–770

Chen X, Levine JD (2005) Epinephrine-induced excitation and sensitization of rat C-fiber nociceptors. J Pain 6:439–446

Chen XH, Geller EB, Rogers TJ, Adler MW (2007) The chemokine CX3CL1/fractalkine interferes with the antinociceptive effect induced by opioid agonists in the periaqueductal grey of rats. Brain Res 1153:52–57

Cheng JK, Chou RCC, Hwang LL, Chiou LC (2003) Antiallodynic effects of intrathecal orexins in a rat model of postoperative pain. J Pharmacol Exp Ther 307:1065–1071

Chiang C-Y, Zhuo M (1989) Evidence for the involvement of a descending cholinergic pathway in systemic morphine analgesia. Brain Res 478:293–300

Chiari A, Eisenach JC (1999) Intrathecal adenosine: interactions with spinal clonidine and neostigmine in rat models of acute nociception and postoperative hypersensitivity. Anesthesiology 90:1413–1421

Chronwall BM, Zukowska Z (2004) Neuropeptide Y, ubiquitous and elusive. Peptides 25:359–363

Chu CJ, Huang SM, De Petrocellis L, Bisogno T, Ewing SA, Miller JD, Zipkin RE, Daddario N, Appandino G, Di Marzo V, Walker JS (2003) N-Oleoyldopamine, a novel endogenous capsaicin-like lipid that produces hyperalgesia. J Biol Chem 278:13633–13639

Chu YC, Guan Y, Skinner J, Raja SN, Johns RA, Tao YX (2005) Effect of genetic knockout or pharmacologic inhibition of neuronal nitric oxide synthase on complete Freund's adjuvant-induced persistent pain. Pain 119:113–123

Chung KM, Song DK, Suh HW, Lee MH, Kim YH (1994) Effects of intrathecal or intracerebroventricular pretreatment with pertussis toxin on antinociception induced by beta-endorphin or morphine administered intracerebroventricularly in mice. Naunyn Schmiedebergs Arch Pharmacol 349:588–593

Chung YH, Miyake H, Kamei C, Tasaka K (1984) Analgesic effect of histamine induced by intracerebral injection into mice. Agents Actions 15:137–142

Ciccarelli R, Ballerini P, Sabatino G, Rathbone MP, D'Onofrio M, Caciagli F, Di Iorio P (2001) Involvement of astrocytes in purine-mediated reparative processes in the brain. Int J Dev Neurosci 19:395–414

Cicero TJ, Nock B, O'Connor L, Meyer ER (2002) Role of steroids in sex differences in morphine-induced analgesia: activational and organizational effects. J Pharmacol Exp Ther 300:695–701

Citera G, Arias MA, Maldonado-Cocco JA, Lazaro MA, Rosemffet MG, Brusco LI, Scheines EJ, Cardinalli DP (2000) The effect of melatonin in patients with fibromyalgia: a pilot study. Clin Rheumatol 19:9–13

Coggeshall RE, Carlton SM (1997) Receptor localization in the mammalian dorsal horn and primary afferent neurons. Brain Res Rev 24:28–66

Commissiong JW, Galli CL, Neff NH (1978) Differentiation of dopaminergic and noradrenergic neurons in rat spinal-cord. J Neurochem 30:1095–1099

Commons KG, Van Bockstaele EJ, Pfaff DW (1999) Frequent colocalization of mu opioid and NMDA-type glutamate receptorst postsynaptic sites in periaqueductal gray neurons. J Comp Neurol 408:549–559

Condes-Lara M, Maie IA, Dickenson AH (2005) Oxytocin actions on afferent evoked spinal cord neuronal activities in neuropathic but not in normal rats. Brain Res 1045:124–133

Conn PJ, Pin J-P (1997) Pharmacology and functions of metabotropic glutamate receptors. Annu Rev Pharmacol Toxicol 37:205–237

Connor M, Schuller A, Pintar JE, Christie MJ (1999) Mu-opioid receptor modulation of calcium channel current in periaqueducal grey neurons from C57B16/J mice and mutant mice lacking MOR-1. Br J Pharmacol 126:1553–1558

Corpechot C, Leclerc P, Baulieu EE, Brazeau P (1985) Neurosteroids - regulatory mechanisms in male-rat brain during heterosexual exposure. Steroids 45:229–234

Costa A, Smeraldi A, Tassorelli C, Greco R, Nappi G (2005) Effects of acute and chronic restraint stress on nitroglycerin-induced hyperalgesia in rats. Neurosci Lett 383:7–11

Costa B, Comelli F, Bettoni I, Colleoni M, Giagnoni G (2008) The endogenous fatty acid amide, palmitoylethanolamide, has anti-allodynic and anti-hyperalgesic effects in a murine model of neuropathic pain: involvement of CB1, TRPV1 and PPAR gamma receptors and neurotrophic factors. Pain 139:541–550

Covey DF, Nathan D, Kalkbrenner M, Nilsson KR, Hu Y, Yorumski CF, Evers AS (2000) Enantioselectivity of pregnanolone-induced gamma-aminobutyric acid(A) receptor modulation and anesthesia. J Pharmacol Exp Ther 293:1009–1116

Cowley MA, Smith RG, Diano S, Tschop M, Pronchuk N, Grove KL, Strasburger CJ, Bidlingmaier M, Esterman M, Heiman ML, Garcia-Segura LM, Nillni EA, Mendez P, Low MJ, Sotonyi P, Friedman JM, Liu HY, Pinto S, Colmers WF, Cone RD, Horvath TL (2003) The distribution and mechanism of action of ghrelin in the CNS demonstrates a novel hypothalamic circuit regulating energy homeostasis. Neuron 37:649–661

Crain SM, Shen K-F (1992) After chronic opioid exposure sensory neurons become supersensitive to the excitatory effects of opioid agonists and antagonists as occurs after acute elevation of GM1 ganglioside. Brain Res 575:13–24

Crain SM, Shen K-F (2000) Enhanced analgesic potency and reduced tolerance of morphine in 129/SvEv mice: evidence for a deficiency in GM1 ganglioside-regulated excitatory opioid receptor function. Brain Res 856:227–235

Cravatt BF, Giang DK, Mayfield SP, Boger DL, Lerner RA, Gilula NB (1996) Molecular characterization of an enzyme that degrades neuromodulatory fatty-acid amides. Nature 384:83–87

Cravatt BF, Prosperogarcia O, Siuzdak G, Gilula NB, Henriksen SJ, Boger DL, Lerner RA (1995) Chemical characterization of a family of brain lipids that induce sleep. Science 268: 1506–1509

Cridland RA, Henry JL (1988a) Effects of intrathecal administration of neuropeptides on a spinal nociceptive reflex in the rat: VIP, galanin, CGRP, TRH, somatostatin and angiotensin II. Neuropeptides 11:23–32

Cridland RA, Henry JL (1988b) Intrathecal administration of substance P in the rat: spinal transection or morphine blocks the behavioural responses but not the facilitation of the tail flick reflex. Neurosci Lett 84:203–208

Cridland RA, Henry JL (1992) Bombesin, neuromedin C and neuromedin B given intrathecally facilitate the tail flick reflex in the rat. Brain Res 584:163–168

Cristino L, De Petrocellis L, Pryce G, Baker D, Guglielmotti V, Di Marzo V (2006) Immunohistochemical localization of cannabinoid type 1 and vanilloid transient receptor potential vanilloid type 1 receptors in the mouse brain. Neuroscience 139:1405–1415

Cronin JN, Bradbury EJ, Lidierth M (2004) Laminar distribution of GABAA- and glycine-receptor mediated tonic inhibition in the dorsal horn of the rat lumbar spinal cord: effects of picrotoxin and strychnine on expression of Fos-like immunoreactivity. Pain 112:156–163

Crosby G, Marota JJA, Huang PL (1995) Intact nociception-induced neuroplasticity in transgenic mice deficient in neuronal nitric oxide synthase. Neuroscience 69:1013–1017

Csullog E, Joo G, Toth G, Dobos I, Benedek G, Horvath G (2001) Antinociceptive effect of continuous intrathecal administration of endomorphin-1. Pain 94:31–38

Cui M, Honore P, Zhong C, Gauvin D, Mikusa J, Hernandez G, Chandran P, Gomtsyan A, Brown B, Bayburt EK, Marsh K, Bianchi B, McDonald H, Niforatos W, Neelands TR, Moreland RB, Decker MW, Lee CH, Sullivan JP, Faltynek CR (2006) TRPV1 receptors in the CNS play a key role in broad-spectrum analgesia of TRPV1 antagonists. J Neurosci 26:9385–9393

Cunha TM, Dal Secco D, Verri WAJ, Guerrero AT, Souza GR, Vieira SM, Lotufo CM, Neto AF, Ferreira SH, Cunha FQ (2008) Dual role of hydrogen sulfide in mechanical inflammatory hypernociception. Eur J Pharmacol 590:127–135

D'Amico M, Berrino L, Marione S, Filippelli A, De Novellis V, Rossi F (1996) Endothelin-1 in periaqueducal gray of mice induces analgesia via glutamatergic receptors. Pain 65:205–209

Da Cunha JM, Rae GA, Ferreira SH, Cunha Fde Q (2004) Endothelins induce ETB receptor-mediated mechanical hypernociception in rat hindpaw: roles of cAMP and protein kinase C. Eur J Pharmacol 501:87–94

Da Motta PG, Veiga APC, Francischi JN, Tatsuo MAKF (2004) Evidence for participation of GABA(A) receptors in a rat model of secondary hyperalgesia. Eur J Pharmacol 483:233–239

Dafny N, Dong WQ, Prieto-Gomez C, Reyes-Vazquez C, Stanford J, Qiao JT (1996) Lateral hypothalamus: site involved in pain modulation. Neuroscience 70:449–460

Daftary SS, Gore AC (2005) IGF-1 in the brain as a regulator of reproductive neuroendocrine function. Exp Biol Med 230:292–306

Daher JB, Souza GEP, D'Orleans-Juste P, Rae GA (2004) Endothelin ETB receptors inhibit articular nociception and priming induced by carrageenan in the rat knee-joint. Eur J Pharmacol 496:77–85

Dale CS, Pagano RdeL, Rioli V, Hyslop S, Giorgi R, Ferro ES (2005) Antinociceptive action of hemopressin in experimental hyperalgesia. Peptides 26:431–436

Damaj MI, Meyer EM, Martin BR (2000) The antinociceptive effects of alpha7 nicotinic agonists in an acute pain model. Neuropharmacology 39:2785–2791

Danysz W, Parsons CG (1998) Glycine and N-methyl-D-aspartate receptors: physiological significance and possible therapeutic applications. Pharmacol Rev 50:597–664

Dautzenberg FM, Hauger RL (2002) The CRF peptide family and their receptors: yet more partners discovered. Trends Pharmacol Sci 23:71–77

Davis JB, Gray J, Gunthorpe MJ, Hatcher JP, Davey PT, Overend P, Harries MH, Latcham J, Clapham C, Atkinson K, Hughes SA, Rance K, Grau E, Harper AJ, Pugh PL, Rogers DC, Bingham S, Randall A, Sheardown SA (2000) Vanilloid receptor-1 essential for inflammatory thermal hyperalgesia. Nature 405:183–187

Davis TP, Gillespie TJ, Porreca F (1989) Peptide-fragments derived from the beta-chain of hemoglobin (hemorphins) are centrally active in vivo. Peptides 10:747–751

Davis-Taber R, Baker S, Lehto SG, Zhong C, Surowy CS, Faltynek CR, Scott VE, Honore P (2008) Central pituitary adenylate cyclase 1 receptors modulate nociceptive behaviors in both inflammatory and neuropathic pain states. J Pain 9:449–456

Dawson-Basoa M, Gintzler AR (1996) Estrogen and progesterone activate spinal kappa-opiate receptor analgesic mechanisms. Pain 64:608–615

Dawson-Basoa M, Gintzler AR (1997) Involvement of spinal cord delta opiate receptors in the antinociception of gestation and its hormonal simulation. Brain Res 757:37–42

Dawson-Basoa M, Gintzler AR (1998) Gestational and ovarian sex steroid antinociception: synergy between spinal kappa and delta opioid systems. Brain Res 794:61–67

De Novellis V, Palazzo E, Rossi F, De Petrocellis L, Petrosino S, Guida F, Luongo L, Migliozzi A, Cristino L, Marabese I, Starowicz K, Di Marzo V, Maione S (2008) The analgesic effect of N-arachidonoyl-serotonin, a FAAH inhibitor and TRPV1 receptor antagonist, associated with changes in rostral ventromedial medulla and locus coeruleus cell activity in rats. Neuropharmacology 55:1105–1113

De Petrocellis L, Chu CJ, Moriello AS, Kellner JC, Walker JM, Di Marzo V (2004) Actions of two naturally occurring saturated N-acyldopamines on transient receptor potential vanilloid 1 (TRPV1) channels. Br J Pharmacol 143:251–256

De Petrocellis L, Starowicz K, Moriello AS, Vivese M, Orlando P, Di Marzo V (2007) Regulation of transient receptor potential channels of melastatin type 8 (TRPM8): effect of cAMP, cannabinoid CB1 receptors and endovanilloids. Exp Cell Res 313:1911–1920

De Petrocellis L, Bisogno T, Davis JB, Pertwee RG, Di Marzo V (2000) Overlap between the ligand recognition properties of the anandamide transporter and the VR1 vanilloid receptor: inhibitors of anandamide uptake with negligible capsaicin-like activity. FEBS Lett 483: 52–56

De Souza EB, Insel TR, Perrin MH, Rivier C, Vale WW, Kuhar MJ (1985) Corticotropin-releasing factor receptors are widely distributed within the rat central nervous system: an autoradiographic study. J Neurosci 5:3189–3203

Delcroix JD, Valletta JS, Wu C, Hunt SJ, Kowal AS, Mobley WC (2003) NGF signaling in sensory neurons: evidence that early endosomes carry NGF retrograde signals. Neuron 39:69–84

DeLeon M, Covenas R, Chadi G, Narvaez JA, Fuxe K, Cintra A (1994) Subpopulations of primary sensory neurons show coexistence of neuropeptides and glucocorticoid receptors in the rat spinal and trigeminal ganglia. Brain Res 636:338–342

Delfs JM, Zhu Y, Druhan JP, Aston-Jones G (2000) Noradrenaline in the ventral forebrain is critical for opiate withdrawal-induced aversion. Nature 403:430–434

Delgado M, Abad C, Martinez C, Juarranz MG, Leceta J, Ganea D, Gomariz RP (2003) PACAP in immunity and inflammation. Ann N Y Acad Sci 992:141–157

Demuth DG, Molleman A (2006) Cannabinoid signalling. Life Sci 78:549–563

Desroches J, Guindon J, Lambert C, Beaulieu P (2008) Modulation of the anti-nociceptive effects of 2-arachidonoyl glycerol by peripherally administered FAAH and MGL inhibitors in a neuropathic pain model. Br J Pharmacol 155:913–924

Detweiler DJ, Eisenach JC, Tong C, Jackson C (1993) A cholinergic interaction in alpha2 adrenoceptor-mediated antinociception in sheep. J Pharmacol Exp Ther 265:536–542

Devane WA, Hanus L, Breuer A, Pertwee RG, Stevenson LA, Griffin G, Gibson D, Mandelbaum A, Mechoulam R (1992) Isolation and structure of a brain constituent that binds to the cannabionid receptor. Science 258:1946–1949

Dhawan BN, Cesselin F, Raghubir R, Reisine T, Bradley PB, Portoghese PS, Hamon M (1996) International union of pharmacology. XII. Classification of opioid receptors. Pharmacol Rev 48:567–592

Di Marzo V, Breivogel C, Bisogno T, Melck D, Patrick G, Tao Q, Szallasi A, Razdan RK, Martin BR (2000a) Neurobehavioral activity in mice of N-vanillyl-arachidonyl-amide. Eur J Pharmacol 406:363–374

Di Marzo V, Breivogel CS, Tao Q, Bridgen DT, Razdan RK, Zimmer AM, Zimmer A, Martin BR (2000b) Levels, metabolism, and pharmacological activity of anandamide in CB_1 cannabinoid receptor knockout mice. Evidence for non-CB_1, non-CB_2 receptor-mediated actions of anandamide in mouse brain. J Neurochem 75:2434–2444

Di Marzo V, Melck D, Bisogno T, De Petrocellis L (1998) Endocannabinoids: endogenous cannabinoid receptor ligands with neuromodulatory action. Trends Neurosci 21:521–528

Dickinson T, Fleetwood-Walker SM (1999) VIP and PACAP: very important in pain? Trends Pharmacol Sci 20:324–329

Dinh TP, Carpenter D, Leslie FM, Freund TF, Katona I, Sensi SL, Kathuria S, Piomelli D (2002) Brain monoglyceride lipase participating in endocannabinoid inactivation. Proc Natl Acad Sci U S A 99:10819–10824

Dinulescu DM, Cone RD (2000) Agouti and agouti-related protein: analogies and contrasts. J Biol Chem 275:6695–6698

Dirig DM, Yaksh TL (1995) Intrathecal baclofen and muscimol, but not midazolam, are antinociceptive using the rat-formalin model. J Pharmacol Exp Ther 275:219–227

Dirig DM, Yaksh TL (1999) In vitro prostanoid release from spinal cord following peripheral inflammaton: effects of substance P, NMDA and capsaicin. Br J Pharmacol 126: 1333–1340

Distrutti E, Sediari L, Mencarelli A, Renga B, Orlandi S, Antonelli E, Roviezzo F, Morelli A, Cirino G, Wallace JL, Fiorucci S (2006a) Evidence that hydrogen sulfide exerts antinociceptive effects in the gastrointestinal tract by activating K-ATP channels. J Pharmacol Exp Ther 316:325–335

Distrutti E, Sediari L, Mencarelli A, Renga B, Orlandi S, Russo G, Caliendo G, Santagada V, Cirino G, Wallace JL, Fiorucci S (2006b) 5-amino-2-hydroxybenzoic acid 4-(5-thioxo-5H-[1,2]dithiol-3yl)-phenyl ester (ATB-429), a hydrogen sulfide-releasing derivative of mesalamine, exerts antinociceptive effects in a model of postinflammatory hypersensitivity. J Pharmacol Exp Ther 319:447–458

Dixit VD, Schaffer EM, Pyle RS, Collins GD, Sakthivel SK, Palaniappan R, Lillard JW, Taub DD (2004) Ghrelin inhibits leptin- and activation-induced proinflammatory cytokine expression by human monocytes and T cells. J Clin Invest 114:57–66

Dmitrieva N, Shelton D, Rice ASC, McMahon SB (1997) The role of nerve growth factor in a model of visceral inflammation. Neuroscience 78:449–459

Doak GJ, Sawynok J (1997) Formalin-induced nociceptive behavior and edema: involvement of multiple peripheral 5-hydroxytryptamine receptor subtypes. Neuroscience 80:939–949

Dobner PR (2006) Neurotensin and pain modulation. Peptides 27:2405–2414

Dobolyi A, Ueda H, Uchida H, Palkovits M, Usdin TB (2002) Anatomical and physiological evidence for involvement of tuberoinfundibular peptide of 39 residues in nociception. Proc Natl Acad Sci U S A 99:1651–1656

Dogrul A, Gul H, Akar A, Yildiz O, Bilgin F, Guzeldemir E (2003) Topical cannabinoid antinociception: synergy with spinal sites. Pain 105:11–16

Dogrul A, Seyrek M (2006) Systemic morphine produce antinociception mediated by spinal 5-HT7, but not 5-HT1A and 5-HT2 receptors in the spinal cord. Br J Pharmacol 149: 498–505

Donnerer J, Schuligoi R, Stein C (1992) Increased content and transport of substance P and calcitonin gene-related peptide in sensory nerves innervating inflamed tissue: evidence for a regulatory function of nerve growth factor in vivo. Neuroscience 49:693–698

Dray A (1995) Inflammatory mediators of pain. Br J Anaesth 75:125–131

Drew GM, Siddall PJ, Duggan AW (2004) Mechanical allodynia following contusion injury of the rat spinal cord is associated with loss of GABAergic inhibition in the dorsal horn. Pain 109:379–388

Duarte IDG, Dos Santos IR, Lorenzetti BB, Ferreira SH (1992) Analgesia by direct antagonism of nociceptor sensitization involves the arginine-nitric oxide-cGMP pathway. Eur J Pharmacol 217:225–227

Dubocovich ML, Rivera-Bermudez MA, Gerdin MJ, Masana MI (2003) Molecular pharmacology, regulation and function of mammalian melatonin receptors. Front Biosci 8:1093–1108

Dubois A, Savasta M, Curet O, Scatton B (1986) Autoradiographic distribution of the D1 agonist [H-3] Skf 38393, in the rat-brain and spinal cord - comparison with the distribution of D2 dopamine-receptors. Neuroscience 19:125–137

Dubuc I, Sarret P, Labbe-Jullie C, Botto JM, Honore E, Bourdel E, Martinez J, Costentin J, Vincent JP, Kitabgi P, Mazella J (1999) Identification of the receptor subtype involved in the analgesic effect of neurotensin. J Neurosci 19:503–510

Dumont Y, Martel J-C, Fournier A, St-Pierre S, Quirion R (1992) Neuropeptide Y and neuropeptide Y receptor subtypes in brain and peripheral tissues. Prog Neurobiol 38:125–167

Duncan M, Millns P, Smart D, Wright JE, Kendall DA, Ralevic V (2004) Noladin ether, a putative endocannabinoid, attenuates sensory neurotransmission in the rat isolated mesenteric arterial bed via a non-CB1/CB(2)G(i/o) linked receptor. Br J Pharmacol 142:509–518

Duttaroy A, Gomeza J, Gan J-W, Siddiqui N, Basile AS, Harman D, Smith PL, Felder CC, Levey AI, Wess J (2002) Evaluation of muscarinic agonist-induced analgesia in muscarinic acetylcholine receptor knockout mice. Mol Pharmacol 62:1084–1093

Dworkin RH (1994) Pain insensitivity in schizophrenia : a neglected phenomenon and some implications. Schizophr Bull 20:235–248

Eaton MJ, Blits B, Ruitenberg MJ, Verhaagen J, Oudega M (2002) Amelioration of chronic neuropathic pain after nerve injury by adeno-associated viral (AAV) vector-mediated overexpresion of BDNF in the rat spinal cord. Gene Ther 9:1387–1395

Eaton MJ, Karmally S, Martinez MA, Plunkett JA, Lopez T, Cejas PJ (1999) Lumbar transplant of neurons genetically modified to secrete galanin reverse pain-like behaviors after partial sciatic nerve injury. J Peripher Nerv Syst 4:245–257

Egan TM, Henderson G, North RA, Williams JT (1983) Noradrenaline-mediated synaptic inhibition in rat locus coeruleus neurones. J Physiol (Lond) 345:477–488

Eguchi N, Minami T, Shirafuji N, Kanaoka Y, Tanaka T, Nagata A, Yoshida N, Urade Y, Ito S, Hayaishi O (1999) Lack of tactile pain (allodynia) in lipocalin-type prostaglandin D synthase-deficient mice. Proc Natl Acad Sci U S A 96:726–730

Eisenach JC, De Kock M, Klimscha W (1996a) Alpha2-adrenergic agonists for regional anesthesia. A clinical review of clonidine (1984–1995). Anesthesiology 85:655–674

Eisenach JC, Detweiler DJ, Tong C, D'Angelo R, Hood DD (1996b) Cerebrospinal fluid norepinephrine and acetylcholine concentrations during acute pain. Anesth Analg 82: 621–626

Eisenach JC, Gebhart GF (1995) Intrathecal amitriptyline - antinociceptive interactions with intravenous morphine and intrathecal clonidine, neostigmine and carbamylcholine in rats. Anesthesiology 83:1036–1045

Eisenach JC, Hood DD, Curry R (2002) Preliminary efficacy assessment of intrathecal injection of an American formulation of adenosine in humans. Anesthesiology 96:29–34

El Shenawy SM, Abdel-Salam OM, Baiuomy AR, El-Batran S, Arbid MS (2002) Studies on the anti-inflammatory and anti-nociceptive effects of melatonin in the rat. Pharmacol Res 46: 235–243

Ellington HC, Cotter MA, Cameron NA, Ross RA (2002) The effect of cannabinoids on capsaicin-evoked calcitonin gene-related peptide (CGRP) release from the isolated paw skin of diabetic and non-diabetic rats. Neuropharmacology 42:966–975

Endres-Becker J, Heppenstall PA, Mousa SA, Labuz D, Oksche A, Schafer M, Stein C, Zollner C (2007) Mu-opioid receptor activation modulates transient receptor potential vanilloid 1

(TRPV1) currents in sensory neurons in a model of inflammatory pain. Mol Pharmacol 71:12–18

Engstrom M, Brandt A, Wurster S, Savola JM, Panula P (2003) Prolactin releasing peptide has high affinity and efficacy at neuropeptide FF2 receptors. J Pharmacol Exp Ther 305:825–832

Enna SJ, McCarson KE (2006) The role of GABA in the mediation and perception of pain. Adv Pharmacol 54:1–27

Faber ESL, Chamber JP, Evans RH (1998) Depression of NMDA receptor-mediated synaptic transmission by four alpha2-adrenoceptor agonists on the in vitro rat spinal cord preparation. Br J Pharmacol 124:507–512

Fairbanks CA, Schreiber KL, Brewer KL, Yu C-G, Stone LS, Kitto KF, Nguyen HO, Grocholski BM, Shoeman DW, Kehl LJ, Regunathan S, Reis DJ, Yezierski RP, Wilcox GL (2000) Agmatine reverses pain induced by inflammation, neuropathy and spinal cord injury. Proc Natl Acad Sci U S A 97:10584–10589

Farrell EK, Merkler DJ (2008) Biosynthesis, degradation and pharmacological importance of the fatty acid amides. Drug Discov Today 13:558–568

Farzin D, Asghari L, Nowrouzi M (2002) Rodent antinociception following acute treatment with different histamine receptor agonists and antagonists. Pharmacol Biochem Behav 72:751–760

Fedorova I, Hashimoto A, Fecik RA, Hedrick MP, Hanus LO, Boger DL, Rice KC, Basile AS (2001) Behavioral evidence for the interaction of oleamide with multiple neurotransmitter systems. J Pharmacol Exp Ther 299:332–342

Fernandes JC, Martel-Pelletier J, Pelletier JP (2002) The role of cytokines in osteoarthritis pathophysiology. Biorheology 39:237–246

Ferreira SH, Cunha FQ, Lorenzetti BB, Michelin MA, Perretti M, Flower RJ, Poole S (1997) Role of lipocortin-1 in the anti-hyperalgesic actions of dexamethasone. Br J Pharmacol 121:883–888

Ferreira SH, Duarte IDG, Lorenzetti BB (1991) The molecular mechanism of action of peripheral morphine analgesia: stimulation of the cGMP system via nitric oxide release. Eur J Pharmacol 201:121–122

Ferreira SH, Lorenzetti BB, Faccioli LH (1992) Blockade of hyperalgesia and neurogenic oedema by topical application of nitroglycerin. Eur J Pharmacol 217:207–209

Fields H (2004) State-dependent opioid control of pain. Nat Rev Neurosci 5:565–575

Fields HL, Basbaum AI (1999) Central nervous system mechanisms of pain modulation. In: Wall, PD Melzack, R(eds) Textbook of pain, 4th edn. Churchill Livingstone, Edinburgh, pp 309–329

Fields HL, Heinricher MM, Mason P (1991) Neurotransmitters in nociceptive modulatory circuits. Annu Rev Neurosci 14:219–245

Finnegan TF, Chen SR, Pan HL (2005) Effect of the mu opioid on excitatory and inhibitory synaptic inputs to periaqueductal gray-projecting neurons in the amygdala. J Pharmacol Exp Ther 312:441–448

Finnegan TF, Chen SR, Pan HL (2006) Mu opioid receptor activation inhibits GABAergic inputs to basolateral amygdala neurons through Kv1.1/1.2 channels. J. Neurophysiol 95:2032–2041

Fiorentino DF, Zlotnik A, Mosmann TR, Howard M, Ogarra A (1991) Il-10 inhibits cytokine production by activated macrophages. J Immunol 147:3815–3822

Fleetwood-Walker SM, Hope PJ, Mitchell R (1988) Antinociceptive actions of descending dopaminergic tracts on cat and rat dorsal horn somatosensory neurons. J Physiol 399:335–348

Frances B, Lahlou H, Zajac J-M (2001) Cholera and pertussis toxin inhibit differently hypothermic and anti-opioid effects of neuropeptide FF. Regul Pept 98:13–18

Franek M, Vaculin S, Rokyta R (2004) GABA(B) receptor agonist baclofen has non-specific antinociceptive effect in the model of peripheral neuropathy in the rat. Physiol Res 53:351–355

Fredholm BB, IJzerman AP, Jacobson KA, Klotz K-N, Linden J (2001) International union of pharmacology. XXV. nomenclature and classification of adenosine receptors. Pharmacol Rev 53:527–552

Frizzo ME, Antunes Soares FA, Dall'Onder LP, Lara DR, Swanson RA, Souza DO (2003) Extracellular conversion of guanine-based purines to guanosine specifically enhances astrocyte glutamate uptake. Brain Res 972:84–89

Fromm C, DeLeo J, Coombs DW, Colburn RW, Twitchell BB (1993) The ganglioside GM1 decrease autotomy but not substance P depletion in a peripheral mononeuropathy rat model. Anesth Analg 77:501–506

Fu CY, Zhao YL, Dong L, Chen Q, Ni JM, Wang R (2008) In vivo characterization of the effects of human hemokinin-1 and human hemokinin-1(4–11), mammalian tachykinin peptides, on the modulation of pain in mice. Brain Behav Immun 22:850–860

Fu J, Gaetani S, Oveisi F, Lo Verme J, Serrano A, De Fonseca FR, Rosengarth A, Luecke H, Di Giacomo B, Tarzia G, Piomelli D (2003) Oleylethanolamide regulates feeding and body weight through activation of the nuclear receptor PPAR-alpha. Nature 425:90–93

Fuji K, Senba E, Ueda Y, Tohyama M (1983) Vasoactive intestinal polipeptide /VIP/-containing neurons in the spinal cord of the rat and their projections. Neurosci Lett 37:51–55

Fujimoto J, Arts KS, Rady J, Tseng L (1990) Intracerebroventricular analgesia enhanced by intrathecal (IT) dynorphin A (1–17) antibody. Prog Clin Biol Res 328:433–436

Fukui M, Nakagawa T, Minami M, Satoh M (2001) Antinociceptive effects of intracerebroventricularly administered P2 purinoceptor agonists in the rat. Eur J Pharmacol 419:25–31

Fukui M, Nakagawa T, Minami M, Satoh M, Kaneko S (2006) Inhibitory role of supraspinal P2X3/P2X2/3 subtypes on nociception in rats. Mol Pain 2:19

Fukui M, Takishita A, Zhang NN, Nakagawa T, Minami M, Satoh M (2004) Involvement of locus coeruleus noradrenergic neurons in supraspinal antinociception by alpha,beta-methylene-ATP in rats. J Pharmacol Sci 94:153–160

Fukusumi S, Fujii R, Hinuma S (2006) Recent advances in mammalian RFamide peptides: the discovery and functional analyses of PrRP, RFRPs and QRFP. Peptides 27:1073–1086

Furst S (1999) Transmitters involved in antinociception in the spinal cord. Brain Res Bull 48: 129–141

Gagnier JJ (2001) The therapeutic potential of melatonin in migraines and other headaches types. Altern Med Rev 6:383–389

Gamse R, Leeman SE, Holzer P, Lembeck F (1981) Differential effects of capsaicin on the content of somatostatin, substance P, and neurotensin in the nervous system of the rat. Naunyn Schmiedebergs Arch Pharmacol 317:140–148

Ganong AH, Lanthorn TH, Cotman CW (1983) Kynurenic acid inhibits synaptic and acidic amino acid-induced responses in the rat hippocampus and spinal cord. Brain Res 273: 170–174

Gao L, Yu LC (2004) Involvement of opioid receptors in the oxytocin-induced antinociception in the central nervous system of rats. Regul Pept 120:53–58

Gardell LR, Ibrahim M, Wang R, Wang Z, Ossipov MH, Malan TP Jr, Porreca F, Lai J (2004) Mouse strains that lack spinal dynorphin upregulation after peripheral nerve injury do not develop neuropathic pain. Neuroscience 123:43–52

Garraway SM, Anderson AJ, Mendell LM (2005) BDNF-induced facilitation of afferent-evoked responses in lamina II neurons is reduced after neonatal spinal cord contusion injury. J Neurophysiol 94:1798–1804

Gauldie SD, McQueen DS, Pertwee R, Chessell IP (2001) Anandamide activates peripheral nociceptors in normal and arthritic rat knee joints. Br J Pharmacol 132:617–621

Gaumann DM, Brunet PC, Jirounek P (1992) Clonidine enhances the effects of lidocaine on C-fiber action potential. Anesth Analg 74:719–725

Gear RW, Aley KO, Levine JD (1999) Pain-induced analgesia mediated by mesolimbic reward circuits. J Neurosci 19:7175–7181

Genzen JR, McGehee DS (2005) Nicotinic modulation of GABAergic synaptic transmission in the spinal cord dorsal horn. Brain Res 1031:229–237

Gerevich Z, Borvendeg SJ, Schroder W, Franke H, Wirkner K, Norenberg W, Furst S, Gillen C, Illes P (2004) Inhibition of N-type voltage-activated calcium channels in rat dorsal root ganglion neurons by P2Y receptors is a possible mechanism of ADP-induced analgesia. J Neurosci 24:797–807

Gerevich Z, Illes P (2004) P2Y receptors and pain transmission. Purinergic Signal 1:3–10

Gergen KA, Zadina JE, Kastin AJ, Paul D (1996) Intrathecal Tyr-W-MIF-1 produces potent, naloxone-reversible analgesia modulated by alpha 2-adrenoceptors. Eur J Pharmacol 298: 235–239

Gibbs JL, Flores CM, Hargreaves KM (2006a) Attenuation of capsaicin-evoked mechanical allodynia by peripheral neuropeptide Y Y1 receptors. Pain 124:167–174

Gibbs JL, Hargreaves KM (2008) Neuropeptide Y Y1 receptor effects on pulpal nociceptors. J Dent Res 87:948–952

Gibbs TT, Russek SJ, Farb DH (2006b) Sulfated steroids as endogenous neuromodulators. Pharmacol Biochem Behav 84:555–567

Gibson DA, Harris BR, Rogers DT, Littleton JM (2002) Radioligand binding studies reveal agmatine is a more selective antagonist for a polyamine-site on the NMDA receptor than arcaine or ifenprodil. Brain Res 952:71–77

Gilbert AK, Franklin KBJ (2001) GABAergic modulation of descending inhibitory systems from the rostral ventromedial medulla (RVM). Dose-response analysis of nociception and neurological deficits. Pain 90:25–36

Gilbert SD, Clark TM, Flores CM (2001) Antihyperalgesic activity of epibatidine in the formalin model of facial pain. Pain 89:159–165

Gimpl G, Fahrenholz F (2001) The oxytocin receptor system: structure, function, and regulation. Physiol Rev 81:629–683

Giordano J, Rogers L (1989) Peripherally administered serotonin 5-HT3 receptor antagonists reduce inflammatory pain in rats. Eur J Pharmacol 170:83–86

Glick SD, Crane LA (1978) Opiate-like and abstinence-like effects of intracerebral histamine administration in rats. Nature 273:547–549

Goettl VM, Lindsey AE, Neff NH, Hadjiconstantinou M (2000) GM1 ganglioside restores abnormal responses to acute thermal and mechanical stimuli in aged rats. Brain Res 858:380–385

Gomes JA, Li X, Pan H-L, Eisenach JC (1999) Intrathecal adenosine interacts with a spinal noradrenergic system to produce antinociception in nerve-injured rats. Anesthesiology 91: 1072–1079

Gomeza J, Shannon H, Kostenis E, Felder C, Zhang L, Brodkin J, Grinberg A, Sheng H, Wess J (1999a) Pronounced pharmacologic deficits in M2 muscarinic acetylcholine receptor knockout mice. Proc Natl Acad Sci U S A 96:1692–1697

Gomeza J, Zhang L, Kostenis E, Felder C, Bymaster F, Brodkin J, Shannon H, Xia B, Deng C-X, Wess J (1999b) Enhancement of D1 dopamine receptor-mediated locomotor stimulation in M_4 muscarinic acetylcholine receptor knockout mice. Proc Natl Acad Sci U S A 96:10483–10488

Gouarderes C, Sutak M, Zajac J-M, Jhamandas K (2000) Role of adenosine in the spinal antinociceptive and morphine modulatory actions of neuropeptide FF analogs. Eur J Pharmacol 406:391–401

Goudet C, Chapuy E, Alloui A, Acher F, Pin JP, Eschalier A (2008) Group III metabotropic glutamate receptors inhibit hyperalgesia in animal models of inflammation and neuropathic pain. Pain 137:112–124

Grass S, Crawley JN, Xu XJ, Wiesenfeld-Hallin Z (2003a) Reduced spinal cord sensitization to C-fibre stimulation in mice over-expressing galanin. Eur J Neurosci 17:1829–1832

Grass S, Jacoby AS, Iismaa TP, Crawley JN, Xu X-J, Wiesenfeld-Hallin Z (2003b) Flexor reflex excitability in mice lacking galanin receptor galanin-R1. Neurosci Lett 345:153–156

Grazia Cascio M, Minassi A, Ligresti A, Appendino G, Burstein S, Di Marzo V (2004) A structure-activity relationship study on N-arachidonoyl-amino acids as possible endogenous inhibitors of fatty acid amide hydrolase. Biochem Biophys Res Commun 314:192–196

Green PG, Strausbaugh HJ, Levine JD (1998) Annexin 1 is a local mediator in neural-endocrine feedback control of inflammation. J Neurophysiol 80:3120–3126

Grill M, Heinemann A, Hoefler G, Peskar BA, Schuligoi R (2008) Effect of endotoxin treatment on the expression and localization of spinal cyclooxygenase, prostaglandin synthases, and PGD(2) receptors. J Neurochem 104:1345–1357

Guan Y, Terayama R, Dubner R, Ren K (2002) Plasticity in excitatory amino acid receptor-mediated descending pain modulation after inflammation. J Pharmacol Exp Ther 300:513–520

Gue M, Del Rio-Lacheze C, Eutamene H, Theodorou V, Fioramonti J, Bueno L (1997) Stress-induced visceral hypersensitivity to rectal distension in rats: role of CRF and mast cells. Neurogastroenterol Motil 9:271–279

Gui X, Carraway RE, Dobner PR (2004) Endogenous neurotensin facilitates visceral nociception and is required for stress-induced antinociception in mice and rats. Neuroscience 126: 1023–1032

Guimaraes AP, Guimaraes FS, Prado WA (2000) Modulation of carbachol-induced antinociception from the rat periaqueductal gray. Brain Res Bull 51:471–478

Guindon J, Desroches J, Beaulieu P (2007) The antinociceptive effects of intraplantar injections of 2-arachidonoyl glycerol are mediated by cannabinoid CB2 receptors. Br J Pharmacol 150: 693–701

Guindon J, Hohmann AG (2007) Cannabinoid CB2 receptors: a therapeutic target for the treatment of inflammatory and neuropathic pain. Br J Pharmacol 153:319–334

Guindon J, Lo Verme J, De Lean A, Piomelli D, Beaulieu P (2006a) Synergistic antinociceptive effects of anandamide, an endocannabinoid, and nonsteroidal anti-inflammatory drugs in peripheral tissue: a role for endogenous fatty-acid ethanolamides? Eur J Pharmacol 550:68–77

Guindon J, De Lean A, Beaulieu P (2006b) Local interactions between anandamide, an endocannabinoid, and ibuprofen, a nonsteroidal anti-inflammatory drug, in acute and inflammatory pain. Pain 121:85–93

Guitart X, Codony X, Monroy X (2004) Sigma receptors: biology and therapeutic potential. Psychopharmacology 174:301–319

Gulati A, Srimal RC (1992) Endothelin mechanisms in the central nervous system: a target for drug development. Drug Dev Res 26:361–387

Guneli E, Kazikdas KC, Kolatan E (2007) Ghrelin may attenuate proinflammatory cytokine-mediated neuropathic pain. Med Hypotheses 69:356–360

Guo H, Zhao Z-Q (2000) Inhibition of nociceptive withdrawal reflex by microinjection of interleukin 2 into rat locus coeruleus. Neuropeptides 34:216–220

Guo TZ, Jiang JY, Buttermann AE, Maze M (1996) Dexmedetomidine injection into the locus ceruleus produces antinociception. Anesthesiology 84:873–881

Haas H, Panula P (2003) The role of histamine and the tuberomamillary nucleus in the nervous system. Nat Rev Neurosci 4:121–130

Haberberger RV, Bernardini N, Kress M, Hartmann P, Lips KS, Kummer W (2004) Nicotinic acetylcholine receptor subtypes in nociceptive dorsal root ganglion neurons of the adult rat. Auton Neurosci 113:32–42

Hajos M, Engberg G (1990) Kynurenic acid blocks chemogenic nociception. J Pharm Pharmacol 42:373–374

Hajos M, Jancso G, Engberg G (1987) Capsaicin-induced excitation of locus coeruleus neurons. Acta Physiol Scand 129:415–420

Hamon M, Bourgoin S (1999) Serotonin and its receptors in pain control. In: Sawynok, J Cowan, A (eds) Novel aspects of pain management: opioids and beyond. Wiley-Liss Inc, New York, NY, pp 203–228

Hansen HS, Artmann A (2008) Endocannabinoids and nutrition. J Neuroendocrinol 20:94–99

Hansson HA, Rozell B, Skottner A (1987) Rapid axoplasmic transport of insulin-like growth factor I in the sciatic nerve af adult rats. Cell Tissue Res 247:241–247

Hanus L, Abu-Lafi S, Fride E, Breuer A, Vogel Z, Shalev DE, Kustanovich I, Mechoulam R (2001) 2-Arachidonyl glyceryl ether, an endogenous agonist of the cannabinoid CB1 receptor. Proc Natl Acad Sci U S A 98:3662–3665

Hao SL, Takahata O, Iwasaki H (1999) Isobolographic analysis of interaction between spinal endomorphin-1, a newly isolated endogenous opioid peptide, and lidocaine in the rat formalin test. Neurosci Lett 276:177–180

Hart PH, Vitti GF, Burgess DR, Whitty GA, Piccoli DS, Hamilton JA (1989) Potential antiinflammatory effects of interleukin-4 - suppression of human monocyte tumor necrosis factor-alpha, interleukin-1, and prostaglandin-E2. Proc Natl Acad Sci U S A 86:3803–3807

Harte SE, Hoot MR, Borszcz GS (2004) Involvement of the intralaminar parafascicular nucleus in muscarinic-induced antinociception in rats. Brain Res 1019:152–161

Harvey RJ, Depner UB, Wassle H, Ahmadi S, Heindl C, Reinold H, Smart TG, Harvey K, Schutz B, Abo-Salem OM, Zimmer A, Poisbeau P, Welzl H, Wolfer DP, Betz H, Zeilhofer HU, Muller U (2004) GlyR alpha 3: an essential target for spinal PGE(2)-mediated inflammatory pain sensitization. Science 304:884–887

Harvey S, Hayer S (1993) Parathyroid hormone binding sites in the brain. Peptides 14:1187–1191

Hata AN, Zent R, Breyer MD, Breyer RM (2003) Expression and molecular pharmacology of the mouse CRTH2 receptor. J Pharmacol Exp Ther 306:463–470

Haulica I, Neamtu C, Stratone A, Petrescu G, Branisteanu D, Rosca V, Slatineanu S (1986) Evidence for the involvement of cerebral renin-angiotensin system (RAS) in stress analgesia. Pain 27:237–245

Haws CM, Heinricher MM, Fields HL (1990) Alpha-adrenergic receptor agonists, but not antagonists, alter the tail-flick latency when microinjected into the rostral ventromedial medulla of the lightly anesthetized rat. Brain Res 533:192–195

Hayes RL, Mao J, Price DD, Germano A, D'Avella D, Fiori M, Mayer DJ (1992) Pretreatment with gangliosides reduces abnormal nociceptive responses associated with a rodent peripheral neuropathy. Pain 48:391–396

Heimann AS, Gomes L, Dale CS, Pagano RL, Gupta A, de Souza LL, Luchessi AD, Castro LM, Giorgi R, Rioli V, Ferro ES, Devi LA (2007) Hemopressin is an inverse agonist of CB1 cannabinoid receptors. Proc Natl Acad Sci U S A 104:20588–20593

Heinricher MM, McGaraughty S (1998) Analysis of excitatory amino acid transmission within the rostral ventromedial medulla: implications for circuitry. Pain 75:247–255

Heinricher MM, McGaraughty S, Farr DA (1999) The role of excitatory amino acid transmission within the rostral ventromedial medulla in the antinociceptive actions of systemically administered morphine. Pain 81:57–65

Hejazi N, Zhou C, Oz M, Sun H, Ye JH, Zhang L (2006) Delta9-tetrahydrocannabinol and endogenous cannabinoid anandamide directly potentiate the function of glycine receptors. Mol Pharmacol 69:991–997

Helyes Z, Pozsgai G, Borzsei R, Nemeth J, Bagoly T, Mark L, Pinter E, Toth G, Elekes K, Szolcsanyi J, Reglodi D (2007) Inhibitory effect of PACAP-38 on acute neurogenic and non-neurogenic inflammatory processes in the rat. Peptides 28:1847–1855

Helyes Z, Than M, Oroszi G, Pinter E, Nemeth J, Keri S, Szolcsanyi J (2000) Anti-nociceptive effect induced by somatostatin released from sensory nerve teminals and by synthetic somatostatin analogues in the rat. Neurosci Lett 278:185–188

Hermanns H, Muth-Selbach U, Williams R, Krug S, Lipfert P, Werdehausen R, Braun S, Bauer I (2008) Differential effects of spinally applied glycine transporter inhibitors on nociception in a rat model of neuropathic pain. Neurosci Lett 445:214–219

Hernandez-Pacheco A, Araiza-Saldana CI, Granados-Soto V, Mixcoatl-Zecuatl T (2008) Possible participation of the nitric oxide-cyclic GMP-protein kinase G-K+ channels pathway in the peripheral antinociception of melatonin. Eur J Pharmacol 596:70–76

Hilmas C, Pereira EF, Alkondon M, Rassoulpour A, Schwarcz R, Albuquerque EX (2001) The brain metabolite kynurenic acid inhibits alpha7 nicotinic receptor activity and increases non-alpha7 nicotinic receptor expression: physiopathological implications. J Neurosci 12: 7463–7473

Hobson SA, Holmes FE, Kerr NCH, Pope RJP, Wynick D (2006) Mice deficient for galanin receptor 2 have decreased neurite outgrowth from adult sensory neurons and impaired pain-like behaviour. J Neurochem 99:1000–1010

Hohmann AG (2002) Spinal and peripheral mechanisms of cannabinoid antinociception: behavioral, neurophysiological and neuroanatomical perspectives. Chem Phys Lipids 121:173–190

Hohmann AG (2007) Inhibitors of monoacylglycerol lipase as novel analgesics. Br J Pharmacol 150:673–675

Hohmann AG, Suplita RL, Bolton NM, Neely MH, Fegley D, Mangieri R, Krey JF, Walker JM, Holmes PV, Crystal JD, Duranti A, Tontini A, Mor M, Tarzia G, Piomelli D (2005) An enocannabinoid mechanism for stress-induced analgesia. Nature 435:1108–1112

Holden JE, Pizzi JA (2008) Lateral hypothalamic-induced antinociception may be mediated by a substance P connection with the rostral ventromedial medulla. Brain Res 1214:40–49

Holden JE, Van Poppel AY, Thomas S (2002) Antinociception from lateral hypothalamic stimulation may be mediated by NK(1) receptors in the A7 catecholamine cell group in rat. Brain Res 953:195–204

Holland PR, Akerman S, Goadsby PJ (2005) Orexin 1 receptor activation attenuates neurogenic dural vasodilation in an animal model of trigeminovascular nociception. J Pharmacol Exp Ther 315:1380–1385

Holmberg K, Kuteeva E, Brumovsky P, Kahl U, Karlstrom H, Lucas GA, Rodriguez J, Westerblad H, Hilke S, Theodorsson E, Berge OG, Lendahl U, Bartfai T, Hokfelt T (2005) Generation and phenotypic characterization of a galanin overexpressing mouse. Neuroscience 133:59–77

Holmes C, Landry D, Granton J (2003) Science review: vasopressin and the cardiovascular system part 1 - receptor physiology. Crit Care 7:427–434

Holmes FE, Arnott N, Vanderplank PA, Kerr NCH, Longbrake EE, Popovich PG, Imai T, Combadiere C, Murphy PM, Wynick D (2008) Intra-neural administration of fractalkine attenuates neuropathic pain-related behaviour. J Neurochem 106:640–649

Honda M, Okutsu H, Matsuura T, Miyagi T, Yamamoto Y, Hazato T, Ono H (2001) Spinorphin, and endogenous inhibitor of enkephalin-degrading enzymes, potentiates Leu-enkephalin-induced anti-allodynic and antinociceptive effects in mice. Jpn J Pharmacol 87:261–267

Honda M, Uchida K, Tanabe M, Ono H (2006) Fluvoxamine, a selective serotonin reuptake inhibitor, exerts its antiallodynic effects on neuropathic pain in mice via 5-HT2A/2C receptors. Neuropharmacology 51:866–872

Hood DD, Mallak KA, James RL, Tuttle R, Eisenach JC (1997) Enhancement of analgesia from systemic opioid in humans by spinal cholinesterase inhibition. J Pharmacol Exp Ther 282:86–92

Hornfeldt CS, Smullin DH, Schamber CD, Sun X, Larson AA (1992) Antinociceptive effects of intrathecal taurine and calcium in the mouse. Life Sci 50:1925–1934

Horvath G (2000) Endomorphin-1 and endomorphin-2. Pharmacology of the selective endogenous mu-opioid receptor agonists. Pharmacol Ther 88:437–463

Horvath G, Agil A, Joo G, Dobos I, Benedek G, Baeyens JM (2003) Evaluation of endomorphin-1 on the activity of Na^{+},K^{+}-ATPase using in vitro and in vivo studies. Eur J Pharmacol 458: 291–297

Horvath G, Farkas I, Tuboly G, Benedek G (2008a) The antinociceptive effects of intrathecal injection of N-arachidonoyl-dopamine (NADA) are mediated by cannabinoid receptors. Clin Neorosci 61(S1):44–45

Horvath G, Kekesi G (2006) Interaction of endogenous ligands mediating antinociception. Brain Res Rev 52:69–92

Horvath G, Kekesi G, Tuboly G, Benedek G (2007) Antinociceptive interactions of triple and quadruple combinations of endogenous ligands at the spinal level. Brain Res 1155:42–48

Horvath G, Kovacs M, Szikszay M, Benedek G (1994) Mydriatic and antinociceptive effects of intrathecal dexmedetomidine in conscious rats. Eur J Pharmacol 253:61–66

Horvath G, Szikszay M, Tomboly C, Benedek G (1999) Antinociceptive effects of intrathecal endomorphin-1 and -2 in rats. Life Sci 65:2635–2641

Horvath G, Kekesi G, Nagy E, Benedek G (2008b) The role of TRPV1 receptors in the antinociceptive effect of anandamide at spinal level. Pain 134:277–284

Hou S-W, Qi J-S, Zhang Y, Qiao J-T (2003) Spinal antinociceptive effect of agmatine and tentative analysis of involved receptors: study in an electrophysiological model of rats. Brain Res 968:277–280

Houghton LA, Lea R, Jackson N, Whorwell PJ (2002) The menstrual cycle affects rectal sensitivity in patients with irritable bowel syndrome but not healthy volunteers. Gut 50:471–474

Howlett AC, Breivogel CS, Childers SR, Deadwyler SA, Hampson RE, Porrino LJ (2004) Cannabinoid physiology and pharmacology: 30 years of progress. Neuropharmacology 47:345–358

Hu LF, Wong PTH, Moore PK, Bian JS (2007) Hydrogen sulfide attenuates lipopolysaccharide-induced inflammation by inhibition of p38 mitogen-activated protein kinase in microglia. J Neurochem 100:1121–1128

Hua X-Y, Boublik JH, Spicer MA, Rivier JE, Brown MR, Yaksh TL (1991) The antinociceptive effects of spinally administered neuropeptide Y in the rat: systematic studies on structure-activity relationship. J Pharmacol Exp Ther 258:243–248

Hua XY, Hayes CS, Hofer A, Fitzsimmons B, Kilk K, Langel U, Bartfai T, Yaksh TL (2004) Galanin acts at GalR1 receptors in spinal antinociception: synergy with morphine and AP-5. J Pharmacol Exp Ther 308:574–582

Huang EJ, Reichardt LF (2003) Trk receptors: roles in neuronal signal transduction. Annu Rev Biochem 72:609–642

Huang J, Wang YY, Wang W, Li YQ, Tamamaki N, Wu SX (2008) 5-HT3A receptor subunit is expressed in a subpopulation of GABAergic and enkephalinergic neurons in the mouse dorsal spinal cord. Neurosci Lett 441:1–6

Huang SM, Bisogno T, Petros TJ, Chang SY, Zavitsanos PA, Zipkin RE, Sivakumar R, Coop A, Maeda DY, De Petrocellis L, Burstein S, Di Marzo V, Walker JM (2001) Identification of a new class of molecules, the arachidonyl amino acids, and characterization of one member that inhibits pain. J Biol Chem 276:42639–42644

Huang SM, Bisogno T, Trevisani M, Al-Hayani A, De Petrocellis L, Fezza F, Tognetto M, Petros TJ, Krey JF, Chu CJ, Miller JD, Davies SN, Geppetti P, Walker JM, Di Marzo V (2002) An endogenous capsaicin-like substance with high potency at recombinant and native vanilloid VR1 receptors. Proc Natl Acad Sci U S A 99:8400–8405

Hucho TB, Dina OA, Kuhn J, Levine JD (2006) Estrogen controls PKC epsilon-dependent mechanical hyperalgesia through direct action on nociceptive neurons. Eur J Neurosci 24:527–534

Hughes J, Smith TW, Kosterlitz HW (1975) Identification of two related pentapeptides from the brain with potent opiate agonist activity. Nature 258:577–579

Hunter JC, Atwal P, Woodruff GN, Singh L (1994) Differential modulation of kappa and mu opioid antinociception by the glycine/NMDA receptor agonist D-serine. Br J Pharmacol 112: 1002–1003

Hurley RW, Banfor P, Hammond DL (2003) Spinal pharmacology of antinociception produced by microinjection of mu or delta opioid receptor agonists in the ventromedial medulla of the rat. Neuroscience 118:789–796

Hussey MJ, Clarke GD, Ledent C, Hourani SMO, Kitchen I (2007) Reduced response to the formalin test and lowered spinal NMDA glutamate receptor binding in adenosine A2A receptor knockout mice. Pain 129:287–294

Hwang JH, Yaksh TL (1997) The effect of spinal GABA receptor agonists on tactile allodynia in a surgically-induced neuropathic pain model in the rat. Pain 70:15–22

Ibrahim MM, Porreca F, Lai J, Albrecht PJ, Rice FL, Khodorova A, Davar G, Makriyannis A, Vanderah TW, Mata HP, Malan TP Jr (2005) CB2 cannabinoid receptor activation produces antinociception by stimulating peripheral release of endogenous opioids. Proc Natl Acad Sci U S A 102:3093–3098

Illes P, Finta EP, Nieber K (1993) Neuropeptide Y potentiates via Y2-receptors the inhibitory effect of noradrenaline in rat locus cooeruleus neurones. Naunyn Schmiedebergs Arch Pharmacol 348:546–548

Inoue K, Tsuda M, Tozaki-Saitoh H (2007) Modification of neuropathic pain sensation through microglial ATP receptors. Purinergic Signal 3:311–316

Inoue M, Kawashima T, Allen RG, Ueda H (2003) Nocistatin and prepro-nociceptin/orphanin FQ 160–187 cause nociception through activation of $G_{i/o}$ in capsaicin-sensitive and of G_s in capsaicin-insensitive nociceptors, respectively. J Pharmacol Exp Ther 306: 141–146

Inoue M, Nakayamada H, Tokuyama S, Ueda H (1997) Peripheral non-opioid analgesic effects of kyotorphin in mice. Neurosci Lett 236:60–62

Inoue N, Hatanaka H (1982) Nerve growth factor induces specific enkephalin binding sites in a nerve cell line. J Biol Chem 257:9238–9241

Intondi AB, Dahlgren MN, Eilers MA, Taylor BK (2008) Intrathecal neuropeptide Y reduces behavioral and molecular markers of inflammatory or neuropathic pain. Pain 137: 352–365

Ishikawa T, Marsala M, Sakabe T, Yaksh TL (2000) Characterization of spinal amino acid release and touch-evoked allodynia produced by spinal glycine or $GABA_A$ receptor antagonist. Neuroscience 95:781–786

Issa PC, Lever IJ, Michael GJ, Bradbury EJ, Malcangio M (2001) Intrathecally delivered glial cell line-derived neurotrophic factor produces electrically evoked release of somatostatin in the dorsal horn of the spinal cord. J Neurochem 78:221–229

Ito K, Yoshikawa M, Maeda M, Jin XL, Takahashi S, Matsuda M, Tamaki R, Kobayashi H, Suzuki T, Hashimoto A (2008) Midazolam attenuates the antinociception induced by D-serine or morphine at the supraspinal level in rats. Eur J Pharmacol 586:139–144

Ito S, Okuda-Ashitaka E, Minami T (2001) Central and peripheral roles of prostaglandins in pain and their interactions with novel neuropeptides nociceptin and nocistatin. Neurosci Res 41: 299–332

Iwamoto ET, Marion L (1993) Characterization of the antinociception produced by intrathecally administered muscarinic agonists in rats. J Pharmacol Exp Ther 266:329–338

Jackson KC, St Onge EL (2003) Antidepressant pharmacotherapy: considerations for the pain clinician. Pain Pract 3:135–143

Jacquet YF (1988) The NMDA receptor: central role in pain inhibition in rat periaqueductal gray. Eur J Pharmacol 154:271–276

Jaggar SI, Hasnie FS, Sellaturay S, Rice ASC (1998) The anti-hyperalgesic actions of the cannabinoid anandamide and the putative CB2 receptor agonist palmitoylethanolamide in visceral and somatic inflammatory pain. Pain 76:189–199

Jancso G, Hokfelt T, Lundberg JM, Kiraly E, Halasz N, Nilsson G, Terenius L, Rehfeld J, Steinbusch H, Verhofstad A, Elde R, Said A, Brown M (1981) Immunohistochemical studies on the effect of capsaicin on spinal and medullary peptide and monoamine neurons using antisera to substance P, gastrin/CCK, somatostatin, VIP, enkephalin, neurotensin and 5-hydroxytryptamine. J Neurocytol 10:963–980

Jancso G, Jancso-Gabor A (1980) Effect of capsaicin on morphine analgesia- possible involvement of hypothalamic structures. Naunyn Schmiedebergs Arch Pharmacol 311: 285–288

Jasmin L, Rabkin SD, Granato A, Boudah A, Ohara PT (2003) Analgesia and hyperalgesia from GABA-mediated modulation of the cerebral cortex. Neuroscience 23:641–647

Jensen RT, Battey JF, Spindel ER, Benya RV (2008) International Union of Pharmacology. LXVIII. mammalian bombesin receptors: nomenclature, distribution, pharmacology, signaling, and functions in normal and disease states. Pharmacol Rev 60:1–42

Jensen TS, Yaksh TL (1984) Effects of an intrathecal dopamine agonist, apomorphine, on thermal and chemical evoked noxious responses in rats. Brain Res 296:285–293

Jensen TS, Yaksh TL (1989) Comparison of the antinociceptive effect of morphine and glutamate at coincidental sites in the periaqueductal gray and medial medulla in rats. Brain Res 476:1–9

Jeong CY, Choi JI, Yoon MH (2004) Roles of serotonin receptor subtypes for the antinociception of 5-HT in the spinal cord of rats. Eur J Pharmacol 502:205–211

Ji G, Neugebauer V (2008) Pro- and anti-nociceptive effects of corticotropin-releasing factor (CRF) in central amygdala neurons are mediated through different receptors. J Neurophysiol 99:1201–1212

Ji R, Zhang Q, Bedecs K, Arvidsson J, Zhang X, Xu X, Wiesenfeld-Hallin Z, Bartfai T, Hokfelt T (1994a) Galanin antisense oligonucleotides reduce galanin levels in dorsal root ganglia and induce autotomy in rats after axotomy. Proc Natl Acad Sci U S A 91:12540–12543

Ji RR, Zhang X, Wiesenfeldhallin Z, Hokfelt T (1994b) Expression of neuropeptide Y and neuropeptide Y (Y1) receptor mRNA in rat spinal cord and dorsal root ganglia following peripheral tissue inflammation. J Neurosci 14:6423–6434

Ji Y, Tang B, Traub RJ (2005) Modulatory effects of estrogen and progesterone on colorectal hyperalgesia in the rat. Pain 117:433–442

Jiang CL, Xu D, Lu CL, Wang YX, You ZD, Liu XY (2000a) Interleukin-2: structural and biological relatedness to opioid peptides. Neuroimmunomodulation 8:20–24

Jiang CL, You ZD, Lu CL, Xu D, Wang AJ, Wang YX, Liu XY (2000b) Leu-enkephalin induced by IL-2 administration mediates analgesic effect of IL-2. NeuroReport 11:1483–1485

Jiang P, Yang CX, Wang YT, Xu TL (2006) Mechanisms of modulation of pregnanolone on glycinergic response in cultured spinal dorsal horn neurons of rat. Neuroscience 141: 2041–2050

Jimenez-Andrade JM, Lundstrom L, Sollenberg UE, Langel U, Castaneda-Hernandez G, Carlton SM (2006) Activation of peripheral galanin receptors: differential effects on nociception. Pharmacol Biochem Behav 85:273–280

Jimenez-Andrade JM, Zhou S, Du J, Yamani A, Grady JJ, Castaneda-Hernandez G, Carlton SM (2004) Pro-nociceptive role of peripheral galanin in inflammatory pain. Pain 110:10–21

Joels M (1997) Steroid hormones and excitability in the mammalian brain. Front Neuroendocrinol 18:2–48

Johansson O, Hokfelt T, Pernow B, Jeffcoate SL, White N, Steinbusch HWM, Verhofstad AAJ, Emson PC, Spindel E (1981) Immunohistochemical support for three putative transmitters in one neuron: coexistence of 5-hydroxytryptamine, substance P- and thyrotropin releasing hormone-like immunoreactivity in medullary neurons projecting to the spinal cord. Neuroscience 6:1857–1881

Johnston IN, Milligan ED, Wieseler-Frank J, Frank MG, Zapata V, Campisi J, Langer S, Martin D, Green P, Fleshner M, Leinwand L, Maier SF, Watkins LR (2004) A role for proinflammatory cytokines and fractalkine in analgesia, tolerance, and subsequent pain facilitation induced by chronic intrathecal morphine. J Neurosci 24:7353–7365

Jones PG, Dunlop J (2007) Targeting the cholinergic system as a therapeutic strategy for the treatment of pain. Neuropharmacology 53:197–206

Jongsma H, Pettersson LME, Zhang Y, Reimer MK, Kanje M, Waldenstrom A, Sundler F, Danielsen N (2001a) Markedly reduced chronic nociceptive response in mice lacking the PAC_I receptor. NeuroReport 12:2215–2219

Jongsma WH, Danielsen N, Johnson JM, Gratto KA, Karchewski LA, Verge VM (2001b) Exogenous NT-3 and NGF differentially modulate PACAP expression in adult sensory neurons, suggesting distinct roles in injury and inflammation. Eur J Neurosci 14:267–282

Jordan BA, Cvejic S, Devi LA (2000) Opioids and their complicated receptor complex. Neuropsychopharmacology 23:S5–S18

Joseph T, Lee TL, Li C, Siau C, Nishiuchi Y, Kimura T, Tachibana S (2007) Levels of neuropeptides nocistatin, nociceptin/orphanin FQ and their precursor protein in a rat neuropathic pain model. Peptides 28:1433–1440

Kajiyama S, Kawamoto M, Shiraishi S, Gaus S, Matsunaga A, Suyama H, Yuge O (2005) Spinal orexin-1 receptors mediate anti-hyperalgesic effects of intrathecally-administered orexins in diabetic neuropathic pain model rats. Brain Res 1044:76–86

Kalamida D, Poulas K, Avramopoulou V, Fostieri E, Lagoumintzis G, Lazaridis K, Sideri A, Zouridakis M, Tzartos SJ (2007) Muscle and neuronal nicotinic acetylcholine receptors - Structure, function and pathogenicity. FEBS J 274:3799–3845

Kalliomaki J, Pertovaara A, Brandt A, Wei H, Pietila P, Kalmari J, Xu M, Kalso E, Panula P (2004) Prolactin-releasing peptide affects pain, allodynia and autonomic reflexes through medullary mechanisms. Neuropharmacology 46:412–424

Kalso EA, Poyhia R, Rosenberg PH (1991) Spinal antinociception by dexmedetomidine, a highly selective alpha2- adrenergic agonist. Pharmacol Toxicol 68:140–143

Kamei J, Hitosugi H, Kawashima N, Misawa M (1993a) Effects of diabetes on the antinociceptive effects of intrathecally administered endothelin-1. Res Commun Chem Pathol Pharmacol 79:209–217

Kamei J, Hitosugi H, Kawashima N, Misawa M, Kasuya Y (1993b) Antinociceptive effects of intrathecally administered endothelin-1 in mice. Neurosci Lett 153:69–72

Kanjhan R, Housley GD, Burton LD, Christie DL, Kippenberger A, Thorne PR, Luo L, Ryan AF (1999) Distribution of the P2X(2) receptor subunit of the ATP-gated ion channels in the rat central nervous system. J Comp Neurol 407:11–32

Karper PE, Nazarian A, Crawford CA, Drago J, McDougall SA (2000) Role of dopamine D(1) receptors for kappa-opioid-mediated locomotor activity and antinociception during the preweaning period: a study using D(1) receptor knockout mice. Physiol Behav 68: 585–590

Kastin AJ, Stephens E, Ehrensing RH, Fischman AJ (1984) Tyr-MIF-1 acts as an opiate antagonist in the tail-flick test. Pharmacol Biochem Behav 21:937–941

Kavaliers M (1987) MIF-1 and Tyr-MIF-1 antagonize morphine and opioid but not non-opioid stress-induced analgesia in the snail, Cepaea nemoralis. Peptides 8:1–5

Kavaliers M, Wiebe JP (1987) Analgesic effects of the progesterone metabolite, 3alpha-hydroxy-5alpha- 20-one, and possible modes of action in mice. Brain Res 415:393–398

Kawabata A, Ishiki T, Nagasawa K, Yoshida S, Maeda Y, Takahashi T, Sekiguchi F, Wada T, Ichida S, Nishikawa H (2007) Hydrogen sulfide as a novel nociceptive messenger. Pain 132:74–81

Kawabata A, Manabe S, Manabe Y, Takagi H (1994a) Effect of topical administration of L-arginine on formalin -induced nociception in the mouse: a dual role of peripherally formed NO in pain modulation. Br J Pharmacol 112:547–550

Kawabata A, Manabe S, Takagi H (1994b) Comparison of antinociception induced by supraspinally administered L-arginine and kyotorphin. Br J Pharmacol 112:817–822

Kawabata A, Tanaka M, Muguruma H, Takagi H (1995) NG-nitro-L-arginine methyl ester and alpha-methyl-L-ornithine inhibit kyotorphin synthetase from rat brain. Peptides 16: 1317–1319

Kawabata A, Umeda N, Takagi H (1992) L-Arginine exerts a dual role in nociceptive processing in the brain: involvement of the kyotorphin-Met-enkephalin pathway and NO-cyclic GMP pathway. Br J Pharmacol 109:73–79

Kawamata T, Omote K, Toriyabe M, Kawamata M, Namiki A (2002) Intracerebroventricular morphine produces antinociception by evoking gamma-aminobutyric acid release through activation of 5-hydroxytryptamine 3 receptors in the spinal cord. Anesthesiology 96:1175–1182

Kayser V, Berkley K, Keita H, Gautron M, Guilbaud G (1996) Estrous and sex variations in vocalization thresholds to hindpaw and tail pressure stimulation in the rat. Brain Res 742:352–354

Kedzierski RM, Yanagisawa M (2001) Endothelin system: the double-edged sword in health and disease. Annu Rev Pharmacol Toxicol 41:851–876

Kekesi G, Dobos I, Benedek G, Horvath G (2004) The antinociceptive potencies and interactions of endogenous ligands during continuous intrathecal administration: adenosine, agmatine, and endomorphin-1. Anesth Analg 98:420–426

Kekesi G, Joo G, Csullog E, Dobos I, Klimscha W, Toth K, Benedek G, Horvath G (2002) The antinociceptive effect of intrathecal kynurenic acid and its interaction with endomorphin-1 in rats. Eur J Pharmacol 445:93–96

Keller AF, Breton JD, Schlichter R, Poisbeau P (2004) Production of 5alpha-reduced neurosteroids is developmentally regulated and shapes GABA(A) miniature IPSCs in lamina II of the spinal cord. J Neurosci 24:907–915

Kerr BJ, Bradbury EJ, Bennett DLH, Trivedi PM, Dassan P, French J, Shelton DB, McMahon SB, Thompson SWN (1999) Brain-derived neurotrophic factor modulates nociceptive sensory inputs and NMDA-evoked responses in the rat spinal cord. J Neurosci 19:5138–5148

Kerr BJ, Cafferty WBJ, Gupta Y, Bacon A, Wynick D, McMahon SB, Thompson SWN (2000) Galanin knockout mice reveal nociceptive deficits following peripheral nerve injury. Eur J Neurosci 12:793–802

Kerr BJ, Gupta Y, Pope R, Thompson SWN, Wynick D, McMahon SB (2001) Endogenous galanin potentiates spinal nociceptive processing following inflammation. Pain 93:267–277

Kesim M, Duman EN, Kadioglu M, Yaris E, Kalyoncu NI, Erciyes N (2005) The different roles of 5-HT2 and 5-HT3 receptors on antinociceptive effect of paroxetine in chemical stimuli in mice. J Pharmacol Sci 97:61–66

Khan I, Osaka H, Stanislaus S, Calvo RM, Deerinck T, Yaksh TL, Taylor P (2003) Nicotinic acetylcholine receptor distribution in relation to spinal neurotransmission pathways. J Comp Neurol 467:44–59

Khan IM, Buerkle H, Taylor P, Yaksh TL (1998) Nociceptive and antinociceptive responses to intrathecally administered nicotinic agonists. Neuropharmacology 37:1515–1525

Khasar SG, Burkham J, Dina OA, Brown AS, Bogen O, Alessandri-Haber N, Green PG, Reichling DB, Levine JD (2008) Stress induces a switch of intracellular signaling in sensory neurons in a model of generalized pain. J Neurosci 28:5721–5730

Khasar SG, McCarter G, Levine JD (1999) Epinephrine produces a beta-adrenergic receptor-mediated mechanical hyperalgesia and in vitro sensitization of rat nociceptors. J Neurophysiol 81:1104–1112

Khasar SG, Wang J-F, Taiwo YO, Heller PH, Green PG, Levine JD (1995) Mu-opioid agonist enhancement of prostaglandin-induced hyperalgesia in the rat: a G-protein beta gamma subunit-mediated effect? Neuroscience 67:189–195

Kiang JG, Wei ET (1985) CRF: an inhibitor of neurogenic plasma extravasation produced by saphenous nerve stimulation. Eur J Pharmacol 114:111–112

Kibaly C, Meyer L, Patte-Mensah C, Mensah-Nyagan AG (2008) Biochemical and functional evidence for the control of pain mechanisms by dehydroepiandrosterone endogenously synthesized in the spinal cord. FASEB J 22:93–104

Kim SJ, Calejesan AA, Zhuo M (2002) Activation of brainstem metabotropic glutamate receptors inhibits spinal nociception in adult rats. Pharmacol Biochem Behav 73:429–437

Kirouac GJ, Li S, Mabrouk G (2004) GABAergic projection from the ventral tegmental area and substantia nigra to the periaqueductal gray region and the dorsal raphe nucleus. J Comp Neurol 469:170–184

Kitabgi P (2002) Targeting neurotensin receptors with agonists and antagonists for therapeutic purposes. Curr Opin Drug Discov Devel 5:764–776

Kiyosawa A, Katsurabayashi S, Akaike N, Pang ZP (2001) Nicotine facilitates glycine release in the rat spinal dorsal horn. J Physiol 536:101–110

Ko MCH, Willmont KJ, Burritt A, Hruby VJ, Woods JH (2000) Local inhibitory effects of dynorphin A-(1–17) on capsaicin-induced thermal allodynia in rhesus monkeys. Eur J Pharmacol 402:69–76

Kohno M, Hasegawa H, Inoue A, Muraoka M, Miyazaki T, Oka K, Yasukawa M (2006) Identification of N-arachidonylglycine as the endogenous ligand for orphan G-protein-coupled receptor GPR18. Biochem Biophys Res Commun 347:827–832

Kojima M, Kangawa K (2005) Ghrelin: structure and function. Physiol Rev 85:495–522

Kolesnikov Y, Jain S, Pasternak GW (1996) Modulation of opioid analgesia by agmatine. Eur J Pharmacol 296:17–22

Kontinen VK, Kalso EA (1995) Differential modulation of alpha(2)-adrenergic and mu-opioid spinal antinociception by neuropeptide FF. Peptides 16:973–977

Korosi A, Kozicz T, Richter J, Veening JG, Olivier B, Roubos EW (2007) Corticotropin-releasing factor, urocortin 1, and their receptors in the mouse spinal cord. J Comp Neurol 502:973–989

Kostenis E, Ulven T (2006) Emerging roles of DP and CRTH2 in allergic inflammation. Trends Mol Med 12:148–158

Kovacs GL (1986) Oxytocin and behavior. In: Ganten, D and Pfaff, D (eds) Neurobiology of oxytocin. Springer, Berlin/Heidelberg, pp 91–128

Krukowski JA, Hood DD, Eisenach JC, Mallak KA, Parker RL (1997) Intrathecal neostigmine for post-cesaren section analgesia: dose response. Anesth Analg 84:1269–1275

Kukkonen JP, Holmqvist T, Ammoun S, Akerman KEO (2002) Functions of the orexinergic/hypocretinergic system. Am J Physiol Cell Physiol 283:C1567–C1591

Kuphal KE, Solway B, Pedrazzini T, Taylor BK (2008) Y1 receptor knockout increases nociception and prevents the anti-allodynic actions of NPY. Nutrition 24:885–891

Kuraishi Y, Hirota N, Satoh M, Takagi H (1985) Antinociceptive effects of intrathecal opioids, noradrenaline and serotonin in rats: mechanical and thermal algesic tests. Brain Res 326: 168–171

Kwiat GC, Basbaum AI (1992) The origin of brain-stem noradrenergic and serotonergic projections to the spinal-cord dorsal horn in the rat. Somatosens Mot Res 9:157–173

LaBuda CJ, Usdin TB (2004) Tuberoinfundibular peptide of 39 residues decreases pain-related affective behavior. NeuroReport 15:1779–1782

Labuz D, Berger S, Mousa SA, Zollner C, Rittner HL, Shaqura MA, Segovia-Silvestre T, Przewlocka B, Stein C, Machelska H (2006) Peripheral antinociceptive effects of exogenous and immune cell-derived endomorphins in prolonged inflammatory pain. J Neurosci 26:4350–4358

LaGraize SC, Borzan J, Peng YB, Fuchs PN (2006) Selective regulation of pain affect following activation of the opioid anterior cingulate cortex system. Exp Neurol 197:22–30

Lai J, Luo MC, Chen Q, Porreca F (2008) Pronociceptive actions of dynorphin via bradykinin receptors. Neurosci Lett 437:175–179

Lambert DG (2008) The nociceptin/orphanin FQ receptor: a target with broad therapeutic potential. Nat Rev Drug Discov 7:694–710

Lamberti C, Bartolini A, Ghelardini C, Malmberg-Aiello P (1996) Investigation into the role of histamine receptors in rodent antinociception. Pharmacol Biochem Behav 53:567–574

Lang PM, Burgstahler R, Sippel W, Irnich D, Schlotter-Weigel B, Grafe P (2003) Characterization of neuronal nicotinic acetylcholine receptors in the membrane of unmyelinated human C-fiber axons by in vitro studies. J Neurophysiol 90:3295–3303

Laurent P, Becker JAJ, Valverde O, Ledent C, D'Exaerde AD, Schiffmann SN, Maldonado R, Vassart G, Parmentier M (2005) The prolactin-releasing peptide antagonizes the opioid system through its receptor GPR10. Nat Neurosci 8:1735–1741

Lauretti GR, Oliveira A-PM, Juliao M-CC, Reis MP, Pereira NL (2000) Transdermal nitroglycerine enhances spinal neostigmine postoperative analgesia following gynecological surgery. Anesthesiology 93:943–946

Lavand'homme P, Eisenach JC (1999) Exogenous and endogenous adenosine enhance the spinal antiallodynic effects of morphine in a rat model of neuropathic pain. Pain 80:31–36

Lawand NB, Willis WD, Westlund KN (1997) Blockade of joint inflammation and secundary hyperalgesia by L-NAME, a nitric oxide synthase inhibitor. NeuroReport 8:595–899

Ledent C, Vaugeois J-M, Schiimann SN, Pedrazzini T, El Yacoubi M, Vanderhaeghen J-J, Costentin J, Heath JK, Vassart G, Parmentier M (1997) Aggressiveness, hypoalgesia and high blood pressure in mice lacking the adenosine A2 receptor. Nature 388:674–682

Lee AT, Shah JJ, Li L, Cheng Y, Moore PK, Khanna S (2008) A nociceptive-intensity-dependent role for hydrogen sulphide in the formalin model of persistent inflammatory pain. Neuroscience 152:89–96

Lee NM, Leybin L, Chang JK, Loh HH (1980) Opiate and peptide interaction: effect of enkephalins on morphine analgesia. Eur J Pharmacol 68:181–185

Lee T-L, Fung FMY, Chen F-G, Chou N, Okuda-Ashitaka E, Ito S, Nishiuchi Y, Kimura T, Tachibana S (1999) Identification of human, rat and mouse nocistatin in brain and human cerebrospinal fluid. NeuroReport 10:1537–1541

Legendre P (2001) The glycinergic inhibitory synapse. Cell Mol Life Sci 58:760–793

Leone M, D'Amico D, Moschiano F, Fraschini F, Bussone G (1996) Melatonin versus placebo in the prophylaxis of cluster headache: a double-blind pilot study with parallel groups. Cephalalgia 16:494–496

Lever I, Cunningham J, Grist J, Yip PK, Malcangio M (2003) Release of BDNF and GABA in the dorsal horn of neuropathic rats. Eur J Neurosci 18:1169–1174

Lewin GR, Rueff A, Mendell LM (1994) Peripheral and central mechanisms of NGF-induced hyperalgesia. Eur J Neurosci 6:1903–1912

Li D, Ren Y, Xu X, Zou X, Fang L, Lin Q (2008) Sensitization of primary afferent nociceptors induced by intradermal capsaicin involves the peripheral release of calcitonin gene-related peptide driven by dorsal root reflexes. J Pain 9:1155–1168

Li DP, Chen SR, Pan YZ, Levey AI, Pan HL (2002a) Role of presynaptic muscarinic and GABAB receptors in spinal glutamate release and cholinergic analgesia in rats. J Physiol (Lond) 543:807–818

Li G, Regunathan S, Barrow CJ, Eshraghi J, Cooper R, Reis DJ (1994a) Agmatine: an endogenous clonidine-displacing substance in the brain. Science 263:966–969

Li H, Lang B, Kang JF, Li YQ (2000) Serotonin potentiates the response of neurons of the superficial laminae of the rat spinal dorsal horn to gamma-aminobutyric acid. Brain Res Bull 52:559–565

Li JJ, Zhou X, Yu LC (2005a) Involvement of neuropeptide Y and Y1 receptor in antinociception in the arcuate nucleus of hypothalamus, an immunohistochemical and pharmacological study in intact rats and rats with inflammation. Pain 118:232–242

Li P, Tong C, Eisenach JC, Figueroa JP (1994b) NMDA causes release of nitric oxide from rat spinal cord in vitro. Brain Res 637:287–291

Li SR, Wang T, Wang R, Dai X, Chen Q, Li RD (2005b) Melatonin enhances antinociceptive effects of delta -, but not mu-opioid agonist in mice. Brain Res 1043:132–138

Li WG, Gavrila D, Liu X, Wang L, Gunnlaugsson S, Stoll LL, McCormick ML, Sigmund CD, Tang C, Weintraub NL (2004) Ghrelin inhibits proinflammatory responses and nuclear factor-kappaB activation in human endothelial cells. Circulation 109:2221–2226

Li X, Clark JD (2001) Spinal cord nitric oxide synthase and heme oxygenase limit morphine induced analgesia. Mol Brain Res 95:96–102

Li X, Clark JD (2002) Spinal cord heme oxygenase participates in glutamate-induced pain-related behaviors. Eur J Pharmacol 450:43–48

Li X, Clark JD (2003) Heme oxygenase type 2 participates in the development of chronic inflammatory and neuropathic pain. J Pain 4:101–107

Li X, Eisenach JC (2001) Alpha2A-adrenoceptor stimulation reduces capsaicin-induced glutamate release from spinal cord synaptosomes. J Pharmacol Exp Ther 299:939–944

Li X, Eisenach JC (2002) Nicotinic acetylcholine receptor regulation of spinal norepinephrine release. Anesthesiology 96:1450–1456

Li Y, Li JJ, Yu LC (2002b) Anti-nociceptive effect of neuropeptide Y in the nucleus accumbens of rats: an involvement of opioid receptors in the effect. Brain Res 940:69–78

Liang D, Li X, Lighthall G, Clark JD (2003) Heme oxygenase type 2 modulates behavioral and molecular changes during chronic exposure to morphine. Neuroscience 121: 999–1005

Liang D-Y, Li X, Clark JD (2004) Formalin-induced spinal cord calcium/calmodulin-dependent protein kinase IIa expression is modulated by heme oxygenase in mice. Neurosci Lett 360: 61–64

Liebmann C, Schrader U, Brantl V (1989) Opioid receptor affinities of the blood-derived tetrapeptides hemorphin and cytochrophin. Eur J Pharmacol 166:523–526

Lin S, Boey D, Herzog H (2004) NPY and Y receptors: lessons from transgenic and knockout models. Neuropeptides 38:189–200

Lindfors PH, Voikar V, Rossi J, Airaksinen MS (2006) Deficient nonpeptidergic epidermis innervation and reduced inflammatory pain in glial cell line-derived neurotrophic factor family receptor alpha2 knock-out mice. J Neurosci 26:1953–1960

Lippton H, Lin B, Gumusel B, Witriol N, Wasserman A, Knight M (2006) Hemopressin, a hemoglobin fragment, dilates the rat systemic vascular bed through release of nitric oxide. Peptides 27:2284–2288

Liu EH, Nishiuchi Y, Kimura T, Tachibana S (2006) Supraspinal nocistatin and its amide derivative antagonize the hyperalgesic effects of nociceptin in mice. Neurosci Lett 397:59–63
Liu HX, Hokfelt T (2002) The participation of galanin in pain processing at the spinal level. Trends Pharmacol Sci 23:468–474
Lo Verme J, Fu J, Astarita G, La Rana G, Russo R, Calignano A, Piomelli D (2005) The nuclear receptor peroxisome proliferator-activated receptor-alpha mediates the anti-inflammatory actions of palmitoylethanolamide. Mol Pharmacol 67:15–19
Lo Verme J, Russo R, La Rana G, Fu J, Farthing J, Mattace-Raso G, Meli R, Hohmann A, Calignano A, Piomelli D (2006) Rapid broad-spectrum analgesia through activation of peroxisome proliferator-activated receptor-alpha. J Pharmacol Exp Ther 319:1051–1061
Loh HH, Tseng LF, Wei E, Li CH (1976) Beta-endorphin is a potent analgesic agent. Proc Natl Acad Sci U S A 73:2895–2898
Lopes SCDN, Soares CM, Baptista AM, Goormaghtigh E, Cabral BJC, Casthanho MARB (2006) Conformational and orientational guidance of the analgesic dipeptide kyotorphin induced by lipidic membranes:putative correlation toward receptor docking. J Phys Chem 110: 3385–3394
Lorenzetti BB, Ferreira SH (1996) Activation of the arginine-nitric oxide pathway in primary sensory neurons contributes to dipyrone-induced spinal and peripheral analgesia. Inflamm Res 45:308–311
Losel R, Wehling M (2003) Nongenomic actions of steroid hormones. Nat Rev Mol Cell Biol 4:46–56
Lozovaya N, Yatsenko N, Beketov A, Tsintsadze T, Burnashev N (2005) Glycine receptors in CNS neurons as a target for nonretrograde action of cannabinoids. J Neurosci 25:7499–7506
Lynch JW (2009) Native glycine receptor subtypes and their physiological roles. Neuropharmacology 56:303–309
Ma F, Xie H, Dong ZQ, Wang YQ, Wu GC (2003) Effect of intrathecal nocistatin on nociceptin/orphanin FQ analgesia in chronic constriction injury rat. Brain Res 988:189–192
Ma QP, Bleasdale C (2002) Modulation of brain stem monoamines and gamma-aminobutyric acid by NKI receptors in rats. NeuroReport 13:1809–1812
Ma W, Eisenach JC (2007) Neuronal nitric oxide synthase is upregulated in a subset of primary sensory afferents after nerve injury which are necessary for analgesia from alpha2-adrenoceptor stimulation. Brain Res 1127:52–58
Mabuchi T, Matsumura S, Okuda-Ashitaka E, Kitano T, Kojima H, Nagano T, Minami T, Ito S (2003) Attenuation of neuropathic pain by the nociceptin/orphanin FQ antagonist JTC-801 is mediated by inhibition of nitric oxide production. Eur J Neurosci 17:1384–1392
Mabuchi T, Shintani N, Matsumura S, Okuda-Ashitaka E, Hashimoto H, Muratani T, Minami T, Baba A, Ito S (2004) Pituitary adenylate cyclase-activating polypeptide is required for the development of spinal sensitization and induction of neuropathic pain. J Neurosci 24: 7283–7291
Macaluso A, McCoy D, Ceriani G, Watanabe T, Biltz T, Catania A, Lipton JM (1994) Antiinflammatory influences of alpha-MSH molecules: central neurogenic and peripheral actions. J Neurosci 14:2377–2382
Mackie K (2008) Cannabinoid receptors: where they are and what they do. J Neuroendocrinol 20:10–14
Macsai M, Pataki I, Toth G, Szabo Gy (2002) The effects of pituitary adenylate cyclase-activating polypeptide on acute and chronic morphine actions in mice. Regul Pept 109:57–62
Macsai M, Szabo G, Telegdy G (1998) Vasoactive intestinal polypeptide induces analgesia and impairs the antinociceptive effect of morphine in mice. Neuropeptides 32:557–562
Maeda Y, Aoki Y, Sekiguchi F, Matsunami M, Takahashi T, Nishikawa H, Kawabata A (2009) Hyperalgesia induced by spinal and peripheral hydrogen sulfide: evidence for involvement of Cav3.2 T-type calcium channels. Pain 142:127–132
Maggi CA, Patacchini R, Rovero P, Giachetti A (1993) Tachykinin receptors and tachykinin receptor antagonists. J Auton Pharmacol 13:23–93

Mahinda TB, Taylor BK (2004) Intrathecal neuropeptide Y inhibits behavioral and cardiovascular responses to noxious inflammatory stimuli in awake rats. Physiol Behav 80:703–711

Mahmoudi M, Zarrindast MR (2002) Effect of intracerebroventricular injection of GABA receptor agents on morphine-induced antinociception in the formalin test. J Psychopharmacol 16:85–91

Maingret F, Patel AJ, Lazdunski M, Honore E (2001) The endocannabinoid anandamide is a direct and selective blocker of the background K(+) channel TASK-1. EMBO J 20:47–54

Maione S, Bisogno T, De Novellis V, Palazzo E, Cristino L, Valenti M, Petrosino S, Guglielmotti V, Rossi F, Marzo VD (2006) Elevation of endocannabinoid levels in the ventrolateral periaqueductal grey through inhibition of fatty acid amide hydrolase affects descending nociceptive pathways via both cannabinoid receptor type 1 and transient receptor potential vanilloid type-1 receptors. J Pharmacol Exp Ther 316:969–982

Majewska MD (1992) Neurosteroids: endogenous bimodal modulators of the GABAA receptor. Mechanism of action and physiological significance. Prog Neurobiol 38:379–395

Malan TP, Mata HP, Porreca F (2002) Spinal GABA(A) and GABA(B) receptor pharmacology in a rat model of neuropathic pain. Anesthesiology 96:1161–1167

Malcangio M, Lessmann V (2003) A common thread for pain and memory synapses? Brain-derived neurotrophic factor and trkB receptors. Trends Pharmacol Sci 24:116–121

Maldonado R, Valverde O, Turcaud S, Fournie-Zaluski M-C, Roques BP (1994) Antinociceptive response induced by mixed inhibitors of enkephalin catabolism in peripheral inflammation. Pain 58:77–83

Malin SA, Molliver DC, Koerber HR, Cornuet P, Frye R, Albers KM, Davis BM (2006) Glial cell line-derived neurotrophic factor family members sensitize nociceptors in vitro and produce thermal hyperalgesia in vivo. J Neurosci 26:8588–8599

Malkmus S, Lu X, Bartfai T, Yaksh TL, Hua XY (2005) Increased hyperalgesia after tissue injury and faster recovery of allodynia after nerve injury in the GalR1 knockout mice. Neuropeptides 39:217–221

Malmberg AB, Yaksh TL (1993) Spinal nitric oxide synthesis inhibition block NMDA-induced thermal hyperalgesia and produces antinociception in the formalin test in rats. Pain 54: 291–300

Malmberg-Aiello P, Lamberti C, Ipponi A, Bartolini A, Schunack W (1998) Evidence for hypernociception induction following histamine H1 receptor activation in rodents. Life Sci 63:463–476

Malmberg-Aiello P, Lamberti C, Ghelardini C, Giotti A, Bartolini A (1994) Role of histamine in rodent antinociception. Br J Pharmacol 111:1269–1279

Mameli M, Carta M, Partridge LD, Valenzuela CF (2005) Neurosteroid-induced plasticity of immature synapses via retrograde modulation of presynaptic NMDA receptors. J Neurosci 25:2285–2294

Mansikka H, Idanpaan-Heikkila J-J, Pertovaara A (1996) Different roles of alpha2-adrenoceptors of the medulla versus the spinal cord in modulation of mustard oil-induced central hyperalgesia in rats. Eur J Pharmacol 297:19–26

Mansikka H, Pertovaara A (1995) The role of alpha 2-adrenoceptors of the medullary lateral reticular nucleus in spinal antinociception in rats. Brain Res Bull 37:633–638

Mantovani M, Kaster MP, Pertile R, Calixto JB, Rodrigues ALS, Santos ARS (2006) Mechanisms involved in the antinociception caused by melatonin in mice. J Pineal Res 41:382–389

Mantovani M, Pertile R, Calixto JB, Santos ARS, Rodrigues AL (2003) Melatonin exerts an antidepressant-like effect in the tail suspension test in mice: evidence for involvement of N-methyl-D-aspartate receptors and the L-arginine-nitric oxide pathway. Neurosci Lett 343:1–4

Mao J, Price DD, Hayes RL, Mayer DJ (1992) Intrathecal GM1 ganglioside and local nerve anesthesia reduce nociceptive behaviors in rats with experimental peripheral mononeuropathy. Brain Res 584:28–35

Marcus JN, Aschkenasi CJ, Lee CE, Chemelli RM, Saper CB, Yanagisawa M, Elmquist JK (2001) Differential expression of orexin receptors 1 and 2 in the rat brain. J Comp Neurol 435:6–25

Marek P, Ben-Eliyahu S, Gold M, Liebeskind JC (1991) Excitatory amino acid antagonists (kynurenic acid and MK-801) attenuate the development of morphine tolerance in the rat. Brain Res 574:77–81

Marinelli S, Vaughan CW, Christie MJ, Connor M (2002) Capsaicin activation of glutamatergic synaptic transmission in the rat locus coeruleus in vitro. J Physiol 543:531–540

Martin BR, Mechoulam R, Razdan RK (1999) Discovery and characterization of endogenous cannabinoids. Life Sci 65:573–595

Martin M, Otto C, Santamarta MT, Torrecilla M, Pineda J, Schutz G, Maldonado R (2003) Morphine withdrawal is modified in pituitary adenylate cyclase-activating polypeptide type I-receptor-deficient mice. Mol Brain Res 110:109–118

Martinez V, Wang L, Million M, Rivier J, Tache Y (2004) Urocortins and the regulation of gastrointestinal motor function and visceral pain. Peptides 25:1733–1744

Martire M, Altobelli D, Maurizi S, Preziosi P, Fuxe K (2000) K+-evoked [(3)H]D-aspartate release in rat spinal cord synaptosomes: modulation by neuropeptide Y and calcium channel antagonists. J Neurosci Res 62:722–729

Masocha W, Horvath G, Agil A, Ocana M, Del Pozo E, Szikszay M, Baeyens JM (2003) Role of Na^{+}, K^{+}-ATPase in morphine-induced antinociception. J Pharmacol Exp Ther 306:1122–1128

Mathers DA, Grewal A, Wang YH (1989) Membrane channels activated by taurine in cultured mouse spinal-cord neurons. Neurosci Lett 98:229–233

Matsumoto M, Xie W, Inoue M, Ueda H (2007) Evidence for the tonic inhibition of spinal pain by nicotinic cholinergic transmission through primary afferents. Mol Pain 3:41

Matsunami M, Tarui T, Mitani K, Nagasawa K, Fukushima O, Okubo K, Yoshida S, Takemura M, Kawabata A (2009) Luminal hydrogen sulfide plays a pro-nociceptive role in mouse colon. Gut 58(6):751–761

Mazella J (2001) Sortilin/neurotensin receptor-3: a new tool to investigate neurotensin signaling and cellular trafficking? Cell Signal 13:1–6

Mazella J, Vincent JP (2006) Functional roles of the NTS2 and NTS3 receptors. Peptides 27: 2469–2475

Mazzari S, Canella R, Petrelli L, Marcolongo G, Leon A (1996) N-(2-hydroxyethil) hexadecanamide is orally active in reducing edema formation and inflammatory hyperalgesia by down-modulating mast cell activation. Eur J Pharmacol 300:227–236

McCarson KE, Duric V, Reisman SA, Winter M, Enna SJ (2006) GABA(B) receptor function and subunit expression in the rat spinal cord as indicators of stress and the antinociceptive response to antidepressants. Brain Res 1068:109–117

McDonald HA, Neelands TR, Kort M, Han P, Vos MH, Faltynek CR, Moreland RB, Puttfarcken PS (2008) Characterization of A-425619 at native TRPV1 receptors: a comparison between dorsal root ganglia and trigeminal ganglia. Eur J Pharmacol 596:62–69

McDougall JJ, Larson SEM (2006) Nociceptin/orphanin FQ evokes knee joint pain in rats via a mast cell independent mechanism. Neurosci Lett 398:135–138

McDougall JJ, Watkins L, Li Z (2006) Vasoactive intestinal peptide (VIP) is a modulator of joint pain in a rat model of osteoarthritis. Pain 123:98–105

McKinley MJ, Albiston AL, Allen AM, Mathai ML, May CN, McAllen RM, Oldfield BJ, Mendelsohn FAO, Chai SY (2003) The brain renin-angiotensin system: location and physiological roles. Int J Biochem Cell Biol 35:901–918

Mechoulam R, Ben Shabat S, Hanus L, Ligumsky M, Kaminski NE, Schatz AR, Gopher A, Almog S, Martin BR, Compton DR (1995) Identification of an endogenous 2-monoglyceride, present in canine gut, that binds to cannabinoid receptors. Biochem Pharmacol 50:83–90

Mechoulam R, Fride E, Di Marzo V (1998) Endocannabinoids. Eur J Pharmacol 359:1–18

Mechoulam R, Fride E, Hanus L, Sheskin T, Bisogno T, DiMarzo V, Bayewitch M, Vogel Z (1997) Anandamide may mediate sleep induction. Nature 389:25–26

Mecs L, Tuboly G, Nagy E, Benedek G, Horvath G (2009) Peripheral antinociceptive effects of endomorphin-1 and kynurenic acid in the rat inflamed joint model. Anesth Analg 109(4): 1297–1304

Medvedeva YV, Kim MS, Usachev YM (2008) Mechanisms of prolonged presynaptic Ca2+ signaling and glutamate release induced by TRPV1 activation in rat sensory neurons. J Neurosci 28:5295–5311

Meller ST, Cummings CP, Traub RJ, Gebhart GF (1994a) The role of nitric oxide in the development and maintenance of the hyperalgesia produced by intraplantar injection of carrageenan in the rat. Neuroscience 60:367–374

Meller ST, Dykstra C, Gebhart GF (1992a) Production of endogenous nitric oxide and activation of soluble guanylate cyclase are required for N-methyl-D-aspartate-produced facilitation of the nociceptive tail-flick reflex. Eur J Pharmacol 214:93–96

Meller ST, Dykstra C, Gryzbycki D, Murphy S, Gebhart GF (1994b) The possible role of glia in nociceptive processing and hyperalgesia in the spinal cord of the rat. Neuropharmacology 33:1471–1478

Meller ST, Dykstra CL, Gebhart GF (1994c) Investigations of the possible role for carbon monoxide (CO) in thermal and mechanical hyperalgesia in the rat. NeuroReport 5: 2337–2341

Meller ST, Pechman PS, Gebhart GF, Maves TJ (1992b) Nitric oxide mediates the thermal hyperalgesia produced in a model of neuropathic pain in the rat. Neuroscience 50:7–10

Menetrey D, Basbaum AI (1987) The distribution of substance P-, enkephalin- and dynorphin-immunoreactive neurons in the medulla of the rat and their contribution to bulbospinal pathways. Neuroscience 23:173–187

Merchenthaler I, Lane MV, Numan S, Dellovade TL (2004) Distribution of estrogen receptor alpha and beta in the mouse central nervous system: in vivo autoradiographic and immunocytochemical analyses. J Comp Neurol 473:270–291

Mezey E, Toth ZE, Corthright DN, Arzubi MK, Krause JE, Elde R, Guo A, Blumberg PM, Szallasi A (2000) Distribution of mRNA for vanilloid receptor subtype 1 (VR1), and VR1-like immunorectivity, in the central nervous system of the rat and human. Proc Natl Acad Sci U S A 97:3655–3660

Mi WL, Mao-Ying QL, Liu Q, Wang XW, Li X, Wang YQ, Wu GC (2009) The distribution of kisspeptin and its receptor GPR54 in rat dorsal root ganglion and up-regulation of its expression after CFA injection. Brain Res Bull 78:254–260

Michael GJ, Averill S, Nitkunan A, Rattray M, Bennett DL, Yan Q, Priestley JV (1997) Nerve growth factor treatment increases brain-derived neurotrophic factor selectively in TrkA-expressing dorsal root ganglion cells and in their central terminations within the spinal cord. J Neurosci 17:8476–8490

Millan MJ (2002) Descending control of pain. Prog Neurobiol 66:355–474

Millan MJ, Millan MH, Pilcher CW, Czlonkowski A, Herz A, Colpaert FC (1985) Spinal cord dynorphin may modulate nociception via a kappa-opioid receptor in chronic arthritic rats. Brain Res 340:156–159

Million M, Wang L, Wang Y, Adelson DW, Yuan PQ, Maillot C, Coutinho SV, Mcroberts JA, Bayati A, Mattsson H, Wu V, Wei JY, Rivier J, Vale W, Mayer EA, Tache Y (2006) CRF2 receptor activation prevents colorectal distension induced visceral pain and spinal ERK1/2 phosphorylation in rats. Gut 55:172–181

Milner TA, Drake CT, Aicher SA (2002) C1 adrenergic neurons are contacted by presynaptic profiles containing DELTA-opioid receptor immunoreactivity. Neuroscience 110:691–701

Minami T, Okuda-Ashitaka E, Mori H, Ito S, Hayaishi O (1996) Prostaglandin D2 inhibits prostaglandin E2-induced allodynia in conscious mice. J Pharmacol Exp Ther 278: 1146–1152

Minami T, Okuda-Ashitaka E, Nishizawa M, Mori H, Ito S (1997) Inhibition of nociceptin-induced allodynia in conscious mice by prostaglandin D2. Br J Pharmacol 122:605–610

Minami T, Uda R, Horiguchi S, Ito S, Hyodo M, Hayaishi O (1994) Allodynia evoked by intrathecal administration of prostaglandin E2 to conscious mice. Pain 57:217–223

Missale C, Nash SR, Robinson SW, Jaber M, Caron MG (1998) Dopamine receptors: from structure to function. Physiol Rev 78:189–225

Miyata A, Arimura A, Dahl RR, Minamino N, Uehara A, Jiang L, Culler MD, Coy DH (1989) Isolation of a novel 38 residue-hypothalamic polypeptide which stimulates adenylate cyclase in pituitary cells. Biochem Biophys Res Commun 164:567–574

Mizoguchi H, Narita M, Kampine JP, Tseng LF (1997) [Met5]enkephalin and delta2-opioid receptors in the spinal cord are involved in the cold water swimming-induced antinociception in the mouse. Life Sci 61:PL81–PL86

Mizutani T, Layon AJ (1996) Clinical applications of nitric oxide. Chest 110:506–524

Mobarakeh JI, Sakurada S, Hayashi T, Orito T, Okuyama K, Sakurada T, Kuramasu A, Watanabe T, Watanabe T, Yanai K (2002) Enhanced antinociception by intrathecally-administered morphine in histamine H1 receptor gene knockout mice. Neuropharmacology 42:1079–1088

Mobarakeh JI, Sakurada S, Katsuyama S, Kutsuwa M, Kuramasu A, Lin ZY, Watanabe T, Hashimoto Y, Watanabe T, Yanai K (2000) Role of histamine H(1) receptor in pain perception: a study of the receptor gene knockout mice. Eur J Pharmacol 391:81–89

Mobarakeh JI, Takahashi K, Sakurada S, Nishino S, Watanabe H, Kato M, Naghdi N, Yanai K (2005) Enhanced antinociception by intracerebroventricularly administered orexin A in histamine H1 or H2 receptor gene knockout mice. Pain 118:254–262

Mobarakeh JI, Takahashi K, Sakurada S, Kuramasu A, Yanai K (2006) Enhanced antinociceptive effects of morphine in histamine H2 receptor gene knockout mice. Neuropharmacology 51:612–622

Mogil JS, Pasternak GW (2001) The molecular and behavioral pharmacology of the orphanin FQ/nociceptin peptide and receptor family. Pharmacol Rev 53:381–415

Mogil JS, Wilson SG, Chesler EJ, Rankin AL, Nemmani KVS, Lariviere WR, Groce MK, Wallace MR, Kaplan L, Staud R, Ness TJ, Glover TL, Stankova M, Mayorov A, Hruby VJ, Grisel JE, Fillingim RB (2003) The melanocortin-1 receptor gene mediates female-specific mechanisms of analgesia in mice and humans. Proc Natl Acad Sci U S A 100:4867–4872

Monnet FP, Mahe V, Robel P, Baulieu E-E (1995) Neurosteroids, via sigma receptors, modulate the [3H]norepinephrine release evoked by N-methyl-D-aspartate in the rat hippocampus. Proc Natl Acad Sci U S A 92:3774–3778

Moon DE, Lee DH, Han HC, Xie J, Coggeshall RE, Chung JM (1999) Adrenergic sensitivity of the sensory receptors modulating mechanical allodynia in a rat neuropathic pain model. Pain 80:589–595

Moore KA, Kohno T, Karchewski LA, Scholz J, Baba H, Woolf CJ (2002) Partial peripheral nerve injury promotes a selective loss of GABAergic inhibition in the superficial dorsal horn of the spinal cord. J Neurosci 22:6724–6731

Morciano M, Ortinau S, Zimmermann H (2004) Guanine nucleotides inhibit NMDA and kainate-induced neurotoxicity in cultured rat hippocampal and neocortical neurons. Neurochem Int 45:95–101

Morgan MJ, Franklin KBJ (1991) Dopamine receptor subtypes and formalin test analgesia. Pharmacol Biochem Behav 40:317–322

Morgan PJ, Barrett P, Howell HE, Helliwell R (1994) Melatonin receptors: localization, molecular pharmacology and physiological significance. Neurochem Int 24:101–146

Moroni F, Russi P, Lombardi G, Beni M, Carla V (1988) Presence of kynurenic acid in the mammalian brain. J Neurochem 51:177–180

Morteau O, Lu B, Gerard C, Gerard NP (2001) Hemokinin 1 is a full agonist at the substance P receptor. Nat Immunol 2:1088

Morton CR, Hutchison WD, Hendry IA (1991) Intraspinal release of substance P and calcitonin gene-related peptide during opiate dependence and withdrawal. Neuroscience 43:593–600

Mouchet P, Manier M, Dietl M, Feuerstein C, Berod A, Arluison M, Denoroy L, Thibault J (1986) Immunohistochemical study of catecholaminergic cell bodies in the rat spinal cord. Brain Res Bull 16:341–353

Mousa SA, Cheppudira BP, Shaqura M, Fischer O, Hofmann J, Hellweg R, Schafer M (2007) Nerve growth factor governs the enhanced ability of opioids to suppress inflammatory pain. Brain 130:502–513

Mousa SA, Machelska H, Schäfer M, Stein C (2002) Immunhistochemical localization of endomorphin-1 and endomorphin-2 in immune cells and spinal cord in a model of inflammatory pain. J Neuroimmunol 126:5–15

Murphy RM, Zemlan FP (1990) Selective serotonin1A/1B agonists differentially affect spinal nociceptive reflexes. Neuropharmacology 29:463–468

Muth-Selbach U, Dybek E, Kollosche K, Stegmann JU, Holthusen H, Lipfert P, Zeilhofer HU (2004) The spinal antinociceptive effect of nocistatin in neuropathic rats is blocked by D-serine. Anesthesiology 101:753–758

Nagano M, Sakai A, Takahashi N, Umino M, Yoshioka K, Suzuki M (2003) Decreased expression of glial cell line-derived neurotrophic factor signaling in rat models of neuropathic pain. Br J Pharmacol 140:1252–1260

Naguib M, Yaksh TL (1997) Characterization of muscarinic receptor subtypes that mediate antinociception in the rat spinal cord. Anesth Analg 85:847–853

Nagy I, Santha P, Jancso G, Urban L (2004) The role of the vanilloid (capsaicin) receptor (TRPV1) in physiology and pathology. Eur J Pharmacol 500:351–369

Nahin RL, Byers MR (1994) Adjuvant-induced inflammation of rat paw is associated with altered calcitonin gene-related peptide immunoreactivity within cell bodies and peripheral endings of primary afferent neurons. J Comp Neurol 349:475–485

Nakagawa T, Kaneko M, Inamura S, Satoh M (1999) Intracerebroventricular administration of nocistatin reduces inflammatory hyperalgesia in rats. Neurosci Lett 265:64–66

Nakamura M, Ferreira SH (1988) Peripheral analgesic action of clonidine: mediation by release of endogenous enkephalin-like substances. Eur J Pharmacol 146:223–228

Nakano H, Minami T, Abe K, Arai T, Tokumura M, Ibii N, Okuda-Ashitaka E, Mori H, Ito S (2000) Effect of intrathecal nocistatin on the formalin-induced pain in mice versus that of nociceptin/orphanin FQ. J Pharmacol Exp Ther 292:331–336

Naono R, Nakayama T, Ikeda T, Matsushima O, Nishimori T (2007) Leucine at the carboxyl-terminal of endokinins C and D contributes to elicitation of the antagonistic effect on substance P in rat pain processing. Brain Res 1165:71–80

Narita M, Dun SL, Dun NJ, Tseng LF (1996) Hyperalgesia induced by pituitary adenylate cyclase-activating polypeptide in the mouse spinal cord. Eur J Pharmacol 311:121–126

Nascimento CG, Branco LG (2007) Role of the peripheral heme oxygenase-carbon monoxide pathway on the nociceptive response of rats to the formalin test: evidence for a cGMP signaling pathway. Eur J Pharmacol 556:55–61

Nascimento CG, Branco LG (2008) Role of the spinal cord heme oxygenase-carbon monoxide-cGMP pathway in the nociceptive response of rats. Eur J Pharmacol 581:71–76

Nayebi AR, Ahmadiani A (1999) Involvement of the spinal serotonergic system in analgesia produced by castration. Pharmacol Biochem Behav 64:467–471

Nayebi AR, Hassapour M, Rezazadeh H (2001) Effect of chronic and acute administration of fluoxetine and its additive effect with morphine on the behavioural response in the formalin test in rats. J Pharm Pharmacol 53:219–225

Nayebi AR, Rezazadeh H (2004) Involvement of serotoninergicic mechanism in analgesia by castration and flutamide, a testosterone antagonist, in the rat formalin test. Pharmacol Biochem Behav 77:9–14

Nemeroff CB, Osbahr AJ, Manberg PJ, Ervin GN, Prange AJ (1979) Alterations in nociception and body temperature after intracisternal administration of neurotensin, beta-endorphin, other endogenous peptides, and morphine. Proc Natl Acad Sci U S A 76:5368–5371

Nemeth H, Toldi J, Vecsei L (2005) Role of kynurenines in the central and peripherial nervous systems. Curr Neurovasc Res 2:249–260

Nemeth J, Reglodi D, Pozsgai G, Szabo A, Elekes K, Pinter E, Szolcsanyi J, Helyes Z (2006) Effect of pituitary adenylate cyclase activating polypeptide-38 on sensory neuropeptide release and neurogenic inflammation in rats and mice. Neuroscience 143:223–230

Netti C, Guidobono F, Sibilia V, Villa I, Cazzamalli E, Pecile A (1988) Central effects of histamine H2-receptor agonists and antagonists on nociception in the rat. Agents Actions 23:247–249

Neugebauer V, Li W, Bird GC, Han JS (2004) The amygdala and persistent pain. Neuroscientist 10:221–234

Nicholson R, Dixon AK, Spanswick D, Lee K (2005) Noradrenergic receptor mRNA expression in adult rat superficial dorsal horn and dorsal root ganglion neurons. Neurosci Lett 380: 316–321

Niederberger E, Schmidtko A, Gao W, Kuhlein H, Ehnert C, Geisslinger G (2007) Impaired acute and inflammatory nociception in mice lacking the p50 subunit of NF-kappaB. Eur J Pharmacol 559:55–60

Nijenhuis WA, Oosterom J, Adan RA (2001) AgRP(83–132) acts as an inverse agonist on the human-melanocortin-4 receptor. Mol Endocrinol 15:164–171

Nijsen M, Ongenae N, Meulemans A, Coulie B (2005) Divergent role for CRF1 and CRF2 receptors in the modulation of visceral pain. Neurogastroenterol Motil 17:423–432

Nishimura K, Hazato T (1993) Isolation and identification of an endogenous inhibitor of enkephalin-degrading enzymes from bovine spinal cord. Biochem Biophys Res Commun 194:713–719

Nishimura S, Fukushima O, Ishikura H, Takahashi T, Matsunami M, Tsujiuchi T, Sekiguchi F, Naruse M, Kamanaka Y, Kawabata A (2009) Hydrogen sulfide as a novel mediator for pancreatic pain in rodents. Gut 58(6):762–770; Epub

Nyberg F, Sanderson K, Glamsta EL (1997) The hemorphins: a new class of opioid peptides derived from the blood protein hemoglobin. Biopolymers 43:147–156

Oatway MA, Chen Y, Weaver LC (2004) The 5-HT3 receptor facilitates at-level mechanical allodynia following spinal cord injury. Pain 110:259–268

Obara I, Przewlocki R, Przewlocka B (2004) Local peripheral effects of mu-opioid receptor agonists in neuropathic pain in rats. Neurosci Lett 360:85–89

Obara I, Przewlocki R, Przewlocka B (2005) Spinal and local peripheral antiallodynic activity of Ro64-6198 in neuropathic pain in the rat. Pain 116:17–25

Ochi T, Motoyama Y, Goto T (2000) The spinal antinociceptive effect of FR140423 is mediated through kyotorphin receptors. Life Sci 66:2239–2245

Ochi T, Ohkubo Y, Mutoh S (2002) Blockade of the antinociceptive effect of spinally administered kyotorphin by naltrindole in mice. Neurosci Lett 322:95–98

Oh DY, Yoon JM, Moon MJ, Hwang J-I, Choe H, Lee JY, Kim JI, Kim S, Rhim H, O'Dell DK, Walker JM, Na HS, Lee MG, Kwon HB, Kim K, Seong JY (2008) Indentification of farnesyl pyrophosphate and N-arachidonylglycine as endogenous ligands for GPR92. J Biol Chem 283:21054–21064

Ohsawa M, Brailoiu GC, Shiraki M, Dun NJ, Paul K, Tseng LF (2002) Modulation of nociceptive transmission by pituitary adenylate cyclase activating polypeptide in the spinal cord of the mouse. Pain 100:27–34

Ohshiro H, Tonai-Kachi H, Ichikawa K (2008) GPR35 is a functional receptor in rat dorsal root ganglion neurons. Biochem Biophys Res Commun 365:344–348

Oja SS, Saransaari P (2000) Modulation of taurine release by glutamate receptors and nitric oxide. Prog Neurobiol 62:407–425

Okada M, Nakagawa T, Minami M, Satoh M (2002) Analgesic effects of intrathecal administration of P2Y nucleotide receptor agonists UTP and UDP in normal and neuropathic pain model rats. J Pharmacol Exp Ther 303:66–73

Okamoto K, Bereiter DF, Thompson R, Tashiro A, Bereiter DA (2008) Estradiol replacement modifies c-fos expression at the spinomedullary junction evoked by temporomandibular joint stimulation in ovariectomized female rats. Neuroscience 156:729–736

Okamoto Y, Wang J, Morishita J, Ueda N (2007) Biosynthetic pathways of the endocannabinoid anandamide. Chem Biodivers 4:1842–1857

Okuda-Ashitaka E, Ito S (2000) Nocistatin: a novel neuropeptide encoded by the gene for the nociceptin/orphanin FQ precursor. Peptides 21:1101–1109

Okuda-Ashitaka E, Minami T, Tachibana S, Yoshihara Y, Nishiuchi Y, Kimura T, Ito S (1998) Nocistatin, a peptide that blocks nociceptin action in pain transmission. Nature 392:286–289

Olgiati VR, Guidobono F, Netti C, Pecile A (1983) Localization of calcitonin binding sites in rat central nervous system: evidence of its neuroactivity. Brain Res 265:209–215

Oliver KR, Keyvan-Fouladi N (2000) Calcitonin gene-related peptide family receptor localization in the rat central nervous system. Eur J Neurosci 12:S472

Oluyomi AO, Hart SL (1991) Involvement of histamine in naloxone-resistant and naloxone-sensitive models of swim stress-induced antinociception in the mouse. Neuropharmacology 30:1021–1027

Omote K, Hazama K, Kawamata T, Kawamata M, Nakayaka Y, Toriyabe M, Namiki A (2001) Peripheral nitric oxide in carrageenan-induced inflammation. Brain Res 912:171–175

Ortiz JP, Close LN, Heinricher MM, Selden NR (2008) Alpha(2)-noradrenergic antagonist administration into the central nucleus of the amygdala blocks stress-induced hypoalgesia in awake behaving rats. Neuroscience 157:223–228

Osbahr AJ, Nemeroff CB, Luttinger D, Mason GA, Prange AJ (1981) Neurotensin-induced antinociception in mice: antagonism by thyrotropin-releasing hormone. J Pharmacol Exp Ther 217:645–651

Ossipov MH, Gebhart GF (1983) Absence of antinociceptive effect of alpha-2 agonists microinjected in the periaqueductal gray of the rat. Brain Res 289:349–354

Ossipov MH, Kovelowski CJ, Wheeler-Aceto H, Cowan A, Hunter JC, Lai J, Malan TP, Porreca F (1996) Opioid antagonists and antisera to endogenous opioids increase the nociceptive response to formalin: demonstration of an opioid kappa and delta inhibitory tone. J Pharmacol Exp Ther 277:784–788

Ossipov MH, Zhang ET, Carvajal C, Gardell L, Quirion R, Dumont Y, Lai J, Porreca F (2002) Selective mediation of nerve injury-induced tactile hypersensitivity by neuropeptide Y. J Neurosci 22:9858–9867

Oz M (2006) Receptor-independent actions of cannabinoids on cell membranes: focus on endocannabinoids. Pharmacol Ther 111:114–144

Page NM (2004) Hemokinins and endokinins. Cell Mol Life Sci 61:1652–1663

Palazzo E, De Novellis V, Marabese I, Cuomo D, Rossi F, Berrino L, Rossi F, Maione S (2002) Interaction between vanilloid and glutamate receptors in the central modulation of nociception. Eur J Pharmacol 439:69–75

Paldyova E, Bereczki E, Santha M, Wenger T, Borsodi A, Benyhe S (2008) Noladin ether, a putative endocannabinoid, inhibits mu-opioid receptor activation via CB2 cannabinoid receptors. Neurochem Int 52:321–328

Palkovits M (1984) Distribution of neuropeptides in the central nervous system: a review of biochemical mapping studies. Prog Neurobiol 23:151–189

Palkovits M (2000) Stress-induced expression of co-localized neuropeptides in hypothalamic and amygdaloid neurons. Eur J Pharmacol 405:161–166

Palkovits M, Mezey E, Eskay RL (1987) Pro-opiomelanocortin-derived peptides (ACTH/beta-endorphin/ alpha-MSH) in brainstem baroreceptor areas of the rat. Brain Res 436: 323–338

Palmer JA, Higuera ES, Chang L, Chaplan SR (2008) Fatty acid amide hydrolase inhibition enhances the anti-allodynic actions of endocannabinoids in a model of acute pain adapted for the mouse. Neuroscience 154:1554–1561

Pan HL, Chen SR, Eisenach JC (1998) Role of spinal NO in antiallodynic effect of intrathecal clonidine in neuropathic rats. Anesthesiology 89:1518–1523

Pan HL, Wu ZZ, Zhou HY, Chen SR, Zhang HM, Li DP (2008) Modulation of pain transmission by G-protein-coupled receptors. Pharmacol Ther 117:141–161

Pan YZ, Li DP, Chen SR, Pan HL (2002) Activation of delta-opioid receptors excites spinally projecting locus coeruleus neurons through inhibition of GABAergic Inputs. J Neurophysiol 88:2675–2683

Panocka I, Sadowski B, Marek P (1987) Adrenalectomy and dexamethasone differentially affect postswim antinociception in mice selectively bred for high and low stress-induced analgesia. Physiol Behav 40:597–601

Panula P, Aarnisalo AA, Wasowicz K (1996) Neuropeptide FF, a mammalian neuropeptide with multiple functions. Prog Neurobiol 48:461–487

Parolaro D, Patrini G, Massi P, Mainardi P, Giagnoni G, Sala M, Gori E (1989) Histamine as a central modulator of rat intestinal transit. J Pharmacol Exp Ther 249:324–328

Paszcuk AF, Gadotti VM, Tibola D, Quintao NL, Rodrigues AL, Calixto JB, Santos AR (2007) Anti-hypernociceptive properties of agmatine in persistent inflammatory and neuropathic models of pain in mice. Brain Res 1159:124–133

Patacchini R, Santicioli P, Giuliani S, Maggi CA (2004) Hydrogen sulfide (H2S) stimulates capsaicin-sensitive primary afferent neurons in the rat urinary bladder. Br J Pharmacol 142:31–34

Patacchini R, Maggi CA (2001) Peripheral tachykinin receptors as targets for new drugs. Eur J Pharmacol 429:13–21

Patel S, Naeem S, Kesingland A, Froestl W, Capogna M, Urban L, Fox A (2001) The effects of GABAB agonists and gabapentin on mechanical hyperalgesia in models of neuropathic and inflammatory pain in the rat. Pain 90:217–226

Pawlak D, Takada Y, Urano T, Takada A (2000) Serotonergic and kynurenic pathways in rats exposed to foot shock. Brain Res Bull 52:197–205

Pedersen LH, Scheel-Kruger J, Blackburn-Munro G (2007) Amygdala GABA-A receptor involvement in mediating sensory-discriminative and affective-motivational pain responses in a rat model of peripheral nerve injury. Pain 127:17–26

Pelegrini-da-Silva A, Martins AR, Prado WA (2005) A new role for the renin-angiotensin system in the rat periaqueductal gray matter: angiotensin receptor-mediated modulation of nociception. Neuroscience 132:453–463

Pellicer F, Lopez-Avila A, Coffeen U, Manuel Ortega-Legaspi J, Angel RD (2007) Taurine in the anterior cingulate cortex diminishes neuropathic nociception: a possible interaction with the glycine(A) receptor. Eur J Pain 11:444–451

Pelligrino DA, Laurito CE, VadeBoncouer TR (1996) Nitric oxide synthase inhibition modulates the ventilatory depressant and antinociceptive actions of fourth ventricular infusions of morphine in the awake dog. Anesthesiology 85:1367–1377

Peng HY, Chang HM, Chang SY, Tung KC, Lee SD, Chou D, Lai CY, Chiu CH, Chen GD, Lin TB (2008) Orexin-A modulates glutamatergic NMDA-dependent spinal reflex potentiation via inhibition of NR2B subunit. Am J Physiol Endocrinol Metab 295:E117–E129

Perrin MH, Vale WW (1999) Corticotropin releasing factor receptors and their ligand family. Ann N Y Acad Sci 885:312–328

Pert A, Moody TW, Pert CB, Dewald LA, Rivier J (1980) Bombesin: receptor distribution in brain and effects on nociception and locomotor activity. Brain Res 193:209–220

Pertovaara A (2006) Noradrenergic pain modulation. Prog Neurobiol 80:53–83

Pertovaara A, Ostergard M, Anko ML, Lehti-Koivunen S, Brandt A, Hong W, Korpi ER, Panula P (2005) RFamide-related peptides signal through the neuropeptide FF receptor and regulate pain-related responses in the rat. Neuroscience 134:1023–1032

Pertwee RG (2001) Cannabinoid receptors and pain. Prog Neurobiol 63:569–611

Pertwee RG (2007) GPR55: a new member of the cannabinoid receptor clan? Br J Pharmacol 152:984–986

Pettibone DJ, Hess JF, Hey PJ, Jacobson MA, Leviten M, Lis EV, Mallorga PJ, Pascarella DM, Snyder MA, Williams JB, Zeng Z (2002) The effects of deleting the mouse neurotensin receptor NTR1 on central and peripheral responses to neurotensin. J Pharmacol Exp Ther 300: 305–313

Pezet S, Malcangio M, McMahon SB (2002) BDNF: a neuromodulator in nociceptive pathways? Brain Res Rev 40:240–249

Pham T, Carrega L, Sauze N, Fund-Saunier O, Devaux C, Peragut JC, Saadjian A, Guieu R (2003) Supraspinal antinociceptive effects of mu and delta agonists involve modulation of adenosine uptake. Anesthesiology 98:459–464

Pieretti S, Di Giannuario A, De Felice M, Perretti M, Cirino G (2004) Stimulus-dependent specificity for annexin 1 inhibition of the inflammatory nociceptive response: the involvement of the receptor for formylated peptides. Pain 109:52–63

Pilcher WH, Joseph SA, Mcdonald JV (1988) Immunocytochemical localization of pro-opiomelanocortin neurons in human-brain areas subserving stimulation analgesia. J Neurosurg 68:621–629

Pinter E, Helyes Z, Szolcsanyi J (2006) Inhibitory effect of somatostatin on inflammation and nociception. Pharmacol Ther 112:440–456

Pinthong D, Wright IK, Hanmer C, Millns P, Mason R, Kendall DA, Wilson VG (1995) Agmatine recognizes alpha 2-adrenoceptor binding sites but neither activates nor inhibits alpha 2-adrenoceptors. Naunyn Schmiedebergs Arch Pharmacol 351:10–16

Piovezan AP, D'Orleans-Juste P, Tonussi CR, Rae GA (1997) Endothelins potentiate formalin-induced nociception and paw edema in mice. Can J Physiol Pharmacol 75:596–600

Pitcher MH, Price TJ, Entrena JM, Cervero F (2007) Spinal NKCC1 blockade inhibits TRPV1-dependent allodynia. Mol Pain 3:17

Pittman QJ, Blume HW, Renaud LP (1981) Connections of the hypothalamic paraventricular nucleus with the neurohypophysis, median-eminence, amygdala, lateral septum and midbrain periaqueductal gray - an electro-physiological study in the rat. Brain Res 215:15–28

Pivonello R, Ferone D, Lombardi G, Colao A, Lamberts SW, Hofland LJ (2007) Novel insights in dopamine receptor physiology. Eur J Endocrinol 156:S13–S21

Placzek EA, Okamoto Y, Ueda N, Barker EL (2008) Mechanisms for recycling and biosynthesis of endogenous cannabinoids anandamide and 2-arachidonylglycerol. J Neurochem 107:987–1000

Poisbeau P, Patte-Mensah C, Keller AF, Barrot M, Breton JD, Luis-Delgado OE, Freund-Mercier MJ, Mensah-Nyagan AG, Schlichter R (2005) Inflammatory pain upregulates spinal inhibition via endogenous neurosteroid production. J Neurosci 25:11768–11776

Poole S, Cunha FQ, Selkirk S, Lorenzetti BB, Ferreira SH (1995) Cytokine-mediated inflammatory hyperalgesia limited by interleukin-10. Br J Pharmacol 115:684–688

Poon A, Sawynok J (1998) Antinociception by adenosine analogs and inhibitors of adenosine metabolism in an inflammatory thermal hyperalgesia model in the rat. Pain 74:235–245

Porter AC, Sauer JM, Knierman MD, Becker GW, Berna MJ, Bao JQ, Nomikos GG, Carter P, Bymaster FP, Leese AB, Felder CC (2002) Characterization of a novel endocannabinoid, virodhamine, with antagonist activity at the CB1 receptor. J Pharmacol Exp Ther 301:1020–1024

Portugal-Santana P, Doretto MC, Tatsuo MAKF, Duarte IDG (2004) Involvement of prolactin, vasopressin and opioids in post-ictal antinociception induced by electroshock in rats. Brain Res 1003:1–8

Prado WA, Pelegrini-da-Silva A, Martins AR (2003) Microinjection of renin-angiotensin system peptides in discrete sites within the rat periaqueductal gray matter elicits antinociception. Brain Res 972:207–215

Prell GD, Green JP (1986) Histamine as a neuroregulator. Annu Rev Neurosci 9:209–254

Price J, Mudge AW (1983) A subpopulation of rat dorsal-root ganglion neurons is catecholaminergic. Nature 301:241–243

Price TJ, Patwardhan A, Akopian AN, Hargreaves KM, Flores CM (2004) Modulation of trigeminal sensory neuron activity by the dual cannabinoid-vanilloid agonists anandamide, N-arachidonoyl-dopamine and arachidonyl-2-chloroethylamide. Br J Pharmacol 141:1118–1130

Proudfit HK (1988) Pharmacologic evidence for the modulation of nociception by noradrenergic neurons. Prog Brain Res 77:357–370

Proudfit HK, Clark FM (1991) The projections of locus coeruleus neurons to the spinal cord. Prog Brain Res 88:123–141

Prusakiewicz JJ, Kingsley PJ, Kozak KR, Marnett LJ (2002) Selective oxygenation of N-arachidonylglycine by cyclooxygenase-2. Biochem Biophys Res Commun 296:612–617

Przewlocka B, Mika J, Labuz D, Toth G, Przewlocki R (1999) Spinal analgesic action of endomorphins in acute, inflammatory and neuropathic pain in rats. Eur J Pharmacol 367:189–196

Przewlocki R, Labuz D, Mika J, Przewlocka B, Tomboly C, Toth G (1999) Pain inhibition by endomorphins. Ann N Y Acad Sci 897:154–164

Qu K, Lee SW, Bian JS, Low CM, Wong PTH (2008) Hydrogen sulfide: neurochemistry and neurobiology. Neurochem Int 52:155–165

Qu ZX, Isaac L (1993) Dynorphin A (1–13) potentiates dynorphin A (1–17) on loss of the tail flick reflex intrathecal injection in the rat. Brain Res 610:340–343

Quartara L, Maggi CA (1998) The tachykinin NK1 receptor. Part II: distribution and pathophysiological roles. Neuropeptides 32:1–49

Quartilho A, Mata HP, Ibrahim MM, Vanderah TW, Porreca F, Kakriyannis A, Malan TP Jr (2003) Inhibition of inflammatory hyperalgesia by activation of peripheral CB2 cannabinoid receptors. Anesthesiology 99:955–960

Raasch W, Regunathan S, Li G, Reeis DJ (1995) Agmatine is widely and unequally distributed in rat organs. Ann N Y Acad Sci 763:330–334

Racz I, Bilkei-Gorzo A, Markert A, Stamer F, Gothert M, Zimmer A (2008) Anandamide effects on 5-HT3 receptors in vivo. Eur J Pharmacol 596:98–101

Rady JJ, Fujimoto JM (2002) Nociceptin and dynorphin A(1–17) produce antianalgesia trough independent systems in mice. Life Sci 70:1085–1093

Raffa RB, Schupsky JJ, Lee DKH, Jacoby HI (1996) Characterization of endothelin-induced nociception in mice: evidence for a mechanistically distinct analgesic model. J Pharmacol Exp Ther 278:1–7

Raffa RB, Stone DJ, Hipp SJ (1999) Differential cholera-toxin sensitivity of supraspinal antinociception induced by the cannabinoid agonists delta9-THC, WIN 55,212-2 and anandamide in mice. Neurosci Lett 263:29–32

Raghavendra V, Agrewala JN, Kulkarni SK (2000) Melatonin reversal of lipopolysacharides-induced thermal and behavioral hyperalgesia in mice. Eur J Pharmacol 395:15–21

Raghavendra V, Chopra K, Kulkarni SK (1999) Brain renin angiotensin system (RAS) in stress-induced analgesia and impaired retention. Peptides 20:335–342

Raghavendra V, Kulkarni SK (1999) Reversal of morphine tolerance and dependence by melatonin: possible role of central and peripheral benzodiazepine receptors. Brain Res 834:178–181

Raghavendra V, Kulkarni SK (2000) Possible mechanisms of action in melatonin reversal of morphine tolerance and dependence in mice. Eur J Pharmacol 409:279–289

Raigorodsky G, Urca G (1990) Spinal antinociceptive effects of excitatory amino acid antagonists: quisqualate modulates the action of N-methyl-D-aspartate. Eur J Pharmacol 182:37–47

Rainville P, Duncan GH, Price DD, Carrier B, Bushnell MC (1997) Pain affect encoded in human anterior cingulate but not somatosensory cortex. Science 277:968–971

Ramaswamy S, Viswanatham S, Bapna JS (1985) Prolactin analgesia: tolerance and cross-tolerance with morphine. Eur J Pharmacol 112:123–125

Rane K, Sollevi A, Segerdahl M (2000) Intrathecal adenosine administration in abdominal hysterectomy lacks analgesic effect. Acta Anaesthesiol Scand 44:868–872

Rashid MH, Furue H, Yoshimura M, Ueda H (2006) Tonic inhibitory role of alpha4beta2 subtype of nicotinic acetylcholine receptors on nociceptive transmission in the spinal cord in mice. Pain 125:125–135

Rashid MH, Ueda H (2002) Neuropathy-specific analgesic action of intrathecal nicotinic agonists and its spinal GABA-mediated mechanism. Brain Res 953:53–62

Raymond JR, Mukhin YV, Gelasco A, Turner J, Collinsworth G, Gettys TW, Grewal JS, Garnovskaya MN (2001) Multiplicity of mechanisms of serotonin receptor signal transduction. Pharmacol Ther 92:179–212

Reddy DS, Kulkarni SK (1997) Chronic neurosteroid treatment prevents the development of morphine tolerance and attenuates abstinence behavior in mice. Eur J Pharmacol 337:19–25

Regaya I, Pham T, Andreotti N, Sauze N, Carrega L, Martin-Eauclaire MF, Jouirou B, Peragut JC, Vacher H, Rochat H, Devaux C, Sabatier JM, Guieu R (2004) Small conductance calcium-activated K+ channels, SkCa, but not voltage-gated K+ (Kv) channels, are implicated in the antinociception induced by CGS21680, a A2A adenosine receptor agonist. Life Sci 76:367–377

Reimann W, Schlutz H, Selve N (1999) The antinociceptive effects of morphine, desipramine, and serotonin and their combinations after intrathecal injection in the rat. Anesth Analg 88:141–145

Reinscheid RK, Nothacker H-P, Bourson A, Ardati A, Henningsen RA, Bunzow JR, Grandy DK, Langen H, Monsma FJ Jr, Civelli O (1995) Orphanin FQ: a neuropeptide that activates an opioidlike G protein-coupled receptor. Science 270:792–794

Reinscheid RK, Nothacker H-P, Civelli O (2000) The orphanin FQ/nociceptin gene: structure, tissue distribution of expression and functional implications obtained from knockout mice. Peptides 21:901–906

Reis DJ, Regunathan S (2000) Is agmatine a novel neurotransmitter in brain? Trends Pharmacol Sci 21:187–193

Reis GM, Duarte ID (2007) Involvement of chloride channel coupled GABA(C) receptors in the peripheral antinociceptive effect induced by GABA(C) receptor agonist cis-4-aminocrotonic acid. Life Sci 80:1268–1273

Reiter RJ (1991) Pineal melatonin: cell biology of its synthesis and of its physiological interactions. Endocr Rev 12:151–180

Ren K, Wei F, Dubner R, Murphy A, Hoffman GE (2000) Progesterone attenuates persistent inflammatory hyperalgesia in female rats: involvement of spinal NMDA receptor mechanisms. Brain Res 865:272–277

Ren X, Noda Y, Mamiya T, Nagai T, Nabeshima T (2004) A neuroactive steroid, dehydroepiandrosterone sulfate, prevents the development of morphine dependence and tolerance via c-fos expression linked to the extracellular signal-regulated protein kinase. Behav Brain Res 152:243–250

Renganathan M, Cummins TR, Waxman SG (2002) Nitric oxide blocks fast, slow, and persistent Na+ channels in C-type DRG neurons by S-nitrosylation. J Neurophysiol 87:761–775

Reny-Palasse V, Monier C, Rips R (1989) Opioid involvement in TRH-induced antinociception in the rat following intracerebral administration. Pain 38:193–201

Reppert SM, Weaver DR, Godson C (1996) Melatonin receptors step into the light: cloning and classification of subtypes. Trends Pharmacol Sci 17:100–102

Riley RC, Trafton JA, Chi SI, Basbaum AI (2001) Presynaptic regulation of spinal cord tachykinin signaling via GABA(B) but not GABA(A) receptor activation. Neuroscience 103:725–737

Rimmerman N, Bradshaw HB, Hughes HV, Chen JS, Hu SS, McHugh D, Vefring E, Jahnsen JA, Thompson EL, Masuda K, Cravatt BF, Burstein S, Vasko MR, Prieto AL, O'Dell DK, Walker JM (2008) N-Palmitoyl glycine, a novel endogenous lipid that acts as a modulator of calcium influx and nitric oxide production in sensory neurons. Mol Pharmacol 74:213–224

Rioli V, Gozzo FC, Heimann AS, Linardi A, Krieger JE, Shida CS, Almeida PC, Hyslop S, Eberlin MN, Ferro ES (2003) Novel natural peptide substrates for endopeptidase 24.15, neurolysin, and angiotensin-converting enzyme. J Biol Chem 278:8547–8555

Rittner HL, Brack A, Stein C (2008) Pain and the immune system. Br J Anaesth 101:40–44

Robbins MT, Ness TJ (2008) Footshock-induced urinary bladder hypersensitivity: role of spinal corticotropin-releasing factor receptors. J Pain 9:991–998

Roerig SC (2003) Spinal and supraspinal agmatine activate different receptors to enhance spinal morphine antinociception. Ann N Y Acad Sci 1009:116–126

Roh DH, Kim HW, Yoon SY, Seo HS, Kwon YB, Kim KW, Han HJ, Beitz AJ, Na HS, Lee JH (2008) Intrathecal injection of the sigma(1) receptor antagonist BD1047 blocks both mechanical allodynia and increases in spinal NR1 expression during the induction phase of rodent neuropathic pain. Anesthesiology 109:879–889

Rokaeus A, Melander T, Hokfelt T, Lundberg JM, Tatemoto K, Carlquist M, Mutt V (1984) A galanin-like peptide in the central nervous system and intestine of the rat. Neurosci Lett 47: 161–166

Romano GJ, Mobbs CV, Howells RD, Pfaff DW (1989) Estrogen regulation of proenkephain gene-expression in the ventromedial hypothalamus of the rat - temporal qualities and synergism with progesterone. Mol Brain Res 5:51–58

Romero MT, Cooper ML, Komisaruk BR, Bodnar RJ (1988) Gender-specific and gonadectomy-specific effects upon swim analgesia: role of steroid replacement therapy. Physiol Behav 44:257–265

Romero-Sandoval A, Eisenach JC (2007) Clonidine reduces hypersensitivity and alters the balance of pro- and anti-inflammatory leukocytes after local injection at the site of inflammatory neuritis. Brain Behav Immun 21:569–580

Rosa AO, Egea J, Lorrio S, Rojo AI, Cuadrado A, Lopez MG (2008) Nrf2-mediated haeme oxygenase-1 up-regulation induced by cobalt protoporphyrin has antinociceptive effects against inflammatory pain in the formalin test in mice. Pain 137:332–339

Rosen A, Zhang YX, Lund I, Lundeberg T, Yu LC (2004) Substance P microinjected into the periaqueductal gray matter induces antinociception and is released following morphine administration. Brain Res 1001:87–94

Rossier J, French ED, Rivier C, Ling N, Guillemin R, Bloom F (1977) Foot-shock induced stress increases beta-endorphin levels in blood but not brain. Nature 270:618–620

Roumy M, Zajac J-M (1998) Neuropeptide FF, pain and analgesia. Eur J Pharmacol 345:1–11

Rousselot P, Papadopoulos G, Merighi A, Poulain DA, Theodosis DT (1990) Oxytocinergic innervation of the rat spinal cord. An electron microscopic study. Brain Res 529: 178–184

Roussy G, Dansereau MA, Dore-Savard L, Belleville K, Beaudet N, Richelson E, Sarret P (2008) Spinal NTS1 receptors regulate nociceptive signaling in a rat formalin tonic pain model. J Neurochem 105:1100–1114

Rubanyi GM, Polokoff MA (1994) Endothelins: molecular biology, biochemistry, pharmacology, physiology, and pathophysiology. Pharmacol Rev 46:325–394

Ruiz G, Ceballos D, Banos J-E (2004) Behavioral and histological effects of endoneurial administration of nerve growth factor: possible implications in neuropathic pain. Brain Res 1011:1–6

Rupprecht R (2003) Neuroactive steroids: mechanisms of action and neuropsychopharmacological properties. Psychoneuroendocrinology 28:139–168

Ryberg E, Larsson N, Sjogren S, Hjorth S, Hermansson NO, Leonova J, Elebring T, Nilsson K, Drmota T, Geasley PJ (2007) The orphan receptor GPR55 is a novel cannabinoid receptor. Br J Pharmacol 152:1092–1101

Saario SM, Laitinen JT (2007) Therapeutic potential of endocannabinoid-hydrolysing enzyme inhibitors. Basic Clin Pharmacol Toxicol 101:287–293

Sakagawa T, Okuyama S, Kawashima N, Hozumi S, Nakagawasai O, Tadano T, Kisara K, Ichiki T, Inagami T (2000) Pain threshold, learning and formation of brain edema in mice lacking the angiotensin II type 2 receptor. Life Sci 67:2577–2585

Sakai A, Asada M, Seno N, Suzuki H (2008) Involvement of neural cell adhesion molecule signaling in glial cell line-derived neurotrophic factor-induced analgesia in a rat model of neuropathic pain. Pain 137:378–388

Sakurada S, Hayashi T, Yuhki M, Fujimura T, Murayama K, Yonezawa A, Sakurada C, Takeshita M, Zadina JE, Kastin AJ, Sakurada T (2000) Differential antagonism of endomorphin-1 and endomorphin-2 spinal nociception by naloxonazine and 3-methoxynaltraxone. Brain Res 881:1–8

Sakurada T, Watanabe C, Okuda K, Sugiyama A, Moriyama T, Sakurada C, Tan-No K, Sakurada S (2002) Intrathecal high-dose morphine induces spinally-mediated behavioral responses through NMDA receptors. Mol Brain Res 98:111–118

Sanchez-Blazquez P, Boronat MA, Olmos G, Garcia-Sevilla JA, Garzon J (2000) Activation of I2-imidazoline receptors channels supraspinal morphine analgesia in mice: a model to detect agonist and antagonist activites at these receptors. Br J Pharmacol 130:146–152

Sanderson K, Nyberg F, Khalil Z (1998) Modulation of peripheral inflammation by locally administered hemorphin-7. Inflamm Res 47:49–55

Sandkuhler J (1996) The organization and function of endogenous antinociceptive system. Prog Neurobiol 50:49–81

Sandman CA, Kastin AJ (1981) Intraventricular administration of MSH induces hyperalgesia in rats. Peptides 2:231–233

Sandor K, Bolcskei K, McDougall JJ, Schuelert N, Reglodi D, Elekes K, Petho G, Pinter E, Szolcsanyi J, Helyes Z (2009) Divergent peripheral effects of pituitary adenylate cyclase-activating polypeptide-38 on nociception in rats and mice. Pain 141:143–150

Sandor K, Elekes K, Szabo A, Pinter E, Engstrom M, Wurster S, Szolcsanyi J, Helyes Z (2006) Analgesic effects of the somatostatin sst4 receptor selective agonist J-2156 in acute and chronic pain models. Eur J Pharmacol 539:71–75

Sann H, McCarthy PW, Mader M, Schemann M (1995) Choline acetyltransferase-like immunoreactivity in small diameter neurones of the rat dorsal root ganglion. Neurosci Lett 198: 17–20

Santicioli P, Del Bianco E, Maggi CA (1993) Adenosine A1 receptors mediate the presynaptic inhibition of calcitonin gene-related peptide release by adenosine in the rat spinal cord. Eur J Pharmacol 231:139–142

Sarret P, Esdaile MJ, Perron A, Martinez J, Stroh T, Beaudet A (2005) Potent spinal analgesia elicited through stimulation of NTS2 neurotensin receptors. J Neurosci 25:8188–8196

Sasamura T, Kuraishi Y (1999) Peripheral and central actions of capsaicin and VR1 receptor. Jpn J Pharmacol 80:275–280

Sawchenko PE, Swanson LW (1982) Immunohistochemical identification of neurons in the paraventricular nucleus of the hypothalamus that project to the medulla or to the spinal-cord in the rat. J Comp Neurol 205:260–272

Sawynok J (1998) Adenosine receptor activation and nociception. Eur J Pharmacol 347:1–11

Sawynok J, Liu XJ (2003) Adenosine in the spinal cord and periphery: release and regulation of pain. Prog Neurobiol 69:313–340

Sawynok J, Sweeney MI, White TD (1989) Adenosine release may mediate spinal analgesia by morphine. Trends Pharmacol Sci 10:186–189

Schafer M, Mousa SA, Zhang Q, Carter L, Stein C (1996) Expression of corticotropin-releasing factor in inflamed tissue is required for intrinsic peripheral opioid analgesia. Proc Natl Acad Sci U S A 93:6096–6100

Schepers RJ, Mahoney JL, Gehrke BJ, Shippenberg TS (2008) Endogenous kappa-opioid receptor systems inhibit hyperalgesia associated with localized peripheral inflammation. Pain 138: 423–439

Schlichter R, Keller AF, De Roo M, Breton LD, Inquimbert P, Poisbeau P (2006) Fast nongenomic effects of steroids on synaptic transmission and role of endogenous neurosteroids in spinal pain pathways. J Mol Neurosci 28:33–51

Schmidt AP, Bohmer AE, Leke R, Schallenberger C, Antunes C, Pereira MS, Wofchuk ST, Elisabetsky E, Souza DO (2008) Antinociceptive effects of intracerebroventricular administration of guanine-based purines in mice: evidences for the mechanism of action. Brain Res 1234:50–58

Schmidt C, Peyroux J, Noble F, Fournie-Zaluski M-C, Roques BP (1991) Analgesic responses elicited by endogenous enkephalins (protected by mixed peptidase inhibitors) in a variety of morphine-sensitive noxious tests. Eur J Pharmacol 192:253–262

Scholz J, Woolf CJ (2007) The neuropathic pain triad: neurons, immune cells and glia. Nat Neurosci 10:1361–1368

Schulte G, Robertson B, Fredholm BB, DeLander GE, Shortland P, Molander C (2003) Distribution of antinociceptive adenosine A1 receptors in the spinal cord dorsal horn, and relationship to primary afferents and neuronal subpopulations. Neuroscience 121:907–916

Schwarcz R, Pellicciari R (2002) Manipulation of brain kynurenines: glial targets, neuronal effects, and clinical opportunities. J Pharmacol Exp Ther 303:1–10

Schwartz JC, Arrang JM, Garbarg M, Pollard H, Ruat M (1991) Histaminergic transmission in the mammalian brain. Physiol Rev 71:1–51

Scoto GM, Santangelo N, Parenti C (2005) Effect of supraspinal nocistatin on nociceptin/orphanin FQ antagonism of selective opioid analgesia. Neurosci Lett 387:126–129

Senogles SE (2000) The D2s dopamine receptor stimulates phospholipase D activity: a novel signaling pathway for dopamine. Mol Pharmacol 58:455–462

Serrano MI, Serrano JS, Fernandez A, Asadi I, Serrano-Martino MC (1998) GABAB receptors and opioid mechanisms involved in homotaurine-induced analgesia. Gen Pharmacol 30:411–415

Sherman SE, Loomis CW (1996) Strychnine-sensitive modulation is selective for non-noxious somatosensory input in the spinal cord of the rat. Pain 66:321–330

Shi TJ, Zhang X, Berge OG, Erickson JC, Palmiter RD, Hokfelt T (1998) Effect of peripheral axotomy on dorsal root ganglion neuron phenotype and autotomy behaviour in neuropeptide Y-deficient mice. Regul Pept 75–76:161–173

Shimada S, Inagaki S, Kubota Y, Kito S, Funaki H, Takagi H (1989) Light and electron-microscopic studies of calcitonin gene-related peptide-like immunoreactive terminals in the central nucleus of the amygdala and the bed nucleus of the stria terminalis of the rat. Exp Brain Res 77:217–220

Shimizu T, Katahira M, Sugawara H, Inoue K, Miyata A (2004) Diverse effects of intrathecal pituitary adenylate cyclase-activating polypeptide on nociceptive transmission in mice spinal cord. Regul Pept 123:117–122

Shin MS (2005) Vasoactive intestinal peptide in the amygdala inhibits tail flick reflexes in rats. Brain Res 1040:197–201

Shiomi H, Kuraishi Y, Ueda H, Harada Y, Amano H, Takagi H (1981) Mechanism of kyotorphin-induced release of Met-enkephalin from guinea pig striatum and spinal cord. Brain Res 221:161–169

Shoemaker JL, Joseph BK, Ruckle MB, Mayeux PR, Prather PL (2005) The endocannabinoid noladin ether acts as a full agonist at human CB2 cannabinoid receptors. J Pharmacol Exp Ther 314:868–875

Shukla VK, Bansinath M, Dumont M, Lemaire S (1992) Selective involvement of kappa opioid and phencyclidine receptors in the analgesic and motor effects of dynorphin-A-(1–13)-Tyr-Leu-Phe-Asn-Gly-Pro. Brain Res 591:176–180

Sibilia V, Lattuada N, Rapetti D, Pagani F, Vincenza D, Bulgarelli I, Locatelli V, Guidobono F, Netti C (2006) Ghrelin inhibits inflammatory pain in rats: involvement of the opioid system. Neuropharmacology 51:497–505

Sim LJ, Liu Q, Childers SR, Selley DE (1998) Endomorphin-stimulated [35S]GTPgammaS binding in rat brain: evidence for partial agonist activity at mu-opioid receptors. J Neurochem 70:1567–1576

Simon GS, Dewey WL (1981) Narcotics and diabetes. I. The effects of streptozotocin-induced diabetes on the antinociceptive potency of morphine. J Pharmacol Exp Ther 218: 318–323

Simpson DA, Headley PM, Lumb BM (2008) Selective inhibition from the anterior hypothalamus of C- versus A-fibre mediated spinal nociception. Pain 136:305–312

Singh VP, Jain NK, Kulkarni SK (2001) On the antinociceptive effect of fluoxetine, a selective serotonin reuptake inhibitor. Brain Res 915:218–226

Sipe JC, Arbour N, Gerber A, Beutler E (2005) Reduced endocannabinoid immune modulation by a common cannabinoid 2 (CB2) receptor gene polymorphism: possible risk for autoimmune disorders. J Leukoc Biol 78:231–238

Skagerberg G, Lindvall O (1985) Organization of diencephalic dopamine neurones projecting to the spinal cord in the rat. Brain Res 342:340–351

Skilling SR, Smullin DH, Larson AA (1990) Differential-effects of C-terminal and N-terminal substance P metabolites on the release of amino-acid neurotransmitters from the spinal cord: potential role in nociception. J Neurosci 10:1309–1318

Skofitsch G, Jacobowitz DM (1985a) Calcitonin gene-related peptide coexists with substance P in capsaicin sensitive neurons and sensory ganglia of the rat. Peptides 6:747–754

Skofitsch G, Jacobowitz DM (1985b) Galanin-like immunoreactivity in capsaicin sensitive sensory neurons and ganglia. Brain Res Bull 15:191–195

Smart D (1999) Orexins: a new family of neuropeptides. Br J Anaesth 83:695–697

Smith DJ, Perrotti JM, Crisp T, Cabral MEY, Long JT, Scalzitti JM (1988) The mu opiate receptor is responsible for descending pain inhibition originating in the periaqueductal gray region of the rat brain. Eur J Pharmacol 156:47–54

Smith PA, Moran TD, Abdulla F, Tumber KK, Taylor BK (2007) Spinal mechanisms of NPY analgesia. Peptides 28:464–474

Smith PB, Compton DR, Welch SP, Razdan RK, Mechoulam R, Martin BR (1994) The pharmacological activity of anandamide, a putative endogenous cannabinoid, in mice. J Pharmacol Exp Ther 270:219–227

Smullin DH, Skilling SR, Larson AA (1990) Interactions between substance P, calcitonin gene-related peptide, taurine and excitatory amino-acids in the spinal cord. Pain 42: 93–101

Sofroniew MV, Weindl A, Schrell U, Wetzstein R (1981) Immunohistochemistry of vasopressin, oxytocin and neurophysin in the hypothalamus and extrahypothalamic regions of the human and primate brain. Acta Histochem S24:79–95

Sollevi A, Belfrage M, Lundeberg T, Segerdahl M, Hansson P (1995) Systemic adenosine infusion: a new treatment modality to alleviate neuropathic pain. Pain 61:155–158

Song GH, Leng PH, Gwee KA, Moochhala SM, Ho KY (2005) Melatonin improves abdominal pain in irritable bowel syndrome patients who have sleep disturbances: a randomised, double blind, placebo controlled study. Gut 54:1402–1407

Song HK, Pan HL, Eisenach JC (1998) Spinal nitric oxide mediates antinociception from intravenous morphine. Anesthesiology 89:215–221

Song P, Zhao Z (2000) Interleukin 2-induced antinociception partially coupled with mu receptor. Cytokine 12:1240–1242

Sorkin LS, Yu AL, Junger H, Doom CM (2002) Antibody directed against GD2 produces mechanical allodynia, but not thermal hyperalgesia when administered systemically or intrathecally despite its dependence on capsaicin sensitive afferents. Brain Res 930:67–74

Sorrells SF, Sapolsky RM (2007) An inflammatory review of glucocorticoid actions in the CNS. Brain Behav Immun 21:259–272

Sousa AM, Prado WA (2001) The dual effect of a nitric oxide donor in nociception. Brain Res 897:9–19

Starowicz K, Cristino L, Di Marzo V (2008) TRPV1 receptors in the central nervous system: potential for previously unforeseen therapeutic applications. Curr Pharm Des 14:42–54

Starowicz K, Maione S, Cristino L, Palazzo E, Marabese I, Rossi F, De Novellis V, Di Marzo V (2007) Tonic endovanilloid facilitation of glutamate release in brainstem descending antinociceptive pathways. J Neurosci 27:13739–13749

Starowicz K, Przewlocki R, Gispen WH, Przewlocka B (2002) Modulation of melanocortin-induced changes in spinal nociception by mu-opioid receptor agonist and antagonist in neuropathic rats. NeuroReport 13:2447–2452

Stein C, Gramsch C, Herz A (1990) Intrinsic mechanisms of antinociception in inflammation: local opioid receptors and beta-endorphin. J Neurosci 10:1292–1298

Steiner AA, Branco LGS, Cunha FQ, Ferreira SH (2001) Role of the haeme oxygenase/carbon monoxide pathway in mechanical nociceptor hypersensitivity. Br J Pharmacol 132: 1673–1682

Stone EA, Lin Y, Rosengarten H, Kramer HK, Quartermain D (2003) Emerging evidence for a central epinephrine-innervated alpha 1-adrenergic system that regulates behavioral activation and is impaired in depression. Neuropsychopharmacology 28:1387–1399

Stone TW (1993) Neuropharmacology of quinolinic and kynurenic acids. Pharmacol Rev 45: 309–379

Stone TW (2000) Development and therapeutic potential of kynurenic acid and kynurenine derivatives for neuroprotection. Trends Pharmacol Sci 21:149–154

Suardiaz M, Estivill-Torrus G, Goicoechea C, Bilbao A, Rodriguez de Fonseca F (2007) Analgesic properties of oleoylethanolamide (OEA) in visceral and inflammatory pain. Pain 133: 99–110

Suarez C, Tornadu IG, Cristina C, Vela J, Iglesias AG, Libertun C, Diaz-Torga G, Becu-Villalobos D (2002) Angiotensin and calcium signaling in the pituitary and hypothalamus. Cell Mol Neurobiol 22:315–333

Succar R, Mitchell VA, Vaughan CW (2007) Actions of N-arachidonyl-glycine in a rat inflammatory pain model. Mol Pain 3:24

Sugden D (1983) Psychopharmacological effects of melatonin in mouse and rat. J Pharmacol Exp Ther 227:587–591

Sugiura T, Kishimoto S, Oka S, Gokoh M (2006) Biochemistry, pharmacology and physiology of 2-arachidonoylglycerol, an endogenous cannabinoid receptor ligand. Prog Lipid Res 45: 405–446

Suh HW, Song DK, Choi YS, Kim YH (1996) Effects of intrathecally injected histamine receptor antagonists on the antinociception induced by morphine, beta-endorphin, and U50, 488H administered intrathecally in the mouse. Neuropeptides 30:485–490

Suh HW, Song D-K, Choi YS, Cheon S-H, Kim YH (1994) Differential effects of intrathecally injected galanin on antinociception induced by beta-endorphin and morphine administered intracerebroventricularly in mice. Neuropeptides 26:297–303

Sullivan AF, Kalso EA, McQuay HJ, Dickenson AH (1991) FLFQPQRF-amide modulates alpha 2-adrenergic antinociception in the rat dorsal horn in vivo. Brain Res 562:327–328

Sun YG, Chen Z-F (2007) A gastrin-releasing peptide receptor mediates the itch sensation in the spinal cord. Nature 448:700–703

Sun YG, Lundeberg T, Yu LC (2003) Involvement of endogenous beta-endorphin in antinociception in the arcuate nucleus of hypothalamus in rats with inflammation. Pain 104:55–63

Sun YG, Yu LC (2005) Interactions of galanin and opioids in nociceptive modulation in the arcuate nucleus of hypothalamus in rats. Regul Pept 124:37–43

Suplita II RL, Gutierrez T, Fegley D, Piomelli D, Hohmann AG (2006) Endocannabinoids at the spinal level regulate, but do not mediate, nonopioid stress-induced analgesia. Neuropharmacology 50:372–379

Suplita RL, Farthing JN, Gutierrez T, Hohmann AG (2005) Inhibition of fatty-acid amide hydrolase enhances cannabinoid stress-induced analgesia: sites of action in the dorsolateral periaqueductal gray and rostral ventromedial medulla. Neuropharmacology 49:1201–1209

Sutton LC, Fleshner M, Mazzeo R, Maier SF, Watkins LR (1994) A permissive role of corticosterone in an opioid form of stress-induced analgesia: blockade of opiate analgesia is not due to stress-induced hormone release. Brain Res 663:19–29

Suzuki R, Rahman W, Hunt SP, Dickenson AH (2004) Descending facilitatory control of mechanically evoked responses is enhanced in deep dorsal horn neurones following peripheral nerve injury. Brain Res 1019:68–76

Swanson LW, McKellar S (1979) Distribution of oxytocin-stained and neurophysin-stained fibers in the spinal-cord of the rat and monkey. J Comp Neurol 188:87–106

Sweeney MI, White TD, Sawynok J (1987) Involvement of adenosine in the spinal antinociceptive effects of morphine and noradrenaline. J Pharmacol Exp Ther 243:657–665

Szabo B (2008) Pharmacology of cannabinoid receptors. Biotrend Rev 2:1–13

Szabo C (2007) Hydrogen sulphide and its therapeutic potential. Nat Rev Drug Discov 6:917–935

Szeto HH (2003) Dynorphin and the hypothalamo-pituitary-adrenal axis during fetal development. Life Sci 73:749–758

Szolcsanyi J (2000) Are cannabinoids endogenous ligands for the VR1 capsaicin receptor? Trends Pharmacol Sci 21:41–42

Szolcsanyi J (2004) Forty years in capsaicin research for sensory pharmacology and physiology. Neuropeptides 38:377–384

Szolcsanyi J, Sandor Z, Petho G, Varga A, Bolcskei K, Almasi R, Riedl Z, Hajos G, Czeh G (2004) Direct evidence for activation and desensitization of the capsaicin receptor by *N*-oleyldopamine on TRPV1-transfected cell, line in gene deleted mice and in the rat. Neurosci Lett 361:155–158

Taguchi H, Oishi K, Sakamoto S, Shingu K (2007) Intrathecal betamethasone for cancer pain in the lower half of the body: a study of its analgesic efficacy and safety. Br J Anaesth 98:385–389

Taiwo OB, Taylor BK (2002) Antihyperalgesic effects of intrathecal neuropeptide Y during inflammation are mediated by Y1 receptors. Pain 96:353–363

Taiwo YO, Levine JD (1992) Serotonin is a directly-acting hyperalgesic agent in the rat. Neuroscience 48:485–490

Takada M, Li ZK, Hattori T (1988) Single thalamic dopaminergic neurons project to both the neocortex and spinal cord. Brain Res 455:346–352

Takagi H, Shiomi H, Ueda H, Amano H (1979) A novel analgesic dipeptide from bovine brain is a possible Met-enkephalin releaser. Nature 282:410–412

Takeda D, Nakatsuka T, Papke R, Gu JG (2003) Modulation of inhibitory synaptic activity by a non-alpha4beta2, non-alpha7 subtype of nicotinic receptors in the substantia gelatinosa of adult rat spinal cord. Pain 101:13–23

Takemori AE, Portoghese PS (1993) Enkephalin antinociception in mice is mediated by delta1- and delta2-opioid receptors in the brain and spinal cord, respectively. Eur J Pharmacol 242:145–150

Takeshita N, Yamaguchi I (1997) Insulin attenuates formalin-induced nociceptive response in mice through a mechanism that is deranged by diabetes mellitus. Mol Brain Res 63:180–183

Tanabe M, Takasu K, Yamaguchi S, Kodama D, Ono H (2008) Glycine transporter inhibitors as a potential therapeutic strategy for chronic pain with memory impairment. Anesthesiology 108:929–937

Tanabe M, Tokuda Y, Takasu K, Ono K, Honda M, Ono H (2007) The synthetic TRH analogue taltirelin exerts modality-specific antinociceptive effects via distinct descending monoaminergic systems. Br J Pharmacol 150:403–414

Tang B, Ji Y, Traub RJ (2008) Estrogen alters spinal NMDA receptor activity via a PKA signaling pathway in a visceral pain model in the rat. Pain 137:540–549

Tao F, Tao YX, Zhao C, Dore S, Liaw WJ, Raja SN, Johns RA (2004) Differential roles of neuronal and endothelial nitric oxide synthases during carrageenan-induced inflammatory hyperalgesia. Neuroscience 128:421–430

Tao F, Tao Y-X, Mao P, Zhao C, Li D, Liaw W-J, Raja SN, Johns RA (2003) Intact carrageenan-induced thermal hyperalgesia in mice lacking inducible nitric oxide synthase. Neuroscience 120:847–854

Tashev R, Belcheva S, Milenov K, Belcheva I (2001) Antinociceptive effect of somatostatin microinjected into caudate putamen. Peptides 22:1079–1083

Tata AM, De Stefano ME, Srubek Tomassy G, Vilaro MT, Levey AI, Biagioni S (2004) Subpopulations of rat dorsal root ganglion neurons express active vesicular acetylcholine transporter. J Neurosci Res 75:194–202

Taylor BK, Joshi C, Uppal H (2003) Stimulation of dopamine D2 receptors in the nucleus accumbens inhibits inflammatory pain. Brain Res 987:135–143

Telleria-Diaz A, Ebersberger A, Vasquez E, Schache F, Kahlenbach J, Schaible HG (2008) Different effects of spinally applied prostaglandin D2 on responses of dorsal horn neurons with knee input in normal rats and in rats with acute knee inflammation. Neuroscience 156:184–192

Terashvili M, Tseng LF, Wu H, Narayanan J, Hart LM, Falck JR, Pratt PF, Harder DR (2008) Antinociception produced by 14,15-epoxyeicosatrienoic acid is mediated by the activation of beta-endorphin and Met-enkephalin in the rat ventrolateral periaqueductal gray. J Pharmacol Exp Ther 326:614–622

Than M, Nemeth J, Szilvassy Z, Pinter E, Helyes Z, Szolcsanyi J (2000) Systemic anti-inflammatory effect of somatostatin released from capsaicin-sensitive vagal and sciatic sensory fibers of the rat and guinea-pig. Eur J Pharmacol 399:251–258

Theodosiou M, Rush RA, Zhou XF, Hu D, Walker JS, Tracey DJ (1999) Hyperalgesia due to nerve damage: role of nerve growth factor. Pain 81:245–255

Thoburn KK, Hough LB, Nalwalk JW, Mischler SA (1994) Histamine-induced modulation of nociceptive responses. Pain 58:29–37

Thompson SW, Bennett DLH, Kerr BJ, Bradbury EJ, McMahon SB (1999) Brain-derived neurotrophic factor is an endogenous modulator of nociceptive responses in the spinal cord. Proc Natl Acad Sci U S A 96:7714–7718

Thurston CL, Campbell IG, Culhane ES, Carstens E, Watkins LR (1992) Characterization of intrathecal vasopressin-induced antinociception, scratching behavior, and motor suppression. Peptides 13:17–25

Todd AJ, Spike RC, Russell G, Johnston HM (1992) Immunohistochemical evidence that Met-enkephalin and GABA coexist in some neurones in rat dorsal horn. Brain Res 584:149–156

Tokunaga A, Saika M, Senba E (1998) 5-HT2A receptor subtype is involved in the thermal hyperalgesic mechanism of serotonin in the periphery. Pain 76:349–355

Tozaki-Saitoh H, Tsuda M, Miyata H, Ueda K, Kohsaka S, Inoue K (2008) P2Y12 receptors in spinal microglia are required for neuropathic pain after peripheral nerve injury. J Neurosci 28:4949–4956

Traversa U, Bombi G, Camaioni E, Macchiarulo A, Costantino G, Palmieri C, Caciagli F, Pellicciari R (2003) Rat brain guanosine binding site. Biological studies and pseudo-receptor construction. Bioorg Med Chem 11:5417–5425

Tribollet E, Duboisdauphin M, Dreifuss JJ, Barberis C, Jard S (1992) Oxytocin receptors in the central-nervous-system - distribution, development, and species-differences. Ann N Y Acad Sci 652:29–38

Trivedi P, Yu H, MacNeil DJ, Van der Ploeg LHT, Guan XM (1998) Distribution of orexin receptor mRNA in the rat brain. FEBS Lett 438:71–75

Tseng LF, Collins KA (1993) Spinal involvement of both dynorphin A and Met-enkephalin in the antinociception induced by intracerebroventriculary administered bremazocine but not morphine in the mouse. J Pharmacol Exp Ther 266:1430–1438

Tseng L-F, Narita M, Suganuma C, Mizoguchi H, Ohsawa M, Nagase H, Kampine JP (2000) Differential antinociceptive effects of endomorphin-1 and endomorphin-2 in the mouse. J Pharmacol Exp Ther 292:576–583

Tsuda M, Inoue K, Salter MW (2005) Neuropathic pain and spinal microglia: a big problem from molecules in 'small' glia. Trends Neurosci 28:101–107

Tu Y, Sun RQ, Willis WD (2004) Effects of intrathecal injections of melatonin analogs on capsaicin-induced secondary mechanical allodynia and hyperalgesia in rats. Pain 109:340–350

Tuboly G, Benedek G, Horvath G (2009) Selective disturbance of pain sensitivity after social isolation. Physiol Behav 96:18–22

Turski WA, Schwarcz R (1988) On the disposition of intrahippocampally injected kynurenic acid in the rat. Exp Brain Res 71:563–567

Uchida H, Mizuno K, Yoshida A, Ueda H (2003) Neurosteroid-induced hyperalgesia through a histamine release is inhibited by progesterone and p,p′-DDE, an endocrine disrupting chemical. Neurochem Int 42:401–407

Ueda H, Shiomi H, Takagi H (1980) Regional distribution of a novel analgesic dipeptide kyotorphin (Tyr-Arg) in the rat brain and spinal cord. Brain Res 198:460–464

Ueda H, Yoshihara Y, Misawa H, Fukushima N, Katada T, Ui M, Takagi H, Satoh M (1989) The kyotorphin (tyrosine-arginine) receptor and a selective reconstitution with purified Gi, measured with GTPase and phospholipase C assays. J Biol Chem 264:3732–3741

Ulugol A, Dokmeci D, Guray G, Sapolyo N, Ozyigit F, Tamer M (2006) Antihyperalgesic, but not antiallodynic, effect of melatonin in nerve-injured neuropathic mice: possible involvements of the L-arginine-NO pathway and opioid system. Life Sci 78:1592–1597

Urban MO, Coutinho SV, Gebhart GF (1999) Biphasic modulation of visceral nociception by neurotensin in rat rostral ventromedial medulla. J Pharmacol Exp Ther 290:207–213

Urbanska EM, Chmielewski M, Kocki T, Turski WA (2000) Formation of endogenous glutamatergic receptors antagonist kynurenic acid - differences between cortical and spinal cord slices. Brain Res 878:210–212

Usdin TB, Dobolyi A, Ueda H, Palkovits M (2003) Emerging functions for tuberoinfundibular peptide of 39 residues. Trends Endocrinol Metab 14:14–19

Vaccarino F, Guidotti A, Costa E (1987) Ganglioside inhibition of glutamate-mediated protein kinase C translocation in primary cultures of cerebellar neurons. Proc Natl Acad Sci U S A 84:8707–8711

Vale ML, Marques JB, Moreira CA, Rocha FA, Ferreira SH, Poole S, Cunha FQ, Ribeiro RA (2003) Antinociceptive effects of interleukin-4, -10, and -13 on the writhing response in mice and zymosan-induced knee joint incapacitation in rats. J Pharmacol Exp Ther 304:102–108

Vallone D, Picetti R, Borrelli E (2000) Structure and function of dopamine receptors. Neurosci Biobehav Rev 24:125–132

Van den Pol AN (1999) Hypothalamic hypocretin (orexin): robust innervation of the spinal cord. J Neurosci 19:3171–3182

Van Huysse JW, Leenen FH (1998) Role of endogenous brain "ouabain" in the sympathoexcitatory and pressor effects of sodium. Clin Exp Hypertens 20:657–667

Van Sickle MD, Duncan M, Kingsley PJ, Mouihate A, Urbani P, Mackie K, Stella N, Makriyannis A, Piomelli D, Davison JS, Marnett LJ, Di Marzo V, Pittman QJ, Patel KD, Sharkey KA (2005) Identification and functional characterization of brainstem cannabinoid CB2 receptors. Science 310:329–332

Van Steenwinckel J, Brisorgueil MJ, Fischer J, Verge D, Gingrich JA, Bourgoin S, Hamon M, Bernard R, Conrath M (2008) Role of spinal serotonin 5-HT2A receptor in 2′,3′-dideoxycytidine-induced neuropathic pain in the rat and the mouse. Pain 137:66–80

Vanderah TW, Laughlin T, Lashbrook JM, Nichols ML, Wilcox GL, Ossipov MH, Malan TP, Porreca F (1996) Single intrathecal injections of dynorphin A or des-Tyr-dynorphins produce long-lasting allodynia in rats: blockade by MK-801 but not naloxone. Pain 68:275–281

Vanecek J (1998) Cellular mechanisms of melatonin action. Physiol Rev 78:687–721

Vanegas H, Schaible HG (2001) Prostaglandins and cyclooxygenases in the spinal cord. Prog Neurobiol 64:327–363

Vasquez E, Bar KJ, Ebersberger A, Klein B, Vanegas H, Schaible HG (2001) Spinal prostaglandins are involved in the development but not the maintenance of inflammation-induced spinal hyperexcitability. J Neurosci 21:9001–9008

Vaught JL, Pelley K, Costa LG, Setler P, Enna SJ (1985) A comparison of the antinociceptive responses to the GABA-receptor agonists THIP and baclofen. Neuropharmacology 24:211–216

Vecsei L, Beal MF (1991) Comparative behavioral and pharmacological studies with centrally administered kynurenine and kynurenic acid in rats. Eur J Pharmacol 196:239–246

Velligan DI, Ritch JL, Sui D, DiCocco M, Huntzinger CD (2002) Frontal systems behavior scale in schizophrenia: relationships with psychiatric symptomatology, cognition and adaptive function. Psychiatry Res 113:227–236

Vergnano AM, Ferrini F, Salio C, Lossi L, Baratta M, Merighi A (2008) The gastrointestinal hormone ghrelin modulates inhibitory neurotransmission in deep laminae of mouse spinal cord dorsal horn. Endocrinology 149:2306–2312

Vergnano AM, Salio C, Merighi A (2004) NK1 receptor activation leads to enhancement of inhibitory neurotransmission in spinal substantia gelatinosa neurons of mouse. Pain 112:37–47

Vergnano AM, Schlichter R, Poisbeau P (2007) PKC activation sets an upper limit to the functional plasticity of GABAergic transmission induced by endogenous neurosteroids. Eur J Neurosci 26:1173–1182

Verma A, Hirsch DJ, Glatt CE, Ronnett GV, Snyder SH (1993) Carbon-monoxide - a putative neural messenger. Science 259:381–384

Vincler M, Eisenach JC (2004) Plasticity of spinal nicotinic acetylcholine receptors following spinal nerve ligation. Neurosci Res 48:139–145

Vit JP, Clauw DJ, Moallem T, Boudah A, Ohara PT, Jasmin L (2006) Analgesia and hyperalgesia from CRF receptor modulation in the central nervous system of Fischer and Lewis rats. Pain 121:241–260

Vitte PA, Harthe C, Pevet P, Claustrat B (1990) Brain autoradiographic study in the golden hamster after intracarotid injection of [14C]melatonin. Neurosci Lett 110:1–5

Von Heijne M, Hao JX, Sollevi A, Xu XJ (1999) Intrathecal adenosine does not relieve allodynia-like behavior in spinally injured rats. NeuroReport 10:3247–3251

Vorwerk CK, Bonheur J, Kreutz MR, Dreyer EB, Laev H (1999) GM1 ganglioside administration protects spinal neurons after glutamate excitotoxicity. Restor Neurol Neurosci 14:47–51

Vrinten DH, Adan RAH, Groen GJ, Gispen WH (2001) Chronic blockade of melanocortin receptors alleviates allodynia in rats with neuropathic pain. Anesth Analg 93:1572–1577

Vrinten DH, Gispen WH, Groen GJ, Adan RAH (2000) Antagonism of the melanocortin system reduces cold and mechanical allodynia in mononeuropathic rats. J Neurosci 20:8131–8137

Vuckovic S, Tomic M, Stepanovic-Petrovic R, Ugresic N, Prostran M, Boskovic B (2006) Peripheral antinociception by carbamazepine in an inflammatory mechanical hyperalgesia model in the rat: a new target for carbamazepine? J Pharmacol Sci 100:310–314

Vuong LAQ, Mitchell VA, Vaughan CW (2008) Actions of N-arachidonyl-glycine in a rat neuropathic pain model. Neuropharmacology 54:189–193

Walker JM, Hohmann AG (2005) Cannabinoid mechanisms of pain suppression. Handb Exp Pharmacol 168:509–554

Walker JM, Krey JF, Chen JS, Vefring E, Jahnsen JA, Bradshaw H, Huang SM (2005) Targeted lipidomics: fatty acid amides and pain modulation. Prostaglandins Other Lipid Mediat 77: 35–45

Walter L, Franklin A, Witting A, Moller T, Stella N (2002) Astrocytes in culture produce anandamide and other acylethanolamides. J Biol Chem 277:20869–20876

Walter L, Stella N (2003) Endothelin-1 increases 2-arachidonoyl glycerol (2-AG) production in astrocytes. Glia 44:85–90

Walter L, Stella N (2004) Cannabinoids and neuroinflammation. Br J Pharmacol 141:775–785

Wang D, Lundeberg T, Yu L-C (2000) Antinociceptive role of galanin in periaqueductal grey of rats with experimentally induced mononeuropathy. Neuroscience 96:767–771

Wang HR, Ng K, Hayes D, Gao XR, Forster G, Blaha C, Yeomans J (2004) Decreased amphetamine-induced locomotion and improved latent inhibition in mice mutant for the M5 muscarinic receptor gene found in the human 15q schizophrenia region. Neuropsychopharmacology 29:2126–2139

Wang JH, Simonavicius N, Wu XS, Swaminath G, Reagan J, Tian H, Ling L (2006a) Kynurenic acid as a ligand for orphan G protein-coupled receptor GPR35. J Biol Chem 281:22021–22028

Wang JL, Zhu CB, Cao XD, Wu GC (1999) Distinct effect of intracerebroventricular and intrathecal injections of nociceptin/orphanin FQ in the rat formalin test. Regul Pept 79:159–163

Wang JZ, Lundeberg T, Yu LC (2001a) Anti-nociceptive effect of neuropeptide Y in periaqueductal grey in rats with inflammation. Brain Res 893:264–267

Wang R, Guo W, Ossipov MH, Vanderah TW, Porreca F, Lai J (2003a) Glial cell line-derived neurotrophic factor normalizes neurochemical changes in injured dorsal root ganglion neurons and prevents the expression of experimental neuropathic pain. Neuroscience 121:815–824

Wang T, Li SR, Dai X, Peng YL, Chen Q, Wang R (2006b) Effects of melatonin on orphanin FQ/nociceptin-induced hyperalgesia in mice. Brain Res 1085:43–48

Wang X, Miyares RL, Ahern GP (2005a) Oleoylethanolamide excites vagal sensory neurones, induces visceral pain and reduces short-term food intake in mice via capsaicin receptor TRPV1. J Physiol 564:541–547

Wang XL, Zhang HM, Li DP, Hen SR, An HL (2006c) Dynamic regulation of glycinergic input to spinal dorsal horn neurones by muscarinic receptor subtypes in rats. J Physiol 571:403–413

Wang Y, Pei G, Cai YC, Zhao ZQ, Wang JB, Jiang CL, Zheng ZC, Liu XY (1996) Human interleukin-2 could bind to opioid receptor and induce corresponding signal transduction. NeuroReport 8:11–14

Wang Y, Su DM, Wang RH, Liu Y, Wang H (2005b) Antinociceptive effects of choline against acute and inflammatory pain. Neuroscience 132:49–56

Wang YY, Wu SX, Liu XY, Wang W, Li YQ (2003b) Effects of c-fos antisense oligodeoxynucleotide on 5-HT-induced upregulation of preprodynorphin, preproenkephalin, and glutamic acid decarboxylase mRNA expression in cultured rat spinal dorsal horn neurons. Biochem Biophys Res Commun 309:631–636

Wang Z, Gardell LR, Ossipov MH, Vanderah TW, Brennan MB, Hochgeschwender U, Hruby VJ, Malan TP Jr, Lai J, Porreca F (2001b) Pronociceptive actions of dynorphin maintain chronic neuropathic pain. J Neurosci 21:1779–1786

Watanabe S, Kuwaki T, Yanagisawa M, Fukuda Y, Shimoyama M (2004) Persistent pain and stress activate pain-inhibitory orexin pathways. NeuroReport 16:5–8

Webb TI, Lynch JW (2007) Molecular pharmacology of the glycine receptor chloride channel. Curr Pharm Des 13:2350–2367

Webster VAD, Griffiths EC, Slater P (1983) Antinociceptive effects of thyrotrophin-releasing hormone and its analogues in the rat periaqueductal grey region. Neurosci Lett 42:67–70

Wei H, Panula P, Pertovaara A (1998) A differential modulation of allodynia, hyperalgesia and nociception by neuropeptide FF in the periaqueductal gray of neuropathic rats: interactions with morphine and naloxone. Neuroscience 86:311–319

Wei H, Panula P, Pertovaara A (2001) Modulation of pain by [1DMe]NPYF, a stable analogue of neuropeptide FF, in neuropathic rats. Brain Res 900:234–243

Weil-Fugazza J, Godefroy F (1993) Dorsal and ventral dopaminergic innervation of the spinal cord: functional implications. Brain Res Bull 30:319–324

Welch SP, Cooper CW, Dewey WL (1986) Antinociceptive activity of salmon calcitonin injected intraventricularly in mice: modulation of morphine antinociception. J Pharmacol Exp Ther 237:54–58

Welch SP, Dewey WL (1990) The activity of several peptide fragments of parathyroid hormone, alone and in combination with salmon calcitonin and morphine, in antinociceptive tests in the mouse. J Pharmacol Exp Ther 252:140–146

Welch SP, Huffman JW, Lowe J (1998) Differential blockade of the antinociceptive effects of centrally administered cannabinoids by SR14176A. J Pharmacol Exp Ther 286:1301–1308

Wen HL, Mehal ZD, Ong BH, Ho WKK, Wen DYK (1985) Intrathecal administration of beta-endorphin and dynorphin-(1–13) for the treatment of intractable pain. Life Sci 37:1213–1220

Wess J (1996) Molecular biology of muscarinic acetylcholine receptors. Crit Rev Neurobiol 10:69–99

Wess J, Duttaroy A, Gomeza J, Zhang W, Yamada M, Felder CC, Bernardini N, Reeh PW (2003) Muscarinic receptor subtypes mediating central and peripheral antinociception studied with muscarinic receptor knockout mice: a review. Life Sci 72:2047–2054

Wess J, Eglen RM, Gautam D (2007) Muscarinic acetylcholine receptors: mutant mice provide new insights for drug development. Nat Rev Drug Discov 6:721–733

Wiesenfeld-Hallin Z (1987) Intrathecal vasoactive intestinal polypeptide modulates spinal reflex excitability primarily to cutaneous thermal stimuli in rats. Neurosci Lett 80:293–297

Wiesenfeld-Hallin Z, Persson A (1984) Subarachnoid injection of salmon calcitonin does not induce analgesia in rats. Eur J Pharmacol 104:375–377

Wiesenfeld-Hallin Z, Xu XJ, Crawley JN, Hokfelt T (2005) Galanin and spinal nociceptive mechanisms: recent results from transgenic and knock-out models. Neuropeptides 39:207–210

Wiesenfeld-Hallin Z, Xu X-J, Langel Ü, Bedecs K, Bartfai T, Hokfelt T, Bartfai T (1992) Galanin-mediated control of pain: enhanced role after nerve injury. Proc Natl Acad Sci U S A 89: 3334–3337

Wikberg JE, Mutulis F (2008) Targeting melanonortin receptors: an approach to treat weight disorders and sexual dysfunction. Nat Rev Drug Discov 7:307–323

Wiklund L, Behzadi G, Kalen P, Headley PM, Nicolopoulos LS, Parsons CG, West DC (1988) Autoradiographic and electrophysiological evidence for excitatory amino acid transmission in the periaqueductal gray projection to nucleus raphe magnus in the rat. Neurosci Lett 93: 158–163

Wiles AL, Pearlman RJ, Rosvall M, Aubrey KR, Vandenberg RJ (2006) N-Arachidonyl-glycine inhibits the glycine transporter, GLYT2a. J Neurochem 99:781–786

Williams DW, Lipton JM, Giesecke AH (1986) Influence of centrally administered peptides on ear withdrawal from heat in the rabbit. Peptides 7:1095–1100

Williams FG, Beitz AJ (1990) Ultrastructural morphometric analysis of GABA-immunoreactive terminals in the ventrocaudal periaqueductal gray: analysis of the relationship of GABA terminals and the GABAA receptor to periaqueductal gray-raphe magnus projection neurons. J Neurocytol 19:686–696

Willingale HL, Gardiner NJ, McLymont N, Giblett S, Grubb BD (1997) Prostanoids synthesized by cyclo-oxygenase isoforms in rat spinal cord and their contribution to the development of neuronal hyperexcitability. Br J Pharmacol 122:1593–1604

Willis WD Jr (1988) Dorsal horn neurophysiology of pain. Ann N Y Acad Sci 531:76–89

Willis WD, Westlund KN (1997) Neuroanatomy of the pain system and of the pathways that modulate pain. J Clin Neurophysiol 14:2–31

Wilson-Gerwing TD, Dmyterko MV, Zochodne DW, Johnston JM, Verge VMK (2005) Neurotrophin-3 suppresses thermal hyperalgesia associated with neuropathic pain and attenuates transient receptor potential vanilloid receptor-1 expression in adult sensory neurons. J Neurosci 25:758–767

Wilson-Gerwing TD, Verge VMK (2006) Neurotrophin-3 attenuates galanin expression in the chronic constriction injury model of neuropathic pain. Neuroscience 141:2075–2085

Wirkner K, Sperlagh B, Illes P (2007) P2X(3) receptor involvement in pain states. Mol Neurobiol 36:165–183

Wise LE, Shelton CC, Cravatt BF, Martin BR, Lichtman AH (2007) Assessment of anandamide's pharmacological effects in mice deficient of both fatty acid amide hydrolase and cannabinoid CB1 receptors. Eur J Pharmacol 557:44–48

Wittau N, Grosse R, Kalkbrenner F, Gohla A, Schultz G, Gudermann T (2000) The galanin receptor type 2 initiates multiple signaling pathways in small cell lung cancer cells by coupling to G(q), G(i) and G(12) proteins. Oncogene 19:4199–4209

Wood PB (2008) Role of central dopamine in pain and analgesia. Expert Rev Neurother 8: 781–797

Woolf CJ, Costigan M (1999) Transcriptional and posttranslational plasticity and the generation of inflammatory pain. Proc Natl Acad Sci U S A 96:7723–7730

Wu D, Doods H, Arndt K, Schindler M (2002) Development and potential of non-peptide antagonists for calcitonin-gene-related peptide (CGRP) receptors: evidence for CGRP receptor heterogeneity. Biochem Soc Trans 30:468–473

Wu FS, Yang YC, Tsai JJ (1999a) Melatonin potentiates the GABAA receptor-mediated current in cultured chick spinal cord neurons. Neurosci Lett 260:177–180

Wu GY, Morris SM (1998) Arginine metabolism: nitric oxide and beyond. Biochem J 336:1–17

Wu X, Li HD, Ruan HZ, Wang J (1999b) Effects of intra-hippocampal injection of interleukin-2 on pain threshold and formaldehyde-induced substance P-like immunoreactivity in periaqueductal gray and spinal cord. Acta Pharmacol Sin 20:839–843

Xu H, Zhou KQ, Huang YN, Chen L, Xu TL (2004a) Taurine activates strychnine-sensitive glycine receptors in neurons of the rat inferior colliculus. Brain Res 1021:232–240

Xu IS, Grass S, Xu X-J, Wiesenfeld-Hallin Z (1998) On the role of galanin in mediating spinal flexor reflex excitability in inflammation. Neuroscience 85:827–835

Xu IS, Hao J-X, Xu X-J, Hokfelt T, Wiesenfeld-Hallin Z (1999a) The effect of intrathecal selective agonists on Y1 and Y2 neuropeptide Y receptors on the flexor reflex in normal axomized rats. Brain Res 833:251–257

Xu IS, Hashemi M, Calo G, Regoli D, Wiesenfeld-Hallin Z, Yacoub MH (1999b) Effects of intrathecal nocistatin on the flexor reflex and its interaction with orphanin FQ nociceptin. NeuroReport 10:3681–3684

Xu M, Petraschka M, McLaughlin JP, Westenbroek RE, Caron MG, Lefkowitz RJ, Czyzyk TA, Pintar JE, Terman GW, Chavkin C (2004b) Neuropathic pain activates the endogenous kappa opioid system in mouse spinal cord and induces opioid receptor tolerance. J Neurosci 24: 4576–4584

Xu W, Lundeberg T, Wang YT, Li Y, Yu L-C (2003) Antinociceptive effect of calcitonin gene-related peptide in the central nucleus of amygdala: activating opioid receptors through amygdala-periaqueductal gray pathway. Neuroscience 118:1015–1022

Xu XJ, Wiesenfeldhallin Z (1991) An analog of growth-hormone releasing-factor (GRF), (Ac-Try1, D-Phe2)-GRF-(1–29), specifically antagonizes the facilitation of the flexor reflex induced by intrathecal vasoactive-intestinal-peptide in rat spinal-cord. Neuropeptides 18:129–135

Xu Z, Chen S-R, Eisenach JC, Pan H-L (2000) Role of spinal muscarinic and nicotinic receptors in clonidine-induced nitric oxide release in a rat model of neuropathic pain. Brain Res 861: 390–398

Xu Z, Tong C, Eisenach JC (1996a) Acetylcholine stimulates the release of nitric oxide from rat spinal cord. Anesthesiology 85:107–111

Xu Z, Tong C, Pan H-L, Cerda SE, Eisenach JC (1997) Intravenous morphine increases release of nitric oxide from spinal cord by an alpha-adrenergic and cholinergic mechanism. J Neurophysiol 78:2072–2078

Xu ZQ, Shi TJ, Landry M, Hokfelt T (1996b) Evidence for galanin receptors in primary sensory neurones and effect of axotomy and inflammation. NeuroReport 8:237–242

Yagi J, Sumino R (1998) Inhibition of a hyperpolarization-activated current by clonidine in rat dorsal root ganglion neurons. J Neurophysiol 80:1094–1104

Yaksh TL (1989) Behavioral and autonomic correlates of the tactile evoked allodynia produced by spinal glycine inhibition: effects of modulatory receptor systems and excitatory amino acid antagonists. Pain 37:111–123

Yaksh TL, Farb DH, Leeman SE, Jessell TM (1979) Intrathecal capsaicin depletes substance P in the rat spinal cord and produces prolonged thermal analgesia. Science 206:481–483

Yamada K, Santo-Yamada Y, Wada K (2003) Stress-induced impairment of inhibitory avoidance learning in female neuromedin B receptor-deficient mice. Physiol Behav 78:303–309

Yamamoto T, Nozaki-Taguchi N, Chiba T (2002a) Analgesic effect of intrathecally administered orexin-A in the rat formalin test and in the rat hot plate test. Br J Pharmacol 137: 170–176

Yamamoto T, Saito O, Shono K, Aoe T, Chiba T (2003) Anti-mechanical allodynic effect of intrathecal and intracerebroventricular injection of orexin-A in the rat neuropathic pain model. Neurosci Lett 347:183–186

Yamamoto T, Sakashita Y (1999) Effect of nocistatin and its interaction with nociceptin/orphanin FQ on the rat formalin test. Neurosci Lett 262:179–182

Yamamoto T, Tatsuno I (1995) Antinociceptive effect of intrathecally administered pituitary adenylate cyclase activating peptide (PACAP) on the rat formalin test. Neurosci Lett 184:32–35

Yamamoto T, Wada T, Miyazaki R (2008) Analgesic effects of intrathecally administered 26RFa, an intrinsic agonist for GPR103, on formalin test and carrageenan test in rats. Neuroscience 157:214–222

Yamamoto T, Yaksh TL (1992) Spinal pharmacology of thermal hyperesthesia induced by constriction injury of sciatic nerve. Excitatory amino acid antagonists. Pain 49:121–128

Yamamoto Y, Ono H, Ueda A, Shimamura M, Nishimura K, Hazato T (2002b) Spinorphin as an endogenous inhibitor of enkephalin-degrading enzymes: roles in pain and inflammation. Curr Protein Pept Sci 3:587–599

Yamazaki N, Umeno H, Kuraishi Y (1999) Involvement of brain serotonergic terminals in the antinociceptive action of peripherally applied calcitonin. Jpn J Pharmacol 81:367–374

Yan T, Liu BG, Du DP, Eisenach JC, Tong C (2007) Estrogen amplifies pain responses to uterine cervical distension in rats by altering transient receptor potential-1 function. Anesth Analg 104:1246–1250

Yanai K, Mobarakeh JI, Kuramasu A, Sakurada S (2003) Roles of histamine receptors in pain perception: a study using receptors gene knockout mice. Nippon Yakurigaku Zasshi 122: 391–399

Yang HY, Fratta W, Majane EA, Costa E (1985) Isolation, sequencing, synthesis, and pharmacological characterization of two brain neuropeptides that modulate the action of morphine. Proc Natl Acad Sci U S A 82:7757–7761

Yang HY, Iadarola MJ (2006) Modulatory roles of the NPFF system in pain mechanisms at the spinal level. Peptides 27:943–952

Yang J (1994) Intrathecal administration of oxytocin-induced analgesia in low-back-pain involving the endogenous opiate peptide system. Spine 19:867–871

Yang J, Chen JM, Liu WY, Song CY, Lin BC (2006) Through V2, not V1 receptor relating to endogenous opiate peptides, arginine vasopressin in periaqueductal gray regulates antinociception in the rat. Regul Pept 137:156–161
Yang J, Yang Y, Chen JM, Xu HT, Liu WY, Wang CH, Lin BC (2007a) Arginine vasopressin is an important regulator in antinociceptive modulation of hypothalamic paraventricular nucleus in the rat. Neuropeptides 41:165–176
Yang J, Yang Y, Xu HT, Chen JM, Liu WY, Lin BC (2007b) Arginine vasopressin induces periaqueductal gray release of enkephalin and endorphin relating to pain modulation in the rat. Regul Pept 142:29–36
Yang K, Ma WL, Feng YP, Dong YX, Li YQ (2002) Origins of GABA(B) receptor-like immunoreactive terminals in the rat spinal dorsal horn. Brain Res Bull 58:499–507
Yang X-C, Reis DJ (1999) Agmatine selectively blocks the N-methyl-D-aspartate subclass of glutamate receptor channels in rat hippocampal neurons. J Pharmacol Exp Ther 288: 544–549
Yang YH, Morand EF, Getting SJ, Paul-Clark M, Liu DL, Yona S, Hannon R, Buckingham JC, Perretti M, Flower RJ (2004) Modulation of inflammation and response to dexamethasone by annexin 1 in antigen-induced arthritis. Arthritis Rheum 50:976–984
Yang Z, Aubrey KR, Alroy I, Harvey RJ, Vandenberg RJ, Lynch JW (2009) Subunit-specific modulation of glycine receptors by cannabinoids and N-arachidonyl-glycine. Biochem Pharmacol 76:1014–1023
Yao MZ, Gu JF, Wang JH, Sun LY, Lang MF, Liu J, Zhao ZQ, Liu XY (2002a) Interleukin-2 gene therapy of chronic neuropathic pain. Neuroscience 112:409–416
Yao MZ, Wang JH, Gu JF, Sun LY, Liu H, Zhao ZQ, Liu XY (2002b) Interleukin-2 gene has superior antinociceptive effects when delivered intrathecally. NeuroReport 13:791–794
Yeomans DC, Onyuksel H, Dagar S, Ikezaki H, Lu Y, Rubinstein I (2003) Conformation-dependent effects of VIP on nociception in rats. Peptides 24:617–622
Yesilyurt O, Dogrul A, Gul H, Seyrek M, Kusmez O, Ozkan Y, Yildiz O (2003) Topical cannabinoid enhances topical morphine antinociception. Pain 105:303–308
Yien HW, Chan JYH, Tsai HF, Lee TY, Chan SHH (1993) Participation of nucleus-reticularis gigantocellularis in the antinociceptive effect of angiotensin-III in the rat. Neurosci Lett 159:9–12
Yokogawa F, Kiuchi Y, Ishikaw Y, Otsuka N, Masuda Y, Oguchi K, Hosoyamada A (2002) An investigation of monoamine receptors involved in antinociceptive effects of antidepressants. Anesth Analg 95:163–168
Yonehara N, Yoshimura M (1999) Effect of nitric oxide on substance P release from the peripheral endings primary afferent neurons. Neurosci Lett 271:199–201
Yoshikawa M, Ito K, Maeda M, Akahori K, Takahashi S, Jin XL, Matsuda M, Suzuki T, Oka T, Kobayashi H, Hashimoto A (2007) Activation of supraspinal NMDA receptors by both D-serine alone or in combination with morphine leads to the potentiation of antinociception in tail-flick test of rats. Eur J Pharmacol 565:89–97
Young T, Wittenauer S, McIntosh JM, Vincler M (2008) Spinal alpha3beta2∗ nicotinic acetylcholine receptors tonically inhibit the transmission of nociceptive mechanical stimuli. Brain Res 1229:118–124
Yu CX, Zhu CB, Xu SF, Cao XD, Wu GC (2000a) Selective MT(2) melatonin receptor antagonist blocks melatonin-induced antinociception in rats. Neurosci Lett 282:161–164
Yu C-X, Zhu C-B, Xu S-F, Cao X-D, Wu G-C (2000b) The analgesic effects of peripheral and central administration of melatonin in rats. Eur J Pharmacol 403:49–53
Yu LC, Weng XH, Wang JW, Lundeberg T (2003) Involvement of calcitonin gene-related peptide and its receptor in anti-nociception in the periaqueductal grey of rats. Neurosci Lett 349:1–4
Yu LC, Xu SL, Xiong W, Lundeberg T (2001) The effect of galanin on wide-dynamic range neuron activity in the spinal dorsal horn of rats. Regul Pept 101:179–182
Yu Y, Jawa A, Pan W, Kastin AJ (2004) Effects of peptides, with emphasis on feeding, pain, and behavior. A 5-year (1999–2003) review of publications in Peptides. Peptides 25:2257–2289

Zadina JE, Hackler L, Ge LJ, Kastin AJ (1997) A potent and selective endogenous agonist for the mu-opiate receptor. Nature 386:499–502

Zadina JE, Kastin AJ, Ge L-J, Hackler L (1994) Mu, delta, and kappa opiate receptor binding of Tyr-MIF-1 and of Tyr-W-MIF-1, its active fragments, and two potent analogs. Life Sci 55:PL461–PL466

Zadina JE, Kastin AJ, Kenigs V, Bruno C, Hackler L (1993) Prolonged analgesia after intracerebroventricular Tyr-W-MIF-1 (Tyr-Pro-Trp-Gly-Nh2). Neurosci Lett 155:220–222

Zadina JE, Kastin AJ, Kersh D, Wyatt A (1992) Tyr-MIF-1 and hemorphin can act as opiate agonists as well as antagonists in the guinea pig ileum. Life Sci 51:869–885

Zadina JE, Paul D, Gergen KA, Ge L-J, Hackler L, Kastin AJ (1996) Binding of Tyr-W-MIF-1 (Tyr-Pro-Trp-Gly-NH2) and related peptides to mu1 and mu 2 opiate receptors. Neurosci Lett 215:65–69

Zahn PK, Lansmann T, Berger E, Speckmann EJ, Musshoff U (2003) Gene expression and functional characterisation of melatonin receptors in the spinal cord of the rat: implications for pain modulation. J Pineal Res 35:24–31

Zamfirova R, Bocheva A, Dobrinova Y, Todorov S (2007) Study on the antinociceptive action of Tyr-K-MIF-1, a peptide from the MIF family. Auton Autacoid Pharmacol 27:93–98

Zeilhofer HU (2005) The glycinergic control of spinal pain processing. Cell Mol Life Sci 62: 2027–2035

Zeilhofer HU, Muth-Selbach U, Guhring H, Erb K, Ahmadi S (2000) Selective suppression of inhibitory synaptic transmission by nocistatin in the rat spinal cord dorsal horn. J Neurosci 20:4922–4929

Zeitz KP, Guy N, Malmberg AB, Dirajlal S, Martin WJ, Sun L, Bonhaus DW, Stucky CL, Julius D, Basbaum AI (2002) The 5-HT3 subtype of serotonin receptor contributes to nociceptive processing via a novel subset of myelinated and unmyelinated nociceptors. J Neurosci 22: 1010–1019

Zeller CB, Marchase RB (1992) Gangliosides as modulators of cell function. Am J Physiol 262(6 Pt 1):C1341–C1355

Zeng Q, Wang S, Lim G, Yang L, Mao J, Sung B, Chang Y, Lim JA, Guo G, Mao J (2008) Exacerbated mechanical allodynia in rats with depression-like behavior. Brain Res 1200: 27–38

Zeng W, Dohi S, Shimonaka H, Asano T (1999) Spinal antinociceptive action of Na^{+}-K^{+} pump inhibitor ouabain and its interaction with morphine and lidocain in rats. Anesthesiology 90:500–508

Zhang J, Chen C (2008) Endocannabinoid 2-arachidonoylglycerol protects neurons by limiting COX-2 elevation. J Biol Chem 283:22601–22611

Zhang J, Hoffert C, Vu HK, Groblewski T, Ahmad S, O'Donnell D (2003a) Induction of CB2 receptor expression in the rat spinal cord of neuropathic but not inflammatory chronic pain models. Eur J Neurosci 17:2750–2754

Zhang XC, Zhang YQ, Zhao ZQ (2005) Involvement of nitric oxide in long-term potentiation of spinal nociceptive responses in rats. NeuroReport 16:1197–1201

Zhang Y, Malmberg AB, Sjolund B, Yaksh TL (1996) The effect of pituitary adenylate cyclase activating peptide (PACAP) on the nociceptive formalin test. Neurosci Lett 207: 187–190

Zhang YP, Lundeberg T, Yu LC (2000a) Interactions of galanin and morphine in the spinal antinociception in rats with mononeuropathy. Brain Res 852:485–487

Zhang YQ, Ji GC, Wu GC, Zhao ZQ (2003b) Kynurenic acid enhances electroacupuncture analgesia in normal and carrageenan-injected rats. Brain Res 966:300–307

Zhang YX, Lundeberg T, Yu LC (2000b) Involvement of neuropeptide Y and Y1 receptor in antinociception in nucleus raphe magnus of rats. Regul Pept 95:109–113

Zhao CS, Li BS, Zhao GY, Liu HX, Luo F, Wang Y, Tian JH, Chang JK, Han JS (1999) Nocistatin reverses the effect of orphanin FQ/nociceptin in antagonizing morphine analgesia. NeuroReport 10:297–299

Zhao J, Seereeram A, Nassar MA, Levato A, Pezet S, Hathaway G, Morenilla-Palao C, Stirling C, Fitzgerald M, McMahon SB (2006) Nociceptor-derived brain-derived neurotrophic factor regulates acute and inflammatory but not neuropathic pain. Mol Cell Neurosci 31:539–548

Zhao M, Wang JY, Jia H, Tang JS (2007) Roles of different subtypes of opioid receptors in mediating the ventrolateral orbital cortex opioid-induced inhibition of mirror-neuropathic pain in the rat. Neuroscience 144:1486–1494

Zheng L, Li XC (1995) Effect of intracerebroventricular injection of somatostatin or gaba on pain threshold and contents of GABA or somatostatin in rat-brain. Acta Pharmacol Sin 16:329–332

Zheng W, Xie W, Zhang J, Strong JA, Wang L, Yu L, Xu M, Lu L (2003) Function of gamma-aminobutyric acid receptor/channel rho 1 subunits in spinal cord. J Biol Chem 278: 48321–48329

Zhou XF, Deng YS, Xian CJ, Zhong JH (2000) Neurotrophins from dorsal root ganglia trigger allodynia after spinal nerve injury in rats. Eur J Neurosci 12:100–105

Zhuo M, Gebhart GF (1990) Spinal cholinergic and monoaminergic receptors mediate descending inhibition from the nuclei reticularis gigantocellularis and gigantocellularis pars alpha in the rat. Brain Res 535:67–78

Zhuo M, Gebhart GF (1991) Tonic cholinergic inhibition of spinal mechanical transmission. Pain 46:211–222

Zollner C, Shaqura MA, Bopaiah CP, Mousa S, Stein C, Schafer M (2003) Painful inflammation-induced increase in mu-opioid receptor binding and G-protein coupling in primary afferent neurons. Mol Pharmacol 64:202–210

Zubrzycka M, Fichna J, Janecka A (2005) Inhibition of trigemino-hypoglossal reflex in rats by oxytocin is mediated by mu and kappa opioid receptors. Brain Res 1035:67–72

Zubrzycka M, Janecka A (2007) Effect of tooth pulp stimulation on oxytocin and vasopressin release into the cerebrospinal fluid and fluid perfusing the cerebral ventricles in rats. Endocr Regul 41:149–154

Zubrzycka M, Janecka A (2008) Interactions of galanin with endomorphin-2, vasopressin and oxytocin in nociceptive modulation of the trigemino-hypoglossal reflex in rats. Physiol Res 57:769–776

Biology of Demyelinating Diseases

Danielle Pham-Dinh and Nicole Baumann

Abstract Demyelinating diseases are those in which myelin is the primary target of damage on the basis of neuroradiological, neuropatholgical, neurochemical, and genetic studies. This review describes the morphological aspects of the myelin sheath which is the most abundant membrane structure in the vertebrate nervous system. It is made of oligodendrocytes in the CNS and Schwann cells in the PNS. It comprises four distinct regions: the node of Ranvier which contains voltage-gated Na+ channels, paranodal loops which are major sites of myelin-axon adhesion, the juxtaparanode, and the internode which is the part of the axon which is ensheathed by a segment of myelin. Demyelination is segmental in the peripheral nervous system and focal in the central nervous system. Myelin is necessary for nerve conduction velocity. Dys- and demyelination can involve specific constituents of the CNS and the PNS both for genetic (leukodystrophies) or acquired diseases. Numerous components are different and differently involved in CNS and PNS myelin, both among proteins and lipids (sphingolipids). Outside of the abnormalities of specific myelin components leading to genetic diseases, experimental models of demyelination (experimental autoimmune encephalomyelitis, cuprizone intoxication, lysolecithin-induced demyelination, and ethidium bromide treatment are also described). During myelin repair, a thinner myelin sheath is produced with shorter internodes and efficient nerve conduction is produced. Dysfunction of astrocytes may be involved in some genetic diseases of myelin. There are many growth factors and transcription factors involved in the process of myelination and demyelination among which eukaryotic initiation factor 2B (elf2B) leading to vanishing white matter disease (CACH). The role of hormones and sexual dimorphism of oligodendrocytes and myelin anre also described. New areas of research are being developed showing the involvement of myelin deficiency in psychiatric diseases and cognition.

N. Baumann (✉)
Laboratoire de Neurochimie, Hôpital de la Salpêtrière, F-75013 Paris, France; Université Paris 6, F-75000 Paris, France
e-mail: nicole.baumann@upmc.fr

J.P. Blass (ed.), *Neurochemical Mechanisms in Disease*,
Advances in Neurobiology 1, DOI 10.1007/978-1-4419-7104-3_16,

Keywords Myelin proteins · Myelin lipids · Demyelination · Leukodystrophies · Multiple sclerosis

Contents

1 Introduction

Different myelin disorders affect several million people in the world, whereby myelin is the primary target of damage on the basis of neuroradiological, neuropathological, neurochemical, and genetic studies. There are primary genetic diseases in which the targets are specific components of myelin or myelin control. Interestingly, even in the latter which are leukodystrophies in CNS or peripheral neuropathies in PNS, there are wide variations according to each individual, and in most cases there are no clear relations between genotype and phenotype. This indicates that the factors involved may be multiple. Myelin disorders also include acquired diseases such as multiple sclerosis (MS) in the central nervous system (CNS), and Guillain–Barré syndrome (already recognized as having multiple causes) in the peripheral nervous system (PNS), although genetic susceptibility may be involved.

The myelin sheath is one of the most abundant membrane structures in the vertebrate nervous system. It is produced by two types of glial cells, oligodendrocytes in the CNS and Schwann cells in the PNS. The myelin sheath is formed by the spiral wrapping of glial plasma membrane extensions around the axon, followed by the extrusion of cytoplasm and the compaction of the stacked myelin bilayers (Simons and Trajkovic, 2006). The myelination process is dependent on neuronal activity (Demerens et al., 1996; Lubetzki and Stankoff, 2000) and has recently been found to be relayed in the CNS by the nonmyelinating cells, the astrocytes (Ishibashi et al., 2006; Spiegel and Peles, 2006). Myelin structure and composition differ somehow in both CNS and PNS (Baumann and Pham-Dinh, 2001), but the function of myelin is the same: to allow rapid nerve conduction through nerve fibers, especially those integrating the motor and sensory functions in vertebrates (Waxman and Bangalore, 2003). Moreover, in mammals, and especially in higher primates in which myelination is a long-lasting process (Yakovlev and Lecours, 1966), particularly in association areas, it is now clear that myelin is involved in cognitive functions such as language (Aslin and Schlaggar, 2006; Pujol et al., 2006) and behavior (Beckman, 2004; Seldon, 2007); recently it has been shown that alteration of CNS myelin may be involved in some psychiatric diseases (Stewart and Davis, 2004; Kubicki et al., 2005; Regenold et al., 2007) and dementia (Filley, 1998).

The symptoms characteristic of myelin disorders may be caused by abnormal formation of myelin (i.e., dysmyelination) or damage to myelin (i.e., demyelination; Baumann and Pham-Dinh, 2001). In fact, it may be sometimes difficult to separate the two aspects, as some of those diseases, even the genetic ones linked to alterations of myelin constituents, may appear only at an adult age. Thus it is necessary to consider both aspects. Furthermore, conduction abnormalities are not due only to changes in the electrical properties related to myelin loss, but also to modifications in electrogenic properties related to alterations of the molecular organization within the axonal membrane. It is not always easy to understand what phenomenon is at the origin of the disease. Nevertheless, we focus here on dys- and demyelinating diseases in which myelin modifications appear to be the primary events. This includes diseases involving the cells that build and preserve myelin such as oligodendrocytes in the CNS and Schwann cells in the PNS, and alterations of neuron–glia interactions (involving astrocytes in the CNS and possibly microglia).

In this review, we do not speak about secondary demyelinations in which the cause is related to abnormalities of cerebral vasculature such as vascular dementia, stroke, CADASIL (cerebral autosomal dominant arteriopathy with subcortical infarcts and leukoencephalopathy), or Fabry's disease, or abnormalities of mitochondrial functions such as MELAS (mitochondrial encephalomyopathy, lactic acidosis, and stroke-like episodes). Also, clinically significant white matter changes can be seen on MRI (magnetic resonance imaging) in some genetic and inflammatory diseases in which the most prominent neuropathological abnormalities are in the grey matter. For instance, tumors with the potential to infiltrate the white matter may also cause dramatic changes in white matter that are evident in MRI.

We do not cite here many references that have been mentioned extensively in a review on the biology of myelination in the central nervous system by Baumann and Pham-Dinh (2001) especially because this review is freely available on the Internet.

Extensive reviews on different aspects of myelin biology and disorders are available in the book by R.A. Lazzarini (Lazzarini, 2004).

Before determining the chemical factors involved in myelin disorders, it is useful to first recall the morphological particularities of myelin and the Ranvier node, in CNS and PNS, as these are the major targets of demyelination.

2 Morphological Aspects

2.1 Myelin Structure in the Central and Peripheral Nervous System

Myelin is a spiral structure made up of extensions of the plasma membrane from the myelinating glial cells, namely the oligodendrocytes in the CNS and the Schwann cells in the PNS. These cells send out sail-like extensions of their cytoplasmic membrane, each of which forms a segment of sheathing around an axon, which then constitutes the myelin sheath.

The structure of myelinated axons is similar in both the CNS and PNS. There are four distinct regions: the node of Ranvier, the paranode, the juxtaparanode, and the internode. The internode is the part of the axon that is ensheathed by a segment of myelin; it ends at both sides in the juxtaparanode; then paranodal loops border the node of Ranvier, a portion of the axon that is exposed to the extracellular milieu. Several structural features characterize myelin in electron microscopy: a periodic structure, with alternating concentric electron-dense and light layers. The major dense line (dark layer) forms as the cytoplasmic surfaces of the expanding myelinating processes of the oligodendrocyte or the Schwann cell are brought into close apposition; the fused two outer leaflets (extracellular apposition) form the intraperiodic lines or minor dense lines. The periodicity of compact CNS and PNS myelin differs only slightly, however, CNS and PNS myelin can be easily distinguished by the presence of a basal lamina around PNS but not CNS myelin. Also, the endoneural or extracellular space of peripheral nerves is particularly abundant. PNS and CNS myelin internodes are separated from the axon by a 12–14-nm periaxonal space.

The regularly spaced unmyelinated gaps that constitute the node of Ranvier are critical to the proper function of CNS and PNS myelin. Both the structure and the molecular organization of the nodal region are dependent on the formation of the appropriate axo–glial interactions. These interactions establish the spacing of adjacent myelin segments (internodes) and ultimately determine the position and length of the nodal gaps. Compact myelin constitutes the internodes whereas juxtaparanode and paranodal loops are mostly formed by noncompact myelin. The juxtaparanode is just under the compact myelin sheath beyond the innermost paranodal junction and may therefore be considered as a specialized portion of the internode (Peles and Salzer, 2000).

2.2 Node of Ranvier

The nodal axolemma is a functional part of myelin formation as it contains voltage-gated Na+ channels and thereby is directly responsible for saltatory conduction. This architecture is accompanied by a selective disposition of different types of ion exchangers and channels that allow the saltatory conduction of nervous influx and are responsible for its high velocity, without increasing axonal diameter. A high density of voltage-gated Na+ channels is found at the node of Ranvier, whereas K+ channels are found at the juxtaparanodal loops.

The node of Ranvier, paranodal junctions and the adjacent juxtaparanodal regions each contain distinct protein complexes. Each of these membrane domains contains a specific set of cell adhesion molecules that are stabilized and retained through interactions with cytoskeletal and scaffolding proteins. These in turn recruit and stabilize the appropriate channels (Coman et al., 2005; Simons and Trajkovic, 2006).

Although PNS and CNS nodes can appear similar by transmission electronic microscopy (TEM), each is surrounded by distinctly different immediate environments. The nodal axolemma is surrounded by Schwann cell microvilli in the PNS and by glial cell processes in the CNS. The microvilli extending from the outer border of the adjacent myelin internodes are best visualized in cross-sections. The Schwann cell basal lamina is continuous across the node but does not surround individual microvilli. The glial cell processes surrounding CNS nodes originate from a recently described cell population: the NG2 cells (Butt et al., 2002). These CNS glial processes do not, however, closely adhere to all regions of the nodal axolemma.

2.3 Paranode

The cytoplasmic channels or paranodal loops at the lateral end of the internode are a major site of myelin–axon adhesion. The membrane of the inner or adaxonal surface of the myelin sheath is in direct contact with the axons. Their cytoplasmic channels may transmit axonal signals that regulate myelin formation and help determine the length and thickness of the myelin internode. These channels contain microtubules and other cytoskeletal components for transport and stability and mitochondria for energy. Also, in some areas, they contain smooth endoplasmic reticulum and free polysomes for the synthesis of local membrane components. In addition, membranes of noncompact myelin serve special functions that are reflected by unique molecular composition.

2.4 Myelination

As mentioned previously, myelination is carried out by highly specialized glial cells, oligodendrocytes in the CNS and Schwann cells in the PNS. These cells differentiate from precursor cells of different origin during development: the neural crest for Schwann cells, and neural tube for oligodendrocytes. Myelin characteristics differ morphologically in the CNS and in the PNS, and differ also inside the CNS. Thus regulation of myelination and most probably demyelination takes place according to different criteria. Oligodendrocytes differ from Schwann cells in that they have the ability to form multiple myelin internodes and do so by each extending multiple processes to form an internode on several partner axons at a distance from the cell body (Baumann and Pham-Dinh, 2001). In the PNS, a single Schwann cell synthesizes only one internode. Myelin sheath thickness (Graham and Lantos, 1997), internodal length, and width of nodes show a constant relationship to axonal diameter in normal tissue. In the CNS, nodal length is related to the diameter of the axon and can vary from less than 1 μm in the small fibers of the optic nerve to more than 5 μm in the large fibers of the spinal cord. In the CNS the majority of axons over 0.6 μm in diameter are surrounded by a myelin sheath. In the PNS myelin internodes are 0.5 μm long. CNS internodes surrounding small diameter axons are shorter and thinner than those surrounding large diameter axons. The number of internodes synthesized by one oligodendrocyte depends on the location in the CNS, for example, up to 50 in the optic nerve where axons are of small diameter, but much less in the spinal cord comprising large-diameter axons. In both CNS and PNS, an external layer of noncompact myelin still containing cytoplasm lines the exterior of the internodes. In addition to this structure, internodes of PNS myelin comprise specific channels, the Schmidt–Lanterman incisures that connect outer and inner regions of the internode. This structure is not found in CNS internode myelin.

In the CNS, there is also a radial component formed by points of focal adhesion between sheaths which helps maintain CNS myelin integrity.

Interestingly, different areas of the CNS are myelinated at different times (Yakovlev and Lecours, 1966). In the spinal cord, there is a caudo–rostral progression but even within a tract system, all the axons are not myelinated simultaneously.

The process of myelination represents one of the clearest examples of cell–cell cooperation. Neither axons nor myelin-forming cells can functionally differentiate to completion without each other (Trapp et al., 2004), although oligodendrocytes produce significant amounts of myelin-like membrane in neuron-free culture.

The Schwann cells that myelinate PNS behave differently. Once they reach their final destination, each Schwann cell surrounds several small-diameter axons in a polyaxonal pocket. As axons mature, the Schwann cell segregates a single axon from the polyaxonal pocket. Thus Schwann cells in peripheral nerves have two major phenotypes: those that ensheath multiple axons (unmyelinated fibers) and those that myelinate single axons. The production of a basal lamina is a prerequisite for myelination.

Our knowledge of the molecular composition of myelin internodes is substantial, however, we know little about the molecular mechanism responsible for spiral wrapping of myelin membranes or for axon–myelin forming cell communication.

2.5 Demyelination

One of the particulars of demyelination in the CNS and the PNS is the fact that they are focal for the CNS and segmental for the PNS.

2.5.1 Primary Demyelination and Hypomyelination in the CNS. Prospects for Remyelination in MS and leukodystrophies

One must distinguish active and chronic demyelination. In active primary demyelination (Graham and Lantos, 1997), the pathology is often focal and the removal of myelin sheaths is accompanied by a florid infiltration of macrophages that quickly accumulate myelin debris and become transformed in fat-filled macrophages; there is also a marked astrocytic hypertrophy and hyperplasia. Areas of chronic demyelination appear as areas of astrocytosis devoid of myelin in which demyelinated axons can be shown to be in continuity with normally myelinated axons in the surrounding white matter. The identification of remyelination, partial demyelination, and hypomyelination can prove difficult. All these changes appear as axons enveloped by myelin sheaths too thin for the axons they surround. Too thin myelin sheaths and/or too short internodes indicate hypomyelination if present throughout the white matter and remyelination if found in otherwise normal white matter in the adult. In partial demyelination, the dimensions of myelin sheaths are irregularly reduced; internodes of normal length may be too thin for the diameter of the axon they enclose, or short, thinly myelinated axons are interposed between normal size internodes. In fact this formulation is too schematic, as some genetic demyelinating diseases which start at adolescence and/or adulthood show normal areas of myelination, and focal and diffuse demyelination, in relation to areas that are myelinated at later stages. In most instances, selective loss of whole internodes of myelin results from the death of the oligodendrocytes. The paranodal junctions attaching the myelin sheath to the axonal surface may represent privileged targets at the onset of demyelination (Coman et al., 2005).

Leukodystrophies are specific diseases affecting the white matter of the CNS (brain, optic nerve, and spinal cord); they are genetic diseases affecting different components of the myelinating glial cells, the oligodendrocytes, or of myelin itself. Structure proteins, growth factors, or transcription factors may be involved. Recently, the nonmyelinating glial cells of the CNS, the astrocytes, have been shown to be the direct target of a myelin disease. Numerous animal models have been engineered to mimic and study human diseases; they include overexpressing transgenic mice or rats, knock-out mice in which a gene has been invalidated, and knock-in animals bearing a mutation pathogenic in a human gene. All these models have been invaluable tools to study genetic diseases of myelin.

Multiple sclerosis (MS) is probably not a monogenic disease, and may be caused by a deregulation of the immune system, with more or less specificity for several myelin constituents, proteins or lipids. In general, multiple sclerosis begins in early adulthood and has two phases. There is a relapsing-remitting phase which often lasts 5–10 years; 30% of individuals enter a secondary chronic progressive state. Occasionally, clinical disability begins with this progressive phase in which case the disease is called "primary progressive MS" (Steinman, 2001). Evidence indicates that the earlier phase of the disease, characterized by distinct attacks followed by remission may be mediated by an autoimmune inflammatory reaction. The subsequent chronic phase of the disease is caused by degeneration of both the myelin sheath and the underlying axon. Axon loss in the spinal cord and spinal cord atrophy correlate most strongly with the inability to walk and with paralysis. With the discovery of early and widespread loss of axons in the disease, new emphasis has been put on the role of axon–oligodendrocyte interactions in MS (Williams et al., 2007a). Myelin repair and neuroprotection represent major goals in strategies for MS and represent very active fields of research therapy (Lubetzki et al., 2005). Glial scars composed of astrocytes may prevent remyelinating cells from gaining access to demyelinated axons. However astrocytes also produce a wide range of signaling molecules that support recruitment, and so it is not clear whether astrocytes are friends or foes (Williams et al., 2007a). The signaling environment of the plaque is crucial for the success of remyelination, and it appears that inflammation may play a beneficial role (Bradl and Hohfeld, 2003).

2.5.2 Primary Demyelination and Hypomyelination in the PNS: Charcot-Marie Tooth Diseases (CMT)

Primary segmental demyelination is a disturbance of Schwann cell function or of myelin itself. The initial changes are observed at the nodes of Ranvier. There is myelin retraction and widening of the node. There is paranodal demyelination or the process extends to the whole of the internode leaving a denuded axon. The nerve is invaded by macrophages that engulf the myelin debris. There can be a primary demyelination related to abnormality of Schwann cell function. Demyelination may also be related to an axonal disease. Repeated demyelination and remyelination is encountered in a wide range of disorders. An excess of Schwann cells after demyelination creates an onion bulb formation in chronic neuropathies. Possibly the presence of unmyelinated axons is required for supernumerary Schwann cells to persist (Graham and Lantos, 1997).

CMT diseases are the most frequent hereditary sensory-motor neuropathies. They are distinguished from other types of genetic neuropathies, either purely motor, mainly distal and dysautonomous neuropathies which mainly alter sensory and sympathetic fibers of the peripheral nerves. We only deal here with CMT diseases and among the many genetic causes, those that give rise to primary demyelinating diseases of the peripheral nervous system (PNS).

The accepted classification of CMT relies on the type of genetic transmission as well as on the electroneuromyographic (ENMG) criteria. Measurement of nerve conduction velocity allows a distinction between demyelinating and axonal forms, because conduction velocity is decreased when there is demyelination. The classification may be discussed in relation to the degree of nerve conduction velocity reduction. We adopt the classification of Dubourg (2004). There are truly demyelinating forms with a nerve conduction velocity for the median nerve of 30 m/s, axonal forms with a nerve conduction velocity for the median nerve above 40 m/s and intermediary forms. The molecular analysis confirms the data of the EMG with a very good genotype/phenotype correlation but the penetrance can be variable in the same families. The autosomal dominant form is the most frequent, classically called CMT1, with several subtypes related to the mutations of different proteins.

We do not discuss the axonal forms of CMT, although they may involve secondary demyelination. Axonal loss is also observed in the demyelinating type 1 CMT (Bjartmar et al., 1999). There are some rare autosomal recessive forms called CMT4 which are also demyelinating.

Acute autoimmune demyelinating polyneuropathies such as Guillain–Barré syndrome and chronic polyneuropathies may involve some glycolipid antigens, as described later.

2.5.3 Ion Channels and Demyelination

Voltage-gated ion channels of the Na_V 1.6 type are normally localized at the Ranvier node.

In experimental models of demyelination as well as in MS lesions, a diffuse distribution of Na_V channels along the naked demyelinated axon has been reported. In addition, there is a reversion of the Na_V channel from a mature Na_V 1.6 expression to an immature Na_V 1.2 isoform and this may limit axonal injury (Craner et al., 2004a, b). However, in addition to this diffuse distribution of Na_V channels, loose clusters of Na_V channels persist on some denuded axons. In demyelinated plaques, nodal, paranodal, and juxtanodal axonal molecules are diffusely distributed along the naked axons. Potassium channels Kv1.1 and Kv1.2 are normally confined to the juxtaparanodes. Demyelination can give rise to dispersion of these channels.

During myelin repair, a thinner myelin sheath is produced with shorter internodes, but efficient nerve conduction is produced (Smith et al., 1979). The aggregation of nodal, paranodal, and juxtaparanodal axonal molecules recapitulates the pattern observed during development with the initial step being Na_V channel clustering (Coman et al., 2005; Ogawa et al., 2006).

2.5.4 Astrocytes and Demyelination

Astrocytes are situated in key positions among microvessels, neurons, and oligodendrocytes where they participate in a wide range of functions during brain construction and maintenance. A pattern of early and active astrocyte involvement in several neurodegenerative disorders is emerging. The detrimental role of activated astrocytes during neuroinflammation was recently demonstrated in vivo (Vesce et al., 2007). Thus, astrocytes are now considered as important regulators of many neurophysiological processes and neuropathological conditions (Baumann and Pham-Dinh, 2002; Giaume et al., 2007; Oberheim et al., 2006; Tian et al., 2005, 2007). Moreover, they have also been recognized as new players in CNS myelination via the secretion of molecules signaling oligodendrocytes to myelinate (Ishibashi et al., 2006). As a consequence, myelin could be the target of numerous types of afflictions, from inflammatory/autoimmune (multiple sclerosis) to genetic (leukodystrophies) diseases, which can involve astrocyte dysfunction.

In further support of the important role of astrocytes in neurodegenerative disorders, the causative genes implicated in three leucodystrophies were recently characterized at the genetic level. Indeed the genes responsible for Alexander disease, CACH/VWM syndrome, and "megalencephalic leukoencephalopathy with subcortical cysts" (MLC1), have been identified (GFAP gene, the 5 eIF2B (eukaryotic initiation factor 2B) genes and MLC1, respectively). These three diseases share some similarities: cavitated white matter lesions; sensitivity to different forms of stress, such as febrile episodes or head trauma, triggering episodes of rapid neurological deterioration; and astrocytes as primary target cells. GFAP expression is specific to nonmyelinating glial lineages, as is the MLC gene (Teijido et al., 2007), but EIF2B is ubiquitously expressed. However, the eIF2B mutations are specifically deleterious in astrocytes (Dietrich et al., 2005).

2.5.5 Experimental Models of Demyelination in the CNS

Experimental autoimmune encephalomyelitis (EAE) is the most important animal model of MS. EAE is induced by injection (active EAE) in presence of immune stimulating adjuvants, of CNS tissue, whole spinal cord, purified myelin, or myelin proteins or peptides in susceptible animals (reviewed in Bradl and Linington, 1996; Lassmann, 2004). Specific myelin proteins have been used as immunogens (see below). The onset, severity, and nature of the disease (demyelinating or predominantly inflammatory) are extremely variable and depend on the genetic background of the animal (species and strain differences) and on environmental factors such as the dose and nature of the sensitizing antigen and adjuvant. The initial model was generated to understand acute disseminated encephalomyelitis. Later versions of more chronic EAE have been developed with pathology including demyelination, axonal damage, and clinical events such as relapsing and remitting episodes of paralysis, all of which are features common to MS (Steinman and Zamvil, 2006). EAE has led to therapies approved for use in MS.

Cuprizone intoxication. The presence of the copper chelator cuprizone (biscyclohexanone oxalydihydrazone) in the diet of young adult mice produces a massive demyelination of certain brain regions (Suzuki and Kikkawa, 1969; Blakemore, 1973a; Ludwin, 1978, 1994). Mice exhibit neurological symptoms in the late stages of exposure to cuprizone. Remyelination occurs if the metabolic insult is removed before the end of six weeks treatment (Blakemore, 1973b). After a period of recovery of a few weeks on a normal diet, animals appear normal neurologically and myelin is regained (Matsushima and Morell, 2001). The corpus callosum is one area that is preferentially affected by exposure to cuprizone as well as the cerebellar peduncles (Blakemore, 1973b). Although there is rapid regeneration of the oligodendrocyte population following an acute lesion, most of these newly regenerated cells undergo apoptosis if mice remain on a cuprizone diet. Interestingly, even if the mice are returned to a normal diet following 12 weeks of exposure to cuprizone, remyelination and oligodendrocyte regeneration do not occur (Mason et al., 2004). The mechanisms of the selective effect of cuprizone on oligodendrocytes in certain areas of the CNS need to be elucidated.

Lysolecithin-induced demyelination. Injection of lysolecithin into the spinal cord causes a dramatic decrease in the levels of some myelin protein transcripts (MBP and PLP/DM20). Myelin protein gene expression associated with myelinogenesis during remyelination follows a similar pattern to that of myelinogenesis during development (Woodruff and Franklin, 1999). One week after injection, very little extracellular myelin debris is detected and remyelination has begun. Remyelination progresses rapidly so that almost all axons are engulfed by myelin sheaths by the end of the third week. Remyelination is accompanied by a prominent astrocytosis.

Ethidium bromide treatment. Intracisternal injection of ethidium bromide induces spongiosis with prominent degenerative changes in oligodendroglia in the subpial regions of the rat CNS. Chronologic investigation of the lesions has revealed that status spongiosus results in myelin degeneration, and by the sixth day postinjection many axons are demyelinated. Vesicular transformation of myelin is the common degenerative change. In the demyelinated areas, oligodendroglial cells disappear completely. By the twelfth day postinjection, remyelination is apparent and numerous active oligodendroglia appear in association with thinly myelinated axons. Locally produced IGF could partly be involved in some of the mechanisms underlying remyelination in the mouse spinal cord (Fushimi and Shirabe, 2004).

3 Dys- and Demyelination in Relation to Specific Constituents

Myelin disorders can have a genetic, toxic, or infectious origin and some such as MS even an immunological component. For most of these diseases, there are animal models. Whatever the causes, the targets are structural or metabolic elements necessary for intact myelin. Diseases involving myelin are often related to specific myelin constituents whether common to CNS and PNS, or different. One can believe that these specialized compounds play significant roles in the pathology of myelin. Thus there are diseases or experimental models that involve only CNS or PNS, and others that involve both systems.

To understand the role of the different constituents of oligodendrocytes, Schwann cells, and myelin, and the cellular interactions, it is necessary to consider each constituent and the abnormalities that occur in experimental models and in human pathology. Each specific compound of myelin, and many factors involved in myelination, can give rise to genetic diseases or autoimmune diseases. There are many animal models and human diseases that help in understanding the demyelination process.

It is clear that specific components are particular targets for diseases, whether these are genetic or acquired. These components, although they may be present in other cell types or tissues, must be viewed as modular elements that, with other elements, generate complex and unique surface patterns in the nervous system.

3.1 CNS Myelin Proteins

Major myelin proteolipid proteins, PLP and DM-20. In 1951 Folch and Lees discovered that a substantial amount of proteins from brain white matter could be extracted by organic solvent techniques. They were given the generic name of proteolipids (PLP), as these proteins were lipid–protein complexes. The responsible gene is PLP1. The PLP1 gene encodes two major products, PLP itself, and DM-20 encoded by an alternatively spliced transcript that lacks 35 residues from the cytoplasmic domain of PLP. PLP represents about half the protein mass of CNS myelin (reviewed in Garbern, 2005). PLP is also present in the PNS; however, it constitutes only a small fraction of myelin proteins in the periphery (Pham-Dinh et al., 1993). PLP is a highly hydrophobic transmembrane protein. In addition to a high content of hydrophobic amino acids, PLP is anchored to the lipid bilayer by fatty acylation of several cysteine residues. PLP and DM-20 are both acylated by covalent linkage of mainly palmitic, oleic, and stearic acids to cysteine residues in regions of the PLP/DM-20 proteins localized on the cytoplasmic side of the myelin membrane (Weimbs and Stoffel, 1992). DM-20 is present at about 10% of the level of PLP in the CNS whereas in the PNS they are present in approximately equivalent amounts.

It is worth noting that there is a 100% sequence identity between rodent and human PLP proteins (Macklin et al., 1987), which is a rather striking conservation. The PLP1 gene is located on the X chromosome (Xq22 in humans) (Mattei et al., 1986).

There are a great number of spontaneous myelin PLP1 gene mutants (Duncan, 2005). Myelin mutants are often named according to their phenotypes. Among the mutations are the jimpy, jimpy^{msd} and jimpy 4 J mice. There is also a myelin-deficient rat (md rat). A long-lived md rat has been described (Duncan et al., 1995) that has recently allowed new therapeutic approaches (Espinosa-Jeffrey et al., 2006). A canine shaking pup has been also described. The rumpshaker mouse and the paralytic tremor rabbit (pt) have a normal life span.

Null mutations are informative for inferring the normal biological functions of a gene product. Despite its abundance in normal CNS myelin, complete lack of PLP and DM20 results in a surprisingly mild phenotype in mice and humans; PLP appears not to be essential for oligodendrocyte differentiation and myelination. The most striking pathology in both mice and humans with PLP null mutations is a relatively late and progressive degeneration of axons (Garbern et al., 2002). Interestingly, although PLP/DM20 represents less than 0.01% of peripheral nerve myelin, complete lack of these proteins causes a demyelinating peripheral neuropathy (Garbern et al., 1997). In the PLP-null CNS, axons large enough to be myelinated often lack myelin entirely or are surrounded by abnormally thin sheaths. Short stretches of cytoplasm persist in many myelin lamellae. In thinly myelinated fibers, there are interlamellar spaces across the full width of the sheaths. In thick myelin sheaths, the spaces appear irregular but diffuse. These spaces constitute a spiral pathway through which ions and other extracellular agents may penetrate gradually, possibly contributing to the axonal damage known to occur in this mutant, especially in thinly myelinated fibers, where the spiral path length is shortest. The "radial component" of myelin is distorted in the mutant ("diagonal component"), extending across the sheaths at 45° instead of 90°. These observations indicate a direct or indirect role for PLP in maintaining myelin compaction along the external surfaces of the lamellae and to a limited extent along the cytoplasmic surface as well, and also in maintaining the normal alignment of the radial component (Rosenbluth et al., 2006). Absence of PLP may give rise to abnormal axonal transport (Edgar et al., 2004).

In some mutants and transgenic mice overexpressing PLP1, there is an accumulation of PLP in the endoplasmic reticulum of the oligodendrocyte, which may eventually trigger apoptosis. The latter is related to an unfolded protein response, UPR (Gow and Sharma, 2003). In these forms, increasing the level of PLP may exacerbate pathology (Gow et al., 1998).

Increased PLP1 gene dosage affects expression of other myelin proteins, particularly MBP which is lower in homozygotes in both myelin and early myelinating oligodendrocytes (Karim et al., 2007). As shown in the jimpy mouse, there is a drastic decrease of the 14 kDa isoform of MBP (Campagnoni et al., 1984).

Pelizaeus–Merzbacher disease (PMD) in an X-linked leukodystrophy primarily associated with the duplication, deletion, or mutation of the PLP1 gene. The principal effect of many mutations in the coding region of PLP1 is to disrupt the highly ordered structure of the resulting protein isoforms DM20 and PLP. This leads to their accumulation in the endoplasmic reticulum of oligodendrocytes and ultimately to diminished biosynthetic capacity or survival of these cells. The UPR protein response modulates disease severity in Pelizaeus–Merzbacher disease (Southwood et al., 2002). PMD patients in which PLP1 is deleted or functionally null may well benefit from gene replacement (Gow et al., 1998). A spastic paraplegia type 2 (SPG-2) phenotype is also caused by a PLP1 mutation. This milder form resembles that in the rumpshaker mutant mouse.

As noted above, experimental autoimmune encephalomyelitis (EAE) has been studied for decades as an experimental model for MS (Steinman and Zamvil, 2006). A PLP peptide (139–151) induces a chronic relapsing disease; combinations of antigens are also used (Kuerten et al., 2007). The strain of mouse influences the expression of EAE (Kuerten et al., 2007). As gene knock-out and knock-in mice are also becoming increasingly indispensable for mechanism-oriented studies, animal models in the mouse have been increasingly useful. EAE has also been provoked using a proteolipid suppressor of cytokine signaling 1 (PLP/SOCS1) transgenic mouse line that displays suppressed oligodendrocyte responsiveness to interferon-gamma; mice under these conditions develop an accelerated onset and increased oligodendrocyte apoptosis (Balabanov et al., 2007).

Myelin basic proteins (MBP). Basic proteins are abundant both in CNS and PNS myelin, where they are associated with negatively charged lipids. They are assumed to be involved in myelin compaction on the cytoplasmic side of the membrane bilayer. A spontaneous MBP mutant, the shiverer mouse, is devoid of the major dense line of myelin in the CNS (Dupouey et al., 1979). Another such mutant, the Long Evans Shaker (LES) rat, shows major changes in spinal cord white matter, with dispersed labeling of Kv1.1 and Kv1.2 K+ channel subunits as well as of Caspr, a molecule normally confined to paranodes along LES rat spinal axons (Eftekharpour et al., 2005).

MBP constitutes about 30% of the protein content of myelin. In fact, the MBPs constitute a family of proteins comprising many isoforms (reviewed in Campagnoni and Skoff, 2001). The molecular weight of the major forms are 21.5, 20.2, 18.5, and 17.2 kDa in man, and 21.5, 18.5, 17, and 14 kDa in mouse. In the adult, two major isoforms constitute about 95% of the MBPs; they are the 18.5 and 17.2 kDa isoforms in humans, and the 18.5 and 14 kDa in mouse (Staugaitis et al., 1990). The MBP isoforms are coded by alternative transcripts generated from the MBP gene which consists of 7 exons (Roach et al., 1983). Subsequent studies have found that the classical MBP gene is contained within another and huge transcription unit called the Golli-MBP gene (golli for "gene expressed in the oligodendrocyte lineage"). It is 195 kb in mice and 179 kb in humans, and produces a number of alternative transcripts from three possible transcription start sites, most of them containing the MBP sequences. The Golli-MBP gene contains three additional specific exons located 5′ to the seven constituting the classical MBP gene. The Golli-MBP gene is located on chromosome 18 in mouse and human (18q23).

The MBP Golli transcripts and proteins are also found in immune system cells (reviewed in Feng, 2007): they have recently been found to directly regulate T-cell activation, thus modulating EAE induction. MBP mRNAs are transported to glial processes to be translated on free ribosomes. Posttranslational modifications can occur on the MBPs, including phosphorylation, methylation, citrullination, and N-terminal acetylation. Moreover, the presence of exon 6-containing MBPs in the nucleus suggests a regulatory role in myelination for these MBPs isoforms (reviewed in Baumann and Pham-Dinh, 2001; Campagnoni and Macklin, 1988). In the PNS, MBPs represent 5–20% of the total PNS myelin protein content and are located on the intracellular side of the myelin. Contrary to what occurs in the CNS, the absence of MBP isoforms does not alter the major dense line of myelin in the PNS.

The exon-6 containing MBP isoforms of 17 and 21.5 kDa in the mouse, and 20.2 and 21.5 kDa in humans, are mainly expressed during myelinogenesis. They are re-expressed in chronic lesions of MS, and their re-expression correlates with remyelination (Capello et al., 1997). Some isoforms are modified as consequences of mutations on other myelin proteins, especially PLP (see above).

MBP or MBP peptides are very commonly used to induce EAE. The MBP-induced disease in some strains of mice is often monophasic with inflammation and no demyelination. The mice recover completely after a single episode of a short and acute disease and become resistant to reinduction of EAE. Therefore, a combination of antigens involving proteins or peptides (Kuerten et al., 2006) or sphingoglycolipids (Raine et al., 1981) are often used to induce demyelination.

A genetic disease, the 18q-syndrome is a rare leukodystrophy presenting a genomic deletion that includes the MBP gene. Proton magnetic resonance data indicate demyelination or increased myelin turnover rather than dysmyelination (Hausler et al., 2005).

Recent evidence obtained via magnetic resonance imaging and spectroscopy techniques supports the view that the normal-appearing white matter (NAWM) in the MS brain is altered. Several biochemical changes in NAWM have been determined. These include the cationicity of myelin basic protein (MBP) as a result of peptidyl arginine–deiminase (PAD) activity converting arginyl residues to citrulline. The accompanying loss of positive charges renders myelin susceptible to vesiculation and MBP more susceptible to proteolytic activity. An increase of MBP autocatalysis in the MS brain might also contribute to the generation of immunodominant epitopes (Mastronardi and Moscarello, 2005).

OSP/claudin-11 (Oligodendrocyte-specific protein). OSP/claudin-11, a 22-kDa protein, is the third most abundant protein in CNS myelin, after PLP/DM20 and MBP. It accounts for about 7% of the protein content. OSP is related to PMP-22 found in PNS myelin, with which it has 48% amino acid similarity and 21% identity. OSP was recognized as a previously known tight junction protein, claudin-11. The gene for OSP is located on chromosome 3 in the mouse and the 3q26.2-26.3 region of chromosome 3 in humans. Myelin compaction is not significantly disrupted in the knock-out mouse. This is not the case for a double knock-out OSP/Claudin 11 and PLP1/DM20, indicating that these proteins have essential structural functions in maintaining myelin compaction, but that there is redundancy in their functions (Chow et al., 2005). K+ channel Kv3.1 associates with OSP/claudin 11 and regulates oligodendrocyte development (Tiwari-Woodruff et al., 2006).

MAL (*myelin and lymphocyte protein*) (formerly rMAL for rat MAL) was the first cloned member of a new myelin–oligodendrocyte proteolipid protein family (MAL family) including MVP17 (myelin vesicular protein of 17 kDa) and Plasmolipin. MAL is a tetraspan raft-associated proteolipid predominantly expressed by oligodendrocytes and Schwann cells. Genetic ablation of MAL leads to reverted paranodal loops away from the axon, with a marked reduction of contactin-associated protein/paranodin, neurofascin 155, and the potassium channel Kv1.2, whereas nodal clusters of sodium channels remain unaltered. MAL has a critical role in the maintenance of CNS paranodes, likely by controlling the trafficking and/or sorting of NF155 and other membrane components in oligodendrocytes (Schaeren-Wiemers et al., 2004). MAL is modified in neurological mutants affecting myelination through a defect in the catabolism of sphingoglycolipids (Saravanan et al., 2004).

Connexins (Cx) 32 and 47 are part of a family of gap junction proteins, which form channels, generally between adjacent cells. These specialized channels span two plasma membranes of adjacent cells and allow the passage of ions, amino acids, second messenger molecules, and small metabolites. In general, six connexins oligomerize to form a homomeric or heteromeric connexon (hemichannel), and a functional gap junction pathway between two cells is formed by homotypic or heterotypic interactions of two connexons. In myelin, Cx32 and 47 may form channels between adjacent layers of the myelin sheath in regions where myelin is not compacted.

Cx32 is a ubiquitous protein, also found in CNS myelin. Cx32 has a molecular weight of 32 kDa and contains four transmembrane domains, two extracellular loops, and three intracellular domains. It is expressed widely in a number of tissues, particularly in the liver, and also in PNS and CNS myelin. Cx32 is found in noncompact regions (paranodes and Schmidt–Lantermann incisures) of the PNS myelin sheath (Scherer et al., 1995). The human Cx32 gene is located on chromosome

Xq13.1. Its structure is similar to all connexin genes, that is, a large exon containing the coding sequence within one uninterrupted block, which is separated by an intron from a small noncoding exon located on the 5′-flanking region. Three alternative promoters, that appear to be activated in a cell-type manner, regulate the tissue-specific expression of Cx32. Its presence in Schwann cells was discovered when Cx32 mutations were associated with Charcot–Marie–Tooth disease of the CMTX type (see § PNS). This gap-junction molecule allows the direct passage of ions and small molecules through the myelin sheath in the paranodal regions. There are subtle abnormalities of the myelin sheath and of the Ranvier node (Hahn et al., 2001).

In humans, *Cx47* is expressed specifically in oligodendrocytes, where it is partially colocalized with Cx32. The gene encoding Cx47 is regulated in parallel with myelin genes. Mice lacking either Cx47 or Cx32 are viable. However, animals lacking both connexins die by postnatal week 6 with profound anomalies in central myelin, characterized by thin or absent myelin sheaths, vacuolation, enlarged periaxonal collars, oligodendrocyte cell death, and axonal loss. Thus gap-junction communication is crucial for normal central myelination (Menichella et al., 2003). Connexin 47 (gap junction protein alpha 12) mutations cause a Pelizaeus–Merzbacher-like disease (Orthmann-Murphy et al., 2007). Connexin 47 is also involved in peripheral myelination in humans (Uhlenberg et al., 2004).

Tetraspanin 2 has recently been identified in cells of the oligodendrocyte lineage. Expressed after birth in rodents, tetraspanin may play a role in signaling in oligodendrocytes at the early stages of their terminal differentiation into myelin-forming glia; it is also hypothesized that it may function in stabilizing the mature sheath.

Myelin-associated/oligodendrocyte basic protein (MOBP) is a small highly basic protein. Alternative splicing generates three isoforms of 8.2, 9.7, and 11.7 kDa. Like MBP, MOBPs are located in the major dense line of myelin where they could play a role similar to that played by MBP in myelin compaction. MOBP transcripts are less abundant than PLP1 but more than those for the CNP (cyclic nucleotide phosphodiesterase). The MOBP mRNA is initially located in the cell bodies of the oligodendrocytes, and moves distally into the processes when myelination occurs, as do MBP mRNAs. The gene for MOBP has been mapped to chromosome 9 in the mouse, in a region syntenic with the human chromosome 3 (3p22).

Myelin-associated glycoprotein (MAG) has an apparent molecular weight of 100 kDa, of which 30% is carbohydrate; MAG bears the L2/HNK1 epitope, a glycosylated epitope also present on glycolipids of the PNS. It is a minor constituent, representing 1% of the total protein content in CNS myelin and 0.1% in PNS myelin. It has been extensively reviewed recently (Quarles, 2007). Two MAG isoforms have been identified, large MAG (L-MAG) and small MAG (S-MAG), corresponding to polypeptides of 72 and 67 kDa, respectively, in the absence of glycosylation. MAG proteins have both a membrane-spanning domain and an extracellular region containing 5 immunoglobulin domains. The MAG gene includes 13 exons (from which exons 1, 2, and 3 are noncoding); the isoforms differ only in their cytoplasmic domains, resulting from alternative splicing. Exon 12, present in S-MAG, contains an alternative stop codon. MAG is located on chromosome 7 in mice and 19 in humans (19q13.1). The 72 kDa L-MAG can be phosphorylated and acylated. MAG is found in the periaxonal space in CNS and PNS; it derives from the Schwann cell and oligodendroglial membrane. CNS findings suggest that the absence of MAG causes oligodendrocytes to form myelin less efficiently during development and to become dystrophic with aging. MAG, together with the proteins Nogo 66 and Omgp, inhibits axonal regeneration. MAG binds to the Nogo 66 receptor called NgR; this activates a P75 neurotrophin receptor (p75NTR) and the transduction of the resulting signal activates the small GTPase Rho leading to inhibition of axonal growth following injury (Spencer et al., 2003; Filbin, 2003). Although MAG binds to gangliosides, sialic acid binding is unnecessary for MAG to exert inhibition (Cao et al., 2007). MAG is also currently well known as an antigen for IgM monoclonal antibodies that cause demyelinating peripheral neuropathies (reviewed also in Quarles, 2007). It comprises the HNK-1 epitope present on the sulfated glycolipid with glucuronic acid SGPG, that is, sulfated glucuronylparagloboside (see below). Interestingly the sensory-motor neuropathies with monoclonal IgM that react with both lipid and proteic antigens are demyelinating, and the neuropathies with IgM monoclonal antibodies that react with SGPG and not MAG are axonal (Chassande et al., 1998).

Myelin-oligodendrocyte glycoprotein (MOG) was first identified as the antigen responsible for the demyelination observed in animals injected with whole CNS homogenate; it was later identified as a minor glycoprotein specific for CNS myelin. MOG was further characterized by immunological methods using a mouse monoclonal antibody obtained against glycoproteins of rat cerebellum. MOG is a minor protein of 25 kDa with some glycosylation resulting in doublets of 26–28 kDa on SDS page, which can form dimers of 52–54 kDa. MOG is only present in mammalian species. In humans, MOG expresses the L2/HNK1 epitope. The amino-terminal, extracellular domain of MOG has characteristics of an Ig-variable domain and is 46% identical with the amino-terminus of bovine butyrophilin protein expressed in the mammary gland, and chick histocompatibility BG antigens. Although MOG contains two highly hydrophobic regions, only one is a truly transmembrane domain, thus MOG presents the same topology as other members of the Ig-superfamily. The human MOG gene is encoded by 11 exons that exhibit a complex pattern of alternative splicing (Pham-Dinh et al., 2004; Delarasse et al., 2006). Complex alternative splicing of MOG is unique to human and nonhuman primates (Delarasse et al., 2006). The MOG gene is located in the distal part of the major histocompatibility complex (MHC) in the class Ib region on chromosome 6p22-p21.3 in humans and 17 in rodents. MOG is specific to the CNS and localized on the outer surfaces of myelin sheaths and oligodendrocytes where it is accessible to components in the external environment, such as complement and antibodies. MOG is a highly encephalitogenic autoantigen and a target for aggressive autoimmune responses in CNS inflammatory demyelinating diseases (Delarasse et al., 2003). Autoantibody responses against conformational epitopes of MOG have the power to destroy myelin, as demonstrated in the marmoset model of human MS (von Budingen et al., 2004, 2006). Controversy exists regarding the pathogenic or predictive role of anti-MOG antibodies in patients with MS (Lalive et al., 2006; Pittock et al., 2007).

2′3′-Cyclic nucleotide-3′-phosphodiesterase (CNP) represents 4% of total CNS myelin proteins. In vitro, this protein hydrolyzes artificial substrates, 2′3′-cyclic nucleotides into their 2′ derivatives. However, the biological role of this enzymatic activity is obscure because 2′3′ nucleotides have not been detected in the brain. Overexpression of CNP in transgenic mice disturbs myelin formation and creates aberrant oligodendrocyte membrane expansion. CNP appears on SDS-page as a doublet of two peptides, with molecular weights of 48 and 46 kDa, referred as CNP2 and CNP1, respectively. The two CNP isoforms are produced by alternative use of two transcription start sites. The CNP gene is located on chromosome 17 (17q21) in human and chromosome 11 in mouse. CNP mRNAs are detected in mouse spinal cord during embryonic stages. CNP is present in the cytoplasm of noncompacted oligodendroglial ensheathment of axons and in the paranodal loops of myelin internodes. The protein is posttranslationally modified, acylated, and phosphorylated. CNP (mainly CNP1) is associated by isoprenylation to the cytoplasmic plasma membrane of the oligodendrocyte. In double knock-out mice, inactivation of both CNP1 and FGF-2 lead to hyperactivity, starting around two weeks of age. When hyperactive mice receive dopamine receptor antagonists or catecholamine synthesis inhibitors, their behavior reverts to normal, suggesting that their symptoms are caused by a dysregulation in the dopaminergic system. The molecular mechanisms leading to hyperactivity have not yet been elucidated. This mouse model supports the evidence cited above that oligodendrocytes and myelin may be involved in the genesis of neuropsychiatric disorders, and also that it is almost impossible to predict the impact of genetic interactions on the behavior of transgenic animals (Kaga et al., 2006).

Nogo proteins, formerly named NI-35/250 proteins, are membrane-bound proteins highly enriched in mammalian CNS myelin and oligodendrocytes (reviewed in Bandtlow and Schwab, 2000; Goldberg and Barres, 2000). Nogo comprises three isoforms, Nogo-A, -B, and -C. It is predominantly associated with the endoplasmic reticulum of the oligodendrocyte. Following injury, Nogo would become exposed to the extracellular environment. MAG, Nogo, and OMgp share the same functional receptor (Spencer et al., 2003; Filbin, 2003). As mentioned previously, they inhibit axonal growth following injury. Antibodies against Nogo or Nogo-blocking peptides enhance sprouting of damaged axons after a partial spinal cord section in a primate, the marmoset, and this is associated with clinical improvement (Fouad et al., 2004). Nogo receptor-interacting protein (LRR and Ig domain-containing Nogo receptor-interacting protein, LINGO-1) is a negative regulator of oligodendrocyte differentiation and myelination. Antagonism of LINGO-1 or its pathway is a promising approach for treatment of demyelinating diseases in the CNS (Mi et al., 2005, 2007).

Enzymes. Many enzyme activities have been found in myelin: neuraminidase, cholesterol ester hydrolase, lipid synthesizing and catabolizing enzymes, proteases, protein kinases, and phosphatases. Two of them have been especially characterized.

1. UDP-Galactose: ceramide galactosyltransferase (CGT) is found mainly in myelinating glia where it parallels expression of MBP and PLP1. The CGT gene spans about 70 kb, comprises 5 exons and has been mapped to mouse chromosome 3 bands E3-F1 and to the human chromosome 4 band q26. Rodent and human CGT sequences are strongly conserved. More details on CGT are in the paragraph related to myelin lipids, below.

2. Fyn kinase is a nonreceptor-type tyrosine kinase that has been proposed to act as a signaling molecule downstream of MAG. MAG and Fyn are coexpressed in oligodendrocytes, and can be coimmunoprecipited by biochemical methods. Fyn-null mice show an important reduction (about 50%) of CNS myelin, whereas myelination is quite normal in the MAG-null mice (see above). Double-deficient mice present a massive hypomyelination, associated with behavioral deficits. These data indicate the importance of both molecules in the initiation of myelination; however, they could also mean that MAG and Fyn act in concert or independently in initiating myelination (Biffiger et al., 2000).

3.2 PNS Myelin Proteins

PNS myelin proteins include two abundant constituents: glycoprotein zero (P0) and MBPs (already discussed in III.1), and a set of minor ones including PMP22, MAG, and connexin 32 (already discussed in III.1).

Protein zero (P0) is the most abundant glycoprotein of peripheral myelin; it represents 50–70% of the total myelin protein content. P0 is specific for the PNS. It is predominantly localized to compacted regions of myelin. P0 is expressed constitutively in neural crest and embryonic nerves; its expression is strongly upregulated in myelinating Schwann cells. P0 has a molecular weight of 28 kDa, and is composed of an Ig-like extracellular domain by which P0 can form tetramers, a single highly hydrophobic transmembrane domain and an intracellular cytoplasmic domain that contains abundant positive-charged amino acids that could stabilize negative-charged lipid heads. The gene encoding the P0 protein, called MPZ for myelin protein zero, contains 6 exons and is on chromosome 1 in humans (1q22). Interestingly, P0 is essential for normal spacing of PNS compact myelin (reviewed in Trapp et al., 2004). CMT1B is an autosomal dominant demyelinating hereditary neuropathy involving the MPZ gene, that is, a CMT disease (Dubourg, 2004; Shy, 2004). Other mutations of this gene cause severe neuropathies of infancy (Dejerine–Sottas disease), and still others lead to disability with a late onset (Shy et al., 2004). MPZ mutations disrupt the tertiary structure of P0 protein, interfering with P0-mediated adhesion during myelination and with myelin compaction. In contrast, late onset neuropathies result from mutations that allow myelination but chronically disrupt Schwann cell–axonal interactions. A genotype/phenotype correlation is clear even though penetrance can vary within single families.

Peripheral myelin protein 22 (PMP22) has a molecular weight of 22 kDa and represents 2–5% of the total myelin protein content. Despite its name, PMP22 is not specific to the PNS inasmuch as it is expressed, albeit at low levels, in other tissues such as lung, gut, heart, and some neurons. It contains four transmembrane domains with two extracellular loops and one short intracellular loop and is preferentially localized in compacted regions of myelin. The human PMP22 gene is about 40 kb long and is located on chromosome 17p12-p11.2. It contains 6 exons (4 coding exons and 2 untranslated exons in the 5′ flanking region) and its expression is regulated by two alternative promoters. Charcot–Marie–Tooth 1A (CMT1A), the most frequent genetic demyelinating neuropathy, is a clinical expression of an autosomal dominant mutation in this gene (Dubourg, 2004; Shy, 2004). The most frequent mutation is a duplication of the PMP-22 gene. There is a spontaneous mouse model of PMP-22 mutation, the trembler-J mouse, which leads to failure of myelination and continuous Schwann cell proliferation; this implies trophic support by Schwann cells (Friedman et al., 1996). Therapies reducing PMP22 overexpression in rodent models of CMT1A offer potential treatments. Progesterone is known to increase PMP22 messenger RNA expression in cultured Schwann cells. On this basis, Sereda and colleagues (Sereda et al., 2003) in Nave's laboratory, using a progesterone receptor antagonist, onapristone, have been able to reduce the expression of PMP22 in a transgenic rat, thus opening ways for symptomatic treatment of this form of the disease. In another mouse model overexpressing a human PMP 22 gene, ascorbic acid reduces PMP22 levels, improving the phenotype of this CMT1A model (Passage et al., 2004).

Lipopolysaccharide-induced tumor necrosis factor (LITAF) is encoded on a gene located on 16p13.1-p12.3 in humans. It is a putative degradation protein also called SIMPLE (for small integral membrane protein of the lysosome/late endosome). Although SIMPLE is expressed in many cell types, when mutated it seems to cause only a demyelinating neuropathy, which suggests that the disease specificity may come from the impaired degradation of specific Schwann cell proteins (Shy, 2004). These mutations give rise to a rare form of CMT called CMT1C.

Periaxin represents about 5% of the total PNS myelin protein content; it was given this name because of its specific localization in the periaxonal membranes of myelinated Schwann cells (Gillespie et al., 1994). Two periaxin isoforms have been identified: L-periaxin (147 kDa), localized to the plasma membrane of myelinating Schwann cells, and S-periaxin (16 kDa). The latter occurs diffusely throughout Schwann cell cytoplasm as a cytoskeletal component, and also in the nucleus. The human periaxin gene is located on chromosome 19q13.13-q13.2. Mutations in the periaxin gene cause a form of demyelinating CMT, CMT4F (Dubourg, 2004; Meyer zu Horste et al., 2006; Shy, 2004). These mutations alter myelin and also involve onion bulb formation with chronic processes of demyelination and remyelination.

As noted in the preceding section, *Cx32* is a ubiquitous protein that is also found in CNS myelin. Its presence in Schwann cells was discovered when Cx32 mutations were associated with Charcot–Marie–Tooth disease of the CMTX type. CMTX is an X-linked demyelinating neuropathy. The molecule is located in the paranodes. There are subtle anomalies of the myelin sheath and the Ranvier node (Hahn et al., 2001). Mutations in Cx32 gap junction protein compromise Schwann cell functions and lead to impaired Schwann cell–axon interactions with subsequent pathology in both myelin and axons.

Cx47 (description in § CNS) is another protein involved in peripheral myelination in humans (Uhlenberg et al., 2004). Diseases associated with mutations of the gene encoding this protein have not been described, at least as yet.

Mutations in several other genes are known to give rise to demyelinating CMT (reviewed in Dubourg, 2004; Meyer zu Horste et al., 2006).

A gene, located on 10q21.1-q22.1, can cause early demyelination and CMT1D when mutated. The gene codes for EGR2 (early growth response 2 gene) which is a transcription factor also termed *Krox-20*. It allows the regulation of key genes coding for myelin proteins such as PMP22, P0, and Cx32, and gives rise to CMT4E when mutated. There are also mutations involving the gene *NEFL* (neurofilament light chain) which in some cases can give rise to demyelinating neuropathies.

Mutations in the CMT4 genes lead to hypomyelination and onion bulb formation. The gene for CMT4A is located on 8q13.q21.1 and codes for a protein called *GDAP1*, that is, ganglioside-induced differentiation-associated protein 1. Another mutated protein may be involved in CMT4B1: MTMR2 (myotubularin-related protein 2). MTMR2 is a dual specific phosphatase that participates in the dephosphorylation of membrane phospholipids involved in the regulation of intracellular membrane trafficking. Mutations in CMT4B2 are related to mutations in MTMR13. MTMR13 is also known as set-binding factor (SBF2). Mutations in CMT4B1 and CMT4B2 lead to characteristic misfolding and redundant loops of myelin. CMT4C is related to a defect in the gene coding for protein *KIAA* and CMT4D with mutations coding for the gene *NDRG1* (N-myc downstream regulated gene). All these mutations alter myelin and also involve onion bulb formation with chronic processes of demyelination and remyelination.

Epithelial cadherin (E-cadherin) has a molecular weight of 130 kDa and is found in PNS myelin. E-cadherin is a protein of the superfamily of calcium-dependent cell adhesion molecules that can usually form adherent junctions. This protein has an N-terminal extracellular domain, a short transmembrane domain and a C-terminal intracellular domain.

Basic protein P2 has a molecular weight of 15 kDa and is a member of a family of cytoplasmic lipid-binding proteins. Unlike other myelin proteins, its quantitative expression varies greatly according to species, ranging from less than 1% in rodent sciatic nerves up to 5–14% in human, bovine, and rabbit nerve. P2 is located on the cytoplasmic side of compacted regions of myelin.

3.3 *Proteins and Specific Lipids of the Node of Ranvier and the Paranodal and Juxtaparanodal Areas*

The nodes of Ranvier are critical for the proper function of the CNS and PNS. The composition of the myelin node and the modifications observed in natural mutant mice, transgenic mice, or knock-out mice have aided understanding of the function of these specialized molecules. Voltage-gated sodium channels (see § Ion Channels and demyelination), ankyrin G, NrCAM (NgCAM-related CAM; i.e., neuron–glia-related CAM) are highly enriched at the node (Simons and Trajkovic, 2006). Other constituents are also present and detailed below because they are more involved in demyelination.

Oligodendrocyte-myelin glycoprotein (OMgp) is clustered at the nodes of Ranvier in both CNS and PNS. It regulates nodal formation through an unidentified mechanism. In the CNS, its abundance is closely linked with axonal size, and OMgp is undetectable in a subset of smaller axons. In transgenic mice in which expression of OMgp is downregulated, myelin thickness diminishes, lateral oligodendrocyte loops at the node–paranode junction are less compacted, and there are shortened nodal gaps (Nie et al., 2006). Transgenic mice, in which OMgp is reduced by 50–70%, show a significant abnormality in the node–paranode junction. Disorganized lateral glial loops at the node–paranode junctions and shortening of nodal space in the OMgp mutants strongly indicate that OMgp has a role in attaching lateral loops and in demarcation of the node–paranode junctions. Thus, this oligodendrocyte protein is involved in regulation of both myelin development and nodal formation. Nodal location of OMgp does not occur along demyelinated axons of either the shiverer natural mutant mice or PLP transgenic mice. Omgp is also a myelin-associated inhibitor of axonal regeneration as a GPI-linked Omgp (Spencer et al., 2003).

The 186 kDa neuron specific isoform of the adhesion molecule neurofascin (NF186) is required for the clustering of voltage-gated channels at the node (Howell et al., 2006). Its expression is disrupted following demyelination.

The myelin protein cyclic nucleotide phosphodiesterase (CNP) is required for maintenance of axon–glia interactions at the node of Ranvier. It also maintains the integrity of the paranodes (Rasband et al., 2005).

In the PNS, both laminin and Schwann cell dystroglycan are necessary for the proper clustering of sodium channels at nodes of Ranvier (Occhi et al., 2005). Gliomedin is also necessary (Eshed et al., 2005).

At the paranodes, myelin loops are anchored to axons through septate-like junctions characterized by the enrichment of paranodin/Caspr (contactin-associated protein) and the GPI-anchored cell adhesion molecule contactin. The paranodin/Caspr–contactin complex interacts with the 155 kDa isoform of neurofascin NF155 that is expressed on the oligodendroglial membrane. NF155 is essential for the tight interaction between myelin and axon. It is a member of the L1 family of cell adhesion molecules (L1-CAM; reviewed in Maier et al., 2006). Changes to NF155 expression accompany inflammation and demyelination and contribute to the destruction of the neurofascin 186/sodium channel complex vital to successful neurotransmission in the CNS (Howell et al., 2006).

The tetraspanin protein CD9 is a novel paranodal component regulating paranodal junctional formation (Ishibashi et al., 2004).

Myelin galactolipids are essential for the proper formation of axo–glial interactions. Disruption of these interactions results in profound abnormalities in the molecular organization of the paranodal axolemma (Dupree et al., 1999). Mice incapable of synthesizing the abundant galactolipids of myelin exhibit disrupted paranodal axo–glial interactions in both the CNS and PNS. Whereas the clustering of the nodal proteins, sodium channels, ankyrin, and neurofascin are only slightly affected, the distribution of potassium channels and paranodin proteins is dramatically altered. The potassium channels, which are normally concentrated in the juxtaparanode, are no longer restricted to this region but are detected throughout the internode in the mutant. The paranodin/contactin-associated protein Caspr, a paranodal protein, is not concentrated in the paranodal region, but diffusely distributed along the internodal regions.

The paranodal junction also contains specialized cytoskeletal components that may be important in stabilizing axon–glia interactions (Ogawa et al., 2006; Voas et al., 2007). A significant alteration in NF155 paranodal structures occurs within and adjacent to actively demyelinating white matter regions that are associated with damaged axons (Howell et al., 2006).

The juxtaparanode is just under the compact myelin sheath beyond the innermost paranodal junction and may therefore be considered a specialized portion of the internode (Peles and Salzer, 2000). Potassium channels aggregate in the juxtaparanode. Potassium channels Kv1.1 and Kv1.2 are normally confined to this area. Demyelination can lead to the dispersion of these channels.

The juxtaparanodal region, just beyond the innermost paranodal junction is enriched in shaker-type potassium (K_V) channels, in association with Caspr2, a second member of the Caspr family, as well as in the cell adhesion molecule TAG-1 (for review see Coman et al., 2006). Kv channels are in the juxtaparanode area (Howell et al., 2006). In demyelinating white matter, shaker-type Kv1.2 channels move and precede alterations at the node itself, in relation with NF155 disruption. TAG1 and Caspr 2 are essential for the molecular organization of the juxtaparanodal region of myelinated fibers (Traka et al., 2003).

During myelin repair, NF 155 is an early marker of myelin damage (Howell et al., 2006). During myelin repair a thinner myelin sheath is produced with shorter internodes, but efficient nerve conduction is nevertheless produced (Smith et al., 1979). The aggregation process of nodal, paranodal, and juxtaparanodal axonal molecules recapitulate development, with the initial step being Na_V channel clustering (Coman et al., 2006; Ogawa et al., 2006).

3.4 Dys- or Demyelination in the CNS and/or the PNS Related to Myelin Lipid Compounds

One of the major characteristics of myelin lipids is their richness in sphingoglycolipids (SGLs) particularly the galactosphingolipids galactosylceramides (GalC) and their sulfated derivatives sulfogalactosylceramides, that is, sulfatides (Baumann and Pham-Dinh, 2001; Colsch et al., 2004). SGLs are present on virtually all mammalian cell plasma membranes. They are amphipathic molecules consisting of a ceramide lipid moiety embedded in the outer leaflet of the membrane, linked to an oligosaccharide structure oriented externally. Both the lipid moiety and the oligosaccharide structure show huge structural diversity. SGLs are very abundant in the nervous system, with different constituents in the CNS and the PNS and within different cell types in these tissues. It is not clear yet whether each SGL isoform has a specific location or function, although the extreme diversity of their constitution possibly contributes to the diversity of stereospecific recognition at the surface of the cellular membranes. SGLs in myelin contain very long chain fatty acids. The fact that they are on the external surface of the cell favors their involvement in the modulation of protein receptors and favors their acting as signaling molecules. These galactolipids are important for the activity and maintenance of myelin and myelin-producing cells (oligodendrocytes) and for the constitution of the Ranvier node area (Dupree et al., 1999; Honke et al., 2002). They are essential to the proper formation of axo–glial interactions and a disruption of these interactions results in profound abnormalities in the molecular organization of the paranodal axolemma (Dupree et al., 1999).

Galactosylceramide (Galactocerebroside) synthesis and degradation in CNS and PNS. The inactivation of the CGT (ceramide galactosyl transferase) gene has allowed the analysis of galactolipid function (Coetzee et al., 1998). The mutants cannot synthesize major myelin lipids, galactosylceramide (GalC), and sulfogalactosylceramide (sulfatide). As explained in Section 3.3 myelin galactolipids are essential for the proper formation of axo–glial interactions and demonstrate that a disruption of these interactions results in profound abnormalities in the molecular organization of the paranodal axolemma.

Many brain abnormalities are related to a defect in the catabolism of galactosylceramide, caused by galactocerebrosidase deficiency. There are similarities in mouse and human diseases, for example, the twitcher mouse and Krabbe's disease (Suzuki, 2003). Krabbe's disease is a recessive autosomal disease caused by a deficiency in galactosylceramidase (galactocerebrosidase). It leads to demyelination in CNS and often in PNS. Globoid cells, of macrophagic origin, are typical of this disease, and contain the undegraded substrate, galactosylceramide. There is a very early disappearance of oligodendrocytes, due to the accumulation of galactosylsphingosine (psychosine), a cytotoxic metabolite. Krabbe's disease generally presents clinically in early childhood (~6 months of age). The most common clinical manifestation is the onset of a paralysis in the four limbs (tetraplegia), but other neurological signs may occur. Peripheral nerve conduction velocity is also reduced. Decerebration follows rapidly, with total degradation of mental capabilities. Death occurs related to brain stem alterations. However, variant forms of this disorder are known. Isolated demyelinating peripheral neuropathies with no CNS involvement for a long period have been described, as well as Krabbe's disease starting in adolescence and adulthood (Baumann and Turpin, 2000); in the latter, MRI shows that white matter involvement predominates symmetrically in the periventricular parieto-occipital regions. Up to now, there is no explanation for this late onset form with reduced brain regional alterations. The mouse model, the twitcher mouse, when on a mixed genetic background, gives rise to myelin alterations and also to neuronal death, especially in the hippocampus. Thus, some sphingolipids may have functions in the hippocampal neuronal organization and maintenance (Tominaga et al., 2004), possibly in relation with modulator genes.

Sulfogalactosylceramide (Sulfatide) synthesis and degradation in CNS and PNS. Regional and cellular abnormalities may also be related to defects in the synthesis of sulfatides. Sulfatides, coded by the galactosylceramide sulfotransferase gene, are essential for maintenance of Na^+ ion channels on myelinated axons but are not required for initial cluster formation (Ishibashi et al., 2002). Mice deficient for this gene are unable to synthesize sulfatides. They display abnormal paranodal junctions in the CNS and PNS, whereas their compact myelin is preserved (Honke et al., 2002). Recent work has evidenced that although sulfatide appears to play a limited role in myelin development in comparison to galactocerebrosides, this lipid is essential for myelin maintenance, as the prevalence of redundant, uncompacted, and degenerating myelin sheaths as well as deteriorating nodal/paranodal structures is increased significantly in aged sulfatide-null mice as compared with wild-type littermates. The role played by sulfatide in CNS is not limited to the myelin sheath as axonal caliber is significantly altered in aged sulfatide-null mice (Marcus et al., 2006).

Metachromatic leukodystrophy (MLD) is a disease involving a defect in sulfatide catabolism. The cause is deficiency of the catabolic lysosomal enzyme arylsulfatase A There is essentially a myelin deficiency, with an excess of sulfatides in myelinating cells, and also in neurons and macrophages (Gieselmann et al., 2003). In an experimental model, an arylsulfatase A-deficient mouse, the lysosomal sulfatide storage disease affects the lipid composition of myelin itself and the amount and localization of specific myelin membrane-associated proteins, particularly the protein MAL (Saravanan et al., 2004). Sulfatiduria is an important element in the diagnosis of MLD. Sulfatides in urine can be identified by thin-layer chromatography with immunodetection (Colsch et al., 2008), and quantitated by mass spectrometry (Cui et al., 2008). Clinically MLD is usually a disease of childhood, with clinical abnormalities developing at about the age of walking. It can, however, first manifest itself clinically at a later age. In some cases it presents clinically in adult life. In these patients, behavioral abnormalities may constitute the first symptoms and the only symptoms to occur for many years. These patients can receive a diagnosis of schizophrenia (Rauschka et al., 2006). The role of sulfatide in brain cognitive functions is certainly important as the first Alzheimer case may have been a metachromatic leukodystrophy (Amaducci et al., 1991). Interestingly, advances in imaging techniques associated with genetic findings suggest that white matter abnormalities are present in schizophrenia (Kubicki et al., 2005; Stewart and Davis, 2004).

Previous work, which has not been revisited, showed that sulfatides are necessary for the optimal function of enzymes such as sodium–potassium-dependent ATPase, and the sulfatide content seems to be directly related to the activity of the enzyme (Karlsson et al., 1974). Sulfatides may also be involved in the functioning of certain opiate receptors (Craves et al., 1980) and in chloride transport systems (Zalc et al., 1978). Implantation in the spinal cord of a hybridoma secreting specific antisulfatide antibodies has been shown to cause demyelination of the CNS in the rat (Rosenbluth et al., 2003). Antisulfatide antibodies have been found in HIV patients with distal sensory neuropathies (Lopate et al., 2005), in MS (Ilyas et al., 2003; Kanter et al., 2006) and in diabetic neuropathies (Buschard et al., 2005). These observations suggest that autoimmune responses directed against sulfatides can contribute to the pathogenesis of some of these diseases. The mechanism of demyelination still remains obscure.

Very long chain fatty acids (VLCFA) of cerebrosides and sulfatides. VLCFA are major constituents of the ceramide part of galactocerebrosides and sulfatides. These are unbranched fatty acids with a chain length of 24 or more carbon atoms. They accumulate in the peroxisomal X-linked genetic disease adrenoleukodystrophy (ALD) because of an impaired beta-oxidation in peroxisomes. A variety of clinical presentations can occur in a single kindred with this disorder (Turpin et al., 1985). The cerebral demyelinating form of ALD mainly affects boys between 5 and 12 years of age (40% of ALD cases) and leads to a vegetative stage or death within 2–5 years. The adult form, adrenomyeloneuropathy (AMN), which represents 40% of ALD cases, mainly affects the spinal cord and leads to spastic paraplegia often complicated by cerebral demyelination (35%) (Aubourg and Dubois-Dalcq, 2000). The ALD gene encodes an ATP-binding cassette (ABC) half-transporter of 75 KDa (ALDP) that must dimerize with itself or a related partner to exert its function within the peroxisomal membrane (Mosser et al., 1993). The disease affects the CNS and PNS. Understanding the mechanism of demyelination in ALD remains a major challenge. The disease shows wide phenotypical variation that is not predictable and is probably under the influence of both genetic and environmental factors (Aubourg and Dubois-Dalcq, 2000). Although the cloning of the ALD gene has allowed the generation of an ALD model, these mice do not show any neurological symptoms and therefore do not reveal how VLCFA accumulation can lead to demyelination.

Specific SGLs of PNS and demyelination. The major SGLs in the PNS are different. The major SGLs of the PNS are the sialosylparagloboside LM1 (also called SPG), the sulfoglucuronylparagloboside (SGPG) and the sulfoglucuronyl-lactosaminylparagloboside (SGLPG). SGPG has a terminal trisaccharide sequence, N-acetylglucosaminyl-galactosyl-glucuronylsulfate. This sequence is the HNK-1 epitope, common to the SGPG and its derivatives with two lactosaminyl residues, and to several adhesion molecules and myelin proteins including P0, the major protein of PNS, and MAG (Baumann, 2000; Willison and Yuki, 2002).

Chronic polyneuropathies associated with IgM gammapathies are mostly sensory demyelinating neuropathies. A small proportion of cases are nondemyelinating and have the characteristics of axonal neuropathies (Chassande et al., 1998). Interestingly, the typical sensory demyelinating neuropathies have anti-MAG and anti-SGPG antibodies whereas axonal neuropathies, often predominantly motor, present only monoclonal IgM anti-SGPG activity with no anti-MAG reactivity. Thus, the fine structure of the epitope recognized by the IgM may be involved.

Gangliosides. Gangliosides are mainly present in neurons, except for ganglioside GM1 and GM4. The latter is a sialylated derivative of galactosylceramide that is present in the myelin of mice and primates including humans. Ganglioside GM1 can be a target for an autoimmune demyelination process or motor conduction blocks in the PNS (Willison and Yuki, 2002).

Molecular mimicry between microbial and self-components is postulated as the mechanism accounting for the antigen and tissue specificity of immune responses in postinfectious autoimmune diseases. Guillain–Barré syndrome, the most frequent cause of acute neuromuscular paralysis, can occur 1–2 weeks after various infections, in particular *Campylobacter jejuni enteritis* (*C. jejuni*). Carbohydrate mimicry between the bacterial lipo-oligosaccharide and human GM1 ganglioside is relevant to the pathogenesis of Guillain–Barré syndrome, as documented by Yuki (Yuki et al., 2004). Upon sensitization with *C. jejuni* lipo-oligosaccharide, rabbits develop anti-GM1 IgG antibodies and flaccid limb weakness. Paralyzed rabbits have pathological changes in their peripheral nerves identical with those present in Guillain–Barré syndrome. Immunization of mice with the lipo-oligosaccharide generates a mAb that reacts with GM1 and binds to human peripheral nerves. The mAb and anti-GM1 IgG from patients with Guillain–Barré syndrome did not induce paralysis but blocked muscle action potentials in a muscle–spinal cord coculture, indicating that anti-GM1 antibody can cause muscle weakness. These findings show that carbohydrate mimicry is an important cause of autoimmune neuropathy that can involve demyelination.

Cholesterol and phospholipids. Most lipids found in myelin are common to other cellular membranes. Cholesterol content is high and cholesterol esters are not present in normal myelin. Phospholipids are also common to other cellular membranes, except for the great quantity of ethanolamine phosphoglycerides in the plasmalogen form. The synthesis of plasmalogens is modified in Zellweger syndrome which is a peroxisomal syndrome that also increases VLCFA. This syndrome and other peroxisomal diseases may cause demyelination (Powers, 2005).

4 Other Glial Cell Types and Factors Involved in Myelination and Demyelination in the CNS

4.1 Astrocytes and Mutations in GFAP

Alexander disease (AxD) is a leukodystrophy caused by dominant mutations in GFAP, the main intermediate filament protein of astrocytes. This neurodegenerative disease is characterized by dystrophic astrocytes containing intermediate filament aggregates associated with myelin abnormalities. Overexpressing human GFAP in mice leads to a toxic gain of function induced by aggregates of GFAP and small heat shock proteins. However, GFAP-null mice also display some myelin abnormalities and blood–brain barrier dysfunction (reviewed in Mignot et al., 2004). AxD was the first human disease to be described in which astrocyte dysfunction induces myelin destruction (Brenner et al., 2001; Rodriguez et al., 2001). A transgenic mouse expressing the R239H mutation presented aggregates (Tanaka et al., 2007), as did KI mice expressing R239C or R79H (Hagemann et al., 2006). This abnormality was correlated in both cases to an overexpression of the mutated allele. Diverse hypotheses have been put forward on the impact of GFAP mutations on myelination (Mignot et al., 2004). Recent works have focused on abnormal formation of the astrocytic intermediate network (Der Perng et al., 2006; Quinlan et al., 2007; Tanaka et al., 2007; Tian et al., 2006), or on the fate of pathogenic astrocytes: cell death versus survival (Mignot et al., 2007). Rare radiological and pathological tumor-like lesions have already been reported in AXD patients. Enlargement of the optic chiasm is a rare feature of AXD, possibly linked to abnormal astrocytic proliferation (Mignot et al., 2009). Taken together, these data highlight pathological astrocytes as key players and valuable therapeutic targets in neurological disorders, in particular myelin diseases. The mechanisms leading to myelin pathogenicity by astrocytes (2.5.4) are undefined, and so far, there is no cure for such diseases.

4.2 Oligodendrocyte Precursors

Myelination requires sequential steps in the maturation of the oligodendrocyte lineage with a co-ordinated change in the expression of cell surface antigens; these antigens can be recognized by monoclonal antibodies. Dys- and demyelination may also act at an early stage of development of the oligodendrocyte lineage, although the roles of oligodendrocyte precursors in human pathology are often not clearly defined. Transcription factors, growth factors, neurotransmitters, and modifications of cell surface components can be involved. Myelination requires a tightly regulated balance between the disappearance of inhibitory signals and the induction of positive signals, some of which are mediated by neuronal electrical activity (Demerens et al., 1996). Cocultures of oligodendrocyte progenitor cells (OPCs) and neurons in the presence of highly specific neurotoxins, which can either block (tetrodotoxin) or increase (alpha-scorpion toxin) the firing of neurons, demonstrate that myelinogenesis is dependent on the electrical activity of neurons (Demerens et al., 1996). We only develop here what occurs in relation to CNS myelination and demyelination (Collarini et al., 1991).

Oligodendrocytes descend from a progenitor cell (OPC) which originates in specialized regions of the subventricular zone. Early OPCs express the platelet-derived growth factor receptor alpha (PDGFαR) and the sulphated proteoglycan NG2 (Nishiyama et al., 1999). Early OPCs differentiate into a late OPC stage (or oligodendroblasts) that appears to be committed to oligodendrogenesis. Late OPCs express the tetraspan protein CD9 (Terada et al., 2002) and the POA antigen. Although uncharacterized, the POA antigen may be recognized by the monoclonal antibody O4 that targets sulfatides in mature cells (Bansal et al., 1992). Once oligodendrogenesis is complete, NG2/PDGFαR positive cells remain as a major glial component of the adult mammalian CNS, apparently providing a pool of quiescent progenitors that can be tapped later for repair of demyelinated axons (Menn et al., 2006; Wilson et al., 2006). Premyelinating oligodendrocytes extend multiple processes. They express many but not all myelin proteins. Present are DM20, MAG, CNP, MBP, and CD; PLP and MOG are not detected by current methods. As myelination begins, the oligodendrocyte targets myelin proteins to specific membrane domains: MAG is selectively targeted to periaxonal membranes, PLP to compact myelin, CNP to noncompact regions of the myelin internode, and MBP mRNA to oligodendrocyte processes (Trapp et al., 2004). As described by Yakovlev and Lecours (1966), myelination takes place until adult age and occurs at different ages according to the area, the latest being the prefrontal area and the associative areas. Thus leukodystrophies may start clinically in adulthood (Baumann and Turpin, 2000).

Axonal damage secondary to myelin loss is a major cause of sensory, motor, and cognitive disabilities in adult MS (Bjartmar and Trapp, 2001). The lack of myelin recovery may be due primarily to deficiency in the genesis of OPCs and in their maturation in the adult CNS (Franklin, 2002; Stangel and Hartung, 2002). Limited myelin regeneration is observed in early demyelinating lesions in MS (Wolswijk, 1998).

Possible explanations for remyelination failure in MS (Franklin, 2002; Stangel and Hartung, 2002) can be the inadequate recruitment of OPCs (Keirstead et al., 1998) or the inability of OPCs to turn into myelinating oligodendrocytes. Thus, studies aiming at identifying factors involved in OPC differentiation during remyelination are of great interest. Guidance molecules Semaphorin 3A and 3F, already known to direct oligodendroglial migration during development, may also be active in controlling OPC migration in MS and may determine the ability of plaques to remyelinate (Williams et al., 2007b).

It is conceivable that the process of remyelination mimics that of myelination during development, but the key factors affecting the differentiation and maturation of OPCs into myelinating oligodendrocytes do not perfectly trigger remyelination in the adult brain.

4.3 Biochemical Factors

As noted above, myelination requires a tightly regulated balance between the disappearance of inhibitory signals, and the induction of positive signals.

Adhesion molecules. The downregulation of the polysialylated neuronal cell adhesion molecule (PSA-NCAM) from the axonal surface (Charles et al., 2000) is a necessary prerequisite to render the axon permissive to myelination (Charles et al., 2002; Coman et al., 2005). L1, another adhesion molecule expressed at the axonal surface, promotes myelination (Coman et al., 2005).

In demyelination such as MS, PSA-NCAM is expressed on denuded axons and might act as an inhibitor of remyelination, whereas the myelinated part outside the plaque is PSA-NCAM negative (Charles et al., 2000; Coman et al., 2005). On the other hand, in MS a two- to threefold increase in OPC density and proliferation was found in the subventricular zone (SVZ), which correlated with enhanced numbers of PSA-NCAM(+) cells (Nait-Oumesmar et al., 2007). EAE in rodents is another important example of the activation of the SVZ and the involvement of progenitor cells expressing the polysialylated form of neural cell adhesion molecule (PSA-NCAM) in the repair process (Picard-Riera et al., 2002).

4.3.1 Growth Factors and Transcription Factors

PDGF alpha and laminin. Laminin-2 deficient mice demonstrate the crucial role of laminin-2 in CNS myelination (Chun et al., 2003). Survival of oligodendrocytes that contact axons requires laminin. In the absence of laminin, the concentration of survival factors such as PDGF is too low to promote survival of newly formed oligodendrocytes. Oligodendrocytes that contact laminin on axon tracts initiate integrin signaling that amplifies the survival response by PDGF. Oligodendrocytes that contact axons are then able to survive and myelinate (Colognato et al., 2005). When α6β1 integrin on the oligodendrocyte binds axonal laminin, Fyn (a member of the Src family kinase) is activated, promoting oligodendrocyte differentiation. Fyn knock-out and α2 laminin knock-out exhibit similar region-specific phenotypes, with a severe myelin deficit in the forebrain in contrast to normal appearing myelin in the spinal cord (Camara and Ffrench-Constant, 2007; Chun et al., 2003). Dystroglycan is a second laminin receptor in oligodendrocytes that expresses and uses this receptor to regulate myelin formation. Blocking the function of dystroglycan receptors leads oligodendrocytes to fail to produce complex myelin membrane sheets and to initiate myelinating segments when cocultured with dorsal ganglion neurons (Colognato et al., 2007).

Jagged is developmentally regulated in neurons and activates the Notch pathway in OPCs, which inhibits their differentiation into oligodendrocytes (Givogri et al., 2002). Because Jagged decreases with a time course that parallels myelination, it is likely that neurons help to regulate the timing of myelination. In the demyelinating brain, the inappropriate upregulation of molecules, including those of the Jagged-1-Notch-1 signal transduction pathway, affects OPC differentiation (Mastronardi and Moscarello, 2005). The importance of communication between astrocytes and oligodendrocytes was also demonstrated in MS in which the abnormal expression of Jagged 1 by reactive astrocytes could be responsible for the failure of myelin repair following myelin destruction caused by inhibition of progenitor differentiation (John et al., 2002). However, in the mouse model, remyelination can proceed to completion despite widespread Notch–Jagged expression; thus Notch–Jagged signaling is not a rate-limiting determinant of remyelination in rodent models of demyelination (Stidworthy et al., 2004).

The *neuregulins* (NRGs) constitute a family of proteins containing an epidermal growth factor (EGF)-like domain that activates the membrane associated ErbB2, ErbB3, ErbB4 receptor tyrosine kinases. NRGs activate ErbBs on oligodendrocytes in the developing CNS. In the absence of ERBb signaling, oligodendrocytes fail to undergo terminal differentiation and to ensheath axons (Park et al., 2001a, b). Loss of erbB signaling, by expression of a dominant negative erbB receptor transgene, in oligodendrocytes alters myelin and dopaminergic function (Roy et al., 2007). These transgenic mice have increased levels of dopamine receptors and transporters, and exhibit behavioral alterations consistent with neuropsychiatric disorders. These results indicate that defects in white matter can cause alterations in dopaminergic function and behavior relevant to neuropsychiatric disorders (Roy et al., 2007).

There are several subgroups of NRG among which are NRG1 type III. Axonal NRG1 regulates myelin sheath thickness in the PNS (Michailov et al., 2004). NRG1 type III, independent of axon diameter, provides a key instructive signal that determines the ensheathment fate of axons (Taveggia et al., 2005). Ensheathed axons express low levels whereas myelinated fibers express high levels of NRG1 type III. Type III is the sole NRG1 isoform retained at the axon surface and activates phosphatidylinositol 3-kinase, which is required for Schwann cell myelination.

Oligodendrocytes also respond to *insulin growth factor* IGF-1 that stimulates oligodendrocyte growth and prevents oligodendrocyte apoptosis. Overexpression of IgF-1 increases the percentage of myelinated axons and the thickness of myelin sheaths. IGF type 1 receptor is required for normal in vivo development and myelination (Zeger et al., 2007). The association of transferring and IGF-1 favors remyelination in the myelin-deficient rat (Espinosa-Jeffrey et al., 2006).

Olig1 and *Olig2* encode basic helix–loop–helix (bHLH) transcription factors that are expressed in both the developing and mature CNS. Expression of Olig in human brain tumors and demyelinating lesions suggest the possibility of additional functions in a variety of neurological diseases (Ligon et al., 2006; Zhao et al., 2005). Mice lacking a functional Olig1 gene develop severe neurological deficits and die in the third postnatal week. In the brains of these mice, expression of myelin-specific genes is abolished, whereas the formation of OPCs is not affected. Furthermore, multilamellar wrapping of myelin membranes around axons does not occur, despite recognition and contact of axons by oligodendrocytes, and Olig1-null mice develop widespread progressive axonal degeneration and gliosis. In contrast, myelin sheaths are formed in the spinal cord, although the extent of myelination is severely reduced. At the molecular level, Olig1 regulates transcription of the major myelin-specific genes, MBP, PLP1, and MAG, and suppresses expression of a major astrocyte-specific gene, Gfap. Thus Olig1 is a central regulator of oligodendrocyte myelinogenesis in brain, and axonal recognition and myelination by oligodendrocytes are distinct processes (Xin et al., 2005).

Eukaryotic initiation factor 2B (elF2B) is a five-subunit guanine nucleotide exchange factor that exchanges GDP for GTP to form the elF2B-GTP complex. e1F2B mutations lead to an abnormal control of protein translation that predominantly affects glial cells. Mutations in elF2B (Leegwater et al., 2001) cause one of the most common leukodystrophies: childhood ataxia with CNS hypomyelination/vanishing white matter disease or CACH/VWM (reviewed in Schiffmann and Elroy-Stein, 2006). Astrocytes are affected (Dietrich et al., 2005), oligodendrocytes are overcrowded (Rodriguez et al., 1999) and become foamy, and neurons are spared. The disease is autosomal dominant. There is a cystic breakdown of white matter or "cavitation" and no gadolinium enhancement of the lesions on MRI. The disease can be caused by mutations in any of the five subunits of elF2B.

Qkl (quaking). The quaking viable (qkv) is a spontaneous recessive mutation in the mouse that deletes an enhancer of the qkI gene and causes diminished qkI transcription, specifically in myelin-producing cells. The qkv mice provide a unique animal model linking RNA binding proteins to defects in oligodendrocyte cell fate and myelination (Larocque and Richard, 2005). The qkI gene encodes RNA binding proteins that are involved in the transport of myelin-specific RNAs, such as those encoding myelin basic proteins (MBP), to specific cellular locations for translation. Schizophrenia, a severe mental disorder, comprising social and cognitive defects may be linked to a qk susceptibility locus (Aberg et al., 2006; Lindholm et al., 2001). QKI, which is essential for myelination, is decreased in schizophrenia (McInnes and Lauriat, 2006). Downregulation of QK1 might be among the primary causes of downregulation of myelin-related genes in schizophrenia (Karoutzou et al., 2007).

The major cognitive disturbances in schizophrenia may result from a deficit of myelination in relevant neuronal structures, such as the corpus callosum, involved in connectivity between both hemispheres; the resulting decrease of electrical conduction in fiber tracks linking different parts of the brain may affect behavior and perception (Haroutunian and Davis, 2007; Haroutunian et al., 2007).

Transferrin (Tf), the iron transport glycoprotein found in the biological fluids of vertebrates, is also synthesized by oligodendrocytes in the CNS. Overexpressing Tf in the brain of transgenic mice accelerates oligodendrocyte maturation, early maturation of the cerebellum and spinal cord, and myelination in the corpus callosum (Sow et al., 2006). The association of IGF-1 and transferrin favors remyelination in the myelin deficient rat (Espinosa-Jeffrey et al., 2006).

Neurotransmitters. Numerous neurotransmitters affect the development of oligodendrocytes. AMPA (alpha-amino-3-hydroxy-5-methyl-4-isoxazolepropionic acid) and kaïnate receptors are expressed on oligodendrocytes. Glutamate has an inhibitory role in the proliferation of oligodendrocytes, especially in relation to AMPA and NMDA receptors (Karadottir and Attwell, 2007). Indeed, glutamate can be toxic to white matter oligodendrocytes through AMPA, kainate glutamate receptor, and N-methyl-D-aspartate receptors (NMDA) (Matute, 2006). Drugs that interact with glutamate receptors in experimental models of MS can contribute to a more favorable outcome (Bolton and Paul, 2006). Dopamine D3 and D2 receptors are also present as well as GABAa receptors. Their roles are not yet elucidated. In experimental models of demyelinating diseases (Theiler's virus) cannabinoids reduce microglial activation, abrogate major histocompatibility complex Class II antigen expression, and decrease the number of CD4+ infiltrating T cells (Arevalo-Martin et al., 2003). N-acetyl aspartate is synthesized from aspartate and acetyl coenzyme A in neurons. The NAA-degrading enzyme is N-aspartoacylase (ASPA). ASPA cleaves the acetate moiety for use in fatty acid and steroid derivatives. Mutations in the gene coding for ASPA result in Canavan disease, a fatal leukodystrophy (Moffett et al., 2007).

Second messengers: Adenosine, ATP, and LIF. Adenosine regulates proliferation and differentiation of OPCs (Stevens et al., 2002), whereas ATP affects mature oligodendrocytes. ATP does not act directly on oligodendrocytes but rather on astrocytes, causing the release of LIF (leukaemia inhibitory factor) by these cells, which in turn triggers the myelination process by promyelinating oligodendrocytes (Ishibashi et al., 2006; Simons and Trajkovic, 2006). However, in LIF-deficient animals, myelin may be formed in the absence of LIF (Bugga et al., 1998), indicating that other factors/cytokines, may complement for that function.

By contrast to oligodendrocytes (Lubetzki et al., 1993), Schwann cells absolutely need the presence of neurons to differentiate and myelinate in vitro (Jessen and Mirsky, 1991; Owens and Bunge, 1989). As for oligodendrocytes in the CNS, calcium imaging in glia in the PNS revealed that purinergic receptors allow premyelinating Schwann cells to detect action potential firing, due to ATP released from axons (Stevens and Fields, 2000; Stevens et al., 2004). Different purinergic receptors (Fields, 2006) are expressed on both types of glia resulting, however, in opposite effects of impulse activity on differentiation of Schwann cells and OPC. In the PNS, ATP regulates early development and myelination by Schwann cells, whereas it inhibits differentiation and myelination (Jessen and Mirsky, 1991), in striking contrast to what happens in the CNS (Stevens, 2006). Both ATP and adenosine inhibit proliferation of Schwann cells induced by PDGF. Unlike ATP, adenosine failed to inhibit differentiation of Schwann cells, in contrast with its role in oligodendrocyte differentiation in the CNS (Stevens et al., 2004).

Hormones. It is well established that thyroid hormone (TH) is required for the normal timing of OPCs differentiation and maturation (Rogister et al., 1999). Also, normal cell-cycle progression mechanisms and terminal differentiation and maturation require TH (Durand and Raff, 2000). Studies of myelination in hypo- and hyperthyroid animals (Jagannathan et al., 1998) have provided strong evidence that TH plays an important role in regulating oligodendrocyte lineage and maturation in vivo and that the TH receptor $\alpha 1$ seems to be responsible for this process (Billon et al., 2002). The administration of TH during the acute phase of experimental allergic encephalomyelitis (EAE) in rats, a commonly used experimental model for MS, is able to generate oligodendroglial cells (Calza et al., 2002).

Steroid hormones: Androgens. Interestingly, a sexual dimorphism of oligodendrocytes and myelin has been demonstrated in rodents. The density of oligodendrocytes in corpus callosum, fornix, and spinal cord is 20–40% greater in males compared with females, independent of age, strain, and species of rodent. This is associated with an elevated level of PLP and CAII (carbonic anhydrase 2). Moreover, oligodendrogenesis and apoptosis of glia are two times greater in female corpus callosum, indicating that the lifespan of oligodendrocytes is shorter in females than in males. Castration of males produces a female phenotype characterized by fewer oligodendrocytes and increased generation of new glia (Cerghet et al., 2006). In EAE castration of males increased the severity of the disease (Bebo et al., 1998) whereas in MS, the lowest levels of serum testosterone in affected women correlates with the severity of the disease, again indicating that androgens are protective (Tomassini et al., 2005), possibly more than estrogens. Altogether, these data indicate that exogenous androgens differentially affect the lifespan of male and female oligodendrocytes, and can override the endogenous production of neurosteroids. These data imply that the turnover of myelin is greater in females than in males, a process that may account for more myelin breakdown products in females. These findings have a potential significance for MS, a sexually dimorphic disease, whose progression is altered by exogenous hormones (Cerghet et al., 2006).

The steroid hormones progesterone and derivatives promote the viability of neurons in the CNS and play an important role in developmental myelination and in myelin repair. The hormone may promote neuroregeneration by several different actions—reducing inflammation, swelling, and apoptosis—thereby increasing the survival of neurons, and promoting the formation of new myelin sheaths. Recognition of the important pleiotropic effects of progesterone opens novel perspectives for the treatment of brain lesions and diseases of the nervous system. Exogenous administration of progesterone or some of its metabolites can be successfully used to treat traumatic brain and spinal cord injury, as well as ischemic stroke (reviewed in Schumacher et al., 2007). Progesterone can be synthesized by neurons and by glial cells within the nervous system, as neurosteroids (Jung-Testas et al., 1999). This finding opens the way for the use of pharmacological agents, such as ligands of TSPO (translocator protein), the peripheral benzodiazepine receptor, to locally increase the synthesis of steroids with neuroprotective and neuroregenerative properties (reviewed in Schumacher et al., 2007).

Prolactin. Motherhood has been shown to attenuate the age-related decline in learning and memory in the rat (Gatewood et al., 2005). Remission of MS during pregnancy led to the hypothesis that remyelination is enhanced in the maternal brain. In MS, the elevated prolactin levels during pregnancy may allow myelin repair, during a temporal window when there is a shift from proinflammatory Th1 to anti-inflammatory Th2-mediated immunity. Using animal models, it has been shown that prolactin treatment promotes myelin repair in female mice (Gregg et al., 2007), mimicking the regenerative effect of pregnancy on white matter damage. Prolactin induces changes early in pregnancy: increased oligodendrogenesis, MBP expression, and the number of myelinated axons. Remarkably, pregnant mice have an enhanced ability to remyelinate white matter lesions. The hormone prolactin regulates oligodendrocyte precursor proliferation and mimics the regenerative effects of pregnancy.

5 Conclusion

Our knowledge of myelin constituents has greatly increased, as well as the role of a bidirectional dialogue between glial cells and neurons in myelination and demyelination; but, little is known of the mechanisms responsible for myelin repair. Why is remyelination incomplete with less myelin and shorter internodes?

Many mysteries remain about the timing of myelination and demyelination, as many genetic diseases become manifest only in adulthood. There is a time and regional control of myelination and demyelination as, for instance, in the cuprizone model. That only implicates certain brain areas, but we know very little about it.

A variety of pathogenic mechanisms has been shown to be at work in myelin diseases: point mutations, recombination events leading to deletions, and duplication of genomic regions including myelin genes. The exquisite sensibility to gene dosage of myelinating glial cells has been pointed out in human myelin diseases as in genetically modified animal models. Nevertheless, there is not always a phenotype/genotype relationship, indicating that many factors involved still remain unknown in human demyelinating diseases. New areas of research are being developed showing the involvement of myelin deficiency in psychiatric diseases and cognition.

Although the roles of major constituents of myelin in relation to pathological experimental models are clear, the specific mechanisms in many human diseases still need to be investigated.

Acknowledgments Drs Saïd Ghandour and Jean-Claude Turpin are gratefully acknowledged for their help and advice during the redaction of this manuscript, and Eric Noe for the careful reading of text and references. The research work of the authors is supported by grants from ELA Foundation (European Leukodystrophy Association), Association Jerome Lejeune, and INSERM to DP-D, and ARNC (association pour la recherche en neurochimie) to NB.

References

Aberg K, Saetre P, Jareborg N, Jazin E (2006) Human QKI, a potential regulator of mRNA expression of human oligodendrocyte-related genes involved in schizophrenia. Proc Natl Acad Sci U S A 103:7482–7487

Amaducci L, Sorbi S, Piacentini S, Bick KL (1991) The first Alzheimer disease case: a metachromatic leukodystrophy? Dev Neurosci 13:186–187

Arevalo-Martin A, Vela JM, Molina-Holgado E, Borrell J, Guaza C (2003) Therapeutic action of cannabinoids in a murine model of multiple sclerosis. J Neurosci 23:2511–2516

Aslin RN, Schlaggar BL (2006) Is myelination the precipitating neural event for language development in infants and toddlers? Neurology 66:304–305

Aubourg P, Dubois-Dalcq M (2000) X-linked adrenoleukodystrophy enigma: how does the ALD peroxisomal transporter mutation affect CNS glia? Glia 29:186–190

Balabanov R, Strand K, Goswami R, McMahon E, Begolka W, Miller SD, Popko B (2007) Interferon-gamma-oligodendrocyte interactions in the regulation of experimental autoimmune encephalomyelitis. J Neurosci 27:2013–2024

Bandtlow CE, Schwab ME (2000) NI-35/250/nogo-a: a neurite growth inhibitor restricting structural plasticity and regeneration of nerve fibers in the adult vertebrate CNS. Glia 29:175–181

Bansal R, Stefansson K, Pfeiffer SE (1992) Proligodendroblast antigen (POA), a developmental antigen expressed by A007/O4-positive oligodendrocyte progenitors prior to the appearance of sulfatide and galactocerebroside. J Neurochem 58:2221–2229

Baumann N (2000) Specificity of antiglycolipid antibodies. Clin Rev Allergy Immunol 19:31–40

Baumann N, Pham-Dinh D (2001) Biology of oligodendrocyte and myelin in the mammalian central nervous system. Physiol Rev 81:871–927

Baumann N, Pham-Dinh D (2002) The Astrocyte. In: Ramachandran VS (ed) Encyclopaedia of the Human Brain. Academic Press/Elsevier Science, New York, NY, pp 251–268

Baumann N, Turpin JC (2000) Adult-onset leukodystrophies. J Neurol 247:751–759

Bebo BF Jr, Zelinka-Vincent E, Adamus G, Amundson D, Vandenbark AA, Offner H (1998) Gonadal hormones influence the immune response to PLP 139–151 and the clinical course of relapsing experimental autoimmune encephalomyelitis. J Neuroimmunol 84:122–130

Beckman M (2004) Crime, culpability and the adolescent brain. Science 305:596–599

Biffiger K, Bartsch S, Montag D, Aguzzi A, Schachner M, Bartsch U (2000) Severe hypomyelination of the murine CNS in the absence of myelin-associated glycoprotein and fyn tyrosine kinase. J Neurosci 20:7430–7437

Billon N, Jolicoeur C, Tokumoto Y, Vennstrom B, Raff M (2002) Normal timing of oligodendrocyte development depends on thyroid hormone receptor alpha 1 (TRalpha1). EMBO J 21:6452–6460

Bjartmar C, Trapp BD (2001) Axonal and neuronal degeneration in multiple sclerosis: mechanisms and functional consequences. Curr Opin Neurol 14:271–278

Bjartmar C, Yin X, Trapp BD (1999) Axonal pathology in myelin disorders. J Neurocytol 28: 383–395

Blakemore WF (1973a) Remyelination of the superior cerebellar peduncle in the mouse following demyelination induced by feeding cuprizone. J Neurol Sci 20:73–83

Blakemore WF (1973b) Demyelination of the superior cerebellar peduncle in the mouse induced by cuprizone. J Neurol Sci 20:63–72

Bolton C, Paul C (2006) Glutamate receptors in neuroinflammatory demyelinating disease. Mediators Inflamm 2006:93684

Bradl M, Hohfeld R (2003) Molecular pathogenesis of neuroinflammation. J Neurol Neurosurg Psychiatry 74:1364–1370

Bradl M, Linington C (1996) Animal models of demyelination. Brain Pathol 6:303–311

Brenner M, Johnson AB, Boespflug-Tanguy O, Rodriguez D, Goldman JE, Messing A (2001) Mutations in GFAP, encoding glial fibrillary acidic protein, are associated with Alexander disease. Nat Genet 27:117–120

Bugga L, Gadient RA, Kwan K, Stewart CL, Patterson PH (1998) Analysis of neuronal and glial phenotypes in brains of mice deficient in leukemia inhibitory factor. J Neurobiol 36: 509–524

Buschard K, Fredman P, Bog-Hansen E, Blomqvist M, Hedner J, Rastam L, Lindblad U (2005) Low serum concentration of sulfatide and presence of sulfated lactosylceramid are associated with Type 2 diabetes. The Skaraborg Project. Diabet Med 22:1190–1198

Butt AM, Kiff J, Hubbard P, Berry M (2002) Synantocytes: new functions for novel NG2 expressing glia. J Neurocytol 31:551–565

Calza L, Fernandez M, Giuliani A, Aloe L, Giardino L (2002) Thyroid hormone activates oligodendrocyte precursors and increases a myelin-forming protein and NGF content in the spinal cord during experimental allergic encephalomyelitis. Proc Natl Acad Sci U S A 99:3258–3263

Camara J, Ffrench-Constant C (2007) Lessons from oligodendrocyte biology on promoting repair in multiple sclerosis. J Neurol 254(Suppl 1):I15–I22

Campagnoni AT, Campagnoni CW, Bourre JM, Jacque C, Baumann N (1984) Cell-free synthesis of myelin basic proteins in normal and dysmyelinating mutant mice. J Neurochem 42:733–739

Campagnoni AT, Macklin WB (1988) Cellular and molecular aspects of myelin protein gene expression. Mol Neurobiol 2:41–89

Campagnoni AT, Skoff RP (2001) The pathobiology of myelin mutants reveals novel biological functions of the MBP and PLP-1 genes. Brain Pathol 11:74–91

Cao Z, Qiu J, Domeniconi M, Hou J, Bryson JB, Mellado W, Filbin MT (2007) The inhibition site on myelin-associated glycoprotein is within Ig-domain 5 and is distinct from the sialic acid binding site. J Neurosci 27:9146–9154

Capello E, Voskuhl RR, McFarland HF, Raine CS (1997) Multiple sclerosis: re-expression of a developmental gene in chronic lesions correlates with remyelination. Ann Neurol 41: 797–805

Cerghet M, Skoff RP, Bessert D, Zhang Z, Mullins C, Ghandour MS (2006) Proliferation and death of oligodendrocytes and myelin proteins are differentially regulated in male and female rodents. J Neurosci 26:1439–1447

Charles P, Hernandez MP, Stankoff B, Aigrot MS, Colin C, Rougon G, Zalc B, Lubetzki C (2000) Negative regulation of central nervous system myelination by polysialylated-neural cell adhesion molecule. Proc Natl Acad Sci U S A 97:7585–7590

Charles P, Reynolds R, Seilhean D, Rougon G, Aigrot MS, Niezgoda A, Zalc B, Lubetzki C (2002) Re-expression of PSA-NCAM by demyelinated axons: an inhibitor of remyelination in multiple sclerosis? Brain 125:1972–1979

Chassande B, Leger JM, Younes-Chennoufi AB, Bengoufa D, Maisonobe T, Bouche P, Baumann N (1998) Peripheral neuropathy associated with IgM monoclonal gammopathy: correlations between M-protein antibody activity and clinical/electrophysiological features in 40 cases. Muscle Nerve 21:55–62

Chow E, Mottahedeh J, Prins M, Ridder W, Nusinowitz S, Bronstein JM (2005) Disrupted compaction of CNS myelin in an OSP/Claudin-11 and PLP-1/DM20 double knockout mouse. Mol Cell Neurosci 29:405–413

Chun SJ, Rasband MN, Sidman RL, Habib AA, Vartanian T (2003) Integrin-linked kinase is required for laminin-2-induced oligodendrocyte cell spreading and CNS myelination. J Cell Biol 163:397–408

Coetzee T, Suzuki K, Popko B (1998) New perspectives on the function of myelin galactolipids. Trends Neurosci 21:126–130

Collarini EJ, Pringle N, Mudhar H, Stevens G, Kuhn R, Monuki ES, Lemke G, Richardson WD (1991) Growth factors and transcription factors in oligodendrocyte development. J Cell Sci Suppl 15:117–123

Colognato H, Ffrench-Constant C, Feltri ML (2005) Human diseases reveal novel roles for neural laminins. Trends Neurosci 28:480–486

Colognato H, Galvin J, Wang Z, Relucio J, Nguyen T, Harrison D, Yurchenco PD, Ffrench-Constant. C (2007) Identification of dystroglycan as a second laminin receptor in oligodendrocytes, with a role in myelination. Development 134:1723–1736

Colsch B, Afonso C, Popa I, Portoukalian J, Fournier F, Tabet JC, Baumann N (2004) Characterization of the ceramide moieties of sphingoglycolipids from mouse brain by ESI-MS/MS: identification of ceramides containing sphingadienine. J Lipid Res 45: 281–286

Colsch B, Baumann N, Ghandour S (2008) Generation and characterization of the binding epitope of a novel monoclonal antibody to sulfatide (sulfogalactosylceramide) OL-2: applications of antigen immunodetections in brain tissues and urinary samples. J Neuroimmunol 193:52–58

Coman I, Aigrot MS, Seilhean D, Reynolds R, Girault JA, Zalc B, Lubetzki C (2006) Nodal, paranodal and juxtaparanodal axonal proteins during demyelination and remyelination in multiple sclerosis. Brain 129:3186–3195

Coman I, Barbin G, Charles P, Zalc B, Lubetzki C (2005) Axonal signals in central nervous system myelination, demyelination and remyelination. J Neurol Sci 233:67–71

Craner MJ, Hains BC, Lo AC, Black JA, Waxman SG (2004a) Co-localization of sodium channel Nav1.6 and the sodium-calcium exchanger at sites of axonal injury in the spinal cord in EAE. Brain 127:294–303

Craner MJ, Newcombe J, Black JA, Hartle C, Cuzner ML, Waxman SG (2004b) Molecular changes in neurons in multiple sclerosis: altered axonal expression of Nav1.2 and Nav1.6 sodium channels and Na+/Ca2+ exchanger. Proc Natl Acad Sci U S A 101:8168–8173

Craves FB, Zalc B, Leybin L, Baumann N, Loh HH (1980) Antibodies to cerebroside sulfate inhibit the effects of morphine and beta-endorphin. Science 207:75–76

Cui Y, Colsch B, Afonso C, Baumann N, Tabet JC, Mallet JM, Zhang Y (2008) Synthetic sulfogalactosylceramide (sulfatide) and its use for the mass spectrometric quantitative urinary determination in metachrompatic leukodystrophies. Glycoconj J 25:147–155

Delarasse C, Daubas P, Mars LT, Vizler C, Litzenburger T, Iglesias A, Bauer J, Della Gaspera B, Schubart A, Decker L, Dimitri D, Roussel G, Dierich A, Amor S, Dautigny A, Liblau R, Pham-Dinh D (2003) Myelin/oligodendrocyte glycoprotein-deficient (MOG-deficient) mice reveal lack of immune tolerance to MOG in wild-type mice. J Clin Invest 112:544–553

Delarasse C, Della Gaspera B, Lu CW, Lachapelle F, Gelot A, Rodriguez D, Dautigny A, Genain C, Pham-Dinh D (2006) Complex alternative splicing of the myelin oligodendrocyte glycoprotein gene is unique to human and non-human primates. J Neurochem 98:1707–1707

Demerens C, Stankoff B, Logak M, Anglade P, Allinquant B, Couraud F, Zalc B, Lubetzki C (1996) Induction of myelination in the central nervous system by electrical activity. Proc Natl Acad Sci U S A 93:9887–9892

Der Perng M, Su M, Wen SF, Li R, Gibbon T, Prescott AR, Brenner M, Quinlan RA (2006) The Alexander disease-causing glial fibrillary acidic protein mutant, R416W, accumulates into Rosenthal fibers by a pathway that involves filament aggregation and the association of alpha B-crystallin and HSP27. Am J Hum Genet 79:197–213

Dietrich J, Lacagnina M, Gass D, Richfield E, Mayer-Proschel M, Noble M, Torres C, Proschel C (2005) EIF2B5 mutations compromise GFAP+ astrocyte generation in vanishing white matter leukodystrophy. Nat Med 11:277–283

Dubourg O (2004) Charcot-Marie-Tooth disease: from phenotype to genotype. Rev Neurol (Paris) 160:1221–1229

Duncan ID (2005) The PLP-1 mutants from mouse to man. J Neurol Sci 228:204–205

Duncan ID, Nadon NL, Hoffman RL, Lunn KF, Csiza C, Wells MR (1995) Oligodendrocyte survival and function in the long-lived strain of the myelin deficient rat. J Neurocytol 24:745–762

Dupouey P, Jacque C, Bourre JM, Cesselin F, Privat A, Baumann N (1979) Immunochemical studies of myelin basic protein in shiverer mouse devoid of major dense line of myelin. Neurosci Lett 12:113–118

Dupree JL, Girault JA, Popko B (1999) Axo-glial interactions regulate the localization of axonal paranodal proteins. J Cell Biol 147:1145–1152

Durand B, Raff M (2000) A cell-intrinsic timer that operates during oligodendrocyte development. Bioessays 22:64–71

Edgar JM, McLaughlin M, Yool D, Zhang SC, Fowler JH, Montague P, Barrie JA, McCulloch MC, Duncan ID, Garbern J, Nave KA, Griffiths IR (2004) Oligodendroglial modulation of fast axonal transport in a mouse model of hereditary spastic paraplegia. J Cell Biol 166:121–131

Eftekharpour E, Karimi-Abdolrezaee S, Sinha K, Velumian AA, Kwiecien JM, Fehlings MG (2005) Structural and functional alterations of spinal cord axons in adult Long Evans Shaker (LES) dysmyelinated rats. Exp Neurol 193:334–349

Eshed Y, Feinberg K, Poliak S, Sabanay H, Sarig-Nadir O, Spiegel I, Bermingham JR Jr, Peles E (2005) Gliomedin mediates Schwann cell-axon interaction and the molecular assembly of the nodes of Ranvier. Neuron 47:215–229

Espinosa-Jeffrey A, Zhao P, Awosika W, Wu N, Macias F, Cepeda C, Levine M, De Vellis J (2006) Activation, proliferation and commitment of enogenous stem/progenitor cells to the oligodendrocyte lineage by TS1 in a rat model of dysmyelination. Dev Neurosci 28: 488–498

Feng JM (2007) Minireview: expression and function of golli protein in immune system. Neurochem Res 32:273–278

Fields RD (2006) Nerve impulses regulate myelination through purinergic signalling. Novartis Found Symp 276:148–158; discussion 158–61, 233–7, 275–281

Filbin MT (2003) Myelin-associated inhibitors of axonal regeneration in the adult mammalian CNS. Nat Rev Neurosci 4:703–713

Filley C (1998) The behavioral neurology of cerebral white matter. Neurology 50:1535–1545

Folch J, Lees. M (1951) Proteolipides, a new type of tissue lipoproteins; their isolation from brain. J Biol Chem 191:807–817

Fouad K, Klusman I, Schwab ME (2004) Regenerating corticospinal fibers in the Marmoset (Callitrix jacchus) after spinal cord lesion and treatment with the anti-Nogo-A antibody IN-1. Eur J Neurosci 20:2479–2482

Franklin RJ (2002) Why does remyelination fail in multiple sclerosis? Nat Rev Neurosci 3: 705–714

Friedman HC, Jelsma TN, Bray GM, Aguayo AJ (1996) A distinct pattern of trophic factor expression in myelin-deficient nerves of Trembler mice: implications for trophic support by Schwann cells. J Neurosci 16:5344–5350

Fushimi S, Shirabe T (2004) Expression of insulin-like growth factors in remyelination following ethidium bromide-induced demyelination in the mouse spinal cord. Neuropathology 24: 208–218

Garbern JY (2005) Pelizaeus-Merzbacher disease: pathogenic mechanisms and insights into the roles of proteolipid protein 1 in the nervous system. J Neurol Sci 228:201–203

Garbern JY, Cambi F, Tang XM, Sima AA, Vallat JM, Bosch EP, Lewis R, Shy M, Sohi J, Kraft G, Chen KL, Joshi I, Leonard DG, Johnson W, Raskind W, Dlouhy SR, Pratt V, Hodes ME, Bird T, Kamholz J (1997) Proteolipid protein is necessary in peripheral as well as central myelin. Neuron 19:205–218

Garbern JY, Yool DA, Moore GJ, Wilds IB, Faulk MW, Klugmann M, Nave KA, Sistermans EA, van der Knaap MS, Bird TD, Shy ME, Kamholz JA, Griffiths IR (2002) Patients lacking the major CNS myelin protein, proteolipid protein 1, develop length-dependent axonal degeneration in the absence of demyelination and inflammation. Brain 125: 551–561

Gatewood JD, Morgan MD, Eaton M, McNamara IM, Stevens LF, Macbeth AH, Meyer EA, Lomas LM, Kozub FJ, Lambert KG, Kinsley CH (2005) Motherhood mitigates aging-related decrements in learning and memory and positively affects brain aging in the rat. Brain Res Bull 66:91–98

Giaume C, Kirchhoff F, Matute C, Reichenbach A, Verkhratsky A (2007) Glia: the fulcrum of brain diseases. Cell Death Differ 14:1324–1335

Gieselmann V, Franken S, Klein D, Mansson JE, Sandhoff R, Lullmann Rauch R, Hartmann D, Saravanan VP, De Deyn PP, D'Hooge R, Van Der Linden AM, Schaeren-Wiemers N (2003) Metachromatic leukodystrophy: consequences of sulphatide accumulation. Acta Paediatr Suppl 92:74–79; discussion 45

Gillespie CS, Sherman DL, Blair GE, Brophy PJ (1994) Periaxin, a novel protein of myelinating Schwann cells with a possible role in axonal ensheathment. Neuron 12:497–508

Givogri MI, Costa RM, Schonmann V, Silva AJ, Campagnoni AT, Bongarzone ER (2002) Central nervous system myelination in mice with deficient expression of Notch1 receptor. J Neurosci Res 67:309–320

Goldberg JL, Barres BA (2000) Nogo in nerve regeneration. Nature 403:369–370

Gow A, Sharma R (2003) The unfolded protein response in protein aggregating diseases. Neuromolecular Med 4:73–94

Gow A, Southwood CM, Lazzarini RA (1998) Disrupted proteolipid protein trafficking results in oligodendrocyte apoptosis in an animal model of Pelizaeus-Merzbacher disease. J Cell Biol 140:925–934

Graham DI, Lantos PA (1997) Greenfield's Neuropathology, 6th edn. Arnold, London

Gregg C, Shikar V, Larsen P, Mak G, Chojnacki A, Yong VW, Weiss S (2007) White matter plasticity and enhanced remyelination in the maternal CNS. J Neurosci 27:1812–1823

Hagemann TL, Connor JX, Messing A (2006) Alexander disease-associated glial fibrillary acidic protein mutations in mice induce Rosenthal fiber formation and a white matter stress response. J Neurosci 26:11162–11173

Hahn AF, Ainsworth PJ, Bolton CF, Bilbao JM, Vallat JM (2001) Pathological findings in the x-linked form of Charcot-Marie-Tooth disease: a morphometric and ultrastructural analysis. Acta Neuropathol (Berl) 101:129–139

Haroutunian VP, Davis KL (2007) Introduction to the special section: Myelin and oligodendrocyte abnormalities in schizophrenia. Int J Neuropsychopharmacol 10:499–502

Haroutunian V, Katsel P, Dracheva S, Stewart DG, Davis KL (2007) Variations in oligodendrocyte-related gene expression across multiple cortical regions: implications for the pathophysiology of schizophrenia. Int J Neuropsychopharmacol 10:565–573

Hausler M, Anhuf D, Schuler H, Ramaekers VT, Thron A, Zerres K, Moller-Hartmann W (2005) White-matter disease in 18q deletion (18q-) syndrome: magnetic resonance spectroscopy indicates demyelination or increased myelin turnover rather than dysmyelination. Neuroradiology 47:83–86

Honke K, Hirahara Y, Dupree J, Suzuki K, Popko B, Fukushima K, Fukushima J, Nagasawa T, Yoshida N, Wada Y, Taniguchi N (2002) Paranodal junction formation and spermatogenesis require sulfoglycolipids. Proc Natl Acad Sci U S A 99:4227–4232

Howell OW, Palser A, Polito A, Melrose S, Zonta B, Scheiermann C, Vora AJ, Brophy PJ, Reynolds R (2006) Disruption of neurofascin localization reveals early changes preceding demyelination and remyelination in multiple sclerosis. Brain 129:3173–3185

Ilyas AA, Chen ZW, Cook SD (2003) Antibodies to sulfatide in cerebrospinal fluid of patients with multiple sclerosis. J Neuroimmunol 139:76–80

Ishibashi T, Dakin KA, Stevens B, Lee PR, Kozlov SV, Stewart CL, Fields RD (2006) Astrocytes promote myelination in response to electrical impulses. Neuron 49:823–832

Ishibashi T, Ding L, Ikenaka K, Inoue Y, Miyado K, Mekada E, Baba H (2004) Tetraspanin protein CD9 is a novel paranodal component regulating paranodal junctional formation. J Neurosci 24:96–102

Ishibashi T, Dupree JL, Ikenaka K, Hirahara Y, Honke K, Peles E, Popko B, Suzuki K, Nishino H, Baba H (2002) A myelin galactolipid, sulfatide, is essential for maintenance of ion channels on myelinated axon but not essential for initial cluster formation. J Neurosci 22:6507–6514

Jagannathan NR, Tandon N, Raghunathan P, Kochupillai N (1998) Reversal of abnormalities of myelination by thyroxine therapy in congenital hypothyroidism: localized in vivo proton magnetic resonance spectroscopy (MRS) study. Brain Res Dev Brain Res 109:179–186

Jessen KR, Mirsky R (1991) Schwann cell precursors and their development. Glia 4:185–194

John GR, Shankar SL, Shafit-Zagardo B, Massimi A, Lee SC, Raine CS, Brosnan CF (2002) Multiple sclerosis: re-expression of a developmental pathway that restricts oligodendrocyte maturation. Nat Med 8:1115–1121

Jung-Testas I, Do Thi A, Koenig H, Desarnaud F, Shazand K, Schumacher M, Baulieu EE (1999) Progesterone as a neurosteroid: synthesis and actions in rat glial cells. J Steroid Biochem Mol Biol 69:97–107

Kaga Y, Shoemaker WJ, Furusho M, Bryant M, Rosenbluth J, Pfeiffer SE, Oh L, Rasband M, Lappe-Siefke C, Yu K, Ornitz DM, Nave KA, Bansal R (2006) Mice with conditional inactivation of fibroblast growth factor receptor-2 signaling in oligodendrocytes have normal myelin but display dramatic hyperactivity when combined with Cnp1 inactivation. J Neurosci 26:12339–12350

Kanter JL, Narayana S, Ho PP, Catz I, Warren KG, Sobel RA, Steinman L, Robinson WH (2006) Lipid microarrays identify key mediators of autoimmune brain inflammation. Nat Med 12: 138–143

Karadottir R, Attwell D (2007) Neurotransmitter receptors in the life and death of oligodendrocytes. Neuroscience 145:1426–1438

Karim SA, Barrie JA, McCulloch MC, Montague P, Edgar JM, Kirkham D, Anderson TJ, Nave KA, Griffiths IR, McLaughlin M (2007) PLP-1 overexpression perturbs myelin protein composition and myelination in a mouse model of Pelizaeus-Merzbacher disease. Glia 55:341–351

Karlsson KA, Samuelsson BE, Steen GO (1974) The lipid composition and Na+-K+-dependent adenosine-triphosphatase activity of the salt (nasal) gland of eider duck and herring gull. A role for sulphatides in sodium-ion transport. Eur J Biochem 46:243–258

Karoutzou G, Emrich HM, Dietrich DE (2007) The myelin-pathogenesis puzzle in schizophrenia: a literature review. Mol Psychiatry 13(3):245–260

Keirstead HS, Levine JM, Blakemore WF (1998) Response of the oligodendrocyte progenitor cell population (defined by NG2 labelling) to demyelination of the adult spinal cord. Glia 22: 161–170

Kubicki M, McCarley RW, Shenton ME (2005) Evidence for white matter abnormalities in schizophrenia. Curr Opin Psychiatry 18:121–134

Kuerten S, Kostova-Bales DA, Frenzel LP, Tigno JT, Tary-Lehmann M, Angelov DN, Lehmann PV (2007) MP4- and MOG:35–55-induced EAE in C57BL/6 mice differentially targets brain, spinal cord and cerebellum. J Neuroimmunol 189:31–40

Kuerten S, Lichtenegger FS, Faas S, Angelov DN, Tary-Lehmann M, Lehmann PV (2006) MBP-PLP-1 fusion protein-induced EAE in C57BL/6 mice. J Neuroimmunol 177:99–111

Lalive PH, Menge T, Delarasse C, Della Gaspera B, Pham-Dinh D, Villoslada P, von Budingen HC, Genain CP (2006) Antibodies to native myelin oligodendrocyte glycoprotein are serologic markers of early inflammation in multiple sclerosis. Proc Natl Acad Sci U S A 103:2280–2285

Larocque D, Richard S (2005) QUAKING KH domain proteins as regulators of glial cell fate and myelination. RNA Biol 2:37–40

Lassmann H (2004) Experimental autoimmune encephalomyelitis. In: LazzariniRA(ed) Myelin Biology and Disorders. Elsevier Academic Press, San Diego, CA, pp 1039–1071

Lazzarini R (2004) Myelin biology and disorders. Elsevier Academic Press, San Diego, CA

Leegwater PA, Vermeulen G, Konst AA, Naidu S, Mulders J, Visser A, Kersbergen P, Mobach D, Fonds D, van Berkel CG, Lemmers RJ, Frants RR, Oudejans CB, Schutgens RB, Pronk JC, van der Knaap MS (2001) Subunits of the translation initiation factor eIF2B are mutant in leukoencephalopathy with vanishing white matter. Nat Genet 29:383–388

Ligon KL, Fancy SP, Franklin RJ, Rowitch DH (2006) Olig gene function in CNS development and disease. Glia 54:1–10

Lindholm E, Ekholm B, Shaw S, Jalonen P, Johansson G, Pettersson U, Sherrington R, Adolfsson R, Jazin E (2001) A schizophrenia-susceptibility locus at 6q25, in one of the world's largest reported pedigrees. Am J Hum Genet 69:96–105

Lopate G, Pestronk A, Evans S, Li L, Clifford D (2005) Anti-sulfatide antibodies in HIV-infected individuals with sensory neuropathy. Neurology 64:1632–1634

Lubetzki C, Demerens C, Anglade P, Villarroya H, Frankfurter A, Lee VM, Zalc B (1993) Even in culture, oligodendrocytes myelinate solely axons. Proc Natl Acad Sci U S A 90: 6820–6824

Lubetzki C, Stankoff B (2000) Role of axonal signals in myelination of the central nervous system. Pathol Biol (Paris) 48:63–69

Lubetzki C, Williams A, Stankoff B (2005) Promoting repair in multiple sclerosis: problems and prospects. Curr Opin Neurol 18:237–244

Ludwin SK (1978) Central nervous system demyelination and remyelination in the mouse: an ultrastructural study of cuprizone toxicity. Lab Invest 39:597–612

Ludwin SK (1994) Central nervous system remyelination: studies in chronically damaged tissue. Ann Neurol 36(Suppl):S143–S145

Macklin WB, Campagnoni CW, Deininger PL, Gardinier MV (1987) Structure and expression of the mouse myelin proteolipid protein gene. J Neurosci Res 18:383–394

Maier O, van der Heide T, Johnson R, de Vries H, Baron W, Hoekstra D (2006) The function of neurofascin155 in oligodendrocytes is regulated by metalloprotease-mediated cleavage and ectodomain shedding. Exp Cell Res 312:500–511

Marcus J, Honigbaum S, Shroff S, Honke K, Rosenbluth J, Dupree JL (2006) Sulfatide is essential for the maintenance of CNS myelin and axon structure. Glia 53:372–381

Mason JL, Toews A, Hostettler JD, Morell P, Suzuki K, Goldman JE, Matsushima GK (2004) Oligodendrocytes and progenitors become progressively depleted within chronically demyelinated lesions. Am J Pathol 164:1673–1682

Mastronardi FG, Moscarello MA (2005) Molecules affecting myelin stability: a novel hypothesis regarding the pathogenesis of multiple sclerosis. J Neurosci Res 80:301–308

Matsushima GK, Morell P (2001) The neurotoxicant, cuprizone, as a model to study demyelination and remyelination in the central nervous system. Brain Pathol 11:107–116

Mattei MG, Alliel PM, Dautigny A, Passage E, Pham-Dinh D, Mattei JF, Jolles P (1986) The gene encoding for the major brain proteolipid (PLP-1) maps on the q-22 band of the human X chromosome. Hum Genet 72:352–353

Matute C (2006) Oligodendrocyte NMDA receptors: a novel therapeutic target. Trends Mol Med 12:289–292

McInnes LA, Lauriat TL (2006) RNA metabolism and dysmyelination in schizophrenia. Neurosci Biobehav Rev 30:551–561

Menichella DM, Goodenough DA, Sirkowski E, Scherer SS, Paul DL (2003) Connexins are critical for normal myelination in the CNS. J Neurosci 23:5963–5973

Menn B, Garcia-Verdugo JM, Yaschine C, Gonzalez-Perez O, Rowitch D, Alvarez-Buylla A (2006) Origin of oligodendrocytes in the subventricular zone of the adult brain. J Neurosci 26: 7907–7918

Meyer zu Horste G, Prukop T, Nave KA, Sereda MW (2006) Myelin disorders: causes and perspectives of Charcot-Marie-Tooth neuropathy. J Mol Neurosci 28:77–88

Mi S, Hu B, Hahm K, Luo Y, Kam Hui ESZ, Yuan Q, Wong WM, Wang L, Su H, Chu TH, Guo J, Zhang W, So KF, Perpinsky B, Shao Z, Graff C, Garber E, Jung V, Wu EX, Wu W (2007) Lingo-1 antagnonist promotes spinal cord remyelination and axonal integrity in MOG-induced experimental auto-immune encephalomyelitis. Nat Med 13:1228–3123

Mi S, Miller RH, Lee X, Scott ML, Shulag-Morskaya S, Shao Z, Chang J, Thill G, Levesque M, Zhang M, Hession C, Sah D, Trapp B, He Z, Jung V, McCoy JM, Pepinsky RB (2005) LINGO-1 negatively regulates myelination by oligodendrocytes. Nat Neurosci 8: 745–751

Michailov GV, Sereda MW, Brinkmann BG, Fischer TM, Haug B, Birchmeier C, Role L, Lai C, Schwab MH, Nave KA (2004) Axonal neuregulin-1 regulates myelin sheath thickness. Science 304:700–703

Mignot C, Boespflug-Tanguy O, Gelot A, Dautigny A, Pham-Dinh D, Rodriguez D (2004) Alexander disease: putative mechanisms of an astrocytic encephalopathy. Cell Mol Life Sci 61:369–385

Mignot C, Delarasse C, Escaich S, Della Gaspera B, Noe E, Colucci-Guyon E, Babinet C, Pekny M, Vicart P, Boespflug-Tanguy O, Dautigny A, Rodriguez D, Pham-Dinh D (2007) Dynamics of mutated GFAP aggregates revealed by real-time imaging of an astrocyte model of Alexander disease. Exp Cell Res 313:2766–2779

Mignot C, Desguerre I, Burglen L, Hertz-Pannier L, Renaldo F, Gadisseux JF, Gallet S, Pham-Dinh D, Boespflug-Tanguy O, Rodriguez D (2009) Tumor-like enlargement of the optic chiasm in an infant with Alexander disease. Brain Dev 31:244–247

Moffett JR, Ross B, Arun P, Madhavarao CN, Namboodiri AM (2007) N-Acetylaspartate in the CNS: from neurodiagnostics to neurobiology. Prog Neurobiol 81:89–131

Mosser J, Douar AM, Sarde CO, Kioschis P, Feil R, Moser H, Poustka AM, Mandel JL, Aubourg P (1993) Putative X-linked adrenoleukodystrophy gene shares unexpected homology with ABC transporters. Nature 361:726–730

Nait-Oumesmar B, Picard-Riera N, Kerninon C, Decker L, Seilhean D, Hoglinger GU, Hirsch EC, Reynolds R, Baron-Van Evercooren A (2007) Activation of the subventricular zone in multiple sclerosis: evidence for early glial progenitors. Proc Natl Acad Sci U S A 104:4694–4699

Nie DY, Ma QH, Law JW, Chia CP, Dhingra NK, Shimoda Y, Yang WL, Gong N, Chen QW, Xu G, Hu QD, Chow PK, Ng YK, Ling EA, Watanabe K, Xu TL, Habib AA, Schachner M, Xiao ZC (2006) Oligodendrocytes regulate formation of nodes of Ranvier via the recognition molecule OMgp. Neuron Glia Biol 2:151–164

Nishiyama A, Chang A, Trapp BD (1999) NG2+ glial cells: a novel glial cell population in the adult brain. J Neuropathol Exp Neurol 58:1113–1124

Oberheim NA, Wang X, Goldman S, Nedergaard M (2006) Astrocytic complexity distinguishes the human brain. Trends Neurosci 29:547–553

Occhi S, Zambroni D, Del Carro U, Amadio S, Sirkowski EE, Scherer SS, Campbell KP, Moore SA, Chen ZL, Strickland S, Di Muzio A, Uncini A, Wrabetz L, Feltri ML (2005) Both laminin and Schwann cell dystroglycan are necessary for proper clustering of sodium channels at nodes of Ranvier. J Neurosci 25:9418–9427

Ogawa Y, Schafer DP, Horresh I, Bar V, Hales K, Yang Y, Susuki K, Peles E, Stankewich MC, Rasband MN (2006) Spectrins and ankyrinB constitute a specialized paranodal cytoskeleton. J Neurosci 26:5230–5239

Orthmann-Murphy JL, Enriquez AD, Abrams CK, Scherer SS (2007) Loss-of-function GJA12/Connexin47 mutations cause Pelizaeus-Merzbacher-like disease. Mol Cell Neurosci 34:629–641

Owens GC, Bunge RP (1989) Evidence for an early role for myelin-associated glycoprotein in the process of myelination. Glia 2:119–128

Park SK, Miller R, Krane I, Vartanian T (2001a) The erbB2 gene is required for the development of terminally differentiated spinal cord oligodendrocytes. J Cell Biol 154:1245–1258

Park SK, Solomon D, Vartanian T (2001b) Growth factor control of CNS myelination. Dev Neurosci 23:327–337

Passage E, Norreel JC, Noack-Fraissignes P, Sanguedolce V, Pizant J, Thirion X, Robaglia-Schlupp A, Pellissier JF, Fontes M (2004) Ascorbic acid treatment corrects the phenotype of a mouse model of Charcot-Marie-Tooth disease. Nat Med 10:396–401

Peles E, Salzer JL (2000) Molecular domains of myelinated axons. Curr Opin Neurobiol 10: 558–565

Pham-Dinh D, Dautigny A, Linington C (2004) The myelin/oligodendrocyte glycoprotein (MOG): structure and physiopathological functions. In: LazzariniRA(ed) Myelin Biology and Disorders. Elsevier Academic Press, San Diego, CA, pp 469–498

Pham-Dinh D, Mattei MG, Nussbaum JL, Roussel G, Pontarotti P, Roeckel N, Mather IH, Arzt K, Lindahl KF, Dautigny A (1993) Myelin/oligodendrocyte glycoprotein is a member of a subset of the immunoglobulin superfamily encoded within the major histocompatibility complex. Proc Natl Acad Sci U S A 90:7990–7994

Picard-Riera N, Decker L, Delarasse C, Goude K, Nait-Oumesmar B, Liblau R, Pham-Dinh D, Barron van Evercooren A (2002) Experimental autoimmune encephalomyelitis mobilizes neural progenitors from the subventricular zone to undergo oligodendrogenesis in adult mice. Proc Natl Acad Sci U S A 99:13211–13216

Pittock SJ, Reindl M, Achenbach S, Berger T, Bruck W, Konig F, Morales Y, Lassmann H, Bryant S, Moore SB, Keegan BM, Lucchinetti CF (2007) Myelin oligodendrocyte glycoprotein antibodies in pathologically proven multiple sclerosis: frequency, stability and clinicopathologic correlations. Mult Scler 13:7–16

Powers JM (2005) Demyelination in peroxisomal diseases. J Neurol Sci 228:206–207

Pujol JC, Soriano-Mas H, Ortiz N, Sebastian-Galles JM, Losilla JM, Deus J (2006) Myelination of language-related areas in the developing brain. Neurology 66:339–343

Quarles RH (2007) Myelin-associated glycoprotein (MAG): past, present and beyond. J Neurochem 100:1431–1448

Quinlan RA, Brenner M, Goldman JE, Messing A (2007) GFAP and its role in Alexander disease. Exp Cell Res 313:2077–2087

Raine CS, Traugott U, Farooq M, Bornstein MB, Norton WT (1981) Augmentation of immune-mediated demyelination by lipid haptens. Lab Invest 45:174–182

Rasband MN, Tayler J, Kaga Y, Yang Y, Lappe-Siefke C, Nave KA, Bansal R (2005) CNP is required for maintenance of axon-glia interactions at nodes of Ranvier in the CNS. Glia 50: 86–90

Rauschka H, Colsch B, Baumann N, Wevers R, Schmidbauer M, Krammer M, Turpin JC, Lefevre M, Olivier C, Tardieu S, Krivit W, Moser H, Moser A, Gieselmann V, Zalc B, Cox T, Reuner U, Tylki-Szymanska A, Aboul-Enein F, LeGuern E, Bernheimer H, Berger J (2006) Late-onset metachromatic leukodystrophy: genotype strongly influences phenotype. Neurology 67: 859–863

Regenold WT, Phatak P, Marano CM, Gearhart L, Viens CH, Hisley KC (2007) Myelin staining of deep white matter in the dorsolateral prefrontal cortex in schizophrenia, bipolar disorder, and unipolar major depression. Psychiatry Res 151:179–188

Roach A, Boylan K, Horvath S, Prusiner SB, Hood LE (1983) Characterization of cloned cDNA representing rat myelin basic protein: absence of expression in brain of shiverer mutant mice. Cell 34:799–806

Rodriguez D, Gauthier F, Bertini E, Bugiani M, Brenner M, N'Guyen S, Goizet C, Gelot A, Surtees R, Pedespan JM, Hernandorena X, Troncoso M, Uziel G, Messing A, Ponsot G, Pham-Dinh D, Dautigny A, Boespflug-Tanguy O (2001) Infantile Alexander disease: spectrum of GFAP mutations and genotype-phenotype correlation. Am J Hum Genet 69:1134–1140

Rodriguez D, Gelot A, Della Gaspera B, Robain O, Ponsot G, Sarlieve LL, Ghandour S, Pompidou A, Dautigny A, Aubourg P, Pham-Dinh D (1999) Increased density of oligodendrocytes in childhood ataxia with diffuse central hypomyelination (CACH) syndrome: neuropathological and biochemical study of two cases. Acta Neuropathol (Berl) 97:469–480

Rogister B, Ben-Hur T, Dubois-Dalcq M (1999) From neural stem cells to myelinating oligodendrocytes. Mol Cell Neurosci 14:287–300

Rosenbluth J, Nave KA, Mierzwa A, Schiff R (2006) Subtle myelin defects in PLP-1-null mice. Glia 54:172–182

Rosenbluth J, Schiff R, Liang WL, Dou W (2003) Antibody-mediated CNS demyelination II. Focal spinal cord lesions induced by implantation of an IgM antisulfatide-secreting hybridoma. J Neurocytol 32:265–276

Roy K, Murtie JC, El-Khodor BF, Edgar N, Sardi SP, Hooks BM, Benoit-Marand M, Chen C, Moore H, O'Donnell P, Brunner D, Corfas G (2007) Loss of erbB signaling in oligodendrocytes alters myelin and dopaminergic function, a potential mechanism for neuropsychiatric disorders. Proc Natl Acad Sci U S A 104:8131–8136

Saravanan K, Schaeren-Wiemers N, Klein D, Sandhoff R, Schwarz A, Yaghootfam A, Gieselmann V, Franken S (2004) Specific downregulation and mistargeting of the lipid raft-associated protein MAL in a glycolipid storage disorder. Neurobiol Dis 16: 396–406

Schaeren-Wiemers N, Bonnet A, Erb M, Erne B, Bartsch U, Kern F, Mantei N, Sherman D, Suter U (2004) The raft-associated protein MAL is required for maintenance of proper axon–glia interactions in the central nervous system. J Cell Biol 166:731–742

Scherer SS, Deschenes SM, Xu YT, Grinspan JB, Fischbeck KH, Paul DL (1995) Connexin32 is a myelin-related protein in the PNS and CNS. J Neurosci 15:8281–8294

Schiffmann R, Elroy-Stein O (2006) Childhood ataxia with CNS hypomyelination/vanishing white matter disease – a common leukodystrophy caused by abnormal control of protein synthesis. Mol Genet Metab 88:7–15

Schumacher M, Guennoun R, Stein DG, De Nicola AF (2007) Progesterone: therapeutic opportunities for neuroprotection and myelin repair. Pharmacol Ther 116:77–106

Seldon HL (2007) Extended neocortical maturation time encompasses speciation, fatty acid and lateralization theories of the evolution of schizophrenia and creativity. Med Hypotheses 69(5):1085–1089

Sereda MW, Meyer zu Horste G, Suter U, Uzma N, Nave KA (2003) Therapeutic administration of progesterone antagonist in a model of Charcot-Marie-Tooth disease (CMT-1A). Nat Med 9:1533–1537

Shy ME (2004) Charcot-Marie-Tooth disease: an update. Curr Opin Neurol 17:579–585

Shy ME, Jani A, Krajewski K, Grandis M, Lewis RA, Li J, Shy RR, Balsamo J, Lilien J, Garbern JY, Kamholz J (2004) Phenotypic clustering in MPZ mutations. Brain 127: 371–384

Simons M, Trajkovic K (2006) Neuron-glia communication in the control of oligodendrocyte function and myelin biogenesis. J Cell Sci 119:4381–4389

Smith EJ, Blakemore WF, McDonald WI (1979) Central remyelination restores secure conduction. Nature 280:395–396

Southwood CM, Garbern J, Jiang W, Gow A (2002) The unfolded protein response modulates disease severity in Pelizaeus-Merzbacher disease. Neuron 36:585–596

Sow A, Lamant M, Bonny JM, Larvaron P, Piaud O, Lecureuil C, Fontaine I, Saleh MC, Garcia Otin AL, Renou JP, Baron B, Zakin M, Guillou F (2006) Oligodendrocyte differentiation is increased in transferrin transgenic mice. J Neurosci Res 83:403–414

Spencer T, Domeniconi M, Cao Z, Filbin MT (2003) New roles for old proteins in adult CNS axonal regeneration. Curr Opin Neurobiol 13:133–139

Spiegel I, Peles E (2006) A new player in CNS myelination. Neuron 49:777–778

Stangel M, Hartung HP (2002) Remyelinating strategies for the treatment of multiple sclerosis. Prog Neurobiol 68:361–376

Staugaitis SM, Smith PR, Colman DR (1990) Expression of myelin basic protein isoforms in nonglial cells. J Cell Biol 110:1719–1727

Steinman L (2001) Blockade of gamma interferon might be beneficial in MS. Mult Scler 7: 275–276

Steinman L, Zamvil SS (2006) How to successfully apply animal studies in experimental allergic encephalomyelitis to research on multiple sclerosis. Ann Neurol 60:12–21

Stevens B (2006) Cross-talk between growth factor and purinergic signalling regulates Schwann cell proliferation. Novartis Found Symp 276:162–175; discussion 175–80, 233–237, 275–81

Stevens B, Fields RD (2000) Response of Schwann cells to action potentials in development. Science 287:2267–2271

Stevens B, Ishibashi T, Chen JF, Fields RD (2004) Adenosine: an activity-dependent axonal signal regulating MAP kinase and proliferation in developing Schwann cells. Neuron Glia Biol 1:23–34

Stevens B, Porta S, Haak LL, Gallo V, Fields RD (2002) Adenosine: a neuron-glial transmitter promoting myelination in the CNS in response to action potentials. Neuron 36:855–868

Stewart DG, Davis KL (2004) Possible contributions of myelin and oligodendrocyte dysfunction to schizophrenia. Int Rev Neurobiol 59:381–424

Stidworthy MF, Genoud S, Li WW, Leone DP, Mantei N, Suter U, Franklin RJ (2004) Notch1 and Jagged1 are expressed after CNS demyelination, but are not a major rate-determining factor during remyelination. Brain 127:1928–1941

Suzuki K (2003) Globoid cell leukodystrophy (Krabbe's disease): update. J Child Neurol 18: 595–603

Suzuki K, Kikkawa Y (1969) Status spongiosus of CNS and hepatic changes induced by cuprizone (biscyclohexanone oxalyldihydrazone). Am J Pathol 54:307–325

Tanaka KF, Takebayashi H, Yamazaki Y, Ono K, Naruse M, Iwasato T, Itohara S, Kato H, Ikenaka K (2007) Murine model of Alexander disease: analysis of GFAP aggregate formation and its pathological significance. Glia 55:617–631

Taveggia C, Zanazzi G, Petrylak A, Yano H, Rosenbluth J, Einheber S, Xu X, Esper RM, Loeb JA, Shrager P, Chao MV, Falls DL, Role L, Salzer JL (2005) Neuregulin-1 type III determines the ensheathment fate of axons. Neuron 47:681–694

Teijido O, Casaroli-Marano R, Kharkovets T, Aguado F, Zorzano A, Palacin M, Soriano E, Martinez A, Estevez R (2007) Expression patterns of MLC1 protein in the central and peripheral nervous systems. Neurobiol Dis 26:532–545

Terada N, Baracskay K, Kinter M, Melrose S, Brophy PJ, Boucheix C, Bjartmar C, Kidd G, Trapp BD (2002) The tetraspanin protein, CD9, is expressed by progenitor cells committed to oligodendrogenesis and is linked to beta1 integrin, CD81, and Tspan-2. Glia 40: 350–359

Tian GF, Azmi H, Takano T, Xu Q, Peng W, Lin J, Oberheim N, Lou N, Wang X, Zielke HR, Kang J, Nedergaard M (2005) An astrocytic basis of epilepsy. Nat Med 11:973–981

Tian DS, Dong Q, Pan DJ, He Y, Yu ZY, Xie MJ, Wang W (2007) Attenuation of astrogliosis by suppressing of microglial proliferation with the cell cycle inhibitor olomoucine in rat spinal cord injury model. Brain Res 1154:206–214

Tian R, Gregor M, Wiche G, Goldman JE (2006) Plectin regulates the organization of glial fibrillary acidic protein in Alexander disease. Am J Pathol 168:888–897

Tiwari-Woodruff S, Beltran-Parrazal L, Charles A, Keck T, Vu T, Bronstein J (2006) K+ channel KV3.1 associates with OSP/claudin-11 and regulates oligodendrocyte development. Am J Physiol Cell Physiol 291:C687–C698

Tomassini V, Onesti E, Mainero C, Giugni E, Paolillo A, Salvetti M, Nicoletti F, Pozzilli C (2005) Sex hormones modulate brain damage in multiple sclerosis: MRI evidence. J Neurol Neurosurg Psychiatry 76:272–275

Tominaga K, Matsuda J, Kido M, Naito E, Yokota I, Toida K, Ishimura K, Suzuki K, Kuroda Y (2004) Genetic background markedly influences vulnerability of the hippocampal neuronal organization in the "twitcher" mouse model of globoid cell leukodystrophy. J Neurosci Res 77:507–516

Traka M, Goutebroze L, Denisenko N, Bessa M, Nifli A, Havaki S, Iwakura Y, Fukamauchi F, Watanabe K, Soliven B, Girault JA, Karagogeos D (2003) Association of TAG-1 with Caspr2

is essential for the molecular organization of juxtaparanodal regions of myelinated fibers. J Cell Biol 162:1161–1172
Trapp BD, Pfeiffer SE, Anitei M, Kidd G (2004) Cell biology of myelin assembly. In: Lazzarini RA et al (eds) Myelin biology and disorders. Elsevier Academic Press, New York, NY, pp 29–55
Turpin JC, Jouas M, Sereni C, Pluot M, Baumann N (1985) Révélation à l'âge adulte d'un cas d'adrénoleucodystrophie familiale. Rev Neurol 141:289–295
Uhlenberg B, Schuelke M, Ruschendorf F, Ruf N, Kaindl AM, Henneke M, Thiele H, Stoltenburg-Didinger G, Aksu F, Topaloglu H, Nurnberg P, Hubner C, Weschke B, Gartner J (2004) Mutations in the gene encoding gap junction protein alpha 12 (connexin 46.6) cause Pelizaeus-Merzbacher-like disease. Am J Hum Genet 75:251–260
Vesce S, Rossi D, Brambilla L, Volterra A (2007) Glutamate release from astrocytes in physiological conditions and in neurodegenerative disorders characterized by neuroinflammation. Int Rev Neurobiol 82:57–71
Voas MG, Lyons DA, Naylor SG, Arana N, Rasband MN, Talbot WS (2007) alphaII-spectrin is essential for assembly of the nodes of Ranvier in myelinated axons. Curr Biol 17: 562–568
von Budingen HC, Hauser SL, Ouallet JC, Tanuma N, Menge T, Genain CP (2004) Frontline: Epitope recognition on the myelin/oligodendrocyte glycoprotein differentially influences disease phenotype and antibody effector functions in autoimmune demyelination. Eur J Immunol 34:2072–2083
von Budingen HC, Menge T, Hauser SL, Genain CP (2006) Restrictive and diversifying elements of the anti-myelin/oligodendrocyte glycoprotein antibody response in primate experimental allergic encephalomyelitis. Immunogenetics 58:122–128
Waxman SG, Bangalore L (2003) Electrophysiological consequences of myelination. In: Lazzarini, RA et al (eds) Myelin Biology and Disorders. Academic Press, New York, NY, pp 117–142
Weimbs T, Stoffel W (1992) Proteolipid protein (PLP-1) of CNS myelin: positions of free, disulfide-bonded, and fatty acid thioester-linked cysteine residues and implications for the membrane topology of PLP-1. Biochemistry 31:12289–12296
Williams A, Piaton G, Aigrot MS, Belhadi A, Théaudin M, Petermann F, Thomas JL, Zalc B, Lubetzki C (2007b) Semaphorin A and 3F, key players in myelin repair in multiple sclerosis? Brain 130:2554–2565
Williams A, Piaton G,, Lubetzki C (2007a) Astrocytes-friends or foes in multiple sclerosis? Glia 55:1300–1312
Willison HJ, Yuki N (2002) Peripheral neuropathies and anti-glycolipid antibodies. Brain 125:2591–2625
Wilson HC, Scolding NJ, Raine CS (2006) Co-expression of PDGF alpha receptor and NG2 by oligodendrocyte precursors in human CNS and multiple sclerosis lesions. J Neuroimmunol 176:162–173
Wolswijk G (1998) Chronic stage multiple sclerosis lesions contain a relatively quiescent population of oligodendrocyte precursor cells. J Neurosci 18:601–609
Woodruff RH, Franklin RJ (1999) Demyelination and remyelination of the caudal cerebellar peduncle of adult rats following stereotaxic injections of lysolecithin, ethidium bromide, and complement/anti-galactocerebroside: a comparative study. Glia 5:216–228
Xin M, Yue T, Ma Z, Wu FF, Gow A, Lu QR (2005) Myelinogenesis and axonal recognition by oligodendrocytes in brain are uncoupled in Olig1-null mice. J Neurosci 25:1354–1365
Yakovlev PI, Lecours LA (1966) The myelinogenic cycles of regional maturation of the brain. BlacKwell, Oxford
Yuki N, Susuki K, Koga M, Nishimoto Y, Odaka M, Hirata K, Taguchi K, Miyatake T, Furukawa K, Kobata T, Yamada M (2004) Carbohydrate mimicry between human ganglioside GM1 and Campylobacter jejuni lipooligosaccharide causes Guillain-Barre syndrome. Proc Natl Acad Sci U S A 101:11404–11409
Zalc B, Helwig JJ, Ghandour MS, Sarlieve L (1978) Sulfatide in the kidney: how is this lipid involved in sodium chloride transport? FEBS Lett 92:92–96

Zeger M, Popken G, Zhang J, Xuan S, Lu QR, Schwab MH, Nave KA, Rowitch D, D'Ercole AJ, Ye P (2007) Insulin-like growth factor type 1 receptor signaling in the cells of oligodendrocyte lineage is required for normal in vivo oligodendrocyte development and myelination. Glia 55:400–411

Zhao C, Fancy SP, Magy L, Urwin JE, Franklin RJ (2005) Stem cells, progenitors and myelin repair. J Anat 207:251–258

Brain Protein Oxidation and Modification for Good or for Bad in Alzheimer's Disease

Rukhsana Sultana and D. Allan Butterfield

Abstract Alzheimer's disease (AD) is the most common cause of dementia in the elderly and is characterized by senile plaques, neurofibrillary tangles, synapse loss, and progressive neuronal deficits. There is an abundance of evidence suggesting that oxidative stress is involved in the pathogenesis of Alzheimer's disease. Several investigations have revealed the presence of oxidation products of proteins, lipids, and DNA in postmortem tissue from AD patients, indices that are indicative of increased oxidative stress. In the present review we discuss the role of protein oxidation in the brain of subjects with AD and MCI.

Keywords Alzheimer's disease · Protein oxidation · Protein carbonyl · 3-Nitrotyrosine · Mild cognitive impairment

Contents

D.A. Butterfield (✉)
Department of Chemistry, Center of Membrane Sciences, Sanders-Brown Center on Aging, University of Kentucky, Lexington, KY 40506, USA
e-mail: dabcns@uky.edu

J.P. Blass (ed.), *Neurochemical Mechanisms in Disease*,
Advances in Neurobiology 1, DOI 10.1007/978-1-4419-7104-3_17,

1 Introduction

Alzheimer's disease (AD) is the most common cause of dementia in the elderly (Evans et al., 1989). AD is characterized by senile plaques, neurofibrillary tangles, synapse loss, and progressive neuronal deficits (Hardy and Selkoe, 2002; McGeer and McGeer, 2003). In addition to the formation of amyloid plaques and NFTs, gliosis, chronic inflammatory reactions, excitotoxic damage, and oxidative stress all appear to contribute to the progression of AD. The proximate cause(s) of the progressive cell death and loss of memory and cognitive functions resulting in profound dementia are still poorly understood. Increased risk factors for AD include stroke, hypertension, diabetes, atherosclerosis, and hypercholesterolemia (Kalaria, 2000; Iadecola, 2003; Casserly and Topol, 2004; Messier and Teutenberg, 2005), and the presence of multiple factors further increases the risk of AD (Luchsinger et al., 2005). These pathologies are associated with chronic inflammation and altered blood vessel responsiveness (Akiyama et al., 2000; Iadecola, 2003; McGeer and McGeer, 2003).

Oxidative stress occurs due to an imbalance in the oxidant and antioxidant levels (Butterfield and Stadtman, 1997). Oxidants can damage virtually all biological molecules: DNA, RNA, cholesterol, lipids, carbohydrates, proteins, and antioxidants. In AD brain, the levels of antioxidants were found to be decreased with an associated increase in protein oxidation (indexed by protein carbonyls and 3-nitrotyrosine), lipid peroxidation, DNA oxidation, advanced glycation end-products, and reactive oxygen species (ROS) formation, among other indices, strongly suggesting a role for oxidative stress in the pathogenesis of AD (Markesbery, 1997; Smith et al., 1997; Butterfield et al., 2001; Lovell and Markesbery, 2001; Butterfield, 2002; Butterfield and Lauderback, 2002; Castegna et al., 2003; Smith et al., 2004; Sultana et al., 2006d, 2006a). Moreover, the use of vitamin E in cell culture diminishes Aβ(1-42)-induced toxicity, further consistent with a role of oxidative damage in AD pathology (Behl, 1999; Yatin et al., 2000; Butterfield and Lauderback, 2002; Boyd-Kimball et al., 2004; Butterfield and Boyd-Kimball, 2005). In addition, Aβ(1-42) can bind to receptors on neuronal and glial cells, for example, the α7-nicotinic acetylcholine receptor, neurotrophin p75 receptor, the N-methyl-D-aspartate receptor, the receptor for advanced glycation end-products (RAGE) (Wang et al., 2000; Verdier and Penke, 2004), and others, forming calcium and potassium channels in cell membranes (Arispe et al., 1993; Etcheberrigaray et al., 1994; Engstrom et al., 1995), decreasing glucose transport across brain endothelial cells (Blanc et al., 1997), and activating the release of chemokines (Fiala et al., 1998) and cytokines (Akama and Van Eldik, 2000). In this review we discuss the role of brain protein oxidation in AD pathology.

2 Role of Aβ(1-42) in Oxidative Stress

Amyloid β-peptide, particularly the 42-mer Aβ(1-42), is thought to be central to the pathogenesis of AD (Selkoe, 2001). In vitro and in vivo studies suggest that Aβ promotes oxidative stress and lipid peroxidation in synaptosomes and neuronal cultures (Butterfield et al., 1994; Keller et al., 1997; Mark et al., 1997; Yatin et al., 1999; Butterfield et al., 2001; Drake et al., 2003; Mohmmad Abdul et al., 2004; Boyd-Kimball et al., 2005a, b; Mattson et al., 1998). Altogether, these studies favor the suggestion that Aβ plays a central role in the pathogenesis of AD as a mediator of oxidative stress. Although senile plaques contain deposited Aβ, the toxic species of Aβ is likely small oligomeric species (Drake et al., 2003; Klein et al., 2004; Ashe, 2005; Walsh et al., 2005).

Aβ peptides (39–43 amino acids) are derived from the amyloid precursor protein (APP) present in both neurons and glial cells in the brain. Although the cellular function of APP has not been completely elucidated, this protein is comprised of a hydrophobic membrane-spanning domain, N-glycosylation sites, and sites for binding Zn^{2+} and Cu^{2+} with high affinity (Hesse et al., 1994). Copper is present at substantial levels in the brain and its release as a result of synaptic activation can reach mM concentrations in the synaptic cleft (Kardos et al., 1989). The interaction between APP and Cu^{2+} may result in the reduction of Cu^{2+} to Cu^{+} and the formation of an intramolecular disulfide bond (Mucke et al., 1994). In the presence of oxidants such as H_2O_2, the APP–Cu interaction may cause APP fragmentation and increase in the production of Aβ. Under normal conditions, Aβs appear to be normal products of APP metabolism and are present in cerebrospinal fluid (CSF) and plasma. Studies by Huang et al. (Huang et al., 1999) reported evidence for a direct interaction between Aβ and Fe^{3+}/Cu^{2+} to create a strong positive formal reduction potential, which can rapidly reduce Fe^{3+} and Cu^{2+} ions and trap molecular oxygen to generate H_2O_2. Transition metals are highly enriched in senile plaques, where they are likely to be bound to amyloid-β (Bush, 2000). Chelation of transition metals efficiently resolves aggregated amyloid-β and senile plaques in vitro (Cherny et al., 2000). The more pathological Aβ(1-42), which has a greater ability to aggregate than Aβ(1-40), has a greater affinity for metals when compared to the latter peptide (Atwood et al., 2000).

The neurotoxic properties of Aβ have also been shown to be associated with methionine at residue 35 of Aβ (Met35) (Butterfield and Boyd-Kimball, 2005). The substitution of methionine by norleucine from Aβ abolishes free radical production, protein oxidation, and toxicity to hippocampal neurons (Butterfield and Boyd-Kimball, 2005). In addition, substitution of a carbon atom for the S atom of methionine completely abrogates Aβ(1-42) neurotoxicity (Yatin et al., 1999; Butterfield and Kanski, 2002), and in vivo studies indicate methionine residue 35 is central for Aβ-induced oxidative damage (Yatin et al., 1999). Studies from our laboratory (Varadarajan et al., 2000) and others (Curtain et al., 2001) showed that Cu^{+2} bound to Aβ(1-42) interacts with Met35 residue to produce free radicals; in the absence of methionine in Aβ(1-42) redox metal ions play no role in the oxidative stress and neurotoxicity induced by the peptide (Varadarajan et al., 2000, 2001). Taken together these results are consonant with the notion that Aβ-induced protein oxidation may in part account for neurodegeneration in AD brain (Butterfield and Boyd-Kimball, 2005).

3 Protein Oxidation in AD

Reactive nitrogen species (RNS) and reactive oxygen species (ROS) can react with proteins leading to oxidatively modified proteins indexed by protein carbonyls and 3-nitrotyrosine (3-NT). Hence, measuring the levels of protein carbonyls and 3-NT reflects the level of protein oxidation in neurons. Increased oxidative stress during aging or in neurodegenerative diseases may lead to accumulation of oxidized proteins. As a consequence, proteins become resistant to proteasomal degradation. On the other hand, oxidative stress is also known to enhance protein turnover accompanied by a specific removal of oxidized proteins. The removal of oxidized proteins has been considered to be impaired in AD due to decreased proteasomal activity (Ding et al., 2006; Halliwell, 2006). Therefore, increased oxidative stress may lead to alterations in both oxidative modification and turnover of proteins and may further trigger immune response. Oxidative damage has been associated with aggregation of proteins, energy dysfunction, calcium dysregulation, mitochondrial malfunction, chronic inflammation, altered antioxidant function, and accumulation of redox-active metals (Butterfield, 2002).

3.1 Protein Carbonyls in AD Brain

Protein carbonyls are formed by backbone fragmentation (Fig. 1), hydrogen atom abstraction at alpha carbons, attack on several amino acid side-chains (Lys, Arg, His, Pro, Thr, etc.), and by the formation of Michael adducts between His, Lys, and Cys residues and reactive alkenals (e.g., hydroxynonenal (HNE)) (Fig. 2). Furthermore, glycation/glycoxidation of Lys amino groups, forming advance glycation end-products (AGEs) (Berlett and Stadtman, 1997; Butterfield and Stadtman, 1997; Dalle-Donne et al., 2003a,b; Stadtman and Levine, 2003), can also lead to protein carbonyl formation. In addition, a number of reactions of protein radicals

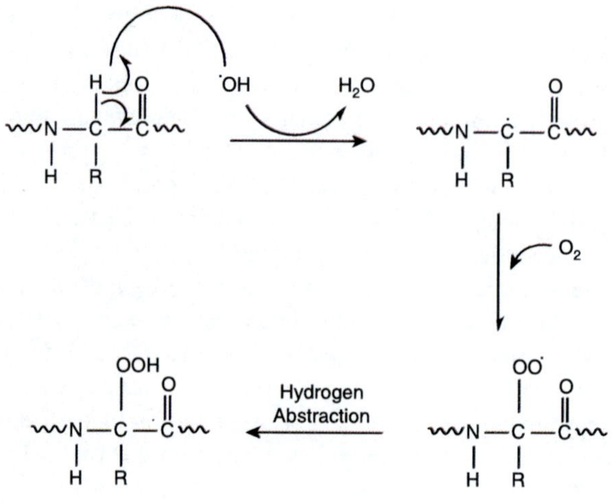

Fig. 1 Mechanism of carbonyl formation from peptide backbone scission

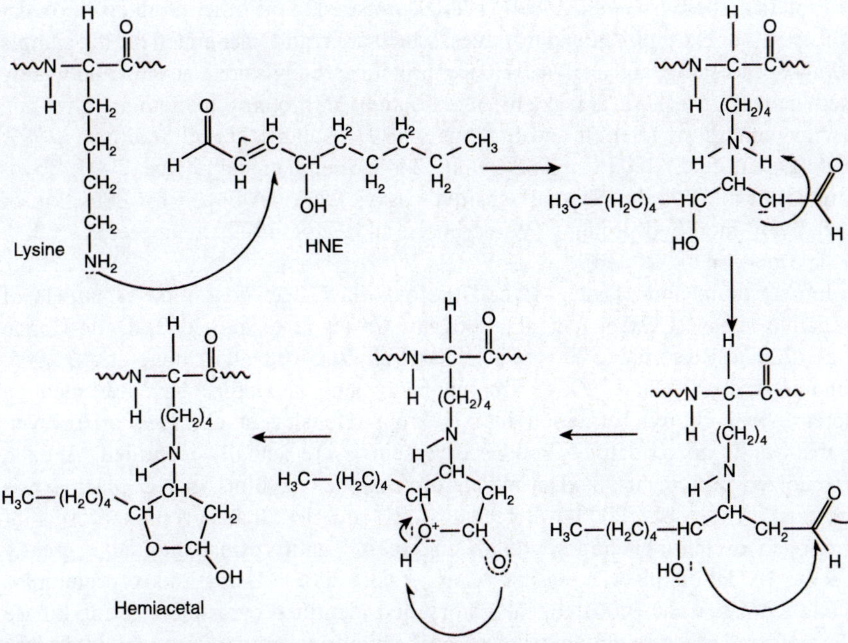

Fig. 2 Mechanism of Michael addition of HNE to lysine

can give rise to other radicals, which can cause damage to other biomolecules. Protein carbonylation leads to oxidation of side-chain hydroxyls, converting them into ketone or aldehyde derivatives, backbone fragmentation, formation of new reactive species (peroxides, DOPA), release of further radicals and occurrence of chain reactions. Most protein damage is irreparable and could lead to a wide range of downstream functional consequences, such as dimerization or aggregation, unfolding or conformational changes to expose more hydrophobic residues to an aqueous environment, loss of structural or functional activity, alterations in cellular handling/turnover, effects on gene regulation and expression, modulation of cell signaling, induction of apoptosis and necrosis, and so on. These phenomena indicate that protein oxidation has physiological and pathological significance (Butterfield and Stadtman, 1997).

Certain oxidation products of proteins, such as oxidation of Cys to cystine, and Met residues to methionine sulfoxide can be repaired by enzymes such as glutathione reductase and methionine sulfoxide (Gabbita et al., 1999; Moskovitz et al., 2002). Other enzymatic means of removing protein carbonyls involve carbonyl reductase and aldehyde dehydrogenase. The majority of the oxidized proteins are catabolized by proteasomal and lysosomal pathways, but some materials appear to be poorly degraded and accumulate within the cell (Dean et al., 1997; Grune et al., 2003). The accumulation of such damaged material may contribute to a range of human pathologies.

Protein carbonyls are chemically stable compared to the other products of oxidative stress, for example, F2 isoprostanes, which are readily generated during sample storage, processing, and analysis. Hence, protein carbonyls are a general and widely used marker to assess the extent of oxidation of proteins in both in vivo and in vitro conditions (Berlett and Stadtman, 1997; Butterfield and Stadtman, 1997; Dalle-Donne et al., 2003a, b; Drake et al., 2003; Stadtman and Levine, 2003; Boyd-Kimball et al., 2005b). Several sensitive assays were developed for detection of oxidatively modified proteins (Winterbourn and Buss, 1999; Levine et al., 2000; Dalle-Donne et al., 2003b).

In AD brain and plasma, several proteins have been identified as targets of oxidative stress (Castegna et al., 2002a,b; Castegna et al., 2003; Dalle-Donne et al., 2003a; Butterfield, 2004; Castegna et al., 2004b; Dalle-Donne et al., 2005; Sultana et al., 2006d, 2006a). Immunohistochemical studies have revealed an increase in carbonyl formation in AD brain (Hensley et al., 1995). However, in that study, no oxidatively modified proteins were actually identified. A band of oxidized protein at 78 kDa on one-dimensional oxyblots in AD plasma was observed (Yu et al., 2003). A recent study revealed that several isoforms of fibrinogen α-chain precursor protein and of α-1-antitrypsin exhibited a greater specific oxidation in AD plasma (Choi et al., 2002). Using redox proteomics (Dalle-Donne et al., 2006) our laboratory first identified cytosolic creatine kinase BB isoform, β-actin, glutamine synthase, ubiquitin carboxy-terminal hydrolase L-1, dihydropyrimidinase-related protein 2, alpha-enolase, phosphoglycerate mutase 1 (PGM1), gamma-soluble *N*-ethylmaleimide-sensitive factor attachment protein (SNAP), peptidyl-prolyl *cis/trans* isomerase 1 (Pin1), triosephosphate isomerase, ATP synthase alpha chain, and carbonic anhydrase 2 as targets of protein oxidation in AD brain, and further studies showed that the oxidatively modified proteins are generally functionally inactive [see Table 1] (Castegna et al., 2002a,b; Castegna et al., 2003; Butterfield, 2004; Sultana et al., 2006c, 2006a). Thus, identification of carbonylated proteins should be followed by functional assessment of the protein, whether it is enzyme or structural protein. These functional studies may identify metabolic or structural defects caused by oxidative modification.

These data support the notion that protein carbonylation perturbs energy metabolism, pH regulation, cell cycle, and mitochondrial functions, as well as Aβ production (Pastorino et al., 2006).

3.2 Protein Nitration in AD

Another marker for protein oxidation is nitration of tyrosine residues, and numerous previous studies support the notion that nitrosative stress also contributes to neurodegeneration in AD (Smith et al., 1997; Tohgi et al., 1999; Castegna et al., 2003; Sultana et al., 2006b). A number of mechanisms for tyrosine nitration of protein have been proposed, and the two widely believed to exist in vivo involve formation of peroxynitrite or mediation via hemeperoxidases (Brennan et al., 2002). These mechanisms involve NO or its by-products that react with ROS (Beckman et al.,

Table 1 Carbonylated and nitrated proteins in AD brain

Protein Functions	Carbonylated Proteins	Nitrated Proteins	References
Energy-related enzymes	CK, Enolase, TPI, PGM1	Enolase, TPI, LDH, GAPDH	Castegna et al. (2002a, b, 2003) Aksenova et al. (2002) Sultana et al. (2006c, d)
Neurotransmitter-related proteins	GS	–	Butterfield et al. (1997) Castegna et al. (2002b)
Proteasome-related proteins	UCHL1 HSC 71	–	Castegna et al. (2002b) Sultana et al. (2005b)
Cholinergic system		Neuropoly-peptide h3-	Castegna et al. (2003)
PH regulation-protein	CA2	CA2	Sultana et al. (2005b)
Structural proteins	DRP2	β-actin	Castegna et al. (2002a) Sultana et al. (2005b)
Cell cycle	Pin 1		Sultana et al. (2005a, b)
Synaptic abnormalities and LTP	Gamma-SNAP		Sultana et al. (2005b)
Mitochondrial abnormalities	–	ATP synthase alpha chain VDAC-1	Sultana et al. (2006b)

CK, creatine kinase BB; TPI, triose phosphate isomerise; PGM1, phosphoglycerate mutase 1; LDH, lactate dehydrogenase; CA2, carbonic anhydrase 2; GS, glutamine synthase; UCHL1, ubiquitin carboxy-terminal hydrolase L-1; HSC 71, heat shock cognate 71; DRP2, dihydropyrimidinase-related protein 2; Pin1, peptidyl-prolyl cis/trans isomerase; Gamma-SNAP, gamma-Soluble NSF-attachment proteins; VDAC, voltage dependent anion channel protein.

2001). NO reacts with superoxide, a toxic anion produced in the cells, to form peroxynitrite (a potent protein nitrating agent) especially after reacting with carbon dioxide (Radi et al., 1999) (Fig. 3). The level of superoxide is kept low in the cell by superoxide dismutase, an antioxidant enzyme. Increased levels of superoxide may be caused either by the overproduction of NAD(P)H oxidases and NOSs or by processes that produce ROSs, such as the electron transport chain (ETC) in mitochondria or by xanthine oxidase (Beckman, 1996; Xia et al., 1996; Ischiropoulos, 1998).

The AD brain has been reported to show mitochondrial abnormalities (Beal, 1998) which could lead to leakage of $O_2^{-\cdot}$ leading to the production of peroxynitrite at diffusion-controlled rates. Peroxynitrite is highly reactive with a very short half-life, and therefore it would react with the proteins, lipids, and carbohydrates near the site of generation and might be involved in the neuronal deterioration observed in AD. The amino acids cysteine, methionine, phenylalanine, and tyrosine are particularly susceptible to nitration. A second mechanism of tyrosine nitration is via hemeproteins. Nitrite, a breakdown product of NO, reacts with hydrogen peroxide to generate nitrogen dioxide. Nitrogen dioxide is also highly unstable and reacts close to the site of its generation as does peroxynitrite.

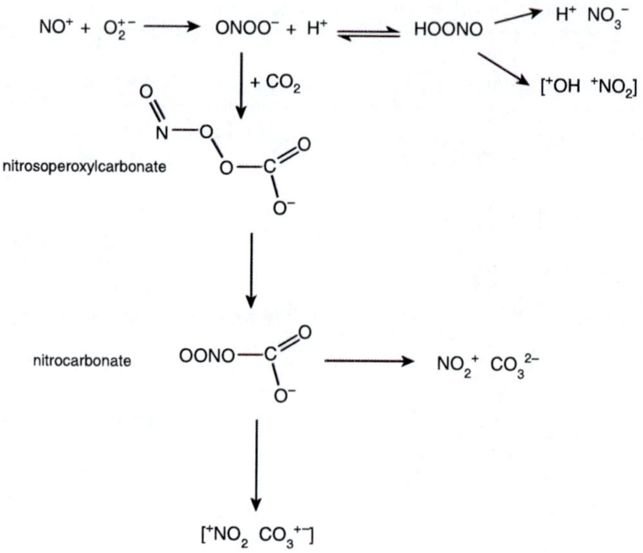

Fig. 3 Formation of peroxynitrite from nitric oxide and superoxide anion and reaction products with carbon dioxide

RNS generated within a physiologically relevant concentration by Ca^{+2}-activated constitutive nitric oxide synthase (NOS) are not toxic; rather RNS so generated are relatively specific in their cellular targets (Lafon-Cazal et al., 1993). NO is generated from three isoforms of NOS: neuronal NOS (nNOS), inducible (iNOS), and endothelial NOS (eNOS) (Stuehr, 1999; Alderton et al., 2001). The role of NO in cell physiology is unclear and its role in disease is controversial with both beneficial and detrimental effects. NO may have direct effects such as signaling by interacting with soluble guanylate cyclase leading to vasodilation and causing alterations in gene expression and also may interact with other molecules to generate more reactive species (Ignarro et al., 1987; Beckman, 1996; Ischiropoulos, 1998; Zamora et al., 2002). One such reactivity is to generate potent nitrating species that result in the formation of 3-nitrotyrosine in proteins. The mechanism, regulation, and role of protein tyrosine nitration are controversial (Beckman and Koppenol, 1996; Ischiropoulos, 1998).

Recently, several studies suggested that protein nitration could be a cellular signaling mechanism and is often a reversible and selective process, similar to protein phosphorylation (Aulak et al., 2004; Koeck et al., 2004). In addition, modified proteins are believed to be either degraded or subject to processes that could lead to enzymatic “denitration” (Gow et al., 1996; Kamisaki et al., 1998; Irie et al., 2003). The latter possibility is intriguing because this would allow the process of tyrosine nitration to be reversible and thus enable a more dynamic physiological role. Protein nitration is observed under normal conditions in all tissues. In AD brain levels of nitrated proteins were found to be increased compared to that of control (Smith et al.,

1997; Castegna et al., 2003; Sultana et al., 2006b). Ubiquitin carboxyl-terminal hydrolase L-1 (UCH L-1), one of the components of the proteosomal pathway, was identified as an oxidized protein in the inferior parietal and hippocampal regions of AD, further suggesting a role for nitration in protein accumulation (Castegna et al., 2003; Sultana et al., 2006b).

Tyrosine residues are crucial in redox cell signaling and oxidative inflammatory injury, due to the fact that nitration has been shown to alter protein function, including modulation of catalytic activity, cell signaling, and cytoskeletal organization (Schopfer et al., 2003). Addition of nitrite to the protein at the 3-position of tyrosine residues (Fig. 4) sterically hinders the phosphorylation of the tyrosine OH moiety, a prominent functional regulation site of proteins. Thus, 3-NT could, potentially, render a protein dysfunctional. Decreased tyrosine phosphorylation could lead to cell death (Lafon-Cazal et al., 1993; Butterfield and Stadtman, 1997). Nitration of proteins may also lead to irreversible damage to the proteins and also affect the energy status of neurons by inactivating key enzymes (Ischiropoulos, 1998; Aulak et al., 2004; Koeck et al., 2004). This widespread occurrence of oxidative alterations not only decreases or eliminates the normal functions of these macromolecules, but also may activate an inflammatory response (the complement cascade, cytokines, acute phase reactants, and proteases) in the AD brain (Meda et al., 1995; Fiala et al., 1998).

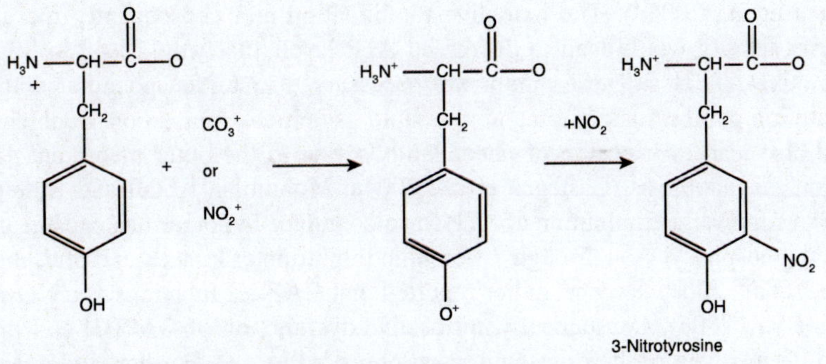

Fig. 4 Mechanism of 3-NT formation

A number of previous studies showed that dityrosine and 3-NT levels were elevated in the hippocampus, IPL, and neocortical regions of the AD brain and in ventricular cerebrospinal fluid (VF) (Smith et al., 1997; Hensley et al., 1998; Tohgi et al., 1999; Castegna et al., 2003; Sultana et al., 2006b). The increased 3-NT protein adducts in CSF of AD subjects probably reflect increased leakage of mitochondrial electron equivalents and protein nitrating agents, with resultant and increased protein nitration in brain tissue. Furthermore, recent work demonstrates that $ONOO^-$ can induce α-synuclein oligomerization through covalent 3,3′-dityrosine cross-linking and may facilitate the misfolding and deposition of select proteins through nitrosative and/or oxidative modification. Horiguchi et al. (Horiguchi et al., 2003) demonstrated the presence of nitrated tau in pretangles,

tangles, and tau inclusions in AD brain. The expression of nitration was robust in pretangles of early AD cases compared to those of more advanced cases, suggesting that tau nitration may be an early event in AD.

Using a redox proteomics approach we reported specific nitration of alpha enolase, gamma-enolase, L-lactate dehydrogenase, triosephosphate isomerase, glyceraldehyde-3-phosphate dehydrogenase (GAPDH), ATP synthase alpha chain, voltage-dependent anion channel protein 1, and carbonic anhydrase 2 in AD brain (Castegna et al., 2003; Sultana et al., 2006b) (Table 1). These data support the notion that nitration of specific proteins perturbs energy metabolism, pH regulation, and mitochondrial functions, which could contribute to the mechanisms for the onset and progression of AD.

The brain depends on glucose as a source of energy, and AD brain has been reported to have altered glucose utilization and consequently altered energy metabolism assessed by PET scanning (Geddes et al., 1996; Messier and Gagnon, 1996; Vanhanen and Soininen, 1998; Rapoport, 1999). The identification of CK, ENO1, TPI, GAPDH, PGM1, and α-ATPase as oxidized proteins using redox proteomics suggest that these proteins are involved directly or indirectly in ATP production (Aksenova et al., 2002; Castegna et al., 2002a,b; Castegna et al., 2003; Sultana et al., 2006d, 2006a). Furthermore, the activity of these identified oxidized proteins (e.g., CK, enolase, PGM1, GAPDH, and ATPase activities), were reportedly diminished in AD brain (Hensley et al., 1995; Aksenova et al., 2002; Sultana et al., 2006d). The oxidative modification and consequently the altered enzyme activity would lead to decreased ATP levels that would lead to impaired ion-motive ATPases. These pumps are necessary to maintain potential gradients, operate ion pumps, maintain membrane lipid asymmetry, and so on. Such changes could also lead to exposure of phosphatidylserine to the outer membrane leaflet, a signal for apoptosis (Castegna et al., 2004a; Mohmmad Abdul and Butterfield, 2005). Moreover, a diminution of ATP can also induce hypothermia, causing abnormal tau phosphorylation through differential inhibition of kinases and phosphatases (Planel et al., 2004). Recent studies reported that GAPDH functions as a NO sensor (Hara et al., 2006). Consequently, nitrosative dysfunction of GAPDH conceivably could be involved in the excess nitration observed in AD. Further studies will be necessary to clarify this point.

In AD brain the ubiquitin–proteasome pathway was found to be dysfunctional (Castegna et al., 2002a; Choi et al., 2004; Sultana et al., 2006d), and the identification of UCHL-1 as an oxidized protein with reduced activity could inhibit the process of the degradation of damaged, excess, or altered proteins and may further promote the aggregation of proteins that could lead to synaptic degeneration in AD brain (Castegna et al., 2002a; Choi et al., 2004; Healy et al., 2004; Sultana et al., 2006d). Recent in vitro studies showed that HNE, a lipid peroxidation product, decreased the activity of recombinant UCH-L1 (Okada et al., 1999; Shringarpure et al., 2001; Hyun et al., 2002), suggesting oxidative modification of UCH-L1 inactivates its hydrolase activity. Proteomics identification of UCH-L1 as an oxidatively modified protein in AD (Castegna et al., 2002a; Sultana et al., 2006d) was recently confirmed by others (Choi et al., 2004). Taken together, these different lines of evidence support a role for dysfunction of the ubiquitin–proteasome pathway in the pathogenesis of AD. Others showed that diminished proteasome function could lead to neurodegeneration (Halliwell, 2002) and oxidative stress (Ding et al., 2003). On the other hand, oxidative stress leads to proteasome dysfuntion (Halliwell, 2002, 2006), suggesting a vicious feedforward cycle of oxidative stress, proteasome dysfunction, and neurodegeneration.

Neuropolypeptide h3 (NPH3), a phosphatidyloethanolamine-binding protein [PEBP] or cholinergic neurostimulating peptide, may play an important role in regulating choline acetyl transferase (ChAT) and maintaining phospholipid asymmetry, a process that is important to normal mitochondrial and plasma membrane function (Castegna et al., 2004a; Mohmmad Abdul and Butterfield, 2005). Oxidation of this protein could lead to impaired cholinergic properties, mitochondria function, and apoptosis in AD.

β-actin (ACT) and dihydropyrimidinase-related protein 2 (DRP2) were found to be downregulated and oxidatively modified in AD brain (Coleman and Flood, 1987; Lubec et al., 1999; Castegna et al., 2002a,b, 2003). Alterations in the structure of proteins induced due to oxidation could be one of the contributing factors involved in the observed loss of interneuronal connections, neuronal repair, and shortened dendritic lengths in AD brain (Coleman and Flood, 1987), conceivably leading to memory impairment and synapse loss, clearly important for AD.

Another important protein that is found to have reduced expression and is also oxidized and has reduced activity in AD brain is peptidyl-prolyl *cis/trans* isomerase (Pin1). This protein is colocalized with phosphorylated tau (Holzer et al., 2002; Kurt et al., 2003; Ramakrishnan et al., 2003; Sultana et al., 2006c, d). Pin1 is a chaperone enzyme that recognizes phosphorylated Ser-Pro and phosphorylated Thr-Pro motifs in proteins, and alters the conformation of proteins from *cis* to *trans* between a given amino acid and a proline (Schutkowski et al., 1998). One of the target proteins of Pin1 is a protein that removes phosphate moieties from tau (Shen et al., 1998). Oxidative modification of Pin1 may lead to hyperphosphorylation of tau, and entry into a cell cycle eventually leading to tangle formation and apoptosis (Nagy et al., 1997; Zhou et al., 2000; Smith et al., 2004). In addition to a role of Pin1 in neurofibrillary tangle formation, recent studies suggest that Pin1 plays a role in APP processing, and therefore, in Aβ levels in brain (Pastorino et al., 2006). Thus, Pin1 is involved in two of the major pathological hallmarks of AD. Pin1 is oxidatively modified and dysfunctional in mild cognitive impairment (MCI), a precursor condition to AD (Butterfield, 2006). Further studies are required to understand the role of Pin1 in the disease progression.

Soluble N-ethylmaleimidesensitive factor (NSF) attachment protein (γ-SNAP) is another protein found to be oxidatively modified in AD brain, and this protein is important in vesicular transport for neurotransmitter release, hormone secretion, and mitochondrial integrity. Hence, oxidation may lead to an altered neurotransmission system and impaired learning and memory in AD (Masliah et al., 1994; Scheff and Price, 2003; Sultana et al., 2006d).

The pH of the cell is crucial for the normal functioning of the cells. Carbonic anhydrase 2 (CA2) regulates cellular pH, CO_2, and HCO_3^- transport, and maintains H_2O and electrolyte balance (Sly and Hu, 1995) by reversible hydration of CO_2 in normal cells. This protein has been reported to be oxidized in AD brain and also showed a decrease in activity (Meier-Ruge et al., 1984; Poon et al., 2004; Sultana et al., 2006d). Functionally inactive CA2 could induce changes in buffering systems in the brain, which could consequently lead to protein aggregation. Protein aggregation is more pronounced in AD brain, and, because cellular pH could be altered, altered mitochondrial production of ATP could be affected.

The voltage-dependent anion channel (VDAC) is identified as one of the oxidized proteins in AD brain (Sultana et al., 2006a,b). The oxidation of this protein in AD suggests an alteration in the function of the mitochondrial permeability transition pore (MPTP) leading to mitochondrial depolarization and altered signal transduction pathways, which could be crucial in synaptic transmission and plasticity. Moreover, alterations in the MPTP could lead to apoptotic processes. In addition, dysfunction of mitochondria recently has been reported to alter APP metabolism, enhancing the intraneuronal accumulation of amyloid β-peptide and enhancing neuronal vulnerability (Busciglio et al., 2002).

Overall, from the data presented above it is clear that oxidation of specific brain proteins alters the structure and thereby function of the proteins. Such changes could be important in AD pathology.

4 Is Protein Oxidation an Early or Late Event in AD Pathogenesis?

In recent years, the clinical stages preceding AD presenting memory impairment but without overt dementia have attained increased attention in the AD clinical and research fields. Patients with MCI are subjects with memory or other cognitive complaints but who do not fulfill the dementia criteria (Visser et al., 2001). Persons with MCI represent a heterogeneous group of patients with several possible explanations for the cognitive deficits. A high proportion of MCI patients are probably early AD subjects, although other diagnoses are also included in this diagnostic entity. Biochemical markers for AD should reflect the pathogenesis of the disorder.

Both in MCI and AD patients, mean plasma levels of nonenzymatic antioxidants and activity of antioxidant enzymes appeared to be lower than in controls, with no parallel induction of antioxidant enzymes (Keller et al., 2005). In order to explain these results it has been suggested that the increased free radical production in MCI might lead to a rapid consumption of plasma antioxidants without a simultaneous activation of new molecules of antioxidant enzymes. Individuals with MCI, and subsequently with AD, are likely to have an inadequate antioxidant enzymatic activity, unable to counteract the increased production of free radicals during the pathogenesis of the disease.

Subjects with MCI have increased protein oxidation in hippocampus and IPL (Butterfield et al., 2006a) and superior and medial temporal gyri (Keller et al., 2005). Additionally, using redox proteomics we identified three specific proteins, that is, enolase, glutamine synthase, and Pin1 as common targets of protein oxidation between MCI and AD which suggests that protein oxidation of these selected proteins could be important in initial events involved in AD pathogenesis (Fig. 5). Furthermore, several gene mutations associated with AD have been observed in subjects with MCI including mutations in apolipoprotein E, presenilin 1, and the amyloid precursor protein (Traykov et al., 2002; Nacmias et al., 2004). Increased levels of lipid peroxidation have been reported in the brain of persons with MCI (Keller et al., 2005; Markesbery et al., 2005; Butterfield et al., 2006b). Thus, increased levels of protein and lipid peroxidation could be implicated as early events in AD pathophysiology and also suggest that pharmacological intervention to prevent protein and lipid peroxidation at the MCI stage or earlier may be a promising therapeutic strategy to delay or prevent progression to AD.

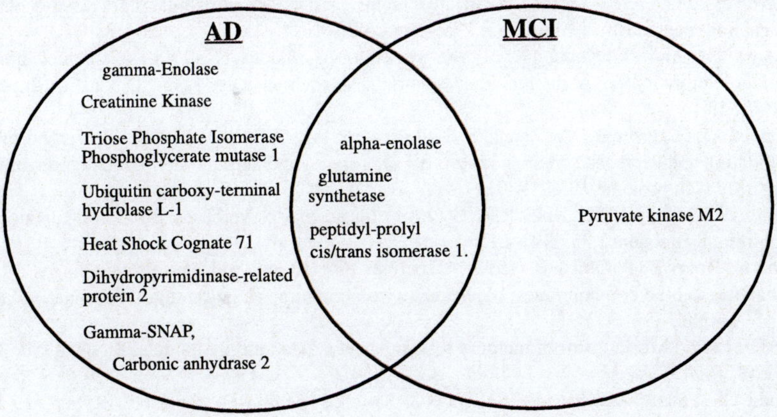

Fig. 5 Comparison between the MCI and AD brain to see the common targets of protein oxidation

Very recent studies reported increased oxidative damage in nuclear and mitochondrial DNA in MCI, as indexed by increased levels of 8-hydroxyguanosine (8-OHdG), 2,6-diamino-4-hydroxy-5-formamidopyrimidine (fapyguanine), 8-hydroadenine, 4,6-diamino-5-formamidopyrimidine (fapyadenine), and 5-hydroxycytosine (Wang et al., 2006). Due to the crucial role that DNA plays in cells, high levels of oxidation, particularly early in the progression of AD, may result in a decline of normal cell function through altered transcription, changes in protein expression, or cross-linking with proteins. Taken together, these results suggest that oxidative damage is one of the factors involved in the pathogenesis of neurodegeneration in AD and is not simply a late effect of the neurodegenerative process.

5 Conclusions

Protein oxidation may be an early event in AD pathogenesis as supported by the data from the MCI brain. With exceptions related to oxidative signaling or other beneficial processes, excessive oxidation of proteins has been reported to decrease the functionality of most proteins, which suggests that protein oxidation is harmful for cell survival. Further studies are in progress to understand the role of protein oxidation and its abrogation in AD by using in vivo and in vitro models of AD centered around Aβ(1-42) (Sultana et al., 2006a).

Acknowledgments This work was supported in part by NIH grants [AG-10836; AG-05119].

References

Akama KT, Van Eldik LJ (2000) Beta-amyloid stimulation of inducible nitric-oxide synthase in astrocytes is interleukin-1beta- and tumor necrosis factor-alpha (TNFalpha)-dependent, and involves a TNFalpha receptor-associated factor- and NFkappaB-inducing kinase-dependent signaling mechanism. J Biol Chem 275:7918–7924

Akiyama H, Arai T, Kondo H, Tanno E, Haga C, Ikeda K (2000) Cell mediators of inflammation in the Alzheimer disease brain. Alzheimer Dis Assoc Disord 14(Suppl 1): S47–S53

Aksenova M, Butterfield DA, Zhang SX, Underwood M, Geddes JW (2002) Increased protein oxidation and decreased creatine kinase BB expression and activity after spinal cord contusion injury. J Neurotrauma 19:491–502

Alderton WK, Cooper CE, Knowles RG (2001) Nitric oxide synthases: structure, function and inhibition. Biochem J 357:593–615

Arispe N, Rojas E, Pollard HB (1993) Alzheimer disease amyloid beta protein forms calcium channels in bilayer membranes: blockade by tromethamine and aluminum. Proc Natl Acad Sci U S A 90:567–571

Ashe KH (2005) Mechanisms of memory loss in Abeta and tau mouse models. Biochem Soc Trans 33:591–594

Atwood CS, Huang X, Khatri A, Scarpa RC, Kim YS, Moir RD, Tanzi RE, Roher AE, Bush AI (2000) Copper catalyzed oxidation of Alzheimer Abeta. Cell Mol Biol (Noisy-le-grand) 46:777–783

Aulak KS, Koeck T, Crabb JW, Stuehr DJ (2004) Dynamics of protein nitration in cells and mitochondria. Am J Physiol Heart Circ Physiol 286:H30–H38

Beal MF (1998) Mitochondrial dysfunction in neurodegenerative diseases. Biochim Biophys Acta 1366:211–223

Beckman JS (1996) Oxidative damage and tyrosine nitration from peroxynitrite. Chem Res Toxicol 9:836–844

Beckman JS, Estevez AG, Crow JP, Barbeito L (2001) Superoxide dismutase and the death of motoneurons in ALS. Trends Neurosci 24:S15–S20

Beckman JS, Koppenol WH (1996) Nitric oxide, superoxide, and peroxynitrite: the good, the bad, and ugly. Am J Physiol 271:C1424–C1437

Behl C (1999) Vitamin E and other antioxidants in neuroprotection. Int J Vitam Nutr Res 69: 213–219

Berlett BS, Stadtman ER (1997) Protein oxidation in aging, disease, and oxidative stress. J Biol Chem 272:20313–20316

Blanc EM, Kelly JF, Mark RJ, Waeg G, Mattson MP (1997) 4-Hydroxynonenal, an aldehydic product of lipid peroxidation, impairs signal transduction associated with muscarinic acetylcholine and metabotropic glutamate receptors: possible action on G alpha(q/11). J Neurochem 69:570–580

Boyd-Kimball D, Castegna A, Sultana R, Poon HF, Petroze R, Lynn BC, Klein JB, Butterfield DA (2005a) Proteomic identification of proteins oxidized by Abeta(1-42) in synaptosomes: implications for Alzheimer's disease. Brain Res 1044:206–215

Boyd-Kimball D, Sultana R, Mohmmad-Abdul H, Butterfield DA (2004) Rodent Abeta(1-42) exhibits oxidative stress properties similar to those of human Abeta(1-42): Implications for proposed mechanisms of toxicity. J Alzheimers Dis 6:515–525

Boyd-Kimball D, Sultana R, Poon HF, Lynn BC, Casamenti F, Pepeu G, Klein JB, Butterfield DA (2005b) Proteomic identification of proteins specifically oxidized by intracerebral injection of amyloid beta-peptide (1-42) into rat brain: implications for Alzheimer's disease. Neuroscience 132:313–324

Brennan ML, Wu W, Fu X, Shen Z, Song W, Frost H, Vadseth C, Narine L, Lenkiewicz E, Borchers MT, Lusis AJ, Lee JJ, Lee NA, Abu-Soud HM, Ischiropoulos H, Hazen SL (2002) A tale of two controversies: defining both the role of peroxidases in nitrotyrosine formation in vivo using eosinophil peroxidase and myeloperoxidase-deficient mice, and the nature of peroxidase-generated reactive nitrogen species. J Biol Chem 277:17415–17427

Busciglio J, Pelsman A, Wong C, Pigino G, Yuan M, Mori H, Yankner BA (2002) Altered metabolism of the amyloid beta precursor protein is associated with mitochondrial dysfunction in Down's syndrome. Neuron 33:677–688

Bush AI (2000) Metals and neuroscience. Curr Opin Chem Biol 4:184–191

Butterfield DA (2002) Amyloid beta-peptide (1-42)-induced oxidative stress and neurotoxicity: implications for neurodegeneration in Alzheimer's disease brain. A review. Free Radic Res 36:1307–1313

Butterfield DA (2004) Proteomics: a new approach to investigate oxidative stress in Alzheimer's disease brain. Brain Res 1000:1–7

Butterfield DA (2006) Oxidative stress in neurodegenerative disorders. Antioxidant Redox Signal 8:1971–1974

Butterfield DA, Boyd-Kimball D (2005) The critical role of methionine 35 in Alzheimer's amyloid beta-peptide (1-42)-induced oxidative stress and neurotoxicity. Biochim Biophys Acta 1703:149–156

Butterfield DA, Drake J, Pocernich C, Castegna A (2001) Evidence of oxidative damage in Alzheimer's disease brain: central role for amyloid beta-peptide. Trends Mol Med 7:548–554

Butterfield DA, Hensley K, Harris M, Mattson M, Carney J (1994) beta-Amyloid peptide free radical fragments initiate synaptosomal lipoperoxidation in a sequence-specific fashion: implications to Alzheimer's disease. Biochem Biophys Res Commun 200:710–715

Butterfield DA, Hensley K, Cole P, Subramaniam R, Aksenov M, Aksenova M, Bummer PM, Haley BE, Carney JM (1997) Oxidatively-induced structural alteration of glutamine synthetase assessed by analysis of spin labeled incorporation kinetics: relevance to Alzheimer's disease. J Neurochem 68:2451–2457

Butterfield DA, Kanski J (2002) Methionine residue 35 is critical for the oxidative stress and neurotoxic properties of Alzheimer's amyloid beta-peptide 1-42. Peptides 23:1299–1309

Butterfield DA, Lauderback CM (2002) Lipid peroxidation and protein oxidation in Alzheimer's disease brain: potential causes and consequences involving amyloid beta-peptide-associated free radical oxidative stress. Free Radic Biol Med 32:1050–1060

Butterfield DA, Poon HF, St Clair D, Keller JN, Pierce WM, Klein JB, Markesbery WR (2006a) Redox proteomics identification of oxidatively modified hippocampal proteins in mild cognitive impairment: Insights into the development of Alzheimer's disease. Neurobiol Dis 22(2):223–232

Butterfield DA, Reed T, Perluigi M, De Marco C, Coccia R, Cini C, Sultana R (2006b) Elevated protein-bound levels of the lipid peroxidation product, 4-hydroxy-2-nonenal, in brain from persons with mild cognitive impairment. Neurosci Lett 397:170–173

Butterfield DA, Stadtman ER (1997) Protein oxidation processes in aging brain. Adv Cell Aging Gerontol vol. 2:pp. 161–191

Casserly I, Topol E (2004) Convergence of atherosclerosis and Alzheimer's disease: inflammation, cholesterol, and misfolded proteins. Lancet 363:1139–1146

Castegna A, Aksenov M, Aksenova M, Thongboonkerd V, Klein JB, Pierce WM, Booze R, Markesbery WR, Butterfield DA (2002a) Proteomic identification of oxidatively modified proteins in Alzheimer's disease brain. Part I: creatine kinase BB, glutamine synthase, and ubiquitin carboxy-terminal hydrolase L-1. Free Radic Biol Med 33:562–571

Castegna A, Aksenov M, Thongboonkerd V, Klein JB, Pierce WM, Booze R, Markesbery WR, Butterfield DA (2002b) Proteomic identification of oxidatively modified proteins in Alzheimer's disease brain. Part II: dihydropyrimidinase-related protein 2, alpha-enolase and heat shock cognate 71. J Neurochem 82:1524–1532

Castegna A, Lauderback CM, Mohmmad-Abdul H, Butterfield DA (2004a) Modulation of phospholipid asymmetry in synaptosomal membranes by the lipid peroxidation products, 4-hydroxynonenal and acrolein: implications for Alzheimer's disease. Brain Res 1004:193–197

Castegna A, Thongboonkerd V, Klein JB, Lynn B, Markesbery WR, Butterfield DA (2003) Proteomic identification of nitrated proteins in Alzheimer's disease brain. J Neurochem 85:1394–1401

Castegna A, Thongboonkerd V, Klein J, Lynn BC, Wang YL, Osaka H, Wada K, Butterfield DA (2004b) Proteomic analysis of brain proteins in the gracile axonal dystrophy (gad) mouse, a syndrome that emanates from dysfunctional ubiquitin carboxyl-terminal hydrolase L-1, reveals oxidation of key proteins. J Neurochem 88:1540–1546

Cherny RA, Barnham KJ, Lynch T, Volitakis I, Li QX, McLean CA, Multhaup G, Beyreuther K, Tanzi RE, Masters CL, Bush AI (2000) Chelation and intercalation: complementary properties in a compound for the treatment of Alzheimer's disease. J Struct Biol 130:209–216

Choi J, Levey AI, Weintraub ST, Rees HD, Gearing M, Chin LS, Li L (2004) Oxidative modifications and down-regulation of ubiquitin carboxyl-terminal hydrolase L1 associated with idiopathic Parkinson's and Alzheimer's diseases. J Biol Chem 279:13256–13264

Choi J, Malakowsky CA, Talent JM, Conrad CC, Gracy RW (2002) Identification of oxidized plasma proteins in Alzheimer's disease. Biochem Biophys Res Commun 293: 1566–1570

Coleman PD, Flood DG (1987) Neuron numbers and dendritic extent in normal aging and Alzheimer's disease. Neurobiol Aging 8:521–545

Curtain CC, Ali F, Volitakis I, Cherny RA, Norton RS, Beyreuther K, Barrow CJ, Masters CL, Bush AI, Barnham KJ (2001) Alzheimer's disease amyloid-beta binds copper and zinc to generate an allosterically ordered membrane-penetrating structure containing superoxide dismutase-like subunits. J Biol Chem 276:20466–20473

Dalle-Donne I, Giustarini D, Colombo R, Rossi R, Milzani A (2003a) Protein carbonylation in human diseases. Trends Mol Med 9:169–176

Dalle-Donne I, Rossi R, Giustarini D, Milzani A, Colombo R (2003b) Protein carbonyl groups as biomarkers of oxidative stress. Clin Chim Acta 329:23–38

Dalle-Donne I, Scaloni A, Butterfield DA (2006) Redox proteomics: from protein modifications to cellualr dysfunction and disease. New York, NY, Wiley

Dalle-Donne I, Scaloni A, Giustarini D, Cavarra E, Tell G, Lungarella G, Colombo R, Rossi R, Milzani A (2005) Proteins as biomarkers of oxidative/nitrosative stress in diseases: the contribution of redox proteomics. Mass Spectrom Rev 24:55–99

Dean RT, Fu S, Stocker R, Davies MJ (1997) Biochemistry and pathology of radical-mediated protein oxidation. Biochem J 324(Pt 1):1–18

Ding Q, Dimayuga E, Keller JN (2006) Proteasome regulation of oxidative stress in aging and age-related diseases of the CNS. Antioxid Redox Signal 8:163–172

Ding Q, Reinacker K, Dimayuga E, Nukala V, Drake J, Butterfield DA, Dunn JC, Martin S, Bruce-Keller AJ, Keller JN (2003) Role of the proteasome in protein oxidation and neural viability following low-level oxidative stress. FEBS Lett 546:228–232

Drake J, Link CD, Butterfield DA (2003) Oxidative stress precedes fibrillar deposition of Alzheimer's disease amyloid beta-peptide (1-42) in a transgenic Caenorhabditis elegans model. Neurobiol Aging 24:415–420

Engstrom I, Ronquist G, Pettersson L, Waldenstrom A (1995) Alzheimer amyloid beta-peptides exhibit ionophore-like properties in human erythrocytes. Eur J Clin Invest 25:471–476

Etcheberrigaray R, Ito E, Kim CS, Alkon DL (1994) Soluble beta-amyloid induction of Alzheimer's phenotype for human fibroblast K+ channels. Science 264:276–279

Evans DA, Funkenstein HH, Albert MS, Scherr PA, Cook NR, Chown MJ, Hebert LE, Hennekens CH, Taylor JO (1989) Prevalence of Alzheimer's disease in a community population of older persons. Higher than previously reported. J Am Med Assoc 262:2551–2556

Fiala M, Zhang L, Gan X, Sherry B, Taub D, Graves MC, Hama S, Way D, Weinand M, Witte M, Lorton D, Kuo YM, Roher AE (1998) Amyloid-beta induces chemokine secretion and monocyte migration across a human blood–brain barrier model. Mol Med 4: 480–489

Gabbita SP, Aksenov MY, Lovell MA, Markesbery WR (1999) Decrease in peptide methionine sulfoxide reductase in Alzheimer's disease brain. J Neurochem 73:1660–1666

Geddes JW, Pang Z, Wiley DH (1996) Hippocampal damage and cytoskeletal disruption resulting from impaired energy metabolism. Implications for Alzheimer disease. Mol Chem Neuropathol 28:65–74

Gow AJ, Duran D, Malcolm S, Ischiropoulos H (1996) Effects of peroxynitrite-induced protein modifications on tyrosine phosphorylation and degradation. FEBS Lett 385:63–66

Grune T, Merker K, Sandig G, Davies KJ (2003) Selective degradation of oxidatively modified protein substrates by the proteasome. Biochem Biophys Res Commun 305:709–718

Halliwell B (2002) Hypothesis: proteasomal dysfunction: a primary event in neurogeneration that leads to nitrative and oxidative stress and subsequent cell death. Ann N Y Acad Sci 962: 182–194

Halliwell B (2006) Proteasomal Dysfunction: A Common Feature of Neurodegenerative Diseases? Implications for the Environmental Origins of Neurodegeneration. Antioxid Redox Signal 8(11–12):2007–2019

Hara MR, Cascio MB, Sawa A (2006) GAPDH as a sensor of NO stress. Biochim Biophys Acta 1762:502–509

Hardy J, Selkoe DJ (2002) The amyloid hypothesis of Alzheimer's disease: progress and problems on the road to therapeutics. Science 297:353–356

Healy DG, Abou-Sleiman PM, Wood NW (2004) Genetic causes of Parkinson's disease: UCHL-1. Cell Tissue Res 318:189–194

Hensley K, Hall N, Subramaniam R, Cole P, Harris M, Aksenov M, Aksenova M, Gabbita SP, Wu JF, Carney JM, et al. (1995) Brain regional correspondence between Alzheimer's disease histopathology and biomarkers of protein oxidation. J Neurochem 65:2146–2156

Hensley K, Maidt ML, Yu Z, Sang H, Markesbery WR, Floyd RA (1998) Electrochemical analysis of protein nitrotyrosine and dityrosine in the Alzheimer brain indicates region-specific accumulation. J Neurosci 18:8126–8132

Hesse L, Beher D, Masters CL, Multhaup G (1994) The beta A4 amyloid precursor protein binding to copper. FEBS Lett 349:109–116

Holzer M, Gartner U, Stobe A, Hartig W, Gruschka H, Bruckner MK, Arendt T (2002) Inverse association of Pin1 and tau accumulation in Alzheimer's disease hippocampus. Acta Neuropathol (Berl) 104:471–481

Horiguchi T, Uryu K, Giasson BI, Ischiropoulos H, LightFoot R, Bellmann C, Richter-Landsberg C, Lee VM, Trojanowski JQ (2003) Nitration of tau protein is linked to neurodegeneration in tauopathies. Am J Pathol 163:1021–1031

Huang X, Atwood CS, Hartshorn MA, Multhaup G, Goldstein LE, Scarpa RC, Cuajungco MP, Gray DN, Lim J, Moir RD, Tanzi RE, Bush AI (1999) The A beta peptide of Alzheimer's disease directly produces hydrogen peroxide through metal ion reduction. Biochemistry 38:7609–7616

Hyun DH, Lee MH, Halliwell B, Jenner P (2002) Proteasomal dysfunction induced by 4-hydroxy-2,3-trans-nonenal, an end-product of lipid peroxidation: a mechanism contributing to neurodegeneration? J Neurochem 83:360–370

Iadecola C (2003) Cerebrovascular effects of amyloid-beta peptides: mechanisms and implications for Alzheimer's dementia. Cell Mol Neurobiol 23:681–689

Ignarro LJ, Byrns RE, Wood KS (1987) Endothelium-dependent modulation of cGMP levels and intrinsic smooth muscle tone in isolated bovine intrapulmonary artery and vein. Circ Res 60:82–92

Irie Y, Saeki M, Kamisaki Y, Martin E, Murad F (2003) Histone H1.2 is a substrate for denitrase, an activity that reduces nitrotyrosine immunoreactivity in proteins. Proc Natl Acad Sci U S A 100:5634–5639

Ischiropoulos H (1998) Biological tyrosine nitration: a pathophysiological function of nitric oxide and reactive oxygen species. Arch Biochem Biophys 356:1–11

Kalaria RN (2000) The role of cerebral ischemia in Alzheimer's disease. Neurobiol Aging 21: 321–330

Kamisaki Y, Wada K, Bian K, Balabanli B, Davis K, Martin E, Behbod F, Lee YC, Murad F (1998) An activity in rat tissues that modifies nitrotyrosine-containing proteins. Proc Natl Acad Sci U S A 95:11584–11589

Kardos J, Kovacs I, Hajos F, Kalman M, Simonyi M (1989) Nerve endings from rat brain tissue release copper upon depolarization. A possible role in regulating neuronal excitability. Neurosci Lett 103:139–144

Keller JN, Mark RJ, Bruce AJ, Blanc E, Rothstein JD, Uchida K, Waeg G, Mattson MP (1997) 4-Hydroxynonenal, an aldehydic product of membrane lipid peroxidation, impairs glutamate transport and mitochondrial function in synaptosomes. Neuroscience 80:685–696

Keller JN, Schmitt FA, Scheff SW, Ding Q, Chen Q, Butterfield DA, Markesbery WR (2005) Evidence of increased oxidative damage in subjects with mild cognitive impairment. Neurology 64:1152–1156

Klein WL, Stine WB Jr., Teplow DB (2004) Small assemblies of unmodified amyloid beta-protein are the proximate neurotoxin in Alzheimer's disease. Neurobiol Aging 25:569–580

Koeck T, Levison B, Hazen SL, Crabb JW, Stuehr DJ, Aulak KS (2004) Tyrosine nitration impairs mammalian aldolase A activity. Mol Cell Proteomics 3:548–557

Kurt MA, Davies DC, Kidd M, Duff K, Howlett DR (2003) Hyperphosphorylated tau and paired helical filament-like structures in the brains of mice carrying mutant amyloid precursor protein and mutant presenilin-1 transgenes. Neurobiol Dis 14:89–97

Lafon-Cazal M, Culcasi M, Gaven F, Pietri S, Bockaert J (1993) Nitric oxide, superoxide and peroxynitrite: putative mediators of NMDA-induced cell death in cerebellar granule cells. Neuropharmacology 32:1259–1266

Levine RL, Wehr N, Williams JA, Stadtman ER, Shacter E (2000) Determination of carbonyl groups in oxidized proteins. Methods Mol Biol 99:15–24

Lovell MA, Markesbery WR (2001) Ratio of 8-hydroxyguanine in intact DNA to free 8-hydroxyguanine is increased in Alzheimer disease ventricular cerebrospinal fluid. Arch Neurol 58:392–396

Lubec G, Nonaka M, Krapfenbauer K, Gratzer M, Cairns N, Fountoulakis M (1999) Expression of the dihydropyrimidinase related protein 2 (DRP-2) in Down syndrome and Alzheimer's disease brain is downregulated at the mRNA and dysregulated at the protein level. J Neural Transm Suppl 57:161–177

Luchsinger JA, Reitz C, Honig LS, Tang MX, Shea S, Mayeux R (2005) Aggregation of vascular risk factors and risk of incident Alzheimer disease. Neurology 65:545–551

Mark RJ, Lovell MA, Markesbery WR, Uchida K, Mattson MP (1997) A role for 4-hydroxynonenal, an aldehydic product of lipid peroxidation, in disruption of ion homeostasis and neuronal death induced by amyloid beta-peptide. J Neurochem 68:255–264

Markesbery WR (1997) Oxidative stress hypothesis in Alzheimer's disease. Free Radic Biol Med 23:134–147

Markesbery WR, Kryscio RJ, Lovell MA, Morrow JD (2005) Lipid peroxidation is an early event in the brain in amnestic mild cognitive impairment. Ann Neurol 58:730–735

Masliah E, Mallory M, Hansen L, DeTeresa R, Alford M, Terry R (1994) Synaptic and neuritic alterations during the progression of Alzheimer's disease. Neurosci Lett 174:67–72

Mattson MP, Parin J, Begley JG (1998) Amyloid beta-peptide induces apoptosis-related events in synapses and dendrites. Brain Res 807:167–176

McGeer EG, McGeer PL (2003) Inflammatory processes in Alzheimer's disease. Prog Neuropsychopharmacol Biol Psychiatry 27:741–749
Meda L, Cassatella MA, Szendrei GI, Otvos L Jr., Baron P, Villalba M, Ferrari D, Rossi F (1995) Activation of microglial cells by beta-amyloid protein and interferon-gamma. Nature 374: 647–650
Meier-Ruge W, Iwangoff P, Reichlmeier K (1984) Neurochemical enzyme changes in Alzheimer's and Pick's disease. Arch Gerontol Geriatr 3:161–165
Messier C, Gagnon M (1996) Glucose regulation and cognitive functions: relation to Alzheimer's disease and diabetes. Behav Brain Res 75:1–11
Messier C, Teutenberg K (2005) The role of insulin, insulin growth factor, and insulin-degrading enzyme in brain aging and Alzheimer's disease. Neural Plast 12:311–328
Mohmmad Abdul H, Butterfield DA (2005) Protection against amyloid beta-peptide (1-42)-induced loss of phospholipid asymmetry in synaptosomal membranes by tricyclodecan-9-xanthogenate (D609) and ferulic acid ethyl ester: implications for Alzheimer's disease. Biochim Biophys Acta 1741:140–148
Mohmmad Abdul H, Wenk GL, Gramling M, Hauss-Wegrzyniak B, Butterfield DA (2004) APP and PS-1 mutations induce brain oxidative stress independent of dietary cholesterol: implications for Alzheimer's disease. Neurosci Lett 368:148–150
Moskovitz J, Yim MB, Chock PB (2002) Free radicals and disease. Arch Biochem Biophys 397:354–359
Mucke L, Masliah E, Johnson WB, Ruppe MD, Alford M, Rockenstein EM, Forss-Petter S, Pietropaolo M, Mallory M, Abraham CR (1994) Synaptotrophic effects of human amyloid beta protein precursors in the cortex of transgenic mice. Brain Res 666:151–167
Nacmias B, Piccini C, Bagnoli S, Tedde A, Cellini E, Bracco L, Sorbi S (2004) Brain-derived neurotrophic factor, apolipoprotein E genetic variants and cognitive performance in Alzheimer's disease. Neurosci Lett 367:379–383
Nagy Z, Esiri MM, Cato AM, Smith AD (1997) Cell cycle markers in the hippocampus in Alzheimer's disease. Acta Neuropathol (Berl) 94:6–15
Okada K, Wangpoengtrakul C, Osawa T, Toyokuni S, Tanaka K, Uchida K (1999) 4-Hydroxy-2-nonenal-mediated impairment of intracellular proteolysis during oxidative stress. Identification of proteasomes as target molecules. J Biol Chem 274:23787–23793
Pastorino L, Sun A, Lu PJ, Zhou XZ, Balastik M, Finn G, Wulf G, Lim J, Li SH, Li X, Xia W, Nicholson LK, Lu KP (2006) The prolyl isomerase Pin1 regulates amyloid precursor protein processing and amyloid-beta production. Nature 440:528–534
Planel E, Miyasaka T, Launey T, Chui DH, Tanemura K, Sato S, Murayama O, Ishiguro K, Tatebayashi Y, Takashima A (2004) Alterations in glucose metabolism induce hypothermia leading to tau hyperphosphorylation through differential inhibition of kinase and phosphatase activities: implications for Alzheimer's disease. J Neurosci 24:2401–2411
Poon HF, Joshi G, Sultana R, Farr SA, Banks WA, Morley JE, Calabrese V, Butterfield DA (2004) Antisense directed at the Abeta region of APP decreases brain oxidative markers in aged senescence accelerated mice. Brain Res 1018:86–96
Radi R, Denicola A, Freeman BA (1999) Peroxynitrite reactions with carbon dioxide-bicarbonate. Methods Enzymol 301:353–367
Ramakrishnan P, Dickson DW, Davies P (2003) Pin1 colocalization with phosphorylated tau in Alzheimer's disease and other tauopathies. Neurobiol Dis 14:251–264
Rapoport SI (1999) In vivo PET imaging and postmortem studies suggest potentially reversible and irreversible stages of brain metabolic failure in Alzheimer's disease. Eur Arch Psychiatry Clin Neurosci 249(Suppl 3):46–55
Scheff SW, Price DA (2003) Synaptic pathology in Alzheimer's disease: a review of ultrastructural studies. Neurobiol Aging 24:1029–1046
Schopfer FJ, Baker PR, Freeman BA (2003) NO-dependent protein nitration: a cell signaling event or an oxidative inflammatory response? Trends Biochem Sci 28:646–654

Schutkowski M, Bernhardt A, Zhou XZ, Shen M, Reimer U, Rahfeld JU, Lu KP, Fischer G (1998) Role of phosphorylation in determining the backbone dynamics of the serine/threonine-proline motif and Pin1 substrate recognition. Biochemistry 37:5566–5575

Selkoe DJ (2001) Alzheimer's disease results from the cerebral accumulation and cytotoxicity of amyloid beta-protein. J Alzheimers Dis 3:75–80

Shen M, Stukenberg PT, Kirschner MW, Lu KP (1998) The essential mitotic peptidyl-prolyl isomerase Pin1 binds and regulates mitosis-specific phosphoproteins. Genes Dev 12: 706–720

Shringarpure R, Grune T, Davies KJ (2001) Protein oxidation and 20S proteasome-dependent proteolysis in mammalian cells. Cell Mol Life Sci 58:1442–1450

Sly WS, Hu PY (1995) Human carbonic anhydrases and carbonic anhydrase deficiencies. Annu Rev Biochem 64:375–401

Smith PD, O'Hare MJ, Park DS (2004) Emerging pathogenic role for cyclin dependent kinases in neurodegeneration. Cell Cycle 3:289–291

Smith MA, Richey Harris PL, Sayre LM, Beckman JS, Perry G (1997) Widespread peroxynitrite-mediated damage in Alzheimer's disease. J Neurosci 17:2653–2657

Stadtman ER, Levine RL (2003) Free radical-mediated oxidation of free amino acids and amino acid residues in proteins. Amino Acids 25:207–218

Stuehr DJ (1999) Mammalian nitric oxide synthases. Biochim Biophys Acta 1411:217–230

Sultana R, Boyd-Kimball D, Poon HF, Cai J, Pierce WM, Klein JB, Markesbery WR, Zhou XZ, Lu KP, Butterfield DA (2006c) Oxidative modification and down-regulation of Pin1 in Alzheimer's disease hippocampus: A redox proteomics analysis. Neurobiol Aging 27(7): 918–925

Sultana R, Boyd-Kimball D, Poon HF, Cai J, Pierce WM, Klein JB, Merchant M, Markesbery WR, Butterfield DA (2006d) Redox proteomics identification of oxidized proteins in Alzheimer's disease hippocampus and cerebellum: An approach to understand pathological and biochemical alterations in AD. Neurobiol Aging 27(11):1564–1576

Sultana R, Perluigi M, Butterfield DA (2006a) Redox proteomics identification of oxidatively modified proteins in Alzheimer's disease brain and in vivo and in vitro models of AD centered around Abeta(1-42). J Chromatogr B Analyt Technol Biomed Life Sci 833:3–11

Sultana R, Poon HF, Cai J, Pierce WM, Merchant M, Klein JB, Markesbery WR, Butterfield DA (2006b) Identification of nitrated proteins in Alzheimer's disease brain using a redox proteomics approach. Neurobiol Dis 22:76–87

Tohgi H, Abe T, Yamazaki K, Murata T, Ishizaki E, Isobe C (1999) Alterations of 3-nitrotyrosine concentration in the cerebrospinal fluid during aging and in patients with Alzheimer's disease. Neurosci Lett 269:52–54

Traykov L, Rigaud AS, Baudic S, Smagghe A, Boller F, Forette F (2002) Apolipoprotein E epsilon 4 allele frequency in demented and cognitively impaired patients with and without cerebrovascular disease. J Neurol Sci 203–204:177–181

Vanhanen M, Soininen H (1998) Glucose intolerance, cognitive impairment and Alzheimer's disease. Curr Opin Neurol 11:673–677

Varadarajan S, Kanski J, Aksenova M, Lauderback C, Butterfield DA (2001) Different mechanisms of oxidative stress and neurotoxicity for Alzheimer's A beta(1–42) and A beta(25–35). J Am Chem Soc 123:5625–5631

Varadarajan S, Yatin S, Aksenova M, Butterfield DA (2000) Review: Alzheimer's amyloid beta-peptide-associated free radical oxidative stress and neurotoxicity. J Struct Biol 130:184–208

Verdier Y, Penke B (2004) Binding sites of amyloid beta-peptide in cell plasma membrane and implications for Alzheimer's disease. Curr Protein Pept Sci 5:19–31

Visser PJ, Verhey FR, Ponds RW, Jolles J (2001) Diagnosis of preclinical Alzheimer's disease in a clinical setting. Int Psychogeriatr 13:411–423

Walsh DM, Klyubin I, Shankar GM, Townsend M, Fadeeva JV, Betts V, Podlisny MB, Cleary JP, Ashe KH, Rowan MJ, Selkoe DJ (2005) The role of cell-derived oligomers of Abeta in Alzheimer's disease and avenues for therapeutic intervention. Biochem Soc Trans 33: 1087–1090

Wang HY, Lee DH, D'Andrea MR, Peterson PA, Shank RP, Reitz AB (2000) beta-Amyloid(1-42) binds to alpha7 nicotinic acetylcholine receptor with high affinity. Implications for Alzheimer's disease pathology. J Biol Chem 275:5626–5632

Wang J, Markesbery WR, Lovell MA (2006) Increased oxidative damage in nuclear and mitochondrial DNA in mild cognitive impairment. J Neurochem 96:825–832

Winterbourn CC, Buss IH (1999) Protein carbonyl measurement by enzyme-linked immunosorbent assay. Methods Enzymol 300:106–111

Xia Y, Dawson VL, Dawson TM, Snyder SH, Zweier JL (1996) Nitric oxide synthase generates superoxide and nitric oxide in arginine-depleted cells leading to peroxynitrite-mediated cellular injury. Proc Natl Acad Sci U S A 93:6770–6774

Yatin SM, Varadarajan S, Butterfield DA (2000) Vitamin E Prevents Alzheimer's Amyloid beta-Peptide (1-42)-Induced Neuronal Protein Oxidation and Reactive Oxygen Species Production. J Alzheimers Dis 2:123–131

Yatin SM, Varadarajan S, Link CD, Butterfield DA (1999) In vitro and in vivo oxidative stress associated with Alzheimer's amyloid beta-peptide (1-42). Neurobiol Aging 20:325–330; discussion 339–342

Yu HL, Chertkow HM, Bergman H, Schipper HM (2003) Aberrant profiles of native and oxidized glycoproteins in Alzheimer plasma. Proteomics 3:2240–2248

Zamora R, Vodovotz Y, Aulak KS, Kim PK, Kane JM 3rd, Alarcon L, Stuehr DJ, Billiar TR (2002) A DNA microarray study of nitric oxide-induced genes in mouse hepatocytes: implications for hepatic heme oxygenase-1 expression in ischemia/reperfusion. Nitric Oxide 7:165–186

Zhou XZ, Kops O, Werner A, Lu PJ, Shen M, Stoller G, Kullertz G, Stark M, Fischer G, Lu KP (2000) Pin1-dependent prolyl isomerization regulates dephosphorylation of Cdc25C and tau proteins. Mol Cell 6:873–883

Oxidative Stress and Alzheimer Disease: Mechanisms and Therapeutic Opportunities

Siddhartha Mondragón-Rodríguez, Francisco García-Sierra, Gemma Casadesus, Hyoung-gon Lee, Robert B. Petersen, George Perry, Xiongwei Zhu, and Mark A. Smith

Abstract Oxidative stress is an early event in the development of Alzheimer disease (AD), preceding classic fibril formation which eventually deposits as amyloid-β senile plaques and neurofibrillary tangles composed of tau protein. Mitochondrial and metallic abnormalities are likely precursors of oxidative stress during the early stages of AD and, under degenerative conditions, the capacity of neurons to maintain redox balance decreases and results in mitochondrial dysfunction, a critical organelle involved in AD progression. Fibril formation, including amyloid-β production and tau phosphorylation, can be explained as a compensatory mechanism that may eventually enhance oxidative stress by increasing reactive oxygen species levels among many other free radicals. In this scenario, deposition of Aβ in the extracellular environment and tau protein in the intracellular environment can be explained as a redox imbalance with tragic consequences. If this hypothesis is correct, pharmacological treatments directed against amyloid-β or tau may not provide a benefit. In contrast, antioxidant strategies may be helpful in treating AD symptoms, although significant extended benefits have not been realized to date. In sum, the damage observed in the brain tissue of AD patients may be minimized with a healthy daily diet, exercise, and intellectual activities, factors that all reduce oxidative stress.

Keywords Alzheimer disease · Amyloid-beta · Antioxidants · Fibrils · Mitochondria · Oxidative stress

Contents

M.A. Smith (✉)
Department of Pathology, Case Western Reserve University, Cleveland, OH 44106, USA
e-mail: mark.smith@case.edu

J.P. Blass (ed.), *Neurochemical Mechanisms in Disease*,
Advances in Neurobiology 1, DOI 10.1007/978-1-4419-7104-3_18,

1 Introduction

Alzheimer disease (AD) is defined by insoluble filamentous aggregates known as senile plaques and neurofibrillary tangles (NFT), of which the major components are amyloid-β (Aβ) and tau protein, respectively (Wood et al., 1986; Arriagada et al., 1992; Goedert et al., 1998). These lesions accumulate in regions that are responsible for cognitive functions (Ball et al., 1988) and contribute towards the declining activities of daily living as well as the neuropsychiatric symptoms and behavioral changes seen in patients with AD. Definitive risk factors for AD include genetic predisposition (i.e., apolipoprotein E (ApoE), amyloid-β protein precursor (AβPP), presenilins (PS)), medical conditions, environment, and lifestyle (Smith, 1998; Wang et al., 2004a,b; Williamson and LaRusse, 2004; Lemos et al., 2009; Zawia et al., 2009).

Oxidative stress has also been strongly associated with the development of AD (Smith et al., 1997a, 1998, 1999; Paola et al., 2000; Smith et al., 2000a,b; Nunomura et al., 2001; Zhu et al., 2007). Therefore, it is not surprising that oxidative stress is related to the neurodegenerative process, inasmuch as it is known to affect, and result in, metabolic dysfunction, mitochondrial dysfunction, dysregulation of metal homeostasis, and alterations in the cell cycle, all of which contribute to the classical fibril aggregations, Aβ plaques and NFT. These structures, counter-intuitively, may be compensatory responses mounted to combat such oxidative stress (Lee et al., 2005; Nunomura et al., 2006; Zhu et al., 2006; Nunomura et al., 2007a; Lee et al., 2009a; Lu et al., 2009). If this scenario is correct, pharmacological treatments reducing fibril aggregation may be detrimental (Perry et al., 2000; Smith et al., 2002a).

The aim of this chapter is to evaluate the relationship between AD and oxidative stress and to consider how this knowledge may dictate treatment options for the clinical symptoms of the disease.

2 Oxidative Stress, Genetics, and Alzheimer Disease Pathology

Behavioral and cognitive decline in AD is accompanied by pathological accumulations of Aβ-containing senile plaques and tau-containing NFTs (Van Hoesen and

Hyman, 1990; Van Hoesen et al., 1991). AβPP and PS1 mutations result in heterogeneity in the clinical expression of neurological features during disease progression compared to sporadic AD, suggesting a genetic influence (Zekanowski et al., 2006; Larner and Doran, 2009). Although an accurate cascade that charts the effect of mutations through to the progression of dementia at the end-stage is an area of extensive study, one common feature in all AD cases is Aβ deposition resulting from the cleavage of AβPP (Younkin, 1994). Mutations are not just limited to AβPP and the PS genes. Recently, mutations in the tau gene, found in familial cases of frontotemporal dementia, which are characterized by an intracellular accumulation of polymerized tau as the primary cause of neurodegeneration, also exhibit increased $A\beta_{40}$ and $A\beta_{42}$ deposition. These data bring new support for a relationship between tau gene mutations and Aβ deposition (Vitali et al., 2004).

Despite all of the evidence that links AD to a genetic component, a detailed mechanism leading from one event to the other remains elusive. In this regard, some authors have suggested that early life exposure to the xenobiotic metal lead (Pb) enhanced the expression of genes associated with AD, repressed the expression of others, and result in an increased burden of oxidative DNA damage in the aged brain; the mechanism acts through either hypomethylation or hypermethylation of DNA (Zawia et al., 2009).

The association between genetics and neurodegeneration is also supported by predisposing risk factors such as the ε allele of the ApoE gene. In particular, the ε4 allele has been strongly correlated with increased risk of AD, whereas the ε3 allele is not (Basurto-Islas et al., 2008). In addition, the ε4 allele of the ApoE gene has been associated with increased vascular Aβ deposition, whereas, in contrast, the ε2 allele is associated with cerebral amyloid angiopathy (CAA) related to intracerebral hemorrhage (Hamaguchi and Yamada, 2008). However, the genetic association is not just the result of coding sequence changes; different levels of gene expression may also be involved, that is, the hypo- or hypermethylation of DNA, as mentioned earlier, that also contribute to the disease (de Carvalho et al., 2000; Speranca et al., 2008).

Interestingly, deficient or altered energy metabolism that could change the overall oxidative microenvironment in neurons during the pathogenesis and progression of AD, leading to alterations in mitochondrial enzymes and in glucose metabolism in AD brain tissue, has been found in AD patients that also carried the ApoE ε4 allele (Mosconi et al., 2008).

On the other hand, oxidative stress has also been found to induce PS1 transcription, thereby promoting production of pathological levels of Aβ in AD (Tamagno et al., 2008). Indeed, it has been proposed that the pathophysiology of oxidative stress is reflected in damage to tissue biomolecules, including lipids, proteins, and DNA by free radicals (Migliore and Coppede, 2009).

Clearly, although not fully understood, the genetic component and oxidative stress may act synergistically or cooperatively, creating a pathological condition that contributes to the protein deposition seen in AD, although the precise mechanisms involved are unknown.

2.1 Fibrillary Aggregates and Oxidative Stress

Senile plaques and NFTs are also present in a considerable percentage of elderly non-AD brains. Although these markers constitute the criteria for the diagnosis of AD, they do not always correlate with cognitive decline (Davis et al., 1999; Lee et al., 2007; Castellani et al., 2008). In fact, studies have appropriately raised the question whether senile plaque deposition has any relationship to the cognitive decline observed in AD (Dickson et al., 1992) inasmuch as Aβ deposition shows no correlation with neuronal loss (Gomez-Isla et al., 1997). In contrast, cognitive decline correlates well with NFT density (Garcia-Sierra et al., 2001), although the degree of neuronal loss greatly exceeds the amount of NFTs (Gomez-Isla et al., 1997). Nonetheless, neuronal loss has been halted and memory defects reversed in transgenic models of tau mutations by turning off mutant tau expression (Santacruz et al., 2005).

Despite the debate over the relationship of fibril deposition and clinical symptoms, their role in neurodegeneration remains as strong as ever. Regarding such a relationship, it has been found that Aβ-induced nitro-oxidative damage promotes the nitrotyrosination of the glycolytic enzyme triosephosphate isomerase in human neuroblastoma cells, suggesting an oxidative stress pathway as the molecular mediator (Guix et al., 2009). Indeed, it has been reported that behavioral stress aggravates AD pathology via generation of metabolic oxidative stress and MMP-2 downregulation in AD mouse models (Lee et al., 2009b).

Clearly, oxidative stress plays a crucial role in neurodegeneration (Nunomura et al., 2007b; Sajad et al., 2009), however, the mechanism by which amyloid deposition causes oxidative stress is the subject of extensive study. In vitro studies have shown that monomeric Aβ1-40 and Aβ1-42 exhibit antioxidant activity in cultured neurons (Zou et al., 2002). Furthermore, Aβ was found to be one of the most important antioxidants in cerebrospinal fluid (CSF). Indeed, recent reports suggest that the fibrillary forms typically observed in senile plaques and NFTs may actually be neuroprotective, because Aβ seems to inhibit oxidation by chelating metal ions, a function that may also equally apply to tau protein (Kontush et al., 1996; Kontush, 2001; Smith et al., 2002b; Caughey and Lansbury, 2003; Walsh and Selkoe, 2004). Nevertheless, this fibrillar formation seems to be accompanied by further compensatory changes that ultimately result in additional oxidative insult during the disease (Lee et al., 2004, 2005).

Based on these observations, it is clear that a close and highly complicated relationship exists between fibril deposition and oxidative stress during neurodegeneration.

2.2 Amyloid-β Peptide

The Aβ plaque was one of the first identified hallmarks of AD. It has been proposed that these structures are the main cause of AD inasmuch as they appear in the limbic area, which is affected in AD (Giannakopoulos et al., 2003), and it is thought that

Aβ binds to neurons activating apoptotic pathways that eventually contribute to neurodegeneration (Gamblin et al., 2003). Furthermore, Aβ1-42 has been determined to be the most toxic form to cultured neurons, because the Aβ1-42 oligomer was able to activate the apoptotic pathway leading to caspase activation (Yankner et al., 1990; Dahlgren et al., 2002; Gamblin et al., 2003; Yao et al., 2005). However, numerous studies support the idea that an oxidative event is critical for Aβ toxicity (Pratico et al., 2002): (1) Aβ staging does not distinguish between cognitive changes and dementia; and (2) Aβ shows overlap among the various clinical dementias (Gold et al., 2001). Along these lines, plaques have been further classified into subtypes such as senile, diffuse, and neuritic, and, in this regard, it is generally accepted that diffuse plaques (Aβ deposits without cores or a neuritic component) are merely decorative in nature, having little impact, if any, on cognitive function, whereas neuritic plaques are more pathogenic. It is thought that diffuse plaques appear during the early preclinical stages of the disease and eventually mature into a defined structure, the neuritic plaque. However, this hypothesis and the role of plaque maturation during AD pathogenesis remain controversial. Here again, no clinical correlation has been found between plaques and the degree of cognitive decline (Arriagada et al., 1992).

In addition to being a pathological hallmark of AD, Aβ plays an important role in normal cell development and maintenance (Atwood et al., 2003). Some propose that diffuse amyloid plaques may be a compensatory response aimed at reducing oxidative stress, because there is a negative correlation between Aβ deposition and oxidative damage in Down syndrome patients, as well as AD patients (Nunomura et al., 2000; Smith et al., 2000b; Nunomura et al., 2001). Aβ is also highly involved in the neurodegenerative process, although its contribution remains debatable. Some authors have suggested that Aβ leads to depletion of cellular choline stores and consequently contributes to the selective vulnerability of cholinergic neurons in AD (Allen et al., 1997). Indeed, a pathological role has been attributed to Aβ accumulation in the brain of AD patients (Wang et al., 2007b). In point of fact, the majority of efforts for developing AD therapeutics have been directed towards eliminating the fibrillar form of aggregated Aβ (Asuni et al., 2006; Matsuda et al., 2009), although there is debate about the potential efficacy of this approach (Lee et al., 2006; Shah et al., 2008). Why should we doubt the effectiveness? The answer is far from simple, however, growing evidence supports a nonpathological role for the Aβ peptide (Lee et al., 2007). In fact, it has been proposed that Aβ deposition may be a primary antioxidant defense indicating that Aβ expression is an adaptive response rather than a cause of AD (Castellani et al., 2006; Lee et al., 2007; Castellani et al., 2008).

Chelation of metals by Aβ may play a role in oxidative stress (Dong et al., 2003; House et al., 2004). Specifically, the methionine at residue 35 of the Aβ sequence can scavenge free radicals and also reduce metals to their high-activity, low-valency form, showing both pro- and antioxidant properties (Cuajungco et al., 2005). On the other hand, Aβ is able to initiate oxidation of different biomolecules; for example, Aβ induces the peroxidation of membrane lipids and lipoproteins, which generates H_2O_2 and hydroxynonenal in neurons, and damages DNA (Huang et al., 1999a; Xu et al., 2001; Butterfield et al., 2002).

Despite the controversy over the toxicity of AβPP and Aβ, it can be inferred from the current data that Aβ may be playing, as proposed, a compensatory role that eventually becomes pathological by activating oxidative stress pathways, although more data are needed to support this hypothesis.

2.3 Tau Protein

The main function of the tau protein is to stabilize microtubules; moreover, tau may also be involved in signal transduction, organelle transport, and cell growth, as well as anchoring of enzymes (Johnson and Hartigan, 1999; Johnson and Jenkins, 1999; Sontag et al., 1999). Furthermore, tau has a role in modulating axonal morphology and polarity (Buee et al., 2000). Therefore, it is not surprising that microtubule abnormalities and tau phosphorylation are also associated with AD because cell cycle re-entry, an early feature in AD (Webber et al., 2005; Evans et al., 2007; Lee et al., 2009a), and candidate tau kinases that have been implicated in cell cycle control such as Cdk2, Cdk5, Cdc2, and MAPK are all increased in AD in a topographical manner that overlaps with hyperphosphorylated tau (Vincent et al., 1997; Swatton et al., 2004; Wang et al., 2004c, 2007a), which has also been proposed as an early event in tau-mediated pathology (Mondragon-Rodriguez et al., 2008a,b). A tau transgenic mouse line, THY-Tau22, expressing a mutated human tau protein that has been linked to frontotemporal dementia with Parkinsonism linked to chromosome-17 displays increased neurogenesis associated with tau hyperphosphorylation. Later, cell cycle events, abnormal tau phosphorylation, and tau aggregation occur preceding neuronal death and neurodegeneration (Schindowski et al., 2008).

Tau is highly phosphorylated; at least 30 phosphorylation sites have been described, the majority are Ser-Pro and Thr-Pro motifs. Some of these motifs seem to be crucial to the development of AD. For example, phosphorylation at Ser262 mediated by P70S6 kinase dramatically reduces the affinity of tau for microtubules in vivo (Hamdane et al., 2003; Zhou et al., 2008b). Furthermore, phosphorylation at Ser202 appears to enhance tau polymerization; and phosphorylation at two sites (Ser202-Thr205) makes filament formation more sensitive to small changes in tau concentration (Rankin et al., 2005). Thus, phosphorylation outside the microtubule-binding domains, such as Ser202 and Thr205, may strongly influence tubulin assembly by modifying the affinity between microtubules and tau as well as tau itself (Alonso et al., 1996, 2001). Moreover, phosphorylation has been found to regulate axonal transport by controlling tau binding to kinesin (Cuchillo-Ibanez et al., 2008).

Regarding phosphorylation, members of the stress-activated protein kinase (SAPK) family have been shown to phosphorylate tau in vitro. SAPKIγ (or Jun N-Terminal kinase, JNK1), SAPK2a (p38), SAPK2β (p38 β), SAPK3 (p38γ), and SAPK4 can phosphorylate tau, although SAPK3 and SAPK4 are the most efficient in vitro (Goedert et al., 1997; Reynolds et al., 1997). In AD, hyperphosphorylated tau accumulates in neurons, and being a constitutive element of NFTs, eventually leads to degeneration (Alonso et al., 2001; Garcia and Cleveland, 2001).

The abnormal phosphorylation of tau associated with AD may be related to either an increase in kinase activity (glycogen synthase kinase 3β, cyclin-dependent kinase-5, p42/44 MAP kinase, p38 MAPK, stress-activated protein kinases, mitotic protein kinases) or a decrease in phosphatase activity (protein phosphatases 1, 2a, 2b), suggesting soluble tau as a cause of neuronal degeneration (Buee et al., 2000; Tian and Wang, 2002; Chen et al., 2008; Liu et al., 2008; Yang et al., 2008; Zhou et al., 2008a).

Hyperphosphorylation of tau has also been proposed to be protective (Lee et al., 2005); phosphorylation may prevent advanced tau processing, that is, cleavage of tau at site Asp421, an event that enhances fibril formation (Guillozet-Bongaarts et al., 2006). Phosphorylation plays a pivotal role in redox balance, so it is perhaps not surprising that oxidative stress, through activation of MAP kinase pathways, leads to phosphorylation of tau (Zhu et al., 2000, 2001a,b). MAP kinase activation and heme oxygenase (HO-1) induction may be but a few of the many responses that result from lipid peroxidation. Consequently, oxidative damage can no longer be considered an end-stage event, but rather a signal of an underlying change of state that is related to the phosphorylation of tau.

3 Oxidative Stress and Metabolism

Oxidative damage has been found in several entities that are critical for neuronal structure and functional integrity. It is possible that under degenerative conditions the capacity of cells to maintain redox balance decreases resulting in mitochondrial dysfunction, a critical organelle involved in AD progression (Cash et al., 2002; Zhu et al., 2006). A significant body of evidence supports the hypothesis that mitochondrial and metallic abnormalities are direct precursors of oxidative stress during the early stages of AD (Halliwell, 1999; Nunomura et al., 2000, 2001; Atamna, 2004; Zhu et al., 2004b, 2007). Also, increased intracellular iron may promote oxidative stress/free radical damage in vulnerable neurons (Casadesus et al., 2004; Zhu et al., 2004a, 2007; Dwyer et al., 2009). Interestingly, the loss of iron homeostasis directly affects mitochondrial function (Lu et al., 2009) and the proximal causes of mitochondrial abnormalities likely involve re-entry into the cell cycle (Cash et al., 2002; Zhu et al., 2006). Recent studies have also shown that reactive oxygen species generated by mitochondria regulate p53 activity, which in turn regulates cell-cycle progression and DNA repair and, in cases of irreparable DNA damage, executes programmed cell death (Holley and St Clair, 2009).

Oxidative stress increases during aging, in parallel with the increased susceptibility to several neurodegenerative diseases including AD. In AD, NFT accumulation within the neuronal cytoplasm is associated with impaired axonal transport of mitochondria between the cell nucleus and synapse, which leads to severe energy dysfunction and an imbalance in the generation of reactive oxygen species (ROS) and reactive nitrogen species (RNS) (Smith et al., 1997b; Rapoport, 2003). DNA and RNA oxidation are marked by increased levels of 8-hydroxy-2-deoxyguanosine (8OHdG) and 8-hydroxyguanosine (8OHG), respectively (Nunomura et al., 1999,

2000, 2001, 2004, 2007b). Meanwhile protein oxidation is marked by elevated levels of protein carbonyls and nitration of tyrosine residues (Smith et al., 1995). Modification of sugars by glycation and glycoxidation is another component of the disease, although the levels of these modifications decrease as the disease progresses to advanced AD (Smith et al., 1994; Castellani et al., 2001; Perry et al., 2002). These data support the hypothesis that increased oxidative damage is an early event in the progression AD.

3.1 Energy Utilization

The role of mitochondrial and redox metal ions as potential neuronal compensatory responses against oxidative stress remains unclear, nevertheless, both are important elements of the energy metabolism deficiency in AD (Blass and Gibson, 1999; Blass et al., 2000). Glucose and oxygen are the primary sources of energy in neurons (Erecinska and Silver, 1989) and there is reduced glucose metabolism in the tempoparietal and posterior cingulate cortex in AD (Drzezga et al., 2003). Furthermore, reduced glucose metabolism in limbic and associative areas of the brain have been reported in AD cases with ApoE ε4, reflecting its genetic influence (Kamino et al., 2000; Mosconi et al., 2004). Other features of AD are increased oxygen consumption (Hoyer, 1998), atrophy in the vasculature (Praprotnik et al., 1996; Perry et al., 1998, 2003), and reduced cerebral glucose transport activity (Kalaria et al., 1988; Perry et al., 2003). The reduction of ATP production from glucose by approximately 50% at the onset of sporadic AD is further evidence of the glucose metabolism imbalance (Hoyer, 1992). All these data support the involvement of altered glucose metabolism in the early pathophysiology of AD. Furthermore, the activity of many enzymes involved in metabolism is decreased in AD, such as glutamine synthetase, creatine kinase, and pyruvate dehydrogenase (Sorbi et al., 1983; Gibson et al., 2000). On the other hand, the activities of succinate dehydrogenase (complex II) and malate dehydrogenase have been reported to increase, suggesting a coordinated alteration of metabolic activity in the mitochondria (Bubber et al., 2005).

3.2 Mitochondria

Due to the high oxygen consumption rate and relative paucity of antioxidant enzymes compared with other organs, the brain is especially vulnerable to free radical damage (Floyd and Hensley, 2002; Mattson et al., 2002). The major source of free radicals [hydrogen peroxide (H_2O_2), hydroxyl ($^{\cdot}OH$) and superoxide ($O_2^{\cdot-}$)] is oxidative phosphorylation (Wallace, 1999). The reactive oxygen species generated by mitochondria have many targets such as lipids, proteins, RNA, DNA, and mitochondrial DNA (mtDNA), which due to the lack of histones, becomes a vulnerable target of oxidative stress. Indeed, nucleic acid oxidation is also deemed a hallmark of AD (Moreira et al., 2008).

These findings establish a link between oxidative stress and mitochondria, creating a pathological feedback loop, although in other cell types such as astrocytes, which are known to regulate glutathione availability, mitochondria are unaffected (Pope et al., 2008).

Mitochondria are also susceptible to apoptotic pathways, mediated through members of (a) Bcl-2 family (Bid, Bad, and Bax), (b) death receptor pathway, and (c) endoplasmic-reticulum-specific pathway. The final result of these three pathways is the activation of caspases (Ferri and Kroemer, 2001). It has been reported that the neurons exhibiting increased oxidative damage in AD are coincident with striking and significant increase in cytochrome oxidases and mtDNA (Hirai et al., 2001). Cytochrome oxidase is found in the neuronal cytoplasm and mtDNA in vacuoles associated with lipofuscin. Furthermore, lipoic acid antisera specifically mark lipofuscin in AD, suggesting increased autophagocytosis (Moreira et al., 2007). Also, a significant reduction of intact mitochondria, as well as a reduction in microtubules, is found in AD (Cash et al., 2003). Oxidative stress markers, mtDNA deletion, and abnormalities in mitochondrial structure in the vascular walls of AD cases are also increased (Aliev et al., 2004). In addition to changes in mitochondrial enzymes, mitochondrial structure, localization, and mobility are all changed in AD. Specifically, markers for mitochondrial fission and fusion are altered in models of AD and are affected by Aβ oligomers, which impair mitochondrial function, leading to energy hypometabolism and elevated reactive oxygen species production (Wang et al., 2007b, 2008a,b).

In sum, it is apparent that mitochondria and oxidative stress are closely related and together, through different pathways, contribute to the neurodegenerative process.

3.3 Metals

The other important source of free radicals comes from redox-active metals. Strikingly, NFT and Aβ plaques were found coincident with overaccumulation of iron in the hippocampus, cerebral cortex, and basal nucleus of Meynert (Lovell et al., 1998). Iron is an important cause of oxidative stress in AD through the Fenton reaction. Aβ is also a substrate for hydroxyl radicals and, in the presence of iron, it has been reported that cleavage and synthesis of Aβ are promoted (Atwood et al., 1999; Rogers et al., 2002).

Copper can also participate in the Fenton reaction to generate ROS (Huang et al., 1999b; Finefrock et al., 2003) and is affected by Aβ (Hayashi et al., 2007; Nakamura et al., 2007). Furthermore, iron and copper in their redox competent states are bound to NFT and Aβ deposits (Smith et al., 1997a; Sayre et al., 2000). Nevertheless, the exact role of copper and iron in their redox competent states remains to be elucidated. In this regard, it is been suggested that mitochondrial dysfunction, acting in concert with cytoskeletal pathology, serves to increase redox-active heavy metals and initiates a cascade of abnormal events culminating in AD pathology (Castellani et al., 2004).

4 Current and Future Pharmacological Treatments for Alzheimer Disease

An enormous effort has been devoted to developing treatments for the clinical symptoms of AD. Cholinesterase inhibitors, antiglutamatergic treatment, β- and γ-secretase inhibitors, cholesterol-lowering drugs, Aβ immunotherapy, nonsteroidal anti-inflammatory drugs, Aβ channel blockers, hormonal replacement therapy, and antioxidant therapies are the current pharmacological options for treatment of AD (Bartus et al., 1982; Schenk et al., 1999; Pratico and Trojanowski, 2000; Dovey et al., 2001; Golde and Eckman, 2001; Nunomura et al., 2001; Cholerton et al., 2002; Farlow et al., 2003; Perry et al., 2003; Moreira et al., 2006; Diaz et al., 2009; Lopez-Bastida et al., 2009; Moriguchi et al., 2009).

Cholinergic transmission plays a fundamental role in cognitive function. Due to a reported decrease of cholinergic function in the neocortex and hippocampus, several acetylcholinesterase inhibitors have been used to treat AD in the past few years, such as tacrine, donepezil, rivastigmine, and galactine (Lleo et al., 2005; Moriguchi et al., 2009). Acetylcholinesterase inhibitors can delay cognitive impairment for at least six months (Takeda et al., 2006). Indeed, these drugs have been found to possess some antioxidant properties as well as neuroprotective properties (Fernandez-Bachiller et al., 2009). However, they were recently shown to have opposing effects on blood pressure and cerebral perfusion (Claassen et al., 2009).

The most important excitatory neurotransmitter in the central nervous system (CNS) is glutamate, reported to regulate Ca^{2+} accumulation through excessive activation of NMDA receptors. Memantine is an NMDA antagonist that has been used to treat neurological syndromes and cognitive dysfunction (Farlow et al., 2003). A small beneficial effect of memantine was observed at six months of treatment in moderate to severe AD (Areosa et al., 2005).

Numerous studies have supported the idea that an oxidative event is critical in AD. It is thought that Aβ is capable of generating reactive oxygen species. However, the source of Aβ toxicity has yet to be established (Rottkamp et al., 2001). Although deposition of Aβ into senile plaques is by no means specific to AD, several treatments against Aβ deposition are currently in use. In this context β- and γ-secretase inhibitors have been used therapeutically. Although the main goal is to block the production of Aβ (Josien, 2002), γ-secretase inhibitors also block the proteolytic processing and function of Notch, which is essential for brain morphogenesis (Louvi et al., 2004). In contrast, no side effects have been found with β-secretase inhibitors in knock-out mice (Dominguez et al., 2005). Indeed, novel therapeutic strategies contemplate the use of dual effectors, such as the new dual inhibitor of acetylcholinesterase and β-secretase (Zhu et al., 2009). The preliminary data in transgenic mice looks promising.

Cholesterol has been reported to negatively regulate α-secretase, whereas β- and γ-secretase activities are positively regulated by cholesterol (Golde and Eckman, 2001). Disappointingly, a three-year trial with pravastatin (a cholesterol-lowering drug) showed no significant effect on cognitive function in elderly individuals (Shepherd et al., 2002).

Aβ immunization is a novel approach to AD treatment. Simple immunization with Aβ42 in transgenic mice blocked deposition of amyloid and cleared existing amyloid (Schenk et al., 1999; Boche et al., 2008). Recently, increased dendritic spine formation in PDAPP transgenic mice was found after amyloid clearance, suggesting functional recovery of neural circuits (Spires-Jones et al., 2009). However, active vaccination with Aβ in patients with mild to moderate AD in a Phase II trial showed a CNS inflammatory response (Monsonego et al., 2003). The meningoencephalitis that occurred in some of the patients was reported to be unrelated to the anti-Aβ antibody titer (Hock et al., 2003; Wilcock and Colton, 2008), but rather to the involvement of a specific T-cell inflammatory response (Ferrer et al., 2004). Others have speculated that the adverse effects were due to external contamination during lumbar punctures, which were required as part of the protocol (Dodel et al., 2003). Furthermore, Fox and colleagues (Fox et al., 2005) showed, by standard volumetry, that the brains of vaccinated individuals lost tissue and gained ventricular volume faster than did brains of individuals vaccinated with placebo. If, as we suspect, Aβ is functioning as an antioxidant, removal of Aβ by immunization and/or other methods may actually exacerbate disease progression.

Extensive epidemiological data suggest a gender-based predisposition that is specific to AD such that there is a higher prevalence (Jorm et al., 1987; Breitner et al., 1988; Rocca et al., 1991; McGonigal et al., 1993) and incidence (Jorm and Jolley, 1998) of AD in women. Hormone replacement therapy protection against AD is restricted to administration during a "critical period" that constitutes the climacteric years. Efficacy is variable when administered after such time (Rapp et al., 2003) or during the latent preclinical stage of AD, which usually occurs much later in life (Resnick and Henderson, 2002; Craig and Murphy, 2009; Henderson, 2009; Hogervorst et al., 2009). The mechanism(s) relating hormones and pathology is (are) being actively pursued; it has been proposed that luteinizing hormone may contribute to AD pathology through an amyloid-dependent mechanism (Casadesus et al., 2006; Webber et al., 2007; Berry et al., 2008), that again underlies the bias towards females developing AD. In this regard, previous reports demonstrate a twofold increase in the gonadotropin LH in AD patients compared to age-matched control subjects (Bowen et al., 2000; Short et al., 2001; Webber et al., 2007). Thus, the presence of functional LH receptors was, at least in part, responsible for the cognitive decline seen in transgenic mice (Casadesus et al., 2007). Additionally, it has been reported that increased serum LH, rather than lower serum-free testosterone, is associated with the accumulation of Aβ in plasma (Verdile et al., 2008). Therefore, therapeutic strategies that are targeted towards decreasing LH may prove successful in treatment of AD (Webber et al., 2006, 2007; Casadesus et al., 2008).

Because chronic inflammation is associated with AD, several anti-inflamatory drugs have been tested including celecoxib (sc-125; sc-560), r-flurbiprofen, naproxen, and rocoxib (Szekely et al., 2004). Unfortunately, no consistent improvement in AD symptoms after treatment has been reported.

The involvement of oxidative stress in AD has opened a new door for potential therapeutic targets. In this regard, several antioxidants are currently in clinical trials such as Idebenone, α-Lipoic acid, acetyl-L-carnitine (ALC), vitamin E, vitamin C, flavonoids, β-carotene, gingko biloba, and metal-chelating agents. Idebenone is a metabolic antioxidant and is normally synthesized as part of the mitochondrial oxidative phosphorylation system. Improvements in clinical status after treatment with idebenone have been shown in a dose-dependent manner compared to placebo and tacrine (Thal et al., 2003).

α-Lipoic acid is another metabolic antioxidant that can recycle other antioxidants such as glutathione. Patients treated with α-Lipoic acid exhibited stabilization of cognitive measures (Hager et al., 2001). Acetyl-L-carnitine (ALC) is another metabolic antioxidant that acts as an intracellular carrier of acetyl groups across the inner mitochondrial membrane. Treatment with ALC showed a 38% response rate, and 50% when combined with acetyl cholinesterase inhibitors (Montgomery et al., 2003).

Vitamins, flavonoids, and terpenoids are examples of direct antioxidants (McShea et al., 2008; Ramiro-Puig et al., 2009). Vitamin E and selegiline appear to delay the time of progression to severe dementia in AD patients (Sano et al., 1997; Grundman, 2000). Vitamin E is the most important lipid-soluble chain-breaking natural antioxidant in mammalian cells and is able to cross the blood–brain barrier and accumulate at therapeutic levels in the brain, where it reduces lipid peroxidation (Veinbergs et al., 2000). In a cross-sectional study of 4809 elderly, decreasing serum levels of vitamin E per unit of cholesterol were consistently associated with decreasing cognitive function, whereas serum levels of vitamins A and C, β-carotene, and selenium were not associated with poor memory performance (Perkins et al., 1999). The Chicago Health and Aging Project with samples of 2889 community residents aged 65–102 years found that supplementary or dietary intake of vitamin E, but not vitamin C or carotenes, was inversely related to cognitive decline (Morris et al., 2002b). However, data from prospective studies relating intake of vitamin E and risk of AD are conflicting. The Chicago Health and Aging Project found that dietary, but not supplementary, intake of vitamin E was associated with a lowered risk of AD only among noncarriers of the ApoE ε4 allele (Morris et al., 2002a). Furthermore, in the Washington Heights–Inwood Columbia Aging Project, no association was found between dietary or supplementary intake of vitamin E and a decreased risk of AD (Luchsinger et al., 2003). The lack of efficacy of vitamin E in preventing the progression from MCI to AD indicates that single supplementary vitamin treatment has no significant affect in the secondary prevention of AD, which is consistent with the previous cohort studies on the progression to AD from the cognitively normal elderly. Gingko biloba contains, among other things, flavonoids and terpenoids. No differences in soluble Aβ and hippocampal Aβ were found in mice treated with Gingko biloba, although they showed improved spatial memory retention (Stackman et al., 2003).

Thus, re-examination of metal-chelating agents, classical indirect antioxidants, is warranted. NFT and senile plaques have been shown to contain redox-active transition metals (Smith et al., 1997a; Sayre et al., 2000) and may exert pro-oxidant or maybe antioxidant activities, depending on the local microenviroment. Aβ transgenic mice exhibited a 40% decrease in Aβ deposition after a nine-week treatment with clioquinol, a metal–protein chelating agent (Cherny et al., 2001). In two familial AD patients, increases in cerebral glucose metabolism were present after extended clioquinol treatment (Ibach et al., 2005).

5 Conclusion

As discussed in this chapter, oxidative stress plays an important role in AD, but more importantly, oxidative stress seems to be an early event, preceding classic fibril formation (Aβ plaques and NFT). Fibril formation can be explained as a compensatory mechanism that eventually enhances oxidative stress by increasing ROS levels among many other free radicals. In this scenario, deposition of Aβ in the extracellular environment and tau protein in the intracellular environment can be explained as an imbalance and tragic consequence. If this hypothesis is current, the current pharmacological treatments will not provide a solution, because the majority are directed against the fibril structures. In contrast, antioxidant strategies may be helpful in treating AD symptoms, although significant extended benefits have not been realized to date.

In sum, the damage observed in the brain tissue of AD patients that is enhanced by the fibril structures may be minimized with a healthy daily diet, exercise, and intellectual activities as the best option so far.

References

Aliev G, Smith MA, de la Torre JC, Perry G (2004) Mitochondria as a primary target for vascular hypoperfusion and oxidative stress in Alzheimer's disease. Mitochondrion 4:649–663

Allen DD, Galdzicki Z, Brining SK, Fukuyama R, Rapoport SI, Smith QR (1997) Beta-amyloid induced increase in choline flux across PC12 cell membranes. Neurosci Lett 234:71–73

Alonso AC, Grundke-Iqbal I, Iqbal K (1996) Alzheimer's disease hyperphosphorylated tau sequesters normal tau into tangles of filaments and disassembles microtubules. Nat Med 2:783–787

Alonso A, Zaidi T, Novak M, Grundke-Iqbal I, Iqbal K (2001) Hyperphosphorylation induces self-assembly of tau into tangles of paired helical filaments/straight filaments. Proc Natl Acad Sci U S A 98:6923–6928

Areosa SA, Sherriff F, McShane R (2005) Memantine for dementia. Cochrane Database Syst Rev 19(2):CD003154

Arriagada PV, Growdon JH, Hedley-Whyte ET, Hyman BT (1992) Neurofibrillary tangles but not senile plaques parallel duration and severity of Alzheimer's disease. Neurology 42:631–639

Asuni AA, Boutajangout A, Scholtzova H, Knudsen E, Li YS, Quartermain D, Frangione B, Wisniewski T, Sigurdsson EM (2006) Vaccination of Alzheimer's model mice with Abeta derivative in alum adjuvant reduces Abeta burden without microhemorrhages. Eur J Neurosci 24:2530–2542

Atamna H (2004) Heme, iron, and the mitochondrial decay of ageing. Ageing Res Rev 3:303–318

Atwood CS, Huang X, Moir RD, Tanzi RE, Bush AI (1999) Role of free radicals and metal ions in the pathogenesis of Alzheimer's disease. Met Ions Biol Syst 36:309–364

Atwood CS, Obrenovich ME, Liu T, Chan H, Perry G, Smith MA, Martins RN (2003) Amyloid-beta: a chameleon walking in two worlds: a review of the trophic and toxic properties of amyloid-beta. Brain Res Brain Res Rev 43:1–16

Ball MJ, Griffin-Brooks S, MacGregor JA, Nagy B, Ojalvo-Rose E, Fewster PH (1988) Neuropathological definition of Alzheimer disease: multivariate analyses in the morphometric distinction between Alzheimer dementia and normal aging. Alzheimer Dis Assoc Disord 2:29–37

Bartus RT, Dean RL 3rd, Beer B, Lippa AS (1982) The cholinergic hypothesis of geriatric memory dysfunction. Science 217:408–414

Basurto-Islas G, Luna-Munoz J, Guillozet-Bongaarts AL, Binder LI, Mena R, Garcia-Sierra F (2008) Accumulation of aspartic acid421- and glutamic acid391-cleaved tau in neurofibrillary tangles correlates with progression in Alzheimer disease. J Neuropathol Exp Neurol 67: 470–483

Berry A, Tomidokoro Y, Ghiso J, Thornton J (2008) Human chorionic gonadotropin (a luteinizing hormone homologue) decreases spatial memory and increases brain amyloid-beta levels in female rats. Horm Behav 54:143–152

Blass JP, Gibson GE (1999) Cerebrometabolic aspects of delirium in relationship to dementia. Dement Geriatr Cogn Disord 10:335–338

Blass JP, Sheu RK, Gibson GE (2000) Inherent abnormalities in energy metabolism in Alzheimer disease. Interaction with cerebrovascular compromise. Ann N Y Acad Sci 903:204–221

Boche D, Zotova E, Weller RO, Love S, Neal JW, Pickering RM, Wilkinson D, Holmes C, Nicoll JA (2008) Consequence of Abeta immunization on the vasculature of human Alzheimer's disease brain. Brain 131:3299–3310

Bowen RL, Isley JP, Atkinson RL (2000) An association of elevated serum gonadotropin concentrations and Alzheimer disease? J Neuroendocrinol 12:351–354

Breitner JC, Silverman JM, Mohs RC, Davis KL (1988) Familial aggregation in Alzheimer's disease: comparison of risk among relatives of early-and late-onset cases, and among male and female relatives in successive generations. Neurology 38:207–212

Bubber P, Haroutunian V, Fisch G, Blass JP, Gibson GE (2005) Mitochondrial abnormalities in Alzheimer brain: mechanistic implications. Ann Neurol 57:695–703

Buee L, Bussiere T, Buee-Scherrer V, Delacourte A, Hof PR (2000) Tau protein isoforms, phosphorylation and role in neurodegenerative disorders. Brain Res Brain Res Rev 33:95–130

Butterfield DA, Castegna A, Lauderback CM, Drake J (2002) Evidence that amyloid beta-peptide-induced lipid peroxidation and its sequelae in Alzheimer's disease brain contribute to neuronal death. Neurobiol Aging 23:655–664

Casadesus G, Garrett MR, Webber KM, Hartzler AW, Atwood CS, Perry G, Bowen RL, Smith MA (2006) The estrogen myth: potential use of gonadotropin-releasing hormone agonists for the treatment of Alzheimer's disease. Drugs R D 7:187–193

Casadesus G, Milliken EL, Webber KM, Bowen RL, Lei Z, Rao CV, Perry G, Keri RA, Smith MA (2007) Increases in luteinizing hormone are associated with declines in cognitive performance. Mol Cell Endocrinol 269:107–111

Casadesus G, Rolston RK, Webber KM, Atwood CS, Bowen RL, Perry G, Smith MA (2008) Menopause, estrogen, and gonadotropins in Alzheimer's disease. Adv Clin Chem 45: 139–153

Casadesus G, Smith MA, Zhu X, Aliev G, Cash AD, Honda K, Petersen RB, Perry G (2004) Alzheimer disease: evidence for a central pathogenic role of iron-mediated reactive oxygen species. J Alzheimers Dis 6:165–169

Cash AD, Aliev G, Siedlak SL, Nunomura A, Fujioka H, Zhu X, Raina AK, Vinters HV, Tabaton M, Johnson AB, Paula-Barbosa M, Avila J, Jones PK, Castellani RJ, Smith MA, Perry G (2003) Microtubule reduction in Alzheimer's disease and aging is independent of tau filament formation. Am J Pathol 162:1623–1627

Cash AD, Perry G, Ogawa O, Raina AK, Zhu X, Smith MA (2002) Is Alzheimer's disease a mitochondrial disorder? Neuroscientist 8:489–496

Castellani RJ, Harris PL, Sayre LM, Fujii J, Taniguchi N, Vitek MP, Founds H, Atwood CS, Perry G, Smith MA (2001) Active glycation in neurofibrillary pathology of Alzheimer disease: N(epsilon)-(carboxymethyl) lysine and hexitol-lysine. Free Radic Biol Med 31:175–180

Castellani RJ, Honda K, Zhu X, Cash AD, Nunomura A, Perry G, Smith MA (2004) Contribution of redox-active iron and copper to oxidative damage in Alzheimer disease. Ageing Res Rev 3:319–326

Castellani RJ, Lee HG, Perry G, Smith MA (2006) Antioxidant protection and neurodegenerative disease: the role of amyloid-beta and tau. Am J Alzheimers Dis Other Demen 21:126–130

Castellani RJ, Lee HG, Zhu X, Perry G, Smith MA (2008) Alzheimer disease pathology as a host response. J Neuropathol Exp Neurol 67:523–531

Caughey B, Lansbury PT (2003) Protofibrils, pores, fibrils, and neurodegeneration: separating the responsible protein aggregates from the innocent bystanders. Annu Rev Neurosci 26:267–298

Chen S, Li B, Grundke-Iqbal I, Iqbal K (2008) I1PP2A affects tau phosphorylation via association with the catalytic subunit of protein phosphatase 2A. J Biol Chem 283(16):10513–10521

Cherny RA, Atwood CS, Xilinas ME, Gray DN, Jones WD, McLean CA, Barnham KJ, Volitakis I, Fraser FW, Kim Y, Huang X, Goldstein LE, Moir RD, Lim JT, Beyreuther K, Zheng H, Tanzi RE, Masters CL, Bush AI (2001) Treatment with a copper-zinc chelator markedly and rapidly inhibits beta-amyloid accumulation in Alzheimer's disease transgenic mice. Neuron 30: 665–676

Cholerton B, Gleason CE, Baker LD, Asthana S (2002) Estrogen and Alzheimer's disease: the story so far. Drugs Aging 19:405–427

Claassen JA, Van Beek AH, Olde Rikkert MG (2009) Short review: acetylcholinesterase-inhibitors in Alzheimer's disease have opposing effects on blood pressure and cerebral perfusion. J Nutr Health Aging 13:231–233

Craig MC, Murphy DG (2009) Alzheimer's disease in women. Best Pract Res Clin Obstet Gynaecol 23:53–61

Cuajungco MP, Frederickson CJ, Bush AI (2005) Amyloid-beta metal interaction and metal chelation. Subcell Biochem 38:235–254

Cuchillo-Ibanez I, Seereeram A, Byers HL, Leung KY, Ward MA, Anderton BH, Hanger DP (2008) Phosphorylation of tau regulates its axonal transport by controlling its binding to kinesin. FASEB J 22(9):3186–3195

Dahlgren KN, Manelli AM, Stine WB Jr., Baker LK, Krafft GA, LaDu MJ (2002) Oligomeric and fibrillar species of amyloid-beta peptides differentially affect neuronal viability. J Biol Chem 277:32046–32053

Davis DG, Schmitt FA, Wekstein DR, Markesbery WR (1999) Alzheimer neuropathologic alterations in aged cognitively normal subjects. J Neuropathol Exp Neurol 58:376–388

de Carvalho CV, Payao SL, Smith MA (2000) DNA methylation, ageing and ribosomal genes activity. Biogerontology 1:357–361

Diaz JC, Simakova O, Jacobson KA, Arispe N, Pollard HB (2009) Small molecule blockers of the Alzheimer Abeta calcium channel potently protect neurons from Abeta cytotoxicity. Proc Natl Acad Sci U S A 106:3348–3353

Dickson DW, Crystal HA, Mattiace LA, Masur DM, Blau AD, Davies P, Yen SH, Aronson MK (1992) Identification of normal and pathological aging in prospectively studied nondemented elderly humans. Neurobiol Aging 13:179–189

Dodel RC, Hampel H, Du Y (2003) Immunotherapy for Alzheimer's disease. Lancet Neurol 2: 215–220

Dominguez D, Tournoy J, Hartmann D, Huth T, Cryns K, Deforce S, Serneels L, Camacho IE, Marjaux E, Craessaerts K, Roebroek AJ, Schwake M, D'Hooge R, Bach P, Kalinke U, Moechars D, Alzheimer C, Reiss K, Saftig P, De Strooper B (2005) Phenotypic and biochemical analyses of BACE1- and BACE2-deficient mice. J Biol Chem 280:30797–30806

Dong J, Atwood CS, Anderson VE, Siedlak SL, Smith MA, Perry G, Carey PR (2003) Metal binding and oxidation of amyloid-beta within isolated senile plaque cores: Raman microscopic evidence. Biochemistry (Mosc) 42:2768–2773

Dovey HF, John V, Anderson JP, Chen LZ, de Saint Andrieu P, Fang LY, Freedman SB, Folmer B, Goldbach E, Holsztynska EJ, Hu KL, Johnson-Wood KL, Kennedy SL, Kholodenko D, Knops JE, Latimer LH, Lee M, Liao Z, Lieberburg IM, Motter RN, Mutter LC, Nietz J, Quinn KP, Sacchi KL, Seubert PA, Shopp GM, Thorsett ED, Tung JS, Wu J, Yang S, Yin CT, Schenk DB, May PC, Altstiel LD, Bender MH, Boggs LN, Britton TC, Clemens JC, Czilli DL, Dieckman-McGinty DK, Droste JJ, Fuson KS, Gitter BD, Hyslop PA, Johnstone EM, Li WY, Little SP, Mabry TE, Miller FD, Audia JE (2001) Functional gamma-secretase inhibitors reduce beta-amyloid peptide levels in brain. J Neurochem 76:173–181

Drzezga A, Lautenschlager N, Siebner H, Riemenschneider M, Willoch F, Minoshima S, Schwaiger M, Kurz A (2003) Cerebral metabolic changes accompanying conversion of mild cognitive impairment into Alzheimer's disease: a PET follow-up study. Eur J Nucl Med Mol Imaging 30:1104–1113

Dwyer BE, Zacharski LR, Balestra DJ, Lerner AJ, Perry G, Zhu X, Smith MA (2009) Getting the iron out: Phlebotomy for Alzheimer's disease? Med Hypotheses 72:504–509

Erecinska M, Silver IA (1989) ATP and brain function. J Cereb Blood Flow Metab 9:2–19

Evans TA, Raina AK, Delacourte A, Aprelikova O, Lee HG, Zhu X, Perry G, Smith MA (2007) BRCA1 may modulate neuronal cell cycle re-entry in Alzheimer disease. Int J Med Sci 4: 140–145

Farlow MR, Tariot P, Grossberg GT, Gergel I, Grahm S, Jin J (2003) Memantine/donepezil dual therapy is superior to placebo/donepezil therapy for treatment of moderate to severe Alzheimer's disease. Neurology 60:A412

Fernandez-Bachiller MI, Perez C, Campillo NE, Paez JA, Gonzalez-Munoz GC, Usan P, Garcia-Palomero E, Lopez MG, Villarroya M, Garcia AG, Martinez A, Rodriguez-Franco MI (2009) Tacrine-Melatonin Hybrids as Multifunctional Agents for Alzheimer's Disease, with Cholinergic, Antioxidant, and Neuroprotective Properties. ChemMedChem 4(5):828–841

Ferrer I, Boada Rovira M, Sanchez Guerra ML, Rey MJ, Costa-Jussa F (2004) Neuropathology and pathogenesis of encephalitis following amyloid-beta immunization in Alzheimer's disease. Brain Pathol 14:11–20

Ferri KF, Kroemer G (2001) Organelle-specific initiation of cell death pathways. Nat Cell Biol 3:E255–E263

Finefrock AE, Bush AI, Doraiswamy PM (2003) Current status of metals as therapeutic targets in Alzheimer's disease. J Am Geriatr Soc 51:1143–1148

Floyd RA, Hensley K (2002) Oxidative stress in brain aging. Implications for therapeutics of neurodegenerative diseases. Neurobiol Aging 23:795–807

Fox NC, Black RS, Gilman S, Rossor MN, Griffith SG, Jenkins L, Koller M (2005) Effects of Abeta immunization (AN1792) on MRI measures of cerebral volume in Alzheimer disease. Neurology 64:1563–1572

Gamblin TC, Chen F, Zambrano A, Abraha A, Lagalwar S, Guillozet AL, Lu M, Fu Y, Garcia-Sierra F, LaPointe N, Miller R, Berry RW, Binder LI, Cryns VL (2003) Caspase cleavage of tau: linking amyloid and neurofibrillary tangles in Alzheimer's disease. Proc Natl Acad Sci U S A 100:10032–10037

Garcia ML, Cleveland DW (2001) Going new places using an old MAP: tau, microtubules and human neurodegenerative disease. Curr Opin Cell Biol 13:41–48

Garcia-Sierra F, Wischik CM, Harrington CR, Luna-Munoz J, Mena R (2001) Accumulation of C-terminally truncated tau protein associated with vulnerability of the perforant pathway in early stages of neurofibrillary pathology in Alzheimer's disease. J Chem Neuroanat 22:65–77

Giannakopoulos P, Herrmann FR, Bussiere T, Bouras C, Kovari E, Perl DP, Morrison JH, Gold G, Hof PR (2003) Tangle and neuron numbers, but not amyloid load, predict cognitive status in Alzheimer's disease. Neurology 60:1495–1500

Gibson GE, Park LC, Sheu KF, Blass JP, Calingasan NY (2000) The alpha-ketoglutarate dehydrogenase complex in neurodegeneration. Neurochem Int 36:97–112

Goedert M, Cuenda A, Craxton M, Jakes R, Cohen P (1997) Activation of the novel stress-activated protein kinase SAPK4 by cytokines and cellular stresses is mediated by SKK3

(MKK6); comparison of its substrate specificity with that of other SAP kinases. EMBO J 16: 3563–3571

Goedert M, Spillantini MG, Davies SW (1998) Filamentous nerve cell inclusions in neurodegenerative diseases. Curr Opin Neurobiol 8:619–632

Gold G, Kovari E, Corte G, Herrmann FR, Canuto A, Bussiere T, Hof PR, Bouras C, Giannakopoulos P (2001) Clinical validity of A beta-protein deposition staging in brain aging and Alzheimer disease. J Neuropathol Exp Neurol 60:946–952

Golde TE, Eckman CB (2001) Cholesterol modulation as an emerging strategy for the treatment of Alzheimer's disease. Drug Discov Today 6:1049–1055

Gomez-Isla T, Hollister R, West H, Mui S, Growdon JH, Petersen RC, Parisi JE, Hyman BT (1997) Neuronal loss correlates with but exceeds neurofibrillary tangles in Alzheimer's disease. Ann Neurol 41:17–24

Grundman M (2000) Vitamin E and Alzheimer disease: the basis for additional clinical trials. Am J Clin Nutr 71:630S–636S

Guillozet-Bongaarts AL, Cahill ME, Cryns VL, Reynolds MR, Berry RW, Binder LI (2006) Pseudophosphorylation of tau at serine 422 inhibits caspase cleavage: in vitro evidence and implications for tangle formation in vivo. J Neurochem 97:1005–1014

Guix FX, Ill-Raga G, Bravo R, Nakaya T, de Fabritiis G, Coma M, Miscione GP, Villa-Freixa J, Suzuki T, Fernandez-Busquets X, Valverde MA, de Strooper B, Munoz FJ (2009) Amyloid-dependent triosephosphate isomerase nitrotyrosination induces glycation and tau fibrillation. Brain 132(Pt 5):1335–1345

Hager K, Marahrens A, Kenklies M, Riederer P, Munch G (2001) Alpha-lipoic acid as a new treatment option for Azheimer type dementia. Arch Gerontol Geriatr 32:275–282

Halliwell B (1999) Antioxidant defence mechanisms: from the beginning to the end (of the beginning). Free Radic Res 31:261–272

Hamaguchi T, Yamada M (2008) [Genetic factors for cerebral amyloid angiopathy]. Brain Nerve 60:1275–1283

Hamdane M, Delobel P, Sambo AV, Smet C, Begard S, Violleau A, Landrieu I, Delacourte A, Lippens G, Flament S, Buee L (2003) Neurofibrillary degeneration of the Alzheimer-type: an alternate pathway to neuronal apoptosis? Biochem Pharmacol 66:1619–1625

Hayashi T, Shishido N, Nakayama K, Nunomura A, Smith MA, Perry G, Nakamura M (2007) Lipid peroxidation and 4-hydroxy-2-nonenal formation by copper ion bound to amyloid-beta peptide. Free Radic Biol Med 43:1552–1559

Henderson VW (2009) Menopause, cognitive ageing and dementia: practice implications. Menopause Int 15:41–44

Hirai K, Aliev G, Nunomura A, Fujioka H, Russell RL, Atwood CS, Johnson AB, Kress Y, Vinters HV, Tabaton M, Shimohama S, Cash AD, Siedlak SL, Harris PL, Jones PK, Petersen RB, Perry G, Smith MA (2001) Mitochondrial abnormalities in Alzheimer's disease. J Neurosci 21:3017–3023

Hock C, Konietzko U, Streffer JR, Tracy J, Signorell A, Muller-Tillmanns B, Lemke U, Henke K, Moritz E, Garcia E, Wollmer MA, Umbricht D, de Quervain DJ, Hofmann M, Maddalena A, Papassotiropoulos A, Nitsch RM (2003) Antibodies against beta-amyloid slow cognitive decline in Alzheimer's disease. Neuron 38:547–554

Hogervorst E, Yaffe K, Richards M, Huppert FA (2009) Hormone replacement therapy to maintain cognitive function in women with dementia. Cochrane Database Syst Rev (1):CD003799

Holley AK, St Clair DK (2009) Watching the watcher: regulation of p53 by mitochondria. Future Oncol 5:117–130

House E, Collingwood J, Khan A, Korchazkina O, Berthon G, Exley C (2004) Aluminium, iron, zinc and copper influence the in vitro formation of amyloid fibrils of Abeta42 in a manner which may have consequences for metal chelation therapy in Alzheimer's disease. J Alzheimers Dis 6:291–301

Hoyer S (1992) Oxidative energy metabolism in Alzheimer brain. Studies in early-onset and late-onset cases. Mol Chem Neuropathol 16:207–224

Hoyer S (1998) Risk factors for Alzheimer's disease during aging. Impacts of glucose/energy metabolism. J Neural Transm Suppl 54:187–194

Huang X, Atwood CS, Hartshorn MA, Multhaup G, Goldstein LE, Scarpa RC, Cuajungco MP, Gray DN, Lim J, Moir RD, Tanzi RE, Bush AI (1999a) The A beta peptide of Alzheimer's disease directly produces hydrogen peroxide through metal ion reduction. Biochemistry (Mosc) 38:7609–7616

Huang X, Cuajungco MP, Atwood CS, Hartshorn MA, Tyndall JD, Hanson GR, Stokes KC, Leopold M, Multhaup G, Goldstein LE, Scarpa RC, Saunders AJ, Lim J, Moir RD, Glabe C, Bowden EF, Masters CL, Fairlie DP, Tanzi RE, Bush AI (1999b) Cu(II) potentiation of alzheimer abeta neurotoxicity. Correlation with cell-free hydrogen peroxide production and metal reduction. J Biol Chem 274:37111–37116

Ibach B, Haen E, Marienhagen J, Hajak G (2005) Clioquinol treatment in familiar early onset of Alzheimer's disease: a case report. Pharmacopsychiatry 38:178–179

Johnson GV, Hartigan JA (1999) Tau protein in normal and Alzheimer's disease brain: an update. J Alzheimers Dis 1:329–351

Johnson GV, Jenkins SM (1999) Tau protein in normal and Alzheimer's disease brain. J Alzheimers Dis 1:307–328

Jorm AF, Jolley D (1998) The incidence of dementia: a meta-analysis. Neurology 51: 728–733

Jorm AF, Korten AE, Henderson AS (1987) The prevalence of dementia: a quantitative integration of the literature. Acta Psychiatr Scand 76:465–479

Josien H (2002) Recent advances in the development of gamma-secretase inhibitors. Curr Opin Drug Discov Devel 5:513–525

Kalaria RN, Gravina SA, Schmidley JW, Perry G, Harik SI (1988) The glucose transporter of the human brain and blood-brain barrier. Ann Neurol 24:757–764

Kamino K, Nagasaka K, Imagawa M, Yamamoto H, Yoneda H, Ueki A, Kitamura S, Namekata K, Miki T, Ohta S (2000) Deficiency in mitochondrial aldehyde dehydrogenase increases the risk for late-onset Alzheimer's disease in the Japanese population. Biochem Biophys Res Commun 273:192–196

Kontush A (2001) Amyloid-beta: an antioxidant that becomes a pro-oxidant and critically contributes to Alzheimer's disease. Free Radic Biol Med 31:1120–1131

Kontush A, Finckh B, Karten B, Kohlschutter A, Beisiegel U (1996) Antioxidant and prooxidant activity of alpha-tocopherol in human plasma and low density lipoprotein. J Lipid Res 37: 1436–1448

Larner AJ, Doran M (2009) Genotype-Phenotype Relationships of Presenilin-1 Mutations in Alzheimer's Disease: An Update. J Alzheimers Dis 17(2):259–265

Lee HG, Casadesus G, Zhu X, Castellani RJ, McShea A, Perry G, Petersen RB, Bajic V, Smith MA (2009a) Cell cycle re-entry mediated neurodegeneration and its treatment role in the pathogenesis of Alzheimer's disease. Neurochem Int 54:84–88

Lee HG, Casadesus G, Zhu X, Takeda A, Perry G, Smith MA (2004) Challenging the amyloid cascade hypothesis: senile plaques and amyloid-beta as protective adaptations to Alzheimer disease. Ann N Y Acad Sci 1019:1–4

Lee KW, Kim JB, Seo JS, Kim TK, Im JY, Baek IS, Kim KS, Lee JK, Han PL (2009b) Behavioral stress accelerates plaque pathogenesis in the brain of Tg2576 mice via generation of metabolic oxidative stress. J Neurochem 108:165–175

Lee HG, Perry G, Moreira PI, Garrett MR, Liu Q, Zhu X, Takeda A, Nunomura A, Smith MA (2005) Tau phosphorylation in Alzheimer's disease: pathogen or protector? Trends Mol Med 11:164–169

Lee HG, Zhu X, Castellani RJ, Nunomura A, Perry G, Smith MA (2007) Amyloid-beta in Alzheimer disease: the null versus the alternate hypotheses. J Pharmacol Exp Ther 321: 823–829

Lee HG, Zhu X, Nunomura A, Perry G, Smith MA (2006) Amyloid-beta vaccination: testing the amyloid hypothesis? heads we win, tails you lose! Am J Pathol 169:738–739

Lemos RR, Castelletti C, Lima Filho JL, Marques ET, Oliveira JR (2009) *In Silico* Identification of New Genetic Variations as Potential Risk Factors for Alzheimer's Disease in a Microarray-oriented Simulation. J Mol Neurosci 39(1–2):242–247

Liu R, Zhou XW, Tanila H, Bjorkdahl C, Wang JZ, Guan ZZ, Cao Y, Gustafsson JA, Winblad B, Pei JJ (2008) Phosphorylated PP2A (tyrosine 307) is associated with Alzheimer neurofibrillary pathology. J Cell Mol Med 12(1):241–257

Lleo A, Greenberg SM, Growdon JH (2005) Current Pharmacotherapy for Alzheimer's Disease. Annu Rev Med 57:513–533

Lopez-Bastida J, Hart W, Garcia-Perez L, Linertova R (2009) Cost-effectiveness of donepezil in the treatment of mild or moderate Alzheimer's disease. J Alzheimers Dis 16:399–407

Louvi A, Sisodia SS, Grove EA (2004) Presenilin 1 in migration and morphogenesis in the central nervous system. Development 131:3093–3105

Lovell MA, Robertson JD, Teesdale WJ, Campbell JL, Markesbery WR (1998) Copper, iron and zinc in Alzheimer's disease senile plaques. J Neurol Sci 158:47–52

Lu Z, Nie G, Li Y, Soe-Lin S, Tao Y, Cao Y, Zhang Z, Liu N, Ponka P, Zhao B (2009) Overexpression of Mitochondrial Ferritin Sensitizes Cells to Oxidative Stress Via an Iron-Mediated Mechanism. Antioxid Redox Signal 11(8):1791–1803

Luchsinger JA, Tang MX, Shea S, Mayeux R (2003) Antioxidant vitamin intake and risk of Alzheimer disease. Arch Neurol 60:203–208

Matsuda JI, Kaminaka K, Nozaki C (2009) Amyloid beta peptides with an additional cysteine residue can enhance immunogenicity and reduce the amyloid beta burden in an Alzheimer's disease mouse model. Biochem Biophys Res Commun 382(1):149–152

Mattson MP, Chan SL, Duan W (2002) Modification of brain aging and neurodegenerative disorders by genes, diet, and behavior. Physiol Rev 82:637–672

McGonigal G, Thomas B, McQuade C, Starr JM, MacLennan WJ, Whalley LJ (1993) Epidemiology of Alzheimer's presenile dementia in Scotland, 1974–88. BMJ (Clinical research ed) 306:680–683

McShea A, Ramiro-Puig E, Munro SB, Casadesus G, Castell M, Smith MA (2008) Clinical benefit and preservation of flavonols in dark chocolate manufacturing. Nutr Rev 66:630–641

Migliore L, Coppede F (2009) Environmental-induced oxidative stress in neurodegenerative disorders and aging. Mutat Res 674:73–84

Mondragon-Rodriguez S, Basurto-Islas G, Santa-Maria I, Mena R, Binder LI, Avila J, Smith MA, Perry G, Garcia-Sierra F (2008a) Cleavage and conformational changes of tau protein follow phosphorylation during Alzheimer's disease. Int J Exp Path 89(2):81–90

Mondragon-Rodriguez S, Mena R, Binder LI, Smith MA, Perry G, Garcia-Sierra F (2008b) Conformational changes and cleavage of tau in Pick bodies parallel the early processing of tau found in Alzheimer pathology. Neuropathol Appl Neurobiol 34:62–75

Monsonego A, Zota V, Karni A, Krieger JI, Bar-Or A, Bitan G, Budson AE, Sperling R, Selkoe DJ, Weiner HL (2003) Increased T cell reactivity to amyloid beta protein in older humans and patients with Alzheimer disease. J Clin Invest 112:415–422

Montgomery SA, Thal LJ, Amrein R (2003) Meta-analysis of double blind randomized controlled clinical trials of acetyl-L-carnitine versus placebo in the treatment of mild cognitive impairment and mild Alzheimer's disease. Int Clin Psychopharmacol 18:61–71

Moreira PI, Nunomura A, Nakamura M, Takeda A, Shenk JC, Aliev G, Smith MA, Perry G (2008) Nucleic acid oxidation in Alzheimer disease. Free Radic Biol Med 44(8): 1493–1505

Moreira PI, Siedlak SL, Wang X, Santos MS, Oliveira CR, Tabaton M, Nunomura A, Szweda LI, Aliev G, Smith MA, Zhu X, Perry G (2007) Autophagocytosis of mitochondria is prominent in Alzheimer disease. J Neuropathol Exp Neurol 66:525–532

Moreira PI, Zhu X, Nunomura A, Smith MA, Perry G (2006) Therapeutic options in Alzheimer's disease. Expert Rev Neurother 6:897–910

Moriguchi S, Zhao X, Marszalec W, Yeh JZ, Fukunaga K, Narahashi T (2009) Nefiracetam and galantamine modulation of excitatory and inhibitory synaptic transmission via stimulation of neuronal nicotinic acetylcholine receptors in rat cortical neurons. Neuroscience 160(2): 484–491

Morris MC, Evans DA, Bienias JL, Tangney CC, Bennett DA, Aggarwal N, Wilson RS, Scherr PA (2002a) Dietary intake of antioxidant nutrients and the risk of incident Alzheimer disease in a biracial community study. J Am Med Assoc 287:3230–3237

Morris MC, Evans DA, Bienias JL, Tangney CC, Wilson RS (2002b) Vitamin E and cognitive decline in older persons. Arch Neurol 59:1125–1132

Mosconi L, Nacmias B, Sorbi S, De Cristofaro MT, Fayazz M, Tedde A, Bracco L, Herholz K, Pupi A (2004) Brain metabolic decreases related to the dose of the ApoE e4 allele in Alzheimer's disease. J Neurol Neurosurg Psychiatry 75:370–376

Mosconi L, Pupi A, De Leon MJ (2008) Brain glucose hypometabolism and oxidative stress in preclinical Alzheimer's disease. Ann N Y Acad Sci 1147:180–195

Nakamura M, Shishido N, Nunomura A, Smith MA, Perry G, Hayashi Y, Nakayama K, Hayashi T (2007) Three histidine residues of amyloid-beta peptide control the redox activity of copper and iron. Biochemistry (Mosc) 46:12737–12743

Nunomura A, Castellani RJ, Zhu X, Moreira PI, Perry G, Smith MA (2006) Involvement of oxidative stress in Alzheimer disease. J Neuropathol Exp Neurol 65:631–641

Nunomura A, Chiba S, Lippa CF, Cras P, Kalaria RN, Takeda A, Honda K, Smith MA, Perry G (2004) Neuronal RNA oxidation is a prominent feature of familial Alzheimer's disease. Neurobiol Dis 17:108–113

Nunomura A, Moreira PI, Lee HG, Zhu X, Castellani RJ, Smith MA, Perry G (2007a) Neuronal death and survival under oxidative stress in Alzheimer and Parkinson diseases. CNS Neurol Disord Drug Targets 6:411–423

Nunomura A, Moreira PI, Takeda A, Smith MA, Perry G (2007b) Oxidative RNA damage and neurodegeneration. Curr Med Chem 14:2968–2975

Nunomura A, Perry G, Aliev G, Hirai K, Takeda A, Balraj EK, Jones PK, Ghanbari H, Wataya T, Shimohama S, Chiba S, Atwood CS, Petersen RB, Smith MA (2001) Oxidative damage is the earliest event in Alzheimer disease. J Neuropathol Exp Neurol 60:759–767

Nunomura A, Perry G, Pappolla MA, Friedland RP, Hirai K, Chiba S, Smith MA (2000) Neuronal oxidative stress precedes amyloid-beta deposition in Down syndrome. J Neuropathol Exp Neurol 59:1011–1017

Nunomura A, Perry G, Pappolla MA, Wade R, Hirai K, Chiba S, Smith MA (1999) RNA oxidation is a prominent feature of vulnerable neurons in Alzheimer's disease. J Neurosci 19: 1959–1964

Paola D, Domenicotti C, Nitti M, Vitali A, Borghi R, Cottalasso D, Zaccheo D, Odetti P, Strocchi P, Marinari UM, Tabaton M, Pronzato MA (2000) Oxidative stress induces increase in intracellular amyloid beta-protein production and selective activation of betaI and betaII PKCs in NT2 cells. Biochem Biophys Res Commun 268:642–646

Perkins AJ, Hendrie HC, Callahan CM, Gao S, Unverzagt FW, Xu Y, Hall KS, Hui SL (1999) Association of antioxidants with memory in a multiethnic elderly sample using the Third National Health and Nutrition Examination Survey. Am J Epidemiol 150:37–44

Perry G, Nunomura A, Raina AK, Aliev G, Siedlak SL, Harris PL, Casadesus G, Petersen RB, Bligh-Glover W, Balraj E, Petot GJ, Smith MA (2003) A metabolic basis for Alzheimer disease. Neurochem Res 28:1549–1552

Perry G, Nunomura A, Raina AK, Smith MA (2000) Amyloid-beta junkies. Lancet 355:757

Perry G, Smith MA, McCann CE, Siedlak SL, Jones PK, Friedland RP (1998) Cerebrovascular muscle atrophy is a feature of Alzheimer's disease. Brain Res 791:63–66

Perry G, Taddeo MA, Nunomura A, Zhu X, Zenteno-Savin T, Drew KL, Shimohama S, Avila J, Castellani RJ, Smith MA (2002) Comparative biology and pathology of oxidative stress in Alzheimer and other neurodegenerative diseases: beyond damage and response. Comp Biochem Physiol C Toxicol Pharmacol 133:507–513

Pope S, Land JM, Heales SJ (2008) Oxidative stress and mitochondrial dysfunction in neurodegeneration; cardiolipin a critical target? Biochim Biophys Acta 1777:794–799

Praprotnik D, Smith MA, Richey PL, Vinters HV, Perry G (1996) Filament heterogeneity within the dystrophic neurites of senile plaques suggests blockage of fast axonal transport in Alzheimer's disease. Acta Neuropathol (Berl) 91:226–235

Pratico D, Trojanowski JQ (2000) Inflammatory hypotheses: novel mechanisms of Alzheimer's neurodegeneration and new therapeutic targets? Neurobiol Aging 21:441–445; discussion 451–443

Pratico D, Uryu K, Sung S, Tang S, Trojanowski JQ, Lee VM (2002) Aluminum modulates brain amyloidosis through oxidative stress in APP transgenic mice. FASEB J 16:1138–1140

Ramiro-Puig E, Casadesus G, Lee HG, Zhu X, McShea A, Perry G, Perez-Cano FJ, Smith MA, Castell M (2009) Neuroprotective effect of cocoa flavonoids on in vitro oxidative stress. Eur J Nutr 48:54–61

Rankin CA, Sun Q, Gamblin TC (2005) Pseudo-phosphorylation of tau at Ser202 and Thr205 affects tau filament formation. Brain Res Mol Brain Res 138:84–93

Rapoport SI (2003) Coupled reductions in brain oxidative phosphorylation and synaptic function can be quantified and staged in the course of Alzheimer disease. Neurotox Res 5:385–398

Rapp SR, Espeland MA, Shumaker SA, Henderson VW, Brunner RL, Manson JE, Gass ML, Stefanick ML, Lane DS, Hays J, Johnson KC, Coker LH, Dailey M, Bowen D (2003) Effect of estrogen plus progestin on global cognitive function in postmenopausal women: the Women's Health Initiative Memory Study: a randomized controlled trial. J Am Med Assoc 289:2663–2672

Resnick SM, Henderson VW (2002) Hormone therapy and risk of Alzheimer disease: a critical time. J Am Med Assoc 288:2170–2172

Reynolds CH, Nebreda AR, Gibb GM, Utton MA, Anderton BH (1997) Reactivating kinase/p38 phosphorylates tau protein in vitro. J Neurochem 69:191–198

Rocca WA, Hofman A, Brayne C, Breteler MM, Clarke M, Copeland JR, Dartigues JF, Engedal K, Hagnell O, Heeren TJ, Jonker C, Lindesay J, Lobo A, Mann AH, Molsa PK, Morgan K, O'Connor DW, da Silva Droux A, Sulkava R, Kay DWK, Amaducci L; EURODEM-Prevalence Research Group (1991) Frequency and distribution of Alzheimer's disease in Europe: a collaborative study of 1980–1990 prevalence findings. The EURODEM-Prevalence Research Group. Ann Neurol 30:381–390

Rogers JT, Randall JD, Cahill CM, Eder PS, Huang X, Gunshin H, Leiter L, McPhee J, Sarang SS, Utsuki T, Greig NH, Lahiri DK, Tanzi RE, Bush AI, Giordano T, Gullans SR (2002) An iron-responsive element type II in the 5′-untranslated region of the Alzheimer's amyloid precursor protein transcript. J Biol Chem 277:45518–45528

Rottkamp CA, Raina AK, Zhu X, Gaier E, Bush AI, Atwood CS, Chevion M, Perry G, Smith MA (2001) Redox-active iron mediates amyloid-beta toxicity. Free Radic Biol Med 30:447–450

Sajad M, Zargan J, Chawla R, Umar S, Sadaqat M, Khan HA (2009) Hippocampal neurodegeneration in experimental autoimmune encephalomyelitis (EAE): potential role of inflammation activated myeloperoxidase. Mol Cell Biochem 328(1–2):183–188

Sano M, Ernesto C, Thomas RG, Klauber MR, Schafer K, Grundman M, Woodbury P, Growdon J, Cotman CW, Pfeiffer E, Schneider LS, Thal LJ (1997) A controlled trial of selegiline, alpha-tocopherol, or both as treatment for Alzheimer's disease. The Alzheimer's Disease Cooperative Study. N Engl J Med 336:1216–1222

Santacruz K, Lewis J, Spires T, Paulson J, Kotilinek L, Ingelsson M, Guimaraes A, DeTure M, Ramsden M, McGowan E, Forster C, Yue M, Orne J, Janus C, Mariash A, Kuskowski M, Hyman B, Hutton M, Ashe KH (2005) Tau suppression in a neurodegenerative mouse model improves memory function. Science 309:476–481

Sayre LM, Perry G, Harris PL, Liu Y, Schubert KA, Smith MA (2000) In situ oxidative catalysis by neurofibrillary tangles and senile plaques in Alzheimer's disease: a central role for bound transition metals. J Neurochem 74:270–279

Schenk D, Barbour R, Dunn W, Gordon G, Grajeda H, Guido T, Hu K, Huang J, Johnson-Wood K, Khan K, Kholodenko D, Lee M, Liao Z, Lieberburg I, Motter R, Mutter L, Soriano F, Shopp G, Vasquez N, Vandevert C, Walker S, Wogulis M, Yednock T, Games D, Seubert P

(1999) Immunization with amyloid-beta attenuates Alzheimer-disease-like pathology in the PDAPP mouse. Nature 400:173–177

Schindowski K, Belarbi K, Bretteville A, Ando K, Buee L (2008) Neurogenesis and cell cycle-reactivated neuronal death during pathogenic tau aggregation. Genes Brain Behav 7 (Suppl 1):92–100

Shah RS, Lee HG, Xiongwei Z, Perry G, Smith MA, Castellani RJ (2008) Current approaches in the treatment of Alzheimer's disease. Biomed Pharmacother 62:199–207

Shepherd J, Blauw GJ, Murphy MB, Bollen EL, Buckley BM, Cobbe SM, Ford I, Gaw A, Hyland M, Jukema JW, Kamper AM, Macfarlane PW, Meinders AE, Norrie J, Packard CJ, Perry IJ, Stott DJ, Sweeney BJ, Twomey C, Westendorp RG (2002) Pravastatin in elderly individuals at risk of vascular disease (PROSPER): a randomised controlled trial. Lancet 360:1623–1630

Short RA, Bowen RL, O'Brien PC, Graff-Radford NR (2001) Elevated gonadotropin levels in patients with Alzheimer disease. Mayo Clin Proc 76:906–909

Smith MA (1998) Alzheimer disease. Int Rev Neurobiol 42:1–54

Smith MA, Atwood CS, Joseph JA, Perry G (2002a) Predicting the failure of amyloid-beta vaccine. Lancet 359:1864–1865

Smith MA, Casadesus G, Joseph JA, Perry G (2002b) Amyloid-beta and tau serve antioxidant functions in the aging and Alzheimer brain. Free Radic Biol Med 33:1194–1199

Smith MA, Harris PL, Sayre LM, Perry G (1997a) Iron accumulation in Alzheimer disease is a source of redox-generated free radicals. Proc Natl Acad Sci U S A 94:9866–9868

Smith MA, Hirai K, Hsiao K, Pappolla MA, Harris PL, Siedlak SL, Tabaton M, Perry G (1998) Amyloid-beta deposition in Alzheimer transgenic mice is associated with oxidative stress. J Neurochem 70:2212–2215

Smith MA, Nunomura A, Zhu X, Takeda A, Perry G (2000a) Metabolic, metallic, and mitotic sources of oxidative stress in Alzheimer disease. Antioxid Redox Signal 2:413–420

Smith MA, Petot GJ, Perry G (1999) Diet and oxidative stress: a novel synthesis of epidemiological data on Alzheimer's disease. J Alzheimers Dis 1:203–206

Smith MA, Richey Harris PL, Sayre LM, Beckman JS, Perry G (1997b) Widespread peroxynitrite-mediated damage in Alzheimer's disease. J Neurosci 17:2653–2657

Smith MA, Rottkamp CA, Nunomura A, Raina AK, Perry G (2000b) Oxidative stress in Alzheimer's disease. Biochim Biophys Acta 1502:139–144

Smith MA, Rudnicka-Nawrot M, Richey PL, Praprotnik D, Mulvihill P, Miller CA, Sayre LM, Perry G (1995) Carbonyl-related posttranslational modification of neurofilament protein in the neurofibrillary pathology of Alzheimer's disease. J Neurochem 64:2660–2666

Smith MA, Taneda S, Richey PL, Miyata S, Yan SD, Stern D, Sayre LM, Monnier VM, Perry G (1994) Advanced Maillard reaction end products are associated with Alzheimer disease pathology. Proc Natl Acad Sci U S A 91:5710–5714

Sontag E, Nunbhakdi-Craig V, Lee G, Brandt R, Kamibayashi C, Kuret J, White CL 3rd, Mumby MC, Bloom GS (1999) Molecular interactions among protein phosphatase 2A, tau, and microtubules. Implications for the regulation of tau phosphorylation and the development of tauopathies. J Biol Chem 274:25490–25498

Sorbi S, Bird ED, Blass JP (1983) Decreased pyruvate dehydrogenase complex activity in Huntington and Alzheimer brain. Ann Neurol 13:72–78

Speranca MA, Batista LM, Lourenco Rda S, Tavares WM, Bertolucci PH, Rigolin Vde O, Payao SL,, Smith Mde A (2008) Can the rDNA methylation pattern be used as a marker for Alzheimer's disease? Alzheimers Dement 4:438–442

Spires-Jones TL, Mielke ML, Rozkalne A, Meyer-Luehmann M, de Calignon A, Bacskai BJ, Schenk D, Hyman BT (2009) Passive immunotherapy rapidly increases structural plasticity in a mouse model of Alzheimer disease. Neurobiol Dis 33:213–220

Stackman RW, Eckenstein F, Frei B, Kulhanek D, Nowlin J, Quinn JF (2003) Prevention of age-related spatial memory deficits in a transgenic mouse model of Alzheimer's disease by chronic Ginkgo biloba treatment. Exp Neurol 184:510–520

Swatton JE, Sellers LA, Faull RL, Holland A, Iritani S, Bahn S (2004) Increased MAP kinase activity in Alzheimer's and Down syndrome but not in schizophrenia human brain. Eur J Neurosci 19:2711–2719

Szekely CA, Thorne JE, Zandi PP, Ek M, Messias E, Breitner JC, Goodman SN (2004) Nonsteroidal anti-inflammatory drugs for the prevention of Alzheimer's disease: a systematic review. Neuroepidemiology 23:159–169

Takeda A, Loveman E, Clegg A, Kirby J, Picot J, Payne E, Green C (2006) A systematic review of the clinical effectiveness of donepezil, rivastigmine and galantamine on cognition, quality of life and adverse events in Alzheimer's disease. Int J Geriatr Psychiatry 21:17–28

Tamagno E, Guglielmotto M, Aragno M, Borghi R, Autelli R, Giliberto L, Muraca G, Danni O, Zhu X, Smith MA, Perry G, Jo DG, Mattson MP, Tabaton M (2008) Oxidative stress activates a positive feedback between the gamma- and beta-secretase cleavages of the beta-amyloid precursor protein. J Neurochem 104:683–695

Thal LJ, Grundman M, Berg J, Ernstrom K, Margolin R, Pfeiffer E, Weiner MF, Zamrini E, Thomas RG (2003) Idebenone treatment fails to slow cognitive decline in Alzheimer's disease. Neurology 61:1498–1502

Tian Q, Wang J (2002) Role of serine/threonine protein phosphatase in Alzheimer's disease. Neurosignals 11:262–269

Van Hoesen GW, Hyman BT (1990) Hippocampal formation: anatomy and the patterns of pathology in Alzheimer's disease. Prog Brain Res 83:445–457

Van Hoesen GW, Hyman BT, Damasio AR (1991) Entorhinal cortex pathology in Alzheimer's disease. Hippocampus 1:1–8

Veinbergs I, Mallory M, Sagara Y, Masliah E (2000) Vitamin E supplementation prevents spatial learning deficits and dendritic alterations in aged apolipoprotein E-deficient mice. Eur J Neurosci 12:4541–4546

Verdile G, Yeap BB, Clarnette RM, Dhaliwal S, Burkhardt MS, Chubb SA, De Ruyck K, Rodrigues M, Mehta PD, Foster JK, Bruce DG, Martins RN (2008) Luteinizing hormone levels are positively correlated with plasma amyloid-beta protein levels in elderly men. J Alzheimers Dis 14:201–208

Vincent I, Jicha G, Rosado M, Dickson DW (1997) Aberrant expression of mitotic cdc2/cyclin B1 kinase in degenerating neurons of Alzheimer's disease brain. J Neurosci 17: 3588–3598

Vitali A, Piccini A, Borghi R, Fornaro P, Siedlak SL, Smith MA, Gambetti P, Ghetti B, Tabaton M (2004) Soluble amyloid beta-protein is increased in frontotemporal dementia with tau gene mutations. J Alzheimers Dis 6:45–51

Wallace DC (1999) Mitochondrial diseases in man and mouse. Science 283:1482–1488

Walsh DM, Selkoe DJ (2004) Oligomers on the brain: the emerging role of soluble protein aggregates in neurodegeneration. Protein Pept Lett 11:213–228

Wang JZ, Grundke-Iqbal I, Iqbal K (2007a) Kinases and phosphatases and tau sites involved in Alzheimer neurofibrillary degeneration. Eur J Neurosci 25:59–68

Wang YP, Song JF, Rao ZR (2004c) [CDK5 and neurodegenerative disease]. Sheng Li Ke Xue Jin Zhan 35:45–48

Wang X, Su B, Fujioka H, Zhu X (2008a) Dynamin-like protein 1 reduction underlies mitochondrial morphology and distribution abnormalities in fibroblasts from sporadic Alzheimer's disease patients. Am J Pathol 173(2):470–482

Wang X, Su B, Perry G, Smith MA, Zhu X (2007b) Insights into amyloid-beta-induced mitochondrial dysfunction in Alzheimer disease. Free Radic Biol Med 43:1569–1573

Wang X, Su B, Siedlak SL, Moreira PI, Fujioka H, Wang Y, Casadesus G, Zhu X (2008b) Amyloid-beta overproduction causes abnormal mitochondrial dynamics via differential modulation of mitochondrial fission/fusion proteins. Proc Natl Acad Sci U S A 105:19318–19323

Wang PN, Yang CL, Lin KN, Chen WT, Chwang LC, Liu HC (2004b) Weight loss, nutritional status and physical activity in patients with Alzheimer's disease. A controlled study. J Neurol 251:314–320

Wang B, Zhang C, Zheng W, Lu Z, Zheng C, Yang Z, Wang L, Jin F (2004a) Association between a T/C polymorphism in intron 2 of cholesterol 24S-hydroxylase gene and Alzheimer's disease in Chinese. Neurosci Lett 369:104–107

Webber KM, Casadesus G, Zhu X, Obrenovich ME, Atwood CS, Perry G, Bowen RL, Smith MA (2006) The cell cycle and hormonal fluxes in Alzheimer disease: a novel therapeutic target. Curr Pharm Des 12:691–697

Webber KM, Perry G, Smith MA, Casadesus G (2007) The contribution of luteinizing hormone to Alzheimer disease pathogenesis. Clin Med Res 5:177–183

Webber KM, Raina AK, Marlatt MW, Zhu X, Prat MI, Morelli L, Casadesus G, Perry G, Smith MA (2005) The cell cycle in Alzheimer disease: a unique target for neuropharmacology. Mech Ageing Dev 126:1019–1025

Wilcock DM, Colton CA (2008) Anti-amyloid-beta immunotherapy in Alzheimer's disease: relevance of transgenic mouse studies to clinical trials. J Alzheimers Dis 15:555–569

Williamson J, LaRusse S (2004) Genetics and genetic counseling: recommendations for Alzheimer's disease, frontotemporal dementia, and Creutzfeldt-Jakob disease. Curr Neurol Neurosci Rep 4:351–357

Wood JG, Mirra SS, Pollock NJ, Binder LI (1986) Neurofibrillary tangles of Alzheimer disease share antigenic determinants with the axonal microtubule-associated protein tau (tau). Proc Natl Acad Sci U S A 83:4040–4043

Xu J, Chen S, Ahmed SH, Chen H, Ku G, Goldberg MP, Hsu CY (2001) Amyloid-beta peptides are cytotoxic to oligodendrocytes. J Neurosci 21:RC118

Yang W, Leystra-Lantz C, Strong MJ (2008) Upregulation of GSK3beta expression in frontal and temporal cortex in ALS with cognitive impairment (ALSci). Brain Res 1196:131–139

Yankner BA, Duffy LK, Kirschner DA (1990) Neurotrophic and neurotoxic effects of amyloid beta protein: reversal by tachykinin neuropeptides. Science 250:279–282

Yao M, Nguyen TV, Pike CJ (2005) Beta-amyloid-induced neuronal apoptosis involves c-Jun N-terminal kinase-dependent downregulation of Bcl-w. J Neurosci 25:1149–1158

Younkin SG (1994) The amyloid beta protein precursor mutations linked to familial Alzheimer's disease alter processing in a way that fosters amyloid deposition. Tohoku J Exp Med 174: 217–223

Zawia NH, Lahiri DK, Cardozo-Pelaez F (2009) Epigenetics, oxidative stress, and Alzheimer disease. Free Radic Biol Med 46(9):1241–1249

Zekanowski C, Golan MP, Krzysko KA, Lipczynska-Lojkowska W, Filipek S, Kowalska A, Rossa G, Peplonska B, Styczynska M, Maruszak A, Religa D, Wender M, Kulczycki J, Barcikowska M, Kuznicki J (2006) Two novel presenilin 1 gene mutations connected with frontotemporal dementia-like clinical phenotype: genetic and bioinformatic assessment. Exp Neurol 200: 82–88

Zhou XW, Gustafsson JA, Tanila H, Bjorkdahl C, Liu R, Winblad B, Pei JJ (2008a) Tau hyperphosphorylation correlates with reduced methylation of protein phosphatase 2A. Neurobiol Dis 31(3):386–394

Zhou XW, Tanila H, Pei JJ (2008b) Parallel increase in p70 kinase activation and tau phosphorylation (S262) with Abeta overproduction. FEBS Lett 582:159–164

Zhu X, Perry G, Moreira PI, Aliev G, Cash AD, Hirai K, Smith MA (2006) Mitochondrial abnormalities and oxidative imbalance in Alzheimer disease. J Alzheimers Dis 9: 147–153

Zhu X, Raina AK, Lee HG, Casadesus G, Smith MA, Perry G (2004a) Oxidative stress signalling in Alzheimer's disease. Brain Res 1000:32–39

Zhu X, Raina AK, Rottkamp CA, Aliev G, Perry G, Boux H, Smith MA (2001a) Activation and redistribution of c-jun N-terminal kinase/stress activated protein kinase in degenerating neurons in Alzheimer's disease. J Neurochem 76:435–441

Zhu X, Rottkamp CA, Boux H, Takeda A, Perry G, Smith MA (2000) Activation of p38 kinase links tau phosphorylation, oxidative stress, and cell cycle-related events in Alzheimer disease. J Neuropathol Exp Neurol 59:880–888

Zhu X, Rottkamp CA, Hartzler A, Sun Z, Takeda A, Boux H, Shimohama S, Perry G, Smith MA (2001b) Activation of MKK6, an upstream activator of p38, in Alzheimer's disease. J Neurochem 79:311–318

Zhu X, Smith MA, Perry G, Aliev G (2004b) Mitochondrial failures in Alzheimer's disease. Am J Alzheimers Dis Other Demen 19:345–352

Zhu X, Su B, Wang X, Smith MA, Perry G (2007) Causes of oxidative stress in Alzheimer disease. Cell Mol Life Sci 64:2202–2210

Zhu Y, Xiao K, Ma L, Xiong B, Fu Y, Yu H, Wang W, Wang X, Hu D, Peng H, Li J, Gong Q, Chai Q, Tang X, Zhang H, Shen J (2009) Design, synthesis and biological evaluation of novel dual inhibitors of acetylcholinesterase and beta-secretase. Bioorg Med Chem 17:1600–1613

Zou K, Gong JS, Yanagisawa K, Michikawa M (2002) A novel function of monomeric amyloid beta-protein serving as an antioxidant molecule against metal-induced oxidative damage. J Neurosci 22:4833–4841

Tau and Tauopathies

R. Mathew, G. Srinivas, and P.S. Mathuranath

Abstract Tau protein is a neuronal microtubule associated protein, which localizes primarily in the axon. It plays a major role in promoting microtubule assembly, stabilizing microtubules and maintaining the normal morphology of the neurons. Structurally tau is a heterogenous molecule due to several posttranslational modifications. Tauopathies are a group of disorders that are the consequence of abnormal tau phosphorylation, abnormal levels of tau, abnormal tau splicing, or mutations in the tau gene. These disorders are characterized not only by neuronal, but also oligodendroglial and astrocytic filamentous tau inclusions. Tauopathies are the commonest among the neurodegerative diseases with filamentous inclusions. Tauopathies include frontotemporal dementia, Parkinsonism plus syndromes, neuromuscular disorders, and certain genetic and metabolic syndromes. The occurrence of neurofibrillary tangles in a wide range of conditions, including Alzheimer's disease, initially led to the suggestion that tau deposition may be an incidental nonspecific finding associated with cell death or cellular dysfunction. Later the discovery of close to 20 different mutations in tau in frontotemporal dementia with Parkinsonism linked to chromosome-17 (FTDP-17) clearly showed that dysfunction of tau protein causes neurodegeneration and dementia. Among the tauopathies, the most studied is Alzheimer's disease. Frontotemporal dementia, progressive supranuclear palsy, and corticobasal ganglionic degeneration are some of the other common tauopathies that have been extensively studied. Overlap of clinical and histopathological features occurs between various tauopathies. The role of CSF tau in the diagnosis of dementias is under investigation. The measures of total tau as well as species of phospho-tau detected by antibodies in CSF correlates best with a diagnosis of AD. The discovery of a tau transgenic mouse model has paved the way for testing various therapeutic models for targeting tau.

P.S. Mathuranath (✉)
Cognition and Behavioural Neurology Section (CBNC), Department of Neurology, Sree Chitra Tirunal Institute for Medical Sciences and Technology (SCTIMST), Thiruvananthapuram 695011, Kerala, India
e-mail: mathu@sctimst.ac.in

J.P. Blass (ed.), *Neurochemical Mechanisms in Disease*,
Advances in Neurobiology 1, DOI 10.1007/978-1-4419-7104-3_19,

Keywords Tau · Tauopathy · Alzheimer's disease · Frontotemporal dementia · Parkinsonism · Exon · Microtubule · Structural · Protein · Phosphorylation

Contents

1 Introduction

Tau protein is a neuronal microtubule associated protein (MAP) that localizes primarily in the axon (Leger et al., 1994). It is one of the major and most-studied MAPs in the central nervous system (Alonso et al., 2001). Tau has been recognized to play major roles in promoting microtubule assembly, stabilizing microtubules and maintaining the normal morphology of the neurons. Tau has been the focus of intense research for more than a decade after it was discovered to be a key component of neurofibrillary tangles in Alzheimer's disease (AD).

Tauopathies are a group of disorders that are the consequence of abnormal tau phosphorylation, abnormal levels of tau, abnormal tau splicing, or mutations in the tau gene. These disorders are characterized not only by neuronal, but also oligodendroglial and astrocytic filamentous tau inclusions (Avila, 2000; Ulloa et al., 1994).

In some tauopathies such as AD the tau pathology is associated with other cerebral changes (Avila et al., 2004). That a presumably neuronal protein was also a component of glial lesions in a host of non-Alzheimer degenerative diseases was unexpected and offered an entirely new perspective on neurodegenerative disorders. The discovery of mutations in the tau gene on chromosome 17 in frontotemporal dementias with Parkinsonism (FTDP-17) added to the importance of tauprotein in cognitive neuroscience (Dickson, 1999). Tau pathology is not restricted to the central nervous system. Clusters ("tangles") of paired-helical filaments containing phosphorylated tau are one of the characteristic features of certain myopathies such as the inclusion body myositis (Askanas and Engel, 1998) and myotonic dystrophy (Tolnay and Probst, 1999). Some adults with myotonic dystrophy type 1 (DM1) have been seen to develop dementia with aging, agreeing with recent studies documenting an abnormal tau-protein expression in the brain tissues of patients with DM1 (Modoni et al., 2004). A possible distinct subclass of peripheral tauopathy has been postulated based on immunoblot studies (Maurage et al., 2004). This chapter, however, focuses on the vast majority of tauopathies that are well recognized. In most of them it so happens that cognitive and or motor impairments are the core clinical manifestation.

2 Biochemistry and Molecular Biology of Tau

Microtubules (MT) are a major component of neuronal cell processes involved in maintaining the cell shape and axonal transport (Buee et al., 2000). It is probable that microtubule-associated proteins play a major role in this function (Wang and Liu, 2008).

2.1 Tau Gene

Tau proteins are translated from a single gene located over 100 kilobases on the long arm of chromosome 17 (17q21.1) and consisting of 16 exons (Fig. 1a) (Kosik, 1993; Neve et al., 1986; Andreadis, 2006). Expression of human tau complex regulation and regarding its control of gene expression, the presence of specific transcription factors (such as AP2/SP1) are involved, even though RNA-based control has been recently proposed that awaits more evidence (D'Souza and Schellenberg, 2006). The tau gene is transcribed into nuclear RNA, which by alternative splicing yields different RNA species (Fig. 1b) (Goedert et al., 1989b). Tau's interactions with microtubules are mediated by the tubulin-binding domains/repeat at the C-terminal region (Fig. 1c) as detailed below.

In vitro and transgenic animal models have demonstrated that different mutations impair protein function, promote tau fibrilization, or perturb tau gene splicing, leading to aberrant and distinct tau aggregates (Cairns et al., 2004). The mutations in the autosomal dominant tauopathies are of two types: intronic mutations that disrupt the splicing of tau and missense mutations that alter the function of tau. The splicing of tau is tightly regulated so as to maintain the relative proportion of the 3R-tau and 4R-tau isoforms. The function of tau is normally tightly regulated through phosphorylation. It is likely that loss of this normal regulation somehow results in tau aggregation, although it should be noted that, in vitro, the mutations also increase tau aggregation itself. Transgenic mice carrying tau mutations have been shown to exhibit behavioural and neuropathological correlates of the disease process. This indicates that tau aggregates are a sign of primary pathology. Tau aggregation without amyloid pathology is sufficient to cause dementia in mice and in

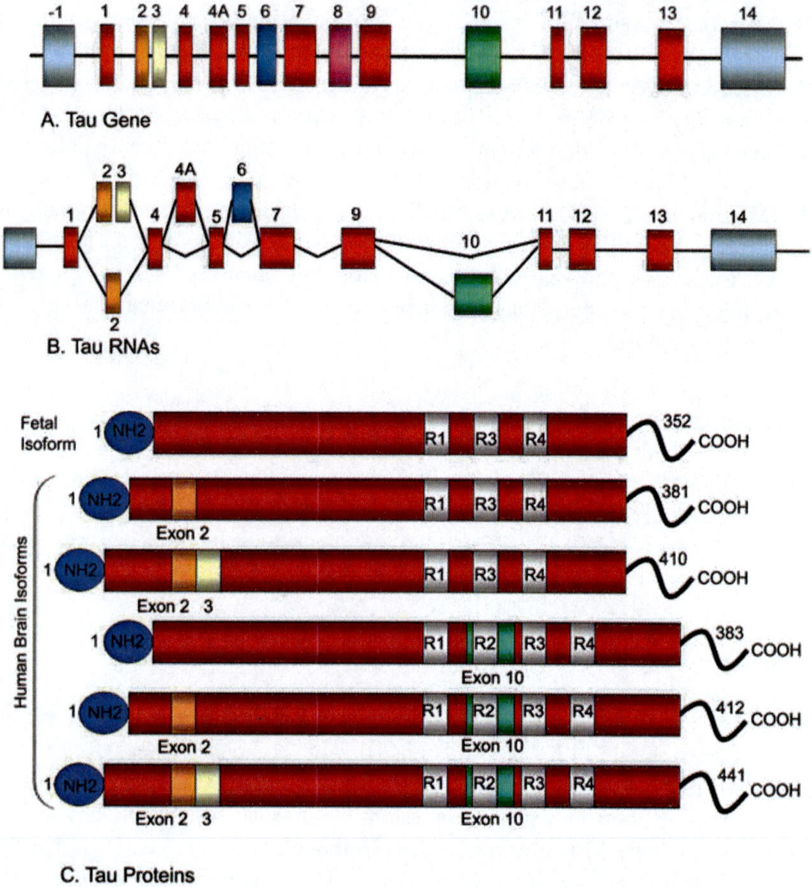

Fig. 1 Schematic representation of the human tau gene (**a**), RNA (**b**), and brain tau isoforms (**c**). Alternatively spliced exons (2, 3, and 10) are shown in *orange*, *cream*, and *green*, respectively. The tandem repeats (3 or 4) are shown in *white bars*. The number of aminoacids is indicated on the *right-hand side* (modified from Spillantini and Goedert, 2000)

humans and hence is likely to be a pathogenic protein (Lovestone and McLoughlin, 2002).

2.2 *Structure, Cellular Localization, and Putative Functions of Tau Protein*

Tau protein was first discovered as an acid- and heat-stable protein essential for microtubule assembly. It was identified as a factor that lowered the concentration at which tubulin polymerizes into microtubules in the brain. Tau is one such neuronal MAP, which localizes primarily in the axon with a molecular tau weight of approximately 50,000–64,000 Da. When purified from the brain it has very little secondary structure (Kosik, 1993). Because of its enormous molecular weights and meager tendency to form highly ordered 3D crystal lattices, tau had long evaded high-resolution structure determination (Margittai and Langen, 2004). Nevertheless, it is now known that several specific proteins serve to stabilize microtubules and tau is one such MAP. Tau organizes MTs into evenly spaced parallel assemblies known as MT-bundles. It is also well recognized that tau plays an important role in the assembly of tubulin monomers into MT to constitute the neuronal microtubule network, maintains the MT structure,(Alonso et al., 2001), and establishes links between microtubules and other cytoskeletal elements and proteins (Buee et al., 2000). It has been proposed that, in vivo, tau induces the bundling and stabilization of cellular microtubules, promotes neurite outgrowth, and establishes and maintains neuronal cell polarity.

Tau fulfills several functions critical for neuronal formation and health. It discharges its functions by producing multiple isoforms via intricately regulated alternative splicing. These isoforms modulate tau function in the normal brain by altering the domains of the protein, thereby influencing its conformation and post-translational modifications and hence its affinity for microtubules and other ligands. Disturbances in tau expression result in disruption of the neuronal cytoskeleton and formation of pathological tau structures (e.g., neurofibrillary tangles in brains of patients with Alzheimer's disease; Andreadis, 2006). It is not clear, however, how tau's ability to decrease the dynamic instability of microtubules directly relates to these changes in microtubule organization and cell morphology (Leger et al., 1994). They are expressed predominantly in axons of central nervous system (CNS) neurons, and also are found in axons of peripheral nervous system (PNS) neurons, but are barely detectable in CNS astrocytes and oligodendrocytes (Trojanowski and Lee, 2002). It is one of the major and most studied MAPs in the central nervous system.

Tau has been shown to be a highly asymmetric protein, compatible with the long rod structure, when observed by electron microscopy (Hirokawa et al., 1996). Earlier studies had suggested that it is a hydrophilic protein having a random coil structure. Attempts to crystallize tau alone or tau associated with microtubules have been unsuccessful and under normal conditions, tau demonstrates the properties of a highly soluble natively unfolded molecule, essentially devoid of secondary or tertiary structural elements (Mandelkow et al., 1996; Friedhoff et al., 2000). The above findings are in agreement with the detailed information of the structure of tau protein in its soluble state obtained through conventional methods such as electron microscopy, spectroscopy, and X-ray diffraction (Crowther et al., 1989, 1992, 1994; Schweers et al., 1994; Wille et al., 1992). However, tau molecule adopts specific secondary and tertiary structures that interact in an orderly fashion to produce the highly regular filaments in Alzheimer's disease and many other neurodegenerative disorders (Gamblin et al., 2000). It is unlikely that random aggregation of tau in the disease state could lead to creation of these highly ordered structures. It can be understood that tau is partially folded when interacting with microtubules by a combined cryoelectron microscopy and tomographic 3-D analysis with freeze-drying and high-resolution unidirectional surface shadowing.

In the adult human brain, six brain-specific isoforms are generated by alternative mRNA splicing of 11 exons (Fig. 1b; Buee et al., 2000). Alternative splicing of exons 2 (E2), 3 (E3), and 10 (E10) give rise to six tau isoforms that range from 352 to 441 amino acids (Fig. 1c; Goedert et al., 1989a). The isoforms differ in whether they contain three (tau-3L, tau-3S, or tau-3: collectively 3R) or four (tau-4L, tau-4S, or tau-4: collectively 4R) tubulin-binding domains/repeats (R) of 31 or 32 amino acids each at the C-terminal (i.e., the presence or absence of a fourth 31-amino acid repeat, coded by exon 10). They also differ on whether they have two (tau-3L, tau-4L), one (tau-3S, tau-4S), or no (tau-3, tau-4) repeats of 29 amino acids each in the N-terminal portion of the molecule (Fig. 1c; Trojanowski and Lee, 2002). Thus, exons 2 and 3 are alternatively spliced cassettes; exon 2 exists alone, but exon 3 never appears independent of exon 2 (Andreadis, 2006). As seen above, the isoforms are designated according to the number of MBDs they possess at the C-terminal. Each of these repeats can be divided into two parts, one composed of an 18-residue sequence that contains the minimal region with tubulin-binding capacity and the less conserved domain called the interrepeat. The proportion of these tau isoforms, as well as their phosphorylation status, changes during development (Kosik, 1990a, b; Kosik et al., 1986; Goedert et al., 1989a, b; Buee et al., 2000). In fact, in the adult human brain, the proportion of 3R-tau to 4R-tau isoforms is about 50% each, but that of tau-3L (or 4L), tau-3S (or 4S), and tau-3 (or 4) is about 54, 37, and 9%, respectively. Tau 4R binds microtubules with a greater affinity and can displace the previously bound tau 3R from microtubules that may produce physiological consequences in cells. As tau is developmentally regulated, only the shortest tau isoform (tau-3) is expressed in fetal brain, but all six isoforms are seen in the adult human brain (Trojanowski and Lee, 2002). Moreover, different neurons seem to have different tau isoforms. In the peripheral nervous system too, there is a high molecular weight tau isoform expressing the exon 4A, whose product forms a protein known as big tau with an approximate size of 100 kDa (Couchie et al., 1990; Goedert et al., 1992).

In summary, brain tau isoforms have been divided into two large domains such as projection domains (containing the amino terminal two-thirds of the molecule) and the microtubule binding domain (containing the carboxy terminal one-third of the molecule) (Avila et al., 2004). The projection domain has been further divided into two regions: the amino terminal region with a high proportion of acidic residues and the proline-rich region. The microtubule binding domain has also been subdivided into the basic, true tubulin-binding region and the acidic carboxy terminal region. Several distinct roles have been proposed for the projection domain including that of determining the spacing between axonal microtubules, interactions with other cytoskeletal proteins, or cation binding due to the presence of the acidic residues. The proline-rich domain plays an important role in interaction with proteins with SH3 domains, facilitating the binding of tau to the plasma membrane proteins (Brandt and Lee, 1993, 1994).

A structure function relationship study by (Gamblin et al., 2000) points out that tau contains very few predicted structural elements, but these small structural units, whether predicted or measured through biochemical/biophysical methods, are likely

Table 1 Predictable Secondary Structure Forming Motifs in Tau Protein

	Sequence	Possible Secondary Structure and Possible Implications
1	^{7}EFEVME^{12}	α-helix; tau aggregation is accompanied by a dramatic conformational change that brings the amino-terminus in close proximity to the microtubule-binding repeats.
2	^{31}MH^{32}	β-strand; imparts structural rigidity?
3	^{117}EAAGHVTQ^{124}	α-helix; region is adjacent to the second most hydrophobic region of the molecule; helical wheel analysis shows amphipathic making it a candidate to interact with the microtubule-binding repeats or the carboxy-terminus of tau.
4	^{226}VAVVR^{230}	β-strand; direct interaction of this region with the microtubule-binding repeats to strengthen the interaction of tau with microtubules.
5	^{275}VQII^{278} and ^{306}VQIVY^{310}	β-strands; core structural elements for filament elongation.
6	^{315}LSKVTSKCGSL^{325}	α-helix; amphipathic in nature structural element that participates in tau–tau interactions in the aggregated state.
7	^{338}EVK^{340} and ^{361}TH^{362}	β-strands; positioning in MTBR4 suggests that they could contribute to microtubule binding.
8	^{426}ATLADEVSASLA^{437}	α-helix; structural element can directly bind to some other element of tau and prevent the aggregation of the molecules.

responsible for the normal and abnormal functions of tau by providing sites for specific interactions either with microtubules or other tau molecules. The study identifies certain structural elements that have a potential for adopting secondary structures which have a role in normal and pathological conditions (Table 1).

The microtubule-associated tau protein participates in the organization and integrity of the neuronal cytoskeleton. Even though tau protein is mainly a neuronal MAP, localizing primarily in the axon, it has been demonstrated that tau is present within the somatodentritic compartment of neurons (Migheli et al., 1988). Tau that is present in the somatodendritic compartment is phosphorylated mainly in its proline-rich region, whereas when this region becomes dephosphorylated, it can be found principally in the distal region of the axon (Mandell and Banker, 1996). Additionally, presence of nuclear tau isoforms has been identified in human neuroblastoma cells. Nuclear tau was found to be associated with both the fibrillar regions of interphase nucleoli and the nucleolar organizer regions of mitotic chromosomes; recent studies have also shown that nuclear tau is mainly present at the internal periphery of nucleoli, partially colocalizing with the nucleolar protein nucleolin and human AT-richα-satellite DNA sequences organized as constitutive heterochromatin (Sjoberg et al., 2006). The import of tau into the nucleus is possibly either by interacting with other nuclear proteins containing a nuclear import sequence or catalyzed by the basic 3/4 repeat MBDs. Because nuclear tau has also been found in neurons from patients with AD, aberrant nuclear tau could affect the nucleolar organization during the course of AD. Recent studies suggest that binding of tau to DNA was in an aggregation-dependent, and a phosphorylation-independent, manner. Tau has also been seen to localize on ribosomes. One possible function for this nonmicrotubule, ribosome-associated tau is to target ribosomes to microtubules for transport into the somatodendritic compartment of neurons to facilitate local protein synthesis. The localization and the function of the different tau isoforms are regulated by its posttranslational modifications.

2.3 Posttranslational Modifications of Tau

Heterogeneity of tau is due to several posttranslational modifications including phosphorylation, glycosylation, ubiquitinylation, oxidation, nitration, cross-linking, deamidation, glycation, truncation by protein cleavage (Avila et al., 2004), prolyl isomerization, association with heparan sulfate proteoglycan, and modification by advanced glycation end-products (Chen et al., 2004). Phosphorylation is of functional and clinical importance and therefore has been the most studied of all posttranslational modifications.

2.3.1 Phosphorylation

Tau is a phosphoprotein and its biological activity is regulated by phosphorylation (Feijoo et al., 2005; Ihara et al., 1986; Grundke-Iqbal et al., 1986). Tau phosphorylation is developmentally regulated and fetal tau is more highly phosphorylated in the embryonic compared to the adult CNS. The degree of phosphorylation of the six adult tau isoforms decreases with age. The tau phosphorylation sites are clustered in regions flanking the microtubule binding repeats. Phosphorylation at these sites has been reported in normal tau, however, the phosphorylation negatively regulates microtubule binding. Although the relative importance of individual sites for regulating the binding of tau to microtubules is unclear, phosphorylation of some sites such as Serine-262 and 396 has been reported to play a dominant role in reducing the binding of tau to microtubules. Both sites are phosphorylated in fetal tau and they are hyperphosphorylated in all six adult human brain tau isoforms that form paired helical filaments (PHFs) in Alzheimer's disease. Other potentially important phosphate acceptor sites also have been described and it is possible that phosphorylation at multiple phosphate acceptor sites regulates the binding of tau to microtubules (Trojanowski and Lee, 2002).

Hyperphosporylation dislodges tau from the microtubule surface, resulting in compromised axonal integrity and accumulation of toxic tau peptides (Drewes, 2004). The definite role of tau in the neurodegenerative process is still not very clear. Recent studies suggest that, before forming fibrils but after becoming hyperphosphorylated, tau actively contributes to neurodgeneration (Takashima, 2008).

There are 85 putative serine or threonine phosphorylation sites on the longest CNS tau isoform. Phosphorylation sites were characterized by phospho-dependent tau antibodies, phopho-peptide mapping, mass spectrometry, and NMR. Most of the phosphorylation sites surround the microtubule-binding domains in the proline-rich region of the C terminal region of tau. A large number of serine/threonine protein kinases have been suggested to play a role in regulating tau functions in vivo, however, this aspect of tau biology remains controversial. The major candidate tau kinases include mitogen-activated protein kinase, glycogen synthase kinase 3β, cyclin-dependent kinase 2 (cdk2), cyclin-dependent kinase 5, cAMP-dependent protein kinase, Ca^{2+}/calmodulin-dependent protein kinase II, microtubule-affinity regulating kinase, and stress-activated protein kinases (Trojanowski and Lee, 2002). The available evidence points to glycogen synthase kinase-3 being the predominant tau kinase in the brain, although other kinases also phosphorylate tau (Lovestone and McLoughlin, 2002). Protein phosphatases counterbalance the effects of tau kinases, although their role in vivo is unclear. In vitro experiments showed that inhibition of protein phosphatases by okadaic acid in cultured human neurons was followed by increased tau phosphorylation, decreased tau binding to microtubules, selective destruction of stable microtubules and rapid axonal degeneration (Trojanowski and Lee, 2002). Today we know more than 20 protein kinases that can phosphorylate the tau protein and they are grouped into several different types as follows.

1. The proline-directed protein kinases (PDPK), which phosphorylate tau on serine or threonine residues that are followed by a proline residue. This group includes tau protein kinase II (cdk5), cdk2, MAP kinase (p38), JNK, and other SAPKs (Baumann et al., 1993; Holzer et al., 1994; Goedert et al., 1997).
2. The nonproline-directed protein kinases (NPDPKs) such as tau-tubulin kinase 1 and 2, protein kinase A (PKA), protein kinase C (PKC), PKB/AKT, calmodulin (CaM) kinase II, MARK kinases, or CK I and II that modify residues close to acidic residues mainly in exons 2 and 3. NPDPKs modify Ser or Thr residues that are not followed by prolines (Kitano-Takahashi et al., 2007; Sergeant et al., 2005).
3. The protein kinases that phosphorylates tau on serine or threonine residues, sometimes but not always, followed by a proline residue which includes tau protein kinase I (glycogen synthase kinase 3, GSK3; Hanger et al., 1992).
4. The tyrosine protein kinases such as Src kinases and c-abl.

Along with tau kinases, several phosphatases, such as protein phosphatase [PP1, PP2A, PP2B (calcineurin), and PP2C] regulate the extent of tau phosphorylation (Goedert et al., 1992; Yamamoto et al., 1995) However, only PP1, PP2, and PP2B have been shown to dephosphorylate abnormally hyperphosphorylated tau (Gong et al., 1994a,b,c). PP2A binds to tau through its tubulin binding region (Sontag et al., 1999). Mutations in this region could decrease the capacity of PP2A to bind to tau and, as a consequence, produce an increase in tau phosphorylation, a feature that has been observed in some FTDP-17 patients bearing such mutations.

The Physiological Role of Tau Phosphorylation

The phosphorylation of tau at specific sites is the predominant mechanism by which tau function is regulated. Dynamic, site-specific phosphorylation of tau is essential for its proper functioning. Interestingly, phosphorylation at different sites could take place in different tau isoforms. This could be due to the different cellular localization or subcellular compartmentalization of the different tau isoforms, or the fact that different kinases or phosphatases can modulate tau phosphorylation in many different ways (Avila et al., 2004) as described below. A tentative figure explaining the role of tau phosphorylation is depicted in Fig. 2.

Microtubule Binding

The ability to bind and stabilize microtubules is a hallmark of tau, and it is becoming obvious that phosphorylation of a few specific sites plays a significant role in regulating tau–microtubule interactions. Phosphorylation of the KXGS motifs within the microtubule-binding repeats of tau strongly reduces the binding of tau to microtubules in vitro and probably in vivo (Biernat and Mandelkow, 1999; Drewes et al., 1995). The NPDPK phosphorylation mainly occurs at the tubulin-binding region of the tau molecule. Therefore, it has been suggested that this type of modification could result in a decrease in the binding of tau to microtubules whereas modification of tau by PDPK mainly affects tau self-aggregation.

In vitro studies have shown that phosphorylation of Ser262 alone is sufficient to attenuate significantly the ability of tau to bind microtubules in vitro. GSK3 plays an important role in regulating tau phosphorylation under normal and pathological conditions. Two types of GSK3 phosphorylation have been proposed: primed (following prior phosphorylation of the substrate by another kinase) or unprimed phosphorylation. Primed phosphorylation appears to occur at Thr231 and affects microtubule binding (which means that Ser235 must be phosphorylated first to get efficient phosphorylation of Thr231), whereas unprimed phosphorylation can take place at serine-396 or -404 and does not appear to affect microtubule binding (Goedert et al., 1994). Although the relative importance of individual sites for regulating the binding of tau to microtubules is unclear, phosphorylation of some sites such as Serine-262 and 396 has been reported to play a dominant role in reducing the binding of tau to microtubules. In addition to serine/threonine modifications, phosphorylation at tau tyrosines has been also reported. It is known that fetal tau is more extensively phosphorylated than adult tau and promotes microtubule assembly less efficiently than the latter.

Altered Intracellular Trafficking/Polarity

This occurs by two major mechanisms explained below.

(a) *Neurite outgrowth*: Neuritic extension is essential for maintaining synaptic plasticity and in CNS repair. Evidence regarding the role of tau was obtained from earlier studies on cultured cerebellar neurons using antisense oligonucleotides (Kosik, 1990a,b). Primary cultures of hippocampal neurons lacking tau exhibit decreased rates of neurite extension and inhibited neuronal polarization (i.e., the development of axons and dendrites; Dawson et al., 2007), during axonogenesis, tau function appears to be locally regulated by phosphorylation. Tau mRNA may also play a role in the determination of polarity, inasmuch as it is

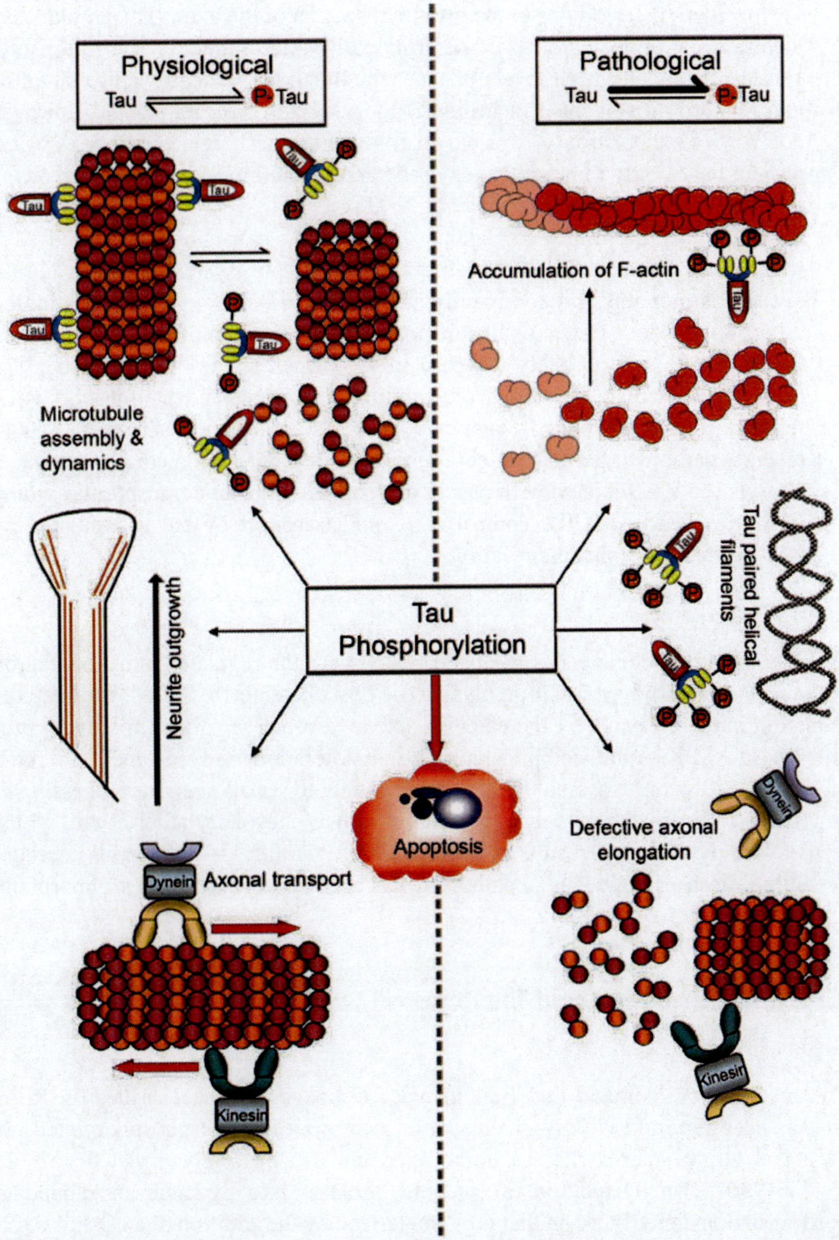

Fig. 2 Schematic representation of the physiological and pathological functions of tau phosphorylation. When the phosphorylation state of tau is appropriately coordinated, it plays a role in regulating neurite outgrowth, axonal transport, and microtubule stability and dynamics. However, in pathological conditions in which there is an imbalance in the phosphorylation/dephosphorylation of tau, aberrant tau phosphorylation can cause tau/actin filament formation, disrupt microtubule-based processes owing to decreased microtubule binding, and perhaps even increase cell death (modified from Johnson et al., 2004 (129))

localized on microtubules in the proximal section of the axon. This implies that local translation of tau at a site determined by microtubular organization sites can lead to locally high levels of tau, which can theoretically cause bundling and forward movement of neurites (Billingsley and Kincaid, 1997). However, contrary to expectation, it was surprising to find that targeted deletions of tau protein led to only minor changes in the axonal calibre of small-fibre axons in restricted brain regions (Harada et al., 1994).

(b) *Axonal transport*: Tau also regulates axonal transport. In mouse models in which tau are overexpressed in the central nervous system, there is almost always axonopathy, predominantly in spinal cord neurons. Tau can inhibit kinesin-dependent fast axonal transport in cell culture models (Ebneth et al., 1998), and this is probably the case in vivo when tau is overexpressed. The primary mechanism by which tau inhibits kinesin-dependent transport is by reducing the attachment frequency of the motors. Tatebayashi et al. (2004) recently demonstrated that, in cell culture models, GSK3β-mediated tau phosphorylation is associated with proper anterograde organelle transport providing further evidence that the control of axonal transport by tau is regulated by GSK3β-mediated phosphorylation.

Altered Proteolysis

One key structural change that has been linked with the regulated phosphorylation of tau is altered turnover and proteolysis. The best characterized effect has been the reduction in tau cleavage by the calcium-activated protease calpain following protein kinase A (PKA)-induced phosphorylation. Alterations in lysosomal trafficking of tau and/or loss of lysosomal function are thought to set off aberrant processing of tau. There is increasing evidence that inappropriate phosphorylation of tau, which leads to tau dysfunction, results in decreased cell viability. Indeed, in all neurodegenerative diseases in which tau pathology has been observed, the tau is abnormally phosphorylated.

2.3.2 Other Modifications of Tau Proteins

Glycosylation

The presence of *N*-linked and (mucin-type) *O*-linked oligosaccharides on PHF-tau has been reported with *N*-glycosylation occurring in hyperphosphorylated tau (Wang et al., 1996a,b) whereas unmodified tau can be *O*-glycosylated (Arnold et al., 1996). O-glycosylation of cytosolic proteins is a dynamic and abundant posttranslational modification that is characterized by the addition of an O-linked N-acetylglucosamine (O-GlcNAc) in the serine or threonine in the vicinity of proline residues by an O-GlcNAc transferase. This relationship between phosphorylation and *O*GlcNAc glycosylation of tau proteins may play a role in transcriptional regulation, cell cycle regulation, protein degradation, cell activation, and the correct assembly of multimeric protein complexes and in the nuclear localization of tau.

Ubiquitnylation

In addition to phosphorylation, tau is also subjected to ubiquitinylation. Ubiquitin is a 76 amino acid protein that flags the proteins to be degraded in an ATP-dependent manner. Ubiquinated tau has been found in inclusion bodies found in Pick's disease or Parkinson's disease or in PHFs in AD.

Glycation: Proteins with slow turnover rates can be modified at lysine residues by nonenzymatic reactions involving the condensation of a sugar aldehyde or ketone group with the NH2-groups of the lysines. The products of this reaction can undergo irreversible changes to form the advanced glycation end-products that can result in the cross-linking of the modified proteins. Tau isolated from PHF is glycated, and this glycation might be involved in the loss of microtubule binding and aggregation of PHF into more complex aggregates possibly through receptors for AGE products (RAGE) present on microglial cells and subsets of neurons (Ledesma et al., 1994; Yan et al., 1994). AGE-tau conjugates may also more likely to form covalent cross-links with each other.

Oxidation

The presence of one or two cysteines in the tau isoforms lacking or containing exon 10 has raised the possibility of tau forming dimers through the formation of inter-molecular S–S bonds (Schweers et al., 1995). In this case, the oxidation of tau could result in its aberrant aggregation.

Truncation

Tau truncation has been defined as the cleavage of tau that occurs at the glutamic acid residue 391 (Wischik et al., 1995). This modification could facilitate aberrant tau aggregation.

Deamidation

The deamidation of tau at asparagine (or glutamine) residues has also been described and could also play a role in tau aggregation.

Prolyl Isomerization

The peptidyl-prolyl *cis/trans* isomerase Pin1 isomerizes the peptide bond of a phosphorylated serine or phosphorylated threonine followed by a proline. Through isomerization of pSer-Pro or PThr-Pro, Pin1 regulates a number of proteins. Together with its ability to regulate phosphorylation and conformation of tau proteins, Pin1 is considered a potential neuroprotective function against AD.

2.4 Turnover of Tau Protein

Intracellularly, proteins are constantly being synthesized, modified, and after the specific function is over, they are routed for degradation. Major factors that determine the half-life of proteins are the presence of signals that control its degradation and stabilization. Among the signals for degradation, the presence of a specific amino-terminal residue, the PEST sequence (a PEST sequence is a peptide sequence which is rich in proline (P), glutamic acid (E), serine (S), and threonine (T)), and the destruction box must be considered. PEST sequences are present in the tau molecule, whereas the tubulin-binding region is a glycine-rich sequence. However, little is known about the implication of these two regions in the stability of the tau protein. Of the stabilization signals, amino acid repeats containing polyglutamine, glycine, or alanine residues are among the most common.

The existence of ubiquitin-independent proteosomal degradation of tau protein has been reported based on the fact that tau protein is degraded by the 20S proteasome in vitro (David et al., 2002). However, there is strong evidence of tau degradation by the ubiquitin–proteasome system (UPS) after ubiquitylation of the protein. The degradation of tau by different proteases has been studied including cleavage by caspases. Tau is also a substrate of the calcium-activated protease calpain, although phosphorylated tau is more resistant to proteolysis by calpain degradation than unphosphorylated tau. Cathepsin D has been shown to cleave tau proteins, generating fragments similar to those found in NFT. HSP-27 is a protein that facilitates degradation of pathogenic hyperphosphorylated tau by an unknown mechanism.

3 Tauopathies

Neurodegenerative diseases with filamentous inclusions can be classified into these groups: (i) tauopathies, (ii) alfa-synucleinopathies, (iii) polyglutamine disorders, and (iv) ubiquitin disorders. Tauopathy is the most common group (Tolnay and Probst, 1999). All these diseases have in common the presence of aberrant tau aggregates. Tau was first implicated in the pathogenesis of Alzheimer's disease when it was discovered to be a major component of the neurofibrillary tangle (Lovestone and McLoughlin, 2002). Subsequently the occurrence of neurofibrillary tangles in a wide range of conditions led to the suggestion that tau deposition may be an incidental nonspecific finding associated with cell death or cellular dysfunction. Later the discovery of close to 20 different mutations in tau in FTDP-17 clearly showed that dysfunction of tau protein causes neurodegeneration and dementia (Spillantini and Goedert, 2000). Table 2 gives a list of diseases grouped under tauopathies. Overlap of some clinical and histopathological features occurs between tauopathies. For example, neurofibrillary tangles (NFT) can be seen in AD, FTDP-17, progressive supranuclear palsy (PSP) and neuropil threads can be seen in AD, cortico-basalganglionic degeneration (CBD), FTDP-17, and PSP. Silver impregnation technique usually detects most of the tau inclusions. Immunohistochemistry with monoclonal antibodies against phosphorylated or nonphosphorylated epitopes of tau, however, are invaluable for detecting the full extent of tau deposition.

Immunohistochemical studies have also revealed tau-positive glial inclusions in both oligodendrocytes and astrocytes in most, although not in all tauopathies

Table 2 Neurodegenerative disorders with abundant filamentous tau inclusions

Dementia Syndromes	Parkinsonism Plus Syndromes	Neuromuscular Disorders	Genetic/Metabolic Disorders
Alzheimer's disease Tangle-only dementia Argyrophilic grain disease Pick's disease Presenile dementia with tangles and calcification Progressive subcortical gliosis Familial frontotemporal dementia and Parkinsonism linked to chromosome 17 Subacute sclerosing panencephalitis Gerstmann–Straussler–Scheinker disease with tangles Prion protein amyloid angiopathy	Corticobasal degeneration Progressive supranuclear palsy Multiple system atrophy Postencephalitic Parkinsonism Dementia pugilistica	Amyotrophic lateral sclerosis with Parkinsonism Dementia complex of Guam Non-Guamanian motor neuron disease with neurofibrillary tangles Myotonic dystrophy Inclusion body myositis	Down's syndrome Hallervorden–Spatz disease Niemann–Pick disease type C

Modified from Tolnay and Probst (1999), Trojanowski and Lee (2002), and Robert and Mathuranath (2007).

(Tolnay and Probst, 1999). Among the tauopathies, the most studied is AD. The analyses of other types of dementia with tau pathology have usually been performed in comparison with AD (Spillantini and Goedert, 2000). Based on electrophoretic pattern, several classes of tau aggregation are presently described. (1) AD and Parkinsonism dementia complex (six tau isoforms); (2) PSP and CBD (three isoforms with exon 10 corresponding sequence); (3) Pick's disease (PiD) (three isoforms without exon 10), and (4) myotonic dystrophy-the shortest tau isoform (Caparros-Lefebvre et al., 2002).

3.1 Frontotemporal Dementia

3.1.1 Spectrum

Frontotemporal dementia (FTD) is one of the common forms of primary degenerative dementias after Alzheimer's disease and can affect presenile individuals. It is a clinically heterogeneous disorder characterized by alterations in language and/or behavior. Rarely it may be associated with Parkinsonism or amyotrophy. Depending on the initial and core clinical feature FTD is further classified into primary progressive aphasia (PPA) when language impairment is the initial and core feature, frontal variant FTD (fvFTD) when behaviour and personality are the initial and core features, and FTD with Parkinsonism (FTDP) when Parkinsonism is an associated initial and core feature in fvFTD and FTD with amyotrophy when muscle amyotrophy attributable to spinal motor neuron involvement is an associated initial and core feature in fvFTD. PPA is further subclassified as progressive nonfluent aphasia (PNFA) if initial language impairment is expressive aphasia and as semantic dementia (SD) if the initial language impairment is sensory aphasia.

3.1.2 Clinical Features

FTD occurs most commonly between the ages 45 and 65 years. There is an equal incidence in men and women. The mean duration of illness is 8 years ranging from 2 to 20 years. A family history of dementia is present in about half of cases. In fvFTD the salient clinical characteristic is a profound alteration in character and social conduct. The behavioural disorders consist of decline in personal hygiene and grooming, mental rigidity and inflexibility, distractability and impersistence, hyperorality and dietary changes, perseveration and stereotypy, and utilization behaviour. There will be early emotional blunting and loss of insight. Language abnormality consists of altered speech output, stereotypy of speech, echolalia, perseveration, and mutism (Snowden et al., 2002). On cognitive testing the patients show attentional deficits, poor abstraction abilities, difficulty shifting mental set, and a tendency to perseveration. Other cognitive domains including memory, praxis, and visuospatial skills remain relatively preserved early in the disease. In PNFA, speech production is effortful with phonological and grammatical errors and word retrieval difficulties. In SD there is a naming and word comprehension problem although speech expression remains fluent, effortless, and grammatical. Behavioral changes occur in all subtypes of FTD as the disease progresses. Clinical diagnosis of various types of FTD is based on consensus criteria (Neary et al., 1998).

3.1.3 Neuropathology

The typical changes seen in all subtypes of FTD are atrophy of the prefrontal and anterior temporal neocortex. The subtype determines the distribution of the pathology. In FTD there is prominent bilateral and usually symmetrical involvement of the frontal lobes. In PNFA, atrophy is asymmetric, involving chiefly the left frontotemporal lobes, concentrated in Broca's area. In SD, atrophy is typically bilateral and is most marked in the anterior temporal neocortex, with inferior and middle temporal gyri being predominantly affected (Snowden et al., 2002).

Histology shows neuronal loss from superficial cortical layers with spongiosis and variable degrees of gliosis with or without Pick cells or Pick bodies (Mathuranath et al., 2000).

Pathologically, FTD is heterogeneous; some cases may show tau- or ubiquitin-positive inclusions, or rarely they may lack distinctive histological features. Sensitive methods for detecting tau abnormalities and ubiquitin are essential in the neuropathological evaluation of FTD (Josephs, 2007).

3.1.4 Neurochemistry and Neurobiology

Tau mutations have been well characterized in FTDP-17. The mutations described are missense, deletion, or silent mutation in the coding region or intronic mutation located close to exon 10. Coding region mutations are located in the microtubule-binding repeat region or close to it. Mutation in exon 10 affects only 4R isoforms whereas mutation in exons 9, 12, and 13 affects all isoforms. Coding region mutations reduce the ability of tau to interact with microtubules. Intronic mutation leads to a net increase in 4R isoforms. This leads to filamentous tau pathology (Spillantini and Goedert, 2000). Bird et al. described three separate families with frontotemporal dementia, having the same molecular mutation in exon 10 of the tau gene (*P301L*). However, differences were seen in clinical features as well as pathological findings among diseased members of the family, in spite of the same mutation in all. This led to the suggestion that in addition to the mutation, there are other environmental and/or genetic factors also influencing the phenotype (Bird et al., 1999).

FTD occurs in familial and sporadic forms; frontotemporal dementia and Parkinsonism linked to chromosome 17 (FTDP-17) is one type characterized by mutations in MAPT (gene encoding tau), which are inherited in an autosomal dominant pattern. Its neuropathological hallmark is an abundance of hyperphosphorylated tau protein and degeneration of neurons and glia. In many cases of sporadic frontotemporal dementia (FTD) and in FTD caused by tau mutations (FTDP-17) there is disruption of the normal splicing of tau leading to the aberrant expression of tau isoforms and neurodegeneration. This suggests a central role for tau in the pathogenesis of FTD. However, more than half the cases of sporadic FTD show no tau deposition (Sutherland et al., 2007; Van Deerlin et al., 2007) or distinctive molecular pathological features involving the deposition of 3R tau protein. However, there may be further tau polymorphisms that remain to be identified, outside the standard sequenced regions, which may have a role in the pathogenesis of PiD. Furthermore, PiD can be distinguished immunohistochemically from other tauopathies by the deposition of abnormally hyperphosphorylated tau, and by the absence of phosphorylation of tau Ser262, which is specifically recognized by the antitau antibody 12-E8 (Morris et al., 2002).

3.2 *Alzheimer's Disease*

3.2.1 Clinical Features

Alzheimer's disease (AD) is characterized initially by progressive loss of recent memory and orientation. As the disease progresses, language, visuospatial, and executive function can also be impaired with neurobehavioural abnormalities developing in the late stages. Typical cases have prominent early memory disturbances and subsequent other cognitive abnormalities occur. Behavioural and psychological symptoms develop during midstage of the disease and include delusions and hallucinations, anxiety, sleep disturbances, and depression. Presentation and clinical course are variable. In the majority of patients neurological examination other than mental status examination will be normal. As disease progresses mild abnormalities in tone and reflexes such as grasp, root, and suck reflexes are seen and at the end-stage patient is mute, incontinent, and bedridden with swallowing difficulty and flexion deformity. There are no diagnostic markers on blood examination. Diagnosis can be established antimortem using established criteria with an accuracy of 90% (Bradley et al., 2004).

3.2.2 Neuropathology

The outstanding gross pathology feature is hippocampal atrophy which is even picked up on neuroimaging. Histopathology is characterized by reduced synaptic density and neuronal loss in selected brain areas. The two molecular pathologies that coexist in AD are amyloidosis and tau pathology. Amyloidosis consists of extracellular aggregation of Aβ peptides into amyloid plaques. Tau pathology is characterized by the intraneuronal aggregation of tau proteins into abnormal filaments forming neurofibrillay tangles deposition that begins in the *trans* entorhinal and entorhinal cortex and spreads from there to hippocampus, temporal neocortex, and beyond that (Bradley et al., 2004). Amyloidosis is closely related to etiology and tau pathology is strongly correlated to the clinical expression of the disease. Little is known about the relationship between amyloid-β precursor protein (APP) and tau pathologies (Delacourte et al., 2002). The quantification of Aβ in the different brain areas demonstrates that the spreading pathway of tau pathology remains constant, whatever the cortical distribution of Aβ aggregates (Delacourte et al., 2002).

3.2.3 Neurochemistry and Neurobiology

Although there is a consensus that Aβ is upstream of tau in the pathological cascade in AD there is little knowledge on either the nature of the interaction or how direct it is. The relationship between Aβ toxicity and tau protein was identified in a study where neurons knocked out of the MAPT gene were less toxic to Aβ neurotoxicity. As more and more of the relationship between amyloid-β precursor protein (APP) and tau pathologies is emerging, one of the missing links in our fully understanding AD is being unravelled. The present section deals exclusively with the neurofibrillary pathology in AD and related disorders.

Tau Hyperphosphorylation, Assembly of P-Tau, and Tau Pathology

Tau pathology corresponds to the intraneuronal aggregation of microtubule-associated tau proteins into abnormal filaments. Paired helical filaments (PHF) are the most characteristic cytoskeletal alterations affecting numerous neurons in AD. Using a combined immunocytochemical and biochemical approach (Iqbal et al., 1989) demonstrated for the first time that the microtubule-associated protein tau, a normal brain cytoskeletal protein, is a component of the PHF. The authors also indicated for the first time that posttranslational modification of tau such as phosphorylation might occur which would allow it to assemble either alone or together with other components of PHF (Iqbal et al., 1989). However, once the pathological process is initiated, several factors join in making this disease course torturous to follow.

The hyperphosphorylation hypothesis of AD was derived from the sequential discoveries that the PHF-related proteins from AD brain are, in fact, persistently phosphorylated forms of tau protein (Iqbal et al., 1989; Kosik et al., 1986). In AD Thr-231, Ser-396 (adjacent to microtubule binding region), and Ser-262 (in the microtubule-binding region) of tau protein are phosphorylated in PHF (Biernat and Mandelkow, 1999; Biernat et al., 1992). Tau loses its biological activity such as tubulin binding on hyperphosphorylation and becomes resistant to proteolytic degradation. This stability of PHF tau is conferred by increased levels of transglutaminase-induced $\epsilon\gamma$ glutamyl lysine bonds which lead hyperphosphorylated tau aggregated into a high molecular weight polymeric complex. In AD, two major classes of tau kinases are involved: tau kinase I (GSK3) and tau kinase II (Cdk5). Prior phosphorylation by tau kinase I is followed by tau kinase II emphasising the role of tau kinase II (Cdk5) on tau phosphorylation in AD. There is evidence that a dysregulation of Cdk5 as a result of proteolytic cleavage of its regulatory subunit p35 by calpain yields a fragment p25 that allows constitutive activation of this kinase, which appears to underlie NFT formation in AD.

The phosphorylation of tau may promote a conformational change, possibly resulting in an increase in the number of α helices in the secondary structure of tau which is confirmed by the observation that the content of α helices is greater in tau isolated from PHF. This conformational change could involve the binding of the amino terminal region of the tau molecule to its microtubule-binding region (Avila et al., 2004). This facilitates its aggregation into PHF.

In a healthy neuron, phosphorylated tau is transported back (retrograde) to the cell body from the axonal compartment where it is ubiquitinated and undergoes proteolysis by the cell's quality control machinery. Hyperphosphorylated tau cannot be dephosphorylated by phosphatases such as PP2A and PP2B due to steric hindrance caused by the cross-linking. Hyperphosphorylated tau predominant in the axonal compartment thus gets trapped.

In AD, heparan sulphate or other sulphated glycosaminoglycans induce p-tau to dimerise followed by polymerisation, glycation, ubiquitination, and cross-linking which leads to the formation of insoluble PHF. A high concentration of free p-tau is required for assembly; often a deamidation facilitates polymerisation. The sulfoglycosaminoglycans (sGAG, polyanions), present along with tau in NFT accelerate polymerisation of more p-tau. Sulphated glycosaminoglycans stimulate phosphorylation of tau by a number of protein kinases, and prevent tau from binding to microtubules and from promoting microtubule assembly (Hasegawa et al., 1997). Earlier studies have shown that heparan sulphate and hyperphosphorylated tau colocalize in so-called "pretangle" neurones in AD brains, which suggests that the accumulation of heparan sulphate precedes the hyperphosphorylation of tau (Bancher et al., 1987; Braak et al., 1994). Other polyanions such as glutamate-rich regions present in the C-terminal region of tubulin also facilitate aggregation, this requiring the third tubulin binding motif of the tau molecule; hence 3R isoforms only give rise to PHF (Perez et al., 1996). Moreover, the tau found in PHFs are often ubiquitinated and glycated in AD (Ledesma et al., 1994)

Tau isoforms with three repeats assemble into PHFs, whereas tau isoforms with four repeats assemble into SFs. It is proposed that oxidation of Cys to produce disulfide cross-linking favors tau assembly into PHF in tau3R molecules whereas tau4R having 2 Cys may permit the formation of intramolecular S–S bonds, and thus assemble into straight filaments (SFs; Barghorn and Mandelkow, 2002). However, the interplay among phosphorylation as a targeting event, tau glycation, oxidation, and heparan-mediated PHF formation remains to be elucidated. Excess calcium influx seems one reasonable downstream mediator for Aβ-induced toxicity that could explain the activation of certain kinases such as calcium–calmodulin-dependent protein kinase II and transglutaminase that induce phosphorylation of tau, inducing their aggregation into PHF (Johnson et al., 1997)

Neurobiology of NFT

It has been proposed that NFTs are an independent feature accumulating slowly with age within the median temporal lobes. However, under the influence of altered amyloid metabolism, which leads to the formation of βAmyloid plaques during the initial stages of the disease, there is an acceleration of NFT formation that spreads further to neocortex (Price and Morris, 2004). However, quantification of Aβ in the different brain areas demonstrates that the spreading pathway of tau pathology remains constant, whatever the cortical distribution of Aβ aggregates (Delacourte et al., 2002). In AD brain, tau is abnormally hyperphosphorylated, cleaved, and conformationally changed and is present mostly as PHF (Komori, 1999). Conformational changes in tau have been proposed to be among the earliest neurobiological changes in AD. Unlike normal tau, which contains two or three phosphate groups, the cytosolic hyperphosphorylated tau from AD brain (AD P-tau) contains 5–9 mol of phosphate/mol of the protein (Alonso et al., 2001). The finding that phosphorylated tau fails to bind microtubules led to the hypothesis that phosphorylated PHF tau was microtubule-assembly incompetent, leading to the destabilisation of neuronal cytoskeleton and cellular demise (Mandelkow et al., 1996)

In AD neurons, there are three main types of neurofibrillary lesions (NFLs) according to their intracellular localization; neurofibrillary tangles (NFTs) inside the cell body and apical dendrites, and neuropil threads (NThs) in distal dendrites and dysrophic neurites, associated with senile plaques. Morphologically, three major subtypes of NFTs can be distinguished, corresponding to different evolutionary stages of these lesions.

1. Pretangle stage: The earliest stage is characterized by the accumulation of hyperphosphorylated tau in the somatodendritic domain of affected neurons without forming any PHF or SF. Pretangle neurons are nonargyrophilic and therefore, only detectable with antitau antibodies (Bancher et al., 1987; Braak et al., 1994)

2. Classical NFT: In AD tau is hyperphosphorylated, cleaved, and conformationally changed (Mondragon-Rodriguez et al., 2008). It is present mostly as paired helical filaments (PHF). Conformational changes in tau have been proposed to be among the earliest neurobiological changes in AD (Haroutunian et al., 2007). Unlike normal tau, which contains two or three phosphate groups, the cytosolic hyperphosphorylated tau from AD brain (AD P-tau) contains 5–9 mol of phosphate/mol of the protein (Alonso et al., 2001). The criteria for AD diagnosis have been revised to include the presence of tau pathology for diagnosing definite AD. Neuropathologically, AD is defined by the accumulation of two types of insoluble fibrous material–extracellular amyloid protein in the shape of senile plaques and intracellular neurofibrillary lesions made of abnormally and hyperphophorylated tau protein. The NFL consists of neurofibrillary tangles, neuropil threads and dystrophic neurites associated with senile plaques. Ultrastructurally NFL contains PHL as a major fibrous component and SFs as a minor component. Both types are formed of the six brain tau isoforms that are hyperphosphorylated. The mechanism of NFL formation in AD is only now beginning to be understood. Tau is first phosphorylated, accumulates in cytoplasm, and then dimers form followed by polymers. Polymers form the globular particles. As the concentration of globular particles increases, tau fibrils, PHFs, and SFs appear (Fig. 2). Tau 3R isoforms assemble and give rise to twisted PHFs with a diameter of 20 nm and Tau 4R isoforms give rise to SFs with a diameter of 15 nm. In the ultrastructure of PHFs, the three microtubule-binding regions are found in the core of the structure, making it inaccessible to antibodies raised against tubulin-binding region (Goedert and Klug, 1999).

   Although globular tau particles were found in non-AD brain, their concentration was lower and there were no filaments. This suggests that the trigger converting a non-AD brain to an AD brain is the concentration of globular tau particles. Unlike other tauopathies, glial tau pathology is only a minor feature of AD (Tolnay and Probst, 1999). Typical NFTs comprising the PHFs and SFs are identified by means of the silver staining technique (using Bielschowsky and Gallyas stains).

3. Ghost tangles (tombstones): These comprise the extracellular residua of NFTs after the degeneration of neurons.

Generally, the degree of dementia correlates with the sites and severity of tau-based NFT accumulation (Holzer et al., 1994). Based on the manner of NFT formation on various brain areas of the AD patient, during the course of disease development, six neuropathological stages are defined.

Stages I and II (Transentorhinal stages) – Earliest NFTs are observed in the entorhinal cortex and in this stage, patients are cognitively unimpaired.

Stages III and IV (Limbic stages): Characterized by more extensive formation of NFTs in the entorhinal cortex and CA1 region of the hippocampus and in this stage patients experience mild cognitive impairment (MCI, the preclinical stage of AD)

Stages V and VI (Isocortical stage): Development of abundant NFTs in the neocortex and the patients are diagnosed as AD in this stage

Interaction of ApoE and Tau

It is hypothesized that apoE isoforms may differentially influence tau pathology. ApoE is the apolipoprotein constituent of chylomicron, which functions in the transport and internalization of cholesterol and triglyceride-rich lipoproteins. ApoEϵ4 isoform is the most important known risk factor of AD. Apo Eϵ3, which is considered to be neuroprotective can bind to the microtubule binding region of nonphosphorylated tau. In vitro studies have shown that phosphorylation of tau prevents this interaction supporting that apoEϵ3 prevents hyperphosphorylation of tau and thus reduces AD risk. The C-terminal bioactive fragments of apoEϵ4 stimulate hyperphosphorylation of tau. This microtubule destabilization impedes cellular transport leading to functional deficits and ultimately results in neuronal loss (Goedert, 1993; Strittmatter et al., 1994).

ApoEϵ4 can assume the "molten globule" conformation inside the late endosome due to the low pH as it possesses Arg residue in its 112 and 158 position. In this state, the molecule attains increased permeability through the membrane of late endosome and escapes into the cytosol, where it undergoes proteolysis by AECE (Apo E cleaving Enzyme) generating truncated bioactive neurotoxic apoEϵ4 fragments. These fragments interact with p-tau and neurofilament protein of high molecular weight thus leading to the formation of pre-NFT-like filaments which are deposited near the mitochondria. The toxic apoEϵ4 fragments from the pre-NFT complex bind on the F0–F1 ATPase of mitochondria resulting in mitochondrial dysfunction and thus disrupting the energy metabolism of neuron (Huang and Stultz, 2007).

Cell Cycle Re-Entry Hypothesis – Role of Tau

The latest and inevitably the most controversial theory for the pathogenesis of AD is the "cell cycle re-entry hypothesis" which postulates that the formation of amyloid plaques and NFT is due to the reactivation of cell division like phenomena in the aging neurons. When the cellular signals controlling the connectivity/synaptic plasticity of a healthy neuron which is integrated in a synaptic network is lost, it re-enters the cell cycle and attempts to proliferate, but G1 arrest halts its progression through the cell cycle, thus resulting in cell death, which could be the mechanism by which age-related neuronal death occurs in the CNS.

In AD neurons, the G1/S regulatory mechanisms seem to be absent/fail and these neurons are allowed to progress through DNA replication into the G2 phase of the cell cycle. One of the prominent features of the G2 phase is the activation of kinases such as MAP Kinase and Cdk2&5, required for pushing the cell through division. The MAP Kinase initiates mitotic re-entry and Cdk2&5 ensure the cell cycle progression. The activation of these kinases is associated with a downregulation of phosphatases and a gradual destabilisation of the microtubule system in an attempt to prepare the cell for cytokinesis. This destabilisation of microtubules results in an increased amount of free tau in the neuron. The concomitant activation of kinases will inevitably lead to the phosphorylation of this pool of tau which is not bound to tubulin. This in turn prevents the further reattachment of tau to the microtubules and favors the formation of PHF typical of the NFTs.

Neither cytokinesis is possible for the terminally differentiated neurons, nor are the cells apoptosed due to the lack of downstream caspases which leads to a state of drawn-out agony termed *Aposklesis* instead of apoptosis. As a consequence, neurons survive for longer periods in the G2 phase of the cell cycle and the aberrant cellular mechanisms will alter neuronal metabolism generating reactive ions and free radicals which bring about oxidative injury to the neuron. It is now known that oxidative stress alone causes no increase in tau phosphorylation, but subtly changes the pattern of tau phosphorylation. The G2 arrest also upregulates the cell cycle inhibitor proteins such as GSK3 and CDK inhibitors. The combination of oxidative stress plus tau phosphorylation (by GSK3) may lead to significant inhibition of tau degradation. This brings about more tau phosphorylation and aggregation (Reynolds et al., 1997).

The Genomic Instability Model of AD – The Tau Connection

The microtubule disassembly in AD neurons gives rise to aneuploidy by causing defects in the organization of the mitotic spindle. Because the chromosomes carrying tau and APP (Chr 17 and Chr 21) are relatively smaller, hence they are subjected to telomere shortening and undergo breakage fusion bridge cycles (BFB) resulting in genomic instability due to overexpression of these genes.

With all the compelling evidence accumulated thus far, the criteria for AD diagnosis have been revised to include the presence of tau pathology for diagnosing definite AD.

3.2.4 The Tau Isoforms in AD

Tau gene expression is developmentally regulated by an alternative splicing mechanism and six different isoforms exists in the adult human brain such as Tau3L, Tau3S, Tau3, Tau4L, Tau4S, and Tau4 (Buee et al., 2000). Distinct sets of tau isoforms expressed in different neuronal populations could lead to different pathologies. In AD brain, all six isoforms are present as a part of NFTs in their hyperphosphorylated state. PHF tau consists of a characteristic triplet of peptides that run as three major bands at 60, 64, and 68 kDa (hence leading to the name A68 protein) and a minor band of 72 kDa apparent molecular mass (Flament et al., 1989). The slowed electrophoretic mobility of A68 proteins was restored by treatment with alkaline phosphatase at elevated temperature (Lee, 1990).

Fetal tissue has only one tau isoform, namely tau3. Interestingly, fetal tau is highly phosphorylated. This similar pattern of phosphorylation at AD-related sites on tau led to the idea that fetal tau phosphorylation was recapitulated at AD (Goedert, 1993). Even though fetal tau is highly phosphorylated, there is no evidence that it begins to self-aggregate. There are developmental factors which account for this finding. It can be suggested that something more than stoichiometry and the site of phosphorylation play a role in PHF formation. Trafficking and compartmentalisation may be of prime importance.

3.3 Progressive Supranuclear Palsy

3.3.1 Clinical Features

The onset of symptoms in PSP is insidious, and the evolution of symptoms may vary. The mean age at PSP onset ranges from 55 to 70 years, with a few cases beginning as early as 45 years of age. The first symptoms of PSP are usually postural instability and falls, which occur either at the onset of the disease or during the first year. Cognitive or behavioural changes usually begin in the first year, but rarely occur at the disease onset. After postural instability, dysarthria is the second most common symptom of PSP and bradykinesia is the third most common problem. A variety of oculomotor abnormalities are seen of which vertical gaze palsy is considered the hallmark of the disease. In general the early symptoms and signs of PSP steadily worsen. The mean duration of illness is 5–7 years. The most common cause of death is pneumonia (Litvan et al., 1996).

3.3.2 Neuropathology

PSP is characterised by subcortical pathology of destruction in the globus pallidus, subthalamic nucleus, midbrain/pontine reticular formation, and homogeneous depletion of substantia nigra pars reticulate. The brain stem involvement typically consists of damage to the supranuclear eye-movement control areas: the interstitial nucleus of cajal, the rostral interstitial nucleus of the medial longitudinal fasciculus, and the nucleus of Darkschewitsch. There is no biological marker for the diagnosis of PSP. Neuropathological examination is the gold standard. The neuropathological characteristics of PSP include a high density of neurofibrillary tangles and neuropil threads in the basal ganglia and brainstem with a characteristic distribution. Neuropil threads are filamentous structures scattered throughout the neuropil, occurring independently of neurofibrillary tangles. Tau-positive astrocytes or processes in areas of involvement help to confirm the diagnosis. Nonspecific neuronal loss and gliosis are also seen (Litvan et al., 1996).

PSP is the most extensively studied disease in this group. A similar propensity for damage of the globus pallidus as well as substantia nigra has been demonstrated in CBD, Parkinsonism dementia complex of Guam and postencephalitic Parkinsonism. The clinical observation that many of these diseases affect supranuclear control of gaze further suggests that these diseases, all of which involve tau protein deposition, share similarities in their topographic pathology.

3.3.3 Neurochemistry and Neurobiology

The NFT in PSP is made of straight filaments and predominantly 4R tau. In vitro experiments have confirmed that 4R tau forms into straight filament NFTs. It

Table 3 Tau Neurodegenerative Diseases Classified by the Isoform of Tau Expressed

	All Isoforms	4R Tau	3R Tau
Familial	Alzheimer's disease Frontotemporal dementia	Frontotemporal dementia	Myotonic dystrophy
Sporadic	Alzheimer's disease Parkinsonism dementia complex of Guam Postencephalitic Parkinsonism	Progressive supranuclear palsy Corticobasal degeneration	Frontotemporal dementia (Pick's disease)

Modified from Morris et al. (1999) and Robert and Mathuranath (2007).

remains unclear if these differences in tau protein deposition reflect a topographically restricted pattern determined by tau gene expression. If true, then one could postulate that cortical neurons when damaged in AD express all six isoforms of the tau gene whereas substantia nigra neurons preferentially express 4R isoforms. Alternatively the tau protein expression may represent a more fundamental aspect of the disease as well. The isoform(s) of tau expressed in various diseases is (are) shown in Table 3. Interestingly, the morphology of tangles varies with the isoform. Thus when all six isoforms are expressed, as in AD, they are paired helical filaments, whereas in 4R diseases they are either twisted ribbon filaments (as in PiD) or straight filaments (as in PSP) (Morris et al., 1999). In addition, in many of these diseases, tau pathology has been described in glial cells as well. This contrasts with the findings in AD, where tau pathology is largely restricted to the neurons (Komori, 1999).

3.4 Corticobasalganglionic Degeneration

3.4.1 Clinical Features

This is a rare entity. The disease has an insidious onset above 40 years of age. The initial symptoms are often subtle. Patient may complain of inexplicable loss of motor skill in one limb. This is typically and predominantly a motor disorder. The affected limb shows rigidity, slowness of movement, apraxia, and often stimulus-sensitive myoclonus. A striking finding is the presence of a fixed dystonic posture of the hand and fingers with flexion of some fingers into the palm, combined with extension of others. Some of the subjects exhibit alien limb behaviour, although less frequently than the above motor signs. Here the subject fails to recognize his limb as part of his body. This results in wide ranges of abnormalities starting from a tendency of the limb to drift away and assume abnormal postures especially when the eyes are closed or attention is diverted to involuntary unwanted movements of the limb which the patient just can't control. With the passage of time, motor symptoms spread to affect other limbs, most often the ipsilateral arm or leg, with later spread to the contralateral limbs. Gait difficulties eventually emerge. The disease is progressive, with increasing manual and locomotor disability, and usually with a progressive dysarthria. Eventually the patients lose their ability to walk or to use their arms and the disease finally leads to a bedridden state (Riley et al., 1990; Rinne et al., 1994).

3.4.2 Neuropathology

Pathologically CBD is characterised by asymmetric cortical atrophy predominantly in the perirolandic area. Histology is the gold standard for the diagnosis of CBD. It is distinguished by achromatic ballooned neurons that stain positive with tau-immunohistochemical stains. Tau-positive inclusions are also found in glial cells in the atrophic areas of the cortex (Fig. 3). Neuronal loss and gliosis with basophilic inclusions are prominent in the substantia nigra and pallidum. The other subcortical structures involved include the caudate, putamen, subthalamus, and the dentate nucleus (Mathuranath et al., 2000).

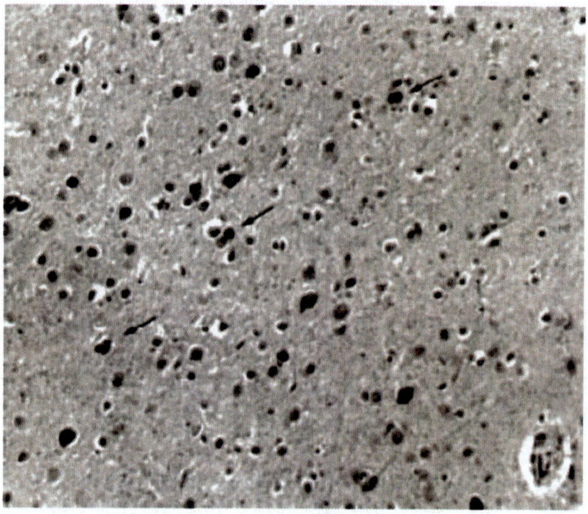

Fig. 3 Photomicrograph of temporal neocortex showing cytoplasmic inclusions in glial cells within subcortical white matter (Mathuranath et al., 2000)

3.5 Multiple System Atrophy (MSA)

3.5.1 Clinical Features

Multiple system atrophy (MSA) is a progressive sporadic neurodegenerative disease of undetermined etiology that causes Parkinsonism and cerebellar, autonomic, and pyramidal dysfunction in varying combinations (Osaki et al., 2002). Historically, it has been described in three different ways: olivopontocerebellar atrophy (OPCA), striatonigral degeneration (SND),and the Shy–Drager syndrome (SDS). The considerable clinicopathological overlap among these three subgroups led Graham and Oppenheimer to introduce MSA as an umbrella term (Wenning, 2000).

3.5.2 Neuropathology

Macroscopically, the brain in MSA shows varying degrees of atrophy of the cerebellum, cerebellar peduncles (especially the middle and inferior peduncles), pons, medulla, and also the posterolateral putamen. There may be loss of pigment in the substantia nigra and also discolouration of the striatum (notably the putamen). Excessive iron accumulation has been demonstrated within the striatum to account for this pigmentary change. The histopathological hallmark is the formation of alpha-synuclein-positive glial cytoplasmic inclusions (GCIs) in oligodendroglia. Alpha-synuclein aggregation is also found in glial nuclear inclusions, neuronal cytoplasmic inclusions (NCIs), neuronal nuclear inclusions (NNIs), and dystrophic neuritis (Burn and Jaros, 2001).

4 Future Direction

4.1 Tau as a Diagnostic Marker

The role of CSF tau in the diagnosis of dementias is being studied quite extensively. The most commonly used assay for tau is the ELISA (Schooneboom et al., 2006). Tau is readily measured in cerebrospinal fluid (CSF) by ELISA, and is one of the most extensively studied AD biomarkers, as thousands of patients with AD, as well as various normal and diseased control subjects, have been studied. Measures of total tau as well as species of phospho-tau detected by antibodies in CSF correlate best with a diagnosis of AD. Total tau is two- to threefold higher in CSF of patients with AD compared with normal controls. The release of tau and species of phospho-tau from degenerating neurons harbouring NFTs and dystrophic neurites in AD is thought to account for the increase in CSF levels of these proteins. So, the effects of therapies that ameliorate tau-mediated neurodegeneration and the further accumulation of species of pathological tau could be reflected in CSF tau biomarker assays (Shaw et al., 2007).

Tau is one of the components of the core neuropathologic changes in AD that can be measured in CSF and has been frequently studied as a candidate diagnostic biomarker. It has been shown in one of the studies that with the use of a cutoff value of 234 pg/ml, CSF tau demonstrated a sensitivity of 85%, specificity of 84%, positive predictive value of 87%, and positive likelihood ratio of 5.3 in distinguishing patients with AD from cognitively normal controls. CSF tau was also useful in distinguishing AD from frontotemporal dementia and diffues Lewy body dementia, although the positive likelihood ratio of correct identification was only 3:1. It has been proposed that CSF tau may also be helpful in differentiating AD from vascular dementia (Leszek et al., 2003). Recently phosphorylated tau level in CSF has been found to be useful as a biological marker of AD (Hampel et al., 2004). Correlation also has been found between impairment of cerebral metabolism, estimated throughout FDG-PET, and CSF Tau protein levels (Ceravolo et al., 2008).

The source of CSF tau remains unclear but most likely is related to the degeneration of neurofibrillary tangle-laden neurons. The protein has not been well characterised in the CSF and may exist in fragmented forms. Although a report has indicated that it may require 3–5 months for elevated CSF tau levels to return to normal after an acute stroke, the clearance rate of tau from the CSF in patients with neurodegenerative dementia remains unknown. Elevated CSF tau has also been reported in mild cognitive impairment (Schonknecht et al., 2007), CBD, FTD, and in many patients with Creutzfeldt–Jakob disease (CJD) (Clark et al., 2003). It has been shown that a higher amount of phosphorylated tau in CSF in sporadic CJD is associated with a rapid progression of the disease to akinetic mutism (Van Everbroeck et al., 2002). Increased CSF Tau has been reported in nondegenerative conditions such as Wernicke's encephalopathy (Matsushita et al., 2008), neurosyphilis (Paraskevas et al., 2007), and multiple sclerosis (Terzi et al., 2007).

4.2 Tau as a Therapeutic Target

In tau transgenic mice, mutations have been discovered which precipitate authentic forms of neurofibrillary degeneration. So these animal models very much resemble AD even though Tau mutation is not described in AD. The discovery of this model has paved the way for testing various therapeutic models targeting tau (Roder and Hutton, 2007). There is good evidence indicating that targeting the kinases responsible for tau protein hyperphosphorylation should be able to arrest and maybe even reverse the degeneration. Because the main tau phosphorylation kinases regulate many other physiological functions apart from tau phosphorylation, it will be important to design kinase inhibitors that minimize the potential toxicity arising from inhibition of such off-target functions, while maximizing effective suppression of tau hyperphosphorylation. On the basis of the current body of knowledge about relevant protein kinase inhibitors, molecules with varying kinase inhibition selectivity profiles and appropriate bioavailability properties might be designed as pharmacological tools and, it is hoped, drug candidates for the treatment of neurodegenerative tauopathies (Mazanetz and Fischer, 2007). Recently using a triple transgenic model it has been shown that Aβ immunotherapy leads to the clearance of early tau pathology. The clearance of the tau pathology is mediated by the proteasome and is dependent on the phosphorylation state of tau, as hyperphosphorylated tau aggregates are unaffected by the Aβ antibody treatment (Oddo et al., 2004). It has been shown that the inhibition of the proteasome leads to a bidirectional degradation of Tau (Delobel et al., 2005).

4.3 Research Avenues

The relation between tau protein and α-synneuclein and amyloid has to be delineated. This will further clarify the role of tau protein in Parkinsonism as well as amyloidosis As well as understanding the tau neurobiology in CBD and PSP.

References

Alonso AD, Zaidi T, Novak M, Barra HS, Grundke-Iqbal I, Iqbal K (2001) Interaction of tau isoforms with Alzheimer's disease abnormally hyperphosphorylated tau and in vitro phosphorylation into the disease-like protein. J Biol Chem 276:37967–37973

Andreadis A (2006) Misregulation of tau alternative splicing in neurodegeneration and dementia. Prog Mol Subcell Biol 44:89–107

Arnold CS, Johnson GV, Cole RN, Dong DL, Lee M, Hart GW (1996) The microtubule-associated protein tau is extensively modified with O-linked N-acetylglucosamine. J Biol Chem 271:28741–28744

Askanas V, Engel WK (1998) Sporadic inclusion-body myositis and its similarities to Alzheimer disease brain. Recent approaches to diagnosis and pathogenesis, and relation to aging. Scand J Rheumatol 27:389–405

Avila J (2000) Tau aggregation into fibrillar polymers: taupathies. FEBS Lett 476:89–92

Avila J, Lucas JJ, Perez M, Hernandez F (2004) Role of tau protein in both physiological and pathological conditions. Physiol Rev 84:361–384

Bancher C, Lassmann H, Budka H, Grundke-Iqbal I, Iqbal K, Wiche G et al (1987) Neurofibrillary tangles in Alzheimer's disease and progressive supranuclear palsy: antigenic similarities and differences. Microtubule-associated protein tau antigenicity is prominent in all types of tangles. Acta Neuropathol 74:39–46

Barghorn S, Mandelkow E (2002) Toward a unified scheme for the aggregation of tau into Alzheimer paired helical filaments. Biochemistry 41:14885–14896

Baumann K, Mandelkow EM, Biernat J, Piwnica-Worms H, Mandelkow E (1993) Abnormal Alzheimer-like phosphorylation of tau-protein by cyclin-dependent kinases cdk2 and cdk5. FEBS Lett 336:417–424

Biernat J, Mandelkow EM (1999) The development of cell processes induced by tau protein requires phosphorylation of serine 262 and 356 in the repeat domain and is inhibited by phosphorylation in the proline-rich domains. Mol Biol Cell 10:727–740

Biernat J, Mandelkow EM, Schroter C, Lichtenberg-Kraag B, Steiner B, Berling B, et al. (1992) The switch of tau protein to an Alzheimer-like state includes the phosphorylation of two serine-proline motifs upstream of the microtubule binding region. EMBO J 11:1593–1597

Billingsley ML, Kincaid RL (1997) Regulated phosphorylation and dephosphorylation of tau protein: effects on microtubule interaction, intracellular trafficking and neurodegeneration. Biochem J 323(Pt 3):577–591

Bird TD, Nochlin D, Poorkaj P, Cherrier M, Kaye J, Payami H et al (1999) A clinical pathological comparison of three families with frontotemporal dementia and identical mutations in the tau gene (P301L). Brain 122(Pt 4):741–756

Braak H, Braak E, Strothjohann M (1994) Abnormally phosphorylated tau protein related to the formation of neurofibrillary tangles and neuropil threads in the cerebral cortex of sheep and goat. Neurosci Lett 171:1–4

Bradley WG, Daroff RB, Fenichel GM, Jankovic J (2004) Neurology in clinical Practice, vol 2. Butterworth Heineman, Philadelphia

Brandt R, Lee G (1993) Functional organization of microtubule-associated protein tau. Identification of regions which affect microtubule growth, nucleation, and bundle formation in vitro. J Biol Chem 268:3414–3419

Brandt R, Lee G (1994) assembly Orientation, and stability of microtubule bundles induced by a fragment of tau protein. Cell Motil Cytoskeleton 28:143–154

Buee L, Bussiere T, Buee-Scherrer V, Delacourte A, Hof PR (2000) Tau protein isoforms, phosphorylation and role in neurodegenerative disorders. Brain Res Brain Res Rev 33:95–130

Burn DJ, Jaros E (2001) Multiple system atrophy: cellular and molecular pathology. J Clin Pathol Mol Pathol 54:419–426

Cairns NJ, Lee VM, Trojanowski JQ (2004) The cytoskeleton in neurodegenerative diseases. J Pathol 204:438–449

Caparros-Lefebvre D, Sergeant N, Lees A, Camuzat A, Daniel S, Lannuzel A et al (2002) Guadeloupean parkinsonism: a cluster of progressive supranuclear palsy-like tauopathy. Brain 125:801–811

Ceravolo R, Borghetti D, Kiferle L, Tognoni G, Giorgetti A, Neglia D, et al. (2008) CSF phosporylated TAU protein levels correlate with cerebral glucose metabolism assessed with PET in Alzheimer's disease. Brain Res Bull 76:80–84

Chen F, David D, Ferrari A, Gotz J (2004) Posttranslational modifications of tau–role in human tauopathies and modeling in transgenic animals. Curr Drug Targets 5:503–515

Clark CM, Xie S, Chittams J, Ewbank D, Peskind E, Galasko D et al (2003) Cerebrospinal fluid tau and beta-amyloid: how well do these biomarkers reflect autopsy-confirmed dementia diagnoses? Arch Neurol 60:1696–1702

Couchie D, Legay F, Guilleminot J, Lebargy F, Brion JP, Nunez J (1990) Expression of Tau protein and Tau mRNA in the cerebellum during axonal outgrowth. Exp Brain Res 82:589–596

Crowther T, Goedert M, Wischik CM (1989) The repeat region of microtubule-associated protein tau forms part of the core of the paired helical filament of Alzheimer's disease. Ann Med 21:127–132

Crowther RA, Olesen OF, Jakes R, Goedert M (1992) The microtubule binding repeats of tau protein assemble into filaments like those found in Alzheimer's disease. FEBS Lett 309: 199–202

Crowther RA, Olesen OF, Smith MJ, Jakes R, Goedert M (1994) Assembly of Alzheimer-like filaments from full-length tau protein. FEBS Lett 337:135–138

David DC, Layfield R, Serpell L, Narain Y, Goedert M, Spillantini MG (2002) Proteasomal degradation of tau protein. J Neurochem 83:176–185

Dawson HN, Cantillana V, Chen L, Vitek MP (2007) The tau N279K exon 10 splicing mutation recapitulates frontotemporal dementia and parkinsonism linked to chromosome 17 tauopathy in a mouse model. J Neurosci 27:9155–9168

Delacourte A, Sergeant N, Champain D, Wattez A, Maurage CA, Lebert F, et al. (2002) Nonoverlapping but synergetic tau and APP pathologies in sporadic Alzheimer's disease. Neurology 59:398–407

Delobel P, Leroy O, Hamdane M, Sambo AV, Delacourte A, Buee L (2005) Proteasome inhibition and Tau proteolysis: an unexpected regulation. FEBS Lett 579:1–5

Dickson DW (1999) Tau and synuclein and their role in neuropathology. Brain Pathol 9:657–661

Drewes G (2004) MARKing tau for tangles and toxicity. Trends Biochem Sci 29:548–555

Drewes G, Trinczek B, Illenberger S, Biernat J, Schmitt-Ulms G, Meyer HE et al (1995) Microtubule-associated protein/microtubule affinity-regulating kinase (p110mark). A novel protein kinase that regulates tau-microtubule interactions and dynamic instability by phosphorylation at the Alzheimer-specific site serine 262. J Biol Chem 270:7679–7688

D'Souza I, Schellenberg GD (2006) Arginine/serine-rich protein interaction domain-dependent modulation of a tau exon 10 splicing enhancer: altered interactions and mechanisms for functionally antagonistic FTDP-17 mutations Delta280K AND N279K. J Biol Chem 281: 2460–2469

Ebneth A, Godemann R, Stamer K, Illenberger S, Trinczek B, Mandelkow E (1998) Overexpression of tau protein inhibits kinesin-dependent trafficking of vesicles, mitochondria, and endoplasmic reticulum: implications for Alzheimer's disease. J Cell Biol 143:777–794

Feijoo C, Campbell DG, Jakes R, Goedert M, Cuenda A (2005) Evidence that phosphorylation of the microtubule-associated protein Tau by SAPK4/p38delta at Thr50 promotes microtubule assembly. J Cell Sci 118:397–408

Flament S, Delacourte A, Hemon B, Defossez A (1989) Characterization of two pathological tau protein, variants in Alzheimer brain cortices. J Neurol Sci 92:133–141

Friedhoff P, von Bergen M, Mandelkow EM, Mandelkow E (2000) Structure of tau protein and assembly into paired helical filaments. Biochim Biophys Acta 1502:122–132

Gamblin TC, King ME, Dawson H, Vitek MP, Kuret J, Berry RW, et al. (2000) In vitro polymerization of tau protein monitored by laser light scattering: method and application to the study of FTDP-17 mutants. Biochemistry 39:6136–6144

Goedert M (1993) Tau protein and the neurofibrillary pathology of Alzheimer's disease. Trends Neurosci 16:460–465

Goedert M, Cohen ES, Jakes R, Cohen P (1992) p42 MAP kinase phosphorylation sites in microtubule-associated protein tau are dephosphorylated by protein phosphatase 2A1. Implications for Alzheimer's disease [corrected]. FEBS Lett 312:95–99

Goedert M, Hasegawa M, Jakes R, Lawler S, Cuenda A, Cohen P (1997) Phosphorylation of microtubule-associated protein tau by stress-activated protein kinases. FEBS Lett 409:57–62

Goedert M, Jakes R, Crowther RA, Cohen P, Vanmechelen E, Vandermeeren M, et al. (1994) Epitope mapping of monoclonal antibodies to the paired helical filaments of Alzheimer's disease: identification of phosphorylation sites in tau protein. Biochem J 301(Pt 3): 871–877

Goedert M, Klug A (1999) Tau protein and the paired helical filament of Alzheimer's disease. Brain Res Bull 50:469–470

Goedert M, Spillantini MG, Jakes R, Rutherford D, Crowther RA (1989a) Multiple isoforms of human microtubule-associated protein tau: sequences and localization in neurofibrillary tangles of Alzheimer's disease. Neuron 3:519–526

Goedert M, Spillantini MG, Potier MC, Ulrich J, Crowther RA (1989b) Cloning and sequencing of the cDNA encoding an isoform of microtubule-associated protein tau containing four tandem repeats: differential expression of tau protein mRNAs in human brain. EMBO J 8:393–399

Gong CX, Grundke-Iqbal I, Damuni Z, Iqbal K (1994a) Dephosphorylation of microtubule-associated protein tau by protein phosphatase-1 and -2C and its implication in Alzheimer disease. FEBS Lett 341:94–98

Gong CX, Grundke-Iqbal I, Iqbal K (1994b) Dephosphorylation of Alzheimer's disease abnormally phosphorylated tau by protein phosphatase-2A. Neuroscience 61:765–772

Gong CX, Singh TJ, Grundke-Iqbal I, Iqbal K (1994c) Alzheimer's disease abnormally phosphorylated tau is dephosphorylated by protein phosphatase-2B (calcineurin). J Neurochem 62:803–806

Grundke-Iqbal I, Iqbal K, Tung YC, Quinlan M, Wisniewski HM, Binder LI (1986) Abnormal phosphorylation of the microtubule-associated protein tau (tau) in Alzheimer cytoskeletal pathology. Proc Natl Acad Sci U S A 83:4913–4917

Hampel H, Buerger K, Zinkowski R, Teipel SJ, Goernitz A, Andreasen N et al (2004) Measurement of phosphorylated tau epitopes in the differential diagnosis of Alzheimer disease: a comparative cerebrospinal fluid study. Arch Gen Psychiatry 61:95–102

Hanger DP, Mann DM, Neary D, Anderton BH (1992) Tau pathology in a case of familial Alzheimer's disease with a valine to glycine mutation at position 717 in the amyloid precursor protein. Neurosci Lett 145:178–180

Harada A, Oguchi K, Okabe S, Kuno J, Terada S, Ohshima T et al (1994) Altered microtubule organization in small-calibre axons of mice lacking tau protein. Nature 369:488–491

Haroutunian V, Davies P, Vianna C, Buxbaum JD, Purohit DP (2007) Tau protein abnormalities associated with the progression of alzheimer disease type dementia. Neurobiol Aging 28:1–7

Hasegawa M, Crowther RA, Jakes R, Goedert M (1997) Alzheimer-like changes in microtubule-associated protein Tau induced by sulfated glycosaminoglycans. Inhibition of microtubule binding, stimulation of phosphorylation, and filament assembly depend on the degree of sulfation. J Biol Chem 272:33118–33124

Hirokawa N, Funakoshi T, Sato-Harada R, Kanai Y (1996) Selective stabilization of tau in axons and microtubule-associated protein 2C in cell bodies and dendrites contributes to polarized localization of cytoskeletal proteins in mature neurons. J Cell Biol 132:667–679

Holzer M, Holzapfel HP, Zedlick D, Bruckner MK, Arendt T (1994) Abnormally phosphorylated tau protein in Alzheimer's disease: heterogeneity of individual regional distribution and relationship to clinical severity. Neuroscience 63:499–516

Huang A, Stultz CM (2007) Conformational sampling with implicit solvent models: application to the PHF6 peptide in tau protein. Biophys J 92:34–45

Ihara Y, Nukina N, Miura R, Ogawara M (1986) Phosphorylated tau protein is integrated into paired helical filaments in Alzheimer's disease. J Biochem 99:1807–1810

Iqbal K, Smith AJ, Zaidi T, Grundke-Iqbal I (1989) Microtubule-associated protein tau. Identification of a novel peptide from bovine brain. FEBS Lett 248:87–91

Johnson GV, Seubert P, Cox TM, Motter R, Brown JP, Galasko D (1997) The tau protein in human cerebrospinal fluid in Alzheimer's disease consists of proteolytically derived fragments. J Neurochem 68:430–433

Johnson GV, Stoothoff WH (2004) Tau phosphorylation in neuronal cell function and dysfunction. J Cell Sci 117:5721–5729

Josephs KA (2007) Frontotemporal lobar degeneration. Neurol Clin 25:683–694

Kitano-Takahashi M, Morita H, Kondo S, Tomizawa K, Kato R, Tanio M, et al. (2007) Expression, purification and crystallization of a human tau-tubulin kinase 2 that phosphorylates tau protein. Acta Crystallogr Sect F Struct Biol Cryst Commun 63:602–604

Komori T (1999) Tau-positive glial inclusions in progressive supranuclear palsy, corticobasal degeneration and Pick's disease. Brain Pathol 9:663–679

Kosik KS (1990a) Tau protein and Alzheimer's disease. Curr Opin Cell Biol 2:101–104

Kosik KS (1990b) Tau protein and neurodegeneration. Mol Neurobiol 4:171–179

Kosik KS (1993) The molecular and cellular biology of tau. Brain Pathol 3:39–43

Kosik KS, Joachim CL, Selkoe DJ (1986) Microtubule-associated protein tau (tau) is a major antigenic component of paired helical filaments in Alzheimer disease. Proc Natl Acad Sci U S A 83:4044–4048

Ledesma MD, Bonay P, Colaco C, Avila J (1994) Analysis of microtubule-associated protein tau glycation in paired helical filaments. J Biol Chem 269:21614–21619

Lee G (1990) Tau protein: an update on structure and function. Cell Motil Cytoskeleton 15: 199–203

Leger JG, Brandt R, Lee G (1994) Identification of tau protein regions required for process formation in PC12 cells. J Cell Sci 107(Pt 12):3403–3412

Leszek J, Malyszczak K, Janicka B, Kiejna A, Wiak A (2003) Total tau in cerebrospinal fluid differentiates Alzheimer's disease from vascular dementia. Med Sci Monit 9:CR484–CR488

Litvan I, Agid Y, Calne D (1996) et alClinical research criteria for the diagnosis of progressive supranuclear palsy (Steele-Richardson-Olszewski syndrome): report of the NINDS-SPSP international workshop. Neurology 47(1):1–9

Lovestone S, McLoughlin DM (2002) Protein aggregates and dementia: is there a common toxicity? J Neurol Neurosurg Psychiatry 72:152–161

Mandelkow EM, Schweers O, Drewes G, Biernat J, Gustke N, Trinczek B et al (1996) Structure, microtubule interactions, and phosphorylation of tau protein. Ann N Y Acad Sci 777:96–106

Mandell JW, Banker GA (1996) A spatial gradient of tau protein phosphorylation in nascent axons. J Neurosci 16:5727–5740

Margittai M, Langen R (2004) Template-assisted filament growth by parallel stacking of tau. Proc Natl Acad Sci U S A 101:10278–10283

Mathuranath PS, Xuereb JH, Bak T, Hodges JR (2000) Corticobasal ganglionic degeneration and/or frontotemporal dementia? A report of two overlap cases and review of literature. J Neurol Neurosurg Psychiatry 68:304–312

Matsushita S, Miyakawa T, Maesato H, Matsui T, Yokoyama A, Arai H et al (2008) Elevated cerebrospinal fluid tau protein levels in Wernicke's encephalopathy. Alcohol Clin Exp Res 32:1091–1095

Maurage CA, Bussiere T, Sergeant N, Ghesteem A, Figarella-Branger D, Ruchoux MM et al (2004) Tau aggregates are abnormally phosphorylated in inclusion body myositis and have an immunoelectrophoretic profile distinct from other tauopathies. Neuropathol Appl Neurobiol 30:624–634

Mazanetz MP, Fischer PM (2007) Untangling tau hyperphosphorylation in drug design for neurodegenerative diseases. Nat Rev Drug Discov 6:464–479

Migheli A, Butler M, Brown K, Shelanski ML (1988) Light and electron microscope localization of the microtubule-associated tau protein in rat brain. J Neurosci 8:1846–1851

Modoni A, Silvestri G, Pomponi MG, Mangiola F, Tonali PA, Marra C (2004) Characterization of the pattern of cognitive impairment in myotonic dystrophy type 1. Arch Neurol 61: 1943–1947

Mondragon-Rodriguez S, Basurto-Islas G, Santa-Maria I, Mena R, Binder LI, Avila J, et al. (2008) Cleavage and conformational changes of tau protein follow phosphorylation during Alzheimer's disease. Int J Exp Pathol 89:81–90

Morris HR, Baker M, Yasojima K, Houlden H, Khan MN, Wood NW, et al. (2002) Analysis of tau haplotypes in Pick's disease. Neurology 59:443–445

Morris HR, Lees AJ, Wood NW (1999) Neurofibrillary tangle parkinsonian disorders–tau pathology and tau genetics. Mov Disord 14:731–736

Neary D, Snowden JS, Gustafson L, Passant U, Stuss D, Black S et al (1998) Frontotemporal lobar degeneration:A consensus on clinical diagnostic criteria. Neurology 51:1546–1554

Neve RL, Harris P, Kosik KS, Kurnit DM, Donlon TA (1986) Identification of cDNA clones for the human microtubule-associated protein tau and chromosomal localization of the genes for tau and microtubule-associated protein 2. Brain Res 387:271–280

Oddo S, Billings L, Kesslak JP, Cribbs DH, LaFerla FM (2004) Abeta immunotherapy leads to clearance of early, but not late, hyperphosphorylated tau aggregates via the proteasome. Neuron 43:321–332

Osaki Y, Wenning GK, Daniel SE, Hughes A, Lees AJ, Mathias CJ (2002) Do published criteria improve clinical diagnostic accuracy in multiple system atrophy? Neurology 59(10): 1486–1491

Paraskevas GP, Kapaki E, Kararizou E, Mitsonis C, Sfagos C, Vassilopoulos D (2007) Cerebrospinal fluid tau protein is increased in neurosyphilis: a discrimination from syphilis without nervous system involvement? Sex Transm Dis 34:220–223

Perez M, Valpuesta JM, Medina M (1996) Montejo de Garcini E, Avila J. Polymerization of tau into filaments in the presence of heparin: the minimal sequence required for tau-tau interaction. J Neurochem 67:1183–1190

Price JL, Morris JC (2004) So what if tangles precede plaques? Neurobiol Aging 25:721–723; discussion 743–6

Reynolds CH, Utton MA, Gibb GM, Yates A, Anderton BH (1997) Stress-activated protein kinase/c-jun N-terminal kinase phosphorylates tau protein. J Neurochem 68:1736–1744

Riley DE, Lang AE, Lewis A, Resch L, Ashby P, Hornykiewicz O et al (1990) Cortical-basal ganglionic degeneration. Neurology 40:1203–1212

Rinne JO, Lee MS, Thompson PD, Marsden CD (1994) Corticobasal degeneration. A clinical study of 36 cases. Brain 117(Pt 5):1183–1196

Robert M, Mathuranath PS (2007) Tau and tauopathies. Neurol India 55:11–16

Roder HM, Hutton ML (2007) Microtubule-associated protein tau as a therapeutic target in neurodegenerative disease. Expert Opin Ther Targets 11:435–442

Schonknecht P, Pantel J, Kaiser E, Thomann P, Schroder J (2007) Increased tau protein differentiates mild cognitive impairment from geriatric depression and predicts conversion to dementia. Neurosci Lett 416:39–42

Schooneboom NHH, Scheltens P, Leon M (2006) Cerebrospinal fluid markers for the diagnosis of Alzheimer's disease. In: Gauthier SSP, Cummings JL (eds) Alzheimer's Disese and related Disorders Annual, Vol 5. Taylor and Francis, Oxford, pp 17–33

Schweers O, Mandelkow EM, Biernat J, Mandelkow E (1995) Oxidation of cysteine-322 in the repeat domain of microtubule-associated protein tau controls the in vitro assembly of paired helical filaments. Proc Natl Acad Sci U S A 92:8463–8467

Schweers O, Schonbrunn-Hanebeck E, Marx A, Mandelkow E (1994) Structural studies of tau protein and Alzheimer paired helical filaments show no evidence for beta-structure. J Biol Chem 269:24290–24297

Sergeant N, Delacourte A, Buee L (2005) Tau protein as a differential biomarker of tauopathies. Biochim Biophys Acta 1739:179–197

Shaw LM, Korecka M, Clark CM, Lee VM, Trojanowski JQ (2007) Biomarkers of neurodegeneration for diagnosis and monitoring therapeutics. Nat Rev Drug Discov 6:295–303

Sjoberg MK, Shestakova E, Mansuroglu Z, Maccioni RB, Bonnefoy E (2006) Tau protein binds to pericentromeric DNA: a putative role for nuclear tau in nucleolar organization. J Cell Sci 119:2025–2034

Snowden JS, Neary D, Mann DM (2002) Frontotemporal dementia. Br J Psychiatry 180:140–143

Sontag E, Nunbhakdi-Craig V, Lee G, Brandt R, Kamibayashi C, Kuret J, et al. (1999) Molecular interactions among protein phosphatase 2A, tau, and microtubules. Implications for the regulation of tau phosphorylation and the development of tauopathies. J Biol Chem 274:25490–25498

Spillantini MG, Goedert M (2000) Tau mutations in familial frontotemporal dementia. Brain 123(Pt 5):857–859

Strittmatter WJ, Saunders AM, Goedert M, Weisgraber KH, Dong LM, Jakes R, et al. (1994) Isoform-specific interactions of apolipoprotein E with microtubule-associated protein tau: implications for Alzheimer disease. Proc Natl Acad Sci U S A 91:11183–11186

Sutherland GT, Nowak G, Halliday GM, Kril JJ (2007) Tau isoform expression in frontotemporal dementia without tau deposition. J Clin Neurosci 14:1182–1185

Takashima A (2008) Hyperphosphorylated tau is a cause of neuronal dysfunction in tauopathy. J Alzheimers Dis 14:371–375

Tatebayashi Y, Haque N, Tung YC, Iqbal K, Grundke-Iqbal I (2004) Role of tau phosphorylation by glycogen synthase kinase-3beta in the regulation of organelle transport. J Cell Sci 117:1653–1663

Terzi M, Birinci A, Cetinkaya E, Onar MK (2007) Cerebrospinal fluid total tau protein levels in patients with multiple sclerosis. Acta Neurol Scand 115:325–330

Tolnay M, Probst A (1999) REVIEW: tau protein pathology in Alzheimer's disease and related disorders. Neuropathol Appl Neurobiol 25:171–187

Trojanowski JQ, Lee VM (2002) The role of tau in Alzheimer's disease. Med Clin North Am 86:615–627

Ulloa L, Montejo de Garcini E, Gomez-Ramos P, Moran MA, Avila J (1994) Microtubule-associated protein MAP1B showing a fetal phosphorylation pattern is present in sites of neurofibrillary degeneration in brains of Alzheimer's disease patients. Brain Res Mol Brain Res 26:113–122

Van Deerlin VM, Forman MS, Farmer JM, Grossman M, Joyce S, Crowe A et al (2007) Biochemical and pathological characterization of frontotemporal dementia due to a Leu266Val mutation in microtubule-associated protein tau in an African American individual. Acta Neuropathol 113:471–479

Van Everbroeck B, Green AJ, Vanmechelen E, Vanderstichele H, Pals P, Sanchez-Valle R, et al. (2002) Phosphorylated tau in cerebrospinal fluid as a marker for Creutzfeldt-Jakob disease. J Neurol Neurosurg Psychiatry 73:79–81

Wang JZ, Grundke-Iqbal I, Iqbal K (1996a) Glycosylation of microtubule-associated protein tau: an abnormal posttranslational modification in Alzheimer's disease. Nat Med 2:871–875

Wang JZ, Grundke-Iqbal I, Iqbal K (1996b) Restoration of biological activity of Alzheimer abnormally phosphorylated tau by dephosphorylation with protein phosphatase-2A, -2B and -1. Brain Res Mol Brain Res 38:200–208

Wang JZ, Liu F (2008) Microtubule-associated protein tau in development, degeneration and protection of neurons. Prog Neurobiol 85:148–175

Wenning GK, S YB-, Hughes A, Daniel SE, Lees A, Quinn, NP (2000) What clinical features are most useful to distinguish definite multiple system atrophy from Parkinson's disease? J Neurol Neurosurg Psychiatry 68:434–440

Wille H, Drewes G, Biernat J, Mandelkow EM, Mandelkow E (1992) Alzheimer-like paired helical filaments and antiparallel dimers formed from microtubule-associated protein tau in vitro. J Cell Biol 118:573–584

Wischik CM, Edwards PC, Lai RY, Gertz HN, Xuereb JH, Paykel ES et al (1995) Quantitative analysis of tau protein in paired helical filament preparations: implications for the role of tau protein phosphorylation in PHF assembly in Alzheimer's disease. Neurobiol Aging 16:409–417; discussion 418–31

Yamamoto H, Hasegawa M, Ono T, Tashima K, Ihara Y, Miyamoto E (1995) Dephosphorylation of fetal-tau and paired helical filaments-tau by protein phosphatases 1 and 2A and calcineurin. J Biochem 118:1224–1231

Yan SD, Chen X, Schmidt AM, Brett J, Godman G, Zou YS, et al. (1994) Glycated tau protein in Alzheimer disease: a mechanism for induction of oxidant stress. Proc Natl Acad Sci U S A 91:7787–7791

Zinc and Zinc Transport and Sequestration Proteins in the Brain in the Progression of Alzheimer's Disease

Mark A. Lovell

Abstract Multiple studies over the past 25 years have demonstrated alterations of zinc (Zn) in the brain in Alzheimer's disease (AD), although the potential fole of these alterations in the pathogenesis of AD remains unclear. The following examines normal and abnormal roles of Zn and Zn transport (ZIP and ZnT) proteins in brain and the potential effects of their alterations in the pathogenesis of AD.

Keywords Zinc · Early Alzheimer's disease · Mild cognitive impairment · Zinc transporter proteins · Amyloid beta peptide · Neurodegeneration

Contents

M.A. Lovell (✉)
Department of Chemistry, Sanders-Brown Center on Aging and Alzheimer's Disease Center, University of Kentucky, Lexington, KY 40536, USA
e-mail: malove2@uky.edu

J.P. Blass (ed.), *Neurochemical Mechanisms in Disease*,
Advances in Neurobiology 1, DOI 10.1007/978-1-4419-7104-3_20,

1 Introduction

1.1 Clinical Parameters of Mild Cognitive Impairment (MCI), Early AD (EAD), and Late Stage AD (LAD)

Alzheimer's disease (AD), the fourth leading cause of death in the United States, affected 4.5 million Americans in 2000 and may affect as many as 14 million by 2040 (Hebert et al., 2003). Current estimates suggest ~3% of Americans between ages 65 and 74, 19% ages 75–84, and 47% over age 85 are victims of the disease (Evans et al., 1989) with ~60% of nursing home patients over age 65 suffering from AD. Alzheimer's disease is characterized clinically by a progressive decline in multiple cognitive functions and is thought to begin with amnestic mild cognitive impairment (MCI), widely considered to be a transition between normal aging and dementia. Recent studies suggest conversion from MCI to dementia occurs at a rate of 10–15% per year (Petersen and Morris, 2003) with a conversion rate of ~80% by the sixth year of followup. Of the remaining MCI subjects ~5% remain stable or convert back to normal (Bennett et al., 2002; DeCarli, 2003). Clinically, MCI is diagnosed based on the Petersen et al. criteria and is characterized by: (a) memory complaints, (b) objective memory impairment for age and education, (c) intact general cognitive function, (4) intact activities of daily living (ADLs), and (5) the subject is not demented (Petersen et al., 1999). Objective memory test impairment is based on a score of $\leq$1.5 standard deviations from the mean of controls on the CERAD Word List Learning Task (Morris et al., 1989) and corroborated in some cases with the Free and Cued Selective Reminding Test. As the disease progresses patients are classified as early AD (EAD), patients and are clinically characterized by (a) a decline in cognitive function from a previous higher level, (b) decline in one or more areas of cognition in addition to memory, (c) a clinical dementia rating scale score of 0.5–1, (d) impaired ADLs, and (e) a clinical evaluation that excludes other causes of dementia. Disease progression ultimately leads to late stage AD (LAD) which is characterized clinically by impairment of recent memory, language disturbances, and alterations of abstract reasoning, concentration, and thought sequencing (executive function) (American Psychiatric Association, 2000). Diagnosis of probable AD is based on criteria from the National Institute of Neurological and Communicative Diseases and Stroke/Alzheimer's Disease and Related Disorders Association (NINCDS-ADRDA) and is made when patients demonstrate (a) dementia established by clinical examination and documented by mental status tests, (b) deficits in two or more areas of cognition, (c) progressive worsening, (d) no disturbance in consciousness, (e) onset between age 40 and 90, and (f) no systemic or other brain diseases that could account for the progressive deficits (National Institute on Aging and Reagan Institute Working Group, 1997). The mean length of life following diagnosis is 8.5 years with a range of 1–25 years (Jost and Grossberg, 1995).

1.2 Pathological Characterization of MCI, EAD, and LAD

Pathological examination of the AD brain shows an abundance of neurofibrillary tangles (NFT), senile plaques (SP), increased neuropil thread formation, increased neuron and synapse loss and proliferation of reactive astrocytes, primarily in the hippocampus, amygdala, entorhinal cortex, and neocortex. Neurofibrillary tangles are intracellular lesions consisting of paired helical filaments composed primarily of hyperphosphorylated tau. Senile plaques are extracellular lesions and are present in two forms: (a) diffuse plaques (DP) composed of amorphous extracellular deposits of Aβ lacking neurites, and (b) neuritic plaques (NP) composed of extracellular deposits of insoluble Aβ surrounded by dystrophic neurites, reactive astrocytes, and activated microglia. In addition to insoluble Aβ present in SP, recent studies suggest soluble Aβ oligomers are present in the AD brain and may represent the main toxic form of Aβ, thus implicating them in the disease process (Glabe, 2006; Klein, 2002; Walsh et al., 2002).

Senile plaques and NFT are the hallmark pathological lesions employed for the histopathologic diagnosis of AD based on the National Institute on Aging-Reagan Institute (NIA-RI) criteria (The National Institute on Aging, 1997). The NIA-RI criteria combine NP scores used by the Consortium to Establish a Registry for Alzheimer's Disease (CERAD) with Braak staging scores to provide classifications of low, intermediate, and high likelihood for the diagnosis of AD. The CERAD criteria use NP densities in three neocortical regions (frontal, temporal, and parietal) to provide an age-related NP score that is used in conjunction with the clinical history to reach a diagnosis of possible, probable, or definite AD. Braak staging scores (Braak and Braak, 1994) are based on the observation that NFT pathology progresses in a topographically predictable manner from the transentorhinal (stages I and II) to entorhinal, hippocampus, amygdala, and adjacent temporal cortex (limbic stages III and IV) and then to the isocortex (stages V and VI).

Pathologically, MCI subjects show significant increases in neocortical NP and NFT densities in entorhinal cortex, hippocampus, and amygdala compared to normal control subjects (Markesbery et al., 2006) with Braak staging scores ranging from III to IV. Subjects with EAD generally meet NIA-RI high likelihood criteria for the histopathological diagnosis of AD with Braak staging scores of V but have less severe overall NFT and NP formation than observed in LAD.

Multiple risk factors have been identified for AD and include age (Evans et al., 1989), a variety of genetic factors including mutations of presenilin 1 (PS1) and 2 (PS2), and the amyloid precursor protein (APP) (Levy-Lahad et al., 1995; St. George-Hyslop, 1994). In addition, single nucleotide polymorphisms in ubiquilin-1 (Bertram et al., 2005), a genetic locus on chromosome 10 that includes the insulin-degrading enzyme (Bertram et al., 2000; Ertekin-Taner et al., 2000; Myers et al., 2000) that may interact with and degrade Aβ, inherited variants in SORL1 (Rogaeva et al., 2007), and the presence of apolipoprotein E4 alleles (Corder et al., 1993) are associated with the risk of AD. Additional risk factors for AD include head injury (Mortimer et al., 1991), diabetes (Chan et al., 1999; Leibson et al., 1997; Peila et al., 2002), hyperlipidemia (Jick et al., 2000), hypertension (Skoog et al., 1996), heart disease (Kleineke and Brand, 1997), smoking (McMahon and Cousins, 1998; Merchant et al., 1999), elevated plasma homocysteine (Seshadri et al., 2002), obesity (Gustafson et al., 2003), and low educational attainment and low linguistic ability early in life (Snowdon et al., 2000, 1996).

Despite considerable research, the major barrier to treating and eventually preventing AD is a lack of understanding of the cause and mechanisms of neuron degeneration and loss. Because of the complexity of the disease, AD is likely a heterogeneous disease of multiple, probably interrelated, etiologic/pathogenic factors. Numerous etiologic/pathogenic mechanisms have been suggested for the cause of AD including genetic defects (St. George-Hyslop et al., 1987; St. George-Hyslop, 1994), the amyloid cascade hypothesis (reviewed in Sommer, 2002), the oxidative stress hypothesis (Coyle and Puttfarcken, 1993), mitochondrial defects (Wallace, 1992), trace element (including Zn) toxicity (reviewed in Markesbery and Ehmann, 1994), or a combination of the above. One hypothesis receiving renewed interest is the potential role of alterations of Zn homeostasis in the pathogenesis of AD.

2 Zinc and Zinc Homeostasis

Zinc is an essential trace (μg/g) element (Prasad et al., 1963) in human health and biology. Although Zn is present in all organs ~90% of total body zinc (1–2 g) is associated with bones and skeletal muscle (Sturniolo et al., 2000). Most dietary Zn is absorbed from the jejunum through passive diffusion and specific transporter proteins (Sturniolo et al., 2000). Once absorbed, Zn is transported in the plasma bound largely to albumin (Smith et al., 1979). Circulating Zn is transported into the brain via the blood/brain and blood/cerebrospinal fluid (CSF) barriers (Nunomura et al., 2001) where brain capillary endothelial cells respond to changes in Zn status by increasing or decreasing Zn uptake (Lehmann et al., 2002). Once transferred to the CSF, Zn is quite mobile and is taken up by the brain in processes that are not completely understood but likely involve transporters of the Zrt-Irt (ZIP) family, zinc transporter (ZnT) family, or through a variety of other specific gated Zn permeable channels. At the cellular level Zn is redox inert and has structural, catalytic, and regulatory roles (Bettger and O'Dell, 1981; Golden, 1989; Vallee and Falchuk, 1993). Zinc is a crucial component in over 300 enzymes and transcription factors where it serves as an essential cofactor for catalytic activity (Frederickson, 1989) or by conferring structural stability to Zn finger domains of DNA binding proteins (Colvin et al., 2003) including stimulating protein-1 (sp-1), a transcription factor responsible for ~30% of APP transcription (Bittel et al., 1998; Dalton et al., 1997, 1996). Additionally, recent studies suggest free Zn may possess important signaling functions including modulation of protein kinase C (PKC) signaling pathways (Korichneva et al., 2002), modulation of p53 mediated DNA repair through stabilization of p53/genomic DNA interactions (Mocchegiani et al., 2005), inhibition of gamma aminobutyric acid (GABA-ergic) neurotransmission (Haase and Beyersmann, 2002), and modulation of glycogen synthase kinase 3β (An et al., 2005; Ilouz et al., 2002).

2.1 Zinc Transport and Sequestration

In the brain, Zn is distributed in discrete pools: (a) a membrane-bound metalloprotein, or protein–metal complex pool involved in metabolic reactions and nonmetabolic functions such as biomembrane structure and support; (b) a vesicular pool present in nerve terminal synaptic vesicles; and (c) a cytoplasmic pool of free or loosely bound ions (Frederickson, 1989). The easily chelated vesicular pool may be the most important (Danscher et al., 1985; Frederickson et al., 1983; Haug, 1967; Perez-Clausell and Danscher, 1986) because it is released during neurotransmission and may reach neurotoxic levels of 300 μM in the synapse. Without immediate uptake and sequestration these Zn gradients could potentially induce neurodegeneration. Mean brain Zn concentrations are highest in the hippocampus, amygdala, and neocortex and are relatively low in cerebellum (Danscher et al., 1997; Frederickson et al., 2005), a pattern that mirrors the distribution of pathological features in AD. These Zn concentrations range between 150 and 200 μM (Ebadi et al., 1995; Price and Joshi, 1982) and are ~10 times serum Zn levels (Takeda, 2000). At the cellular level Zn concentrations range from nanomolar levels in the cytoplasm of most neurons to millimolar concentrations in vesicles of mossy fiber terminals (Frederickson et al., 1983; Williams, 1989).

Although Zn is critical for normal brain function, in vitro and in vivo studies show high concentrations of Zn are toxic to neurons (Choi et al., 1988; Duncan et al., 1992; Yokoyama et al., 1986; Chuah et al., 1995; Cuajungco and Lees, 1996; Koh et al., 1996) resulting in increased oxidative stress, and necrotic and apoptotic cell death occurring in as little as 30 min (Choi et al., 1988; Gaskin and Kress, 1977; Manev et al., 1997; Kim et al., 1999). Although elevated Zn can be neurotoxic, the exact mechanism of Zn-induced cell death remains unclear. One possible mechanism by which Zn mediates neurotoxicity is through the potentiation of glutamate (Beaulieu et al., 1992; Bramham et al., 1990; Danscher et al., 1985; Frederickson et al., 1983; Kesslak et al., 1987; Stengaard-Pedersen et al., 1983), AMPA (Buschke et al., 1999; Choi et al., 1988; Freund and Reddig, 1994; Koh and Choi, 1987), or kainic acid (Choi et al., 1988; Shore et al., 1984; Yin and Weiss, 1995) toxicity. In addition, Zn has been shown to play a role in mitochondrial dysfunction by inhibiting the transfer of an electron between coenzyme Q and cytochrome b of the bc_1 complex (Blennow et al., 1995; Hunter and Ford, 1955; Kleiner and von Jagow, 1972), thus inhibiting the initial step of respiration. At high Zn concentrations levels of complex I and II and cytochrome oxidase are inhibited (Skhulachev et al., 1967), although Yamaguchi et al. (1982) demonstrated increased mitochondrial function in rat liver after a single low dose of Zn. Later studies (Canzoniero et al., 1999; Ho et al., 2000; Krotkiewska and Banas, 1992) showed nM–μM concentrations of Zn can inhibit a number of enzymes required for mitochondrial respiration and glycolysis. Zinc-mediated dysfunction in oxidative phosphorylation and the resultant increase of free radical generation could in turn lead to release of Zn from MT and further increased intracellular concentrations of Zn (Fliss and Menard, 1992). Zinc at relatively low concentrations can inhibit sodium/potassium ATPase (Na^+K^+ ATPase) activity in isolated protein, inhibit glutamate and GABA uptake in mice synaptosomes (Gabrielsson et al., 1986), and glutamate transport by human excitatory amino acid transporter (EAAT) 1 in Xenopus laevis oocytes (Vandenberg et al., 1998).

Zinc is hypothesized to influence assembly and disassembly of tubulin (Eagle et al., 1983; Gaskin and Kress, 1977; Gaskin et al., 1978) and several microtubule associated proteins in vitro (Backstrom et al., 1992; Gaskin and Kress, 1977; Gaskin et al., 1978; Kress et al., 1981) contributing to structural abnormalities. In addition, Zn may mediate tau phosphorylation through modulation of P13/AKT, ERK1/2, and p38/MAPK signaling cascades (An et al., 2005). Influx of Zn through NMDA receptor channels may lead to neuronal depolarization and an increase of intracellular calcium (Ca) that could further activate second messenger systems via PKC-mediated phosphorylation of receptor ion channels or voltage-dependent gene expression (Atar et al., 1995; Murakami et al., 1987; Rubin and Koide, 1973). Calcium homeostasis may be further disrupted by Zn binding to calmodulin (Baudier et al., 1983) and the inhibition of calmodulin-complexed Ca ATPase (Brewer et al., 1979). Chelatable Zn has been shown to accumulate in the cell perikarya of apoptotic neurons before and during degeneration following ischemia insult (Kress et al., 1981; Tonder et al., 1990) or seizure activity (Frederickson, 1989), and is suggested to play a pathological role in neuron death. More recent studies demonstrated increased intracellular Zn as an early event in the apoptotic pathway that occurs in the absence of exogenous Zn and is consistent with a release of Zn from intracellular stores (Zalewski et al., 1994). Because of the essential but potentially toxic qualities of Zn it is imperative that cells regulate Zn levels through control of influx and efflux and through chelation to Zn sequestering proteins.

3 Maintenance of Zinc Homeostasis

In general, Zn homeostasis is maintained by three families of proteins: (a) metallothioneins (MT) that quickly bind, sequester, and hold Zn after influx into the cytoplasm, (b) Zrt–Irt-like (ZIP) proteins that likely mediate Zn influx into the cell, and (c) zinc transporter (ZnT) proteins that mediate efflux of cytoplasmic Zn to the extracellular space or sequestration in intracellular organelles.

3.1 Metallothioneins

Metallothioneins are a superfamily of nonenzymatic low molecular weight (6–7 kDa) single polypeptide chains of 61–68 amino acids, 25–30% of which are cysteines. Metallothioneins display high Zn binding affinity ($K_{Zn} = 3.2 \times 10^{-13}$ M^{-1} at pH 7.4), can bind 7 atoms of Zn per molecule, and function to sequester Zn immediately after uptake by cells to prevent toxicity (Palmiter, 1998). These proteins are ubiquitous and expression can be induced by metals including mercury and cadmium, glucocorticoids, proinflammatory cytokines, oxidative stress, electrophilic compounds, and xenobiotics (Kagi and Schaffer, 1988; Palmiter, 1998; Vallee, 1995). In mammals, four subfamilies of MT exist (MT-I, MT-II, MT-III, and MT-IV) with three functional isoforms expressed in brain including MT-I and II, which are expressed in astrocytes, the perivascular space, and pia mater (Penkowa et al., 1999) in a Zn-dependent manner (Atar et al., 1995; Durnam and Palmiter, 1981) and MT-III, which is most abundant in neurons that sequester Zn in synaptic vesicles (Bush et al., 1994). In general MT are thought to be largely intracellular with localization in the cytoplasm, lysosomes, and mitochondria (Penkowa, 2006) and are regularly translocated to the nucleus during cell division and under oxidative stress (Cherian and Apostolova, 2000; Klaassen et al., 1999; Maret, 2002; Trayhurn et al., 2000). Because of their small size MT-I and MT-II are able to diffuse into the nucleus through nuclear pore complexes where they are retained by nuclear factors (Penkowa, 2006). Metallothioneins have a low redox potential (–366 mV) that allows mild oxidation to decrease Zn binding and facilitates release of Zn for binding to Zn finger and other transcription factors that modulate DNA binding efficiency and expression of antioxidant genes during periods of oxidative stress (Mocchegiani et al., 2005).

3.2 ZIP Proteins

Although the mechanism of transport of Zn from brain extracellular environments to intracellular compartments in neurons and glia is not completely understood, it is thought to involve members of the ZIP family of proteins (Chromy et al., 2003). The ZIP family of proteins was initially identified based on their functional and structural similarity to the ZRT yeast family (Eide, 1998) and the IRT transporters of Arabidopsis thaliana (Grotz et al., 1998). ZIP proteins are predicted to have 8 transmembrane domains with a histidine-rich intracellular loop between domains 3 and 4 (Huang et al., 2005) and are part of the plasma membrane or membranes of intracellular organelles. Using mouse and human sequence analysis 14 mammalian ZIP proteins that elevate intracellular Zn by increasing Zn uptake (ZIP 1–5; 7–15) or by releasing Zn from intracellular stores when Zn is deficient (ZIP 6 and 7) have been identified. ZIPs have no ATP binding sites or ATPase domains and function in an energy-independent manner (Gaither and Eide, 2000, 2001). ZIP-1 mRNA is expressed ubiquitously (Gaither and Eide, 2001) whereas ZIP-2 is specific to spleen, small intestine, and bone marrow (Gaither and Eide, 2000). Similarly, ZIP-3 expression is high in bone marrow and spleen (Gaither and Eide, 2000). ZIP-4 expression is associated primarily with small intestine, and kidney (Wang et al., 2002b) and is increased during periods of Zn deficiency (Cousins et al., 2003; Dufner-Beattie et al., 2003). Human ZIP-5 is highest in intestine, liver, kidney, and pancreas (Wang et al., 2004) whereas ZIP-6 is associated with prostate and placenta (Taylor et al., 2003). In humans ZIP-7 is ubiquitously expressed and is subcellularly localized to the Golgi apparatus where it functions to release Zn to the cytoplasm during periods of low intracellular Zn (Huang et al., 2005). The remaining ZIP proteins (8–15) have been identified by database searches but are yet to be localized. In addition to ZIP proteins, neuronal Zn uptake may also be mediated by a variety of Zn-permeable membrane spanning channels including Ca^{2+} permeable AMPA/kainate channels (Jia et al., 2002), voltage-gated L-type Ca^{2+} (Colvin et al., 2003), N-methyl-D aspartate (NMDA) receptor gated (Koh and Choi, 1994), and Na^{+}/Zn^{2+} exchangers (Cheng and Reynolds, 1998).

3.3 ZnT Proteins

Zinc transport (ZnT) proteins serve as a counterpoint to ZIP proteins and function in the export of cytoplasmic Zn to the extracellular space or the sequestration of Zn in intracellular organelles. ZnT proteins are members of the cation diffusion facilitator family of proteins and are predicted to have 6 transmembrane domains with a histidine-rich loop between transmembrane domains 4 and 5. Presently, eight ZnT proteins have been described (reviewed in Eide, 2006) with two additional ZnT genes (ZNT-9 and ZNT-10) predicted based on analysis of the mouse and human genome (Seve et al., 2004; Sim and Chow, 1999). ZnT-1 is located at the plasma membrane, whereas the other ZnT proteins are expressed at the membrane of intracellular organelles. ZnT-1 is present in multiple organs including brain (Palmiter, 1995) and is induced in the presence of elevated cytoplasmic Zn through direct binding of Zn to the Zn-finger domain of metal response element-binding transcription factor-1 (MTF-1; reviewed in Andrews, 2001). After binding Zn, MTF-1 translocates to the nucleus where it binds the metal response element (MRE) in genes for ZnT-1, MT, and gamma glutamylcysteine synthetase heavy chain which controls the rate-limiting step in glutathione synthesis (reviewed in Andrews, 2001). Initial in vitro studies of ZnT-1 showed overexpression in baby hamster kidney cells conferred resistance to increased Zn with the rate of Zn efflux increasing as extracellular Zn concentrations increased suggesting Zn efflux mediated by ZnT-1 is an energy-dependent process and argues against ZnT-1 being a channel or facilitated transporter (Palmiter, 1995). Later studies demonstrated that ZnT-1 reduces Zn influx through the L-type calcium channels (LTCC) without increasing Zn efflux (Nolte et al., 2004; Ohana et al., 2006; Segal et al., 2004). In addition, in vivo studies (Chowanadisai et al., 2005) showed rats provided a Zn-deficient diet demonstrated decreased brain ZnT-1, suggesting low systemic Zn could decrease ZnT-1 to maintain or increase brain Zn stores which is consistent with studies of Takeda et al. (2001) who found rats on a Zn-deficient diet showed increased brain Zn. Studies from our laboratory show ZnT-1 protein expression and function can be inactivated by HNE (2006aSmith et al.,), a neurotoxic aldehydic marker of lipid peroxidation present in MCI and LAD brain (Lovell et al., 1997; Williams et al., 2005).

ZnT-2, a component of vesicular acid intracellular compartments, is predominantly expressed in intestine, kidney, and testis and is scarcely detected in brain in mice (Palmiter et al., 1996). Overexpression of ZnT-2 in baby hamster kidney cells conferred resistance to elevated Zn with sequestration into acidic compartments at higher concentrations (Palmiter et al., 1996). In contrast, coexpression of ZnT-1 suppressed ZnT-2 mediated transport into acidic vesicles suggesting ZnT-2 has a relatively low affinity for Zn and functions only under excessive elevations of Zn as a second line of defense when other ZnTs fail to function properly (Palmiter et al., 1996). ZnT-3 sequesters Zn in vesicles and has expression limited to brain and testis (Palmiter et al., 1996). In mouse brain, ZnT-3 is associated with hippocampal dentate granule cells, pyramidal, and intraneurons as evidenced by levels of mRNA (Palmiter et al., 1996).

ZnT-4 exhibits considerable homology with ZnT-2 and 3 and has expression in mammary gland and brain (Huang and Gitschier, 1997). Functionally, ZnT-4 sequesters Zn in acidic vesicles and is involved in the transport of Zn^{2+} into milk during lactation (Kelleher and Lonnerdal, 2002). In contrast to other ZnT proteins, ZnT-5 is predicted to have 15 membrane spanning domains and is less than twice the size of other ZnT proteins (Colvin et al., 2003).

In mice ZnT-5 mRNA is found in most organs although the highest protein expression is in the pancreas where is it associated with Zn-enriched secretory granules in insulin containing β cells. ZnT-5 is scarcely detected in brain (Kambe et al., 2004) in mice although more recent studies observed ZnT-5 immunostaining in SP of AD brain (Zhang et al., 2008a).

In mice ZnT-6 mRNA is present in multiple organs including brain and sequesters cytoplasmic Zn in the trans-Golgi network (TGN) and vesicular compartments (Huang et al., 2002). ZnT-6 mRNA is present in multiple organs including brain. Similarly, ZnT-7 sequesters Zn in the TGN but has expression limited to lung and small intestine (Kirschke and Huang, 2003). ZnT-8 has been characterized and is primarily associated with secretory granules of pancreatic β cells (Kleineke and Brand, 1997; Rivlin et al., 1999) where it likely plays a role in insulin transport. In mice ZnT-8 has limited expression in brain.

4 Zinc, Zinc Transport, Alzheimer's Disease, and Mouse Models of AD

The potential role of Zn in the pathogenesis of AD has been of interest since 1981 when Burnet (Burnet, 1981) proposed that Zn deficiencies led to dementia. Initial studies of AD and control brain showed significantly decreased Zn in the hippocampus, inferior parietal lobule, and occipital cortex of LAD subjects (Andrasi et al., 1990, 1993; Corrigan et al., 1993; Deng et al., 1994). In contrast, later studies using short postmortem interval tissue specimens from well-characterized LAD and control subjects showed significant elevations of Zn in LAD hippocampus, amygdala, and multiple neocortical areas (Cornett et al., 1998; Danscher et al., 1997; Deibel et al., 1996; Ehmann et al., 1986; Samudralwar et al., 1995; Wenstrup et al., 1990). The use of formalin-fixed tissues in some of the earlier studies, which could have led to mobilization and loss of Zn, has been suggested to account for the observed differences in these studies. In addition, earlier studies may have also included control subjects that were not prospectively evaluated. Although multiple studies show alterations of Zn in LAD, there are few reports of Zn concentrations in brain in earlier stages of the disease.

Although several studies have quantified changes in Zn at the bulk level, changes in the cellular localization of Zn in the progression of AD remains unclear. Studies of Zn at the microprobe level have primarily focused on the association of Zn with SP. Initial studies using microparticle induced X-ray emission (micro-PIXE), showed increased Zn in SP compared to adjacent neuropil and an elevation of Zn in LAD neuropil compared to age-matched normal control (NC) subjects (Lovell et al., 1998). Subsequent studies confirmed those findings in AD (Cherny et al., 1999; Frederickson et al., 2000; Miller et al., 2006; Stoltenberg et al., 2005) and in amyloid plaques of Tg2576 transgenic mice expressing mutant APP (Friedlich et al., 2004; Lee et al., 1999). Using Raman microscopy to evaluate the structure and composition of isolated senile plaques Dong et al. (2003) showed Zn^{2+} and Cu^{2+} were specifically coordinated with histidine residues in Aβ. Despite considerable study of Zn in SP, there have been relatively few studies that measure Zn in individual neurons in AD.

Although the subject of extensive study over the past 25 years, the reasons for elevated brain Zn in AD are unclear. Several studies have attempted to relate changes in peripheral Zn to elevated brain levels, although results have been contradictory. Haines et al. (1991), Molina et al. (1998), and Shore et al. (1984) showed no significant differences between AD and control serum Zn, whereas Jeandel et al. (1989) showed a significant decrease in Zn and other nutrients and antioxidant properties in AD serum, although the AD group may have contained malnourished subjects. The study of Haines et al. (1991) may also be questioned because it included control subjects whose Mini Mental Status Examination scores were considered cognitively impaired. In contrast, Rulon et al. (2000) and Gonzales et al. (1999) showed significant elevations of Zn in AD serum. Additionally, Gonzales et al. (1999) showed that serum Zn correlated with the presence of APOE4 alleles and concluded that of the indices analyzed in their study, only serum Zn appeared to be an independent risk factor associated with the development of AD. In a subsequent study of serum Zn in the progression of AD, we showed a statistically significant decrease of serum Zn in men with MCI compared to women with MCI or age-matched normal control men (Dong et al., 2008). In contrast, there were no significant differences in serum Zn between well-characterized LAD subjects and cognitively normal control subjects. The observation of decreased serum Zn in MCI is of interest in light of previous in vivo rat studies that showed systemic Zn deficiencies led to diminished ZnT-1 levels and increased brain Zn (Chowanadisai et al., 2005; Nunomura et al., 2001; Takeda et al., 2001). These data support the hypothesis that elevated brain Zn in AD may be due to increased Zn uptake by brain under conditions of diminished extraparenchymal Zn in MCI.

Similar to serum studies, measures of CSF Zn levels have also been inconsistent. Molina et al. (1998) showed decreased Zn in AD CSF compared to age-matched control subjects whereas Basun et al. (1991) showed no significant changes. In addition, recent studies (Gerhardsson et al., 2008; Strozyk et al., 2007) showed there is an inverse relationship between Zn and copper concentrations and levels of $A\beta_{1-42}$ in CSF of LAD subjects and that degradation of soluble Aβ is normally promoted by physiological concentrations of both Cu and Zn (Strozyk et al., 2007). Although the potential variation of Zn through the progression of AD is of interest there have been no published studies of CSF levels of Zn in MCI subjects.

Despite considerable interest in the mechanism by which Zn accumulates in the brain in AD, there has been relatively little study of proteins responsible for Zn influx and efflux. In the first study of ZnT-1 in AD, we used Western blot analyses to show significantly decreased ZnT-1 levels in the hippocampus/parahippocampal gyri (HPG) of MCI, but significant elevations in EAD and LAD (Lovell and Markesbery, 2005). In studies of multiple ZnT proteins in SP in AD brain, Zhang et al. (2008a) used confocal microscopy and double immunolabeling to show colocalization of ZnT-1, ZnT-3, ZnT-4, ZnT-5, ZnT-6, and ZnT-7 with amyloid in SP in AD. Although all six ZnT proteins were present to varying degrees in SP, ZnT-5 demonstrated the most pronounced immunostaining in SP whereas ZnT-3 immunostaining was more pronounced in amyloid angiopathic vessels. These data are similar to those observed in our studies of ZnT-4 and ZnT-6 in the progression of AD which showed significantly elevated ZnT-4 in the HPG and superior and middle temporal gyri (SMTG) of EAD and LAD subjects compared to age-matched controls (Smith et al., 2006b) and significantly increased ZnT-6 in the HPG of EAD and LAD subjects compared to normal control subjects and a trend toward a significant elevation in MCI (Smith et al., 2006b). We also observed a striking association of ZnT-6 with NFT-bearing neurons identified using the modified Bielschowsky stain in LAD and in neurons positive for MC-1, a marker of early NFT formation in MCI (Lovell et al., 2006).

In studies of transgenic mouse models of amyloid deposition, Zhang et al. (2008b) used Western blot analysis to show significant elevations of ZnT-1, ZnT-3, ZnT-4, ZnT-6, and ZnT-7 in the hippocampus and neocortex of mice expressing mutant APP and PS1 (APPSwePS1dE9). Immunolocalization showed that most amyloid plaques of APP/PS1 mice were immunopositive for ZnT-1 and ZnT-4 whereas ZnT-3, ZnT-5, and ZnT-6 were mainly associated with degenerating neurites at the plaque periphery. Levels of ZnT-1 were increased 300% in hippocampus and 200% in neocortex of APP/PS1 mice compared to wild-type (WT) mice of the same age. Levels of ZnT-6 and ZnT-7 showed the smallest increase in APP/PS1 hippocampus and neocortex with levels ~150% those of WT mice. Levels of ZnT-5 were also elevated in APP/PS1 mice but did not reach statistical significance. Of the proteins studied, ZnT-3 showed the most pronounced changes in hippocampus and neocortex of APP/PS1 mice (400 and 200%) compared to WT mice providing further support for the studies of Gosavi et al. (2002) who showed that crossing mice expressing mutant APP with ZnT-3-null mice led to diminished Aβ deposition. More recently, Friedlich et al. (2004) showed that these mice also demonstrate reduced cerebral amyloid angiopathy that is hypothesized to be due to diminished Zn concentrations in the perivascular space of ZnT-3-null mice. In additional studies, Stoltenberg et al. (2007) showed that providing APP/PS1 mice a Zn-deficient diet from 9 to 12 months of age led to increased Aβ deposition but no significant changes in autometallographic staining of Zn or ZnT immunostaining. Although the mechanism by which Zn deficiencies would lead to increased Aβ deposition, but not alterations in ZnT proteins, is unclear, the data do support the hypothesis that alterations of Zn may contribute to the pathogenic changes in AD.

5 Zinc and Amyloid Beta (Aβ) Peptide Processing and Aggregation

Although considerable evidence suggests there are alterations of Zn homeostasis in the AD brain, direct evidence for its role in the pathogenesis of AD has been lacking. Although Zn may play a role in multiple pathways relevant to AD, to date the most widely studied has been the possible role of Zn in processing of APP and aggregation of Aβ. APP synthesis is regulated by Zn-containing transcription factors, NF-κβ and sp1, and although Zn is essential for their activity (Yang et al., 1995; Zabel et al., 1991; Zeng et al., 1991), it is unclear whether the activity in vivo is regulated by Zn availability. In addition to the potential influence of Zn on APP expression, it may also affect proteolytic processing of the protein. Normal (nonamyloidogenic) processing of APP by α-secretase cleavage in the Golgi complex leads to formation of sAPP, a neurotrophic factor (Wilquet and De Strooper, 2004). In contrast, proteolytic processing of APP by β-secretase (BACE) at the β-cleavage site (Andrasi et al., 2000; Calingasan et al., 1999; Hussain et al., 1999; Sinha et al., 1999; Vassar et al., 1999) occurs in endosomes (Kinoshita et al., 2003; Koo and Squazzo, 1994), where acidic pH necessary for β-secretase activity is possible (Wilquet and De Strooper, 2004) and coupled with further processing by the γ-secretase complex at the plasma membrane (reviewed in Sisodia and St. George-Hyslop, 2002) leads to formation of Aβ, a 40 or 42 amino acid peptide that is the major component of SP in AD (Selkoe, 1999) (amyloidogenic pathway). Additionally, APP contains a ligand-binding site for Zn spanning the α-secretase position (Bush et al., 1993, 1994). Zn concentrations less than 50 μM inhibit α-secretase-mediated sAPP formation and increase generation of Aβ (Bush et al., 1994) perhaps through altered protein conformation. In addition, high Zn concentrations can inhibit matrix metalloproteinase-2 (MMP-2) (Backstrom et al., 1992) an enzyme that partially degrades soluble $A\beta_{1\text{-}42}$ in vitro (Bergeron et al., 1996) which could lead to increased amyloidogenic Aβ levels. Most APP molecules are transported through the TGN where α-secretase cleavage likely occurs leading to formation of secreted APP (Wilquet and De Strooper, 2004). Because ZnT-6-mediated accumulation of Zn in the TGN could initially diminish α-secretase cleavage of APP, Zn could significantly modulate APP processing leading to increased Aβ production. In addition, the presence of elevated Zn in endosomes mediated by ZnT-2 or ZnT-4 or both could further enhance β-secretase activity through modulation of pH.

Once generated, several reports indicate that Zn at low physiological concentrations induces Aβ aggregation (Bush et al., 1996, 1994; Bush et al., 1995; Mantyh et al., 1993), although later studies indicate that higher Zn concentrations are required (Clements et al., 1996; Esler et al., 1996) for significant aggregation (fibril formation). A subsequent study using atomic force and transmission electron microscopy and $A\beta_{13\text{-}21}$ shows Zn^{2+} specifically controls the rate of fibril assembly and regulates fibril morphology via specific coordination sites (Dong et al., 2006).

Multiple studies show that treatment of cortical neuron cultures with Aβ leads to increased levels of reactive oxygen species, increased lipid peroxidation, protein oxidation, mitochondrial dysfunction, caspase activation, and neuron death (Butterfield, 2003; Canzoniero et al., 1999; Keller et al., 2005; Yatin et al., 1999). In addition, several transgenic models of AD including those with mutant APP, mutant APP/PS1, or mutant APP/PS1 and tau show increased Aβ deposition (Gotz et al., 2001; Lewis et al., 2001; Oddo et al., 2004). Although Aβ deposits are associated with AD, the specific Aβ species responsible for neurodegeneration are unclear. Fibrillar Aβ, the predominant component of insoluble amyloid plaques, is neurotoxic (Lorenzo and Yankner, 1994; Pike et al., 1993). However, in vivo, insoluble Aβ deposits do not accurately predict the severity of dementia in AD subjects (Cherny et al., 1999). In addition, studies of transgenic mice including those with APP mutations show cognitive dysfunction and synaptic damage that precede amyloid plaque deposition and neuron loss (Irizarry et al., 1997; Kumar-Singh et al., 2000; Moechars et al., 1996; Mucke et al., 2000; Westerman et al., 2002), leading to the suggestion that soluble oligomeric or protofibril Aβ species may the most toxic.

In vitro studies of synthetic Aβ show monomeric Aβ aggregates in a time-dependent manner that may be accelerated by Zn leading to oligomeric species, which may eventually form fibrils (Chromy et al., 2003; Pike et al., 1991; Walsh et al., 1997). Increasing evidence suggests that these soluble oligomeric species are the predominant neurotoxic species for neurons (Demuro et al., 2005; Klein, 2002), leading to inhibition of long-term potentiation in synaptic hippocampal slices (Lambert et al., 1998; Wang et al., 2002a), calcium dysregulation, and membrane dysfunction (Demuro et al., 2005; Kayed et al., 2004). Although the exact Aβ species responsible for mediating neurodegeneration in AD is unclear, several lines of evidence support a role for Zn in their formation.

6 Zinc as a Therapeutic Target in AD

Because of the potential role of Zn and Cu in the deposition of Aβ in AD brain, there has been considerable interest in the use of metal chelation to decrease amyloid pathology (Bush, 2003). In vitro studies show clioquinol (CQ), an 8-OH quinoline inhibits Aβ aggregation mediated by Cu and Zn (Cherny et al., 2001). In vivo studies show transgenic mouse models of amyloid deposition (Tg 2576) treated with CQ for 9 weeks showed significantly reduced amyloid plaque burdens (Cherny et al., 2001). In initial, phase-2 double-blind placebo-controlled clinical trials, CQ significantly slowed cognitive decline in AD patients compared to placebo controls (Ritchie et al., 2003). More recently, PBT2, an 8-hydroxy quinoline with increased blood–brain barrier permeability has been developed (Adlard et al., 2008) and in a 12-week phase-IIa clinical trial of AD subjects reversed frontal lobe functional deficits and significantly decreased $A\beta_{1-42}$ levels in CSF (Lannfelt et al., 2008). Together, these data suggest modulation of Zn may be an effective potential therapeutic target in AD.

7 Conclusions and Future Directions

Although considerable evidence suggests a link between alterations in Zn and Zn transport and sequestration proteins in the progression of AD, further in-depth study is needed particularly early in the progression of AD (MCI) when therapeutic interventions would have greater efficacy. In particular Zn levels in CSF of subjects with MCI and EAD need to be quantified and correlated with brain ZnT, ZIP, and Zn levels. Based on in vivo studies, it is tempting to hypothesize that low extraparenchymal Zn early in disease progression may lead to decreased ZnT-1 levels and a concomitant elevation of intracellular Zn that leads to increased levels of ZnT-2, ZnT-4, and ZnT-6 and increased localization of Zn in subcellular organelles in which Aβ processing occurs. As the disease progresses and extraparenchymal Zn levels normalize, the resulting alterations in multiple ZnT proteins could further promote Aβ aggregation and SP formation.

Acknowledgments Supported by NIH grants 1R01-AG16269, 5-P01-AG05119, and 1P30-AG028383, and by a grant from the Abercrombie Foundation. The author thanks Ms. Paula Thomason for editorial assistance.

References

Adlard PA, Cherny RA, Finkelstein DI, Gautier E, Robb E, Cortes M, Volitakis I, Liu X, Smith JP, Perez K, Laughton K, Li QX, Charman SA, Nicolazzo JA, Wilkins S, Deleva K, Lynch T, Kok G, Ritchie CW, Tanzi RE, Cappai R, Masters CL, Barnham KJ, Bush AI (2008) Rapid restoration of cognition in Alzheimer's transgenic mice with 8-hydroxy quinoline analogs is associated with decreased interstitial Abeta. Neuron 59:43–55

American Psychiatric Association. (2000) Diagnostic and statistical manual of mental disorders DSM-IV-TR, 4th edn. American Psychiatric Publishing, Arlington, VA

An WL, Bjorkdahl C, Liu R, Cowburn RF, Winblad B, Pei JJ (2005) Mechanism of zinc-induced phosphorylation of p70 S6 kinase and glycogen synthase kinase 3beta in SH-SY5Y neuroblastoma cells. J Neurochem 92:1104–1115

Andrasi E, Farkas E, Gawlik D, Rosick U, Bratter P (2000) Brain Iron and Zinc Contents of German Patients with Alzheimer Disease. J Alzheimers Dis 2:17–26

Andrasi E, Nadasdi J, Molnar Z, Bezur L, Ernyei L (1990) Determination of main and trace element contents in human brain by NAA and ICP-AES methods. Biol Trace Elem Res 26–27:691–698

Andrasi E, Suhajda M, Saray I, Bezur L, Ernyei L, Reffy A (1993) Concentration of elements in human brain: glioblastoma multiforme. Sci Total Environ 139–140:399–402

Andrews GK (2001) Cellular zinc sensors: MTF-1 regulation of gene expression. Biometals 14:223–237

Atar D, Backx PH, Appel MM, Gao WD, Marban E (1995) Excitation-transcription coupling mediated by zinc influx through voltage-dependent calcium channels. J Biol Chem 270:2473–2477

Backstrom JR, Miller CA, Tokes ZA (1992) Characterization of neutral proteinases from Alzheimer-affected and control brain specimens: identification of calcium-dependent metalloproteinases from the hippocampus. J Neurochem 58:983–992

Basun H, Forssell LG, Wetterberg L, Winblad B (1991) Metals and trace elements in plasma and cerebrospinal fluid in normal aging and Alzheimer's disease. J Neural Transm Park Dis Dement Sect 3:231–258

Baudier J, Haglid K, Haiech J, Gerard D (1983) Zinc ion binding to human brain calcium binding proteins, calmodulin and S100b protein. Biochem Biophys Res Commun 114:1138–1146

Beaulieu C, Dyck R, Cynader M (1992) Enrichment of glutamate in zinc-containing terminals of the cat visual cortex. NeuroReport 3:861–864

Bennett DA, Wilson RS, Schneider JA, Evans DA, Beckett LA, Aggarwal NT, Barnes LL, Fox JH, Bach J (2002) Natural history of mild cognitive impairment in older persons. Neurology 59:198–205

Bergeron C, Petrunka C, Weyer L (1996) Copper/zinc superoxide dismutase expression in the human central nervous system. Correlation with selective neuronal vulnerability. Am J Pathol 148:273–279

Bertram L, Blacker D, Mullin K, Keeney D, Jones J, Basu S, Yhu S, McInnis MG, Go RC, Vekrellis K, Selkoe DJ, Saunders AJ, Tanzi RE (2000) Evidence for genetic linkage of Alzheimer's disease to chromosome 10q. Science 290:2302–2303

Bertram L, Hiltunen M, Parkinson M, Ingelsson M, Lange C, Ramasamy K, Mullin K, Menon R, Sampson AJ, Hsiao MY, Elliott KJ, Velicelebi G, Moscarillo T, Hyman BT, Wagner SL, Becker KD, Blacker D, Tanzi RE (2005) Family-based association between Alzheimer's disease and variants in UBQLN1. N Engl J Med 352:884–894

Bettger WJ, O'Dell BL (1981) A critical physiological role of zinc in the structure and function of biomembranes. Life Sci 28:1425–1438

Bittel D, Dalton T, Samson SL, Gedamu L, Andrews GK (1998) The DNA binding activity of metal response element-binding transcription factor-1 is activated in vivo and in vitro by zinc, but not by other transition metals. J Biol Chem 273:7127–7133

Blennow K, Wallin A, Agren H, Spenger C, Siegfried J, Vanmechelen E (1995) Tau protein in cerebrospinal fluid: a biochemical marker for axonal degeneration in Alzheimer disease? Mol Chem Neuropathol 26:231–245

Braak H, Braak E (1994) Pathology of Alzheimer's Disease. In: Calne D (ed) Neurodegnerative Diseases. WB Saunders Co, Philadelphia, PA, pp 585–613

Bramham CR, Torp R, Zhang N, Storm-Mathisen J, Ottersen OP (1990) Distribution of glutamate-like immunoreactivity in excitatory hippocampal pathways: a semiquantitative electron microscopic study in rats. Neuroscience 39:405–417

Brewer GJ, Aster JC, Knutsen CA, Kruckeberg WC (1979) Zinc inhibition of calmodulin: a proposed molecular mechanism of zinc action on cellular functions. Am J Hematol 7:53–60

Burnet FM (1981) A possible role of zinc in the pathology of dementia. Lancet 1:186–188

Buschke H, Kuslansky G, Katz M, Stewart WF, Sliwinski MJ, Eckholdt HM, Lipton RB (1999) Screening for dementia with the memory impairment screen. Neurology 52:231–238

Bush AI (2003) The metallobiology of Alzheimer's disease. Trends Neurosci 26:207–214

Bush AI, Atwood CS, Huang L (1996) Abnormal homeostasis of ph1 and zinc: the prelunde for cerebral Aß amyloid formation. Eur Neuropsychopharmacol 6(suppl 3):S2–S4

Bush AI, Multhaup G, Moir RD, Williamson TG, Small DH, Rumble B, Pollwein P, Beyreuther K, Masters CL (1993) A novel zinc(II) binding site modulates the function of the beta A4 amyloid protein precursor of Alzheimer's disease. J Biol Chem 268:16109–16112

Bush AI, Pettingell WH, Multhaup G, d Paradis M, Vonsattel JP, Gusella JF, Beyreuther K, Masters CL, Tanzi RE (1994) Rapid induction of Alzheimer A beta amyloid formation by zinc. Science 265:1464–1467

Bush AI, Pettingell WH, Paradis MD, Tanzi R (1995) Zinc and Alzheimer's disease. Science 268:1921–1923

Butterfield DA (2003) Amyloid beta-peptide [1-42]-associated free radical-induced oxidative stress and neurodegeneration in Alzheimer's disease brain: mechanisms and consequences. Curr Med Chem 10:2651–2659

Calingasan NY, Uchida K, Gibson GE (1999) Protein-bound acrolein: a novel marker of oxidative stress in Alzheimer's disease. J Neurochem 72:751–756

Canzoniero LM, Turetsky DM, Choi DW (1999) Measurement of intracellular free zinc concentrations accompanying zinc-induced neuronal death. J Neurosci 19:RC31

Chan SL, Griffin WS, Mattson MP (1999) Evidence for caspase-mediated cleavage of AMPA receptor subunits in neuronal apoptosis and Alzheimer's disease. J Neurosci Res 57:315–323

Cheng C, Reynolds IJ (1998) Calcium-sensitive fluorescent dyes can report increases in intracellular free zinc concentration in cultured forebrain neurons. J Neurochem 71:2401–2410

Cherian MG, Apostolova MD (2000) Nuclear localization of metallothionein during cell proliferation and differentiation. Cell Mol Biol (Noisy-le-grand) 46:347–356

Cherny RA, Atwood CS, Xilinas ME, Gray DN, Jones WD, McLean CA, Barnham KJ, Volitakis I, Fraser FW, Kim Y, Huang X, Goldstein LE, Moir RD, Lim JT, Beyreuther K, Zheng H, Tanzi RE, Masters CL, Bush AI (2001) Treatment with a copper-zinc chelator markedly and rapidly inhibits beta-amyloid accumulation in Alzheimer's disease transgenic mice. Neuron 30:665–676

Cherny RA, Legg JT, McLean CA, Fairlie DP, Huang X, Atwood CS, Beyreuther K, Tanzi RE, Masters CL, Bush AI (1999) Aqueous dissolution of Alzheimer's disease Abeta amyloid deposits by biometal depletion. J Biol Chem 274:23223–23228

Choi DW, Yokoyama M, Koh J (1988) Zinc neurotoxicity in cortical cell culture. Neuroscience 24:67–79

Chowanadisai W, Kelleher SL, Lonnerdal B (2005) Zinc deficiency is associated with increased brain zinc import and LIV-1 expression and decreased ZnT-1 expression in neonatal rats. J Nutr 135:1002–1007

Chromy BA, Nowak RJ, Lambert MP, Viola KL, Chang L, Velasco PT, Jones BW, Fernandez SJ, Lacor PN, Horowitz P, Finch CE, Krafft GA, Klein WL (2003) Self-assembly of Abeta(1-42) into globular neurotoxins. Biochemistry 42:12749–12760

Chuah MI, Tennent R, Jacobs I (1995) Response of olfactory Schwann cells to intranasal zinc sulfate irrigation. J Neurosci Res 42:470–478

Clements A, Allsop D, Walsh DM, Williams CH (1996) Aggregation and metal-binding properties of mutant forms of the amyloid A beta peptide of Alzheimer's disease. J Neurochem 66: 740–747

Colvin RA, Fontaine CP, Laskowski M, Thomas D (2003) Zn2+ transporters and Zn2+ homeostasis in neurons. Eur J Pharmacol 479:171–185

Corder EH, Saunders AM, Strittmatter WJ, Schmechel DE, Gaskell PC, Small GW, Roses AD, Haines JL, Pericak-Vance MA (1993) Gene dose of apolipoprotein E type 4 allele and the risk of Alzheimer's disease in late onset families. Science 261:921–923

Cornett CR, Markesbery WR, Ehmann WD (1998) Imbalances of trace elements related to oxidative damage in Alzheimer's disease brain. Neurotoxicology 19:339–345

Corrigan FM, Reynolds GP, Ward NI (1993) Hippocampal tin, aluminum and zinc in Alzheimer's disease. Biometals 6:149–154

Cousins RJ, Blanchard RK, Popp MP, Liu L, Cao J, Moore JB, Green CL (2003) A global view of the selectivity of zinc deprivation and excess on genes expressed in human THP-1 mononuclear cells. Proc Natl Acad Sci U S A 100:6952–6957

Coyle JT, Puttfarcken P (1993) Oxidative stress, glutamate, and neurodegenerative disorders. Science 262:689–695

Cuajungco MP, Lees GJ (1996) Prevention of zinc neurotoxicity in vivo by N,N,N′,N′-tetrakis (2-pyridylmethyl) ethylene-diamine (TPEN). NeuroReport 7:1301–1304

Dalton TP, Bittel D, Andrews GK (1997) Reversible activation of mouse metal response element-binding transcription factor 1 DNA binding involves zinc interaction with the zinc finger domain. Mol Cell Biol 17:2781–2789

Dalton TP, Li Q, Bittel D, Liang L, Andrews GK (1996) Oxidative stress activates metal-responsive transcription factor-1 binding activity. Occupancy in vivo of metal response elements in the metallothionein-I gene promoter. J Biol Chem 271:26233–26241

Danscher G, Howell G, Perez-Clausell J, Hertel N (1985) The dithizone, Timm's sulphide silver and the selenium methods demonstrate a chelatable pool of zinc in CNS. A proton activation (PIXE) analysis of carbon tetrachloride extracts from rat brains and spinal cords intravitally treated with dithizone. Histochemistry 83:419–422

Danscher G, Jensen KB, Frederickson CJ, Kemp K, Andreasen A, Juhl S, Stoltenberg M, Ravid R (1997) Increased amount of zinc in the hippocampus and amygdala of Alzheimer's diseased brains: a proton-induced X-ray emission spectroscopic analysis of cryostat sections from autopsy material. J Neurosci Methods 76:53–59

DeCarli C (2003) Mild cognitive impairment: prevalence, prognosis, aetiology, and treatment. Lancet neurol 2:15–21

Deibel MA, Ehmann WD, Markesbery WR (1996) Copper, iron, and zinc imbalances in severely degenerated brain regions in Alzheimer's disease: possible relation to oxidative stress. J Neurol Sci 143:137–142

Demuro A, Mina E, Kayed R, Milton SC, Parker I, Glabe CG (2005) Calcium dysregulation and membrane disruption as a ubiquitous neurotoxic mechanism of soluble amyloid oligomers. J Biol Chem 280:17294–17300

Deng QS, Turk GC, Brady DR, Smith QR (1994) Evaluation of brain element composition in Alzheimer's disease using inductively coupled plasma mass spectrometry. Neurobiol Aging S113:464

Dong J, Atwood CS, Anderson VE, Siedlak SL, Smith MA, Perry G, Carey PR (2003) Metal binding and oxidation of amyloid-beta within isolated senile plaque cores: Raman microscopic evidence. Biochemistry 42:2768–2773

Dong J, Robertson JD, Markesbery WR, Lovell MA (2008) Serum Zinc in the Progression of Alzheimer's Disease. J Alzheimers Dis 15(3):443–450

Dong J, Shokes JE, Scott RA, Lynn DG (2006) Modulating amyloid self-assembly and fibril morphology with Zn(II). J Am Chem Soc 128:3540–3542

Dufner-Beattie J, Wang F, Kuo YM, Gitschier J, Eide D, Andrews GK (2003) The acrodermatitis enteropathica gene ZIP4 encodes a tissue-specific, zinc-regulated zinc transporter in mice. J Biol Chem 278:33474–33481

Duncan MW, Marini AM, Watters R, Kopin IJ, Markey SP (1992) Zinc, a neurotoxin to cultured neurons, contaminates cycad flour prepared by traditional guamanian methods. J Neurosci 12:1523–1537

Durnam DM, Palmiter RD (1981) Transcriptional regulation of the mouse metallothionein-I gene by heavy metals. J Biol Chem 256:5712–5716

Eagle GR, Zombola RR, Himes RH (1983) Tubulin-zinc interactions: binding and polymerization studies. Biochemistry 22:221–228

Ebadi M, Iversen PL, Hao R, Cerutis DR, Rojas P, Happe HK, Murrin LC, Pfeiffer RF (1995) Expression and regulation of brain metallothionein. Neurochem Int 27:1–22

Ehmann WD, Markesbery WR, Alauddin M, Hossain TI, Brubaker EH (1986) Brain trace elements in Alzheimer's disease. Neurotoxicology 7:195–206

Eide DJ (1998) The molecular biology of metal ion transport in Saccharomyces cerevisiae. Annu Rev Nutr 18:441–469

Eide DJ (2006) Zinc transporters and the cellular trafficking of zinc. Biochim Biophys Acta 1763:711–722

Ertekin-Taner N, Graff-Radford N, Younkin LH, Eckman C, Baker M, Adamson J, Ronald J, Blangero J, Hutton M, Younkin SG (2000) Linkage of plasma Abeta42 to a quantitative locus on chromosome 10 in late-onset Alzheimer's disease pedigrees. Science 290: 2303–2304
Esler WP, Stimson ER, Jennings JM, Ghilardi JR, Mantyh PW, Maggio JE (1996) Zinc-induced aggregation of human and rat beta-amyloid peptides in vitro. J Neurochem 66:723–732
Evans DA, Funkenstein HH, Albert MS, Scherr PA, Cook NR, Chown MJ, Hebert LE, Hennekens CH, Taylor JO (1989) Prevalence of Alzheimer's disease in a community population of older persons. Higher than previously reported. J Am Med Assoc 262:2551–2556
Fliss H, Menard M (1992) Oxidant-induced mobilization of zinc from metallothionein. Arch Biochem Biophys 293:195–199
Frederickson CJ (1989) Neurobiology of zinc and zinc-containing neurons. Int Rev Neurobiol 31:145–238
Frederickson CJ, Klitenick MA, Manton WI, Kirkpatrick JB (1983) Cytoarchitectonic distribution of zinc in the hippocampus of man and the rat. Brain Res 273:335–339
Frederickson CJ, Koh JY, Bush AI (2005) The neurobiology of zinc in health and disease. Nat Rev Neurosci 6:449–462
Frederickson CJ, Suh SW, Silva D, Thompson RB (2000) Importance of zinc in the central nervous system: the zinc-containing neuron. J Nutr 130:1471S–1483S
Freund WD, Reddig S (1994) AMPA/Zn(2+)-induced neurotoxicity in rat primary cortical cultures: involvement of L-type calcium channels. Brain Res 654:257–264
Friedlich AL, Lee JY, van Groen T, Cherny RA, Volitakis I, Cole TB, Palmiter RD, Koh JY, Bush AI (2004) Neuronal zinc exchange with the blood vessel wall promotes cerebral amyloid angiopathy in an animal model of Alzheimer's disease. J Neurosci 24:3453–3459
Gabrielsson B, Robson T, Norris D, Chung SH (1986) Effects of divalent metal ions on the uptake of glutamate and GABA from synaptosomal fractions. Brain Res 384:218–223
Gaither LA, Eide DJ (2000) Functional expression of the human hZIP2 zinc transporter. J Biol Chem 275:5560–5564
Gaither LA, Eide DJ (2001) The human ZIP1 transporter mediates zinc uptake in human K562 erythroleukemia cells. J Biol Chem 276:22258–22264
Gaskin F, Kress Y (1977) Zinc ion-induced assembly of tubulin. J Biol Chem 252:6918–6924
Gaskin F, Kress Y, Brosnan C, Bornstein M (1978) Abnormal tubulin aggregates induced by zinc sulfate in organotypic cultures of nerve tissue. Neuroscience 3:1117–1128
Gerhardsson L, Lundh T, Minthon L, Londos E (2008) Metal Concentrations in Plasma and Cerebrospinal Fluid in Patients with Alzheimer's Disease. Dement Geriatr Cogn Disord 25:508–515
Glabe CG (2006) Common mechanisms of amyloid oligomer pathogenesis in degenerative disease. Neurobiol Aging 27:570–575
Golden BE (1989) Zinc in cell division and tissue growth: Physiological aspects. In: Mills CF (ed) Zinc in Human Biology. Springer, London, pp 119–127
Gonzalez C, Martin T, Cacho J, Brenas MT, Arroyo T, Garcia-Berrocal B, Navajo JA, Gonzalez-Buitrago JM (1999) Serum zinc, copper, insulin and lipids in Alzheimer's disease epsilon 4 apolipoprotein E allele carriers. Eur J Clin Invest 29:637–642
Gosavi N, Lee HJ, Lee JS, Patel S, Lee SJ (2002) Golgi fragmentation occurs in the cells with prefibrillar alpha-synuclein aggregates and precedes the formation of fibrillar inclusion. J Biol Chem 277:48984–48992
Gotz J, Chen F, van Dorpe J, Nitsch RM (2001) Formation of neurofibrillary tangles in P301l tau transgenic mice induced by Abeta 42 fibrils. Science 293:1491–1495
Grotz N, Fox T, Connolly E, Park W, Guerinot ML, Eide D (1998) Identification of a family of zinc transporter genes from Arabidopsis that respond to zinc deficiency. Proc Natl Acad Sci U S A 95:7220–7224
Gustafson D, Rothenberg E, Blennow K, Steen B, Skoog I (2003) An 18-year follow-up of overweight and risk of Alzheimer disease. Arch Intern Med 163:1524–1528

Haase H, Beyersmann D (2002) Intracellular zinc distribution and transport in C6 rat glioma cells. Biochem Biophys Res Commun 296:923–928

Haines A, Iliffe S, Morgan P, Dormandy T, Wood B (1991) Serum aluminium and zinc and other variables in patients with and without cognitive impairment in the community. Clin Chim Acta 198:261–266

Haug FM (1967) Electron microscopical localization of the zinc in hippocampal mossy fibre synapses by a modified sulfide silver procedure. Histochemie 8:355–368

Hebert LE, Scherr PA, Bienias JL, Bennett DA, Evans DA (2003) Alzheimer disease in the US population: prevalence estimates using the 2000 census. Arch Neurol 60:1119–1122

Ho LH, Ratnaike RN, Zalewski PD (2000) Involvement of intracellular labile zinc in suppression of DEVD-caspase activity in human neuroblastoma cells. Biochem Biophys Res Commun 268:148–154

Huang L, Gitschier J (1997) A novel gene involved in zinc transport is deficient in the lethal milk mouse. Nat Genet 17:292–297

Huang L, Kirschke CP, Gitschier J (2002) Functional characterization of a novel mammalian zinc transporter, ZnT6. J Biol Chem 277:26389–26395

Huang L, Kirschke CP, Zhang Y, Yu YY (2005) The ZIP7 gene (Slc39a7) encodes a zinc transporter involved in zinc homeostasis of the Golgi apparatus. J Biol Chem 280:15456–15463

Hunter FEJ, Ford L (1955) Inactivation of oxidative and phosphorylative systems in mitochondria by preincubation with phosphate and other. J Biol Chem 216:357–369

Hussain I, Powell D, Howlett DR, Tew DG, Meek TD, Chapman C, Gloger IS, Murphy KE, Southan CD, Ryan DM, Smith TS, Simmons DL, Walsh FS, Dingwall C, Christie G (1999) Identification of a novel aspartic protease (Asp 2) as beta-secretase. Mol Cell Neurosci 14:419–427

Ilouz R, Kaidanovich O, Gurwitz D, Eldar-Finkelman H (2002) Inhibition of glycogen synthase kinase-3beta by bivalent zinc ions: insight into the insulin-mimetic action of zinc. Biochem Biophys Res Commun 295:102–106

Irizarry MC, McNamara M, Fedorchak K, Hsiao K, Hyman BT (1997) APPSw transgenic mice develop age-related A beta deposits and neuropil abnormalities, but no neuronal loss in CA1. J Neuropathol Exp Neurol 56:965–973

Jeandel C, Nicolas MB, Dubois F, Nabet-Belleville F, Penin F, Cuny G (1989) Lipid peroxidation and free radical scavengers in Alzheimer's disease. Gerontology 35:275–282

Jia Y, Jeng JM, Sensi SL, Weiss JH (2002) Zn2+ currents are mediated by calcium-permeable AMPA/kainate channels in cultured murine hippocampal neurones. J Physiol 543:35–48

Jick H, Zornberg GL, Jick SS, Seshadri S, Drachman DA (2000) Statins and the risk of dementia. Lancet 356:1627–1631

Jost BC, Grossberg GT (1995) The natural history of Alzheimer's disease: a brain bank study. J Am Geriatr Soc 43:1248–1255

Kagi JH, Schaffer A (1988) Biochemistry of metallothionein. Biochemistry 27:8509–8515

Kambe T, Yamaguchi-Iwai Y, Sasaki R, Nagao M (2004) Overview of mammalian zinc transporters. Cell Mol Life Sci 61:49–68

Kayed R, Sokolov Y, Edmonds B, McIntire TM, Milton SC, Hall JE, Glabe CG (2004) Permeabilization of lipid bilayers is a common conformation-dependent activity of soluble amyloid oligomers in protein misfolding diseases. J Biol Chem 279:46363–46366

Kelleher SL, Lonnerdal B (2002) Zinc transporters in the rat mammary gland respond to marginal zinc and vitamin A intakes during lactation. J Nutr 132:3280–3285

Keller JN, Schmitt FA, Scheff SW, Ding Q, Chen Q, Butterfield DA, Markesbery WR (2005) Evidence of increased oxidative damage in subjects with mild cognitive impairment. Neurology 64:1152–1156

Kesslak JP, Frederickson CJ, Gage FH (1987) Quantification of hippocampal noradrenaline and zinc changes after selective cell destruction. Exp Brain Res 67:77–84

Kim EY, Koh JY, Kim YH, Sohn S, Joe E, Gwag BJ (1999) Zn2+ entry produces oxidative neuronal necrosis in cortical cell cultures. Eur J Neurosci 11:327–334

Kinoshita A, Fukumoto H, Shah T, Whelan CM, Irizarry MC, Hyman BT (2003) Demonstration by FRET of BACE interaction with the amyloid precursor protein at the cell surface and in early endosomes. J Cell Sci 116:3339–3346

Kirschke CP, Huang L (2003) ZnT7, a novel mammalian zinc transporter, accumulates zinc in the Golgi apparatus. J Biol Chem 278:4096–4102

Klaassen CD, Liu J, Choudhuri S (1999) Metallothionein: an intracellular protein to protect against cadmium toxicity. Annu Rev Pharmacol Toxicol 39:267–294

Klein WL (2002) Abeta toxicity in Alzheimer's disease: globular oligomers (ADDLs) as new vaccine and drug targets. Neurochem Int 41:345–352

Kleineke JW, Brand IA (1997) Rapid changes in intracellular Zn2+ in rat hepatocytes. J Pharmacol Toxicol Methods 38:181–187

Kleiner D, von Jagow G (1972) On the inhibition of mitochondrial electron transport by Zn(2+) ions. FEBS Lett 20:229–232

Koh JY, Choi DW (1987) Quantitative determination of glutamate mediated cortical neuronal injury in cell culture by lactate dehydrogenase efflux assay. J Neurosci Methods 20:83–90

Koh JY, Choi DW (1994) Zinc toxicity on cultured cortical neurons: involvement of N-methyl-D-aspartate receptors. Neuroscience 60:1049–1057

Koh JY, Suh SW, Gwag BJ, He YY, Hsu CY, Choi DW (1996) The role of zinc in selective neuronal death after transient global cerebral ischemia. Science 272:1013–1016

Koo EH, Squazzo SL (1994) Evidence that production and release of amyloid beta-protein involves the endocytic pathway. J Biol Chem 269:17386–17389

Korichneva I, Hoyos B, Chua R, Levi E, Hammerling U (2002) Zinc release from protein kinase C as the common event during activation by lipid second messenger or reactive oxygen. J Biol Chem 277:44327–44331

Kress Y, Gaskin F, Brosnan CF, Levine S (1981) Effects of zinc on the cytoskeletal proteins in the central nervous system of the rat. Brain Res 220:139–149

Krotkiewska B, Banas T (1992) Interaction of Zn2+ and Cu2+ ions with glyceraldehyde-3-phosphate dehydrogenase from bovine heart and rabbit muscle. Int J Biochem 24: 1501–1505

Kumar-Singh S, Dewachter I, Moechars D, Lubke U, De Jonghe C, Ceuterick C, Checler F, Naidu A, Cordell B, Cras P, Van Broeckhoven C, Van Leuven F (2000) Behavioral disturbances without amyloid deposits in mice overexpressing human amyloid precursor protein with Flemish (A692G) or Dutch (E693Q) mutation. Neurobiol Dis 7:9–22

Lambert MP, Barlow AK, Chromy BA, Edwards C, Freed R, Liosatos M, Morgan TE, Rozovsky I, Trommer B, Viola KL, Wals P, Zhang C, Finch CE, Krafft GA, Klein WL (1998) Diffusible, nonfibrillar ligands derived from Abeta1-42 are potent central nervous system neurotoxins. Proc Natl Acad Sci U S A 95:6448–6453

Lannfelt L, Blennow K, Zetterberg H, Batsman S, Ames D, Harrison J, Masters CL, Targum S, Bush AI, Murdoch R, Wilson J, Ritchie CW (2008) Safety, efficacy, and biomarker findings of PBT2 in targeting Abeta as a modifying therapy for Alzheimer's disease: a phase IIa, double-blind, randomised, placebo-controlled trial. Lancet Neurol 7(9):779–786

Lee JY, Mook-Jung I, Koh JY (1999) Histochemically reactive zinc in plaques of the Swedish mutant beta-amyloid precursor protein transgenic mice. J Neurosci 19:RC10

Lehmann HM, Brothwell BB, Volak LP, Bobilya DJ (2002) Zinc status influences zinc transport by porcine brain capillary endothelial cells. J Nutr 132:2763–2768

Leibson CL, Rocca WA, Hanson VA, Cha R, Kokmen E, O'Brien PC, Palumbo PJ (1997) Risk of dementia among persons with diabetes mellitus: a population-based cohort study. Am J Epidemiol 145:301–308

Levy-Lahad E, Wasco W, Poorkaj P, Romano DM, Oshima J, Pettingell WH, Yu CE, Jondro PD, Schmidt SD, Wang K et al (1995) Candidate gene for the chromosome 1 familial Alzheimer's disease locus. Science 269:973–977

Lewis J, Dickson DW, Lin WL, Chisholm L, Corral A, Jones G, Yen SH, Sahara N, Skipper L, Yager D, Eckman C, Hardy J, Hutton M, McGowan E (2001) Enhanced neurofibrillary degeneration in transgenic mice expressing mutant tau and APP. Science 293:1487–1491

Lorenzo A, Yankner BA (1994) Beta-amyloid neurotoxicity requires fibril formation and is inhibited by congo red. Proc Natl Acad Sci U S A 91:12243–12247

Lovell MA, Ehmann WD, Mattson MP, Markesbery WR (1997) Elevated 4-hydroxynonenal in ventricular fluid in Alzheimer's disease. Neurobiol Aging 18:457–461

Lovell MA, Markesbery WR (2005) Ectopic expression of Musashi-1 in Alzheimer disease and Pick disease. J Neuropathol Exp Neurol 64:675–680

Lovell MA, Robertson JD, Teesdale WJ, Campbell JL, Markesbery WR (1998) Copper, iron and zinc in Alzheimer's disease senile plaques. J Neurol Sci 158:47–52

Lovell MA, Smith JL, Markesbery WR (2006) Elevated zinc transporter-6 in mild cognitive impairment, Alzheimer disease, and pick disease. J Neuropathol Exp Neurol 65:489–498

Manev H, Kharlamov E, Uz T, Mason RP, Cagnoli CM (1997) Characterization of zinc-induced neuronal death in primary cultures of rat cerebellar granule cells. Exp Neurol 146:171–178

Mantyh PW, Ghilardi JR, Rogers S, DeMaster E, Allen CJ, Stimson ER, Maggio JE (1993) Aluminum, iron, and zinc ions promote aggregation of physiological concentrations of beta-amyloid peptide. J Neurochem 61:1171–1174

Maret W (2002) Optical methods for measuring zinc binding and release, zinc coordination environments in zinc finger proteins, and redox sensitivity and activity of zinc-bound thiols. Methods Enzymol 348:230–237

Markesbery WR, Ehmann WD (1994) Trace element alterations in Alzheimer's disease. In: Terry RD, Katzman R (eds) Alzheimer's disease. Raven Press, New York, NY

Markesbery WR, Schmitt FA, Kryscio RJ, Davis DG, Smith CD, Wekstein DR (2006) Neuropathologic substrate of mild cognitive impairment. Arch Neurol 63:38–46

McMahon RJ, Cousins RJ (1998) Regulation of the zinc transporter ZnT-1 by dietary zinc. Proc Natl Acad Sci U S A 95:4841–4846

Merchant C, Tang MX, Albert S, Manly J, Stern Y, Mayeux R (1999) The influence of smoking on the risk of Alzheimer's disease. Neurology 52:1408–1412

Miller LM, Wang Q, Telivala TP, Smith RJ, Lanzirotti A, Miklossy J (2006) Synchrotron-based infrared and X-ray imaging shows focalized accumulation of Cu and Zn co-localized with beta-amyloid deposits in Alzheimer's disease. J Struct Biol 155:30–37

Mocchegiani E, Bertoni-Freddari C, Marcellini F, Malavolta M (2005) Brain, aging and neurodegeneration: role of zinc ion availability. Prog Neurobiol 75:367–390

Moechars D, Lorent K, De Strooper B, Dewachter I, Van Leuven F (1996) Expression in brain of amyloid precursor protein mutated in the alpha-secretase site causes disturbed behavior, neuronal degeneration and premature death in transgenic mice. EMBO J 15:1265–1274

Molina JA, Jimenez-Jimenez FJ, Aguilar MV, Meseguer I, Mateos-Vega CJ, Gonzalez-Munoz MJ, de Bustos F, Porta J, Orti-Pareja M, Zurdo M, Barrios E, Martinez-Para MC (1998) Cerebrospinal fluid levels of transition metals in patients with Alzheimer's disease. J Neural Transm 105:479–488

Morris JC, Heyman A, Mohs RC, Hughes JP, van Belle G, Fillenbaum G, Mellits ED, Clark C (1989) The Consortium to Establish a Registry for Alzheimer's Disease (CERAD). Part I. Clinical and neuropsychological assessment of Alzheimer's disease. Neurology 39: 1159–1165

Mortimer JA, van Duijn CM, Chandra V, Fratiglioni L, Graves AB, Heyman A, Jorm AF, Kokmen E, Kondo K, Rocca WA et al (1991) Head trauma as a risk factor for Alzheimer's disease: a collaborative re-analysis of case-control studies. EURODEM Risk Factors Research Group. Int J Epidemiol 20(Suppl 2):S28–S35

Mucke L, Yu GQ, McConlogue L, Rockenstein EM, Abraham CR, Masliah E (2000) Astroglial expression of human alpha(1)-antichymotrypsin enhances alzheimer-like pathology in amyloid protein precursor transgenic mice. Am J Pathol 157:2003–2010

Murakami K, Whiteley MK, Routtenberg A (1987) Regulation of protein kinase C activity by cooperative interaction of Zn2+ and Ca2+. J Biol Chem 262:13902–13906

Myers A, Holmans P, Marshall H, Kwon J, Meyer D, Ramic D, Shears S, Booth J, DeVrieze FW, Crook R, Hamshere M, Abraham R, Tunstall N, Rice F, Carty S, Lillystone S, Kehoe P, Rudrasingham V, Jones L, Lovestone S, Perez-Tur J, Williams J, Owen MJ, Hardy J, Goate

AM (2000) Susceptibility locus for Alzheimer's disease on chromosome 10. Science 290: 2304–2305
National institute on aging and Reagan institute working group (1997) Diagnostic criteria for the neuropathological assessment of Alzheimer's disease. Neurobiol Aging 18(S4):S
Nolte C, Gore A, Sekler I, Kresse W, Hershfinkel M, Hoffmann A, Kettenmann H, Moran A (2004) ZnT-1 expression in astroglial cells protects against zinc toxicity and slows the accumulation of intracellular zinc. Glia 48:145–155
Nunomura A, Perry G, Aliev G, Hirai K, Takeda A, Balraj EK, Jones PK, Ghanbari H, Wataya T, Shimohama S, Chiba S, Atwood CS, Petersen RB, Smith MA (2001) Oxidative damage is the earliest event in Alzheimer disease. J Neuropathol Exp Neurol 60:759–767
Oddo S, Billings L, Kesslak JP, Cribbs DH, LaFerla FM (2004) Abeta immunotherapy leads to clearance of early, but not late, hyperphosphorylated tau aggregates via the proteasome. Neuron 43:321–332
Ohana E, Sekler I, Kaisman T, Kahn N, Cove J, Silverman WF, Amsterdam A, Hershfinkel M (2006) Silencing of ZnT-1 expression enhances heavy metal influx and toxicity. J Mol Med 84:753–763
Palmiter RD (1995) Constitutive expression of metallothionein-III (MT-III), but not MT-I, inhibits growth when cells become zinc deficient. Toxicol Appl Pharmacol 135:139–146
Palmiter RD (1998) The elusive function of metallothioneins. Proc Natl Acad Sci U S A 95: 8428–8430
Palmiter RD, Cole TB, Findley SD (1996) ZnT-2, a mammalian protein that confers resistance to zinc by facilitating vesicular sequestration. EMBO J 15:1784–1791
Peila R, Rodriguez BL, Launer LJ (2002) Type 2 diabetes, APOE gene, and the risk for dementia and related pathologies: The Honolulu-Asia Aging Study. Diabetes 51: 1256–1262
Penkowa M (2006) Metallothioneins are multipurpose neuroprotectants during brain pathology. FEBS J 273:1857–1870
Penkowa M, Nielsen H, Hidalgo J, Bernth N, Moos T (1999) Distribution of metallothionein I + II and vesicular zinc in the developing central nervous system: correlative study in the rat. J Comp Neurol 412:303–318
Perez-Clausell J, Danscher G (1986) Release of zinc sulphide accumulations into synaptic clefts after in vivo injection of sodium sulphide. Brain Res 362:358–361
Petersen RC, Morris JC (2003) Clinical features. In: Petersen RC (ed) Mild Cognitive Impairment. Oxford University, New York, NY, pp 15–39
Petersen RC, Smith GE, Waring SC, Ivnik RJ, Tangalos EG, Kokmen E (1999) Mild cognitive impairment: clinical characterization and outcome. Arch Neurol 56:303–308
Pike CJ, Burdick D, Walencewicz AJ, Glabe CG, Cotman CW (1993) Neurodegeneration induced by beta-amyloid peptides in vitro: the role of peptide assembly state. J Neurosci 13: 1676–1687
Pike CJ, Walencewicz AJ, Glabe CG, Cotman CW (1991) In vitro aging of beta-amyloid protein causes peptide aggregation and neurotoxicity. Brain Res 563:311–314
Prasad AS, Miale A Jr., Farid Z, Sandstead HH, Schulert AR (1963) Zinc metabolism in patients with the syndrome of iron deficiency anemia, hepatosplenomegaly, dwarfism, and hypognadism. J Lab Clin Med 61:537–549
Price D, Joshi JG (1982) Ferritin: a zinc detoxicant and a zinc ion donor. Proc Natl Acad Sci U S A 79:3116–3119
Ritchie CW, Bush AI, Mackinnon A, Macfarlane S, Mastwyk M, MacGregor L, Kiers L, Cherny R, Li QX, Tammer A, Carrington D, Mavros C, Volitakis I, Xilinas M, Ames D, Davis S, Beyreuther K, Tanzi RE, Masters CL (2003) Metal-protein attenuation with iodochlorhydroxyquin (clioquinol) targeting Abeta amyloid deposition and toxicity in Alzheimer disease: a pilot phase 2 clinical trial. Arch Neurol 60:1685–1691
Rivlin AA, Chan YL, Wool IG (1999) The contribution of a zinc finger motif to the function of yeast ribosomal protein YL37a. J Mol Biol 294:909–919

Rogaeva E, Meng Y, Lee JH, Gu Y, Kawarai T, Zou F, Katayama T, Baldwin CT, Cheng R, Hasegawa H, Chen F, Shibata N, Lunetta KL, Pardossi-Piquard R, Bohm C, Wakutani Y, Cupples LA, Cuenco KT, Green RC, Pinessi L, Rainero I, Sorbi S, Bruni A, Duara R, Friedland RP, Inzelberg R, Hampe W, Bujo H, Song YQ, Andersen OM, Willnow TE, Graff-Radford N, Petersen RC, Dickson D, Der SD, Fraser PE, Schmitt-Ulms G, Younkin S, Mayeux R, Farrer LA, St. George-Hyslop P (2007) The neuronal sortilin-related receptor SORL1 is genetically associated with Alzheimer disease. Nat Genet 39:168–177

Rubin H, Koide T (1973) Stimulation of DNA synthesis and 2-deoxy-D-glucose transport in chick embryo cultures by excessive metal concentrations and by a carcinogenic hydrocarbon. J Cell Physiol 81:387–396

Rulon LL, Robertson JD, Lovell MA, Deibel MA, Ehmann WD, Markesber WR (2000) Serum zinc levels and Alzheimer's disease. Biol Trace Elem Res 75:79–85

Samudralwar DL, Diprete CC, Ni BF, Ehmann WD, Markesbery WR (1995) Elemental imbalances in the olfactory pathway in Alzheimer's disease. J Neurol Sci 130:139–145

Segal D, Ohana E, Besser L, Hershfinkel M, Moran A, Sekler I (2004) A role for ZnT-1 in regulating cellular cation influx. Biochem Biophys Res Commun 323: 1145–1150

Selkoe DJ (1999) Translating cell biology into therapeutic advances in Alzheimer's disease. Nature 399:A23–A31

Seshadri S, Beiser A, Selhub J, Jacques PF, Rosenberg IH, D'Agostino RB, Wilson PW, Wolf PA (2002) Plasma homocysteine as a risk factor for dementia and Alzheimer's disease. N Engl J Med 346:476–483

Seve M, Chimienti F, Devergnas S, Favier A (2004) In silico identification and expression of SLC30 family genes: an expressed sequence tag data mining strategy for the characterization of zinc transporters' tissue expression. BMC Genomics 5:32

Shore D, Henkin RI, Nelson NR, Agarwal RP, Wyatt RJ (1984) Hair and serum copper, zinc, calcium, and magnesium concentrations in Alzheimer-type dementia. J Am Geriatr Soc 32:892–895

Sim DL, Chow VT (1999) The novel human HUEL (C4orf1) gene maps to chromosome 4p12-p13 and encodes a nuclear protein containing the nuclear receptor interaction motif. Genomics 59:224–233

Sinha S, Anderson JP, Barbour R, Basi GS, Caccavello R, Davis D, Doan M, Dovey HF, Frigon N, Hong J, Jacobson-Croak K, Jewett N, Keim P, Knops J, Lieberburg I, Power M, Tan H, Tatsuno G, Tung J, Schenk D, Seubert P, Suomensaari SM, Wang S, Walker D, Zhao J, McConlogue L, John V (1999) Purification and cloning of amyloid precursor protein beta-secretase from human brain. Nature 402:537–540

Sisodia SS, St. George-Hyslop PH (2002) gamma-Secretase, Notch, Abeta and Alzheimer's disease: where do the presenilins fit in? Nat Rev Neurosci 3:281–290

Skhulachev VP, Christyakov VV, Jasaitis AA, Smirnova EG (1967) Inhibition of the respiratory chain by zinc ions. Biochem Biophys Res Commun 26:1–6

Skoog I, Lernfelt B, Landahl S, Palmertz B, Andreasson LA, Nilsson L, Persson G, Oden A, Svanborg A (1996) 15-year longitudinal study of blood pressure and dementia. Lancet 347:1141–1145

Smith KT, Failla ML, Cousins RJ (1979) Identification of albumin as the plasma carrier for zinc absorption by perfused rat intestine. Biochem J 184:627–633

Smith JL, Xiong S, Lovell MA (2006a) 4-Hydroxynonenal disrupts zinc export in primary rat cortical cells. Neurotoxicology 27:1–5

Smith JL, Xiong S, Markesbery WR, Lovell MA (2006b) Altered expression of zinc transporters-4 and -6 in mild cognitive impairment, early and late Alzheimer's disease brain. Neuroscience 140:879–888

Snowdon DA, Greiner LH, Markesbery WR (2000) Linguistic ability in early life and the neuropathology of Alzheimer's disease and cerebrovascular disease. Findings from the Nun Study. Ann N Y Acad Sci 903:34–38

Snowdon DA, Kemper SJ, Mortimer JA, Greiner LH, Wekstein DR, Markesbery WR (1996) Linguistic ability in early life and cognitive function and Alzheimer's disease in late life. Findings from the Nun Study. J Am Med Assoc 275:528–532

Sommer B (2002) Alzheimer's disease and the amyloid cascade hypothesis: ten years on. Curr Opin Pharmacol 2:87–92

St. George-Hyslop PH, Tanzi RE, Polinsky RJ, Haines JL, Nee L, Watkins PC, Myers RH, Feldman RG, Pollen D, Drachman D et al (1987) The genetic defect causing familial Alzheimer's disease maps on chromosome 21. Science 235:885–890

St. George-Hyslop PH (1994) The molecular genetics of Alzheimer's disease. In: Terry RD, Katzman R, Bick LL (eds) Alzheimer's disease. Raven Press, New York, NY, pp 345–352

Stengaard-Pedersen K, Fredens K, Larsson LI (1983) Comparative localization of enkephalin and cholecystokinin immunoreactivities and heavy metals in the hippocampus. Brain Res 273:81–96

Stoltenberg M, Bruhn M, Sondergaard C, Doering P, West MJ, Larsen A, Troncoso JC, Danscher G (2005) Immersion autometallographic tracing of zinc ions in Alzheimer beta-amyloid plaques. Histochem Cell Biol 123:605–611

Stoltenberg M, Bush AI, Bach G, Smidt K, Larsen A, Rungby J, Lund S, Doering P, Danscher G (2007) Amyloid plaques arise from zinc-enriched cortical layers in APP/PS1 transgenic mice and are paradoxically enlarged with dietary zinc deficiency. Neuroscience 150:357–369

Strozyk D, Launer LJ, Adlard PA, Cherny RA, Tsatsanis A, Volitakis I, Blennow K, Petrovitch H, White LR, Bush AI (2007) Zinc and copper modulate Alzheimer Abeta levels in human cerebrospinal fluid. Neurobiol Aging 30(7):1069–1077

Sturniolo GC, Mestriner C, D'Inca R (2000) Trace Element and Mineral Nutrition in Gastrointestinal Disease. In: Bogden JD, Klevay LM (eds) Clinical Nutrition of the Essential Trace Elements and Minerals: The Guide for Health Professionals. Humana Press, Totowa, NJ

Takeda A (2000) Movement of zinc and its functional significance in the brain. Brain Res Brain Res Rev 34:137–148

Takeda A, Minami A, Takefuta S, Tochigi M, Oku N (2001) Zinc homeostasis in the brain of adult rats fed zinc-deficient diet. J Neurosci Res 63:447–452

Taylor KM, Morgan HE, Johnson A, Hadley LJ, Nicholson RI (2003) Structure-function analysis of LIV-1, the breast cancer-associated protein that belongs to a new subfamily of zinc transporters. Biochem J 375:51–59

The National Institute on Aging, and Reagan Institute Working Group on Diagnostic Criteria for the Neuropathological Assessment of Alzheimer's Disease (1997) Consensus recommendations for the postmortem diagnosis of Alzheimer's disease. The National Institute on Aging, and Reagan Institute Working Group on Diagnostic Criteria for the Neuropathological Assessment of Alzheimer's Disease. Neurobiol Aging 18:S

Tonder N, Johansen FF, Frederickson CJ, Zimmer J, Diemer NH (1990) Possible role of zinc in the selective degeneration of dentate hilar neurons after cerebral ischemia in the adult rat. Neurosci Lett 109:247–252

Trayhurn P, Duncan JS, Wood AM, Beattie JH (2000) Regulation of metallothionein gene expression and secretion in rat adipocytes differentiated from preadipocytes in primary culture. Horm Metab Res 32:542–547

Vallee BL (1995) The function of metallothionein. Neurochem Int 27:23–33

Vallee BL, Falchuk KH (1993) The biochemical basis of zinc physiology. Physiol Rev 73:79–118

Vandenberg RJ, Mitrovic AD, Johnston GA (1998) Molecular basis for differential inhibition of glutamate transporter subtypes by zinc ions. Mol Pharmacol 54:189–196

Vassar R, Bennett BD, Babu-Khan S, Kahn S, Mendiaz EA, Denis P, Teplow DB, Ross S, Amarante P, Loeloff R, Luo Y, Fisher S, Fuller J, Edenson S, Lile J, Jarosinski MA, Biere AL, Curran E, Burgess T, Louis JC, Collins F, Treanor J, Rogers G, Citron M (1999) Beta-secretase cleavage of Alzheimer's amyloid precursor protein by the transmembrane aspartic protease BACE. Science 286:735–741

Wallace DC (1992) Mitochondrial genetics: a paradigm for aging and degenerative diseases? Science 256:628–632

Walsh DM, Klyubin I, Fadeeva JV, Cullen WK, Anwyl R, Wolfe MS, Rowan MJ, Selkoe DJ (2002) Naturally secreted oligomers of amyloid beta protein potently inhibit hippocampal long-term potentiation in vivo. Nature 416:535–539

Walsh DM, Lomakin A, Benedek GB, Condron MM, Teplow DB (1997) Amyloid beta-protein fibrillogenesis. Detection of a protofibrillar intermediate. J Biol Chem 272:22364–22372

Wang F, Kim BE, Petris MJ, Eide DJ (2004) The mammalian Zip5 protein is a zinc transporter that localizes to the basolateral surface of polarized cells. J Biol Chem 279:51433–51441

Wang HW, Pasternak JF, Kuo H, Ristic H, Lambert MP, Chromy B, Viola KL, Klein WL, Stine WB, Krafft GA, Trommer BL (2002a) Soluble oligomers of beta amyloid (1-42) inhibit long-term potentiation but not long-term depression in rat dentate gyrus. Brain Res 924:133–140

Wang K, Zhou B, Kuo YM, Zemansky J, Gitschier J (2002b) A novel member of a zinc transporter family is defective in acrodermatitis enteropathica. Am J Hum Genet 71:66–73

Wenstrup D, Ehmann WD, Markesbery WR (1990) Trace element imbalances in isolated subcellular fractions of Alzheimer's disease brains. Brain Res 533:125–131

Westerman MA, Cooper-Blacketer D, Mariash A, Kotilinek L, Kawarabayashi T, Younkin LH, Carlson GA, Younkin SG, Ashe KH (2002) The relationship between Abeta and memory in the Tg2576 mouse model of Alzheimer's disease. J Neurosci 22:1858–1867

Williams RJP (1989) An introduction to the biochemistry of zinc. In: Mills CF (ed) Zinc in Human Biology. Springer, London, pp 15–31

Williams TI, Lynn BC, Markesbery WR, Lovell MA (2005) Increased levels of 4-hydroxynonenal and acrolein, neurotoxic markers of lipid peroxidation, in the brain in Mild Cognitive Impairment and early Alzheimer's disease. Neurobiol Aging 27(8):1094–1099

Wilquet V, De Strooper B (2004) Amyloid-beta precursor protein processing in neurodegeneration. Curr Opin Neurobiol 14:582–588

Yamaguchi M, Kura M, Okada S (1982) Role of zinc as an activator of mitochondrial function in rat liver. Biochem Pharmacol 31:1289–1293

Yang JP, Merin JP, Nakano T, Kato T, Kitade Y, Okamoto T (1995) Inhibition of the DNA-binding activity of NF-kappa B by gold compounds in vitro. FEBS Lett 361:89–96

Yatin SM, Varadarajan S, Link CD, Butterfield DA (1999) In vitro and in vivo oxidative stress associated with Alzheimer's amyloid beta-peptide (1-42). Neurobiol Aging 20:325–330; discussion 339–342

Yin HZ, Weiss JH (1995) Zn(2+) permeates Ca(2+) permeable AMPA/kainate channels and triggers selective neural injury. NeuroReport 6:2553–2556

Yokoyama M, Koh J, Choi DW (1986) Brief exposure to zinc is toxic to cortical neurons. Neurosci Lett 71:351–355

Zabel U, Schreck R, Baeuerle PA (1991) DNA binding of purified transcription factor NF-kappa B. Affinity, specificity, Zn2+ dependence, and differential half-site recognition. J Biol Chem 266:252–260

Zalewski PD, Forbes IJ, Seamark RF, Borlinghaus R, Betts WH, Lincoln SF, Ward AD (1994) Flux of intracellular labile zinc during apoptosis (gene-directed cell death) revealed by a specific chemical probe, Zinquin. Chem Biol 1:153–161

Zeng J, Heuchel R, Schaffner W, Kagi JH (1991) Thionein (apometallothionein) can modulate DNA binding and transcription activation by zinc finger containing factor Sp1. FEBS Lett 279:310–312

Zhang LH, Wang X, Stoltenberg M, Danscher G, Huang L, Wang ZY (2008a) Abundant expression of zinc transporters in the amyloid plaques of Alzheimer's disease brain. Brain Res Bull 77:55–60

Zhang LH, Wang X, Zheng ZH, Ren H, Stoltenberg M, Danscher G, Huang L, Rong M, Wang ZY (2008b) Altered expression and distribution of zinc transporters in APP/PS1 transgenic mouse brain. Neurobiol Aging 31(1):74–87

The Genetics of Alzheimer's Disease and Parkinson's Disease

Lynn M. Bekris, Chang-En Yu, Thomas D. Bird, and Debby Tsuang

Abstract Alzheimer's disease (AD) is the most common neurodegenerative disorder. It is characterized by progressive loss of memory and other cognitive domains along with functional decline that can occur in the third to eighth decades. The early onset (<60 years old) familial forms of AD have an autosomal dominant inheritance linked to three causative genes: *APP*, *PSEN1*, and *PSEN2*. The most common sporadic form of AD occurs after the age of 60 and is associated with the *APOE* gene. The mechanistic contribution of these genes in AD pathogenesis has been studied extensively but is still unclear, suggesting that other AD associated genes remain to be elucidated. Parkinson's disease (PD) is the second most common neurodegenerative disorder. Idiopathic PD is the most frequent form of Parkinsonism, although rare forms of PD in which genetic factors dominate exist. Family studies have identified 13 causative genetic loci linked to PD of which 8 genes have been described: four autosomal dominant (*SNCA*, *LRRK2*, *UCHL1*, and *HTRA2*) and four autosomal recessive (*PRKN*, *DJ1*, *PINK1*, and *ATP13A2*). In addition, another gene has recently been described as a possible risk factor for PD (*GBA*). The function of these genes and their contribution to PD pathogenesis remains to be fully elucidated. Like AD, other genes that contribute to PD risk likely exist. The prevalence, incidence, clinical manifestations, and genetic components of these two neurodegenerative disorders, AD and PD, are discussed in this chapter.

Keywords Alzheimer's disease · Parkinson's disease · Presenilin · Amyloid precursor protein · Apolipoprotein E · Synuclein · Parkin · LRRK2 · PINK1 · neurodegeneration

Contents

D. Tsuang (✉)
Department of Psychiatry and Behavioral Sciences, University of Washington School of Medicine, Seattle, WA, USA; Mental Illness Research, Education and Clinical Center, Veterans Affairs Puget Sound Health Care System, Seattle, WA, USA
e-mail: dwt1@u.washington.edu

J.P. Blass (ed.), *Neurochemical Mechanisms in Disease*,
Advances in Neurobiology 1, DOI 10.1007/978-1-4419-7104-3_21,

1 Alzheimer's Disease

1.1 Introduction

1.1.1 Prevalence and Incidence

Alzheimer's disease (AD) (OMIM #104300) is the most common irreversible, progressive brain disease. It is characterized by a gradual loss of memory and cognitive skills. AD accounts for over 50% of all dementia cases, and presently affects more than 24 million people worldwide, with over 5 million new cases each year, a figure that is likely to increase as a greater proportion of the population ages (Ferri et al., 2005).

Age is the largest known risk factor, with AD prevalence increasing significantly with age. AD incidence increases from 2.8 per 1000 person-years when 65–69 years and to 56.1 per 1000 person-years when older than 90 years (Kukull et al., 2002). Approximately 10% of persons older than 70 years have significant memory loss and more than half of these individuals have probable AD. An estimated 25–45% of persons older than 85 years have dementia (Bird, 2008). The duration of disease is typically 8–10 years, with a range from 2 to 25 years after diagnosis. The disease is divided into two subtypes based on the age of onset: early-onset AD (EOAD) and late-onset AD (LOAD). EOAD accounts for approximately 1–6% of all cases and ranges roughly from 30 years to 60 or 65 years. On the other hand, the most common form of AD, LOAD, is defined as an age-at-onset later than 60 or 65 years. Both EOAD and LOAD may have a positive family history of AD. With the exception of a few autosomal dominant families that are single-gene disorders (see below), most AD appears to be a complex disorder that is likely to involve multiple susceptibility genes and environmental factors (Bertram and Tanzi, 2004b; Bird, 2008; Kamboh, 2004; Roses, 2006; Serretti et al., 2005). Approximately 60% of EOAD is familial, with multiple cases of AD within a family. Thirteen percent of these familial cases are inherited in an autosomal dominant manner with at least three generations affected (Brickell et al., 2006; Campion et al., 1999). Early-onset cases can also occur in families with late-onset disease (Bird, 2008).

1.1.2 Clinical Symptoms

Both EOAD and LOAD present clinically as dementia that begins with a gradual decline of memory which slowly increases in severity until symptoms eventually become incapacitating. Other common symptoms are confusion, poor judgment, language disturbance, agitation, withdrawal, and hallucinations. Rare symptoms include seizures, Parkinsonism, increased muscle tone, myoclonus, incontinence, and mutism. Death commonly occurs from general inanition, malnutrition, and pneumonia (Bird, 2008). Treatment of AD with cholinesterase inhibitors and memantine may have some improvement in cognitive decline in mild to moderate dementia cases but overall there is clinically marginal improvement in measures of cognition and global assessment of dementia (Raina et al., 2008; Raschetti et al., 2007).

1.1.3 Clinical Diagnosis

Currently, the diagnosis of AD is based on clinical history and neuropsychological tests. The *Diagnostic and Statistical Manual of Mental Disorders, 4th Edition (DSM-IV)* criteria for diagnosing dementia requires loss of two or more of the following: memory, language, calculation, orientation, or judgment (Kawas, 2003). The Mini-Mental State Examination (MMSE) helps to evaluate changes in a patient's cognitive abilities. In addition, a diagnosis of probable AD necessitates the exclusion of other degenerative disorders associated with dementia, such as frontotemporal dementia (including frontotemporal dementia with Parkinsonism-17 and Pick's disease), Parkinson's disease, diffuse Lewy body disease, Creutzfeldt–Jakob disease, and cerebral autosomal dominant arteriopathy with subcortical infarcts and leukoencephalopathy (CADASIL) (Rogan and Lippa, 2002). Discrimination from other forms of dementia is usually based on clinical history and neuroimaging (Bird, 2008). In addition, other possible causes of dementia also need to be excluded, especially the treatable forms of cognitive impairment, such as that due to depression, chronic drug intoxication, chronic central nervous system infection, thyroid disease, vitamin deficiencies (i.e., B_{12} and thiamine), central nervous system angitis, and normal-pressure hydrocephalus (Bird, 2008). Individuals who do not meet these criteria but have short-term memory loss and have only minimal impairment in other cognitive abilities and are not functionally impaired at work or at home are considered to have "mild cognitive impairment" (Petersen et al., 2001).

1.1.4 Neuropathological Diagnosis

A definitive diagnosis of AD requires not only the presence of severe dementia in life but also postmortem confirmation, with the presence of two histopathological features: neurofibrillary tangles and amyloid plaques (Braak and Braak, 1997; Goedert and Spillantini, 2006; Nussbaum and Ellis, 2003). The clinical diagnosis of AD, before autopsy confirmation, is correct about 80–90% of the time by expert clinicians (Kaye, 1998). Even though plaques and tangles are often also found in cognitively normal age-matched controls, the density and distribution are more severe in patients with AD, according to standardized histological assessments (Braak and Braak, 1997). Amyloid plaques are extracellular with a cross-beta structure and characteristic dye-binding (neuritic amyloid plaques contain thioflavin S and Congo red-positive fibrillar deposits with both Aβ40 and Aβ42 present; Kidd,

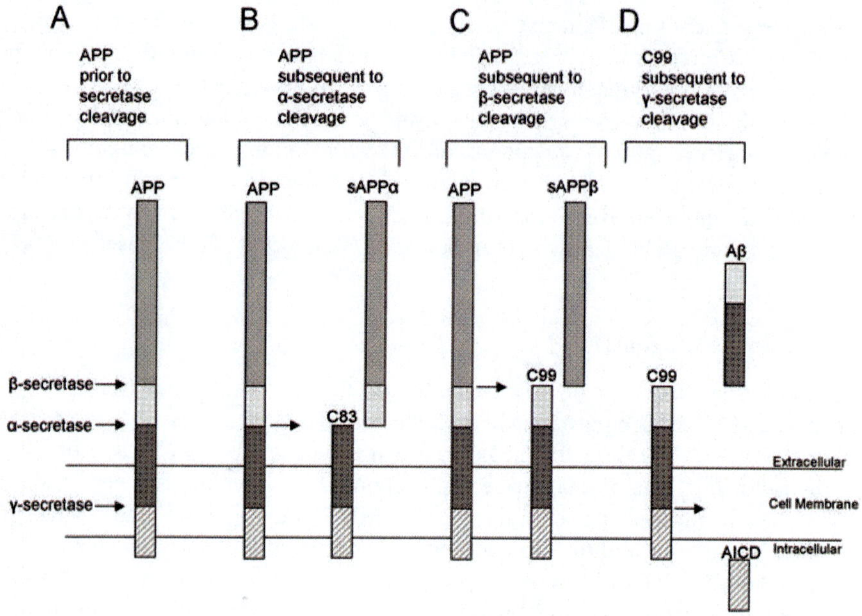

Fig. 1 APP cleavage. The APP protein can be cleaved by three different secretases: α, β, or γ (**panel a**). Subsequent to "normal" α-secretase cleavage, sAPPα is produced and released into the extracellular space and the C83 peptide remains in the cell membrane (**panel b**). Subsequent to β-secretase cleavage, sAPPβ is produced and released into the extracellular space and the C99 peptide remains in the cell membrane (**panel c**). Subsequent to β-secretase cleavage, the C99 peptide is "abnormally" cleaved by γ-secretase to yield an Aβ peptide and the AICD peptide (**panel d**). Scale is approximate

1963; Terry et al., 1964). The major component of amyloid plaques is amyloid-beta (Aβ), which can be stained and detected using Aβ antibodies (Glenner et al., 1984; Iwatsubo et al., 1994). The most common form of Aβ is 40 amino acids long and is called Aβ40. A 42 amino acid long fragment, Aβ42, is less abundant and differs only by having two additional amino acid residues at the C-terminus. Aβ42 is associated with AD (Bentahir et al., 2006). Aβ is derived from the amyloid precursor protein (APP) by the action of two aspartyl proteases. First α-secretase (nonneurotoxic "normal" cleavage) or β-secretase (potential neurotoxic "abnormal" cleavage) cleaves APP (Fig. 1). Second γ-secretases cleave APP (Haass et al., 1992; Seubert et al., 1992; Shoji et al., 1992). Upon cleavage by α-secretase, a large ectodomain referred to as soluble APP alpha (sAPPα) is released and a C-terminal 83 amino acid fragment (C83) remains membrane bound. Upon cleavage by β-secretase, APP sheds a large ectodomain referred to as soluble APP beta (s*APP*β) and leaves a membrane-bound C-terminal fragment (Cai et al., 2001; Vassar et al., 1999). This 99 amino acid fragment (C99) is membrane bound and is subsequently cleaved by γ-secretase to release Aβ and the APP intracellular domain (AICD) (De Strooper, 2000; Schroeter et al., 2003) (Fig. 1). Thus two main forms of Aβ are produced depending on the point of cleavage by γ-secretase; producing either 40 or 42 amino acid residues. The proportion of Aβ42 to Aβ40 formed is particularly important in AD because Aβ42 is far more prone to oligomerize and form fibrils than the more abundantly produced Aβ40 peptide. In a small number of individuals an increased proportion of Aβ42 appears sufficient to cause EOAD even though it appears that the production of Aβ isoforms is a normal process of unknown function (Goedert and Spillantini, 2006; Irvine et al., 2008).

Neurons bearing neurofibrillary tangles containing hyperphosphorylated tau are frequently found in AD brain (Kosik et al., 1986; Wood et al., 1986), and their temporal and spatial appearance more closely reflects disease severity than does the presence of amyloid plaques (Braak and Braak, 1991; Thal et al., 2006). However, neurofibrillary tangles are not specific to AD, are found in other disorders (such as frontotemporal dementia and progressive supranuclear palsy), and are not necessarily associated with the cognitive dysfunction and memory impairment typical of AD, and mutations in the gene that encodes the tau protein (*MAPT*) have not been genetically linked to AD (Iwatsubo et al., 1994).

1.2 Genetics of Alzheimer's Disease

1.2.1 Introduction

To date autosomal dominant early-onset familial AD (EOFAD) is associated with three genes: the *APP* gene, the presenilin 1 gene (*PSEN1*), and the presenilin 2 gene (*PSEN2*) (Goedert and Spillantini, 2006). However, it is likely that other genes will be identified as a cause of EOFAD because there are still kindreds with autosomal-dominant EOFAD with no known mutations in these three genes (Bird, 2008; Cruts and Van Broeckhoven, 1998; Raux et al., 2005). Despite evidence from family studies that genetic mutations cause EOFAD, more than 90% of AD cases appear to be sporadic, without a family history, and have a later age-at-onset of 60–65 years (Bertram and Tanzi, 2004a). The only gene consistently found to be associated with sporadic LOAD, across multiple studies, is the apolipoprotein E gene (*APOE*) (Coon et al., 2007; Couzin, 2008; Roses et al., 1995; Schellenberg, 1995; Selkoe, 2001) (Table 1). Although twin studies support the existence of a genetic component in LOAD, no causative gene has been yet identified. The age-at-onset of LOAD is significantly more variable for dizygotic twins than for monozygotic twins, suggesting that both genetic and environmental factors play a role in the disease (Gatz et al., 2006). The *APOE* gene is the only well-validated gene strongly associated with LOAD risk (Coon et al., 2007; Couzin, 2008; Roses et al., 1995; Schellenberg, 1995; Selkoe, 2001). However, many carriers of the *APOE* risk allele (ϵ4) live into their 90s, suggesting the existence of other LOAD genetic and/or environmental risk factors yet to be identified. Several other genetic variants have been reported and suggest that there may be five to seven major LOAD susceptibility genes, but most are without replication among studies (Bird, 2008; Chai, 2007; Daw et al., 2000). For a catalogue of candidate gene association studies, please refer to the AlzGene online database (http://www.alzforum.org/res/com/gen/alzgene/default.asp).

Table 1 Alzheimer's disease and Parkinson's disease genes. Alzheimer's disease genes; AD1–4 (panel A) and Parkinson's disease genes; PARK1–13 (panel B)

A

AD loci	Gene symbol	Gene name	Chromosome	Inheritance
AD1	APP	Amyloid precursor protein	21q21	Autosomal dominant
AD2	APOE	ApolipoproteinE	19q13.32	Sporadic
AD3	PSEN1	Presenilin 1	14q24.2	Autosomal dominant
AD4	PSEN2	Presenilin 2	1q42.13	Autosomal dominant

B

PARK loci	Gene symbol	Gene name	Chromosome	Inheritance
PARK1	SNCA	α-Synuclein	4q22.1	Autosomal dominant
PARK2	PRKN	Parkin	6q26	Autosomal recessive
PARK3	Unknown	Unknown	2p13	Autosomal dominant
PARK4	SNCA	α-Synuclein	4q22.1	Autosomal dominant; sporadic
PARK5	UCHL1	Ubiquitin carboxy-terminal esterase L1	4p13	Autosomal dominant; sporadic
PARK6	PINK1	Phosphatase and tensin (PTEN) induced kinase 1	1p36.12	Autosomal recessive
PARK7	DJ-1	DJ-1	1p36.23	Autosomal recessive
PARK8	LRRK2	Dardarin; leucine repeat rich kinase 2	12p12	Autosomal dominant; sporadic
PARK9	ATP13A2	ATP13A2	1p36.13	Autosomal recessive
PARK10	Unknown	Unknown	1p32	Sporadic
PARK11	Unknown	Unknown	2q36–37	Autosomal dominant
PARK12	Unknown	Unknown	Xq21–25	Sporadic
PARK13	HTRA2	HtrA serine peptidase 2	2p13.1	Autosomal dominant
Unnamed	GRA	Glucocerebrosidase	1q21	Unknown

1.2.2 Genes Associated with Autosomal Dominant Alzheimer's Disease

AD1: *App*

Inheritance and Clinical Features

The purification of both plaque and vascular amyloid deposits and the isolation of their 40-residue constituent peptide (Aβ) led to the cloning of the APP type I integral membrane glycoprotein from which Aβ is proteolytically derived (Kang et al., 1987). The *APP* gene was mapped to chromosome 21q which accounts for the observation that Down syndrome patients (trisomy 21) develop amyloid deposits and the neuropathological features of AD in their 40 s (Giaccone et al., 1989; Iwatsubo et al., 1994; Lemere et al., 1996; Mann et al., 1989). Subsequent searches for autosomal dominant EOAD families with genetic linkage to chromosome 21 resulted in the identification of six different missense mutations in *APP*, five associated with familial AD (Chartier-Harlin et al., 1991a, b; Goate et al., 1991; Mullan, 1992; Murrell et al., 1991), and one with the neuropathologically related syndrome of hereditary cerebral hemorrhage with amyloidosis of the Dutch type (Levy et al., 1990).

Subsequently, over 20 different *APP* missense mutations have been identified in 60 families. Interestingly, most of these mutations are located at exons 16 and 17 where the secretase cleavage sites or the *APP* transmembrane domain are located (Fig. 2). Information regarding *APP* mutations is available in the NCBI database and the Alzheimer Disease Mutation Database (www.molgen.ua.ac.be/ADMutations) (Cruts and Van Broeckhoven, 1998). Mutations within *APP* account for 10–15% of EOFAD (Bird, 2008; Janssen et al., 2003; Raux et al., 2005; Sherrington et al., 1996), appear to be family specific, and do not occur within the

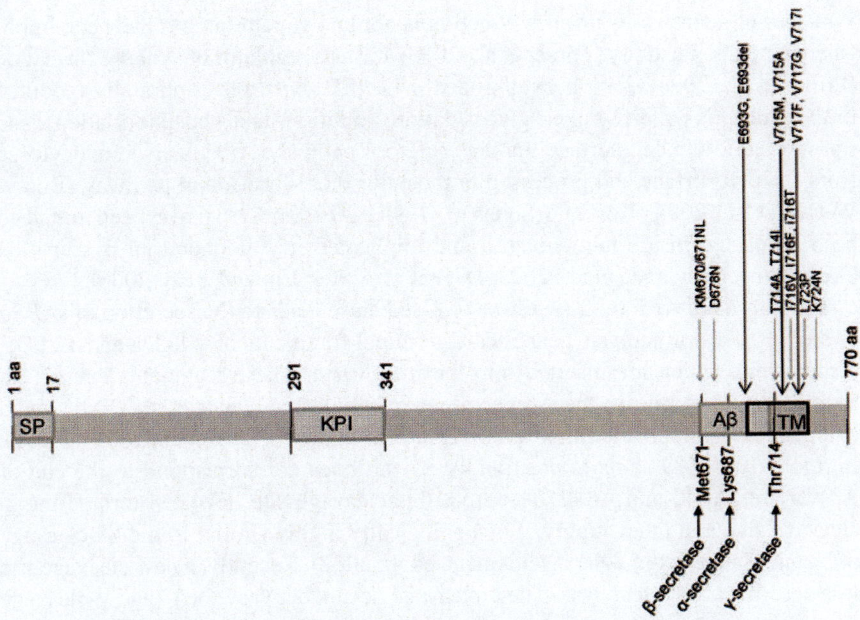

Fig. 2 AD1: *APP* structure and mutations. SP, signal peptide; KPI, Kunitz protease inhibitor domain; Aβ, amyloid beta; TM, transmembrane domain. Scale is approximate

majority of sporadic AD cases. The majority of these EOFAD mutations are located in or adjacent to the Aβ peptide sequence (Fig. 2), the major component of the amyloid plaques (Esler and Wolfe, 2001; Suzuki et al., 1994). Most cases containing *APP* mutations have an age of onset in the mid-40 s and 50 s (Hardy, 2001).

Gene Location and Structure

Sequences encoding *APP* were first cloned by screening cDNA libraries (Kang et al., 1987). The initial full-length cDNA clone encoded a 695 amino acid protein (*APP*695) (Schellenberg, 1995) and consisted of 18 exons. The *APP* gene, located at chromosome 21q21, is alternatively spliced into several products, named according to their length in amino acids (i.e., *APP*695, *APP*714, *APP*751, *APP*770, and *APP*563) and expressed differentially by tissue type whereby three isoforms, most relevant to AD, are restricted to the central nervous system (*APP*695) or expressed in both the peripheral and CNS tissues (*APP*751 and *APP*770) (de Sauvage and Octave, 1989; Golde et al., 1990; Goldgaber et al., 1987; Kang et al., 1987; Kitaguchi et al., 1988; Ponte et al., 1988; Tanzi et al., 1988; Yoshikai et al., 1990).

Gene Function and Expression

APP is a type I integral membrane protein (Kang et al., 1987) that resembles a signal-transduction receptor. It is expressed in many tissues and concentrated in the synapses of neurons. Its primary function is not known, although it has been implicated in neural plasticity (Turner et al., 2003) and as a regulator of synapse formation (Priller et al., 2006). *APP* is synthesized in the ER, posttranscriptionally modified in the Golgi (N- and O-linked glycosylation, sulfation, and phosphorylation), and transported to the cell surface via the secretory pathway. *APP* is also endocytosed from the cell surface and processed in the endosomal–lysosomal pathway (Bossy-Wetzel et al., 2004; Koo and Squazzo, 1994). *APP* and Aβ have been found to be translocated inside mitochondria and implicated in mitochondrial dysfunction (Anandatheerthavarada et al., 2003; Devi et al., 2006; Lin and Beal, 2006).

Proteolysis of *APP* by α-secretase or β-secretase leads to the secretion of s*APP*α or s*APP*β. This proteolysis generates C-terminal fragments of 10 kDa and 12 kDa, respectively, which are inserted into the membrane. These fragments can be cut by γ-secretase to release the Aβ peptide extracellularly (Walter et al., 2001) and a cytoplasmic fragment identified as AICD intracellularly (Sastre et al., 2001) (Fig. 1). Intriguingly, AICD starts at position 49/50 and does not correspond to the end of Aβ variants Aβ40 and Aβ42. Therefore this cleavage site has been termed the ϵ-cleavage site, and interestingly, it is topologically highly similar to the S3 cleavage of Notch (Sastre et al., 2001; Weidemann et al., 2002). Recently, a new cleavage site was described for γ-secretase. The ξ-cleavage occurs between the ϵ- and γ-cleavage sites and generates longer Aβ isoforms within cells and in the brain, including Aβ43, Aβ45, Aβ46, and Aβ48 (Qi-Takahara et al., 2005; Zhao et al., 2004). The majority of EOAD mutations alter this processing of *APP* in such a way that Aβ42 levels relative to other Aβ isoforms are changed (Scheuner et al., 1996; Walker et al., 2005). The function of these *APP* proteolytic fragments is still unclear.

The missense APP "Swedish" mutations (APPSW, APPK670N, and M671L) and the "London" mutations (APPLON and APPV717I) are examples of *APP* mutations that lead to increased Aβ production and development of AD (Goate et al., 1991; Mullan, 1992). Transgenic mouse models of *APP* mutations have been developed such as: PDAPP, Tg2576, APP23, TgCRND8, and J20 (Higgins and Jacobsen, 2003). Each of these transgenic mouse models has different mutations and different promoters that lead to different expression levels and different levels of neuroanatomical abnormalities (Higgins and Jacobsen, 2003; Mineur et al., 2005). For example, the Tg2576 mouse model that carries the "Swedish" mutation has high *APP* levels, high Aβ levels, and cognitive disturbances (Irizarry et al., 1997) that are progressive and start as early as six months of age (Westerman et al., 2002).

Genetic Variation

APP transcripts have been identified in which exons 7, 8, and 15 are alternatively spliced. Exon 7 encodes 57 amino acids with homology to the Kunitz-type protease inhibitor (KPI) domain (Kitaguchi et al., 1988; Ponte et al., 1988; Tanzi et al., 1988) and exon 8 (Kitaguchi et al., 1988; Lemaire et al., 1989). The Aβ peptide is encoded by parts of both exons 16 and 17 (exon and codon numbering based on the *APP*770 splice variant) (Lemaire et al., 1989) (Fig. 2). In neurons, the predominant isoform is *APP*695 (Weidemann et al., 1989), which contains exon 15 but excludes exons

7 and 8. The major isoforms in peripheral tissue (*APP*751 and *APP*770), and also in neurons, encode KPI-containing forms of *APP* (Kitaguchi et al., 1988; Ponte et al., 1988; Sandbrink et al., 1994; Tanzi et al., 1988). Other splice variants have been observed that are missing exon 15 in various combinations with exons 7 and 8 and are referred to as L-*APP*s (Konig et al., 1991; Sandbrink et al., 1994). A number of studies have indicated that alternative splicing of exons 7 and 8 in *APP* mRNAs is changed in the brain during aging and possibly during AD (Johnson et al., 1989; Konig et al., 1991; Neve et al., 1988; Palmert et al., 1988; Sisodia et al., 1990; Tanaka et al., 1988). Even though the function of *APP* and its various splice variants is unknown, differential expression of these splice variants between tissues may imply functional differences. It is important to note that although most of the described splice variants contain Aβ-encoding sequences, two additional rare transcripts, *APP*365 and *APP*563, do not, implicating additional variability in *APP* function (de Sauvage and Octave, 1989; Jacobsen et al., 1991).

The first described and best characterized *APP* mutation (V717I) was identified in a London family and is located within the transmembrane domain near the γ-secretase cleavage site (Goate et al., 1991) (Fig. 2). Subsequently, other substitutions at this site have been identified and many other groups have reported the V717I mutation in other families. Many other mutations have been identified, most of which are located near the gamma-secretase cleavage site and have been associated with modulation of Aβ levels. For example, a C-terminal L723P mutation was identified in an Australian family and is reported to generate an increase of Aβ42 peptide levels in CHO cells (Kwok et al., 2000). The majority of EOAD mutations alter processing of *APP* in such a way that the relative level of Aβ42 is increased, either by increasing Aβ42 or decreasing Aβ40 peptide levels or both (Scheuner et al., 1996; Walker et al., 2005).

AD3: Presenilin 1

Inheritance and Clinical Features

Linkage studies established the presence of an AD3 locus on chromosome 14 (Schellenberg et al., 1992) and positional cloning led to the identification of mutations in the *PSEN1* gene, which encodes a polytopic membrane protein (Sherrington et al., 1995). Presenilins are major components of the atypical aspartyl protease complexes responsible for the γ-secretase cleavage of *APP* (De Strooper et al., 1998; Wolfe et al., 1999b). Mutations in *PSEN1* are the most common cause of EOFAD. *PSEN1* missense mutations account for 18–50% of the autosomal dominant EOFAD (Theuns et al., 2000). *PSEN1* mutations appear to increase the ratio of Aβ42 to Aβ40, and this appears to result in a change in function that leads to reduced γ-secretase activity (Citron et al., 1997). In preclinical cases with *PSEN1* mutations, deposition of Aβ42 may be an early event (Lippa et al., 1998).

Defects in *PSEN1* cause the most severe forms of AD, with complete penetrance and an onset occurring as early as 30 years of age. A second form of *PSEN1*-associated AD has a mean age of onset greater than 58 years. Both are autosomal dominant neurodegenerative disorders characterized by progressive dementia, Parkinsonism, and notch signaling, as well as Aβ intracellular domain generation (Goedert and Spillantini, 2006; Wolfe, 2007). There is considerable phenotypic variability in EOFAD, including some patients with spastic paraparesis and other atypical AD symptoms. Some of these variable clinical phenotypes have been described by specific mutations. Neuropathological studies often confirm the clinical diagnosis of AD with measurement of amyloid plaque and Braak stage (as described above) but vary in other brain areas according to the presence of specific *PSEN1* mutations (Moehlmann et al., 2002; Rudzinski et al., 2008). For example, clinical and neuropathologic features of a Greek family with a *PSEN1* mutation (N135S) include memory loss in their 30 s, as well as variable limb spasticity and seizures. Upon neuropathological examination, the diagnosis of AD was confirmed but in addition, there was histological evidence of corticospinal tract degeneration (Rudzinski et al., 2008). A *PSEN1* mutation (I143M) that lies in a cluster in the second transmembrane domain of the protein has been described in an African family with an age-at-onset in the early 50 s that lasts for 6–7 years. Neuropathologically, these cases were characterized by neuronal loss, abundant Aβ neuritic plaques, and neurofibrillary tangles as well as degeneration extending into the brainstem (Heckmann et al., 2004).

Gene Location and Structure

PSEN1 is located on chromosome 14q24.2 and consists of 12 exons that encode a 467 amino acid protein that is predicted to traverse the membrane 6–10 times; the amino and carboxyl termini are both oriented toward the cytoplasm (Hutton and Hardy, 1997).

Gene Function and Expression

PSEN1 is a polytopic membrane protein that forms the catalytic core of the gamma-secretase complex (De Strooper et al., 1998; Wolfe et al., 1999a). Gamma-secretase is an integral membrane protein found at the cell surface, but it may also be found in the Golgi, endoplasmic reticulum, and mitochondria (Baulac et al., 2003; De Strooper et al., 1998). *PSEN1*, nicastrin (Nct), anterior pharynx defective 1 (Aph-1), and presenilin enhancer 2 (PSENEN) are required for the stability and activity of the γ-secretase complex (Edbauer et al., 2003; Francis et al., 2002; Goutte et al., 2002; Kimberly et al., 2003; Takasugi et al., 2003). This complex cleaves many type I transmembrane proteins including APP and Notch (De Strooper et al., 1999, 1998) in the hydrophobic environment of the phospholipid bilayer of the membrane (Kimberly et al., 2003). Gamma-secretase is biologically and biochemically heterogeneous, consisting of four and potentially more different complexes that result from the mutually exclusive incorporation of *PSEN1*, *PSEN2*, and PSENEN or Aph-1-A and Aph-1-B protein subunits (Kimberly et al., 2003; Serneels et al., 2005). *PSEN1* knock-out mice are not viable (Shen et al., 1997) but a conditional *PSEN1* knock-out mouse model, where the loss of the gene is limited to the postnatal forebrain, shows mild cognitive impairments in long-term spatial reference memory and retention (Yu et al., 2001), suggesting that presenilins play a role in cognitive memory. Knock-in mouse models with missense mutations of the endogenous murine *PSEN1* and high Aβ42 levels perform poorly on the object recognition test (Huang et al., 2003; Janus et al., 2000). Double *PSEN1/APP* transgenics have been developed and suggest that *PSEN1*, *APP*, and mutations within these genes, play a role in the production of Aβ (Holcomb et al., 1998; Mineur et al., 2005).

Genetic Variation

To date, there have been 123 *PSEN1* mutations reported (Fig. 3). A comprehensive list of *PSEN1* mutations is available through the NCBI database (http://www.molgen.ua.ac.be/ADmutations). The majority of these mutations are missense mutations. These missense mutations cause amino acid substitutions throughout the *PSEN1* protein and appear to result in a relative increase in the ratio of the Aβ42 to Aβ40 peptides via either increased Aβ42 or decreased Aβ40 generation, or a combination of both (Scheuner et al., 1996). For example, individuals that carry the *PSEN1* L166P mutation can have an age-at-onset in adolescence,

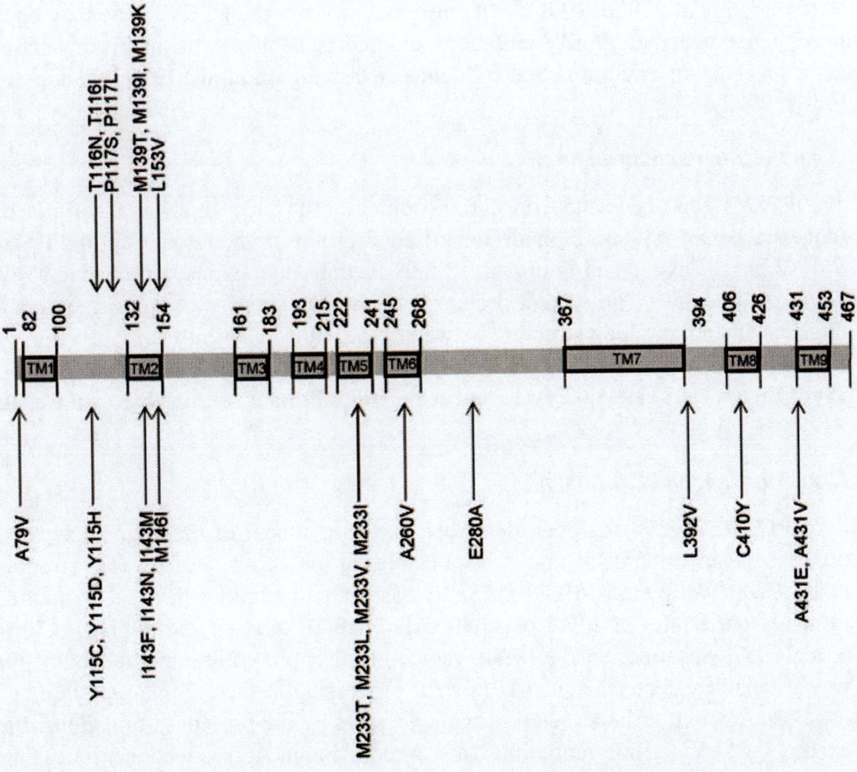

Fig. 3 AD3: *PSEN1* structure and mutations. Thus far, at least 123 mutations in the *PSEN1* gene have been described, of which a few are shown. For a more complete list of *PSEN1* mutations, see http://www.molgen.ua.ac.be/ADMutations. TM, transmembrane domains. Scale is approximate

and in vitro studies indicate that this mutation induces exceptionally high levels of Aβ42 production as well as impairs notch intracellular domain production and notch signaling (Moehlmann et al., 2002).

AD4: Presenilin 2

Inheritance and Clinical Features

A candidate gene for the chromosome 1 AD4 locus was identified in 1995 in a Volga German AD kindred with a high homology to the AD3 locus (*PSEN1*) and was later named presenilin 2 (*PSEN2*) (Levy-Lahad et al., 1995; Rogaev et al., 1995; Sherrington et al., 1996). In contrast to mutations in the *PSEN1* gene, missense mutations in the *PSEN2* gene are a rare cause of EOFAD, at least in Caucasian populations. The age of onset in *PSEN2*-affected families appears to be older (45–88 years) than that observed in *PSEN1*-affected families (25–65 years). Age of onset is highly variable among *PSEN2*-affected family members within the same family, whereas for *PSEN1*-affected families, the age of onset is generally quite similar among affected family members and is even similar among members of different families with the same mutation (Campion et al., 1999; Rogaev et al., 1995; Sherrington et al., 1996, 1995). Missense mutations in the *PSEN2* gene may be of lower penetrance than *PSEN1* mutations and thus be subject to the modifying action of other genes or environmental influences (Sherrington et al., 1996; Tandon and Fraser, 2002).

Gene Location and Structure

The *PSEN2* gene is located on chromosome 1 (1q42.13) and was identified by sequence homology and cloned (Levy-Lahad et al., 1995; Rogaev et al., 1995). *PSEN2* has 12 exons and is organized into 10 translated exons that encode a 448 amino acid peptide. The *PSEN2* protein is predicted to consist of 9 transmembrane domains and a large loop structure between the sixth and seventh domains (Fig. 4). *PSEN2* also displays tissue-specific alternative splicing (ADCG, 1995; Anwar et al., 1996; Hutton et al., 1996; Levy-Lahad et al., 1995; Prihar et al., 1996; Rogaev et al., 1995).

Gene Function and Expression

Like *PSEN1*, *PSEN2* has been described as a component of the atypical aspartyl protease called γ-secretase that is responsible for the cleavage of Aβ (De Strooper et al., 1998; Wolfe et al., 1999b). *PSEN2*-associated mutations have been reported to increase the ratio of Aβ42 to Aβ40 (Aβ42/Aβ40) in mice and humans (Citron et al., 1997; Scheuner et al., 1996), indicating that presenilins might modify the way in which γ-secretase cuts *APP*. *APP* processing at the gamma-secretase site has been reported to be affected in variable ways by the presenilin mutations. For example, *PSEN1*-L166P mutations cause a reduction in Aβ production whereas the *PSEN1*-G384A mutant significantly increases Aβ42. In contrast, *PSEN2* appears to be a less efficient producer of Aβ than *PSEN1* (Bentahir et al., 2006). The functions and biological importance of presenilin splice variants are poorly understood. But

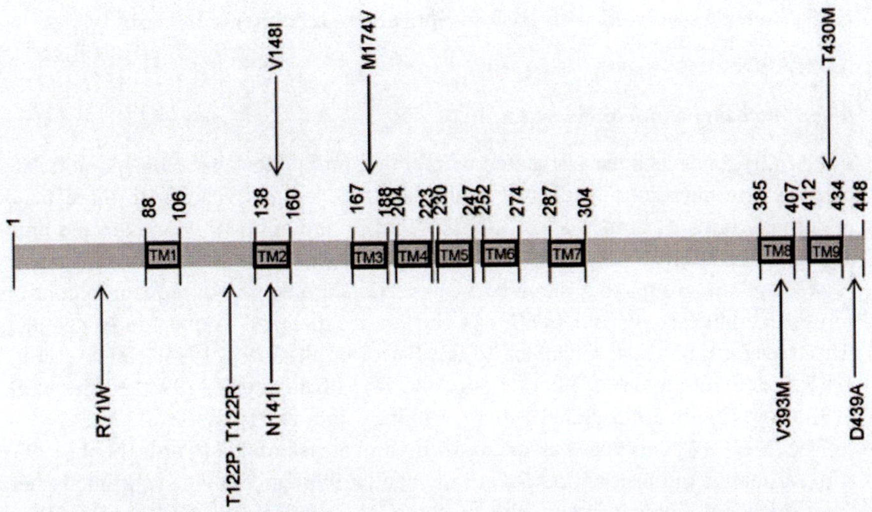

Fig. 4 AD4: *PSEN2* structure and mutations. Thus far, at least 16 mutations in the *PSEN2* gene have been described, of which a few are shown. For a more complete list of *PSEN2* mutations, see http://www.molgen.ua.ac.be/ADMutations. The V393M novel mutation was most recently found in one case (Lindquist et al., 2008). TM, transmembrane domains. Scale is approximate

it appears that differential expression of presenilin isoforms may lead to differential regulation of the proteolytic processing of the *APP* protein. For example, aberrant *PSEN2* transcripts lacking exon 5 increase the rate of production of Aβ peptide (Sato et al., 2001), whereas naturally occurring isoforms without exons 3 and 4 and/or without exon 8 do not affect production of Aβ (ADCG, 1995; Grunberg et al., 1998). *PSEN2* is expressed in a variety of tissues, including the brain where it is expressed primarily in neurons (Kovacs et al., 1996).

Genetic Variation

Mutations in *PSEN2* are a much rarer cause of FAD than are *PSEN1* mutations, having been described in only six families, including the Volga German kindred where a founder effect has been demonstrated (Cruts and Van Broeckhoven, 1998; Levy-Lahad et al., 1995; Rogaev et al., 1995; Sherrington et al., 1996). To date, as many as 16 *PSEN2* mutations have been identified. One of the first mutations to be identified was a point mutation resulting in the substitution of an isoleucine for an asparagine at residues 141 (N141I) located within the second transmembrane domain (Levy-Lahad et al., 1995). Most recently, a V393M mutation located within the seventh transmembrane domain has been described (Lindquist et al., 2008) (Fig. 4). A comprehensive list of *PSEN2* mutations is available through the NCBI database (http://www.molgen.ua.ac.be/ADmutations).

1.2.3 Genes Associated with Risk in Sporadic Alzheimer's Disease

AD2: *APOE*

Inheritance and Clinical Features

The *APOE* gene has been associated with both familial late-onset and sporadic late-onset AD in numerous studies of multiple ethnic groups. There are three major protein isoforms of human apoE (apoE2, apoE3, and apoE4), which are the products of three alleles (ϵ2, ϵ3, and ϵ4). The frequency of the *APOE* ϵ4 allele varies between ethnic groups, but the *APOE* ϵ4– carriers are the most frequent in controls across all ethnic groups and *APOE* ϵ4+ carriers are the most frequent in AD patients (Brousseau et al., 1994; Chauhan, 2003; Farrer et al., 1997, 1995; Hendrie et al., 1995; Liddell et al., 1994; Lucotte et al., 1994; Mayeux et al., 1993; Poirier et al., 1993; Roses et al., 1995; Schellenberg, 1995; Selkoe, 2001; Tsai et al., 1994).

The *APOE* ϵ4 genotype is associated with higher risk of AD (Corder et al., 1993), with earlier age of onset of both AD (Tang et al., 1996) and Down syndrome (where there is an additional copy of chromosome 21 carrying the *APP* gene) (Schupf and Sergievsky, 2002), and also with a worse outcome after head trauma (Nicoll et al., 1995) and stroke, both in humans (Liu et al., 2002b) and in transgenic mice expressing human apoE4 (Horsburgh et al., 2000).

Gene Location and Structure

The *APOE* gene is located on chromosome 19q13.2 and consists of 4 exons that encode a 299 amino acid protein. The *APOE* gene is in a cluster with other apolipoprotein genes: *APOC1*, *APOC2*, and *APOC4*. The *APOE* ϵ4 loci are located within exon 4 of the gene. The three *APOE* ϵ4 alleles (ϵ2, ϵ3, and ϵ4) defined by two single nucleotide polymorphisms, rs429358 and rs7412, encode three protein isoforms (E2, E3, and E4). The most frequent isoform is apoE3, which contains cysteine and arginine at amino acid positions 112 and 158. Both positions contain cysteine residues in apoE2 and arginine residues in apoE4 (Fig. 5). This substitution affects the three-dimensional structure and the lipid-binding properties between isoforms. In apoE4, the amino acid substitution results in a changed structure with the formation of a salt-bridge between an arginine in position 61 and a glutamic acid in 255 that causes this isoform to bind preferentially to VLDL whereas apoE3 and apoE2 bind preferentially to high-density lipoproteins (HDLs) (Mahley et al., 2006).

Gene Function and Expression

The mechanisms that govern apoE toxicity in the brain are not fully understood. Some proposed mechanisms include isoform specific toxicity, apoE E4–mediated amyloid aggregation, and apoE E4–mediated tau hyperphosphorylation (Huang, 2006).

It is known that apoE plays an important role in the distribution and metabolism of cholesterol and triglycerides within many organs and cell types in the human body (Mahley et al., 2006). The apoE polymorphism is unique to humans and has been

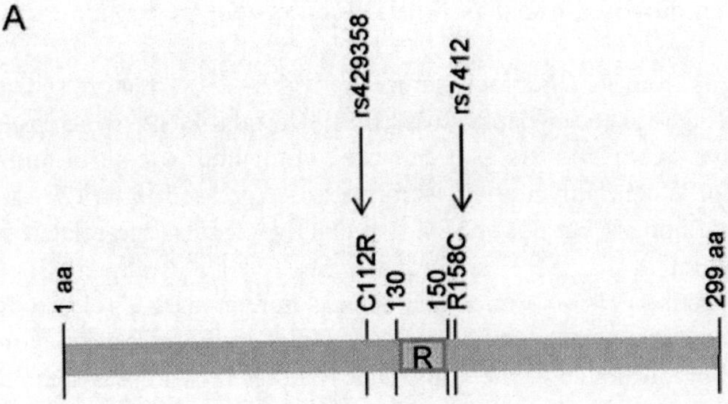

B

Allele	SNP Position (Amino Acid): rs429358 (C112R)	SNP Position (Amino Acid): rs7412 (R158C)
ε2	TGC (C)	TGC (C)
ε3	TGC (C)	CGC (R)
ε4	CGC (R)	CGC (R)

Fig. 5 AD2: *APOE* structure and single nucleotide polymorphisms (SNPs). The general protein structure of apoE is shown (**panel a**). The two SNPs and corresponding protein locations are shown (rs429358 and rs7412; C112R and R158C). The *APOE* ϵ2, ϵ3, ϵ4 haplotype is shown in **panel b**. Receptor binding domain; R. Scale is approximate

proposed to have evolved as a result of adaptive changes to diet (Finch and Stanford, 2004; Mahley and Rall, 1999). Individuals carrying *APOE ϵ4* have higher total and LDL cholesterol (Sing and Davignon, 1985). Neurons, in vitro, have a cholesterol uptake that is lower when the lipid is bound to apoE4 compared to apoE2 and apoE3 (Rapp et al., 2006), and apoE4 appears to be less efficient than the other isoforms in promoting cholesterol efflux from both neurons and astrocytes (Michikawa et al., 2000).

Chylomicron remnants and very low density lipoprotein (VLDL) remnants are rapidly removed from the circulation by receptor-mediated endocytosis. ApoE, the major apolipoprotein of the chylomicron in the brain, binds to a specific receptor and is essential for the normal catabolism of triglyceride-rich lipoprotein constituents. Defects in apolipoprotein E result in familial dysbetalipoproteinemia, or type III hyperlipoproteinemia (HLP III), in which increased plasma cholesterol and triglycerides are the consequence of impaired clearance of chylomicron and VLDL remnants (Mahley et al., 1999). In the brain, lipidated apoE binds aggregated Aβ in a isoform-specific manner, apoE4 being much more effective than the other forms, and has been proposed to enhance deposition of the Aβ peptide (Stratman et al., 2005).

Brain cells from *APOE* knock-out mice (*APOE−/−*) are more sensitive to excitotoxic and age-related synaptic loss (Buttini et al., 1999), whereas Aβ-induced synaptosomal dysfunction is also enhanced compared to control animals (Keller et al., 2000). When human apoE isoforms are expressed in *APOE−/−* mice, the expression of apoE3, but not apoE4, is protective against age-related neurodegeneration (Buttini et al., 1999) and Aβ toxicity (Keller et al., 2000). In addition, astrocytes, from *APOE−/−* mice that express human apoE3, release more cholesterol than those expressing apoE4, suggesting that apoE isoforms may modulate the amount of lipid available for neurons. Other studies report apoE-specific effects on Aβ removal from the extracellular space whereby the apoE3 isoform has a higher Aβ binding capacity than ApoE4 when associated with lipids (Canevari and Clark, 2007; LaDu et al., 1995).

In humans the greatest expression of apoE is found in the liver, followed by the brain. Animal and in vitro models show that in the brain, astrocytes and microglia are the main producers of secreted apoE (Pitas et al., 1987; Uchihara et al., 1995) whereas neurons appear to produce apoE under stress conditions (Aoki et al., 2003; Xu et al., 1999). In a rodent model, moderate injury induces enhancement of apoE levels in clusters of CA1 and CA3 pyramidal neurons (Boschert et al., 1999); in another model, apoE levels increase in response to peripheral nerve injury (Ignatius et al., 1986) whereas apoE secretion in human primary astrocytes can be reduced by a combination cytokines (Baskin et al., 1997).

In addition, individuals carrying apoE4 have higher amyloid and tangle pathology (Nagy et al., 1995), and they have an increase in mitochondrial damage (Gibson et al., 2000) compared to those carrying other forms.

Genetic Variation

The gene dose of *APOE* ϵ4 is a major risk factor for the disease, with many studies reporting an association between gene dose, age-at-onset (Blacker et al., 1997), and cognitive decline (Martins et al., 2005). After age 65, the risk among family members increases depending on the number of ϵ4 alleles present in the affected individual. Risks to family members with the *APOE* 2/2 and 2/3 genotypes are nearly identical at all ages to risks for family members with the *APOE* 3/3 genotype. Among family members with *APOE* 3/3, the lifetime risk for AD by age 90 can be as much as three times greater than the expected proportion of *APOE* ϵ4 carriers, suggesting that factors other than *APOE* contribute to AD risk. In addition, a 44% risk of AD by age 93 among family members of *APOE* 4/4 carriers indicates that as many as 50% of people having at least one e4 allele do not develop AD. There also appears to be a gender modification effect because the risk to male family members with *APOE* 3/4 is similar to that for the *APOE* 3/3 group but significantly less than the risk for the *APOE* 4/4 carriers; whereas among female family members the risk for the *APOE* 3/4 carriers is nearly twice that for the *APOE* 3/3 carriers (Brousseau et al., 1994; Farrer et al., 1997, 1995; Hendrie et al., 1995; Liddell et al., 1994; Lucotte et al., 1994; Mayeux et al., 1993; Poirier et al., 1993; Tsai et al., 1994).

1.3 Summary

AD is characterized by an irreversible progressive loss of memory and cognitive skills that can occur in rare familial cases as early as the third decade. Currently there is no cure for AD, and treatments only slow AD progression slightly in some patients (Raina et al., 2008; Raschetti et al., 2007). The early-onset familial forms of AD have an autosomal dominant inheritance linked to three genes: *APP*, *PSEN1*, and *PSEN2*. The most common sporadic form of AD occurs after the age of 60 and has thus far been consistently, across numerous studies, associated with only one gene, the *APOE* gene. The mechanistic contribution of these genes in AD pathogenesis has been studied extensively but the specific biology involved in the progression of AD remains unclear, suggesting that AD is a genetic and environmentally complex disease.

2 Parkinson's Disease

2.1 Introduction

2.1.1 Prevalence and Incidence

Parkinson's disease (PD) (OMIM #168600) is the second most common neurodegenerative disorder. The incidence is similar worldwide, with the prevalence increasing in proportion to regional increases in population longevity with more than 1% affected over the age of 65 years and more than 4% of the population affected by the age of 85 years (de Rijk et al., 2000). Idiopathic PD is the most frequent form of Parkinsonism and accounts for over 75% of all PD cases, and it usually refers to a syndrome characterized by late-onset, largely non-genetic movement disorder (Gibb and Lees, 1988). Rare forms of PD in which genetic factors dominate, represent 5–10% of all PD patients (Belin and Westerlund, 2008).

2.1.2 Clinical Symptoms

Clinical manifestations that can be detected by neurological examinations include tremor, rigidity, bradykinesia, and postural instability. Disruption of motor abilities is associated with striatal dopamine levels thought to arise from selective and progressive loss of dopaminergic cells within the substantia nigra pars compacta and the locus ceruleus of the midbrain (Tan and Skipper, 2007). Secondary symptoms may involve cognitive dysfunction and subtle language problems. Symptoms can be both chronic and progressive. Levodopa remains the most effective treatment of PD symptoms but its use is complicated by the emergence of motor fluctuations and dyskinesias. Dopamine agonists, catechol-O-methyltransferase inhibitors, and other anti-Parkinsonian drugs may diminish or prevent these complications and possibly exert disease-modifying effects (Jankovic, 2006).

2.1.3 Clinical Diagnosis

Diagnostic clinical criteria of PD include four cardinal symptoms: bradykinesia, rest tremor, rigidity, and postural instability. An additional criterion includes a therapeutic response of tremor to levodopa (Galpern and Singhal, 2006). In addition, other common motor signs and symptoms include loss of automatic motor movements such as loss of arm swing, loss of blinking, and difficulty in performing simultaneous motor acts. Many nonmotor symptoms can also be present in PD, such as cognitive impairment, hallucination, delusion, behavioral abnormalities, clinical depression, disturbances of sleep and wakefulness, loss of smell, pain, and autonomic dysfunctions such as constipation, hypotension, urinary frequency, impotence, and sweating (Mizuno et al., 2008).

2.1.4 Neuropathological Diagnosis

The diagnosis of idiopathic PD may also involve confirmation upon autopsy where neuropatholgical assessment of the amount of neuronal loss and Lewy-related pathology (Lewy bodies and Lewy neurites), in the brainstem and elsewhere in the brain, is performed. Eosinophilic neuronal cytoplasmic inclusions known as Lewy bodies (LBs) are found in PD postmortem brain (Gibb and Lees, 1988). The disease is also characterized by dopamine neuron degeneration and depigmentation of the substantia nigra accompanied by neuronal loss in other brainstem regions including the ventral tegmental area and locus ceruleus (Belin and Westerlund, 2008; Love, 2005).

The principal component of LBs is α-synuclein, and LBs are best visualized immunohistochemically, using an antibody to α-synuclein (Love, 2005). The function of α-synuclein is unknown. It is primarily found in neural tissue in presynaptic terminals. It can also be found in glial cells. It is predominantly expressed in the neocortex, hippocampus, substantia nigra, thalamus, and cerebellum (George, 2002).

LBs are typically found in the substantia nigra and locus ceruleus, where there is substantial neuronal loss and gliosis. LBs may also be found in the dorsal motor nucleus of the vagus where LBs are usually roughly spherical, with an eosinophilic core surrounded by a paler "halo." Within the cerebrum, LBs are usually present in the amygdaloid nuclei, parahippocampal and cingulate gyri, and insula, but they may also be found in other parts of the neocortex. The cholinergic nucleus basalis of Meynert may also be affected. Cortical LBs appear as regions of homogeneous eosinophilic staining of neuronal cytoplasm and eccentric displacement of the nucleus (Love, 2005). Lewy neurites are nerve cell processes that contain aggregates of α-synuclein and are most numerous in the CA2/3 region of the hippocampus and in the substantia nigra (Love, 2005).

2.2 Genetics of Parkinson's Disease

2.2.1 Introduction

Historically, PD was considered to be largely sporadic in nature without genetic origin. However, in the past decade, genetic studies of PD families from different geographical regions worldwide have strengthened the hypothesis that PD has a substantial genetic component. One of the first autosomal dominant inherited forms of PD was identified in an Italian family, and it is named PARK1 (Polymeropoulos et al., 1996). Since then, 13 loci, PARK1–13, have been linked to rare forms of PD: autosomal dominant and autosomal recessive PD (Belin and Westerlund, 2008; Farrer, 2006). Of these 13 loci, eight genes have been described as causing PD: four autosomal dominant (*SNCA, LRRK2, UCHL1*, and *HTRA2*) and four autosomal recessive (*PRKN, DJ1, PINK1*, and *ATP13A2*; Table 1). Mutations in the *SNCA, LRRK2, PRKN,* and *PINK1* genes are the most well-chararacterized as causing PD whereas mutations in the other genes listed do not have as much supporting evidence as causes of PD. Recently, a clinical association has been reported between PD and type-1 Gaucher's disease, which is caused by a glucocerebrosidase deficiency owing to mutations in the glucocerebrosidase gene (*GBA*), and several studies have found an association between *GBA* mutations and PD (Aharon-Peretz et al., 2004; Bras et al., 2007; Clark et al., 2005, 2007; De Marco et al., 2008; Eblan et al., 2006; Gan-Or et al., 2008; Lwin et al., 2004; Sato et al., 2005; Spitz et al., 2008; Tan et al., 2007; Toft et al., 2006; Wu et al., 2007; Ziegler et al., 2007). The *GBA* gene has not yet been named as a PD gene but is described briefly here. Some PD genes where mutations have been linked to familial forms of PD are also candidate genes for sporadic forms of PD, as those genes (*SNCA* and *LRRK2*) may also carry other mutations that merely increase risk (Table 1).

2.2.2 Genes Associated with Autosomal Dominant Parkinson's Disease

PARK1 and PARK4: *SNCA*

Inheritance and Clinical Features

PARK1- and PARK4-linked PD are both of autosomal dominant inheritance, but PARK1 is caused by missense mutations in the α-synuclein gene (*SNCA*) and PARK4, by multiplications of *SNCA*. Affected family members are mostly of juvenile-onset with atypical clinical features including myoclonus and hypoventilation, with rapid progression of symptoms. Three missense mutations, A53T (Polymeropoulos et al., 1996), A30P (Kruger et al., 1998), and E46K (Zarranz et al., 2004); duplications (Chartier-Harlin et al., 2004; Fuchs et al., 2007; Ibanez et al., 2004; Nishioka et al., 2006); and triplications (Farrer et al., 2004; Singleton et al., 2003) of *SNCA* are known (Fig. 6). The A53T substitution was

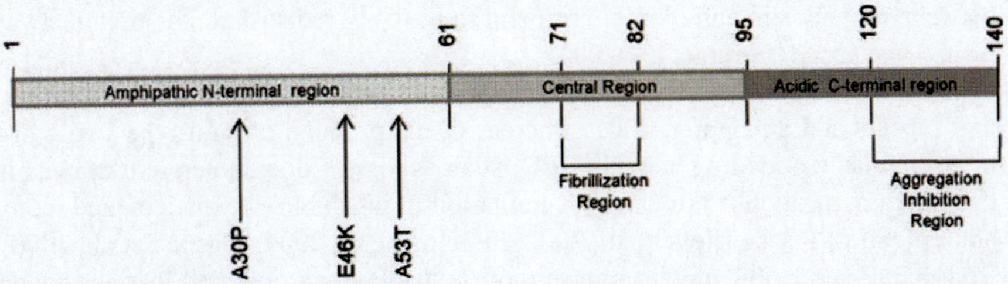

Fig. 6 PARK1 and PARK4: *SNCA* structure and mutations. The general protein structure of α-synuclein is shown (Bisaglia et al., 2008). Scale is approximate

the first mutation identified in a large family with autosomal dominant disease (Polymeropoulos et al., 1996). Later, A30P and E46K substitutions were identified in a German and a Spanish family, respectively, with clinical features described as dementia with LB (Kruger et al., 1998; Zarranz et al., 2004). PARK1 missense mutations and PARK4 multiplications are both extremely rare causes of familial Parkinsonism (Chartier-Harlin et al., 2004; Farrer et al., 2004; Fuchs et al., 2007; Ibanez et al., 2004; Nishioka et al., 2006; Singleton et al., 2003).

Gene Location and Structure

SNCA is located on chromosome 4q22.1, has six exons, and encodes a 140 amino acid protein. The N-terminus consists of an amphipathic α-helical domain that associates with membrane microdomains, known as lipid rafts (Fortin et al., 2004). The central region contains a fibrillization region, and the C-terminus contains an aggregation inhibition region (Fig. 6) (Bisaglia et al., 2008).

Gene Function and Expression

SNCA is expressed throughout the mammalian brain and is enriched in presynaptic nerve terminals (George, 2002). The protein can adopt partially folded structures but in its native form is unfolded and can assume both monomeric and oligomeric alpha helix and beta-sheet conformations, as well as morphologically diverse aggregates, ranging from those that are amorphous to amyloid-like fibrils (Uversky, 2003). These fibrillar moieties are a component of LBs in both familial and idiopathic PD (Spillantini et al., 1997), but it is unclear whether the fibrils themselves, or the oligomeric fibrilization intermediates (protofibrils), are toxic to the cell. Interestingly, *SNCA* genomic multiplications in familial PD are associated with an increase in protein expression (Farrer et al., 2004) and brain samples of triplication mutant carriers show protofibril formation is enhanced with an increase in *SNCA* expression (Miller et al., 2004). In vitro, A30P, A53T, and E46K mutant proteins show an increased propensity for self-aggregation and oligomerization into protofibrils, compared with wild-type protein (Conway et al., 1998; Pandey et al., 2006) that may be related to the membrane permeabilization activity of these protofibrils, which form pore-like and tubular structures (Lashuel et al., 2002). It appears that only A53T and E46K promote formation of the fibrils (Conway et al., 2000; Greenbaum et al., 2005) whereas A30P has been reported to disrupt the interaction between α-synuclein and the lipid raft and to possibly redistribute the protein away from the synapse (Fortin et al., 2004).

A mouse spontaneous deletion strain is viable, fertile, and phenotypically normal (Specht and Schoepfer, 2001) whereas overexpression of wild-type *SNCA* in a mouse model has many features of PD, such as loss of dopaminergic terminals in the striatum, mislocalization and accumulation of insoluble α-synuclein, and motor abnormalities (Rockenstein et al., 2002; Fleming et al., 2004; Masliah et al., 2000). Both A30P and A53T mutant mouse models display neuronal cell loss and motor changes (Melrose et al., 2006).

Increased tendency for oligomer and aggregate formation in *SNCA* mutants has been suggested to be a cause of PARK1-linked PD, and PARK4-linked PD multiplications with an increased amount of normal α-synuclein may predispose neurons to oligomer and aggregate formations (El-Agnaf et al., 1998; Fredenburg et al., 2007). In LBs the α-synuclein protein is phosphorylated at serine 129 (Ser129) which is located within the C-terminal that has been implicated as playing an important role in aggregation (Bisaglia et al., 2008). Interestingly, in a *Drosophila* model, mutation of Ser129 to alanine (which prevents phosphorylation) can suppress dopaminergic neuronal loss (Chen and Feany, 2005).

Genetic Variation

Multiple *SNCA* mutations and multiplications have been described. Genetic variation in *SNCA* appears to contribute to PD phenotype. For example, PARK4 American and European families with *SNCA* triplication show different clinical features from families with the *SNCA* duplication where the phenotype closely resembles idiopathic PD, with late age of onset, slow progression, and no atypical features, suggesting that *SNCA* gene dosage may play a role in disease progression (Chartier-Harlin et al., 2004; Ibanez et al., 2004). *SNCA* duplications are rarely associated with dementia (Fuchs et al., 2007; Nishioka et al., 2006). Multiplications of *SNCA* appear to be slightly more common than missense mutations with the *SNCA* triplication found in a large Iowan family (Singleton et al., 2003). Duplications of *SNCA* (Chartier-Harlin et al., 2004; Ibanez et al., 2004; Nishioka et al., 2006) were reported in a Swedish–American family (Fuchs et al., 2007) with patients in the Swedish branch carriers of the duplication and those in the American branch carriers of the triplication (Farrer et al., 2004), suggesting unequal recombination or crossing over as the potential mechanisms for duplication and triplication, respectively (Fuchs et al., 2007). Both the E46K mutation and the triplication are associated with Parkinsonism and dementia, and the age of onset is younger than the other mutations with diffuse Lewy body disease. A30P mutation is usually not associated with dementia. The A53T mutation has been associated with dementia and the presence of cortical Lewy bodies (Golbe, 1990; Golbe et al., 1990).

In addition, *SNCA* promoter polymorphisms have been associated with idiopathic PD disease risk (Maraganore et al., 2006; Pals et al., 2004; Tan et al., 2004a), and recently *SNCA* polymorphic mutations associated with increased α-synuclein expression have been reported to be significant risk factors for sporadic PD (Mizuta et al., 2006; Mueller et al., 2005).

PARK8: *LRRK2*

Inheritance and Clinical Features

Autosomal dominant PARK8-linked PD was first identified in a Japanese family known as the Samagihara kindred (Funayama et al., 2002). Clinical features were first described in 1978 in a large Japanese family (Nukada et al., 1978) with similar symptoms as sporadic PD with a slightly earlier onset of age, and this linkage has been replicated in Caucasian families (Zimprich et al., 2004). Although affected individuals have clinically typical PD, pathologically the disease appears to be heterogeneous with reports of Lewy body pathology and tau pathology as well as neuronal loss without intracellular inclusions (Nicholl et al., 2002; Wszolek et al., 2004) in addition to motor neuron disease (Zimprich et al., 2004). Dementia is not a common feature but has been described in some families (Zimprich et al., 2004).

Gene Location and Structure

The gene for PARK8 was recently identified as Leucine Repeat Rich Kinase 2 (*LRRK2*; also called *dardarin*, from the Basque word for tremor) in families from the Basque region of Spain, Britain, Western Nebraska, and in an American kindred of German descent (Paisan-Ruiz et al., 2004; Zimprich et al., 2004). It is located on chromosome 12p12 and is a huge gene encompassing 144 kb in the genome, consisting of 51 exons (7449 bp cDNA) and encoding a protein consisting of 2517 amino acids. The *LRRK2* gene contains several functional domains including ANK (ankyrin repeat domain), LRR (leucine-repeat-rich), ROC (Ras of complex proteins), COR (carboxy terminal of ROC), MAPKKK (mitogen-activated protein kinase kinase kinase), and a WD40 domain that is rich in tryptophan and aspartate repeats.

Gene Function and Expression

Function of *LRRK2* is not well known, although it has been identified as a tyrosine kinase-like protein (Mata et al., 2006). The ROC domain is able to bind GTP and is essential for the MAPKKK domain to exert kinase activity but does not have GTPase activity (Ito et al., 2007). Some of the *LRRK2* mutations appear to exert increased kinase activity (Gloeckner et al., 2006; West et al., 2005). Other functional domains are believed to be important in protein–protein interactions (Zimprich et al., 2004). *LRRK2* also interacts with other familial PD proteins. For example, *LRRK2* appears to interact with parkin through the ROC domain; however, the interaction with parkin does not seem to enhance polyubiquitylation of *LRRK2* (Smith et al., 2005). *LRRK2* expression has been described in the central nervous system (cerebral cortex, medulla, cerebellum, spinal cord, putamen, and substantia nigra), heart, kidney, lung, liver, and peripheral leukocytes (Paisan-Ruiz et al., 2004; Zimprich et al., 2004). *LRRK2* protein is found in the cytosol and mitochondrial outer membrane (West et al., 2005), plasma membrane, lysosomes, endosomes, transport vesicles, Golgi apparatus, a cytoskeleton protein microtubule, synaptic vesicles, and lipid rafts (Biskup et al., 2006; Hatano et al., 2007). Interestingly, α-synuclein is also expressed in the presynaptic membranes and lipid rafts (Fortin et al., 2004).

There is currently very limited postmortem data on pathogenic *LRRK2* mutations but it appears that typical LB pathology is seen in most *LRRK2*-related patients. One report of *LRRK2* brain expression shows substantia nigra cell loss, Lewy body formation, and small numbers of cortical Lewy bodies (Khan et al., 2005). In the same study18F-dopa positron emission tomography (PET) in another patient, but not in unaffected family members, showed a pattern of nigrostriatal dysfunction typical of idiopathic PD (Khan et al., 2005). The mechanism that links *LRRK2* protein to

SNCA protein accumulation remains unknown, but evidence suggests that there may be a direct interaction between *LRRK2* and the *SNCA* protein (Silveira-Moriyama et al., 2008; Smith et al., 2005).

Genetic Variation

Over 20 missense or nonsense mutations are concentrated in these functional domains (Fig. 7) (Funayama et al., 2002; Mata et al., 2006; Paisan-Ruiz et al., 2004; Zimprich et al., 2004). Several coding mutations have been identified in the *LRRK2* gene including: Y1699C, R1441C, I1122V, I2020T, and R1369G and a splice site mutation, 3342A. The most frequent mutation is G2019S that accounts for as many as 40% of cases of Arab descent, about 20% of Ashkenazi Jewish patients, and is the most frequent *LRRK2* mutation in a large British kindred (Khan et al., 2005; Lesage et al., 2006, 2005; Zabetian et al., 2006). *LRRK2* mutations have also been reported in some apparently sporadic PD patients (Gilks et al., 2005). One of these polymorphisms, G2385R, is a genetic risk factor for sporadic PD in Asian populations (Di Fonzo et al., 2006; Funayama et al., 2002; Tan and Skipper, 2007). A mutation was also identified in 5 out of 107 sporadic Spanish/Basque PD cases suggesting that this gene may have a reduced penetrance. In addition, the mean age at onset (64 years) in a British kindred and the occurrence of mutations in apparently sporadic PD patients suggests that mutations in this gene may be more widely distributed in the late-onset PD population than the *SCNA* gene (Nicholl et al., 2002).

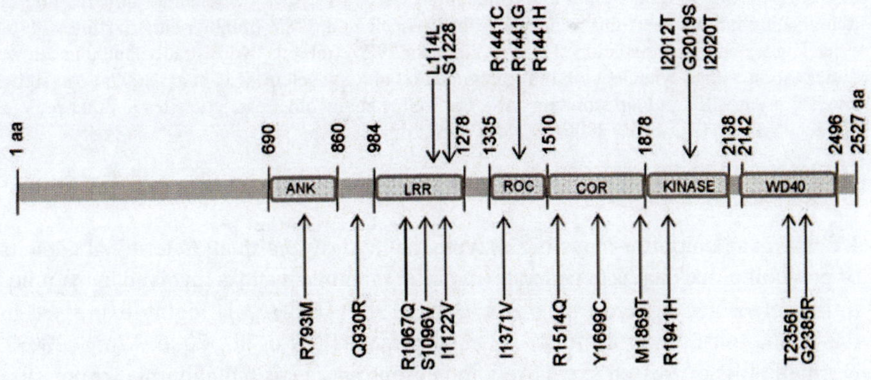

Fig. 7 PARK8: *LRRK2* structure and mutations. ANK, ankyrin repeat region; LRR, leucine-rich repeat domain; ROC, Ras of complex; COR, C-terminal of Ras (GTPase) (Tan et al., 2007). Scale is approximate

PARK5: *UCHL1*

Inheritance and Clinical Features

PARK5-linked PD is an autosomal dominant PD. Clinical features are similar to those of sporadic PD with the age of onset from 49 to 50. A mutation (I93M) has been described in a single German PD family and named as the PARK5 locus (Leroy et al., 1998). However, another study did not find an association with the I93M mutation in familial PD (Harhangi et al., 1999). A S18Y polymorphism in the PARK5 gene has been associated with familial and sporadic PD in some studies but not others, leading some to speculate that the PARK5 locus mutations may confer only a weak effect on risk for the sporadic form of PD (Hutter et al., 2008; Levecque et al., 2001; Maraganore et al., 1999; Mellick and Silburn, 2000; Wintermeyer et al., 2000; Zhang et al., 2008).

Gene Location and Structure

The PARK5-linked PD gene has been reported as the ubiquitin carboxyl-terminal hydrolase-L1 (*UCHL1*) located on chromosome 4p13 (Leroy et al., 1998). *UCHL1* is a protein of 223 amino acids, 9 exons, and a transcript length of 1172 bps (Wilkinson et al., 1989) (Fig. 8).

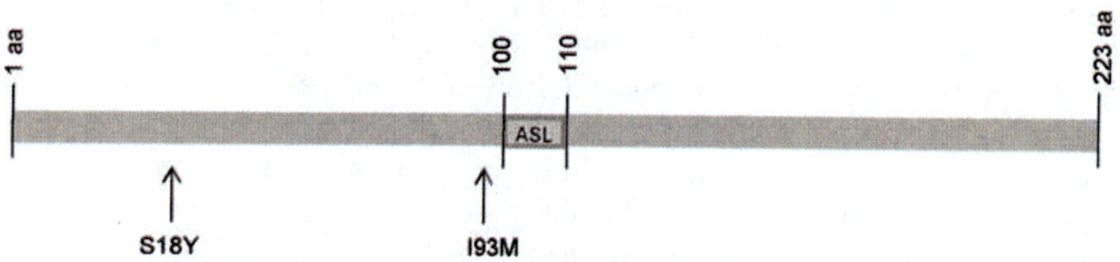

Fig. 8 PARK5: *UCHL1* structure and mutations. In vitro assays demonstrate dimerization or ubiquination inhibition by this molecule with α-synuclein at S18Y (inhibits dimerization with α-synuclein leading to degradation of α-synuclein) and I93M (inhibits hydrolyzation and thus allows dimerization with α-synuclein inhibiting degradation of α-synuclein) (Liu et al., 2002a). An active site is present within a loop structure (ASL) as well as many other catalytic sites in other regions not denoted here (Das et al., 2006). Scale is approximate

Gene Function and Expression

UCHL1 is an ubiquitin-recycling enzyme that hydrolyzes small C-terminal adducts of polyubiquitine chains to generate ubiquitin monomers and is involved in the ubiquitin proteosome system (Wilkinson et al., 1989). *UCHL1* is highly expressed in the brain, constituting up to 2% of total protein (Das et al., 2006; Wing, 2003). Normally it is expressed exclusively in neurons and testis but abnormal expression has been described in many primary lung tumors, lung tumor cell lines, and colorectal cancer (Hibi et al., 1998; Sasaki et al., 2001; Yamazaki et al., 2002). Neuronal functions include dimerization-dependent ubiquitin ligase activity (Liu et al., 2002a; Wilkinson et al., 1989) and the maintenance of ubiquitin homeostasis by promoting ubiquitin monomer stability (Osaka et al., 2003). There is also evidence that *UCHL1* may modulate tubulin polymerization (Kabuta et al., 2008).

Postmortmem studies indicate that the *UCHL1* protein is found in LBs of sporadic PD cases and that it can promote the accumulation of SNCA protein (Leroy et al., 1998; Liu et al., 2002a). Therefore, further study of the pathogenesis and potential role of *UCHL1* in PD pathology may be warranted.

Genetic Variation

The heterozygous I93M amino acid substitution in *UCHL1* was identified in a sibling pair, both affected by PD, and the transmitting parent was asymptomatic. Functional studies show that *UCHL1* I93M mutant protein has a 50% decrease in hydrolytic activity in vitro (Leroy et al., 1998; Nishikawa et al., 2003) with catalytic activity of half of the wild-type enzyme (Leroy et al., 1998), suggesting that the supply of ubiquitin for 26S proteasome may be reduced with this mutation. The I93M site is located within the hydrophobic core holding the structure of the right lobe together, and even though subtle perturbation caused by substituting methionine for isoleucine is unlikely to have significant structural consequences, it has been suggested that any movement of this residue could possibly distort the geometry of the catalytic triad located there (Das et al., 2006) (Fig. 8). It has also been suggested that the I93M mutation inhibits hydrolyzation and thus allows dimerization with α-synuclein inhibiting degradation of α-synuclein (Leroy et al., 1998; Liu et al., 2002a). At 20 weeks of age, high-expressing I93M Tg mice show a significant reduction in dopamine neurons in the substantia nigra and in dopamine content in the striatum compared to the non-Tg mice. In addition, high-expressing I93M Tg mice have an increased amount of insoluble *UCHL1* in the midbrain, suggesting a toxic gain of function (Setsuie et al., 2007). Deletion of *UCHL1* exons 7 and 8 in a mouse model causes gracile axonal dystrophy (gad mouse); this is an autosomal recessive condition characterized by axonal degeneration and formation of spheroid bodies in motor and sensory nerve terminals (Saigoh et al., 1999). *UCHL1* binds to and stabilizes monoubiquitin in neurons (Osaka et al., 2003). The I93M mutation in *UCHL1* alters the conformation of *UCHL1* (Naito et al., 2006; Nishikawa et al., 2003). An incidental mutant, the gracile axonal dystrophy mouse lacks functional *UCHL1* due to an intragenic exonic deletion. In these mice, *UCHL1* dysfunction appears to disturb the reuse of free ubiquitin, which results in the accumulation of abnormal proteins in the brain. Mice deficient in *UCHL1* do not exhibit obvious dopaminergic cell loss, in contrast to *UCHL1*I93M-transgenic mice (Osaka et al., 2003; Saigoh et al., 1999; Setsuie et al., 2007), suggesting that a loss or decrease in the level of *UCHL1* is not the main cause of PD and that I93M-associated PD is caused by an acquired toxicity. Thus, although the hydrolase activity of I93M is decreased (Leroy et al., 1998; Nishikawa et al., 2003), this decreased activity may not be a major cause of PD.

An *UCHL1* S18Y variant has been described that may be associated with a decreased risk of idiopathic PD (Healy et al., 2004b; Maraganore et al., 2004; Tan et al., 2006a). Although close analysis of consequences of the S18Y on protein structure did not yield any insight on the functional impact of this mutation (Das et al., 2006), in vitro studies indicate that the S18Y variant has reduced ligase activity and possibly increased hydrolase activity compared with wild-type enzyme (Liu et al., 2002a; Nishikawa et al., 2003). It has been suggested that the S18Y site may inhibit dimerization with α-synuclein leading to the degradation of α-synuclein and thus less accumulation in brain (Liu et al., 2002a; Setsuie and Wada, 2007).

PARK13: *HTRA2*

Inheritance and Clinical Features

The PARK13 locus was identified in a German idiopathic PD case-control study (Strauss et al., 2005). The PARK13 gene encodes HtrA serine peptidase 2 (*HTRA2*) and is located on chromosome 2p13.1 (Strauss et al., 2005). Two gene variants, A141S and G399S, were identified in the German PD case-control study, and these variants resulted in a defective activation of protease activity and mitochondrial dysfunction in vitro (Strauss et al., 2005). However, further studies are necessary to determine if this is a new familial PD-inducing protein.

Gene Location and Structure

A cDNA of *HTRA2* was first isolated by Faccio et al. (2000a,b). It has 8 exons with a transcript length of 2367 bps that encodes a protein called *HTRA2* consisting of 458 amino acids and homology to bacterial HtrA endoprotease with a PDZ domain (Faccio et al., 2000a, b; Vande Walle et al., 2008) (Fig. 9).

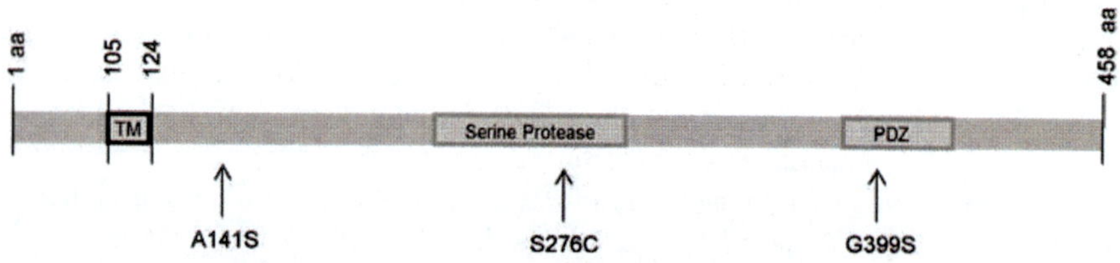

Fig. 9 PARK13: *HTRA2* structure and mutations. The PDZ domain helps anchor transmembrane proteins to the cytoskeleton. Mutations A141S, S276C, and G399S are shown (Strauss et al., 2005; Vande Walle et al., 2008). TM, transmembrane domain. Scale is approximate

Gene Function and Expression

HTRA2 is a nuclear-encoded protein located in the intermembrane space of the mitochondria and released into the cytosol during apoptosis. It has a serine protease domain that interacts with inhibitor apoptosis proteins (IAPs) to enhance the progression of apoptosis (Althaus et al., 2007; Hegde et al., 2002; Li et al., 2002; Martins et al., 2002; Srinivasula et al., 2003; Suzuki et al., 2001; Takahashi et al., 1998; Verhagen et al., 2002; Yang et al., 2003). Upon receiving various apoptotic stimuli it is released from the mitochondrial intermembrane space into the cytosol, where it is thought to induce apoptotic cell death by binding to IAPs. The binding of *HTRA2* to IAP appears to block the caspase-inhibitory activities of IAPs (Althaus et al., 2007; Hegde et al., 2002; Li et al., 2002; Martins et al., 2002; Srinivasula et al., 2003; Suzuki et al., 2001; Takahashi et al., 1998; Verhagen et al., 2002; Yang et al., 2003). *HTRA2* also enhances caspase activity by contributing to permeabilization of the mitochondrial outer membrane, which leads to the release of cytochrome c (Suzuki et al., 2001).

Mouse models with loss of *HTRA2* activity show a Parkinsonian phenotype with striatum-specific neuronal loss (Jones et al., 2003; Martins et al., 2004) and may suggest that other stress response proteins, in addition to IAPs, are involved in *HTRA2* associated neuronal loss, including the transcription factor CHOP (Moisoi et al., 2008).

Genetic Variation

Three *HTRA2* mutations have been reported: A141S, G399S, and S276C (Fig. 9). The *HTRA2* G399S mutation induces mitochondrial dysfunction in vitro and is associated with altered mitochondrial morphology whereby cells overexpressing G399S mutant *HTRA2* are more susceptible to stress-induced cell death than wild-type (Strauss et al., 2005; Vande Walle et al., 2008). The S276C mutation is the cause of the mouse mutant mnd2 (motor neuron degeneration 2) phenotype, which exhibits muscle wasting, neurodegeneration, involution of the spleen and thymus, and death by 40 days of age. Striatal neuron degeneration, with astrogliosis and microglia activation, begins at around three weeks of age, and other neurons are affected at later stages (Jones et al., 2003).

2.2.3 Genes Associated with Autosomal Recessive Parkinson's Disease

PARK2: *PRKN*

Inheritance and Clinical Features

Clinical features of the autosomal recessive young onset PARK2-linked PD include an age of onset between 20 and 40, but the age of onset can be earlier than 10 years and above 60 years (Yamamura et al., 1973).When the age of onset is young, dystonia is a characteristic symptom and patients are levodopa responsive. Motor fluctuations soon develop. Pathologically, the substantia nigra undergoes severe neuronal loss and gliosis whereas the locus ceruleus is much less severely involved and usually no Lewy bodies are seen (Mori et al., 1998; Takahashi et al., 1994), although rare Lewy body positive cases have been reported (Farrer et al., 2001).

Gene Location and Structure

Linkage analysis of several PD families mapped the PARK2 disease locus to chromosome 6q26, near the sod2 locus (Jones et al., 1998; Matsumine et al., 1997; Tassin et al., 1998). By screening a BAC library using the D6S305 marker at this region a cDNA was cloned consisting of the open reading frame of the novel gene *PRKN* (Kitada et al., 1998). PRKN protein belongs to the RING-IBR-RING family, which is a subgroup of RING finger type E3 ubiquitin ligase. The PRKN protein is 465 amino acids, with 12 exons and a 1395 bp open reading frame. It contains two RING finger domains at the carboxyl (C) terminus. RING stands for rare interesting gene and RING-like structures have been found in proteins with ubiquitin ligase activity (Lorick et al., 1999). Similar to other RING finger proteins, the PRKN protein has been found to function as an E3 ubiquitin ligase (Shimura et al., 2000).

Gene Function and Expression

PRKN appears to be a cytosolic protein normally, but it may also colocalize to synaptic vesicles, the Golgi complex, endoplasmic reticulum, and the mitochondrial outer membrane (Darios et al., 2003; Kubo et al., 2001; Mouatt-Prigent et al., 2004; Shimura et al., 2000). Many of the single amino acid substitutions appear to alter wild-type PRKN cellular localization, solubility, and propensity to aggregate (Cookson et al., 2003; Gu et al., 2003; Wang et al., 2005). It has been reported that parkin binds tubulin and associates with microtubules (Ren et al., 2003). However, PD-linked mutations, including those that impair E3 activity (Matsuda et al., 2006), appear not to affect this binding activity (Yang et al., 2005). Many ubiquitination substrates have been proposed including the aminoacyl-tRNA synthetase cofactor, p38, and a rare, 22-kDa glycosylated form of α-synuclein (Corti et al., 2003; Shimura et al., 2001; von Coelln et al., 2004a). Some mutations appear to result in *PRKN* loss of function, although *PRKN* knock-out mice have only subtle behavior and glutaminergic transmission alterations and do not suffer nigral neuronal degeneration or clinical manifestations of Parkinsonism (Goldberg et al., 2003; Itier et al., 2003) and reduced numbers of noradrenergic neurons in the locus ceruleus were reported in one strain (Von Coelln et al., 2004b). Accumulation of p38 leading to catecholaminergic cell death has been shown in one strain as well as in PARK2-linked PD and idiopathic PD brain (Ko et al., 2005).

PRKN knock-out mice also have reduced numbers of mitochondrial oxidative phosphorylation proteins, a decrease in mitochondrial respiratory capacity, and age-dependent increases in oxidative damage (Palacino et al., 2004). Mitochondrial defects have also been reported in parkin knock-out *Drosophila*, suggesting that PRKN ubiquination dysfunction may be secondary in the course of pathogenic events (Greene et al., 2003; Pesah et al., 2004). In vitro studies of a *PRKN* knock-down SH-SY5Y cell line showed apoptotic cell death and an increase in the auto-oxidized forms of levodopa and dopamine, implicating parkin antioxidative properties (Machida et al., 2005).

PARK2-linked recessive, loss-of-function mutations do not usually exhibit the classical Lewy body pathology seen in idiopathic disease, although this is not the case for some mutations that reduce but not completely ablate PRKN activity. In addition, it appears that PRKN need not be mutated to participate in the pathogenic process because it is also found in the Lewy bodies of idiopathic disease brain (Schlossmacher et al., 2002). Interestingly, Lewy bodies have been reported in a patient carrying an R275W substitution and an exon 3 deletion (Farrer et al., 2001), and an autopsy in a 73-year-old patient carrying a deletion of exon 7 as well as the del1072T point deletion showed PD-type cell loss, reactive gliosis, and SNCA-positive Lewy bodies (Pramstaller et al., 2005).

Genetic Variation

Reported mutations in parkin now exceed 100 including missense and nonsense mutations as well as exonic deletions, rearrangements, and duplications (Abbas et al., 1999; Hattori et al., 1998; Hedrich et al., 2002; Kann et al., 2002; Klein

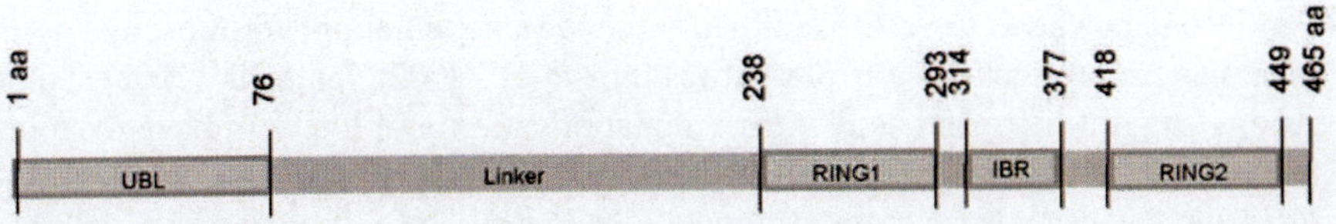

Fig. 10 PARK2: *PRKN* structure. The general protein structure of parkin is shown (Schlehe et al., 2008). More than 100 mutations have been identified and are not shown here. Five common alterations account for 35% of all *PRKN* mutations: (1) deletions of exon 4 ($n = 28$), (2) deletions of exon 3 ($n = 27$), (3) deletions of exons 3–4 ($n = 23$), (4) a point mutation in exon 7 (924C>T; $n = 38$), and (5) a single base pair deletion in exon 2 (255/256delA; $n = 17$). Hotspots for common parkin mutations appear to be concentrated in exons 2–7, whereas hotspots for exon rearrangements are more likely to occur in introns 2–4 (Hedrich et al., 2004). Scale is approximate

et al., 2003, 2000). Exonic deletions in the parkin gene were first identified in Japanese families with autosomal recessive juvenile Parkinsonism (Kitada et al., 1998). Parkin mutations have been found to account for about 50% of familial cases and about 70% of sporadic cases with age of onset of 20 years depending on the ethnicity of the population sample (Lucking et al., 2000; Mata et al., 2004; Periquet et al., 2003). Parkin mutational frequency in late-onset PD is lower than early-onset cases, accounting for between 0 and 11% depending largely on whether the sample is familial or sporadic (Foroud et al., 2003; Mata et al., 2004; Oliveri et al., 2001).

Many *PRKN* mutations including deletion mutations and point mutations have been detected in the *PRKN* gene of PARK2-linked patients. The site of these mutations spans almost all regions including the N-terminal UBL domain and the RING-IBR-RING domain (Fig. 10) but there seems to be very little difference in symptom phenotypes among these mutation sites, suggesting that the entire region of the PRKN protein may be essential for exerting its physiology. *PRKN*-linked PD has been initially characterized as a recessively inherited disease with a deleterious alteration on both alleles with the presumption that heterozygous carriers are unaffected. However, evidence now suggests that a large proportion of the total number of cases identified with a parkin mutation have only a single heterozygous mutation (Lucking et al., 2000; Mata et al., 2004; Periquet et al., 2003). Thus patients having a loss of one *PRKN* allele may suffer from haploinsufficiency as a consequence of a reduced *PRKN* expression or enzymatic activity. Indeed, one brain imaging study showed that some asymptomatic heterozygous *PRKN* carriers show significant striatal dopaminergic dysfunction, suggesting there may be a gene dosage effect (Kann et al., 2002). In addition, other reports suggest that *PRKN* heterozygous mutation carrier status significantly influences age at onset of PD (Foroud et al., 2003) with many of the heterozygous mutations in the first RING finger domain associated with later age of onset (Oliveira et al., 2003). A common polymorphism in the *PRKN* promoter has been associated with late-onset idiopathic disease (Tan et al., 2005a; West et al., 2002).

Differential expression of a *PRKN* splice variant where the RING domains have been deleted has also been shown to modulate the risk of sporadic PD (Tan et al., 2005b). In one family, patients with a recessive pattern of inheritance for the *PRKN* Ex3-Δ40 mutation manifest symptoms of early-onset levodopa-responsive

Parkinsonism whereas in other families with the same mutation, an autosomal dominant pattern of inheritance is present (Munoz et al., 2002; Tan et al., 2003). Some reports suggest that carriers of *PRKN* mutations are more likely to have dystonia and symmetric symptoms than noncarriers, and some may even have atypical features such as psychiatric manifestations (Khan et al., 2003; Lucking et al., 2000). However, a wide overlap of Parkinsonian symptoms between some groups suggests that no specific diagnostic clinical feature can be demonstrated (Munhoz et al., 2004). Thus the influence of the *PRKN* gene on PD risk may involve a complex interplay between environment and gene dose that manifests varying phenotypes.

In summary, over 100 different mutations have been identified in the parkin gene including, but not limited to, 40 exon rearrangements (26 deletions and 14 multiplications), 43 single base pair substitutions, and 12 small deletions or insertions of one or several base pairs. The most common mutations appear to be (1) deletions of exon 4 ($n = 28$), (2) deletions of exon 3 ($n = 27$), (3) deletions of exons 3–4 ($n = 23$), (4) a point mutation in exon 7 (924C>T; $n = 38$), and (5) a single base pair deletion in exon 2 (255/256delA; $n = 17$). These five common alterations account for 35% of all parkin mutations. Hotspots for common parkin mutations appear to be concentrated in exons 2 and 7, whereas hotspots for exon rearrangements are more likely to occur in introns 2 through 4 (Hedrich et al., 2004) (Fig. 10).

PARK7: *DJ1*

Inheritance and Clinical Features

DJ1 recessively inherited missense and exonic deletion mutations were first identified in two European families with an age of onset of 20–40 years (Bonifati et al., 2003). PARK7-linked PD appears to be very rare (Bonifati et al., 2003; Hague et al., 2003; Hering et al., 2004). Very few *DJ1* patients have been reported in the literature, thus clinical features and correlations with *DJ1* mutations are still difficult to determine. Some clinical features such as psychiatric symptoms (Dekker et al., 2003), short stature, and brachydactyly (Dekker et al., 2004) have been reported. *DJ1* mutations rarely associate with PD but some missense, splice-site, and exonic deletion mutations have been identified, accounting for less than 1% of early-onset PD (Clark et al., 2004; Hering et al., 2004; Lockhart et al., 2004; Tan et al., 2004b).

Gene Location and Structure

DJ1 has been cloned and is located on chromosome 1p36.23. It has a transcript length of 949 bps with seven exons (Nagakubo et al., 1997). It encodes a protein consisting of 189 amino acids.

Gene Function and Expression

DJ1 is a homodimer that belongs to the peptidase C56 family of proteins (Moore et al., 2003). It is a cytoplasmic protein, but it can also translocate into the mitochondria (Zhang et al., 2005) and it appears to act as an antioxidant (Abou-Sleiman et al., 2003; Canet-Aviles et al., 2004; Moore et al., 2005; Nagakubo et al., 1997).

Its antioxidant properties may depend on a cysteine residue at 106, which upon oxidation forms a disulphide bond (Canet-Aviles et al., 2004). *DJ1* may act as either a redox sensor protein that can prevent the aggregation of α-synuclein or an antioxidant (Batelli et al., 2008; Canet-Aviles et al., 2004; Mitsumoto and Nakagawa, 2001; Mitsumoto et al., 2001; Zhou and Freed, 2005; Zhou et al., 2006). *DJ1* may also act as a reactive oxygen species scavenger through auto-oxidation (Taira et al., 2004). Thus, it has been proposed that inasmuch as substantia nigral neurons are exposed to high oxidative stress owing to the presence of dopamine, *DJ1*may be acting as a strong antioxidative protein.

Expression of *DJ1* is ubiquitous and abundant in most mammalian tissues including the brain, where it is found in both neuronal and glial cells (Bandopadhyay et al., 2004). Downregulation of endogenous *DJ1* protein of the neuronal cell line by siRNA enhances oxidative stress-induced cell death, ER stress, and proteasome inhibition, but not by proapoptotic stimulus (Taira et al., 2004; Yokota et al., 2003). The L166P mutant protein has a reduced antioxidative activity (Takahashi-Niki et al., 2004). Mutant *DJ1* appears to interact with parkin (Moore et al., 2005) whereby parkin acts as an E3 ligase to remove mutated *DJ1*. *DJ1*-null mice are sensitive to oxidative stress and MPTP (Kim et al., 2005). DJ1 protein expression is increased upon oxidative stress induced by paraquat (Mitsumoto et al., 2001). Other *DJ1* knock-out strains show normal numbers of dopaminergic neurons but also sensitivity to the PD associated environmental toxins paraquat and rotenone (Goldberg et al., 2005; Meulener et al., 2005).

DJ1 does not appear to be an essential component of LBs in sporadic cases (Bandopadhyay et al., 2004). *DJ1* mutations are rare in sporadic PD but recent studies suggest that *DJ1* may play an important role in common forms of the disease. Sporadic PD brain exhibits *DJ1* with oxidative damage (Choi et al., 2006). Sporadic PD patients also demonstrate a significant increase in total cerebrospinal fluid *DJ1* protein levels compared to normal controls (Waragai et al., 2006).

Genetic Variation

In general, *DJ1* mutations are found in the homozygous or compound heterozygous state, putatively resulting in a loss of protein function. The L166P mutation causes destabilization through unfolding of the C-terminus, inhibiting dimerization, and enhancing degradation by the proteasome (Miller et al., 2003; Moore et al., 2003; Olzmann et al., 2004). In addition, probably consequential to instability, L166P reduces the neuroprotective function of *DJ1* (Taira et al., 2004). Reduced nuclear localization, in favor of the mitochondria, is also seen for L166P, as well as for the M26I and D149A mutations (Bonifati et al., 2003; Xu et al., 2005). In addition, there appear to be structural perturbations associated with *DJ1*mutations L166P, E64D, M26I, A104T, and D149A, which can lead to global destabilization, unfolding of the protein structure, heterodimer formation, or reduced antioxidant activity, implicating these mutations in pathogenicity associated with *DJ1* (Anderson and Daggett, 2008; Malgieri and Eliezer, 2008; Takahashi-Niki et al., 2004) (Fig. 11).

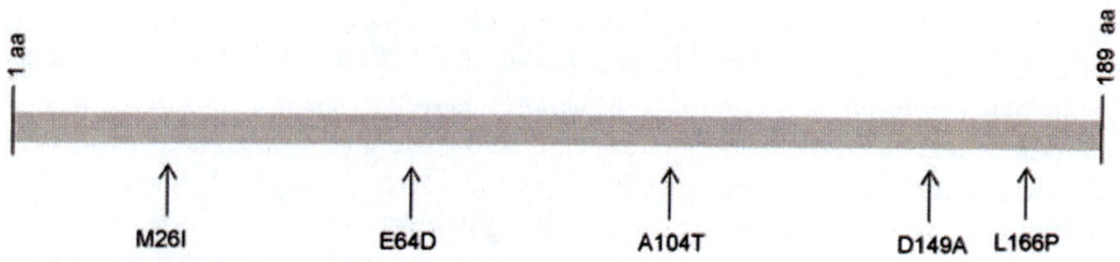

Fig. 11 PARK7: *DJ1* structure and mutations. *DJ1* mutations are found in both the homozygous or heterozygous state, putatively resulting in a loss of protein function. Scale is approximate

PARK6: *PINK1*

Inheritance and Clinical Features

Mutations in the phosphatase and Tensin (PTEN) Induced Kinase 1 gene (*PINK1*) were first identified in patients with recessive young-onset autosomal recessive PD designated as PARK6. The age of onset is from 32 to 48 years (Valente et al., 2001). *PINK1* mutations account for approximately 1–7% of autosomal recessive PD in Caucasians (Healy et al., 2004a; Rohe et al., 2004; Valente et al., 2001), about 8.9% of autosomal recessive PD in Japanese autosomal recessive PD families (Li et al., 2005), and 2–3% of sporadic and familial PD in individuals of Chinese origin (Tan et al., 2006b, 2005c). Clinical features of PARK6 are similar to late-onset PD, with rare features such as dystonia at onset, sleep benefit, and psychiatric disturbances (Hatano et al., 2004b; Tan et al., 2006b; Valente et al., 2001). Clinical characteristics of *PINK1* are similar to *PRKN* with dystonia at onset and increased reflexes that were originally thought to be related only to parkin (Ibanez et al., 2006). However, age at onset is earlier in *PINK1* mutational carriers than in those with *PRKN* mutations (Leutenegger et al., 2006; Valente et al., 2001). Brain imaging of PARK6 carriers indicates a 20–30% reduction of the caudate and putamen (Khan et al., 2002).

Gene Location and Structure

PINK1 is located on chromosome 1p36.12 and has eight exons and cDNA that spans 1.8 kb. It encodes a protein with 581 amino acids. It has a serine/threonine protein kinase domain. However, its function is not known (Valente et al., 2001). It is a mitochondrial protein located in the matrix and the intermembrane space that is ubiquitously expressed in the brain and systemic organs and contains a mitochondrial-targeting motif and a conserved serine/threonine kinase domain (Silvestri et al., 2005).

Gene Function and Expression

Functional studies have shown that *PINK1* can be localized to mitochondria both in vitro and in vivo (Gandhi et al., 2006). Wild-type *PINK1* appears to be important in neuroprotection against mitochondrial dysfunction and proteasome-induced apoptosis whereas the G309D mutation impairs this protective effect, possibly by interfering with adenosine diphosphate (ADP) binding and thus inhibiting kinase activity (Valente et al., 2004, 2001). E240K and L489P mutants disrupt *PINK1*'s protectivity by either enhancing the instability of the protein or disrupting the kinase activity of the protein (Petit et al., 2005). In vitro studies indicate that cells transfected with *PINK1* mutants have disrupted mitochondrial membrane potential under stressful conditions (Abou-Sleiman et al., 2006). Knock-out models of the *Drosophila PINK1* orthologue have defects in mitochondrial morphology and increased sensitivity to oxidative stress and appear to be rescued by human parkin (Clark et al., 2006).

PINK1 haploinsufficiency may be sufficient to cause disease because *PINK1* is detected in some LBs in sporadic PD, as well as in samples carrying only one mutant *PINK1* allele, which are clinically and pathologically indistinguishable from sporadic cases (Gandhi et al., 2006).

Genetic Variation

The first mutations discovered were the G309D missense and a W437X truncating mutation found in families of Italian and Spanish descent (Valente et al., 2004, 2001). Several point mutations, frameshifts, and truncating mutants have been identified (Bonifati et al., 2005; Ibanez et al., 2006; Tan et al., 2006b). Interestingly, in contrast to *PRKN*, most of the *PINK1* mutations reported are either missense or nonsense mutations (Hatano et al., 2004a, b; Li et al., 2005; Rohe et al., 2004; Valente et al., 2004). One family with a large deletion mutation involving exons 6–8 homozygotes has been reported (Li et al., 2005).

Japanese and Israeli *PINK1*-linked families and a sporadic PD patient of Chinese ethnicity have a R246X mutation (Tan et al., 2006b). Most of the reported mutations are located in a highly conserved amino acid position in the protein kinase domain and are absent in healthy controls, thus suggesting that these mutations are pathogenic (Abou-Sleiman et al., 2006) and that homozygous mutation carriers appear to be clinically affected whereas heterozygous carriers are not (Hiller et al., 2007). A large kindred from Sudan with early-onset Parkinsonism (ages 9–17 years) is associated with a novel mutation, A217D, in the *PINK1* gene. Phenotypes in this family vary from dopa-responsive dystonia-like to typical early-onset Parkinsonism. A217D is located in the highly conserved adenosine triphosphate orientation site of the *PINK1* kinase domain (Leutenegger et al., 2006; Tan and Skipper, 2007) (Fig. 12).

PARK9: *ATP13A2*

Inheritance and Clinical Features

Homozygous and compound heterozygous mutations in the P-type ATPase gene (*ATP13A2*) have been demonstrated in a Jordanian family (Myhre et al., 2008; Najim al-Din et al., 1994) and a Chilean family (Ramirez et al., 2006) with Kufor–Rakeb syndrome, a form of recessively inherited atypical Parkinsonism that is clinically characterized by very early age of onset (11–16 years), levodopa-responsive Parkinsonism, pyramidal signs, dementia, and a supranuclear gaze palsy (Najim al-Din et al., 1994, Williams et al., 2005). MRIs show significant atrophy of the globus pallidus and the pyramids, as well as generalized brain atrophy in later

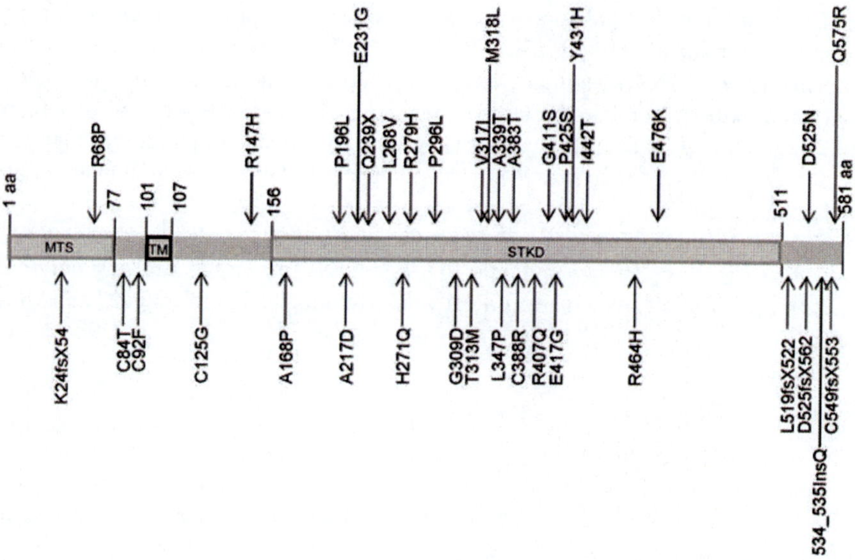

Fig. 12 PARK6: *PINK1* structure and mutations. Many *PINK1* mutations have been described of which a few are shown. For a more complete list of *PINK1* mutations, see Tan et al., (2007). MTS, mitochondrial targeting sequence; TM, transmembrane domain; STKD, serine threonine kinase domain (Tan et al., 2007). Scale is approximate

stages. Some develop facial–faucial–finger minimyoclonus, visual hallucinations, and oculogyric dystonic spasm (Williams et al., 2005).

Gene Location and Structure

The disease locus designated as PARK9 was mapped to 1p36.13 with a maximum LOD score of 3.6.1p36.13 is, a hotspot for autosomal recessive familial PD (Hampshire et al., 2001). The disease gene was subsequently identified as *ATP13A2*. The transcript has 29 exons and is 3854 bps in length. The ATP13A2 protein contains 1180 amino acids and has 10 transmembrane domains.

Gene Function and Expression

ATP13A2 is a lysosomal membrane protein with an ATPase domain (Ramirez et al., 2006). It is a member of the P5 subfamily of ATPases that transports inorganic cations and other substrates. The exact function of the ATP13A2 protein is still unknown. *ATP13A2* is predominantly expressed in brain tissues, and *ATP13A2* mRNA levels are about tenfold higher in the substantia nigra dopaminergic neurons of sporadic patients than control subject brains (Ramirez et al., 2006).

Genetic Variation

All known *ATP13A2* mutations appear to directly or indirectly affect transmembrane domains (Ramirez et al., 2006) (Fig. 13). In vitro evidence indicates that wild-type

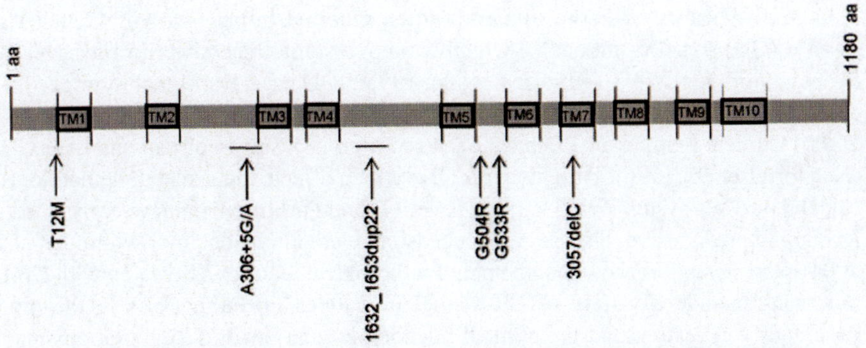

Fig. 13 PARK9: *ATP13A2* structure and mutations. Mutations found in young-onset PD (T12M, G504R, and G533R; Di Fonzo et al., 2007) and Kufor-Rakeb syndrome PD (1306+5G/A, 1632_1653dup22, 3057delC; Ramirez et al., 2006) are shown. TM, transmembrane domain. Scale is approximate

ATP13A2 is localized to the lysosome membrane of transiently transfected cells whereas unstable truncated mutants are retained in the endoplasmic reticulum and degraded by the proteasome (Ramirez et al., 2006). A homozygous missense mutation (G504R) has been identified in one sporadic case from Brazil with juvenile Parkinsonism (Di Fonzo et al., 2007). This patient had symptoms onset at age 12, levodopa-responsive severe akinetic-rigid Parkinsonism, levodopa-induced motor fluctuations and dyskinesias, severe visual hallucinations, supranuclear vertical gaze paresis, and moderate diffuse atrophy but no pyramidal deficit nor dementia. In this same study, two Italian cases with youth-onset PD without atypical features carried a novel missense mutation (T12M, G533R) in a single heterozygous state (Di Fonzo et al., 2007). A rare variant associated with an increased risk of PD among ethnic Chinese in Asia has recently been described that has a clinical phenotype and brain image similar to that seen in idiopathic PD (Lin et al., 2008).

GBA

Inheritance and Clinical Features

An association between mutations within the glucocerebrosidase gene (*GBA*) and PD has been reported in multiple studies including an initial study where a small group of PD postmortem brain samples were found to have a *GBA* mutation frequency of 14% and where an Israeli PD patient sample was found to have a *GBA* mutation frequency of 31.3% (Aharon-Peretz et al., 2004). The frequency for *GBA* mutations appears to be less than 1% in the general population and 6–7% in Ashkenazi Jews (Aharon-Peretz et al., 2004; Bras et al., 2007; Clark et al., 2005, 2007; De Marco et al., 2008; Eblan et al., 2006; Gan-Or et al., 2008; Lwin et al., 2004; Nichols et al., 2009; Sato et al., 2005; Spitz et al., 2008; Tan et al., 2007; Toft et al., 2006; Wu et al., 2007; Ziegler et al., 2007).

A family history of Parkinsonism is often reported in patients with Gaucher's disease (GD) (OMIM #606463), which is an autosomal recessive disorder caused by mutations in *GBA* (Neudorfer et al., 1996). GD is a lysosomal storage disease characterized by an accumulation of glucocerebrosides (Goker-Alpan et al., 2008). Clinical features of PD have been reported in a subset of patients with GD (Neudorfer et al., 1996). Patients with GD (which affects the skeletal, hematological, and nervous systems with varying severity) and Parkinsonism have early-onset, levodopa-unresponsive disease with occasional cognitive decline (Wong et al., 2004). Recent studies show that the neuropathological features associated with *GBA* mutations include a variety of LB synucleinopathies, including LBs in the hippocampus, suggesting that the clinical phenotype of PD with *GBA* mutations may be diverse (Goker-Alpan et al., 2008; Mata et al., 2008).

Gene Location and Structure

GBA is located on chromosome 1 (1q21). The *GBA* cDNA is approximately 2 kb in length (Horowitz et al., 1989; Reiner et al., 1988). A *GBA* pseudogene has a 96% homology to *GBA* and is located approximately 12 kb downstream. There are two in-frame translational start sites in exon 1 and exon 2. Initiation at each exon leads to a different leader sequence but both are processed into a mature functional enzyme of the same length. The protein is cleaved to produce a mature polypeptide of 497 amino acids with a molecular weight of 55.5 kDa. The polypeptide contains five potential glycosylation sites, four of which appear to be glycosylated. The active site of this enzyme resides in the C-terminal half of the molecule at exon 9 and exon 10 (Dinur et al., 1986).

Gene Function and Expression

The *GBA* gene encodes the lysosomal membrane protein, glucocerebrosidase, which cleaves the beta-glucosidic linkage of glycosylceramide, an intermediate in glycolipid metabolism (Dinur et al., 1986). *GBA* mRNA levels vary among cell lines, with high, moderate, low, and negligible levels reported in epithelial, fibroblast, macrophage, and B-cell lines, respectively (Reiner and Horowitz, 1988; Reiner et al., 1987; Wigderson et al., 1989). There appears to be a poor correlation between the levels of mRNA and the amount of identified enzymatic activity (Doll and Smith, 1993; Reiner and Horowitz, 1988) implicating a complex regulatory system for the expression of glucocerebrosidase at the level of transcription, translation, and posttranslational modification (Xu et al., 1995).

Genetic Variation

Over 250 mutations have been reported in *GBA*: 203 missense mutations, 18 nonsense mutations, 36 small insertions or deletions that lead to either frameshifts or in-frame alterations, 14 splice junction mutations, and 13 complex alleles carrying two or more mutations in *cis* (Hruska et al., 2008). Recombination events with a highly homologous pseudogene downstream of the *GBA* locus also have been identified, resulting from gene conversion, fusion, or duplication. Some of the alleles for disease mutations are also found in the pseudogene, making analysis complicated. The *GBA* mutations that influence GD or PD are not necessarily disease-specific. For example, N370S and L444P are the most common mutations associated with both GD and PD whereas, in contrast, R120W is found mainly in PD but not GD, and R463C is found mainly in GD but not PD (Aharon-Peretz et al., 2004; Bras et al., 2007; Clark et al., 2005, 2007; De Marco et al., 2008; Eblan et al., 2006; Gan-Or et al., 2008; Hruska et al., 2008; Lwin et al., 2004; Sato et al., 2005; Spitz et al., 2008; Tan et al., 2007; Toft et al., 2006; Wu et al., 2007; Ziegler et al., 2007) (Fig. 14).

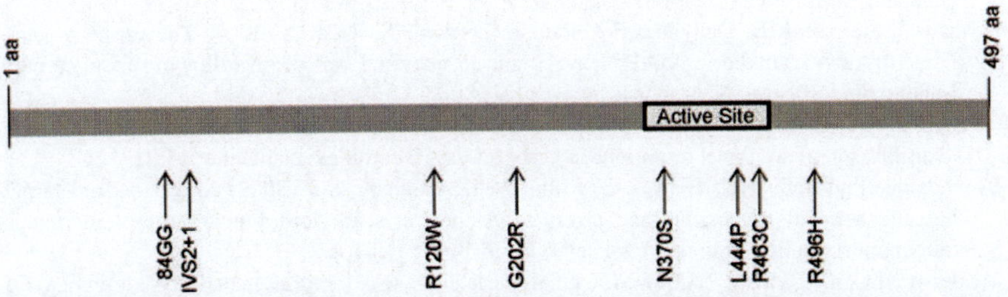

Fig. 14 *GBA* structure and mutations. Over 250 *GBA* mutations have been described (Hruska et al., 2008). Mutations found in PD include R120W, N370S, and L444P. The enzyme's active site is located in the region of mutations; N370S (exon 9) and L444P (exon 10). Scale is approximate

2.3 Summary

Thirteen loci have been linked to PD of which eight genes have been described: four autosomal dominant (*SNCA*, *LRRK2*, *UCHL1*, and *HTRA2*) and four autosomal recessive (*PRKN*, *DJ1*, *PINK1*, and *ATP13A2*). In addition, another gene has recently been described as a robust risk factor for PD (*GBA*). These findings suggest that PD pathology involves a strong genetic component and provides numerous clues to the etiology of the disease. The function of these genes and their contribution to PD pathogenesis remains unclear. However, many of these genes play a role in ubiquitination, oxidative stress, and apoptosis, suggesting that PD, may be a genetically complex and heterogeneous disease. In addition to the link between these genes and familial forms of PD, many are also candidate genes for idiopathic forms of the disease suggesting that some of these genes carry other mutations that simply increase risk.

Acknowledgments This work is supported in part by the Office of Research and Development, Biomedical Laboratory Research Program, U.S. Department of Veterans Affairs; the Mental Illness Research, Education, and Clinical Center, Veterans Affairs Puget Sound Health Care System; the Geriatric Research, Education, and Clinical Center, Veterans Affairs Puget Sound Health Care System; and the University of Washington Alzheimer's Disease Research Center, NIA AG005136.

References

Abbas N, Lucking CB, Ricard S, Durr A, Bonifati V, De Michele G, Bouley S, Vaughan JR, Gasser T, Marconi R, Broussolle E, Brefel-Courbon C, Harhangi BS, Oostra BA, Fabrizio E, Bohme GA, Pradier L, Wood NW, Filla A, Meco G, Denefle P, Agid Y, Brice A (1999) A wide variety of mutations in the parkin gene are responsible for autosomal recessive parkinsonism in Europe. French Parkinson's Disease Genetics Study Group and the European Consortium on Genetic Susceptibility in Parkinson's Disease. Hum Mol Genet 8:567–574

Abou-Sleiman PM, Healy DG, Quinn N, Lees AJ, Wood NW (2003) The role of pathogenic DJ-1 mutations in Parkinson's disease. Ann Neurol 54:283–286

Abou-Sleiman PM, Muqit MM, McDonald NQ, Yang YX, Gandhi S, Healy DG, Harvey K, Harvey RJ, Deas E, Bhatia K, Quinn N, Lees A, Latchman DS, Wood NW (2006) A heterozygous effect for PINK1 mutations in Parkinson's disease? Ann Neurol 60:414–419

Aharon-Peretz J, Rosenbaum H, Gershoni-Baruch R (2004) Mutations in the glucocerebrosidase gene and Parkinson's disease in Ashkenazi Jews. N Engl J Med 351:1972–1977

Althaus J, Siegelin MD, Dehghani F, Cilenti L, Zervos AS, Rami A (2007) The serine protease Omi/HtrA2 is involved in XIAP cleavage and in neuronal cell death following focal cerebral ischemia/reperfusion. Neurochem Int 50:172–180

Alzheimer's Disease Collaborative Group (1995) The structure of the presenilin 1 (S182) gene and identification of six novel mutations in early onset AD families. Nat Genet 11:219–222

Anandatheerthavarada HK, Biswas G, Robin MA, Avadhani NG (2003) Mitochondrial targeting and a novel transmembrane arrest of Alzheimer's amyloid precursor protein impairs mitochondrial function in neuronal cells. J Cell Biol 161:41–54

Anderson PC, Daggett V (2008) Molecular basis for the structural instability of human DJ-1 induced by the L166P mutation associated with Parkinson's disease. Biochemistry 47: 9380–9393

Anwar R, Moynihan TP, Ardley H, Brindle N, Coletta PL, Cairns N, Markham AF, Robinson PA (1996) Molecular analysis of the presenilin 1 (S182) gene in "sporadic" cases of Alzheimer's disease: identification and characterisation of unusual splice variants. J Neurochem 66: 1774–1777

Aoki K, Uchihara T, Nakamura A, Komori T, Arai N, Mizutani T (2003) Expression of apolipoprotein E in ballooned neurons-comparative immunohistochemical study on neurodegenerative disorders and infarction. Acta Neuropathol 106:436–440

Bandopadhyay R, Kingsbury AE, Cookson MR, Reid AR, Evans IM, Hope AD, Pittman AM, Lashley T, Canet-Aviles R, Miller DW, McLendon C, Strand C, Leonard AJ, Abou-Sleiman PM, Healy DG, Ariga H, Wood NW, de Silva R, Revesz T, Hardy JA, Lees AJ (2004) The expression of DJ-1 (PARK7) in normal human CNS and idiopathic Parkinson's disease. Brain 127:420–430

Baskin F, Smith GM, Fosmire JA, Rosenberg RN (1997) Altered apolipoprotein E secretion in cytokine treated human astrocyte cultures. J Neurol Sci 148:15–18

Batelli S, Albani D, Rametta R, Polito L, Prato F, Pesaresi M, Negro A, Forloni G (2008) DJ-1 modulates alpha-synuclein aggregation state in a cellular model of oxidative stress: relevance for Parkinson's disease and involvement of HSP70. PLoS One 3:e1884

Baulac S, LaVoie MJ, Kimberly WT, Strahle J, Wolfe MS, Selkoe DJ, Xia W (2003) Functional gamma-secretase complex assembly in Golgi/trans-Golgi network: interactions among presenilin, nicastrin, Aph1, Pen-2, and gamma-secretase substrates. Neurobiol Dis 14:194–204

Belin AC, Westerlund M (2008) Parkinson's disease: a genetic perspective. FEBS J 275:1377–1383

Bentahir M, Nyabi O, Verhamme J, Tolia A, Horre K, Wiltfang J, Esselmann H, De Strooper B (2006) Presenilin clinical mutations can affect gamma-secretase activity by different mechanisms. J Neurochem 96:732–742

Bertram L, Tanzi RE (2004a) Alzheimer's disease: one disorder, too many genes? Hum Mol Genet 13(Spec No 1):R135–R141

Bertram L, Tanzi RE (2004b) The current status of Alzheimer's disease genetics: what do we tell the patients? Pharmacol Res 50:385–396

Bird TD (2008) Genetic aspects of Alzheimer disease. Genet Med 10:231–239

Bisaglia M, Mammi S, Bubacco L (2008) Structural insights on physiological functions and pathological effects of {alpha}-synuclein. FASEB J 23(2):329–340

Biskup S, Moore DJ, Celsi F, Higashi S, West AB, Andrabi SA, Kurkinen K, Yu SW, Savitt JM, Waldvogel HJ, Faull RL, Emson PC, Torp R, Ottersen OP, Dawson TM, Dawson VL (2006) Localization of LRRK2 to membranous and vesicular structures in mammalian brain. Ann Neurol 60:557–569

Blacker D, Haines JL, Rodes L, Terwedow H, Go RC, Harrell LE, Perry RT, Bassett SS, Chase G, Meyers D, Albert MS, Tanzi R (1997) ApoE-4 and age at onset of Alzheimer's disease: the NIMH genetics initiative. Neurology 48:139–147

Bonifati V, Rizzu P, Squitieri F, Krieger E, Vanacore N, van Swieten JC, Brice A, van Duijn CM, Oostra B, Meco G, Heutink P (2003) DJ-1(PARK7), a novel gene for autosomal recessive, early onset parkinsonism. Neurol Sci 24:159–160

Bonifati V, Rohe CF, Breedveld GJ, Fabrizio E, De Mari M, Tassorelli C, Tavella A, Marconi R, Nicholl DJ, Chien HF, Fincati E, Abbruzzese G, Marini P, De Gaetano A, Horstink MW, Maat-Kievit JA, Sampaio C, Antonini A, Stocchi F, Montagna P, Toni V, Guidi M, Dalla Libera A, Tinazzi M, De Pandis F, Fabbrini G, Goldwurm S, de Klein A, Barbosa E, Lopiano L, Martignoni E, Lamberti P, Vanacore N, Meco G, Oostra BA (2005) Early-onset parkinsonism associated with PINK1 mutations: frequency, genotypes, and phenotypes. Neurology 65:87–95

Boschert U, Merlo-Pich E, Higgins G, Roses AD, Catsicas S (1999) Apolipoprotein E expression by neurons surviving excitotoxic stress. Neurobiol Dis 6:508–514

Bossy-Wetzel E, Schwarzenbacher R, Lipton SA (2004) Molecular pathways to neurodegeneration. Nat Med 10(Suppl):S2-S9

Braak H, Braak E (1991) Demonstration of amyloid deposits and neurofibrillary changes in whole brain sections. Brain Pathol 1:213–216

Braak H, Braak E (1997) Frequency of stages of Alzheimer-related lesions in different age categories. Neurobiol Aging 18:351–357

Bras J, Paisan-Ruiz C, Guerreiro R, Ribeiro MH, Morgadinho A, Januario C, Sidransky E, Oliveira C, Singleton A (2007) Complete screening for glucocerebrosidase mutations in Parkinson disease patients from Portugal. Neurobiol Aging 30(9):1515–15157

Brickell KL, Steinbart EJ, Rumbaugh M, Payami H, Schellenberg GD, Van Deerlin V, Yuan W, Bird TD (2006) Early-onset Alzheimer disease in families with late-onset Alzheimer disease: a potential important subtype of familial Alzheimer disease. Arch Neurol 63:1307–1311

Brousseau T, Legrain S, Berr C, Gourlet V, Vidal O, Amouyel P (1994) Confirmation of the epsilon 4 allele of the apolipoprotein E gene as a risk factor for late-onset Alzheimer's disease. Neurology 44:342–344

Buttini M, Orth M, Bellosta S, Akeefe H, Pitas RE, Wyss-Coray T, Mucke L, Mahley RW (1999) Expression of human apolipoprotein E3 or E4 in the brains of Apoe-/- mice: isoform-specific effects on neurodegeneration. J Neurosci 19:4867–4880

Cai H, Wang Y, McCarthy D, Wen H, Borchelt DR, Price DL, Wong PC (2001) BACE1 is the major beta-secretase for generation of Abeta peptides by neurons. Nat Neurosci 4:233–234

Campion D, Dumanchin C, Hannequin D, Dubois B, Belliard S, Puel M, Thomas-Anterion C, Michon A, Martin C, Charbonnier F, Raux G, Camuzat A, Penet C, Mesnage V, Martinez M, Clerget-Darpoux F, Brice A, Frebourg T (1999) Early-onset autosomal dominant Alzheimer disease: prevalence, genetic heterogeneity, and mutation spectrum. Am J Hum Genet 65: 664–670

Canet-Aviles RM, Wilson MA, Miller DW, Ahmad R, McLendon C, Bandyopadhyay S, Baptista MJ, Ringe D, Petsko GA, Cookson MR (2004) The Parkinson's disease protein DJ-1 is neuroprotective due to cysteine-sulfinic acid-driven mitochondrial localization. Proc Natl Acad Sci U S A 101:9103–9108

Canevari L, Clark JB (2007) Alzheimer's disease and cholesterol: the fat connection. Neurochem Res 32:739–750

Chai CK (2007) The genetics of Alzheimer's disease. Am J Alzheimers Dis Other Demen 22:37–41

Chartier-Harlin MC, Crawford F, Hamandi K, Mullan M, Goate A, Hardy J, Backhovens H, Martin JJ, Broeckhoven CV (1991a) Screening for the beta-amyloid precursor protein mutation (APP717: Val→Ile) in extended pedigrees with early onset Alzheimer's disease. Neurosci Lett 129:134–135

Chartier-Harlin MC, Crawford F, Houlden H, Warren A, Hughes D, Fidani L, Goate A, Rossor M, Roques P, Hardy J et al (1991b) Early-onset Alzheimer's disease caused by mutations at codon 717 of the beta-amyloid precursor protein gene. Nature 353:844–846

Chartier-Harlin MC, Kachergus J, Roumier C, Mouroux V, Douay X, Lincoln S, Levecque C, Larvor L, Andrieux J, Hulihan M, Waucquier N, Defebvre L, Amouyel P, Farrer M, Destee A (2004) Alpha-synuclein locus duplication as a cause of familial Parkinson's disease. Lancet 364:1167–1169

Chauhan NB (2003) Membrane dynamics, cholesterol homeostasis, and Alzheimer's disease. J Lipid Res 44:2019–2029

Chen L, Feany MB (2005) Alpha-synuclein phosphorylation controls neurotoxicity and inclusion formation in a Drosophila model of Parkinson disease. Nat Neurosci 8:657–663

Choi J, Sullards MC, Olzmann JA, Rees HD, Weintraub ST, Bostwick DE, Gearing M, Levey AI, Chin LS, Li L (2006) Oxidative damage of DJ-1 is linked to sporadic Parkinson and Alzheimer diseases. J Biol Chem 281:10816–10824

Citron M, Westaway D, Xia W, Carlson G, Diehl T, Levesque G, Johnson-Wood K, Lee M, Seubert P, Davis A, Kholodenko D, Motter R, Sherrington R, Perry B, Yao H, Strome R, Lieberburg I, Rommens J, Kim S, Schenk D, Fraser P, St George Hyslop P, Selkoe DJ (1997) Mutant presenilins of Alzheimer's disease increase production of 42-residue amyloid beta-protein in both transfected cells and transgenic mice. Nat Med 3:67–72

Clark LN, Afridi S, Mejia-Santana H, Harris J, Louis ED, Cote LJ, Andrews H, Singleton A, Wavrant De-Vrieze F, Hardy J, Mayeux R, Fahn S, Waters C, Ford B, Frucht S, Ottman R, Marder K (2004) Analysis of an early-onset Parkinson's disease cohort for DJ-1 mutations. Mov Disord 19:796–800

Clark IE, Dodson MW, Jiang C, Cao JH, Huh JR, Seol JH, Yoo SJ, Hay BA, Guo M (2006) Drosophila pink1 is required for mitochondrial function and interacts genetically with parkin. Nature 441:1162–1166

Clark LN, Nicolai A, Afridi S, Harris J, Mejia-Santana H, Strug L, Cote LJ, Louis ED, Andrews H, Waters C, Ford B, Frucht S, Fahn S, Mayeux R, Ottman R, Marder K (2005) Pilot association study of the beta-glucocerebrosidase N370S allele and Parkinson's disease in subjects of Jewish ethnicity. Mov Disord 20:100–103

Clark LN, Ross BM, Wang Y, Mejia-Santana H, Harris J, Louis ED, Cote LJ, Andrews H, Fahn S, Waters C, Ford B, Frucht S, Ottman R, Marder K (2007) Mutations in the glucocerebrosidase gene are associated with early-onset Parkinson disease. Neurology 69: 1270–1277

Conway KA, Harper JD, Lansbury PT (1998) Accelerated in vitro fibril formation by a mutant alpha-synuclein linked to early-onset Parkinson disease. Nat Med 4:1318–1320

Conway KA, Harper JD, Lansbury PT Jr. (2000) Fibrils formed in vitro from alpha-synuclein and two mutant forms linked to Parkinson's disease are typical amyloid. Biochemistry 39: 2552–2563

Cookson MR, Lockhart PJ, McLendon C, O'Farrell C, Schlossmacher M, Farrer MJ (2003) RING finger 1 mutations in Parkin produce altered localization of the protein. Hum Mol Genet 12:2957–2965

Coon KD, Myers AJ, Craig DW, Webster JA, Pearson JV, Lince DH, Zismann VL, Beach TG, Leung D, Bryden L, Halperin RF, Marlowe L, Kaleem M, Walker DG, Ravid R, Heward CB, Rogers J, Papassotiropoulos A, Reiman EM, Hardy J, Stephan DA (2007) A high-density whole-genome association study reveals that APOE is the major susceptibility gene for sporadic late-onset Alzheimer's disease. J Clin Psychiatry 68:613–618

Corder EH, Saunders AM, Strittmatter WJ, Schmechel DE, Gaskell PC, Small GW, Roses AD, Haines JL, Pericak-Vance MA (1993) Gene dose of apolipoprotein E type 4 allele and the risk of Alzheimer's disease in late onset families. Science 261:921–923

Corti O, Hampe C, Koutnikova H, Darios F, Jacquier S, Prigent A, Robinson JC, Pradier L, Ruberg M, Mirande M, Hirsch E, Rooney T, Fournier A, Brice A (2003) The p38 subunit of the aminoacyl-tRNA synthetase complex is a Parkin substrate: linking protein biosynthesis and neurodegeneration. Hum Mol Genet 12:1427–1437

Couzin J (2008) Genetics. Once shunned, test for Alzheimer's risk headed to market. Science 319:1022–1023

Cruts M, Van Broeckhoven C (1998) Molecular genetics of Alzheimer's disease. Ann Med 30: 560–565

Darios F, Corti O, Lucking CB, Hampe C, Muriel MP, Abbas N, Gu WJ, Hirsch EC, Rooney T, Ruberg M, Brice A (2003) Parkin prevents mitochondrial swelling and cytochrome c release in mitochondria-dependent cell death. Hum Mol Genet 12:517–526

Das C, Hoang QQ, Kreinbring CA, Luchansky SJ, Meray RK, Ray SS, Lansbury PT, Ringe D, Petsko GA (2006) Structural basis for conformational plasticity of the Parkinson's disease-associated ubiquitin hydrolase UCH-L1. Proc Natl Acad Sci U S A 103:4675–4680

Daw EW, Payami H, Nemens EJ, Nochlin D, Bird TD, Schellenberg GD, Wijsman EM (2000) The number of trait loci in late-onset Alzheimer disease. Am J Hum Genet 66:196–204

De Marco EV, Annesi G, Tarantino P, Rocca FE, Provenzano G, Civitelli D, Ciro Candiano IC, Annesi F, Carrideo S, Condino F, Nicoletti G, Messina D, Novellino F, Morelli M, Quattrone A (2008) Glucocerebrosidase gene mutations are associated with Parkinson's disease in southern Italy. Mov Disord 23:460–463

de Rijk MC, Launer LJ, Berger K, Breteler MM, Dartigues JF, Baldereschi M, Fratiglioni L, Lobo A, Martinez-Lage J, Trenkwalder C, Hofman A (2000) Prevalence of Parkinson's disease in Europe: A collaborative study of population-based cohorts. Neurologic Diseases in the Elderly Research Group. Neurology 54:S21–S23

de Sauvage F, Octave JN (1989) A novel mRNA of the A4 amyloid precursor gene coding for a possibly secreted protein. Science 245:651–653

De Strooper B (2000) Alzheimer's disease. Closing in on gamma-secretase. Nature 405(627):629

De Strooper B, Annaert W, Cupers P, Saftig P, Craessaerts K, Mumm JS, Schroeter EH, Schrijvers V, Wolfe MS, Ray WJ, Goate A, Kopan R (1999) A presenilin-1-dependent gamma-secretase-like protease mediates release of Notch intracellular domain. Nature 398:518–522

De Strooper B, Saftig P, Craessaerts K, Vanderstichele H, Guhde G, Annaert W, Von Figura K, Van Leuven F (1998) Deficiency of presenilin-1 inhibits the normal cleavage of amyloid precursor protein. Nature 391:387–390

Dekker MC, Galjaard RJ, Snijders PJ, Heutink P, Oostra BA, van Duijn CM (2004) Brachydactyly and short stature in a kindred with early-onset parkinsonism. Am J Med Genet A 130A:102–104

Dekker MC, van Swieten JC, Houwing-Duistermaat JJ, Snijders PJ, Boeren E, Hofman A, Breteler MM, Heutink P, Oostra BA, van Duijn CM (2003) A clinical-genetic study of Parkinson's disease in a genetically isolated community. J Neurol 250:1056–1062

Devi L, Prabhu BM, Galati DF, Avadhani NG, Anandatheerthavarada HK (2006) Accumulation of amyloid precursor protein in the mitochondrial import channels of human Alzheimer's disease brain is associated with mitochondrial dysfunction. J Neurosci 26:9057–9068

Di Fonzo A, Chien HF, Socal M, Giraudo S, Tassorelli C, Iliceto G, Fabbrini G, Marconi R, Fincati E, Abbruzzese G, Marini P, Squitieri F, Horstink MW, Montagna P, Libera AD, Stocchi F, Goldwurm S, Ferreira JJ, Meco G, Martignoni E, Lopiano L, Jardim LB, Oostra BA, Barbosa ER, Italian Parkinson Genetics Network, Bonifati V (2007) ATP13A2 missense mutations in juvenile parkinsonism and young onset Parkinson disease. Neurology 68:1557–1562

Di Fonzo A, Wu-Chou YH, Lu CS, van Doeselaar M, Simons EJ, Rohe CF, Chang HC, Chen RS, Weng YH, Vanacore N, Breedveld GJ, Oostra BA, Bonifati V (2006) A common missense variant in the LRRK2 gene, Gly2385Arg, associated with Parkinson's disease risk in Taiwan. Neurogenetics 7:133–138

Dinur T, Osiecki KM, Legler G, Gatt S, Desnick RJ, Grabowski GA (1986) Human acid beta-glucosidase: isolation and amino acid sequence of a peptide containing the catalytic site. Proc Natl Acad Sci U S A 83:1660–1664

Doll RF, Smith FI (1993) Regulation of expression of the gene encoding human acid beta-glucosidase in different cell types. Gene 127:255–260

Eblan MJ, Nguyen J, Ziegler SG, Lwin A, Hanson M, Gallardo M, Weiser R, De Lucca M, Singleton A, Sidransky E (2006) Glucocerebrosidase mutations are also found in subjects with early-onset parkinsonism from Venezuela. Mov Disord 21:282–283

Edbauer D, Winkler E, Regula JT, Pesold B, Steiner H, Haass C (2003) Reconstitution of gamma-secretase activity. Nat Cell Biol 5:486–488

El-Agnaf OM, Jakes R, Curran MD, Middleton D, Ingenito R, Bianchi E, Pessi A, Neill D, Wallace A (1998) Aggregates from mutant and wild-type alpha-synuclein proteins and NAC peptide induce apoptotic cell death in human neuroblastoma cells by formation of beta-sheet and amyloid-like filaments. FEBS Lett 440:71–75

Esler WP, Wolfe MS (2001) A portrait of Alzheimer secretases – new features and familiar faces. Science 293:1449–1454

Faccio L, Fusco C, Chen A, Martinotti S, Bonventre JV, Zervos AS (2000a) Characterization of a novel human serine protease that has extensive homology to bacterial heat shock endoprotease HtrA and is regulated by kidney ischemia. J Biol Chem 275:2581–2588

Faccio L, Fusco C, Viel A, Zervos AS (2000b) Tissue-specific splicing of Omi stress-regulated endoprotease leads to an inactive protease with a modified PDZ motif. Genomics 68:343–347

Farrer MJ (2006) Genetics of Parkinson disease: paradigm shifts and future prospects. Nat Rev Genet 7:306–318

Farrer M, Chan P, Chen R, Tan L, Lincoln S, Hernandez D, Forno L, Gwinn-Hardy K, Petrucelli L, Hussey J, Singleton A, Tanner C, Hardy J, Langston JW (2001) Lewy bodies and parkinsonism in families with parkin mutations. Ann Neurol 50:293–300

Farrer LA, Cupples LA, Haines JL, Hyman B, Kukull WA, Mayeux R, Myers RH, Pericak-Vance MA, Risch N, van Duijn CM (1997) Effects of age, sex, and ethnicity on the association between apolipoprotein E genotype and Alzheimer disease. A meta-analysis. APOE and Alzheimer Disease Meta Analysis Consortium. J Am Med Assoc 278:1349–1356

Farrer LA, Cupples LA, van Duijn CM, Kurz A, Zimmer R, Muller U, Green RC, Clarke V, Shoffner J, Wallace DC et al (1995) Apolipoprotein E genotype in patients with Alzheimer's disease: implications for the risk of dementia among relatives. Ann Neurol 38:797–808

Farrer M, Kachergus J, Forno L, Lincoln S, Wang DS, Hulihan M, Maraganore D, Gwinn-Hardy K, Wszolek Z, Dickson D, Langston JW (2004) Comparison of kindreds with parkinsonism and alpha-synuclein genomic multiplications. Ann Neurol 55:174–179

Ferri CP, Prince M, Brayne C, Brodaty H, Fratiglioni L, Ganguli M, Hall K, Hasegawa K, Hendrie H, Huang Y, Jorm A, Mathers C, Menezes PR, Rimmer E, Scazufca M (2005) Global prevalence of dementia: a Delphi consensus study. Lancet 366:2112–2117

Finch CE, Stanford CB (2004) Meat-adaptive genes and the evolution of slower aging in humans. Q Rev Biol 79:3–50

Fleming SM, Salcedo J, Fernagut PO, Rockenstein E, Masliah E, Levine MS, Chesselet MF (2004) Early and progressive sensorimotor anomalies in mice overexpressing wild-type human alpha-synuclein. J Neurosci 24:9434–9440

Foroud T, Uniacke SK, Liu L, Pankratz N, Rudolph A, Halter C, Shults C, Marder K, Conneally PM, Nichols WC (2003) Heterozygosity for a mutation in the parkin gene leads to later onset Parkinson disease. Neurology 60:796–801

Fortin DL, Troyer MD, Nakamura K, Kubo S, Anthony MD, Edwards RH (2004) Lipid rafts mediate the synaptic localization of alpha-synuclein. J Neurosci 24:6715–6723

Francis R, McGrath G, Zhang J, Ruddy DA, Sym M, Apfeld J, Nicoll M, Maxwell M, Hai B, Ellis MC, Parks AL, Xu W, Li J, Gurney M, Myers RL, Himes CS, Hiebsch R, Ruble C, Nye JS, Curtis D (2002) aph-1 and pen-2 are required for Notch pathway signaling, gamma-secretase cleavage of betaAPP, and presenilin protein accumulation. Dev Cell 3:85–97

Fredenburg RA, Rospigliosi C, Meray RK, Kessler JC, Lashuel HA, Eliezer D, Lansbury PT Jr. (2007) The impact of the E46K mutation on the properties of alpha-synuclein in its monomeric and oligomeric states. Biochemistry 46:7107–7118

Fuchs J, Nilsson C, Kachergus J, Munz M, Larsson EM, Schule B, Langston JW, Middleton FA, Ross OA, Hulihan M, Gasser T, Farrer MJ (2007) Phenotypic variation in a large Swedish pedigree due to SNCA duplication and triplication. Neurology 68:916–922

Funayama M, Hasegawa K, Kowa H, Saito M, Tsuji S, Obata F (2002) A new locus for Parkinson's disease (PARK8) maps to chromosome 12p11.2-q13.1. Ann Neurol 51:296–301

Galpern WR, Singhal AB (2006) Neuroprotection: lessons from a spectrum of neurological disorders. Int J Stroke 1:97–99

Gan-Or Z, Giladi N, Rozovski U, Shifrin C, Rosner S, Gurevich T, Bar-Shira A, Orr-Urtreger A (2008) Genotype-phenotype correlations between GBA mutations and Parkinson disease risk and onset. Neurology 70:2277–2283

Gandhi S, Muqit MM, Stanyer L, Healy DG, Abou-Sleiman PM, Hargreaves I, Heales S, Ganguly M, Parsons L, Lees AJ, Latchman DS, Holton JL, Wood NW, Revesz T (2006) PINK1 protein in normal human brain and Parkinson's disease. Brain 129:1720–1731

Gatz M, Reynolds CA, Fratiglioni L, Johansson B, Mortimer JA, Berg S, Fiske A, Pedersen NL (2006) Role of genes and environments for explaining Alzheimer disease. Arch Gen Psychiatry 63:168–174

George JM (2002) The synucleins. Genome Biol 3:S3002; review

Giaccone G, Tagliavini F, Linoli G, Bouras C, Frigerio L, Frangione B, Bugiani O (1989) Down patients: extracellular preamyloid deposits precede neuritic degeneration and senile plaques. Neurosci Lett 97:232–238

Gibb WR, Lees AJ (1988) A comparison of clinical and pathological features of young- and old-onset Parkinson's disease. Neurology 38:1402–1406

Gibson GE, Haroutunian V, Zhang H, Park LC, Shi Q, Lesser M, Mohs RC, Sheu RK, Blass JP (2000) Mitochondrial damage in Alzheimer's disease varies with apolipoprotein E genotype. Ann Neurol 48:297–303

Gilks WP, Abou-Sleiman PM, Gandhi S, Jain S, Singleton A, Lees AJ, Shaw K, Bhatia KP, Bonifati V, Quinn NP, Lynch J, Healy DG, Holton JL, Revesz T, Wood NW (2005) A common LRRK2 mutation in idiopathic Parkinson's disease. Lancet 365:415–416

Glenner GG, Wong CW, Quaranta V, Eanes ED (1984) The amyloid deposits in Alzheimer's disease: their nature and pathogenesis. Appl Pathol 2:357–369

Gloeckner CJ, Kinkl N, Schumacher A, Braun RJ, O'Neill E, Meitinger T, Kolch W, Prokisch H, Ueffing M (2006) The Parkinson disease causing LRRK2 mutation I2020T is associated with increased kinase activity. Hum Mol Genet 15:223–232

Goate A, Chartier-Harlin MC, Mullan M, Brown J, Crawford F, Fidani L, Giuffra L, Haynes A, Irving N, James L et al (1991) Segregation of a missense mutation in the amyloid precursor protein gene with familial Alzheimer's disease. Nature 349:704–706

Goedert M, Spillantini MG (2006) A century of Alzheimer's disease. Science 314:777–781

Goker-Alpan O, Lopez G, Vithayathil J, Davis J, Hallett M, Sidransky E (2008) The spectrum of parkinsonian manifestations associated with glucocerebrosidase mutations. Arch Neurol 65:1353–1357

Golbe LI (1990) The genetics of Parkinson's disease: a reconsideration. Neurology 40(suppl 7–14):14–16; discussion

Golbe LI, Di Iorio G, Bonavita V, Miller DC, Duvoisin RC (1990) A large kindred with autosomal dominant Parkinson's disease. Ann Neurol 27:276–282

Goldberg MS, Fleming SM, Palacino JJ, Cepeda C, Lam HA, Bhatnagar A, Meloni EG, Wu N, Ackerson LC, Klapstein GJ, Gajendiran M, Roth BL, Chesselet MF, Maidment NT, Levine MS, Shen J (2003) Parkin-deficient mice exhibit nigrostriatal deficits but not loss of dopaminergic neurons. J Biol Chem 278:43628–43635

Goldberg MS, Pisani A, Haburcak M, Vortherms TA, Kitada T, Costa C, Tong Y, Martella G, Tscherter A, Martins A, Bernardi G, Roth BL, Pothos EN, Calabresi P, Shen J (2005) Nigrostriatal dopaminergic deficits and hypokinesia caused by inactivation of the familial Parkinsonism-linked gene DJ-1. Neuron 45:489–496

Golde TE, Estus S, Usiak M, Younkin LH, Younkin SG (1990) Expression of beta amyloid protein precursor mRNAs: recognition of a novel alternatively spliced form and quantitation in Alzheimer's disease using PCR. Neuron 4:253–267

Goldgaber D, Lerman MI, McBride OW, Saffiotti U, Gajdusek DC (1987) Characterization and chromosomal localization of a cDNA encoding brain amyloid of Alzheimer's disease. Science 235:877–880

Goutte C, Tsunozaki M, Hale VA, Priess JR (2002) APH-1 is a multipass membrane protein essential for the Notch signaling pathway in Caenorhabditis elegans embryos. Proc Natl Acad Sci U S A 99:775–779

Greenbaum EA, Graves CL, Mishizen-Eberz AJ, Lupoli MA, Lynch DR, Englander SW, Axelsen PH, Giasson BI (2005) The E46K mutation in alpha-synuclein increases amyloid fibril formation. J Biol Chem 280:7800–7807

Greene JC, Whitworth AJ, Kuo I, Andrews LA, Feany MB, Pallanck LJ (2003) Mitochondrial pathology and apoptotic muscle degeneration in Drosophila parkin mutants. Proc Natl Acad Sci U S A 100:4078–4083

Grunberg J, Walter J, Eckman C, Capell A, Schindzielorz A, Younkin S, Mehta N, Hardy J, Haass C (1998) Truncated presenilin 2 derived from differentially spliced mRNA does not affect the ratio of amyloid beta-peptide 1-42/1-40. NeuroReport 9:3293–3299

Gu WJ, Corti O, Araujo F, Hampe C, Jacquier S, Lucking CB, Abbas N, Duyckaerts C, Rooney T, Pradier L, Ruberg M, Brice A (2003) The C289G and C418R missense mutations cause rapid sequestration of human Parkin into insoluble aggregates. Neurobiol Dis 14:357–364

Haass C, Koo EH, Mellon A, Hung AY, Selkoe DJ (1992) Targeting of cell-surface beta-amyloid precursor protein to lysosomes: alternative processing into amyloid-bearing fragments. Nature 357:500–503

Hague S, Rogaeva E, Hernandez D, Gulick C, Singleton A, Hanson M, Johnson J, Weiser R, Gallardo M, Ravina B, Gwinn-Hardy K, Crawley A, St George-Hyslop PH, Lang AE, Heutink P, Bonifati V, Hardy J (2003) Early-onset Parkinson's disease caused by a compound heterozygous DJ-1 mutation. Ann Neurol 54:271–274

Hampshire DJ, Roberts E, Crow Y, Bond J, Mubaidin A, Wriekat AL, Al-Din A, Woods CG (2001) Kufor-Rakeb syndrome, pallido-pyramidal degeneration with supranuclear upgaze paresis and dementia, maps to 1p36. J Med Genet 38:680–682

Hardy J (2001) The genetic causes of neurodegenerative diseases. J Alzheimers Dis 3:109–116

Harhangi BS, Farrer MJ, Lincoln S, Bonifati V, Meco G, De Michele G, Brice A, Durr A, Martinez M, Gasser T, Bereznai B, Vaughan JR, Wood NW, Hardy J, Oostra BA, Breteler MM (1999) The Ile93Met mutation in the ubiquitin carboxy-terminal-hydrolase-L1 gene is not observed in European cases with familial Parkinson's disease. Neurosci Lett 270:1–4

Hatano T, Kubo S, Imai S, Maeda M, Ishikawa K, Mizuno Y, Hattori N (2007) Leucine-rich repeat kinase 2 associates with lipid rafts. Hum Mol Genet 16:678–690

Hatano Y, Li Y, Sato K, Asakawa S, Yamamura Y, Tomiyama H, Yoshino H, Asahina M, Kobayashi S, Hassin-Baer S, Lu CS, Ng AR, Rosales RL, Shimizu N, Toda T, Mizuno Y, Hattori N (2004a) Novel PINK1 mutations in early-onset parkinsonism. Ann Neurol 56:424–427

Hatano Y, Sato K, Elibol B, Yoshino H, Yamamura Y, Bonifati V, Shinotoh H, Asahina M, Kobayashi S, Ng AR, Rosales RL, Hassin-Baer S, Shinar Y, Lu CS, Chang HC, Wu-Chou YH, Atac FB, Kobayashi T, Toda T, Mizuno Y, Hattori N (2004b) PARK6-linked autosomal recessive early-onset parkinsonism in Asian populations. Neurology 63:1482–1485

Hattori N, Matsumine H, Asakawa S, Kitada T, Yoshino H, Elibol B, Brookes AJ, Yamamura Y, Kobayashi T, Wang M, Yoritaka A, Minoshima S, Shimizu N, Mizuno Y (1998) Point mutations (Thr240Arg and Gln311Stop) [correction of Thr240Arg and Ala311Stop] in the Parkin gene. Biochem Biophys Res Commun 249:754–758

Healy DG, Abou-Sleiman PM, Gibson JM, Ross OA, Jain S, Gandhi S, Gosal D, Muqit MM, Wood NW, Lynch T (2004a) PINK1 (PARK6) associated Parkinson disease in Ireland. Neurology 63:1486–1488

Healy DG, Abou-Sleiman PM, Wood NW (2004b) Genetic causes of Parkinson's disease: UCHL-1. Cell Tissue Res 318:189–194

Heckmann JM, Low WC, de Villiers C, Rutherfoord S, Vorster A, Rao H, Morris CM, Ramesar RS, Kalaria RN (2004) Novel presenilin 1 mutation with profound neurofibrillary pathology in an indigenous Southern African family with early-onset Alzheimer's disease. Brain 127: 133–142

Hedrich K, Eskelson C, Wilmot B, Marder K, Harris J, Garrels J, Meija-Santana H, Vieregge P, Jacobs H, Bressman SB, Lang AE, Kann M, Abbruzzese G, Martinelli P, Schwinger E, Ozelius LJ, Pramstaller PP, Klein C, Kramer P (2004) Distribution, type, and origin of Parkin mutations: review and case studies. Mov Disord 19:1146–1157

Hedrich K, Marder K, Harris J, Kann M, Lynch T, Meija-Santana H, Pramstaller PP, Schwinger E, Bressman SB, Fahn S, Klein C (2002) Evaluation of 50 probands with early-onset Parkinson's disease for Parkin mutations. Neurology 58:1239–1246

Hegde R, Srinivasula SM, Zhang Z, Wassell R, Mukattash R, Cilenti L, DuBois G, Lazebnik Y, Zervos AS, Fernandes-Alnemri T, Alnemri ES (2002) Identification of Omi/HtrA2 as a mitochondrial apoptotic serine protease that disrupts inhibitor of apoptosis protein-caspase interaction. J Biol Chem 277:432–438

Hendrie HC, Hall KS, Hui S, Unverzagt FW, Yu CE, Lahiri DK, Sahota A, Farlow M, Musick B, Class CA et al (1995) Apolipoprotein E genotypes and Alzheimer's disease in a community study of elderly African Americans. Ann Neurol 37:118–120

Hering R, Strauss KM, Tao X, Bauer A, Woitalla D, Mietz EM, Petrovic S, Bauer P, Schaible W, Muller T, Schols L, Klein C, Berg D, Meyer PT, Schulz JB, Wollnik B, Tong L, Kruger R, Riess O (2004) Novel homozygous p.E64D mutation in DJ1 in early onset Parkinson disease (PARK7). Hum Mutat 24:321–329

Hibi K, Liu Q, Beaudry GA, Madden SL, Westra WH, Wehage SL, Yang SC, Heitmiller RF, Bertelsen AH, Sidransky D, Jen J (1998) Serial analysis of gene expression in non-small cell lung cancer. Cancer Res 58:5690–5694

Higgins GA, Jacobsen H (2003) Transgenic mouse models of Alzheimer's disease: phenotype and application. Behav Pharmacol 14:419–438

Hiller A, Hagenah JM, Djarmati A, Hedrich K, Reetz K, Schneider-Gold C, Kress W, Munchau A, Klein C (2007) Phenotypic spectrum of PINK1-associated parkinsonism in 15 mutation carriers from 1 family. Mov Disord 22:145–147

Holcomb L, Gordon MN, McGowan E, Yu X, Benkovic S, Jantzen P, Wright K, Saad I, Mueller R, Morgan D, Sanders S, Zehr C, O'Campo K, Hardy J, Prada CM, Eckman C, Younkin S, Hsiao K, Duff K (1998) Accelerated Alzheimer-type phenotype in transgenic mice carrying both mutant amyloid precursor protein and presenilin 1 transgenes. Nat Med 4:97–100

Horowitz M, Wilder S, Horowitz Z, Reiner O, Gelbart T, Beutler E (1989) The human glucocerebrosidase gene and pseudogene: structure and evolution. Genomics 4:87–96

Horsburgh K, McCulloch J, Nilsen M, Roses AD, Nicoll JA (2000) Increased neuronal damage and apoE immunoreactivity in human apolipoprotein E, E4 isoform-specific, transgenic mice after global cerebral ischaemia. Eur J Neurosci 12:4309–4317

Hruska KS, LaMarca ME, Scott CR, Sidransky E (2008) Gaucher disease: mutation and polymorphism spectrum in the glucocerebrosidase gene (GBA). Hum Mutat 29:567–583

Huang Y (2006) Molecular and cellular mechanisms of apolipoprotein E4 neurotoxicity and potential therapeutic strategies. Curr Opin Drug Discov Devel 9:627–641

Huang XG, Yee BK, Nag S, Chan ST, Tang F (2003) Behavioral and neurochemical characterization of transgenic mice carrying the human presenilin-1 gene with or without the leucine-to-proline mutation at codon 235. Exp Neurol 183:673–681

Hutter CM, Samii A, Factor SA, Nutt JG, Higgins DS, Bird TD, Griffith A, Roberts JW, Leis BC, Montimurro JS, Kay DM, Edwards KL, Payami H, Zabetian CP (2008) Lack of evidence for an association between UCHL1 S18Y and Parkinson's disease. Eur J Neurol 15:134–139

Hutton M, Busfield F, Wragg M, Crook R, Perez-Tur J, Clark RF, Prihar G, Talbot C, Phillips H, Wright K, Baker M, Lendon C, Duff K, Martinez A, Houlden H, Nichols A, Karran E, Roberts

G, Roques P, Rossor M, Venter JC, Adams MD, Cline RT, Phillips CA, Goate A et al (1996) Complete analysis of the presenilin 1 gene in early onset Alzheimer's disease. NeuroReport 7:801–805

Hutton M, Hardy J (1997) The presenilins and Alzheimer's disease. Hum Mol Genet 6:1639–1646

Ibanez P, Bonnet AM, Debarges B, Lohmann E, Tison F, Pollak P, Agid Y, Durr A, Brice A (2004) Causal relation between alpha-synuclein gene duplication and familial Parkinson's disease. Lancet 364:1169–1171

Ibanez P, Lesage S, Lohmann E, Thobois S, De Michele G, Borg M, Agid Y, Durr A, Brice A (2006) Mutational analysis of the PINK1 gene in early-onset parkinsonism in Europe and North Africa. Brain 129:686–694

Ignatius MJ, Gebicke-Harter PJ, Skene JH, Schilling JW, Weisgraber KH, Mahley RW, Shooter EM (1986) Expression of apolipoprotein E during nerve degeneration and regeneration. Proc Natl Acad Sci U S A 83:1125–1129

Irizarry MC, McNamara M, Fedorchak K, Hsiao K, Hyman BT (1997) APPSw transgenic mice develop age-related A beta deposits and neuropil abnormalities, but no neuronal loss in CA1. J Neuropathol Exp Neurol 56:965–973

Irvine GB, El-Agnaf OM, Shankar GM, Walsh DM (2008) Protein aggregation in the brain: the molecular basis for Alzheimer's and Parkinson's diseases. Mol Med 14:451–464

Itier JM, Ibanez P, Mena MA, Abbas N, Cohen-Salmon C, Bohme GA, Laville M, Pratt J, Corti O, Pradier L, Ret G, Joubert C, Periquet M, Araujo F, Negroni J, Casarejos MJ, Canals S, Solano R, Serrano A, Gallego E, Sanchez M, Denefle P, Benavides J, Tremp G, Rooney TA, Brice A, Garcia de Yebenes J (2003) Parkin gene inactivation alters behaviour and dopamine neurotransmission in the mouse. Hum Mol Genet 12:2277–2291

Ito G, Okai T, Fujino G, Takeda K, Ichijo H, Katada T, Iwatsubo T (2007) GTP binding is essential to the protein kinase activity of LRRK2, a causative gene product for familial Parkinson's disease. Biochemistry 46:1380–1388

Iwatsubo T, Odaka A, Suzuki N, Mizusawa H, Nukina N, Ihara Y (1994) Visualization of A beta 42(43) and A beta 40 in senile plaques with end-specific A beta monoclonals: evidence that an initially deposited species is A beta 42(43). Neuron 13:45–53

Jacobsen JS, Muenkel HA, Blume AJ, Vitek MP (1991) A novel species-specific RNA related to alternatively spliced amyloid precursor protein mRNAs. Neurobiol Aging 12: 575–583

Jankovic J (2006) An update on the treatment of Parkinson's disease. Mt Sinai J Med 73:682–689

Janssen JC, Beck JA, Campbell TA, Dickinson A, Fox NC, Harvey RJ, Houlden H, Rossor MN, Collinge J (2003) Early onset familial Alzheimer's disease: Mutation frequency in 31 families. Neurology 60:235–239

Janus C, D'Amelio S, Amitay O, Chishti MA, Strome R, Fraser P, Carlson GA, Roder JC, St George-Hyslop P, Westaway D (2000) Spatial learning in transgenic mice expressing human presenilin 1 (PS1) transgenes. Neurobiol Aging 21:541–549

Johnson SA, Rogers J, Finch CE (1989) APP-695 transcript prevalence is selectively reduced during Alzheimer's disease in cortex and hippocampus but not in cerebellum. Neurobiol Aging 10:755–760

Jones JM, Datta P, Srinivasula SM, Ji W, Gupta S, Zhang Z, Davies E, Hajnoczky G, Saunders TL, Van Keuren ML, Fernandes-Alnemri T, Meisler MH, Alnemri ES (2003) Loss of Omi mitochondrial protease activity causes the neuromuscular disorder of mnd2 mutant mice. Nature 425:721–727

Jones AC, Yamamura Y, Almasy L, Bohlega S, Elibol B, Hubble J, Kuzuhara S, Uchida M, Yanagi T, Weeks DE, Nygaard TG (1998) Autosomal recessive juvenile parkinsonism maps to 6q25.2-q27 in four ethnic groups: detailed genetic mapping of the linked region. Am J Hum Genet 63:80–87

Kabuta T, Setsuie R, Mitsui T, Kinugawa A, Sakurai M, Aoki S, Uchida K, Wada K (2008) Aberrant molecular properties shared by familial Parkinson's disease-associated mutant UCH-L1 and carbonyl-modified UCH-L1. Hum Mol Genet 17:1482–1496

Kamboh MI (2004) Molecular genetics of late-onset Alzheimer's disease. Ann Hum Genet 68: 381–404

Kang J, Lemaire HG, Unterbeck A, Salbaum JM, Masters CL, Grzeschik KH, Multhaup G, Beyreuther K, Muller-Hill B (1987) The precursor of Alzheimer's disease amyloid A4 protein resembles a cell-surface receptor. Nature 325:733–736

Kann M, Jacobs H, Mohrmann K, Schumacher K, Hedrich K, Garrels J, Wiegers K, Schwinger E, Pramstaller PP, Breakefield XO, Ozelius LJ, Vieregge P, Klein C (2002) Role of parkin mutations in 111 community-based patients with early-onset parkinsonism. Ann Neurol 51:621–625

Kawas CH (2003) Clinical practice. Early Alzheimer's disease. N Engl J Med 349:1056–1063

Kaye JA (1998) Diagnostic challenges in dementia. Neurology 51:S45–S52; discussion S65–47

Keller JN, Lauderback CM, Butterfield DA, Kindy MS, Yu J, Markesbery WR (2000) Amyloid beta-peptide effects on synaptosomes from apolipoprotein E-deficient mice. J Neurochem 74:1579–1586

Khan NL, Graham E, Critchley P, Schrag AE, Wood NW, Lees AJ, Bhatia KP, Quinn N (2003) Parkin disease: a phenotypic study of a large case series. Brain 126:1279–1292

Khan NL, Jain S, Lynch JM, Pavese N, Abou-Sleiman P, Holton JL, Healy DG, Gilks WP, Sweeney MG, Ganguly M, Gibbons V, Gandhi S, Vaughan J, Eunson LH, Katzenschlager R, Gayton J, Lennox G, Revesz T, Nicholl D, Bhatia KP, Quinn N, Brooks D, Lees AJ, Davis MB, Piccini P, Singleton AB, Wood NW (2005) Mutations in the gene LRRK2 encoding dardarin (PARK8) cause familial Parkinson's disease: clinical, pathological, olfactory and functional imaging and genetic data. Brain 128:2786–2796

Khan NL, Valente EM, Bentivoglio AR, Wood NW, Albanese A, Brooks DJ, Piccini P (2002) Clinical and subclinical dopaminergic dysfunction in PARK6-linked parkinsonism: an 18F-dopa PET study. Ann Neurol 52:849–853

Kidd M (1963) Paired helical filaments in electron microscopy of Alzheimer's disease. Nature 197:192–193

Kim RH, Smith PD, Aleyasin H, Hayley S, Mount MP, Pownall S, Wakeham A, You-Ten AJ, Kalia SK, Horne P, Westaway D, Lozano AM, Anisman H, Park DS, Mak TW (2005) Hypersensitivity of DJ-1-deficient mice to 1-methyl-4-phenyl-1,2,3,6-tetrahydropyrindine (MPTP) and oxidative stress. Proc Natl Acad Sci U S A 102:5215–5220

Kimberly WT, LaVoie MJ, Ostaszewski BL, Ye W, Wolfe MS, Selkoe DJ (2003) Gamma-secretase is a membrane protein complex comprised of presenilin, nicastrin, Aph-1, and Pen-2. Proc Natl Acad Sci U S A 100:6382–6387

Kitada T, Asakawa S, Hattori N, Matsumine H, Yamamura Y, Minoshima S, Yokochi M, Mizuno Y, Shimizu N (1998) Mutations in the parkin gene cause autosomal recessive juvenile parkinsonism. Nature 392:605–608

Kitaguchi N, Takahashi Y, Tokushima Y, Shiojiri S, Ito H (1988) Novel precursor of Alzheimer's disease amyloid protein shows protease inhibitory activity. Nature 331:530–532

Klein C, Hedrich K, Wellenbrock C, Kann M, Harris J, Marder K, Lang AE, Schwinger E, Ozelius LJ, Vieregge P, Pramstaller PP, Kramer PL (2003) Frequency of parkin mutations in late-onset Parkinson's disease. Ann Neurol 54:415–416; author reply 416–417

Klein C, Pramstaller PP, Kis B, Page CC, Kann M, Leung J, Woodward H, Castellan CC, Scherer M, Vieregge P, Breakefield XO, Kramer PL, Ozelius LJ (2000) Parkin deletions in a family with adult-onset, tremor-dominant parkinsonism: expanding the phenotype. Ann Neurol 48:65–71

Ko HS, von Coelln R, Sriram SR, Kim SW, Chung KK, Pletnikova O, Troncoso J, Johnson B, Saffary R, Goh EL, Song H, Park BJ, Kim MJ, Kim S, Dawson VL, Dawson TM (2005) Accumulation of the authentic parkin substrate aminoacyl-tRNA synthetase cofactor, p38/JTV-1, leads to catecholaminergic cell death. J Neurosci 25:7968–7978

Konig G, Salbaum JM, Wiestler O, Lang W, Schmitt HP, Masters CL, Beyreuther K (1991) Alternative splicing of the beta A4 amyloid gene of Alzheimer's disease in cortex of control and Alzheimer's disease patients. Brain Res Mol Brain Res 9:259–262

Koo EH, Squazzo SL (1994) Evidence that production and release of amyloid beta-protein involves the endocytic pathway. J Biol Chem 269:17386–17389

Kosik KS, Bakalis SF, Selkoe DJ, Pierce MW, Duffy LK (1986) High molecular weight microtubule-associated proteins: purification by electro-elution and amino acid compositions. J Neurosci Res 15:543–551

Kovacs DM, Fausett HJ, Page KJ, Kim TW, Moir RD, Merriam DE, Hollister RD, Hallmark OG, Mancini R, Felsenstein KM, Hyman BT, Tanzi RE, Wasco W (1996) Alzheimer-associated presenilins 1 and 2: neuronal expression in brain and localization to intracellular membranes in mammalian cells. Nat Med 2:224–229

Kruger R, Kuhn W, Muller T, Woitalla D, Graeber M, Kosel S, Przuntek H, Epplen JT, Schols L, Riess O (1998) Ala30Pro mutation in the gene encoding alpha-synuclein in Parkinson's disease. Nat Genet 18:106–108

Kubo SI, Kitami T, Noda S, Shimura H, Uchiyama Y, Asakawa S, Minoshima S, Shimizu N, Mizuno Y, Hattori N (2001) Parkin is associated with cellular vesicles. J Neurochem 78:42–54

Kukull WA, Higdon R, Bowen JD, McCormick WC, Teri L, Schellenberg GD, van Belle G, Jolley L, Larson EB (2002) Dementia and Alzheimer disease incidence: a prospective cohort study. Arch Neurol 59:1737–1746

Kwok JB, Li QX, Hallupp M, Whyte S, Ames D, Beyreuther K, Masters CL, Schofield PR (2000) Novel Leu723Pro amyloid precursor protein mutation increases amyloid beta42(43) peptide levels and induces apoptosis. Ann Neurol 47:249–253

LaDu MJ, Pederson TM, Frail DE, Reardon CA, Getz GS, Falduto MT (1995) Purification of apolipoprotein E attenuates isoform-specific binding to beta-amyloid. J Biol Chem 270: 9039–9042

Lashuel HA, Petre BM, Wall J, Simon M, Nowak RJ, Walz T, Lansbury PT Jr. (2002) Alpha-synuclein, especially the Parkinson's disease-associated mutants, forms pore-like annular and tubular protofibrils. J Mol Biol 322:1089–1102

Lemaire HG, Salbaum JM, Multhaup G, Kang J, Bayney RM, Unterbeck A, Beyreuther K, Muller-Hill B (1989) The PreA4(695) precursor protein of Alzheimer's disease A4 amyloid is encoded by 16 exons. Nucleic Acids Res 17:517–522

Lemere CA, Blusztajn JK, Yamaguchi H, Wisniewski T, Saido TC, Selkoe DJ (1996) Sequence of deposition of heterogeneous amyloid beta-peptides and APO E in Down syndrome: implications for initial events in amyloid plaque formation. Neurobiol Dis 3:16–32

Leroy E, Boyer R, Auburger G, Leube B, Ulm G, Mezey E, Harta G, Brownstein MJ, Jonnalagada S, Chernova T, Dehejia A, Lavedan C, Gasser T, Steinbach PJ, Wilkinson KD, Polymeropoulos MH (1998) The ubiquitin pathway in Parkinson's disease. Nature 395:451–452

Lesage S, Durr A, Tazir M, Lohmann E, Leutenegger AL, Janin S, Pollak P, Brice A (2006) LRRK2 G2019S as a cause of Parkinson's disease in North African Arabs. N Engl J Med 354:422–423

Lesage S, Ibanez P, Lohmann E, Pollak P, Tison F, Tazir M, Leutenegger AL, Guimaraes J, Bonnet AM, Agid Y, Durr A, Brice A (2005) G2019S LRRK2 mutation in French and North African families with Parkinson's disease. Ann Neurol 58:784–787

Leutenegger AL, Salih MA, Ibanez P, Mukhtar MM, Lesage S, Arabi A, Lohmann E, Durr A, Ahmed AE, Brice A (2006) Juvenile-onset Parkinsonism as a result of the first mutation in the adenosine triphosphate orientation domain of PINK1. Arch Neurol 63:1257–1261

Levecque C, Destee A, Mouroux V, Becquet E, Defebvre L, Amouyel P, Chartier-Harlin MC (2001) No genetic association of the ubiquitin carboxy-terminal hydrolase-L1 gene S18Y polymorphism with familial Parkinson's disease. J Neural Transm 108:979–984

Levy E, Carman MD, Fernandez-Madrid IJ, Power MD, Lieberburg I, van Duinen SG, Bots GT, Luyendijk W, Frangione B (1990) Mutation of the Alzheimer's disease amyloid gene in hereditary cerebral hemorrhage, Dutch type. Science 248:1124–1126

Levy-Lahad E, Wijsman EM, Nemens E, Anderson L, Goddard KA, Weber JL, Bird TD, Schellenberg GD (1995) A familial Alzheimer's disease locus on chromosome 1. Science 269:970–973

Li W, Srinivasula SM, Chai J, Li P, Wu JW, Zhang Z, Alnemri ES, Shi Y (2002) Structural insights into the pro-apoptotic function of mitochondrial serine protease HtrA2/Omi. Nat Struct Biol 9:436–441

Li Y, Tomiyama H, Sato K, Hatano Y, Yoshino H, Atsumi M, Kitaguchi M, Sasaki S, Kawaguchi S, Miyajima H, Toda T, Mizuno Y, Hattori N (2005) Clinicogenetic study of PINK1 mutations in autosomal recessive early-onset parkinsonism. Neurology 64:1955–1957

Liddell M, Williams J, Bayer A, Kaiser F, Owen M (1994) Confirmation of association between the e4 allele of apolipoprotein E and Alzheimer's disease. J Med Genet 31:197–200

Lin MT, Beal MF (2006) Alzheimer's APP mangles mitochondria. Nat Med 12:1241–1243

Lin CH, Tan EK, Chen ML, Tan LC, Lim HQ, Chen GS, Wu RM (2008) Novel ATP13A2 variant associated with Parkinson disease in Taiwan and Singapore. Neurology 71:1727–1732

Lindquist SG, Hasholt L, Bahl JM, Heegaard NH, Andersen BB, Norremolle A, Stokholm J, Schwartz M, Batbayli M, Laursen H, Pardossi-Piquard R, Chen F, St George-Hyslop P, Waldemar G, Nielsen JE (2008) A novel presenilin 2 mutation (V393M) in early-onset dementia with profound language impairment. Eur J Neurol 15:1135–1139

Lippa CF, Nee LE, Mori H, St George-Hyslop P (1998) Abeta-42 deposition precedes other changes in PS-1 Alzheimer's disease. Lancet 352:1117–1118

Liu Y, Fallon L, Lashuel HA, Liu Z, Lansbury PT Jr. (2002a) The UCH-L1 gene encodes two opposing enzymatic activities that affect alpha-synuclein degradation and Parkinson's disease susceptibility. Cell 111:209–218

Liu Y, Laakso MP, Karonen JO, Vanninen RL, Nuutinen J, Soimakallio S, Aronen HJ (2002b) Apolipoprotein E polymorphism and acute ischemic stroke: a diffusion- and perfusion-weighted magnetic resonance imaging study. J Cereb Blood Flow Metab 22:1336–1342

Lockhart PJ, Lincoln S, Hulihan M, Kachergus J, Wilkes K, Bisceglio G, Mash DC, Farrer MJ (2004) DJ-1 mutations are a rare cause of recessively inherited early onset parkinsonism mediated by loss of protein function. J Med Genet 41:e22

Lorick KL, Jensen JP, Fang S, Ong AM, Hatakeyama S, Weissman AM (1999) RING fingers mediate ubiquitin-conjugating enzyme (E2)-dependent ubiquitination. Proc Natl Acad Sci U S A 96:11364–11369

Love S (2005) Neuropathological investigation of dementia: a guide for neurologists. J Neurol Neurosurg Psychiatry 76(Suppl 5):v8–v14

Lucking CB, Durr A, Bonifati V, Vaughan J, De Michele G, Gasser T, Harhangi BS, Meco G, Denefle P, Wood NW, Agid Y, Brice A (2000) Association between early-onset Parkinson's disease and mutations in the parkin gene. N Engl J Med 342:1560–1567

Lucotte G, Turpin JC, Landais P (1994) Apolipoprotein E-epsilon 4 allele doses in late-onset Alzheimer's disease. Ann Neurol 36:681–682

Lwin A, Orvisky E, Goker-Alpan O, LaMarca ME, Sidransky E (2004) Glucocerebrosidase mutations in subjects with parkinsonism. Mol Genet Metab 81:70–73

Machida Y, Chiba T, Takayanagi A, Tanaka Y, Asanuma M, Ogawa N, Koyama A, Iwatsubo T, Ito S, Jansen PH, Shimizu N, Tanaka K, Mizuno Y, Hattori N (2005) Common anti-apoptotic roles of parkin and alpha-synuclein in human dopaminergic cells. Biochem Biophys Res Commun 332:233–240

Mahley RW, Huang Y, Rall SC Jr. (1999) Pathogenesis of type III hyperlipoproteinemia (dysbetalipoproteinemia). Questions, quandaries, and paradoxes. J Lipid Res 40:1933–1949

Mahley RW, Rall SC Jr. (1999) Is epsilon4 the ancestral human apoE allele? Neurobiol Aging 20:429–430

Mahley RW, Weisgraber KH, Huang Y (2006) Apolipoprotein E4: a causative factor and therapeutic target in neuropathology, including Alzheimer's disease. Proc Natl Acad Sci U S A 103:5644–5651; Epub 2006 Mar 5627

Malgieri G, Eliezer D (2008) Structural effects of Parkinson's disease linked DJ-1 mutations. Protein Sci 17:855–868

Mann DM, Prinja D, Davies CA, Ihara Y, Delacourte A, Defossez A, Mayer RJ, Landon M (1989) Immunocytochemical profile of neurofibrillary tangles in Down's syndrome patients of different ages. J Neurol Sci 92:247–260

Maraganore DM, de Andrade M, Elbaz A, Farrer MJ, Ioannidis JP, Kruger R, Rocca WA, Schneider NK, Lesnick TG, Lincoln SJ, Hulihan MM, Aasly JO, Ashizawa T, Chartier-Harlin MC, Checkoway H, Ferrarese C, Hadjigeorgiou G, Hattori N, Kawakami H, Lambert JC, Lynch T, Mellick GD, Papapetropoulos S, Parsian A, Quattrone A, Riess O, Tan EK, Van Broeckhoven C (2006) Collaborative analysis of alpha-synuclein gene promoter variability and Parkinson disease. J Am Med Assoc 296:661–670

Maraganore DM, Farrer MJ, Hardy JA, Lincoln SJ, McDonnell SK, Rocca WA (1999) Case-control study of the ubiquitin carboxy-terminal hydrolase L1 gene in Parkinson's disease. Neurology 53:1858–1860

Maraganore DM, Lesnick TG, Elbaz A, Chartier-Harlin MC, Gasser T, Kruger R, Hattori N, Mellick GD, Quattrone A, Satoh J, Toda T, Wang J, Ioannidis JP, de Andrade M, Rocca WA (2004) UCHL1 is a Parkinson's disease susceptibility gene. Ann Neurol 55:512–521

Martins LM, Iaccarino I, Tenev T, Gschmeissner S, Totty NF, Lemoine NR, Savopoulos J, Gray CW, Creasy CL, Dingwall C, Downward J (2002) The serine protease Omi/HtrA2 regulates apoptosis by binding XIAP through a reaper-like motif. J Biol Chem 277:439–444

Martins LM, Morrison A, Klupsch K, Fedele V, Moisoi N, Teismann P, Abuin A, Grau E, Geppert M, Livi GP, Creasy CL, Martin A, Hargreaves I, Heales SJ, Okada H, Brandner S, Schulz JB, Mak T, Downward J (2004) Neuroprotective role of the Reaper-related serine protease HtrA2/Omi revealed by targeted deletion in mice. Mol Cell Biol 24:9848–9862

Martins CA, Oulhaj A, de Jager CA, Williams JH (2005) APOE alleles predict the rate of cognitive decline in Alzheimer disease: a nonlinear model. Neurology 65:1888–1893

Masliah E, Rockenstein E, Veinbergs I, Mallory M, Hashimoto M, Takeda A, Sagara Y, Sisk A, Mucke L (2000) Dopaminergic loss and inclusion body formation in alpha-synuclein mice: implications for neurodegenerative disorders. Science 287:1265–1269

Mata IF, Lockhart PJ, Farrer MJ (2004) Parkin genetics: one model for Parkinson's disease. Hum Mol Genet 13(Spec No 1):R127–R133

Mata IF, Ross OA, Kachergus J, Huerta C, Ribacoba R, Moris G, Blazquez M, Guisasola LM, Salvador C, Martinez C, Farrer M, Alvarez V (2006) LRRK2 mutations are a common cause of Parkinson's disease in Spain. Eur J Neurol 13:391–394

Mata IF, Samii A, Schneer SH, Roberts JW, Griffith A, Leis BC, Schellenberg GD, Sidransky E, Bird TD, Leverenz JB, Tsuang D, Zabetian CP (2008) Glucocerebrosidase gene mutations: a risk factor for Lewy body disorders. Arch Neurol 65:379–382

Matsuda N, Kitami T, Suzuki T, Mizuno Y, Hattori N, Tanaka K (2006) Diverse effects of pathogenic mutations of Parkin that catalyze multiple monoubiquitylation in vitro. J Biol Chem 281:3204–3209

Matsumine H, Saito M, Shimoda-Matsubayashi S, Tanaka H, Ishikawa A, Nakagawa-Hattori Y, Yokochi M, Kobayashi T, Igarashi S, Takano H, Sanpei K, Koike R, Mori H, Kondo T, Mizutani Y, Schaffer AA, Yamamura Y, Nakamura S, Kuzuhara S, Tsuji S, Mizuno Y (1997) Localization of a gene for an autosomal recessive form of juvenile Parkinsonism to chromosome 6q25.2-27. Am J Hum Genet 60:588–596

Mayeux R, Stern Y, Ottman R, Tatemichi TK, Tang MX, Maestre G, Ngai C, Tycko B, Ginsberg H (1993) The apolipoprotein epsilon 4 allele in patients with Alzheimer's disease. Ann Neurol 34:752–754

Mellick GD, Silburn PA (2000) The ubiquitin carboxy-terminal hydrolase-L1 gene S18Y polymorphism does not confer protection against idiopathic Parkinson's disease. Neurosci Lett 293:127–130

Melrose HL, Lincoln SJ, Tyndall GM, Farrer MJ (2006) Parkinson's disease: a rethink of rodent models. Exp Brain Res 173:196–204

Meulener MC, Graves CL, Sampathu DM, Armstrong-Gold CE, Bonini NM, Giasson BI (2005) DJ-1 is present in a large molecular complex in human brain tissue and interacts with alpha-synuclein. J Neurochem 93:1524–1532

Michikawa M, Fan QW, Isobe I, Yanagisawa K (2000) Apolipoprotein E exhibits isoform-specific promotion of lipid efflux from astrocytes and neurons in culture. J Neurochem 74:1008–1016

Miller DW, Ahmad R, Hague S, Baptista MJ, Canet-Aviles R, McLendon C, Carter DM, Zhu PP, Stadler J, Chandran J, Klinefelter GR, Blackstone C, Cookson MR (2003) L166P mutant DJ-1, causative for recessive Parkinson's disease, is degraded through the ubiquitin-proteasome system. J Biol Chem 278:36588–36595

Miller DW, Hague SM, Clarimon J, Baptista M, Gwinn-Hardy K, Cookson MR, Singleton AB (2004) Alpha-synuclein in blood and brain from familial Parkinson disease with SNCA locus triplication. Neurology 62:1835–1838

Mineur YS, McLoughlin D, Crusio WE, Sluyter F, Huynh LX (2005) Genetic mouse models of Alzheimer's disease Social behavior deficits in the Fmr 1 mutant mouse Genetic dissection of learning and memory in mice. Neural Plast 12:299–310

Mitsumoto A, Nakagawa Y (2001) DJ-1 is an indicator for endogenous reactive oxygen species elicited by endotoxin. Free Radic Res 35:885–893

Mitsumoto A, Nakagawa Y, Takeuchi A, Okawa K, Iwamatsu A, Takanezawa Y (2001) Oxidized forms of peroxiredoxins and DJ-1 on two-dimensional gels increased in response to sublethal levels of paraquat. Free Radic Res 35:301–310

Mizuno Y, Hattori N, Kubo S, Sato S, Nishioka K, Hatano T, Tomiyama H, Funayama M, Machida Y, Mochizuki H (2008) Progress in the pathogenesis and genetics of Parkinson's disease. Philos Trans R Soc Lond B Biol Sci 363:2215–2227

Mizuta I, Satake W, Nakabayashi Y, Ito C, Suzuki S, Momose Y, Nagai Y, Oka A, Inoko H, Fukae J, Saito Y, Sawabe M, Murayama S, Yamamoto M, Hattori N, Murata M, Toda T (2006) Multiple candidate gene analysis identifies alpha-synuclein as a susceptibility gene for sporadic Parkinson's disease. Hum Mol Genet 15:1151–1158

Moehlmann T, Winkler E, Xia X, Edbauer D, Murrell J, Capell A, Kaether C, Zheng H, Ghetti B, Haass C, Steiner H (2002) Presenilin-1 mutations of leucine 166 equally affect the generation of the Notch and APP intracellular domains independent of their effect on Abeta 42 production. Proc Natl Acad Sci U S A 99:8025–8030

Moisoi N, Klupsch K, Fedele V, East P, Sharma S, Renton A, Plun-Favreau H, Edwards RE, Teismann P, Esposti MD, Morrison AD, Wood NW, Downward J, Martins LM (2008) Mitochondrial dysfunction triggered by loss of HtrA2 results in the activation of a brain-specific transcriptional stress response. Cell Death Differ 16(3):449–464

Moore DJ, Zhang L, Dawson TM, Dawson VL (2003) A missense mutation (L166P) in DJ-1, linked to familial Parkinson's disease, confers reduced protein stability and impairs homo-oligomerization. J Neurochem 87:1558–1567

Moore DJ, Zhang L, Troncoso J, Lee MK, Hattori N, Mizuno Y, Dawson TM, Dawson VL (2005) Association of DJ-1 and parkin mediated by pathogenic DJ-1 mutations and oxidative stress. Hum Mol Genet 14:71–84

Mori H, Kondo T, Yokochi M, Matsumine H, Nakagawa-Hattori Y, Miyake T, Suda K, Mizuno Y (1998) Pathologic and biochemical studies of juvenile parkinsonism linked to chromosome 6q. Neurology 51:890–892

Mouatt-Prigent A, Muriel MP, Gu WJ, El Hachimi KH, Lucking CB, Brice A, Hirsch EC (2004) Ultrastructural localization of parkin in the rat brainstem, thalamus and basal ganglia. J Neural Transm 111:1209–1218

Mueller JC, Fuchs J, Hofer A, Zimprich A, Lichtner P, Illig T, Berg D, Wullner U, Meitinger T, Gasser T (2005) Multiple regions of alpha-synuclein are associated with Parkinson's disease. Ann Neurol 57:535–541

Mullan M (1992) Familial Alzheimer's disease: second gene locus located. BMJ 305:1108–1109

Munhoz RP, Sa DS, Rogaeva E, Salehi-Rad S, Sato C, Medeiros H, Farrer M, Lang AE (2004) Clinical findings in a large family with a parkin ex3delta40 mutation. Arch Neurol 61:701–704

Munoz E, Tolosa E, Pastor P, Marti MJ, Valldeoriola F, Campdelacreu J, Oliva R (2002) Relative high frequency of the c.255delA parkin gene mutation in Spanish patients with autosomal recessive parkinsonism. J Neurol Neurosurg Psychiatry 73:582–584

Murrell J, Farlow M, Ghetti B, Benson MD (1991) A mutation in the amyloid precursor protein associated with hereditary Alzheimer's disease. Science 254:97–99

Myhre R, Steinkjer S, Stormyr A, Nilsen GL, Zayyad HA, Horany K, Nusier MK, Klungland H (2008) Significance of the parkin and PINK1 gene in Jordanian families with incidences of young-onset and juvenile parkinsonism. BMC Neurol 8:47

Nagakubo D, Taira T, Kitaura H, Ikeda M, Tamai K, Iguchi-Ariga SM, Ariga H (1997) DJ-1, a novel oncogene which transforms mouse NIH3T3 cells in cooperation with ras. Biochem Biophys Res Commun 231:509–513

Nagy Z, Esiri MM, Jobst KA, Johnston C, Litchfield S, Sim E, Smith AD (1995) Influence of the apolipoprotein E genotype on amyloid deposition and neurofibrillary tangle formation in Alzheimer's disease. Neuroscience 69:757–761

Naito S, Mochizuki H, Yasuda T, Mizuno Y, Furusaka M, Ikeda S, Adachi T, Shimizu HM, Suzuki J, Fujiwara S, Okada T, Nishikawa K, Aoki S, Wada K (2006) Characterization of multimetric variants of ubiquitin carboxyl-terminal hydrolase L1 in water by small-angle neutron scattering. Biochem Biophys Res Commun 339:717–725

Najim al-Din AS, Wriekat A, Mubaidin A, Dasouki M, Hiari M (1994) Pallido-pyramidal degeneration, supranuclear upgaze paresis and dementia: Kufor-Rakeb syndrome. Acta Neurol Scand 89:347–352

Neudorfer O, Giladi N, Elstein D, Abrahamov A, Turezkite T, Aghai E, Reches A, Bembi B, Zimran A (1996) Occurrence of Parkinson's syndrome in type I Gaucher disease. QJM 89: 691–694

Neve RL, Finch EA, Dawes LR (1988) Expression of the Alzheimer amyloid precursor gene transcripts in the human brain. Neuron 1:669–677

Nicholl DJ, Vaughan JR, Khan NL, Ho SL, Aldous DE, Lincoln S, Farrer M, Gayton JD, Davis MB, Piccini P, Daniel SE, Lennox GG, Brooks DJ, Williams AC, Wood NW (2002) Two large British kindreds with familial Parkinson's disease: a clinico-pathological and genetic study. Brain 125:44–57

Nichols WC, Pankratz N, Marek DK, Pauciulo MW, Elsaesser VE, Halter CA, Rudolph A, Wojcieszek J, Pfeiffer RF, Foroud T (2009) Mutations in GBA are associated with familial Parkinson disease susceptibility and age at onset. Neurology 72:310–316

Nicoll JA, Roberts GW, Graham DI (1995) Apolipoprotein E epsilon 4 allele is associated with deposition of amyloid beta-protein following head injury. Nat Med 1: 135–137

Nishikawa K, Li H, Kawamura R, Osaka H, Wang YL, Hara Y, Hirokawa T, Manago Y, Amano T, Noda M, Aoki S, Wada K (2003) Alterations of structure and hydrolase activity of parkinsonism-associated human ubiquitin carboxyl-terminal hydrolase L1 variants. Biochem Biophys Res Commun 304:176–183

Nishioka K, Hayashi S, Farrer MJ, Singleton AB, Yoshino H, Imai H, Kitami T, Sato K, Kuroda R, Tomiyama H, Mizoguchi K, Murata M, Toda T, Imoto I, Inazawa J, Mizuno Y, Hattori N (2006) Clinical heterogeneity of alpha-synuclein gene duplication in Parkinson's disease. Ann Neurol 59:298–309

Nukada H, Kowa H, Saitoh T, Tazaki Y, Miura S (1978) [A big family of paralysis agitans (author's transl)]. Rinsho Shinkeigaku 18:627–634

Nussbaum RL, Ellis CE (2003) Alzheimer's disease and Parkinson's disease. N Engl J Med 348:1356–1364

Oliveira SA, Scott WK, Martin ER, Nance MA, Watts RL, Hubble JP, Koller WC, Pahwa R, Stern MB, Hiner BC, Ondo WG, Allen FH Jr., Scott BL, Goetz CG, Small GW, Mastaglia F, Stajich JM, Zhang F, Booze MW, Winn MP, Middleton LT, Haines JL, Pericak-Vance MA, Vance JM (2003) Parkin mutations and susceptibility alleles in late-onset Parkinson's disease. Ann Neurol 53:624–629

Oliveri RL, Zappia M, Annesi G, Annesi F, Spadafora P, Pasqua AA, Tomaino C, Nicoletti G, Bosco D, Messina D, Logroscino G, Manobianca G, Epifanio A, Morgante L, Savettieri G, Quattrone A (2001) The parkin gene is not a major susceptibility locus for typical late-onset Parkinson's disease. Neurol Sci 22:73–74

Olzmann JA, Brown K, Wilkinson KD, Rees HD, Huai Q, Ke H, Levey AI, Li L, Chin LS (2004) Familial Parkinson's disease-associated L166P mutation disrupts DJ-1 protein folding and function. J Biol Chem 279:8506–8515

Osaka H, Wang YL, Takada K, Takizawa S, Setsuie R, Li H, Sato Y, Nishikawa K, Sun YJ, Sakurai M, Harada T, Hara Y, Kimura I, Chiba S, Namikawa K, Kiyama H, Noda M, Aoki S, Wada K (2003) Ubiquitin carboxy-terminal hydrolase L1 binds to and stabilizes monoubiquitin in neuron. Hum Mol Genet 12:1945–1958

Paisan-Ruiz C, Jain S, Evans EW, Gilks WP, Simon J, van der Brug M, Lopez de Munain A, Aparicio S, Gil AM, Khan N, Johnson J, Martinez JR, Nicholl D, Carrera IM, Pena AS, de Silva R, Lees A, Marti-Masso JF, Perez-Tur J, Wood NW, Singleton AB (2004) Cloning of the gene containing mutations that cause PARK8-linked Parkinson's disease. Neuron 44: 595–600

Palacino JJ, Sagi D, Goldberg MS, Krauss S, Motz C, Wacker M, Klose J, Shen J (2004) Mitochondrial dysfunction and oxidative damage in parkin-deficient mice. J Biol Chem 279:18614–18622

Palmert MR, Golde TE, Cohen ML, Kovacs DM, Tanzi RE, Gusella JF, Usiak MF, Younkin LH, Younkin SG (1988) Amyloid protein precursor messenger RNAs: differential expression in Alzheimer's disease. Science 241:1080–1084

Pals P, Lincoln S, Manning J, Heckman M, Skipper L, Hulihan M, Van den Broeck M, De Pooter T, Cras P, Crook J, Van Broeckhoven C, Farrer MJ (2004) alpha-Synuclein promoter confers susceptibility to Parkinson's disease. Ann Neurol 56:591–595

Pandey N, Schmidt RE, Galvin JE (2006) The alpha-synuclein mutation E46K promotes aggregation in cultured cells. Exp Neurol 197:515–520

Periquet M, Latouche M, Lohmann E, Rawal N, De Michele G, Ricard S, Teive H, Fraix V, Vidailhet M, Nicholl D, Barone P, Wood NW, Raskin S, Deleuze JF, Agid Y, Durr A, Brice A (2003) Parkin mutations are frequent in patients with isolated early-onset parkinsonism. Brain 126:1271–1278

Pesah Y, Pham T, Burgess H, Middlebrooks B, Verstreken P, Zhou Y, Harding M, Bellen H, Mardon G (2004) Drosophila parkin mutants have decreased mass and cell size and increased sensitivity to oxygen radical stress. Development 131:2183–2194

Petersen RC, Doody R, Kurz A, Mohs RC, Morris JC, Rabins PV, Ritchie K, Rossor M, Thal L, Winblad B (2001) Current concepts in mild cognitive impairment. Arch Neurol 58:1985–1992

Petit A, Kawarai T, Paitel E, Sanjo N, Maj M, Scheid M, Chen F, Gu Y, Hasegawa H, Salehi-Rad S, Wang L, Rogaeva E, Fraser P, Robinson B, St George-Hyslop P, Tandon A (2005) Wild-type PINK1 prevents basal and induced neuronal apoptosis, a protective effect abrogated by Parkinson disease-related mutations. J Biol Chem 280:34025–34032

Pitas RE, Boyles JK, Lee SH, Foss D, Mahley RW (1987) Astrocytes synthesize apolipoprotein E and metabolize apolipoprotein E-containing lipoproteins. Biochim Biophys Acta 917:148–161

Poirier J, Davignon J, Bouthillier D, Kogan S, Bertrand P, Gauthier S (1993) Apolipoprotein E polymorphism and Alzheimer's disease. Lancet 342:697–699

Polymeropoulos MH, Higgins JJ, Golbe LI, Johnson WG, Ide SE, Di Iorio G, Sanges G, Stenroos ES, Pho LT, Schaffer AA, Lazzarini AM, Nussbaum RL, Duvoisin RC (1996) Mapping of a gene for Parkinson's disease to chromosome 4q21-q23. Science 274:1197–1199

Ponte P, Gonzalez-DeWhitt P, Schilling J, Miller J, Hsu D, Greenberg B, Davis K, Wallace W, Lieberburg I, Fuller F (1988) A new A4 amyloid mRNA contains a domain homologous to serine proteinase inhibitors. Nature 331:525–527

Pramstaller PP, Schlossmacher MG, Jacques TS, Scaravilli F, Eskelson C, Pepivani I, Hedrich K, Adel S, Gonzales-McNeal M, Hilker R, Kramer PL, Klein C (2005) Lewy body Parkinson's disease in a large pedigree with 77 Parkin mutation carriers. Ann Neurol 58:411–422

Prihar G, Fuldner RA, Perez-Tur J, Lincoln S, Duff K, Crook R, Hardy J, Philips CA, Venter C, Talbot C, Clark RF, Goate A, Li J, Potter H, Karran E, Roberts GW, Hutton M, Adams MD (1996) Structure and alternative splicing of the presenilin-2 gene. NeuroReport 7: 1680–1684

Priller C, Bauer T, Mitteregger G, Krebs B, Kretzschmar HA, Herms J (2006) Synapse formation and function is modulated by the amyloid precursor protein. J Neurosci 26:7212–7221

Qi-Takahara Y, Morishima-Kawashima M, Tanimura Y, Dolios G, Hirotani N, Horikoshi Y, Kametani F, Maeda M, Saido TC, Wang R, Ihara Y (2005) Longer forms of amyloid beta protein: implications for the mechanism of intramembrane cleavage by gamma-secretase. J Neurosci 25:436–445

Raina P, Santaguida P, Ismaila A, Patterson C, Cowan D, Levine M, Booker L, Oremus M (2008) Effectiveness of cholinesterase inhibitors and memantine for treating dementia: evidence review for a clinical practice guideline. Ann Intern Med 148:379–397

Ramirez A, Heimbach A, Grundemann J, Stiller B, Hampshire D, Cid LP, Goebel I, Mubaidin AF, Wriekat AL, Roeper J, Al-Din A, Hillmer AM, Karsak M, Liss B, Woods CG, Behrens MI, Kubisch C (2006) Hereditary parkinsonism with dementia is caused by mutations in ATP13A2, encoding a lysosomal type 5 P-type ATPase. Nat Genet 38:1184–1191

Rapp A, Gmeiner B, Huttinger M (2006) Implication of apoE isoforms in cholesterol metabolism by primary rat hippocampal neurons and astrocytes. Biochimie 88:473–483

Raschetti R, Albanese E, Vanacore N, Maggini M (2007) Cholinesterase inhibitors in mild cognitive impairment: a systematic review of randomised trials. PLoS Med 4:e338

Raux G, Guyant-Marechal L, Martin C, Bou J, Penet C, Brice A, Hannequin D, Frebourg T, Campion D (2005) Molecular diagnosis of autosomal dominant early onset Alzheimer's disease: an update. J Med Genet 42:793–795

Reiner O, Horowitz M (1988) Differential expression of the human glucocerebrosidase-coding gene. Gene 73:469–478

Reiner O, Wigderson M, Horowitz M (1988) Structural analysis of the human glucocerebrosidase genes. DNA 7:107–116

Reiner O, Wilder S, Givol D, Horowitz M (1987) Efficient in vitro and in vivo expression of human glucocerebrosidase cDNA. DNA 6:101–108

Ren Y, Zhao J, Feng J (2003) Parkin binds to alpha/beta tubulin and increases their ubiquitination and degradation. J Neurosci 23:3316–3324

Rockenstein E, Mallory M, Hashimoto M, Song D, Shults CW, Lang I, Masliah E (2002) Differential neuropathological alterations in transgenic mice expressing alpha-synuclein from the platelet-derived growth factor and Thy-1 promoters. J Neurosci Res 68:568–578

Rogaev EI, Sherrington R, Rogaeva EA, Levesque G, Ikeda M, Liang Y, Chi H, Lin C, Holman K, Tsuda T et al (1995) Familial Alzheimer's disease in kindreds with missense mutations in a gene on chromosome 1 related to the Alzheimer's disease type 3 gene. Nature 376:775–778

Rogan S, Lippa CF (2002) Alzheimer's disease and other dementias: a review. Am J Alzheimers Dis Other Demen 17:11–17

Rohe CF, Montagna P, Breedveld G, Cortelli P, Oostra BA, Bonifati V (2004) Homozygous PINK1 C-terminus mutation causing early-onset parkinsonism. Ann Neurol 56:427–431

Roses AD (2006) On the discovery of the genetic association of Apolipoprotein E genotypes and common late-onset Alzheimer disease. J Alzheimers Dis 9:361–366

Roses AD, Saunders AM, Alberts MA, Strittmatter WJ, Schmechel D, Gorder E, Pericak-Vance MA (1995) Apolipoprotein E E4 allele and risk of dementia. J Am Med Assoc 273:374–375; author reply 375–376

Rudzinski LA, Fletcher RM, Dickson DW, Crook R, Hutton ML, Adamson J, Graff-Radford NR (2008) Early onset familial Alzheimer Disease with spastic paraparesis, dysarthria, and seizures and N135S mutation in PSEN1. Alzheimer Dis Assoc Disord 22:299–307

Saigoh K, Wang YL, Suh JG, Yamanishi T, Sakai Y, Kiyosawa H, Harada T, Ichihara N, Wakana S, Kikuchi T, Wada K (1999) Intragenic deletion in the gene encoding ubiquitin carboxy-terminal hydrolase in gad mice. Nat Genet 23:47–51

Sandbrink R, Masters CL, Beyreuther K (1994) Similar alternative splicing of a non-homologous domain in beta A4-amyloid protein precursor-like proteins. J Biol Chem 269:14227–14234

Sasaki H, Yukiue H, Moriyama S, Kobayashi Y, Nakashima Y, Kaji M, Fukai I, Kiriyama M, Yamakawa Y, Fujii Y (2001) Expression of the protein gene product 9.5, PGP9.5, is correlated with T-status in non-small cell lung cancer. Jpn J Clin Oncol 31:532–535

Sastre M, Steiner H, Fuchs K, Capell A, Multhaup G, Condron MM, Teplow DB, Haass C (2001) Presenilin-dependent gamma-secretase processing of beta-amyloid precursor protein at a site corresponding to the S3 cleavage of Notch. EMBO Rep 2:835–841

Sato N, Imaizumi K, Manabe T, Taniguchi M, Hitomi J, Katayama T, Yoneda T, Morihara T, Yasuda Y, Takagi T, Kudo T, Tsuda T, Itoyama Y, Makifuchi T, Fraser PE, St George-Hyslop P, Tohyama M (2001) Increased production of beta-amyloid and vulnerability to endoplasmic reticulum stress by an aberrant spliced form of presenilin 2. J Biol Chem 276:2108–2114

Sato C, Morgan A, Lang AE, Salehi-Rad S, Kawarai T, Meng Y, Ray PN, Farrer LA, St George-Hyslop P, Rogaeva E (2005) Analysis of the glucocerebrosidase gene in Parkinson's disease. Mov Disord 20:367–370

Schlehe JS, Lutz AK, Pilsl A, Lämmermann K, Grgur K, Henn IH, Tatzelt J, Winklhofer KF (2008) Aberrant folding of pathogenic Parkin mutants: aggregation versus degradation. J Biol Chem 283:13771–13779

Schellenberg GD (1995) Genetic dissection of Alzheimer disease, a heterogeneous disorder. Proc Natl Acad Sci U S A 92:8552–8559

Schellenberg GD, Bird TD, Wijsman EM, Orr HT, Anderson L, Nemens E, White JA, Bonnycastle L, Weber JL, Alonso ME et al (1992) Genetic linkage evidence for a familial Alzheimer's disease locus on chromosome 14. Science 258:668–671

Scheuner D, Eckman C, Jensen M, Song X, Citron M, Suzuki N, Bird TD, Hardy J, Hutton M, Kukull W, Larson E, Levy-Lahad E, Viitanen M, Peskind E, Poorkaj P, Schellenberg G, Tanzi R, Wasco W, Lannfelt L, Selkoe D, Younkin S (1996) Secreted amyloid beta-protein similar to that in the senile plaques of Alzheimer's disease is increased in vivo by the presenilin 1 and 2 and APP mutations linked to familial Alzheimer's disease. Nat Med 2:864–870

Schlossmacher MG, Frosch MP, Gai WP, Medina M, Sharma N, Forno L, Ochiishi T, Shimura H, Sharon R, Hattori N, Langston JW, Mizuno Y, Hyman BT, Selkoe DJ, Kosik KS (2002) Parkin localizes to the Lewy bodies of Parkinson disease and dementia with Lewy bodies. Am J Pathol 160:1655–1667

Schroeter EH, Ilagan MX, Brunkan AL, Hecimovic S, Li YM, Xu M, Lewis HD, Saxena MT, De Strooper B, Coonrod A, Tomita T, Iwatsubo T, Moore CL, Goate A, Wolfe MS, Shearman M, Kopan R (2003) A presenilin dimer at the core of the gamma-secretase enzyme: insights from parallel analysis of Notch 1 and APP proteolysis. Proc Natl Acad Sci U S A 100:13075–13080

Schupf N, Sergievsky GH (2002) Genetic and host factors for dementia in Down's syndrome. Br J Psychiatry 180:405–410

Selkoe DJ (2001) Alzheimer's disease: genes, proteins, and therapy. Physiol Rev 81:741–766

Serneels L, Dejaegere T, Craessaerts K, Horre K, Jorissen E, Tousseyn T, Hebert S, Coolen M, Martens G, Zwijsen A, Annaert W, Hartmann D, De Strooper B (2005) Differential contribution of the three Aph1 genes to gamma-secretase activity in vivo. Proc Natl Acad Sci U S A 102:1719–1724

Serretti A, Artioli P, Quartesan R, De Ronchi D (2005) Genes involved in Alzheimer's disease, a survey of possible candidates. J Alzheimers Dis 7:331–353

Setsuie R, Wada K (2007) The functions of UCH-L1 and its relation to neurodegenerative diseases. Neurochem Int 51:105–111

Setsuie R, Wang YL, Mochizuki H, Osaka H, Hayakawa H, Ichihara N, Li H, Furuta A, Sano Y, Sun YJ, Kwon J, Kabuta T, Yoshimi K, Aoki S, Mizuno Y, Noda M, Wada K (2007) Dopaminergic neuronal loss in transgenic mice expressing the Parkinson's disease-associated UCH-L1 I93M mutant. Neurochem Int 50:119–129

Seubert P, Vigo-Pelfrey C, Esch F, Lee M, Dovey H, Davis D, Sinha S, Schlossmacher M, Whaley J, Swindlehurst C et al (1992) Isolation and quantification of soluble Alzheimer's beta-peptide from biological fluids. Nature 359:325–327

Shen J, Bronson RT, Chen DF, Xia W, Selkoe DJ, Tonegawa S (1997) Skeletal and CNS defects in Presenilin-1-deficient mice. Cell 89:629–639

Sherrington R, Froelich S, Sorbi S, Campion D, Chi H, Rogaeva EA, Levesque G, Rogaev EI, Lin C, Liang Y, Ikeda M, Mar L, Brice A, Agid Y, Percy ME, Clerget-Darpoux F, Piacentini

S, Marcon G, Nacmias B, Amaducci L, Frebourg T, Lannfelt L, Rommens JM, George-Hyslop PH St (1996) Alzheimer's disease associated with mutations in presenilin 2 is rare and variably penetrant. Hum Mol Genet 5:985–988

Sherrington R, Rogaev EI, Liang Y, Rogaeva EA, Levesque G, Ikeda M, Chi H, Lin C, Li G, Holman K et al (1995) Cloning of a gene bearing missense mutations in early-onset familial Alzheimer's disease. Nature 375:754–760

Shimura H, Hattori N, Kubo S, Mizuno Y, Asakawa S, Minoshima S, Shimizu N, Iwai K, Chiba T, Tanaka K, Suzuki T (2000) Familial Parkinson disease gene product, parkin, is a ubiquitin-protein ligase. Nat Genet 25:302–305

Shimura H, Schlossmacher MG, Hattori N, Frosch MP, Trockenbacher A, Schneider R, Mizuno Y, Kosik KS, Selkoe DJ (2001) Ubiquitination of a new form of alpha-synuclein by parkin from human brain: implications for Parkinson's disease. Science 293:263–269

Shoji M, Golde TE, Ghiso J, Cheung TT, Estus S, Shaffer LM, Cai XD, McKay DM, Tintner R, Frangione B et al (1992) Production of the Alzheimer amyloid beta protein by normal proteolytic processing. Science 258:126–129

Silveira-Moriyama L, Guedes LC, Kingsbury A, Ayling H, Shaw K, Barbosa ER, Bonifati V, Quinn NP, Abou-Sleiman P, Wood NW, Petrie A, Sampaio C, Ferreira JJ, Holton J, Revesz T, Lees AJ (2008) Hyposmia in G2019S LRRK2-related parkinsonism: clinical and pathologic data. Neurology 71:1021–1026

Silvestri L, Caputo V, Bellacchio E, Atorino L, Dallapiccola B, Valente EM, Casari G (2005) Mitochondrial import and enzymatic activity of PINK1 mutants associated to recessive parkinsonism. Hum Mol Genet 14:3477–3492

Sing CF, Davignon J (1985) Role of the apolipoprotein E polymorphism in determining normal plasma lipid and lipoprotein variation. Am J Hum Genet 37:268–285

Singleton AB, Farrer M, Johnson J, Singleton A, Hague S, Kachergus J, Hulihan M, Peuralinna T, Dutra A, Nussbaum R, Lincoln S, Crawley A, Hanson M, Maraganore D, Adler C, Cookson MR, Muenter M, Baptista M, Miller D, Blancato J, Hardy J, Gwinn-Hardy K (2003) alpha-Synuclein locus triplication causes Parkinson's disease. Science 302:841

Sisodia SS, Koo EH, Beyreuther K, Unterbeck A, Price DL (1990) Evidence that beta-amyloid protein in Alzheimer's disease is not derived by normal processing. Science 248: 492–495

Smith WW, Pei Z, Jiang H, Moore DJ, Liang Y, West AB, Dawson VL, Dawson TM, Ross CA (2005) Leucine-rich repeat kinase 2 (LRRK2) interacts with parkin, and mutant LRRK2 induces neuronal degeneration. Proc Natl Acad Sci U S A 102:18676–18681

Specht CG, Schoepfer R (2001) Deletion of the alpha-synuclein locus in a subpopulation of C57BL/6 J inbred mice. BMC Neurosci 2:11

Spillantini MG, Schmidt ML, Lee VM, Trojanowski JQ, Jakes R, Goedert M (1997) Alpha-synuclein in Lewy bodies. Nature 388:839–840

Spitz M, Rozenberg R, Pereira Lda V, Reis Barbosa E (2008) Association between Parkinson's disease and glucocerebrosidase mutations in Brazil. Parkinsonism Relat Disord 14:58–62

Srinivasula SM, Gupta S, Datta P, Zhang Z, Hegde R, Cheong N, Fernandes-Alnemri T, Alnemri ES (2003) Inhibitor of apoptosis proteins are substrates for the mitochondrial serine protease Omi/HtrA2. J Biol Chem 278:31469–31472

Stratman NC, Castle CK, Taylor BM, Epps DE, Melchior GW, Carter DB (2005) Isoform-specific interactions of human apolipoprotein E to an intermediate conformation of human Alzheimer amyloid-beta peptide. Chem Phys Lipids 137:52–61

Strauss KM, Martins LM, Plun-Favreau H, Marx FP, Kautzmann S, Berg D, Gasser T, Wszolek Z, Müller T, Bornemann A, Wolburg H, Downward J, Riess O, Schulz JB, Krüger R (2005) Loss of function mutations in the gene encoding Omi/HtrA2 in Parkinson's disease. Hum Mol Genet 14:2099–2111

Suzuki N, Cheung TT, Cai XD, Odaka A, Otvos L Jr., Eckman C, Golde TE, Younkin SG (1994) An increased percentage of long amyloid beta protein secreted by familial amyloid beta protein precursor (beta APP717) mutants. Science 264:1336–1340

Suzuki Y, Imai Y, Nakayama H, Takahashi K, Takio K, Takahashi R (2001) A serine protease, HtrA2, is released from the mitochondria and interacts with XIAP, inducing cell death. Mol Cell 8:613–621

Taira T, Saito Y, Niki T, Iguchi-Ariga SM, Takahashi K, Ariga H (2004) DJ-1 has a role in antioxidative stress to prevent cell death. EMBO Rep 5:213–218

Takahashi R, Deveraux Q, Tamm I, Welsh K, Assa-Munt N, Salvesen GS, Reed JC (1998) A single BIR domain of XIAP sufficient for inhibiting caspases. J Biol Chem 273:7787–7790

Takahashi-Niki K, Niki T, Taira T, Iguchi-Ariga SM, Ariga H (2004) Reduced anti-oxidative stress activities of DJ-1 mutants found in Parkinson's disease patients. Biochem Biophys Res Commun 320:389–397

Takahashi H, Ohama E, Suzuki S, Horikawa Y, Ishikawa A, Morita T, Tsuji S, Ikuta F (1994) Familial juvenile parkinsonism: clinical and pathologic study in a family. Neurology 44: 437–441

Takasugi N, Tomita T, Hayashi I, Tsuruoka M, Niimura M, Takahashi Y, Thinakaran G, Iwatsubo T (2003) The role of presenilin cofactors in the gamma-secretase complex. Nature 422: 438–441

Tan EK, Chai A, Teo YY, Zhao Y, Tan C, Shen H, Chandran VR, Teoh ML, Yih Y, Pavanni R, Wong MC, Puvan K, Lo YL, Yap E (2004a) Alpha-synuclein haplotypes implicated in risk of Parkinson's disease. Neurology 62:128–131

Tan EK, Puong KY, Chan DK, Yew K, Fook-Chong S, Shen H, Ng PW, Woo J, Yuen Y, Pavanni R, Wong MC, Puvan K, Zhao Y (2005a) Impaired transcriptional upregulation of Parkin promoter variant under oxidative stress and proteasomal inhibition: clinical association. Hum Genet 118:484–488

Tan EK, Puong KY, Fook-Chong S, Chua E, Shen H, Yuen Y, Pavanni R, Wong MC, Puvan K, Zhao Y (2006a) Case-control study of UCHL1 S18Y variant in Parkinson's disease. Mov Disord 21:1765–1768

Tan EK, Shen H, Tan JM, Lim KL, Fook-Chong S, Hu WP, Paterson MC, Chandran VR, Yew K, Tan C, Yuen Y, Pavanni R, Wong MC, Puvan K, Zhao Y (2005b) Differential expression of splice variant and wild-type parkin in sporadic Parkinson's disease. Neurogenetics 6:179–184

Tan EK, Skipper LM (2007) Pathogenic mutations in Parkinson disease. Hum Mutat 28: 641–653

Tan EK, Tan C, Zhao Y, Yew K, Shen H, Chandran VR, Teoh ML, Yih Y, Pavanni R, Wong MC (2004b) Genetic analysis of DJ-1 in a cohort Parkinson's disease patients of different ethnicity. Neurosci Lett 367:109–112

Tan LC, Tanner CM, Chen R, Chan P, Farrer M, Hardy J, Langston JW (2003) Marked variation in clinical presentation and age of onset in a family with a heterozygous parkin mutation. Mov Disord 18:758–763

Tan EK, Tong J, Fook-Chong S, Yih Y, Wong MC, Pavanni R, Zhao Y (2007) Glucocerebrosidase mutations and risk of Parkinson disease in Chinese patients. Arch Neurol 64:1056–1058

Tan EK, Yew K, Chua E, Puvan K, Shen H, Lee E, Puong KY, Zhao Y, Pavanni R, Wong MC, Jamora D, de Silva D, Moe KT, Woon FP, Yuen Y, Tan L (2006b) PINK1 mutations in sporadic early-onset Parkinson's disease. Mov Disord 21:789–793

Tan EK, Yew K, Chua E, Shen H, Jamora RD, Lee E, Puong KY, Zhao Y, Pavanni R, Wong MC, Puvan K, Yih Y, Tan LC (2005c) Analysis of PINK1 in Asian patients with familial parkinsonism. Clin Genet 68:468–470

Tanaka S, Nakamura S, Ueda K, Kameyama M, Shiojiri S, Takahashi Y, Kitaguchi N, Ito H (1988) Three types of amyloid protein precursor mRNA in human brain: their differential expression in Alzheimer's disease. Biochem Biophys Res Commun 157:472–479

Tandon A, Fraser P (2002) The presenilins. Genome Biol 3:s3014; review

Tang MX, Maestre G, Tsai WY, Liu XH, Feng L, Chung WY, Chun M, Schofield P, Stern Y, Tycko B, Mayeux R (1996) Relative risk of Alzheimer disease and age-at-onset distributions, based on APOE genotypes among elderly African Americans, Caucasians, and Hispanics in New York City. Am J Hum Genet 58:574–584

Tanzi RE, McClatchey AI, Lamperti ED, Villa-Komaroff L, Gusella JF, Neve RL (1988) Protease inhibitor domain encoded by an amyloid protein precursor mRNA associated with Alzheimer's disease. Nature 331:528–530

Tassin J, Durr A, de Broucker T, Abbas N, Bonifati V, De Michele G, Bonnet AM, Broussolle E, Pollak P, Vidailhet M, De Mari M, Marconi R, Medjbeur S, Filla A, Meco G, Agid Y, Brice A (1998) Chromosome 6-linked autosomal recessive early-onset Parkinsonism: linkage in European and Algerian families, extension of the clinical spectrum, and evidence of a small homozygous deletion in one family. The French Parkinson's Disease Genetics Study Group, and the European Consortium on Genetic Susceptibility in Parkinson's Disease. Am J Hum Genet 63:88–94

Terry RD, Gonatas NK, Weiss M (1964) Ultrastructural Studies in Alzheimer's Presenile Dementia. Am J Pathol 44:269–297

Thal DR, Capetillo-Zarate E, Del Tredici K, Braak H (2006) The development of amyloid beta protein deposits in the aged brain. Sci Aging Knowledge Environ 2006(6) re1

Theuns J, Del-Favero J, Dermaut B, van Duijn CM, Backhovens H, Van den Broeck MV, Serneels S, Corsmit E, Van Broeckhoven CV, Cruts M (2000) Genetic variability in the regulatory region of presenilin 1 associated with risk for Alzheimer's disease and variable expression. Hum Mol Genet 9:325–331

Toft M, Pielsticker L, Ross OA, Aasly JO, Farrer MJ (2006) Glucocerebrosidase gene mutations and Parkinson disease in the Norwegian population. Neurology 66:415–417

Tsai MS, Tangalos EG, Petersen RC, Smith GE, Schaid DJ, Kokmen E, Ivnik RJ, Thibodeau SN (1994) Apolipoprotein E: risk factor for Alzheimer disease. Am J Hum Genet 54: 643–649

Turner PR, O'Connor K, Tate WP, Abraham WC (2003) Roles of amyloid precursor protein and its fragments in regulating neural activity, plasticity and memory. Prog Neurobiol 70:1–32

Uchihara T, Duyckaerts C, He Y, Kobayashi K, Seilhean D, Amouyel P, Hauw JJ (1995) ApoE immunoreactivity and microglial cells in Alzheimer's disease brain. Neurosci Lett 195:5–8

Uversky VN (2003) A protein-chameleon: conformational plasticity of alpha-synuclein, a disordered protein involved in neurodegenerative disorders. J Biomol Struct Dyn 21:211–234

Valente EM, Abou-Sleiman PM, Caputo V, Muqit MM, Harvey K, Gispert S, Ali Z, Del Turco D, Bentivoglio AR, Healy DG, Albanese A, Nussbaum R, Gonzalez-Maldonado R, Deller T, Salvi S, Cortelli P, Gilks WP, Latchman DS, Harvey RJ, Dallapiccola B, Auburger G, Wood NW (2004) Hereditary early-onset Parkinson's disease caused by mutations in PINK1. Science 304:1158–1160

Valente EM, Bentivoglio AR, Dixon PH, Ferraris A, Ialongo T, Frontali M, Albanese A, Wood NW (2001) Localization of a novel locus for autosomal recessive early-onset parkinsonism, PARK6, on human chromosome 1p35-p36. Am J Hum Genet 68:895–900

Vande Walle L, Lamkanfi M, Vandenabeele P (2008) The mitochondrial serine protease HtrA2/Omi: an overview. Cell Death Differ 15:453–460

Vassar R, Bennett BD, Babu-Khan S, Kahn S, Mendiaz EA, Denis P, Teplow DB, Ross S, Amarante P, Loeloff R, Luo Y, Fisher S, Fuller J, Edenson S, Lile J, Jarosinski MA, Biere AL, Curran E, Burgess T, Louis JC, Collins F, Treanor J, Rogers G, Citron M (1999) Beta-secretase cleavage of Alzheimer's amyloid precursor protein by the transmembrane aspartic protease BACE. Science 286:735–741

Verhagen AM, Silke J, Ekert PG, Pakusch M, Kaufmann H, Connolly LM, Day CL, Tikoo A, Burke R, Wrobel C, Moritz RL, Simpson RJ, Vaux DL (2002) HtrA2 promotes cell death through its serine protease activity and its ability to antagonize inhibitor of apoptosis proteins. J Biol Chem 277:445–454

von Coelln R, Dawson VL, Dawson TM (2004a) Parkin-associated Parkinson's disease. Cell Tissue Res 318:175–184

Von Coelln R, Thomas B, Savitt JM, Lim KL, Sasaki M, Hess EJ, Dawson VL, Dawson TM (2004b) Loss of locus coeruleus neurons and reduced startle in parkin null mice. Proc Natl Acad Sci U S A 101:10744–10749

Walker LC, Ibegbu CC, Todd CW, Robinson HL, Jucker M, LeVine H 3rd, Gandy S (2005) Emerging prospects for the disease-modifying treatment of Alzheimer's disease. Biochem Pharmacol 69:1001–1008

Walter J, Kaether C, Steiner H, Haass C (2001) The cell biology of Alzheimer's disease: uncovering the secrets of secretases. Curr Opin Neurobiol 11:585–590

Wang C, Tan JM, Ho MW, Zaiden N, Wong SH, Chew CL, Eng PW, Lim TM, Dawson TM, Lim KL (2005) Alterations in the solubility and intracellular localization of parkin by several familial Parkinson's disease-linked point mutations. J Neurochem 93:422–431

Waragai M, Wei J, Fujita M, Nakai M, Ho GJ, Masliah E, Akatsu H, Yamada T, Hashimoto M (2006) Increased level of DJ-1 in the cerebrospinal fluids of sporadic Parkinson's disease. Biochem Biophys Res Commun 345:967–972

Weidemann A, Eggert S, Reinhard FB, Vogel M, Paliga K, Baier G, Masters CL, Beyreuther K, Evin G (2002) A novel epsilon-cleavage within the transmembrane domain of the Alzheimer amyloid precursor protein demonstrates homology with Notch processing. Biochemistry 41:2825–2835

Weidemann A, Konig G, Bunke D, Fischer P, Salbaum JM, Masters CL, Beyreuther K (1989) Identification, biogenesis, and localization of precursors of Alzheimer's disease A4 amyloid protein. Cell 57:115–126

West AB, Maraganore D, Crook J, Lesnick T, Lockhart PJ, Wilkes KM, Kapatos G, Hardy JA, Farrer MJ (2002) Functional association of the parkin gene promoter with idiopathic Parkinson's disease. Hum Mol Genet 11:2787–2792

West AB, Moore DJ, Biskup S, Bugayenko A, Smith WW, Ross CA, Dawson VL, Dawson TM (2005) Parkinson's disease-associated mutations in leucine-rich repeat kinase 2 augment kinase activity. Proc Natl Acad Sci U S A 102:16842–16847

Westerman MA, Cooper-Blacketer D, Mariash A, Kotilinek L, Kawarabayashi T, Younkin LH, Carlson GA, Younkin SG, Ashe KH (2002) The relationship between Abeta and memory in the Tg2576 mouse model of Alzheimer's disease. J Neurosci 22:1858–1867

Wigderson M, Firon N, Horowitz Z, Wilder S, Frishberg Y, Reiner O, Horowitz M (1989) Characterization of mutations in Gaucher patients by cDNA cloning. Am J Hum Genet 44:365–377

Wilkinson KD, Lee KM, Deshpande S, Duerksen-Hughes P, Boss JM, Pohl J (1989) The neuron-specific protein PGP 9.5 is a ubiquitin carboxyl-terminal hydrolase. Science 246:670–673

Williams DR, Hadeed A, al-Din AS, Wreikat AL, Lees AJ (2005) Kufor Rakeb disease: autosomal recessive, levodopa-responsive parkinsonism with pyramidal degeneration, supranuclear gaze palsy, and dementia. Mov Disord 20:1264–1271

Wing SS (2003) Deubiquitinating enzymes – the importance of driving in reverse along the ubiquitin-proteasome pathway. Int J Biochem Cell Biol 35:590–605

Wintermeyer P, Kruger R, Kuhn W, Muller T, Woitalla D, Berg D, Becker G, Leroy E, Polymeropoulos M, Berger K, Przuntek H, Schols L, Epplen JT, Riess O (2000) Mutation analysis and association studies of the UCHL1 gene in German Parkinson's disease patients. NeuroReport 11:2079–2082

Wolfe MS (2007) When loss is gain: reduced presenilin proteolytic function leads to increased Abeta42/Abeta40. Talking Point on the role of presenilin mutations in Alzheimer disease. EMBO Rep 8:136–140

Wolfe MS, De Los Angeles J, Miller DD, Xia W, Selkoe DJ (1999a) Are presenilins intramembrane-cleaving proteases? Implications for the molecular mechanism of Alzheimer's disease. Biochemistry 38:11223–11230

Wolfe MS, Xia W, Ostaszewski BL, Diehl TS, Kimberly WT, Selkoe DJ (1999b) Two transmembrane aspartates in presenilin-1 required for presenilin endoproteolysis and gamma-secretase activity. Nature 398:513–517

Wong K, Sidransky E, Verma A, Mixon T, Sandberg GD, Wakefield LK, Morrison A, Lwin A, Colegial C, Allman JM, Schiffmann R (2004) Neuropathology provides clues to the pathophysiology of Gaucher disease. Mol Genet Metab 82:192–207

Wood JG, Mirra SS, Pollock NJ, Binder LI (1986) Neurofibrillary tangles of Alzheimer disease share antigenic determinants with the axonal microtubule-associated protein tau (tau). Proc Natl Acad Sci U S A 83:4040–4043

Wszolek ZK, Pfeiffer RF, Tsuboi Y, Uitti RJ, McComb RD, Stoessl AJ, Strongosky AJ, Zimprich A, Muller-Myhsok B, Farrer MJ, Gasser T, Calne DB, Dickson DW (2004) Autosomal dominant parkinsonism associated with variable synuclein and tau pathology. Neurology 62:1619–1622

Wu YR, Chen CM, Chao CY, Ro LS, Lyu RK, Chang KH, Lee-Chen GJ (2007) Glucocerebrosidase gene mutation is a risk factor for early onset of Parkinson disease among Taiwanese. J Neurol Neurosurg Psychiatry 78:977–979

Xu PT, Gilbert JR, Qiu HL, Ervin J, Rothrock-Christian TR, Hulette C, Schmechel DE (1999) Specific regional transcription of apolipoprotein E in human brain neurons. Am J Pathol 154:601–611

Xu YH, Wenstrup R, Grabowski GA (1995) Effect of cellular type on expression of acid beta-glucosidase: implications for gene therapy in Gaucher disease. Gene Ther 2:647–654

Xu J, Zhong N, Wang H, Elias JE, Kim CY, Woldman I, Pifl C, Gygi SP, Geula C, Yankner BA (2005) The Parkinson's disease-associated DJ-1 protein is a transcriptional co-activator that protects against neuronal apoptosis. Hum Mol Genet 14:1231–1241

Yamamura Y, Sobue I, Ando K, Iida M, Yanagi T (1973) Paralysis agitans of early onset with marked diurnal fluctuation of symptoms. Neurology 23:239–244

Yamazaki T, Hibi K, Takase T, Tezel E, Nakayama H, Kasai Y, Ito K, Akiyama S, Nagasaka T, Nakao A (2002) PGP9.5 as a marker for invasive colorectal cancer. Clin Cancer Res 8:192–195

Yang QH, Church-Hajduk R, Ren J, Newton ML, Du C (2003) Omi/HtrA2 catalytic cleavage of inhibitor of apoptosis (IAP) irreversibly inactivates IAPs and facilitates caspase activity in apoptosis. Genes Dev 17:1487–1496

Yang F, Jiang Q, Zhao J, Ren Y, Sutton MD, Feng J (2005) Parkin stabilizes microtubules through strong binding mediated by three independent domains. J Biol Chem 280:17154–17162

Yokota T, Sugawara K, Ito K, Takahashi R, Ariga H, Mizusawa H (2003) Down regulation of DJ-1 enhances cell death by oxidative stress, ER stress, and proteasome inhibition. Biochem Biophys Res Commun 312:1342–1348

Yoshikai S, Sasaki H, Doh-ura K, Furuya H, Sakaki Y (1990) Genomic organization of the human amyloid beta-protein precursor gene. Gene 87:257–263

Yu H, Saura CA, Choi SY, Sun LD, Yang X, Handler M, Kawarabayashi T, Younkin L, Fedeles B, Wilson MA, Younkin S, Kandel ER, Kirkwood A, Shen J (2001) APP processing and synaptic plasticity in presenilin-1 conditional knockout mice. Neuron 31:713–726

Zabetian CP, Hutter CM, Yearout D, Lopez AN, Factor SA, Griffith A, Leis BC, Bird TD, Nutt JG, Higgins DS, Roberts JW, Kay DM, Edwards KL, Samii A, Payami H (2006) LRRK2 G2019S in families with Parkinson disease who originated from Europe and the Middle East: evidence of two distinct founding events beginning two millennia ago. Am J Hum Genet 79:752–758

Zarranz JJ, Alegre J, Gomez-Esteban JC, Lezcano E, Ros R, Ampuero I, Vidal L, Hoenicka J, Rodriguez O, Atares B, Llorens V, Gomez Tortosa E, del Ser T, Munoz DG, de Yebenes JG (2004) The new mutation, E46K, of alpha-synuclein causes Parkinson and Lewy body dementia. Ann Neurol 55:164–173

Zhang ZJ, Burgunder JM, An XK, Wu Y, Chen WJ, Zhang JH, Wang YC, Xu YM, Gou YR, Yuan GG, Mao XY, Peng R (2008) Lack of evidence for association of a UCH-L1 S18Y polymorphism with Parkinson's disease in a Han-Chinese population. Neurosci Lett 442:200–202

Zhang L, Shimoji M, Thomas B, Moore DJ, Yu SW, Marupudi NI, Torp R, Torgner IA, Ottersen OP, Dawson TM, Dawson VL (2005) Mitochondrial localization of the Parkinson's disease related protein DJ-1: implications for pathogenesis. Hum Mol Genet 14:2063–2073

Zhao G, Mao G, Tan J, Dong Y, Cui MZ, Kim SH, Xu X (2004) Identification of a new presenilin-dependent zeta-cleavage site within the transmembrane domain of amyloid precursor protein. J Biol Chem 279:50647–50650

Zhou W, Freed CR (2005) DJ-1 up-regulates glutathione synthesis during oxidative stress and inhibits A53T alpha-synuclein toxicity. J Biol Chem 280:43150–43158

Zhou W, Zhu M, Wilson MA, Petsko GA, Fink AL (2006) The oxidation state of DJ-1 regulates its chaperone activity toward alpha-synuclein. J Mol Biol 356:1036–1048

Ziegler SG, Eblan MJ, Gutti U, Hruska KS, Stubblefield BK, Goker-Alpan O, LaMarca ME, Sidransky E (2007) Glucocerebrosidase mutations in Chinese subjects from Taiwan with sporadic Parkinson disease. Mol Genet Metab 91:195–200

Zimprich A, Muller-Myhsok B, Farrer M, Leitner P, Sharma M, Hulihan M, Lockhart P, Strongosky A, Kachergus J, Calne DB, Stoessl J, Uitti RJ, Pfeiffer RF, Trenkwalder C, Homann N, Ott E, Wenzel K, Asmus F, Hardy J, Wszolek Z, Gasser T (2004) The PARK8 locus in autosomal dominant parkinsonism: confirmation of linkage and further delineation of the disease-containing interval. Am J Hum Genet 74:11–19

Nicotinic Receptors in Brain Diseases

Jerry A. Stitzel

Abstract The existence of neuronal nicotinic acetylcholine receptor (nAChRs) expression in the brain was discovered 30 years ago. Although the relevance of neuronal nAChRs at the time of their discovery was debated, it is now clear that nAChRs are expressed throughout the brain where they mainly serve a modulatory role. Neuronal nAChRs increasingly have become of interest due to the many observations that various nAChR subtypes exhibit abnormal expression or function in a wide assortment of neurological diseases. In this review, the putative role of nAChRs in brain disease is discussed in several broad categories: (1) diseases associated with a loss of nAChRs, (2) diseases associated with innate differences in the expression of nAChRs, (3) diseases associated with genetic variability in genes that code for nAChR subunit proteins, and (4) diseases in which nAChRs are implicated based on the observation that nicotine has a therapeutic effect.

Keywords Receptors, nicotinic · Parkinson's disease · Alzheimer's disease · Schizophrenia · Autism · Autosomal dominant nocturnal frontal lobe epilepsy (ADNFLE) · CHRNA5 · CHRNA3 · Nicotine dependence · Tourette's syndrome · Down syndrome

Contents

J.A. Stitzel (✉)
Department of Integrative Physiology, Institute for Behavioral Genetics, University of Colorado, Boulder, CO 80309, USA
e-mail: stitzel@colorado.edu

J.P. Blass (ed.), *Neurochemical Mechanisms in Disease*,
Advances in Neurobiology 1, DOI 10.1007/978-1-4419-7104-3_22,

1 Introduction

1.1 Brief History

Neuronal nicotinic acetylcholine receptors (nAChRs) are members of the cysteine loop superfamily of ligand gated ion channels that includes ionotropic 5-HT, GABA, and glycine receptors. As their name implies, nAChRs are receptors for the endogenous neurotransmitter acetylcholine in the nicotinic branch of the cholinergic system. The existence of nAChR in the brain was first demonstrated by ligand binding studies in the 1980s. Using radio-ligand binding techniques, several groups established that there were at least two distinct nAChR populations in the rodent brain: one that binds the ligand [125I]-α-bungarotoxin with high affinity (Marks and Collins, 1982; Morley et al., 1979; Oswald and Freeman, 1981) and one that binds the ligands [3H]-L-nicotine or [3H] acetylcholine with high affinity (Abood et al., 1980; Marks and Collins, 1982; Romano and Goldstein, 1980; Schwartz et al., 1982; Sershen et al., 1981). The two binding sites also were found to be expressed in overlapping yet distinct patterns in the brain (Clarke et al., 1985; Marks et al., 1986; Marks and Collins, 1982). At the time of their identification, the functional relevance of these binding sites in the brain was not clear (Abood et al., 1980, 1981; Sershen et al., 1981). However, from the mid-1980s through the early 1990s cDNAs for multiple nAChR subunits were cloned from rat and chicken brain (Boyd, 1997). These studies not only led to the identification of 11 different genes (12 in chickens) that code for neuronal nAChR subunits but also demonstrated that various subunit combinations could form functional nAChRs that could be activated by acetylcholine and nicotine. The subunit genes identified were named α2–α10 (α8 only found in chickens) and β2–β4 based on the presence (α subunit) or absence (β subunit) of vicinal cysteines in the N-terminal extracellular domain and the order in which they were cloned. Neuronal nAChRs, like nAChRs at the neuromuscular junction, also were found to be composed of five subunits that form a pentameric ring around a central cation pore. These early studies also demonstrated that some nAChRs are heteromeric, requiring both an α subunit (α2–α4, α6) and a β subunit (β2 or β4) in order to form a functional receptor in vitro. The most abundant heteromeric nAChR in brain is comprised of the subunits α4 and β2 (Flores et al., 1992; Whiting et al., 1991). The α4β2* (the asterisk indicates that other subunits such as α5 can contribute to α4β2 nAChRs) receptor exhibits high affinity for nicotinic agonists and has been demonstrated to be the [3H]-L nicotine binding site described in the early ligand-binding studies (Flores et al., 1992; Marubio et al., 1999; Picciotto et al., 1995; Whiting et al., 1991). Other nAChR α subunits were identified that could form functional pentameric receptors in vitro without a β subunit. The most prevalent of these so-called homomeric nAChRs in the brain is composed of α7 subunits. Homomeric α7 nAChRs exhibit low affinity for nicotinic agonists and immunological (Chen and Patrick, 1997) and genetic studies (Orr-Urtreger et al., 1997) demonstrated that α7 nAChRs are the previously described [125I]-α-bungarotoxin binding sites in brain.

Although α4β2* nAChRs are the most abundant nAChR expressed in the brain, several other heteromeric nAChR subtypes exist in the brain. For example, within dopamine terminals there are at least five different heteromeric nAChRs composed of anywhere between two and four different subunits (Champtiaux et al., 2002; Cui et al., 2003; Klink et al., 2001; Marubio et al., 2003; Salminen et al., 2004). The nAChRs on dopamine terminals in the striatum include α4β2, α4β2α5, α6β2, α6β2β3, and α4α6β2β3. Data also indicate that the nAChRs in GABAergic terminals are α4β2 and α4β2α5 (Lu et al., 1998; McClure-Begley et al., 2009; Zhu and Chiappinelli, 1999), whereas nAChRs that modulate acetylcholine release in the interpeduncular nucleus are α3β4 and α3β3β4 heteromers (Grady et al., 2001, 2009). A combination of immunoprecipitation experiments and in situ hybridization studies also suggest the existence of additional heteromeric nAChR subtypes (Gotti et al., 2006b), including an α7β2* subtype (Liu et al., 2009) although the functional relevance of these potential nAChR subtypes remains to be determined.

1.2 Activation, Desensitization, and Upregulation

Activation of nAChRs by agonists leads to the opening of a central channel that is permeable to cations including calcium (Mulle et al., 1992). Permeability to calcium is receptor subtype-dependent with α7 nAChRs exhibiting the greatest calcium permeability (Fucile et al., 2003; Fucile, 2004; Ragozzino et al., 1998). Although acute exposure to a nicotinic agonist activates nAChRs, continuous exposure to activating and even subactivating concentrations of agonist leads to receptor desensitization, a state in which the receptors become refractory to activation to agonists (Giniatullin et al., 2005; Quick and Lester, 2002). This property of nAChRs also is subtype-dependent with α7 nAChRs exhibiting the fastest rate of desensitization (Couturier et al., 1990; Seguela et al., 1993). However, due to the low sensitivity of α7 nAChRs to activation by nicotinic agonists, α7 nAChRs appear to remain active at nicotine concentrations in the range found in smokers (Mansvelder et al., 2002; Wooltorton et al., 2003). In contrast, α4β2* nAChRs appear to be desensitized at the same concentrations of nicotine (Mansvelder et al., 2002; Mansvelder and McGehee, 2000). The ability of nicotine to produce long-term desensitization of at least some nAChR subtypes at physiologically relevant concentrations have led some to refer to nicotine as a time-averaged antagonist (Hulihan-Giblin et al., 1990).

Another property of at least some nAChR subtypes is upregulation of receptor numbers in response to long-term nicotine exposure. This phenomenon first was discovered in rodents by Marks et al. (1983) and Schwartz and Kellar (1983). Both of these groups demonstrated that chronic treatment of mice and rats led to upregulation of what we now know are α4β2* nAChRs. High doses of nicotine also led to modest increases in α7 nAChRs (Marks et al., 1983). Upregulation of α4β2* nAChRs also is seen in brain tissue from smokers (Benwell et al., 1988; Breese et al., 1997; Perry et al., 1999). This upregulation of nAChRs has been termed paradoxical because the expected effect of chronic agonist exposure is receptor downregulation (Wonnacott, 1990). However, it has been postulated that the upregulation is due to the time-averaged antagonist property of nicotine. Support for this possibility comes from a study by Marks et al. (2004) that demonstrated that, despite the increase in numbers of α4β2* nAChRs following chronic nicotine treatment, the level of function of α4β2* nAChRs remained unchanged relative to controls. These findings suggest that upregulation serves as a homeostatic mechanism to maintain normal levels of receptor function in the presence of a "time-averaged" antagonist.

Due to the widespread expression of nAChRs throughout the brain and their involvement in modulating the release of many neurotransmitters, it is not surprising that aberrant expression or function of nAChRs might contribute to a wide range of diseases. In the following sections, diseases of the brain in which nAChRs have been implicated are discussed in four broad categories, diseases associated with the loss of nicotinic receptors, diseases associated with innate differences in the expression of nicotinic receptors, diseases related to genetic variants in the genes that code for the nicotinic receptor subunits, and diseases in which nAChRs have been implicated due to a therapeutic effect of nicotine.

2 Diseases Associated with Loss of Brain Nicotinic Receptors

The diseases associated with loss of nAChRs are typically neurodegenerative diseases. The two most studied diseases that are associated with the loss of nicotinic receptors in the brain are Parkinson's and Alzheimer's.

2.1 Parkinson's Disease

In Parkinson's disease (PD), loss of nAChRs occurs in the nigrostriatal pathway (Aubert et al., 1992; Pimlott et al., 2004; Quik et al., 2004) as well as in the basal forebrain and cortex (Aubert et al., 1992; Lange et al., 1993; Perry et al., 1995; Rinne et al., 1991). Early studies indicated that high-affinity nAChRs are preferentially lost in PD. A combination of studies in both rodent and nonhuman primate models of PD suggests that the α6α4β2β3 nAChR is the most labile high affinity nAChR in response to nigrostriatal damage (Kulak et al., 2002; Quik et al., 2005). The loss of this nAChR subtype also closely coincides with the loss of the dopamine transporter (Bordia et al., 2007; Gotti et al., 2006a; Quik et al., 2004). α4β2* nAChRs also appear to be lost in animal models of PD but only when lesions are severe (Bordia et al., 2007; Kulak et al., 2002; Quik et al., 2003). Similarly, α6β2* appear to be lost to a greater extent than α4β2* nAChRs in several brain areas of PD patients (Bohr et al., 2005; Bordia et al., 2007; Quik et al., 2004). In contrast, there appears to be no loss of α7 nAChRs in striatal tissue in both animal models and humans (Guan et al., 2002; Quik et al., 2005; Zoli et al., 2002). However, there may be a loss of α7 nAChRs in cortical regions of Parkinson's disease patients (Banerjee et al., 2000; Burghaus et al., 2003) although this finding is not universal (Guan et al., 2002).

It remains to be determined whether the loss of nAChRs in PD contributes to the development of the disease or simply is a marker of the disease and α6β2* nAChRs are simply present on the neurons most sensitive to damage. Data from knock-out mice demonstrate that the lack of any of the striatal expressed nAChR subunits does not lead to striatal neurodegeneration. Thus, the simple loss of these nAChRs alone is not sufficient to elicit a neurodegenerative state. Nonetheless, a role of nAChRs in PD is supported by epidemiological evidence that clearly demonstrates that there is an inverse relationship between smoking and the development of PD. Although the mechanism through which smoking delays the onset of PD remains to be elucidated, it is generally thought that nicotine acts as a neuroprotective agent via interaction with nAChRs. The ability of nicotine to be neuroprotective in general and in animal models of PD more specifically has been demonstrated in several in vitro and in vivo studies (Quik et al., 2008). Whether nicotine acts as a neuroprotective agent through activating or desensitizing nicotinic receptors is not clear. However, recent studies with mice possessing a hyperactive form of the α4 subunit suggest that heightened activity rather than loss of activity is neurodegenerative in the striatum (Labarca et al., 2001; Schwarz et al., 2006). Based on this and the fact that nAChR knock-out mice show no striatal neurodegeneration suggests that the neuroprotective properties of nicotine may be through desensitization/inactivation of nAChRs rather than through activation.

2.2 Alzheimer's Disease

2.2.1 Altered Expression of nAChRs

Loss of nAChRs also is associated with Alzheimer's disease (AD). The most profoundly affected nAChR subtype in AD is the α4β2* subtype. Results from both receptor-ligand binding assays (Nordberg et al., 1988; Nordberg and Winblad, 1986; Perry et al., 2000; Whitehouse et al., 1986, 1988) and immunological experiments (Burghaus et al., 2000; Gotti et al., 2006a; Guan et al., 2000; Martin-Ruiz et al., 1999; Wevers et al., 1999) indicate that α4β2* nAChRs are reduced by as much as 50% in cortical and hippocampal regions of postmortem brain tissue of AD patients. α4β2* nAChRs begin to decline in the earliest stages of AD (Marutle et al., 1999) and several recent studies have shown a significant correlation between the degree of loss of this nAChR subtype and cognitive deficits in early AD patients (Kadir et al., 2006; Sabri et al., 2008). Other studies have reported a correlation between the level of expression of cortical α4β2* nAChRs and degree of cognitive deficits in AD patients (Nordberg et al., 1995; Perry et al., 2000). However, not all studies have observed a significant correlation between the expression levels of α4β2* nAChRs and cognitive deficits in early AD patients (Ellis et al., 2008, 2009). Nonetheless, a putative role of α4β2* nAChRs in AD-related neurodegeneration is supported by the observation that β2 nAChR-null mutant mice exhibit elevated age-related neurodegeneration in cortical brain areas and hippocampus and increased age-related cognitive deficits (Zoli et al., 1999).

Some studies also have found alterations in the expression of other nAChR subunits, including α3 and α7, in postmortem brain tissue of AD patients (Guan et al., 2000; Mousavi et al., 2003; Wevers et al., 1999) and animal models (Bednar et al., 2002; Jones et al., 2006; Mousavi et al., 2004). However, these findings generally are not consistent. In the case of α7, some studies have found no change in expression of this subunit (Gotti et al., 2006a; Martin-Ruiz et al., 1999), others have found a decrease in expression of this subunit (Burghaus et al., 2000; Engidawork et al., 2001; Guan et al., 2000; Wevers et al., 1999), and a few studies have reported an increase in the expression of this nAChR subunit (Counts et al., 2007; Hellstrom-Lindahl et al., 1999; Teaktong et al., 2003). This apparent disparity in the relationship between α7 expression and AD may be explained by a recent study by Jones et al. (2006) in which α7 expression was assessed in a transgenic mouse possessing a mutant form of the human amyloid precursor protein (APP) that results in familial AD. Results of this study demonstrated that α7 expression increases progressively to levels three- or fourfold higher than normal control brain by 9 months of age. However, by 12 months of age the transgenic mice expressed lower levels of α7 than controls. Therefore, the relationship between α7 expression and AD may be age and/or disease state-dependent. It is of interest to note that despite the fact that several studies have demonstrated a reduction in α4β2* and potentially other nAChRs in AD patients, changes in RNA levels for these receptor subunits generally have not been observed in AD patients (Mousavi et al., 2003; Terzano et al., 1998). This observation suggests that the loss of nAChRs in AD is mediated posttranscriptionally.

2.2.2 Interaction of nAChRs with β Amyloid

nAChRs also have been implicated in the etiology of AD via interactions with amyloid β (Aβ), a 39–43 amino acid polypeptide that is thought to play a critical role in the pathogenesis of AD. Several studies have shown that $A\beta_{1\text{-}42}$ binds with high affinity to both α4β2* and α7 nAChRs. In addition, nicotinic receptors are implicated in neuroprotection from Aβ toxicity by the observations that nicotine reduces Aβ accumulation and neurotoxicity both in vitro (Kihara et al., 1998, 1999, 2001; Liu and Zhao, 2004; Zamani et al., 1997) and in animal models (Gahring et al., 2003; Hellstrom-Lindahl et al., 2004; Nordberg et al., 2002; Zhang et al., 2006). Moreover, the deposition of Aβ is significantly reduced in postmortem brain from AD patients who were smokers (Hellstrom-Lindahl et al., 2004). However, there are conflicting data regarding whether the interaction of Aβ with various nAChR subtypes activates (Chin et al., 2006; Dineley et al., 2001, 2002; Fu and Jhamandas, 2003) or inhibits (Grassi et al., 2003; Lamb et al., 2005; Liu et al., 2001, 2009; Magdesian et al., 2005; Soderman et al., 2008; Tozaki et al., 2002; Wu et al., 2004) the function of the receptors. In addition, a recent paper reported that oligomeric $A\beta_{1\text{-}42}$ at low concentrations (1 nM) selectively inhibits a novel and putatively α7β2 nAChR (Liu et al., 2009). It also has been reported that there is a physical interaction between Aβ and α7 nAChRs (Wang et al., 2000a,b) and that the interaction between Aβ and the α7 nAChR facilitates the internalization of Aβ in neurons (Nagele et al., 2002). This reported internalization of Aβ by α7 nAChRs may explain the observation that Aβ and α7 nAChRs have been found to be colocalized in neurons of AD patients (Nagele et al., 2002; Wang et al., 2000a,b; Wevers et al., 1999). It has been postulated that the excessive intraneuronal accumulation of Aβ via internalization by the α7 nAChR leads to neuronal death (Nagele et al., 2002). However, this hypothesis is not consistent with the observation that α7 nAChRs are not preferentially lost in AD.

3 Diseases Associated with Innate Differences in the Expression of nAChRs

Although there obviously is individual variability in the expression of brain nAChRs in the population, altered expression of some nAChRs relative to healthy controls is associated with several neuropsychiatric disorders. The best-studied example of low nAChR expression in brain and disease is schizophrenia. A second example discussed is autism.

3.1 Schizophrenia

Schizophrenia is characterized by multiple symptoms including, but certainly not limited to, psychosis, apathy, and cognitive impairment (Austin, 2005; Mueser and McGurk, 2004). Another common feature of schizophrenia is poor sensory inhibition including the inability to "filter" repetitive stimuli (Baker et al., 1987; Boutros et al., 1999; Braff et al., 2001; Clementz et al., 1998; Holzman, 2000; Kelley and Bakan, 1999; Lee and Williams, 2000). The inability to filter repetitive stimuli is believed to lead to personality decompensation (Venables, 1964, 1992) and almost certainly contributes to the cognitive deficits associated with schizophrenia (Erwin et al., 1998; Simosky et al., 2002). The first evidence that nicotinic receptors may be involved in schizophrenia was the observation that either smoking or nicotine-normalized deficits in sensory inhibition, as measured by P50 auditory gating, in schizophrenic patients (Adler et al., 1992). In addition, smoking, nicotine, or nicotinic agonists more recently have been shown to improve cognitive performance in schizophrenic patients (Freedman et al., 2008; Harris et al., 2004; Olincy et al., 2006; Sacco et al., 2005). These apparent "beneficial" effects of nicotinic agents in schizophrenics may explain the extremely high rate of smoking in this population. It is estimated that anywhere between 50 and 90% of schizophrenic patients smoke (Dalack et al., 1998; Hughes et al., 1986; Lohr and Flynn, 1992). In contrast, smoking rates in individuals with other mental illnesses are around 25% and the prevalence of smoking in the general population is about 20% (Dalack et al., 1998; Williams and Ziedonis, 2004). Moreover, schizophrenic patients exhibit altered smoking behaviors that allow them to extract significantly more nicotine per cigarette than nonschizophrenic smokers (Tidey et al., 2005).

The first direct evidence that alterations in nicotinic receptor expression might contribute to schizophrenia was from a study by Freedman and colleagues (1995) who demonstrated that schizophrenic patients had lower levels of α7 nAChRs as measured by ^{125}I-αBTX binding and lower levels of α4β2* nAChRs as measured by [3H] cytisine binding in hippocampus relative to controls. The reduced binding was the result of both fewer labeled cells and diminished labeling per cell. In addition to reduced expression in the hippocampus, α7 nAChRs also have been shown to be decreased in other brain areas of schizophrenic subjects including the reticular thalamic nucleus (Court et al., 1999) and multiple cortical regions (Guan et al., 1999; Marutle et al., 1999, 2001). However, the data regarding the expression of high affinity (predominantly α4β2* nAChRs) is less clear. Results suggest that in schizophrenic patients, there is a reduction in high-affinity nAChRs in hippocampus (Breese et al., 2000; Freedman et al., 1995) and no change from controls in thalamus (Breese et al., 2000; Court et al., 1999). However, there are conflicting results on the expression of high-affinity nAChRs in the striatum and cortex of schizophrenic subjects. Some studies indicate that there is an increase in high-affinity receptors in these brain regions of schizophrenic subjects (Court et al., 2000; Martin-Ruiz et al., 2003; Marutle et al., 2001) whereas other reports show that high-affinity receptor binding is lower in these brain regions of schizophrenic patients (Breese et al., 2000; Durany et al., 2000). Despite the equivocal results for high-affinity receptor expression in schizophrenia, there is evidence that regulation of this nAChR population is abnormal. As mentioned previously, smoking generally leads to a significant upregulation of α4β2* nAChRs in brain. However, depending upon brain region, upregulation of high-affinity nAChRs is either absent or substantially reduced in schizophrenic brain relative to controls (Breese et al., 2000).

Support for a role of both α4β2* and α7 nAChRs in schizophrenia also comes from pharmacological and animal model data. For example, the α4β2* selective agonists ABT-418 (Stevens and Wear, 1997) and A-85380 (Wildeboer and Stevens, 2008) and the α7 selective agonist DMXB-A (O'Neill et al., 2003; Simosky et al., 2001; Stevens et al., 1998) improve innate and drug-induced deficits in auditory gating in rodents. DMXB-A also has been shown to improve sensory gating and cognitive function and to reduce negative symptoms in two recent clinical trials (Freedman et al., 2008; Olincy et al., 2006).

Additional support for nAChRs in schizophrenia largely is based on mouse genetic models. Mice heterozygous for a null mutation in Chrna7, the gene that codes for the α7 subunit, exhibit reduced expression of the α7 subunit, poor auditory gating, and other functional deficits in the hippocampus similar to those observed in schizophrenic patients (Adams et al., 2008). In addition, a naturally occurring variant allele of Chrna7 has been identified in mice. This allele is linked to variability in α7 expression in the hippocampus (Stitzel et al., 1996), neuroanatomical distribution of α7 nAChRs in the hippocampus (Adams et al., 2001), developmental expression of α7 nAChRs in the hippocampus (Adams et al., 2006), and auditory gating deficits (Stevens et al., 2001). The fact that the allele of Chrna7 that leads to reduced α7 expression in the hippocampus also leads to impaired auditory gating is consistent with the role of Chrna7 in regulating the auditory gating phenotype in schizophrenics. Recently, Liu et al. (2006) reported that α7 nAChRs are involved in the normal development of the GABAergic system in the hippocampus. Thus, abnormal expression of α7 nAChRs during pre- and perinatal periods of development may have long-term consequences on brain function. Suggestive support for a developmental role of α7 nAChRs in impaired auditory gating comes from two recent studies that have shown that perinatal dietary supplementation with choline, an α7-selective agonist, permanently improves gating in two animal models of impaired auditory gating (Stevens et al., 2008a,b).

3.2 Autism

A second disease where there appears to be altered expression of nAChRs is autism. Studies have shown that high-affinity nicotinic receptors as measured by [3H] epibatidine, α4 RNA and anti-α4 antibodies, are reduced in various cortical regions in autistic subjects (Martin-Ruiz et al., 2004; Perry et al., 2001). Using both [3H] epibatidine and anti-α4 antibodies, Lee et al. (2002) and Martin-Ruiz et al. (2004) also reported that α4 nAChRs are reduced in cerebellar regions in subjects with autism relative to normal controls. The α7 subunit was not found to be altered in expression in cortical regions of autistic patients but was found to be upregulated in cerebellum (Lee et al., 2002; Martin-Ruiz et al., 2004); the binding of [125I] α-bungarotoxin was increased in cerebellum although no change in α7 RNA or α7 immunoreactivity was detected. Finally, in a small sample, α7 and β2 but not α4 immunoreactivity was found to be decreased in the thalamus of individuals with autism (Ray et al., 2005).

In addition to altered levels of nAChRs, there also appear to be increased numbers and enlarged morphology of cholinergic neurons in the cortex of children with autism (Bauman and Kemper, 2005). Based on this observation, it has been hypothesized that the downregulation of nAChRs in the cortex and thalamus in autism is the result of a homeostatic response to hypercholinergic activity in the cortex (Lippiello, 2006). The potential hyperactivity in the cortex of individuals with autism may explain the low level of smoking associated with autism relative to both the general population and other mental diseases (Bejerot and Nylander, 2003; Poirier et al., 2002). Nonetheless, there currently are no pharmacological or animal model data to convincingly implicate nAChRs in the etiology of autism. Therefore, the relevance of the altered expression of nAChRs in this disease remains to be determined.

4 Genetic Variants of nAChR Subunit Genes and Brain Disease

Each nAChR subunit is encoded by a different gene and any mutation in any of these genes that affects the expression or function of an nAChR could lead to disease or contribute to individual differences in risk for disease. In this section, one disease directly caused by mutations in nAChR subunit genes is discussed. In addition, the potential role of genetic variability in nAChR subunit genes in altering risk for disease is summarized.

4.1 Autosomal Dominant Nocturnal Frontal Lobe Epilepsy (ADNFLE)

Autosomal dominant nocturnal frontal lobe epilepsy (ADNFLE) is the only brain disease known to be caused by mutations in genes that code for nicotinic receptor subunits. ADNFLE is a rare, inherited form of epilepsy characterized by hyperkinetic or tonic seizures that tend to occur in clusters. Seizures also are of frontal lobe origin and tend to occur during periods of light sleep (Scheffer et al., 1995). To date, there have been ten mutations in genes that code for nAChR subunits that cause ADNFLE, four in CHRNA4 the gene that encodes the nAChR α4 subunit (Hirose et al., 1999; Leniger et al., 2003; McLellan et al., 2003; Phillips et al., 2000; Saenz et al., 1999; Steinlein et al., 1995, 1997, 2000), five in CHRNB2, the gene that codes the β2 nAChR subunit (Bertrand et al., 2005; De Fusco et al., 2000; Hoda et al., 2008; Phillips et al., 2001), and one in CHRNA2, the gene that encodes the α2 nAChR subunit (Aridon et al., 2006). For more details on these ADNFLE-causing mutations, please see the recent review by Steinlein and Bertrand (2008). A fifth mutation in CHRNA4 recently has been identified that may add to this long list of nAChR subunit gene mutations that cause ADNFLE (Chen et al., 2009b). There also is some debate as to whether the seizure disorder caused by the CHRNA2 mutation is ADNFLE or a related seizure disorder (Hoda et al., 2009). Regardless of whether the seizure disorder caused by the CHRNA2 mutation is ADNFLE or a related disease, it still is an example of a mutation in an nAChR subunit gene that directly causes an inherited disease.

Although mutations in CHRNA2, CHRNA4, and CHRNB2 have been shown to cause ADNFLE or related seizure disorders, how these mutations cause epilepsy remains unknown. However, in vitro functional analysis indicates that a common feature of nAChRs possessing ADNFLE mutations is a gain of function, either by increased sensitivity to acetylcholine or reduced desensitization (Aridon et al., 2006; Bertrand et al., 2002; Hoda et al., 2008, 2009; Leniger et al., 2003; Moulard et al., 2001; Phillips et al., 2001). In addition, two recent studies found that smoking or nicotine treatment decreased seizure frequency in ADNFLE patients with nAChR mutations (Brodtkorb and Picard, 2006; Willoughby et al., 2003). The therapeutic effect of smoking/nicotine presumably was the result of decreasing or inhibiting the function of the hyperactive nAChRs via the well-characterized desensitizing effect of nicotine on α4β2*. A mechanism by which nAChR gain of function mutations might lead to ADNFLE has been suggested by studies using two lines of mice a engineered to possess different ADNFLE mutations in Chrna4 (Klaassen et al., 2006). In these studies, it was found that nicotine was greater than 20 times more potent at activating inhibitory postsynaptic currents in cortical regions of mice with ADNFLE mutations than in their control littermates. In contrast, nicotine had no effect on excitatory postsynaptic currents. Based on these data and the observation that picrotoxin, a use-dependent GABA antagonist, transiently eliminated epileptiform activity in ADNFLE mice, the authors concluded that nAChR-mediated ADNFLE may be caused by hyperactive nAChRs in GABAergic neurons that leads to synchronization of cortical networks.

An interesting feature of ADNFLE-causing nAChR mutants is their differential sensitivity to the antiepileptic drug carbamazepine. Carbamazepine can inhibit nAChR function via open channel blockade and three of the known ADNFLE mutations, two in α4 and one in β2, have substantially increased sensitivity to inhibition by carbamazepine (Bertrand et al., 2005; Hogg and Bertrand, 2004; Picard et al., 1999). For individuals with any of these carbamazepine-sensitive mutations, carbamazepine has proven to be an effective treatment. In contrast, other ADNFLE-causing mutations in CHRNA4 and CHRNA2 either do not show altered sensitivity to carbamazepine or actually show a reduced sensitivity to inhibition by this drug (Bertrand et al., 2005; Hoda et al., 2009; Leniger et al., 2003). Individuals with these mutations apparently do not benefit from carbamazepine treatment. Thus knowing which nAChR mutation a patient carries can be a valuable aid in treatment selection. However, it should be pointed out that mutations in nAChR subunit genes only account for a fraction of total ADNFLE cases and therefore, the predictive power of nAChR subunit gene mutation identification is restricted to a small percentage of ADNFLE patients.

4.2 *Other Genetic Variants in nAChR Subunit Genes and Their Relation to Diseases of the Brain*

A significant number of polymorphisms and rare mutations in nAChR subunit genes have been implicated in various diseases of the brain through linkage and association studies. Diseases thought to be influenced by nAChR subunit gene variants include schizophrenia, Alzheimer's disease, non-ADNFLE epilepsies, various cognitive disorders including attention deficits, and drug addiction-related phenotypes. Because there have been several recent reviews on this topic (Portugal and Gould, 2008; Steinlein and Bertrand, 2008; Stitzel, 2008) it is not extensively reviewed here. However, at the time of these reviews, studies began appearing that implicated the gene cluster on chromosome 15q24 that contains CHRNA5, CHRNA3, and CHRNB4 in various aspects of addiction to nicotine, alcohol, and cocaine. This cluster of genes encodes the α5, α3, and β4 nAChR subunits, respectively. Because this gene cluster repeatedly has been implicated in influencing individual variability in addiction-related measures over the past two years, it warrants some further discussion. The first two reports that implicated this nAChR gene cluster in addiction were published by Bierut et al. (2007) and

Saccone et al. (2007). These two studies identified single nucleotide polymorphisms (SNPs) in both CHRNA5 and CHRNA3 that were associated with nicotine dependence.

Subsequent studies have confirmed the association between the CHRNA5 and CHRNA3 SNPs and nicotine dependence (Baker et al., 2009; Bierut et al., 2008; Caporaso et al., 2009; Chen et al., 2009a; Saccone et al., 2009; Spitz et al., 2008; Stevens et al., 2008; Thorgeirsson et al., 2008; Wang et al., 2009; Weiss et al., 2008) as well as implicated the gene cluster in individual differences in level of smoking (Berrettini et al., 2008; Le et al., 2008), subjective effects of smoking (Sherva et al., 2008), age of initiation of smoking (Schlaepfer et al., 2008), cocaine addiction (Grucza et al., 2008), and alcohol dependence (Joslyn et al., 2008; Schlaepfer et al., 2008; Wang et al., 2008). The same SNPS also have been associated with risk for lung cancer (Amos et al., 2008; Hung et al., 2008; Liu et al., 2008; Shiraishi et al., 2009; Spitz et al., 2008; Thorgeirsson et al., 2008) and chronic obstructive pulmonary disease (COPD) (Pillai et al., 2009; Young et al., 2008). Whether the association between the CHRNA5-CHRNA3 SNPs and lung cancer or COPD are due to an altered risk for smoking or represent an independent signal remains a matter of debate (Volkow et al., 2008). Although beyond the scope of this review (see Egleton et al. (2008), and Song and Spindel (2008) for recent reviews of this topic), nAChRs, including those that contain the α3 and/or α5 subunit are expressed in pulmonary epithelial cells and lung cancer cells so an independent role of SNPs in CHRNA3 and/or CHRNA5 on risk for these diseases certainly is feasible.

Although the repeated associations between the CHRNA5–CHRNA3–CHRNB4 gene cluster and the mentioned addiction-related measures strongly suggest that there is one or more polymorphism in the cluster that alters risk for drug use and abuse, the identity of the causative SNP or SNPs is not known. However, a strong candidate is an amino-acid-altering SNP in CHRNA5 that changes a highly conserved aspartic acid codon at amino acid position 398 in the α5 subunit to an asparagine codon. Preliminary in vitro data indicate that the amino acid change associated with increased risk for nicotine dependence (asparagine at position 398) reduces the function of α4β2α5 nAChRs (Bierut et al., 2008). Whether this functional effect of the polymorphism is responsible for altered liability to nicotine dependence and if so, by what mechanism does the change in function of α4β2α5 nAChRs alter addiction risk, are questions that remain to be answered.

5 Diseases Where nAChRs Are Implicated by Therapeutic Effects of Nicotine

A putative role for nAChRs in schizophrenia was suggested by the high rate of smoking in schizophrenic patients and the observation that nicotine normalizes neurophysiological deficits associated with the disease. As described elsewhere in this review, subsequent studies provided strong evidence for a role of nAChRs in the etiology of this disease. However, there are some diseases where nicotine has been shown to have therapeutic value although a specific role of nAChRs has yet to be established. Two examples of such diseases are discussed here.

5.1 Tourette Syndrome

Tourette syndrome is a neurological disorder characterized by repetitive, stereotyped, involuntary movements and vocalizations called tics (NINDS, 2008). In cases where the tics interfere with normal functioning, therapeutics such as haloperidol often are used. The first evidence for the role of nicotinic receptors in Tourette syndrome came from a study by Sanberg et al. (1988) that reported that nicotine gum in combination with haloperidol improved symptoms in two patients where haloperidol alone was without effect. Follow-up studies have confirmed that nicotine gum potentiates the therapeutic effects of haloperidol in Tourette syndrome (McConville et al., 1991, 1992; Sanberg et al., 1989). In addition, the use of a transdermal nicotine patch rather than nicotine gum has been shown to have long-lasting potentiation of the effects of neuroleptics on tic frequency and severity (Dursun et al., 1994; Silver et al., 1996, 2001). Dursun et al. (1994) also reported that nicotine alone improved Tourette syndrome symptoms. In studies where it has been assessed, combined nicotine/neuroleptic treatment also improved measures of attention in Tourette syndrome patients relative to neuroleptic treatment alone (Dursun et al., 1994; Howson et al., 2004). Although the mechanism through which nicotine improves symptoms of Tourette syndrome is not known, a relatively recent study demonstrated that nicotine normalized deficits in inhibitory function of motor cortex in Tourette syndrome patients (Orth et al., 2005). Nonetheless, there are no pharmacological data to suggest which nAChR subtypes might be responsible for the therapeutic effects of nicotine in this disease and no postmortem data to evaluate whether there might be abnormalities in nAChR expression that may directly contribute to the disease.

5.2 Down Syndrome

Down syndrome is a genetic disease caused by the inheritance of an extra copy (trisomy) of chromosome 21. In addition to some common physical features and health problems, most subjects with Down syndrome also have mild to moderate mental retardation. Postmortem brain tissue of Down syndrome patients exhibits amyloid plaques (Burger and Vogel, 1973; Ellis et al., 1974) and cholinergic deficits (Yates et al., 1980) similar to those observed in postmortem brain tissue from Alzheimer patients. In addition, studies with primary cultures from Down syndrome patient brain or cell lines derived from a mouse model of Down syndrome (trisomy 16) suggest that there are cholinergic deficiencies in trisomy 21/16 neurons (Allen et al., 2000; Cardenas et al., 2002; Fiedler et al., 1994). Based on the apparent cholinergic deficits in Down syndrome, Lubec and colleagues (Bernert et al., 2001; Seidl et al.,

2000) examined whether transdermal nicotine could improve some of the cognitive deficits associated with Down Syndrome. In both published studies, nicotine was found to improve cognitive performance in the Down syndrome subjects. However, despite the cholinergic deficits and presumably related therapeutic effect of nicotine in Down syndrome, a specific contribution of nAChRs remains to be established for this disease. For example, neither Lee et al. (2002) nor Ray et al. (2005) found any deficits in [3H] epibatidine or [125I] α bungarotoxin binding in postmortem brain of Down syndrome patients. These findings contradict the observation by Engidawork et al. (2001) that the expression of α3 and α7 subunits is altered in Down syndrome. This apparent discrepancy likely is due to the fact that Engidawork et al. (2001) used immunohistochemical methods to detect nAChR subunits. The use of antibodies for standard immunohistochemical detection of nAChR subunits has come under recent scrutiny (Jones and Wonnacott, 2005; Moser et al., 2007).

Another mechanism proposed for the therapeutic effect of nicotine in Down syndrome is that the high levels of β amyloid present in the Down syndrome brain are inhibiting the function of α7 nAChRs essentially as described above in Alzheimer's disease (Deutsch et al., 2003). However, a recent report found no correlation between β amyloid levels and dementia in older Down syndrome subjects (Jones et al., 2009). Therefore, despite the therapeutic effect of nicotine in Down syndrome, the specific role of nAChRs remains elusive.

6 Conclusions

The research summarized in this review suggests that nAChRs contribute to a wide range of neuropathologies. In many cases the combined therapeutic effect of nicotine and/or nicotinic drug in addition to detectable differences in nAChR expression provides compelling evidence for a contribution of nAChRs to neuropathology. However, in the diseases that fall into this category, including Alzheimer's and Parkinson's disease, schizophrenia, and autism (among others), the mechanism responsible for the altered expression of the nAChRs is not known. Moreover, whether the altered expression of nAChRs and these diseases is causal or casual remains to be established. In the case of ADNFLE, identified mutations in nAChR subunit genes clearly define a role of nAChRs in diseases of the brain and animal models provide a plausible mechanism. In contrast, the contribution of genetic variants in genes that code for nAChR subunits in diseases other than ADNFLE is only beginning to emerge. Not surprisingly, very little is known regarding the biological mechanisms responsible for the associations between nAChR subunit gene variation and diseases such as nicotine dependence. Finally, there are several diseases such as Tourette syndrome and Down syndrome where nicotine has therapeutic effects despite the lack of any detectable alterations in nAChR expression or function. In summary, there is substantial evidence that nAChRs contribute to a wide assortment of brain disease. Nonetheless, much work remains to be done to establish the mechanisms through which nAChRs contribute to the etiology of disease.

Acknowedgments This work was supported by grants from the NIH (CA089392, DA022462, MH068582).

References

Abood LG, Reynolds DT, Bidlack JM (1980) Stereospecific 3H-nicotine binding to intact and solubilized rat brain membranes and evidence for its noncholinergic nature. Life Sci 27: 1307–1314

Abood LG, Reynolds DT, Booth H, Bidlack JM (1981) Sites and mechanisms for nicotine's action in the brain. Neurosci Biobehav Rev 5:479–486

Adams CE, Stitzel JA, Collins AC, Freedman R (2001) Alpha7-nicotinic receptor expression and the anatomical organization of hippocampal interneurons. Brain Res 922:180–190

Adams CE, Yonchek JC, Stitzel JA (2006) Development of hippocampal alpha7 nicotinic receptors in C3H and DBA/2 congenic mice. Brain Res 1122:27–35

Adams CE, Yonchek JC, Zheng L, Collins AC, Stevens KE (2008) Altered hippocampal circuit function in C3H alpha7 null mutant heterozygous mice. Brain Res 1194:138–145

Adler LE, Hoffer LJ, Griffith J, Waldo MC, Freedman R (1992) Normalization by nicotine of deficient auditory sensory gating in the relatives of schizophrenics. Biol Psychiatry 32: 607–616

Allen DD, Martin J, Arriagada C, Cardenas AM, Rapoport SI, Caviedes R, Caviedes P (2000) Impaired cholinergic function in cell lines derived from the cerebral cortex of normal and trisomy 16 mice. Eur J Neurosci 12:3259–3264

Amos CI, Wu X, Broderick P, Gorlov IP, Gu J, Eisen T, Dong Q, Zhang Q, Gu X, Vijayakrishnan J, Sullivan K, Matakidou A, Wang Y, Mills G, Doheny K, Tsai YY, Chen WV, Shete S, Spitz MR, Houlston RS (2008) Genome-wide association scan of tag SNPs identifies a susceptibility locus for lung cancer at 15q25.1. Nat Genet 40:616–622

Aridon P, Marini C, Di Resta C, Brilli E, De Fusco M, Politi F, Parrini E, Manfredi I, Pisano T, Pruna D, Curia G, Cianchetti C, Pasqualetti M, Becchetti A, Guerrini R, Casari G (2006) Increased sensitivity of the neuronal nicotinic receptor alpha 2 subunit causes familial epilepsy with nocturnal wandering and ictal fear. Am J Hum Genet 79:342–350

Aubert I, Araujo DM, Cecyre D, Robitaille Y, Gauthier S, Quirion R (1992) Comparative alterations of nicotinic and muscarinic binding sites in Alzheimer's and Parkinson's diseases. J Neurochem 58:529–541

Austin J (2005) Schizophrenia: an update and review. J Genet Couns 14:329–340

Baker N, Adler LE, Franks RD, Waldo M, Berry S, Nagamoto H, Muckle A, Freedman R (1987) Neurophysiological assessment of sensory gating in psychiatric inpatients: comparison between schizophrenia and other diagnoses. Biol Psychiatry 22:603–617

Baker TB, Weiss RB, Bolt D, von Niederhausern A, Fiore MC, Dunn DM, Piper ME, Matsunami N, Smith SS, Coon H, McMahon WM, Scholand MB, Singh N, Hoidal JR, Kim SY, Leppert MF, Cannon DS (2009) Human neuronal acetylcholine receptor A5-A3-B4 haplotypes are associated with multiple nicotine dependence phenotypes. Nicotine Tob Res 11(7):785–796

Banerjee C, Nyengaard JR, Wevers A, de Vos RA, Jansen Steur EN, Lindstrom J, Pilz K, Nowacki S, Bloch W, Schroder H (2000) Cellular expression of alpha7 nicotinic acetylcholine receptor protein in the temporal cortex in Alzheimer's and Parkinson's disease – a stereological approach. Neurobiol Dis 7:666–672

Bauman ML, Kemper TL (2005) Neuroanatomic observations of the brain in autism: a review and future directions. Int J Dev Neurosci 23:183–187

Bednar I, Paterson D, Marutle A, Pham TM, Svedberg M, Hellstrom-Lindahl E, Mousavi M, Court J, Morris C, Perry E, Mohammed A, Zhang X, Nordberg A (2002) Selective nicotinic receptor consequences in APP(SWE) transgenic mice. Mol Cell Neurosci 20:354–365

Bejerot S, Nylander L (2003) Low prevalence of smoking in patients with autism spectrum disorders. Psychiatry Res 119:177–182

Benwell ME, Balfour DJ, Anderson JM (1988) Evidence that tobacco smoking increases the density of (-)-[3H]nicotine binding sites in human brain. J Neurochem 50:1243–1247

Bernert G, Sustrova M, Sovcikova E, Seidl R, Lubec G (2001) Effects of a single transdermal nicotine dose on cognitive performance in adults with Down syndrome. J Neural Transm Suppl (61):237–245

Berrettini W, Yuan X, Tozzi F, Song K, Francks C, Chilcoat H, Waterworth D, Muglia P, Mooser V (2008) Alpha-5/alpha-3 nicotinic receptor subunit alleles increase risk for heavy smoking. Mol Psychiatry 13:368–373

Bertrand D, Elmslie F, Hughes E, Trounce J, Sander T, Bertrand S, Steinlein OK (2005) The CHRNB2 mutation I312M is associated with epilepsy and distinct memory deficits. Neurobiol Dis 20:799–804

Bertrand D, Picard F, Le Hellard S, Weiland S, Favre I, Phillips H, Bertrand S, Berkovic SF, Malafosse A, Mulley J (2002) How mutations in the nAChRs can cause ADNFLE epilepsy. Epilepsia 43(Suppl 5):112–122

Bierut LJ, Madden PA, Breslau N, Johnson EO, Hatsukami D, Pomerleau OF, Swan GE, Rutter J, Bertelsen S, Fox L, Fugman D, Goate AM, Hinrichs AL, Konvicka K, Martin NG, Montgomery GW, Saccone NL, Saccone SF, Wang JC, Chase GA, Rice JP, Ballinger DG (2007) Novel genes identified in a high-density genome wide association study for nicotine dependence. Hum Mol Genet 16:24–35

Bierut LJ, Stitzel JA, Wang JC, Hinrichs AL, Grucza RA, Xuei X, Saccone NL, Saccone SF, Bertelsen S, Fox L, Horton WJ, Breslau N, Budde J, Cloninger CR, Dick DM, Foroud T, Hatsukami D, Hesselbrock V, Johnson EO, Kramer J, Kuperman S, Madden PA, Mayo K, Nurnberger J Jr, Pomerleau O, Porjesz B, Reyes O, Schuckit M, Swan G, Tischfield JA, Edenberg HJ, Rice JP, Goate AM (2008) Variants in nicotinic receptors and risk for nicotine dependence. Am J Psychiatry 165:1163–1171

Bohr IJ, Ray MA, McIntosh JM, Chalon S, Guilloteau D, McKeith IG, Perry RH, Clementi F, Perry EK, Court JA, Piggott MA (2005) Cholinergic nicotinic receptor involvement in movement disorders associated with Lewy body diseases. An autoradiography study using [(125)I]alpha-conotoxinMII in the striatum and thalamus. Exp Neurol 191:292–300

Bordia T, Grady SR, McIntosh JM, Quik M (2007) Nigrostriatal damage preferentially decreases a subpopulation of alpha6beta2* nAChRs in mouse, monkey, and Parkinson's disease striatum. Mol Pharmacol 72:52–61

Boutros NN, Belger A, Campbell D, D'Souza C, Krystal J (1999) Comparison of four components of sensory gating in schizophrenia and normal subjects: a preliminary report. Psychiatry Res 88:119–130

Boyd RT (1997) The molecular biology of neuronal nicotinic acetylcholine receptors. Crit Rev Toxicol 27:299–318

Braff DL, Geyer MA, Swerdlow NR (2001) Human studies of prepulse inhibition of startle: normal subjects, patient groups, and pharmacological studies. Psychopharmacology (Berl) 156: 234–258

Breese CR, Lee MJ, Adams CE, Sullivan B, Logel J, Gillen KM, Marks MJ, Collins AC, Leonard S (2000) Abnormal regulation of high affinity nicotinic receptors in subjects with schizophrenia. Neuropsychopharmacology 23:351–364

Breese CR, Marks MJ, Logel J, Adams CE, Sullivan B, Collins AC, Leonard S (1997) Effect of smoking history on [3H]nicotine binding in human postmortem brain. J Pharmacol Exp Ther 282:7–13

Brodtkorb E, Picard F (2006) Tobacco habits modulate autosomal dominant nocturnal frontal lobe epilepsy. Epilepsy Behav 9:515–520

Burger PC, Vogel FS (1973) The development of the pathologic changes of Alzheimer's disease and senile dementia in patients with Down's syndrome. Am J Pathol 73:457–476

Burghaus L, Schutz U, Krempel U, Lindstrom J, Schroder H (2003) Loss of nicotinic acetylcholine receptor subunits alpha4 and alpha7 in the cerebral cortex of Parkinson patients. Parkinsonism Relat Disord 9:243–246

Burghaus L, Schutz U, Krempel U, de Vos RA, Jansen Steur EN, Wevers A, Lindstrom J, Schroder H (2000) Quantitative assessment of nicotinic acetylcholine receptor proteins in the cerebral cortex of Alzheimer patients. Brain Res Mol Brain Res 76:385–388

Caporaso N, Gu F, Chatterjee N, Sheng-Chih J, Yu K, Yeager M, Chen C, Jacobs K, Wheeler W, Landi MT, Ziegler RG, Hunter DJ, Chanock S, Hankinson S, Kraft P, Bergen AW (2009) Genome-wide and candidate gene association study of cigarette smoking behaviors. PLoS ONE 4:e4653

Cardenas AM, Arriagada C, Allen DD, Caviedes R, Cortes JF, Martin J, Couve E, Rapoport SI, Shimahara T, Caviedes P (2002) Cell lines derived from hippocampal neurons of the normal and trisomy 16 mouse fetus (a model for Down syndrome) exhibit neuronal markers, cholinergic function, and functional neurotransmitter receptors. Exp Neurol 177:159–170

Champtiaux N, Han ZY, Bessis A, Rossi FM, Zoli M, Marubio L, McIntosh JM, Changeux JP (2002) Distribution and pharmacology of alpha 6-containing nicotinic acetylcholine receptors analyzed with mutant mice. J Neurosci 22:1208–1217

Chen X, Chen J, Williamson VS, An SS, Hettema JM, Aggen SH, Neale MC, Kendler KS (2009a) Variants in nicotinic acetylcholine receptors alpha5 and alpha3 increase risks to nicotine dependence. Am J Med Genet B Neuropsychiatr Genet 150B(7):926–933

Chen D, Patrick JW (1997) The alpha-bungarotoxin-binding nicotinic acetylcholine receptor from rat brain contains only the alpha7 subunit. J Biol Chem 272:24024–24029

Chen Y, Wu L, Fang Y, He Z, Peng B, Shen Y, Xu Q (2009b) A novel mutation of the nicotinic acetylcholine receptor gene CHRNA4 in sporadic nocturnal frontal lobe epilepsy. Epilepsy Res 83:152–156

Chin JH, Tse FW, Harris K, Jhamandas JH (2006) Beta-amyloid enhances intracellular calcium rises mediated by repeated activation of intracellular calcium stores and nicotinic receptors in acutely dissociated rat basal forebrain neurons. Brain Cell Biol 35:173–186

Clarke PB, Schwartz RD, Paul SM, Pert CB, Pert A (1985) Nicotinic binding in rat brain: autoradiographic comparison of [3H]acetylcholine [3H]nicotine, and [125I]-alpha-bungarotoxin. J Neurosci 5:1307–1315

Clementz BA, Geyer MA, Braff DL (1998) Poor P50 suppression among schizophrenia patients and their first-degree biological relatives. Am J Psychiatry 155:1691–1694

Counts SE, He B, Che S, Ikonomovic MD, Dekosky ST, Ginsberg SD, Mufson EJ (2007) Alpha7 nicotinic receptor up-regulation in cholinergic basal forebrain neurons in Alzheimer disease. Arch Neurol 64:1771–1776

Court JA, Piggott MA, Lloyd S, Cookson N, Ballard CG, McKeith IG, Perry RH, Perry EK (2000) Nicotine binding in human striatum: elevation in schizophrenia and reductions in dementia with Lewy bodies, Parkinson's disease and Alzheimer's disease and in relation to neuroleptic medication. Neuroscience 98:79–87

Court J, Spurden D, Lloyd S, McKeith I, Ballard C, Cairns N, Kerwin R, Perry R, Perry E (1999) Neuronal nicotinic receptors in dementia with Lewy bodies and schizophrenia: alpha-bungarotoxin and nicotine binding in the thalamus. J Neurochem 73:1590–1597

Couturier S, Bertrand D, Matter JM, Hernandez MC, Bertrand S, Millar N, Valera S, Barkas T, Ballivet M (1990) A neuronal nicotinic acetylcholine receptor subunit (alpha 7) is developmentally regulated and forms a homo-oligomeric channel blocked by alpha-BTX. Neuron 5:847–856

Cui C, Booker TK, Allen RS, Grady SR, Whiteaker P, Marks MJ, Salminen O, Tritto T, Butt CM, Allen WR, Stitzel JA, McIntosh JM, Boulter J, Collins AC, Heinemann SF (2003) The beta3 nicotinic receptor subunit: a component of alpha-conotoxin MII-binding nicotinic acetylcholine receptors that modulate dopamine release and related behaviors. J Neurosci 23:11045–11053

Dalack GW, Healy DJ, Meador-Woodruff JH (1998) Nicotine dependence in schizophrenia: clinical phenomena and laboratory findings. Am J Psychiatry 155:1490–1501

De Fusco M, Becchetti A, Patrignani A, Annesi G, Gambardella A, Quattrone A, Ballabio A, Wanke E, Casari G (2000) The nicotinic receptor beta 2 subunit is mutant in nocturnal frontal lobe epilepsy. Nat Genet 26:275–276

Deutsch SI, Rosse RB, Mastropaolo J, Chilton M (2003) Progressive worsening of adaptive functions in Down syndrome may be mediated by the complexing of soluble Abeta peptides with the alpha 7 nicotinic acetylcholine receptor: therapeutic implications. Clin Neuropharmacol 26:277–283

Dineley KT, Westerman M, Bui D, Bell K, Ashe KH, Sweatt JD (2001) Beta-amyloid activates the mitogen-activated protein kinase cascade via hippocampal alpha7 nicotinic acetylcholine receptors: in vitro and in vivo mechanisms related to Alzheimer's disease. J Neurosci 21: 4125–4133

Dineley KT, Xia X, Bui D, Sweatt JD, Zheng H (2002) Accelerated plaque accumulation, associative learning deficits, and up-regulation of alpha 7 nicotinic receptor protein in transgenic mice co-expressing mutant human presenilin 1 and amyloid precursor proteins. J Biol Chem 277:22768–22780

Durany N, Zochling R, Boissl KW, Paulus W, Ransmayr G, Tatschner T, Danielczyk W, Jellinger K, Deckert J, Riederer P (2000) Human post-mortem striatal alpha4beta2 nicotinic acetylcholine receptor density in schizophrenia and Parkinson's syndrome. Neurosci Lett 287:109–112

Dursun SM, Reveley MA, Bird R, Stirton F (1994) Longlasting improvement of Tourette's syndrome with transdermal nicotine. Lancet 344:1577

Egleton RD, Brown KC, Dasgupta P (2008) Nicotinic acetylcholine receptors in cancer: multiple roles in proliferation and inhibition of apoptosis. Trends Pharmacol Sci 29:151–158

Ellis WG, McCulloch JR, Corley CL (1974) Presenile dementia in Down's syndrome. Ultrastructural identity with Alzheimer's disease. Neurology 24:101–106

Ellis JR, Nathan PJ, Villemagne VL, Mulligan RS, Saunder T, Young K, Smith CL, Welch J, Woodward M, Wesnes KA, Savage G, Rowe CC (2009) Galantamine-induced improvements in cognitive function are not related to alterations in alpha(4)beta (2) nicotinic receptors in early Alzheimer's disease as measured in vivo by 2-[(18)F]Fluoro-A-85380 PET. Psychopharmacology (Berl) 202:79–91

Ellis JR, Villemagne VL, Nathan PJ, Mulligan RS, Gong SJ, Chan JG, Sachinidis J, O'Keefe GJ, Pathmaraj K, Wesnes KA, Savage G, Rowe CC (2008) Relationship between nicotinic receptors and cognitive function in early Alzheimer's disease: a 2-[18F]fluoro-A-85380 PET study. Neurobiol Learn Mem 90:404–412

Engidawork E, Gulesserian T, Balic N, Cairns N, Lubec G (2001) Changes in nicotinic acetylcholine receptor subunits expression in brain of patients with Down syndrome and Alzheimer's disease. J Neural Transm Suppl (61):211–222

Erwin RJ, Turetsky BI, Moberg P, Gur RC, Gur RE (1998) P50 abnormalities in schizophrenia: relationship to clinical and neuropsychological indices of attention. Schizophr Res 33: 157–167

Fiedler JL, Epstein CJ, Rapoport SI, Caviedes R, Caviedes P (1994) Regional alteration of cholinergic function in central neurons of trisomy 16 mouse fetuses, an animal model of human trisomy 21 (Down syndrome). Brain Res 658:27–32

Flores CM, Rogers SW, Pabreza LA, Wolfe BB, Kellar KJ (1992) A subtype of nicotinic cholinergic receptor in rat brain is composed of alpha 4 and beta 2 subunits and is up-regulated by chronic nicotine treatment. Mol Pharmacol 41:31–37

Freedman R, Hall M, Adler LE, Leonard S (1995) Evidence in postmortem brain tissue for decreased numbers of hippocampal nicotinic receptors in schizophrenia. Biol Psychiatry 38:22–33

Freedman R, Olincy A, Buchanan RW, Harris JG, Gold JM, Johnson L, Allensworth D, Guzman-Bonilla A, Clement B, Ball MP, Kutnick J, Pender V, Martin LF, Stevens KE, Wagner BD, Zerbe GO, Soti F, Kem WR (2008) Initial phase 2 trial of a nicotinic agonist in schizophrenia. Am J Psychiatry 165:1040–1047

Fu W, Jhamandas JH (2003) Beta-amyloid peptide activates non-alpha7 nicotinic acetylcholine receptors in rat basal forebrain neurons. J Neurophysiol 90:3130–3136

Fucile S (2004) Ca2+ permeability of nicotinic acetylcholine receptors. Cell Calcium 35:1–8

Fucile S, Renzi M, Lax P, Eusebi F (2003) Fractional Ca(2+) current through human neuronal alpha7 nicotinic acetylcholine receptors. Cell Calcium 34:205–209

Gahring LC, Meyer EL, Rogers SW (2003) Nicotine-induced neuroprotection against N-methyl-D-aspartic acid or beta-amyloid peptide occur through independent mechanisms distinguished by pro-inflammatory cytokines. J Neurochem 87:1125–1136

Giniatullin R, Nistri A, Yakel JL (2005) Desensitization of nicotinic ACh receptors: shaping cholinergic signaling. Trends Neurosci 28:371–378

Gotti C, Moretti M, Bohr I, Ziabreva I, Vailati S, Longhi R, Riganti L, Gaimarri A, McKeith IG, Perry RH, Aarsland D, Larsen JP, Sher E, Beattie R, Clementi F, Court JA (2006a) Selective nicotinic acetylcholine receptor subunit deficits identified in Alzheimer's disease, Parkinson's disease and dementia with Lewy bodies by immunoprecipitation. Neurobiol Dis 23: 481–489

Gotti C, Zoli M, Clementi F (2006b) Brain nicotinic acetylcholine receptors: native subtypes and their relevance. Trends Pharmacol Sci 27:482–491

Grady SR, Meinerz NM, Cao J, Reynolds AM, Picciotto MR, Changeux JP, McIntosh JM, Marks MJ, Collins AC (2001) Nicotinic agonists stimulate acetylcholine release from mouse interpeduncular nucleus: a function mediated by a different nAChR than dopamine release from striatum. J Neurochem 76:258–268

Grady SR, Moretti M, Zoli M, Marks MJ, Zanardi A, Pucci L, Clementi F, Gotti C (2009) Rodent habenulo-interpeduncular pathway expresses a large variety of uncommon nAChR subtypes, but only the alpha3beta4* and alpha3beta3beta4* subtypes mediate acetylcholine release. J Neurosci 29:2272–2282

Grassi F, Palma E, Tonini R, Amici M, Ballivet M, Eusebi F (2003) Amyloid beta(1-42) peptide alters the gating of human and mouse alpha-bungarotoxin-sensitive nicotinic receptors. J Physiol 547:147–157

Grucza RA, Wang JC, Stitzel JA, Hinrichs AL, Saccone SF, Saccone NL, Bucholz KK, Cloninger CR, Neuman RJ, Budde JP, Fox L, Bertelsen S, Kramer J, Hesselbrock V, Tischfield J, Nurnberger JI Jr, Almasy L, Porjesz B, Kuperman S, Schuckit MA, Edenberg HJ, Rice JP, Goate AM, Bierut LJ (2008) A risk allele for nicotine dependence in CHRNA5 is a protective allele for cocaine dependence. Biol Psychiatry 64:922–929

Guan ZZ, Nordberg A, Mousavi M, Rinne JO, Hellstrom-Lindahl E (2002) Selective changes in the levels of nicotinic acetylcholine receptor protein and of corresponding mRNA species in the brains of patients with Parkinson's disease. Brain Res 956:358–366

Guan ZZ, Zhang X, Blennow K, Nordberg A (1999) Decreased protein level of nicotinic receptor alpha7 subunit in the frontal cortex from schizophrenic brain. Neuroreport 10: 1779–1782

Guan ZZ, Zhang X, Ravid R, Nordberg A (2000) Decreased protein levels of nicotinic receptor subunits in the hippocampus and temporal cortex of patients with Alzheimer's disease. J Neurochem 74:237–243

Harris JG, Kongs S, Allensworth D, Martin L, Tregellas J, Sullivan B, Zerbe G, Freedman R (2004) Effects of nicotine on cognitive deficits in schizophrenia. Neuropsychopharmacology 29:1378–1385

Hellstrom-Lindahl E, Mousavi M, Ravid R, Nordberg A (2004) Reduced levels of Abeta 40 and Abeta 42 in brains of smoking controls and Alzheimer's patients. Neurobiol Dis 15:351–360

Hellstrom-Lindahl E, Mousavi M, Zhang X, Ravid R, Nordberg A (1999) Regional distribution of nicotinic receptor subunit mRNAs in human brain: comparison between Alzheimer and normal brain. Brain Res Mol Brain Res 66:94–103

Hirose S, Iwata H, Akiyoshi H, Kobayashi K, Ito M, Wada K, Kaneko S, Mitsudome A (1999) A novel mutation of CHRNA4 responsible for autosomal dominant nocturnal frontal lobe epilepsy. Neurology 53:1749–1753

Hoda JC, Gu W, Friedli M, Phillips HA, Bertrand S, Antonarakis SE, Goudie D, Roberts R, Scheffer IE, Marini C, Patel J, Berkovic SF, Mulley JC, Steinlein OK, Bertrand D (2008) Human nocturnal frontal lobe epilepsy: pharmocogenomic profiles of pathogenic nicotinic acetylcholine receptor beta-subunit mutations outside the ion channel pore. Mol Pharmacol 74:379–391

Hoda JC, Wanischeck M, Bertrand D, Steinlein OK (2009) Pleiotropic functional effects of the first epilepsy-associated mutation in the human CHRNA2 gene. FEBS Lett 583(10):1599–1604

Hogg RC, Bertrand D (2004) Nicotinic acetylcholine receptors as drug targets. Curr Drug Targets CNS Neurol Disord 3:123–130

Holzman PS (2000) Eye movements and the search for the essence of schizophrenia. Brain Res Brain Res Rev 31:350–356

Howson AL, Batth S, Ilivitsky V, Boisjoli A, Jaworski M, Mahoney C, Knott VJ (2004) Clinical and attentional effects of acute nicotine treatment in Tourette's syndrome. Eur Psychiatry 19:102–112

Hughes JR, Hatsukami DK, Mitchell JE, Dahlgren LA (1986) Prevalence of smoking among psychiatric outpatients. Am J Psychiatry 143:993–997

Hulihan-Giblin BA, Lumpkin MD, Kellar KJ (1990) Effects of chronic administration of nicotine on prolactin release in the rat: inactivation of prolactin response by repeated injections of nicotine. J Pharmacol Exp Ther 252:21–25

Hung RJ, McKay JD, Gaborieau V, Boffetta P, Hashibe M, Zaridze D, Mukeria A, Szeszenia-Dabrowska N, Lissowska J, Rudnai P, Fabianova E, Mates D, Bencko V, Foretova L, Janout V, Chen C, Goodman G, Field JK, Liloglou T, Xinarianos G, Cassidy A, McLaughlin J, Liu G, Narod S, Krokan HE, Skorpen F, Elvestad MB, Hveem K, Vatten L, Linseisen J, Clavel-Chapelon F, Vineis P, Bueno-de-Mesquita HB, Lund E, Martinez C, Bingham S, Rasmuson T, Hainaut P, Riboli E, Ahrens W, Benhamou S, Lagiou P, Trichopoulos D, Holcatova I, Merletti F, Kjaerheim K, Agudo A, Macfarlane G, Talamini R, Simonato L, Lowry R, Conway DI, Znaor A, Healy C, Zelenika D, Boland A, Delepine M, Foglio M, Lechner D, Matsuda F, Blanche H, Gut I, Heath S, Lathrop M, Brennan P (2008) A susceptibility locus for lung cancer maps to nicotinic acetylcholine receptor subunit genes on 15q25. Nature 452:633–637

Jones EL, Hanney M, Francis PT, Ballard CG (2009) Amyloid beta concentrations in older people with Down syndrome and dementia. Neurosci Lett 451:162–164

Jones IW, Westmacott A, Chan E, Jones RW, Dineley K, O'Neill MJ, Wonnacott S (2006) Alpha7 nicotinic acetylcholine receptor expression in Alzheimer's disease: receptor densities in brain regions of the APP(SWE) mouse model and in human peripheral blood lymphocytes. J Mol Neurosci 30:83–84

Jones IW, Wonnacott S (2005) Why doesn't nicotinic ACh receptor immunoreactivity knock out? Trends Neurosci 28:343–345

Joslyn G, Brush G, Robertson M, Smith TL, Kalmijn J, Schuckit M, White RL (2008) Chromosome 15q25.1 genetic markers associated with level of response to alcohol in humans. Proc Natl Acad Sci U S A 105:20368–20373

Kadir A, Almkvist O, Wall A, Langstrom B, Nordberg A (2006) PET imaging of cortical 11C-nicotine binding correlates with the cognitive function of attention in Alzheimer's disease. Psychopharmacology (Berl) 188:509–520

Kelley MP, Bakan P (1999) Eye tracking in normals: spem asymmetries and association with schizotypy. Int J Neurosci 98:27–81

Kihara T, Shimohama S, Akaike A (1999) Effects of nicotinic receptor agonists on beta-amyloid beta-sheet formation. Jpn J Pharmacol 79:393–396

Kihara T, Shimohama S, Sawada H, Honda K, Nakamizo T, Shibasaki H, Kume T, Akaike A (2001) alpha 7 nicotinic receptor transduces signals to phosphatidylinositol 3-kinase to block A beta-amyloid-induced neurotoxicity. J Biol Chem 276:13541–13546

Kihara T, Shimohama S, Urushitani M, Sawada H, Kimura J, Kume T, Maeda T, Akaike A (1998) Stimulation of alpha4beta2 nicotinic acetylcholine receptors inhibits beta-amyloid toxicity. Brain Res 792:331–334

Klaassen A, Glykys J, Maguire J, Labarca C, Mody I, Boulter J (2006) Seizures and enhanced cortical GABAergic inhibition in two mouse models of human autosomal dominant nocturnal frontal lobe epilepsy. Proc Natl Acad Sci U S A 10350:19152–19157

Klink R, de Kerchove dA, Zoli M, Changeux JP (2001) Molecular and physiological diversity of nicotinic acetylcholine receptors in the midbrain dopaminergic nuclei. J Neurosci 21: 1452–1463

Kulak JM, Musachio JL, McIntosh JM, Quik M (2002) Declines in different beta2* nicotinic receptor populations in monkey striatum after nigrostriatal damage. J Pharmacol Exp Ther 303:633–639

Labarca C, Schwarz J, Deshpande P, Schwarz S, Nowak MW, Fonck C, Nashmi R, Kofuji P, Dang H, Shi W, Fidan M, Khakh BS, Chen Z, Bowers BJ, Boulter J, Wehner JM, Lester HA (2001) Point mutant mice with hypersensitive alpha 4 nicotinic receptors show dopaminergic deficits and increased anxiety. Proc Natl Acad Sci U S A 98:2786–2791

Lamb PW, Melton MA, Yakel JL (2005) Inhibition of neuronal nicotinic acetylcholine receptor channels expressed in Xenopus oocytes by beta-amyloid1-42 peptide. J Mol Neurosci 27: 13–21

Lange KW, Wells FR, Jenner P, Marsden CD (1993) Altered muscarinic and nicotinic receptor densities in cortical and subcortical brain regions in Parkinson's disease. J Neurochem 60: 197–203

Le ML, Derby KS, Murphy SE, Hecht SS, Hatsukami D, Carmella SG, Tiirikainen M, Wang H (2008) Smokers with the CHRNA lung cancer-associated variants are exposed to higher levels of nicotine equivalents and a carcinogenic tobacco-specific nitrosamine. Cancer Res 68: 9137–9140

Lee M, Martin-Ruiz C, Graham A, Court J, Jaros E, Perry R, Iversen P, Bauman M, Perry E (2002) Nicotinic receptor abnormalities in the cerebellar cortex in autism. Brain 125:1483–1495

Lee KH, Williams LM (2000) Eye movement dysfunction as a biological marker of risk for schizophrenia. Aust N Z J Psychiatry 34(Suppl):S91–S100

Leniger T, Kananura C, Hufnagel A, Bertrand S, Bertrand D, Steinlein OK (2003) A new Chrna4 mutation with low penetrance in nocturnal frontal lobe epilepsy. Epilepsia 44:981–985

Lippiello PM (2006) Nicotinic cholinergic antagonists: a novel approach for the treatment of autism. Med Hypotheses 66:985–990

Liu Q, Huang Y, Xue F, Simard A, DeChon J, Li G, Zhang J, Lucero L, Wang M, Sierks M, Hu G, Chang Y, Lukas RJ, Wu J (2009) A novel nicotinic acetylcholine receptor subtype in basal forebrain cholinergic neurons with high sensitivity to amyloid peptides. J Neurosci 29:918–929

Liu Q, Kawai H, Berg DK (2001) beta -Amyloid peptide blocks the response of alpha 7-containing nicotinic receptors on hippocampal neurons. Proc Natl Acad Sci U S A 98: 4734–4739

Liu Z, Neff RA, Berg DK (2006) Sequential interplay of nicotinic and GABAergic signaling guides neuronal development. Science 314:1610–1613

Liu P, Vikis HG, Wang D, Lu Y, Wang Y, Schwartz AG, Pinney SM, Yang P, de AM, Petersen GM, Wiest JS, Fain PR, Gazdar A, Gaba C, Rothschild H, Mandal D, Coons T, Lee J, Kupert E, Seminara D, Minna J, Bailey-Wilson JE, Wu X, Spitz MR, Eisen T, Houlston RS, Amos CI, Anderson MW, You M (2008) Familial aggregation of common sequence variants on 15q24-25.1 in lung cancer. J Natl Cancer Inst 100:1326–1330

Liu Q, Zhao B (2004) Nicotine attenuates beta-amyloid peptide-induced neurotoxicity, free radical and calcium accumulation in hippocampal neuronal cultures. Br J Pharmacol 141:746–754

Lohr JB, Flynn K (1992) Smoking and schizophrenia. Schizophr Res 8:93–102

Lu Y, Grady S, Marks MJ, Picciotto M, Changeux JP, Collins AC (1998) Pharmacological characterization of nicotinic receptor-stimulated GABA release from mouse brain synaptosomes. J Pharmacol Exp Ther 287:648–657

Magdesian MH, Nery AA, Martins AH, Juliano MA, Juliano L, Ulrich H, Ferreira ST (2005) Peptide blockers of the inhibition of neuronal nicotinic acetylcholine receptors by amyloid beta. J Biol Chem 280:31085–31090

Mansvelder HD, Keath JR, McGehee DS (2002) Synaptic mechanisms underlie nicotine-induced excitability of brain reward areas. Neuron 33:905–919

Mansvelder HD, McGehee DS (2000) Long-term potentiation of excitatory inputs to brain reward areas by nicotine. Neuron 27:349–357

Marks MJ, Burch JB, Collins AC (1983) Effects of chronic nicotine infusion on tolerance development and nicotinic receptors. J Pharmacol Exp Ther 226:817–825

Marks MJ, Collins AC (1982) Characterization of nicotine binding in mouse brain and comparison with the binding of alpha-bungarotoxin and quinuclidinyl benzilate. Mol Pharmacol 22: 554–564

Marks MJ, Rowell PP, Cao JZ, Grady SR, McCallum SE, Collins AC (2004) Subsets of acetylcholine-stimulated 86Rb+ efflux and [125I]-epibatidine binding sites in C57BL/6 mouse brain are differentially affected by chronic nicotine treatment. Neuropharmacology 46: 1141–1157

Marks MJ, Stitzel JA, Romm E, Wehner JM, Collins AC (1986) Nicotinic binding sites in rat and mouse brain: comparison of acetylcholine, nicotine, and alpha-bungarotoxin. Mol Pharmacol 30:427–436

Martin-Ruiz CM, Court JA, Molnar E, Lee M, Gotti C, Mamalaki A, Tsouloufis T, Tzartos S, Ballard C, Perry RH, Perry EK (1999) Alpha4 but not alpha3 and alpha7 nicotinic acetylcholine receptor subunits are lost from the temporal cortex in Alzheimer's disease. J Neurochem 73:1635–1640

Martin-Ruiz CM, Haroutunian VH, Long P, Young AH, Davis KL, Perry EK, Court JA (2003) Dementia rating and nicotinic receptor expression in the prefrontal cortex in schizophrenia. Biol Psychiatry 54:1222–1233

Martin-Ruiz CM, Lee M, Perry RH, Baumann M, Court JA, Perry EK (2004) Molecular analysis of nicotinic receptor expression in autism. Brain Res Mol Brain Res 123:81–90

Marubio LM, Gardier AM, Durier S, David D, Klink R, Arroyo-Jimenez MM, McIntosh JM, Rossi F, Champtiaux N, Zoli M, Changeux JP (2003) Effects of nicotine in the dopaminergic system of mice lacking the alpha4 subunit of neuronal nicotinic acetylcholine receptors. Eur J Neurosci 17:1329–1337

Marubio LM, Mar Arroyo-Jimenez M, Cordero-Erausquin M, Lena C, Le Novere N, de Kerchove dA, Huchet M, Damaj MI, Changeux JP (1999) Reduced antinociception in mice lacking neuronal nicotinic receptor subunits. Nature 398:805–810

Marutle A, Warpman U, Bogdanovic N, Lannfelt L, Nordberg A (1999) Neuronal nicotinic receptor deficits in Alzheimer patients with the Swedish amyloid precursor protein 670/671 mutation. J Neurochem 72:1161–1169

Marutle A, Zhang X, Court J, Piggott M, Johnson M, Perry R, Perry E, Nordberg A (2001) Laminar distribution of nicotinic receptor subtypes in cortical regions in schizophrenia. J Chem Neuroanat 22:115–126

McClure-Begley TD, King NM, Collins AC, Stitzel JA, Wehner JM, Butt CM (2009) Acetylcholine-stimulated [3H]GABA release from mouse brain synaptosomes is modulated by alpha4beta2 and alpha4alpha5beta2 nicotinic receptor subtypes. Mol Pharmacol 75:918–926

McConville BJ, Fogelson MH, Norman AB, Klykylo WM, Manderscheid PZ, Parker KW, Sanberg PR (1991) Nicotine potentiation of haloperidol in reducing tic frequency in Tourette's disorder. Am J Psychiatry 148:793–794

McConville BJ, Sanberg PR, Fogelson MH, King J, Cirino P, Parker KW, Norman AB (1992) The effects of nicotine plus haloperidol compared to nicotine only and placebo nicotine only in reducing tic severity and frequency in Tourette's disorder. Biol Psychiatry 31:832–840

McLellan A, Phillips HA, Rittey C, Kirkpatrick M, Mulley JC, Goudie D, Stephenson JB, Tolmie J, Scheffer IE, Berkovic SF, Zuberi SM (2003) Phenotypic comparison of two Scottish families with mutations in different genes causing autosomal dominant nocturnal frontal lobe epilepsy. Epilepsia 44:613–617

Morley BJ, Kemp GE, Salvaterra P (1979) Alpha-Bungarotoxin binding sites in the CNS. Life Sci 24:859–872

Moser N, Mechawar N, Jones I, Gochberg-Sarver A, Orr-Urtreger A, Plomann M, Salas R, Molles B, Marubio L, Roth U, Maskos U, Winzer-Serhan U, Bourgeois JP, Le Sourd AM, De Biasi M, Schroder H, Lindstrom J, Maelicke A, Changeux JP, Wevers A (2007) Evaluating the suitability of nicotinic acetylcholine receptor antibodies for standard immunodetection procedures. J Neurochem 102:479–492

Moulard B, Picard F, Le Hellard S, Agulhon C, Weiland S, Favre I, Bertrand S, Malafosse A, Bertrand D (2001) Ion channel variation causes epilepsies. Brain Res Brain Res Rev 36: 275–284

Mousavi M, Bednar I, Nordberg A (2004) Selective changes in expression of different nicotinic receptor subtypes in brain and adrenal glands of mice carrying human mutated gene for APP or over-expressing human acetylcholinestrase. Int J Dev Neurosci 22:545–549

Mousavi M, Hellstrom-Lindahl E, Guan ZZ, Shan KR, Ravid R, Nordberg A (2003) Protein and mRNA levels of nicotinic receptors in brain of tobacco using controls and patients with Alzheimer's disease. Neuroscience 122:515–520

Mueser KT, McGurk SR (2004) Schizophrenia. Lancet 363:2063–2072

Mulle C, Choquet D, Korn H, Changeux JP (1992) Calcium influx through nicotinic receptor in rat central neurons: its relevance to cellular regulation. Neuron 8:135–143

Nagele RG, D'Andrea MR, Anderson WJ, Wang HY (2002) Intracellular accumulation of beta-amyloid(1-42) in neurons is facilitated by the alpha 7 nicotinic acetylcholine receptor in Alzheimer's disease. Neuroscience 110:199–211

NINDS (2008) Tourette syndrome fact sheet. http://www.ninds.nih.gov/disorders/tourette/detail_tourette.htm

Nordberg A, Adem A, Hardy J, Winblad B (1988) Change in nicotinic receptor subtypes in temporal cortex of Alzheimer brains. Neurosci Lett 86:317–321

Nordberg A, Hellstrom-Lindahl E, Lee M, Johnson M, Mousavi M, Hall R, Perry E, Bednar I, Court J (2002) Chronic nicotine treatment reduces beta-amyloidosis in the brain of a mouse model of Alzheimer's disease (APPsw). J Neurochem 81:655–658

Nordberg A, Lundqvist H, Hartvig P, Lilja A, Langstrom B (1995) Kinetic analysis of regional (S)(–)11C-nicotine binding in normal and Alzheimer brains – in vivo assessment using positron emission tomography. Alzheimer Dis Assoc Disord 9:21–27

Nordberg A, Winblad B (1986) Reduced number of [3H]nicotine and [3H]acetylcholine binding sites in the frontal cortex of Alzheimer brains. Neurosci Lett 72:115–119

Olincy A, Harris JG, Johnson LL, Pender V, Kongs S, Allensworth D, Ellis J, Zerbe GO, Leonard S, Stevens KE, Stevens JO, Martin L, Adler LE, Soti F, Kem WR, Freedman R (2006) Proof-of-concept trial of an alpha7 nicotinic agonist in schizophrenia. Arch Gen Psychiatry 63:630–638

Orr-Urtreger A, Goldner FM, Saeki M, Lorenzo I, Goldberg L, De Biasi M, Dani JA, Patrick JW, Beaudet AL (1997) Mice deficient in the alpha7 neuronal nicotinic acetylcholine receptor lack alpha-bungarotoxin binding sites and hippocampal fast nicotinic currents. J Neurosci 17: 9165–9171

Orth M, Amann B, Robertson MM, Rothwell JC (2005) Excitability of motor cortex inhibitory circuits in Tourette syndrome before and after single dose nicotine. Brain 128:1292–1300

Oswald RE, Freeman JA (1981) Alpha-bungarotoxin binding and central nervous system nicotinic acetylcholine receptors. Neuroscience 6:1–14

O'Neill HC, Rieger K, Kem WR, Stevens KE (2003) DMXB, an alpha7 nicotinic agonist, normalizes auditory gating in isolation-reared rats. Psychopharmacology (Berl) 169:332–339

Perry DC, Davila-Garcia MI, Stockmeier CA, Kellar KJ (1999) Increased nicotinic receptors in brains from smokers: membrane binding and autoradiography studies. J Pharmacol Exp Ther 289:1545–1552

Perry EK, Lee ML, Martin-Ruiz CM, Court JA, Volsen SG, Merrit J, Folly E, Iversen PE, Bauman ML, Perry RH, Wenk GL (2001) Cholinergic activity in autism: abnormalities in the cerebral cortex and basal forebrain. Am J Psychiatry 158:1058–1066

Perry E, Martin-Ruiz C, Lee M, Griffiths M, Johnson M, Piggott M, Haroutunian V, Buxbaum JD, Nasland J, Davis K, Gotti C, Clementi F, Tzartos S, Cohen O, Soreq H, Jaros E, Perry R, Ballard C, McKeith I, Court J (2000) Nicotinic receptor subtypes in human brain ageing, Alzheimer and Lewy body diseases. Eur J Pharmacol 393:215–222

Perry EK, Morris CM, Court JA, Cheng A, Fairbairn AF, McKeith IG, Irving D, Brown A, Perry RH (1995) Alteration in nicotine binding sites in Parkinson's disease, Lewy body dementia and Alzheimer's disease: possible index of early neuropathology. Neuroscience 64:385–395

Phillips HA, Favre I, Kirkpatrick M, Zuberi SM, Goudie D, Heron SE, Scheffer IE, Sutherland GR, Berkovic SF, Bertrand D, Mulley JC (2001) CHRNB2 is the second acetylcholine receptor subunit associated with autosomal dominant nocturnal frontal lobe epilepsy. Am J Hum Genet 68:225–231

Phillips HA, Marini C, Scheffer IE, Sutherland GR, Mulley JC, Berkovic SF (2000) A de novo mutation in sporadic nocturnal frontal lobe epilepsy. Ann Neurol 48: 264–267

Picard F, Bertrand S, Steinlein OK, Bertrand D (1999) Mutated nicotinic receptors responsible for autosomal dominant nocturnal frontal lobe epilepsy are more sensitive to carbamazepine. Epilepsia 40:1198–1209

Picciotto MR, Zoli M, Lena C, Bessis A, Lallemand Y, LeNovere N, Vincent P, Pich EM, Brulet P, Changeux JP (1995) Abnormal avoidance learning in mice lacking functional high-affinity nicotine receptor in the brain. Nature 374:65–67

Pillai SG, Ge D, Zhu G, Kong X, Shianna KV, Need AC, Feng S, Hersh CP, Bakke P, Gulsvik A, Ruppert A, Lodrup Carlsen KC, Roses A, Anderson W, Rennard SI, Lomas DA, Silverman EK, Goldstein DB (2009) A genome-wide association study in chronic obstructive pulmonary disease (COPD): identification of two major susceptibility loci. PLoS Genet 5:e1000421

Pimlott SL, Piggott M, Owens J, Greally E, Court JA, Jaros E, Perry RH, Perry EK, Wyper D (2004) Nicotinic acetylcholine receptor distribution in Alzheimer's disease, dementia with lewy bodies, Parkinson's disease, and vascular dementia: in vitro binding study using 5-[(125)i]-a-85380. Neuropsychopharmacology 29:108–116

Poirier MF, Canceil O, Bayle F, Millet B, Bourdel MC, Moatti C, Olie JP, Attar-Levy D (2002) Prevalence of smoking in psychiatric patients. Prog Neuropsychopharmacol Biol Psychiatry 26:529–537

Portugal GS, Gould TJ (2008) Genetic variability in nicotinic acetylcholine receptors and nicotine addiction: converging evidence from human and animal research. Behav Brain Res 193:1–16

Quick MW, Lester RA (2002) Desensitization of neuronal nicotinic receptors. J Neurobiol 53: 457–478

Quik M, Bordia T, Forno L, McIntosh JM (2004) Loss of alpha-conotoxinMII- and A85380-sensitive nicotinic receptors in Parkinson's disease striatum. J Neurochem 88:668–679

Quik M, O'Leary K, Tanner CM (2008) Nicotine and Parkinson's disease: implications for therapy. Mov Disord 23:1641–1652

Quik M, Sum JD, Whiteaker P, McCallum SE, Marks MJ, Musachio J, McIntosh JM, Collins AC, Grady SR (2003) Differential declines in striatal nicotinic receptor subtype function after nigrostriatal damage in mice. Mol Pharmacol 63:1169–1179

Quik M, Vailati S, Bordia T, Kulak JM, Fan H, McIntosh JM, Clementi F, Gotti C (2005) Subunit composition of nicotinic receptors in monkey striatum: effect of treatments with 1-methyl-4-phenyl-1, 2, 3, 6-tetrahydropyridine or L-DOPA. Mol Pharmacol 67:32–41

Ragozzino D, Barabino B, Fucile S, Eusebi F (1998) Ca2+ permeability of mouse and chick nicotinic acetylcholine receptors expressed in transiently transfected human cells. J Physiol 507(Pt 3):749–757

Ray MA, Graham AJ, Lee M, Perry RH, Court JA, Perry EK (2005) Neuronal nicotinic acetylcholine receptor subunits in autism: an immunohistochemical investigation in the thalamus. Neurobiol Dis 19:366–377

Rinne JO, Myllykyla T, Lonnberg P, Marjamaki P (1991) A postmortem study of brain nicotinic receptors in Parkinson's and Alzheimer's disease. Brain Res 547:167–170

Romano C, Goldstein A (1980) Stereospecific nicotine receptors on rat brain membranes. Science 210:647–650

Sabri O, Kendziorra K, Wolf H, Gertz HJ, Brust P (2008) Acetylcholine receptors in dementia and mild cognitive impairment. Eur J Nucl Med Mol Imaging 35(Suppl 1):S30–S45

Sacco KA, Termine A, Seyal A, Dudas MM, Vessicchio JC, Krishnan-Sarin S, Jatlow PI, Wexler BE, George TP (2005) Effects of cigarette smoking on spatial working memory and attentional deficits in schizophrenia: involvement of nicotinic receptor mechanisms. Arch Gen Psychiatry 62:649–659

Saccone SF, Hinrichs AL, Saccone NL, Chase GA, Konvicka K, Madden PA, Breslau N, Johnson EO, Hatsukami D, Pomerleau O, Swan GE, Goate AM, Rutter J, Bertelsen S, Fox L, Fugman D, Martin NG, Montgomery GW, Wang JC, Ballinger DG, Rice JP, Bierut LJ (2007) Cholinergic nicotinic receptor genes implicated in a nicotine dependence association study targeting 348 candidate genes with 3713 SNPs. Hum Mol Genet 16:36–49

Saccone NL, Saccone SF, Hinrichs AL, Stitzel JA, Duan W, Pergadia ML, Agrawal A, Breslau N, Grucza RA, Hatsukami D, Johnson EO, Madden PA, Swan GE, Wang JC, Goate AM, Rice JP, Bierut LJ (2009) Multiple distinct risk loci for nicotine dependence identified by dense coverage of the complete family of nicotinic receptor subunit (CHRN) genes. Am J Med Genet B Neuropsychiatr Genet 150B(4):453–466

Saenz A, Galan J, Caloustian C, Lorenzo F, Marquez C, Rodriguez N, Jimenez MD, Poza JJ, Cobo AM, Grid D, Prud'homme JF, Lopez dM (1999) Autosomal dominant nocturnal frontal lobe epilepsy in a Spanish family with a Ser252Phe mutation in the CHRNA4 gene. Arch Neurol 56:1004–1009

Salminen O, Murphy KL, McIntosh JM, Drago J, Marks MJ, Collins AC, Grady SR (2004) Subunit composition and pharmacology of two classes of striatal presynaptic nicotinic acetylcholine receptors mediating dopamine release in mice. Mol Pharmacol 65: 1526–1535

Sanberg PR, Fogelson HM, Manderscheid PZ, Parker KW, Norman AB, McConville BJ (1988) Nicotine gum and haloperidol in Tourette's syndrome. Lancet 1:592

Sanberg PR, McConville BJ, Fogelson HM, Manderscheid PZ, Parker KW, Blythe MM, Klykylo WM, Norman AB (1989) Nicotine potentiates the effects of haloperidol in animals and in patients with Tourette syndrome. Biomed Pharmacother 43:19–23

Scheffer IE, Bhatia KP, Lopes-Cendes I, Fish DR, Marsden CD, Andermann E, Andermann F, Desbiens R, Keene D, Cendes F, (1995) Autosomal dominant nocturnal frontal lobe epilepsy. A distinctive clinical disorder. Brain 118(Pt 1):61–73

Schlaepfer IR, Hoft NR, Collins AC, Corley RP, Hewitt JK, Hopfer CJ, Lessem JM, McQueen MB, Rhee SH, Ehringer MA (2008) The CHRNA5/A3/B4 gene cluster variability as an important determinant of early alcohol and tobacco initiation in young adults. Biol Psychiatry 63: 1039–1046

Schwartz RD, Kellar KJ (1983) Nicotinic cholinergic receptor binding sites in the brain: regulation in vivo. Science 220:214–216

Schwartz RD, McGee R Jr, Kellar KJ (1982) Nicotinic cholinergic receptors labeled by [3H]acetylcholine in rat brain. Mol Pharmacol 22:56–62

Schwarz J, Schwarz SC, Dorigo O, Stutzer A, Wegner F, Labarca C, Deshpande P, Gil JS, Berk AJ, Lester HA (2006) Enhanced expression of hypersensitive alpha4* nAChR in adult mice increases the loss of midbrain dopaminergic neurons. FASEB J 20:935–946

Seguela P, Wadiche J, Dineley-Miller K, Dani JA, Patrick JW (1993) Molecular cloning, functional properties, and distribution of rat brain alpha 7: a nicotinic cation channel highly permeable to calcium. J Neurosci 13:596–604

Seidl R, Tiefenthaler M, Hauser E, Lubec G (2000) Effects of transdermal nicotine on cognitive performance in Down's syndrome. Lancet 356:1409–1410

Sershen H, Reith ME, Lajtha A, Gennaro J Jr (1981) Noncholinergic, saturable binding of (+/-)-[3H]nicotine to mouse brain. J Recept Res 2:1–15

Sherva R, Wilhelmsen K, Pomerleau CS, Chasse SA, Rice JP, Snedecor SM, Bierut LJ, Neuman RJ, Pomerleau OF (2008) Association of a single nucleotide polymorphism in neuronal acetylcholine receptor subunit alpha 5 (CHRNA5) with smoking status and with 'pleasurable buzz' during early experimentation with smoking. Addiction 103:1544–1552

Shiraishi K, Kohno T, Kunitoh H, Watanabe S, Goto K, Nishiwaki Y, Shimada Y, Hirose H, Saito I, Kuchiba A, Yamamoto S, Yokota J (2009) Contribution of nicotine acetylcholine receptor polymorphisms to lung cancer risk in a smoking-independent manner in the Japanese. Carcinogenesis 30:65–70

Silver AA, Shytle RD, Philipp MK, Sanberg PR (1996) Case study: long-term potentiation of neuroleptics with transdermal nicotine in Tourette's syndrome. J Am Acad Child Adolesc Psychiatry 35:1631–1636

Silver AA, Shytle RD, Philipp MK, Wilkinson BJ, McConville B, Sanberg PR (2001) Transdermal nicotine and haloperidol in Tourette's disorder: a double-blind placebo-controlled study. J Clin Psychiatry 62:707–714

Simosky JK, Stevens KE, Freedman R (2002) Nicotinic agonists and psychosis. Curr Drug Targets CNS Neurol Disord 1:149–162

Simosky JK, Stevens KE, Kem WR, Freedman R (2001) Intragastric DMXB-A, an alpha7 nicotinic agonist, improves deficient sensory inhibition in DBA/2 mice. Biol Psychiatry 50:493–500

Soderman A, Thomsen MS, Hansen HH, Nielsen EO, Jensen MS, West MJ, Mikkelsen JD (2008) The nicotinic alpha7 acetylcholine receptor agonist ssr180711 is unable to activate limbic neurons in mice overexpressing human amyloid-beta1-42. Brain Res 1227:240–247

Song P, Spindel ER (2008) Basic and clinical aspects of non-neuronal acetylcholine: expression of non-neuronal acetylcholine in lung cancer provides a new target for cancer therapy. J Pharmacol Sci 106:180–185

Spitz MR, Amos CI, Dong Q, Lin J, Wu X (2008) The CHRNA5-A3 region on chromosome 15q24-25.1 is a risk factor both for nicotine dependence and for lung cancer. J Natl Cancer Inst 100:1552–1556

Steinlein OK, Bertrand D (2008) Neuronal nicotinic acetylcholine receptors: from the genetic analysis to neurological diseases. Biochem Pharmacol 76:1175–1183

Steinlein OK, Magnusson A, Stoodt J, Bertrand S, Weiland S, Berkovic SF, Nakken KO, Propping P, Bertrand D (1997) An insertion mutation of the CHRNA4 gene in a family with autosomal dominant nocturnal frontal lobe epilepsy. Hum Mol Genet 6:943–947

Steinlein OK, Mulley JC, Propping P, Wallace RH, Phillips HA, Sutherland GR, Scheffer IE, Berkovic SF (1995) A missense mutation in the neuronal nicotinic acetylcholine receptor alpha 4 subunit is associated with autosomal dominant nocturnal frontal lobe epilepsy. Nat Genet 11:201–203

Steinlein OK, Stoodt J, Mulley J, Berkovic S, Scheffer IE, Brodtkorb E (2000) Independent occurrence of the CHRNA4 Ser248Phe mutation in a Norwegian family with nocturnal frontal lobe epilepsy. Epilepsia 41:529–535

Stevens KE, Adams CE, Mellott TJ, Robbins E, Kisley MA (2008a) Perinatal choline deficiency produces abnormal sensory inhibition in Sprague-Dawley rats. Brain Res 1237:84–90

Stevens KE, Adams CE, Yonchek J, Hickel C, Danielson J, Kisley MA (2008b) Permanent improvement in deficient sensory inhibition in DBA/2 mice with increased perinatal choline. Psychopharmacology (Berl) 198:413–420

Stevens VL, Bierut LJ, Talbot JT, Wang JC, Sun J, Henrichs AL, Thun MJ, Goate A, Calle EE (2008) Nicotinic Receptor Gene Variants Influence Susceptibility to Heavy Smoking. Cancer Epidemiol Biomarkers Prev 17:3517–3525

Stevens KE, Kem WR, Mahnir VM, Freedman R (1998) Selective alpha7-nicotinic agonists normalize inhibition of auditory response in DBA mice. Psychopharmacology (Berl) 136: 320–327

Stevens KE, Stitzel JA, Jimenez M, Collins AC 2001 Transferring the alpha7 nicotinic receptor between different mouse backgrounds alters auditory gating. Society for Neuroscience Annual Meeting Abstract viewer/Itinerary planner, San Diego, CA, Program Number: 145.4.

Stevens KE, Wear KD (1997) Normalizing effects of nicotine and a novel nicotinic agonist on hippocampal auditory gating in two animal models. Pharmacol Biochem Behav 57:869–874

Stitzel JA (2008) Naturally occurring genetic variability in the nicotinic acetylcholine receptor alpha4 and alpha7 subunit genes and phenotypic diversity in humans and mice. Front Biosci 13:477–491

Stitzel JA, Farnham DA, Collins AC (1996) Linkage of strain-specific nicotinic receptor alpha 7 subunit restriction fragment length polymorphisms with levels of alpha-bungarotoxin binding in brain. Brain Res Mol Brain Res 43:30–40

Teaktong T, Graham A, Court J, Perry R, Jaros E, Johnson M, Hall R, Perry E (2003) Alzheimer's disease is associated with a selective increase in alpha7 nicotinic acetylcholine receptor immunoreactivity in astrocytes. Glia 41:207–211

Terzano S, Court JA, Fornasari D, Griffiths M, Spurden DP, Lloyd S, Perry RH, Perry EK, Clementi F (1998) Expression of the alpha3 nicotinic receptor subunit mRNA in aging and Alzheimer's disease. Brain Res Mol Brain Res 63:72–78

Thorgeirsson TE, Geller F, Sulem P, Rafnar T, Wiste A, Magnusson KP, Manolescu A, Thorleifsson G, Stefansson H, Ingason A, Stacey SN, Bergthorsson JT, Thorlacius S, Gudmundsson J, Jonsson T, Jakobsdottir M, Saemundsdottir J, Olafsdottir O, Gudmundsson LJ, Bjornsdottir G, Kristjansson K, Skuladottir H, Isaksson HJ, Gudbjartsson T, Jones GT, Mueller T, Gottsater A, Flex A, Aben KK, de Vegt F, Mulders PF, Isla D, Vidal MJ, Asin L, Saez B, Murillo L, Blondal T, Kolbeinsson H, Stefansson JG, Hansdottir I, Runarsdottir V, Pola R, Lindblad B, van Rij AM, Dieplinger B, Haltmayer M, Mayordomo JI, Kiemeney LA, Matthiasson SE, Oskarsson H, Tyrfingsson T, Gudbjartsson DF, Gulcher JR, Jonsson S, Thorsteinsdottir U, Kong A, Stefansson K (2008) A variant associated with nicotine dependence, lung cancer and peripheral arterial disease. Nature 452:638–642

Tidey JW, Rohsenow DJ, Kaplan GB, Swift RM (2005) Cigarette smoking topography in smokers with schizophrenia and matched non-psychiatric controls. Drug Alcohol Depend 80:259–265

Tozaki H, Matsumoto A, Kanno T, Nagai K, Nagata T, Yamamoto S, Nishizaki T (2002) The inhibitory and facilitatory actions of amyloid-beta peptides on nicotinic ACh receptors and AMPA receptors. Biochem Biophys Res Commun 294:42–45

Venables PH (1964) Input dysfunction in schizophrenia. Prog Exp Pers Res 72:1–47

Venables PH (1992) Hippocampal function and schizophrenia. Experimental psychological evidence. Ann N Y Acad Sci 658:111–127

Volkow N, Rutter J, Pollock JD, Shurtleff D, Baler R (2008) One SNP linked to two diseases-addiction and cancer: a double whammy? Nicotine addiction and lung cancer susceptibility. Mol Psychiatry 13:990–992

Wang JC, Cruchaga C, Saccone NL, Bertelsen S, Liu P, Budde JP, Duan W, Fox L, Grucza RA, Kern J, Mayo K, Reyes O, Rice J, Saccone SF, Spiegel N, Steinbach JH, Stitzel JA, Anderson MW, You M, Stevens VL, Bierut LJ, Goate AM (2009) Risk for nicotine dependence and lung cancer is conferred by mRNA expression levels and amino acid change in CHRNA5. Hum Mol Genet 18(16):3125–3135

Wang JC, Grucza R, Cruchaga C, Hinrichs AL, Bertelsen S, Budde JP, Fox L, Goldstein E, Reyes O, Saccone N, Saccone S, Xuei X, Bucholz K, Kuperman S, Nurnberger J Jr, Rice JP, Schuckit M, Tischfield J, Hesselbrock V, Porjesz B, Edenberg HJ, Bierut LJ, Goate AM (2008) Genetic variation in the CHRNA5 gene affects mRNA levels and is associated with risk for alcohol dependence. Mol Psychiatry 14(5):501–510

Wang HY, Lee DH, D'Andrea MR, Peterson PA, Shank RP, Reitz AB (2000a) beta-Amyloid(1-42) binds to alpha7 nicotinic acetylcholine receptor with high affinity. Implications for Alzheimer's disease pathology. J Biol Chem 275:5626–5632

Wang HY, Lee DH, Davis CB, Shank RP (2000b) Amyloid peptide Abeta(1-42) binds selectively and with picomolar affinity to alpha7 nicotinic acetylcholine receptors. J Neurochem 75: 1155–1161

Weiss RB, Baker TB, Cannon DS, von Niederhausern A, Dunn DM, Matsunami N, Singh NA, Baird L, Coon H, McMahon WM, Piper ME, Fiore MC, Scholand MB, Connett JE, Kanner RE, Gahring LC, Rogers SW, Hoidal JR, Leppert MF (2008) A candidate gene approach identifies the CHRNA5-A3-B4 region as a risk factor for age-dependent nicotine addiction. PLoS Genet 4:e1000125

Wevers A, Monteggia L, Nowacki S, Bloch W, Schutz U, Lindstrom J, Pereira EF, Eisenberg H, Giacobini E, de Vos RA, Steur EN, Maelicke A, Albuquerque EX, Schroder H (1999) Expression of nicotinic acetylcholine receptor subunits in the cerebral cortex in Alzheimer's disease: histotopographical correlation with amyloid plaques and hyperphosphorylated-tau protein. Eur J Neurosci 11:2551–2565

Whitehouse PJ, Martino AM, Antuono PG, Lowenstein PR, Coyle JT, Price DL, Kellar KJ (1986) Nicotinic acetylcholine binding sites in Alzheimer's disease. Brain Res 371:146–151

Whitehouse PJ, Martino AM, Wagster MV, Price DL, Mayeux R, Atack JR, Kellar KJ (1988) Reductions in [3H]nicotinic acetylcholine binding in Alzheimer's disease and Parkinson's disease: an autoradiographic study. Neurology 38:720–723

Whiting PJ, Schoepfer R, Conroy WG, Gore MJ, Keyser KT, Shimasaki S, Esch F, Lindstrom JM (1991) Expression of nicotinic acetylcholine receptor subtypes in brain and retina. Brain Res Mol Brain Res 10:61–70

Wildeboer KM, Stevens KE (2008) Stimulation of the alpha4beta2 nicotinic receptor by 5-I A-85380 improves auditory gating in DBA/2 mice. Brain Res 1224:29–36

Williams JM, Ziedonis D (2004) Addressing tobacco among individuals with a mental illness or an addiction. Addict Behav 29:1067–1083

Willoughby JO, Pope KJ, Eaton V (2003) Nicotine as an antiepileptic agent in ADNFLE: an N-of-one study. Epilepsia 44:1238–1240

Wonnacott S (1990) The paradox of nicotinic acetylcholine receptor upregulation by nicotine. Trends Pharmacol Sci 11:216–219

Wooltorton JR, Pidoplichko VI, Broide RS, Dani JA (2003) Differential desensitization and distribution of nicotinic acetylcholine receptor subtypes in midbrain dopamine areas. J Neurosci 23:3176–3185

Wu J, Kuo YP, George AA, Xu L, Hu J, Lukas RJ (2004) beta-Amyloid directly inhibits human alpha4beta2-nicotinic acetylcholine receptors heterologously expressed in human SH-EP1 cells. J Biol Chem 279:37842–37851

Yates CM, Simpson J, Maloney AF, Gordon A, Reid AH (1980) Alzheimer-like cholinergic deficiency in Down syndrome. Lancet 2:979

Young RP, Hopkins RJ, Hay BA, Epton MJ, Black PN, Gamble GD (2008) Lung cancer gene associated with COPD: triple whammy or possible confounding effect? Eur Respir J 32: 1158–1164

Zamani MR, Allen YS, Owen GP, Gray JA (1997) Nicotine modulates the neurotoxic effect of beta-amyloid protein(25–35) in hippocampal cultures. Neuroreport 8:513–517

Zhang J, Liu Q, Chen Q, Liu NQ, Li FL, Lu ZB, Qin C, Zhu H, Huang YY, He W, Zhao BL (2006) Nicotine attenuates beta-amyloid-induced neurotoxicity by regulating metal homeostasis. FASEB J 20:1212–1214

Zhu PJ, Chiappinelli VA (1999) Nicotine modulates evoked GABAergic transmission in the brain. J Neurophysiol 82:3041–3045

Zoli M, Moretti M, Zanardi A, McIntosh JM, Clementi F, Gotti C (2002) Identification of the nicotinic receptor subtypes expressed on dopaminergic terminals in the rat striatum. J Neurosci 22:8785–8789

Zoli M, Picciotto MR, Ferrari R, Cocchi D, Changeux JP (1999) Increased neurodegeneration during ageing in mice lacking high-affinity nicotine receptors. EMBO J 18:1235–1244

Lysosomal Storage Diseases

Gregory M. Pastores

Abstract The lysosomal storage diseases (LSD) are a heterogeneous group of disorders, characterized by the progressive accumulation of various substrates in multiple cell types, as a consequence of defects in the degradation of by-products of cellular turnover. Several subtypes are associated with neurodegenerative features, which present as a major therapeutic challenge. Although the causal gene defects and corresponding enzyme, cofactor, or transport deficiency have been delineated, there remains incomplete understanding of the downstream pathways leading to organ dysfunction and clinical symptomatology. Recent studies suggest that several processes, including inflammation, apoptosis, and defects of autophagy, may be involved. Therapy remains palliative for most LSDs, although enzyme replacement therapy is available for several disorders that are caused by a deficiency in a soluble hydrolase. Novel strategies, which involve the use of small molecular agents that inhibit substrate synthesis or act as pharmacological chaperones to rescue the mutant protein, are current subjects of investigation. In addition, gene therapy and stem cell therapy are being evaluated. The multifactorial basis of LSDs will likely necessitate a combination of approaches to optimize therapeutic outcome. Meanwhile, preimplantation genetic diagnosis and prenatal detection are being offered as an option to families at risk. Newborn screening and carrier detection in populations at risk is also being undertaken, to enable early diagnosis, appropriate counseling, and timely intervention.

Keywords Enzyme replacement therapy · Enzyme deficiency · Lysosomal storage disease · Substrate reduction therapy

Contents

G.M. Pastores (✉)
Neurology and Pediatrics, New York University School of Medicine, New York, NY 10016, USA;
Neurogenetics Laboratory, New York University School of Medicine, New York, NY 10016, USA
e-mail: gregory.pastores@med.nyu.edu

J.P. Blass (ed.), *Neurochemical Mechanisms in Disease*,
Advances in Neurobiology 1, DOI 10.1007/978-1-4419-7104-3_23,

1 Introduction

The lysosomal storage diseases (LSD) are a heterogeneous group of disorders resulting from an inherited defect in the metabolism of by-products of cellular turnover (Reuser and Drost, 2006). As a consequence, there is progressive accumulation of one or more substrates within the lysosome, eventually leading to multiple organ/system dysfunction. Although individual disorders may be rarely encountered, collectively 1 in 5000 children will be found to have an LSD, caused primarily by a deficiency of a lysosomal hydrolase or its cofactor. Given the frequent delays in diagnosis, and the introduction of therapies for certain subtypes, several groups have advocated for screening programs of newborns or high-risk populations (i.e., based on ethnic group or defined clinical groups).

There are at least 50 distinct LSDs, grouped according to the biochemical composition of the storage material into the sphingolipidoses, mucopolysaccharidoses (MPS), oligosaccharidoses, and so on. Several LSDs have also been given an eponymous designation (e.g., Gaucher disease, Fabry disease, Tay–Sachs) in recognition of the physician/scientist who played a role in their seminal description (Table 1). The majority of LSDs are inherited in an autosomal recessive fashion, except for three disorders: Fabry disease, Dannon disease, and Hunter syndrome (MPS type II).

The diagnosis in suspected LSD cases can be confirmed by biochemical and/or molecular assays, which can be applied for prenatal and presymptomatic diagnosis (Meikle et al., 2004). Although most LSDs have onset in childhood, several subtypes have later-onset of disease, with symptoms that may not be evident until adulthood. The latter individuals often suffer from delayed diagnosis, unless there is a prior family history. However, there can be heterogeneity in clinical expression, and even siblings can have a distinct clinical course of disease, particularly among those with a chronic variant. Except for null alleles, which are often associated with the "classic phenotype," studies of the correlation between genotype and phenotype suggest a role for factors that modifies disease expression.

2 Modes of Clinical Presentation

A significant proportion of patients with an LSD have neurological involvement, which can be manifested as developmental delay and behavioral changes (Table 2). The presence of specific findings may suggest the diagnosis (e.g., leucodystrophy in patients with metachromatic leucodystrophy or globoid cell leucodystrophy; cherry

Table 1 The lysosomal storage disorders classified according to relevant substrate involved

Stored Substrate	Disease	Enzyme Deficiency	Gene Locus
A. Sphingolipids			
GM_2-gangliosides, glycolipids, globoside oligosaccharides	Tay–Sachs	α-subunit β-hexosaminidase	15q23-4
	GM_2 gangliosidosis (three types)		
	Sandhoff disease GM_2 gangliosidosis	β-subunit β-hexosaminidase	5q13
	GM_2 gangliosidosis, AB variant	G_{M2} activator	5q32-33
GM_1-gangliosides, oligosaccharides, keratan sulfate, glycolipids	GM_1 gangliosidosis (three types)	β-galactosidase	3p21-3pter
Sulphatides	Metachromatic leukodystrophy (MLD)	Arylsulphatase A (galactose-3-sulphatase)	22q13.31-qter
GM_1-gangliosides, sphingomyelin, glycolipids, sulphatide	MLD variant	Saposin B activator	10q21
Galactosylceramides	Krabbe disease	Galactocerebrosidase	14q31
α-galactosyl-sphingolipids, oligosaccharides	Fabry disease	α-galactosidase A	Xq22
Glucosylceramide, globosides	Gaucher disease (GD) (three types)	β-glucosidase	1q2l
Glucosylceramide, globosides	GD (variant)	Saposin C	10q21
Ceramide	Farber disease (seven types)	Acid ceramidase	8p22-21.2
Sphingomyelin	Niemann–Pick disease types A and B	Sphingomyelinase	11p15.1-15.4
B.Mucopolysaccharidoses (glycosaminoglycans)			
	MPS I, Hurler, Scheie	α-L-iduronidase	4p16.3
Dermatan sulfate (DS) and Heparan sulfate (HS)	MPS II, Hunter	Iduronate-2-sulphatase	Xq27.3-28

Table 1 (continued)

Stored Substrate	Disease	Enzyme Deficiency	Gene Locus
HS	MPS IIIA, Sanfilippo A	Sulfamidase	17q25.3
	MPS IIIB, Sanfilippo B	α-N-acetylglucosaminidase	17q21.1
	MPS IIIC, Sanfilippo C	Acetyl CoA:α-glucosaminide-N-acetyltransferase	1412q14
	MPS IIID, Sanfilippo D	N-acetylglucosamine-6-sulfatase	16q24.3
	MPS IVA, Morquio A	Galactosamine-6-sulphatase	3p21.33
Keratan sulfate (KS)	MPS IVB, Morquio B	β-D-galactosidase	5q13-14
DS	MPS VI, Maroteaux–Lamy	N-acetylgalactosamine-4-sulfatase	7q21.1-22
DS and HS	MPS VII, Sly	β-D-glucuronidase	3p21.3
Hyaluronan	MPS IX Natowicz	Hyaluronidase	17q25
C. Glycogen			
Glycogen	Pompe, GSD IIA	α-D glucosidase	
Glycogen	Danon disease	Lysosomal associated membrane protein-2 (LAMP-2)	Xq24
D. Oligosaccharides/ Glycopeptides			
α-mannoside	α-mannosidosis	α-mannosidase	19p13.2-q12
β-mannoside	β-mannosidosis	β-mannosidase	4q22-25
α-—fucosides, glycolipids	α-fucosidosis	α-fucosidase	1p34.1-36.1
α-N-acetylgalactosaminide	Schindler/Kanzaki disease	α-N-acetylgalactosaminidase	22q13.1-13.2
sialyloligosaccharides	Sialidosis	α-neuraminidase	6p21.3
aspartylglucosamine	Aspartylglucosaminuria	Aspartylglucosaminidase	4q34-35
E. Multiple Enzyme deficiencies			
Glycolipids, oligosaccharides			
	Mucolipidosis II (I-cell disease); mucolipidosis III (pseudo-Hurler polydystrophy) – three complementation groups	N-acetylglucosamine-1-phosphotransferase	4q21-q23; ML-III subtype C (gamma subunit mutations-16p)

Table 1 (continued)

Stored Substrate	Disease	Enzyme Deficiency	Gene Locus
Sulphatides, glycolipids, glycosaminoglycans	Galactosialidosis (protective protein/cathepsin A)	protective protein/cathepsin A	20
	Multiple sulfatases (Austin disease)	SUMF-1	3p26
F. Lipids			
Cholesterol esters	Wolman disease, CESD (cholesterol ester storage disease)	Acid lipase	10q23.2-q23.3
Cholesterol, sphingomyelin, GM_2-gangliosides	Niemann–Pick disease type C	NPC1; HE1	18q11-12; 14q24.3
G. Monosaccraides/amino acid/monomers			
Sialic acid, glucuronic acid	Salla, ISSD	Sialin	6q14-15
cystine	Cystinosis	Cystinosis	17p13
H. Peptides			
Bone proteins	Pycnodysostosis	Cathepsin K	1q21
S-acylated proteins			
Palmitoylated proteins	Infantile neuronal ceroid lipofuscinosis (NCL)	Palmitoyl-protein thioesterase	1p32
Pepstatin-insensitive lysosomal peptidase	Late-infantile NCL	Pepstatin-insensitive lysosomal peptidase	11p15.5
Cathepsin D	Congenital NCL	Lysosomal cysteine protease	11p15.5

Table 2 Neurological features encountered in patients with an LSD

Cherry-red spot, Optic atrophy, Visual loss*
- Galactosialidosis
- G_{M1}-gangliosidosis
- Infantile free sialic acid storage disease (ISSD)
- Mucolipidosis II (I-cell disease)
- Mucopolysaccharidosis types IV (MPS IV) and VII (MPS VII)
- Neuronal ceroid lipofuscinosis
- Niemann-Pick disease type A
- Sialidosis type 1
- Sandhoff disease
- Tay–Sachs disease

Retinitis pigmentosa
- Neuronal ceroid lipofuscinosis

Corneal opacities (clouding)
- I-cell disease (ML II)
- Mucolipidosis IV (MLIV)
- MPS I, IV, VI
- Oligosaccharidosis (late-onset α-mannosidosis)
- Fabry disease

Lenticular opacities (cataracts)
- Oligosaccharidosis (sialidosis, α-mannosidosis)
- Fabry disease

Ophthalmoplegia (Abnormal eye movements), Nystagmus
- Gaucher disease 3
- Niemann–Pick C

Leukosdystrophy
- Krabbe disease
- MLD
- Fabry disease*

Myoclonic seizures
- Galactosialidosis
- Gaucher disease III
- G_{M2}-gangliosidosis
- Neuronal ceroid lipofuscinosis
- Niemann-Pick C
- Oligosaccharidosis (α-N-acetylgalactosaminidase deficiency, fucosidosis, Sialidosis type 1)

Deafness
- Fabry disease
- Galactosialidosis
- Gaucher disease type 2
- I-cell disease
- MPS I, II, IV
- Oligosaccharidosis (α- and β-mannosidosis)
- Metachromatic leukodystrophy
- Infantile Pompe disease

Macrocephaly
- Tay–Sachs disease
- Sandhoff disease
- Krabbe disease

Peripheral neuropathy
- Krabbe disease
- MLD (spastic paraplegia)
- Multiple sulfatase deficiency

Cortical atrophy
- Late stage of G_{M1}- and G_{M2}-gangliosidosis (cerebellar atrophy)
- MLD
- I-cell disease
- Neuronal ceroid lipofuscinosis

Cerebrovascular or strokelike episodes and other vascular events (e.g., Raynaud's phenomenon)
- Fabry disease

Ataxia
- Galactosialidosis
- Gaucher disease III
- G_{M1}-gangliosidosis
- Late-onset G_{M2}-gangliosidosis (cerebellar hypoplasia)
- Krabbe disease
- MLIV
- MLD
- Neuronal ceroid lipofucinosis
- Niemann-Pick C
- Salla disease
- Sialidosis I

Extrapyramidal signs
- Gaucher disease 3
- G_{M1}-gangliosidosis (adult form)
- Late-onset G_{M2}-gangliosidosis
- Krabbe disease
- Niemann–Pick C
- Oligosaccharidosis

Dementia, Psychosis
- Fabry disease
- Gaucher disease 3
- G_{M1}-gangliosidosis
- Late-onset G_{M2}-gangliosidosis
- Krabbe disease
- MLD
- MPS III (Sanfilippo disease)
- Neuronal ceroid lipofuscinosis
- Niemann–Pick C

red spot in Tay–Sachs disease, G_{M1}-gangliosidosis, Niemann–Pick type A and Sialidosis; ophthalmoplegia in Gaucher disease type 3 and Niemann–Pick disease type C).

Extraneurological features that should lead to consideration of an LSD diagnosis include hepatosplenomegaly, short stature, joint contractures, and cardiomyopathy.

An early age of symptom onset often portends a rapidly progressive course, although each LSD subtype is associated with chronic subtypes, with a clinical course that can run into decades.

In general, null alleles are associated with the classic early-onset phenotype, whereas missense mutations which lead to defective proteins that exhibit residual enzyme activity lead to attenuated phenotypes (Froissart et al., 2002). However, studies of genotype–phenotype correlation have revealed a lack of perfect concordance, which suggests other factors may be involved that influence disease outcome (Froissart et al., 2002). At present, the putative factors that modify LSD-phenotypes among patients with identical genotypes remain obscure.

3 Diagnostic Confirmation

Diagnosis of an LSD is critical for several reasons: (1) it focuses attention on the needs of the patient, and the potential to intervene in subtypes for which treatment is available; (2) the inherited basis implies a risk of recurrence during future pregnancies, and as prenatal diagnosis is available for most, families are given the opportunity to plan accordingly; (3) although treatment is available for certain subtypes, early diagnosis is essential as current approaches are unlikely to restore organ function when there is considerable pre-existing pathology at the time of initiation.

For disorders characterized by an underlying enzyme deficiency (e.g., Gaucher disease, Fabry disease, Tay–Sachs, Hurler syndrome), assays of enzyme activity in blood and/or tissues is generally available (Meikle et al., 2004). Mutation analysis is also available, particularly for populations in whom the common disease alleles are known (e.g., mutations among Ashkenazi Jews for Gaucher, Tay–Sachs, Niemann–Pick type A, and mucolipidosis type IV; Ostrer, 2001). In other cases, analysis of the gene defect responsible for rare subtypes is available through specialized laboratories.

Examination of skin or other tissues (e.g., liver, bone marrow) may suggest the presence of lysosomal storage, however, this involves invasive procedures and requires expertise in interpretation of the findings (Alroy and Ucci, 2006). Analysis of urine for excess substrates (e.g., glycosaminoglycans in the Mucopolysaccharidoses, globotriaolsylceramide in Fabry disease) may also suggest the presence of an LSD. In any case, all patients suspected to have an LSD should have diagnostic confirmation by means of biochemical and/or molecular genetic analysis.

4 Pathophysiological Mechanisms

Intralysosomal substrate storage represents the initial insult to cells; by-products of intermediary metabolism (e.g., psychosine in globoid cell leukodytrorphy), a disruption of normal lysosome function, and/or the consequent deficiency in recycling of certain substrates are putative disease events (Ballabio and Gieselmann, 2009).

In disorders characterized by primary (e.g., Tay–Sachs disease) or secondary (e.g., Niemann–Pick type C) ganglioside storage, neuronal cells develop ectopic dendritogenesis and meganeurite formation (Walkley, 2009). These changes may be associated with a disturbance in neuronal signal transmission and/or the transport of trophic factors along the length of the axon; partly accounting for the neurodegenerative features seen in these conditions.

There is incomplete understanding of the disease mechanism beyond substrate storage, although several processes (such as inflammation, apoptosis, defects of autophagy and activation of the ER-stress response) may have a contributory role (Ballabio and Gieselmann, 2009).

In globoid cell leucodystrophy the accumulation of galactosylsphingosine (psychosine) is believed to promote energy depletion, loss of oligodendrocytes, and the induction of gliosis and aberrant inflammation by astrocytes in the central nervous system (CNS) (Suzuki, 1998). Recently, psychosine has also been shown to down-regulate AMP-activated protein kinase (AMPK), the "cellular energy switch" in oligodendrocytes and astrocytes (Giri et al., 2008). In an oligodendrocyte cell line (MO3.13) and primary astrocytes, psychosine accumulation increased the biosynthesis of lipids, including cholesterol and free fatty acid. These findings delineate an explicit role for AMPK in psychosine-induced inflammation in astrocytes, without directly affecting the cell death of oligodendrocytes.

In the brain obtained from the mucopolysaccharidosis type IIIB mouse model, the accumulating substrate—heparan sulfate oligosaccharides—activated microglial cells by signaling through the Toll-like receptor 4 and the adaptor protein MyD88 (Ausseil et al., 2008). Although intrinsic to the disease, the observed phenomenon was deemed not to be a major determinant of the neurodegenerative process, with a possibly greater role for inflammatory changes in the later stages of the disease (Ausseil et al., 2008).

In multiple sulphatase deficiency and mucopolysaccharidosis type IIIA, studies in the respective mouse models suggest defects in autophagy; a lysosomal-dependent catabolic pathway through which long-lived cytosolic proteins and organelles (such as mitochondria) are sequestered by double-membrane vesicles and ultimately degraded after fusion with lysosomes. In affected cells, reduced colocalization of the lysosomal membrane protein LAMP-1 with the autophagosome marker LC3 have been observed; indicative of an impairment of lysosome/autophagosome fusion (Settembre et al., 2008). Accumulation of autophagic vacuoles in the heart and skeletal muscle are hallmarks of Danon disease (Yang and Vatta, 2007). LAMP2, which is defective in Danon disease, is believed to be normally involved in lysosome/autophagosome fusion, and may have a role in dynein-based centripetal motility. In Niemann–Pick type C, there is increased expression of Beclin-1 and LC3-II; the Purkinje neuron cell death encountered in this disorder is believed to be dependent on autophagy (Pacheco and Lieberman, 2008.). A disturbance of autophagy has also been found in the mouse model of Pompe disease; which interestingly has been linked to a deficiency in the trafficking/processing of recombinant enzyme along the endocytic pathway (Raben et al., 2008).

Several endeavors are being directed towards identifying biomarkers that can serve as a surrogate indicator of disease severity, in terms of either overall disease burden or involvement of a particular organ/system in patients with an LSD. In mucopolysaccharidosis type I, the analysis of the levels of oligosaccharides derived from GAGs in cultured fibroblasts (as measured by electrospray ionization tandem mass spectrometry) combined with the residual α-L-iduronidase activity have been shown to distinguish patients with and without CNS involvement (Fuller et al., 2005). The practical application of these techniques in the final assignment of disease subtype remains to be determined, but may be relevant when combined with genotype information in the selection of appropriate therapy for diagnosed patients with mucopolysaccharidosis type I. Meanwhile, ongoing efforts, employing proteomic-based screening tools (such as SELDI-TOF-MS), are anticipated to reveal markers that will help with prediction of disease severity and that may also be useful in monitoring of therapy (Hendriks et al., 2007). Protein profiling provides an opportunity to identify and analyze multiple markers, and enables a systems biology approach to ascertain the impact of the primary deficiency in lysosomal function.

It is likely that one or more of these pathological events may promote cellular dysfunction or tissue damage in the LSDs (Table 3). At present, it is uncertain which of the processes that have been identified plays a dominant role. Certain mechanisms may also be cell-type-specific, but this remains to be clarified.

Table 3 Putative mechanisms of disease

- Altered trafficking of molecules through the endolysosomal network, including sequestration of membrane rafts, leading to a disruption in signaling
- Aberrant inflammatory response, either through activation of resident microglia and/or recruitment and activation of peripheral monocytes
- Oxidative stress and activation of ER-stress response
- Disruption of autophagy
- Initiation of apoptosis

5 Therapeutic Approaches

The management of patients with an LSD is mainly palliative, particularly for subtypes associated with neurological involvement. Commonly encountered problems include seizures, altered sleep–wake cycles, and behavioral problems such as hyperactivity and aggression. Attempts at controlling or modifying these problems may help improve the quality of life of an affected individual and their relatives.

Observations of metabolic cross-correction provided the rationale for cellular replacement, achieved primarily through allogeneic hematopoietic stem cell or bone transplantation (HSCT) (Prasad and Kurtzberg, 2008). More recently, the use of neural stem cells (NSC) implanted in the brain of patients with late-infantile neuronal ceroid lipofuscinosis has been contemplated (Pierret et al., 2008) but there are no reports as yet of its potential efficacy. Within the central nervous system there must be proper integration of donor cells, and differentiation into appropriate cell types. As specialized cell types within the nervous system elaborate neurotransmitters and are involved with conducting electrical impulses, functional differentiation may be a major hurdle for the neurodegenerative LSDs.

Increasingly, donor material is isolated from umbilical cord blood (UCB); these cells are deemed to have greater potential for transdifferentiation into appropriate cell types, and thus may have greater facility for tissue-specific regeneration or repair (Gluckman and Rocha, 2009). In addition, the incidence of graft versus host disease appears to decrease following the use of UCB cells, potentially resulting in decreased morbidity.

HSCT has been performed in several disorders associated with primary CNS involvement (e.g., globoid cell leukodystrophy, Hurler syndrome, α-mannosidosis) (Prasad and Kurtzberg, 2008). The justification has been based on the presence of monocytes in the donor pool, which can traverse the blood–brain barrier (BBB) and differentiate into microglia; serving as the source of functional enzyme. The replacement of endogenous microglia by donor cells is estimated to take at least six to nine months, during which time pathogenic influences may remain; this may explain the potential limitations of HSCT, particularly in cases where the diagnosis is delayed. In globoid cell leukodystrophy, over 80% of infantile cases subjected to HSCT in the first few weeks of life develop gross motor problems after the age two years; often requiring assistance with ambulation (Prasad and Kurtzberg, 2008).

Enzyme replacement therapy is available for several subtypes associated with a soluble hydrolase deficiency; this therapeutic approach has been shown to modify disease course, primarily features of the disorder resulting from extraneurological involvement (Grabowski, 2008). Unfortunately, the ultimate prognosis is not significantly altered in patients with neurodenegerative features, likely because the intravenously administered enzyme does not gain sufficient access across the blood–brain barrier (Pastores, 2003). In addition, therapeutic response is limited in patients with an advanced disease stage, wherein organ function may not be fully restored in cases with significant tissue damage from fibrosis or necrosis. Varying proportions of patients given recombinant enzymes have developed antibodies, which can lead to neutralization of enzyme activity and/or altered tissue distribution (Pastores, 2003). The significance of these observations on long-term outcome remains to be established.

Substrate reduction therapy (SRT) involves the inhibition of substrate synthesis to a level where the load falls within the capacity of the mutant enzyme that exhibits residual function (Platt and Jeyakumar, 2008). Thus this approach, as in the case of pharmacological chaperones, may be dependent on the type of mutation responsible for disease in an individual.

The imino sugar miglustat has been shown to lead to partial glycosphingolipid synthesis inhibition and modification of disease course in treated patients with Gaucher disease (Pastores and Barnett, 2003). As miglustat can gain access to the CNS and inhibit the formation of G_{M2}-gangliosides, its potential use was explored in patients with late-onset Tay–Sachs disease, Gaucher disease type III, and Niemann–Pick type C. In patients with Niemann–Pick type C, miglustat has been shown to improve saccadic eye movements and swallowing difficulties (Patterson et al., 2007). Unfortunately, the miglustat trials in patients with late-onset Tay–Sachs disease and Gaucher disease type III failed to show any measurable benefit, perhaps because of the advanced stage of disease suffered by the study subjects (Shapiro Schiffmann et al., 2008; Shapiro et al., 2009). Surprisingly, there have been minimal side effects (e.g., diarrhea, weight loss) with the use of miglustat; although long-term studies are required to ascertain safety and benefit of its use (Pastores and Barnett, 2003).

Enzyme enhancement therapy involves the use of pharmacological chaperones, which in cell culture and animal studies have been demonstrated to increase residual enzyme activity of the mutant enzyme by preventing its premature degradation within the endoplasmic reticulum (Fan, 2008). Several studies have shown that deficient lysosomal hydrolysis may in the majority of cases be due to mutations that promote protein misfolding and failure of its delivery to the lysosome; as opposed to mutations involving the catalytic site that inactivates enzyme activity altogether (Steet et al., 2007; Sugawara et al., 2009). This approach is currently in clinical trials; its effectiveness in substantially clearing tissue deposits and clinical efficacy in modifying disease phenotype when used as a singular approach remains to be established. As the drugs (isofogamine for Gaucher disease and the imino sugar N-deoxygalactonojirimycin in Fabry disease) currently under study are also inhibitors of enzyme activity, a particular challenge with the use of pharmacological chaperones relates to determination of the appropriate dose and frequency of drug administration, to ensure enzyme enhancement has the upper hand (Fan, 2008).

Gene therapy and stem cell therapies are other approaches that have been explored, primarily in mouse models of various LSDs (Sands and Haskins, 2008). Although results of various experiments have been promising, the application of these techniques in human patients awaits further preclinical studies, ideally involving large animal models of disease (i.e., in dog, cats, and sheep), in which a larger brain size and higher level of complexity may provide greater insights into the challenge of these therapeutic strategies in humans (Haskins, 2009).

6 Summary

The clinical features of most LSDs likely have a multifactorial basis, and several processes, such as inflammation and apoptosis, contribute to disease development. However, the downstream events triggered by substrate storage in the lysosome are incompletely understood. Research in this area is motivated by the hope of discovering markers that can serve as a surrogate for tissue substrate storage, and avoid the need for invasive procedures. Furthermore, the discovery of disease mechanisms may lead to the identification of putative therapeutic targets.

The LSD are defined by regulatory agencies as "orphan" disorders, that is, affecting individuals numbering <200,000 in the United States, or no more than 5/10,000 in Europe (Graul, 2009). In the United States, therapeutic options for the LSDs have and are being developed, pursuant to two pieces of landmark legislation: the Bayh–Dole Act (BDA, 1980) and the Orphan Drug Act (ODA, 1983). Essentially, these Acts of Congress enabled universities to patent their discoveries and license them to private corporations (BDA); in turn, the biotech companies have received several incentives (including the potential for fast-track approval and subsequent marketing exclusivity) to stimulate development of medical drugs and devices for rare disorders (ODA). Patient support and advocacy groups have played a major part in upholding the enactment of these and related pieces of legislation, including the more recent Genetic Information Nondiscrimination Act (GINA) of 2007–2008.

Several disease-based registries, sponsored by the drug manufacturers, have been established; primarily for disorders in which there is commercially available treatment. Guidelines for the monitoring and treatment of the affected individual are being formulated under the auspices of various experts involved in these surveillance efforts (Martin et al., 2008; Muenzer et al., 2009).

References

Alroy J, Ucci AA (2006) Skin biopsy: a useful tool in the diagnosis of lysosomal storage diseases. Ultrastruct Pathol 30(6):489–503

Ausseil J, Desmaris N, Bigou S, Attali R, Corbineau S, et al. (2008) Early neurodegeneration progresses independently of microglial activation by heparan sulfate in the brain of mucopolysaccharidosis IIIB mice. PLoS ONE 3(5):e2296

Ballabio A, Gieselmann V (2009) Lysosomal disorders: from storage to cellular damage. Biochim Biophys Acta 1793(4):684–696

Fan JQ (2008) A counterintuitive approach to treat enzyme deficiencies: use of enzyme inhibitors for restoring mutant enzyme activity. Biol Chem 389(1):1–11

Froissart R, Moreira da Silva I, Guffon N, Bozon D, Maire I (2002) Mucopolysaccharidosis type II–genotype/phenotype aspects. Acta Paediatr Suppl 91(439):82–87

Fuller M, Brooks DA, Evangelista M, Hein LK, Hopwood JJ, Meikle PJ (2005) Prediction of neuropathology in mucopolysaccharidosis I patients. Mol Genet Metab 84(1):18–24

Giri S, Khan M, Nath N, Singh I, Singh AK (2008) The role of AMPK in psychosine mediated effects on oligodendrocytes and astrocytes: implication for Krabbe disease. J Neurochem 105(5):1820–1833

Gluckman E, Rocha V (2009) Cord blood transplantation: state of the art. Haematologica 94(4):451–454

Grabowski GA (2008) Treatment perspectives for the lysosomal storage diseases. Expert Opin Emerg Drugs 13(1):197–211

Graul AI (2009) Promoting, improving and accelerating the drug development and approval processes. Drug News Perspect 22(1):30–38

Haskins M (2009) Gene therapy for lysosomal storage diseases (LSDs) in large animal models. ILAR J 50(2):112–121

Hendriks MM, Smit S, Akkermans WL, Reijmers TH, Eilers PH, et al. (2007) How to distinguish healthy from diseased? Classification strategy for mass spectrometry-based clinical proteomics. Proteomics 7(20):3672–3680

Martin R, Beck M, Eng C, Giugliani R, Harmatz P, Muñoz V, Muenzer J (2008) Recognition and diagnosis of mucopolysaccharidosis II (Hunter syndrome). Pediatrics 121(2):e377–e386

Meikle PJ, Fietz MJ, Hopwood JJ (2004) Diagnosis of lysosomal storage disorders: current techniques and future directions. Expert Rev Mol Diagn 4(5):677–691

Muenzer J, Wraith JE, Clarke LA International Consensus Panel on Management and Treatment of Mucopolysaccharidosis I (2009) Mucopolysaccharidosis I: management and treatment guidelines. Pediatrics123(1):19–29

Ostrer H (2001) A genetic profile of contemporary Jewish populations. Nat Rev Genet 2(11): 891–898

Pacheco CD, Lieberman AP (2008) The pathogenesis of Niemann-Pick type C disease: a role for autophagy? Expert Rev Mol Med 10:e26

Pastores GM (2003) Enzyme therapy for the lysosomal storage disorders: principles, patents, practice and prospects. Expert Opin Ther Patents 13(8):1157–1172

Pastores GM, Barnett NL (2003) Substrate reduction therapy: miglustat as a remedy for symptomatic patients with Gaucher disease type 1. Expert Opin Investig Drugs 12(2):273–281

Patterson MC, Vecchio D, Prady H, Abel L, Wraith JE (2007) Miglustat for treatment of Niemann-Pick C disease: a randomised controlled study. Lancet Neurol 6(9):765–772

Pierret C, Morrison JA, Kirk MD (2008) Treatment of lysosomal storage disorders: focus on the neuronal ceroid-lipofuscinoses. Acta Neurobiol Exp (Wars) 68(3):429–442

Platt FM, Jeyakumar M (2008) Substrate reduction therapy. Acta Paediatr Suppl 97(457):88–93

Prasad VK, Kurtzberg J (2008) Emerging trends in transplantation of inherited metabolic diseases. Bone Marrow Transplant 41(2):99–108

Raben N, Hill V, Shea L, Takikita S, Baum R, et al. (2008) Suppression of autophagy in skeletal muscle uncovers the accumulation of ubiquitinated proteins and their potential role in muscle damage in Pompe disease. Hum Mol Genet 17(24):3897–3908

Reuser AJ, Drost MR (2006) Lysosomal dysfunction, cellular pathology and clinical symptoms: basic principles. Acta Paediatr Suppl 95(451):77–82

Sands MS, Haskins ME (2008) CNS-directed gene therapy for lysosomal storage diseases. Acta Paediatr Suppl 97(457):22–27

Schiffmann R, Fitzgibbon EJ, Harris C, DeVile C, Davies EH, et al. (2008) Randomized, controlled trial of miglustat in Gaucher's disease type 3. Ann Neurol 64(5):514–522

Settembre C, Fraldi A, Jahreiss L, Spampanato C, Venturi C, et al. (2008) A block of autophagy in lysosomal storage disorders. Hum Mol Genet 17(1):119–129

Shapiro BE, Pastores GM, Gianutsos J, Luzy C, Kolodny EH (2009) Miglustat in late-onset Tay-Sachs disease: a 12-month, randomized, controlled clinical study with 24 months of extended treatment. Genet Med [Epub ahead of print]

Steet R, Chung S, Lee WS, Pine CW, Do H, Kornfeld S (2007) Selective action of the iminosugar isofagomine, a pharmacological chaperone for mutant forms of acid-beta-glucosidase. Biochem Pharmacol 73(9):1376–1383

Sugawara K, Tajima Y, Kawashima I, Tsukimura T, Saito S, et al. (2009) Molecular interaction of imino sugars with human alpha-galactosidase: insight into the mechanism of complex formation and pharmacological chaperone action in Fabry disease. Mol Genet Metab 96(4):233–238

Suzuki K (1998) Twenty five years of the "psychosine hypothesis": a personal perspective of its history and present status. Neurochem Res 23(3):251–259

Walkley SU (2009) Pathogenic cascades in lysosomal disease-Why so complex? J Inherit Metab Dis 32(2):181–189

Yang Z, Vatta M (2007) Danon disease as a cause of autophagic vacuolar myopathy. Congenit Heart Dis 2(6):404–409

Genetic Signaling in Glioblastoma Multiforme (GBM): A Current Overview

Walter J. Lukiw and Frank Culicchia

Abstract Cancers of the brain comprise a genetically and morphologically heterogeneous class of proliferating neural cells derived from incompletely differentiated brain tumor stem cells (BTSCs). The molecular and genetic mechanisms that contribute to their development and propagation are incompletely understood, however, current research is expanding our knowledge as to what specific gene activation and deactivation mechanisms are triggered during the onset of brain cell neoplasia. Apparently, only relatively small populations of BTSCs are capable of driving the proliferative and invasive nature of these cancers, and the intrinsic ability to reinitiate and propagate aberrant cell growth at any metabolic cost. This chapter provides a current overview of gene expression patterns in glioma and glioblastoma multiforme (GBM), with special emphasis on messenger RNA (mRNA) and micro RNA (miRNA) speciation and abundance, and how our recent understanding of specific mRNA–miRNA interactions have increased our comprehension of this insidious neoplastic process.

Keywords Amyloid beta peptides · Brain tumor stem cells · Caspase-3 · Cyclin-dependent kinase · Glioblastoma · Micro RNA · Pentraxin

Abbreviations

Aβ peptides	amyloid beta peptides
ATCC	American tissue culture collection
Bapp	beta amyloid precursor protein
BDC	brain differentiated cell
BTSC	brain tumor stem cell
CD133	neuronal precursor cell surface marker prominin-1
CDKN2A	cyclin-dependent kinase inhibitor 2A
CRL-1690	an experimental glioblastoma (GBM) cell line; also known as T98G (ATCC)

W.J. Lukiw (✉)
LSU Neuroscience Center of Excellence, Louisiana State University Health Science Center, New Orleans, LA 70112, USA
e-mail: wlukiw@lsuhsc.edu

J.P. Blass (ed.), *Neurochemical Mechanisms in Disease*,
Advances in Neurobiology 1, DOI 10.1007/978-1-4419-7104-3_24,

EGFR	epidermal growth factor receptor
NBSCs	normal brain stem cells
GBM	glioblastoma multiforme
HTB-138	an experimental glioma cell line; also known as Hs683 (ATCC)
LOH	loss of heterozygosity
MBAD	metal-based anticancer drugs
miRNA	micro RNA
mRNA	messenger RNA
NHA	normal human astrocytes
NPX2	neuronal pentraxin-2
NSC	neural stem cell
NV	neovascularization
PDGFα	platelet-derived growth factor-alpha
PDGFR	platelet-derived growth factor receptor
Rb	retinoblastoma
SAP	serum amyloid P component
TMZ	temozolomide
VEGF	vascular endothelial growth factor
WHO	World Health Organization

Contents

1 Introduction

Brain cancers constitute a genetically and phenotypically diverse class of proliferative neoplasms derived from incompletely differentiated neuroglial stem cells, sometimes referred to as brain tumor stem cells (BTSCs). Early pathogenetic events appear to differ between glioma and glioblastoma multiforme (GBM), and whether the glioma-to-GBM transition is a developmental attribute or is related to brain tumor progression, is not well understood. The molecular–genetic mechanisms and pathological neurobiology of glioma and GBM remain unclear, however, current oncological research into molecular alterations observed in tumors over time are expanding our understanding as to what brain-enriched genes and induction mechanisms are specifically activated during the onset and propagation of the neoplastic process. Although brain cancer cells are pathologically heterogeneous, only a small population of BTSCs appears to drive the invasive neoplastic phenotype, and the intrinsic ability repeatedly to reinitiate and propagate cancer cell growth at any metabolic cost (Fig. 1 and 2). Several interrelated alterations in gene expression are common among different tumor cell types, especially those that target cell-cycle regulation and growth-promoting pathways, resulting, ultimately, in angiogenesis, apoptosis, necrosis, and deregulated mitotic proliferation. The molecular, genetic, and cellular heterogeneity of glioma and GBM may well underlie the basis for each type of brain cancer's highly variable resistance to current pharmacotherapeutic treatment strategies. The scope of this chapter is to provide a current overview concerning gene expression patterns in glioma and GBM with special emphasis on specific messenger RNA (mRNA), micro RNA (miRNA) interactions, and the contribution of altered miRNA-mRNA-directed signaling pathways to this currently incurable neoplastic process.

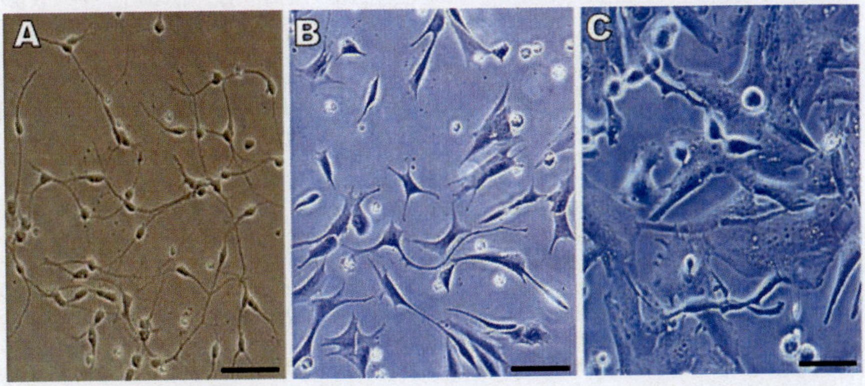

Fig. 1 Cultured human neurons and glia can be differentially viewed or stained to study the contribution of each cell type to brain cell morphology, growth, cell type, drug interaction, and gene expression (see e.g., Lukiw et al., 2005) Control (normal) human neuronal-glial (HNG) cells in primary co-culture exhibit complex, small diameter neuritic extensions and extensive (**a**); cultured HTB-138 glioma cells (American Type Tissue Collection, Bethesda, MD) (**b**); cultured CRL-1690 glioblastoma cells (**c**); all cultured brain cells are about 15–20% confluent, photographed using phase contrast light microscopy (Lukiw et al., 2009); note cell-contact avoidance in the HTB-138 glioma culture, lack of small diameter extensions and altered, flattened, and diverse morphologies of glioma, and especially of GBM cells (**b**), (**c**) when compared to control (**a**); 1 week of culture; bar = 25 μm

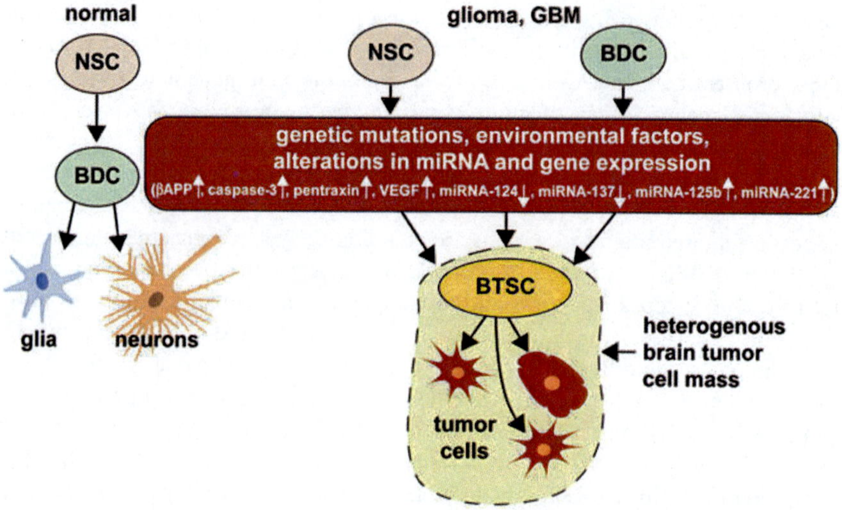

Fig. 2 Highly schematicized representation of normal neural stem cell (NSC), brain differentiated cell (BDC), and brain tumor stem cell (BTSC) development into glioma and gliobalstoma muliforme (GBM) tumor cells. NSCs have an intrinsic property for long-term self-renewal and are pluripotent, that is, have an intrinsic capability to give rise to multiple types of differentiated progeny. In the normal condition (*left*), NSCs differentiate into BDCs such as neurons, glia, and neuroglial subspecies. In contrast, in glioma and GBM, genetic mutations, environmental factors, and alterations in miRNA signaling and pathogenic gene expression trigger the development of BTSCs from both NSCs and BDCs. BTSCs, that make up only a relatively small fraction of the entire heterogeneous tumor cell mass, give rise to a series of genotypically and phenotypically heterogeneous tumor cells and a proliferating and invasive tumor cell mass (see Singh et al., 2004; Xie and Chin, 2008; Godlewski et al., 2008; Hide et al., 2008; Yadirgi and Marino, 2009). Recent evidence suggests the participation of the polytopic membrane protein beta-amyloid precursor protein (βAPP), the apoptosis effector protein caspase-3, the cell contact and synaptic remodeling protein pentraxin-2, and vascular endothelial growth factor (VEGF), the most potent vascular substance known in driving brain oncogenesis. More recently, specific micro RNA (miRNA; miRNA-124 and miRNA-137) downregulation has been shown to affect cellular proliferation and/or induce unscheduled differentiation of BTSCs (Gurdon and Melton, 2008; Silber et al., 2008). GBM is further associated with an upregulation in miRNA-125b and miRNA-221. miRNA-125b is upregulated in IL-6-stressed normal human astrocytes (NHA), a treatment known to induce astrogliosis, and in vitro, anti-miRNA-125b added exogenously to IL-6-stressed NHA cultures attenuated both glial cell proliferation and increased the expression of CDKN2A, a predicted miRNA-125b target and negative regulator of cell growth (Pogue et al., 2010). GBM-up-regulated miRNA-221 appears to target the cell growth suppressive cyclin-dependent kinase inhibitors p27 and p57, linking the cell cycle checkpoint at S phase initiation with growth factors, which may be another trigger for tumor cell proliferation (Li et al., 1999; le Sage et al., 2007; Mellai et al., 2008; Medina et al., 2008; Lukiw et al., 2009)

2 Brain Cancer Etiology—Glioma and GBM

Tumors are classified by their tissue of origin. Astrocytomas fall into the largest category of tumors of neuroepithelial tissue. Neoplastic neuroepithelial tumors of the central nervous system (CNS) are categorized by the World Health Organization

(WHO) rating as being pilocytic, and having circumscribed growth that tends to respect anatomic boundaries because they do not invade (WHO grade I). The more diffuse (WHO grade II) tumors demonstrate slow growth, moderate hypercellularity, occasional nuclear atypia, and diffuse infiltration of neighboring brain cell structures. These lesions have a tendency for malignant transformation, possibly dedifferentiating all the way to glioblastoma multiforme (GBM) and included in this group are protoplasmic, gemistocytic, fibrillary, and mixed variants. Anaplastic (WHO grade III) tumors demonstrate hypercellularity, moderate nuclear atypia, prominent mitotic activity, and diffuse infiltration. These tumors are most often the result of dedifferentiation of a grade II astrocytoma. Glioblastoma multiforme (WHO grade IV) demonstrate marked nuclear atypia, high mitotic activity, microvascular proliferation, and areas of coagulative necrosis. This group includes GBM and two variants: giant-cell glioblastoma and gliosarcoma. Although a glioblastoma may represent a dedifferentiated grade II or III astrocytoma, most are primary glioblastomas and do not derive from a less malignant precursor. Primary GBMs often manifest de novo; without clinical or histopathological evidence of a pre-existing, less-malignant precursor lesion. These tumors are identified in patients after a short clinical history of usually less than three months. Primary GBM accounts for the vast majority of cases (60%) in adults older than 50 years of age and secondary GBMs (40%) typically develop in younger patients (<45 years) through malignant progression from a low-grade astrocytoma (WHO grade II) or anaplastic astrocytoma (WHO grade III). The time required for this progression varies considerably, ranging from less than 1 year to more than 10 years, the mean interval being 4–5 years. These classifications provide the standard for communication between different medical institutions in the United States and around the world, and are based on the premise that each type of tumor results from the abnormal growth from a specific CNS cell class (Lopes et al., 1993; Louis, 2006; Rosell et al., 2008; Rueger et al., 2008; Tatter, 2005; Fuller et al., 2002).

Of the estimated 17,000 primary brain tumors diagnosed in the United States each year, gliomas account for more than 75% of all brain tumors and are the most common supratentorial tumor in all age groups. These tumors comprise a heterogeneous group of neoplasms that differ in location within the CNS, in age and sex distribution, in growth potential, in extent of invasiveness, in morphological features, in tendency for progression, and in response to treatments. WHO grade IV GBMs are the most frequent and malignant histological brain tumor cell type (Ohgaki and Kleihues, 2005; Ohgaki, 2009). There is a tendency toward a higher incidence of gliomas in Caucasians of the more highly developed, industrialized societies (Ohgaki and Kleihues, 2005; Fisher et al., 2007; Ohgaki, 2009). The epidemiology of GBM as a spontaneously occurring malignant neoplasm remains largely unknown. Familial gliomas account for about 5% or less of all malignant gliomas, and less than 1% of gliomas are associated with known genetic syndromes such as tuberous sclerosis, neurofibromatosis, Turcot syndrome, Li–Fraumeni, von Hippel–Lindau, or related neurological syndromes (Fisher et al., 2007; Farrell and Plotkin, 2007). About 95% of all brain cancers are of idiopathic, sporadic, or unknown origin (Louis, 2006; Fisher et al., 2007; Ohgaki, 2009).

Recent concerns regarding the association between GBM and head injury, labile nitrogenated and nitroso-compounds, exogenous or endogenous genomic alkylating factors, occupational hazards, and electromagnetic field exposure including cell phone use have been inconclusive and to date no hard and fast rules apply (Fisher et al., 2007; Ohgaki, 2009). GBM most often occurs in the subcortical white matter of the cerebral hemispheres, and in one recent epidemiological study of 987 cases of GBM, the most frequently affected sites were the temporal (31%), parietal (24%), frontal (23%), and occipital (16%) lobes (Ohgaki and Kleihues, 2005; Ohgaki, 2009). The prognosis of patients diagnosed with GBM remains dismal, and the median survival time of patients with this most common form of malignant glioma currently averages less than one year. Some of the newer treatment strategies and novel pharmacological approaches are further described in the sections below.

3 Brain Tumors—Subpopulations of Brain Tumor Stem Cells

An evolving concept in the neuro-oncological mechanism driving glioma and GBM is that brain tumor stem cells, which represent a relatively minor population of the entire tumor mass, constitute the essential "functional core" of the tumor that drives neoplastic proliferation. As do normal brain stem cells (NBSCs), BTSCs exhibit two defining properties including the capability for long-term self-renewal, and pluripotency, that is, the capability to give rise to multiple types of differentiated progeny. In normal brain stem cells the balanced coordination of these two defining properties is essential for brain development and functional homeostasis, yet these same two parameters are fundamentally altered in brain tumor development (Gurdon and Melton, 2008; Yadirgi and Marino, 2009). In brain cancers, variable populations of BTSCs have been detected in glioblastoma, medulloblastoma, and ependymoma (Singh et al., 2003, 2004; Xie et al., 2008). In the framework of this brain cancer stem cell hypothesis, genes important for normal neural stem cell homeostatic function appear also to be essential to support their pathological development into BTSCs. This concept of nuclear reprogramming, describing a switch in gene expression from one kind of cell to that of another unrelated cell type, may be central to oncogenesis (Fig. 2; Gurdon and Melton, 2008). BTSCs appear to incompletely differentiate in vivo, and their neoplastic potential depends on the balance between their replicative index and the degree of terminal differentiation that these minority brain cell populations achieve.

A specific oncogenic family of genes may be involved in triggering BTSC development, proliferation, and pathogenic functions, including polytopic surface sensor proteins such as the neural precursor cell surface marker prominin-1 (CD133), beta-amyloid precursor protein (βAPP), several neural-enriched pentraxin species, vascular endothelial growth factor (VEGF), caspase-3 and other potentially oncogenic genes associated with growth rate, cell cycle regulation, angiogenesis, apoptosis and/or necrosis, and deregulated mitotic proliferation (Singh et al., 2004; Xie et al., 2008; Bauer et al., 2008; Culicchia et al., 2008). Uncovering the molecular mechanism of how these individual genes are activated, if their expression is in any way coordinated, the individual or coordinated contribution of BTSCs to tumor cell proliferation, and how BTSC-specific proteins interact with each other and with chemoactive, antineoplastic agents should be of use not only in expanding our understanding of how brain cancers develop, but also in the design of future neurotherapeutic approaches and multimodal treatment strategies.

4 Gene Expression in the Human Brain

Mammalian brain cells have an intrinsically higher index of genetic output and complexity compared to gene transcription profiles in other cells and tissue cell systems (Lukiw et al., 2000; Sutcliffe, 2001; Colangelo et al., 2002; Mattick and Makunin, 2005). Furthermore, complex biological behaviors and functional neurochemical mechanisms that accompany aging and neuropathology, including neuro-oncological change, are probably not controlled by single genes but rather by groups of functionally related genes sometimes referred to as gene families. The use of DNA array technologies is currently capable of interrogating 33,000 genes on a single DNA array, or the levels of all expressed genes in a single brain biopsy sample (Affymetrix Corporation, Santa Clara, CA; Lukiw, 2004; Lukiw et al., 2005; Macdonald et al., 2007). Although some pathogenically related upregulated oncogenic genes thus far identified may share common overlapping functions such as stress response and adaptive processes that support several related aspects of brain inflammation, apoptosis and/or angiogenesis, glioma and GBM may represent some of the most heterogeneous gene expression patterns of any neurological disease known (Kavsan et al., 2006; Idbaih et al., 2007; Juric et al., 2007; Culicchia et al., 2008).

5 Gene Expression in Brain Cancers

Molecular–genetic and population-based studies have identified several gene mutations that associate with brain tumor development (Tatter, 2005; Nicholas et al., 2006; Van den Bent and Kros, 2007; Juric et al., 2007; Kavsan et al., 2007; Johansson Swartling, 2008). These may be linked to an ordered accumulation of multiple genetic mutations on multiple chromosomes and sequential, temporally mediated pathogenic interactions. For example, in about one third of all cases, the transition from healthy astroglial cells to astrocytoma has been associated with p53 gene mutations at chromosome 17p, and mutations in p53, a tumor suppressor gene, were among the first genetic alterations ever to be identified in astrocytic brain tumors (Tatter, 2005; Nicholas et al., 2006; Van den Bent and Kros, 2007). Deletion or alteration of the p53 gene appears to be present in approximately 25–40% of all GBMs, and p53 immunoreactivity also appears to be associated with tumors that arise in younger patients. Brain cancers also often exhibit loss of heterozygosity (LOH), and LOH at chromosome 10q is the most frequent gene alteration for both primary and secondary glioblastomas, ranging from 60 to 90% of all cases. This mutation appears to be specific for GBM, is found rarely in other tumor grades, and is associated with poor survival. LOH at 10q plus 1 or 2 of the additional gene mutations appear to be frequent alterations and are most likely major players in the development of GBM. LOH leads in almost one half of all subsequent cases to anaplastic astrocytoma associated with the retinoblastoma (Rb) gene at chromosome 13q. Involvement of additional mutations at chromosomes 9p and 19q, followed by GBM development associated with chromosome 10 mutations and amplification of the epidermal growth factor receptor gene is further postulated in this highly complex pathway of oncogenic development (see, e.g., Tatter, 2005; Nicholas et al., 2006; Van den Bent and Kros, 2007; Johansson Swartling, 2008). The involvement of multiple gene loci and reduced or incomplete penetrance of these gene mutations indicate that the resulting altered developmental or oncogenic pathways induce tumors possessing a highly variable phenotype and heterogeneous morphology.

Brain cancers are also genetically associated with homeostatic disturbances in the epidermal growth factor receptor (EGFR), MDM2, platelet-derived growth factor-alpha (PDGFα) and PTEN genes. The EGFR gene is involved in the control of cell proliferation and EGFR overexpression and mutant EGFR expression occurs in approximately 50% of patients with GBM (Nicholas et al., 2006; Voelzke et al., 2008). In fact multiple genetic mutations in EGFR are apparent, including overexpression of the receptor as well as rearrangements that result in truncated isoforms (Voelzke et al., 2008). However, all the clinically relevant EGFR mutations appear to generate a similar phenotype resulting in increased EGFR activity. Amplification or overexpression of MDM2 constitutes an alternative mechanism to escape from p53-regulated control of cell growth by binding to p53 and blunting its activity. Overexpression of MDM2 is the second most common gene mutation in GBMs and is observed in 10–15% of patients. Some studies show that this mutation associates with a poor prognosis. The PDGF gene acts as a major mitogen for glial cells by binding to the PDGF receptor (PDGFR) and amplification or overexpression of PDGFR is typical (60%) in the pathway leading to secondary glioblastomas. PTEN (also known as MMAC and TEP1) encodes a tyrosine phosphatase located at chromosome 10q23.3 that functions as a cellular phosphatase, turning off signaling pathways, and is consistent with possible tumor-suppression activities. When phosphatase activity is lost because of genetic mutation, signaling pathways can become activated constitutively, resulting in aberrant proliferation. PTEN mutations have been found in as many as 30% of all glioblastoma cases studied (Koul, 2008; Cheng et al., 2009).

Current DNA array technologies and statistical and comparative bioinformatics analysis enable a comprehensive examination of the expression of all genes associated with brain health and disease. Genomewide gene expression patterns of neoplastic brain cells in their various developmental stages provide a fascinating reflection of the physiological and pathological status of those pathogenic brain cells. Robust gene expression analyses have been applied to whole tumor cells and some recurrent themes, besides such variables as patient age, sex, affected lobe and disease onset, duration, and other clinical parameters, are emerging for specific pathology-related genes. Interestingly, the progression from low-grade glioma to high-grade GBM may be associated with distinct molecular–genetic changes that vary according to WHO grade (MacDonald et al., 2007; Juric et al., 2007). Several excellent reviews of DNA array analysis of glioma and GBM have recently appeared and the material in them is not reiterated here (Boudreau et al., 2005; Belda-Iniesta et al., 2006; Kavsan et al., 2005; Tso et al., 2006; Faury et al., 2007; Idbaih et al., 2007; Juric et al., 2007; MacDonald et al., 2007; Johansson Swartling, 2008). Rather we focus on some current observations on increases in the expression of glioma and GBM of several altered markers involved in brain cell contact, cell cycle, cell death, and vascular proliferation markers at the level of gene expression in virtually all brain tumors examined: beta-amyloid precursor protein (βAPP), the apoptosis effector caspase-3, a cell–cell contact neuronal-enriched protein pentraxin-2, and the angiogenesis promoting vascular endothelial growth factor (VEGF).

5.1 Beta-Amyloid Precursor Protein (βAPP)

Beta amyloid precursor protein (βAPP), a brain-abundant trans-membrane glycoprotein "sensor" implicated in neuronal–glial intercellular contact and progressive apoptotic and necrotic brain cell death appears to be part of a pathogenic gene family that associates with glial cell proliferation in glioma, GBM, and neurodegeneration (Colangelo et al., 2002; Lukiw, 2004; Fuso et al., 2007; Culicchia et al., 2008). In fact most of the original studies on the role of βAPP in neurobiology and neurodegeneration were first performed in transformed human glioblastoma cell lines (Lahiri et al., 1997; Paris et al., 2005; Fuso et al., 2007; Culicchia et al., 2008). Cancer-affected glial cells are characterized by highly unusual and diverse morphology that often correlates to the grade of the neoplasm and display noncontact inhibited cells, lack of cell–cell adhesion and connectivity, and a highly varied range of cellular morphology (Fig. 1) (Hyun Huang et al., 2007; Caltagarone et al., 2007; Culicchia et al., 2008). Bizarre glial cell morphology in malignant gliomas and GBM have been correlated with the depletion of cytoskeletal-matrix actin-bundling proteins and alterations in integrin-mediated communication between the extracellular matrix and the actin cytoskeleton (Venezia et al., 2007; Caltagarone et al., 2007; Young-Pearse et al., 2007). Although the abundant cytoskeletal protein β-actin itself is not upregulated, β-actin-associated cytoskeletal proteins and integrins that further drive glial cell division and proliferation processes during the cell cycle have been implicated in brain cancer development. Interestingly βAPP structural orientation within the membrane, βAPP trafficking, and intracellular signaling functions are associated with β-actin-associated proteins and β-actin-mediated cellular shape. It is difficult to rationalize whether βAPP upregulation in glioma and GBM is either a consequence of, or contributory factor to altered microfilament of microtubule cytoarchitecture, or increased cellular proliferation, or both (Venezia et al., 2007; Caltagarone et al., 2007; Young-Pearse et al., 2007; Culicchia et al., 2008).

Although βAPP is known to be variably upregulated in chronic neurodegenerative disease, depending on the stage of the disease, the observation of upregulated proinflammatory and amyloidogenic neural degenerative markers in glioma and

GBM is a relatively recent one (Lukiw, 2004; Lukiw et al., 2005; Fuso et al., 2007; Sin et al., 2008; Culicchia et al., 2008). Excessive βAPP-mediated signaling is thought to be responsible, in part, for driving neural inflammation, glial cell growth and expansion, and brain cell degeneration events such as apoptosis (Melhorn et al., 2000; Radde et al., 2007; Venezia et al., 2007; Herber et al., 2007). Ischemic brain damage is also known to induce inflammatory cytokine and βAPP over-expression that further induce widespread brain cell death via apoptosis or necrosis (Pluta, 2002; Bates et al., 2002). Chronic gliosis is, in addition, associated with altered processing of βAPP in vivo, and thus may trigger pathological changes associated with aberrant interneural communication between brain cells, thus contributing to progressive alterations in glial cell morphology (Pluta, 2002; Bates et al., 2002; Young-Pearse et al., 2007).

Increased upregulation of βAPP expression in glioma and GBM further suggests that unscheduled proliferative events of brain cells are accompanied by the significant elevation of integral transmembrane receptors that are pathogenic markers for neurodegeneration. βAPP appears to be part of a poorly understood cell-contact signaling pathway whose disruption induces cell-cycle signaling, mitotic abnormalities, and glial cell expansion (Paris et al., 2005; Venezia et al., 2007; Young-Pearse et al., 2007; Fuso et al., 2007; Lukiw et al., 2008). Increased βAPP expression has long been associated with gliosis, the localized expansion of astrocyte populations, and the production of dense fibrous networks of neuroglia in the area of a pathogenic lesion. Similar gliosis-related increases in the expression of βAPP in glioma and GBM and in the neurodegenerating brain tissues support some underlying commonality in disorganized interneural signaling, aberrant cytoarchitecture and neural cell shape characteristic of both neurological conditions. Interestingly, Alzheimer's disease and GBM have similar age-specific incidence rates and accumulation of senile plaque deposits consisting of amyloid beta (Aβ) peptides derived from the secretase cleavage of βAPP holoprotein. About one third of all GBM cases exhibit age-related plaque scores indicative or suggestive of AD; and progressive neurodegenerative pathology is present in about half of all cases of GBM (Nelson, 2002; Lukiw et al., 2008).

5.2 Caspase-3

Whether brain cell death in the malignant neoplasms, triggered by hypoxia, lack of nutritive support or other pathogenic factors, is driven by apoptosis or necrosis is not well understood. In fact both neural-destructive processes may be operating simultaneously. Apoptosis and necrosis appear to lie at either end of a spectrum of functional brain cell dysfunction and progressive cell loss spanning from programmed cell death (apoptosis; internucleosomal DNA fragmentation) at one end to induced and premature cell death (necrosis; randomized DNA fragmentation) of brain cells at the other. Cysteine–aspartic acid protease-3 (caspase-3), a key member of a family of 11 human cysteine proteases, plays key essential effector roles in both apoptosis and necrosis and in neuroinflammatory aspects of neurodegeneration and brain tumor growth. There is evidence for both caspase-3 upregulation (Ray et al., 2002; Lukiw et al., 2009) and caspase-3 downregulation in human brain tumors (Stegh et al., 2008). Nonhomeostatic levels of caspase-3 indicate alterations in the cellular balance of both pro-apoptotic and anti-apoptotic signals (Takuma et al., 2004; Lefranc et al., 2007). The increased expression of the pro-apoptotic Bax protein, upregulation of calpain and caspase-3, and occurrence of internucleosomal DNA fragmentation indicate that one mechanism of cell death in malignant brain tumors is apoptosis (Ray et al., 2002; Lukiw et al., 2009). These results may be explained by the fact that the apoptotic process only approaches the stage of caspase-3 activation, followed by a subsequent variable activation of the apoptotic cascade and "programmed" cell death mechanism, resulting in apoptotic blockage and an accumulation of brain cell mass.

5.3 *Pentraxin-2 (NP2; NPTX2)*

Pentraxins are a family of pentameric calcium-dependent ligand-binding proteins bearing a highly distinctive structure similar to that of the ring-shaped lectins (Emsley et al., 1994). Pentraxins represent a novel neuronal uptake pathway that functions during intercellular and extracellular signaling, synapse formation and clustering, remodeling, and cell–cell contact (Gerrow and El-Husseini, 2007). "Short" pentraxins include the inflammation-related serum amyloid P component (SAP) and C reactive protein (CRP) and the "long" pentraxins include PTX3 (a cytokine-modulated molecule) and several prominent brain-enriched pentraxins such as neuronal pentraxin-2 (NP2; NPTX2). Interestingly, NPTX2, normally expressed in the CNS, is a member of a family of proteins related to CRP and other acute-phase inflammatory mediators, and has been found to be correlated with glioma and GBM edema, the swelling of soft tissues as the result of loss of brain water balance and excessive water accumulation. Increased NPTX2 are in turn strongly associated with poorer survival rates in tumors with the highest levels of edema (Hsu and Perin, 1995; Goodman et al., 1996). Several gene expression studies have shown NPTX2 to be consistently and significantly upregulated in glioma and GBM (Carlson et al., 2007; Pope et al., 2008; unpublished observations). It is important to note that the NPTX2 upregulation associated with angiogenic- and edema-related signaling is often coregulated with the simultaneous upregulation of vascular endothelial growth factor (VEGF) and the proliferation of neovascularization.

5.4 *Vascular Endothelial Growth Factor (VEGF)*

Angiogenesis, the proliferation of vascular growth that provides nutritive support to the expanding tumor cell mass, is one of the hallmarks of all cancers.

Vascular endothelial growth factor stands out as a key mediator of tumor-associated angiogenesis among a complex signaling system involving pro- and antiangiogenic factors (Chamberlain, 2008; Grothey et al., 2008; Pope et al., 2008). The upregulation of VEGF, originally described as a vascular permeability factor in brain tumors, has often been proposed to be the major cause of both vasogenic edema in gliomas and neovascularization (NV) (Bruce et al., 1987; Buie and Valgus, 2008; Norden et al., 2008). A consistent observation in brain cancer biology is that malignant gliomas invariably express vast amounts of VEGF, now regarded as an important pathogenic marker of angiogenesis and NV, essential for the proliferation and the survival at any cost for malignant glioma cells. NV is orchestrated by the coordinate induction of a family of growth-factor genes and most prominently by VEGF which also possesses endothelial cell-specific mitogenic effects that closely correlate with NV during embryonic development and normal systemic physiology, fetal anemia, in retinal NV, in models of hypoxic ischemia, and in malignant tumors. These combined observations are suggestive of VEGF's key role in vascular proliferation in growth, health, and disease. Hypoxia is thought to be one crucial physiological stimulus for VEGF upregulation that precedes NV, and low cellular oxygen tension rapidly induces a number of transient genetic signals through which this is accomplished (Larrivee and Karsan, 2000; Hasan and Jayson, 2001; Giles, 2001; L. Lukiw et al., 2003; Norden et al., 2008). The multiple roles of VEGF in brain tumor development and proliferation and anti-VEGF-based therapies have been recently examined in the last year in several excellent reviews and interested readers are encouraged to refer to these thoughtful works and the published papers referenced within (Brandsma et al., 2008; Chamberlain, 2008; Grothey and Ellis, 2008; Pope et al., 2008; Reardon et al., 2008; Wong and Brem, 2008).

6 Specific Alterations in the Expression of Brain-Enriched Genes

Neurological disorders including glioblastoma involve a highly complex pathogenesis with multiple etiological factors, and this is reflected in the expression of brain genes in this disease. Several glioma and GBM tumor cell lines have been immortalized and "standardized brain tumor cell cultures" are available to oncology researchers through government-funded sources such as the American Type Tissue Collection (ATCC, Bethesda MD). Commonly used human brain cell cultures include glioma cell line HTB-138 (Hs683) and glioblastoma tumor cell lines CRL-2020 (DBTRG-05MG), CRL-1690 (T98G), CRL-2365 (M059K), and CRL-2366 (M059J). The majority of these standardized neoplastic, immortalized cell lines develop as a mixture of floating and adherent cells growing as heterogeneous clusters of neuroblastic cells with multiple, short, fine cell processes (neurites) that often aggregate, forming clumps, detach from solid surface, and float within the cell culture medium (Fig. 1). Total DNA, RNA, and protein can be effectively and efficiently isolated from these archived cell lines and are subsequently used for gene expression analysis and downstream molecular–genetic investigations. Recent studies in these cell lines have indicated increases in the integral membrane β-amyloid precursor protein (βAPP) as a proinflammatory, neurodegenerative, and proliferative pathogenic marker (Culicchia et al., 2008). Indeed, from the perspective of dysregulated pathogenic gene expression, glioma and glioblastoma multiforme display significant upregulation of disease markers such as βAPP and caspase-3 with features of rapid-onset, progressive, glial cell proliferating, degenerative brain disease. The known disease-related functions of these inflammatory and neurodegenerative markers may further contribute to the pathogenic phenotype and unscheduled misregulated propagation of glial cells in the brain. The important point is that brain tumors consist of a spectrum of tumors of varying differentiation, malignancy grades, and gene expression profiles. Despite the fact that all tumors have an initially invasive phenotype, early genetic events appear to differ between astrocytic and oligodendroglial tumors, and this may form, in part, the molecular genetic basis for variation in brain cancer cell composition that complicates more effective therapies. Knowledge of malignant glioma genetics has already affected clinical management of these tumors, and researchers and clinicians can only hope that further knowledge of the evolution of the molecular pathology of malignant gliomas will result in novel therapies that employ multiple, multimodal treatment strategies.

7 Micro RNAs (miRNAs): Specific miRNA and mRNA Alterations in Human Brain Cancer

Micro RNAs (miRNAs) are small RNA polymerase II and III transcribed, noncoding RNA molecules that play important posttranscriptional regulatory roles by recognizing and binding to the 3′ untranslated region (3′UTR) of mature messenger RNAs (mRNAs). By doing so, miRNAs repress translation and expression of their particular mRNA targets (Mattick and Makunin, 2005; Cao et al., 2006; Lukiw, 2007; Lukiw and Pogue, 2007; Cho, 2007; Amaral et al., 2008; Dogini et al., 2008; Zeng 2009). Transcription of protein-encoding genes and miRNAs by RNA polymerase II and III and their interrelated functions in the modulation of gene expression suggests the possibility of some coordinated mode of interaction, possibly through miRNA interaction with specific transcription factors (Mattick and Makunin, 2005; Hobert, 2008; Lukiw et al., 2008; Makeyev and Maniatis, 2008; Williams et al., 2008; Amaral et al., 2008). Interestingly, small signaling molecules such as miRNA may transfer epigenetic information not only within cells but also between cells and organ systems as part of a dynamic RNA-mediated interplay between the environment and the genome (Zhao et al., 2006; Hill et al., 2009; Mattick et al., 2009). Such novel genetic mechanisms may explain, in part, cancer invasiveness and metastasis throughout cells, organs, and tissue systems (Louis, 2006; Hyun Hwang et al., 2008).

To date about 911 miRNAs in the human brain have been identified (Lukiw, 2007; Lukiw and Pogue, 2007). The miRNA-mediated regulation of messenger RNA (mRNA) complexity in the human central nervous system is evolving as a critical and determining factor in regulating CNS-specific gene expression during development, plasticity, aging, and disease (Hobert, 2008; Makeyev and Maniatis, 2008; Williams et al., 2008). Several excellent recent reviews on miRNA speciation in the CNS and specific examples in brain tumors have recently appeared in the literature and the authors would encourage interested researchers, clinicians, and medical and graduate students to read them over (Mattick and Makunin, 2005; Ciafrè et al., 2005; Zhang et al., 2007; Mourelatos, 2008; Nicoloso and Calin, 2008; Papagiannakopoulos and Kosik, 2008; Silber et al., 2008; Hobert, 2008; Makeyev and Maniatis, 2008; Williams et al., 2008; Zeng, 2009, Lukiw et al., 2009).

Current studies indicate that specific miRNAs may function at multiple hierarchical levels in gene regulatory networks, from targeting hundreds of effector genes to controlling the levels of global regulators of transcription and alternative pre-mRNA splicing (Cao et al., 2006; Makeyev and Maniatis, 2008; Silber et al., 2008). An expanding number of miRNAs have been reported to be altered in abundance in glioma and GBM, and largely because of their disease-related expression and selection of specific mRNA targets in the brain, these miRNAs are strongly implicated as important regulatory controls in neoplastic onset and evolution. In general, abrogation of global miRNA-mediated mRNA processing and homeostatic control is associated with accelerated cellular transformation and tumorigenesis, and some specific examples are given below (Lukiw, 2004; Pogue and Lukiw, 2004; Lukiw and Bazan, 2006; Kumar et al., 2007; Lukiw and Bazan, 2008; Lukiw, 2009; Zeng, 2009).

Several decreased or increased miRNA species implicated in miRNA-mediated brain cell tumor growth, oncogenesis, apoptosis, and survival (sometimes referred to as oncomirs) are miRNA-124 and miRNA-137 (Ciafrè et al., 2005; Cho, 2007; Silber et al., 2008; Papagiannakopoulos and Kosik, 2008; unpublished observations). In one recent study the expression levels of miRNA-124 and miRNA-137 were found to be significantly decreased in anaplastic astrocytoma (WHO grade III) and GBM (WHO grade IV) relative to nonneoplastic control tissue (Silber et al., 2008; Papagiannakopoulos and Kosik, 2008). Interestingly, when miRNA-124 was introduced into nonneuronal mammalian cells a preferential reduction in the amounts of nonneuronal mRNAs, including those encoding protein required for cell proliferation or neural stem cell function was observed, and promotion of a neuronal-like mRNA profile (Conaco et al., 2006; Makeyev et al., 2007). Conversely, a depletion of miRNA-124 from primary neurons accumulated a number of nonneuronal mRNA targets, suggesting that miRNA-124 ensures that progenitor genes are posttranscriptionally inhibited in neurons (Makeyev et al., 2007; Cao et al., 2007). Such evidence suggests the roles of miRNAs are in controlling cell fate and the proliferating capacity of brain cells. That miRNA-124 and miRNA-137 induce differentiation of adult neural stem cells, oligodendroglioma-derived stem cells, and human GBM-derived stem cells and induce cell cycle arrest in GBM suggests that targeted delivery of these highly soluble and mobile small RNAs to glioma and GBM cells may provide an efficacious and novel therapeutic treatment strategy for containing the growth of cancerous brain cells (Silber et al., 2008; unpublished observations).

Another miRNA species implicated in cell tumor growth, oncogenesis, apoptosis, and survival is miRNA-221 (Ciafrè et al., 2005; Gillies and Lorimer, 2007; Medina et al., 2008; Lukiw et al., 2009). Support for the pathogenic role of miRNA-221 in tumor growth comes from the recent observations that upregulated miRNA-221 targets the cell growth suppressive cyclin-dependent kinase inhibitors p27 and p57, thus linking a cell-cycle checkpoint at S phase initiation with growth factors that trigger cell proliferation (Li et al., 1999; le Sage et al., 2007; Mellai et al., 2008; Medina et al., 2008; Lukiw et al., 2009). Other recent work reported a selective upregulation of miRNA-221 and downregulation of a miRNA-221 messenger RNA target encoding the survivin-1 homologue BIRC1, a neuronal inhibitor of apoptosis protein and brain cell marker for neural degeneration (Lukiw et al., 2009). In these later studies the expression of BIRC5 (survivin-1) and caspase-3 was found to be significantly upregulated, particularly in the more advanced stages of GBM. It is important to note that paracrine signaling between adjacent brain cells may contribute to significant positive feedback regulation and the progressive intercellular proliferation of pathogenic signaling in both degenerating brain cells and brain tumors (Zhao et al., 2006; Culicchia et al., 2008). Indeed, tumor invasion occurs not only through dysfunction of the adhesive properties of tumor cells but also in their pathogenic secretion of small lysosomal proteolytic enzymes such as cathepsin-L (Levicar et al., 2002, 2003). Use of online accessible miRNA–mRNA database searches, other miRNA-221-targeted components of apoptotic signaling in glioma and GBM, and interactions with the Bcl-2 protein family of apoptosis include anti-apoptotic protein Bcl-2-binding component 3 and other Bcl-2-modifying factors (Sanger mirBase version 10.1; http://microrna.sanger.ac.uk/cgi-bin), hence miRNA-221 may further modulate apoptotic signaling via quenching or augmentation of the expression of a number of alternate antiapoptotic mRNA targets, such as additional Bcl-2-modifing factors. Again the small size and high solubility of specific brain-enriched miRNA species suggests that they may perform ancillary intracellular and extracellular signaling roles involved in the spreading and propagation of tumor cell growth and associated metastatic events (Lukiw, 2007; Lukiw and Pogue, 2007; Felicetti et al., 2008; Mattick et al., 2009). More recently, miRNA-125b has been shown to be upregulated in interleukin-6 (IL-6)-stressed normal human astrocytes (NHA), a treatment known to induce astrogliosis, and in vitro, anti-miRNA-125b added exogenously to IL-6-stressed NHA cultures attenuated both glial cell proliferation and increased the expression of the cyclin-dependent kinase inhibitor 2A (CDKN2A), a miRNA-125b target, and negative regulator of cell growth (Pogue et al., 2010).

8 Therapeutic Strategies for the Clinical Management of Glioma and GBM

Current treatment strategies for GBM are multimodal and typically involve surgical resection followed by radiation therapy and chemotherapy. Upon tumor recurrence repeat resection or stereotactic radiosurgery followed by additional radiotherapy and chemotherapy may improve outcome in certain cases; several new strategies have been developed to optimize designer therapies for GBM (Sathornsumetee and Rich, 2008; Sathornsumetee and Reardon, 2009; Tentori and Graziani, 2009). The most commonly used chemotherapeutic drug for GBM, temozolomide (TMZ), typically administered both during and after radiotherapy, is a potent DNA methylating agent that generates a wide spectrum of random methyl adducts in the genome. The antitumor activity of TMZ and related alkylating agents has been mainly attributed to the production of O(6)-methylguanine as a potent cytotoxic and antimitotic (Tentori and Graziani, 2009). TMZ also promotes autophagic cell death, a caspase-independent process characterized by the accumulation of cytoplasmic autophagic vacuoles and accompanied by extensive degradation of polyribosomes, the endoplasmic reticulum, and the Golgi apparatus that precedes the destruction of the nucleus (Lefranc et al., 2007). As brain neoplasms are generally associated with altered βAPP, pentraxin, caspase-3, VEGF expression, and the kinases that modify these effector molecules, co-ordinated inhibition of these oncogenic markers might be an effective therapeutic strategy. These kinds of treatment approaches have recently been reviewed (Anderson et al., 2008; Lakka and Rao, 2008; Hide et al., 2008; Norden et al., 2008). Anti-inflammatory, anti-βAPP, and antiamyloid pharmacologic strategies directed at neurodegenerative processes may also have some therapeutic value in the treatment of glioma- or glioblastoma-affected brain cells (Nelson, 2002; Lukiw and Bazan, 2006; Lukiw and Bazan, 2008; Tschape and Hartmann, 2008; Culicchia et al., 2008).

Unfortunately, chemotherapeutic drug resistance occurs relatively often and effective drug delivery to cancer targets remains an accessory concern affecting the clinical response in brain cancer patients. Because malignant gliomas are highly vascularized tumors that produce VEGF, a key mediator of angiogenesis, and given the fact that angiogenesis is essential for the proliferation and survival of malignant glioma cells, angiogenesis antagonists such as angiostatin, endostatin, and vasostatin may provide yet another specifically targeted, therapeutic strategy. Recent studies have investigated the use of bevacizumab—a humanized monoclonal antibody against VEGF—for patients with recurrent malignant glioma, however, the results have been inconsistent, and larger, randomized clinical trials are needed to determine the magnitude of the benefit (Buie and Valgus, 2008; Norden et al., 2008). Moreover, angiogenesis antagonists have numerous unwanted side effects in interfering with normal wound healing, bleeding, and blood clotting, and are associated with heart, immune, and reproductive dysfunction (Norden et al., 2008). Interestingly, gamma- and beta-secretases that act on βAPP processing appear to play an essential role during angiogenesis, and inhibitors of these secretases may constitute a novel evolving class of antiangiogenic and antitumoral compounds (Paris et al., 2005).

Just as for angiogenesis antagonists, toxic metal-based anticancer drugs (MBADs), including cisplatin, carboplatin, and oxaliplatin, and other arsenic-, cadmium-, copper-, gallium-, lanthanum-, platinum-, ruthenium-, or titanium-containing antitumor drug complexes have adverse effects on physiological systems outside of the CNS (Zhang and Lippard, 2003). In fact, despite the synthesis of thousands of compounds over the last decade only very few novel neurotoxic MBADs have successfully reached the clinical development stage in brain cancer chemotherapy (Zhang and Lippard, 2003; Heffeter et al., 2008).

Anti-miRNA therapeutic strategies remain attractive in that single miRNAs may interact with the expression of a relatively large number of dysregulated pathogenic genes in neurological disease processes (Corsten et al., 2007; Lukiw et al., 2009). For example miRNA-21 levels have been reported to be elevated in glioma and their knock-down is associated with increased apoptotic activity. The use of anti-miRNA-21 oligonucleotides in vitro shows that suppression of miRNA-21 leads to a synergistic increase in caspase-3 activity and decreased cell viability (Corsten et al., 2007). Similar effects on the use of miRNA-based therapies on the stem-cell-like characteristics of glioma have been suggested to have considerable therapeutic potential (Godlewski et al., 2008; Hide et al., 2008). The development of advanced combinatorial therapies involving surgery, radiotherapy, antiangiogenics, MBADs, miRNA antisense strategies, and chemotherapeutics remain as attractive and evolving strategies in the future clinical management of glioma and GBM.

9 Summary

Glioma and glioblastoma multiforme constitute highly complex, progressive, and insidious neoplastic disorders of the human CNS. Those treated with optimal therapy, including surgical resection, radiation therapy, and chemotherapy, have a median survival of approximately 12 months, with fewer than 25% of patients surviving up to two years and fewer than 10% of patients surviving up to five years. Whether the prognosis of patients with secondary glioblastoma is better than, or similar to, those patients with primary glioblastoma remains controversial. Glioma and GBM each exhibit significantly heterogeneous gene expression profiles, and spontaneous, dysregulated, and highly proliferative invasive cell growth. Although individual genetic signaling patterns are variable, increases in the expression of glioma and GBM markers, such as beta-amyloid precursor protein, caspase-3, pentraxin-2, and vascular endothelial growth factor, indicate upregulated expression of cell–cell contact, cell cycle, vascular proliferation, and apoptotic–necrotic markers at the level of gene expression in virtually all brain tumors examined. The heterogeneous genotypic and phenotypic nature of human brain neoplasms further confounds their molecular and genetic signature as well as pharmacological and therapeutic treatment strategies.

Surgical resection followed by aggressive radiotherapy and chemotherapy using genomic methylating agents, such as temozolomide (TMZ), and tailored to each individual case, currently represents the best treatment options available. Surgical and multimodal radiotherapeutic approaches combined with chemotherapeutic agents, each with independent and sometimes synergistic mechanisms of action, are currently providing the greatest clinical benefit with improved quality of life in many cases. Recent discoveries on the regulation of miRNA-124, miRNA-125b, miRNA-137 and miRNA-221 expression are uncovering another layer of genetic control in neoplastic brain cells, and should provide yet another therapeutic approach, and treatment opportunity, for advanced clinical intervention. The design and application of novel micro-RNA-based therapeutic strategies are highly attractive because a single miRNA may be able to quench the expression of entire families of interrelated neoplastic or oncogenic genes. Several of these novel approaches have been proven to be effective in vitro, however, miRNA and drug delivery systems in vivo remain an imposing biophysical, medical, and clinical research challenge. In the future, combinatorial surgical, radiotherapeutic, and pharmacological strategies, employing several genomic structure and function modifiers simultaneously, appear to hold the most promise for advancing the clinical management of brain cancer and improvement in the prognosis for both the glioma and GBM patient.

Acknowledgments The work in this manuscript was supported by a Translational Research Initiative (TRI) Grant entitled "*Gene expression patterns in glioblastoma multiforme (GBM)*" by the Louisiana State University Board of Regents.

References

Amaral PP, Dinger ME, Mercer TR, Mattick JS (2008) The eukaryotic genome as an RNA machine. Science 319:1787–1789

Anderson JC, McFarland BC, Gladson CL (2008) New molecular targets in angiogenic vessels of glioblastoma tumours. Expert Rev Mol Med 10:e23

Bates KA, Fonte J, Robertson TA, Martins RN, Harvey AR (2002) Chronic gliosis triggers Alzheimer's disease-like processing of amyloid precursor protein. Neuroscience 113:785–796

Bauer N, Fonseca AV, Florek M, Freund D, Jászai J, Bornhäuser M, Fargeas CA, Corbeil D (2008) New insights into the cell biology of hematopoietic progenitors by studying prominin-1 (CD133). Cells Tissues Organs 188:127–138

Belda-Iniesta C, de Castro Carpeño J, Casado Sáenz E, Cejas Guerrero P, Perona R, González Barón M (2006) Molecular biology of malignant gliomas. Clin Transl Oncol 8:635–641

Boudreau CR, Yang I, Liau LM (2005) Gliomas: advances in molecular analysis and characterization. Surg Neurol 64:286–294

Brandsma D, Stalpers L, Taal W, Sminia P, van den Bent MJ (2008) Clinical features, mechanisms, and management of pseudoprogression in malignant gliomas. Lancet Oncol 9:453–461

Bruce JN, Criscuolo GR, Merrill MJ, Moquin RR, Blacklock JB, Oldfield EH (1987) Vascular permeability induced by protein product of malignant brain tumors: inhibition by dexamethasone. J Neurosurg 67:880–884

Buie LW, Valgus J (2008) Bevacizumab: a treatment option for recurrent glioblastoma multiforme. Ann Pharmacother 42:1486–1490

Caltagarone J, Jing Z, Bowser R (2007) Focal adhesions regulate amyloid signaling and cell death in Alzheimer's disease. Biochim Biophys Acta 1772:438–445

Cao X, Pfaff SL, Gage FH (2007) A functional study of miR-124 in the developing neural tube. Genes Dev 21:531–536

Cao X, Yeo G, Muotri AR, Kuwabara T, Gage FH (2006) Noncoding RNAs in the mammalian central nervous system. Annu Rev Neurosci 29:77–103

Carlson MR, Pope WB, Horvath S, Braunstein JG, Nghiemphu P, Tso CL, Mellinghoff I, Lai A, Liau LM, Mischel PS, Dong J, Nelson SF, Cloughesy TF (2007) Relationship between survival and edema in malignant gliomas: role of vascular endothelial growth factor and neuronal pentraxin 2. Clin Cancer Res 13:2592–2598

Chamberlain MC (2008) Antiangiogenic blockage: a new treatment for glioblastoma. Expert Opin Biol Ther 8:1449–1453

Cheng CK, Fan QW, Weiss WA (2009) PI3K signaling in glioma – animal models and therapeutic challenges. Brain Pathol 19:112–120

Cho WC (2007) OncomiRs: the discovery and progress of microRNAs in cancers. Mol Cancer 6:60

Ciafrè SA, Galardi S, Mangiola A, Ferracin M, Liu CG, Sabatino G, Negrini M, Maira G, Croce CM, Farace MG (2005) Extensive modulation of a set of microRNAs in primary glioblastoma. Biochem Biophys Res Commun 334:1351–1358

Colangelo V, Schurr J, Ball MJ, Palacios Pelaez R, Bazan NG, Lukiw WJ (2002) Gene expression profiling of 12,633 genes in Alzheimer hippocampal CA1. J Neurosci Res 70:462–473

Conaco C, Otto S, Han JJ, Mandel G (2006) Reciprocal actions of REST and a microRNA promote neuronal identity. Proc Natl Acad Sci U S A 103:2422–2427

Corsten MF, Miranda R, Kasmieh R, Krichevsky AM, Weissleder R, Shah K (2007) MicroRNA-21 knockdown disrupts glioma growth in vivo and displays synergistic cytotoxicity with neural precursor cell delivered S-TRAIL in human gliomas. Cancer Res 67:8994–9000

Culicchia F, Cui JG, Li YY, Lukiw WJ (2008) Upregulation of beta-amyloid precursor protein (βAPP) expression in glioblastoma multiforme. NeuroReport 19:981–985

Dogini DB, Ribeiro PA, Rocha C, Pereira TC, Lopes-Cendes I (2008) MicroRNA expression profile in murine central nervous system development. J Mol Neurosci 35:331–337

Emsley J, White HE, O'Hara BP, Oliva G, Srinivasan N, Tickle IJ, Blundell TL, Pepys MB, Wood SP (1994) Structure of pentameric human serum amyloid P component. Nature 367:338–345

Farrell CJ, Plotkin SR (2007) Genetic causes of brain tumors: neurofibromatosis, tuberous sclerosis, von Hippel-Lindau, and other syndromes. Neurol Clin 25:925–946

Faury D, Nantel A, Dunn SE, Guiot MC, Haque T, Hauser P (2007) Molecular profiling identifies prognostic subgroups of pediatric glioblastoma and shows increased YB-1 expression in tumors. J Clin Oncol 25:1196–1208

Felicetti F, Errico MC, Bottero L, Segnalini P, Stoppacciaro A, Biffoni M, Felli N, Mattia G, Petrini M, Colombo MP, Peschle C, Carè A (2008) The promyelocytic leukemia zinc finger-microRNA-221/-222 pathway controls melanoma progression through multiple oncogenic mechanisms. Cancer Res 68:2745–2754

Fisher JL, Schwartzbaum JA, Wrensch M, Wiemels JL (2007) Epidemiology of brain tumors. Neurol Clin 25:867–890

Fuller GN, Hess KR, Rhee CH, Yung WK, Sawaya RA, Bruner JM, Zhang W (2002) Molecular classification of human diffuse gliomas by multidimensional scaling analysis of gene expression profiles parallels morphology-based classification, correlates with survival, and reveals clinically-relevant novel glioma subsets. Brain Pathol 12:108–116

Fuso A, Cavallaro RA, Zampelli A, D'Anselmi F, Piscopo P, Confaloni A, Scarpa S (2007) gamma-secretase is differentially modulated by alterations of homocysteine cycle in neuroblastoma and glioblastoma cells. J Alzheimers Dis 11:275–290

Gerrow K, El-Husseini A (2007) Receptors look outward: revealing signals that bring excitation to synapses. Sci STKE 408:56

Giles FJ (2001) The vascular endothelial growth factor (VEGF) signaling pathway: a therapeutic target in patients with hematologic malignancies. Oncologist 6:32–39

Gillies JK, Lorimer IA (2007) Regulation of p27Kip1 by miRNA 221/222 in glioblastoma. Cell Cycle 6:2005–2009

Godlewski J, Nowicki MO, Bronisz A, Williams S, Otsuki A, Nuovo G, Raychaudhury A, Newton HB, Chiocca EA, Lawler S (2008) Targeting of the Bmi-1 oncogene/stem cell renewal factor by microRNA-128 inhibitsglioma proliferation and self-renewal. Cancer Res 68: 9125–9130

Goodman AR, Cardozo T, Abagyan R, Altmeyer A, Wisniewski HG, Vilcek J (1996) Long pentraxins: an emerging group of proteins with diverse functions. Cytokine Growth Factor Rev 7:191–202

Grothey A, Ellis LM (2008) Targeting angiogenesis driven by vascular endothelial growth factors using antibody-based therapies. Cancer J 14:170–177

Grothey E, Gibson T, Sartor O, Bukowski R (2008) Highlights from: the 2008 meeting of the American Society of Clinical Oncology, May 30–June 3, 2008, Chicago, IL. Clin Genitourin Canc 6:65–72

Gurdon JB, Melton DA (2008) Nuclear reprogramming in cells. Science 322:1811–1815

Hasan J, Jayson GC (2001) VEGF antagonists. Expert Opin Biol Ther 1:703–718

Heffeter P, Jungwirth U, Jakupec M, Hartinger C, Galanski M, Elbling L, Micksche M, Keppler B, Berger W (2008) Resistance against novel anticancer metalcompounds: differences and similarities. Drug Resist Updat 11:1–16

Herber DL, Mercer M, Roth LM, Symmonds K, Maloney J, Wilson N (2007) Microglial activation is required for Abeta clearance after intracranial injection of lipopolysaccharide in APP transgenic mice. J Neuroimmune Pharmacol 2:222–231

Hide T, Takezaki T, Nakamura H, Kuratsu J, Kondo T (2008) Brain tumor stem cells as research and treatment targets. Brain Tumor Pathol 25:67–72

Hill JM, Zhao Y, Clement C, Neumann DM, Lukiw WJ (2009) HSV-1 infection of human brain cells induces miRNA-146a and Alzheimer-type inflammatory signaling. Neuroreport 20:1500–1505

Hobert O (2008) Gene regulation by transcription factors and microRNAs. Science 319:1785–1786

Hsu YC, Perin MS (1995) Human neuronal pentraxin II (NPTX2): conservation, genomic structure, and chromosomal localization. Genomics 28:220–227

Huang PH, Cavenee WK, Furnari FB, White FM (2007) Uncovering therapeutic targets for glioblastoma: a systems biology approach. Cell Cycle 6:2750–2754

Hyun Hwang J, Smith CA, Salhia B, Rutka JT (2008) The role of fascin in the migration and invasiveness of malignant glioma cells. Neoplasia 10:149–159

Idbaih A, Omuro A, Ducray F, Hoang-Xuan K (2007) Molecular genetic markers as predictors of response to chemotherapy in gliomas. Curr Opin Oncol 19:606–611

Johansson Swartling F (2008) Identifying candidate genes involved in brain tumor formation. Ups J Med Sci 113:1–38

Juric D, Bredel C, Sikic BI, Bredel M (2007) Integrated high-resolution genome-wide analysis of gene dosage and expression in human brain tumors. Methods Mol Biol 377:187–202

Kavsan VM, Dmitrenko VV, Shostak KO, Bukreieva TV, Vitak NY, Simirenko OE, Malisheva TA, Shamayev MI, Rozumenko VD, Zozulya YA (2007) Comparison of microarray and sage techniques in gene expression analysis of human glioblastoma. Tsitol Genet 41: 36–55

Kavsan V, Shostak K, Dmitrenko V, Zozulya Y, Rozumenko V, Demotes-Mainard J (2005) Characterization of genes with increased expression in human glioblastomas. Tsitol Genet 39:37–49

Koul D (2008) PTEN signaling pathways in glioblastoma. Cancer Biol Ther 7:1321–1325

Kumar MS, Lu J, Mercer KL, Golub TR, Jacks T (2007) Impaired microRNA processing enhances cellular transformation and tumorigenesis. Nat Genet 39:673–677

Lahiri DK, Farlow MR, Nurnberger JI Jr, Greig NH (1997) Effects of cholinesterase inhibitors on the secretion of beta-amyloid precursor protein in cell cultures. Ann N Y Acad Sci 826: 416–421

Lakka SS, Rao JS (2008) Antiangiogenic therapy in brain tumors. Expert Rev Neurother 8: 1457–1473

Larrivee B, Karsan A (2000) Signaling pathways induced by vascular endothelial growth factor (review). Int J Mol Med 5:447–456

Lefranc F, Facchini V, Kiss R (2007) Proautophagic drugs: a novel means to combat apoptosis-resistant cancers, with a special emphasis on glioblastomas. Oncologist 12: 1395–1403

le Sage C, Nagel R, Egan DA, Schrier M, Mesman E, Mangiola A, Anile C, Maira G, Mercatelli N, Ciafrè SA, Farace MG, Agami R (2007) Regulation of the p27(Kip1) tumor suppressor by miR-221 and miR-222 promotes cancer cell proliferation. EMBO J 26:3699–3708

Levicar N, Dewey RA, Daley E, Bates TE, Davies D, Kos J, Pilkington GJ, Lah TT (2003) Selective suppression of cathepsin L by antisense cDNA impairs human brain tumor cell invasion in vitro and promotes apoptosis. Cancer Gene Ther 10:141–151

Levicar N, Strojnik T, Kos J, Dewey RA, Pilkington GJ, Lah TT (2002) Lysosomal enzymes, cathepsins in brain tumour invasion. J Neurooncol 58:21–32

Li F, Ackermann EJ, Bennett CF (1999) Pleiotropic cell-division defects and apoptosis induced by interference with survivin function. Nat Cell Biol 1:461–466

Lopes MBS, VandenBerg SR, Scheithauer BW (1993) The World Health Organization classification of nervous system tumors in experimental neuro-oncology. In: AJ Levine and HH Schmidek, eds Molecular Genetics of Nervous System Tumors. Wiley-Liss, New York, NY, pp 1–36

Louis DN (2006) Molecular pathology of malignant gliomas. Annu Rev Pathol 1:97–117

Lukiw WJ (2004) Gene expression profiling in fetal, aged, and Alzheimer hippocampus: a continuum of stress-related signaling. Neurochem Res 29:1287–1297

Lukiw WJ (2007) Micro-RNA speciation in fetal, adult and Alzheimer's disease hippocampus. NeuroReport 18:297–300

Lukiw WJ (2009) Docosahexaenoic acid and amyloid-beta peptide signaling in Alzheimer's disease. World Rev Nutr Diet 99:55–70

Lukiw WJ, Bazan NG (2006) Survival signaling in Alzheimer's disease. Biochem Soc Trans 34:1277–1282

Lukiw WJ, Bazan NG (2008) Docosahexaenoic acid and the aging brain. J Nutr 138:2510–2514

Lukiw WJ, Carver LA, LeBlanc HJ, Bazan NG (2000) Analysis of 1184 gene transcript levels in AD CA1 hippocampus: synaptic signaling and transcription factor deficits and upregulation of pro-inflammatory pathways. Alzheimers Rep 3:161–167

Lukiw WJ, Cui JG, Li YY, Culicchia F (2009) Up-regulation of micro-RNA-221 (miRNA-221; chr Xp11.3) and caspase-3 accompanies down-regulation of the survivin-1 homolog BIRC1 (NAIP) in glioblastoma multiforme(GBM). J Neurooncol 91:27–32

Lukiw WJ, Cui JG, Marcheselli VL, Bodker M, Botkjaer A, Gotlinger K (2005) A role for docosahexaenoic acid-derived neuroprotectin D1 in neural cell survival and Alzheimer disease. J Clin Invest 115:2774–2783

Lukiw WJ, Cui JG, Zhao Y, Cui JG (2008) An NF-kB-sensitive microRNA-146a-mediated inflammatory circuit in Alzheimer's disease and in stressed human brain cells, J. Biol Chem 283:31315–31322

Lukiw WJ, Ottlecz A, Lambrou G, Grueninger M, Finley J, Thompson HW, Bazan NG (2003) Coordinate activation of HIF-1 and NF-kappaB DNA binding and COX-2 and VEGF expression in retinal cells by hypoxia. Invest Ophthalmol Vis Sci 44:4163–4170

Lukiw WJ, Pogue AI (2007) Induction of specific micro RNA (miRNA) species by ROS-generating metal sulfates in primary human brain cells. J Inorg Biochem 101:1265–1269

MacDonald TJ, Pollack IF, Okada H, Bhattacharya S, Lyons-Weiler J (2007) Progression-associated genes in astrocytoma identified by novel microarray gene expression data reanalysis. Methods Mol Biol 377:203–222

Makeyev EV, Maniatis T (2008) Multilevel regulation of gene expression by microRNAs. Science 319:1789–1790

Makeyev EV, Zhang J, Carrasco MA, Maniatis T (2007) The MicroRNA miR-124 promotes neuronal differentiation by triggering brain-specific alternative pre-mRNA splicing. Mol Cell 27:435–448

Mattick JS, Amaral PP, Dinger ME, Mercer TR, Mehler MF (2009) RNA regulation of epigenetic processes. Bioessays 31:51–59

Mattick JS, Makunin IV (2005) Small regulatory RNAs in mammals. Hum Mol Genet 14:121–132

Medina R, Zaidi SK, Liu CG, Stein JL, van Wijnen AJ, Croce CM, Stein GS (2008) MicroRNAs 221 and 222 bypass quiescence and compromise cell survival. Cancer Res 68: 2773–2780

Mehlhorn G, Hollborn M, Schliebs R (2000) Induction of cytokines in glial cells surrounding cortical beta-amyloid plaques in transgenic Tg2576 mice. Int J Dev Neurosci 18:423–431

Mellai M, Caldera V, Patrucco A, Annovazzi L, Schiffer D (2008) Survivin expression in glioblastomas correlates with proliferation, but not with apoptosis. Anticancer Res 28:109–118

Mourelatos Z (2008) MicroRNAs: biology and roles in neurodegeneration and brain tumours. Introduction and historical background. Brain Pathol 18:110–112

Nelson JS (2002) Alzheimer pathology in elderly patients with glioblastoma multiforme. Arch Pathol Lab Med 126:1515–1517

Nicholas MK, Lukas RV, Jafri NF, Faoro L, Salgia R (2006) Epidermal growth factor receptor – mediated signal transduction in thedevelopment and therapy of gliomas. Clin Cancer Res 12:7261–7270

Nicoloso MS, Calin GA (2008) Micro RNA involvement in brain tumors: from bench to bedside. Brain Pathol 18:122–129

Norden AD, Drappatz J, Wen PY (2008) Novel anti-angiogenic therapies for malignant gliomas. Lancet Neurol 7:1152–1160

Ohgaki H (2009) Epidemiology of brain tumors. Methods Mol Biol 472:323–342

Ohgaki H, Kleihues P (2005) Population-based studies on incidence, survival rates, and genetic alterations in astrocytic and oligodendroglial gliomas. J Neuropathol Exp Neurol 64:479–489

Papagiannakopoulos T, Kosik KS (2008) MicroRNAs: regulators of oncogenesis and stemness. BMC Med 6:15

Paris D, Quadros A, Patel N, DelleDonne A, Humphrey J, Mullan M (2005) Inhibition of angiogenesis and tumor growth by beta and gamma-secretase inhibitors. Eur J Pharmacol 514:1–15

Pluta R (2002) Astroglial expression of the beta-amyloid in ischemia-reperfusion brain injury. Ann N Y Acad Sci 977:102–108

Pogue AI, Cui JG, Li YY, Zhao Y, Culicchia F, Lukiw WJ (2010) Micro RNA-125b (miRNA-125b) function in astrogliosis and glial cell proliferation. Neurosci Lett 476(1):18–22

Pogue AI, Lukiw WJ (2004) Angiogenic signaling in Alzheimer's disease. Neuroreport 15:1507–1510

Pope WB, Chen JH, Dong J, Carlson MR, Perlina A, Cloughesy TF, Liau LM, Mischel PS, Nghiemphu P, Lai A, Nelson SF (2008) Relationship between gene expression and enhancement in glioblastoma multiforme: exploratory DNA microarray analysis. Radiology 249:268–277

Radde R, Bolmont T, Kaeser SA, Coomaraswamy J, Lindau D, Stoltze L (2007) Abeta42-driven cerebral amyloidosis in transgenic mice reveals early and robust pathology. EMBO Rep 7: 940–946

Ray SK, Patel SJ, Welsh CT, Wilford GG, Hogan EL, Banik NL (2002) Molecular evidence of apoptotic death in malignant brain tumors includingglioblastoma multiforme: upregulation of calpain and caspase-3. J Neurosci Res 69:197–206

Reardon DA, Wen PY, Desjardins A, Batchelor TT, Vredenburgh JJ (2008) Glioblastoma multiforme: an emerging paradigm of anti-VEGF therapy. Expert Opin Biol Ther 8:541–553

Rosell R, de Las Peñas R, Balaña C, Santarpia M, Salazar F, de Aguirre I, Reguart N, Villa S, Wei J, Ramirez JL, Molina MA, Ramon y, Cajal S, Jablons D, Taron M (2008) Translational research in GBM: molecular criteria for patient selection. Future Oncol 4:219–228

Rueger MA, Winkeler A, Thomas AV, Kracht LW, Jacobs AH (2008) Molecular imaging-guided gene therapy of gliomas. Handb Exp Pharmacol 185:341–359

Sathornsumetee S, Reardon DA (2009) Targeting multiple kinases in glioblastoma multiforme. Expert Opin Investig Drugs 18:277–292

Sathornsumetee S, Rich JN (2008) Designer therapies for glioblastoma multiforme. Ann N Y Acad Sci 1142:108–132

Silber J, Lim DA, Petritsch C, Persson AI, Maunakea AK, Yu M, Vandenberg SR, Ginzinger DG, James CD, Costello JF, Bergers G, Weiss WA, Alvarez-Buylla A, Hodgson JG (2008) miR-124 and miR-137 inhibit proliferation of glioblastoma multiforme cells andinduce differentiation of brain tumor stem cells. BMC Med 6:14

Sin AH, Gonzalez-Toledo E, Fowler M, Minagar A, Nanda A (2008) Amyloidoma presenting as a butterfly glioma on positron emission tomography scan and magnetic resonance-spectroscopy: a case report and review of the literature. J La State Med Soc 160:44–50

Singh SK, Clarke ID, Hide T, Dirks PB (2004) Cancer stem cells in nervous system tumors. Oncogene 23:7267–7273

Singh SK, Clarke ID, Terasaki M, Bonn VE, Hawkins C, Squire J, Dirks PB (2003) Identification of a cancer stem cell in human brain tumors. Cancer Res 63:5821–5828

Stegh AH, Kesari S, Mahoney JE, Jenq HT, Forloney KL, Protopopov A, Louis DN, Chin L, DePinho RA (2008) Bcl2L12-mediated inhibition of effector caspase-3 and caspase-7 via distinctmechanisms in glioblastoma. Proc Natl Acad Sci U S A 105:10703–10708

Sutcliffe JG (2001) Open-system approaches to gene expression in the CNS. J Neurosci 21: 8306–8309

Takuma K, Baba A, Matsuda T (2004) Astrocyte apoptosis: implications for neuroprotection. Prog Neurobiol 72:111–127

Tatter SB 2005. The new WHO classification of tumors affecting the human central nervous system. http://neurosurgery.mgh.harvard.edu/newwhobt.htm#Mutations

Tentori L, Graziani G (2009) Recent approaches to improve the antitumor efficacy of temozolomide. Curr Med Chem 16:245–257

Tschape JA, Hartmann T (2008) Therapeutic perspectives in Alzheimer's disease. Recent Pat CNS Drug Discov 1:119–127

Tso CL, Freije WA, Day A, Chen Z, Merriman B, Perlina A (2006) Distinct transcription profiles of primary and secondary glioblastoma subgroups. Cancer Res 66:159–167

van den Bent MJ, Kros JM (2007) Predictive and prognostic markers in neuro-oncology. J Neuropathol Exp Neurol 66:1074–1081

Venezia V, Nizzari M, Carlo P, Corsaro A, Florio T, Russo C (2007) Amyloid precursor protein and presenilin involvement in cell signaling. Neurodegener Dis 4:101–111

Voelzke WR, Petty WJ, Lesser GJ (2008) Targeting the epidermal growth factor receptor in high-grade astrocytomas. Curr Treat Options Oncol 9:23–31

Williams AE, Perry MM, Moschos SA, Larner-Svensson HM, Lindsay MA (2008) Role of miRNA-146a in the regulation of the innate immune response and cancer. Biochem Soc Trans 36:1211–1215

Wong ET, Brem S (2008) Antiangiogenesis treatment for glioblastoma multiforme: challenges and opportunities. J Natl Compr Canc Netw 6:515–522

Xie Z, Chin LS (2008) Molecular and cell biology of brain tumor stem cells: lessons from neuralprogenitor/stem cells. Neurosurg Focus 24:E25

Xie Q, Thompson R, Hardy K, DeCamp L, Berghuis B, Sigler R, Knudsen B, Cottingham S, Zhao P, Dykema K, Cao B, Resau J, Hay R, Vande Woude GF (2008) A highly invasive human glioblastoma pre-clinical model for testing therapeutics. J Transl Med 6:77–97

Yadirgi G, Marino S (2009) Adult neural stem cells and their role in brain pathology. J Pathol 217:242–253

Young-Pearse TL, Bai J, Chang R, Zheng JB, LoTurco JJ, Selkoe DJ (2007) A critical function for beta-amyloid precursor protein in neuronal migration. J Neurosci 27:14459–14469

Zeng Y (2009) Regulation of the mammalian nervous system by microRNAs. Mol Pharmacol 75:259–264

Zhang CX, Lippard SJ (2003) New metal complexes as potential therapeutics. Curr Opin Chem Biol 7:481–489

Zhang B, Pan X, Cobb GP, Anderson TA (2007) microRNAs as oncogenes and tumor suppressors. Dev Biol 302:1–12

Zhao Y, Cui JG, Lukiw WJ (2006) Natural secretory products of human neural and microvessel endothelial cells: Implications in pathogenic "spreading" of neurodegenerative disease. Mol Neurobiol 34:181–192

Index

J.P. Blass (ed.), *Neurochemical Mechanisms in Disease*,
Advances in Neurobiology 1, DOI 10.1007/978-1-4419-7104-3,

T